LANDOLT-BÖRNSTEIN

Numerical Data and Functional Relationships in Science and Technology

New Series

Editor in Chief: K.-H. Hellwege

Group II: Atomic and Molecular Physics

Volume 14

Supplement to Volumes II/4 and II/6

Molecular Constants

mostly from Microwave, Molecular Beam, and Electron Resonance Spectroscopy

Subvolume a

J. Demaison · A. Dubrulle · W. Hüttner · E. Tiemann

Editors: K.-H. Hellwege and A. M. Hellwege

Springer-Verlag Berlin · Heidelberg · New York 1982

LANDOLT-BÖRNSTEIN

Zahlenwerte und Funktionen aus Naturwissenschaften und Technik

Neue Serie

Gesamtherausgabe: K.-H. Hellwege

Gruppe II: Atom- und Molekularphysik

Band 14

Ergänzung der Bände II/4 und II/6

Molekelkonstanten

vorwiegend aus Messungen der Mikrowellen-, Molekularstrahl- und Elektronenresonanz-Spektroskopie

Teilband a

J. Demaison · A. Dubrulle · W. Hüttner · E. Tiemann

Herausgeber: K.-H. Hellwege und A. M. Hellwege

Springer-Verlag Berlin · Heidelberg · New York 1982

CIP-Kurztitelaufnahme der Deutschen Bibliothek

Zahlenwerte und Funktionen aus Naturwissenschaften und Technik / Landolt-Börnstein. – Berlin; Heidelberg; New York: Springer.
Parallelt.: Numerical data and functional relationships in science and technology

NE: Landolt, Hans [Begr.]; PT. N.S./Gesamthrsg.: K.-H. Hellwege. Gruppe 2, Atom- und Molekularphysik. Bd. 14. Molekelkonstanten: vorwiegend aus Messungen d. Mikrowellen-, Molekularstrahl- u. Elektronenresonanz-Spektroskopie; Erg. d. Bd. II/4 u. II/6. Teilbd. a. J. Demaison ... Hrsg.: K.-H. Hellwege u. A. M. Hellwege. – 1982.

ISBN 3-540-11365-7 (Berlin, Heidelberg, New York)
ISBN 0-387-11365-7 (New York, Heidelberg, Berlin)

NE: Hellwege, Karl-Heinz [Hrsg.]; Demaison, Jean [Mitverf.]

Printed in Germany

Typesetting, printing and bookbinding: Universitätsdruckerei H. Stürtz AG Würzburg

2163/3020 – 543210

Contributors

Jean Demaison

Université des Sciences et Techniques de Lille, France
UER de Physique Fondamentale

Alain Dubrulle

Université des Sciences et Techniques de Lille, France
UER de Physique Fondamentale

Wolfgang Hüttner

Abteilung Chemie-Physik
Universität Ulm, Germany

Eberhard Tiemann

Institut für Atom- und Molekülphysik
Universität Hannover, Germany

Vorwort

Hiermit wird der erste Teilband, II/14a, des Bandes II/14 „Molekelkonstanten vorwiegend aus Messungen der Mikrowellen-, Molekularstrahl- und Elektronenresonanz-Spektroskopie" vorgelegt, der ein Ergänzungsband von II/6 und damit der zweite Ergänzungsband von II/4 ist.

Die drei Bände II/4, II/6 und II/14 bilden zusammen eine vollständige, kritische Sammlung aller bis 1982 veröffentlichten Daten von Molekelkonstanten, die mit Hilfe der verschiedensten klassischen und neuen spektroskopischen Methoden gewonnen wurden.

Der Band II/14 wurde aus technischen Gründen geteilt. Der vorliegende Teilband II/14a enthält die allgemeine Einleitung und die Tabellen für die diamagnetischen Molekeln. Der Teilband II/14b, der die Strukturdiagramme der im Teilband a behandelten Molekeln, die Tabellen für die freien Radikale und das Substanzenverzeichnis enthält, wird anschließend erscheinen.

Für die zweiatomigen Molekeln sind wieder die Kernabstände zusammen mit den Molekelkonstanten angegeben. Für die Struktur der mehratomigen Molekeln sei – neben den im Teilband II/14b wieder zusammengestellten qualitativen Strukturzeichnungen, die die Richtung der verschiedenen tabellierten Größen illustrieren – auf den Band II/7 „Strukturdaten freier, mehratomiger Molekeln" hingewiesen, der in einem der nächsten Jahre ergänzt werden wird.

Ein detailliertes Substanzenverzeichnis für alle drei Bände II/4, II/6 und II/14a, b am Ende des letzen Teilbandes, II/14b, ermöglicht das schnelle Auffinden der speziellen gesuchten Daten einer Substanz in einem dieser Bände.

Unser Dank gilt wieder den Autoren für ihre gründliche und schnelle Arbeit und für die angenehme Zusammenarbeit, sowie der „Sektion für Strukturdokumentation der Universität Ulm", die ihr vollständiges Dokumentationsmaterial zur Verfügung stellte, aber auch der Landolt-Börnstein-Redaktion, insbesondere Frau R. Lettmann, für ihre zuverlässige Hilfe und dem Verlag für die übliche Sorgfalt bei der Fertigstellung des Bandes.

Auch dieser Band wird ohne finanzielle Hilfe von anderer Seite veröffentlicht.

Darmstadt, Oktober 1982 **Die Herausgeber**

Preface

This is the first subvolume, II/14a, of volume II/14 "Molecular Constants, mostly from Microwave, Molecular Beam and Electron Resonance Spectroscopy", which is the supplement to volume II/6 and thus the second supplement to volume II/4.

The three volumes II/4, II/6 and II/14 together present a complete compilation of the critically evaluated data on molecular constants from measurements with classical and modern spectroscopy which have been published until early 1982.

Volume II/14 had to be divided for technical reasons. The present subvolume contains the general introduction and the tables on diamagnetic molecules. Subvolume II/14b, containing the structure diagrams of the molecules treated in subvolume a, the tables on the free radicals, and the index of substances will appear subsequently.

The internuclear distances of diatomic molecules are supplemented together with the constants. The qualitative structure diagrams of polyatomic molecules illustrate the direction of various constants listed in the tables, while the complete tables of structure data may be found in volume II/7, "Structure Data of Free Polyatomic Molecules", which will be supplemented in one of the next years.

A detailed index of substances treated in the three volumes II/4, II/6, and II/14a, b is given at the end of subvolume II/14b and will help the user to locate quickly any special information on individual molecules in one of these volumes.

Our thanks are again due to the authors for their careful and quick work and for the agreeable cooperation, to the "Sektion für Strukturdokumentation der Universität Ulm" for making available their complete documentation material, and also to the editorial office, especially Mrs. R. Lettmann, for their reliable help and Springer Verlag for their usual care during the completion of the volume.

This subvolume is published without any outside financial support.

Darmstadt, October 1982 **The Editors**

Survey

Subvolume a

Subvolume b

Table of contents

Subvolume a

Subvolume b

1 Introduction

1.1 General remarks

The present volume II/14 (published as two subvolumes II/14a and II/14b) is a supplement to and an extension of volume II/6 published in 1974 and has been prepared on the understanding that it will be used in combination with II/4 and II/6. The preliminary remarks of II/4 apply also to this supplement and are not repeated. The remarks relating to each table and the definitions of the listed constants, however, are rewritten in the present volume. Data on a particular substance can be found through the common index at the end of subvolume II/14b.

Volume II/14 brings the spectroscopic data on molecules up to date considering the publications up to and partly including 1981. The spectroscopic information collected in this volume has been obtained principally from gas phase microwave measurements. In addition, gas phase data have been included derived from methods related to microwave spectroscopy by employing a coherent radiation source. These are molecular beam techniques, radio frequency spectroscopy, electron resonance spectroscopy, laser spectroscopy, and double resonance techniques. Some other methods are considered if the accuracy of the derived molecular parameters is comparable to that of microwave spectroscopy and no microwave data are available. Examples would be Fourier infrared spectroscopy or electric deflection method.

Internuclear distances are listed in the tables only for diatomic molecules and for some small open shell molecules. For all other polyatomic molecules the literature giving structural information has been cited.

A compilation of structural data has been published in 1976 in volume II/7 of the New Series of Landolt-Börnstein and will be supplemented in the near future.

1.2 Review articles and tables

Some books treating recent developments in microwave spectroscopy are listed in 1.7, [1–7].

The Journal of Physical and Chemical Reference Data has published a series of tables of line frequencies and molecular constants for diatomic molecules [8], triatomic molecules [9] and for molecules of particular interest in astrophysics.

A bibliography of experimental microwave spectroscopy (1945–1975) has been compiled in [10].

Since 1973 the Chemical Society (London) has regularly published a review on microwave spectroscopy [11].

1.3 Arrangement of tables, substances and parameters

1.3.1 Arrangement of chapters and sections

The arrangement of Vols. II/4 and II/6 is retained as far as possible. The data on molecules are listed in chapter 2 and those on radicals in chapter 4. In each section of chapter 2, a set of related constants and physical quantities is tabulated for all of the molecules included, whereas in chapter 4 all measured constants and physical quantities of each radical are listed together in individual tables.

In section 2.2 through 2.5 the molecules (radicals are included with references to chapter 4) are ordered according to the type of their respective rotational spectrum as follows: Diatomic molecules (2.2), linear molecules (2.3) symmetric top molecules (2.4), and asymmetric top molecules (2.5). Molecules which are asymmetric only due to isotopic substitution are listed together with their parent species in 2.4. The tables include rotational constants, centrifugal distortion constants, rotation-vibration interaction constants, and *l*-type doubling constants. Some additional molecular constants obtained by microwave type methods have been listed as well. References to publications concerning the molecular structure are cited separately.

Tables 2.6 through 2.9 contain the dipole moments (2.6), nuclear quadrupole coupling constants (2.7), constants of hindered rotation (2.8) and magnetic interaction constants (2.9) of the molecules.

In chapter 3, structural diagrams of various molecules are to be found.

The literature used for chapter 2 is contained in separate sections of chapter 2, collected by the various authors and concerning their own contributions. For the reader's convenience the page where to find the appropriate reference section is indicated at the top of each page of the tables. The references are ordered according to the publication year followed by the first author's name and in few cases, by an additional running number. For each year the references are ordered alphabetically.

In chapter 4 the data from investigations of radicals (non-$^1\Sigma$ electronic states of molecules) have been compiled. The diatomic radicals are listed in 4.1, the polyatomic ones in 4.2.

Chapter 5 consists of additional data which came to our attention after the page proof of the tables.

Chapter 6 gives the index of substances. It lists the running numbers for each molecule in all the tables in which the molecule appears both in the present volume II/14 and in the previous volumes II/4 and II/6. Inorganic (not containing carbon) and organic (containing carbon) substances are listed in separate sections and arranged alphabetically according to their alphabetically ordered gross formula.

Chapters 1 and 2 are contained in the present subvolume II/14a.

Chapters 3, 4, 5 and 6 will be published subsequently in subvolume II/14b.

1.3.2 Arrangement within the sections of chapter 2

The arrangement of the columns within the sections of chapter 2 is the same as in chapter 2 of Vols. II/4 and II/6. The explanation of the symbols used in the tables is found under the subsection "Preliminary remarks" for each individual table. In order to favour comparison between similar molecules, the ordering of molecules in a particular table follows a specific pattern. A brief explanation or survey at the end of the subsection "Preliminary remarks" of each table gives the ordering procedure adopted for the compounds. A particular molecule can easily be found with the help of the index of substances, chapter 6 (Subvolume II/14b).

1.3.3 Explanation of the columns of the tables in chapter 2

Column 1 gives the running number of each molecule in the table. The numbers in the corresponding table of Vols II/4 and II/6 are given below the running number

Column 2 gives the structural formula of the molecule. The isotopic species are labeled with the isotopic numbers with the exception of the most abundant species, where the labels have been omitted: $C={}^{12}C$, $O={}^{16}O$, $S={}^{32}S$, $N={}^{14}N$, etc. The vibrational state of the molecule for which the listed constants were obtained is added (not in tables 2.9.2 and 2.9.3).

The next columns contain the values of the constants with which the table is concerned, and their references. Where several references are given, the first reference listed is the publication from which the numerical values were taken.

The last column of each table contains general remarks and values if necessary. This column also gives references to tables or diagrams which contain further information on the respective molecule. In tables 2.2 through 2.5 references are given in this column to all following tables and chapters. In tables 2.6 through 2.9 references are given only to tables 2.2 through 2.5. Thus all information given for each molecule may reliably be found by use of either tables 2.2 through 2.5, or the index of substances in chapter 6 (Subvolume II/14b).

1.3.4 Arrangements within the sections of chapter 4

The data are compiled in individual tables for each radical. The diatomic radicals are treated in section 4.1, the polyatomic ones in 4.2.

In sections 4.1 and 4.2 radicals are arranged in subsections according to the electronic states with running numbers in each subsection. The data for each molecule are contained in two tables. The first table lists the observed frequencies and their assignment to specific transitions as indicated by the various quantum numbers (electronic, vibrational, rotational, hyperfine, and parity). The second table gives all constants and parameters which were available.

The constants and parameters used in the tables are defined in a short introduction preceding the tables of each subsection. The literature is given for each radical separately.

1.3.5 Errors

The error in a tabulated value is written as defined by the following equations (examples taken from LB, NS, Vol. II/7):

$$3.478\,(21)\,\text{Å} = (3.478 \pm 0.021)\,\text{Å}$$

$$13.4\,(21)\,\text{Å} = (13.4 \pm 2.1)\,\text{Å}$$

$$119.3\,(2)\,\text{deg} = (119.3 \pm 0.2)\,\text{Å}$$

Evidently the error given in parentheses on the left side applies to the last significant digits.

1.4 Selection of data

For a molecule which was studied by several authors, the data of those authors are listed whose work (a) was the most complete (comparison of the data of a particular molecule), (b) was the most recent and (c) appeared to be the most reliable one. The work of other authors is cited in the references, but only the most recent paper of a particular research group is given. Data from dissertations and conference research reports were only included when no other publication could be located.

1.5 Abbreviations used for experimental methods

BMS	beam maser spectroscopy
ED	electric deflection method
EPR	electron paramagnetic resonance
FIR	far infrared spectroscopy [Sections 2.3, 2.4, 2.5, 2.6.2, 2.6.3, 2.7.2, 2.7.3, 2.8]
	Fourier infrared spectroscopy [Sections 2.2, 2.6.1, 2.7.1, 2.9.1]
IR	infrared spectroscopy with or without laser
IRIRDR	infrared-infrared double resonance
IRMWDR	infrared-microwave double resonance
La	laser
LaSt	laser Stark spectroscopy
LC	level crossing spectroscopy
LIF	laser induced fluorescence
LMR	laser magnetic resonance
LRMW	low resolution microwave spectroscopy
MB	molecular beam electric / magnetic resonance (sometimes for the state preparation and detection lasers are used)
MBE (or MBER)	molecular beam electric resonance
MBM	molecular beam magnetic resonance
MBRF	molecular beam radiofrequency spectroscopy
MBMW	molecular beam microwave and mm-wave spectroscopy
MODR	microwave optical double resonance
MOMRIE	microwave optical magnetic resonance induced by electrons
MW	microwave spectroscopy in the cm- and mm-wavelength region
OS	optical spectroscopy
QB	quantum beat spectroscopy
RF/DR	radio frequency optical double resonance
RFIRDR	radio frequency infrared double resonance
RFMWDR	radio frequency microwave double resonance
SLS	saturated laser spectroscopy

1.6 Selected fundamental constants and conversion factors

The calculation of molecular constants from the measured frequencies of spectral lines needs best values of the fundamental constants. The "best values" improve with advancing techniques of measurement and changes in valuation. The recommended values listed in the following table are taken from the publication by Cohen and Taylor [12].

Recommended physical constants

Quantity	Symbol	Value	Units	
			SI	cgs
Velocity of light	c	2.99792458 (12)	10^8 m sec^{-1}	10^{10} cm sec^{-1}
Fine-structure constant	α	7.2973506 (60)	10^{-3}	10^{-3}
	α^{-1}	137.03604 (11)		
Electron charge	e	1.6021892 (46)	10^{-19} C	10^{-20} emu
		4.803242 (14)		10^{-10} esu
Planck's constant	h	6.626176 (36)	10^{-34} J · sec	10^{-27} erg · sec
	$\hbar = h/2\pi$	1.0545887 (57)	10^{-34} J · sec	10^{-27} erg · sec
Avogadro's number	N	6.022045 (31)	10^{26} kmol^{-1}	10^{23} mol^{-1}
Atomic mass unit	amu	1.6605655 (86)	10^{-27} kg	10^{-24} g
Electron rest mass	m_{el}	9.109534 (47)	10^{-31} kg	10^{-28} g
		5.4858026 (21)	10^{-4} amu	10^{-4} amu
Proton rest mass	m_p	1.6726485 (86)	10^{-27} kg	10^{-24} g
		1.007276470 (11)	amu	amu
Neutron rest mass	m_n	1.6749543 (86)	10^{-27} kg	10^{-24} g
		1.008665012 (37)	amu	amu

(continued)

Recommended physical constants (continued)

Quantity	Symbol	Value	Units	
			SI	cgs
Electron charge to mass ratio	e/m_{el}	1.7588047 (49)	10^{11} C kg^{-1}	10^{7} emu g^{-1}
		5.272764 (15)		10^{17} esu g^{-1}
Rydberg constant	R_{∞}	1.097373177 (83)	10^{7} m^{-1}	10^{5} cm^{-1}
Bohr radius	a_0	5.2917706 (44)	10^{-11} m	10^{-9} cm
Classical electron radius	r_0	2.8179380 (70)	10^{-15} m	10^{-13} cm
Electron magnetic moment in Bohr magnetons	μ_{el}/μ_{B}	1.0011596567 (35)		
Bohr magneton	μ_{B}	9.274078 (36)	10^{-24} J T^{-1}	10^{-21} erg Gauss^{-1}
Nuclear magneton	μ_{n}	5.050824 (20)	10^{-27} J T^{-1}	10^{-24} erg Gauss^{-1}
Electron magnetic moment	μ_{el}	9.284832 (36)	10^{-24} J T^{-1}	10^{-21} erg Gauss^{-1}
Proton magnetic moment in Bohr magnetons	μ_{p}/μ_{B}	1.521032209 (16)	10^{-3}	10^{-3}
Proton magnetic moment	μ_{p}	1.4106171 (55)	10^{-26} J T^{-1}	10^{-23} erg Gauss^{-1}

These values yield the conversion factor $I \cdot B$:

$$I\,[\text{amu Å}^2] \cdot B\,[\text{MHz}] = 5.0537905(85) \cdot 10^5 \text{ amu Å}^2 \text{ MHz}.$$

In fact a slightly different value of this conversion factor is very often used:

$$I\,B = 505376 \text{ amu Å}^2 \text{ MHz}.$$

The following table for conversion between different energy scales may be used (uncertainties are all about 3 ppm, if needed, more accurate values may be calculated from the preceding table):

Energy Conversion factors

	J	erg	eV	cm^{-1}	cal · mol^{-1}	Hz
J	1.0	10^{7}	$6.24146 \cdot 10^{18}$	$5.03404 \cdot 10^{22}$	$1.43834 \cdot 10^{23}$	$1.50916 \cdot 10^{33}$
erg	10^{-7}	1.0	$6.24146 \cdot 10^{11}$	$5.03404 \cdot 10^{15}$	$1.43834 \cdot 10^{16}$	$1.50916 \cdot 10^{26}$
eV	$1.60219 \cdot 10^{-19}$	$1.60219 \cdot 10^{-12}$	1.0	8065.48	$2.30450 \cdot 10^{4}$	$2.41797 \cdot 10^{14}$
cm^{-1}	$1.98648 \cdot 10^{-23}$	$1.98648 \cdot 10^{-16}$	$1.23985 \cdot 10^{-4}$	1.0	2.85724	$2.99793 \cdot 10^{10}$
cal · mol^{-1}	$6.95246 \cdot 10^{-24}$	$6.95246 \cdot 10^{-17}$	$4.33934 \cdot 10^{-5}$	$3.49989 \cdot 10^{-1}$	1.0	$1.04851 \cdot 10^{10}$
1 Hz	$6.62620 \cdot 10^{-34}$	$6.62620 \cdot 10^{-27}$	$4.13571 \cdot 10^{-15}$	$3.33564 \cdot 10^{-11}$	$9.53731 \cdot 10^{-11}$	1.0

1.7 References

1 Bates, D.R., Estermann, I.: "Advances in Atomic and Molecular Physics" Vol. 9, Academic Press, New York, **1973**.

2 Buckingham, A.D.: "MTP International Review of Science Physical chemistry, Series 2. Vol. 2: Molecular Structure and Properties", Butterworths, London, **1975**.

3 Buckingham, A.D., Ramsay, D.A.: "MTP International Review of Science. Physical chemistry, Series 2. Vol. 3: Spectroscopy", Butterworths, London, **1976**.

4 Chantry, G.W.: "Modern Aspects of Microwave Spectroscopy", Academic Press, London, **1979**.

5 Kroto, H.W.: "Molecular Rotation Spectra", John Wiley and Sons, New York, **1975**.

6 Lide, D.R., Paul, M.A.: "Critical Evaluation of Chemical and Physical Structural Information", National Academy of Sciences, Washington, D.C. **1974**.

7 Rao, K.N.: "Molecular Spectroscopy: Modern Research, Vol. II", Academic Press, New York, **1976**.

8 Lovas, F.J., Tiemann, E.: J. Phys. Chem. Ref. Data **3** (1974) 609.

9 Lovas, F.J.: J. Phys. Chem. Ref. Data **7** (1978) 1445.

10 Starck, B., Mutter, R., Spreter, C., Ketteman, K., Boggs, A., Botskor, M., Jones, M.: "Bibliography of Microwave Spectroscopy 1945–1975". Zentralstelle für Atomkernenergie-Dokumentation. Eggenstein-Leopoldshafen, 9-1, **1977**.

11 "Molecular Spectroscopy. Specialist Periodical Reports". The Chemical Society, London. Vol. 1 (1973); Vol. 2 (1974); Vol. 3 (1975): R.F. Barrow, D.A. Long, and D.J. Millen, eds.; Vol. 4 (1976); Vol. 5 (1978); Vol. 6 (1979): R.F. Barrow, D.A. Long, and J. Sheridan, eds.

12 Cohen, E.R., Taylor, B.N.: J. Phys. Chem. Ref. Data **2** (1973) 663.

2 Constants of diamagnetic molecules

2.1 Survey

2.2 Diatomic molecules

Rotational constants, centrifugal distortion constants, rotation-vibration interaction constants, Dunham energy coefficients, parameters of the breakdown of the Born-Oppenheimer approximation, and equilibrium internuclear separations

2.2.1 Preliminary remarks

1. Introduction

This table contains the molecular parameters which describe the rotational energy in a specific electronic state and in different vibrational states. Values are given only for molecules the electronic state of which is assigned to $^1\Sigma^\pm$ or $O^\pm$. For other electronic states of the same molecules or for radicals references to tables 2.6, 2.7 and 4 are given.

Most of the molecular constants used are explained in Section 2.2.1 of Vol. II/6 Landolt-Börnstein, New Series. Only a brief supplement will be presented here to introduce some new parameters.

In all disciplines of molecular spectroscopy, the rotational-vibrational energy is described by a power series of $J(J+1)$ and $(v+1/2)$, where J and v are the rotational and vibrational quantum numbers, respectively. As parameters, the Dunham notation Y_{lk} or the original spectroscopic notation $B_e^{spec}, \alpha_e^{spec}, \ldots, \omega_e^{spec}, \ldots$ are in use. But in both notations, the values of these parameters are obtained by fitting the same formula to the measured spectral line positions. Therefore both types of constants are the same and only disagree by the signs $\pm$ which were originally defined for the series $B_e, \alpha_e, \ldots$. The following relations are given as examples:

$$Y_{10} = \omega_e^{spec}$$
$$-Y_{20} = \omega_e x_e^{spec}$$
$$Y_{01} = B_e^{spec}$$
$$-Y_{11} = \alpha_e^{spec} \quad \text{etc.}$$

The upper script "spec" is written in the case of the phenomenological parameters in order to distinguish between these and the physically defined constants B_e and ω_e:

$$B_e^{cor} = \frac{h}{8\pi^2 \mu_{red} (r_e^{cor})^2} \quad \text{rotational constant for equilibrium configuration e}$$

$$\omega_e = 2\sqrt{a_0 \cdot B_e^{cor}} \quad \text{harmonic vibrational constant}$$

which are both related to the potential $U(r)$ for the motion of the two atoms within the Born-Oppenheimer approximation and where

h Planck's constant

μ_{red} reduced atomic mass of the molecule

r_e^{cor} equilibrium internuclear distance corrected for the breakdown of the Born-Oppenheimer approximation

$a_0 \cdot 2$ second derivative of $U(r)$ with respect to $\xi = \frac{r - r_e}{r_e}$ at $\xi = 0$.
a_0 is proportional to the harmonic force constant.

As reviewed by Watson [1] the breakdown of the Born-Oppenheimer approximation can be accounted for by the following formula in the case of the rotational constant Y_{01} for a molecule AB:

$$Y_{01} = B_e^{cor}\left[1 - 2m_e\left(\frac{d_A}{M_A} + \frac{d_B}{M_B}\right)\right] \tag{1}$$

where

m_e mass of the electron

$M_{A,B}$ mass of a specific isotope of atom A or B

$d_{A,B}$ phenomenological parameters for the adiabatic and non-adiabatic Born-Oppenheimer corrections of atom A or B including the usual Dunham correction.

Similar relations are valid for the other Dunham parameters Y_{lk} (compare Watson [1] and Bunker [2]) but are not used to describe the observed spectra. The IR-spectrum of CO is the only exception. The parameters d_A, d_B can be determined by isotopic substitution of both atoms separately. An overall fit of all molecular isotopes observed will give the most reliable set of parameters, which is included in the tables and from which molecular constants of all isotopes can be calculated with high precision by using simple mass relations (compare [3]). If data exist only for individual isotopes of one molecule in the literature, it is all included in the tables.

Sometimes the Born-Oppenheimer correction is described with δ-parameters (compare the definition in Vol. II/6 and [4]), which do not take into account the different behavior of atom A and B. Therefore this data should be recalculated to get comparable information.

2. Explanations of the columns in the tabulation

Column	Parameters		Definitions
1	Nr.		serial number of the diatomic molecule used in all tables of this book
		6–Nr.	the second number represents the serial number in table 2.2 of Vol. II/6 Landolt-Börnstein, New Series
2	B [MHz]		rotational constant
		B_e	for the equilibrium configuration e, compare column 3
		$B_{0,1,2,\dots}$	for the vibrational state $v=0, 1, 2, \dots$, compare column 3
	Y_{01} [MHz]		Dunham's rotational constant
3	D [kHz]		centrifugal distortion constant
		D_e	for the equilibrium configuration e
		$D_{0,1,2,\dots}$	for the vibrational state $v=0, 1, 2, \dots$,
	Y_{02} [kHz]		Dunham's centrifugal distortion constant
		β_e [kHz]	higher-order rotation-vibration interaction constant
		Y_{12} [kHz]	Dunham's higher-order rotation-vibration interaction constant
	H [kHz]	H_e	higher-order centrifugal distortion constant
	Y_{03} [kHz]		Dunham's higher-order centrifugal distortion constant
4	α, γ [MHz]	α_e, γ_e	rotation-vibration interaction constants
	Y_{11} [MHz], Y_{21} [MHz], Y_{31} [MHz]		Dunham energy coefficients for vibration-rotation interaction
5	Ref.		references for columns 2...4, 8, 9
6	Meth.		method of measurement used in the given reference corresponding to the preceding quoted values
		MW	microwave spectroscopy
		MB	molecular beam resonance experiment-electric or magnetic resonance
		IR	infrared with lasers
		MODR	microwave optical double resonance
7	$a_0, a_1, a_2, a_3, \dots$		potential constants for a power series of $U(r)$ according to $\xi = \frac{r - r_e}{r_e}$
	d_A, d_B; δ_1, δ_2		parameters of the breakdown of the Born-Oppenheimer approximation
	$Y_{10}, Y_{20}, \dots$		vibrational parameters determined from MW-measurements or high precision IR-spectroscopy

Column	Parameters		Definitions
8	r [Å]		internuclear distance; see also 3.
		r_e	equilibrium value of the internuclear distance calculated from $Y_{01} = h/(8\pi^2 \mu_{red} r_e^2)$
		$\tilde{r}_e$	equilibrium value of the internuclear distance calculated from B_e, subtracted the Dunham-correction δ_3, compare Eq. (4) of Vol. II/6 Landolt-Börnstein, New Series
		r_e^{cor}	equilibrium value of the internuclear distance calculated from B_e^{cor}, compare Eq. (1)
9			references to following tables

All measured or derived values Y_{ik}–or the equivalent constants as defined in Vol. II/6 Landolt-Börnstein, New Series–, which are listed in the quoted references, are also tabulated, even when they are not indicated in the column headings of the table.

List of abbreviations for indexes:

cor	corrected	= corrected to allow for a treatment of the molecule different from the Born-Oppenheimer approximation
der	derived	= derived from the measured values by simple interrelations
e	equilibrium	= symbol representing the equilibrium configuration
e	electron	
el	electron, electronic	
ev	evaluated	= evaluated with the use of experimental values from other spectroscopic regions
n	nuclear	
p	proton	
red	reduced.	

3. Conversion factors

For the calculation of the internuclear distance different conversion factors $h/(8\pi^2 m_0)$ (m_0 = atomic mass unit) are used in the literature. Throughout these tables the value

$$B\,[\text{MHz}] \cdot I\,[\text{amu Å}^2] = 505391\ \text{MHz amu Å}^2$$

will be adopted. For any theoretical model interpretation, it seems unimportant which of the existing conversion factors is used at the moment, because the different values will change the r_e-parameters by an amount of about $2 \cdot 10^{-5}$ only. The errors shown for r_e are the experimental errors neglecting the uncertainty of the conversion factor. The relative atomic masses for the calculation of the Born-Oppenheimer approximation are usually taken from latest published atomic mass tables [5]. But there are several exceptions because the authors of the original publication used other values for the masses. In order to get full consistency for all data the whole fitting procedure would have to be repeated. But this would lead to slight changes only in the molecular parameters and will not alter the physical interpretation. On the other hand for predicting spectral lines with high reliability the reader should use the originally adopted mass values and is advised to refer to the original work.

For conversion of cm^{-1} to Hz the value 299792458 m/s is used for the speed of light.

4. Arrangement of substances

The same arrangement of the molecules by the number of valence electrons is used as in Vol. II/6, Landolt-Börnstein, New Series.

Nr. in 2.2.2	Group of molecules	Example	Number of valence electrons	Ground state	For rotational constants see table
1	H_2	H_2	2	$^1\Sigma_g^+$	2.2.2
2···3	Group Ia - hydrides	LiH	2	$^1\Sigma^+$	2.2.2
4···6	Group Ia/group Ia	NaK	2	$^1\Sigma^+$	2.2.2
7	He_2	He_2	4	nonbonding	4.1.3 ex. state
8	Group IVa - hydrides	CH	5	$^2\Pi_r$	4.1.5
9···10	Group Va - hydrides	PH	6	$^3\Sigma^-$	4.1.3
11···13	Group VIa - hydrides	OH	7	$^2\Pi_i$	4.1.5
14	Group VIIa - hydrides ion	HBr^+	7	$^2\Pi_i$	4.1.5
15···18	Group VIIa - hydrides	HCl	8	$^1\Sigma^+$	2.2.2
19···29	Group Ia/group VIIa	NaCl	8	$^1\Sigma^+$	2.2.2
30···33	Group Ib/group VIIa	CuBr	8	$^1\Sigma^+$	2.2.2
34···37	Group IIa/group VIa	SrO	8	$^1\Sigma^+$	2.2.2
38	Group Ia/group VIIIa	KAr	9	$^2\Sigma^+$	4.1.2
39···44	Group IIa/group VIIa	CaF	9	$^2\Sigma^+$	4.1.2
45	Group IVa/group VIa ion	CO^+	9	$^2\Sigma^+$	4.1.2
46	Group IVa/group Va	CN	9	$^2\Sigma^+$	4.1.2
47···58	Group IIIa/group VIIa	AlBr	10	$^1\Sigma^+$	2.2.2
59···75	Group IVa/group VIa	CSe	10	$^1\Sigma^+$	2.2.2
76···77	Group Va/group Va	PN	10	$^1\Sigma^+$	2.2.2
78···79	Group IVa/group VIIa	CF	11	$^2\Pi_r$	4.1.5
80···81	Group Va/group VIa	NS	11	$^2\Pi_r$	4.1.5
82···86	Group Va/group VIIa	NF	12	$^3\Sigma^-$	4.1.3
87···90	Group VIa/group VIa	S_2	12	$^3\Sigma^-$	4.1.3
91···95	Group VIa/group VIIa	OF	13	$^2\Pi_i$	4.1.5
96···103	Group VIIa/group VIIa	ClF	14	$^1\Sigma^+$	2.2.2

5. References

1 Watson, J.K.G.: J. Mol. Spectrosc. **45** (1973) 99.
Watson, J.K.G.: J. Mol. Spectrosc. **80** (1980) 411.
2 Bunker, P.R.: J. Mol. Spectrosc. **35** (1970) 306
Bunker, P.R.: J. Mol. Spectrosc. **68** (1977) 367.
3 Townes, C.H., Schawlow, A.L.: Microwave Spectroscopy, McGraw-Hill Book Comp., New York (1955).
4 Gordy, W., Cook, R.: Microwave Moleculare Spectra, John Wiley and Sons, New York (1970).
5 Wapstra, A.H., Bos, K.: Atomic Data and Nuclear Data Tables **19** (1977) 177.

For references in the table, see 2.2.3.

2.2.2 Data

Nr.	Molecule	B [1]) MHz	D [1]) kHz	α, γ [1]) MHz	Ref.	Meth.	Derived constants	r_e [1]) Å	See Tab.
1 *6-2*	$^2D_2$	$B_0 = 896775\,(24)$ $B_1 = 865150\,(27)$	$D_0 = 345.1\,(24)$ MHz $H_0 = 207\,(69)$ $D_1 = 337.6\,(21)$ MHz $H_1 = 192\,(45)$		78McK	IR [2])	$\nu_0 = 2993.614\,(4)\ \mathrm{cm}^{-1}$ band origin of $v=1-v=0$		2.9.1 4.1.3 4.1.6
	$^2D_2$ ortho	electronic state G(3d) $^1\Sigma_g^+$: Data from level anticrossing spectroscopy with r(4d) $^3\Pi_g$-state, compare 4.1.6							
2 *6-3*	LiH								2.6.1 2.9.1
3	NaH								2.6.1
4 *6-5*	Na_2								2.7.1 2.9.1
5 *6-6b*	$^{23}Na^{39}K$	$Y_{01} = 2854.021\,(78)$	$Y_{02} = -6.613\,(50)$ $Y_{12} = -53.8\,(22)$ Hz	$Y_{11} = -13.480\,(19)$ $Y_{21} = -54.7\,(21)$ kHz $Y_{31} = -3.09\,(11)$ kHz	81WOR [3]) 81WOR [3]) 81WOR [3])	MODR MODR MODR			2.6.1
4 *6-5* 5 *6-6b* 6 *6-8*	Na_2 NaK Cs_2	From laser induced fluorescence studies rotational constants are derived which have nearly the accuracy of microwave-spectroscopy standards [75DEM, 76DEM, 78BRE, 79BRE, 79EIS, 79HÖN, 79RAA]							
7	He_2								4.1.3 4.1.7

[1]) Or related terms.
[2]) Pure rotational-vibrational quadrupole transitions.
[3]) Results from a combined fit of optical and microwave data; higher order Y_{lk} are calculated in [81WOR].

Nr.	Molecule	B [1]) MHz	D [1]) kHz	α, γ [1]) MHz	Ref.	Meth.	Derived constants	r_e [1]) Å	See Tab.
8 *6-8b*	CH								2.6.1 4.1.5
9 *6-8c*	NH								2.6.1 4.1.3
10	PH								4.1.3 4.1.7
11 *6-9*	OH								2.6.1 2.7.1 4.1.2 4.1.5
12 *6-10*	SH								2.6.1 2.7.1 4.1.5
13 *6-11*	SeH								4.1.5
14	HBr^+								4.1.5
15 *6-13*	$^1H^{19}F$	$B_0 = 616365.59\,(42)$	$D_0 = 63569.5\,(84)$		76GUE1	IR	$\nu_0 = 3961.4229\,(3)$		2.9.1
			$H_0 = 5.037\,(60)$		76GUE1	IR	cm^{-1} band origin of		
			L_0 [2]) $= -0.72\,(15)$ Hz		76GUE1	IR	$v=1-v=0$		
		$B_1 = 593213.67\,(42)$	$D_1 = 61876.9\,(84)$		76GUE1	IR	$\nu_0 = 7750.7949\,(15)$		
			$H_1 = 4.827\,(60)$		76GUE1	IR	cm^{-1} band origin of		
		$B_2 = 570652.88\,(96)$	$D_2 = 60239\,(23)$		76GUE1	IR	$v=2-v=0$		
			$H_2 = 4.44\,(15)$		76GUE1	IR			
16 *6-14*	$^1H^{35}Cl$	$Y_{01} = 317580.97\,(23)$	$Y_{02} = -15945.8\,(15)$	$Y_{11} = -9208.96\,(57)$	81GUE2	IR	$Y_{10} = 2990.9664\,(39)\ cm^{-1}$		2.6.1 2.7.1 2.9.1
			$Y_{12} = 225.49\,(87)$	$Y_{21} = 53.04\,(28)$	81GUE2	IR	$Y_{20} = -52.8364\,(29)\ cm^{-1}$		
Cont.			$Y_{22} = -11.97\,(21)$	$Y_{31} = -3.585\,(36)$	81GUE2	IR	$Y_{30} = 0.2302\,(8)\ cm^{-1}$		

[1]) Or related terms.

[2]) Centrifugal correction: $L \cdot J^4(J+1)^4$.

Nr.	Molecule	B [1]) MHz	D [1]) kHz	α, γ [1]) MHz	Ref.	Meth.	Derived constants	r_e [1]) Å	See Tab.
16 *6–14* Cont.	$^1H^{35}Cl$		Y_{03} = 520.7 (30) Hz		81GUE2	IR	$Y_{40} = -1.2822(73) \cdot 10^{-2}\ cm^{-1}$		
			$Y_{13} = -19.44(39)$ Hz		81GUE2	IR			
			$Y_{04} = -31.0(21) \cdot 10^{-3}$ Hz		81GUE2	IR			
		$B_e^{cor} = 317558.67(42)$			73WAT/ 71LUC	IR/MW			
					76GUE1	IR			
	$^1H^{37}Cl$	$Y_{01} = 317100.68(33)$	$Y_{02} = -15881.6(30)$	$Y_{11} = -9188.97(78)$	81GUE2	IR	$Y_{10} = 2988.7033(42)\ cm^{-1}$		
			Y_{12} = 216.2 (10)	Y_{21} = 53.39 (23)	81GUE2	IR	$Y_{20} = -52.7569(24)\ cm^{-1}$		
			Y_{03} = 475.2 (81) Hz	[2])	81GUE2	IR	$Y_{30} = 0.22970(39)\ cm^{-1}$		
			[2])		76GUE1	IR	[2])		
	$^2D^{35}Cl$	$Y_{01} = 163351.09(27)$	$Y_{02} = -4211.7(28)$	$Y_{11} = -3396.56(63)$	81GUE2	IR	$Y_{10} = 2145.1523(67)\ cm^{-1}$		
			Y_{12} = 42.12 (60)	Y_{21} = 13.98 (21)	81GUE2	IR	$Y_{20} = -27.1736(37)\ cm^{-1}$		
			Y_{03} = 45 (11) Hz	[2])	81GUE2	IR	$Y_{30} = 0.08390(58)\ cm^{-1}$		
			[2])				[2])		
					77NIA	IR			
	$^2D^{37}Cl$	$Y_{01} = 162870.62(28)$	$Y_{02} = -4187.0(23)$	$Y_{11} = -3381.75(66)$	81GUE2	IR	$Y_{10} = 2142.032(10)\ cm^{-1}$		
			Y_{12} = 42.05 (48)	Y_{21} = 13.93 (22)	81GUE2	IR	$Y_{20} = -27.1143(58)\ cm^{-1}$		
			Y_{03} = 54 (10) Hz	[2])	81GUE2	IR	$Y_{30} = 0.08667(90)\ cm^{-1}$		
Cont.			[2])				[2])		

[1]) Or related terms.
[2]) Other parameters are calculated by mass relations from the results of $^1H^{35}Cl$.

Nr.	Molecule	B [1]) MHz	D [1]) kHz	α, γ [1]) MHz	Ref.	Meth.	Derived constants	r_e [1]) Å	See Tab.
16 *6–14* Cont.	HCl				67WEB		$a_0 = 211129\ cm^{-1}$ $a_1 = -2.36425$ $a_2 = 3.66281$ $a_3 = -4.70624$ $a_4 = 5.2131$ $a_5 = -5.5269$ $a_6 = 8.3664$		
					73WAT/ 71LUC		$d_H = -0.06446\ (14)$ $d_{Cl} = 0.075\ (43)$	$r_e^{cor} = 1.2746149(9)$	
17 *6–15*	$^1H^{81}Br$	$Y_{01} = 253770.6\ (11)$	$Y_{02} = -10366.5\ (54)$	$Y_{11} = -6992.8\ (26)$	76STO	IR	$Y_{10} = 2649.3698\ (88)\ cm^{-1}$		2.6.1 2.7.1 2.9.1
			$Y_{12} = 119.6\ (72)$	$Y_{21} = 25.7\ (16)$	76STO	IR	$Y_{20} = -45.5795\ (70)\ cm^{-1}$		
			$Y_{22} = -11.4\ (19)$	$Y_{31} = -3.50\ (26)$	76STO	IR	$Y_{30} = 0.1245\ (20)\ cm^{-1}$		
			$Y_{03} = 228.7\ (57)$ Hz		76STO	IR	$Y_{40} = -1.518\ (18) \cdot 10^{-2}\ cm^{-1}$		
			$Y_{13} = -16.8\ (39)$ Hz		76STO	IR			
	$^2D^{81}Br$	$Y_{01} = 128538.08\ (21)$	$Y_{02} = -2661.3\ (12)$	$Y_{11} = -2519.70\ (54)$	76BER	MW/IR	$Y_{10} = 1885.5435\ (30)\ cm^{-1}$		
			$Y_{12} = 22.21\ (75)$	$Y_{21} = 6.42\ (24)$	76BER	MW/IR	$Y_{20} = -23.0527\ (20)\ cm^{-1}$		
			$Y_{22} = -1.77\ (24)$	$Y_{31} = -0.630\ (30)$	76BER	MW/IR	$Y_{30} = 0.0370\ (5)\ cm^{-1}$		
			$Y_{03} = 32.1\ (21)$ Hz		76BER	MW/IR	$Y_{40} = -0.00334\ (4)\ cm^{-1}$		
			$Y_{13} = -1.95\ (27)$ Hz		76BER	MW/IR			
	$^2D^{79}Br$	$Y_{01} = 128617.14\ (21)$	$Y_{02} = -2664.6\ (12)$	$Y_{11} = -2522.03\ (54)$	76BER	MW/IR	$Y_{10} = 1886.1231\ (30)\ cm^{-1}$		
			$Y_{12} = 22.24\ (75)$	$Y_{21} = 6.43\ (24)$	76BER	MW/IR	$Y_{20} = -23.0668\ (20)\ cm^{-1}$		
			$Y_{22} = -1.78\ (24)$	$Y_{31} = -0.630\ (30)$	76BER	MW/IR	$Y_{30} = 0.0370\ (5)\ cm^{-1}$		
Cont.									

[1]) Or related terms.

Nr.	Molecule	B [1]) MHz	D [1]) kHz	α, γ [1]) MHz	Ref.	Meth.	Derived constants	r_e [1]) Å	See Tab.
17 *6–15* Cont.	$^{2}D^{79}Br$		$Y_{03} =$ 32.1 (21) Hz		76BER	MW/IR	$Y_{40} =$ −0.00334 (4) cm^{-1}		
			$Y_{13} =$ −1.95 (27) Hz		76BER	MW/IR			
	HBr				76STO	IR	$a_1 =$ −2.43694 (55) $a_2 =$ 3.8336 (35) $a_3 =$ −4.985 (22) $a_4 =$ 5.57 (17) $a_5 =$ −6.64 (88) $a_6 =$ 13.6 (43)		
18 *6–16*	$^{1}H^{127}I$	$Y_{01} =$ 195213.69 (39)	$Y_{02} =$ −6213.5 (13)	$Y_{11} =$ −5107.6 (11)	81GUE1	IR	$Y_{10} =$ 2308.9941 (77) cm^{-1}		2.7.1
			$Y_{12} =$ 24.9 (25)	$Y_{21} =$ −8.53 (69)	81GUE1	IR	$Y_{20} =$ −39.6229 (77) cm^{-1}		
			$Y_{03} =$ 90.9 (45) Hz	$Y_{31} =$ −1.21 (16)	81GUE1	IR	$Y_{30} =$ −0.0272 (36) cm^{-1}		
				$Y_{41} =$ −259 (11) kHz	81GUE1	IR	$Y_{40} =$ −0.01609 (87) cm^{-1}		
					78NIA	IR			
	$^{2}D^{127}I$	$Y_{01} =$ 98452.78 (26)	$Y_{02} =$ −1582.5 (10)	$Y_{11} =$ −1830.59 (72)	81GUE1	IR	$Y_{10} =$ 1639.7927 (35) cm^{-1}		
			$Y_{12} =$ 4.76 (60)	$Y_{21} =$ −1.30 (48)	81GUE1	IR	$Y_{20} =$ −19.9906 (22) cm^{-1}		
			$Y_{22} =$ −0.98 (11)	$Y_{31} =$ −0.41 (12)	81GUE1	IR	$Y_{30} =$ −0.00834 (52) cm^{-1}		
			$Y_{03} =$ 18.0 (27) Hz	$Y_{41} =$ −20.4 (99) kHz	81GUE1	IR	$Y_{40} =$ −0.004055 (43) cm^{-1}		
					79NIA	IR			
	HI				81GUE1	IR	$a_1 =$ −2.54612 (24) $a_2 =$ 4.0511 (14) $a_3 =$ −5.4561 (61) $a_4 =$ 7.058 (75) $a_5 =$ −9.58 (36)		

[1]) Or related terms.

Nr.	Molecule	B [1]) MHz	D [1]) kHz	α, γ [1]) MHz	Ref.	Meth.	Derived constants	r_e [1]) Å	See Tab.
19 *6–17*	$^{6}Li^{19}F$				73CUP	MB/ MW rot. tran- sition $J=1\rightarrow 0$			2.6.1 2.7.1 2.9.1
20 *6–18*	$^{6}Li^{35}Cl$	$B_e^{cor}=24116.37\,(20)$			73WAT/ 69PEA	MW	$d_{Li}=-0.196\,(17)$ $d_{Cl}=+0.87\,(25)$	$r_e^{cor}=2.020700\,(8)$	2.9.1
21 *6–19*	LiBr								2.9.1
22 *6–20*	LiI								2.7.1 2.9.1
23 *6–28*	KI								2.6.1 2.7.1
24 *6–31*	$^{85}Rb^{79}Br$	$Y_{01}=1424.85302\,(64)$	$Y_{02}=-0.44847\,(7)$	$Y_{11}=-5.57482\,(73)$ $Y_{21}=6.41\,(25)$ kHz	77TIE1	MW			2.6.1 2.7.1
25 *6–23*	RbI								2.6.1 2.7.1
26 *6–33*	$^{133}Cs^{19}F$	$Y_{01}=5527.2643\,(19)$	$Y_{02}=-6.0463\,(16)$ $Y_{12}=-9.2\,(19)$ Hz $Y_{03}^{der}=-213.0\,(12)\cdot 10^{-6}$ Hz	$Y_{11}=-35.2443\,(30)$ $Y_{21}=35.37\,(98)$ kHz $Y_{31}=532\,(83)$ Hz	74HON2 74HON2 74HON2	MW MW MW	 $\omega_e=352.556\,(46)$ cm^{-1} $\omega_e x_e=1.615\,(14)$ cm^{-1}		2.6.1
	CsF				74HON2	MW	$a_0=168542\,(44)$ cm^{-1} $a_1=-3.03220\,(19)$ $a_2=5.655\,(49)$ $a_3=-7.32\,(35)$		

[1]) Or related terms.

Nr.	Molecule	B [1]) MHz	D [1]) kHz	α, γ [1]) MHz	Ref.	Meth.	Derived constants	r_e [1]) Å	See Tab.
27 *6–34*	$^{133}Cs^{35}Cl$	$Y_{01} = 2161.2485\,(12)$	$Y_{02} = -0.97957\,(21)$	$Y_{11} = -10.11969\,(93)$	74HON2	MW			2.7.1 2.9.1
			$Y_{12} = -1.15\,(11)$ Hz	$Y_{21} = 10.25\,(18)$ kHz	74HON2	MW	$\omega_e = 214.165\,(23)$ cm^{-1}		
			$Y_{03}^{der} = -141.335\,(71) \cdot 10^{-6}$ Hz	$Y_{31} = 53\,(10)$ Hz	74HON2	MW	$\omega_e x_e = 0.7308\,(30)$ cm^{-1}		
	CsCl				74HON2	MW	$a_0 = 159058\,(34)$ cm^{-1} $a_1 = -3.31834\,(19)$ $a_2 = 7.006\,(27)$ $a_3 = -10.92\,(20)$		
28 *6–35*	$^{133}Cs^{79}Br$	$Y_{01} = 1081.32897\,(84)$	$Y_{02} = -251.229\,(60)$ Hz	$Y_{11} = -3.71777\,(59)$	74HON2	MW			2.6.1
			$Y_{12} = -0.191\,(26)$ Hz	$Y_{21} = 3.057\,(97)$ kHz	74HON2	MW	$\omega_e = 149.662\,(18)$ cm^{-1}		
			$Y_{03}^{der} = -22.043\,(19) \cdot 10^{-6}$ Hz	$Y_{31} = 9.7\,(42)$ Hz	74HON2	MW	$\omega_e x_e = 0.3740\,(19)$ cm^{-1}		
	CsBr				74HON2	MW	$a_0 = 155248\,(37)$ cm^{-1} $a_1 = -3.37765\,(31)$ $a_2 = 7.347\,(36)$ $a_3 = -12.05\,(26)$		
29 *6–36*	$^{133}Cs^{127}I$	$Y_{01} = 708.33033\,(46)$	$Y_{02} = -111.362\,(14)$ Hz	$Y_{11} = -2.04646\,(21)$	74HON2	MW			2.6.1
			$Y_{12} = -0.0689\,(51)$ Hz	$Y_{21} = 1.467\,(24)$ kHz	74HON2	MW	$\omega_e = 119.1776\,(76)$ cm^{-1}		
			$Y_{03}^{der} = -7.5078\,(39) \cdot 10^{-6}$ Hz	$Y_{31} = 3.43\,(76)$ Hz	74HON2	MW	$\omega_e x_e = 0.2505\,(7)$ cm^{-1}		
	CsI				74HON2	MW	$a_0 = 150285\,(19)$ cm^{-1} $a_1 = -3.42882\,(19)$ $a_2 = 7.627\,(20)$ $a_3 = -12.97\,(14)$		
30 *6–37*	CuF								2.7.1 2.9.1

[1]) Or related terms.

Nr.	Molecule	B [1]) MHz	D [1]) kHz	α, γ [1]) MHz	Ref.	Meth.	Derived constants	r_e [1]) Å	See Tab.
31	$^{63}Cu^{35}Cl$	$Y_{01} = 5343.72266\,(100)$	$Y_{02} = -3.89432\,(77)$	$Y_{11} = -30.37285\,(120)$	75MAN3	MW			2.7.1
		$B_e^{cor} = 5343.768\,(20)$	$Y_{12} = 0.00234\,(47)$	$Y_{21} = 0.06021\,(46)$	75MAN3	MW	$\omega_e = 417.599\,(42)\ cm^{-1}$		
			$Y_{03}^{der} = -0.6220\,(10) \cdot 10^{-3}$ Hz	$Y_{31} = -0.061\,(50)$ kHz	75MAN3	MW	$\omega_e x_e = 1.617\,(6)\ cm^{-1}$		
	$^{63}Cu^{37}Cl$	$Y_{01} = 5158.15294\,(100)$	$Y_{02} = -3.62854\,(77)$	$Y_{11} = -28.80453\,(120)$	75MAN3	MW			
		$B_e^{cor} = 5158.195\,(20)$	$Y_{12} = 0.00214\,(47)$	$Y_{21} = 0.05610\,(46)$	75MAN3	MW	$\omega_e = 410.284\,(42)\ cm^{-1}$		
			$Y_{03}^{der} = -0.5594\,(10) \cdot 10^{-3}$ Hz	$Y_{31} = -0.056\,(50)$ kHz	75MAN3	MW	$\omega_e x_e = 1.561\,(6)\ cm^{-1}$		
	$^{65}Cu^{35}Cl$	$Y_{01} = 5284.98184\,(100)$	$Y_{02} = -3.80917\,(77)$	$Y_{11} = -29.87340\,(120)$	75MAN3	MW			
		$B_e^{cor} = 5285.026\,(20)$	$Y_{12} = 0.00227\,(47)$	$Y_{21} = 0.05889\,(46)$	75MAN3	MW	$\omega_e = 415.297\,(42)\ cm^{-1}$		
			$Y_{03}^{der} = -0.6017\,(10) \cdot 10^{-3}$ Hz	$Y_{31} = -0.059\,(50)$ kHz	75MAN3	MW	$\omega_e x_e = 1.599\,(6)\ cm^{-1}$		
	$^{65}Cu^{37}Cl$	$Y_{01} = 5099.41148\,(110)$	$Y_{02} = -3.54636\,(77)$	$Y_{11} = -28.31388\,(120)$	75MAN3	MW			
		$B_e^{cor} = 5099.453\,(20)$	$Y_{12} = 0.00208\,(47)$	$Y_{21} = 0.05483\,(46)$	75MAN3	MW	$\omega_e = 407.941\,(42)\ cm^{-1}$		
			$Y_{03}^{der} = -0.5405\,(10) \cdot 10^{-3}$ Hz	$Y_{31} = -0.054\,(50)$ kHz	75MAN3	MW	$\omega_e x_e = 1.543\,(6)\ cm^{-1}$		
	CuCl				75MAN3	MW	$a_0 = 244588\,(48)\ cm^{-1}$ $a_1 = -3.21936\,(32)$ $a_2 = 6.908\,(26)$ $a_3 = -11.68\,(23)$	$r_e^{cor} = 2.051207\,(8)$	
32	$^{63}Cu^{79}Br$	$Y_{01} = 3055.67223\,(40)$	$Y_{02} = -1.28122\,(10)$	$Y_{11} = -13.55482\,(38)$	75MAN2	MW			2.7.1
		$B_e^{cor} = 3055.707\,(16)$	$Y_{12} = 0.00056\,(4)$	$Y_{21} = 0.02031\,(14)$	75MAN2	MW	$\omega_e = 314.816\,(12)\ cm^{-1}$		
			$Y_{03}^{der} = -0.15211\,(10) \cdot 10^{-3}$ Hz	$Y_{31} = -0.029\,(12)$ kHz	75MAN2	MW	$\omega_e x_e = 0.955\,(1)\ cm^{-1}$		
	$^{63}Cu^{81}Br$	$Y_{01} = 3022.19991\,(36)$	$Y_{02} = -1.25331\,(10)$	$Y_{11} = -13.33271\,(38)$	75MAN2	MW			
		$B_e^{cor} = 3022.234\,(16)$	$Y_{12} = 0.00055\,(4)$	$Y_{21} = 0.01987\,(14)$	75MAN2	MW	$\omega_e = 313.087\,(12)\ cm^{-1}$		
			$Y_{03}^{der} = -0.14717\,(10) \cdot 10^{-3}$ Hz	$Y_{31} = -0.028\,(12)$ kHz	75MAN2	MW	$\omega_e x_e = 0.945\,(1)\ cm^{-1}$		
Cont.									

[1]) Or related terms.

Nr.	Molecule	B [1]) MHz	D [1]) kHz	α, γ [1]) MHz	Ref.	Meth.	Derived constants	r_e [1]) Å	See Tab.
32 Cont	$^{65}Cu^{79}Br$	$Y_{01} = 3003.35282\,(40)$ $B_e^{cor} = 3003.386\,(16)$	$Y_{02} = -1.23772\,(10)$ $Y_{12} = 0.00054\,(4)$ $Y_{03}^{der} = -0.14443\,(10) \cdot 10^{-3}$ Hz	$Y_{11} = -13.20818\,(38)$ $Y_{21} = 0.01962\,(14)$ $Y_{31} = -0.028\,(12)$ kHz	75MAN2 75MAN2 75MAN2	MW MW MW	$\omega_e = 312.110\,(12)$ cm^{-1} $\omega_e x_e = 0.939\,(1)$ cm^{-1}		
	$^{65}Cu^{81}Br$	$Y_{01} = 2969.88036\,(40)$ $B_e^{cor} = 2969.913\,(16)$	$Y_{02} = -1.21029\,(10)$ $Y_{12} = 0.00052\,(4)$ $Y_{03}^{der} = -0.13966\,(10) \cdot 10^{-3}$ Hz	$Y_{11} = -12.98799\,(38)$ $Y_{21} = 0.01918\,(14)$ $Y_{31} = -0.027\,(12)$ kHz	75MAN2 75MAN2 75MAN2	MW MW MW	$\omega_e = 310.365\,(12)$ cm^{-1} $\omega_e x_e = 0.928\,(1)$ cm^{-1}		
	CuBr				75MAN2	MW	$a_0 = 243091\,(20)$ cm^{-1} $a_1 = -3.28353\,(14)$ $a_2 = 7.229\,(8)$ $a_3 = -12.77\,(9)$	$r_e^{cor} = 2.173467\,(6)$	
33	$^{63}Cu^{127}I$	$Y_{01} = 2197.10172\,(98)$	$Y_{02} = -0.67269\,(10)$ $Y_{12} = 0.031\,(34)$ Hz $Y_{03}^{der} = -0.06858\,(10) \cdot 10^{-3}$ Hz	$Y_{11} = -8.51120\,(66)$ $Y_{21} = 0.00934\,(14)$ $Y_{31} = -0.0375\,(88)$ kHz	75MAN1 75MAN1 75MAN1	MW MW MW	$\omega_e = 264.897\,(18)$ cm^{-1} $\omega_e x_e = 0.715\,(2)$ cm^{-1}		2.7.1
	$^{65}Cu^{127}I$	$Y_{01} = 2151.89989\,(94)$	$Y_{02} = -0.64530\,(10)$ $Y_{12} = 0.030\,(34)$ Hz $Y_{03}^{der} = -0.06443\,(10) \cdot 10^{-3}$ Hz	$Y_{11} = -8.24990\,(66)$ $Y_{21} = 0.00896\,(14)$ $Y_{31} = -0.0356\,(88)$ kHz	75MAN1 75MAN1 75MAN1	MW MW MW	$\omega_e = 262.157\,(18)$ cm^{-1} $\omega_e x_e = 0.700\,(2)$ cm^{-1}		
	CuI				75MAN1	MW	$a_0 = 239366\,(34)$ cm^{-1} $a_1 = -3.33365\,(36)$ $a_2 = 7.391\,(16)$ $a_3 = -13.20\,(18)$	2.3383516 (10)	
34	$^{40}Ca^{16}O$	$Y_{01} - \frac{3}{4} Y_{21} = 13324.3578\,(32)$	$Y_{02} = -19.623\,(33)$ $Y_{12} = -0.203\,(30)$	$Y_{11} + 2 Y_{21} = -99.3110\,(28)$	78HOC/ 77CRE 78HOC/ 77CRE	MW MW	$Y_{10} = 732.48\,(62)$ cm^{-1} $Y_{20} = -4.71\,(14)$ cm^{-1}	1.822225 (68)	
35 *6-41*	$^{88}Sr^{16}O$	$Y_{01} = 10132.17765\,(84)$	$Y_{02} = -10.8344\,(32)$ $Y_{12} = -0.1359\,(30)$	$Y_{11} = -64.7187\,(11)$ $Y_{21} = -0.53939\,(50)$	81SCH1/ 81SCH3 81SCH1/ 81SCH3 78HOC 78HOC	MW MW MW MW	$Y_{10} = 653.30\,(33)$ cm^{-1} $Y_{20} = -3.74\,(14)$ cm^{-1}	1.919832 (95)	

[1]) Or related terms.

Nr.	Molecule	B [1]) MHz	D [1]) kHz	α, γ [1]) MHz	Ref.	Meth.	Derived constants	r_e [1]) Å	See Tab.
36 *6-42*	$^{138}Ba^{16}O$	$Y_{01} = 9371.9442\,(16)$	$Y_{02} = -8.1671\,(34)$	$Y_{11} = -41.7581\,(25)$	81SCH1/ 81SCH3	MW			2.6.1
			$Y_{12} = -2.2\,(19)$ Hz	$Y_{21} = -\ 0.1202\,(12)$	81SCH1/ 81SCH3	MW			
				$Y_{31} = -0.70\,(16)$ kHz	81SCH1/ 81SCH3	MW			
	BaO				74TIE	MW	$a_0 = 3.5879\,(10) \cdot 10^5\ cm^{-1}$ $a_1 = -2.59069\,(35)$ $a_2 = 4.009\,(24)$ $a_3 = -6.45\,(20)$		
	$^{138}Ba^{16}O$	$B_e^{cor} = 9372.508\,(16)$			78HOC	MW	$Y_{10} = 670.24\,(28)\ cm^{-1}$ $Y_{20} = -2.26\,(10)\ cm^{-1}$	$r_e^{cor} = 1.939653\,(7)$	
36 *6-42*	$^{138}Ba^{16}O$ $A^1\Sigma^+$	$B_e = 7746.94\,(60)$ $B_0 = 7729.55\,(30)$ $B_1 = 7690.28\,(30)$ $B_2 = 7664.49\,(30)$ $B_3 = 7623.42\,(30)$ $B_4 = 7591.94\,(30)$ $B_5 = 7566.46\,(30)$ $B_7 = 7481.92\,(30)$		$\alpha_e = 34.506\,(90)$	73FIE	MODR			2.6.1
37 *6-43*	$^{138}Ba^{32}S$	$Y_{01} = 3097.28216\,(26)$	$Y_{02} = -0.918568\,(63)$	$Y_{11} = -9.44620\,(33)$	80HEL/ 76TIE3	MW	$Y_{10} = 379.42\ cm^{-1}$		2.6.1
			$Y_{12} = -1.554\,(73)$ Hz	$Y_{21} = -13.323\,(66)$ kHz	80HEL/ 76TIE3	MW	$Y_{20} = -0.8710\ cm^{-1}$		
	$^{137}Ba^{32}S$	$Y_{01} = 3101.53884\,(83)$	$Y_{02} = -0.92130\,(24)$	$Y_{11} = -9.4681\,(11)$	80HEL	MW			
			$Y_{12} = -1.75\,(21)$ Hz	$Y_{21} = -12.40\,(29)$ kHz	80HEL	MW			
	$^{136}Ba^{32}S$	$Y_{01} = 3105.8631\,(12)$	$Y_{02} = -0.92370\,(45)$	$Y_{11} = -9.4864\,(14)$	80HEL	MW			
			$Y_{12} = -1.68\,(33)$ Hz	$Y_{21} = -12.90\,(40)$ kHz	80HEL	MW			
	$^{135}Ba^{32}S$	$Y_{01} = 3110.2478\,(14)$	$Y_{02} = -0.92792\,(56)$	$Y_{11} = -9.5134\,(28)$	80HEL	MW			
Cont.				$Y_{21} = -11.7\,(11)$ kHz	80HEL	MW			

[1]) Or related terms.

Nr.	Molecule	B [1]) MHz	D [1]) kHz	α, γ [1]) MHz	Ref.	Meth.	Derived constants	r_e [1]) Å	See Tab.
37 *6-43* Cont.	$^{134}Ba^{32}S$ $^{138}Ba^{34}S$ BaS	$Y_{01} = 3109.9304\,(23)$ $Y_{01} = 2945.1775\,(62)$	$Y_{02} = -0.92965\,(22)$ $Y_{02} = -0.8370\,(60)$		80HEL 80HEL 76TIE3	MW MW MW	$a_0 = 3.481\,(10) \cdot 10^5\ cm^{-1}$ $a_1 = -2.8574\,(30)$ $a_2 = 5.120\,(30)$ $a_3 = -9.50\,(23)$	2.5073508 (5)	
38	KAr								4.1.2
39	CaF								4.1.2
40	CaCl								4.1.2
41	SrF								4.1.2
42	SrCl								4.1.2
43	BaF								4.1.2
44	BaCl								2.7.1 4.1.2
45	CO^+								4.1.2 4.1.5
46 *6-44*	CN								2.6.1 2.7.1 4.1.2 4.1.5
47 *6-45*	AlF								2.7.1 2.9.1
48 *6-45a*	AlCl								2.7.1
49 *6-45b*	AlBr								2.7.1

[1]) Or related terms.

Nr.	Molecule	B [1]) MHz	D [1]) kHz	α, γ [1]) MHz	Ref.	Meth.	Derived constants	r_e [1]) Å	See Tab.
50 *6-46*	$^{69}Ga^{19}F$	$Y_{01}=10778.0181\,(19)$	$Y_{02}=-14.459\,(31)$	$Y_{11}=-85.8665\,(24)$	77ARN	MW			2.7.1
			$Y_{12}=75\,(20)$ Hz	$Y_{21}=304.67\,(65)$ kHz	77ARN	MW			2.9.1
			$Y_{03}^{der}=-5.680\,(24)\cdot10^{-3}$ Hz		77ARN	MW			
	$^{71}Ga^{19}F$	$Y_{01}=10712.3766\,(19)$			77ARN	MW			
	GaF				77ARN	MW	$a_0=2.6800\,(58)\cdot10^5$ cm^{-1} $a_1=-3.2928\,(25)$ $a_2=7.906\,(75)$ $a_3=-17.3\,(15)$	1.774362	
51	$^{69}Ga^{79}Br$	$Y_{01}=2482.0846\,(20)$	$Y_{02}=-0.95708\,(31)$	$Y_{11}=-9.8280\,(12)$	81NAI3	MW	$Y_{10}=266.661\,(43)$ cm^{-1}		2.7.1
			$Y_{12}=2.07\,(20)$ Hz	$Y_{21}=15.393\,(44)$ kHz	81NAI3	MW	$Y_{20}=-0.658\,(7)$ cm^{-1}		
			$Y_{03}^{der}=-46.32\,(10)\cdot10^{-6}$ Hz		81NAI3	MW			
	$^{69}Ga^{81}Br$	$Y_{01}=2453.5144\,(21)$	$Y_{02}=-0.93539\,(32)$	$Y_{11}=-9.6585\,(5)$	81NAI3	MW			
			$Y_{12}=1.99\,(23)$ Hz	$Y_{21}=15.046\,(45)$ kHz	81NAI3	MW			
	$^{71}Ga^{79}Br$	$Y_{01}=2444.7403\,(23)$	$Y_{02}=-0.92852\,(36)$	$Y_{11}=-9.60745\,(53)$ $Y_{21}=15.13\,(18)$ kHz	81NAI3	MW			
	$^{71}Ga^{81}Br$	$Y_{01}=2416.1688\,(29)$	$Y_{02}=-0.90704\,(43)$	$Y_{11}=-9.4378\,(21)$	81NAI3	MW			
			$Y_{12}=1.86\,(31)$ Hz	$Y_{21}=14.33\,(12)$ kHz	81NAI3	MW			
	GaBr				81NAI3	MW	$a_0=214695\,(83)$ cm^{-1} $a_1=-3.12534\,(32)$ $a_2=6.905\,(70)$ $a_3=-13.44\,(52)$	2.3524907 (82)	
52	$^{69}Ga^{127}I$	$Y_{01}=1706.89645\,(83)$	$Y_{02}=-0.472713\,(60)$	$Y_{11}=-5.68714\,(53)$	80NAI	MW	$Y_{10}=216.38$ cm^{-1}		2.7.1
			$Y_{12}=0.472\,(38)$ Hz	$Y_{21}=6.329\,(43)$ kHz	80NAI	MW	$Y_{20}=-0.471$ cm^{-1}		2.9.1
			$Y_{03}^{der}=-1.446\,(20)\cdot10^{-5}$ Hz		80NAI	MW			
Cont.									

[1]) Or related terms.

Nr.	Molecule	B [1]) MHz	D [1]) kHz	α, γ [1]) MHz	Ref.	Meth.	Derived constants	r_e [1]) Å	See Tab.
52 Cont.	$^{71}Ga^{127}I$	$Y_{01}=1675.72004\,(71)$	$Y_{02}=-0.455700\,(51)$	$Y_{11}=-5.53277\,(57)$	80NAI	MW			
			$Y_{12}=\ 0.522\,(40)$ Hz	$Y_{21}=\ 5.995\,(34)$ kHz	80NAI	MW			
			$Y_{03}^{\rm der}=-1.369\,(15)\cdot 10^{-5}$ Hz		80NAI	MW			
	GaI				80NAI	MW	$a_0=205565\,(35)\ \rm cm^{-1}$ $a_1=\ -3.11045\,(120)$ $a_2=\ 6.602\,(38)$ $a_3=-11.81\,(32)$	2.574667 (12)	
53	$^{115}In^{79}Br$	$Y_{01}=1670.1436\,(52)$	$Y_{02}^{\rm ev}=-0.419$	$Y_{11}=-5.7040\,(47)$	76TIE2	MW		2.5432103 (97)	2.7.1
54 *6-49*	$^{115}In^{127}I$	$Y_{01}=1105.24569\,(11)$	$Y_{02}=-0.191627\,(23)$	$Y_{11}=-3.121148\,(11)$	77ARN	MW			
			$Y_{12}=\ 0.0427\,(38)$ Hz	$Y_{21}=\ 2.5323\,(34)$ kHz	77ARN	MW	$Y_{10}=177.080\,(11)\ \rm cm^{-1}$		
			$Y_{03}^{\rm der}=-8.66\,(3)\cdot 10^{-6}$ Hz	$Y_{31}=-1.00\,(16)$ Hz	77ARN	MW	$Y_{20}=-0.34293\,(44)\ \rm cm^{-1}$		
	$^{113}In^{127}I$	$Y_{01}=1115.51948\,(11)$			77ARN	MW			
	InI				77ARN	MW	$a_0=2.12637\,(26)\cdot 10^5\ \rm cm^{-1}$ $a_1=\ -3.26066\,(15)$ $a_2=\ 7.0886\,(74)$ $a_3=-12.506\,(61)$	2.753642	
55 *6-50*	$^{205}Tl^{19}F$	$Y_{01}=6689.87302\,(31)$	$Y_{02}=-5.8580\,(22)$	$Y_{11}=-45.08276\,(60)$	77ARN	MW	$Y_{10}=476.963\,(91)\ \rm cm^{-1}$		
			$Y_{12}=-2.5\,(14)$ Hz	$Y_{21}=\ 93.52\,(31)$ kHz	77ARN	MW	$Y_{20}=-2.276\,(15)\ \rm cm^{-1}$		
			$Y_{03}^{\rm der}=-2.05449\,(83)\cdot 10^{-3}$ Hz	$Y_{31}=\ 120\,(46)$ Hz	77ARN	MW	$d_{\rm Tl}=9.6\,(17)$		
	$^{203}Tl^{19}F$	$Y_{01}=6695.46707\,(33)$			77ARN	MW			
	TlF				77ARN	MW	$a_0=2.54836\,(97)\cdot 10^5\ \rm cm^{-1}$ $a_1=\ -3.40052\,(47)$ $a_2=\ 7.656\,(42)$ $a_3=-13.53\,(35)$	2.084429	
Cont.									

[1]) Or related terms.

Nr.	Molecule	B [1]) MHz	D [1]) kHz	α, γ [1]) MHz	Ref.	Meth.	Derived constants	r_e [1]) Å	See Tab.
55 Cont.	$^{205}Tl^{19}F$				80HIN1	MB	Search for electric dipole hyperfine structure, negative result		
					80HIN2	MB			
56 *6-51*	$^{205}Tl^{35}Cl$	$Y_{01}=2740.0135(19)$	$Y_{02}=-1.1286(24)$	$Y_{11}=-11.9251(22)$	77ARN	MW	$d_{Tl}=10.36(54)$		
			$Y_{12}=-3.2(28)$ Hz	$Y_{21}=11.65(17)$ kHz	77ARN	MW	$d_{Cl}=0.609(17)$		
			$Y_{03}^{der}=-0.12107(77)\cdot10^{-3}$ Hz	$Y_{31}=36.6(44)$ Hz	77ARN	MW			
	$^{205}Tl^{37}Cl$	$Y_{01}=2613.5627(18)$			77ARN	MW			
	$^{203}Tl^{35}Cl$	$Y_{01}=2743.9510(18)$			77ARN	MW			
	$^{203}Tl^{37}Cl$	$Y_{01}=2617.5001(18)$			77ARN	MW			
	TlCl				77ARN	MW	$a_0=2.2134(33)\cdot10^5$ cm^{-1} $a_1=-3.2585(12)$ $a_2=7.14(33)$ $a_3=-13.3(26)$	2.484815	
57 *6-52*	$^{205}Tl^{79}Br$	$Y_{01}=1293.88633(4)$	$Y_{02}=-0.256871(15)$	$Y_{11}=-3.932042(30)$	77ARN	MW	$d_{Tl}=8.23(38)$ $d_{Br}=0.737(58)$		
			$Y_{12}=0.0060(48)$ Hz	$Y_{21}=2.6780(90)$ kHz	77ARN	MW	$Y_{10}=193.7296(55)$ cm^{-1}		
			$Y_{03}^{der}=0.013950(2)\cdot10^{-3}$ Hz	$Y_{31}=3.07(77)$ Hz	77ARN	MW	$Y_{20}=-0.40921(45)$ cm^{-1}		
			$Y_{13}^{der}=0.076(16)\cdot10^{-6}$ Hz		77ARN	MW			
	$^{205}Tl^{81}Br$	$Y_{01}=1270.81958(4)$			77ARN	MW			
	$^{203}Tl^{79}Br$	$Y_{01}=1297.43362(4)$			77ARN	MW			
	$^{203}Tl^{81}Br$	$Y_{01}=1274.36689(4)$			77ARN	MW			
	TlBr				77ARN	MW	$a_0=2.17399(12)\cdot10^5$ cm^{-1} $a_1=-3.27348(7)$ $a_2=7.0736(67)$ $a_3=-12.480(54)$	2.618166	

[1]) Or related terms.

Nr.	Molecule	B [1]) MHz	D [1]) kHz	α, γ [1]) MHz	Ref.	Meth.	Derived constants	r_e [1]) Å	See Tab.
58	$^{205}Tl^{127}I$	$Y_{01}=814.46788\,(4)$	$Y_{02}=-0.105878\,(6)$	$Y_{11}=-1.995220\,(14)$	77ARN	MW	$d_{Tl}=7.79\,(92)$		
6–53			$Y_{12}=0.0013\,(11)$ Hz	$Y_{21}=0.9384\,(24)$ kHz	77ARN	MW	$Y_{10}=150.7015\,(41)$ cm^{-1}		
			$Y_{03}^{der}=-3.64\,(1)\cdot 10^{-6}$ Hz	$Y_{31}=1.33\,(14)$ Hz	77ARN	MW	$Y_{20}=-0.25663\,(20)$ cm^{-1}		
			$Y_{13}^{der}=-0.0254\,(24)\cdot 10^{-6}$ Hz		77ARN	MW			
	$^{203}Tl^{127}I$	$Y_{01}=817.53952\,(4)$			77ARN	MW			
	TlI				77ARN	MW	$a_0=2.08988\,(12)\cdot 10^5$ cm^{-1}	2.813658	
							$a_1=-3.26485\,(4)$		
							$a_2=7.0261\,(46)$		
							$a_3=-12.503\,(37)$		
59	$^{12}C^{16}O$ [2]) [3])	$Y_{01}=57898.3448\,(69)$	$Y_{02}=-183.502\,(21)$	$Y_{11}=-524.7657\,(42)$	79DAL [4]) [5])	IR/MW	$Y_{10}=2169.81386\,(11)$ cm^{-1}		2.6.1
6–54			$Y_{12}=32.2\,(15)$ Hz	$Y_{21}=16.6\,(12)$ kHz	79DAL [4]) [5])	IR/MW	$Y_{20}=-13.288392\,(63)$ cm^{-1}		2.9.1
			$Y_{22}=5.45\,(18)$ Hz	$Y_{31}=818\,(57)$ Hz	79DAL [4]) [5])	IR/MW	$Y_{30}=0.0105185\,(97)$ cm^{-1}		4.1.6
			$Y_{03}=0.1712\,(54)$ Hz		79DAL [4]) [5])	IR/MW	$Y_{40}=5.712\,(67)\cdot 10^{-5}$ cm^{-1}		
			$Y_{13}=-4.90\,(19)\cdot 10^{-3}$ Hz		79DAL [4]) [5])	IR/MW	$Y_{50}=9.89\,(21)\cdot 10^{-7}$ cm^{-1}		
					79DAL [4]) [5])	IR/MW	$Y_{60}=-3.168\,(24)\cdot 10^{-8}$ cm^{-1}		
		$B_e^{cor}=57907.98\,(6)$			73WAT	MW	$d_C=1.022\,(6)$	$r_e^{cor}=1.1282427\,(6)$	
							$d_O=1.058\,(12)$		

[1]) Or related terms.
[2]) Measurements on $^{12}C^{16}O$ also in: [75MAN4, 79GUE, 74ROH, 74LOV1, 74KIL].
[3]) Measurements on other CO-isotopes in: [74ROS, 79GUE, 74LOV1, 79DAL].
[4]) Combined fit of all existing data, very extensive study of higher order mass corrections.
[5]) Comparable data set in [74KIL].

Nr.	Molecule	B [1]) MHz	D [1]) kHz	α, γ [1]) MHz	Ref.	Meth.	Derived constants	r_e [1]) Å	See Tab.
60 6–55	$^{12}C^{32}S$ [2]) [3])	$Y_{01}=24584.303\,(16)$	$Y_{02}=-40.24\,(11)$ [79TOD]	$Y_{11}=-177.452\,(10)$	79BUS/ 74LOV1	MW			2.9.1
				$Y_{21}=\ -21.3\,(17)$ kHz	79BUS/ 74LOV1	MW			
				$Y_{31}=\ -1.298\,(80)$ kHz	79BUS/ 74LOV1	MW			
	$^{12}C^{34}S$ [2]) [3])	$Y_{01}=24190.125\,(22)$	$Y_{02}=-38.91\,(29)$ [79TOD]	$Y_{11}=-173.158\,(35)$	79BUS/ 74LOV1	MW			
				$Y_{21}=\ -31\,(14)$ kHz	79BUS/ 74LOV1	MW			
				$Y_{31}=\ -0.5\,(16)$ kHz	79BUS/ 74LOV1	MW			
	$^{13}C^{32}S$ [3])	$Y_{01}=23205.234\,(31)$	$Y_{02}=-36.54\,(20)$ [79TOD]	$Y_{11}=-162.729\,(16)$	79BUS/ 74LOV1	MW			
	$^{12}C^{32}S$ [3])				79TOD	IR	$Y_{10}=1285.1537\,(31)$ [5]) cm^{-1} $Y_{20}=\ -6.5021\,(16)$ cm^{-1}		
					79TOD	IR	$Y_{30}=\ 3.77\,(21)\cdot 10^{-3}\ cm^{-1}$		
					79TOD	IR	$a_0=503515.1\,(24)$ cm^{-1} $a_1=-2.88583\,(54)$ $a_2=\ 5.1241\,(42)$ $a_3=-7.086\,(68)$ $a_4=\ 8.1\,(13)$	1.535152 (12) [4])	
61	$^{12}C^{80}Se$	$B_0=17182.77\,(2)$	$Y_{02}^{ev}=-21.25\,(5)$		74LOV2/ 73McG1	MW		$r_e^{cor}=1.676086(64)$	2.6.1 2.9.1
Cont.	$^{12}C^{78}Se$	$B_0=17240.20\,(2)$	$Y_{02}^{ev}=-21.39\,(5)$		74LOV2/ 73McG1	MW			

[1]) Or related terms.
[2]) Measurements on these isotope-combinations also in [77TOD].
[3]) Measurements on these combinations and on $^{12}C^{33}S$ in [79TOD, 74LOV1, 79YAM].
[4]) Value calculated using nuclear masses.
[5]) Calibration accuracy 0.002 cm^{-1}.

Nr.	Molecule	B [1]) MHz	D [1]) kHz	α, γ [1]) MHz	Ref.	Meth.	Derived constants	r_e [1]) Å	See Tab.
61 Cont.	$^{12}C^{77}Se$	$B_0 = 17270.00\,(2)$	$Y_{02}^{ev} = -21.47\,(5)$		74LOV2/ 73McG1	MW			
	$^{12}C^{76}Se$	$B_0 = 17300.62\,(3)$	$Y_{02}^{ev} = -21.54\,(3)$		74LOV2/ 73McG1	MW			
	$^{12}C^{74}Se$	$B_0 = 17364.35\,(2)$	$Y_{02}^{ev} = -21.70\,(5)$		74LOV2/ 73McG1	MW			
	$^{12}C^{82}Se$	$B_0 = 17128.09\,(2)$	$Y_{02}^{ev} = -21.11\,(3)$		74LOV2/ 73McG1	MW			
	$^{13}C^{80}Se$	$B_0 = 16032.00\,(6)$	$Y_{02}^{ev} = -18.49\,(3)$		74LOV2/ 73McG1	MW			
62 6-56	$^{28}Si^{16}O$ [2])	$Y_{01} = 21787.4809\,(43)$	$Y_{02} = -29.749\,(58)$	$Y_{11} = -151.0322\,(49)$	81LOV/ 77MAN	IR/MW	Y_{10}[3]) $= 1241.54304\,(51)\ cm^{-1}$		2.9.1
			$Y_{12} = -0.0219\,(36)$	$Y_{21} = 71.6\,(11)$ kHz	81LOV/ 77MAN	IR/MW	$Y_{20} = -5.97391\,(26)\ cm^{-1}$		
			$Y_{03}^{der} = 1.08 \cdot 10^{-3}$ Hz		81LOV/ 77MAN	IR/MW	$Y_{30} = 0.005931\,(33)\ cm^{-1}$		
	$^{29}Si^{16}O$ [2])	$Y_{01} = 21514.046\,(13)$	$Y_{02} = -28.64\,(13)$	$Y_{11} = -148.193\,(11)$	77MAN/ 68TÖR	MW			
				$Y_{21} = 68.7\,(24)$ kHz	77MAN/ 68TÖR	MW			
	$^{39}Si^{16}O$ [2])	$Y_{01} = 21259.494\,(13)$	$Y_{02} = -27.97\,(13)$	$Y_{11} = -145.570\,(11)$	77MAN/ 68TÖR	MW			
				$Y_{21} = 67.1\,(24)$ kHz	77MAN/ 68TÖR	MW			
	SiO				77MAN	MW	$a_0 = 5.390\,(22) \cdot 10^5\ cm^{-1}$; $a_1 = -2.9899\,(41)$; $a_2 = 5.7\,(7)$; $a_3 = -9.0\,(5)$	1.509755 (40)	

[1]) Or related terms.
[2]) Calculations on this isotope also in: 74LOV1.
[3]) Systematic calibration uncertainty is $\pm 0.002\ cm^{-1}$.

Nr.	Molecule	B [1]) MHz	D [1]) kHz	α, γ [1]) MHz	Ref.	Meth.	Derived constants	r_e [1]) Å	See Tab.
63 *6–57*	$^{28}Si^{32}S$	$B_e^{cor} = 9100.089\ (20)$			76TIE4/ 76STI1	MW	$d_{Si} = 0.713\ (30)$ $d_S = 0.953\ (38)$	$r_e^{cor} = 1.929286(15)$	2.9.1
64 *6–59*	$^{74}Ge^{16}O$	$Y_{01} = 14560.872\ (16)$		$Y_{11} = -92.306\ (27)$ $Y_{21} = 41.8\ (87)$ kHz	73HON2 73HON2	MW MW			2.6.1 2.9.1
65 *6–60*	$^{74}Ge^{32}S$	$Y_{01} = 5593.10069\ (38)$ $B_e^{cor} = 5593.343\ (10)$	$Y_{02} = -2.3632\ (26)$ $Y_{12} = -3.0\ (11)$ Hz	$Y_{11} = -22.45753\ (44)$ $Y_{21} = -1.31\ (19)$ kHz $Y_{31} = -85\ (24)$ Hz	76STI2 76STI2 76STI2	MW MW MW	$d_{Ge} = 0.736\ (18)$ $d_S = 0.947\ (10)$	$r_e^{cor} = 2.012067(10)$	2.9.1
66 *6–61*	$^{74}Ge^{80}Se$	$Y_{01} = 2888.21579\ (17)$ $B_e^{cor} = 2888.2886\ (30)$	$Y_{02} = -0.66166\ (34)$ $Y_{12} = -0.57\ (20)$ Hz	$Y_{11} = -8.66519\ (18)$ $Y_{21} = -1.028\ (91)$ kHz $Y_{31} = -34\ (12)$ Hz	76STI2 76STI2 76STI2	MW MW MW	$d_{Ge} = 0.788\ (23)$ $d_{Se} = 0.966\ (31)$	$r_e^{cor} = 2.134628(10)$	2.9.1
67 *6–62*	GeTe								2.9.1
68 *6–63*	SnO								2.9.1
69 *6–64*	$^{120}Sn^{32}S$	$Y_{01} = 4103.00274\ (29)$ $B_e^{cor} = 4103.1658\ (60)$	$Y_{02} = -1.2974\ (13)$ $Y_{12} = -2.15\ (55)$ Hz	$Y_{11} = -15.15868\ (35)$ $Y_{21} = -5.80\ (14)$ kHz $Y_{31} = -89\ (15)$ Hz	76STI1 76STI1 76STI1	MW MW MW	$d_{Sn} = 0.855\ (24)$ $d_S = 0.921\ (7)$	$r_e^{cor} = 2.209008(10)$	2.9.1
70 *6–65*	$^{120}Sn^{80}Se$	$Y_{01} = 1948.58813\ (14)$ $B_e^{cor} = 1948.6299\ (10)$	$Y_{02} = -0.304572\ (99)$ $Y_{12} = -0.389\ (23)$ Hz	$Y_{11} = -5.125841\ (62)$ $Y_{21} = -1.691\ (15)$ kHz $Y_{31} = -10.5\ (13)$ Hz	76STI1 76STI1 76STI1 76STI1	MW MW MW MW	$d_{Sn} = 0.802\ (42)$ $d_{Se} = 1.028\ (24)$	$r_e^{cor} = 2.325601\ (10)$	2.9.1

[1]) Or related terms.

Nr.	Molecule	B [1]) MHz	D [1]) kHz	α, γ [1]) MHz	Ref.	Meth.	Derived constants	r_e [1]) Å	See Tab.
71 *6-66*	$^{120}Sn^{130}Te$	Y_{01} = 1273.49067 (10) B_e^{cor} = 1273.5104 (10)	Y_{02} = −0.138178 (27) Y_{12} = −0.183 (13) Hz Y_{03} = −4.1 · 10^{-6} Hz	Y_{11} = −2.861490 (73) Y_{21} = −1.233 (23) kHz Y_{31} = −10.2 (25) Hz	76STI1 76STI1 76STI1 76STI1	MW MW MW MW	d_{Sn} = 0.934 (45) d_{Te} = 0.776 (53)	r_e^{cor} = 2.522827 (10)	2.9.1
72 *6-67*	PbO								2.9.1
73 *6-68*	$^{208}Pb^{32}S$	Y_{01} = 3487.14630 (36) B_e^{cor} = 3487.387 (12)	Y_{02} = −1.02393 (93) Y_{12} = −3.97 (39) Hz	Y_{11} = −13.04377 (39) Y_{21} = −15.93 (17) kHz Y_{31} = −149 (20) Hz	75TIE1/ 76STI1 75TIE1/ 76STI1 75TIE/ 76STI1 75TIE1/ 76STI1	MW MW MW MW	d_S = 1.009 (30) d_{Pb} = 6.03 (77)	r_e^{cor} = 2.286811 (10)	2.9.1
74 *6-69*	$^{208}Pb^{80}Se$	Y_{01} = 1516.93622 (21) B_e^{cor} = 1517.0048 (40)	Y_{02} = −0.200497 (88) Y_{12} = −0.409 (22) Hz Y_{03}^{der} = −9.36 (2) · 10^{-6} Hz	Y_{11} = −3.893574 (74) Y_{21} = −3.405 (12) kHz Y_{31} = −21.72 (81) Hz	76STI1 76STI1 76STI1 76STI1	MW MW MW MW	d_{Pb} = 6.07 (42) d_{Se} = 1.007 (38)	r_e^{cor} = 2.402207 (10)	2.9.1
75 *6-70*	$^{208}Pb^{130}Te$	Y_{01} = 938.58181 (6) B_e^{cor} = 938.6174 (30)	Y_{02} = −0.081540 (11) Y_{12} = −0.1448 (75) Hz Y_{03}^{der} = −3.083 (1) · 10^{-6} Hz Y_{13}^{der} = −75 (12) · 10^{-9} Hz	Y_{11} = −2.021054 (41) Y_{21} = −1.7673 (58) kHz Y_{31} = −8.95 (34) Hz	76STI1 76STI1 76STI1 76STI1 76STI1	MW MW MW MW MW	d_{Pb} = 5.69 (54) d_{Te} = 0.90 (7)	r_e^{cor} = 2.594958 (10)	2.9.1
76 *6-71*	N_2								2.7.1 4.1.3

[1]) Or related terms.

Nr.	Molecule	B [1]) MHz	D [1]) kHz	α, γ [1]) MHz	Ref.	Meth.	Derived constants	r_e [1]) Å	See Tab.
77 *6-71a*	$^{31}P^{14}N$	$Y_{01} = 23578.2459(61)$	$Y_{02} = -32.717(32)$	$Y_{11} = -165.975(13)$	81MAK	IR/MW	Y_{10}[2]) $= 1336.99230(66)$ cm^{-1}		
			$Y_{12} = -88.4(41)$ Hz	$Y_{21} = -190.6(67)$ kHz	81MAK	IR/MW	$Y_{20} = -6.91192(40)$ cm^{-1}		
			$Y_{22} = -5.4(10)$ Hz	$Y_{31} = -7.85(99)$ kHz	81MAK	IR/MW	$Y_{30} = -0.004238(62)$ cm^{-1}		
			$Y_{03}^{der} = 0.25 \cdot 10^{-3}$ Hz		81MAK	IR/MW			
78 *6-72*	CF								2.6.1 4.1.5
79	CCl								4.1.5
80 *6-73*	NO								2.6.1 2.7.1 2.9.1 4.1.2 4.1.5
81 *6-74*	NS								4.1.5
82 *6-74a*	NF								2.6.1 2.7.1 4.1.7
83	BiF								2.7.1 4.1.3
84	BiCl								2.7.1 4.1.3
85	BiBr								4.1.3
86	BiI								2.7.1 4.1.3

[1]) Or related terms.
[2]) Systematic calibration uncertainty is ± 0.002 cm^{-1}.

Nr.	Molecule	B [1]) MHz	D [1]) kHz	α, γ [1]) MHz	Ref.	Meth.	Derived constants	r_e [1]) Å	See Tab.
87 *6-75*	O_2								4.1.3
88 *6-76*	SO								2.7.1 4.1.3 4.1.7
89 *6-77*	SeO								4.1.3
90 *6-75a*	S_2								4.1.3
91	OF								4.1.5
92 *6-78*	OCl								2.7.1 4.1.5
93 *6-79*	OBr								2.7.1 4.1.5
94 *6-80*	OI								2.7.1 4.1.5
95 *6-81*	SF								2.6.1 4.1.5
96 *6-84*	$^{35}Cl^{19}F$	Y_{01} = 15483.2702 (36)	Y_{02} = −26.91 (2)	Y_{11} = −129.6497 (50)	79WIL/ 80WIL	MW	ω_e = 783.553 (30) cm^{-1}		
			Y_{12} = −140 (20) Hz	Y_{21} = −520.3 (17) kHz	79WIL/ 80WIL	MW	$\omega_e x_e$ = 5.045 (87) cm^{-1}		
		B_e^{cor} = 15484.4304 (70)	Y_{03}^{der} = −5.47 (3) $\cdot 10^{-3}$ Hz		79WIL/ 80WIL	MW			
Cont.					73McG3	MW			

[1]) Or related terms.

Nr.	Molecule	B [1]) MHz	D [1]) kHz	α, γ [1]) MHz	Ref.	Meth.	Derived constants	r_e [1]) Å	See Tab.
96 *6–84* Cont.	$^{37}Cl^{19}F$	Y_{01} = 15188.8247 (30)	Y_{02} = −25.90 (2)	Y_{11} = −125.9687 (49)	79WIL/ 80WIL	MW	ω_e = 776.067 (30) cm^{-1}		
			Y_{12} = −130 (20) Hz	Y_{21} = −500.7 (16) kHz	79WIL/ 80WIL	MW	$\omega_e x_e$ = 5.026 (84) cm^{-1}		
			Y_{03}^{der} = −5.16 (3) $\cdot 10^{-3}$ Hz		79WIL/ 80WIL	MW			
					73McG3	MW			
	ClF				79WIL/ 80WIL	MW	a_0 = 2.9712 (25) $\cdot 10^5$ cm^{-1} a_1 = −3.1171 (9) a_2 = 5.63 (11) a_3 = −8.8 (8)	r_e^{cor} = 1.628341 (4)	
97 *6–84a*	$^{79}Br^{19}F$	Y_{01} = 10667.4502 (35)	Y_{02} = −12.049 (10)	Y_{11} = −77.8629 (79)	79WIL/ 80WIL	MW	ω_e = 669.679 (28) cm^{-1}		2.6.1 2.7.1 2.9.1
		B_e^{cor} = 10668.0970 (69)	Y_{12} = −66 (11) Hz	Y_{21} = −201.6 (45) kHz	79WIL/ 80WIL	MW	$\omega_e x_e$ = 3.863 (76) cm^{-1}		
			Y_{03}^{der} = −3.937 (7) $\cdot 10^{-3}$ Hz	Y_{31} = −6.2 (8) kHz	79WIL/ 80WIL	MW			
					79NAI2	MW			
	$^{81}Br^{19}F$	Y_{01} = 10616.3476 (36)	Y_{02} = −11.934 (10)	Y_{11} = −77.3040 (78)	79WIL/ 80WIL	MW	ω_e = 668.67 (28) cm^{-1}		
			Y_{12} = −66 (11) Hz	Y_{21} = −199.7 (45) kHz	79WIL/ 80WIL	MW	$\omega_e x_e$ = 3.844 (78) cm^{-1}		
			Y_{03}^{der} = −3.881 (7) $\cdot 10^{-3}$ Hz	Y_{31} = −6.1 (8) kHz	79WIL/ 80WIL	MW			
					79NAI2	MW			
	BrF				79WIL/ 80WIL	MW	a_0 = 3.1503 (26) $\cdot 10^5$ cm^{-1} a_1 = −3.2893 (10) a_2 = 6.29 (15) a_3 = −9.5 (11)	r_e^{cor} = 1.758987 (4)	

[1]) Or related terms.

Nr.	Molecule	B [1]) MHz	D [1]) kHz	α, γ [1]) MHz	Ref.	Meth.	Derived constants	r_e [1]) Å	See Tab.
98 *6–84b*	$^{127}I^{19}F$	$Y_{01} = 8385.520\,(30)$	$Y_{02}^{ev} = -7.095$	$Y_{11} = -56.163\,(50)$	73TIE2	MW			2.6.1 2.9.1
				$Y_{21} = -80\,(20)$ kHz	73TIE2	MW			
					73McG2	MW			
99	$^{79}Br^{35}Cl$	$Y_{01} = 4570.9687\,(13)$	$Y_{02} = -2.1527\,(7)$	$Y_{11} = -23.1426\,(29)$	79WIL/ 80WIL	MW	$\omega_e = 444.322\,(62)$ cm^{-1}		2.6.1 2.7.1 2.9.1
		$B_e^{cor} = 4571.0295\,(26)$	$Y_{12} = -10.7\,(8)$ Hz	$Y_{21} = -61.9\,(18)$ kHz	79WIL/ 80WIL	MW	$\omega_e x_e = 1.856\,(20)$ cm^{-1}		
			$Y_{03}^{der} = -0.4656\,(3) \cdot 10^{-3}$ Hz	$Y_{31} = -0.9\,(3)$ kHz	79WIL/ 80WIL	MW			
					76TIE5	MW			
	$^{81}Br^{35}Cl$	$Y_{01} = 4536.3140\,(13)$	$Y_{02} = -2.1202\,(7)$	$Y_{11} = -22.8800\,(29)$	79WIL/ 80WIL	MW	$\omega_e = 442.634\,(62)$ cm^{-1}		
			$Y_{12} = -10.5\,(8)$ Hz	$Y_{21} = -61.0\,(18)$ kHz	79WIL/ 80WIL	MW	$\omega_e x_e = 1.849\,(20)$ cm^{-1}		
			$Y_{03}^{der} = -0.4551\,(3) \cdot 10^{-3}$ Hz	$Y_{31} = -0.9\,(3)$ kHz	79WIL/ 80WIL	MW			
					76TIE5	MW			
	$^{79}Br^{37}Cl$	$Y_{01} = 4399.8516\,(13)$	$Y_{02} = -1.9945\,(7)$	$Y_{11} = -21.8553\,(27)$	79WIL/ 80WIL	MW	$\omega_e = 435.926\,(61)$ cm^{-1}		
			$Y_{12} = -9.7\,(8)$ Hz	$Y_{21} = -57.4\,(17)$ kHz	79WIL/ 80WIL	MW	$\omega_e x_e = 1.787\,(19)$ cm^{-1}		
			$Y_{03}^{der} = -0.4152\,(3) \cdot 10^{-3}$ Hz	$Y_{31} = -0.8\,(3)$ kHz	79WIL/ 80WIL	MW			
					76TIE5	MW			
	$^{81}Br^{37}Cl$	$Y_{01} = 4365.1979\,(12)$	$Y_{02} = -1.9632\,(7)$	$Y_{11} = -21.5976\,(27)$	79WIL/ 80WIL	MW	$\omega_e = 434.206\,(61)$ cm^{-1}		
			$Y_{12} = -9.5\,(8)$ Hz	$Y_{21} = -56.5\,(16)$	79WIL/ 80WIL	MW	$\omega_e x_e = 1.771\,(2)$ cm^{-1}		
			$Y_{03}^{der} = -0.4055\,(3) \cdot 10^{-3}$ Hz	$Y_{31} = -0.8\,(3)$ kHz	79WIL/ 80WIL	MW			
Cont.						MW			

[1]) Or related terms.

Nr.	Molecule	B [1]) MHz	D [1]) kHz	α, γ [1]) MHz	Ref.	Meth.	Derived constants	r_e [1]) Å	See Tab.
99 Cont.	BrCl				79WIL/ 80WIL	MW	$a_0 = 3.2369\,(9)$ $\cdot 10^5\ cm^{-1}$ $a_1 = -3.4590\,(5)$ $a_2 = 6.85\,(8)$ $a_3 = -11.2\,(7)$	$r_e^{cor} = 2.136124\,(5)$	
100 6–85	$^{127}I^{35}Cl$	$Y_{01} = 3422.3638\,(20)$	$Y_{02} = -1.2081\,(6)$	$Y_{11} = -15.9751\,(43)$	79WIL/ 80WIL	MW	$\omega_e = 384.290\,(96)\ cm^{-1}$		2.6.1 2.7.1 4.1.6
		$B_e^{cor} = 3422.4275\,(38)$	$Y_{12} = -5.9\,(6)$ Hz	$Y_{21} = -35.2\,(26)$ kHz	79WIL/ 80WIL	MW	$\omega_e x_e = 1.501\,(26)\ cm^{-1}$		
			$Y_{03}^{der} = -0.2639\,(3)$ $\cdot 10^{-3}$ Hz	$Y_{31} = -0.6\,(4)$ kHz	79WIL/ 80WIL	MW			
	$^{127}I^{37}Cl$	$Y_{01} = 3277.4173\,(19)$	$Y_{02} = -1.108\,(6)$	$Y_{11} = -14.9710\,(40)$	79WIL/ 80WIL	MW	$\omega_e = 376.064\,(94)\ cm^{-1}$		
			$Y_{12} = -5.5\,(6)$ Hz	$Y_{21} = -32.3\,(24)$ kHz	79WIL/ 80WIL	MW	$\omega_e x_e = 1.437\,(25)\ cm^{-1}$		
			$Y_{03}^{der} = -0.2318\,(3)$ $\cdot 10^{-3}$ Hz	$Y_{31} = -0.5\,(3)$ kHz	79WIL/ 80WIL	MW			
	ICl				79WIL/ 80WIL	MW	$a_0 = 3.2338\,(16)$ $\cdot 10^5\ cm^{-1}$ $a_1 = -3.6188\,(5)$ $a_2 = 7.60\,(15)$ $a_3 = -13.2\,(13)$	$r_e^{cor} = 2.320962\,(6)$	
101	Br_2								2.7.1 2.9.1 4.1.6
102	$^{127}I^{79}Br$	$Y_{01} = 1703.79826\,(30)$	$Y_{02} = -0.30491\,(2)$	$Y_{11} = -5.90413\,(41)$	79WIL/ 80WIL	MW	$\omega_e = 268.69\,(9)\ cm^{-1}$		2.6.1 2.7.1
		$B_e^{cor} = 1703.8172\,(6)$	$Y_{12} = -1.50\,(2)$ Hz	$Y_{21} = -13.66\,(15)$ kHz	79WIL/ 80WIL	MW	$\omega_e x_e = 0.826\,(6)\ cm^{-1}$		
			$Y_{03}^{der} = -39.868\,(11)$ $\cdot 10^{-6}$ Hz	$Y_{31} = -0.14\,(2)$ kHz	79WIL/ 80WIL	MW			
Cont.					75TIE2	MW			

[1]) Or related terms.

Nr.	Molecule	B [1]) MHz	D [1]) kHz	α, γ [1]) MHz	Ref.	Meth.	Derived constants	r_e [1]) Å	See Tab.
102 Cont.	$^{127}I^{81}Br$	$Y_{01} = 1677.85957\,(30)$	$Y_{02} = -0.29570\,(2)$	$Y_{11} = -5.76982\,(40)$	79WIL/ 80WIL	MW	$\omega_e = 266.64\,(9)\ cm^{-1}$		
			$Y_{12} = -1.44\,(2)$ Hz	$Y_{21} = -13.25\,(15)$ kHz	79WIL/ 80WIL	MW	$\omega_e x_e = 0.814\,(6)\ cm^{-1}$		
			$Y_{03}^{der} = -38.075\,(11) \cdot 10^{-6}$ Hz	$Y_{31} = -0.13\,(2)$ kHz	79WIL/ 80WIL	MW			
					75TIE2	MW			
	IBr				79WIL/ 80WIL	MW	$a_0 = 3.17558\,(26) \cdot 10^5\ cm^{-1}$ $a_1 = -3.7305\,(2)$ $a_2 = 7.70\,(3)$ $a_3 = -12.6\,(3)$	$r_e^{cor} = 2.469078\,(6)$	
103 6–86	I_2								2.6.1 2.7.1 2.9.1 4.1.6

[1]) Or related terms.

2.2.3 References for 2.2, 2.6.1, 2.7.1, 2.9.1

67AMA	Amano, E., Hirota, Y., Morino: J. Phys. Soc. Japan **22** (1967) 399.
67WEB	Webb, D.U.: Thesis, Ph.D.; Ohio State University, **1967**
68TÖR	Törring, T.: Z. Naturforsch. **23a** (1968) 777.
69MEC	Melendres, C.A., Hebert, A.J., Street, Jr., K.: J. Chem. Phys. **51** (1969) 855.
69PEA	Pearson, E.F., Gordy, W.: Phys. Rev. **177** (1969) 52.
71LUC	Lucia, F.C. de, Helminger, P., Gordy, W.: Phys. Rev. **A3** (1971) 1849.
72DAG	Dagdigian, P.J., Wharton, L.: J. Chem. Phys. **57** (1972) 1487.
73AMO	Amano, T., Hirota, E.: J. Mol. Spectrosc. **45** (1973) 417.
73CEC	Cecchi, J.L., Ramsey, N.F.: J. Chem. Phys. **60** (1973) 53.
73COO	Cook, T.J., Levy, D.H.: J. Chem. Phys. **59** (1973) 2387.
73CUP	Cupp, R.E., Smith, W.T., Contini, D.A., Woods, D., Gallagher, J.J.: Phys. Lett. **44A** (1973) 305.
73CUR	Curran, A.H., Macdonald, R.G., Stone, A.J., Thrush, B.A.: Proc. Roy. Soc. London **A332** (1973) 355.
73DAB	Dabbousi, O.B., Meerts, W.L., de Leeuw, F.H., Dymanus, A.: Chem. Phys. **2** (1973) 473.
73DEL	De Leeuw, F.H., Dymanus, A.: J. Mol. Spectrosc. **48** (1973) 427.
73DES	De Santis, D., Lurio, A., Miller, T.A., Freund, R.S.: J. Chem. Phys. **58** (1973) 4625.
73FIE	Field, R.W., English, A.D., Tanaka, T., Harris, D.O., Jennings, D.A.: J. Chem. Phys. **59** (1973) 2191.
73HOE	Hoeft, J., Törring, T., Tiemann, E.: Z. Naturforsch. **28A** (1973) 1066.
73HON1	Honerjäger, R., Tischer, R.: Z. Naturforsch. **28A** (1973) 1372.
73HON2	Honerjäger, R., Tischer, R.: Z. Naturforsch **28A** (1973) 1374.
73McG1	McGurk, J.C., Tigelaar, H.L., Rock, S.L., Noris, C.L., Flygare, W.H.: J. Chem. Phys. **58** (1973) 1420.
73McG2	McGurk, J.C., Flygare, W.H.: J. Chem. Phys. **59** (1973) 5742.
73McG3	McGurk, J., Norris, C.L., Tigelaar, H.L., Flygare, W.H.: J. Chem. Phys. **58** (1973) 3118.
73MEE	Meerts, W.L., Dymanus, A.: Chem. Phys. Lett. **23** (1973) 45.
73SAI	Saito, S.: J. Mol. Spectrosc. **48** (1973) 530.
73TIE1	Tiemann, E., El Ali, H., Hoeft, J., Törring, T.: Z. Naturforsch. **28A** (1973) 1058.
73TIE2	Tiemann, E., Hoeft, J., Törring, T.: Z. Naturforsch. **28a** (1973) 1405.
73WAT	Watson, J.K.G.: J. Mol. Spectrosc. **45** (1973) 99.
73WEI	Weinstock, E.M., Zare, R.N.: J. Chem. Phys. **58** (1973) 4319.
74BER	Bergeman, T., Zare, R.N.: J. Chem. Phys. **61** (1974) 4500.
74CUM	Cummings, F.E., Klemperer, W.: J. Chem. Phys. **60** (1974) 2035.
74DAV	Davis, R.E., Muenter, J.S.: J. Chem. Phys. **61** (1974) 2940.
74ENG	Eng, R.S., La Tourrette, J.T.: J. Mol. Spectrosc. **52** (1974) 269.
74FRE	Freeman, R.R., Johnson, D.W., Ramsey, N.F.: J. Chem. Phys. **61** (1974) 3471.
74HON1	Honerjäger, R., Tischer, R.: Z. Naturforsch. **29A** (1974) 1919.
74HON2	Honerjäger, R., Tischer, R.: Z. Naturforsch. **29A** (1974) 819.
74HON3	Honerjäger, R., Tischer, R.: Z. Naturforsch. **29A** (1974) 342.
74HON4	Honerjäger, R., Tischer, R.: Z. Naturforsch **29A** (1974) 1695.
74KIL	Kildal, H., Eng, R.S., Ross, A.H.M.: J. Mol. Spectrosc. **53** (1974) 479.
74LOV1	Lovas, F.J., Krupenie, P.H.: J. Phys. Chem. Ref. Data **3** (1974) 245.
74LOV2	Lovas, F.J., Tiemann, E.: J. Phys. Chem. Ref. Data **3** (1974) 609.
74MEE	Meerts, W.L., Dymanus, A.: Astrophys. J. **187** (1974) L45.
74MEL	Melieres-Marechal, M.-A., Lombardi, M.: J. Chem. Phys. **61** (1974) 2600.
74ROH	Roh, W.B., Rao, K.N.: J. Mol. Spectrosc. **49** (1974) 317.
74ROS	Ross, A.H.M., Eng, R.S., Kildal, H.: Opt. Commun. **12** (1974) 433.
74SCA	Scarl, E.A., Dalby, F.W.: Can. J. Phys. **52** (1974) 1429.
74TIE	Tiemann, E., Bojaschewsky, M., Sauter-Servaes, Ch., Törring, T.: Z. Naturforsch. **29A** (1974) 1692.
74VAN	Van Dijk, F.A., Dymanus, A.: Chem. Phys. **6** (1974) 474.
75BRO	Broyer, M., Lehmann, J.-C., Vigué, J.: J. Phys. (Paris) **36** (1975) 235.
75DEM	Demtröder, W., Stock, M.: J. Mol. Spectrosc. **55** (1975) 476.
75FRE	Freeman, R.R., Jacobson, A.R., Johnson, D.W., Ramsey, N.F.: J. Chem. Phys. **63** (1975) 2597.

75HOY	Hoy, A.R., Johns, J.W.C., McKellar, A.R.W.: Can. J. Phys. **53** (1975) 2029.
75MAN1	Manson, E.L., De Lucia, F.C., Gordy, W.: J. Chem. Phys. **62** (1975) 4796.
75MAN2	Manson, E.L., De Lucia, F.C., Gordy, W.: J. Chem. Phys. **63** (1975) 2724.
75MAN3	Manson, E.L., De Lucia, F.C., Gordy, W.: J. Chem. Phys. **62** (1975) 1040.
75MAN4	Mantz, A.W., Maillard, J.P., Roh, W.B., Rao, K.N.: J. Mol. Spectrosc. **57** (1975) 155.
75MEE	Meerts, W.L., Dymanus, A.: Can. J. Phys. **53** (1975) 2123.
75MUE	Muenter, J.S.: J. Mol. Spectrosc. **55** (1975) 490.
75ROS	Rosner, S.D., Holt, R.A., Gaily, T.D.: Phys. Rev. Lett. **35** (1975) 785.
75TIE1	Tiemann, E., Stieda, W.U., Törring, T., Hoeft, J.: Z. Naturforsch. **30A** (1975) 1606.
75TIE2	Tiemann, E., Möller, T.: Z. Naturforsch. **30A** (1975) 986.
75WIC1	Wicke, B.G., Klemperer, W., Field, R.: J. Chem. Phys. **62** (1975) 3544.
75WIC2	Wicke, B.G., Klemperer, W.: Mol. Phys. **30** (1975) 1021.
75WIC3	Wicke, B.G., Klemperer, W.: J. Chem. Phys. **63** (1975) 3756.
76BER	Bernage, P., Niay, P.: J. Mol. Spectrosc. **63** (1976) 317.
76DEM	Demtröder, W., Stetzenbach, W., Stock, M., Witt, J.: J. Mol. Spectrosc. **61** (1976) 382.
76GER	German, K.R.: J. Chem. Phys. **64** (1976) 4192.
76GUE1	Guelachvili, G.: Opt. Commun. **19** (1976) 150.
76GUE2	Guedard, G., Broyer, M., Vigué, J., Lehmann, J.C.: Chem. Phys. Lett. **43** (1976) 118.
76HEB	Hebert, A.J., Hollowell, C.D.: J. Chem. Phys. **65** (1976) 4327.
76JAC	Jacobson, A.R., Ramsey, N.F.: J. Chem. Phys. **65** (1976) 1211.
76KUI	Kuijpers, P., Törring, T., Dymanus, A.: Chem. Phys. **18** (1976) 401.
76LAN	Landsberg, B.M.: Chem. Phys. Lett. **43** (1976) 102.
76MEE	Meerts, W.L.: Chem. Phys. **14** (1976) 421.
76STI1	Stieda, W.U.: Dissertation Freie Universität Berlin, **1976**.
76STI2	Stieda, W.U., Tiemann, E., Törring, T., Hoeft, J.: Z. Naturforsch. **31A** (1976) 374.
76STO1	Story, T.L., Hebert, A.J.: J. Chem. Phys. **64** (1976) 855.
76STO2	Stocker, R.N., Goldman, A.: J. Quant. Spectros. Radiat. Transfer **16** (1976) 335; **17** (1977) 555.
76TIE1	Tiemann, E., Hoeft, J.: Z. Naturforsch. **31A** (1976) 236.
76TIE2	Tiemann, E., Köhler, U., Hoeft, J.: Z. Naturforsch. **32A** (1976) 6.
76TIE3	Tiemann, E., Ryzlewicz, Ch., Törring, T.: Z. Naturforsch. **31A** (1976) 128.
76TIE4	Tiemann, E.: J. Phys. Chem. Ref. Data **5** (1976) 1147.
76TIE5	Tiemann, E., Wagner, T.: priv. comm., **1976**.
77ARN	Arnst, H.: Dissertation Freie Universität Berlin, **1977**.
77CED	Cederberg, J.: J. Chem. Phys. **66** (1977) 5247.
77CRE	Creswell, R.A., Hocking, W.H., Pearson, E.F.: Chem. Phys. Lett. **48** (1977) 369.
77DIX	Dixon, T.A., Woods, R.C.: J. Chem. Phys. **67** (1977) 3956.
77HEM	Hemminger, J.C., Cavanagh, R., Lisy, J.M., Klemperer, W.: J. Chem. Phys. **67** (1977) 4952.
77HON	Honerjäger, R., Tischer, R.: Z. Naturforsch. **32A** (1977) 1.
77KUI	Kuijpers, P., Dymanus, A.: Chem. Phys. **24** (1977) 97.
77MAN	Manson, E.L., Clark, W.W., De Lucia, F.C., Gordy, W.: Phys. Rev. **A15** (1977) 223.
77MEE	Meerts, W.L., De Leeuw, T.H., Dymanus, A.: Chem. Phys. Lett. **22** (1977) 319.
77NAI	Nair, K.P.R., Tiemann, E., Hoeft, J.: Z. Naturforsch. **32a** (1977) 1053.
77NIA	Niay, P., Coquant, C., Bernage, P., Bocquet, H.: J. Mol. Spectrosc. **65** (1977) 388.
77TIE1	Tiemann, E., Hölzer, B., Hoeft, J.: Z. Naturforsch. **32A** (1977) 123.
77TIE2	Tiemann, E., Hoeft, J.: Z. Naturforsch. **32A** (1977) 1477.
77TIE3	Tiemann, E., Dreyer, A.: Chem. Phys. **23** (1977) 231.
77TOD	Todd, T.R.: J. Mol. Spectrosc. **66** (1977) 162.
77WOR	Wormsbecher, R.F., Laue, S.L., Harris, D.O.: J. Chem. Phys. **66** (1977) 2745.
78BEA	Beaudet, R.A., Poynter, R.L.: J. Phys. Chem. Ref. Data **7** (1978) 311.
78BRE	Breford, E.J., Engelke, F.: Chem. Phys. Lett. **53** (1978) 282.
78CHA	Charifi, M., Gräff, G., Ley, R., Wolf, M.: J. Chem. Phys. **69** (1978) 5020.
78HOC	Hocking, W.H., Pearson, E.F., Creswell, R.A., Winnewisser, G.: J. Chem. Phys. **68** (1978) 1128.
78KAK	Kakar, R.K., Cohen, E.A., Geller, M.: J. Mol. Spectrosc. **70** (1978) 243.
78McK	McKellar, A.R.W., Oka, T.: Can. J. Phys. **56** (1978) 1315.
78NAI	Nair, K.P.R., Hoeft, J., Tiemann, E.: Chem. Phys. Lett. **58** (1978) 153.
78NIA	Niay, P., Bernage, P., Coquant, C.: J. Mol. Spectrosc. **72** (1978) 168.

78PFA	Pfaffe, S., Tiemann, E., Hoeft, J.: Z. Naturforsch **33A** (1978) 1386.
79BRE	Breford, E.J., Engelke, F.: J. Chem. Phys. **71** (1979) 1994.
79BUS	Bustreel, R., Demuynck-Marlière, C., Destombes, J.L., Journel, G.: Chem. Phys. Lett. **67** (1979) 178.
79CHU	Churassy, S., Grenet, G., Gaillard, M.L., Bacis, R.: Opt. Commun. **30** (1979) 41.
79DAG	Dagdigian, P.J.: J. Chem. Phys. **71** (1979) 2328.
79DAL	Dale, R.M., Herman, M., Johns, J.W.C., McKellar, A.R.W., Nagler, S., Strathy, I.K.M.: Can. J. Phys. **57** (1979) 677.
79DOH	Dohnt, G., Hese, A., Renn, A., Schweda, H.S.: Chem. Phys. **42** (1979) 183.
79EIS	Eisel, D., Zevgolis, D., Demtröder, W.: J. Chem. Phys. **71** (1979) 2005.
79FOT	Foth, H.J., Spieweck, F.: Chem. Phys. Lett. **65** (1979) 347.
79GUE	Guelachvili, G.: J. Mol. Spectrosc. **75** (1979) 251.
79HOE	Hoeft, J., Nair, K.P.R.: Z. Naturforsch. **34A** (1979) 1290.
79HÖN	Hönig, H., Czajkowski, M., Stock, M., Demtröder, W.: J. Chem. Phys. **71** (1979) 2138.
79KNÖ	Knöckel, H., Tiemann, E.: Chem. Phys. Lett. **64** (1979) 593.
79LIS1	Lisy, J.M., Klemperer, W.: J. Chem. Phys. **70** (1979) 4967.
79LIS2	Lisy, J.M., Klemperer, W.: J. Chem. Phys. **70** (1979) 228.
79NAI1	Nair, K.P.R., Hoeft, J., Tiemann, E.: Chem. Phys. Lett. **60** (1979) 253.
79NAI2	Nair, K.P.R., Hoeft, J., Tiemann, E.: J. Mol. Spectrosc. **78** (1979) 506.
79NIA	Niay, P., Bernage, P., Bocquet, H., Houdart, R.: J. Mol. Spectrosc. **77** (1979) 133.
79RAA	Raab, M., Höning, G., Castell, R., Demtröder, W.: Chem. Phys. Lett. **66** (1979) 307.
79TOD	Todd, T.R., Olson, W.B.: J. Mol. Spectrosc. **74** (1979) 190.
79WIL	Willis Jr., R.E.: Thesis, Ph.D.; Duke University, **1979**.
79YAM	Yamada, C., Hirota, E.: J. Mol. Spectrosc. **74** (1979) 203.
80BAC	Bacis, R., Broyer, M., Churassy, S., Vergès, J., Vigué, J.: J. Chem. Phys. **73** (1980) 2641.
80BRI	Brieger, M., Hese, A., Renn, A., Sodeik, A.: Chem. Phys. Lett. **76** (1980) 465.
80CAL	Callahan, D., Yokozeki, A., Muenter, J.S.: J. Chem. Phys. **72** (1980) 4791.
80DAG	Dagdigian, P.J.: J. Chem. Phys. **73** (1980) 2049.
80HAC	Hackel, R.P., Hackel, L.A., Ezekiel, S.: Phys. Rev. **A21** (1980) 1342.
80HEL	Helms, D.A., Winnewisser, M., Winnewisser, G.: J. Phys. Chem. **84** (1980) 1758.
80HIN1	Hinds, E.A., Sandars, P.G.H.: Phys. Rev. **A21** (1980) 471.
80HIN2	Hinds, E.A., Sandars, P.G.H.: Phys. Rev. **A21** (1980) 480.
80KNÖ	Knöckel, H.: Dissertation Universität Hannover, **1980**.
80MEE	Meerts, W.L., Veseth, L.: J. Mol. Spectrosc. **82** (1980) 202.
80NAI	Nair, K.P.R., Schütze-Pahlmann, H.-U., Hoeft, J.: Chem. Phys. Lett. **70** (1980) 583.
80SAI	Saito, S., Endo, Y., Hirota, E.: Ann. Rev. of the Institute for Mol. Science, Okazaki, Japan, **1980**.
80STR	Strow, L.L.: Opt. Lett. **5** (1980) 166.
80WIL	Willis, R.E., Clark III, W.C.: J. Chem. Phys. **72** (1980) 4946.
80YOK	Yokozeki, A., Muenter, J.S.: J. Chem. Phys. **72** (1980) 3796.
81BET	Bettin, N., Knöckel, H., Tiemann, E.: Chem. Phys. Lett. **80** (1981) 386.
81BRI	Brieger, M., Hese, A., Renn, A., Sodeik, A.: Chem. Phys. Lett. **78** (1981) 153.
81COH	Cohen, E.A., Pickett, H.M., Geller, M.: J. Mol. Spectrosc. **87** (1981) 459.
81GUE1	Guelachvili, G., Niay, P., Bernage, P.: J. Mol. Spectrosc. **85** (1981) 253.
81GUE2	Guelachvili, G., Niay, P., Bernage, P.: J. Mol. Spectrosc. **85** (1981) 271.
81LOV	Lovas, F.J., Maki, A.G., Olson, W.B.: J. Mol. Spectrosc. **87** (1981) 449.
81MAK	Maki, A.G., Lovas, F.J.: J. Mol. Spectrosc. **85** (1981) 368.
81NAI1	Nair, K.P.R., Hoeft, J.: J. Mol. Spectrosc. **85** (1981) 301.
81NAI3	Nair, K.P.R., Schütze-Pahlmann, H.W., Hoeft, J.: Chem. Phys. Lett. **80** (1981) 349.
81SCH1	Schütze-Pahlmann, H.-U., Hoeft, J., Törring, T.: priv. comm., **1981**.
81SCH2	Schütze-Pahlmann, H.-U., Hoeft, J., Törring, T.: priv. comm., **1981**.
81SCH3	Schütze-Pahlmann, H.-U.: Dissertation Freie Universität Berlin, **1981**.
81TIS	Tischer, R., Möller, K., Törring, T.: Chem. Phys. **62** (1981) 115.
81WOR	Wormsbecher, R.F., Hessel, M.M., Lovas, F.J.: J. Chem. Phys. **74** (1981) 6983.
82TIE	Tiemann, E.: to be published, **1982**.

2.3 Linear molecules

Rotational constants, centrifugal distortion constants, rotation-vibration interaction constants, *l*-type doubling constants, isotopic masses, Fermi resonance interaction constants, and references for structural data

2.3.1 Preliminary remarks

1. Introduction

In favorable cases, the analysis of the rotational spectrum of linear molecules with three or more atoms in vibrational state $v_1 \dots v_j \dots v_{3N-5}$ allows the determination of the constants listed in the table. For determination of the rotation-vibration interaction constants, the analysis of at least two vibrational states of the same normal vibration is necessary. The *l*-type doubling constants can be determined from the splitting of degenerate excited rotation-vibration states as a result of which the rotational lines are split. This splitting is caused by the interaction of the angular momentum of the degenerate vibration with the total angular momentum: each $J \to J+1$ transition has two components which are indicated either as c and d, e and f, or L (lower) and U (upper).

The unperturbed rovibrational energy E_{RV} may be written in the following form [71AMA, 72WIN]

$$E_{RV} = E_V + E_R$$

$$E_V = \sum_{s} \omega_s (v_s + d_s/2) + \sum_{\substack{ss' \\ s \leqq s'}} x_{ss'} (v_s + d_s/2)(v_{s'} + d_{s'}/2) + \sum_{t,t'} g_{l_t l_{t'}} l_t l_{t'} \tag{1}$$

$$E_R = [[B_v + \gamma_{l_t l_{t'}} l_t l_{t'}][J(J+1) - l^2] - D_v [J(J+1) - l^2]^2 + H_v [J(J+1) - l^2]^3 \tag{2}$$

with $l = \sum_t l_t$ and

$$B_v = B_e - \sum_s \alpha_s (v_s + d_s/2) + \sum_{s,s'} \gamma_{ss'} (v_s + d_s/2)(v_{s'} + d_{s'}/2) \tag{3}$$

where s is summed over all normal modes and t over all degenerate modes. For the explanation of the coefficients, see 2.

In order to consider the interaction of the vibrational angular momentum with the rotational angular momentum, and the interaction of the different degenerate modes, the energy matrix for each group of interacting levels must be constructed and diagonalized. There is a separate matrix for each vibrational quantum number v and for each rotational quantum number J. The off-diagonal matrix elements are the following:

A.

$$\langle l_t | H | l_t \pm 2 \rangle = \frac{q_t}{4} \{[J(J+1) - l(l \pm 1)][J(J+1) - (l \pm 1)(l \pm 2)] \cdot (v_t \pm l_t + 2)(v_t \mp l_t)\}^{1/2} \tag{4}$$

with

$$q_t = q_t^{(0)} - q_t^{(1)} \cdot J(J+1) + q_t^{(2)} [J(J+1)]^2 - \cdots. \tag{5}$$

This matrix element found in every excited state $v_t \neq 0$ is responsible for the rotational *l*-type resonance and doubling.

A perturbation calculation gives the following contribution to the energy:

for $v_t = 1$

$$\Delta = \pm \frac{q_t}{2} J(J+1) \tag{6}$$

for $v_t = 2$

$$\Delta = \varepsilon \frac{q_t^2 J(J+1)[J(J+1) - 2]}{4(B_e - g_{l_t l_t})} \qquad \varepsilon = 1 \text{ if } l_t = 0; \quad \varepsilon = -1 \text{ or } 0 \text{ if } l_t = 2 \tag{7}$$

for $v_t > 2$, see [71AMA].

B.

$$\langle l_t, l_{t'} | H | l_t \pm 2, l_{t'} \mp 2 \rangle = \frac{r_{tt'}}{4} [(v_t \pm l_t + 2)(v_t \mp l_t)(v_{t'} \mp l_{t'} + 2)(v_{t'} \pm l_{t'})]^{1/2} \tag{8}$$

with $r_{tt'} = r_{tt'}^0 + r_{tt',J} J(J+1)$.

This matrix element which affects combination states is responsible for the vibrational *l*-type resonance and doubling.

C.

$$\langle l_t | H | l_t \pm 4 \rangle.$$

This matrix element has been proposed to account for experimentally determined higher order effects and may affect states with $v_t \geqq 2$ (coefficient ϱ_t) as well as combination states (coefficient $\varrho_{tt'}$) [72WIN].

2. Explanations of columns in the tabulation

Column	Parameters	Definitions
1	Nr. *6-Nr.* *4-Nr.*	serial number of the molecule in this table the second and third numbers represent respectively the serial number of the molecule in Vol. II/6 and II/4, Landolt-Börnstein, New Series
2	v v_j	vibrational quantum number for j-th normal vibration: $\mathrm{j}=1, \ldots, 3\mathrm{N}-5$ vibrational ground state: $v_1=v_2=\cdots v_j \cdots v_{3N-5}=0;$ in short $v=0$ vibrational state: $v_1=\cdots v_{j-1}=0; \quad v_{j+1}=\cdots v_{3N-5}=0;$ in short $v_j=0$
		for triatomic molecules:
	v_1	vibrational quantum number of the highest energy stretching mode
	v_2	vibrational quantum number of the degenerate bending vibration
	v_3	vibrational quantum number of the lowest energy stretching mode
	l or l_t	angular momentum quantum number which arises in degenerate excited vibration v_t. Compare [55TOW] and [70GOR]
	e	symbol for the equilibrium configuration
3	B [MHz] B_e B_v	rotational constant for the equilibrium configuration e for the vibrational state $v_j=v$ $B_e=\frac{h}{8\pi^2 I_e}; \quad B_v=\frac{h}{8\pi^2 I_v}$ where I_e and I_v are the molecular moments of inertia for the equilibrium configuration e and the vibrational state v_j
	$\alpha_j, \gamma_{ij}, \gamma_{ll}$	rotation-vibration interaction constants conveying the variation of the rotational constants B_v with v_j, v_i or l
4	D [kHz] D_v H [kHz] H_v	centrifugal distortion constants for the vibrational state $v_j=v$
5	q [MHz] q_t	l-type doubling constant for the degenerate excited vibrational state v_t. When q_t is determined for different degenerate vibrational states, the q coefficients are on the same line as the corresponding vibrational quantum number v.
	$q^{(1)}, q^{(2)}, q^{(3)}$	coefficients for centrifugal effects on $q^{(0)}$
6		remarks concerning the previous columns and data of constants which have been determined only for a few molecules in this table:
	σ	standard deviation
	v_s	type, frequency or wave number of the s-th normal vibration.
	ω_s	the harmonic oscillator frequency for the s-th normal mode
	$x_{ss'}, g_{l_t l_{t'}}$	anharmonicity constants
	$r_{tt'}=r^0_{tt'}$ $r_{tt',J}=r^1_{tt'}$	coefficients of the vibrational l-type resonance matrix element, see (8)
Cont.		

Column	Parameters	Definitions
6 Cont.	ϱ_t, $\varrho_{tt'}$	coefficient of the matrix elements $\langle l_t \vert H \vert l_t \pm 4 \rangle$ or $\langle l_t, l_{t'}, l \vert H \vert l_t \pm 2, l_{t'} \pm 2, l \pm 4 \rangle$ respectively
	α_j, γ_{ij}, γ_{ll}	see column 3
	W_{12}	Fermi resonance interaction constant. Compare [55TOW]
	k	force constant
	ζ_{ij}	Coriolis coupling constant
	r_e	equilibrium value of the internuclear separation
7		references to the following tables

2.3.2 Data

Nr.	Molecule	v	B MHz	D kHz	Ref.	q MHz	Ref.	Varia, remarks	See also Tab., Nr.
1 *6-1*	$^{18}OC^{16}O$	$v=0$	11037.892 (14)	3.52 (13)	80END2			for structure, see also Vol. II/7 (3, 43) Uncertainties are 2.5 σ	2.6.2, 1
2 *6-2* *4-1*	OCS	$v=0$	6081.492109 (15)	$D=1.301368$ (81) $H=-8.32$ (129) $\cdot 10^{-8}$	81FAY, 80DUB			for structure, see also Vol. II/7 (3, 44)	2.6.2, 2 2.7.2, 20
		$v_2=1^1$ c	6088.896782 (63)	1.32015 (14)		$q^{(0)}=6.36141271$ (161)	74MAK	$\nu_2=520.62478$ (28) cm^{-1}	2.9.2, 2
		$v_2=1^1$ d	6095.258200 (66)	1.32489 (16)		$q^{(1)}=4.2720$ (177) $\cdot 10^{-6}$ $q^{(2)}=9.3$ (91) $\cdot 10^{-12}$		see also 5, appendix (subvolume b)	
		$v_2=2^0$	6100.19152 (354)	1.09810 (87)	81FAY			$2\nu_2^0=1047.04205$ (3) cm^{-1}	
		$v_2=2^2$ c	6102.55832 (56)	1.57030 (35)				$2\nu_2^2=1042.10312$ (34) cm^{-1}	
		$v_2=2^2$ d	6102.55968 (56)	1.34328 (35)					
		$v_2=3^1$	6108.65350 (720)	1.24594 (167)		$q^{(0)}=6.18845$ (720) $q^{(1)}=0.62$ (167) $\cdot 10^{-3}$	81FAY, 74FAB	$3\nu_2^1=1573.5696$ (3) cm^{-1}	
		$v_2=3^3$	6112.93586 (155)	1.4528 (4)				$3\nu_2^3=1564.4340$ (19) cm^{-1}	
		$v_2=4^0$	6114.8049 (166)	0.5992 (26)				$4\nu_2^0=2104.82772$ (14) cm^{-1}	
		$v_2=4^2$ c	6117.3192 (129)	1.9692 (19)				$4\nu_2^2=2100.3365$ (3) cm^{-1}	
		$v_2=4^2$ d	6117.3235 (129)	1.2657 (19)					
		$v_2=4^4$	6123.2054 (50)	1.46724 (56)				$4\nu_2^4=2087.6162$ (60) cm^{-1}	
		$v_2=5^1$	6121.7233 (262)	1.06660 (425)		$q^{(0)}=$ 6.0577 (262) $q^{(1)}=-4.37$ (425) $\cdot 10^{-3}$	81FAY	$5\nu_2^1=2635.7934$ (3) cm^{-1}	
		$v_2=5^3$	6126.1126 (203)	1.527 (2)				$5\nu_2^3=2627.4337$ (18) cm^{-1}	
		$v_2=5^5$	6133.3655 (127)	1.48641 (67)				$5\nu_2^5=2611.648$ (13) cm^{-1}	
		$v_3=1$	6063.3580 (35)	1.32934 (50)				$\nu_3=858.96691$ (15) cm^{-1}	
		$v_2=1^1, v_3=1$	6075.5045 (78)	1.36183 (43)		$q^{(0)}=6.8420$ (78) $q^{(1)}=9.49$ (43) $\cdot 10^{-3}$		$\nu_2^1+\nu_3=1372.6614$ (4) cm^{-1}	
		$v_2=2^0, v_3=1$	6084.3871 (135)	1.0837 (12)				$2\nu_2^0+\nu_3=1892.22998$ (77) cm^{-1}	
		$v_2=2^2, v_3=1$ c	6087.2308 (127)	1.67137 (91)				$2\nu_2^2+\nu_3=1887.7606$ (25) cm^{-1}	
		$v_2=2^2, v_3=1$ d	6087.2325 (127)	1.39053 (91)					
Cont.		$v_2=3^1, v_3=1$	6093.5398 (199)	1.26728 (240)		$q^{(0)}=6.5482$ (199) $q^{(1)}=3.0$ (24) $\cdot 10^{-3}$		$3\nu_2^1+\nu_3=2412.3276$ (17) cm^{-1}	

Nr.	Molecule	v	B MHz	D kHz	Ref.	q MHz	Ref.	Varia, remarks	See also Tab., Nr.
2	OCS	$v_2=3^3, v_3=1$	6098.6232 (175)	1.5260 (5)	81FAY			$3\nu_2^3+\nu_3=2404.1708\ (72)\ cm^{-1}$	
6-2		$v_3=2$	6044.8748 (92)	1.35984 (89)				$2\nu_3=1710.9764\ (3)\ cm^{-1}$	
4-1		$v_2=1^1, v_3=2$	6058.5529 (160)	1.40486 (81)		$q^{(0)}=\ 7.3323\ (160)$	81FAY	$\nu_2^1+2\nu_3=2218.2303\ (9)\ cm^{-1}$	
Cont.						$q^{(1)}=15.58\ (81)\cdot10^{-3}$			
		$v_1=1$	6045.0624 (88)	1.32318 (64)				$\nu_1=2062.20121\ (19)\ cm^{-1}$	
		$v_1=1, v_2=1^1$	6056.2708 (128)	1.34171 (70)		$q^{(0)}=\ 6.4648\ (128)$		$\nu_1+\nu_2^1=2575.5095\ (15)\ cm^{-1}$	
						$q^{(1)}=13.02\ (70)\cdot10^{-3}$			
		$v_1=1, v_3=1$	6028.9232 (262)	1.54388 (84)	81FAY, 80BOG			$\nu_1+\nu_3=2918.1168\ (36)\ cm^{-1}$	
		$v_1=2$	6008.8715 (195)	1.37545 (134)				$2\nu_1=4101.3870\ (29)\ cm^{-1}$	
	$OC^{33}S$	$v=0$	6004.91536 (57)	1.26992 (61)	80DUB				
		$v_2=1^1$ c	6012.2674 (48)	$D=1.2897\ (60)$	79KRU	$q^{(0)}=6.209878\ (31)$	74REI		
				$H=-0.1\ (25)\cdot10^{-6}$		$q^{(1)}=4.4\ (11)\cdot10^{-6}$			
		$v_2=1^1$ d	6018.4840 (58)	$D=1.3030\ (74)$					
				$H=4.2\ (31)\cdot10^{-6}$					
		$v_2=2^0$	6023.475 (15)	$D=1.066\ (19)$					
				$H=0.2\ (82)\cdot10^{-6}$					
		$v_2=2^2$ c	6025.732 (14)	$D=1.536\ (18)$					
				$H=-3.3\ (76)\cdot10^{-6}$					
		$v_2=2^2$ d	6025.733 (14)	$D=1.314\ (18)$					
				$H=0.4\ (76)\cdot10^{-6}$					
		$v_3=1$	5987.0165 (90)	$D=1.305\ (11)$					
				$H=3.6\ (46)\cdot10^{-6}$					
	$OC^{34}S$	$v=0$	5932.833775 (88)	1.24126 (18)	80DUB				
		$v_2=1^1\ c$	5940.1290 (28)	1.2582 (35)	79KRU	$q^{(0)}=6.0688022\ (100)$	74MAK	$H=-0.7\ (14)$ mHz	
		$v_2=1^1\ d$	5946.1990 (28)	1.2635 (35)		$q^{(1)}=3.926\ (260)\cdot10^{-6}$		$H=-0.2\ (14)$	
		$v_2=2^0$	5951.2603 (45)	1.0455 (56)				$H=\ 2.66\ (22)$	
		$v_2=2^2\ c$	5953.4012 (44)	1.5054 (56)				$H=-1.7\ (22)$	
		$v_2=2^2\ d$	5953.3976 (44)	1.2793 (56)				$H=-1.0\ (22)$	
		$v_2=3^3$	5963.535 (25)	1.385 (29)				$H=-3\ (11)$	
		$v_3=1$	5915.1436 (33)	1.2585 (41)				$H=-2.9\ (17)$	
		$v_2=1^1\ c, v_3=1$	5923.6444 (96)	1.292 (12)				$H=-0.1\ (46)$	
		$v_2=1^1\ d, v_3=1$	5930.1385 (96)	1.297 (12)				$H=-1.7\ (46)$	
Cont.		$v_1=1$	5897.398 (15)	1.19 (25)	80BOG				

Nr.	Molecule	v	B MHz	D kHz	Ref.	q MHz	Ref.	Varia, remarks	See also Tab., Nr.
2	$O^{13}CS$	$v=0$	6061.92381 (13)	1.29773 (25)	80DUB				
6–2		$v_2=1^1$ c	6068.6583 (32)	1.3128 (41)	79KRU	$q^{(0)}=6.5078418\ (52)$	74MAK	$H=-1.3\ (17)$	
4–1		$v_2=1^1$ d	6075.1686 (32)	1.3203 (41)		$q^{(1)}=4.4890\ (843) \cdot 10^{-6}$		$H=\ \ 0.0\ (17)$	
Cont.						$q^{(2)}=2.2\ (48) \cdot 10^{-11}$			
		$v_2=2^0$	6079.250 (5)	1.133 (38)	74MAK				
		$v_2=2^2$	6081.801 (7)	1.363 (46)*)				*) $D(v_2=2^2$ c)	
		$v_2=3^1$	6087.031 (5)			$q=6.287\ (5)$		$=1.597\ (46)$ kHz,	
		$v_2=3^3$	6091.583 (5)					the value given is for $v_2=2^2$ d	
		$v_2=4^0$	6092.200 (17)						
		$v_2=4^2$	6095.070 (8)						
		$v_2=4^4$	6101.262 (9)						
		$v_3=1$	6043.9468 (77)	1.3395 (99)	79KRU			$H=3.9\ (40)$	
		$v_2=1^1, v_3=1$	6055.610 (8)		74MAK	$q=7.066\ (14)$			
		$v_2=2^0, v_3=1$	6063.680 (13)						
		$v_2=2^2, v_3=1$	6066.811 (17)						
		$v_3=2$	6025.643 (12)						
		$v_1=1$	6027.644 (15)	1.31 (25)	80BOG				
	^{18}OCS	$v=0$	5704.85720 (37)	1.13263 (34)	80DUB				
		$v_2=1^1$	5714.995 (4)	1.198 (15)	74MAK	$q^{(0)}=5.6530026\ (101)$			
						$q^{(1)}=3.5940\ (676) \cdot 10^{-6}$			
						$q^{(2)}=3.0\ (30) \cdot 10^{-11}$			
		$v_2=2^0$	5722.924 (5)	1.020 (33)					
		$v_2=2^2$	5725.024 (4)	1.156 (20)**)				**) for $v_2=2^2$ d,	
		$v_2=3^1$	5731.142 (7)			$q=5.504\ (7)$		$D(v_2=2^2$ c$)=1.335\ (20)$ kHz	
		$v_2=3^3$	5734.973 (5)	1.304 (24)					
		$v_2=4^0$	5737.002 (10)						
		$v_2=4^2$	5739.504 (10)						
		$v_2=4^4$	5744.819 (10)	1.284 (71)					
		$v_3=1$	5688.653 (5)	1.170 (25)					
		$v_2=1^1, v_3=1$	5700.182 (6)			$q=6.053\ (11)$			
		$v_2=2^0, v_3=1$	5708.826 (12)						
		$v_2=2^2, v_3=1$	5711.369 (17)	1.386 (81)					
		$v_3=2$	5672.140 (10)						
Cont.		$v_1=1$	5669.854 (20)						

Nr.	Molecule	v	B MHz	D kHz	Ref.	q MHz	Ref.	Varia, remarks	See also Tab., Nr.
2	$O^{13}C^{34}S$	$v=0$	5911.73314 (21)	1.237869 (84)	81BUR				
6–2		$v_2=1^{1c}$	5918.36983 (99)	1.25388 (45)		$q^{(0)}=6.20478\ (130)$	81BUR		
4–1		$v_2=1^d$	5924.57459 (84)	1.25785 (34)		$q^{(1)}=3.96\ (56)\cdot 10^{-6}$			
Cont.		$v_2=2^0$	5928.866 (10)		74MAK				
		$v_2=2^2$	5931.050 (10)						
		$v_3=1$	5894.148 (10)						
	^{17}OCS	$v=0$	5883.67237 (26)	1.211119 (98)	81BUR				
		$v_2=1^1$ c	5891.0279 (13)	1.22805 (48)		$q^{(0)}=5.99600\ (182)$	81BUR		
		$v_2=1^1$ d	5897.0224 (12)	1.23470 (47)		$q^{(1)}=7.02\ (68)\cdot 10^{-6}$			
	$OC^{36}S$	$v=0$	5799.69204 (37)	1.18881 (15)					
		$v_2=1^1$ c	5806.879 (21)	1.205 (11)					
	$^{18}OC^{34}S$	$v=0$	5559.96722 (46)	1.07788 (15)					
		$v_2=1^1$	5569.845 (13)		74MAK	$q=5.393\ (13)$	74MAK		
		$v_2=2^0$	5577.726 (15)						
		$v_3=1$	5544.199 (10)						
	$O^{13}C^{33}S$	$v=0$	5984.56325 (48)	1.26584 (18)	81BUR				
	$^{18}O^{13}CS$	$v=0$	5691.069 (5)		74MAK				
		$v_2=1^1$	5700.627 (20)			$q=5.800\ (20)$			
		$v_3=1$	5674.994 (40)						
	$^{18}O^{13}C^{34}S$	$v=0$	5544.864 (20)						
	$OC^{35}S$	$v=0$	5864.371 (7)						
	$O^{14}CS$	$v=0$	6043.26 (50)						
		$v_2=1^1$	6052.65 (80)			$q=6.70\ (20)$			
3	$OC^{80}Se$	$v=0$	4017.6501 (18)		77MAK1			for structure, see also Vol. II/7 (3, 45)	
6–4		$v_1=1$	3995.8585 (250)						
4–2		$v_2=1^1$	4024.5557 (12)			$q=3.1715\ (8)$	77MAK1	r_e(C—O) = 1.1535 (1) Å	
		$v_2=2^0$	4030.6737 (200)					r_e(C—Se) = 1.7098 (1) Å	
		$v_2=2^2$	4031.3700 (200)						
		$v_3=1$	4004.3887 (200)						
	$^{18}OC^{80}Se$	$v=0$	3722.6821 (10)						
		$v_1=1$	3702.0201 (60)						
		$v_2=1^1$	3729.2448 (40)			$q=2.7530\ (50)$			
		$v_2=2^0$	3735.1216 (60)						
		$v_2=2^2$	3735.7450 (100)						
Cont.		$v_3=1$	3711.0384 (40)						

Nr.	Molecule	v	B MHz	D kHz	Ref.	q MHz	Ref.	Varia, remarks	See also Tab., Nr.
3	$O^{13}C^{80}Se$	$v=0$	3979.9918 (15)		77MAK1				
6-4		$v_1=1$	3959.5582 (60)						
4-2		$v_2=1^1$	3986.4500 (20)			$q=3.1952$ (40)	77MAK1		
Cont.		$v_2=2^0$	3992.1789 (50)						
		$v_2=2^2$	3992.8295 (30)						
		$v_3=1$	3966.7858 (30)						
4	HCN	$v=0$	44315.9757 (4)	87.24 (6)	74MAK			for structure, see	2.6.2, 3
6-6		$v_2=1^{1c}$	44310.381 (13)	88.8 (5)	77LUC	$q^{(0)}=224.47657$ (30)	74MAK	also Vol. II/7 (3, 58)	2.7.2, 1
4-5		$v_2=1^{1d}$	44534.851 (4)	90.2 (2)		$q^{(1)}=$ 2.65865 (292) $\cdot 10^{-3}$			
		$v_2=2^0$	44544.233 (14)	205.4 (6)		$q^{(2)}=$ 3.602 (620) $\cdot 10^{-8}$			
		$v_2=2^{2d}$	44519.895	92.6					
		$v_2=2^{2c}$	44519.563	−24.5					
		$v_2=3^{1c}$	44425.766 (13)	135.0 (5)		$q^{(0)}=231.120$ (4)		$H=-1.32\cdot 10^{-5}$ MHz	
		$v_2=3^{1d}$	44887.974 (50)	139.7 (21)		$q^{(1)}=2.98$ (9) $\cdot 10^{-3}$		$H=$ 1.52 $\cdot 10^{-5}$ MHz	
						$q^{(2)}=3.6\cdot 10^{-8}$			
		$v_2=4^0$	44785.584 (51)	452.97 (34)				Coriolis perturbation between	
		$v_3=1$	44013.803 (6)	88.1 (3)				$v_2=4^0$ and $v_3=1$, $v_2=1^{1c}$:	
		$v_3=2$	43709.846 (14)	87.9 (6)				$\Delta E=78663.1$ (40) MHz	
		$v_1=1$	44003.505 (15)	87.16 (6)				$W_{hi}=$ 112.27 (6) MHz	
		$v_1=2$	43681.998	84.4					
		$v_1=1, v_3=1$	43707.783 (36)	86.3 (14)					
		$v_2=1^{1c}, v_1=1$	44002.465	89.8					
		$v_2=1^{1d}, v_1=1$	44230.503	91.7					
		$v_3=1, v_2=1^{1c}$	44004.714 (37)	88.73 (16)					
		$v_3=1, v_2=1^{1d}$	44229.006	91.2					
		e	44511.62 (3)					r_e(H—C)=1.06549 (24) Å	
								r_e(C—N)=1.15321 (5) Å	
	$H^{13}CN$	$v=0$	43170.140 (20)	83.8 (40)	76PEA1	$q^{(0)}=214.8378$ (2)	78WIN	[76PEA1]	
		$v_2=1^1$				$q^{(1)}=$ 2.435 (2) $\cdot 10^{-3}$			
						$q^{(2)}=$ 4.18 (67) $\cdot 10^{-8}$			
	$HC^{15}N$	$v=0$	43027.644 (20)	81.9 (40)		$q^{(0)}=211.9440$ (2)			
		$v_2=1^1$				$q^{(1)}=$ 2.431 (2) $\cdot 10^{-3}$			
						$q^{(2)}=$ 5.11 (73) $\cdot 10^{-8}$			
Cont.	DCN	$v=0$	36207.4627 (2)	57.83 (4)	74MAK				

Nr.	Molecule	v	B MHz	D kHz	Ref.	q MHz	Ref.	Varia, remarks	See also Tab., Nr.
4	$D^{13}CN$	$v=0$	35587.619 (20)	53.3 (20)	76PEA1	$q^{(0)}=182.2941$ (3)	78WIN		
6–6		$v_2=1^1$				$q^{(1)}=$ 2.075 (4) $\cdot 10^{-3}$			
4–5						$q^{(2)}=$ 3.03 (93) $\cdot 10^{-8}$			
Cont.	$DC^{15}N$	$v=0$	35169.809 (1)	56.0 (2)	80CAZ	$q^{(0)}=176.0807$ (3)			
		$v_2=1^1$				$q^{(1)}=$ 2.008 (4) $\cdot 10^{-3}$			
						$q^{(2)}=$ 2.11 (98) $\cdot 10^{-8}$			
	$H^{13}C^{15}N$	$v=0$	41863.953 (20)	79.0 (40)	76PEA1	$q^{(0)}=202.3556$ (3)			
		$v_2=1^1$				$q^{(1)}=$ 2.216 (4) $\cdot 10^{-3}$			
						$q^{(2)}=$ 4.95 (125) $\cdot 10^{-8}$			
	$D^{13}C^{15}N$	$v=0$	34531.276 (20)	50.7 (20)		$q^{(0)}=172.0455$ (1)			
		$v_2=1^1$				$q^{(1)}=$ 1.896 (1) $\cdot 10^{-3}$			
						$q^{(2)}=$ 4.29 (33) $\cdot 10^{-8}$			
	TCN	$v=0$	30954.080 (8)	43.82 (10)	75LUC1				
		$v_2=1^1$	31080.410 (5)	45.82 (8)		$q^{(0)}=150.822$ (4)	75LUC1, 78WIN		
						$q^{(1)}=$ 1.91 (7) $\cdot 10^{-3}$			
	$T^{13}CN$	$v=0$	30591.036 (7)	42.45 (9)		$q^{(0)}=149.749$ (14)			
		$v_2=1^1$	30709.082 (12)	44.37 (16)		$q^{(1)}=$ 1.41 (19) $\cdot 10^{-3}$			
	$TC^{15}N$	$v=0$	30064.400 (12)	41.19 (14)		$q^{(0)}=142.688$ (15)			
		$v_2=1^1$	30186.560 (8)	43.00 (14)		$q^{(1)}=$ 1.98 (28) $\cdot 10^{-3}$			
5	FCN	$v_2=1^1$	10584.615*)	4.7115 *)	68LAF, 77MAK2	$q^{(0)}=19.68421$ (34)	68LAF	for structure, see also Vol. II/7 (3, 22)	2.6.2, 4
6–7						$q^{(1)}=$ 5.556 (3) $\cdot 10^{-5}$		*) Combined analysis of IR and MW data.	
4–6		$v_2=2^0$	10610.625*)	4.483*)				$\nu_3=1076.492007$ (13) cm^{-1}	
		$v_3=1$	10510.836*)	4.45*)					
		$v_1=1$	10554.186*)	4.483*)					
		$v_2=1^1, v_3=1$	10543.330 (90)$^{+}$)	0.51 $^{++}$)	77MAK2, 79JON1	$q^{(0)}=21.448$ (354)$^{+}$)	77MAK2 79JON1	$^{+}$) Laser Stark	
						$q^{(1)}=$ 2.7 $\cdot 10^{-4++}$)		$^{++}$) two-photon	
		$v_2=2^0, v_3=1$	10567.272*)	4.45	68LAF				
6	$^{35}ClCN$	$v=0$	5970.820 (10)	1.66 (3)	79COG			for structure, see also Vol. II/7 (3, 10)	2.6.2, 5 2.7.2, 10
6–8		$v_2=1^1$				$q^{(0)}=7.467532$ (16)*)	80REI	*) MBER data	
4–7						$q^{(1)}=1.83$ (15) $\cdot 10^{-5}$			
		$v_1=1$	5938.488 (13)	1.77 (7)	77CAZ			r_e(C—N) = 1.1606 (28) Å	
		$v_3=1$	5970.934 (8)	2.11 (2)				r_e(C—Cl) = 1.6290 (24) Å	
Cont.		e	5982.995 (40)					$\alpha_3=24.65$ (5) MHz. [77CAZ]	

Nr.	Molecule	v	B MHz	D kHz	Ref.	q MHz	Ref.	Varia, remarks	See also Tab., Nr.
6	$^{37}ClCN$	$v=0$	5847.246 (8)	1.60 (8)	79COG				
6–8		$v_1=1$	5815.575 (22)	1.38 (13)	77CAZ				
4–7		e	5859.058 (29)						
Cont.									
7	$^{79}BrCN$	$v_1=1$	4099.824 (30)	0.853 (8)	77CAZ			for structure, see also	
6–8a								Vol. II/7 (3, 4)	
4–8	$^{81}BrCN$	$v_1=1$	4076.550 (30)	0.870 (8)					
8	ICN	$v=0$	3225.54263 (77)	0.5923 (33)	77BUR			for structure, see also	2.7.2, 56
6–8b		$v_1=1$	3211.2102 (59)	0.5494 (95)	78CAZ2			Vol. II/7 (3, 36)	2.9.2, 5
4–9		e	3229.159 (6)	0.587 (27)				harmonic force field	
	$I^{13}CN$	$v=0$	3177.0419 (16)	0.5821 (17)				calculated.	
		$v_3=1$	3167.8552 (12)	0.6104 (11)				r_e(I—C) =1.99209 (22) Å	
		$v_2=1^1$	3185.9025 (11)	0.5901 (12)		2.65066 (53)	78CAZ2	r_e(C—N)=1.16044 (33) Å	
		$v_2=2^0$	3192.3160 (13)	0.5374 (14)				[78CAZ2]	
		$v_1=1$	3163.6887 (30)	0.5610 (42)					
		e	3180.674 (6)	0.570 (9)					
9	HNC	$v=0$	45331.999 (40)	101.2 (50)	76PEA1			*) assumed value	2.6.2, 6
		$v_1=1$	45032.70 (6)	100*)	76SAY				2.7.2, 11
		$v_3=1$	44996.81 (4)	100*)					2.9.2, 3
		$v_3=2$	44660.90 (4)	100*)					
	$HN^{13}C$	$v=0$	43545.616 (40)	95.5 (50)	76CRE				
	$H^{15}NC$	$v=0$	44433.052 (40)	97.3 (50)	76PEA1				
	$H^{15}N^{13}C$	$v=0$	42629.642 (40)	90.2 (50)					
	DNC	$v=0$	38153.004 (40)	70.0 (50)	76CRE				
	$DN^{13}C$	$v=0$	36684.003 (25)	62.8 (20)	76PEA1				
	$D^{15}NC$	$v=0$	37643.521 (30)	67.5 (20)					
	$D^{15}N^{13}C$	$v=0$	36155.521 (25)	61.9 (20)					
10	HCP								2.7.2, 2
6–9									2.9.2, 6
4–10									
11	FCP	$v=0$	5257.80 (3)	1.0 (17)	80KRO2				2.6.2, 7
		$v_2=1$	$\alpha_2=-11.95$ (5)			$q=5.971$ (2)	80KRO2	$\nu_2=390$ (10) cm^{-1}.	
	$F^{13}CP$	$v=0$	5250.16					$v_3=1$ satellite measured	

Nr.	Molecule	v	B MHz	D kHz	Ref.	q MHz	Ref.	Varia, remarks	See also Tab., Nr.
12	HBS	$v=0$	19083.006 (4)	21.89 (4)	73PEA1			*) assumed	2.6.2, 8
		$v_2=1^1$	19106.472 (20)	21.9*)		$q=46.45$ (1)	73PEA1		2.7.2, 5
		$v_1=1$	18994.33 (42)		75SAM				2.9.2, 1
	$H^{10}BS$	$v=0$	20080.234 (9)	24.02 (9)	73PEA1				
		$v_2=1^1$	20108.482 (30)	24.0*)		$q=50.80$ (5)			
		$v_1=1$	19979.0 (12)		75SAM				
	$HB^{34}S$	$v=0$	18791.623 (10)	21.25 (10)	73PEA1				
		$v_1=1$	18704.4 (12)		75SAM				
	$H^{10}B^{34}S$	$v=0$	19793.168 (15)	23.18 (25)	73PEA1				
		$v_1=1$	19693.5 (18)		75SAM				
	DBS	$v=0$	15937.934 (7)	14.25 (4)	73PEA1				
		$v_2=1^1$	15970.719 (20)	14.3*)		$q=40.65$ (6)			
	$D^{10}BS$	$v=0$	16563.378 (11)	15.61 (6)					
		$v_2=1^1$	16600.733 (30)	15.6*)		$q=43.16$ (2)			
	$DB^{34}S$	$v=0$	15686.936 (12)	14.02 (8)					
	$D^{10}B^{34}S$	$v=0$	16317.465 (18)	14.97 (14)					
13	$^{35}ClBS$	$v=0$	2796.7796 (7)	0.3440 (80)	80KIR2				2.6.2, 9
		$v_2=1^1$	$\alpha_2=-7.4889$ (7)			$q=1.9942$ (9)	80KIR2	$\gamma_{22}+\gamma_{ll}=-12.0$ (1) kHz	2.7.2, 22
		$v_3=1$	$\alpha_3=3.4620$ (2)					$\nu_2=300$ (40) cm^{-1}	
	$^{35}ClB^{33}S$	$v=0$	2753.2570 (12)	0.341 (15)					
	$^{35}ClB^{34}S$	$v=0$	2712.2219 (7)	0.282 (9)					
		$v_2=1^1$	$\alpha_2=-7.2378$ (3)			$q=1.8786$ (5)			
		$v_3=1$	$\alpha_3=3.4017$ (3)						
	$^{35}Cl^{10}BS$	$v=0$	2796.8613 (14)	0.327 (17)					
		$v_2=1^1$	$\alpha_2=-7.8745$ (15)			$q=1.9160$ (7)		$\gamma_{22}+\gamma_{ll}=-11.5$ (4) kHz	
		$v_3=1$	$\alpha_3=3.7731$ (4)						
	$^{35}Cl^{10}B^{33}S$	$v=0$	2753.2758	0.303					
	$^{37}ClBS$	$v=0$	2722.9999 (6)	0.321 (7)					
		$v_2=1^1$	$\alpha_2=-7.2941$ (15)			$q_2=1.8954$ (7)		$\gamma_{22}+\gamma_{ll}=-11.2$ (4) kHz	
		$v_3=1$	$\alpha_3=3.4755$ (2)						
	$^{37}ClB^{33}S$	$v=0$	2679.7995 (1)	0.316 (1)					
	$^{37}ClB^{34}S$	$v=0$	2639.0735 (24)	0.320 (27)					
	$^{37}Cl^{10}BS$	$v=0$	2723.2127 (17)	0.328 (21)					
		$v_2=1^1$	$\alpha_2=-7.6742$ (17)			$q_2=1.8190$ (7)		$\gamma_{22}+\gamma_{ll}=-11.6$ (5) kHz	
Cont.		$v_3=1$	$\alpha_3=3.6354$ (3)						

Nr.	Molecule	v	B MHz	D kHz	Ref.	q MHz	Ref.	Varia, remarks	See also Tab., Nr.
13 Cont.	$^{37}Cl^{10}B^{33}S$ $^{37}Cl^{10}B^{34}S$	$v=0$ $v=0$	2679.9226 (7) 2639.1327 (32)	0.273 (8) 0.289 (37)	80KIR2				
14	O=C=C=C=O	$\nu_7 \leftarrow 0$	$B'-B''+2D'+3H'$ $=28.55016\,(26)$	$D'-D''+3H'$ $=-27.54\,(58)\cdot 10^{-3}$	79BUR2			for structure, see also Vol. II/7 (3, 343) band origin of ν_7: $\nu_0-B'-D'-H'$ $=545007.524\,(32)$ MHz $H'-H''=-0.00649\,(37)$ Hz	
15	N≡C—C≡P	$v=0$	2704.4803 (19)	0.216 (13)	80COO				2.6.2, 10
16	O=C=C=C=S	$v=0$ $v_5=1^1$ $v_6=1^1$ $v_7=1^1$ $v_7=2$ $v_7=3$	1413.898 1415.171 1415.960 1420.535 1427.140 1433.705	0.046	76WIN	 $q=0.323$ $q=0.382$ $q=1.749$ $q=1.768$	 76WIN	see also 5, appendix (subvolume b)	2.6.2, 11
17 *6–15*	HCNO	$v_3=1$ $v_5=4^0$ $v_5=4^{2+}$ $v_5=4^{2-}$	11433.039 (36) 11530.0800 (93) 11546.9532 (51) 11546.9712 (40)	$-4.41\,(71)$ $H=5.1\,(20)\cdot 10^{-3}$ 10.54 (27) $H=5.1\,(20)\cdot 10^{-3}$ 6.874 (58) 4.033 (45)	75YAM1			for structure, see also Vol. II/7 (3, 62) interaction between $v_4=1$ and $v_5=2$, see [75YAM1]. Fermi resonance between $v_3=1$ and $v_5=4$: $W_a=\sqrt{2}k_{35555}$ $=63000\,(200)$ MHz; $q_5^{(0)}=26.4\,(4)$ MHz (coefficient of matrix element of l-type resonance); $E^0(v_3=1)-E^0(v_5=4^0)$ $=4.939\,(20)$ cm^{-1}; $E^0(v_5=4^0)-E^0(v_5=4^2)$ $=23.07\,(8)$ cm^{-1} [75YAM1]	2.9.2, 8 2.7.2, 12

Nr.	Molecule	v	B MHz	D kHz	Ref.	q MHz	Ref.	Varia, remarks	See also Tab., Nr.
17 *6–15* Cont.	DCNO	$v=0$	10292.48340 (31)	3.5418 (10)	74WIN				
		$v_3=1$	10249.8458 (56)	3.609 (45)	75YAM1				
		$v_4=1^1$	10306.00780 (45)	3.6409 (22)	74WIN				
		$v_4=2$	10318.1833 (20)	3.734 (8)	75WIN1			$g_{tt'}=1.16\,(7)\,cm^{-1}$ $\gamma_{ll}=0.2722\,(5)$ MHz $\varrho_{tt'}=0.037\,(6)$ kHz	
		$v_5=1^1$	10338.65942 (32)	3.6208 (16)	74WIN				
		$v_5=2$	10362.2057 (7)	3.5971 (23)	75WIN1			$g_{tt'}=-6.88\,(33)\,cm^{-1}$ $\gamma_{ll}=2.84792\,(30)$ MHz $\varrho_{tt'}=0.0712\,(31)$ kHz	
		$v_5=3$	10388.9438 (6)	3.6013 (6)		$q^{(0)}=29.27449\,(4)$ $q^{(1)}=0.14000\,(15)\cdot 10^{-3}$ $q^{(2)}=1.5\cdot 10^{-9}$ (fixed)	75WIN1	$g_{tt'}=-4.809\,(33)\,cm^{-1}$ $\gamma_{ll'}=1.44940\,(15)$ MHz	
		$v_4=1, v_5=1$	10351.9895 (7)	3.679 (5)				$g_{tt'}=1.89\,(21)\,cm^{-1}$ $\gamma_{ll}=0.4569\,(5)$ MHz $\varrho_{tt'}=0.017\,(4)$ kHz $r_{45}^{(0)}=-2.91\,(44)\,cm^{-1}$ $r_{45}^{(1)}=0.3951\,(6)$ MHz	
18 *6–14* *4–15*	N_2O	$v=0$	12561.6348 (21)	5.2808 (61) $H=0.8\,(57)\cdot 10^{-6}$	76AND			for structure, see also Vol. II/7 (2, 331) for a detailed analysis of the l-doubling and Fermi resonances, see [76AND] *) MBER data	2.6.2, 12 2.7.2, 13 2.9.2, 7
		$v_1=1$	12458.1706	5.17	75BOG				
		$v_2=1^{1c}$	12566.6411 (21)	5.3539 (61) $H=5.7\,(57)\cdot 10^{-6}$	76AND	$q^{(0)}=23.74375\,(1)$*) $q^{(1)}=32\,(1)\cdot 10^{-6}$	78REI		
		$v_2=1^{1d}$	12590.3906 (21)	5.4002 (61) $H=19.7\,(57)\cdot 10^{-6}$					
		$v_2=2^0$	12588.8913 (21)	7.4403 (61) $H=82.5\,(57)\cdot 10^{-6}$					
		$v_2=2^{2c}$	12595.0614 (21)	3.6281 (61) $H=-70.2\,(57)\cdot 10^{-6}$					
		$v_2=2^{2d}$	12595.0692 (21)	5.4482 (61) $H=2.9\,(57)\cdot 10^{-6}$					
		$v_2=3^{1c}$	12578.8241 (68)	6.340 (19) $H=39\,(18)\cdot 10^{-6}$					
Cont.		$v_2=3^{1d}$	12623.6200 (68)	6.494 (19) $H=-35\,(18)\cdot 10^{-6}$					

Nr.	Molecule	v	B MHz	D kHz	Ref.	q MHz	Ref.	Varia, remarks	See also Tab., Nr.
18 *6-14* *4-15* Cont.	N_2O	$v_2=3^3$	12611.2577 (72)	4.742 (20) $H=-102\,(18)\cdot10^{-6}$	76AND				
		$v_3=1$	12508.9923 (31)	5.1726 (86) $H=1.8\,(77)\cdot10^{-6}$					
		$v_2=1^{1c}, v_3=1$	12515.270 (15)	5.233 (38) $H=-1\,(33)\cdot10^{-6}$					
		$v_2=1^{1d}, v_3=1$	12542.493 (15)	5.120 (38) $H=-23\,(33)\cdot10^{-6}$					
	$^{14}N^{15}NO$	$v=0$	12560.7310 (21)	5.2682 (61) $H=-8.1\,(57)\cdot10^{-6}$					
		$v_2=1^{1c}$	12564.5298 (99)	5.319 (27) $H=-21\,(24)\cdot10^{-6}$		$q^{(0)}=24.2796\,(26)$ $q^{(1)}=28.4\,(34)\cdot10^{-6}$	76AND		
		$v_2=1^{1d}$	12588.8168 (68)	5.368 (19) $H=-3\,(18)\cdot10^{-6}$					
	$^{15}N^{14}NO$	$v=0$	12137.3103 (22)	4.8896 (57) $H=-11.3\,(49)\cdot10^{-6}$					
		$v_2=1^{1c}$	12142.711 (10)	4.960 (25) $H=-4\,(20)\cdot10^{-6}$		$q^{(0)}=22.2914\,(26)$ $q^{(1)}=28.0\,(30)\cdot10^{-6}$			
		$v_2=1^{1d}$	12164.996 (10)	4.972 (25) $H=-16\,(20)\cdot10^{-6}$					
	$N_2{}^{17}O$	$v=0$	12191.719 (10)	4.987 (25) $H=-4\,(20)\cdot10^{-6}$					
	$N_2{}^{18}O$	$v=0$	11859.1097 (22)	4.7274 (57) $H=-11.9\,(49)\cdot10^{-6}$					
		$v_2=1^{1c}$	11864.545 (10)	4.772 (25) $H=-20\,(20)\cdot10^{-6}$		$q^{(0)}=21.3446\,(30)$ $q^{(1)}=29.0\,(34)\cdot10^{-6}$			
		$v_2=1^{1d}$	11885.913 (15)	4.857 (36) $H=23\,(28)\cdot10^{-6}$					
19	NaOH	$v=0$ $v_2=1^1$ $v_3=1$	12567.054 (10) $\alpha_2=97\,(20)$ $B(v_3=1)-B(v=0)$ $=116.3\,(15)$	28.72 (5)	73PEA2 76KUI	$q=51.66\,(3)$	76KUI	$\gamma_{ll}=-10\,(2)$ MHz $\gamma_{22}=9\,(2)$ MHz [76KUI]	

Nr.	Molecule	v	B MHz	D kHz	Ref.	q MHz	Ref.	Varia, remarks	See also Tab., Nr.
20 *6–11*	^{39}KOH	$v=0$	8208.679 (10)	12.19 (6)	73PEA3			uncertainties are 3 σ	2.7.2, 41
		$v_2=1$	8169.03 (2)	12.456 (18)	76PEA2, 75KUI	$q^{(0)}=16.042$ (20) $q^{(1)}=56$ (73) $\cdot 10^{-6}$	76PEA2	$\gamma_{ll}=-2.95$ (2) MHz $\nu_2=300$ cm^{-1}	
		$v_2=2$	8135.190 (28)	12.599 (21)				$\gamma_{ll}=-3.146$ (10) MHz $x_{ll}=$ 8.1 (5) cm^{-1}	
		$v_2=3$	8108.038 (21)	12.811 (51)		$q^{(0)}=16.900$ (9) $q^{(1)}=48$ (36) $\cdot 10^{-6}$		$\gamma_{ll}=-3.337$ (2) MHz $x_{ll}=$ 8.0 (15) cm^{-1}	
		$v_2=4$	8087.009 (78)	12.964 (10)				$\gamma_{ll}=-3.457$ (2) MHz from $l=0$ and 4 $\gamma_{ll}=-3.281$ (30) MHz from $l=0$ and 2 $x_{ll}=$ 8.2 (4) cm^{-1}	
		$v_2=5$	8072.041 (91)	13.082 (223)		$q^{(0)}=17.665$ (8) $q^{(1)}=96$ (37) $\cdot 10^{-6}$		$\gamma_{ll}=-3.317$ (7) MHz $x_{ll}=$ 8.5 (5) cm^{-1}	
		$v_3=1$	8142.656 (8)	12.256 (33)					
		$v_3=2$	8076.698 (18)	12.324 (78)					
		$v_3=3$	8010.823 (12)	12.356 (57)					
		$v_2=1, v_3=1$	8102.47 (2)	12.483 (27)		$q^{(0)}=17.376$ (6) $q^{(1)}=44$ (24) $\cdot 10^{-6}$		$\nu_3=403$ cm^{-1}	
		$v_2=2, v_3=1$	8069.264 (28)	12.666 (51)				$\gamma_{ll}=-3.088$ (10) MHz $x_{ll}=$ 8.0 (6) cm^{-1}	
		$v_2=3, v_3=1$	8043.179 (78)	12.788 (200)		$q^{(0)}=17.793$ (34) $q^{(1)}=49$ (140) $\cdot 10^{-6}$		$\gamma_{ll}=-3.172$ (6) MHz $x_{ll}=$ 7.2 (15) cm^{-1}	
		$v_2=4, v_3=1$	8023.111 (21)	13.007 (42)				$\gamma_{ll}=-3.173$ (7) MHz from $l=0, 2$ $x_{ll}=$ 7.6 (5) cm^{-1}	
		$v_2=1, v_3=2$	8036.10 (2)	12.528 (111)		$q^{(0)}=18.778$ (25)		$\alpha_1=19.7$ MHz (calculated) $\alpha_2=48.94$ (96) MHz $\alpha_3=65.68$ (90) MHz harmonic and cubic force fields calculated. [76PEA2]	
	^{41}KOH	$v=0$	8088.910 (2)	11.852 (57)	76PEA2				
	^{39}KOD	$v=0$	7494.825 (20)	9.455 (57)					
Cont.		$v_2=1$	7483.458 (24)*)	9.652 (36)		$q^{(0)}=16.775$ (24) $q^{(1)}=31$ (63) $\cdot 10^{-6}$		*) corrected for γ_{ll} $\nu_2=223$ cm^{-1}	

Nr.	Molecule	v	B MHz	D kHz	Ref.	q MHz	Ref.	Varia, remarks	See also Tab., Nr.
20 *6-11* Cont.	^{39}KOD	$v_2=2$ $v_2=3$ $v_2=4$ $v_3=1$ $v_2=1, v_3=1$ $v_2=2, v_3=1$	7474.872 (9) 7468.878 (28) 7464.889 (20) 7438.072 (12) 7429.461 (36)*) 7420.653 (17)	9.822 (21) 9.910 (33) 10.00 (fixed) 9.510 (28) 9.788 (40) 9.959 (16)	76PEA2	$q^{(0)}=17.251$ (8) $q^{(1)}=80\,(20)\cdot 10^{-6}$ $q^{(0)}=18.008$ (17) $q^{(1)}=71\,(40)\cdot 10^{-6}$	76PEA2	$\gamma_{ll}=-1.770$ (2) MHz $x_{ll}=$ 4.25 (25) cm^{-1} $\gamma_{ll}=-1.793$ (3) MHz $x_{ll}=$ 4.40 (15) cm^{-1} $\gamma_{ll}=-1.771$ (75) MHz $x_{ll}=$ 4.27 (20) cm^{-1} $\nu_3=399\ cm^{-1}$ *) corrected for γ_{ll} $\gamma_{ll}=-1.921$ (6) MHz $x_{ll}=$ 3.73 (20) cm^{-1} $\alpha_1=24.4$ MHz (calculated) $\alpha_2=14.88$ (100) MHz $\alpha_3=57.19$ (100) MHz r_e(KO) = 2.196 (2) Å r_e(OH) = 0.960 (10) Å [76PEA2]	
21 *6-12*	^{87}RbOH	$v_2=4^2$						for structure, see also Vol. II/7 (2, 291) $E_\Delta-E_\Sigma=41$ (12) cm^{-1} [77KUI]	
22 *6-13*	CsOH	$v_2=4^2$ $v_2=4^2, v_3=1$ $v_2=6^2$						for structure, see also Vol. II/7 (2, 167) $E_\Delta-E_\Sigma=39$ (4) cm^{-1} $E_\Delta-E_\Sigma=42$ (4) cm^{-1} $E_\Delta-E_\Sigma=36$ (4) cm^{-1} [77KUI]	
23 *6-16* *4-11* Cont.	HC≡CH	$v_5=1^1$ $v_5=2^0$	1.178768 (23) $^{\times}$) 1.181008 (14) $^{\times}$)	1.633 (13) $\cdot 10^{-6}$ $^{\times}$) $D=5.190\,(31)\cdot 10^{-6}$ $^{\times}$) $H=1.283\,(36)\cdot 10^{-9}$ $^{\times}$)	80DAS2	$q^{(0)}=0.004711$ (23) $^{\times}$) $q^{(1)}=4.5\,(13)\cdot 10^{-8}$ $^{\times}$)	80DAS2	for structure, see also Vol. II/7 (3, 179) $^{\times}$) diode laser results, values in cm^{-1} $\nu_5^1=730.33281$ (19) cm^{-1}	2.6.2, 13 2.9.2, 9

Nr.	Molecule	v	B MHz	D kHz	Ref.	q MHz	Ref.	Varia, remarks	See also Tab., Nr.
23	HC≡CD	$v=0$	29725.25 (5)	33.6 (32)	69MCK			*) uncertainties are 2.5 σ	
6–16		$v_4=1$	29803.003 (6)*)	$D_c=36.3$ (9)**)	80MAT1	$q^{(0)}=132.993$ (14)*)	80MAT1	$\nu_4=519.372$ cm^{-1}, CCD bend	
4–11				$D_d=30.3$ (12)**)				**) IR value [76BAL]	
Cont.		$v_4=3$, Π state	29969.052 (43)*)	$(D_c+D_d)/2=40$ (7)***)		$q^{(0)}=135.529$ (25)*)		***) assumed value	
				$(D_c-D_d)/2=0$ (6)***)					
		$v_5=1$	29766.750 (4)*)	$D_c=34.2$ (2)+)		$q^{(0)}=105.702$ (8)*)		$\nu_5=678.792$ cm^{-1}, CCH bend	
				$D_d=35.2$ (3)+)				+) IR value [79ANT]	
		$v_5=3$, Π state	29861.189 (27)*)	$D_c=40.2$ (9)**)		$q^{(0)}=108.642$ (28)*)			
				$D_d=45.0$ (12)**)					
	DC≡CD	$v_4=1$	25480.70 (43)++)	24.36 (19)++)	77LAF	$q^{(0)}=97.02$ (18)++)	77LAF	++) combination of IR	
						$q^{(1)}=$ 0.23 (17) $\cdot 10^{-3}$		and MW data;	
		$v_5=1$	25483.24 (46)++)	24.27 (23)++)		$q^{(0)}=98.02$ (22)++)		constants derived for the	
						$q^{(1)}=$ 0.13 (21) $\cdot 10^{-3}$		$\nu_5^1-\nu_4^1$ difference band; 90%	
								confidence intervals	
24	HC≡CF	$v=0$	9706.178 (10)	3.46 (20)	79JON2			for structure, see also	2.6.2, 14
6–17		$v_3=1$	9669.057 (10)	2.7 (5)				Vol. II/7 (3, 173)	2.9.2, 10
4–12		$v_4=1^1$	9715.002 (20)*)	3.57 (10)		$q^{(0)}=12.8202$ (3)	79JON2	+) assumed	
						$q^{(1)}=$ 2.10 (5) $\cdot 10^{-5}$		$\nu_3=1061.4452$ (3) cm^{-1} [76TAN]	
								*) $B_v+\gamma_{ll}$	
		$v_4=2^{0,2}$	9723.27 (5)	3.6+)				$\gamma_{ll}=-0.12$ MHz;	
		$v_5=1^1$	9733.839 (20)*)	3.65 (10)		$q^{(0)}=19.3553$ (2)		$g_{tt}\simeq$ 2.5 GHz	
						$q^{(1)}=$ 6.26 (3) $\cdot 10^{-5}$			
		$v_5=2^{0,2}$	9760.57 (2)	3.7 (5)				$\gamma_{ll}=$ 0.12 MHz;	
		$v_5=3^1$	9786.90 (5)	3.7+)		$q^{(0)}=19.21$ (5)		$g_{tt}=$ 26.5 (10) GHz	
		$v_5=4^{0,2}$	9812.09 (5)	3.8+)				$\gamma_{ll}=$ 0.22 MHz	
		$(v_4=1, v_5=1)^{0,2}$	9742.40 (5)	3.6+)				$\gamma_{ll'}=-0.016$ MHz	
		$(v_4=1, v_5=2)^1$	9769.13 (5)**)	3.6+)		$q^{(0)}=12.84$ (3)		**) $B_v-\gamma_{ll'}$	
		$v_3=1$, $v_4=1$	9679.18 (12)***)	3.47++)	76TAN,	$q^{(0)}=13.2573$ (14)++)	79JON1	***) Laser Stark	
					79JON1	$q^{(1)}=$ 7.00 (41) $\cdot 10^{-5}$		++) two-photon	
						$q^{(2)}=$ 1.93 (29) $\cdot 10^{-8}$			
		$v_3=1$, $v_5=1$	9696.43 (6)***)	3.65++)		$q^{(0)}=19.6115$ (9)++)			
Cont.						$q^{(1)}=$ 6.26 (23) $\cdot 10^{-5}$			

Nr.	Molecule	v	B MHz	D kHz	Ref.	q MHz	Ref.	Varia, remarks	See also Tab., Nr.
24	$H^{13}C{\equiv}CF$	$v=0$	9373.93 (5)	3.5$^+$)	79JON2			$^+$) assumed	
6-17		$v_4=1^1$	9381.37 (5)*)	3.6$^+$)		$q^{(0)}=11.97$ (5)	79JON2	*) $B_v+\gamma_{ll}$	
4-12		$v_5=1^1$	9400.66 (5)*)	3.6$^+$)		$q^{(0)}=18.14$ (5)			
Cont.	$HC{\equiv}^{13}CF$	$v=0$	9700.64 (5)	3.5$^+$)					
		$v_4=1^1$	9709.41 (5)*)	3.6$^+$)		$q^{(0)}=12.78$ (5)			
		$v_5=1^1$	9726.98 (5)*)	3.6$^+$)		$q^{(0)}=19.90$ (5)			
	$DC{\equiv}CF$	$v=0$	8736.028 (10)	2.75 (10)					
		$v_3=1$	8704.86 (3)	3.0 (10)				$\nu_3=1045.9242$ (8) cm^{-1} [80MAT2]	
		$v_4=1^1$	8751.135 (20)*)	2.85 (20)		$q^{(0)}=13.5305$ (2)			
						$q^{(1)}=$ 3.0 (3) $\cdot 10^{-5}$			
		$v_4=2^{0,2}$	8766.42 (5)	2.9$^+$)				$\gamma_{ll}=-0.22$ MHz	
		$v_5=1^1$	8760.580 (10)*)	2.78 (5)		$q^{(0)}=15.6980$ (1)			
						$q^{(1)}=$ 4.26 (10)			
		$v_5=2^{0,2}$	8783.95 (2)	2.5 (5)				$\gamma_{ll}=$ 0.19 MHz;	
								$g_{tt}=$ 35.4 (20) GHz	
		$v_5=3^1$	8807.01 (5)*)	3.0 (5)		$q^{(0)}=15.61$ (5)			
		$v_5=4^{0,2}$	8828.99 (5)	2.8$^+$)				$\gamma_{ll}=$ 0.15 MHz	
		$(v_4=1, v_5=1)^{0,2}$	8775.58 (5)	2.8$^+$)				$\gamma_{ll'}=$ 0.03 MHz	
		$(v_4=1, v_5=2)^1$	8799.20 (5)**)	2.8$^+$)		$q^{(0)}=13.57$ (3)		**) $B_v-\gamma_{ll'}$	
		$(v_4=2, v_5=1)^1$	8790.33 (5)**)	2.8$^+$)		$q^{(0)}=15.69$ (5)			
		$v_3=1, v_4=1, v_5=1$	8744.27 (14)***)	2.8$^+$)	80MAT2			***) Laser Stark	
	$D^{13}C{\equiv}CF$	$v=0$	8486.21 (5)	2.8$^+$)	79JON2				
		$v_4=1^1$	8499.67 (5)*)	2.8$^+$)		$q^{(0)}=12.87$ (5)			
		$v_5=1^1$	8510.16 (5)*)	2.8$^+$)		$q^{(0)}=14.86$ (5)			
	$DC{\equiv}^{13}CF$	$v=0$	8733.90 (5)	2.8$^+$)					
25	$HC{\equiv}C^{35}Cl$	$v=0$	5684.214 (5)	1.39 (2)	78JON			for structure, see also	2.7.2, 23
6-18		$v_3=1$	5665.35 (2)					Vol. II/7 (2, 172)	
4-13		$v_4=1^1$	5687.90 (2)	1.38 (8)		$q=4.30$ (5)	78JON	$\nu_3=756\ cm^{-1}$, C—Cl stretch	
		$v_5=1^1$	5699.328 (10)	1.41 (5)		$q=7.65$ (2)		$\nu_4=604\ cm^{-1}$, C—H bend	
		$v_5=2^2$	5714.32 (3)					$\nu_5=326\ cm^{-1}$, C—Cl bend	
		$v_5=2^0$	5711.94 (3)					Fermi resonance analysis:	
Cont.								$k_{355}=35.2\ cm^{-1}$	

Nr.	Molecule	v	B MHz	D kHz	Ref.	q MHz	Ref.	Varia, remarks	See also Tab., Nr.
25 *6–18* *4–13* Cont.	$HC{\equiv}C^{37}Cl$	$v=0$	5572.355 (5)	1.33 (2)	78JON				
		$v_3=1$	5554.39 (3)					$k_{355}=37.7$ cm^{-1}	
		$v_4=1^1$	5575.92 (3)			$q=4.12$ (5)	78JON		
		$v_5=1^1$	5587.16 (2)	1.25 (10)		$q=7.3$ (4)			
		$v_5=2^2$	5601.87 (3)						
		$v_5=2^0$	5599.17 (3)						
	$DC{\equiv}C^{35}Cl$	$v_3=1$	5171.72 (3)					$\nu_3=742$ cm^{-1}	
		$v_4=1^1$	5194.00 (2)			$q=4.38$ (5)		$\nu_4=472$ cm^{-1}	
		$v_5=1^1$	5200.13 (2)			$q=6.62$ (2)		$\nu_5=312$ cm^{-1}	
		$v_5=2^2$	5213.19 (3)						
		$v_5=2^0$	5211.59 (3)					$k_{355}=34.5$ cm^{-1}	
	$DC{\equiv}C^{37}Cl$	$v_3=1$	5069.32 (3)					$k_{355}=37.0$ cm^{-1}	
		$v_4=1^1$	5091.03 (3)			$q=4.25$ (5)			
		$v_5=1^1$	5097.06 (2)			$q=6.38$ (3)			
		$v_5=2^2$	5109.84 (3)						
		$v_5=2^0$	5108.06 (3)						
26 *6–19*	$HC{\equiv}C^{81}Br$	$v=0$	3978.45		77JON			for structure, see also Vol. II/7 (3, 171)	2.6.2, 15 2.7.2, 44
		$v_3=1$	3976.80						
		$v_4=1^1$	3980.02			$q=2.6$	77JON	$\nu_3=637$ cm^{-1}, C—Br stretch	2.9.2, 11
		$v_5=1^1$	3989.43			$q=4.14$			
		$v_5=2^2$	4000.31					$\nu_4=600$ cm^{-1}, C—H bend	
		$v_5=2^0$	3989.14					$\nu_5=295$ cm^{-1}, C—Br bend	
	$HC{\equiv}C^{79}Br$	$v=0$	4000.08					Fermi resonance between	
		$v_3=1$	3997.50					ν_3 and $2\nu_5$:	
		$v_4=1^1$	4001.69			$q=2.6$		mixing ratio: a/b = 1.45 (151);	
		$v_5=1^1$	4011.10			$q=4.17$		interaction energy:	
		$v_5=2^2$	4022.06					$w_{3,5}=1.313$ (1181) $\delta_{3,5}$ for the ^{81}Br(^{79}Br) species;	
		$v_5=2^0$	4011.28						
	$DC{\equiv}C^{81}Br$	$v=0$	3634.99					$\delta_{3,5}\simeq 25$ cm^{-1} $\rightarrow k_{355}\simeq 44$ cm^{-1}	
		$v_5=1^1$	3644.62			$q=3.62$			
	$DC{\equiv}C^{79}Br$	$v=0$	3655.11						
		$v_5=1^1$	3664.80			$q=3.64$			

Nr.	Molecule	v	B MHz	D kHz	Ref.	q MHz	Ref.	Varia, remarks	See also Tab., Nr.
27 *6–20* *4–14*	HC≡CCN	$v=0$	4549.0583 (8)	0.5444 (15)	78MAL			for structure, see also Vol. II/7 (3, 346)	2.6.2, 16 2.7.2, 14
		$v_4=1$	$\alpha_4=$ 6.255 (4)						
		$v_5=1$	$\alpha_5=$ −1.813 (11)	$\beta_5=$ 0.1067 (470)		$q^{(0)}=2.5811$ (80) $q^{(1)}=8.0$ (32) $\cdot 10^{-5}$	78LAF	$\gamma_{5,7}=-0.093$ (7); $\gamma_{6,6}=-0.019$ (5); $\gamma_{6,7}=$ 0.033 (8); $\gamma_{7,7}=-0.031$ (2); $\gamma_{l_5l_7}=-0.046$ (7); $\gamma_{l_6l_7}=-0.038$ (12); $\gamma_{l_7l_7}=0.0031$ (8) MHz $(q_7^0)^2/g_{77}=2.370$ (25) kHz; $(q_5^0+q_7^0)^2/(r_{57}-2g_{57})$ $=-3.59$ (57) kHz; $(q_5^0-q_7^0)^2/(r_{57}+2g_{57})$ $=-0.60$ (57) kHz; $(q_6^0+q_7^0)^2/(r_{67}-2g_{67})$ $=2.56$ (66) kHz; $(q_6^0-q_7^0)^2/(r_{67}+2g_{67})$ $=-2.05$ (66) kHz; $r_{57J}=-0.044$ (7) MHz; $r_{67J}=$ 0.106 (11) MHz; *) $v_3=2079.30500$ (58) cm^{-1}, diode laser [80YA2]	
		$v_6\geqq 1$	$\alpha_6=$ −9.262 (21)	$\beta_6=-0.0010$ (420)		$q^{(0)}=3.5886$ (43) $q^{(1)}=4.1$ (18) $\cdot 10^{-5}$			
		$v_7\geqq 1$	$\alpha_7=-14.605$ (16)	$\beta_7=$ 0.0176 (32)		$q^{(0)}=6.538254$ (33) $q^{(1)}=1.5983$ (12) $\cdot 10^{-5}$			
		$v_7=3^1$	4592.1931 (43)	0.334 (22)	78LAF	$q^{(0)}=6.58153$ (18) $q^{(1)}=1.134$ (26) $\cdot 10^{-5}$			
		$v_3=1$	0.151212 (10) cm^{-1} *)		80YA2				
		e	4548.091 (48)		78MAL				
	$H^{13}C≡CCN$	$v=0$	4408.438 (3)	0.5444 $^{+}$)	78MAL		78MAL	$^{+}$) fixed, [78LAF] gives 0.5088 (56) kHz, with $B=4408.4423$ (11) MHz and $H=-2.8$ (85) nHz	
		$v_5=1$	$\alpha_5=$ −1.387 (13)			$q=2.428$ (24)			
		$v_6=1$	$\alpha_6=$ −8.960 (11)			$q=3.367$ (22)			
		$v_7=1$	$\alpha_7=-14.067$ (10)			$q=6.175$ (19)			
	$HC≡C^{13}CN$	$v=0$	4530.195 (3)	0.5444 $^{++}$)				$^{++}$) fixed, [78LAF] gives 0.5401 (21) kHz, with $B=4530.19760$ (46) MHz and $H=-0.84$ (242) nHz	
		$v_4=1$	$\alpha_4=$ 6.137 (10)						
		$v_5=1$	$\alpha_5=$ −1.552 (10)			$q=2.516$ (20)			
		$v_6=1$	$\alpha_6=$ −8.983 (10)			$q=3.635$ (20)			
		$v_7=1$	$\alpha_7=-14.184$ (10)			$q=6.529$ (20)			
	$HC≡^{13}CCN$	$v=0$	4529.755 (3)	0.5444 $^{+++}$)				$^{+++}$) fixed, [78LAF] gives 0.5438 (62) kHz with $B=4529.7617$ (13) MHz and $H=0.58$ (896) nHz	
		$v_4=1$	$\alpha_4=$ 6.309 (10)						
		$v_5=1$	$\alpha_5=$ −1.558 (10)			$q=2.519$ (20)			
		$v_6=1$	$\alpha_6=$ −9.075 (10)			$q=3.593$ (20)			
Cont.		$v_7=1$	$\alpha_7=-14.020$ (10)			$q=6.579$ (20)			

Nr.	Molecule	v	B MHz	D kHz	Ref.	q MHz	Ref.	Varia, remarks	See also Tab., Nr.
27 *6-20* *4-14* Cont.	$HC\equiv CC^{15}N$	$v=0$	4416.753 (4)	0.5444**)	78MAL			**) fixed, [78LAF] gives 0.5099 (49) kHz with $B=4416.75278$ (98) MHz and $H=-3.4$ (72) nHz $\gamma_{5,7}=-0.029$ (4); $\gamma_{6,6}=-0.044$ (3); $\gamma_{6,7}=\ 0.091$ (5); $\gamma_{7,7}=-0.002$ (1); $\gamma_{l_7l_7}=-0.011$ (1) MHz $(q_7^0)^2/g_{77}=2.355$ (5) kHz; $(q_5^0+q_7^0)^2/(r_{57}-2g_{57})$ $=-2.52$ (6) kHz; $(q_5^0-q_7^0)^2/(r_{57}+2g_{57})$ $=0.26$ (6) kHz; $(q_6^0+q_7^0)^2/(r_{67}-2g_{67})$ $=3.95$ (6) kHz; $(q_6^0-q_7^0)^2/(r_{67}+2g_{67})$ $=-3.02$ (6) kHz $r_{57J}=-0.062$ (7) MHz; $r_{67J}=\ 0.003$ (7) MHz	
		$v_6=1$	$\alpha_6=-9.044$ (13)			$q=3.394$ (26)	78MAL		
		$v_7=1$	$\alpha_7=-14.009$ (13)			$q=6.205$ (26)			
	$DC\equiv CCN$	$v=0$	4221.5814 (6)	0.4518 (7)					
		$v_4=1$	$\alpha_4=\ 6.019$ (5)						
		$v_5=1$	$\alpha_5=-4.278$ (6)			$q=2.686$ (4)			
		$v_6\geqq 1$	$\alpha_6=-7.844$ (16)	$\beta_6=0.0074$ (43)		$q=3.148$ (4)			
		$v_7\geqq 1$	$\alpha_7=-12.877$ (10)	$\beta_7=0.0179$ (13)		$q=5.894$ (4)			
		$v_7=3$				$q=5.931$ (5)			
		e	4220.103 (52)						
	$D^{13}C\equiv CCN$	$v=0$	4107.205 (3)	0.4518 (fixed)					
		$v_5=1$	$\alpha_5=-4.012$ (10)			$q=2.572$ (20)			
		$v_6=1$	$\alpha_6=-7.438$ (10)			$q=2.981$ (20)			
		$v_7=1$	$\alpha_7=-12.632$ (10)			$q=5.608$ (20)			
	$DC\equiv {}^{13}CCN$	$v=0$	4207.551 (3)	0.4518 (fixed)					
		$v_4=1$	$\alpha_4=\ 6.199$ (10)						
		$v_5=1$	$\alpha_5=-3.876$ (10)			$q=2.684$ (20)			
		$v_6=1$	$\alpha_6=-7.994$ (10)			$q=3.177$ (200)			
		$v_7=1$	$\alpha_7=-12.561$ (10)			$q=5.957$ (20)			
	$DC\equiv C^{13}CN$	$v=0$	4202.515 (3)	0.4518 (fixed)					
		$v_4=1$	$\alpha_4=\ 5.720$ (10)						
		$v_5=1$	$\alpha_5=-4.101$ (10)			$q=2.661$ (20)			
		$v_6=1$	$\alpha_6=-7.682$ (10)			$q=3.185$ (20)			
		$v_7=1$	$\alpha_7=-12.700$ (10)			$q=5.886$ (20)			
	$DC\equiv CC^{15}N$	$v=0$	4100.585 (3)	0.4518 (fixed)				see also 5, appendix (subvolume b)	
		$v_4=1$	$\alpha_4=\ 5.061$ (10)						
		$v_5=1$	$\alpha_5=-4.065$ (10)			$q=2.534$ (20)			
		$v_6=1$	$\alpha_6=-7.662$ (10)			$q=2.981$ (20)			
		$v_7=1$	$\alpha_7=-12.537$ (10)			$q=5.616$ (20)			

Nr.	Molecule	v	B MHz	D kHz	Ref.	q MHz	Ref.	Varia, remarks	See also Tab., Nr.
27a	ClC≡CF							see 5, appendix (subvolume b)	
28	^{35}ClC≡CCN	$v=0$	1382.328 (2)	0.054 (5)	74BJO			for structure, see also	2.6.2, 17
		$v_4=1$	$\alpha=$ 1.0343 (23)					Vol. II/7 (3, 334)	2.7.2, 24
		$v_5=1$	$\alpha=-1.9235$ (4)			$q=0.3582$ (6)	74BJO	$\nu_4=527$ cm^{-1}, C—Cl stretch	
		$v_6=1$	$\alpha=-2.5948$ (14)			$q=0.4444$ (23)		$\nu_5=483$ cm^{-1}, C—C≡N bend	
		$v_7=1$	$\alpha=-3.1905$ (10)			$q=1.0332$ (16)		$\nu_6=333$ cm^{-1}, C—C≡C bend	
								$\nu_7=144$ cm^{-1}, C≡C—Cl bend	
	^{35}Cl^{13}C≡CCN	$v=0$	1381.454 (2)						
	^{35}ClC≡^{13}CCN	$v=0$	1380.357 (2)						
	^{35}ClC≡C^{13}CN	$v=0$	1366.054 (2)						
	^{35}ClC≡CC^{15}N	$v=0$	1343.913 (2)						
	^{37}ClC≡CCN	$v=0$	1350.360 (2)	0.054 (5)					
		$v_4=1$	$\alpha=$ 1.0320 (26)						
		$v_5=1$	$\alpha=-1.8657$ (4)			$q=0.3365$ (7)			
		$v_6=1$	$\alpha=-2.5379$ (12)			$q=0.4131$ (20)			
		$v_7=1$	$\alpha=-3.1109$ (15)			$q=0.9986$ (24)			
	^{37}Cl^{13}C≡CCN	$v=0$	1349.692 (2)						
	^{37}ClC≡^{13}CCN	$v=0$	1348.202 (2)						
	^{37}ClC≡C^{13}CN	$v=0$	1334.089 (2)						
	^{37}ClC≡CC^{15}N	$v=0$	1312.603 (2)						
29	^{79}BrC≡CCN	$v=0$	955.0201 (6)	0.0257 (4)	74BJO			for structure, see also	2.7.2, 45
								Vol. II/7 (3, 329)	
		$v_4=1$	$\alpha=$ 0.8788 (14)					$\nu_4=419$ cm^{-1}, C—Br stretch.	
		$v_5=1$	$\alpha=-1.2035$ (12)			$q=0.1426$ (19)	74BJO	$\nu_5=485$ cm^{-1}, C—C≡N bend	
		$v_6=1$	$\alpha=-1.8808$ (5)			$q=0.2173$ (7)		$\nu_6=312$ cm^{-1}, C—C≡C bend	
		$v_7=1$	$\alpha=-2.0861$ (4)			$q=0.5359$ (8)		$\nu_7=141$ cm^{-1}, C≡C—Br bend	
	^{79}Br^{13}C≡CCN	$v=0$	954.8155 (9)						
	^{79}BrC≡^{13}CCN	$v=0$	950.7730 (9)						
	^{79}BrC≡C^{13}CN	$v=0$	940.0343 (9)						
	^{79}BrC≡CC^{15}N	$v=0$	926.3263 (9)						
	^{81}BrC≡CCN	$v=0$	947.6236 (9)	0.0236 (6)					
		$v_4=1$	$\alpha=$ 0.8805 (14)						
		$v_5=1$	$\alpha=-1.0582$ (17)			$q=0.1453$ (28)			
		$v_6=1$	$\alpha=-1.8681$ (8)			$q=0.2137$ (13)			
Cont.		$v_7=1$	$\alpha=-2.0678$ (2)			$q=0.5391$ (5)			

Nr.	Molecule	v	B MHz	D kHz	Ref.	q MHz	Ref.	Varia, remarks	See also Tab., Nr.
29 Cont.	^{81}Br^{13}C≡CCN	$v=0$	947.3963 (17)		74BJO				
	^{81}BrC≡^{13}CCN	$v=0$	943.3199 (17)						
	^{81}BrC≡C^{13}CN	$v=0$	932.6452 (17)						
	^{81}BrC≡CC^{15}N	$v=0$	919.0699 (17)						
30	IC≡CCN	$v=0$	767.5522 (4)	0.0178 (2)	74BJO			for structure, see also Vol. II/7 (3, 341)	2.7.2, 57
		$v_4=1$	$\alpha=$ 0.8421 (5)					$\nu_4=364$ cm^{-1}, C—I stretch	
		$v_5=1$	$\alpha=-0.8916$ (2)			$q=0.0996$ (4)	74BJO	$\nu_5=496$ cm^{-1}, C—C≡N bend	
		$v_6=1$	$\alpha=-1.5159$ (5)			$q=0.1556$ (9)		$\nu_6=309$ cm^{-1}, C—C≡C bend	
		$v_7=1$	$\alpha=-1.6493$ (2)			$q=0.3842$ (4)		$\nu_7=131$ cm^{-1}, C≡C—I bend	
	I^{13}C≡CCN	$v=0$	766.6689 (2)						
	IC≡^{13}CCN	$v=0$	762.5575 (6)						
	IC≡C^{13}CN	$v=0$	753.9597 (9)						
	IC≡CC^{15}N	$v=0$	743.7545 (5)						
30a	HC≡C—C≡P						see 5, appendix (subvolume b)		2.6.2, 17a
30b	HC≡C—C≡CH						see 5, appendix (subvolume b)		2.6.2, 17b
31	HC≡C—C≡CCN	$v=0$	1331.333 (2)	0.025 (2)	80HUT			see also 5, appendix (subvolume b)	2.6.2, 18
		$v_7=1$				$q=0.215$ (1)	80HUT	$\gamma_{10,11}=3.4$ (2); $\gamma_{11,11}=3.9$ (4);	
		$v_8=1$				$q=0.251$ (1)		$\gamma_{10,10}=3.7$ (4);	
		$v_9=1$				$q=0.325$ (1)		$\gamma_{l_{10},l_{10}}=4.82$ (12); $\gamma_{l_{11},l_{11}}=-3.33$ (2)	
		$v_{10}\geqq1$	$\alpha=2.431$ (2)			$q=0.509$ (2)		$\gamma_{l_{10},l_{11}}=7.33$ (24) kHz.	
		$v_{11}\geqq1$	$\alpha=2.769$ (1)			$q^{(0)}=1.158$ (1)		$g_{11,11}=0.206$ (1) cm^{-1},	
						$q^{(1)}=5.00\,(25)\cdot10^{-3}$		$g_{10,11}=0.247$ (1) cm^{-1}.	
								$\varrho_{10,11}=0.7$ (2) Hz.	
								$r_{10,11}=-0.152$ (1) cm^{-1} +0.30 (16) $J(J+1)$ kHz [80HUT]	
	DC≡C—C≡CCN	$v=0$	1271.057 (1)	0.0244 (20)	76ALE				
	H^{13}C≡C—C≡CCN	$v=0$	1296.677 (2)	0.0299 (20)					
	HC≡^{13}C—C≡CCN	$v=0$	1317.688 (2)	0.0267 (20)					
	HC≡C—^{13}C≡CCN	$v=0$	1329.955 (2)	0.0320 (20)					
	HC≡C—C≡^{13}CCN	$v=0$	1330.115 (2)	0.0269 (20)					
	HC≡C—C≡C^{13}CN	$v=0$	1318.180 (2)	0.0292 (20)					
	HC≡C—C≡CC^{15}N	$v=0$	1298.640 (2)	0.0277 (20)					
32	H(C≡C)$_3$CN	$v=0$	564.00074 (16)	$3.821\,(87)\cdot10^{-3}$	80KIR3				

Nr.	Molecule	v	B MHz	D kHz	Ref.	q MHz	Ref.	Varia, remarks	See also Tab., Nr.
33	HCN···HF	$v=0$	3591.11 (2)	5.2 (7)	80LG1				2.6.2, 19
		$v_\sigma=1$	3529.32 (2)	5.8 (5)				$\nu_\sigma=197\,(15)\,cm^{-1}$,	
		$v_\beta=1^1$	3622.48 (9)	9 (3)		$q=12.91\,(2)$	80LG1	hydrogen bond stretching mode	
		$v_\beta=2^0$	3642.48 (9)	13 (3)				$\nu_\beta=91\,(20)\,cm^{-1}$, hydrogen	
		$v_\beta=2^2$	3653.6 (2)	1 (6)				bond bending mode	
		$v_\beta=3^1$	3673.01 (7)	12 (3)		$q=12.72\,(7)$		harmonic force field	
		$v_\beta=3^3$	3685.53 (6)	8 (4)				calculated	
		$v_\beta=4^0$	3700.4 (4)	17 (14)				dissociation energy:	
		$v_\beta=4^2$	3705.9 (3)	9 (7)				$D_e=26.1\,(16)$ kJ/mol [80LG1]	
		$v_\sigma=1, v_\beta=1^1$	3568.9 (4)	5.5 (fixed)		$q=15.5\,(6)$			
	$HC^{15}N$···HF	$v=0$	3573.65 (4)	7.9 (13)					
		$v_\sigma=1$	3509.92 (12)	6 (4)					
		$v_\beta=1^1$	3603.55 (10)	9 (2)		$q=13.07\,(21)$			
		$v_\beta=2^2$	3633.62 (24)	9 (6)					
	$H^{13}CN$···HF	$v=0$	3496.13 (6)	6.6*)				*) assumed value	
		$v_\sigma=1$	3437.4 (3)	5.9*)					
		$v_\beta=1^1$	3526.95 (11)	9.2*)		$q=12.28\,(15)$			
	DCN···DF	$v=0$	3351.87 (4)	5.5 (8)					
		$v_\sigma=1$	3292.32 (25)	7 (4)					
		$v_\beta=1^1$	3379.86 (11)	8 (3)		$q=12.01\,(22)$		$\nu_\beta=85\,(20)\,cm^{-1}$	
		$v_\beta=2^0$	3403.46 (2)	13*)					
		$v_\beta=2^2$	3408.2 (3)	9*)					
		$v_\beta=3^1$	3430.8 (2)	12*)		$q=11.95\,(9)$			
		$v_\beta=3^3$	3436.7 (3)	8*)					
34	HCN···HCN	$v=0$	1745.80973 (50)	2.133 (30)	81BUX1,				2.7.2,
	HCN···$HC^{15}N$	$v=0$	1700.30190 (30)	1.939 (40)	77LEG				14a
	$HC^{15}N$···HCN	$v=0$	1729.92082 (20)	2.023 (30)					
	$HC^{15}N$···$HC^{15}N$	$v=0$	1684.28825 (25)	1.900 (30)					
	DCN···DCN	$v=0$	1698.37 (20)		77LEG				
34a	NCCN···HF							see 5, appendix (subvolume b)	
35	OC···HF	$v=0$	3063.898 (1)	9.6 (2)	80LG2			see also 5, appendix (subvolume b)	
36	OC···$H^{35}Cl$	$v=0$	1671.7234 (4)	4.802 (9)	80LG2,				2.7.2, 25
	OC···$H^{37}Cl$	$v=0$	1633.3988 (2)	4.574 (6)	81SOP				
	OC···$D^{35}Cl$	$v=0$	1675.8027 (3)	4.507 (6)					
	$O^{13}C$···$H^{35}Cl$	$v=0$	1654.3070 (2)	4.682 (5)					

Nr.	Molecule	v	B MHz	D kHz	Ref.	q MHz	Ref.	Varia, remarks	See also Tab., Nr.
37	OC···H^{79}Br	v = 0	1150.51218 (14)	2.7034 (18)	80KE3				2.7.2, 46 2.9.2, 14
	OC···H^{81}Br	v = 0	1143.23704 (14)	2.6709 (18)					
	OC···D^{79}Br	v = 0	1153.75614 (45)	2.5145 (50)					
	O^{13}C···H^{79}Br	v = 0	1131.18609 (12)	2.6138 (14)					
	^{18}OC···H^{79}Br	v = 0	1079.41084 (48)	2.3460 (47)					
37a	CO_2 · HF							see 5, appendix (subvolume b)	2.6.2, 19b
37b	SCO · HF							see 5, appendix (subvolume b)	2.6.2, 19c
38	Ne · D^{35}Cl	v = 0	2698.1373 (6)	183.07 (5)	80BAR2				2.6.2, 20 2.7.2, 26
	Ne · D^{37}Cl	v = 0	2648.9471 (5)	176.77 (10)					
39	Ar · HF	v = 0	3065.7099 (4)	70.90 (6)	81KEE, 74HAR2				2.6.2, 21 2.7.2, 3
	Ar · DF	v = 0	3039.8529 (4)	59.53 (6)					
40	Ar · H^{35}Cl	v = 0	1678.511 (5)	20.0 (4)	76NOV2				2.6.2, 22 2.7.2, 27
	Ar · H^{37}Cl	v = 0	1631.604	19.0 (4)					
	Ar · D^{35}Cl	v = 0	1657.627	17.1 (4)					
	Ar · D^{37}Cl	v = 0	1611.904	16.4 (4)					
41	Ar · H^{79}Br	v = 0	1106.66951 (17)	12.397 (4)	80KE1, 80JAC				2.6.2, 24 2.7.2, 47 2.9.2, 12
	Ar · H^{81}Br	v = 0	1097.58109 (15)	12.226 (3)					
	Ar · D^{79}Br	v = 0	1087.50889 (27)	8.497 (6)					
	Ar · D^{81}Br	v = 0	1078.59315 (22)	8.352 (5)					
42	Ar · ^{35}ClF	v = 0	1327.113 (5)	4.66 (20)	74HAR1				2.6.2, 23 2.7.2, 28
	Ar · ^{37}ClF	v = 0	1319.650 (5)	4.72 (20)					
43	^{82}Kr · HF	v = 0	2403.4442 (25)	32.31 (15)	81BUX				2.7.2, 55
	^{83}Kr · HF	v = 0	2397.8544 (13)	32.14 (10)					
	^{84}Kr · HF	v = 0	2392.4152 (7)	31.94 (10)					
	^{86}Kr · HF	v = 0	2381.8930 (14)	31.65 (15)					
	^{82}Kr · DF	v = 0	2364.3960 (25)	27.51 (10)					
	^{83}Kr · DF	v = 0	2358.691	27.4					
	^{84}Kr · DF	v = 0	2353.1407 (7)	27.31 (10)					
	^{86}Kr · DF	v = 0	2342.3987 (23)	27.02 (16)					

Nr.	Molecule	v	B MHz	D kHz	Ref.	q MHz	Ref.	Varia, remarks	See also Tab., Nr.
44	$^{82}Kr \cdot H^{35}Cl$	$v=0$	1209.3075 (6)	7.431 (15)	80BAL, 80BAR1			see also 5, appendix (subvolume b)	2.6.2, 25; 2.7.2, 29; 2.9.2, 13
	$^{84}Kr \cdot H^{35}Cl$	$v=0$	1200.6255 (7)	7.389 (16)					
	$^{86}Kr \cdot H^{35}Cl$	$v=0$	1192.3386 (5)	7.273 (12)					
	$^{82}Kr \cdot H^{37}Cl$	$v=0$	1164.7454 (2)	6.935 (5)					
	$^{84}Kr \cdot H^{37}Cl$	$v=0$	1156.0596 (5)	6.795 (12)					
	$^{86}Kr \cdot H^{37}Cl$	$v=0$	1147.7735 (2)	6.691 (4)					
	$^{82}Kr \cdot D^{35}Cl$	$v=0$	1192.4708 (4)	6.612 (8)					
	$^{84}Kr \cdot D^{35}Cl$	$v=0$	1183.7735 (5)	6.530 (12)					
	$^{86}Kr \cdot D^{35}Cl$	$v=0$	1175.4763 (11)	6.489 (23)					
	$^{82}Kr \cdot D^{37}Cl$	$v=0$	1149.1999 (8)	6.126 (19)					
	$^{84}Kr \cdot D^{37}Cl$	$v=0$	1140.5069 (8)	6.026 (17)					
	$^{86}Kr \cdot D^{37}Cl$	$v=0$	1132.2113 (8)	5.909 (18)					
45	$^{82}Kr \cdot H^{79}Br$	$v=0$	692.31400 (14)	2.6579 (14)	80KE1				2.7.2, 48
	$^{84}Kr \cdot H^{79}Br$	$v=0$	684.22952 (15)	2.5973 (11)					
	$^{86}Kr \cdot H^{79}Br$	$v=0$	676.51549 (36)	2.5364 (30)					
	$^{82}Kr \cdot H^{81}Br$	$v=0$	683.71030 (41)	2.6003 (40)					
	$^{84}Kr \cdot H^{81}Br$	$v=0$	675.62442 (14)	2.5266 (11)					
	$^{86}Kr \cdot H^{81}Br$	$v=0$	667.91018 (13)	2.4634 (10)					
	$^{84}Kr \cdot D^{79}Br$	$v=0$	675.84862 (14)	2.1900 (10)					
	$^{84}Kr \cdot D^{81}Br$	$v=0$	667.40217 (20)	2.1384 (15)					
45a	Kr · HCN							see 5, appendix (subvolume b)	
46	$^{82}Kr \cdot {}^{35}ClF$	$v=0$	933.4239 (50)	1.86 (fixed)	75NOV			vibrational force field calculated; see also 5, appendix (subvolume b)	2.6.2, 26; 2.7.2, 30
	$^{84}Kr \cdot {}^{35}ClF$	$v=0$	925.1859 (50)	1.86 (20)					
	$^{86}Kr \cdot {}^{35}ClF$	$v=0$	917.3176 (50)	1.86 (fixed)					
	$^{84}Kr \cdot {}^{37}ClF$	$v=0$	914.4371 (50)	1.86 (fixed)					
46a	Xe · HF							see 5, appendix (subvolume b)	2.6.2, 26a
47	$^{129}Xe \cdot H^{35}Cl$	$v=0$	994.14484 (12)	3.8130 (15)	80KE2, 79CHN			see also 5, appendix (subvolume b)	2.6.2, 27; 2.7.2, 31
	$^{129}Xe \cdot H^{37}Cl$	$v=0$	952.93341 (38)	3.4927 (63)					
	$^{129}Xe \cdot D^{35}Cl$	$v=0$	979.40964 (11)	3.4520 (14)					
	$^{132}Xe \cdot H^{35}Cl$	$v=0$	989.26113 (34)	3.7762 (31)					
	$^{131}Xe \cdot H^{35}Cl$	$v=0$	990.86264 (40)	3.7844 (45)					

2.4 Symmetric top molecules

Rotational constants, centrifugal distortion constants, rotation-vibration interaction constants, isotopic masses, *l*-type doubling constants, Coriolis coupling constants and references for structure data

2.4.1 Preliminary remarks

1. Introduction

In favorable cases, the analysis of the rotational spectrum of symmetric top molecules in vibrational state $v_1 \dots v_j \dots v_{3N-6}$ allows the determination of the constants listed in this table. The rotation-vibration interaction constants are determined by the analysis of at least two vibrational states of the same normal vibration.

The *l*-type doubling constants can be determined from the splitting of degenerate excited rotation-vibration states as a result of which the rotational lines are split. This splitting is caused by the interaction of the angular momentum of the degenerate vibration with the total angular momentum.

For analysis of the spectrum the following Eq. (2) for rotational energy in vibrational state $v_j = v$ is used in most cases.

The special selection rule $\Delta K = 0$ which applies to this type of rotational spectrum does not allow the determination of those constants which are only dependent on K; e.g. D_{Kv} and A_v und C_v.

Note, however, that in some cases "forbidden" $\Delta K = \pm 3$ transitions can be observed [76OKA].

Since the rotational spectrum of a symmetric top molecule generally gives only one rotational constant and therefore provides only one information concerning the molecular structure, the rotational spectrum of isotopic forms of this molecule must also be investigated. In many cases the symmetric top molecule becomes an asymmetric top molecule due to isotope substitution. These asymmetric top molecules appear in this table.

The rovibrational energy E_{RV} may be written in the following form [71AMA, 75BAU2, 75CAE2, 72MIL].

$$E_{RV} = E_V + E_R$$

$$E_V = \sum_s \omega_s \left(v_s + \frac{d_s}{2}\right) + \sum_{\substack{s,s' \\ s \leqq s'}} x_{s,s'} \left(v_s + \frac{d_s}{2}\right)\left(v_{s'} + \frac{d_{s'}}{2}\right) + \sum_{\substack{t,t' \\ t \leqq t'}} x_{l_t l_{t'}} \, l_t \, l_{t'} \tag{1}$$

$$\begin{aligned} E_R = {} & B_V J(J+1) + (A_V - B_V) K^2 - 2(A\zeta_t)_v K l_t \\ & - D_J J^2(J+1)^2 - D_{JK} J(J+1) K^2 - D_K K^4 \\ & + H_J J^3(J+1)^3 + H_{JK} J^2(J+1)^2 K^2 + H_{KJ} J(J+1) K^4 + H_K K^6 \\ & + \eta_J J(J+1) K l_t + \eta_K K^3 l_t \\ & + \Delta_{split} + \Delta_l \end{aligned} \tag{2}$$

with

$$B_V = B_e - \sum_s \alpha_s^B \left(v_s + \frac{d_s}{2}\right) + \sum_{ss'} \gamma_{ss'}^B \left(v_s + \frac{d_s}{2}\right)\left(v_{s'} + \frac{d_{s'}}{2}\right) \tag{3}$$

where s is summed over all normal modes and t over all degenerate modes

J = the rotational angular momentum quantum number

K = the quantum number of the component of the angular momentum in direction of the symmetry axis of the molecule

l_t = the quantum number of the angular momentum which arises in degenerate excited vibrations

$$l = v_t, v_t - 2, \dots, -v_t$$

d_s = degeneracy of the s-th normal vibration.

For the other constants, see explanation in 2.

The formula above are valid for a prolate symmetric-top which is the most common type. To obtain the formular for the oblate symmetric top one replaces the A_v by C_v.

The splitting term Δ_{split} originates from the off-diagonal matrix elements of the centrifugal distortion contribution, and depends on the point group:

- for molecules of threefold or sixfold symmetry, it is non-zero only for the $K = 3$ levels:

$$\Delta(J, 3^\pm) = \pm h_3 J(J+1)[J(J+1)-2][J(J+1)-6] \tag{4}$$

- for molecules of fourfold symmetry, it is non-zero for the $K = 2$ levels:

$$\Delta(J, 2^\pm) = \pm [d_2 + h_2 J(J+1)] J(J+1)[J(J+1)-2]. \tag{5}$$

The l-type doubling term Δ_l comes from the rotational l-type resonance whose matrix element is:

$$\langle K l_t | H | K \pm 2, l_t \pm 2 \rangle = \frac{q_t}{4} \{[J(J+1) - K(K \pm 1)][J(J+1) - (K \pm 1)(K \pm 2)](v_t \pm l_t + 2)(v_t \mp l_t)\}^{1/2}. \quad (6)$$

$$\text{with } q_t = q_0 - q_J\, J(J+1) + \cdots$$

In the particular case $v_t = 1$, a perturbation calculation gives:

$$\Delta_l = \pm \tfrac{1}{2} q_t J(J+1) \quad \text{if } K l_t = 1, \quad (7a)$$

$$\Delta_l = \tfrac{1}{16} q_t^2 \frac{[J(J+1) - K l_t (K l_t - 1)][J(J+1) - (K l_t - 1)(K l_t - 2)]}{(K l_t - 1)(A_e - B_e - A_e \zeta^Z)} \quad \text{if } K l_t \neq 1. \quad (7b)$$

For higher excited states $v_t > 1$ and for the expressions of the other off-diagonal matrix elements, see for instance [71AMA, 75BAU2, 75CAE2].

The rotational energy levels of a spherical top may be obtained by the approximate expression [77WAT]:

$$E(J, \kappa) = BJ(J+1) - DJ^2(J+1)^2 + HJ^3(J+1)^3 + [D_{4T} + H_{4T} J(J+1)]\, f(J, \kappa) + H_{6T}\, g(J, \kappa) \quad (8)$$

where κ is a label specifiying the energy levels, and $f(J, \kappa)$ and $g(J, \kappa)$ are tabulated functions [73KIRS].

2. Explanations of the columns in the tabulation

Column	Parameters		Definitions
1	Nr.		serial number in this table
		6-*Nr.* 4-*Nr.*	the second and third numbers represent respectively the serial number of the molecule in Vol. II/4 and II/6, Landolt-Börnstein, New Series
2	v	v_j	vibrational quantum number of the j-th normal vibration: $j = 1, \ldots, 3N-6$ vibrational ground state: $v_1 = v_2 = \cdots v_j \cdots = v_{3N-6} = 0$; in short $v = 0$ vibrational state: $v_1 = \cdots v_{j-1} = 0$; $v_j = v$; $v_{j+1} = \cdots v_{3N-6} = 0$; in short $v_j = v$
		v_τ	vibrational quantum number of a torsional vibration
		e	this symbol means that the data in the same line of the table are valid for the equilibrium configuration
3	B [MHz]		rotational constant
		B_e	for the equilibrium configuration
		B_v	for the vibrational state $v_j = v$ $B_e = \frac{h}{8\pi^2 I_e}$; $B_v = \frac{h}{8\pi^2 I_{bv}}$
		B^*	for the degenerate excited state $v_t = 1$ $B^* = B_v - D_{JK} + \eta_J - \frac{12(q_{12})^2}{A_e - B_e + 2A_e\zeta^Z}$
	A, B, C [MHz]	A_v, B_v, C_v	rotational constants in vibrational state $v_j = v$ for the asymmetric top molecules which arise from isotope substitution. Constants are defined in 2.5.1 and 2.5.2
		I_{av}, I_{bv}, I_{cv} [amu Å^2], [g cm^2]	principal molecular moments of inertia in vibrational state $v_j = v$ with $I_{cv} \geqq I_{bv} \geqq I_{av}$

Column	Parameters		Definitions
4	D [kHz]		centrifugal distortion constants
		D_J, D_{JK}, D_K	quartic centrifugal distortion constants
		H_J, H_{JK}, H_{KJ}, H_K	sextic centrifugal distortion constants see Eq. (2) and [77WAT]
		$\tau_{\alpha\beta\gamma\delta}$	quartic centrifugal distortion constants according to [36WIL], see also [70GOR]
		η_J, η_K or η_t^J, η_t^K	centrifugal distortion constants for a degenerate vibrational state v_t. See Eq. (2)
		ϱ^*	determinable constant for $v_t=1$ $$\varrho^*=\eta_J-2D_{JK}-\frac{q_0^2}{8(A_e-B_e-A_e\zeta^Z)}-\frac{8(q_{12})^2}{A_e-B_e+2A_e\zeta^Z}$$
		L or G	octic centrifugal distortion constants (coefficients of P^8)
		d_2 or R_6	centrifugal distortion constant, see Eq. (5)
		D_{4T}, H_{4T}, H_{6T}	centrifugal distortion constants of a spherical top, see Eq. (8) and [77WAT]
5	q [MHz]	$q_t^{(0)}$ or $q_t^{(+)}$	l-type doubling constant for a degenerate excited vibrational state v_t. See Eq. (6)
		q_J, q_{JJ}, q_K	coefficients for centrifugal effects in q_t
		q_{12} or r_t	coefficient of the matrix element: $\langle K\,l_t\lvert H\rvert\, K\mp1,\, l_t\pm2\rangle$
		f_{42} or t	coefficient of the matrix element: $\langle K\,l_t\lvert H\rvert\, K\pm4,\, l_t\mp2\rangle$
		g_6	coefficient of the matrix element: $\langle K\,l_t\lvert H\rvert\, K,\, l_t\pm6\rangle$
		$q^{(-)}$	coefficient of the matrix element: $\langle K\,l_t\lvert H\rvert\, K\pm2,\, l\mp2\rangle$
		w	coefficient of the matrix element: $\langle K\,l_t\lvert H\rvert\, K\pm6,\, l_t\rangle$
6			remarks concerning the previous columns and data of constants which have been determined only for a few molecules in this table:
		σ	standard deviation
		ζ_t	Coriolis coupling constant for the t-th degenerate normal vibration
		v_j	type, frequency or wave number of the j-th normal vibration
		$x_{l_t l_t'}$ or $g_{l_t l_t'}$	anharmonicity constant giving the variation of the vibrational energy with l_t, see Eq. (1)
		$\gamma_{l_t l_t'}$	variation of B_v with l_t
		α_s, $\gamma_{s,s'}$	rotation-vibration interaction constants, compare Eq. (3)
		r_e	equilibrium value of the internuclear distance
		k, Φ	force constant
7			references to the following tables; figures in Chapter 3, subvolume b

General remark: Quantities associated with the upper state are indicated by a single prime mark (′) and those associated with the lower state by two prime marks (″).

2.4.2 Data

Nr.	Molecule	v	B MHz	D kHz	Ref.	q MHz	Ref.	Ref. Struct.	Varia, remarks	See also Tab., Nr.
1 *6-1* *4-1*	NH_3	$v=0^s$	298192.85 (18) $C-B$ $=-111471.23$ (72)	D_J $=25468$ (14) $D_{JK}=-47316$ (48) D_K $=30300$ (54) H_J $=7.73$ (13) $H_{JK}=-27.16$ (48) $H_{KJ}=32.32$ (75) H_K $=-6$ (fixed) G_{JJJJ} $=-3.48$ (30) $\cdot 10^{-3}$ G_{JJJK} $=16.9$ (15) $\cdot 10^{-3}$ G_{JJKK} $=-32.1$ (26) $\cdot 10^{-3}$ $G_{JKKK}=27.9$ (26) $\cdot 10^{-3}$ $G_{KKKK}=-1\cdot 10^{-3}$ (fixed)	81URB, 80URB			74HEL1	for structure, see also Vol. II/7 (2, 274) s = symmetric a = antisymmetric diagonal $K=3$ element: η_3 $=\eta_3^0+\eta_3^J J(J+1)$ $+\eta_3^K K^2+\cdots$, η_3^0 $=128.78$ (19) Hz η_3^J $=-0.1670$ (22) Hz $\eta_3^{JJ}=86.6$ (66) $\cdot 10^{-6}$ Hz diagonal $K=6$ element: η_6 $=-0.98$ (22) $\cdot 10^{-9}$ Hz	2.6.2, 28 2.7.2, 15 2.9.2, 15
		$v=0^a$	$\Delta B=-151.5010$ (57) $\Delta(C-B)$ $=211.4379$ (90)	ΔD_J $=-503.29$ (15) *) $\Delta D_{JK}=1388.85$ (39) ΔD_K $=-951.89$ (29) ΔH_J $=-1.1609$ (15) $\Delta H_{JK}=4.7469$ (57) $\Delta H_{KJ}=-6.4434$ (69) ΔH_K $=2.8975$ (36) ΔG_{JJJJ} $=1.8947$ (69) $\cdot 10^{-3}$ ΔG_{JJJK} $=-10.295$ (33) $\cdot 10^{-3}$ ΔG_{JJKK} $=21.010$ (57) $\cdot 10^{-3}$ $\Delta G_{JKKK}=-18.930$ (54) $\cdot 10^{-3}$ $\Delta G_{KKKK}=6.334$ (21) $\cdot 10^{-3}$					$\Delta k=\pm 3$ element: $\alpha=2.401$ (81) MHz component of the inversion doublet: $\Delta E=23785.785$ (72) MHz with $\Delta x=x^{(a)}-x^{(s)}$ *) ΔL_{JJJJJ} $=$ -1.646 (11) ΔL_{JJJJK} $=$ 11.323 (66) ΔL_{JJJKK} $=-31.53$ (16) ΔL_{JJKKK} $=$ 43.46 (22) $\Delta L_{JKKKK}=-29.52$ (15) $\Delta L_{KKKKK}=$ 7.912 (45) mHz inversion spectrum [78YOU]	
Cont.										

Nr.	Molecule	v	B MHz	D kHz	Ref.	q MHz	Ref.	Ref. Struct.	Varia, remarks	See also Tab., Nr.
1 *6-1* *4-1* Cont.	NH_3	$v_2=1^s$	301896.37 (25) $C-B$ $=-119347.62$ (54)	$D_J=33899$ (17) $D_{JK}=-72628$ (45) $D_K=48482$ (57) $H_J=16.72$ (28) $H_{JK}=-66.6$ (11) $H_{KJ}=87.0$ (20) $H_K=-30.3$ (10) $G_{JJJJ}=-7.4$ (12) $\cdot 10^{-3}$ $G_{JJJK}=39.9$ (78) $\cdot 10^{-3}$ $G_{JJKK}=-78$ (21) $\cdot 10^{-3}$ $G_{JKKK}=64$ (22) $\cdot 10^{-3}$ $G_{KKKK}=-10.5$ (78) $\cdot 10^{-3}$	81URB, 80BEL				$\nu_2=932.43384$ (13) cm^{-1} $\eta_3^0=106$ (19) Hz $\eta_3^J=-0.69$ (33) Hz $\eta_3^{JJ}=1.8$ (15) $\cdot 10^{-3}$ Hz $\alpha=3.840$ (11) MHz $\eta_6=0$ (fixed)	
		$v_2=1^a$	$\Delta B=-5400.80$ (12) $\Delta(C-B)$ $=7544.76$ (23)	$\Delta D_J=-13004.4$ (45) $\Delta D_{JK}=35618$ (12) $\Delta D_K=-24170$ (11) $\Delta H_J=-18.269$ (54) $\Delta H_{JK}=73.01$ (22) $\Delta H_{KJ}=-96.08$ (33) $\Delta H_K=41.60$ (20) **)					$\Delta E=35.688107$ (17) cm^{-1} **) $\Delta G_{JJJJ}=12.95$ (21) $\Delta G_{JJJK}=-65.7$ (12) $\Delta G_{JJKK}=121.6$ (27) $\Delta G_{JKKK}=-96.8$ (33) $\Delta G_{KKKK}=27.6$ (17) Hz	
Cont.		$v_4=1$			80URB	$q_0-5q_J-\Delta\eta_J$ $=10361.894$ (4) $f_{42}=1.958$ (40)	80COH		$J=0$ inversion frequency: $\Delta E=32904.98$ (100) MHz [80COH] $f_{42}=K$-type doubling constant	

Nr.	Molecule	v	B MHz	D kHz	Ref.	q MHz	Ref.	Ref. Struct.	Varia, remarks / See also Tab., Nr.
1 *6-1* *4-1* Cont.	$^{15}NH_3$	$v=0$	297392.8 (24) *)	D_J = 25365 (96) *) D_{JK} = −47098 (204) H_J = 8.4 (12) H_{JK} = −32.1 (33) H_{KJ} = 40.2 (41)	80SHI, 80CAR1				*) combined analysis of IR and MW data, inversion spectrum [80SAS]
		$v_2=1$	298598.2 (24) $C'-B'-(C''-B'')$ = −4319.68 (42)	D_J = 27767 (96) D_{JK} = −55813 (207) $D'_K-D''_K$ = 6878.1 (93) H_J = 9.6 (12) H_{JK} = −40.5 (34) H_{KJ} = 53.8 (41) $H'_K-H''_K$ = −6.22 (7)					ν_2 = 945.29533 (15) cm^{-1} ΔE = 34.43729 (30) cm^{-1} inversion parameters (in cm^{-1}): ΔB = −0.177672 (23) $\Delta(C-B)$ = 0.247295 (27) ΔD_J = −4.3221 (49) $\cdot 10^{-4}$ ΔD_{JK} = 1.18310 (100) $\cdot 10^{-3}$ ΔD_K = −8.0231 (62) $\cdot 10^{-4}$ ΔH_J = −5.500 (30) $\cdot 10^{-7}$ ΔH_{JK} = 2.2042 (97) $\cdot 10^{-6}$ ΔH_{KJ} = −2.9067 (110) $\cdot 10^{-6}$ ΔH_K = 1.2594 (45) $\cdot 10^{-6}$ [80SHI]
	$^{15}NH_3$	$v_4=1$				$q_0-5q_J+\Delta\eta_J$ = 10166.022 (4) f_{42} = 2.000 (40)	80COH		$J=0$ inversion frequency: ΔE = 31602.72 (100) MHz
Cont.	NH_2D	$v=0^s$	$\mathscr{A}$ = 290286.53 (21) $\mathscr{B}$ = 192240.15 (42) $\mathscr{C}$ = 140602.50 (42)	Δ_J = 4993.7 (100) Δ_{JK} = 9375.33 (4200) Δ_K = 9789.8 (530) δ_J = 1079.91 (1100) δ_K = 16674 (250) H_{JK} = 48.4054 ± 14 H_{KJ} = −128.660 ± 39 H_K = 114.47 ± 31 h_J = 1.4387 ± 0.46 h_{JK} = −17.97 ± 7.2 h_K = 108.41 ± 32 L_K = −0.1354 (390)	75LUC3				inversion frequency: ΔE = 12169.83 (26) MHz

Nr.	Molecule	v	B MHz	D kHz	Ref.	q MHz	Ref.	Ref. Struct.	Varia, remarks	See also Tab., Nr.
1 *6-1* *4-1* Cont.	NH_2D	$v=0^a$	$\mathscr{A}=290192.14(21)$ $\mathscr{B}=192176.27(42)$ $\mathscr{C}=140625.64(42)$	$\Delta_J=4923.1(100)$ $\Delta_{JK}=9396.26\pm42$ $\Delta_K=9295.3\pm52$ $\delta_J=1052.45\pm11$ $\delta_K=16180\pm240$ $H_{JK}=42.7792\pm14$ $H_{KJ}=-116.117\pm38$ $H_K=105.67\pm30$ $h_J=1.2134\pm0.45$ $h_{JK}=-15.53\pm7.1$ $h_K=96.349\pm31$ $L_K=-0.1094(380)$	75LUC3					
		$v_2=1^s$	$A=294602.7(12)$ $B=192142.0(27)$ $C=138410.0(27)$		80SHI				$v_2=886.2633(6)$ $\Delta E=20.1800(14)\ cm^{-1}$	
		$v_2=1^a$	$A=290767.2(15)$ $B=190694.9(36)$ $C=139351.5(36)$							
Cont.	ND_2H	$v=0^s$	$\mathscr{A}=223222.92(18)$ $\mathscr{B}=160433.96(17)$ $\mathscr{C}=112305.40(17)$	$\Delta_J=4081(20)$ $\Delta_{JK}=-6118.25\pm40$ $\Delta_K=22142.6\pm40$ $\delta_J=1165.38\pm1.5$ $\delta_K=4617.1\pm19$ $H_{JK}=16.4406\pm4$ $H_{KJ}=-44.0645\pm9$ $H_K=43.381\pm6$ $h_J=0.96877\pm0.2$ $h_{JK}=-10.621\pm2$ $h_K=28.025\pm5$	75LUC3				$\Delta E=5118.48(15)$ MHz vibration - rotation interaction constant, coefficient of $\langle v, K, v \mid H \mid J, K+1, v' \rangle$: $D=2.9(1)$ kHz	

Nr.	Molecule	v	B MHz	D kHz	Ref.	q MHz	Ref.	Ref. Struct.	Varia, remarks	See also Tab., Nr.
1 *6-1* *4-1* Cont.	ND_2H	$v=0^a$	$\mathscr{A}=223147.21\ (14)$ $\mathscr{B}=160423.37\ (13)$ $\mathscr{C}=112313.72\ (13)$	$\Delta_J=4075\ (20)$ $\Delta_{JK}=-6030.53\pm30$ $\Delta_K=21493.3\pm30$ $\delta_J=1166.99\pm1.2$ $\delta_K=4523.5\pm15$ $H_{JK}=12.4765\pm3$ $H_{KJ}=-33.3841\pm7$ $H_K=31.281\pm5$ $h_J=0.76366\pm0.2$ $h_{JK}=-7.8206\pm2$ $h_K=22.697\pm4$	75LUC3					
	$^{14}ND_3$	$v=0^s$	154176.16 (6)	$D_J=5945\ (9)$ $D_{JK}=-10496\ (27)$ $H_J=0.717\ (45)$ $H_{JK}=-1.71\ (19)$ $H_{KJ}=1.47\ (33)$	81DEV					
		$v=0^a$	$\Delta B=-5.1975\ (36)$ $\Delta C-\Delta B$ $=7.8360\ (42)$	$\Delta D_J=-6.42\ (51)$ $\Delta D_{JK}=19.6\ (13)$ $\Delta D_K=-14.57\ (84)$ $\Delta H_J=1.08\ (21)\cdot10^{-1}$ $\Delta H_{JK}=-4.62\ (90)\cdot10^{-1}$ $\Delta H_{KJ}=6.8\ (12)\cdot10^{-1}$ $\Delta H_K=-3.33\ (60)\cdot10^{-1}$					$\Delta E=1591.544\ (9)$ MHz with $\Delta x=x^{(a)}-x^{(s)}$ $\eta_3=21.41\ (18)$ Hz	
Cont.		$v_2=1^s$	153569.0 (15) $C'-B'-(C''-B'')$ $=-438.0\ (21)$	$D_J=5888\ (27)$ $D_{JK}=-11047\ (60)$ $H_J=0.63\ (14)$ $H_{JK}=-2.61\ (51)$ $H_{KJ}=3.63\ (60)$					$\nu_2=745.5997\ (7)\ cm^{-1}$ $D'_K-D''_K=746\ (36)$ kHz $H'_K-H''_K=-1.24\ (23)$ kHz	

Nr.	Molecule	v	B MHz	D kHz	Ref.	q MHz	Ref.	Ref. Struct.	Varia, remarks	See also Tab., Nr.
1 *6-1* *4-1* Cont.	$^{14}ND_3$	$v_2=1^a$	$\Delta B=-313.9$ (15) $\Delta C-\Delta B$ $=469.8$ (30)	$\Delta D_J=-486$ (30) $\Delta D_{JK}=1391$ (78) $\Delta D_K=-1022$ (45) $\Delta H_J=-0.38$ (16) $\Delta H_{JK}=1.68$ (57) $\Delta H_{KJ}=-2.64$ (72) $\Delta H_K=1.35$ (33)	81DEV				$\Delta E=106378.4$ (90) MHz $\eta_3=21.41$ Hz (fixed)	
	$^{15}ND_3$	$v=0^s$	153603.40 (12)	$D_J=5927$ (15) $D_{JK}=-10493$ (54) $H_J=0.630$ (90) $H_{JK}=-1.77$ (27) $H_{KJ}=2.07$ (42)						
		$v=0^a$	$\Delta B=-4.80$ (18) $\Delta C-\Delta B$ $=6.69$ (33)	$\Delta D_J=2.8$ (21) $\Delta D_{JK}=-75.6$ (96) $\Delta D_K=0$ (fixed) $\Delta H_J=1.11$ (48) $\cdot 10^{-1}$ $\Delta H_{JK}=-9.0$ (24) $\cdot 10^{-1}$ $\Delta H_{KJ}=12.3$ (27) $\cdot 10^{-1}$ $\Delta H_K=0$ (fixed)					$\Delta E=1432.74$ (15) MHz $\eta_3=21.41$ Hz (fixed)	
		$v_2=1^s$	153060.2 (12) $C'-B'-(C''-B'')$ $=-489.3$ (15)	$D_J=5954$ (18) $D_{JK}=-11191$ (54) $D'_K-D''_K=731$ (21) $H_J=0.824$ (75) $H_{JK}=-3.06$ (30) $H_{KJ}=3.99$ (45) $H'_K-H''_K=-0.45$ (15)					$\nu_2=739.5343$ (9) cm^{-1}	
Cont.		$v_2=1^a$	$\Delta B=-302.2$ (21) $\Delta C-\Delta B$ $=449.4$ (24)	$\Delta D_J=-546$ (36) $\Delta D_{JK}=1553$ (75) $\Delta D_K=-995$ (36) $\Delta H_J=-0.76$ (13) $\Delta H_{JK}=2.40$ (42) $\Delta H_{KJ}=-1.98$ (63) $\Delta H_K=0$ (fixed)					$\Delta E=97295$ (54) MHz $\eta=21.41$ Hz (fixed)	

Nr.	Molecule	v	B MHz	D kHz	Ref.	q MHz	Ref.	Ref. Struct.	Varia, remarks	See also Tab., Nr.
1 *6-1* *4-1* Cont.	NT_3	$v=0$	105565.373 (34)	D_J = 2598.1 (24) $D_{JK} = -4472$ (6)	74HEL1				inversion: $\Delta E = 305.89$ (11) MHz	
2	$N^{35}Cl_3$	$v=0$	3468.603 (3)	D_J = 1.863 (2) $D_{JK} = -3.015$ (6)	74CAZ			74CAZ		2.6.2, 29 2.7.2, 32
	$^{15}N^{35}Cl_3$	$v=0$	3461.311 (4)	D_J = 1.850 (3) $D_{JK} = -2.98$ (1)						
	$N^{37}Cl_3$	$v=0$	3286.97 (2)	D_J = 1.620 (36) $D_{JK} = -2.66$ (9)						
	$N^{35}Cl_2{}^{37}Cl$	$v=0$	$A = 3468.30$ (1) $B = 3347.64$ (1) $C = 1764.7$ (10)							
3 *6-3* *4-3*	PH_3	$v=0$	$B = 133480.129$ (24) $C = 117488.694$ (207)	D_J = 3939.6 (37) $D_{JK} = -5179.0$ (16) D_K = 4127.0 (313) H_J = 545 (132) $\cdot 10^{-3}$ $H_{JK} = -1.4411$ (40) $H_{KJ} = 0.820$ (235) $H_K = -3.984$ (935) *) Δ_3 = 86.3080 (108) $\cdot 10^{-3}$ +)	81BEL, 81BEL1			77HEL	for structure, see also Vol. II/7 (2, 295) $r_e(P-H) = 1.41159$ (60) Å $\measuredangle_e$ HPH = 93.328 (20)° [77HEL] *) with L_{JJJK} = 0.102 (14) Hz, $L_{JJKK} = -0.22$ (14) Hz, L_{JKKK} = 30.51 (673) Hz +) $\Delta_J = -0.009711$ (291) Hz, $\Delta_{JJ} = 0.00000324$ (192) Hz; $\hat{H} = \hat{H}_0 + \frac{1}{2}(\Delta_3 + \Delta_J \hat{P}^2 + \Delta_{JJ} \hat{P}^4) \cdot (\hat{P}_+^6 + \hat{P}_-^6)$	2.6.2, 30 2.9.2, 16
		$v_4=1$				q = 5267.10 (60) $q_J = 9.22$ (6)		81GUA		
	PD_3	$v=0$	$B = 69471.10$ (3) $C = 58974.37$ (5)	D_J = 1021 (2) $D_{JK} = -1312$ (1) D_K = 1023 (2) $\hbar^4 \tau_{xxxz}/4h = 221$ (5) $H_{JJK} = -0.091$ (4) $h_0 = 9.41$ (22) $\cdot 10^{-3}$ $L_{JJJK} = 0.044$ (5) $\cdot 10^{-3}$	77HEL					

Nr.	Molecule	v	B MHz	D kHz	Ref.	q MHz	Ref.	Ref. Struct.	Varia, remarks	See also Tab., Nr.
4 6-4 4-4	PF_3	$v=0$	7819.996 (1)	D_J = 7.8466 (35) D_{JK} = −11.758 (14)	78SMA			78SMA	for structure, see also Vol. II/7 (2, 229) *) constrained value harmonic and anharmonic force fields calculated [78SMA] $\nu_1 = 891.919\ cm^{-1}$ $\nu_2 = 487.718\ cm^{-1}$ $\nu_3 = 859.87\ cm^{-1}$ $\nu_4 = 346.6\ (3)\ cm^{-1}$ Coriolis resonance between ν_1 and ν_3: $\zeta^y_{13} = 0.392$ ζ^c_{33} = 0.4194 [78SMA] ζ^c_{44} = −0.629 [78SMA] r_e(P−F) = 1.561 (1) Å ∢FPF = 97.7 (2)° [77KAW]	2.9.2, 17
		$v_1=1$	7852.393 (2)	D_J = 9.965 (17) D_{JK} = −20.550 (49)						
		$v_2=1$	7811.161 (1)	D_J = 7.714 (4) D_{JK} = −11.473 (28)						
		$v_3=1$	B = 7777.330 (5) $C\zeta_t$ = 2001.8 (48) C = 4784.4*)	D_J = 6.729 (16) D_{JK} = −7.7468 (54) D_K = 5.2*) η_J = 317 (8) τ'_{xxxz} = 0.99 (70)		q = 42.631 (18) q_J = 1.50 (16) · 10^{-3} r = 4.08 (14) t = −6(3) · 10^{-6}	78SMA			
		$v_4=1$	B = 7823.0925 (12) $C\zeta_t$ = −3010.941 (11) C = 4785.965 (10)	D_J = 7.944 (3) D_{JK} = −11.983 (15) D_K = 5.5*) η_J = −39.2 (6) η_K = 35.3 (7)	78SMA 77KAW	q = 29.49308 (5) q_J = 0.16009 (6) · 10^{-3} q_K = −0.293 (9) · 10^{-3} r = 2.9452 (4) t = −0.064807 (16) · 10^{-3} w = 0.0016 · 10^{-6} *)	77KAW			
5 6-5 4-5	$P^{35}Cl_3$	$v_2=1$	2615.277 (1)	D_J = 1.165 (1) D_{JK} = −1.870 (4)	79SMI				for structure, see also Vol. II/7 (2, 148) *) constrained. ζ^c_{44} = −0.738 (12)	
		$v_4=1$	B = 2618.2597 (7) $C\zeta_t$ = −1087.92 (4) C = 1472.98*)	D_J = 1.178 (1) D_{JK} = −1.897 (4) η_J = −4.23 (35) η_K = 4.69 (99)		q = 6.644 (2) q_J = 15 (2) · 10^{-6} r = 0.336 (8)	79SMI			
6 6-7 4-7	AsH_3	$v=0$	B = 112470.596 (23) C = 104884.40 (10)	D_J = 2927.0 (55) D_{JK} = −3723.27 (51) D_K = 3408 (10) H_J = 218 (451) · 10^{-3} H_{JJK} = −1.2778 (46) H_{JKK} = 1.678 (33) H_K = 1.411 (384)*) $\hbar^4 \tau_{xxxz}/4h$ = 679 (3)	80BUR1, 78HEL2			78HEL2	for structure, see also Vol. II/7 (2, 21) *) higher order constants L_J = 5 (11) L_{JJJK} = −0.441 (12) L_{JJKK} = 1.531 (43) L_{JKKK} = −2.02 (30) L_K = −15.0 (34) Hz [80BUR1]	2.6.2, 31 2.7.2, 42 2.9.2, 18

Nr.	Molecule	v	B MHz	D kHz	Ref.	q MHz	Ref.	Ref. Struct.	Varia, remarks See also Tab., Nr.
7 *6–8* *4–8*	AsF_3	$v=0$	5879.0067 (1)	D_J =4.6313 (1) D_{JK}= −6.1757 (3)	78SMI			78SMI	for structure, 2.7.2, 43 see also Vol. II/7 (2, 18) *) constrained harmonic and anharmonic force fields calculated $\nu_1=740.55\ cm^{-1}$ $\nu_2=336.50\ cm^{-1}$ $\nu_3=702.20\ cm^{-1}$ $\nu_4=262.30\ cm^{-1}$ ζ^c_{33}= 0.2434 (11) ζ^c_{44}= −0.5308 (12) r_e=1.7041 (10) Å ∢FAsF=95°46′ (7′) [78SMI]
		$v_1=1$	5874.3896 (1)	D_J =4.5872 (2) D_{JK}= −6.2043 (39)					
		$v_2=1$	5884.1370 (4)	D_J =4.5220 (1) D_{JK}= −5.9095 (7)					
		$v_2=2$	5889.2949 (7)	D_J =4.4026 (36) D_{JK}= −5.6027 (38)					
		$v_3=1$	B=5863.967 (1) $C\zeta_t$=1007.05 (317) C=4137.74 (553)	D_J =4.698 (3) D_{JK}= −6.386 (10) η_J =22.2 (24) τ'_{xxxz}=19.86 (93)		q=6.747 (4) r =2.242 (23)	78SMI		
		$v_4=1$	B=5879.2692 (3) $C\zeta_t$= −2196.07 (11) C=4137.507*)	D_J =4.7186 (9) D_{JK}= −6.3995 (71) η_J = −24.45 (4)		q =10.392 (1) q_J=94.5 (31) · 10^{-6} t =60.8 (16) · 10^{-6}			
8	$As^{35}Cl_3$	$v=0$	2146.773 (2)	D_J =0.7746 (8) D_{JK}= −1.143 (2) H_J =63 (1) · 10^{-7}	78CAZ1			78CAZ1	for structure, see also Vol. II/7 (2, 17) harmonic force field calculated
	$As^{37}Cl_3$	$v=0$	2043.964 (3)	D_J =0.7129 (16) D_{JK}= −1.047 (9) H_J =1.1 (3) · 10^{-6}					
	$As^{37}Cl^{35}Cl_2$	$v=0$	A=2143.595 (1) B=2080.85 (2) C=1314.3 (3)						
	$As^{35}Cl^{37}Cl_2$	$v=0$	A=2108.760 (3) B=2047.13 (2) C=1290.4 (3)						
9	$As^{79}Br_3$ $As^{81}Br_3$	$v=0$ $v=0$	893.269 (5) 873.183 (6)	D_J=0.172 (6) D_J=0.159 (7)	76ROB2			76ROB2	for structure, see also Vol. II/7 (2, 16)

Nr.	Molecule	v	B MHz	D kHz	Ref.	q MHz	Ref.	Ref. Struct.	Varia, remarks	See also Tab., Nr.
10	$Sb^{35}Cl_3$	$v=0$	1753.874 (2)	D_J = 0.5489 (7) D_{JK} = −0.774 (3) H_J = 5.5 (8) · 10^{-7}	76CAZ			76CAZ	for structure, see also Vol. II/7 (2, 153) harmonic force field calculated	
	$^{123}Sb^{35}Cl_3$	$v=0$	1750.362 (2)	D_J = 0.5458 (7) D_{JK} = −0.768 (4) H_J = 6.2 (8) · 10^{-7}						
	$^{121}Sb^{37}Cl_3$	$v=0$	1670.55 (2)	D_J = 0.492 (13)						
	$^{123}Sb^{37}Cl_3$	$v=0$	1667.21 (1)	D_J = 0.493 (7)						
	$^{121}Sb^{35}Cl_2{}^{37}Cl$	$v=0$	A = 1749.478 (3) B = 1702.10 (1) C = 1168.5 (8)							
	$^{123}Sb^{35}Cl_2{}^{37}Cl$	$v=0$	A = 1745.905 (3) B = 1698.71 (1) C = 1171.0 (8)							
11 *6-10*	CH_4	$v=0$		D_T = 132.94341 (71) H_{4T} = −16.9839 (76) · 10^{-3} H_{6T} = 11.0342 (86) · 10^{-3} L_{4T} = 20.27 (24) · 10^{-7} L_{6T} = −26.77 (35) · 10^{-7} L_{8T} = −30.0 (18) · 10^{-7}	75HOL			79HIR2	for structure, see also Vol. II/7 (3, 120) harmonic and anharmonic force fields calculated r_e(C—H) = 1.0870 ± 0.0007 Å [79HIR2]	2.6.2, 32 2.9.2, 19
	$^{13}CH_4$	$v=0$		D_T = 132.98332 (48) *) H_{4T} = −17.0026 (18) · 10^{-3} H_{6T} = 11.1869 (11) · 10^{-3}	76OZ2				*) L_{4T}, L_{6T} and L_{8T} were field fixed at the CH_4 values	
	CH_2D_2	$v=0$	A = 128994.03 *) B = 105103.75 C = 91438.50	Δ_J = 1391.61 (51) *) Δ_{JK} = −473.1 (17) Δ_K = 2188.4 (18) δ_J = 264.93 (27) δ_K = −448.1 (19)	77DER, 75HIR2				*) IR values	
Cont.										

Nr.	Molecule	v	B MHz	D kHz	Ref.	q MHz	Ref.	Ref. Struct.	Varia, remarks	See also Tab., Nr.
11 *6-10* Cont.	CD_4	$v=0$		$D_T=32.6497(17)$ $H_{4T}=-1.9797(44)\cdot 10^{-3}$ $H_{6T}=1.1027(16)\cdot 10^{-3}$	81KRE					
12 *6-11* *4-13*	CH_3F	$v=0$	25536.14942 (70)*)	$D_J=60.235(31)$*) $D_{JK}=439.42(11)$ $H_J=0.88(39)\cdot 10^{-3}$ $H_{JK}=-0.78(55)\cdot 10^{-2}$ $H_{KJ}=0.034(21)$	81HER			76CLA	for structure, see also Vol. II/7 (3, 97) *) combined analysis of IR and MW data, the Coriolis interaction between ν_3 and ν_6 is taken into account $\nu_3=31436560.30(62)$ MHz	2.6.2, 33 2.9.2, 20
		$v_3=1$	$B=25262.819(20)$*) $A'-A''-B'+B''=-18.51(12)$	$D_J=60.43(20)$*) $D_{JK}=400.3(13)$ $D'_K-D''_K=18.7(14)$ $H_J=3.6(7)\cdot 10^{-3}$ $H_{JK}=-2.58(66)\cdot 10^{-2}$ $H_{KJ}=0.158(21)$ $H'_K-H''_K=-0.0662(88)$						
		$v_6=1$	$B=25418.917(47)$ +) $A\zeta_t=46279(33)$ ++) $A=155910(42)$ ++)	$A\zeta_t\eta_J=-562(30)$ +) $A\zeta_t\eta_K=4.8(33)\cdot 10^3$	76HIR2 81HIR	$q=8.70(13)$ +)	76HIR2, 81HIR		$\nu_6=1182.67605(80)\ cm^{-1}$ +) uncertainties are 2.5 σ, D_J and D_{JK} are assumed ++) IR data, uncertainties are 2.5 σ	
		$v_2=1$ $v_5=1$	25432.52 (33)**) $B=25611.60(40)$**) $A\zeta_t=-38772(116)$	$A\zeta_t\eta_J=430(380)$**)	76HIR2	$q=39.58(30)$**) $q_J=+0.036(23)$ $q_{JJ}=0.00028(90)$ $r=76(17)$	76HIR2		**) uncertainties are 2.5 σ, the Coriolis interaction between ν_2 and ν_5 is taken into account: $D=21838.4(82)$ MHz, $\nu_5-\nu_2=252412(112)$ MHz	
Cont.	$^{13}CH_3F$	$v=0$ $v_3=1$ $v_6=1$	24862.785 24542.268 24749.139		75TAN					

Nr.	Molecule	v	B MHz	D kHz	Ref.	q MHz	Ref.	Ref. Struct.	Varia, remarks See also Tab., Nr.
12 *6-11* *4-13* Cont.	$^{13}CH_3F$	$v_2=1$ $v_5=1$	24773.0 (10)***) $B=24925.6$ (17)***) $A\zeta_t=-38.16(60)\cdot 10^3$ (fixed value)	 $A\zeta_t\eta_J=1.0(13)\cdot 10^3$	78HIR3	 q = 28.17 (96) q_J = +0.033 (39) q_{JJ} = 0.0006 (17) r = 87 (55)	 78HIR3		***) uncertainties are 2.5 σ, Coriolis interaction between ν_2 and ν_5 taken into account: $D=21276$ (35) MHz $\nu_5-\nu_2=358.24(87)\cdot 10^3$ MHz
	CD_3F	$v_3=1$ $v_6=1$	20395.776 ± 0.08 20272.47 (4) $A\zeta=0.61950(8)$ cm^{-1} $A'-A''=186.5$ (14)	 $A\zeta_t\eta_J=0.14(3)\cdot 10^3$	75TAN 77DUX	 q = 141.821 (6) q_J = +6.13 (6) · 10^{-3} q_{JJ} = 6.3 (14) · 10^{-7}	 75TAN		 $\nu_6=911.49606(8)$ cm^{-1} [77DUX]
		$v_2=1$ $v_5=1$	20342 (24)$^{+++}$) $B=20552$ (11)$^{+++}$) $A\zeta_t=-22821$ (35)	 $A\zeta_t\eta_J=-1.75(38)\cdot 10^3$	75TAN 75TAN, 81CAL	 q = 31 (23) q_J = −0.0010 (35) q_{JJ} = −4.0 (15) · 10^{-6} r = 58.8 (73)			$^{+++}$) uncertainties are 2.5 σ, Coriolis interaction between ν_2 and ν_5 $\nu_2-\nu_5$ = 1798850 (80873) MHz, $D=15640$ (fixed) MHz
	CH_2DF	$v=0$	A = 119675.0535 (740)*) B = 24043.4415 (720) C = 22959.3732 (720)	Δ_J = 49.371 (110)*) Δ_{JK} = 342.68 (60) Δ_K = 1377.4 (35) δ_J = 2.329 (45) δ_K = 68.7 (360)	76CLA				*) uncertainties are 2 σ
13 *6-12* *4-14*	$CH_3{}^{35}Cl$	$v=0$	13292.87656 (6)	D_J = 18.0931 (10) D_{JK} = 198.794 (6) H_{KJ} = 10.126 (56) · 10^{-3}	81DEM2			76IM	for structure, see also Vol. II/7 (3, 88) 2.6.2, 34; 2.7.2, 33; 2.9.2, 21
Cont.		$v_3=1$	13177.6386 (14)	D_J = 18.120 (12) D_{JK} = 198.949 (73) H_{KJ} = 7.3 (10) · 10^{-3}					

Nr.	Molecule	v	B MHz	D kHz	Ref.	q MHz	Ref.	Ref. Struct.	Varia, remarks See also Tab., Nr.
13	$CH_3{}^{35}Cl$	$v_6=1$	$B^*=13244.1677\,(29)$	$D_J=18.238\,(20)$	81DUB	$q=-14.5474\,(54)$	81DUB		$\nu_6=1018.0685\,(36)\ cm^{-1}$
6–12				$D_{JK}=203.881\,(149)$					[78DER]
4–14				$H_{KJ}=22.3\,(27)\cdot 10^{-3}$					$q_0^2/(A_e-B_e-A_e\zeta)$
Cont.				$\varrho^*=63.80\,(41)$					$=1.60\,(28)$ kHz [81DUB]
				$D'_K-D''_K=79.4\,(66)$*)	75SHI				
				$\eta_J=473.7\,(36)$*)					*) Laser Stark data
			$A\zeta=39273\,(18)$*)	$\eta_K=3.6\,(3)\cdot 10^3$	78DER				
		$v_2=1$	13279.7 (34)	$D_J=17.9$	76IM				$\nu_2=1354.9\ cm^{-1}$
				$D_{JK}=198.2$					$\nu_5=1452.1\ cm^{-1}$
		$v_5=1$	13278.6 (20)	$D_J=17.9$		$q=47.8\,(28)$	76IM		Coriolis interaction
				$D_{JK}=198.2$					between ν_2 and ν_5:
									$D=13138\,(204)$ MHz
	$CH_3{}^{37}Cl$	$v=0$	13088.17043 (29)	$D_J=17.5570\,(34)$	81DEM2				
				$D_{JK}=193.526\,(20)$					
				$H_{KJ}=9.90\,(16)\cdot 10^{-3}$					
		$v_3=1$	12975.8220 (15)	$D_J=17.6497\,(133)$					
				$D_{JK}=193.867\,(81)$					
				$H_{KJ}=12.5\,(11)\cdot 10^{-3}$					
		$v_6=1$	$B^*=13040.2390\,(17)$	$D_J=17.6319\,(115)$	81DUB	$q=-14.3128\,(21)$	81DUB		$\nu_6=1017.6825\,(60)\ cm^{-1}$
				$D_{JK}=198.02\,(43)$					[78DER]
				$H_{KJ}=42\,(27)\cdot 10^{-3}$					$q_0^2/(A_e-B_e-A_e\zeta)$
				$\varrho^*=64.33\,(117)$					$=1.71\,(38)$ kHz [81DUB]
			$A\zeta=39174\,(24)$*)		78DER				
	$^{13}CH_3{}^{35}Cl$	$v=0$	12796.1764 (20)	$D_J=16.904\,(15)$	81DEM2				
				$D_{JK}=189.091\,(94)$					
				$H_{KJ}=9.3\,(16)\cdot 10^{-3}$					
	$^{13}CH_3{}^{37}Cl$	$v=0$	12589.7152 (27)	$D_J=16.437\,(18)$					
				$D_{JK}=184.080\,(153)$					
				$H_{KJ}=12.0\,(31)\cdot 10^{-3}$					
	$CD_3{}^{35}Cl$	$v_3=1$	10752.60 (3)	$D_J=14\,(4)$	76IM				
				$D_{JK}=120\,(12)$					
		$v_6=1$	10814.04 (4)	$D_J=5\,(6)$		$q=1.01\,(4)$	76IM		
				$D_{JK}=100\,(16)$					
Cont.				$A\zeta\eta_J=-210\,(4)$					

Nr.	Molecule	v	B MHz	D kHz	Ref.	q MHz	Ref.	Ref. Struct.	Varia, remarks See also Tab., Nr.
13 *6–12* *4–14* Cont.	$CD_3{}^{35}Cl$	$v_2=1$	$A=78765.20\,(89)$ $B=10805.29\,(26)$	$D_{JK}=145\,(20)$ $D_K=892\,(45)$	77YAM1				$\nu_2=1028.67275\,(15)\,cm^{-1}$ $\nu_5=1059.96970\,(11)\,cm^{-1}$
		$v_5=1$	$A=78030.21\,(109)$ $B=10860.10\,(13)$ $A\,\zeta_t=-25080.77\,(99)$	$D_J=20.9\,(20)$ $D_{JK}=36\,(25)$ $D_K=1197\,(77)$ $A\,\zeta\,\eta_J=-409\,(58)$ $A\,\zeta\,\eta_K=1570\,(190)$		$q=9.42\,(14)$ $q_J=-27.01\,(16)\cdot10^{-3}$ $q_{JJ}=10.05\,(33)\cdot10^{-6}$	77YAM1		Coriolis interaction between ν_2 and ν_5: $D=8756.0\,(43)$ MHz $\zeta^y_{2,5a}=0.57004\,(29)$ $\zeta_5=-0.321424$ [77YAM1]
		$v_3=2$	$B-18\,D_J=10664.43$	$D_{JK}=157$	76IM				
	$^{13}CD_3{}^{35}Cl$	$v_3=1$	$B-18\,D_J=10468.68$						
	$CD_3{}^{37}Cl$	$v_3=1$	10573.14 (3)	$D_J=12\,(4)$ $D_{JK}=90\,(12)$	76IM				
		$v_6=1$	10630.28 (5)	$D_J=10\,(8)$ $D_{JK}=100\,(20)$ $A\,\zeta\,\eta_J=-150\,(4)$		$q=0.54\,(4)$	76IM		
		$v_2=1$	10622.05 (20)	$D_J=10$ $D_{JK}=100$					Coriolis interaction between ν_2 and ν_5: $D=8591\,(6)$ MHz
		$v_5=1$	10676.23 (12)	$D_J=10$ $D_{JK}=100$		$q=9.58\,(20)$			
	$CHD_2{}^{35}Cl$	$v=0$	$Z=95426.0774\,(611)$ $X=11679.7051\,(79)$ $Y=11370.0711\,(80)$	$\Delta_J=12.6698\,(92)$ $\Delta_{JK}=127.356\,(193)$ $\Delta_K=955.26\,(117)$ $\delta_J=0.3102\,(4)$ $\delta_K=59.949\,(223)$ $H_{JK}=0.50\,(2)\cdot10^{-3}$	77MAL1				
	$CHD_2{}^{37}Cl$	$v=0$	$Z=95425.2293\,(1133)$ $X=11485.1293\,(135)$ $Y=11185.5963\,(137)$	$\Delta_J=12.2842\,(123)$ $\Delta_{JK}=123.895\,(214)$ $\Delta_K=958.26\,(139)$ $\delta_J=0.2959\,(3)$ $\delta_K=58.255\,(196)$ $H_{JK}=0.64\,(10)\cdot10^{-3}$					

Nr.	Molecule	v	B MHz	D kHz	Ref.	q MHz	Ref.	Ref. Struct.	Varia, remarks	See also Tab., Nr.
14 *6-13* *4-15*	$CH_3{}^{79}Br$	$v=0$	9568.192831 (44) *)	D_J =9.87337 (39) D_{JK} =128.665 (12) H_J = −0.0041 (12) $\cdot 10^{-3}$ H_{JK}=0.150 (46) $\cdot 10^{-3}$ H_{KJ}=5.911 (66) $\cdot 10^{-3}$	81HER1				for structure, see also Vol. II/7 (3, 85) *) combined analysis of IR and MW data	2.6.2, 35 2.7.2, 49 2.9.2, 22
		$v_6=1$	B=9534.033584(838) *) $A-B-A\zeta_t-2D_K+3H_K$ =111955.794 (106) $A'-B'-A''+B''-6D'_K+15H'_K$ =903.900 (197)	D_J =9.9050 (14) D_{JK} =129.662 (108) D_K-5H_K=2734.9(186) $D'_K-D''_K-15H'_K$ =160.0 (224) η_J =284.588 (128) H_J = −0.0057 (9) $\cdot 10^{-3}$ H_{JK}= −0.958 (227) $\cdot 10^{-3}$ H_{KJ}=11.6 (24) $\cdot 10^{-3}$ H_K =4.425 (948) $H'_K-H''_K$=5.393 (948)		q= −8.95585 (218)	81HER1		$\nu_6+A'-B'-2A\zeta_6-D'_K+H'_K$ =28703473.57 (36) MHz $q_0^2/(A_e-B_e-A_e\zeta)$ =0.725 (26) kHz [81HER1]	
		$v_3=1$	9495.4254 (77)	D_J =9.89716 (1300) D_{JK} =128.528 (37) H_J = −0.01265 (450) $\cdot 10^{-3}$ H_{JK}=0.1363 (120) $\cdot 10^{-3}$ H_{KJ}=5.557 (280) $\cdot 10^{-3}$	77MOS1				$\nu_3=611\ cm^{-1}$	
		$v_3=2$	9423.3612 (140)	D_J =9.96755 (2100) D_{JK} =129.062 (74) H_J = −0.01133 (780) $\cdot 10^{-3}$ H_{JK}=0.1919 (270) $\cdot 10^{-3}$ H_{KJ}=9.219 (870) $\cdot 10^{-3}$						
Cont.										

Nr.	Molecule	v	B MHz	D kHz	Ref.	q MHz	Ref.	Ref. Struct.	Varia, remarks	See also Tab., Nr.
14 *6–13* *4–15* Cont.	$CH_3{}^{81}Br$	$v=0$	9531.830315 (46) *)	D_J = 9.80104 (38) D_{JK} = 127.832 (11) H_J = −0.0056 (10) $\cdot 10^{-3}$ H_{JK} = 0.165 (44) $\cdot 10^{-3}$ H_{KJ} = 5.767 (137) $\cdot 10^{-3}$	81HER1				*) combined analysis of IR and MW data	
		$v_6=1$	B = 9497.812708 (529 *) $A-B-A\zeta_t-2D_K+3H_K$ = 112010.996 (99) $A'-B'-A''+B''-6D_K'+15H_K'$ = 906.002 (61)	D_J = 9.84418 (102) D_{JK} = 128.772 (48) D_K-5H_K = 2417.8 (76) $D_K'-D_K''-15H_K'$ = 79.9 (29) H_J = −0.0057 (6) $\cdot 10^{-3}$ H_{JK} = 0.152 (30) $\cdot 10^{-3}$ H_{KJ} = 4.37 (77) $\cdot 10^{-3}$ η_J = 283.805 (86) H_K = 0.805 (200) $H_K'-H_K''$ = −0.349 (76)		q = −8.89476 (161)	81HER1		$\nu_6+A'-B'-2A\zeta_6-D_K'+H_K'$ = 28701666.03 (25) MHz $q_0^2/(A_e-B_e-A_e\zeta)$ = 0.595 (16) kHz [81HER1]	
		$v_3=1$	9459.5064 (66)	D_J = 9.85485 (1100) D_{JK} = 127.514 (28) H_J = −0.00061 $\cdot 10^{-3}$ H_{JK} = 0.0979 (100) $\cdot 10^{-3}$ H_{KJ} = 5.220 (200) $\cdot 10^{-3}$	77MOS1					
		$v_3=2$	9387.8020 (86)	D_J = 9.855991 (1400) D_{JK} = 127.801 (59) H_J = −0.02253 (520) $\cdot 10^{-3}$ H_{JK} = 0.1250 (180) $\cdot 10^{-3}$ H_{KJ} = 4.443 (670) $\cdot 10^{-3}$						
Cont.										

Nr.	Molecule	v	B MHz	D kHz	Ref.	q MHz	Ref.	Ref. Struct.	Varia, remarks	See also Tab., Nr.
14 *6–13* *4–15* Cont.	$CD_3{}^{79}Br$	$v=0$	7714.6406 (21)	D_J = 5.81462 (200) D_{JK} = 63.788 (35) H_J = 3.79 (57) $\cdot 10^{-6}$ H_{JK} = 0.0562 (7) $\cdot 10^{-3}$ H_{KJ} = 1.472 (23) $\cdot 10^{-3}$	78MOS1					
		$v_3=1$	7666.7073 (3100)	D_J = 6.69519 (13000) D_{JK} = 64.702 (73) H_J = 0.1237 (190) $\cdot 10^{-3}$ H_{JK} = 0.0621 (160) $\cdot 10^{-3}$ H_{KJ} = 1.122 (30) $\cdot 10^{-3}$	77MOS1					
		$v_3=2$	7614.6646 (2300)	D_J = 5.6933 (960) D_{JK} = 66.231 (660) H_J = −0.0229 (140) $\cdot 10^{-3}$ H_{JK} = 0.7666 (1400) $\cdot 10^{-3}$ H_{KJ} = 27.945 (730) $\cdot 10^{-3}$						
	$CD_3{}^{81}Br$	$v=0$	7681.84 (37)	D_J = 6.031 (170) D_{JK} = 63.267 (110) H_J = 42.16 (250) $\cdot 10^{-6}$ H_{JK} = 0.0596 (240) $\cdot 10^{-3}$ H_{KJ} = 1.327 (15) $\cdot 10^{-3}$	78MOS1					
Cont.		$v_3=1$	7631.0958 (1100)	D_J = 5.5354 (450) D_{JK} = 64.036 (63) H_J = −0.0366 (64) $\cdot 10^{-3}$ H_{JK} = 0.0573 (130) $\cdot 10^{-3}$ H_{KJ} = 0.794 (25) $\cdot 10^{-3}$	77MOS1					

Nr.	Molecule	v	B MHz	D kHz	Ref.	q MHz	Ref.	Ref. Struct.	Varia, remarks	See also Tab., Nr.
14 *6–13* *4–15* Cont.	$CD_3{}^{81}Br$	$v_3=2$	7582.2942 (37)	D_J = 5.775 D_{JK} = 64.483 (740) H_J = −0.00254 (47) $\cdot 10^{-3}$ H_{JK} = 0.5357 (1500) $\cdot 10^{-3}$ H_{KJ} = 34.149 (2140) $\cdot 10^{-3}$	77MOS1					
	$^{13}CH_3{}^{79}Br$	$v=0$	9199.4644 (40)	D_J = 9.08519 (580) D_{JK} = 120.713 (27) H_J = 0.001285 $\cdot 10^{-3}$ H_{JK} = 0.1008 (1000) $\cdot 10^{-3}$ H_{KJ} = 5.579 (150) $\cdot 10^{-3}$	78MOS1					
	$^{13}CH_3{}^{81}Br$	$v=0$	9082.8321 (52)	D_J = 9.02932 (780) D_{JK} = 119.986 (40) H_J = 0.005827 (2600) $\cdot 10^{-3}$ H_{JK} = 0.1537 (120) $\cdot 10^{-3}$ H_{KJ} = 6.353 (270) $\cdot 10^{-3}$						
	$^{13}CD_3{}^{79}Br$	$v=0$	7441.2924 (39)	D_J = 5.389 D_{JK} = 60.636 (130) H_J = 0.007421 (210) $\cdot 10^{-3}$ H_{JK} = 0.0018 $\cdot 10^{-3}$ H_{KJ} = 1.237 (110) $\cdot 10^{-3}$						
	$^{13}CD_3{}^{81}Br$	$v=0$	7407.6900 (15)	D_J = 5.39586 (31) D_{JK} = 60.430 (26) H_{JK} = 0.0617 (49) $\cdot 10^{-3}$ H_{KJ} = 1.280 (93) $\cdot 10^{-3}$						

Nr.	Molecule	v	B MHz	D kHz	Ref.	q MHz	Ref.	Ref. Struct.	Varia, remarks	See also Tab., Nr.
15 *6–14* *4–16*	CH_3I	$v=0$	7501.275699 (21)	D_J = 6.3070 (2) D_{JK} = 98.7625 (33) H_J = −0.00506 (61) $\cdot 10^{-3}$ H_{JK} = 0.043 (19) $\cdot 10^{-3}$ H_{KJ} = 4.506 (35) $\cdot 10^{-3}$	78BOU			75MAL	for structure, see also Vol. II/7 (3, 108) harmonic force field calculated [75MAL]	2.6.2, 36 2.7.2, 58 2.9.2, 23
		$v_2=1$ $v_3=1$ $v_3=2$	α_2^B = 41.13 (4) *) α_3^B = 54.42 (4) *) $2\alpha_3^B$ = 108.90 (12) *)		73KAW				*) calculated with B_0 = 7501.25 MHz, uncertainties are 2.5 σ	
		$v_6=1$	B^* = 7477.76697 (63)	D_J = 6.3591 (22) D_{JK} = 98.811 (46) H_{JK} = 0.287 (101) $\cdot 10^{-3}$ H_{KJ} = 2.78 (95) $\cdot 10^{-3}$ ϱ^* = 3.755 (127)	81DUB	q = −5.91920 (88)	81DUB, 80AR2		$\nu_6 + A\zeta^2$ = 882.91083 (31) cm^{-1} **) Diode laser results $q_0^2/(A_e - B_e - A_e\zeta)$ = 0.234 (30) kHz [81DUB]	
			A = 156119.59 (48) **) $A\zeta_t$ = 32891.2 (22) **) B = 7477.7395 (84) **)	η_J = 216.7 (14) **)	80DAS1					
		$v_5=1$	B = 7500.00 (18) ***) $2[A(1-\zeta)-B+\eta_K]$ = 367554.5 (90) $A'-B'-A''+B''+3/2\,\eta_K$ = −1386.84 (90)	D_J = 6.60 (6) ***) D_{JK} = 110.0 (18) $D_K' - 1/4\,\eta_K$ = 4227.1 (60) $D_K'' - 1/4\,\eta_K$ = 4320.0 (90) η_J = −6 (12)	74MAT	q = −0.81 (12) ***)	74MAT		$\nu_5 + A(1-\zeta)^2 - B + 3/4\,\eta_K$ = 1442.376 (1) cm^{-1} ***) combined analysis of IR and MW data, uncertainties are 2.5 σ, the constants given are unperturbed by Fermi and Coriolis interaction.	
Cont.		$v_3=1,\ v_6=1$	B = 7422.86 (15) ***) $2[A(1-\zeta)-B+\eta_K]$ = 234588 (180) $A'-B'-A''+B''+3/2\,\eta_K$ = 1244 (27)	D_J = 6.30 (30) ***) D_{JK} = 108 (27) $D_K' - 1/4\,\eta_K$ = 3598 (300) η_J = 180 (180)	74MAT	q = −5.37 (42)	74MAT		$\nu_{3+6} + A(1-\zeta)^2 - B + 3/4\,\eta_K$ = 1413.76 (1) cm^{-1} k_{356} = 29.6149 (7) cm^{-1}, δ = −1.04 (2) $\cdot 10^{-4}$ cm^{-1}, ξ_{536} = 0.03267 (4) cm^{-1} [74MAT]	

Nr.	Molecule	v	B MHz	D kHz	Ref.	q MHz	Ref.	Ref. Struct.	Varia, remarks See also Tab., Nr.
15 *6-14* *4-16* Cont.	$^{13}CH_3I$	$v=0$	7119.151 (99)	D_J = 5.783 (22) D_{JK} = 86.4 (18)	81DAS				
		$v_6=1$	7097.0348 (84) $2[A(1-\zeta)-B+\eta_K]$ = 231406 (12) $A'-B'-(A''-B'')$ $+3/2\,\eta_K$ = 1060.2 (24)	D_J = 5.825 (23) D_{JK} = 86.4 (18) $D_K-1/4\,\eta_K$ = 1277 (42) η_J = 187.1 (42)					$\nu_6+A(1-\zeta)^2-B+3/4\,\eta_K$ = 881.91830 (67) cm^{-1} with $D'_{JK}=D''_{JK}$ and $D'_K=D''_K$ assumed diode laser results
	CD_3I	$v=0$	6040.293 (4)	D_{JK} = 46.7 (18)	73KUC				
		$v_2=1$	6007.873 (8) $A'-A''$ = 234.0 (11) $^{+})$	D_{JK} = 106.0 (20) $D'_K-D''_K$ = 121 (85) $^{+})$	77KAW2, 78CAL				ν_2 = 949.35995 (3) cm^{-1} $^{+})$ Laser Stark results, uncertainties are 2.5 σ
		$v_3=1$	6003.266 (4)	D_{JK} = 47.1 (20)	73KUC				
		$v_3=2$	5966.373 (10)	D_{JK} = 69.0 (40)					
		$v_5=1$	$B^*-18\,D_J$ = 6051.210 (14)	D_{JK} = 1.5 (100) ϱ^* = −664.0 (140)		q = 17.82 (6)	74MAT		
		$v_6=1$	$B^*-18\,D_J$ = 6025.527 (6)	D_{JK} = 46.2 (40) ϱ^* = −49.8 (80)		q = 4.21 (2)			
		$v_3=1,\ v_6=1$	$B^*-18\,D_J$ = 5988.447 (8)	D_{JK} = 46.3 (24) ϱ^* = −31.0 (52)		q = 4.19 (6)			
	$^{13}CD_3I$	$v=0$	$B-18\,D_J$ = 5801.792 (2) B = 5801.857 (10) $^{++})$	D_{JK} = 44.99 (85)	75MAL				$^{++})$ calculated assuming $D_J(^{13}CD_3I)=D_J(CD_3I)$
	CH_2DI	$v=0$	Z = 118905.6615 (443) X = 6980.6766 (23) Y = 6883.4852 (18)	Δ_J = 5.2045 (309) Δ_{JK} = 73.16 (157) Δ_K = 1801.6 (652) δ_J = 0.0726 (1) δ_K = 11.984 (221)	75MAL				
Cont.		$v''_6=1$	A = 119813.3 (12) B = 6960.446 (57) C = 6865.637 (48)	Δ_J = 5.1645 (66) Δ_{JK} = 83.07 (96) Δ_K = 2644 (54) δ_J = 0.0462 (54) δ_K = −244 (22) H_{KJ} = 0.252 (36)	81DAS				ν''_6 = 862.58230 (19) cm^{-1} diode laser results

Nr.	Molecule	v	B MHz	D kHz	Ref.	q MHz	Ref.	Ref. Struct.	Varia, remarks	See also Tab., Nr.
15 *6-14* 4-*16* Cont.	$^{13}CH_2DI$		Z = 118720 (fixed) X = 6654.2397 (69) Y = 6565.4354 (74)	Δ_J = 4.750 (371) Δ_{JK} = 69.83 (232)$^{+++}$)	75MAL				$^{+++}$) the other centrifugal distortion constants are constrained to values predicted by force field	
16 *6-39* 4-*48*	CH_3CN	$v=0$	9198.9009 (15)	D_J = 3.8075533 ±0.0019 D_{JK} = 177.33849 ±0.018 H_J = −0.90 (76) · 10^{-6} H_{JK} = 0.9874 (71) · 10^{-3} H_{KJ} = 5.756 (52) · 10^{-3}	78MOS2, 80BOU1			79DEM, 78HAL	for structure, see also Vol. II/7 (3; 218) harmonic force field calculated [78HAL]	2.6.2, 37 2.7.2, 16 2.9.2, 32
		$v_4=1$	9152.6361 (18) $A' - A''$ = −212.09 (11)	D_J = 3.79394 (240) D_{JK} = 184.325 (46) H_J = −3.9 (10) · 10^{-6} H_{JK} = 1.3284 (160) · 10^{-3} H_{KJ} = −11.60 (32) · 10^{-3}	78MOS2, 75BAU1, 78ROM				ν_4 = 920.288690 (22) cm^{-1}	
		$v_8=1$	B^* = 9226.6402 (11)	D_J = 3.886 (15) D_{JK} = 177.972 (61) ϱ^* = 28.86 (3)	80BOU1					
	$CH_3C^{15}N$	$v=0$	8922.0402 (18)	D_J = 3.55398 (330) D_{JK} = 169.0387 (500) H_J = −0.0012 · 10^{-3} H_{JK} = 0.9377 (110) · 10^{-3} H_{KJ} = 7.034 (640) · 10^{-3}	77MOS2					
Cont.		$v_4=1$	8878.7778	D_J = 3.560 D_{JK} = 173.1	75BAU1					

Nr.	Molecule	v	B MHz	D kHz	Ref.	q MHz	Ref.	Ref. Struct.	Varia, remarks	See also Tab., Nr.
16 *6–39* *4–48* Cont.	$CH_3C^{15}N$	$v_8=1$	8948.912 $A-A\zeta=19420$	$D_J=3.67$ $D_{JK}=169.7$ $\eta_J=367.2$	75BAU2	$q/4=4.2133$	75BAU2			
		$v_8=2$	8974.706 $x-A=11280$	$D_J=3.73$ $D_{JK}=170.3$ $\eta_J=370.3$						
	$CH_3{}^{13}CN$	$v=0$	9194.3495 (28)	$D_J=3.80228$ (330) $D_{JK}=176.7794$ (580) $H_J=-3.85$ (120) $\cdot 10^{-6}$ $H_{JK}=0.9846$ (200) $\cdot 10^{-3}$ $H_{KJ}=7.614$ (570) $\cdot 10^{-3}$	77MOS2, 79DEM					
	$^{13}CH_3CN$	$v=0$	8933.3075 (62)	$D_J=3.6185$ (120) $D_{JK}=168.3142$ (790) $H_J=-4\cdot 10^{-6}$ $H_{JK}=0.9846$ (700) $\cdot 10^{-3}$ $H_{KJ}=6.177$ (370) $\cdot 10^{-3}$						
	CD_3CN	$v=0$	7857.9689 (43)	$D_J=2.74552$ (510) $D_{JK}=110.63336$ (700) $H_J=-4.30$ (140) $\cdot 10^{-6}$ $H_{JK}=0.58496$ (190) $\cdot 10^{-3}$ $H_{KJ}=2.0659$ (180) $\cdot 10^{-3}$						
Cont.	$CD_3C^{15}N$	$v=0$	7619.2867 (104)	$D_J=2.5539$ (120) $D_{JK}=105.4518$ (300) $H_J=-6.37$ (320) $\cdot 10^{-6}$ $H_{JK}=0.5554$ (110) $\cdot 10^{-3}$ $H_{KJ}=2.007$ (190) $\cdot 10^{-3}$						

Nr.	Molecule	v	B MHz	D kHz	Ref.	q MHz	Ref.	Ref. Struct.	Varia, remarks	See also Tab., Nr.
16 *6-39* *4-48* Cont.	$CD_3{}^{13}CN$	$v=0$	7848.5247 (78)	D_J = 2.7506 (92) D_{JK} = 109.8486 (520) H_J = −3.79 (260) · 10^{-6} H_{JK} = 0.6063 (110) · 10^{-3} H_{KJ} = 2.046 (380) · 10^{-3}	77MOS2, 79DEM					
	$^{13}CD_3CN$	$v=0$	7695.2998 (44)	D_J = 2.64895 (500) D_{JK} = 106.9112 (260) H_J = −1.39 · 10^{-6} H_{JK} = 0.5549 (60) · 10^{-3} H_{KJ} = 2.204 (160) · 10^{-3}						
	CHD_2CN	$v=0$	B_z = 96500.2674 (373) B_x = 8320.0735 (30) B_y = 8164.4282 (30)	D_J = 3.0903 (34) D_{JK} = 126.2140 (808) D_K = 1213.57 (319) d_1 = −0.04786 (14) d_2 = −0.010712 (15) H_{KJ} = 60.14 (418) · 10^{-3} H_K = −1.941 (275) h_1 = −0.40 (3) · 10^{-6} h_2 = 0.26 (1) · 10^{-6}	78HAL					
17 *6-40* *4-49*	CH_3NC	$v=0$	10052.8855 (31)*)	D_J = 4.690 (21)*) D_{JK} = 227.168 (14) H_{KJ} = 7.3 (20) · 10^{-3}	78ROM			78HAL	for structure, see also Vol. II/7 (3, 219) *) uncertainties are 3 σ. harmonic force field calculated [78HAL] ν_4 = 944.903946 (33) cm^{-1}	2.6.2, 38
		$v_4=1$	9992.9474 (74)*) $A'-A''$ = −254.00 (26)	D_J = 4.693 (60) D_{JK} = 225.37 (20) $D'_K-D''_K$ = 63 (19) $H'_K-H''_K$ = 1.38 (36)						
Cont.		$v_8=1$	B^* = 10092.311	D_J = 4.875 D_{JK} = 226.9 ϱ^* = 54.02	77GOD	$q/4$ = 6.9394 $q_J/4$ = 0.043 · 10^{-3}	77GOD			

Nr.	Molecule	v	B MHz	D kHz	Ref.	q MHz	Ref.	Ref. Struct.	Varia, remarks See also Tab., Nr.
17 *6-40* *4-49* Cont.		$v_8=2$	10131.2809	D_J = 5.082 D_{JK} = 227.5 η_J = 588.61	77GOD	$q/4$ = 6.9636 $q_J/4 = 0.047 \cdot 10^{-3}$	77GOD		$\gamma_{ll} = -0.3283$ MHz ζ^z = 0.92723 x_{ll} = 152254 MHz [77GOD]
		$v_8=3$	10169.9594	D_J = 5.282 D_{JK} = 228.1 η_J = 592.6	76BAU2	$q/4$ = 6.9871 $q_J/4 = 0.045 \cdot 10^{-3}$	76BAU2		$\gamma_{ll} = -0.3370$ MHz ζ^z = 0.92753 x_{ll} = 152650 MHz [76BAU2]
		$v_8=4$	10208.3919	D_J = 5.460 D_{JK} = 228.45 η_J = 596.43	77BAU				$\gamma_{ll} = -0.34902$ MHz x_{ll} = 152847 MHz ζ^z = 0.92793 [77BAU]
	$CH_3{}^{15}NC$	$v=0$	10044.3286	D_J = 4.68 D_{JK} = 225.3	74GOD				
		$v_4=1$	9984.4447	D_J = 4.745 D_{JK} = 224.5	75BAU1				
		$v_8=1$	B^* = 10082.297	D_J = 4.873 D_{JK} = 225.4 $\varrho^* = -15.4$	77GOD	$q/4$ = 7.0657 $q_J/4 = 0.037 \cdot 10^{-3}$	77GOD		
		$v_8=2$	10119.8598	D_J = 5.101 D_{JK} = 225.6 η_J = 586.3					$\gamma_{ll} = -0.3387$ MHz x_{ll} = 151193 MHz ζ^z = 0.93087 [77GOD]
		$v_8=3$	10157.1481	D_J = 5.286 D_{JK} = 226.2 η_J = 590.6	76BAU2	$q/4$ = 7.1126 $q_J/4 = 0.033 \cdot 10^{-3}$	76BAU2		$\gamma_{ll} = -0.3488$ MHz x_{ll} = 151525 MHz ζ^z = 0.93117 [76BAU2]
	$CH_3{}^{15}NC$	$v_8=4$	10194.2002	D_J = 5.5415 D_{JK} = 226.02 η_J = 593.90	77BAU				$\gamma_{ll} = -0.35925$ MHz x_{ll} = 151790 MHz ζ^z = 0.93149 g_6 = 0.6 MHz [77BAU]
	$CH_3{}^{15}N^{13}C$	$v=0$	$B - 98D_J$ = 9689.73	D_{JK} = 213	74GOD				
Cont.	$^{13}CH_3{}^{15}NC$	$v=0$	$B - 98D_J$ = 9760.76	D_{JK} = 215					

Nr.	Molecule	v	B MHz	D kHz	Ref.	q MHz	Ref.	Ref. Struct.	Varia, remarks	See also Tab., Nr.
17- *6–40* *4–49* Cont.	CHD_2NC	$v=0$	$B_z=96044.1705\,(335)$ $B_x=9096.6708\,(31)$ $B_y=8910.5106\,(31)$	$D_J=3.7220\,(50)$ $D_{JK}=158.715\,(111)$ $D_K=1183.48\,(286)$ $d_1=-0.06058\,(25)$ $d_2=-0.015157\,(30)$ $H_{KJ}=0.11451\,(343)$ $H_K=-3.352\,(194)$ $h_1=-1.25\,(6)\cdot 10^{-6}$ $h_2=6.5\,(2)\cdot 10^{-7}$	78HAL					
18 *6–15* *4–17*	CF_3H	$v=0$ $v_6=1$	$B-C=4675.36\,(10)$ $B=10352.88\,(3)$*) $C=5663.7\,(8)$ $C\zeta_t=-4558.0\,(8)$	 $\eta_J=-38.4\,(3)$***) $\eta_K=41.7\,(3)$	81MEE 78KAW1	 $q=36.27896\,(7)$*) $q_J=0.17293\,(9)\cdot 10^{-3}$ $q_K=-0.304\,(3)\cdot 10^{-3}$ $r=0.976\,(4)$ $t=-0.02895\,(3)\cdot 10^{-3}$ $w=0.0035\,(3)\cdot 10^{-6}$	 78KAW1 76KAW	78KAW1	for structure, see also Vol. II/7 (3, 55) *) uncertainties are 2 σ r_e(C—H) = 1.091 (14) Å r_e(C—F) = 1.3284 (31) Å ∢$_e$ FCF = 108.58 (34)°	2.6.2, 39 2.7.2, 4 2.9.2, 24
	CF_3D	$v=0$	$B=9921.0318$ $C=5673.23$	$D_K=221.8\,(121)$**)	80IBI					
		$v_5=1$	$B=9911.7382$ $C=5672.48\,(74)$**) $C\zeta_t=4151.97\,(115)$	$D_J=10.2\,(1)$**) $D_{JK}=-13.1\,(25)$ $D_K=331.2\,(112)$ $\eta_J=26.3\,(162)$ $\eta_K=-202.7\,(1163)$		$q=0.6219$	80IBI		$\nu_5=975.4317\,(3)\,cm^{-1}$, **) Laser Stark results, uncertainties are 2.5 σ	
		$v_6=1$	$B=9923.08\,(4)$*) $C=5662.4\,(7)$ $C\zeta_t=-4258.1\,(9)$	$\eta_J=-46.3\,(5)$*)***) $\eta_K=51.4\,(3)$	78KAW1	$q=31.84634\,(19)$*) $q_J=0.13826\,(17)\cdot 10^{-3}$ $q_K=-0.2171\,(17)\cdot 10^{-3}$ $r=1.298\,(6)$ $t=-0.03097\,(3)\cdot 10^{-3}$ $w=0.0032\,(3)\cdot 10^{-6}$	78KAW1 76KAW		***) definition: $\eta_J=-2C\zeta\eta_J$ $\eta_K=-2C\zeta\eta_K$	

Nr.	Molecule	v	B MHz	D kHz	Ref.	q MHz	Ref.	Ref. Struct.	Varia, remarks	See also Tab., Nr.
19	CF_4	$v=0$	5731.59 (3)	D_0 =0.94 (2)	81TAK					2.6.2, 40
6–17				D_{044} =0.1459 (16)						2.9.2, 24
		$v_3=1$	5720.40 (2)	α_{220} =4047 (19)						
			$B\zeta$=4622.77 (3)	F_{110} =75.3 (18)					ν_3=1283.71933 (98) cm^{-1}	
				D =0.73 (16)					ζ_3=0.808119 (8)	
				G_{220} =0.20 (4)						
				α_{224} = −751.2 (2)						
				F_{134} =12.60 (5)						
				D_{044} =0.157 (3)						
				G_{244} = −0.1342 (18)						
				G_{246} =0.0719 (6)						
				H_{044} = −0.024 (3) $\cdot 10^{-3}$						
				H_{066} =0.0436 (12) $\cdot 10^{-3}$						
20	$CF_3{}^{79}Br$	$v=0$	2098.091 (1)	D_{JK}=1.25 (5)	80COX			80COX	for structure,	2.6.2, 41
		$v_1=1$	α_B=6.35		76JON2				see also	2.7.2, 50
		$v_3=1$	2096.56						Vol. II/7 (3, 3)	
		$v_6=1$	2094.47	D_J =0.31		q=3.44	76JON2		ν_1=1085 cm^{-1},	
				D_{JK}=1.33					CF_3 stretch	
		$v_1=1, v_6=1$	2088.13	D_J =0.22					ν_3= 350 cm^{-1},	
				D_{JK}=1.36					C—Br valence	
	$CF_3{}^{81}Br$	$v=0$	2078.529 (2)	D_{JK}=1.26 (3)	80COX				ν_6= 297 cm^{-1}, CF_3 rock	
	$^{13}CF_3{}^{79}Br$	$v=0$	2092.060 (3)	D_{JK}=1.25 (fixed)						
	$^{13}CF_3{}^{81}Br$	$v=0$	2072.402 (4)	D_{JK}=1.26 (fixed)						
21	CF_3I	$v=0$	1523.250 (1)	D_{JK}=1.00 (3)	80COX			80COX	for structure,	2.6.2, 42
		$v_1=1$	1517.50 (5)	D_J =0.1	80KOH				see also	2.7.2, 59
				D_{JK}=0.6 (fixed)					Vol. II/7 (3, 21)	
									ν_1=1075.1915 (10) cm^{-1},	
									CF_3 stretch	
		$v_3=1$	1520.69 (4)	D_J =0.5					ν_3=286 cm^{-1},	
				D_{JK}=1.5					C—I valence	
		$v_5=1$	1516.67 (4)	D_J =0.0					ν_5=540 cm^{-1},	
Cont.				D_{JK}=0.8					CF_3 deformation	

Nr.	Molecule	v	B MHz	D kHz	Ref.	q MHz	Ref.	Ref. Struct.	Varia, remarks	See also Tab., Nr.
21 Cont.	CF_3I	$v_6=1$	1522.32 (4)	$D_J=0.15$ $D_{JK}=0.6$	80KOH				$\nu_6=265\ cm^{-1}$, CF_3 rock	
		$v_6=2$	1521.13 (4)	$D_J=0.1$						
		$v_1=1,\ v_3=1$	1514.90 (20)	$D_J=0.2$ $D_{JK}=0.6$ (fixed)						
		$v_1=1,\ v_5=1$	1512.43 (10)	$D_J=0.0$ $D_{JK}=0.6$ (fixed)						
		$v_1=1,\ v_6=1$	1516.95 (10)	$D_J=0.1$ $D_{JK}=0.6$ (fixed)						
		$v_1=1,\ v_6=2^2$	1515.78 (10)	$D_J=0.03$ $D_{JK}=0.6$ (fixed)						
		$v_x=1$	1516.53 (5)	$D_J=1.3$ $D_{JK}=0.6$ (fixed)					$\nu_x=1075.269\ (10)\ cm^{-1}$	
		$v_y=1$	1520.78 (20)	$D_J=1.3$ $D_{JK}=0.6$ (fixed)					$\nu_y=1074.78\ (10)\ cm^{-1}$	
	$^{13}CF_3I$	$v=0$	1516.227 (6)	$D_{JK}=1.00$ (fixed)	80COX					
22 *6-19*	$^{28}SiH_4$	$v=0$		$D_T=74.7514\ (16)$ $H_{4T}=-6.044\ (10)\cdot 10^{-3}$ $H_{6T}=2.598\ (3)\cdot 10^{-3}$ $L_{4T}=46.5\ (15)\cdot 10^{-8}$ $L_{6T}=-37.9\ (8)\cdot 10^{-8}$ $L_{8T}=-76.6\ (21)\cdot 10^{-8}$	76OZ1				for structure, see also Vol. II/7 (2, 308)	2.6.2, 43 2.9.2, 26
23 *6-20* *4-24*	$^{28}SiH_3F$	$v=0$	14347.575 (6)*)	$D_J=15.6$**) $D_{JK}=214$	76GEO			76ROB3	for structure, see also Vol. II/7 (2, 186) *) uncertainties are 2.5 σ **) calculated from the force field $\nu_3=875\ cm^{-1}$; $\nu_6=730\ cm^{-1}$ Coriolis interaction between ν_2 and ν_5 analyzed: $\nu_2=990\ cm^{-1}$; $\nu_5=960\ cm^{-1}$	2.6.2, 44
		$v_3=1$	14269.65 (2)*)							
		$v_6=1$	14300.85 (3)	$\eta_J=240\ (24)$*)		$q=44.35\ (8)$ *)	76GEO			
		$v_2=1$	14382 (3)							
		$v_5=1$	14355 (2)			$q=-37\ (3)$				
			$A-B-A\zeta_t=81200\ (100)$							
Cont.										

Nr.	Molecule	v	B MHz	D kHz	Ref.	q MHz	Ref.	Ref. Struct.	Varia, remarks	See also Tab., Nr.
23	$^{29}SiH_3F$	$v=0$	14215.994 (10)*)		76GEO				*) uncertainties are 2.5σ	
6–20	$^{30}SiH_3F$	$v=0$	14091.974 (11)*)						**) calculated from the	
4–24	$^{28}SiD_3F$	$v=0$	12253.141 (5)*)	D_J = 10.9**)					force field	
Cont.				D_{JK}=124.7						
		$v_3=1$	12188.35 (2)*)						$\nu_3=889\ cm^{-1}$	
		$v_6=1$	12214.48 (4)	η_J=234 (42)*)		q=33.01 (10)	76GEO		$\nu_6=551\ cm^{-1}$;	
		$v_2=1$	12216 (2)						$\nu_2=710\ cm^{-1}$	
		$v_5=1$	12229 (1)	η_J=200 (40)		q=+25 (2)			Coriolis interaction between ν_2 and ν_5 analyzed:	
			$A-B-A\zeta_t$ =35300 (40)						$\nu_2-\nu_5+(A\zeta)_5$ =246300 (2200) MHz, A_2-A_5=160 (30) MHz, $B_e(\omega_2+\omega_5)\zeta^{x,y}_{2,5}/(2\omega_2\omega_5)^{1/2}$ =9740 (40) MHz [76GEO]	
	$^{29}SiD_3F$	$v=0$	12175.604 (15)*)							
	$^{30}SiD_3F$	$v=0$	12101.972 (11)*)							
	$^{28}SiH_2DF$	$v=0$	A=64842.947 (70)	Δ_J =14.1 $^+$)	76ROB3				$^+$) constrained to the	
			B=13906.428 (11)	Δ_{JK}=180.4 (15)					values calculated from	
			C=13218.515 (11)	Δ_K =340 $^+$)					force field	
				δ_J =0.801 (3)					harmonic force field	
				δ_K =32.1 (5)					calculated [76ROB3]	
	$^{28}SiHD_2F$	$v=0$	A=51740.041 (261)	Δ_J =12.6 $^+$)						
			B=13216.117 (8)	Δ_{JK}=150.2 (20)						
			C=12527.928 (8)	Δ_K =154 $^+$)						
				δ_J =0.669 (3)						
				δ_K =67.4 (9)						
24	$SiH_3{}^{81}Br$	$v=0$	4292.6462 (4)	D_J = 1.81 (1)	77DO3				for structure,	2.6.2, 45
6–22				D_{JK}=29.19 (4)					see also	2.7.2, 51
4–26	$SiH_3{}^{79}Br$	$v=0$	4321.8019 (11)	D_J = 1.96 (6)	79DO1				Vol. II/7 (2, 72)	2.9.2, 27
				D_{JK}=29.58 (14)						
	$SiD_3{}^{79}Br$	$v=0$	3827.6791 (7)	D_J = 1.34 (3)						
				D_{JK}=19.19 (8)						

Nr.	Molecule	v	B MHz	D kHz	Ref.	q MHz	Ref.	Ref. Struct.	Varia, remarks	See also Tab., Nr.
25 *6–26*	SiF_4									2.9.2, 28
26	$H^{28}SiF_3$	$v=0$	7208.05		73HOY			73HOY	for structure, see also	
		$v_2=1$	$\alpha^B=$ 20.79 (20) *)						Vol. II/7 (2, 189)	
		$v_3=1$	$\alpha^B=-4.48$ (3)						harmonic and anharmonic	
		$v_4=1$	$\alpha^B=$ 15.13 (3)	$\eta_J=-20$ (50)		$q=14.53$ (6)	73HOY		force fields calculated	
		$v_5=1$	$\alpha^B=$ 1.80 (3)	$\eta_J=-30$ (50)		$q=19.61$ (2)			*) corrected for the Fermi	
		$v_6=1$	$\alpha^B=-6.81$ (3)	$\eta_J=-30$ (20)		$q=13.16$ (2)			resonance between	
			$C-C\zeta_t-B$						ν_2 and $2\nu_3$:	
			$=-706$ (200)						$\phi_{233}(HSiF_3)=19.5$ (8) cm^{-1}	
									$\phi_{233}(DSiF_3)=23.4$ (8) cm^{-1}	
	$H^{29}SiF_3$	$v=0$	7195.75							
		$v_3=1$	$\alpha^B=-4.83$ (13)						$\nu_2=856.2\ cm^{-1}$	
		$v_6=1$	$\alpha^B=-6.51$ (3)						$\nu_3=425.2\ cm^{-1}$	
	$H^{30}SiF_3$	$v=0$	7183.79						$\nu_4=997.8\ cm^{-1}$	
		$v_3=1$	$\alpha^B=-5.21$ (3)						$\nu_5=843.6\ cm^{-1}$	
		$v_6=1$	$\alpha^B=-6.24$ (6)			$q=12.05$ (20)			$\nu_6=306.2\ cm^{-1}$	
	$D^{28}SiF_3$	$v=0$	6890.10							
		$v_2=1$	$\alpha^B=$ 21.90 (20) *)						$\nu_2=853.8\ cm^{-1}$	
		$v_3=1$	$\alpha^B=-4.54$ (3)						$\nu_3=422.1\ cm^{-1}$	
		$v_4=1$	$\alpha^B=$ 15.20 (10)	$\eta_J=-60$ (50)		$q=13.40$ (6)			$\nu_4=991.8\ cm^{-1}$	
		$v_5=1$	$\alpha^B=-0.45$ (5)	$\eta_J=-10$ (50)		$q=$ 6.83 (5)			$\nu_5=628.4\ cm^{-1}$	
		$v_6=1$	$\alpha^B=-4.95$ (3)	$\eta_J=-10$ (20)		$q=11.54$ (2)			$\zeta_6=-0.56$ (3);	
			$C-C\zeta_t-B$						$\nu_6=302.8\ cm^{-1}$	
			$=-454$ (100)							
	$D^{29}SiF_3$	$v=0$	6880.20						r_e(Si—H) = 1.4468 (5) Å	
		$v_2=1$	$\alpha^B=-4.80$ (15)						r_e(Si—F) = 1.5624 (1) Å	
		$v_6=1$	$\alpha^B=-4.80$ (15)			$q=11.45$ (15)			∡ HSiF = 110.64 (3)°	
	$D^{30}SiF_3$	$v=0$	6870.58							
		$v_3=1$	$\alpha^B=-4.55$ (20)							
		$v_6=1$	$\alpha^B=-4.55$ (10)			$q=11.60$ (6)				
27	$H^{28}Si^{35}Cl_3$	$v=0$	2472.468	$D_J=0.85$	77TA2			77TA2	for structure, see also Vol. II/7 (2, 118)	2.7.2, 34
	$H^{28}Si^{37}Cl_3$	$v=0$	2346.213							
	$H^{29}Si^{35}Cl_3$	$v=0$	2469.279							

Nr.	Molecule	v	B MHz	D kHz	Ref.	q MHz	Ref.	Ref. Struct.	Varia, remarks	See also Tab., Nr.
28 *6-29a*	GeH_4	$v=0$		$D_T=67.77554(86)$ $H_{4T}=-5.3827(55)\cdot10^{-3}$ $H_{6T}=2.9693(19)\cdot10^{-3}$ $L_{4T}=3.996(88)\cdot10^{-7}$ $L_{6T}=-4.122(50)\cdot10^{-7}$ $L_{8T}=-8.01(14)\cdot10^{-7}$	79KAG, 77KRE				for structure, see also Vol. II/7 (2, 258)	2.6.2, 46 2.9.2, 29
29 *6-30b* *4-34*	$^{70}GeH_3{}^{79}Br$ $^{70}GeH_3{}^{81}Br$ $^{72}GeH_3{}^{79}Br$ $^{72}GeH_3{}^{81}Br$ $^{74}GeH_3{}^{79}Br$ $^{74}GeH_3{}^{81}Br$ 	$v=0$ $v_3=1$ $v=0$ $v_3=1$ $v=0$ $v_3=1$ $v=0$ $v_3=1$ $v=0$ $v_3=1$ $v=0$ $v_3=1$	2438.48 (1) $\alpha=8.80(3)$ 2410.11 (1) $\alpha=8.63(3)$ 2406.35 (1) $\alpha=8.62(2)$ 2377.95 (2) $\alpha=8.46(4)$ 2375.82 (1) $\alpha=8.44(3)$ 2347.40 (2) $\alpha=8.29(4)$	$D_{JK}=13(1)$	73KRI				$\nu_3\simeq300\ cm^{-1}$	
30 *6-31*	GeF_4									2.9.2, 30
31 *6-35* *4-37* Cont.	OPF_3	$v=0$	$A=4811.7579(18)$ $B=4594.2624(5)$	$D_J=1.0119(12)$ $D_{JK}=1.2971(7)$ $D_K=-1.114(12)$ $H_J=-6.7(10)\cdot10^{-6}$ $H_{JK}=9.10(27)\cdot10^{-6}$ $H_{KJ}=-6.8(8)\cdot10^{-6}$ $H_K=-73(30)\cdot10^{-6}$	78KAG			78KAG	for structure, see also Vol. II/7 (2, 216) harmonic force field calculated [76SM1].	2.6.2, 47 2.9.2, 31

Nr.	Molecule	v	B MHz	D kHz	Ref.	q MHz	Ref.	Ref. Struct.	Varia, remarks	See also Tab., Nr.
31 *6–35* *4–37*		$v_2=1$	4587.0502 (6)	D_J =1.0166 (11) D_{JK} =1.2962 (34)	76SM1				$\nu_2=872.8\ cm^{-1}$	
		$v_3=1$	4600.1684 (4)	D_J =1.0681 (9) D_{JK} =1.0235 (20)					$\nu_3=473.2\ cm^{-1}$	
		$v_3=2$	4605.940 (4)	*)						
		$v_4=1$	A =4803.06 (16) B =4584.668 (30) $A\zeta_t$ =2429.03 (22)	D_J =1.004 (5) D_{JK} =2.5 (30) D_K = −0.8 (113) η_J =0.8 (68) η_K =26 (24)	78AMA, 76SM1	q =0.347 (3)	76SM1		$\nu_4=992.972904\ (33)\ cm^{-1}$ *) centrifugal distortion constants constrained to ground-state values	
		$v_5=1$	4596.8842 (3) $A\zeta_t$ =524.5 (6)	D_J =0.9972 (5) D_{JK} =1.4490 (91) η_J =8.08 (5) η_K =186 (48) H_{KJ} = −6 (9) · 10^{-6}	76SM1	q_0 =2.535226 (43) q_J =31.8 (8) · 10^{-6} t =10.6 (13) · 10^{-6}	79ME2 76SM1		$\nu_5=483.2\ cm^{-1}$	
		$v_6=1$	A =4797.8 (15) B =4589.8514 (2) $A\zeta_t$ = −710.0 (52)	D_J =1.0156 (3) D_{JK} =1.2927 (29) η_J = −4.53 (1) H_{KJ} =24 (8) · 10^{-6} τ'_{xxxz} =1.834 (17)		q_0 =1.0764528 (33) q_J = −33.46 (5) · 10^{-6} t =24.8 (9) · 10^{-6}	79ME2 76SM1		$\nu_6=335.6\ cm^{-1}$	
		$v_6=2$	A =4803.2 (49) B =4585.535 (1) $A\zeta$ = −729.0 (150)	D_J =0.968 (4) **) D_{JK} =0.943 (42) **) η_J = −14.30 (4) η_K =252 (100) τ_{xxxz} =1.83 ***)	81SMI	q^+ =1.07645 ***) r =1.468 (24) f_{24} =2.2 · 10^{-3} ***)	81SMI		**) $l=0$ values, the $l=2$ values are: D_J =1.038 (2) and D_{JK} =1.370 (19) kHz ***) fixed value γ_{ll} = −16.7 (4) kHz x_{ll} =16134 (390) MHz	
		$v_3=1, v_5=1$	4602.622 (3)	*)	76SM1					
		$v_5=1, v_6=1$	4592.477 (5)	*)						
		$v_3=1, v_6=1$	4595.592 (5)	D_J =1.084 (5)						
	$^{18}OPF_3$	$v=0$	4395.287 (1)	D_J =0.8738 (39) D_{JK} =1.4629 (51)						

Nr.	Molecule	v	B MHz	D kHz	Ref.	q MHz	Ref.	Ref. Struct.	Varia, remarks	See also Tab., Nr.
32	SPF_3	$v=0$	2657.6552 (2)	D_J =0.2742 (3) D_{JK} =1.9047 (11) H_J = −0.4 (1) · 10^{-3} H_{JK} =0.3 (4) · 10^{-3}	76SM2			76SM2	for structure, see also Vol. II/7 (2, 233) harmonic force field calculated [76SM2]	2.7.2, 21
		$v_5=1$	2659.8945 (4) $A\zeta_t$ = −1797 (41)	D_J =0.2769 (5) D_{JK} =1.9181 (8) η_J =4.152 (8)		q_0 =1.3956 (13) q_J = −4.9 (17) · 10^{-6}	76SM2		ν_5 =405 cm^{-1} ζ^z_{55} = −0.367 (15)	
		$v_6=1$	2655.3663 (3) $A\zeta_t$ =941.3 (27)	D_J =0.2736 (4) D_{JK} =1.9031 (8) η_J =0.141 (7)		q_0 =1.6965 (10) q_J = −1.7 (14) · 10^{-6}			ν_6 =274 cm^{-1} ζ^z_{66} =0.192 (5)	
	$^{33}SPF_3$	$v=0$	2617.7569 (7)	D_J =0.2554 (8) D_{JK} =1.8849 (22)						
	$^{34}SPF_3$	$v=0$	2579.7895 (6)	D_J =0.2587 (7) D_{JK} =1.8407 (22)						
		$v_6=1$	2577.6078 (7) $A\zeta_t$ =934.3 (57)	D_J =0.2582 (7) D_{JK} =1.8388 (16) η_J =0.186 (13)		q =1.6596 (7)			ζ^z_{66} =0.191 (6)	
33	NSF_3	$v=0$	4636.2434 (4)	D_J =1.0765 (4) D_{JK} =2.120 (3) H_{JK} =9 (3) · 10^{-3}	79SMA			79SMA	for structure, see also Vol. II/7 (2, 207) harmonic force field calculated [79SMA]	
		$v_3=1$	4634.5785 (5)	D_J =1.073 (1) D_{JK} =2.130 (3)					ν_3 =520 cm^{-1}	
		$v_5=1$	A =5182.2 (5) B =4640.9549 (4) $A\zeta_t$ = −1169.4 (7)	D_J =1.0918 (5) D_{JK} =2.260 (5) η_J =44.0 (4) η_{JJ} =1.12 (8) · 10^{-3} η_{JK} =4.2 (1) · 10^{-3} H_{JK} =0.023 (13) · 10^{-3}		q_0 =6.787 (2) q_J =14 (3) · 10^{-6} r =2.02 (2) t =2.3 (8) · 10^{-6}	79SMA		ν_5 =429 cm^{-1} ζ^z_{55} = −0.2256 (18)	
		$v_6=1$	A =5178.4 (17) B =4628.604 (1) $A\zeta_t$ =812.78 (263)	D_J =1.0672 (5) D_{JK} =2.015 (4) η_J =24.7 (23) τ'_{xxxz} =2.54 (2)		q_0 =0.9901 (14) q_J = −71 (3) · 10^{-6} r = −4.14 (7) t =41.3 (9) · 10^{-6}			ν_6 =342 cm^{-1} ζ^z_{66} =0.1569 (15)	
Cont.										

Nr.	Molecule	v	B MHz	D kHz	Ref.	q MHz	Ref.	Ref. Struct.	Varia, remarks	See also Tab., Nr.
33 Cont.	$N^{34}SF_3$	$v=0$	4630.3771 (8)	D_J =1.081 (2)	79SMA					
				D_{JK} =2.100 (10)						
				H_{JK} =21 (15)·10^{-6}						
		$v_6=1$	4622.7270 (6)	D_J =1.066 (2)		q_0 =1.039 (3)	79SMA		ζ^z_{66} =0.1657 (15)	
			$A\zeta_t$ =858.23 (238)	D_{JK} =1.980 (5)		q_J = −73 (7)·10^{-6}				
				η_J =23.0 (1)		r = −4.14 *)			*) constrained	
				τ'_{xxxz} =2.4 (2)		t =38.6 (19)·10^{-6}				
	$^{15}NSF_3$	$v=0$	4520.222 (1)	D_J =1.002 (2)						
				D_{JK} =2.208 (4)						
34	MnO_3F	$v=0$	4129.141 (9)	D_J =1.12 (29)	76HOG				for structure, see also Vol. II/7 2, 196)	
				D_{JK} =1.87 (86)						
		$v_3=1$	4120.519 (9)							
		$v_5=1$	4141.135 (8)			q =16.005 (10)	76HOG			
		$v_6=1$	4123.099 (8)			q = 8.456 (10)				
35	ClO_3F								for structure, see also Vol. II/7 (2, 97)	2.6.2, 48 2.7.2, 35
36 6–37 4–45	$CD_3{}^{202}Hg^{35}Cl$	$v=0$	1905.82		75WAL			75WAL	for structure, see also Vol. II/7 (3, 89)	
	$CD_3{}^{202}Hg^{37}Cl$	$v=0$	1843.09							
	$^{13}CH_3{}^{202}Hg^{35}Cl$	$v=0$	2034.09							
	$^{13}CH_3{}^{202}Hg^{37}Cl$	$v=0$	1964.96							
37	$CD_3{}^{198}Hg^{79}Br$	$v=0$	1075.10		75WAL			75WAL	for structure, see also Vol. II/7 (3, 86)	
	$CD_3{}^{200}Hg^{79}Br$	$v=0$	1073.87							
	$CD_3{}^{202}Hg^{79}Br$	$v=0$	1072.72							
	$CD_3{}^{198}Hg^{81}Br$	$v=0$	1059.13							
	$CD_3{}^{200}Hg^{81}Br$	$v=0$	1057.91							
	$CD_3{}^{202}Hg^{81}Br$	$v=0$	1056.62							
	$^{13}CH_3{}^{198}Hg^{79}Br$	$v=0$	1125.62							
	$^{13}CH_3{}^{200}Hg^{79}Br$	$v=0$	1124.22							
	$^{13}CH_3{}^{202}Hg^{79}Br$	$v=0$	1122.86							
	$^{13}CH_3{}^{198}Hg^{81}Br$	$v=0$	1108.46							
	$^{13}CH_3{}^{200}Hg^{81}Br$	$v=0$	1107.00							
	$^{13}CH_3{}^{202}Hg^{81}Br$	$v=0$	1105.54							

Nr.	Molecule	v	B MHz	D kHz	Ref.	q MHz	Ref.	Ref. Struct.	Varia, remarks	See also Tab., Nr.
38	$CH_3{}^{198}HgI$	$v=0$	767.43		75WAL			75WAL	for structure, see also Vol. II/7 (3, 107)	
6–38	$CH_3{}^{199}HgI$	$v=0$	766.52							
4–47	$CH_3{}^{200}HgI$	$v=0$	765.67							
	$CH_3{}^{202}HgI$	$v=0$	763.94							
	$CH_3{}^{204}HgI$	$v=0$	762.33							
	$CD_3{}^{198}HgI$	$v=0$	729.58							
	$CD_3{}^{199}HgI$	$v=0$	728.75							
	$CD_3{}^{200}HgI$	$v=0$	727.97							
	$CD_3{}^{202}HgI$	$v=0$	726.48							
	$CD_3{}^{204}HgI$	$v=0$	725.04							
	$^{13}CH_3{}^{198}HgI$	$v=0$	757.50							
	$^{13}CH_3{}^{199}HgI$	$v=0$	756.62							
	$^{13}CH_3{}^{200}HgI$	$v=0$	755.79							
	$^{13}CH_3{}^{202}HgI$	$v=0$	754.69							
	$^{13}CH_3{}^{204}HgI$	$v=0$	752.58							
39	$CF_3{}^{202}HgCH_3$	$v=0$	1022.85 (5)		80GUN			80GUN		
	$^{13}CF_3{}^{202}HgCH_3$	$v=0$	1017.52 (5)							
	$CF_3{}^{202}Hg^{13}CH_3$	$v=0$	1009.29 (5)							
40	$CH_3{}^{11}B{=}S$	$v=0$	4563.2203 (24)	D_J = 0.899 (86)	80KIR1			80KIR1		2.6.2, 49
				D_{JK} = 74.847 (187)						2.7.2, 6
		$v_8 \geqq 1$	$\alpha = -14.311$ (4)	D_J = 0.95 (9)		$q = 5.448$ (2)	80KIR1		$\gamma_{88} = -0.946$ (1) MHz	
				D_{JK} = 75.08 (25)					$\gamma_{l_8 l_8} =$ 0.8737 (5) MHz	
				η_J = 110.97 (54)						
	$CH_3{}^{10}B{=}S$	$v=0$	4572.3034 (29)	D_J = 0.806 (105)						
				D_{JK} = 74.837 (227)						
		$v_8=1$	$\alpha = -12.910$ (1)	D_J = 0.80 (10)		$q = 5.301$ (2)				
				D_{JK} = 74.87 (21)						
				η_J = 123.31 (67)						
	$CH_3{}^{11}B{=}{}^{34}S$	$v=0$	4462.4982 (5) *)	D_{JK} = 70.986 (194)					*) with D_J assumed from parent species	
	$^{13}CH_3{}^{11}B{=}S$	$v=0$	4406.3159 (41) *)							
	$CD_3{}^{11}B{=}S$	$v=0$	3926.9085 (25)	D_J = 0.788 (57)						
				D_{JK} = 49.133 (91)						
		$v_8=1$	$\alpha = -10.204$ (1)	D_J = 0.69 (6)		$q = 4.441$ (2)				
				D_{JK} = 49.14 (13)						
Cont.				η_J = 73.34 (40)						

Nr.	Molecule	v	B MHz	D kHz	Ref.	q MHz	Ref.	Ref. Struct.	Varia, remarks	See also Tab., Nr.
40 Cont.	$CD_3{}^{10}B{=}S$	$v=0$	3930.5933 (52)	D_J = 0.786 (121) D_{JK} = 48.952 (193)	80KIR1					
	$CD_3{}^{11}B{=}{}^{34}S$	$v=0$	3835.9807 (15)	D_J = 0.820 (35) D_{JK} = 46.131 (187)						
	$^{13}CD_3{}^{11}B{=}S$	$v=0$	3823.0898 (75)	D_J = 1.053 (167) D_{JK} = 41.181 (2932)						
	$CH_2D^{11}B{=}S$	$v=0$	B = 4340.6864 (20) C = 4303.0164 (20)	Δ_J = 0.861 **) Δ_{JK} = 58.079 (945)					**) values interpolated from results for CH_3BS and CD_3BS	
	$CHD_2{}^{11}B{=}S$	$v=0$	B = 4131.0469 (146) C = 4092.4485 (145)	Δ_J = 0.825 **) Δ_{JK} = 49.476 (3069)						
41	$CH_3C{\equiv}P$	$v=0$	4991.344 (6)	D_J = 1.06 (22) D_{JK} = 66.61 (45)	79KRO			79KRO		2.6.2, 50
		$v_8 \geqq 1$	$\alpha = -13.004$ (7)	$\alpha(D_J)$ = 0.031 (121) $\alpha(D_{JK}) = -0.76$ (29) η_J = 110.36 (97) $\alpha(\eta_J) = -1.50$ (44)		q = 6.394 (6) $\alpha(q)$ = 0.0755 (25)	79KRO		$\gamma^{B}_{88} = -0.309$ (1) MHz $\gamma_{l_8 l_8}$ = 0.240 (1) MHz g_{88} = 183.1 (258) GHz ζ_8 = 0.906 (9)	
	$^{13}CH_3C{\equiv}P$	$v=0$	4823.651 (12)	D_J = 1.0 (4) D_{JK} = 61.7 (6)						
	$CH_3{}^{13}C{\equiv}P$	$v=0$	4982.628 (22)	D_J = 1.4 (7) D_{JK} = 65.8 (10)						
	$CD_3C{\equiv}P$	$v=0$	4293.634 (3)	D_J = 0 (fixed) D_{JK} = 43.15 (49)						
		$v_8 \geqq 1$	$\alpha = -10.680$ (5)	$\alpha(D_{JK}) = -0.64$ (28) η_J = 73.71 (107) $\alpha(\eta_J) = -1.21$ (41)		q = 5.058 (6)			$\gamma^{B}_{88} = -0.208$ (1) MHz $\gamma^{B}_{l_8 l_8}$ = 0.156 (1) MHz ζ_8 = 0.867 (32)	
	$^{13}CD_3C{\equiv}P$	$v=0$	4182.9400 (4)	D_J = 0.8 (fixed) D_{JK} = 41.10 (7)						
	$CD_3{}^{13}C{\equiv}P$	$v=0$	4290.2190 (7)	D_J = 0.8 (fixed) D_{JK} = 42.85 (14)						
	$CH_2DC{\equiv}P$	$v=0$	B = 4749.172 (7) C = 4704.275 (7)	D_J = 0.79 (24) D_{JK} = 52.25 (57)						
	$CD_2HC{\equiv}P$	$v=0$	B = 4519.339 (9) C = 4473.372 (9)	D_J = 0.81 (32) D_{JK} = 47.72 (76)						

Nr.	Molecule	v	B MHz	D kHz	Ref.	q MHz	Ref.	Ref. Struct.	Varia, remarks	See also Tab., Nr.
42 *6-42* *4-50*	CF_3CN	$v_8=1$	2950.478 (5)	$D_J=0.30$ (1) $D_{JK}=5.78$ (1) $\eta_J=23.0$ (2) $H_J=-1.8\,(9)\cdot 10^{-5}$	74COX	$q/4=0.901$ (3)	74COX		for structure, see also Vol. II/7 (3, 157) $\zeta_8=0.5243$ (4) assumed constants: $A=5700$ MHz $r=0$	2.6.2, 51
43	SiH_3CN	$v=0$	4973.009 (15)	$D_J=1.49$ (10) $D_{JK}=63.4$ (20)	75CAE				for structure, see also Vol. II/7 (3, 119)	2.7.2, 17
		$v_7=1$	$\alpha=2.343$ (30)	$D_{JK}=62.6$ (30) $\eta_J=46.5$ (30)		$q=0.932$ (10)	75CAE		$\nu_7=682\ cm^{-1}$	
		$v_8\geqq 1$	$\alpha=-22.517$ (20)	$\alpha(D_J)=-0.22$ (10) $\alpha(D_{JK})=-2.8$ (10) $\eta_J=180.9$ (15) $\alpha(\eta_J)=-4.1$ (10)		$q_0=7.343$ (5) $q_J=-0.56\,(30)\cdot 10^{-3}$ $\alpha(q)=0.113$ (5)			$\nu_8=235\ cm^{-1}$ $\gamma_{88}^{B}=-0.669$ (5) MHz $\gamma_{l_8 l_8}^{B}=0.571$ (5) MHz $g_{88}=95.4$ (500) GHz $\zeta_8=0.890$ (5)	
		$v_3=1$	4923.979 (20)	$D_J=0.12$ (50)					$\nu_3=920\ cm^{-1}$	
		$v_4=1$	4952.506 (20)	$D_J=1.58$ (30) $D_{JK}=58.8$ (50)					$\nu_4=608\ cm^{-1}$	
	$^{29}SiH_3CN$	$v=0$	4915.184 (15)	$D_J=1.43$ (30) $D_{JK}=62.2$ (50)						
		$v_8=1$	$\alpha=-20.839$ (30)	$D_{JK}=66.7$ (15) $\eta_J=180.8$ (15)		$q=7.074$ (10)			$\zeta_8=0.889$ (5)	
	$^{30}SiH_3CN$	$v=0$	4860.707 (15)	$D_J=1.50$ (3) $D_{JK}=61.3$ (50)						
		$v_8=1$	$\alpha=-20.613$ (30)	$D_{JK}=66.0$ (15) $\eta_J=175.3$ (15)		$q=6.922$ (10)			$\zeta_8=0.893$ (5)	
	$SiH_3{}^{13}CN$	$v=0$	4946.226 (20)	$D_J=1.43$ (30) $D_{JK}=61.3$ (50)						
	$SiH_3C^{15}N$	$v=0$	4805.466 (20)	$D_J=1.57$ (30)						

Nr.	Molecule	v	B MHz	D kHz	Ref.	q MHz	Ref.	Ref. Struct.	Varia, remarks	See also Tab., Nr.
44 *6-43*	SiH_3NCO	$v=0$	2517.932 (3)	D_J =1.30 (5) D_{JK}=642.0 (3) H_{JK}=72 (3) · 10^{-3} H_{KJ}= −16 (8) · 10^{-3}	76DUC			81DUC	for structure, see also Vol. II/7 (3, 116) harmonic force field calculated [76DUC] ν_{10}=22 cm^{-1}, SiNC bend uncertainties are 3 σ	2.7.2, 18
		$v_{10}=1^1$	2542.349 (7) $A-A\zeta-B$=2651 (32)	D_J =1.60 (9) D_{JK}=58.4 (8) H_{JK}= −20 (11) · 10^{-3} H_{KJ}=31 (18) · 10^{-3} η_J =82 (2) θ_{JK} = −1.902 (56) θ_{JJ} =0.444 (32)		q_0=13.663 (19) q_J=0.16 (12) · 10^{-3}	76DUC			
		$v_{10}=2^0$	2543.064 (9)	D_J =0.36 (12) D_{JK}=226.6 (10) H_{JK}=17 (13) · 10^{-3} H_{KJ}= −14 (20) · 10^{-3}						
		$v_{10}=2^2$	2562.764 (10)	D_J =1.21 (12) D_{JK}=64.7 (7) H_{KJ}= −14 (20) · 10^{-3} η_J =106.2 (8) θ_{JK} = −87 (29) · 10^{-3}						
		$v_{10}=3^1$	2565.169 (5) $A-A\zeta-B$=2630 (24)	D_J =0.81 (7) D_{JK}=130.7 (7) H_{KJ}=25 (20) · 10^{-3} η_J =231 (2) θ_{JK} = −255 (59) · 10^{-3} θ_{JJ} =119 (33) · 10^{-3}		q_0=17.906 (18) q_J=0.10 (12) · 10^{-3}				
		$v_{10}=3^3$	2581.126 (7)	D_J =1.17 (8) D_{JK}=61.0 (14) H_{KJ}= −63 (54) · 10^{-3} η_J =105.8 (11) θ_{JK} = −25 (26) · 10^{-3} θ_{JJ} =9 (14) · 10^{-3}						
Cont.		$v_{10}=4^0$	2577.1	D_{JK}=128.1						

Nr.	Molecule	v	B MHz	D kHz	Ref.	q MHz	Ref.	Ref. Struct.	Varia, remarks	See also Tab., Nr.
44 *6-43* Cont.	$^{29}SiH_3NCO$	$v=0$	2484.462 (7)	D_J =1.25 (6) D_{JK}=621.9 (5) H_{JK}=68 (4) · 10^{-3} H_{KJ}=−6 (8) · 10^{-3}	81DUC					
		$v_{10}=1^1$	2508.556 (26) $A-A\zeta-B$=2641 (44)	D_J =1.2 (3) D_{JK}=54 (4) H_{JK}=−54 (39) · 10^{-3} H_{KJ}=45 (55) · 10^{-3} η_J =80 (14) θ_{JK} =−1.759 (233) θ_{JJ} =0.450 (155)		q=13.3 (fixed)	81DUC			
		$v_{10}=2^0$	2509.320 (29)	D_J =0.3 (3) D_{JK}=220.8 (11) H_{JK}=31 (18) · 10^{-3} H_{KJ}=−24 (25) · 10^{-3}						
		$v_{10}=2^2$	2528.747 (27)	D_J =1.2 (3) D_{JK}=63.1 (3) η_J =105 (4) θ_{JK} =−0.050 (47) θ_{JJ} =−0.019 (48)						
		$v_{10}=3^1$	2531.145 (17) $A-A\zeta-B$=2620 (30)	D_J =0.70 (19) D_{JK}=121 (6) H_{KJ}=−51 (68) · 10^{-3} η_J =237.2 (18) θ_{JK} =−0.387 (404)		q=17.446 (6)				
	$^{30}SiH_3NCO$	$v=0$	2542.799 (10)	D_J =1.24 (9) D_{JK}=602.9 (7) H_{JK}=64 (6) · 10^{-3}						
		$v_{10}=1^1$	2476.636 (23) $A-A\zeta-B$=2642 (59)	D_J =1.4 (2) D_{JK}=55.3 (17) H_{JK}=−25 (17) · 10^{-3} H_{KJ}=45 (27) · 10^{-3} η_J =80 (7) θ_{JK} =−1.755 (114) θ_{JJ} =0.432 (82)		q=12.967 (15)				
Cont.		$v_{10}=2^0$	2525.190 (56)	D_J =0.4 (6) D_{JK}=213 (5)						

Nr.	Molecule	v	B MHz	D kHz	Ref.	q MHz	Ref.	Ref. Struct.	Varia, remarks	See also Tab., Nr.
44	$^{28}SiD_3NCO$	$v=0$	2305.799 (66)	D_J = 1.0 (fixed)	81DUC					
6–43				D_{JK} = 516.7 (86)						
Cont.		$v_{10}=1^1$	2353.674 (55)	*)		q = 12.500 (110)	81DUC		*) constants constrained	
		$v_{10}=2^0$	2353.911 (56)	D_J = 0.6 (9)					to the values found for	
				D_{JK} = 186 (4)					$^{30}SiH_3NCO$	
	$^{29}SiD_3NCO$	$v=0$	2305.799 (66)	D_J = 1.0 (fixed)						
				D_{JK} = 516.7 (86)						
	$^{30}SiD_3NCO$	$v=0$	2281.170 (83)	D_J = 1.0 (fixed)						
				D_{JK} = 507.0 (88)						
	$SiH_3{}^{15}NCO$	$v=0$	2517.738 (86)	D_J = 1.1 (9)						
				D_{JK} = 674.0 (67)						
				H_{JK} = 79 (66) · 10^{-3}						
		$v_{10}=1^1$	2541.425 (23)	**)		q = 14.155 (47)			**) constants constrained	
	$^{29}SiH_3{}^{15}NCO$	$v=0$	2484.238 (64)	D_J = 1.0 (fixed)					to the values found for	
				D_{JK} = 650.7 (60)					$^{29}SiH_3NCO$	
				H_{JK} = 70.10^{-3} (fixed)						
	$SiH_3N^{13}CO$	$v=0$	2499.981 (16)	D_J = 1.27 (15)						
				D_{JK} = 630.3 (11)						
				H_{JK} = 67 (10) · 10^{-3}						
		$v_{10}=2^0$	2525.190 (56)	D_J = 0.4 (6)						
				D_{JK} = 213 (5)						
	$SiH_3NC^{18}O$	$v=0$	2387.762 (38)	D_J = 1.0 (fixed)						
				D_{JK} = 602.9 (41)						
				H_{JK} = 60.10^{-3} (fixed)						
		$v_{10}=1^1$	2410.810 (47)	*)		q = 12.541 (93)				
45	SiH_3NCS	$v=0$	1516.040 (1)	D_J = 0.087 (4)	77DO1			79DO2	for structure, see	2.6.2, 52
				D_{JK} = 41.958 (6)					also Vol. II/7	2.7.2, 19
									(3, 118)	2.9.2, 34
		$v_{10}=1^1$	1526.335 (1)	D_J = 0.116 (4)		q = 2.783 (1)	77DO1		ν_{10} = 65 (15) cm^{-1}	
			$A-A\zeta-B$	D_{JK} = 39.184 (6)					ζ_{10} = 0.931 (2)	
			= 4250 (100)	η_J = 52.151 (3)						
		$v_{10}=2^0$	1535.644 (1)	D_J = 0.130 (6)		q = 2.75 (3)			$4[g_{10,10}+A-B-2A\zeta]$	
Cont.				D_{JK} = 37.661 (7)					= −1220 (60) MHz	

Nr.	Molecule	v	B MHz	D kHz	Ref.	q MHz	Ref.	Ref. Struct.	Varia, remarks	See also Tab., Nr.
45 Cont.	SiH_3NCS	$v_{10}=2^2$	1536.124 (2)	D_J = 0.145 (7) D_{JK} = 37.252 (11) η_J = 49.44 (3)	77DO1					
		$v_{10}=3^1$	1544.728 (1)	D_J = 0.162 (4) D_{JK} = 36.458 (7) η_J = 48.69 (3)		$q = 2.657$ (3)	77DO1		$8[g_{10,10}+A-B-2A\zeta]$ = 7100 (400) MHz	
		$v_{10}=3^3$	1545.494 (2)	D_J = 0.135 (9) D_{JK} = 35.863 (33) η_J = 47.80 (3)						
		$v_{10}=4^0$	1553.27	D_{JK} = 36.1						
	$^{29}SiH_3NCS$	$v=0$	1493.640 (1)	D_J = 0.089 (4) D_{JK} = 40.50 (2)						
	$^{30}SiH_3NCS$	$v=0$	1472.411 (3)	D_J = 0.088 (9) D_{JK} = 39.26 (3)						
	$SiH_3NC^{34}S$	$v=0$	1473.556 (2)	D_J = 0.089 (6) D_{JK} = 40.01 (3)						
	$SiH_3{}^{15}NCS$	$v=0$	1514.552 (1)	D_J = 0.086 (6) D_{JK} = 42.30 (1)	77DO2					
	$SiH_3N^{13}CS$	$v=0$	1514.169 (1)	D_J = 0.074 (6) D_{JK} = 41.76 (1)						
	SiD_3NCS	$v=0$	1412.443 (1)	D_J = 0.066 (5) D_{JK} = 34.264 (5)	79DO2					
	$^{30}SiD_3NCS$	$v=0$	1377.342 (1)	D_J = 0.083 (3) D_{JK} = 32.39 (1)						
	$SiD_3NC^{34}S$	$v=0$	1372.529	D_J = 0.076 D_{JK} = 32.1						
45a	$^{70}GeH_3F$	$v=0$	10102.30	$D_J = 7$	81CRA			81CRA		
6-30	$^{72}GeH_3F$	$v=0$	10055.80	$D_J = 7$						
4-32	$^{74}GeH_3F$	$v=0$	10011.60	$D_J = 7$						
	$^{76}GeH_3F$	$v=0$	9969.60	$D_J = 7$						
	$^{70}GeD_3F$	$v=0$	9199.47							
	$^{72}GeD_3F$	$v=0$	9166.04							
	$^{74}GeD_3F$	$v=0$	9134.23							
Cont.	$^{76}GeD_3F$	$v=0$	9103.90							

Nr.	Molecule	v	B MHz	D kHz	Ref.	q MHz	Ref.	Ref. Struct.	Varia, remarks	See also Tab., Nr.
45a	$^{70}GeHD_2F$	$v=0$	$B=9680.81$		81CRA				$A(GeHD_2F)=47164$ MHz	
6-30			$C=9284.80$							
4-32	$^{72}GeHD_2F$	$v=0$	$B=9641.93$							
Cont.			$C=9249.23$							
	$^{74}GeHD_2F$	$v=0$	$B=9604.93$							
			$C=9215.33$							
45b	$^{70}GeH_3{}^{35}Cl$	$v=0$	4401.74	*)	81CRA			81CRA	*) $D_J=1.6$ kHz assumed	
6-30a	$^{74}GeH_3{}^{35}Cl$	$v=0$	4333.94							
4-33	$^{70}GeD_3{}^{35}Cl$	$v=0$	4153.56							
	$^{72}GeD_3{}^{35}Cl$	$v=0$	4126.47							
	$^{74}GeD_3{}^{35}Cl$	$v=0$	4098.21							
	$^{76}GeD_3{}^{35}Cl$	$v=0$	4072.44							
	$^{70}GeHD_2{}^{35}Cl$	$v=0$	$B=4271.51$						$A(GeHD_2Cl)=47585$ MHz	
			$C=4193.39$							
	$^{72}GeHD_2{}^{35}Cl$	$v=0$	$B=4240.61$							
			$C=4163.63$							
	$^{74}GeHD_2{}^{35}Cl$	$v=0$	$B=4211.14$							
			$C=4135.29$							
	$^{74}GeH_3{}^{37}Cl$	$v=0$	4177.93							
	$^{70}GeD_3{}^{37}Cl$	$v=0$	4009.16							
	$^{72}GeD_3{}^{37}Cl$	$v=0$	3980.58							
	$^{74}GeD_3{}^{37}Cl$	$v=0$	3953.34							
	$^{76}GeD_3{}^{37}Cl$	$v=0$	3927.4							
	$^{74}GeHD_2{}^{37}Cl$	$v=0$	$B=4060.19$							
			$C=3989.63$							
45c	$^{70}GeH_3{}^{79}Br$	$v=0$	2438.52	*)	81CRA			81CRA, 81DUR3	for structure, see also Vol. II/7(2, 70)	2.7.2, 51a
6-30b	$^{72}GeH_3{}^{79}Br$	$v=0$	2406.39	*)					*) $D_J=0.5$ kHz assumed	
4-34	$^{74}GeH_3{}^{79}Br$	$v=0$	2375.86	*)						
	$^{70}GeD_3{}^{79}Br$	$v=0$	2316.24							
	$^{72}GeD_3{}^{79}Br$	$v=0$	2288.75							
	$^{74}GeD_3{}^{79}Br$	$v=0$	2262.55	$D_J=0.52(10)$						
				$D_{JK}=9.15(55)$						
	$^{70}GeHD_2{}^{79}Br$	$v=0$	$B=2367.33$	$D_J=0.61(7)$						
Cont.			$C=2343.21$	$D_{JK}=11.36(46)$						

Nr.	Molecule	v	B MHz	D kHz	Ref.	q MHz	Ref.	Ref. Struct.	Varia, remarks	See also Tab., Nr.
45c	$^{72}GeHD_2{}^{79}Br$	$v=0$	$B=2338.08$		81CRA					
6–30b			$C=2314.56$							
4–34	$^{74}GeHD_2{}^{79}Br$	$v=0$	$B=2310.26$						$A(GeHD_2Br)=47682$ MHz	
Cont.			$C=2287.31$							
	$^{74}GeH_2D^{79}Br$	$v=0$	$B=2347.99$ (1)		81DUR3					
			$C=2325.01$ (1)							
	$^{70}GeH_3{}^{81}Br$	$v=0$	2410.15	*)	81CRA				*) $D_J=0.5$ kHz, assumed	
	$^{72}GeH_3{}^{81}Br$	$v=0$	2377.99	*)						
	$^{74}GeH_3{}^{81}Br$	$v=0$	2347.44	*)						
	$^{70}GeD_3{}^{81}Br$	$v=0$	2289.17							
	$^{72}GeD_3{}^{81}Br$	$v=0$	2261.63							
	$^{74}GeD_3{}^{81}Br$		2235.40	$D_J=0.52$ (10)						
				$D_{JK}=9.15$ (55)						
	$^{70}GeHD_2{}^{81}Br$	$v=0$	$B=2339.54$							
			$C=2315.98$							
	$^{72}GeHD_2{}^{81}Br$	$v=0$	$B=2310.28$							
			$C=2287.30$							
	$^{74}GeHD_2{}^{81}Br$	$v=0$	$B=2282.43$	$D_J=0.56$ (3)						
			$C=2260.02$	$D_{JK}=10.5$ (13)						
	$^{74}GeH_2D^{81}Br$	$v=0$	$B=2319.72$ (1)		81DUR3					
			$C=2297.30$ (1)							
45d	$^{70}GeH_3I$	$v=0$	1685.89	*)	81CRA			81CRA	*) $D_J=0.4$ kHz assumed	
6–30c	$^{72}GeH_3I$	$v=0$	1658.38							
	$^{74}GeH_3I$	$v=0$	1632.26							
	$^{76}GeH_3I$	$v=0$	1607.42							
	$^{70}GeD_3I$	$v=0$	1603.42							
	$^{72}GeD_3I$	$v=0$	1579.46							
	$^{74}GeD_3I$	$v=0$	1556.62							
	$^{70}GeHD_2I$	$v=0$	$B=1635.48$						$A(GeHD_2I)=47825$ MHz	
			$C=1623.98$							
	$^{72}GeHD_2I$	$v=0$	$B=1610.23$							
			$C=1599.06$							
	$^{74}GeHD_2I$	$v=0$	$B=1586.09$							
			$C=1575.44$							

Nr.	Molecule	v	B MHz	D kHz	Ref.	q MHz	Ref.	Ref. Struct.	Varia, remarks	See also Tab., Nr.
46	$^{76}GeH_3NCS$	$v_{10}=1$	1016.857	$D_J=0.24$	79DU2			79DU2	$\nu_{10}\simeq 50\ cm^{-1}$	
	$^{74}GeH_3NCS$	$v=0$	1018.386							
		$v_{10}=1$	1026.074	$D_J=0.31$						
	$^{73}GeH_3NCS$	$v_{10}=1$	1030.574							
	$^{72}GeH_3NCS$	$v=0$	1027.915							
		$v_{10}=1$	1035.667	$D_J=0.31$						
	$^{70}GeH_3NCS$	$v=0$	1037.919							
		$v_{10}=1$	1045.736	$D_J=0.31$						
	$^{76}GeH_3{}^{15}NCS$	$v_{10}=1$	1016.247	$D_J=0.29$						
	$^{74}GeH_3{}^{15}NCS$	$v=0$	1018.199							
		$v_{10}=1$	1025.566	$D_J=0.34$						
	$^{73}GeH_3{}^{15}NCS$	$v_{10}=1$	1030.290	$D_J=0.34$						
	$^{72}GeH_3{}^{15}NCS$	$v=0$	1027.750							
		$v_{10}=1$	1035.187	$D_J=0.37$						
	$^{70}GeH_3{}^{15}NCS$	$v=0$	1037.756							
		$v_{10}=1$	1045.280	$D_J=0.34$						
	$^{74}GeD_3NCS$	$v=0$	989.702							
		$v_{10}=1$	996.909	$D_J=0.21$						
	$^{72}GeD_3NCS$	$v=0$	998.219							
		$v_{10}=1$	1005.489	$D_J=0.27$						
	$^{70}GeD_3NCS$	$v=0$	1007.127							
		$v_{10}=1$	1014.476	$D_J=0.33$						
	$^{74}GeH_3N^{13}CS$	$v=0$	1014.124							
		$v_{10}=1$	1021.732	$D_J=0.28$						
47 *6–45* *4–54*	$CH_3C{\equiv}CH$	$v=0$	8545.87712 (6)	$D_J=2.9423\ (7)$ $D_{JK}=163.423\ (10)$ $H_J=9.7\ (20)\cdot 10^{-6}$ *) $H_{JK}=0.935\ (68)\cdot 10^{-3}$ $H_{KJ}=5.23\ (14)\cdot 10^{-3}$	78DUB			78DUB	for structure, see also Vol. II/7 (3, 367) *) The F.I.R. value is more accurate: $H_J=1.6\ (9)\cdot 10^{-13}\ cm^{-1}$ [80MEY] $\nu_5=930.27540\ (7)\ cm^{-1}$	2.6.2, 53 2.9.2, 33
		$v_5=1$	8508.119 (3) $A'-A''=-227.0\ (23)$	$D_J=1.8\ (2)$ $D_{JK}=169\ (1)$ $D'_K-D''_K=-50\ (500)$	80BUR2					
Cont.										

Nr.	Molecule	v	B MHz	D kHz	Ref.	q MHz	Ref.	Ref. Struct.	Varia, remarks	See also Tab., Nr.
47	$CH_3C{\equiv}CH$	$v_{10}=1$	$B^*=8569.9875\,(17)$	$D_J=2.966\,(22)$	79BAU2	$q/4=4.1938\,(6)$	79BAU2		$\zeta_{10}=0.892\,(1)$	
6-45				$D_{JK}=163.780\,(57)$						
4-54				$\varrho^*=1.97\,(22)$						
Cont.	$^{13}CH_3C{\equiv}CH$	$v=0$	8313.24827 (89)	$D_J=2.8122\,(32)$	78DUB					
				$D_{JK}=155.193\,(59)$						
				$H_{JK}=0.813\,(194)\cdot10^{-3}$						
				$H_{KJ}=5.11\,(66)\cdot10^{-3}$						
	$CH_3{}^{13}C{\equiv}CH$	$v=0$	8542.33338 (96)	$D_J=2.9521\,(35)$						
				$D_{JK}=162.845\,(57)$						
				$H_{JK}=1.215\,(204)\cdot10^{-3}$						
				$H_{KJ}=4.76\,(37)\cdot10^{-3}$						
	$CH_3C{\equiv}{}^{13}CH$	$v=0$	8290.24630 (94)	$D_J=2.7586\,(31)$						
				$D_{JK}=155.573\,(59)$						
				$H_{JK}=0.882\,(194)\cdot10^{-3}$						
				$H_{KJ}=5.67\,(61)\cdot10^{-3}$						
48	$CH_3(C{\equiv}C)_2CN$	$v=0$	778.040 (1)	$D_J=0.009\,(1)$	78ALE			78ALE		
				$D_{JK}=4.41\,(2)$						
		$v_{15}=1$	$\alpha=-1.148\,(1)$	$\alpha(D_J)=-0.005\,(1)$		$q=0.235\,(4)$	78ALE		$\zeta_{15}=0.9888\,(6)$	
				$\alpha(D_{JK})=0.10\,(2)$						
				$\eta_J=8.08\,(3)$						
		$v_{16}=1$	$\alpha=-1.3459\,(9)$	$\alpha(D_{JK})=-0.051\,(13)$		$q=0.561\,(3)$			$\zeta_{16}=0.9807\,(5)$	
				$\eta_J=8.59\,(2)$						
49	$CH_3(C{\equiv}C)_3H$	$v=0$	778.2445 (7)	$D_J=0.0095\,(7)$	78ALE			78ALE		
				$D_{JK}=4.441\,(6)$						
		$v_{17}=1$	$\alpha=-1.0238\,(8)$	$\alpha(D_J)=0.0023\,(8)$		$q=0.225\,(3)$	78ALE		$\zeta_{17}=0.9717\,(7)$	
				$\alpha(D_{JK})=-0.031\,(9)$						
				$\eta_J=9.48\,(3)$						
		$v_{18}=1$	$\alpha=-1.2884\,(6)$	$\alpha(D_{JK})=-0.013\,(5)$		$q=0.562\,(2)$			$\zeta_{18}=0.9836\,(2)$	
				$\eta_J=8.14\,(1)$						

Nr.	Molecule	v	B MHz	D kHz	Ref.	q MHz	Ref.	Ref. Struct.	Varia, remarks	See also Tab., Nr.
50 *4-61*	$SiH_3C{\equiv}CH$	$v=0$	4828.66510 (42)	D_J = 1.1737 (5) D_{JK} = 62.724 (17) H_{JK} = 3.0 (1) · 10^{-4} H_{KJ} = 13.3 (14) · 10^{-4}	80CAR, 81CAR				for structure, see also Vol. II/7 (3, 256)	
		$v_{10}=1$	4848.4824 (5) $A\zeta_t$ = 75960 (20)	D_J = 1.226 (3)*) D_{JK} = 63.620 (7) η_J = 172.0 (1) η_K = −0.017 (1)	80CAR	$q/4$ = 1.8940 (3)	80CAR		ν_{10} = 220 cm^{-1} *) H_{JK} and H_{KJ} are fixed at the values obtained for the parent species, $v=0$	
		$v_{10}=2$	4868.0949 (9) $A\zeta_t$ = 76064 (13)	D_J = 1.2828 (9) D_{JK} = 64.171 (15) η_J = 174.56 (8) η_K = −0.0207 (17) H_{JK} = 3 · 10^{-4} (fixed) H_{KJ} = 13 · 10^{-4} (fixed)	81CAR	$q/4$ = 1.8966 (fixed)			x_{ll} = 112190 (100) MHz $x_{10,10}$ = −11.4 cm^{-1}	
	$^{29}SiH_3C{\equiv}CH$	$v=0$	4775.656 (2)	D_J = 1.14 (2)*) D_{JK} = 61.40 (4)	80CAR					
	$^{30}SiH_3C{\equiv}CH$	$v=0$	4725.686 (2)	D_J = 1.13 (2)*) D_{JK} = 60.37 (4)						
	$SiH_3{}^{13}C{\equiv}CH$	$v=0$	4804.021 (2)	D_J = 1.20 (2)*) D_{JK} = 61.60 (4)						
	$SiH_3C{\equiv}{}^{13}CH$	$v=0$	4662.336 (3)	D_J = 1.01 (3)*) D_{JK} = 59.34 (5)						
51	$^{11}BH_3CO$	$v=0$	8657.323 (52)	D_J = 5 (2) D_{JK} = 345 (50)	77VEN		77VEN		for structure, see also Vol. II/7 (3, 84)	2.6.2, 54 2.7.2, 7
	$^{10}BH_3CO$	$v=0$	8980.045 (41)	D_J = 5 (2) D_{JK} = 372 (50)						
	$^{11}BH_3{}^{13}CO$	$v=0$	8656.415 (52)	*)					*) centrifugal distortion constants fixed at the values of the corresponding parent species	
	$^{10}BH_3{}^{13}CO$	$v=0$	8979.975 (41)	*)						
	$^{11}BD_3CO$	$v=0$	7336.655 (41)	D_{JK} = 209 (50)						
	$^{10}BD_3CO$	$v=0$	7530.328 (41)	*)						
	$^{11}BD_3{}^{13}CO$	$v=0$	7332.015 (91)	*)						
	$^{10}BD_3{}^{13}CO$	$v=0$	7526.818 (116)	*)						
	$^{11}BH_3C^{18}O$	$v=0$	8245.397 (52)	*)						
	$^{10}BH_3C^{18}O$	$v=0$	8561.481 (52)	*)						
Cont.	$^{11}BD_3C^{18}O$	$v=0$	6981.751 (91)	*)						

Nr.	Molecule	v	B MHz	D kHz	Ref.	q MHz	Ref.	Ref. Struct.	Varia, remarks	See also Tab., Nr.
51 Cont.	$^{11}BD_2HCO$	$v=0$	$A = 79892.63\ (50)$ $B = 7802.818\ (52)$ $C = 7623.237\ (52)$		77VEN					
	$^{10}BD_2HCO$	$v=0$	$A = 79912.16\ (50)$ $B = 8035.230\ (41)$ $C = 7844.210\ (41)$							
52	$^{11}BF_3CO$	$v=0$	1646.961 (10)	$D_J = 3.8\ (10)$ $D_{JK} = 26\ (10)$	78JAN			78JAN		2.6.2, 55 2.7.2, 8
	$^{10}BF_3CO$ $^{11}BF_3{}^{13}CO$ $^{10}BF_3{}^{13}CO$	$v=0$ $v=0$ $v=0$	1653.015 (10) 1629.301 (15) 1635.426 (15)							
53	$HC(CN)_3$	$v=0$	2865.08 (1)	$D_J = 2.26\ (20)$	77BA1					
54	$(CH_3)_3CH$							73HIL	for structure, see also Vol. II/7 (3, 552)	2.9.2, 35
55	$(CD_3)_2CH_3C^{35}Cl$	$v=0$	$A = 3758.567\ (52)$ $B = 2754.761\ (17)$ $C = 2624.102\ (17)$	$\Delta_J = -0.298\ (1100)$ $\Delta_{JK} = -3.99\ (1390)$ $\Delta_K = 5873\ (29400)$ $\delta_J = -1.55\ (572)$ $\delta_K = 178\ (940)$	74BRA			74BRA	for structure, see also Vol. II/7 (3, 544)	2.7.2, 36
56	$(CH_3)_3C^{81}Br$ $(CH_3)_3C^{79}Br$ $(CH_3)_3{}^{13}C^{81}Br$ $(CH_3)_3{}^{13}C^{79}Br$	$v=0$ $v=0$ $v=0$ $v=0$	2028.369 (9) 2044.237 (7) 2020.292 (10) 2036.331 (9)		79LE2			79LE2		2.7.2, 52
57 *6–51 a* *4–73*	$(CH_3)_3Si^{35}Cl$	$v=0$	2197.382		73ZEI				for structure, see also Vol. II/7 (3, 445)	2.7.2, 37
58	$(CH_3)_3Si^{79}Br$ $(CH_3)_3Si^{81}Br$ $(CH_3)_3{}^{30}Si^{81}Br$	$v=0$ $v=0$ $v=0$	1461.98 (1) 1449.13 (1) 1440.92 (2)		80HAR			80HAR		
59	$(CH_3)_3SiCN$	$v=0$	1995.308 (5)	0.27 (4)	73DUR3					

Nr.	Molecule	v	B MHz	D kHz	Ref.	q MHz	Ref.	Ref. Struct.	Varia, remarks	See also Tab., Nr.
60	$(CH_3)_3SiNC$	$v=0$	2197.32		73DUR3					
61	$(CH_3)_3{}^{70}GeH$	$v=0$	4796.53	$D_J=0.56\,(10)$	73DUR4			73DUR4	for structure, see also Vol. II/7 (3, 445)	
		$v_\tau=1$	4796.15							
	$(CH_3)_3{}^{72}GeH$	$v=0$	4791.10							
		$v_\tau=1$	4790.75							
	$(CH_3)_3{}^{73}GeH$	$v=0$	4788.50							
		$v_\tau=1$	4788.11							
	$(CH_3)_3{}^{74}GeH$	$v=0$	4785.88							
		$v_\tau=1$	4785.52							
	$(CH_3)_3{}^{76}GeH$	$v=0$	4780.83							
		$v_\tau=1$	4780.46							
	$(CH_3)_3{}^{70}GeD$	$v=0$	4657.09							
		$v_\tau=1$	4656.73							
	$(CH_3)_3{}^{72}GeD$	$v=0$	4652.61							
		$v_\tau=1$	4652.28							
	$(CH_3)_3{}^{73}GeD$	$v=0$	4650.42							
		$v_\tau=1$	4649.91							
	$(CH_3)_3{}^{74}GeD$	$v=0$	4648.28							
		$v_\tau=1$	4647.94							
	$(CH_3)_3{}^{76}GeD$	$v=0$	4643.79							
62	$(CH_3)_3{}^{70}Ge^{35}Cl$	$v=0$	1993.113 (4)		75DUR3			75DUR3		
	$(CH_3)_3{}^{72}Ge^{35}Cl$	$v=0$	1991.340 (3)							
	$(CH_3)_3{}^{74}Ge^{35}Cl$	$v=0$	1989.611 (3)							
	$(CH_3)_3{}^{76}Ge^{35}Cl$	$v=0$	1987.929 (2)							
	$(CH_3)_3{}^{70}Ge^{37}Cl$	$v=0$	1942.426 (2)							
	$(CH_3)_3{}^{72}Ge^{37}Cl$	$v=0$	1940.516 (4)							
	$(CH_3)_3{}^{74}Ge^{37}Cl$	$v=0$	1938.622 (2)							
63	$(CH_3)_3{}^{76}Ge^{81}Br$	$v=0$	1226.48		73LI1			73LI1	for structure, see also Vol. II/7 (3, 443)	
		$v_x=1$	1224.66							
	$(CH_3)_3{}^{74}Ge^{81}Br$	$v=0$	1230.388							
		$v_x=1$	1228.67							
Cont.	$(CH_3)_3{}^{73}Ge^{81}Br$	$v=0$	1232.37							

Nr.	Molecule	v	B MHz	D kHz	Ref.	q MHz	Ref.	Ref. Struct.	Varia, remarks	See also Tab., Nr.
63 Cont.	$(CH_3)_3{}^{72}Ge^{81}Br$	$v=0$	1234.396		73LI1					
		$v_x=1$	1232.67							
	$(CH_3)_3{}^{70}Ge^{81}Br$	$v=0$	1238.514							
		$v_x=1$	1236.78							
	$(CH_3)_3{}^{76}Ge^{79}Br$	$v=0$	1240.49							
	$(CH_3)_3{}^{74}Ge^{79}Br$	$v=0$	1244.333							
		$v_x=1$	1242.61							
	$(CH_3)_3{}^{72}Ge^{79}Br$	$v=0$	1248.268							
		$v_x=1$	1246.52							
	$(CH_3)_3{}^{70}Ge^{79}Br$	$v=0$	1252.333							
		$v_x=1$	1250.56							
64	$(CH_3)_3{}^{76}GeCN$	$v=0$	1844.96 (2)		74DUR11			74DUR11	for structure, see also Vol. II/7 (3, 548)	
	$(CH_3)_3{}^{74}GeCN$	$v=0$	1846.06 (3)							
	$(CH_3)_3{}^{72}GeCN$	$v=0$	1847.17 (2)							
	$(CH_3)_3{}^{70}GeCN$	$v=0$	1848.30 (2)							
	$(CH_3)_3{}^{76}Ge^{13}CN$	$v=0$	1826.44 (4)							
	$(CH_3)_3{}^{74}Ge^{13}CN$	$v=0$	1827.69 (4)							
	$(CH_3)_3{}^{72}Ge^{13}CN$	$v=0$	1828.90 (1)							
	$(CH_3)_3{}^{70}Ge^{13}CN$	$v=0$	1830.16 (2)							
	$(CH_3)_3{}^{76}GeC^{15}N$	$v=0$	1793.33 (2)							
	$(CH_3)_3{}^{74}GeC^{15}N$	$v=0$	1794.52 (2)							
	$(CH_3)_3{}^{72}GeC^{15}N$	$v=0$	1795.81 (2)							
	$(CH_3)_3{}^{70}GeC^{15}N$	$v=0$	1797.08 (2)							
65 *6-52* *4-76*	$(CH_3)_3N$								for structure, see also Vol. II/7 (3, 450)	2.8, 108
66	$(CH_3)_3N \cdot ICF_3$	$v=0$	431.95 (20)		75LEG					

Nr.	Molecule	v	B MHz	D kHz	Ref.	q MHz	Ref.	Ref. Struct.	Varia, remarks	See also Tab., Nr.
67	$(CH_3)_3{}^{11}B \cdot N(CH_3)_3$	$v=0$	$I=321.02\,(80)$*)		78KUZ			78KUZ	*) in amu Å^2	
	$(CH_3)_3{}^{10}B \cdot N(CH_3)_3$	$v=0$	$I=320.20\,(80)$*)							
	$(CH_3)_3{}^{11}B \cdot {}^{15}N(CH_3)_3$	$v=0$	$I=321.67\,(80)$*)							
	$(CH_3)_3{}^{10}B \cdot {}^{15}N(CH_3)_3$	$v=0$	$I=320.88\,(80)$*)							
	$(CH_3)_3{}^{11}B \cdot N(CD_3)_3$	$v=0$	$I=359.65\,(85)$*)							
	$(CH_3)_3{}^{10}B \cdot N(CD_3)_3$	$v=0$	$I=358.68\,(85)$*)							
	$(CH_3)_3{}^{11}B \cdot N({}^{13}CH_3)_3$	$v=0$	$I=329.26\,(85)$*)							
	$(CH_3)_3{}^{10}B \cdot {}^{15}N({}^{13}CH_3)_3$	$v=0$	$I=328.52\,(85)$*)							
68	$(CH_3)_3N \cdot {}^{11}BH_3$	$v=0$	4516.43		73DUR6			73DUR6		2.6.2, 56
	$(CH_3)_3N \cdot {}^{10}BH_3$	$v=0$	4627.78					75CAX		2.8, 135
	$(CH_3)_3N \cdot {}^{11}BD_3$	$v=0$	4042.93							
	$(CH_3)_3N \cdot {}^{10}BD_3$	$v=0$	4123.38							
	$(CD_3)_3N \cdot {}^{11}BH_3$	$v=0$	3729.09							
	$(CD_3)_3N \cdot {}^{10}BH_3$	$v=0$	3811.18							
	$(CD_3)_3N \cdot {}^{11}BD_3$	$v=0$	3381.37							
	$(CD_3)_3N \cdot {}^{10}BD_3$	$v=0$	3442.95							
	$(CH_3)_2{}^{13}CH_3N \cdot {}^{11}BH_3$	$v=0$	$A=4640.2$							
			$B=4499.90$							
			$C=4430.02$							
	$(CH_3)_3N \cdot {}^{11}BD_2H$	$v=0$	$A=4616.9$							
			$B=4210.45$							
			$C=4147.65$							
	$(CH_3)_3{}^{15}N \cdot {}^{11}BH_3$	$v=0$	$I=111.8891$ *)						*) in amu Å^2	
	$(CH_3)_3{}^{15}N \cdot {}^{10}BH_3$	$v=0$	$I=109.1940$ *)							
	$(CH_3)_3{}^{15}N \cdot {}^{11}BD_3$	$v=0$	$I=124.9985$ *)							
	$(CH_3)_3{}^{15}N \cdot {}^{10}BD_3$	$v=0$	$I=122.5570$ *)							
69	$({}^{13}CH_3)_3N \cdot {}^{11}BF_3$	$v=0$	$I=296.2155$ *)		77CAX2			77CAX2	*) in amu Å^2	
	$({}^{13}CH_3)_3N \cdot {}^{10}BF_3$	$v=0$	$I=295.5609$ *)							
	$(CD_3)_3N \cdot {}^{11}BF_3$	$v=0$	$I=326.7258$ *)							
	$(CD_3)_3N \cdot {}^{10}BF_3$	$v=0$	$I=325.9573$ *)							

Nr.	Molecule	v	B MHz	D kHz	Ref.	q MHz	Ref.	Ref. Struct.	Varia, remarks	See also Tab., Nr.
70	$(CH_3)_3N \cdot {}^{69}GaH_3$	$v=0$	1875.91 (2)	$D_J=0.50$ (10)	80DUR1			80DUR1		2.8, 136
	$(CH_3)_3N \cdot {}^{71}GaH_3$	$v=0$	1858.66 (1)	$D_J=0.46$ (4)					$\nu(GaD_3$ torsion)	
	$(CH_3)_3{}^{15}N \cdot {}^{69}GaH_3$	$v=0$	1869.24 (1)	$D_J=0.43$ (1)					$\simeq 70\ cm^{-1}$	
	$(CH_3)_3{}^{15}N \cdot {}^{71}GaH_3$	$v=0$	1851.91 (1)	$D_J=0.53$ (1)						
	$(CH_3)_3N \cdot {}^{69}GaD_3$	$v=0$	1816.24 (2)	$D_J=0.43$ (11)						
	$(CH_3)_3N \cdot {}^{71}GaD_3$	$v=0$	1800.94 (1)	$D_J=0.36$ (4)						
70a	$(CH_3)_3N \cdot Ga(CH_3)_3$	$v=0$	1102		81DUR4				harmonic force field calculated	
71	$(CH_3)_3As \cdot {}^{11}BH_3$	$v=0$	2837.97		74DUR6			74DUR6		
	$(CH_3)_3As \cdot {}^{10}BH_3$	$v=0$	2909.23							
	$(CH_3)_3As \cdot {}^{11}BD_3$	$v=0$	2567.43							
	$(CH_3)_3As \cdot {}^{10}BD_3$	$v=0$	2623.21							
72	CH_3CD_3	$v=0$	16474.438 (30)	$D_J=19.6$	77HIR			81HIR1	for structure,	2.6.2, 57
6-54				$D_{JK}=50$					see also	2.8, 1
	${}^{13}CH_3CD_3$	$v=0$	16088.428 (30)						Vol. II/7 (3, 274)	
	$CH_3{}^{13}CD_3$	$v=0$	16251.914 (30)						uncertainties are 2.5 σ	
	CH_3CHD_2	$v=0$	$A=60809.592$ (59) *)	$\Delta_J=23.33$ (45)	79HIR1				*) with $D=-566.319$ MHz,	
			$B=17770.87$ (16)	$\Delta_{JK}=60.2$ (20)					assumed (constant	
			$C=17084.84$ (16)	$\Delta_K=177$ (15)					associated with a cross	
				$\delta_J=0.79$ (16)					product of inertia)	
				$\delta_K=50$ (81)						
		$v_\tau=1$	$A=60847$ (106)	**)	81HIR1				**) ground state	
			$B=17703.167$ (48)						constants assumed	
			$C=17030.976$ (54)							
	CH_3CH_2D	$v=0$	$A=69640.504$ (69)	$\Delta_J=26.61$ (52)					$D=811.64$ MHz assumed	
			$B=18876.23$ (14)	$\Delta_{JK}=67.3$ (13)						
			$C=18209.90$ (13)	$\Delta_K=243$ (16)						
				$\delta_J=1.03$ (17)						
				$\delta_K=-57$ (65)						
	CD_3CH_2D	$v=0$	$A=48650.630$ (58)	$\Delta_J=17.52$ (14)					$D=611.36$ MHz assumed	
			$B=15656.793$ (64)	$\Delta_{JK}=40.93$ (57)						
			$C=15194.324$ (63)	$\Delta_K=100$ (15)						
				$\delta_J=0.576$ (65)						
Cont.				$\delta_K=-50$ (30)						

Nr.	Molecule	v	B MHz	D kHz	Ref.	q MHz	Ref.	Ref. Struct.	Varia, remarks	See also Tab., Nr.
72 *6–54* Cont.	CH_2DCH_2D gauche	$v=0$	$A=61360.344\,(23)^{+})$ $B=17642.0291\,(75)$ $C=17036.3299\,(75)$	$\Delta_J=24.596\,(66)^{+})$ $\Delta_{JK}=49.76\,(10)$ $\Delta_K=196.3\,(35)$ $\delta_J=1.07276\,(91)$ $\delta_K=182.52\,(48)$	81HIR1				$^{+})$ average value	
73	CH_3CF_3	$v=0$	$B=5185.1387\,(24)$ $A=5498.850\,(70)$	$D_J=1.267\,(11)$ $D_{JK}=2.006\,(14)$ $D_K<10$	78MEE				for structure, see also Vol. II/7 (3, 213)	2.6.2, 58 2.8, 2
74	$CD_3C^{35}Cl_3$ $CD_3C^{37}Cl_3$ $CD_3C^{35}Cl_2{}^{37}Cl$ $CD_3C^{35}Cl^{37}Cl_2$	$v=0$ $v=0$ $v=0$ $v=0$	2210.61 2126.42 $A=2208.42\,(2)$ $B=2156.59\,(2)$ $C=1659.08\,(73)$ $A=2180.19\,(3)$ $B=2128.61\,(3)$ $C=1634.02\,(118)$	$D_J=0.2$ $D_J=0.4$	73DUR2			73DUR2	for structure, see also Vol. II/7 (3, 206)	2.8, 3
75 *6–58* *4–86*	$CH_3Si^{35}Cl_3$ $CH_3{}^{29}Si^{35}Cl_3$ $^{13}CH_3Si^{35}Cl_3$	$v=0$ $v=0$ $v=0$	1769.797 1769.228 1741.700	$D_J=0.26$	77TA2			77TA2	for structure, see also Vol. II/7 (3, 95)	
76	$CH_3{}^{70}Ge^{35}Cl_3$ $CH_3{}^{72}Ge^{35}Cl_3$ $CH_3{}^{73}Ge^{35}Cl_3$ $CH_3{}^{74}Ge^{35}Cl_3$ $CH_3{}^{76}Ge^{35}Cl_3$ $CH_3{}^{70}Ge^{37}Cl_3$ $CH_3{}^{72}Ge^{37}Cl_3$ $CH_3{}^{74}Ge^{37}Cl_3$ $CH_3{}^{72}Ge^{35}Cl_2{}^{37}Cl$ $CH_3{}^{74}Ge^{35}Cl_2{}^{37}Cl$	$v=0$ $v=0$ $v=0$ $v=0$ $v=0$ $v=0$ $v=0$ $v=0$ $v=0$ $v=0$	1602.19 1601.42 1601.10 1600.71 1600.02 1537.84 1537.10 1536.36 $A=1597.96$ $B=1559.31$ $C=1203\,(90)$ $A=1597.17$ $B=1558.59$ $C=1207\,(90)$	0.42 0.24 0.42 0.30 0.36	75DUR2			75DUR2		2.8, 67

Nr.	Molecule	v	B MHz	D kHz	Ref.	q MHz	Ref.	Ref. Struct.	Varia, remarks	See also Tab., Nr.
77	$CF_3{}^{70}GeH_3$	$v=0$	1943.97		74KR1					2.8, 66
	$CF_3{}^{72}GeH_3$	$v=0$	1924.269							
		$v_\gamma=1$	1921.767							
		$v_\gamma=2$	1919.317							
		$v_\gamma=3$	1917.024							
		$v_\tau=1$	1925.537							
	$CF_3{}^{73}GeH_3$	$v=0$	1914.75							
	$CF_3{}^{74}GeH_3$	$v=0$	1905.530							
		$v_\gamma=1$	1903.015							
		$v_\gamma=2$	1900.587							
		$v_\gamma=3$	1898.315							
		$v_\tau=1$	1906.811							
	$CF_3{}^{76}GeH_3$	$v=0$	1887.59							
78	SiH_3SiF_3	$v=0$	2075.60_2		75PAS			75PAS		2.6.2, 59
	$^{29}SiH_3SiF_3$	$v=0$	2043.17_7							
	$^{30}SiH_3SiF_3$	$v=0$	2012.38_6							
	$SiH_3{}^{29}SiF_3$	$v=0$	2074.60_3							
	$SiH_3{}^{30}SiF_3$	$v=0$	2073.58_5							
	SiD_3SiF_3	$v=0$	1913.73							
79	$^{11}BH_3 \cdot NH_3$	$v=0$	17517.18 (14)	$D_J=48.6$ (97)	81SUE			81SUE		2.6.2, 60
	$^{10}BH_3 \cdot NH_3$	$v=0$	18054.42 (10)	$D_J=50.1$ (66)						
80	$H_3P \cdot {}^{11}BF_3$	$v=0$	2889.32 (9)	$D_J=70$ (11)	75ODO			75ODO		2.6.2, 61 2.8, 133
	$H_3P \cdot {}^{10}BF_3$	$v=0$	2892.35 (28)	$D_J=51$ (33)						
	$D_3P \cdot {}^{11}BF_3$	$v=0$	2654.14 (22)	$D_J=31$ (21)						
	$D_3P \cdot {}^{10}BF_3$	$v=0$	2657.71 (22)	$D_J=29$ (21)						
81 *6–62a*	$^{11}BH_3PH_3$ *)	$v=0$	10591.31		73DUR5			73DUR5	for structure, see also Vol. II/7 (2, 52) *) staggered conformation	2.6.2, 62 2.8, 132
	$^{10}BH_3PH_3$	$v=0$	11045.78							
	$^{11}BD_3PH_3$	$v=0$	8747.45							
	$^{10}BD_3PH_3$	$v=0$	9011.18							
	$^{11}BD_3PD_3$	$v=0$	7801.40							
	$^{10}BD_3PD_3$	$v=0$	8034.55							
	$^{11}BH_3PD_3$	$v=0$	9357.00							
Cont.	$^{10}BH_3PD_3$	$v=0$	9746.75							

Nr.	Molecule	v	B MHz	D kHz	Ref.	q MHz	Ref.	Ref. Struct.	Varia, remarks	See also Tab., Nr.
81 *6-62a* Cont.	$^{11}BD_2HPH_3$	$v=0$	$B=9393.73$ $C=9136.78$		73DUR5					
	$^{10}BD_2HPH_3$	$v=0$	$B=9714.63$ $C=9438.93$							
	$^{11}BDH_2PH_3$	$v=0$	$B=9993.43$ $C=9746.93$							
	$^{11}BD_2HPD_3$	$v=0$	$B=8342.58$ $C=8140.73$							
	$^{10}BD_2HPD_3$	$v=0$	$B=8622.68$ $C=8406.48$							
	$^{11}BH_3PD_2H$		$B=9879.93$ $C=9572.68$							
82	$CH_2{=}C{=}CD_2$	$v=0$	$(B+C)/2$ $=7846.158(10)$ $[A-(B+C)/2]\,b_p^2$ $=0.133(5)$	$D_J=2.4(5)$ $D_{JK}=130(5)$	73HIR				for structure, see also Vol. II/7 (3, 368)	2.6.2, 63
83	SF_6									2.9.2, 36
84 *6-64* *4-94*	$SF_5{}^{35}Cl$	$v=0$	1824.59248(10)	$D_J=0.11194(25)$ $D_{JK}=0.18684(20)$ $R_6=1.549(10)\cdot10^{-3}$	79BEL			81BAR	for structure, see also Vol. II/7 (2, 100)	2.7.2, 38
	$SF_5{}^{37}Cl$	$v=0$	1783.55904(10)	$D_J=0.10913(25)$ $D_{JK}=0.18536(30)$ $R_6=1.395(10)\cdot10^{-3}$						
	$^{34}SF_5{}^{35}Cl$	$v=0$	1823.8326(10)	$D_J=0.1120(5)$ $D_{JK}=0.1865(5)$ $R_6=1.58(5)\cdot10^{-3}$						
85	$SF_5{}^{81}Br$	$v=0$	1159.7720(1)	$D_J=0.0684(2)$ $D_{JK}=0.1633(3)$ $R_6=0.26(5)\cdot10^{-3}$	81VER				for structure, see also Vol. II/7 (2, 67)	2.7.2, 53

Nr.	Molecule	v	B MHz	D kHz	Ref.	q MHz	Ref.	Ref. Struct.	Varia, remarks	See also Tab., Nr.
86	$^{122}TeF_5{}^{35}Cl$	$v=0$	1396.335		73LEG			73LEG	for structure, see also Vol. II/7 (2, 103)	
	$^{124}TeF_5{}^{35}Cl$	$v=0$	1396.150							
	$^{125}TeF_5{}^{35}Cl$	$v=0$	1396.058							
	$^{126}TeF_5{}^{35}Cl$	$v=0$	1395.964							
	$^{128}TeF_5{}^{35}Cl$	$v=0$	1395.776							
		$v_{11}=1$	1395.006						v_{11}, Te—Cl wagging	
	$^{130}TeF_5{}^{35}Cl$	$v=0$	1395.595							
		$v_{11}=1$	1394.816							
	$^{122}TeF_5{}^{37}Cl$	$v=0$	1363.637							
	$^{124}TeF_5{}^{37}Cl$	$v=0$	1363.418							
	$^{125}TeF_5{}^{37}Cl$	$v=0$	1363.314							
	$^{126}TeF_5{}^{37}Cl$	$v=0$	1363.205							
	$^{128}TeF_5{}^{37}Cl$	$v=0$	1362.987							
		$v_{11}=1$	1362.261							
	$^{130}TeF_5{}^{37}Cl$	$v=0$	1362.772							
		$v_{11}=1$	1362.062							
87	$HC(NO_2)_3$	A *)	1608.410 (9)	$D_J=0.215$ (18)	80CAM2				for structure, see also Vol. II/7 (3, 63)	3, 1
		B	1609.255 (5)	$D_J=0.142$ (10)					*) label of the 6 torsions levels	
		C	1610.087 (3)	$D_J=0.130$ (6)						
		D	1610.916 (10)	$D_J=0.136$ (9)						
		E	1611.791 (3)	$D_J=0.206$ (6)						
		F	1612.663 (1)	$D_J=0.288$ (3)						
	$DC(NO_3)_3$	A *)	1594.409 (5)	$D_J=0.267$ (10)						
		B	1595.226 (8)	$D_J=0.108$ (16)						
		C	1596.037 (4)	$D_J=0.139$ (7)						
		D	1596.846 (9)	$D_J=0.146$ (2)						
		E	1597.681 (4)	$D_J=0.208$ (1)						
		F	1598.519 (3)	$D_J=0.283$ (6)						
88	$^{35}ClF_5$	$v=0$	3550.273 (6)	D_J = 0.764 (6)	76GOU, 76BOD			76BOD	harmonic force field calculated [76GOU]	2.6.2, 64 2.7.2, 39
				$D_{JK}=-0.664$ (5)						
				R_6 = 0.0512 (8)						
	$^{37}ClF_5$	$v=0$	3545.888 (7)	D_J = 0.755 (6)						
				$D_{JK}=-0.586$ (16)						
				R_6 = 0.0512 (40)						

Nr.	Molecule	v	B MHz	D kHz	Ref.	q MHz	Ref.	Ref. Struct.	Varia, remarks / See also Tab., Nr.
89 *6-66*	$^{79}BrF_5$	$v=0$	3009.2612 (8)	D_J = 0.5432 (8) D_{JK} = −0.455 (2) R_6 = 0.078 (2)	73SUZ			78GHE	for structure, 2.7.2, 54 see also Vol. II/7 (2, 65)
		$v_5=1$ $v_9=1$	B = 3094.93 (70) C = 2147.22 (230) B = 3097.97 (35) C = 2146.02 (90)		77BRI, 78GHE	$q^+ = +2.91$ (3) $q^- = -7.45$ (70)	77BRI		Coriolis interaction between ν_5 and ν_9 analyzed: $\nu_5-\nu_9-C_9\zeta_9$ $= -205594$ (1200) MHz $\zeta^y_{5,9}=0.781$ (15) $\zeta_9=-0.251$ (4) [77BRI]
	$^{81}BrF_5$	$v=0$	3096.8196 (8)	D_J = 0.5416 (8) D_{JK} = −0.450 (2) R_6 = 0.078 (2)	73SUZ				
		$v_5=1$ $v_9=1$	B = 3093.19 (70) C = 2148.9 (23) B = 3095.18 (35) C = 2146.08 (110)		77BRI, 78GHE	$q^+ = +2.83$ (3) $q^- = -8.11$ (70)			Coriolis interaction between ν_5 and ν_9: $\zeta^y_{5,9}=0.785$ (15), $\nu_5-\nu_9-C_9\zeta_9$ $= -201664$ (1200) MHz. $\zeta_9=-0.249$ (4) [77BRI]
90 *6-67*	IF_5	$v=0$	2727.4217 (10)	D_J = 0.400 (2) D_{JK} = −0.320 (2) R_6 = 0.0795 (10)	78TR			81BAL	for structure, 2.7.2, 60 see also Vol. II/7 (2, 190)
		$v_5=1$ $v_9=1$	2723.20 (46)*) 2727.70 (23) $C+C\zeta_t-B$ $= -1132.10$ (85) $C-C\zeta_t-B$ $= -470.10$ (34) $C_5-C_9=2.77$ (140) $C_9=1926.70$ (45)		81BAL	$q^+ = 2.259$ (2) *) $q^- = -5.93$ (47)	81BAL		$\nu_5=206.7$ (2) cm^{-1} Coriolis interaction between ν_5 and ν_9: $\nu_5-\nu_9$ $=531475$ (6000) MHz, $R_{5,9}=3046.9$ (360) MHz, $\zeta^y_{5,9}=0.79$ (10) $\zeta_9=-0.172$ (5) *) uncertainties are 2.5 σ

Nr.	Molecule	v	B MHz	D kHz	Ref.	q MHz	Ref.	Ref. Struct.	Varia, remarks	See also Tab., Nr.
91	CHFCHFCHF (ring)	$v=0$	4064.925 (22)	$D_J=$ 3.9 (8)	76GIL1			76GIL1	the uncertainties are 2 σ.	2.6.2, 65
		$v_a=1$	4065.645 (30)	$D_J=$ 16.2 (12)						
		$v_b=1$	4060.911 (4)	$D_J=$ 2.4 (2)						
	CDFCDFCDF (ring)	$v=0$	3750.373 (16)	$D_J=$ 2.4 (4)						
	CHFCHFCDF (ring)	$v=0$	$A=3992.69$ (6)							
			$B=3917.61$ (6)							
			$C=2856.27$ (766)							
	CHFCDFCDF (ring)	$v=0$	$A=3883.47$ (6)							
			$B=3817.44$ (6)							
			$C=2818.81$ (790)							
	^{13}CHFCHFCHF (ring)	$v=0$	$A=4050.98$ (8)							
			$B=4026.50$ (8)							
			$C=2877.98$ (11662)							
92 *6-72* *4-99*	$(CH_2O)_3$	$v_{19}=1$	5276.9304	$D_J=$ 1.381	75COL	$q_0/4=3.2507$	75COL	80COL1	for structure, see also Vol. II/7 (3, 410)	2.9.2, 37
				$D_{JK}=-2.081$		$q_J/4=0.148 \cdot 10^{-4}$				
				$\eta_J=-16.06$		$q_{12}=3.25$			$\zeta^z_{19}=-0.668$	
		$v_{20}=1$	5268.1487	$D_J=$ 1.346		$q_0/4=1.9453$				
				$D_{JK}=-2.012$		$q_{12}=1.8621$			$\zeta^z_{20}=-0.281$	
				$\eta_J=-7.78$						
		$v_{20}=2$	5263.4068	$D_J=$ 1.33		$q_0/4=1.8867$			$\gamma_{ll}=-0.0565$ MHz	
				$D_{JK}=-2.05$		$q_{12}=1.8621$			$x_{ll}=-21330$ MHz	
				$\eta_J=-9.09$		$d_{12}=2.6868$			$\zeta^z_{20}=-0.266$	
		$v_7=1, v_{20}=1$	5256.2403	$D_J=$ 1.285		$q_0/4=2.1211$				
				$D_{JK}=-2.072$		$q_{12}=1.8621$			$\zeta^z_{20}=-0.269$ [75COL]	
				$\eta_J=-4.55$						
		$v_{19}=1, v_{20}=1$	$B=5271.161$ (2)	$D_J=$ 1.40 (1)	80COL2	$q^t_0/4=1.958$ (20)	80COL2		$x_{ll'}=-38620$ (30) MHz,	
			$C_e=2933.95$ (34)	$D_{JK}=-2.067$ (20)		$q^{t'}_0/4=3.218$ (20)			$\gamma_{ll'}=-0.2589$ (20) MHz,	
				$\eta_t=-7.78$		$q^t_{12}=-1.832$ (20)			$\zeta^z_t=-0.281$	
				$\eta_{t'}=-22.6$ (4)		$q^{t'}_{12}=-3.248$ (40)			$\zeta^z_{t'}=-0.665$ [80COL2]	
						$d^{tt'}_{12}=24.43$ (46)				
						$d^{t't}_{12}=3.63$ (53)				
Cont.						$r^{tt'}=897.26$ (1500)				

Nr.	Molecule	v	B MHz	D kHz	Ref.	q MHz	Ref.	Ref. Struct.	Varia, remarks	See also Tab., Nr.
92	$(H_2{}^{13}CO)(H_2CO)_2$	$v=0$	$A=5271.08667\,(256)$	$\Delta_J = 1.3144\,(13)$	80COL1					
6-72			$B=5171.42441\,(242)$	$\Delta_{JK}=-1.9821\,(17)$						
4-99			$C=2904.712\,(155)$							
Cont.	$(H_2C^{18}O)(H_2CO)_2$	$v=0$	$A=5268.0552\,(102)$	$\Delta_J = 1.2947\,(21)$						
			$B=5080.38834\,(894)$	$\Delta_{JK}=-1.9585\,(42)$						
			$C=2874.092\,(222)$	$\delta_J = 0.02937\,(826)$						
93	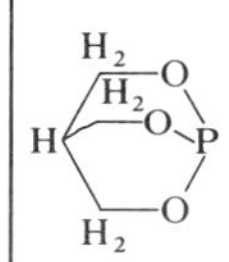	$v=0$	2135.317		74SLA					
		$v_\tau=1$	2140.172						$\nu_\tau=44.2\,(50)\ cm^{-1}$, skeletal torsion	
		$v_\tau=2$	2142.866						$V=0.920\,(200)\ cm^{-1}$	
		$v_\tau=3$	2145.357						[74SLA]	
		$v_\tau=4$	2147.553							
		$v_\tau=5$	2149.514							
		$v_\tau=6$	2151.312							
		$v_\tau=7$	2152.989							
		$v_\tau=8$	2154.535							
		$v_\tau=9$	2155.989							
		$v_\tau=10$	2157.355							
93a	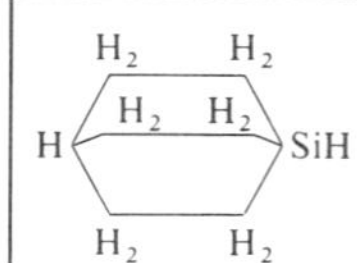	$v=0$	1846.97		80KAW			80KAW	$\nu_\tau=78.6\,(47)\ cm^{-1}$, skeletal torsion	
		$v_\tau=1$	1847.00						equilibrium torsional angle (SiCCC dihedral angle) $=21.3\,(10)°$	
		$v_\tau=2$	1840.87						height of the potential hump $=606\,(40)$ cal/mole	
		$v_\tau=3$	1842.67							
		$v_\tau=4$	1835.54							
		$v_\tau=5$	1841.44							
		$v_\tau=6$	1841.84							
		$v_\tau=7$	1843.76							
		$v_\tau=8$	1845.00							
		$v_\tau=9$	1846.30							
		$v_\tau=10$	1847.28							
	^{29}Si	$v=0$	1835.19							
		$v_\tau=1$	1835.23							
		$v_\tau=2$	1829.17							
		$v_\tau=3$	1830.95							
		$v_\tau=4$	1823.93							
		$v_\tau=5$	1829.75							
Cont.		$v_\tau=6$	1830.14							

Nr.	Molecule	v	B MHz	D kHz	Ref.	q MHz	Ref.	Ref. Struct.	Varia, remarks	See also Tab., Nr.
93a Cont.	^{30}Si	$v=0$	1823.79		80KAW					
		$v_\tau=1$	1823.82							
94	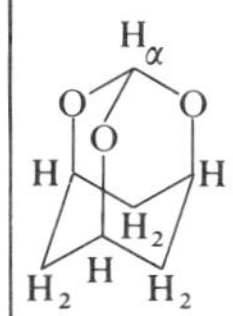	$v=0$	1848.64		73SLA			73SLA	for structure, see also Vol. II/7 (3, 742)	2.6.2, 66
		$v_a=1$	1846.68							
		$v_a=2$	1844.72							
		$v_a=3$	1842.70							
		$v_b=1$	1851.50							
		$v_b=2$	1854.90							
	$^{13}C_\alpha$	$v=0$	1836.11							
95	$C_7H_9{}^{35}Cl$	$v=0$	1406.46 (1)		73LAU			73LAU	for structure, see also Vol. II/7 (3, 733)	2.7.2, 40; 3, 2
		$v_x=1$	1406.15 (1)							
	$C_7H_9{}^{37}Cl$	$v=0$	1371.68 (1)							
		$v_x=1$	1371.38 (1)							
	$^{13}C_1$—^{35}Cl	$v=0$	$B=1395.03$ (1)						numbering of C atoms, see Fig. 2	
			$C=1392.10$ (1)							
	$^{13}C_7$—^{35}Cl	$v=0$	$B=1406.16$ (2)							
			$C=1398.72$ (2)							
96	C_5H_5BeH	$v=0$	6212.619 (8)		75BAR			75BAR	C_{5v} symmetry	2.6.2, 67; 3, 3
		$v_1=1$	6188.492 (17)						relative intensity measurements of ν_1, ν_2, ν_3 and ν_4	
		$v_2=1$	6199.753 (8)							
		$v_3=1$	6207.054 (23)							
		$v_4=1$	6219.952 (10)							
	C_5H_5BeD	$v=0$	5740.264 (36)							
	$^{13}CC_4H_5BeH$	$v=0$	$A=6209.52$ (3)							
			$B=6100.56$ (3)							
			$C=4369.1$ (86)							
	$^{13}CC_4H_5BeD$	$v=0$	$A=5736.56$ (3)							
			$B=5643.51$ (3)							
			$C=4368.0$ (68)							
97	$C_4H_4{}^{56}Fe(Co)_3$	$v=0$	962.41	$D_J=0.415$	73LUCK			73LUCK		

Nr.	Molecule	v	B MHz	D kHz	Ref.	q MHz	Ref.	Ref. Struct.	Varia, remarks	See also Tab., Nr.
98	B_5H_9	$v=0$	7002.88 (7)		77SCH			77SCH	for structure,	3, 4
	$^{10}B_1$	$v=0$	7089.86 (2)						see also Vol. II/7 (2, 46)	
	D_1	$v=0$	6596.98 (9)						C_{4v} symmetry	
	$^{10}B_2$	$v=0$	$A=7169.18$ (2)	$\Delta_{JK}=-1.86$ (2) *)					numbering of the atoms	
			$B=7005.81$ (2)	$\Delta_K=1.1$ (4)					see Fig. 4	
			$C=4968.29$ (2)	$\delta_J=0.018$ (1)					*) representation III^r	
				$\delta_K=0.40$ (7)						
	$^{10}B_2$, $^{10}B_4$	$v=0$	$A=7338.30$ (2)							
			$B=7008.56$ (2)							
			$C=5047.21$ (2)							
	D_6	$v=0$	$A=6901.83$ (4)							
			$B=6724.52$ (4)							
			$C=4805.01$ (5)							
	$^{10}B_1$, D_6	$v=0$	$A=6989.16$ (2)							
			$B=6807.78$ (2)							
			$C=4804.89$ (4)							
	$^{10}B_2$, D_6	$v=0$	$A=7009.31$ (3)							
			$B=6771.08$ (3)							
			$C=4879.50$ (11)							
	$^{10}B_4$, D_6	$v=0$	$A=7014.25$ (3)							
			$B=6776.09$ (3)							
			$C=4881.71$ (10)							
	D_6, D_8	$v=0$	$A=6794.45$ (2)							
			$B=6465.61$ (2)							
			$C=4726.65$ (3)							
	$^{10}B_2$, D_6, D_8	$v=0$	$A=6887.97$ (3)							
			$B=6521.50$ (3)							
			$C=4800.35$ (4)							
	$^{10}B_2$, D_1	$v=0$	$A=6745.53$ (4)							
			$B=6600.80$ (4)							
			$C=4968.13$ (4)							
	D_2	$v=0$	$A=7002.55$ (4)							
			$B=6471.90$ (4)							
Cont.			$C=4624.69$ (5)							

Nr.	Molecule	v	B MHz	D kHz	Ref.	q MHz	Ref.	Ref. Struct.	Varia, remarks	See also Tab., Nr.
98 Cont.	$^{10}B_2, D_3$	$v=0$	$A=7168.78$ (6)		77SCH					
			$B=6474.40$ (6)							
			$C=4694.86$ (16)							
	$^{10}B_2, D_4$	$v=0$	$A=7005.82$ (4)							
			$B=6622.36$ (4)							
			$C=4699.55$ (5)							
	B_4	$v=0$	$A=7260.01$ (3)							
			$B=7092.44$ (3)							
			$C=4968.13$ (3)							
99	$C_{10}H_{10}$	$v=0$	1644.31 (1)	$D_J=0.0787$	75LUK			75LUK	C_{3v} symmetry	3, 5
		$v_a=1$	1644.822		76STI1	$q=0.557$	76STI1		$\nu_a(E)=239$ (2) cm^{-1}	
		$v_a=2$	1645.328						first-order coriolis interaction between ν_a and ν_e [76STI1]	
		$v_a=3$	1645.825							
		$v_b=1$	1644.970						$\nu_b(A_2)=266$ (10) cm^{-1}	
		$v_b=2$	1645.630							
		$v_b=3$	1646.290							
		$v_c=1$	1644.163						$\nu_c(E)=355$ (2) cm^{-1}	
		$v_c=2$	1644.031							
		$v_d=1$	1644.754						$\nu_d(A_1)=443$ (2) cm^{-1}	
		$v_e=1$	1645.142						$\nu_e(E)=586$ (2) cm^{-1}	
100	$C_{11}H_{13}{}^{35}Cl$	$v=0$	865.77 (2)		75LUK			75LUK		3, 6
	$C_{11}H_{13}{}^{37}Cl$	$v=0$	848.07 (2)							
101	$C_{13}H_{13}{}^{35}Cl$	$v=0$	641.22 (2)		75LUK			75LUK		3, 7
	$C_{13}H_{13}{}^{37}Cl$	$v=0$	629.31 (2)							
102	$Ar \cdot {}^{11}BF_3$	$v=0$	1537.261 (10)	$D_J=9.6$ (10)	78JAN			78JAN		2.6.2, 68
				$D_{JK}=74$ (10)						2.7.2, 9
	$Ar \cdot {}^{10}BF_3$	$v=0$	1544.586 (10)							
103	$Ar \cdot SO_3$	$v=0$	1449.089 (5)	$D_J=3.6$ (6)	80BOW			80BOW		2.6.2, 69
	$Ar \cdot {}^{34}SO_3$	$v=0$	1439.090 (5)							
103a	$Kr \cdot SO_3$									2.6.2, 69a
104	$N_2 \cdot {}^{11}BF_3$	$v=0$	1737.0 (10)		78JAN			78JAN		2.6.2, 70
	$N_2 \cdot {}^{10}BF_3$	$v=0$	1743.5 (10)							

Nr.	Molecule	v	B MHz	D kHz	Ref.	q MHz	Ref.	Ref. Struct.	Varia, remarks	See also Tab., Nr.
105	$N_2 \cdot SO_3$									2.6.3, 71
106	$CH_3CN\cdots HF$	$v=0$	1853.37 (4)	D_J = 0.82 (20) D_{JK}=67 (2)	80BEV1			80BEV1	see also 5, appendix (subvolume b) ν_σ=181 (20) cm^{-1}, stretch ν_β=45 (15) cm^{-1}, bend [80BEV1] harmonic force field calculated [80BEV1, 80LG4]	
		$v_\sigma=1$	1845.76 (19)							
		$v_\beta=1$	1866.77 (22)							
		$v_\beta=2$	1880.67 (45)							
	$^{13}CH_3CN\cdots HF$	$v=0$	1808.99 (8)	D_J = 1.0 (fixed) D_{JK}=65 (3)						
		$v_\sigma=1$	1801.37 (13)							
		$v_\beta=1$	1822.08 (8)							
		$v_\beta=2$	1835.96 (13)							
	$CD_3CN\cdots HF$	$v=0$	1686.21 (5)	D_J = 1.00 (20) D_{JK}=50 (4)						
		$v_\sigma=1$	1666.46 (10)							
		$v_\beta=1$	1697.98 (11)							
		$v_\beta=2$	1710.40 (34)							
	$CH_3{}^{13}CN\cdots HF$	$v=0$	1844.64 (8)	D_{JK}=66 (2)*)					*) with D_J=1.0 assumed	
		$v_\sigma=1$	1837.45 (11)							
		$v_\beta=1$	1858.00 (11)							
		$v_\beta=2$	1872.09 (8)							
	$CH_3C^{15}N\cdots HF$	$v=0$	1853.20 (7)	D_{JK}=66 (3)*)						
		$v_\sigma=1$	1843.66 (19)							
		$v_\beta=1$	1866.25 (10)							
		$v_\beta=2$	1880.13 (13)							
	$CH_3CN\cdots DF$	$v=0$	1832.42 (6)	D_{JK}=62 (3)*)						
		$v_\sigma=1$	1811.58 (12)							
		$v_\beta=1$	1845.44 (9)							
		$v_\beta=2$	1859.17 (11)							
107	$CH_3NC \cdot {}^{11}BH_3$	$v=0$	3459.24		77STE			77STE	ν_b= 186 (20) cm^{-1}, bend	2.6.2, 72 2.8, 134
		$v_b=1$	3469.73							
		$v_b=2$	3480.36							
		$v_b=3$	3491.06							
		$v_b=4$	3501.73							
Cont.		$v_b=5$	3512.54							

Nr.	Molecule	v	B MHz	D kHz	Ref.	q MHz	Ref.	Ref. Struct.	Varia, remarks	See also Tab., Nr.
107 Cont.	$CH_3NC \cdot {}^{10}BH_3$	$v=0$	3575.28		77STE					
	${}^{13}CH_3NC \cdot {}^{11}BH_3$	$v=0$	3370.53							
	$CH_3{}^{15}NC \cdot {}^{11}BH_3$	$v=0$	3452.00							
	$CH_3{}^{15}NC \cdot {}^{10}BH_3$	$v=0$	3568.63							
	$CH_3N^{13}C \cdot {}^{11}BH_3$	$v=0$	3450.98							
	$CH_3N^{13}C \cdot {}^{10}BH_3$	$v=0$	3565.22							
	$CD_3NC \cdot {}^{11}BH_3$	$v=0$	3094.75							
	$CD_3NC \cdot {}^{10}BH_3$	$v=0$	3198.24							
	$CD_3NC \cdot {}^{11}BD_3$	$v=0$	2731.94							
	$CD_3NC \cdot {}^{10}BD_3$	$v=0$	2803.29							
	$CH_3NC \cdot {}^{11}BD_3$	$v=0$	3045.41							
	$CH_3NC \cdot {}^{10}BD_3$	$v=0$	3124.27							
	$CH_3N^{13}C \cdot {}^{11}BD_3$	$v=0$	3041.51							
108	$(CH_3)_3CCN \cdots HF$	$v=0$	876.25 (12)		80GEO1			80GEO1	v_σ, H-bond stretching	
		$v_\sigma=1$	$\alpha_\sigma=12.3$ (2)						$v_\beta=55$ (5) cm^{-1},	
		$v_\beta=1$	$\alpha_\beta=-4.3$ (1)						H-bond bending	
	$(CH_3)_3CCN \cdots DF$	$v=0$	862.10 (9)							
		$v_\sigma=1$	$\alpha_\sigma=12.4$ (5)							
		$v_\beta=1$	$\alpha_\beta=-4.1$ (1)							
109	$(CH_3)_3CNC \cdot {}^{11}BH_3$	$v=0$	1424.11 (2)		79BEV			79BEV		
	$(CH_3)_3CNC \cdot {}^{10}BH_3$	$v=0$	1468.15 (2)							
	$(CH_3)_3CN^{13}C \cdot {}^{11}BH_3$	$v=0$	1412.93 (2)							
	$(CH_3)_3CN^{13}C \cdot {}^{10}BH_3$	$v=0$	1455.72 (2)							
110	 · HCl								see 5, appendix (subvolume b)	

2.5 Asymmetric top molecules

Rotational constants, centrifugal distortion constants, rotation-vibration interaction constants, Coriolis coupling constants, equilibrium structure, and references for structural data

2.5.1 Preliminary remarks

1. Introduction

In favorable cases the analysis of the rotational spectrum of asymmetric molecules in the vibrational state $v_1 \cdots v_j \cdots v_{3N-6}$ allows the determination of the constants listed in this table. The vibration-rotation interaction constants must be determined by the analysis of at least two vibrational states of the same normal vibration.

All rotating molecules show the influence of molecular deformation (centrifugal distortion) in their spectra. It is included in Eq. (5). Since there are different theoretical formulations of centrifugal distortion, a variety of centrifugal distortion constants appear in the table (column 4). For these constants the subscript v has been omitted. It is to be understood that the molecule is in the particular vibrational state v_j. In most cases, $v=0$.

For the analysis of the spectrum, the following Eq. (2) for the rotational energy in a vibrational state $v_j=v$ is used in most cases. The selection rules for this type of rotational spectra allows in general a determination of three rotational constants or combinations of them.

The rigid asymmetric rotor Hamiltonian may be written [70GOR]:

$$\mathscr{H}_2 = A_v P_a^2 + B_v P_b^2 + C_v P_c^2 \qquad (1)$$

and the corresponding rotational energy levels may be calculated by using:

$$W_v(J_{K_- K_+})/h = \frac{(A_v + C_v)}{2} J(J+1) + \frac{(A_v - C_v)}{2} E_{J, K_- K_+}(\kappa_v) \qquad (2)$$

and

$$A_v = A_e - \sum_j \alpha_j (v_j + \tfrac{1}{2})$$
$$B_v = B_e - \sum_j \beta_j (v_j + \tfrac{1}{2}) \qquad (3)$$
$$C_v = C_e - \sum_j \gamma_j (v_j + \tfrac{1}{2})$$

where

W_v/h [MHz]	= the rotational energy in vibrational state $v_j=v$
J	= the rotational angular momentum quantum number
K	= the quantum number of the angular momentum component in the direction of the symmetry axis in the limiting case of the symmetric top molecule
$K_- K_+$	= the pseudo-quantum number formed from K_- and $K_+ \cdot K_-$ is the quantum number of the prolate symmetric top limiting case; K_+ is the quantum number of the oblate symmetric top limiting case
A_v, B_v, C_v [MHz]	= the rotational constants. See explanation of column 3
A_e, B_e, C_e [MHz]	= the rotational constants of the equilibrium configuration

$$A_e\,[\text{MHz}] = \frac{h}{8\pi^2 I_{ae}}; \quad B_e\,[\text{MHz}] = \frac{h}{8\pi^2 I_{be}}; \quad C_e\,[\text{MHz}] = \frac{h}{8\pi^2 I_{ce}} \qquad (4)$$

κ	= the asymmetry parameter. See explanation column 3
I_{ae}, I_{be}, I_{ce} [amu Å^2] [g cm^2]	= the molecular principal moments of inertia about the a, b, and c axes for the equilibrium configuration
$E_{J, K_- K_+}(\kappa_v)$	= the reduced energy of an asymmetric top molecule in vibrational state v_j
$\alpha_j, \beta_j, \gamma_j$ [MHz]	= the rotation-vibration interaction constants. See explanation of column 6
P_a, P_b, P_c	= angular momentum components measured in units of $\hbar$

According to Kivelson and Wilson [52KIV, 70GOR] the Hamiltonian for a non-rigid asymmetric rotor may be written:

$$\mathscr{H} = A' P_a^2 + B' P_b^2 + C' P_c^2 + \tfrac{1}{4} \sum_{\alpha, \beta} \tau'_{\alpha\alpha\beta\beta} P_\alpha^2 P_\beta^2 \qquad (5)$$

For non-planar molecules this Hamiltonian has to be "reduced" in order to avoid an indeterminacy [77WAT]. Since there is not a unique unitary transformation which can reduce the Kivelson-Wilson Hamiltonian, several reduced Hamiltonians may be used. Their coefficients are listed in columns 3 and 4.

2. Explanations of the columns in the tabulation

Column	Parameters		Definitions
1	Nr. *6-Nr.* *4-Nr.*		serial number of the molecule in this table the following numbers represent the serial number of the molecule in Vol. II/6, and II/4, Landolt-Börnstein, New Series
2	v	v_j	vibrational quantum number of the j-th normal vibration: $j=1, \dots 3N-6$ vibrational ground state: $v_1=v_2=\cdots v_j \cdots = v_{3N-6}=0$ in short $v=0$ vibrational state: $v_1=\cdots v_{j-1}=0;\quad v_j=v;\quad v_{j+1}=\cdots v_{3N-6}=0,$ in short: $v_j=v$
		v_τ	vibrational quantum number for a torsional vibration
		e	symbol for the equilibrium configuration
3	A, B, C [MHz]		rotational constants:
		A_v, B_v, C_v	for the vibrational state $v_j=v$ $A_v=\frac{h}{8\pi^2 I_{av}};\ B_v=\frac{h}{8\pi^2 I_{bv}};\ C_v=\frac{h}{8\pi^2 I_{cv}}$ (6)
		$I_a, I_b, I_c;\ I_{av}, I_{bv}, I_{cv}$ [amu Å^2], [g cm^2]	molecular principal moments of inertia in vibrational state $v_j=v$ with $I_{cv}>I_{bv}>I_{av}$
		κ	asymmetry parameter according to Ray, compare Eq. (2) $\kappa_v=\frac{2B_v-(A_v+C_v)}{A_v-C_v}$ (7)
		b_p	asymmetry parameter for prolate tops: $b_p=\frac{C_v-B_v}{2A_v-B_v-C_v}$
			rotational constants in presence of centrifugal distortion:
		A', B', C' [MHz]	according to Kivelson-Wilson [52KIV]
		A'', B'', C''	according to Kirchhoff [72KIR]
		$\mathscr{A}, \mathscr{B}, \mathscr{C}$	according to Watson [68WAT]
		B_x^A, B_y^A, B_z^A	Watson's asymmetric top reduction [77WAT]
		B_x^S, B_y^S, B_z^S	Watson's symmetric top reduction [77WAT]

Column	Parameters	Definitions
4	D, H [kHz]	quartic and sextic centrifugal distortion constants:
	$\tau_{\alpha\beta\gamma\delta}$	according to Wilson [36WIL]
	$\tau'_{\alpha\beta\gamma\delta}$	according to Kivelson-Wilson [52KIV], the ′ is often omitted
	$\tau_{aaaa}, \tau_{bbbb}, \tau_{cccc}, \tau_{abab}, \tau_{aabb}$	Kivelson-Wilson parameters for a planar molecule [70GOR]
	$\tau_{aaaa}, \tau_{bbbb}, \tau_{cccc}, \tau_1, \tau_2, \tau_3$	Kirchhoff parameters [72KIR]
	$\Delta_J, \Delta_{JK}, \Delta_K, \delta_J, \delta_K, \phi_J, \phi_{JK}, \phi_{KJ}, \phi_K, \varphi_J, \varphi_{JK}, \varphi_K$	Watson parameters in the asymmetric top reduction [77WAT]. In fact the ϕ's are often noted H
	$D_J, D_{JK}, D_K, d_1, d_2, H_J, H_{JK}, H_{KJ}, H_K, h_1, h_2, h_3$	Watson parameters in the symmetric top reduction [77WAT]
	$D'_J, D'_{JK}, D'_K, \delta_J, R'_6, H'_J, H'_{JK}, H'_{KJ}, H'_K, H'_5, H'_6, H'_{10}$	Van Eijck-Typke parameters [74VAN1, 76TY]
	$d_J, d_{JK}, d_K, d_{WJ}, d_{WK}$	quartic centrifugal distortion constants according to Watson [67WAT] note that d_{WJ}, d_{WK} (or d_{EJ}, d_{EK}) are dimensionless
	L, l, G, g	octic centrifugal distortion constants (coefficients of P^8)
	P, p, F	dectic centrifugal distortion constants (coefficients of P^{10})
5	Ref.	references for columns 3 and 4
6	Ref. struct.	references to structure data. See also the references to Vol. II/7 in the column Varia, remarks, where the numbers following in parentheses indicate the chapter number and the molecule number in II/7
7		remarks concerning the previous columns, and data of constants which have been determined only for a few molecules of this table:
	σ	standard deviation
	$\alpha_j, \beta_j, \gamma_j$ [MHz]	rotation-vibration interaction constants of the j-th normal vibration referred to the ordered principal inertia axes a, b, and c. Compare Eq. (3)
	$\zeta^{(g)}_{jj'}$	Coriolis coupling constants for the j-th and j′-th normal vibration referred to the ordered principal axes ($g = a, b$, and c)
	$\nu_j, \tilde{\nu}_j$	type, frequency or wave number of the j-th normal vibration
	A, B, E a, b, e	classification of the vibrational symmetric species: A, a: nondegenerate, symmetric; B, b: nondegenerate, antisymmetric, E, e: degenerate
8		references to following tables; figures in Chapter 3, subvolume b

2.5.2 Data

2.5.2.1 Inorganic molecules

Nr.	Molecule	v	A, B, C MHz	D, H kHz	Ref.	Ref. struct.	Varia, remarks	See also Tab., Nr.
1a *6-1*	$^{28}SiF_2$	$v_1=1$ $v_3=1$	$A=30651.64$ (17) $B=8768.17$ (6) $C=6797.70$ (90) *) $A=30300.85$ (13) $B=8809.96$ (5) $C=6806.57$ (90) *)		73SHO	73SHO	for structure, see also Vol. II/7 (2, 243) *) constants of the unperturbed state; Coriolis interaction between ν_1 and ν_3: $D=5771.23$ (3900); $F=-63.97$ (230); $E_1-E_3=-461529.16$ (730) MHz harmonic and anharmonic force field calculated r_e(Si—F)=1.5901 (1) Å, θ_e(FSiF)=100° 46′ (1′) [73SHO]	
1b *6-1b*	NH_2Cl	$v=0$					for structure, see also Vol. II/7 (2, 112) separation between the O^+ and O^- levels = 70 MHz upper limit for the barrier of inversion = 4000 cm^{-1} potential function given [73CAZ2]	
2 *6-2* *4-1* Cont.	NH_2CN	$v=0$	$A'=310885$ (450) *) $B'=10130.162$ (30) $C'=9866.422$ (30)	$\tau_1=-1912$ (10) $\tau_2=-74.3$ (74) $\tau_3=23800$ (9600) $\tau_{aaaa}=0$ **) $\tau_{bbbb}=-19.1$ (32) $\tau_{cccc}=-16.6$ (32)	76JOH		*) infrared value **) fixed at 0 $V_{inv}=450$ cm^{-1} inversion angle = 45.5° [74ATT]	3, 8

Nr.	Molecule	v	A, B, C MHz	D, H kHz	Ref.	Ref. struct.	Varia, remarks	See also Tab., Nr.
2 *6-2* *4-1* Cont.	NH_2CN	$v_i=1$	$A'=304589\ (450)$ *) $B'=\ 10113.032\ (67)$ $C'=\ 9866.865\ (67)$	$\tau_1=-1174.1\ (81)$ $\tau_2=-49.7\ (56)$ $\tau_3=20400\ (11600)$ $\tau_{aaaa}=0$ **) $\tau_{bbbb}=-15.8\ (53)$ $\tau_{cccc}=\ -4.7\ (53)$	76JOH		v_i NH_2 inversion *) infrared value **) fixed at 0	
	ND_2CN	$v=0$					$V_{inv}=450\ cm^{-1}$ [74ATT]	
2a	NH_2NC	$v=0$ *)	$A=282152\ (140)$ $B=10757.2791\ (22)$ $C=10525.3389\ (23)$	$D_J\ =5.2845\ (27)$ $D_{JK}=424.36\ (37)$ $d_1\ =-0.1776\ (33)$ $d_2\ =-0.0323\ (21)$ $H_{JK}=4.33\ (37)\cdot 10^{-3}$ $H_{KJ}=-0.516\ (39)$ $L_{KJ}=18.35\ (100)\cdot 10^{-3}$	81SCH		*) O^+ inversion state	
		$v_\tau=1$ **)	$A=282139\ (130)$ $B=10756.7698\ (22)$ $C=10525.2509\ (21)$	$D_J\ =5.2815\ (24)$ $D_{JK}=422.44\ (33)$ $d_1\ =-0.1739\ (33)$ $d_2\ =-0.0309\ (19)$ $H_{JK}=4.15\ (33)\cdot 10^{-3}$ $H_{KJ}=-0.516\ (35)$ $L_{KJ}=18.54\ (87)\cdot 10^{-3}$			**) O^- inversion state	
3 *4-3*	NHF_2	$v=0$	$A''=53019.806\ (45)$ $B''=10895.809\ (9)$ $C''=\ 9307.5468\ (79)$	$\tau_1=84.09\ (401)$ $\tau_2=-17.49\ (66)$ $\tau_3=2183\ (19)$ $\tau_{aaaa}=-3642.89\ (885)$ $\tau_{bbbb}=-78.816\ (401)$ $\tau_{cccc}=-33.51\ (25)$	78KAU4		for structure, see also Vol. II/7 (2, 177)	
4	$HN{=}SOF_2$	$v=0$	$A=5119.28\ (1)$ $B=4836.94\ (1)$ $C=4757.42\ (1)$		77CAX1	77CAX1		2.6.3, 1 2.7.3, 11 3, 9
Cont.	$DN{=}SOF_2$	$v=0$	$A=5003.45\ (1)$ $B=4731.34\ (1)$ $C=4557.24\ (1)$					

Nr.	Molecule	v	A, B, C MHz	D, H kHz	Ref.	Ref. struct.	Varia, remarks	See also Tab., Nr.
4 Cont.	$H^{15}N{=}SOF_2$	$v=0$	$A=5094.79\,(1)$ $B=4745.38\,(1)$ $C=4648.37\,(1)$		77CAX1			
	$HN{=}^{34}SOF_2$	$v=0$	$A=5119.10\,(5)$ $B=4834.14\,(5)$ $C=4754.48\,(5)$					
5 *6–3* *4–7*	HNCO	$v=0$	$A=918504.4\,(154)$ *) $B=11071.00825\,(61)$ $C=10910.57553\,(61)$	$D_J=3.4863\,(29)$ *) $D_{JK}=931.70\,(65)$ $D_K=6065.6\,(191)\cdot 10^3$ $d_1=-72.995\,(83)\cdot 10^{-3}$ $d_2=-36.5\,(30)\cdot 10^{-3}$ $H_{JK}=-2.49\,(16)\cdot 10^{-3}$ $H_{KJ}=32.29\,(30)$ $H_K=287.6\,(41)\cdot 10^3$ **)	80YA1, 81FUS	80YA1, 81FUS	for structure, see also Vol. II/7 (3, 61) *) weighted FIR-MW data **) higher-order constants: $L_{KJ}=3.163\,(44)$ kHz $L_K=14.88\,(36)$ MHz $S_{KJ}=144.4\,(15)$ Hz $S_K=0.493\,(15)$ MHz $T_{KJ}=2.494\,(48)$ Hz $T_K=8.56\,(30)$ kHz $U_K=58.6\,(22)$ Hz	2.6.3, 2 2.7.3, 12 3, 10
Cont.		$v_4=1$	$B(0)=10996.3347\,(6)$ $B(1,U)=11060.1123\,(28)$ $B(1,L)=10987.1398\,(28)$ $B(2,U)=10975.5174\,(49)$ $B(2,L)=10975.5360\,(46)$ $B(3)=10976.9553\,(8)$	$D(0)=4.2087\,(40)$ $D(1,U)=9.822\,(49)$ $H(1,U)=1.72\,(26)\cdot 10^{-3}$ $D(1,L)=2.243\,(49)$ $H(1,L)=0.23\,(26)\cdot 10^{-3}$ $D(2,U)=2.836\,(35)$ $D(2,L)=3.298\,(32)$ $D(3)=3.3922\,(47)$	77YAM4		v_4 in-plane bending, 777.1 cm^{-1}. $B(K,t)$ is an effective constant for the state $(K_a, t)\cdot t$ indicate U (upper) and L (lower) component of the K doublets. Coriolis interaction analyzed	

Nr.	Molecule	v	A, B, C MHz	D, H kHz	Ref.	Ref. struct.	Varia, remarks	See also Tab., Nr.
5 6-3 4-7 Cont.	HNCO	$v_5=1$	$(B+C)/2$ $=11008.70673\,(42)$ $B-C=128.48373\,(39)$	$D_J\ =5.4453\,(14)$ $D_{JK}=-1278.96\,(30)$ $\delta_J\ =\ 0.02530\,(66)$ $H_{JK}=\ 1.1832\,(16)$ $h_J\ =-0.0088\,(11)\cdot10^{-3}$ $L_{JK}=\ 0.0137\,(25)\cdot10^{-3}$	77YAM3		v_5 in-plane bending, 577.5 cm^{-1} energy difference between the $K_a=1$ and 0 levels $=275697.309\,(17)$ MHz	
			$B(2,U)=11010.6472\,(9)$ $B(2,L)=11010.6516\,(8)$ $B(3)\ =11009.1323\,(17)$ $B(4)\ =11006.7900\,(21)$	$D(2,U)=2.1928\,(61)$ $D(2,L)=4.7093\,(57)$ $D(3)\ =3.308\,(12)$ $D(4)\ =3.541\,(14)$	77YAM4			
		$v_6=1$	$B(0)\ =11008.7878\,(11)$ $B(1,U)=11016.6074\,(10)$ $B(1,L)=10989.2826\,(10)$ $B(2,U)=10967.5044\,(37)$ $B(2,L)=10967.5013\,(14)$ $B(3)\ =11008.2329\,(25)$	$D(0)\ =\ 2.9064\,(79)$ $D(1,U)=\ 3.6247\,(61)$ $D(1,L)=\ 3.5297\,(61)$ $D(2,U)=-2.445\,(26)$ $D(2,L)=\ 4.4618\,(98)$ $D(3)\ =\ 3.757\,(16)$			v_6 out-of-plane bending, 659.8 cm^{-1}	
	$H^{15}NCO$	$v=0$	$A=908968.2\,(23)$ *) $B=\ 10737.8304\,(42)$ $C=\ 10585.4653\,(42)$	$D_J\ =3.307\,(19)$ *) $D_{JK}=1020.2\,(61)$ $D_K\ =6065573$ +) $d_1\ =-63.16\,(24)\cdot10^{-3}$ $d_2\ =-36.470\cdot10^{-3}$ +) $H_{JK}=-2.4907\cdot10^{-3}$ +) $H_{KJ}=\ 34.6\,(20)$ $H_K\ =287620$ +) ++)	80YA1, 81FUS		*) weighted FIR-MW data +) fixed value ++) higher-order constants: $L_{KJ}=\ 1.97\,(15)$ kHz $L_K\ =14.8838$ MHz +) $S_K\ =\ 0.49330$ MHz +) $T_K\ =\ 8.5591$ kHz +) $U_K\ =58.571$ Hz +)	
	$HN^{13}CO$	$v=0$	$A=916293.8\,(10)$ *) $B=\ 11071.4798\,(22)$ $C=\ 10910.7307\,(22)$	$D_J\ =3.5038\,(78)$ *) +++) $D_{JK}=933.3\,(26)$ $d_1\ =-73.15\,(25)\cdot10^{-3}$ $d_2\ =-41.4\,(86)\cdot10^{-3}$ $H_{KJ}=\ 35.6\,(11)$ $L_{KJ}=\ 3.647\,(16)$ $S_{KJ}\ =167.8\,(92)\cdot10^{-3}$ $T_{KJ}=\ 2.88\,(17)\cdot10^{-3}$			+++) $D_K, H_{JK}, H_K, L_K, S_K$, T_K and U_K are fixed on the value found for the parent species	
Cont.								

Nr.	Molecule	v	A, B, C MHz	D, H kHz	Ref.	Ref. struct.	Varia, remarks	See also Tab., Nr.
5 *6–3* *4–7* Cont.	$HNC^{18}O$	$v=0$	$A=918416.51\,(34)$ *) $B=\ 10470.98559\,(89)$ $C=\ 10327.24228\,(89)$	$D_J\ =3.1431\,(27)$ *) ×) $D_{JK}=821.91\,(88)$ $d_1\ =-61.414\,(78)\cdot 10^{-3}$ $H_{KJ}=27.56\,(42)$ $L_{KJ}=\ 2.331\,(58)$ $S_{KJ}=66.0\,(22)\cdot 10^{-3}$	80YA1, 81FUS		*) weighted FIR-MW data ×) $D_K, d_2, H_{JK}, H_K, L_K, S_K$, T_K and U_K are fixed on the value found for the parent species	
	DNCO	$v=0$	$A=512472.3\,(57)$ *) $B=\ 10313.71310\,(36)$ $C=\ 10079.67613\,(36)$	$D_J\ =3.2597\,(11)$ *) $D_{JK}=-243.230\,(99)$ $D_K\ =1522.2\,(63)\cdot 10^3$ $d_1\ =-203.92\,(21)\cdot 10^{-3}$ $d_2\ =\ -50.6\,(11)\cdot 10^{-3}$ $H_{JK}=\ 0.41\,(14)\cdot 10^{-3}$ $H_{KJ}=-7.008\,(11)$ $H_K\ =\ 22.03\,(57)\cdot 10^3$ **)			**) higher-order constants: $L_{KJ}=-0.06753\,(39)$ kHz $L_K\ =\ 0.345\,(21)$ MHz $S_{KJ}=-0.4289\,(44)$ Hz $S_K\ =\ 0.00355\,(32)$ MHz $T_K\ =\ 0.0159\,(19)$ kHz harmonic force field calculated [81FUS]	
	$D^{15}NCO$	$v=0$	$B=10044.575\,(50)$ $C=9819.435\,(50)$		75HO1			
6 *4–8*	HNCS	$v=0$	$\tilde{A}=1357250\,(700)$ $\tilde{B}=5883.4627\,(3)$ $\tilde{C}=5845.6113\,(3)$	$D_J\ =1.19393\,(37)$ $D_{JK}=-1025.37\,(28)$ $D_K\ =51.57\,(25)\cdot 10^6$ $d_1\ =-13.7808\,(30)\cdot 10^{-3}$ $d_2\ =\ -4.59\,(19)\cdot 10^{-3}$ $H_{JK}=1.019\,(72)\cdot 10^{-3}$ $H_{KJ}=-185.40\,(11)$ $H_K\ =4002\,(28)\cdot 10^3$ $L_{KJ}=-16.597\,(13)$ $L_K\ =\ 119.7\,(10)\cdot 10^3$ $S_{KJ}=\ -0.53263\,(45)$	79YA	80YA3	for structure, see also Vol. II/7 (3, 64)	2.6.3, 3 2.7.3, 13 3, 11
		$v_4=1$	$B(0)\ =5849.1125\,(8)$ $B(1, U)=5870.9019\,(4)$ $B(1, L)=5856.0814\,(8)$ $B(2, U)=5866.0342\,(8)$	$D(0)\ =0.6659\,(21)$ $D(1, U)=1.1967\,(9)$ $D(1, L)=1.1133\,(19)$ $D(2, U)=1.1780\,(20)$	78YAM		$B(K, t)$ is an effective constant for the state $(K_a, t)\cdot t$ indicate U (upper) and L (lower) component of the K doublets. Coriolis interaction analyzed energy difference between the $K_a=1$ and 0 levels $=-40104.2866\,(71)$ MHz	
Cont.								

Nr.	Molecule	v	A, B, C MHz	D, H kHz	Ref.	Ref. struct.	Varia, remarks	See also Tab., Nr.
6	HNCS	$v_5=1$	$(B+C)/2=5868.6299\,(10)$	$D_J=1.3036\,(28)$	77YAM2			
4–8			$B-C=21.29986\,(26)$	$D_{JK}=-5779.034\,(75)$				
Cont.				$\delta_J=-0.01235\,(27)$				
				$H_{JK}=39.59\,(27)\cdot10^{-3}$				
				$L_{JK}=2.9241\,(77)\cdot10^{-6}$				
				$S_{JK}=-0.05079\,(68)\cdot10^{-9}$				
		$v_6=1$	$B(0)=5875.5299\,(5)$	$D(0)=0.9986\,(11)$	78YAM			
			$B(1, U)=5880.6920\,(10)$	$D(1, U)=1.7340\,(19)$				
			$B(1, L)=5872.3478\,(6)$	$D(1, L)=1.2059\,(12)$				
			$B(2)=5874.1205\,(9)$	$D(2)=1.2262\,(23)$				
	$H^{15}NCS$	$v=0$	$A=1339170.0$ *)	$D_J=1.10858\,(67)$ **)	80YA3		*) fixed from the inertial defect of the normal species	
			$B=5688.69511\,(46)$	$D_{JK}=-786.79\,(49)$			**) D_K, H_K, L_K are fixed on the value found for the parent species	
			$C=5652.98346\,(46)$	$d_1=-11.8997\,(50)\cdot10^{-3}$				
				$d_2=-4.06\,(32)\cdot10^{-3}$				
				$H_{JK}=0.87\,(18)\cdot10^{-3}$				
				$H_{KJ}=-147.15\,(19)$				
				$L_{KJ}=-13.109\,(22)$				
				$S_{KJ}=-0.42210\,(78)$				
	$HN^{13}CS$	$v=0$	$A=1354250.0$ *)	$D_J=1.1917\,(12)$ **)				
			$B=5864.55243\,(60)$	$D_{JK}=-1038.20\,(62)$				
			$C=5826.88808\,(50)$	$d_1=-13.7233\,(59)\cdot10^{-3}$				
				$d_2=-4.25\,(56)\cdot10^{-3}$				
				$H_{JK}=1.37\,(21)\cdot10^{-3}$				
				$H_{KJ}=-187.00\,(24)$				
				$L_{KJ}=-16.821\,(29)$				
				$S_{KJ}=-0.5468\,(10)$				
	$HNC^{34}S$	$v=0$	$A=1357100.0$ *)	$D_J=1.14320\,(94)$ **)				
			$B=5744.8327\,(13)$	$D_{JK}=-1016.46\,(59)$				
			$C=5708.7361\,(13)$	$d_1=-12.2\,(14)\cdot10^{-3}$				
				$d_2=-3.99\,(47)\cdot10^{-3}$				
				$H_{JK}=0.90\,(13)\cdot10^{-3}$				
				$H_{KJ}=-183.22\,(23)$				
				$L_{KJ}=-16.424\,(28)$				
Cont.				$S_{KJ}=-0.52703\,(99)$				

Nr.	Molecule	v	A, B, C MHz	D, H kHz	Ref.	Ref. struct.	Varia, remarks	See also Tab., Nr.
6	DNCS	$v=0$	$A_0=705510.743\,(330)$	$D_J=1.09264\,(83)$	78SZA			
4–8			$B_0=5500.43911\,(53)$	$D_{JK}=-1333.454\,(380)$				
Cont.			$C_0=5445.22521\,(53)$	$D_K=10193\cdot 10^3$ +)			+) fixed value, IR data [68KRA]	
				$d_1=-3.45507\,(140)\cdot 10^{-2}$				
				$d_2=-7.77\,(47)\cdot 10^{-3}$				
				$H_{JK}=3.9920\,(740)\cdot 10^{-3}$				
				$H_{KJ}=-120.4882\,(1700)$				
				$H_K=3.30\cdot 10^5$ +)				
				$L_{JK}=-2.612\,(400)\cdot 10^{-4}$				
				$L_{KJ}=10.69517\,(2600)$				
				$L_K=-5400$ +)				
				$S_{JK}=0.526\,(160)\cdot 10^{-5}$				
				$S_{KJ}=-4.9841\,(150)\cdot 10^{-1}$				
				$S_K=33$ +)				
				$T_{KJ}=8.4987\,(290)\cdot 10^{-3}$				
7	$HNC^{82}Se$ *)	$v=0$	$B=3911.088$		79LAD	79LAD	*) quasilinear molecule	2.6.3, 4; 2.7.3, 14
		$v=1^{1e}$	$B=3909.357$					
		$v=1^{1f}$	$B=3917.043$					
		$v=2^2$	$B=3915.517$					
	$HNC^{80}Se$	$v=0$	$B=3933.242$					
		$v=1^{1e}$	$B=3931.470$					
		$v=1^{1f}$	$B=3939.252$					
		$v=2^2$	$B=3937.690$					
	$HNC^{78}Se$	$v=0$	$B=3956.495$					
		$v=1^{1e}$	$B=3954.688$					
		$v=1^{1f}$	$B=3962.560$					
		$v=2^2$	$B=3960.957$					
	$HNC^{77}Se$	$v=0$	$B=3968.540$					
		$v=1^{1e}$	$B=3966.713$					
		$v=1^{1f}$	$B=3974.638$					
		$v=2^2$	$B=3973.007$					
	$HNC^{76}Se$	$v=0$	$B=3980.928$					
		$v=1^{1e}$	$B=3979.087$					
		$v=1^{1f}$	$B=3987.057$					
Cont.		$v=2^2$	$B=3985.620$					

Nr.	Molecule	v	A, B, C MHz	D, H kHz	Ref.	Ref. struct.	Varia, remarks	See also Tab., Nr.
7 Cont.	$HN^{13}C^{80}Se$	$v=0$	$B=3896.277$		79LAD.			
		$v=1^{1e}$	$B=3894.573$					
		$v=1^{1f}$	$B=3902.230$					
	$DNC^{82}Se$	$v=0$	$B=3640.322$					
		$v=1^{1e}$	$B=3636.795$					
		$v=1^{1f}$	$B=3647.802$					
		$v=2^2$	$B=3645.575$					
	$DNC^{80}Se$	$v=0$	$B=3661.358$					
		$v=1^{1e}$	$B=3657.778$					
		$v=1^{1f}$	$B=3668.910$					
		$v=2^2$	$B=3666.645$					
	$DNC^{78}Se$	$v=0$	$B=3683.438$					
		$v=1^{1e}$	$B=3679.805$					
		$v=1^{1f}$	$B=3691.070$					
		$v=2^2$	$B=3688.753$					
	$DNC^{77}Se$	$v=0$	$B=3694.873$					
		$v=1^{1e}$	$B=3691.208$					
		$v=1^{1f}$	$B=3702.550$					
		$v=2^2$	$B=3700.213$					
	$DNC^{76}Se$	$v=0$	$B=3706.640$					
		$v=1^{1e}$	$B=3702.953$					
		$v=1^{1f}$	$B=3714.360$					
		$v=2^2$	$B=3711.995$					
8 6–4	$^{35}Cl^{15}NCO$	$v=0$	$A=49090.359(52)$	$\tau_{aaaa}=-53151(39)$	75HO2	75HO2	for structure, see also Vol. II/7 (3, 11)	2.7.3, 81
			$B=3130.6330(26)$	$\tau_{bbbb}=-10.610(29)$				
			$C=2936.5686(22)$	$\tau_{aabb}=659.00(78)$				
				$\tau_{abab}=-10.03(52)$				
	$^{37}Cl^{15}NCO$	$v=0$	$A=48771.584(55)$	$\tau_{aaaa}=-52454(29)$				
			$B=3057.5618(23)$	$\tau_{bbbb}=-10.133(28)$				
			$C=2871.0664(38)$	$\tau_{aabb}=641.4(25)$				
				$\tau_{abab}=-11.1(21)$				
	$^{35}ClN^{13}CO$	$v=0$	$A=51513.460(59)$	$\tau_{aaaa}=-59263(40)$				
			$B=3107.0181(28)$	$\tau_{bbbb}=-10.437(29)$				
			$C=2924.0943(30)$	$\tau_{aabb}=690.4(14)$				
Cont.				$\tau_{abab}=-10.7(10)$				

Nr.	Molecule	v	A, B, C MHz	D, H kHz	Ref.	Ref. struct.	Varia, remarks	See also Tab., Nr.
8 *6–4* Cont.	$^{35}ClNC^{18}O$	$v=0$	$A=50689.754(74)$ $B=2957.2339(31)$ $C=2788.5436(29)$	$\tau_{aaaa}=-57524(56)$ $\tau_{bbbb}=-9.569(25)$ $\tau_{aabb}=646.04(92)$ $\tau_{abab}=-9.82(68)$	75HO2			
9	NCNCO	$v=0$	$\tilde{A}=74358.690(81)$ $\tilde{B}=2699.0367(33)$ $\tilde{C}=2597.8663(31)$	$\Delta_J=1.0904(116)$ $\Delta_{JK}=-527.13(71)$ $\Delta_K=86254(38)$ $\delta_J=0.20694(117)$ $\delta_K=26.929(732)$ $H_{KJ}=-1.6024(323)$ $H_J=-0.0535(192)\cdot10^{-5}$ $h_J=1.026(322)\cdot10^{-6}$	76HO2	79BA3		2.6.3, 5 3, 12
		$v_1=1$	$\tilde{A}=82777.808(150)$ $\tilde{B}=2704.3222(57)$ $\tilde{C}=2599.7778(55)$	$\Delta_J=1.2401(177)$ $\Delta_{JK}=-677.12(121)$ $\Delta_K=150581(74)$ $\delta_J=0.21701(94)$ $\delta_K=71.198(1273)$ $H_{KJ}=-3.0275(520)$ $H_J=0.642(312)\cdot10^{-5}$			$v_1=144(40)\ cm^{-1}$ from intensity measurements 1 = lowest frequency vibration	
		$v_1=2$	$\tilde{A}=90431.5(35901)$ $\tilde{B}=2708.320(14)$ $\tilde{C}=2601.244(17)$	$\Delta_J=0.931(180)$ $\Delta_{JK}=-914.9(29)$ $H_{KJ}=-6.02(17)$				
		$v_1=3$	$\tilde{A}=117342.8(47185)$ $\tilde{B}=2711.191(12)$ $\tilde{C}=2601.659(12)$	$\Delta_J=1.333(189)$ $\Delta_{JK}=-1336.1(52)$ $H_{KJ}=-19.20(50)$				
	$NC^{15}NCO$	$v=0$	$B=2699.11$ $C=2594.00$		79BA3			
	$^{15}NCNCO$	$v=0$	$B=2622.41$ $C=2526.52$					
10	NCNCS	$v=0$	$B=1628.09(3)$ $C=1599.37(3)$		80KIN	80KIN	for structures, see also Vol. II/7 (3, 168)	

Nr.	Molecule	v	A, B, C MHz	D, H kHz	Ref.	Ref. struct.	Varia, remarks	See also Tab., Nr.
11	^{35}ClSCN	$v=0$	$A=12109.28\,(10)$ $B=2902.943\,(14)$ $C=2337.285\,(12)$		80RIC	80RIC		2.7.3, 82
	^{37}ClSCN	$v=0$	$A=11975.76\,(28)$ $B=2836.427\,(15)$ $C=2289.116\,(15)$					
12 *6-4a*	HNO *)	$v=0$	$A=18.476055\,(20)$ *) $B=1.41125980\,(68)$ $C=1.30655178\,(68)$	$\Delta_K=4.2790\,(216)\cdot10^{-3}$ *) $\Delta_{JK}=9.379\,(115)\cdot10^{-5}$ $\Delta_J=3.7696\,(294)\cdot10^{-6}$ $\delta_J=3.0581\,(89)\cdot10^{-7}$ $H_K=2.677\,(380)\cdot10^{-6}$	77JO1		for structure, see also Vol. II/7 (2, 276) *) Laser Stark data all values in cm^{-1} the uncertainties are 3σ representation I^r used $\nu_2=1565.345\,(2)\,cm^{-1}$ N—O stretch **) fixed on their ground state values $\nu_3=1500.818\,(3)\,cm^{-1}$ bend Coriolis interaction between ν_2 and ν_3: $G_c=1.0553\,(34)\,cm^{-1}$ $Z_1=8.36\,(330)\cdot10^{-3}\,cm^{-1}$ $Z_2=0.2391\,(24)\,cm^{-1}$ where $G_c=2\,C\,\zeta_{23}^{(c)}$ and Z_1 and Z_2 are the coefficients of the matrix element: $\langle v\,J\,k\,\|H\|\,v'\,J\,k\rangle = Z_1\,J(J+1)+Z_2\,k^2$	2.6.3, 6 2.7.3, 15 3, 13
		$v_2=1$	$A=18.57568\,(80)$ *) $B=1.40058\,(47)$ $C=1.29256\,(44)$	**)				
		$v_3=1$	$A=18.82257\,(113)$ *) $B=1.40836\,(154)$ $C=1.29903\,(55)$	**)				
	DNO *)	$v=0$	$A=10.523867\,(23)$ *) $B=1.292054\,(59)$ $C=1.145827\,(59)$	$\Delta_K=1.4852\,(152)\cdot10^{-3}$ *) $\Delta_{JK}=5.702\,(109)\cdot10^{-5}$ $\Delta_J=3.157\,(137)\cdot10^{-6}$ $\delta_J=4.09\,(153)\cdot10^{-7}$ $\delta_K=4.44\,(294)\cdot10^{-5}$ $H_K=6.12\,(202)\cdot10^{-7}$				
		$v_2=1$	$A=10.454358\,(158)$ *) $B=1.280280\,(195)$ $C=1.136723\,(122)$	$\Delta_K=1.4871\,(68)\cdot10^{-3}$ $\Delta_{JK}=4.828\,(660)\cdot10^{-5}$ $\Delta_J=3.670\,(1467)\cdot10^{-6}$ ***)			$\nu_2=1546.8801\,(7)\,cm^{-1}$ ***) other constants fixed on their ground state values	

Nr.	Molecule	v	A, B, C MHz	D, H kHz	Ref.	Ref. struct.	Varia, remarks	See also Tab., Nr.
13 *6-5* *4-9*	FNO *)	$v=0$	$A=95190.06\,(25)$ *) $B=11844.310\,(19)$ $C=10508.194\,(18)$	$\Delta_K=3933\,(23)$ *) $\Delta_{JK}=-33.18\,(63)$ $\Delta_J=19.698\,(48)$ $\delta_K=106.2\,(19)$ $\delta_J=2.699\,(11)$	78ALL		for structure see also Vol. II/7 (2, 202) *) Laser Stark data the uncertaintics are 3 σ	2.6.3, 7 2.9.3, 1
		$v_1=1$	$A=94796.6\,(14)$ *) $B=11857.0\,(19)$ $C=10519.3\,(18)$	$\Delta_K=3901\,(27)$ *) **) $\Delta_{JK}=-8\,(36)$ $\Delta_J=17.3\,(42)$			$\nu_1=1844.099\,(1)\ cm^{-1}$ **) δ_J and δ_K are fixed on their ground state values	
14 *6-6* *4-10*	$^{35}ClNO$	$v=0$	$A=87374.491\,(14)$ $B=5737.7836\,(8)$ $C=5376.2348\,(1)$	$\Delta_J=6.34368\,(93)$ $\Delta_{JK}=-61.078\,(27)$ $\Delta_K=4394.1\,(27)$ $\delta_J=0.502175\,(67)$ $\delta_K=40.300\,(59)$ $H_J=2.54\,(23)\cdot10^{-6}$ $H_{JK}=4.22\,(21)\cdot10^{-4}$ $H_{KJ}=-3.750\,(33)\cdot10^{-2}$ $H_K=0.872\,(62)$ $h_J=9.86\,(17)\cdot10^{-7}$ $h_{JK}=8.35\cdot10^{-5}$ (fixed) $h_K=4.66\,(58)\cdot10^{-2}$ $L_{KKJ}=6.4\,(12)\cdot10^{-6}$	81CAZ	80MIR	for structure, see also Vol. II/7 (2, 130) harmonic force field calculated [75CAZ] cubic force field calculated [80MIR]	2.7.3, 83
		$v_1=1$	$\alpha^B=-26.73\,(30)$ *) $\alpha^C=-20.95\,(30)$		80MIR		$\nu_1=1799\ cm^{-1}$ *) vibrorotational constants	
		$v_2=1$	$A=88169.8\,(7)$ $B=5715.70\,(2)$ $C=5348.16\,(2)$		78MIR, 75CAZ		$\nu_2=595.8\ cm^{-1}$	
		$v_3=1$	$A=87879.64\,(30)$ $B=5710.53\,(5)$ $C=5349.19\,(3)$		75CAZ		$\nu_3=331.9\ cm^{-1}$	
	$^{37}ClNO$	$v_2=1$	$\alpha^B=21.86\,(30)$ *) $\alpha^C=27.39\,(30)$		80MIR		equilibrium structure: r_e(N—Cl) = 1.976 (4) Å r_e(N—O) = 1.131 (4) Å ∢$_e$(ONCl) = 113° 15′ ± 11′ [80MIR]	
		$v_3=1$	$\alpha^B=26.10\,(30)$ *) $\alpha^C=26.09\,(30)$					
Cont.								

Nr.	Molecule	v	A, B, C MHz	D, H kHz	Ref.	Ref. struct.	Varia, remarks	See also Tab., Nr.
14 *6–6* *4–10* Cont.	$^{35}Cl^{15}NO$	$v=0$	$A=81600$		79EN3,			
			$B=$ 5693.56 (2)		80MIR			
			$C=$ 5322.64 (2)					
		$v_2=1$	$\alpha^B=21.81$ (30) *)		80MIR		*) vibro-rotational constants	
			$\alpha^C=27.63$ (30)					
		$v_3=1$	$\alpha^B=26.58$ (30) *)					
			$\alpha^C=26.53$ (30)					
	$^{37}Cl^{15}NO$	$v=0$	$A=81520$		79EN3,			
			$B=$ 5556.42 (2)		80MIR			
			$C=$ 5201.18 (2)					
		$v_3=1$	$\alpha^B=25.58$ (30) *)		80MIR			
			$\alpha^C=25.50$ (30)					
15 *6–7* *4–11*	$^{81}BrNO$	$v_2=1$	$\alpha_2^B=16.86$ (50) *)		78MIR		for structure, see also Vol. II/7 (2, 79)	
			$\alpha_2^C=20.22$ (50)				*) vibro-rotational constants	
		$v_3=1$	$\alpha_2^B=17.38$ (50) *)				anharmonic force field calculated	
			$\alpha_2^C=16.75$ (50)					
16	NOCN	$v=0$	$A=81216.23$ (6)	$\tau_{aaaa}=-62790$ (100)	78DIC	78DIC	for structure,	2.6.3, 8
			$B=$ 5384.533 (4)	$\tau_{bbbb}=$ −16.02 (9)			see also	2.7.3, 16
			$C=$ 5040.682 (5)	$\tau_{aabb}=$ 742 (6)			Vol. II/7 (3, 40)	3, 14
				$\tau_{abab}=$ −129 (6)			intensities of vibrational satellites measured	
	$^{15}NOCN$		$B=5359.70$	$D_J=$ 4.9				
			$C=5005.37$	$D_{JK}=-199$				
				$\delta_J=$ 0.537				
	$NOC^{15}N$		$B=5205.99$	$D_J=$ 4.4				
			$C=4882.72$	$D_{JK}=-222$				
				$\delta_J=$ 0.427				
	$NO^{13}CN$		$B=5359.48$	$D_J=$ 1.0				
			$C=5018.56$	$D_{JK}=-206$				
				$\delta_J=$ 0.483				
	$N^{18}OCN$		$B=5144.93$	$D_J=$ 2.8				
			$C=4819.26$	$D_{JK}=-236$				
				$\delta_J=$ −0.8				

Nr.	Molecule	v	A, B, C MHz	D, H kHz	Ref.	Ref. struct.	Varia, remarks	See also Tab., Nr.
17 *6–11* *4–14*	$NO_2{}^{35}Cl$	$v=0$	$A=13290.14\,(1)$ $B=5175.51\,(0)$ $C=3719.53\,(0)$	$\tau_{aaaa}=-42.69\,(70)$ $\tau_{bbbb}=-17.30\,(4)$ $\tau_{aabb}=3.83\,(10)$ $\tau_{abab}=-25.25\,(7)$	75FIL	79EN1	for structure, see also Vol. II/7 (2, 131) force field calculated [75FIL]	2.7.3, 84
	$NO_2{}^{37}Cl$	$v=0$	$A=13290.07\,(1)$ $B=5020.61\,(0)$ $C=3638.83\,(0)$	$\tau_{aaaa}=-37.79\,(1000)$ $\tau_{bbbb}=-16.22\,(4)$ $\tau_{aabb}=2.94\,(10)$ $\tau_{abab}=-23.63\,(8)$				
	${}^{15}NO_2{}^{35}Cl$	$v=0$	$A=13298\,(2)$ $B=5156.83\,(6)$ $C=3709.76\,(6)$		79EN1			
	${}^{15}NO_2{}^{37}Cl$	$v=0$	$A=13299\,(25)$ $B=5001.2\,(8)$ $C=3628.6\,(8)$					
18	${}^{35}ClONO$ cis	$v=0$	$A=18230.275\,(30)$ $B=4571.584\,(11)$ $C=3649.963\,(10)$	$\tau_{aaaa}=-258.9\,(15)$ $\tau_{bbbb}=-21.4\,(3)$ $\tau_{aabb}=38.1\,(13)$ $\tau_{abab}=-18.3\,(7)$	79KAW1	79KAW1		2.7.3, 85
	${}^{37}ClONO$ cis	$v=0$	$A=18159.521\,(103)$ $B=4456.947\,(26)$ $C=3573.786\,(31)$	$\tau_{aaaa}=-266.7\,(123)$ $\tau_{bbbb}=-20.7\,(8)$ $\tau_{aabb}=16.4\,(167)$ $\tau_{abab}=-5.0\,(88)$				
19 *6–12* *4–15a* Cont.	trans-HNO_2	$v=0$	$A=92892.0204\,(110)$ $B=12524.9742\,(17)$ $C=11016.6671\,(17)$	$\Delta_J=15.3081\,(47)$ $\Delta_{JK}=51.7709\,(460)$ $\Delta_K=2995.96\,(53)$ $\delta_J=2.06190\,(46)$ $\delta_K=93.120\,(180)$ $H_J=-0.2469\,(490)\cdot10^{-4}$ $H_{JK}=-0.1735\,(240)\cdot10^{-2}$ $H_{KJ}=-0.02297\,(91)$ $H_K=0.2909\,(110)$ $h_{JK}=-0.439\,(190)\cdot10^{-3}$ $h_K=-0.02688\,(2000)$	81BOW1		for structure see also Vol. II/7 (2, 277)	2.6.3, 9

Nr.	Molecule	v	A, B, C MHz	D, H kHz	Ref.	Ref. struct.	Varia, remarks	See also Tab., Nr.
19 *6–12* *4–15a* Cont.	trans-HNO_2	$v_2=1$	$A=92495.6\ (30)$ *) $B=12538.6\ (157)$ $C=10966.2\ (163)$	$\Delta_J=19.48\ (167)$ $\Delta_{JK}=105\ (56)$ $\Delta_K=2906\ (114)$ **)	80ALL		ν_2(N=O stretch) $=1699.760\ (1)\ cm^{-1}$ *) Laser Stark, uncertainties are 3 σ **) δ_J and δ_K are fixed to the ground state values	
	cis-HNO_2	$v=0$	$A=84101.8381\ (110)$ $B=13169.0755\ (19)$ $C=11364.1644\ (19)$	$\Delta_J=17.0502\ (65)$ $\Delta_{JK}=-8.6971\ (760)$ $\Delta_K=1958.59\ (28)$ $\delta_J=2.80810\ (23)$ $\delta_K=84.811\ (58)$ $H_J=-0.3248\ (670)\cdot10^{-4}$ $H_{KJ}=-0.02027\ (110)$ $H_K=0.1496\ (86)$ $h_{JK}=-0.202\ (67)\cdot10^{-3}$ $h_K=0.04024\ (610)$	81BOW1		energy difference between the cis and trans conformers: $\Delta E=(404\pm100)$ cal [76VAR]	
		$v_2=1$	$A=83940.9\ (52)$ *) $B=13146.6\ (41)$ $C=11345.3\ (39)$	***)	80ALL		ν_2(N=O stretch) $=1640.517\ (1)\ cm^{-1}$ ***) centrifugal distortion constants fixed on the ground state values	
20 *4–15* Cont.	HNO_3	$v=0$	$A=13011.0485\ (23)$ $B=12099.8803\ (23)$ $C=6260.6488\ (29)$	$\Delta_J=14.1390\ (31)$ $\Delta_{JK}=-20.18697\ (330)$ $\Delta_K=7.38540\ (250)$ $\delta_J=1.18256\ (45)$ $\delta_K=-20.5601\ (48)$ $H_J=0.3145\ (120)\cdot10^{-4}$ $H_{JK}=-0.010756\ (360)\cdot10^{-2}$ $H_{KJ}=1.1465\ (110)\cdot10^{-4}$ $H_K=-0.3809\ (800)\cdot10^{-4}$ $h_J=-0.9344\ (140)\cdot10^{-5}$ $h_{JK}=-0.1369\ (31)\cdot10^{-3}$ $h_K=0.1089\ (30)\cdot10^{-2}$	81BOW1, 79CAZ		for structure, see also Vol. II/7 (2, 278)	

Nr.	Molecule	v	A, B, C MHz	D, H kHz	Ref.	Ref. struct.	Varia, remarks	See also Tab., Nr.
20 *4-15* Cont.	DNO_3	$v=0$	$A''=12970.623\ (35)$ $B''=11312.627\ (19)$ $C''=6034.9667\ (77)$	$\tau_1=-57.42435\ (2013735)$ $\tau_2=-10.74135\ (698562)$ $\tau_3=1158.07\ (10760)$ $\tau_{aaaa}=-43.61418\ (511908)$ $\tau_{bbbb}=-43.24113\ (580942)$	77KAU4			
21 *6-13* *4-16*	HN_3	$v=0$	$A=610996.2\ (60)$ $B=12034.1465\ (50)$ $C=11781.4512\ (50)$	$\Delta_K=230000\ (6000)$ *) $\Delta_J=4.673\ (35)$ $\Delta_{JK}=791.1858\ (11)$ $\delta_J=0.0888\ (27)$ $H_J=0.088\ (36)\cdot 10^{-3}$ $H_{JK}=40.6\ (86)\cdot 10^{-3}$ $H_{KJ}=-1.210\ (35)$	75BEN	80WIN1	for structure, see also Vol. II/7 (2, 271) *) fixed value, obtained from the IR spectrum [68KRA]	2.6.3, 10 3, 15
	DN_3	$v=0$	$A=344746.589\ (64)$ $B=11350.983\ (16)$ $C=10964.755\ (15)$	$D_K=92242\ (33)$ $D_J=4.281\ (16)$ $D_{JK}=444.51\ (44)$ $\delta_K=365.1\ (77)$ $\delta_J=0.1864\ (41)$ $H_K=59.9\ (96)$ **) $H_{KJ}=-0.308\ (21)$ $H_{JK}=-3.4\ (11)\cdot 10^{-3}$	79BEN		**) fixed value, obtained from the Raman spectrum [77BEN]	
22 *6-14a*	NCN_3	$v=0$	$A=38066.72$ $B=3185.25$ $C=2933.61$		73BLA		for structure see also Vol. II/7 (3, 38)	2.7.3, 17 3, 16
	$^{15}NCN_3$	$v=0$	$A=37595.85$ $B=3093.37$ $C=2852.75$					
23 *6-15* *4-17* Cont.	HNSO	$v=0$	$A=49315.79\ (9)$ $B=9869.78\ (2)$ $C=8205.10\ (2)$	$d_J=-35.1\ (4)$ $d_{JK}=-720\ (10)$ $d_K=-510\ (60)$ $d_{WJ}=0.463\ (8)\cdot 10^{-5}$ $d_{WK}=0.049\ (2)\cdot 10^{-3}$	79DAL	79DAL	for structure see also Vol. II/7 (2, 281)	2.7.3, 18

Nr.	Molecule	v	A, B, C MHz	D, H kHz	Ref.	Ref. struct.	Varia, remarks	See also Tab., Nr.
23 *6–15* *4–17* Cont.	DNSO	$v=0$	$A=42257.50$ (10) $B=$ 9540.94 (2) $C=$ 7765.83 (2)	d_J = −34.0 (4) d_{JK} = −690 (10) d_K = −890 (60) d_{WJ} = 0.479 (6) · 10^{-5} d_{WK} = 0.054 (2) · 10^{-3}	79DAL			
24	$^{35}ClNO_3$	$v=0$	$A''=12106.77$ (40) $B''=$ 2777.026 (20) $C''=$ 2258.114 (10)	$\tau_1=-20.78$ (400) $\tau_2=$ −3.586 (700) $\tau_3=$ 81 (9) $\tau_{aaaa}=-170$ (48) $\tau_{bbbb}=$ −2.593 (200) $\tau_{cccc}=$ −0.908 (300)	77SUE			2.6.3, 11 2.7.3, 86
		$v=1$	$A''=12005.10$ (40) $B''=$ 2776.827 (20) $C''=$ 2262.117 (10)	$\tau_1=-21.45$ (600) $\tau_2=$ −3.839 (1000) $\tau_3=$ 86 (13) $\tau_{aaaa}=-85.1$ (590) $\tau_{bbbb}=$ −2.64 (40) $\tau_{cccc}=$ −1.09 (30)			$\nu=107$ (30) cm^{-1} from relative intensity measurements	
	$^{37}ClNO_3$	$v=0$	$A''=12105.59$ (20) $B''=$ 2700.984 (10) $C''=$ 2207.602 (10)	$\tau_1=-20.41$ (200) $\tau_2=$ −3.747 (500) $\tau_3=$ 88 (6) $\tau_{aaaa}=-85.0$ (380) $\tau_{bbbb}=$ −2.735 (200) $\tau_{cccc}=$ −1.29 (20)				
		$v=1$	$A''=12003.76$ (60) $B''=$ 2700.874 (20) $C''=$ 2211.474 (10)	$\tau_1=-20.15$ (400) $\tau_2=$ −3.843 (600) $\tau_3=$ 87 (11) $\tau_{aaaa}=-92.3$ (1150) $\tau_{bbbb}=$ −2.842 (300) $\tau_{cccc}=$ −1.25 (40)				
25 *4–4*	N_2H_4	$v=0$	$A=143467.6$ $B=$ 24083.0 $C=$ 24070.1		80TSU	76TSU	for structure, see also Vol. II/7 (2, 275)	2.6.3, 12 2.7.3, 19 2.8, 137
Cont.	NHDNHD species 4 *)	$v=0$	$A=102309.0$ $B=$ 21306.7 $C=$ 20356.5		76TSU		*) see Fig. 17 for the numbering	3, 17

Nr.	Molecule	v	A, B, C MHz	D, H kHz	Ref.	Ref. struct.	Varia, remarks	See also Tab., Nr.
25 *4-4* Cont.	NHDNHD species 5	$v=0$	$A=100173.8$ $B=20930.5$ $C=20843.6$		76TSU			
	NHDNHD species 6	$v=0$	$A=98246.3$ $B=21529.1$ $C=20404.8$					
	N_2D_4	$v=0$	$A=74724.3$ $B=18501.9$ $C=18437.5$		78TSU, 77HAR2			
26 *4-5*	N_2F_4	$v=0$	$\tilde{A}=5576.185$ $\tilde{B}=3189.417$ $\tilde{C}=2813.155$	$d_J = -13.89$ $d_{JK}=-256.63$ $d_K=-208.72$ $d_{WJ}=4.7984\cdot10^{-6}$ $d_{WK}=81.164\cdot10^{-6}$	77KAU3		for structure, see also Vol. II/7 (2, 201)	
27 *4-18*	S_2O	$v=0$	$A''=41915.4395\,(100)$ $B''=5059.04706\,(131)$ $C''=4507.15763\,(131)$	$\tau_1=105.983\,(308)$ $\tau_2=4.94026\,(3709)$ $\tau_3=601\,(1)$ $\tau_{aaaa}=-4679.33\,(269)$ $\tau_{bbbb}=-10.3185\,(152)$ $\tau_{cccc}=-4.7746\,(166)$ $H_J=-0.7367\,(4840)\cdot10^{-5}$ $H_{JK}=0.2400\,(1321)\cdot10^{-3}$ $H_{KJ}=-0.6155\,(1234)\cdot10^{-2}$ $H_K=0.1349\,(258)$ $h_J=0.418\,(157)\cdot10^{-5}$ *) $h_{JK}=-0.546\,(292)\cdot10^{-3}$ $h_K=0.3578\,(1459)\cdot10^{-1}$	78LOV	74TIE	for structure, see also Vol. II/7 (2, 343) *) Watson uses $2h_J$ for this parameter	
Cont.		$v_2=1$	$A''=42479.687\,(24)$ $B''=5059.73551\,(471)$ $C''=4500.8588\,(42)$	$\tau_1=97.785\,(3959)$ $\tau_2=3.748\,(464)$ $\tau_3=660\,(20)$ $\tau_{aaaa}=-5033.0\,(170)$ $\tau_{bbbb}=-10.957\,(175)$ $\tau_{cccc}=-5.489\,(187)$			$v_2=387\ cm^{-1}$ bend [73COO]	

Nr.	Molecule	v	A, B, C MHz	D, H kHz	Ref.	Ref. struct.	Varia, remarks	See also Tab., Nr.
27	$^{34}S^{32}SO$	$v=0$	$A=41737.142\,(60)$		74TIE			
4–18			$B=\ 4901.545\,(10)$					
Cont.			$C=\ 4379.740\,(10)$					
	$^{32}S^{34}SO$	$v=0$	$A=40637.232\,(75)$					
			$B=\ 5034.435\,(10)$					
			$C=\ 4472.449\,(10)$					
28	SO_2	$v=0$	$A''=60778.5553\,(56)$	$\tau_1=388.3076\,(1118)$	78LOV,	72KIR	for structure,	2.6.3, 13
6–17			$B''=10317.91220\,(106)$	$\tau_2=\ 31.57071\,(1652)$	81SAT		see also	2.9.3, 2
4–19			$C''=\ 8799.651056\,(927)$	$\tau_3=727.84\,(44)$			Vol. II/7 (2, 344)	
				$\tau_{aaaa}=-9917.148\,(221)$			*) Watson uses $2\,h_J$ for this	
				$\tau_{bbbb}=-40.04591\,(753)$			parameter	
				$\tau_{cccc}=-12.82924\,(624)$			force field calculated [72KIR]	
				$H_J=\ 0.13729\,(1361)\cdot 10^{-4}$				
				$H_{JK}=\ 0.1129\,(291)\cdot 10^{-3}$				
				$H_{KJ}=-0.184810\,(797)\cdot 10^{-1}$				
				$H_K=\ 0.354211\,(587)$				
				$h_J=\ 0.10146\,(241)\cdot 10^{-4}$ *)				
				$h_{JK}=\ 0.49316\,(4775)\cdot 10^{-3}$				
				$h_K=124.11\,(1467)\cdot 10^{-3}$				
		$v_1=1$	$A''=60810.908\,(87)$	$\tau_1=354.762\,(13600)$			$\nu_1=1151.71352\,(31)\ cm^{-1}$	
			$B''=10268.097\,(15)$	$\tau_2=\ 27.035\,(2000)$				
			$C''=\ 8757.3121\,(17)$	$\tau_3=821\,(48)$				
				$\tau_{aaaa}=-10054.6\,(240)$				
				$\tau_{bbbb}=-41.0525\,(8400)$				
				$\tau_{cccc}=-13.630\,(550)$				
				$H_J=\ (0.66187\pm 0.59)\cdot 10^{-3}$				
				$H_{JK}=\ (0.12798\pm 0.17)\cdot 10^{-1}$				
				$H_{KJ}=(-0.51919\pm 0.76)\cdot 10^{-1}$				
				$H_K=\ 0.6008\pm 0.52$				
				$h_J=\ (0.4347\pm 1.66)\cdot 10^{-4}$ *)				
				$h_{JK}=\ (0.1132\pm 0.064)\cdot 10^{-1}$				
				$h_K=\ -0.1522\pm 0.13$				
		$v_2=1$	$A''=61954.7930\,(177)$	$\tau_1=409.4489\,(5437)$	78LOV			
			$B''=10320.2147\,(32)$	$\tau_2=\ 33.61685\,(7973)$				
Cont.			$C''=\ 8783.80517\,(276)$	$\tau_3=699.6\,(21)$				

Nr.	Molecule	v	A, B, C MHz	D, H kHz	Ref.	Ref. struct.	Varia, remarks	See also Tab., Nr.
28 *6-17* *4-19* Cont.	SO_2			$\tau_{aaaa}=-11026.98\,(118)$ $\tau_{bbbb}=-40.13276\,(3566)$ $\tau_{cccc}=-12.7477\,(275)$ $H_J=(0.2793\pm0.78)\cdot10^{-5}$ $H_{JK}=(-0.3777\pm1.64)\cdot10^{-4}$ $H_{KJ}=(-0.21090\pm0.011)\cdot10^{-1}$ $H_K=0.43841\pm0.010$ $h_J=(0.11290\pm0.0080)\cdot10^{-4}$ *) $h_{JK}=(0.1301\pm0.20)\cdot10^{-3}$ $h_K=(0.1509\pm0.048)\cdot10^{-1}$	78LOV		*) Watson uses $2\,h_J$ for this parameter	
		$v_2=2$	$A''=63185.5876\,(656)$ $B''=10322.1482\,(114)$ $C''=8767.77679\,(1004)$	$\tau_1=441.451\pm7.0$ $\tau_2=37.2037\pm1.0$ $\tau_3=635.5\pm24.0$ $\tau_{aaaa}=-12283.9\pm14$ $\tau_{bbbb}=-39.6164\pm0.40$ $\tau_{cccc}=-12.326\pm0.26$ $H_J=(-0.31551\pm0.16)\cdot10^{-3}$ $H_{JK}=(-0.78331\pm0.42)\cdot10^{-2}$ $H_{KJ}=(0.1111\pm0.19)\cdot10^{-1}$ $H_K=(0.26342\pm0.14)$ $h_J=(-0.4662\pm0.29)\cdot10^{-4}$ *) $h_{JK}=(-0.3455\pm0.19)\cdot10^{-2}$ $h_K=0.10331\pm0.038$				
Cont.		$v_3=1$	$A''=60158.405\,(86)$ $B''=10283.0482\,(192)$ $C''=8766.86835\,(1655)$	$\tau_1=481.166\pm23$ $\tau_2=44.7019\pm3.4$ $\tau_3=450\pm80$ $\tau_{aaaa}=-9879.82\pm34$ $\tau_{bbbb}=-36.2180\pm1.4$ $\tau_{cccc}=-11.006\pm0.8$ $H_J=(-0.220903\pm0.079)\cdot10^{-2}$ $H_{JK}=(-0.766512\pm0.23)\cdot10^{-1}$ $H_{KJ}=0.24558\pm0.081$ $H_K=-1.0760\pm0.52$ $h_J=(-0.76972\pm0.23)\cdot10^{-3}$ *) $h_{JK}=(-0.4544\pm0.73)\cdot10^{-2}$ $h_K=(-0.6118\pm1.31)\cdot10^{-1}$				

Nr.	Molecule	v	A, B, C MHz	D, H kHz	Ref.	Ref. struct.	Varia, remarks	See also Tab., Nr.
28 *6-17* *4-19* Cont.	$^{33}SO_2$	$v=0$	$A''=59856.4540\,(681)$ $B''=10318.1328\,(121)$ $C''=\ 8780.08934\,(116)$	$\tau_1=381.875\pm9.2$ $\tau_2=\ 31.6319\pm1.4$ $\tau_3=670\,(32)$ $\tau_{aaaa}=-9628\,(14)$ $\tau_{bbbb}=-39.6980\pm0.56$ $\tau_{cccc}=-12.705\pm0.35$ $H_J=(-0.63683\pm0.36)\cdot10^{-3}$ $H_{JK}=(-0.17436\pm0.089)\cdot10^{-1}$ $H_{KJ}=\ (0.50098\pm0.34)\cdot10^{-1}$ $H_K=\ -0.1257\pm0.25$ $h_J=(-0.1553\pm0.078)\cdot10^{-3}$ *) $h_{JK}=(-0.5277\pm0.46)\cdot10^{-2}$ $h_K=\ 0.1324\pm0.11$	78LOV		*) Watson uses $2\,h_J$ for this parameter	
	$^{34}SO_2$	$v=0$	$A''=58991.1970\,(150)$ $B''=10318.3582\,(28)$ $C''=\ 8761.25810\,(246)$	$\tau_1=367.168\,(717)$ $\tau_2=\ 30.3047\,(1098)$ $\tau_3=654.2\,(24)$ $\tau_{aaaa}=-9338.295\,(1597)$ $\tau_{bbbb}=-40.0718\,(447)$ $\tau_{cccc}=-12.533\,(35)$ $H_J=\ (0.2242\pm0.099)\cdot10^{-4}$ $H_{JK}=\ (0.3456\pm0.63)\cdot10^{-3}$ $H_{KJ}=(-0.21318\pm0.037)\cdot10^{-1}$ $H_K=\ 0.36197\pm0.026$ $h_J=\ (0.672\pm0.82)\cdot10^{-5}$ *) $h_{JK}=\ (0.6369\pm0.59)\cdot10^{-3}$ $h_K=\ (0.7388\pm1.74)\cdot10^{-2}$				
		$v_2=1$	$A''=60129.228\,(339)$ $B''=10320.298\,(63)$ $C''=\ 8745.0711\,(513)$	$\tau_1=379.64\pm29$ $\tau_2=\ 31.482\pm4.3$ $\tau_3=640\pm90$ $\tau_{aaaa}=-10258.0\pm43$ $\tau_{bbbb}=-40.042\pm1.7$ $\tau_{cccc}=-12.357\pm1.2$				

Nr.	Molecule	v	A, B, C MHz	D, H kHz	Ref.	Ref. struct.	Varia, remarks	See also Tab., Nr.
29	$(SO)_2$*)	$v=0$	$A''=12972.958\,(8)$ $B''=3488.9409\,(27)$ $C''=2745.0555\,(29)$	$\tau_1=66.4\,(6)$ $\tau_2=4.98\,(12)$ $\tau_3=227.4\,(20)$ $\tau_{aaaa}=-293.8\,(4)$ $\tau_{bbbb}=-21.85\,(6)$ $\tau_{cccc}=-5.42\,(8)$ $H_J=0.12\,(4)\cdot10^{-3}$ $H_{JK}=-1.2\,(8)\cdot10^{-3}$ $H_{KJ}=5.3\,(24)\cdot10^{-3}$ $H_K=-1.2\,(19)\cdot10^{-3}$ $h_J=-0.08\,(2)\cdot10^{-3}$ $h_{JK}=3.2\,(9)\cdot10^{-3}$ $h_K=-26\,(11)\cdot10^{-3}$	74LOV	74LOV	for structure, see also Vol. II/7 (2, 345) *) planar cis configuration	2.6.3, 14 3, 18
		$v=1$	$A''=13133.251\,(23)$ $B''=3469.541\,(6)$ $C''=2736.198\,(9)$	$\tau_1=66.7\,(12)$ $\tau_2=5.31\,(17)$ $\tau_3=213.5\,(21)$ $\tau_{aaaa}=-303\,(18)$ $\tau_{bbbb}=-20.23\,(15)$ $\tau_{cccc}=-4.77\,(25)$			$\nu\simeq140\ cm^{-1}$ from relative intensity measurements, symmetric in-plane bend or torsion	
	$O^{32}S^{34}SO$	$v=0$	$A''=12845.663\,(42)$ $B''=3441.410\,(25)$ $C''=2709.909\,(20)$	$\tau_1=62.2\,(38)$ $\tau_2=4.5\,(9)$ $\tau_3=222\,(15)$ $\tau_{aaaa}=-266\,(33)$ $\tau_{bbbb}=-20.9\,(6)$ $\tau_{cccc}=-5.05\,(52)$				

Nr.	Molecule	v	A, B, C MHz	D, H kHz	Ref.	Ref. struct.	Varia, remarks	See also Tab., Nr.
30 *6-18* *4-20*	SOF_2	$v=0$	$A=8614.78$ (2) $B=8356.93$ (1) $C=4952.91$ (1)	$\Delta_J = 4.33$ (50) *) $\Delta_{JK} = -1.91$ (5) $\Delta_K = 9.82$ (5) $\delta_J = 1.63$ (1) $\delta_K = 2.02$ (3) $H_J = -762 \cdot 10^{-6}$ $H_{JK} = -4951 \cdot 10^{-6}$ $H_{KJ} = -6732 \cdot 10^{-6}$ $H_K = -2762 \cdot 10^{-6}$ $h_J = 496 \cdot 10^{-6}$ $h_{JK} = -2112 \cdot 10^{-6}$ $h_K = 1550 \cdot 10^{-6}$	75DUB, 73YAM		for structure, see also Vol. II/7 (2, 220) *) representation I^r used	
		$v_4=1$	$A=8576.549$ (15) $B=8387.840$ (13) $C=4859.261$ (16)		73RIM			
		$v_6=1$	$A=8631.524$ (11) $B=8347.391$ (11) $C=5022.696$ (13)					
31 *6-20*	$SOCl_2$						for structure, see also Vol. II/7 (2, 143)	2.7.3, 87
32 *6-21* *4-22*	$SO_2{}^{35}Cl^{37}Cl$	$v=0$	$A=3459.39$ (2) $B=2293.83$ (2) $C=1888.16$ (4)	$\Delta_J = 0.723$ (132) *) $\Delta_{JK} = -0.864$ (99) $\Delta_K = 1.744$ (80) $\delta_J = 0.101$ (10) $\delta_K = 0.218$ (100) $H_J = 356 \cdot 10^{-6}$ $H_{JK} = 117 \cdot 10^{-6}$ $H_{KJ} = -346 \cdot 10^{-6}$ $H_K = 221 \cdot 10^{-6}$ $h_J = 0.5 \cdot 10^{-6}$ $h_{JK} = -70 \cdot 10^{-6}$ $h_K = 410 \cdot 10^{-6}$	74DUB		for structure, see also Vol. II/7 (2, 144) *) representation I^r used	

Nr.	Molecule	v	A, B, C MHz	D, H kHz	Ref.	Ref. struct.	Varia, remarks	See also Tab., Nr.
33	$SO_2{}^{79}BrF$	$v=0$	$A=5069.65$		73RAL1	73RAL1	for structure,	2.7.3, 147
			$B=1773.00$				see also Vol. II/7 (2, 66)	
			$C=1767.81$					
	$SO_2{}^{81}BrF$	$v=0$	$A=5069.66$					
			$B=1754.69$					
			$C=1749.56$					
34	SOF_4	$v=0$	$A-C=974.41\,(30)$		76GRA		for structure, see also	
			$\kappa=-0.83005\,(40)$				Vol. II/7 (2, 224)	
35	SF_2	$v=0$	$A''=26930.5017\,(45)$	$\tau_1=128.603\,(615)$	73KIR3,	79EN2	for structure, see also	
6–22			$B''=9212.04734\,(229)$	$\tau_2=8.0540\,(1504)$	78LOV		Vol. II/7 (2, 238)	
			$C''=6845.87452\,(120)$	$\tau_3=372.0\,(17)$			*) Watson uses $2h_J$ for this	
				$\tau_{aaaa}=-1221.848\,(481)$			parameter	
				$\tau_{bbbb}=-82.3413\,(730)$			**) uncertainties are 2.5 σ	
				$\tau_{cccc}=-15.201\,(78)$			***) fixed value	
				$H_J=(0.36044\pm0.14)\cdot10^{-3}$			harmonic and anharmonic force	
				$H_{JK}=(0.6525\pm1.4)\cdot10^{-3}$			fields calculated [79EN2]	
				$H_{KJ}=(0.1908\pm0.25)\cdot10^{-2}$				
				$H_K=(0.15453\pm0.017)\cdot10^{-1}$			$\nu_1=840\,(20)\,cm^{-1}$	
				$h_J=(-0.3162\pm0.21)\cdot10^{-4}$ *)			$\nu_2=357\,(2)\,cm^{-1}$	
				$h_{JK}=(0.53051\pm0.18)\cdot10^{-2}$			$\nu_3=809\,(10)\,cm^{-1}$	
				$h_K=(-0.8503\pm0.74)\cdot10^{-2}$			$\nu_1-\nu_3=25.72\,(33)\,cm^{-1}$	
		$v_1=1$	$A=26925.973\,(61)$ **)	$\tau_{aaaa}=-1260\,(41)$ **)	79EN2		[79EN2]	
			$B=9162.176\,(20)$	$\tau_{bbbb}=-82.6\,(12)$			$r_e(S{-}F)=1.58745\,(12)\,Å$	
			$C=6860.040\,(20)$	$\tau_{cccc}=-24.1\,(14)$			$\theta_e(FSF)=98.048\,(13)^0$ [79EN2]	
				$\tau_{aabb}=180.4\,(60)$				
				$\tau_{bbcc}=-28.6\,(11)$				
				$\tau_{ccaa}=20.51$ ***)				
		$v_2=1$	$A=27264.264\,(39)$ **)	$\tau_{aaaa}=-1308\,(4)$ **)				
			$B=9188.819\,(14)$	$\tau_{bbbb}=-81.2\,(6)$				
			$C=6819.028\,(12)$	$\tau_{aabb}=227\,(2)$				
Cont.				$\tau_{abab}=-52.7\,(14)$				

Nr.	Molecule	v	A, B, C MHz	D, H kHz	Ref.	Ref. struct.	Varia, remarks	See also Tab., Nr.
35 6–22 Cont.	SF_2	$v_3=1$	$A=26744.451\ (62)$ $B=\ 9191.040\ (25)$ $C=\ 6778.374\ (18)$	$\tau_{aaaa}=-1227\ (22)$ **) +) $\tau_{bbbb}=\ -83.9\ (14)$ $\tau_{cccc}=\ -6.4\ (12)$ $\tau_{aabb}=\ 66.6\ (94)$ $\tau_{bbcc}=\ -35.4\ (13)$	79EN2		**) uncertainties are 2.5 σ +) with $\tau_{ccaa}=62.15$ fixed	
36 6–23 4–23	SF_4	$v=0$	$A=6687.737\ (11)$ $B=4086.689\ (12)$ $C=3219.949\ (12)$	$\tau_{aaaa}=-15.2\ (11)$ $\tau_{bbbb}=\ -7.8\ (11)$ $\tau_{cccc}=\ -4.9\ (10)$ $T_1=-14.7\ (20)$ *) $T_2=\ -5.2\ (16)$ *)	76INO		for structure, see also Vol. II/7 (2, 241) *) $T_1=\tau'_{aabb}+\frac{A-B}{A-C}\tau'_{ccaa}$ $T_2=\tau'_{bbcc}+\frac{B-C}{A-C}\tau'_{ccaa}$ vibrational assignment discussed	
		$v_4=1$	$A=6641.7081\ (84)$ $B=4095.8118\ (89)$ $C=3213.7530\ (90)$	$\tau_{aaaa}=-10.52\ (76)$ $\tau_{bbbb}=\ -4.64\ (63)$ $\tau_{cccc}=\ -1.36\ (66)$ $T_1=-8.5\ (11)$ *) $T_2=-0.24\ (84)$ *)				
		$v_9=1$	$A=6707.7928\ (71)$ $B=4075.9354\ (94)$ $C=3217.9307\ (93)$	$\tau_{aaaa}=-22.5\ (19)$ $\tau_{bbbb}=-13.8\ (19)$ $\tau_{cccc}=\ 4.4\ (43)$ $T_1=-17.8\ (33)$ *) $T_2=-11.1\ (27)$ *)				
37 6–22*a*	$S^{35}Cl_2$	$v=0$	$A=14613.5968\ (52)$ $B=\ 2920.8702\ (11)$ $C=\ 2430.6943\ (10)$	$\Delta_J=\ 1.3249\ (25)$ $\Delta_{JK}=-14.604\ (37)$ $\Delta_K=\ 138.027\ (70)$ $\delta_J=0.33968\ (50)$ $\delta_K=3.927\ (24)$ $H_{JK}=-2.09\ (14)\cdot10^{-4}$ $H_{KJ}=\ 3.46\ (22)\cdot10^{-4}$ $h_J=-2.82\ (33)\cdot10^{-6}$ $h_{JK}=\ 1.98\ (29)\cdot10^{-4}$	77DAV	77DAV	for structure, see also Vol. II/7 (2, 151) harmonic force field determined *) centrifugal distortion assumed to be the same as for the ground state	2.6.3, 15 2.7.3, 88
Cont.		$v_2=1$	$A=14732.830\ (10)$ $B=\ 2918.3174\ (34)$ $C=\ 2426.1032\ (34)$	*)			$\nu_2=208\ cm^{-1}$	

Nr.	Molecule	v	A, B, C MHz	D, H kHz	Ref.	Ref. struct.	Varia, remarks	See also Tab., Nr.
37 *6–22a* Cont.	$S^{35}Cl^{37}Cl$	$v=0$	$A=14490.169\,(21)$ $B=2841.1929\,(47)$ $C=2371.9634\,(35)$	$\Delta_J=1.239\,(13)$ $\Delta_{JK}=-14.40\,(13)$ $\Delta_K=135.7\,(2)$ $\delta_J=0.3168\,(35)$ $\delta_K=3.62\,(21)$ $H_{JK}=-2.56\,(68)\cdot10^{-4}$ $H_{KJ}=3.55\,(125)\cdot10^{-4}$ $h_J=-5.0\,(22)\cdot10^{-6}$ $h_{JK}=2.7\,(10)\cdot10^{-4}$	77DAV			
37a	H_2SO_4	$v=0$	$A''=5160.606\,(36)$ $B''=5024.534\,(27)$ $C''=4881.02\,(12)$	$\tau_1=-8.33\,(75)$ $\tau_2=-2.77\,(25)$ $\tau_3=2.7\,(11)\cdot10^3$ $\tau_{aaaa}=-8.98\,(64)$ $\tau_{bbbb}=-6.78\,(47)$ $\tau_{cccc}=-12.8\,(79)$	81KUC1	81KUC1	conformation with C_2 symmetry	2.6.3, 15a
	$H_2{}^{34}SO_4$	$v=0$	$A''=5159.040\,(77)$ $B''=5023.043\,(57)$ $C''=4881.20\,(30)$	$\tau_1=-6.1\,(25)$ $\tau_2=-2.01\,(84)$ $\tau_3=0.2\,(24)\cdot10^3$ $\tau_{aaaa}=-9.5\,(15)$ $\tau_{bbbb}=-7.19\,(75)$ $\tau_{cccc}=5\,(15)$				
	$HDSO_4$	$v=0$	$A''=5122.307\,(42)$ $B''=4831.917\,(33)$ $C''=4725.747\,(65)$	$\tau_1=-9.26\,(92)$ $\tau_2=-3.11\,(30)$ $\tau_3=2.28\,(33)\cdot10^3$ $\tau_{aaaa}=-7.35\,(80)$ $\tau_{bbbb}=-6.97\,(55)$ $\tau_{cccc}=-9.6\,(22)$				
	D_2SO_4	$v=0$	$A''=5042.420\,(24)$ $B''=4651.802\,(28)$ $C''=4613.252\,(40)$	$\tau_1=-5.27\,(62)$ $\tau_2=-1.80\,(21)$ $\tau_3=4.31\,(28)\cdot10^3$ $\tau_{aaaa}=-11.13\,(41)$ $\tau_{bbbb}=-5.19\,(40)$ $\tau_{cccc}=-6.87\,(82)$				

Nr.	Molecule	v	A, B, C MHz	D, H kHz	Ref.	Ref. struct.	Varia, remarks	See also Tab., Nr.
38 *6–35a*	$(HF)_2$ $(DF)_2$ HFDF	$v=0$ $v=0$ $v=0$	$(B+C)/2=6504.8$ (20) *) $(B+C)/2=6252.194$ (2) *) $(B+C)/2=6500.1$ (1) *)		72DYK	72DYK	for structure, see also Vol. II/7 (2, 175) *) MBER	2.6.3, 16 2.7.3, 2
39	$HF^{35}ClF$ $HF^{37}ClF$ $DF^{35}ClF$	$v=0$ $v=0$ $v=0$	$(B+C)/2=2545.404$ (3) *) $(B+C)/2=2543.188$ (5) *) $(B+C)/2=2443.586$ (5) *)	$D_J=7.27$ (30) *)	76NOV1	76NOV1	*) MBER	2.6.3, 17 2.7.3, 89 3, 19
40	$HFH^{35}Cl$ $HFH^{37}Cl$ $HFD^{35}Cl$	$v=0$ $v=0$ $v=0$	$(B+C)/2=3422.461$ (5) *) $(B+C)/2=3355.897$ (5) *) $(B+C)/2=3424.396$ (5) *)	$D_J=23.8$ (5) *) $D_J=22.0$ (5) *)	77JAN	77JAN	*) MBER	2.6.3, 18 2.7.3, 90 3, 20
41 *6–26a*	HOF DOF	$v=0$ $v=0$	$A''=585631.428$ (159) $B''=26760.6692$ (120) $C''=25515.0321$ (95) $A''=316096.061$ (45) $B''=25593.2238$ (30) $C''=23589.5489$ (27)	$\tau_1=-10785.3$ (20) $\tau_2=-763.4757$ (4589) $\tau_3=202790$ (200) $\tau_{aaaa}=-340301$ (393) $\tau_{bbbb}=-411.2447$ (2611) $\tau_{cccc}=-342.2286$ (5482) $\tau_1=-7884.041$ (1459) $\tau_2=-757.0956$ (1334) $\tau_3=65172$ (9) $\tau_{aaaa}=-106496$ (29) $\tau_{bbbb}=-363.2766$ (463) $\tau_{cccc}=-268.3925$ (417)	78LOV, 72PEA	72PEA	for structure, see also Vol. II/7 (2, 179)	2.6.3, 19 2.9.3, 4
42 *6–27a* Cont.	$F^{35}ClO_2$*)	$v=0$	$A=9635.91$ (3) $B=8275.67$ (3) $C=5019.23$ (3)	$\Delta_J=9.4$ (18) **) $\Delta_{JK}=-4.57$ (7) $\Delta_K=-2.57$ (15) $\delta_J=-1.057$ (5) $\delta_K=23.37$ (4) $\phi_J=0.023$ (3) $\cdot 10^{-3}$ $\phi_K=-2.7$ (3) $\cdot 10^{-3}$	81ROB	81ROB	for structure, see also Vol. II/7 (2, 96) *) C_s symmetry harmonic force field calculated **) representation III^r used	2.6.3, 20 2.7.3, 91

Nr.	Molecule	v	A, B, C MHz	D, H kHz	Ref.	Ref. struct.	Varia, remarks	See also Tab., Nr.
42 *6-27a* Cont.	$F^{37}ClO_2$	$v=0$	$A=9598.43\,(2)$ $B=8239.09\,(2)$ $C=5016.13\,(2)$	$\Delta_J=8.5$ (fixed) **) $\Delta_{JK}=-4.42\,(17)$ $\Delta_K=-2.44\,(37)$ $\delta_J=-1.027\,(12)$ $\delta_K=23.06\,(11)$ $\phi_J=0.027\,(6)\cdot10^{-3}$ $\phi_K=-2.9\,(5)\cdot10^{-3}$	81ROB		**) representation III^r used	
	$F^{35}Cl^{16}O^{18}O$	$v=0$	$A=9178.12\,(2)$ $B=8100.93\,(2)$ $C=4843.10\,(2)$	$\Delta_J=8.8$ (fixed) **) $\Delta_{JK}=-9.76\,(14)$ $\Delta_K=2.67\,(36)$ $\delta_J=-0.856\,(9)$ $\delta_K=14.60\,(10)$ $\phi_J=0.023\,(4)\cdot10^{-3}$ $\phi_K=-2.5\,(4)\cdot10^{-3}$				
43	$SPHF_2$	$v=0$	$A=8336.572\,(6)$ $B=3725.667\,(3)$ $C=2805.302\,(2)$	$\tau_1=-23.2\,(4)$ $\tau_2=-5.2\,(1)$ $\tau_3=129\,(3)$ $\tau_{aaaa}=-39\,(1)$ $\tau_{bbbb}=-4.1\,(1)$ $\tau_{cccc}=-1.35\,(7)$	73NAV	73NAV	for structure, see also Vol. II/7 (2, 184)	2.6.3, 21 3, 21
		$v_6=1$	$A=8344.88\,(3)$ $B=3727.73\,(3)$ $C=2798.75\,(3)$				$\nu_6(A')=344\ cm^{-1}$ deformation FPF	
		$v_5=1$	$A=8336.72\,(3)$ $B=3726.70\,(3)$ $C=2807.56\,(3)$				$\nu_5(A')=419\ cm^{-1}$ in-plane wag PF_2	
	$SPDF_2$	$v=0$	$A=7976.61\,(3)$ $B=3641.52\,(3)$ $C=2798.88\,(3)$					
	$^{34}SPHF_2$	$v=0$	$A=8333.40\,(3)$ $B=3596.86\,(3)$ $C=2731.91\,(3)$					

Nr.	Molecule	v	A, B, C MHz	D, H kHz	Ref.	Ref. struct.	Varia, remarks	See also Tab., Nr.
44	P_2H_4	$v=0$	$A=65433.7\,(28)$ $B=5831.27\,(74)$ $C=5815.19\,(30)$		74DUR2	74DUR2	for structure, see also Vol. II/7 (2, 296) *) isomer of type B=type of the Q branch series. The H is at the near trans position **) isomer of type C=type of the Q branch series. The H is at the near eclipsed position	2.6.3, 22
	$P_2D_3H(B)$ *)	$v=0$	$A=38432.65\,(23)$ $B=5378.13\,(4)$ $C=5303.86\,(5)$					
	$P_2D_3H(C)$ **)	$v=0$	$A=38540.95\,(21)$ $B=5407.92\,(5)$ $C=5278.60\,(4)$					
	P_2D_4		$A=33793.33\,(19)$ $B=5210.16\,(5)$ $C=5184.07\,(4)$					
45	$^{11}B^{35}ClF_2$	$v=0$	$A=10449.32\,(13)$ $B=4705.811\,(20)$ $C=3239.702\,(26)$	$\Delta_J=1.86\,(48)$ $\Delta_{JK}=9.0\,(17)$	77KRO	77KRO	planar molecule *) constrained to the value obtained for $^{11}B^{35}ClF_2$	2.7.3, 92
	$^{11}B^{37}ClF_2$	$v=0$	$A=10449.70\,(15)$ $B=4562.512\,(20)$ $C=3171.021\,(26)$	$\Delta_J=0.93\,(47)$ $\Delta_{JK}=6.81\,(14)$				
	$^{10}B^{35}ClF_2$	$v=0$	$A=10449.32$ *) $B=4713.346\pm0.3$ $C=3242.988\pm0.4$	$\Delta_J=1.86$ *) $\Delta_{JK}=9.0$ *)				
46	$^{11}BF_2NH_2$	$v=0$	$A''=10019.46\,(5)$ $B''=9590.39\,(5)$ $C''=4892.89\,(5)$	$\tau_1=-66\,(1)$ $\tau_2=-15\,(4)$ $\tau_3=3190\,(210)$ $\tau_{aaaa}=-49\,(4)$ $\tau_{bbbb}=-39\,(4)$ $\tau_{cccc}=-6\,(4)$	73LOV	73LOV	for structure see also Vol. II/7 (2, 34) planar molecule	2.6.3, 23 3, 22
	$^{10}BF_2NH_2$	$v=0$	$A''=10018.13\,(5)$ $B''=9589.48\,(5)$ $C''=4892.29\,(5)$	$\tau_1=-54\,(12)$ $\tau_2=-11\,(4)$ $\tau_3=2980\,(23)$ $\tau_{aaaa}=-46\,(4)$ $\tau_{bbbb}=-35\,(4)$ $\tau_{cccc}=-1\,(5)$				
Cont.								

Nr.	Molecule	v	A, B, C MHz	D, H kHz	Ref.	Ref. struct.	Varia, remarks	See also Tab., Nr.
46 Cont.	$^{11}BF_2{}^{15}NH_2$	$v=0$	$A=10019.19$ (15) $B=9234.51$ (15) $C=4798.44$ (10)		73LOV			
	$^{10}BF_2{}^{15}NH_2$	$v=0$	$A=10018.03$ (15) $B=9233.43$ (15) $C=4797.78$ (10)					
	$^{11}BF_2ND_2$	$v=0$	$A=9735.57$ (5) $B=8391.32$ (5) $C=4501.13$ (5)					
	$^{10}BF_2ND_2$	$v=0$	$A=9734.33$ (5) $B=8390.31$ (5) $C=4500.53$ (5)					
47	$^{11}BH_2NH_2$	$v=0$	$A=138212$ (4) $B=27487.83$ (10) $C=22878.44$ (11)	$\tau_{aaaa}=-22(5)\cdot10^3$ $\tau_{bbbb}=-0.23(4)\cdot10^3$ $\tau_{aabb}=1.0(5)\cdot10^3$ $\tau_{abab}=-1.4(2)\cdot10^3$	79SUG		uncertainties are 3σ	2.6.3, 24
	$^{10}BH_2NH_2$	$v=0$	$A=138199$ (6) $B=28420.36$ (11) $C=23520.78$ (12)	$\tau_{aaaa}=-28(9)\cdot10^3$ $\tau_{bbbb}=-0.19(8)\cdot10^3$ $\tau_{aabb}=1.8(10)\cdot10^3$ $\tau_{abab}=-1.8(4)\cdot10^3$				
47a	$^{11}BH_2OH$	$v=0$	$A=172621.1$ (15) *) $B=30470.22$ (27) $C=25812.73$ (24)	$\tau_{aaaa}=-25640(300)$ *) $\tau_{bbbb}=-362(8)$ $\tau_{aabb}=540(50)$ $\tau_{abab}=-1676(13)$	81KAW	81KAW	*) uncertainties are 3σ	2.6.3, 24a 2.7.3, 9b
	$^{10}BH_2OH$	$v=0$	$A=172629.4$ (15) *) $B=31542.88$ (27) $C=26578.55$ (24)	$\tau_{aaaa}=-26040(120)$ $\tau_{bbbb}=-478(210)$ $\tau_{aabb}=80(140)$ $\tau_{abab}=-1677(13)$				
Cont.	$^{11}BH_2{}^{18}OH$	$v=0$	$A=172154.1$ (13) *) $B=29239.67$ (18) $C=24914.16$ (21)	$\tau_{aaaa}=-24500(2100)$ $\tau_{bbbb}=-336(16)$ $\tau_{aabb}=420(120)$ $\tau_{abab}=-1550(60)$				

Nr.	Molecule	v	A, B, C MHz	D, H kHz	Ref.	Ref. struct.	Varia, remarks	See also Tab., Nr.
47a Cont.	$^{11}BD_2OD$	$v=0$	$A=88646.5$ (18) *) $B=23945.64$ (51) $C=18791.50$ (42)	$\tau_{aaaa}=-6940$ (230) $\tau_{bbbb}=-264$ (39) $\tau_{aabb}=370$ (19) $\tau_{abab}=-795$ (31)	81KAW		*) uncertainties are 3 σ	
48	$^{11}BH(OH)_2$	$v=0$	$A=61491$ (372) $B=10281.12$ (27) $C=8804.71$ (27)		78KAW2		$\Delta=0.024$ (51) amu Å^2	2.6.3, 25
	$^{10}BH(OH)_2$	$v=0$	$A=63236$ (675) $B=10279.97$ (27) $C=8841.88$ (27)					
49	$^{11}BFHOH$	$v=0$	$A=70557$ (804) $B=10271.24$ (33) $C=8965.44$ (33)		78KAW2		$\Delta=0.004$ (81) amu Å^2	2.6.3, 26
	$^{10}BFHOH$	$v=0$	$A=72856$ (999) $B=10270.23$ (36) $C=9002.90$ (36)					
50	$^{11}BF(OH)_2$	$v=0$	$A=10210.696$ (48) *) $B=9963.203$ (48) $C=5034.759$ (30)	$\tau_{aaaa}=-33$ (9) *) $\tau_{bbbb}=-27$ (9) $\tau_{aabb}=77$ (28) $\tau_{abab}=-55$ (9)	79KAW2	79KAW2	planar molecule *) error limits are 3 σ	2.6.3, 27 3, 23
	$^{10}BF(OH)_2$	$v=0$	$A=10209.403$ (75) *) $B=9962.117$ (73) $C=5034.124$ (42)	$\tau_{aaaa}=-42$ (19) *) $\tau_{bbbb}=-36$ (19) $\tau_{aabb}=50$ (60) $\tau_{abab}=-46$ (19)				
	$^{11}BF(OD)_2$	$v=0$	$A=9810.776$ (39) *) $B=8972.772$ (39) $C=4680.932$ (27)	$\tau_{aaaa}=-49$ (8) *) $\tau_{bbbb}=-32$ (8) $\tau_{aabb}=35$ (26) $\tau_{abab}=-33$ (9)				
	$^{10}BF(OD)_2$	$v=0$	$A=9809.580$ (63) *) $B=8971.805$ (63) $C=4679.409$ (39)	$\tau_{aaaa}=-51$ (15) *) $\tau_{bbbb}=-33$ (15) $\tau_{aabb}=32$ (48) $\tau_{abab}=-32$ (15)				

Nr.	Molecule	v	A, B, C MHz	D, H kHz	Ref.	Ref. struct.	Varia, remarks	See also Tab., Nr.
51	$^{11}BH(NH_2)_2$	$v=0$	$A=52382.90\,(47)$ *) $B=\ 9093.230\,(85)$ $C=\ 7748.590\,(60)$		78BRI	78BRI	*) $\Delta=-0.0032$ ± 0.001 amu Å^2	2.6.3, 28
	$^{10}BH(NH_2)_2$	$v=0$	$A=53752.60\,(42)$ $B=\ 9092.310\,(74)$ $C=\ 7777.420\,(52)$					
	$^{11}BD(ND_2)_2$	$v=0$	$A=32507.6\,(27)$ $B=\ 7424.80\,(87)$ $C=\ 6045.9\,(12)$					
	$^{10}BD(ND_2)_2$	$v=0$	$A=33035.1\,(12)$ $B=\ 7424.70\,(37)$ $C=\ 6064.40\,(37)$					
52 *6-36* *4-28*	O_3	$v=0$	$A=106536.236\,(4)$ $B=\ 13349.2548\,(7)$ $C=\ 11834.3613\,(6)$	$\Delta_J\ =\ 13.618\,(2)$ $\Delta_{JK}=-55.34\,(4)$ $\Delta_K\ =6345.4\,(3)$ $\delta_J\ =\ 2.0924\,(2)$ $\delta_K\ =96.92\,(3)$ $H_J\ =\ 0.011\,(2)\cdot 10^{-3}$ $H_{JK}=\ -0.18\,(7)\cdot 10^{-3}$ $H_{KJ}=-55.1\,(5)\cdot 10^{-3}$ $H_K\ =1178\,(7)\cdot 10^{-3}$ $h_J\ =\ 0.0053\,(2)\cdot 10^{-3}$ $h_{JK}\ =\ -0.18\,(6)\cdot 10^{-3}$ $h_K\ =\ 67\,(4)\cdot 10^{-3}$ $g_K\ =\ 0.021\,(30)\cdot 10^{-3}$	77DEP	77DEP	for structure, see also Vol. II/7 (2, 337) $r_e\ =1.27156\,(20)$ Å $2\,\theta_e=117°\,47'\,30''\pm 2'$	2.6.3, 29 2.9.3, 3
		$v_1=1$	$A=3.556673$ *) $B=0.442737$ $C=0.392424$	$\Delta_J\ =\ 4.440\cdot 10^{-7}$ *) $\Delta_{JK}=-2.54\cdot 10^{-6}$ $\Delta_K\ =\ 2.179\cdot 10^{-4}$ $\delta_J\ =\ 6.629\cdot 10^{-8}$ $\delta_K\ =\ 3.628\cdot 10^{-6}$	77BAR		*) I.R. data, all values in cm^{-1}, the sextic constants are constrained to the ground state values Coriolis interaction between ν_1 and ν_3: $\nu_1=1103.1332$ cm^{-1} $\nu_3=1042.0846$ cm^{-1} $G_c\ =-0.481990$ cm^{-1}	
		$v_3=1$	$A=3.500549$ *) $B=0.441299$ $C=0.391186$	$\Delta_J\ =\ 4.6176\cdot 10^{-7}$ *) $\Delta_{JK}=-1.842\cdot 10^{-6}$ $\Delta_K\ =\ 2.08327\cdot 10^{-4}$ $\delta_J\ =7.424\cdot 10^{-8}$ $\delta_K\ =2.84\cdot 10^{-6}$				
Cont.								

Nr.	Molecule	v	A, B, C MHz	D, H kHz	Ref.	Ref. struct.	Varia, remarks	See also Tab., Nr.
52 *6-36* *4-28* Cont.	O_3	$v_2=1$	$A=108137.97\,(3)$ $B=13311.443\,(4)$ $C=11765.034\,(3)$	$\Delta_J=13.67\,(1)$ $\Delta_{JK}=-54.5\,(3)$ $\Delta_K=6982\,(2)$ $\delta_J=2.0744\,(9)$ $\delta_K=116.47\,(9)$ $H_{JK}=-59\,(2)\cdot10^{-3}$ **) $H_K=1379\,(44)\cdot10^{-3}$ $h_J=0.0057\,(3)\cdot10^{-3}$ $h_K=77\,(4)\cdot10^{-3}$	78MON		$F_{AB}=-0.0103559\ \mathrm{cm}^{-1}$ where the Coriolis matrix element is: $H_{13}=F_{AB}(P_xP_z+P_zP_x)-iG_cP_y$ [77BAR]. **) the other sextic constants are constrained to the ground state values	
		$v_2=2$	$A=109796.25\,(9)$ $B=13273.153\,(8)$ $C=11694.077\,(9)$	$\Delta_J=13.6\,(2)$ ***) $\Delta_{JK}=-56\,(6)$ $\Delta_K=7700\,(40)$ $\delta_J=2.05\,(1)$ $\delta_K=137.3\,(1)$			***) the sextic constants are constrained to the ground state values	
	$^{16}O^{16}O^{18}O$	$v=0$	$A=104573.161\,(3)$ $B=12591.5331\,(5)$ $C=11212.5057\,(5)$	$\Delta_J=12.198\,(1)$ $\Delta_{JK}=-57.02\,(2)$ $\Delta_K=6112.9\,(2)$ $\delta_J=1.8175\,(2)$ $\delta_K=87.52\,(3)$ $H_J=0.0071\,(6)\cdot10^{-3}$ $H_{JK}=-0.09\,(7)\cdot10^{-3}$ $H_{KJ}=-51.3\,(5)\cdot10^{-3}$ $H_K=1103\,(5)\cdot10^{-3}$ $h_J=0.0048\,(3)\cdot10^{-3}$ $h_{JK}=-0.34\,(8)\cdot10^{-3}$ $h_K=69\,(5)\cdot10^{-3}$	77DEP			
Cont.								

Nr.	Molecule	v	A, B, C MHz	D, H kHz	Ref.	Ref. struct.	Varia, remarks	See also Tab., Nr.
52 *6-36* *4-28* Cont.	$^{16}O^{18}O^{16}O$	$v=0$	$A=98646.687\,(3)$ $B=13352.7323\,(6)$ $C=11731.7677\,(5)$	$\Delta_J = 13.424\,(2)$ $\Delta_{JK} = -39.55\,(3)$ $\Delta_K = 5424.2\,(2)$ $\delta_J = 2.1856\,(3)$ $\delta_K = 93.37\,(4)$ $H_J = 0.0098\,(14)\cdot10^{-3}$ $H_{JK} = -0.09\,(9)\cdot10^{-3}$ $H_{KJ} = -47.5\,(5)\cdot10^{-3}$ $H_K = 939\,(5)\cdot10^{-3}$ $h_J = 0.0056\,(4)\cdot10^{-3}$ $h_{JK} = -0.29\,(10)\cdot10^{-3}$ $h_K = 65\,(5)\cdot10^{-3}$	77DEP			
	$^{16}O^{18}O^{18}O$	$v=0$	$A=96684.920\,(2)$ $B=12593.3499\,(5)$ $C=11115.6935\,(5)$	$\Delta_J = 12.021\,(1)$ $\Delta_{JK} = -41.83\,(2)$ $\Delta_K = 5210.5\,(1)$ $\delta_J = 1.9014\,(2)$ $\delta_K = 84.37\,(3)$ $H_J = 0.0080\,(9)\cdot10^{-3}$ $H_{JK} = -0.19\,(7)\cdot10^{-3}$ $H_{KJ} = -42.6\,(4)\cdot10^{-3}$ $H_K = 875\,(4)\cdot10^{-3}$ $h_J = 0.0046\,(4)\cdot10^{-3}$ $h_{JK} = -0.17\,(9)\cdot10^{-3}$ $h_K = 52\,(5)\cdot10^{-3}$				
Cont.	$^{18}O^{16}O^{18}O$	$v=0$	$A=102579.228\,(3)$ $B= 11865.4883\,(7)$ $C= 10611.8811\,(6)$	$\Delta_J = 10.905\,(2)$ $\Delta_{JK} = -58.23\,(3)$ $\Delta_K = 5883.5\,(3)$ $\delta_J = 1.5747\,(2)$ $\delta_K = 78.86\,(4)$ $H_J = 0.0077\,(12)\cdot10^{-3}$ $H_{JK} = -0.12\,(7)\cdot10^{-3}$ $H_{KJ} = -46.1\,(5)\cdot10^{-3}$ $H_K = 1020\,(8)\cdot10^{-3}$ $h_J = 0.0037\,(3)\cdot10^{-3}$ $h_{JK} = -0.15\,(9)\cdot10^{-3}$ $h_K = 56\,(6)\cdot10^{-3}$				

Nr.	Molecule	v	A, B, C MHz	D, H kHz	Ref.	Ref. struct.	Varia, remarks	See also Tab., Nr.
52 *6-36* *4-28* Cont.	$^{18}O_3$	$v=0$	$A=94689.324\,(4)$ $B=11868.6414\,(6)$ $C=10522.7451\,(6)$	$\Delta_J = 10.745\,(2)$ $\Delta_{JK} = -43.74\,(3)$ $\Delta_K = 5000.0\,(2)$ $\delta_J = 1.6515\,(3)$ $\delta_K = 75.96\,(4)$ $H_J = 0.006\,(1)\cdot 10^{-3}$ $H_{JK} = -0.04\,(8)\cdot 10^{-3}$ $H_{KJ} = -39.2\,(6)\cdot 10^{-3}$ $H_K = 814\,(7)\cdot 10^{-3}$ $h_J = 0.0045\,(4)\cdot 10^{-3}$ $h_{JK} = -0.35\,(12)\cdot 10^{-3}$ $h_K = 55\,(6)\cdot 10^{-3}$	77DEP			
53 *6-37* *4-29*	H_2O^+)	$v=0$	$\mathscr{A}=835840.288\pm0.50$ $\mathscr{B}=435351.717\pm0.45$ $\mathscr{C}=278138.700\pm0.45$	$\Delta_J = 37594.22\pm20$ $\Delta_{JK} = -172912.8\pm170$ $\Delta_K = 973290.52\pm100$ $\delta_J = 15210.40\pm10$ $\delta_K = 41050.2\pm150$ $H_J = 15.6556\pm0.20$ $H_{JK} = -42.081\pm6$ $H_{KJ} = -509.508\pm10$ $H_K = 3733.028\pm8$ $h_J = 7.79579\pm0.09$ $h_{JK} = -25.165\pm1.1$ $h_K = 1097.1\pm30$ $L_{JK} = -3.0647\pm0.46$ $L_{KKJ} = 10.2952\pm1.2$ $L_K = -23.40138\pm0.8$ $l_K = -13.546\pm1.0$ $P_K = (5.19841\pm0.07)\cdot 10^{-2}$ $p_K = (3.7603\pm1.0)\cdot 10^{-2}$	78HEL1	74COO1	for structure, see also Vol. II/7 (2, 287) harmonic force field calculated [74COO1] $r_e = 0.9587$ Å $\theta_e = 103.9°$ [74COO1] $^+$) combined infrared-microwave data	2.6.3, 30 2.7.3, 3 2.9.3, 6
Cont.	$H_2{}^{17}O^+$)	$v=0$	$\mathscr{A}=830282.791\pm1.6$ $\mathscr{B}=435357.289\pm2.6$ $\mathscr{C}=277505.735\pm2.6$	$\Delta_J = 37594.14\pm90$ $\Delta_{JK} = -171853.2\pm300$ $\Delta_K = 960466.34\pm500$ $\delta_J = 15346.63\pm40$ $\delta_K = 41453.8\pm1200$				

Nr.	Molecule	v	A, B, C MHz	D, H kHz	Ref.	Ref. struct.	Varia, remarks	See also Tab., Nr.
53	$H_2{}^{17}O$ $^+$)			H_J = 18.2110 ± 1.2	78HEL1		$^+$) combined infrared-microwave data	
6-37				H_{KJ} = −642.468 ± 260				
4-29				H_K = 3745.123 ± 60				
Cont.				h_J = 10.18803 ± 0.6				
				h_{JK} = 37.859 ± 24				
				h_K = 1149.5 ± 130				
				L_K = −15.17327 ± 1.7				
				l_{JK} = −0.4257 ± 0.26				
				l_K = −3.26 ± 2				
				P_K = (4.71660 ± 1.5) · 10^{-2}				
		v_2 = 1	A = 926698	Δ_J = 38.1 · 10^3	78JOH			
			B = 440193	Δ_{JK} = −235 · 10^3				
			C = 273055	Δ_K = 1693 · 10^3				
				δ_J = 15.5 · 10^3				
				δ_K = 121 · 10^3			*) fixed value	
				ϕ_J = −0.043 · 10^3 *)				
				ϕ_{JK} = 1.02 · 10^3				
				ϕ_{KJ} = −5.6 · 10^3				
				ϕ_K = 12 · 10^3				
				φ_J = −0.0079 · 10^3 *)				
				φ_{JK} = −0.51 · 10^3				
				φ_K = 6.76 · 10^3				
	$H_2{}^{18}O$ $^+$)	v = 0	$\mathscr{A}$ = 825366.844 ± 1.0	Δ_J = 37545.79 ± 17	78HEL1			
			$\mathscr{B}$ = 435356.685 ± 0.8	Δ_{JK} = −171168.3 ± 130				
			$\mathscr{C}$ = 276948.998 ± 0.8	Δ_K = 949882.29 ± 200				
				δ_J = 15231.81 ± 11				
				δ_K = 38885.0 ± 200				
				H_J = 15.3906 ± 0.19				
				H_{JK} = −31.738 ± 4				
				H_{KJ} = −521.898 ± 12				
				H_K = 3624.767 ± 17				
				h_J = 8.04641 ± 0.14				
				h_{JK} = −24.187 ± 2.5				
Cont.				h_K = 1048.8 ± 40				

Nr.	Molecule	v	A, B, C MHz	D, H kHz	Ref.	Ref. struct.	Varia, remarks	See also Tab., Nr.
53	$H_2{}^{18}O$ $^+$)			$L_{JK} = -4.6449 \pm 0.6$	78HEL1		$^+$) combined infrared-	
6-37				$L_{KKJ} = 14.6700 \pm 1.6$			microwave data	
4-29				$L_K = -26.17359 \pm 1.1$				
Cont.				$l_J = (-2.59 \pm 0.8) \cdot 10^{-3}$				
				$l_K = -13.2008 \pm 1.3$				
				$P_K = (5.72244 \pm 0.4) \cdot 10^{-2}$				
		$v_2 = 1$	$A = 921309$	$\Delta_J = 41.6 \cdot 10^3$	78JOH			
			$B = 440195$	$\Delta_{JK} = -232 \cdot 10^3$				
			$C = 272564$	$\Delta_K = 1702 \cdot 10^3$				
				$\delta_J = 17.3 \cdot 10^3$				
				$\delta_K = 198 \cdot 10^3$				
				$\phi_J = 0.018 \cdot 10^3$ *)			*) fixed value	
				$\phi_{JK} = -0.0021 \cdot 10^3$				
				$\phi_{KJ} = -1.4 \cdot 10^3$				
				$\phi_K = 10.1 \cdot 10^3$				
				$\varphi_J = 0.0096 \cdot 10^3$				
				$\varphi_{JK} = -0.029 \cdot 10^3$				
				$\varphi_K = 1.81 \cdot 10^3$				
	D_2O $^+$)	$v = 0$	$\mathscr{A} = 462278.854(52)$ **)	$\tau'_{xxxx} = -66654(11)$ **)	73STE		**) uncertainties are 2 σ	
			$\mathscr{B} = 218038.233(33)$	$\tau'_{yyyy} = -7650(4)$				
			$\mathscr{C} = 145258.022(21)$	$\tau'_{zzzz} = -963936(68)$				
				$\tau'_1 = 71429(76)$				
				$\tau'_2 = -51107(79)$				
				$H_J = 1.74(2)$				
				$H_{JK} = -7.76(41)$				
				$H_{KJ} = -78.62(57)$				
				$H_K = 555.27(42)$				
				$2h_J = 1.81(1)$				
				$h_{JK} = -2.66(17)$				
				$h_K = 118.24(134)$				
				$G_{KJK} = -0.154(16)$				
				$G_{KJ} = 0.612(42)$				
				$G_K = -1.923(29)$				
				$g_K = -0.660(42)$				
				$F_2 = -0.133(27) \cdot 10^{-3}$				
				$F_3 = 4.991(64) \cdot 10^{-3}$				
Cont.				$F_5 = -0.098(2) \cdot 10^{-4}$				

Nr.	Molecule	v	A, B, C MHz	D, H kHz	Ref.	Ref. struct.	Varia, remarks	See also Tab., Nr.
53	D_2O	$v_2=1$	$\mathcal{A}=498730\,(24)$	$\tau'_{xxxx}=-74.1\,(7)\cdot10^3$	73STE			
6-37			$\mathcal{B}=220039\,(14)$	$\tau'_{yyyy}=-7.67\,(24)\cdot10^3$				
4-29			$\mathcal{C}=143611\,(6)$	$\tau'_{zzzz}=-1491.2\,(37)\cdot10^3$				
Cont.				$\tau'_1=101.0\,(32)\cdot10^3$				
				$\tau'_2=-17.5\,(56)\cdot10^3$				
				$H_K=1.05\,(3)\cdot10^3$				
	$D_2{}^{17}O$	$v=0$	$\mathcal{A}=456766.850\,(250)$ **)	$\tau'_{xxxx}=-66219\,(44)$ **)	73BEL		**) uncertainties are 2 σ	
			$\mathcal{B}=218041.011\,(160)$	$\tau'_{yyyy}=-7405\,(18)$				
			$\mathcal{C}=144701.487\,(58)$	$\tau'_{zzzz}=-938124\,(240)$				
				$\tau'_1=72050\,(270)$				
				$\tau'_2=-47660\,(280)$				
				$H_{KJ}=-77.23\,(46)$				
				$H_K=536.55\,(35)$				
				$h_K=116.36\,(63)$				
				$G_K=-1.862\,(4)$			***) the other sextic and higher	
				$F_3=4.867\,(2)\cdot10^{-3}$ ***)			order constants are constrained	
	$D_2{}^{18}O$	$v=0$	$\mathcal{A}=451891.921\,(530)$ **)	$\tau'_{xxxx}=-66653\,(83)$			to the values of $D_2{}^{16}O$	
			$\mathcal{B}=218045.181\,(280)$	$\tau'_{yyyy}=-7537\,(45)$				
			$\mathcal{C}=144201.681\,(140)$	$\tau'_{zzzz}=-920459\,(50)$				
				$\tau'_1=67238\,(550)$				
				$\tau'_2=-49030\,(570)$				
				$H_{KJ}=-70.49\,(120)$				
				$H_K=517.73\,(77)$				
				$2h_J=1.77\,(1)$				
				$h_K=109.05\,(150)$				
				$G_K=-1.834\,(1)$ ***)				
	HTO	$v=0$	$\mathcal{A}=677849.040\pm0.17$ **)	$\Delta_J=5212.023\pm3$ **)	74HEL2			
			$\mathcal{B}=198197.489\,(128)$	$\Delta_{JK}=48522.76\pm20$				
			$\mathcal{C}=150462.412\,(128)$	$\Delta_K=271275.33\pm60$				
				$\delta_J=1414.126\pm0.5$				
Cont.				$\delta_K=51328.33\pm70$				

Nr.	Molecule	v	A, B, C MHz	D, H kHz	Ref.	Ref. struct.	Varia, remarks	See also Tab., Nr.
53	HTO			H_J = 0.519072±0.04	74HEL2			
6-37				H_{JK} = 39.4327±0.6				
4-29				H_{KJ} = −63.4743±1.8				
Cont.				H_K = 896.3777±11				
				h_J = 0.17818±0.008				
				h_{JK} = 17.126±0.2				
				h_K = 538.348±13				
				L_{JK} = −0.542921±0.03				
				L_{KKJ} = 1.98579±0.08				
				L_K = −8.34015±0.7				
				l_{KJ} = (−9.1126±2.0)·10^{-2}				
				l_K = −8.1573±0.7				
				P_K = 0.110422±0.017				
				p_K = (5.95655±1.1)·10^{-2}				
	DTO	v=0	$\mathscr{A}$ = 410174.145 (78) **)	Δ_J = 5199.034±3 **)			**) uncertainties are 2 σ	
			$\mathscr{B}$ = 172101.952 (45)	Δ_{JK} = −15504.11±15				
			$\mathscr{C}$ = 119127.850 (45)	Δ_K = 180310.52±10				
				δ_J = 1941.256±1				
				δ_K = 12889.3±12				
				H_J = 0.59488±0.07				
				H_{JK} = 4.1358±0.8				
				H_{KJ} = −64.4426±1.0				
				H_K = 346.6981±0.9				
				h_J = 0.29687±0.016				
				h_{JK} = 2.962±0.42				
				h_K = 80.8715±1.6				
				L_{JJK} = (1.3294±0.4)·10^{-2}				
				L_{KKJ} = (1.7846±0.2)·10^{-1}				
				L_K = −0.821427±0.02				
				l_K = −0.67028±0.07				
Cont.				P_{JJK} = (−6.0406±1.3)·10^{-5}				
				p_K = (6.1777±0.8)·10^{-3}				

Nr.	Molecule	v	A, B, C MHz	D, H kHz	Ref.	Ref. struct.	Varia, remarks	See also Tab., Nr.
53 *6–37* *4–29* Cont.	T_2O	$v=0$	$\mathcal{A}=338810.923\,(76)$ **) $\mathcal{B}=145665.417\,(44)$ $\mathcal{C}=100259.415\,(44)$	$\Delta_J = 4145.597 \pm 1.4$ **) $\Delta_{JK} = -22038.98 \pm 8$ $\Delta_K = 144137.66 \pm 6$ $\delta_J = 1609.823 \pm 0.8$ $\delta_K = 5440.92 \pm 10$ $H_J = 0.55745 \pm 0.018$ $H_{JK} = -2.4519 \pm 0.18$ $H_{KJ} = -27.3425 \pm 0.34$ $H_K = 203.6106 \pm 0.3$ $h_J = 0.26829 \pm 0.010$ $h_{JK} = 0.2774 \pm 0.24$ $h_K = 36.3685 \pm 0.4$ $L_{JK} = (1.9022 \pm 0.24) \cdot 10^{-2}$ $L_K = -0.377789 \pm 0.004$ $l_J = (7.001 \pm 5.0) \cdot 10^{-5}$ $l_{JK} = (-3.143 \pm 1.4) \cdot 10^{-3}$ $P_{KJ} = (1.6559 \pm 0.6) \cdot 10^{-4}$ $P_K = (5.5117 \pm 0.32) \cdot 10^{-4}$ $p_{KKJ} = (-3.4123 \pm 1.0) \cdot 10^{-4}$	73LUC		**) uncertainties are 2 σ	
54 *6–38* *4–30*	H_2S	$v=0$	$A'=10.361528\,(310)$ *) $B'=9.016079\,(290)$ $C'=4.731214\,(130)$	$D'_J = 1.10011\,(790) \cdot 10^{-3}$ *) $D'_{JK} = -1.9652\,(130) \cdot 10^{-3}$ $D'_K = 0.92695\,(810) \cdot 10^{-3}$ $\delta'_J = 0.20846\,(480) \cdot 10^{-3}$ $R'_6 = -0.27791\,(690) \cdot 10^{-3}$ $H'_J = 0.4244\,(930) \cdot 10^{-6}$ $H'_{JK} = -1.536\,(260) \cdot 10^{-6}$ $H'_{KJ} = 1.866\,(350) \cdot 10^{-6}$ $H'_K = -0.755\,(150) \cdot 10^{-6}$ $H'_5 = 0.325\,(130) \cdot 10^{-6}$ $H'_6 = 0.885\,(160) \cdot 10^{-6}$ $H'_{10} = 0.2412\,(700) \cdot 10^{-6}$	81GI1, 78LOV	75COO	for structure, see Vol. II/7 (2, 300) *) weighted MW-IR analysis, all values are in cm^{-1}, uncertainties are 6 σ other isotopic species also measured harmonic force field calculated [75COO] $r_e = 1.3362$ Å $\theta_e = 92.06°$ [75COO]	

Nr.	Molecule	v	A, B, C MHz	D, H kHz	Ref.	Ref. struct.	Varia, remarks	See also Tab., Nr.
55 *6-39* *4-31*	$H_2{}^{78}Se$	$v=0$	$\mathscr{A}=244259.14\,(119)$ *) $\mathscr{B}=232558.27\,(119)$ $\mathscr{C}=116910.11\,(514)$	$\Delta_J = 35042\,(100)$ *) $\Delta_{JK}=-118184\,(150)$ $\Delta_K = 87908\,(300)$ $\delta_J = -2114.91\,(1800)$ $\delta_K = 458987.2\pm3000$ $H_J = 19.8\,(40)$ $H_{JK}=-1137.58\pm30$ $H_{KJ}= 2499.6\pm200$ $H_K = -868.1\pm200$ $h_J = -5.9680\pm0.2$ $h_{JK} = 207.64\pm10$ $h_K = 8451.90\pm400$ **)	75HEL, 81GI2		for structure, see also Vol. II/7 (2, 305) *) weighted MW-IR analysis. **) higher-order constants (in kHz): $L_{JJK}=1.18122\pm0.17$ $L_{JK} = 135.773\pm8$ $L_{KKJ}=-365.455\pm17$ $L_K = 218.064\pm9$ $l_{JK} = 0.98799\pm0.2$ $l_K = 8.493\pm1.7$ $P_{JK} = -0.02258\pm0.027$ $P_{KJ} = -2.48166\pm0.2$ $P_{KKJ} = 6.47661\pm0.6$ $P_K = -3.96067\pm0.4$	
	$H_2{}^{80}Se$	$v=0$	$\mathscr{A}=244099.89\,(29)$ *) $\mathscr{B}=232561.95\,(29)$ $\mathscr{C}=116874.93\,(176)$	$\Delta_J = 34950\,(30)$ *) $\Delta_{JK}=-118189\,(70)$ $\Delta_K = 88187.3\,(1300)$ $\delta_J = -2120.91\,(1200)$ $\delta_K = 461039.4\pm1000$ $H_J = 21.7\,(14)$ $H_{JK}=-1201.93\pm13$ $H_{KJ}= 2753.5\pm60$ $H_K = -1106.7\pm60$ $h_J = -6.26116\pm0.12$ $h_{JK} = 202.17\pm7$ $h_K = 8276.58\pm120$ ***)			***) higher-order constants (in kHz): $L_{JJK}=1.7430\pm0.04$ $L_{JK} = 135.488\pm2$ $L_{KKJ}=-366.685\pm5$ $L_K = 221.558\pm2$ $l_{JK} = 0.89245\pm0.06$ $l_K = 6.860\pm1.2$ $P_{JK} = -0.043699\pm0.007$ $P_{KJ} = -2.26800\pm0.1$ $P_{KKJ} = 6.04766\pm0.2$ $P_K = -3.73530\pm0.13$	
Cont.	$H_2{}^{76}Se$	$v=0$	$\mathscr{A}=244434.91$ $\mathscr{B}=232546.40$ $\mathscr{C}=116945.72$	$\Delta_J = 35146$ $\Delta_{JK}=-118341$ $\Delta_K = 86561.4$ $\delta_J = -2154.31$ $\delta_K = 452416.9$ $H_{JK}=-1148.09$ $H_{KJ}= 2503.5$ $h_J = -6.10582$ $h_{JK} = 200.83$ $h_K = 8147.09$ $^+$)			$^+$) the distortion constants not shown are fixed at their $H_2{}^{80}Se$ values	

Nr.	Molecule	v	A, B, C MHz	D, H kHz	Ref.	Ref. struct.	Varia, remarks	See also Tab., Nr.
55 *6-39* *4-31* Cont.	$H_2{}^{77}Se$	$v=0$	$\mathcal{A}=244348.74$ $\mathcal{B}=232549.75$ $\mathcal{C}=116927.29$	$\Delta_J = 35122$ $\Delta_{JK} = -118408$ $\Delta_K = 87079.9$ $\delta_J = -2148.38$ $\delta_K = 454192.9$ $H_{JK} = -1169.93$ $H_{KJ} = 2577.8$ $h_J = -6.17029$ $h_{JK} = 198.72$ $h_K = 8128.89$ ⁺)	75HEL, 81GI2			
	$H_2{}^{82}Se$	$v=0$	$\mathcal{A}=243945.78$ $\mathcal{B}=232568.92$ $\mathcal{C}=116841.45$	$\Delta_J = 34906$ $\Delta_{JK} = -118264$ $\Delta_K = 88917.1$ $\delta_J = -2109.03$ $\delta_K = 464572.6$ $H_{JK} = -1238.63$ $H_{KJ} = 2886.24$ $h_J = -6.36121$ $h_{JK} = 199.16$ $h_K = 8274.15$ ⁺)			⁺) the distortion constants not shown are fixed at their $H_2{}^{80}Se$ values	
	$HD^{80}Se$	$v=0$	$A'=7.953039\pm0.00103$ *) $B'=4.017408\pm0.00050$ $C'=2.632836\pm0.00047$	$D'_J = (0.2227\pm0.035)\cdot10^{-4}$ *) $D'_{JK} = (0.9491\pm0.023)\cdot10^{-3}$ $D'_K = (-0.5859\pm0.022)\cdot10^{-3}$ $\delta'_J = (0.1796\pm0.012)\cdot10^{-4}$ $R'_6 = (-0.17200\pm0.0069)\cdot10^{-4}$	80GIL		*) weighted MW-IR analysis, all values are in cm^{-1}, uncertainties are 6 σ other isotopic species also measured	
56 Cont.	$H_2{}^{130}Te$	$v=0$	$A'=187341.11\pm8.74$ $B'=182724.87\pm9.24$ $C'=91012.730\pm0.420$	$\tau'_{aaaa} = -98358\ (186)$ $\tau'_{bbbb} = -85095\ (226)$ $\tau'_{cccc} = -2984\ (118)$ $\tau'_{aabb} = 37113\ (792)$ $\tau'_{aacc} = -26.75\ (1816)\cdot10^3$ $\tau'_{bbcc} = 16.53\ (1740)\cdot10^3$	79BUR1		for structure, see also Vol. II/7 (2, 311)	

Nr.	Molecule	v	A, B, C MHz	D, H kHz	Ref.	Ref. struct.	Varia, remarks	See also Tab., Nr.
56 Cont.	$H_2{}^{128}Te$	$v=0$	$A'=187385.94\pm8.74$ $B'=182725.03\pm9.26$ $C'=\ 91023.564\pm0.420$	$\tau'_{aaaa}=-98457\ (186)$ $\tau'_{bbbb}=-85120\ (228)$ $\tau'_{cccc}=\ -3072\ (118)$ $\tau'_{aabb}=\ 37051\ (802)$ $\tau'_{aacc}=\ -26.87\ (1818)\cdot10^3$ $\tau'_{bbcc}=\ 16.50\ (1742)\cdot10^3$	79BUR1			
	$H_2{}^{126}Te$	$v=0$	$A'=187432.40\pm8.74$ $B'=182725.47\pm9.26$ $C'=\ 91034.820\pm0.424$	$\tau'_{aaaa}=-98693\ (186)$ $\tau'_{bbbb}=-85376\ (228)$ $\tau'_{cccc}=\ -3204\ (118)$ $\tau'_{aabb}=\ 36948\ (812)$ $\tau'_{aacc}=\ -27.08\ (1820)\cdot10^3$ $\tau'_{bbcc}=\ 16.39\ (1744)\cdot10^3$				
57 *6-40* *4-32*	F_2O	$v=0$	$A''=58782.6832\ (151)$ $B''=10896.3521\ (30)$ $C''=\ 9167.37374\ (255)$	$\tau_1=\ 174.809\ (1681)$ $\tau_2=-11.7994\ (2618)$ $\tau_3=2590\ (5)$ $\tau_{aaaa}=-6651.478\ (1426)$ $\tau_{bbbb}=\ -92.78132\ (10292)$ $\tau_{cccc}=\ -36.7054\ (701)$ $H_J\ =(0.59137\pm0.22)\cdot10^{-4}$ $H_{JK}=(0.4706\pm0.38)\cdot10^{-3}$ $H_{KJ}=-0.022290\ (836)$ $H_K\ =\ 0.17668\ (978)$ $h_J\ =\ 0.2282\ (461)\cdot10^{-4}$ *) $h_{JK}\ =\ 0.1664\ (665)\cdot10^{-2}$ $h_K\ =(0.565\pm17.2)\cdot10^{-3}$	78LOV		for structure, see also Vol. II/7 (2, 214) *) Watson uses $2h_J$ for this parameter	
58 *6-41* *4-33*	$^{35}Cl_2O$	$v=0$	$A=42046.336\ (58)$ *) $B=\ 3683.0821\ (53)$ $C=\ 3381.9035\ (70)$	$\tau_{aaaa}=-5400\ (30)$ $\tau_{bbbb}=\ -8.83\ (9)$ $\tau_{aabb}=\ 117\ (4)$ $\tau_{abab}=\ -17\ (3)$	81NAK	81NAK	for structure, see also Vol. II/7 (2, 139) *) uncertainties are 3 σ r_e(O—Cl) =1.69587 (7) Å $\measuredangle_e$ ClOCl =110.886 (6)°	
Cont.	$^{37}Cl_2O$	$v=0$	$A=41621.147\ (35)$ *) $B=\ 3484.2856\ (29)$ $C=\ 3211.0197\ (43)$	$\tau_{aaaa}=-5300\ (20)$ $\tau_{bbbb}=\ -7.97\ (7)$ $\tau_{aabb}=\ 108\ (3)$ $\tau_{abab}=\ -15\ (2)$				

Nr.	Molecule	v	A, B, C MHz	D, H kHz	Ref.	Ref. struct.	Varia, remarks	See also Tab., Nr.
58 *6-41* *4-33* Cont.	$^{35}Cl_2{}^{18}O$	$v=0$	$A=38235.643\,(58)$ *) $B=\ 3683.9040\,(54)$ $C=\ 3355.4739\,(74)$	$\tau_{aaaa}=-4460\,(30)$ $\tau_{bbbb}=\ -8.82\,(12)$ $\tau_{aabb}=\ 109\,(3)$ $\tau_{abab}=\ -20\,(3)$	81NAK		*) uncertainties are 3 σ	
	$^{37}Cl_2{}^{18}O$	$v=0$	$A=37810.686\,(58)$ *) $B=\ 3485.0660\,(47)$ $C=\ 3186.6991\,(77)$	$\tau_{aaaa}=-4370\,(30)$ $\tau_{bbbb}=\ -7.86\,(9)$ $\tau_{aabb}=\ 104\,(4)$ $\tau_{abab}=\ -19\,(3)$				
59	$S_2{}^{35}Cl_2$	$v=0$	$A=5533.19\,(16)$ $B=1393.840\,(5)$ $C=1232.672\,(6)$		79MAR	79MAR	for structure, see also Vol. II/7 (2, 152) *) uncertainties are 2.5 σ $v_x \simeq 80\,(20)\ cm^{-1}$	2.7.3, 93
		$v_x=1$	$A=5540.868\,(48)$ *) $B=1398.765\,(12)$ $C=1234.267\,(12)$	$d_J\ =\ -3.769\,(34)$ *) $d_{JK}\ =\ -72.32\,(211)$ $d_K\ =-147.4\,(69)$ $d_{WJ}=3.300\,(18)\cdot 10^{-6}$ $d_{WK}=4.042\,(165)\cdot 10^{-5}$	79YAM			
	$S^{35}Cl^{37}Cl$	$v=0$	$A=5475.441\,(19)$ $B=1361.852\,(6)$ $C=1205.179\,(8)$		79MAR			
60	$(H_2O)_2$	$v=0;\ E_1$	$(B+C)/2=6160.8$	$D_J\ =46.7$ $D_{JK}=\ 4.01\cdot 10^3$	77DYK	80ODU	MBER E_1 and E_2 are the labels of the tunneling states	2.6.3, 31 3, 24
		$v=0;\ E_2$	$(B+C)/2=6158.3$	$D_J\ =50.1$ $D_{JK}=\ 0.67\cdot 10^3$				
	$(H_2{}^{18}O)_2$	$v=0;\ E_A$	$(B+C)/2=5570.4$	$D_J\ =34.1$ $D_{JK}=\ 3.57\cdot 10^3$				
	$(D_2O)_2$	$v=0;\ E_1$	$(B+C)/2=5432.4$	$D_J\ =33.1$ $D_{JK}=\ 1.74\cdot 10^3$				
		$v=0;\ E_2$	$(B+C)/2=5432.9$	$D_J\ =35.0$ $D_{JK}=\ 1.22\cdot 10^3$				
	$H_2O\cdots DOD$	$v=0$	$(B+C)/2-2D_J$ $=5892.72\,(3)$		80ODU			
	D_3HO_2	$v=0$	$(B+C)/2-2D_J$ $=5674.96\,(3)$					
	D_3HO_2	$v=0$	$(B+C)/2-2D_J$ $=5648.26\,(3)$					

Nr.	Molecule	v	A, B, C MHz	D, H kHz	Ref.	Ref. struct.	Varia, remarks	See also Tab., Nr.
61	$H_2O\cdots HF$	$v=0$	$B=7263.43$ (10) $C=7139.91$ (10)	$D_J=56$ (10) $D_{JK}=1.98$ (5) $\cdot 10^3$	80BEV2	80BEV2	planar C_{2v} equilibrium structure or non planar C_s equilibrium structure with a low barrier to inversion,	2.6.3, 32
	$H_2{}^{18}O\cdots HF$	$v=0$	$B=6916.5$ (4) $C=6804.5$ (4)	$D_J=50$ (fixed) $D_{JK}=1.66$ (15) $\cdot 10^3$				
	$HDO\cdots HF$	$v=0$	$B=6935.3$ (6) $C=6769.7$ (6)	$D_J=50$ (fixed) $D_{JK}=1.54$ (936) $\cdot 10^3$		from relative intensity measurements: ν (H bond stretch) $=198\pm20$ cm^{-1} ν (out of plane bend) $=94\pm20$ cm^{-1} ν (in-plane bend) $=180\pm20$ cm^{-1}		
	$D_2O\cdots DF$	$v=0$	$B=6643.16$ (21) $C=6450.06$ (21)	$D_J=37$ (11) $D_{JK}=1408$ (29)				
	$D_2{}^{18}O\cdots DF$	$v=0$	$B=6377$ (4) $C=6206$ (4)					
62 *4–34*	H_2O_2	$v=0; \tau=1,2$ *)	$A=301874.205$ (37) $B=26212.439$ (46) $C=25098.604$ (46)	$\Delta_J=105.292$ (3) $\Delta_{JK}=1121.80$ (26) $\Delta_K=12034.0$ (120) $\delta_J=-0.258456$ (1400) $\delta_K=9786.31$ (2700) $H_{JK}=0.506307$ (9700) $H_{KJ}=-1.61735$ (8700) $H_K=1.85268$ (97000) $h_J=(0.304918\pm0.022)\cdot10^{-4}$ $h_{JK}=0.172789$ (900) $h_K=44.0464\pm3.8$ $L_{JJK}=(-0.159961\pm0.0045)\cdot10^{-4}$ $L_{JK}=(-0.933110\pm0.051)\cdot10^{-2}$ $L_{KKJ}=0.0218515\pm0.0038$	81BOW2	79CR	for structure, see also Vol. II/7 (2, 289) *) torsional sublevels splitting between the torsional levels 342885.03 (5) MHz [81HEL]	2.6.3, 32a 2.9, 7
		$v=0; \tau=3,4$ *)	$A=301586.074$ (75) $B=26155.639$ (36) $C=25186.461$ (36)	$\Delta_J=99.6169$ (140) $\Delta_{JK}=1152.75$ (51) $\Delta_K=11855.8\pm23$ $\delta_J=0.68678$ (120) $\delta_K=6383.26\pm2.1$ $H_{JK}=0.0594176\pm0.0013$ $H_{KJ}=-0.506399\pm0.073$ $h_J=(-0.515225\pm0.011)\cdot10^{-4}$ $h_{JK}=0.129727$ (1100) $h_K=-63.1712\pm0.28$ $L_{JJK}=(-0.105713\pm0.0071)\cdot10^{-4}$ $L_{JK}=(0.829787\pm0.0077)\cdot10^{-2}$				

Nr.	Molecule	v	A, B, C MHz	D, H kHz	Ref.	Ref. struct.	Varia, remarks	See also Tab., Nr.
63	KCN	$v=0$	$A''=58265.81\,(4)$ $B''=4940.055\,(2)$ $C''=4536.214\,(2)$	$\tau_{aaaa}=-5330\,(10)$ $\tau_{bbbb}=-27.03\,(3)$ $\tau_{cccc}=-18.71\,(3)$ $\tau_1=-1555.7\,(4)$ $\tau_2=-120.83\,(3)$ $H_{JK}=-5.4\,(2)\cdot 10^{-3}$ $H_{KJ}=95\,(1)\cdot 10^{-3}$	80TOR	80TOR	nonlinear, T-shaped structure	

2.5.2.2 Methane, ethane, propane derivatives

Nr.	Molecule	v	A, B, C MHz	D, H kHz	Ref.	Ref. struct.	Varia, remarks	See also Tab., Nr.
64 *6-42a*	CF_2	$v=0$	$A''=88355.0942\,(660)$ $B''=12507.5392\,(100)$ $C''=10932.3151\,(90)$	$\tau_{aaaa}=-11277.8\,(360)$ $\tau_{bbbb}=-69.1387\,(5000)$ $\tau_{cccc}=-34.0714\,(4000)$ $\tau_1=126.447\,(11000)$ $\tau_2=-18.1803\,(15000)$ $\tau_3=3450\,(40)$ $H_J=(0.1162\pm0.2)\cdot 10^{-3}$ $H_{JK}=(-0.0013\pm0.6)\cdot 10^{-2}$ $H_{KJ}=(-0.2555\pm0.5)\cdot 10^{-1}$ $H_K=0.3337\pm0.4$ $h_J=(0.132\pm0.4)\cdot 10^{-4}$ *) $h_{JK}=(0.4325\pm0.3)\cdot 10^{-2}$ $h_K=-0.1580\pm0.14$	73KIR4		for structure, see also Vol. II/7 (3, 20) harmonic force field calculated [73KIR4] *) Watson uses $2h_J$ for this parameter	2.6.3, 33a
65 *6-43* *4-38* Cont.	CH_3OH	$v_{CO}=1$	$B-C=930$ $B+C=48020$ *) $D_{ab}=-343\,(28)$ **)		72LE	76GER	for structure, see also Vol. II/7 (3, 124) ν(CO stretch) $=1033.5\ cm^{-1}$ *) derived on the assumption that $G_v=-5$ MHz **) D_{ab} arises because the axes of quantization are not the principal axes $F_v=-180$ MHz for the definition of the interactions constants see [68LEE]	2.6.3, 33 2.7.3, 4 2.8, 73 2.9.3, 9

Nr.	Molecule	v	A, B, C MHz	D, H kHz	Ref.	Ref. struct.	Varia, remarks See also Tab., Nr.
65 *6–43* *4–38* Cont.	CH_3OH	$v_{CH}=1$			72LE		two transitions measureed by IR MW DR [79TA]
	$^{13}CH_3OH$	$v=0$	$(B+C)/2=23640.058\,(22)$ $B-C=864.18\,(83)$ $D_{ab}=-84.96\,(98)$ **)	$D_J = 48.12\,(57)$ $D_{JK}=282.7\,(23)$	74HAQ		$F_v=-69.583\,(47)$ MHz $G_v=-3.3916\,(61)$ MHz $L_v=-0.022\,(22)$ MHz
	$^{13}CH_3OH$	$v_{CO}=1$	$I_b=35.2553$ *) $I_c=36.6382$		79HEN		*) FIR laser emission, in units of $kg\,m^2\cdot 10^{-47}$ **) D_{ab} arises because the axes of quantization are not the principal axes ν_{CO}=C—O stretch
	$CH_3{}^{18}OH$	$v=0$	$(B+C)/2=23219.002\,(42)$ $B-C=844.20\,(11)$ $D_{ab}=-167.42\,(85)$ **)	$D_J = 46.49\,(77)$ $D_{JK}=218.2\,(30)$	76GER		$F_v=-68.559\,(89)$ MHz $G_v=-3.3391\,(77)$ MHz $L_v=-0.138\,(31)$ MHz
	CH_3OD	$v=0$	$A=110125.46$ ***) $B=23436.49$ $C=21987.31$ $D_{ab}=876.2\,(54)$ **)	$D_J = 47.8\,(80)$ $D_{JK}=196\,(25)$	80KAU		$F_v=-65.14\,(24)$ MHz $G_v=-2.605\,(19)$ MHz $L_v=-0.200\,(25)$ MHz ***) excited torsional states measured
	CH_2DOH	$v=0$	$A_{e_0}=106977\,(1)$ $B_{e_0}=23033\,(1)$ $C_{e_0}=21680\,(1)$ $B_{e_1}=22830\,(1)$ $C_{e_1}=21825\,(1)$ $B_{o_1}=22680\,(1)$ $C_{o_1}=21947\,(1)$		80QUA		e_0 is the substate corresponding to the symmetric configuration e_1 and o_1 are the symmetric and antisymmetric states of the asymmetric form: $\Delta=E_{o_1}-E_{e_1}=101850\,(50)$ MHz Coriolis interaction term: $\varepsilon=44400\,(1000)$ MHz
66 *6–44* *4–39*	CH_3SH	$v=0$	$(B+C)/2=12666.169\,(64)$ $(B-C)/2=276.317\,(20)$ $D_{ab}=-123.4\,(20)$	$D_J = 15.9\,(20)$ $D_{JK}=106.2\,(54)$	80LEE, 80KOJ		for structure, 2.8, 93 see also Vol. II/7 (3, 125) $F_v=-61.744\,(51)$ MHz $G_v=-0.9245\,(21)$ MHz $L_v=-0.0634\,(33)$ MHz see definition of the constants in [68LEE]

Nr.	Molecule	v	A, B, C MHz	D, H kHz	Ref.	Ref. struct.	Varia, remarks	See also Tab., Nr.
67 *6–45*	$CH_3{}^{80}SeH$	$v=0$	$A=93727\,(50)$ $B=\ 9340.36\,(10)$ $C=\ 8982.41\,(10)$		73THO	73THO	for structure, see also Vol. II/7 (3, 126)	2.6.3, 34 2.8, 94
	$CH_3{}^{80}SeD$	$v=0$	$B=9307.58\,(10)$ $C=8635.13\,(10)$					
	$CD_3{}^{80}SeH$	$v=0$	$B=7510.32\,(10)$ $C=7279.71\,(10)$					
	$CD_3{}^{80}SeD$	$v=0$	$B=7482.75\,(10)$ $C=7045.02\,(10)$					
68 *6–45 a* *4–40*	CH_2DNH_2 trans form	$v=0$	$B_{\xi\xi}=\ 20475.318$ $B_{\eta\eta}=\ 20578.934$ $B_{\zeta\zeta}=\ 86051.597$ $B_{\eta\zeta}=-1460.257$	$D_J\ =\ 23.08$ $D_{JK}=\ 43.65$ $D_K\ =412.71$	74TAM3		for structure, see also Vol. II/7 (3, 129) *) ξ, η and ζ are the molecule fixed coordinate axes see theory in [74TAM2] $\Delta E_{gt}=7.06\pm0.03\ cm^{-1}$ energy difference between trans and gauche conformers [74TAM1] gauche-gauche tunneling interactions analyzed [74TAM3]	
	CH_2DNH_2 gauche form	$v=0$	$B_{\xi\xi}=21168.208$ $B_{\eta\eta}=20005.068$ $B_{\zeta\zeta}=86648.841$ $B_{\eta\zeta}=\ 581.371$ $B_{\zeta\xi}=-954.820$ $B_{\xi\eta}=-354.026$	same values as for the trans form				
69	$CH_3NH^{35}Cl$	$v=0$	$A=36204.2\,(6)$ $B=\ 5935.8\,(2)$ $C=\ 5340.4\,(2)$	$d_J\ =-15$ $d_{JK}=\ -4.4\cdot10^3$ $d_K\ =-23\cdot10^3$ $d_{WJ}=\ 3.6\cdot10^{-6}$ $d_{WK}=\ 0.75\cdot10^{-3}$	73MIR	74CA2	for structure, see also Vol. II/7 (3, 121)	2.7.3, 94 2.8, 109 3, 25
		$v_\tau=1$	$A=36126.0\pm2$ $B=\ 5915.09\,(8)$ $C=\ 5329.79\,(9)$	$D_J\ =\ 48\,(9)$ $D_{JK}=-18\,(15)$ $D_K\ =\ 530\,(420)$				
	$CH_3ND^{35}Cl$	$v=0$	$A=31587.0\,(3)$ $B=\ 5878.03\,(3)$ $C=\ 5234.50\,(3)$					
Cont.		$v_\tau=1$	$A=31530.9\,(12)$ $B=\ 5858.25\,(4)$ $C=\ 5224.25\,(13)$	$D_J\ =\ 10\,(2)$ $D_{JK}=-26\,(11)$ $D_K\ =\ 520\,(60)$				

Nr.	Molecule	v	A, B, C MHz	D, H kHz	Ref.	Ref. struct.	Varia, remarks	See also Tab., Nr.
69 Cont.	$CH_3NH^{37}Cl$	$v=0$	$A=36139.9\,(3)$ $B=5808.39\,(4)$ $C=5236.05\,(5)$		73MIR			
	$CH_3ND^{37}Cl$	$v=0$	$A=31530.1\,(3)$ $B=5750.72\,(4)$ $C=5131.63\,(4)$					
	$CH_2D(1)NH^{35}Cl$	$v=0$	$A=32286.97\,(19)$ $B=5720.29\,(8)$ $C=5145.09\,(8)$		74CA2		see Fig. 25 for the numbering of the 3 different positions of the D atom in the methyl group	
	$CH_2D(2)NH^{35}Cl$	$v=0$	$A=32147.82\,(4)$ $B=5725.63\,(2)$ $C=5152.94\,(2)$					
	$CH_2D(3)NH^{35}Cl$	$v=0$	$A=36124.41\,(37)$ $B=5506.14\,(10)$ $C=4989.88\,(10)$					
70	CH_3NHCN	$(a, +)$ *) $v=0$	$A=36090\,(156)$ $B=4977.13\,(7)$ $C=4505.13\,(7)$	$\Delta_J=4.3\,(20)$ $\Delta_{JK}=-17.4\,(53)$	80BAK2	80BAK2	for structure, see also Vol. II/7 (3, 242) *) a = A species, e = E species, $\pm$ = usual notation for inversion	2.6.3, 35 2.8, 110
	CH_3NHCN	$(a, -)$ $v=0$	$A=35380\,(145)$ $B=4982.54\,(7)$ $C=4512.58\,(7)$	$\Delta_J=0.4\,(21)$ $\Delta_{JK}=-176.7\,(52)$				
	CD_3NHCN	$(a, +)$ $v=0$	$A=28190.\,(64)$ $B=4422.69\,(4)$ $C=4033.59\,(4)$	$\Delta_J=2.9\,(13)$ $\Delta_{JK}=89.1\,(25)$				
	CD_3NHCN	$(a, -)$ $v=0$	$A=28390\,(178)$ $B=4428.16\,(11)$ $C=4039.78\,(11)$	$\Delta_J=5.0\,(34)$ $\Delta_{JK}=-242.7\,(67)$				
71 Cont.	CH_3NHOH *)	$v=0$	$A=38930.771\,(5)$ $B=9939.607\,(2)$ $C=8690.716\,(1)$	$\tau_{aaaa}=-1433\,(2)$ $\tau_{bbbb}=-55.6\,(2)$ $\tau_{cccc}=-24.3\,(2)$ $\tau'_{aabb}+\tau'_{aacc}=42\,(2)$ $\tau'_{bbcc}=-38.7\,(2)$	79SUN	79SUN	*) trans conformation	2.6.3, 36 2.7.3, 20 2.8, 111

Nr.	Molecule	v	A, B, C MHz	D, H kHz	Ref.	Ref. struct.	Varia, remarks	See also Tab., Nr.
71 Cont.	CH_3NDOH	$v=0$	$A=34022.903\,(15)$ $B=\ 9847.736\,(5)$ $C=\ 8489.441\,(5)$	$\tau_{aaaa}=\ -1.06\,(9)\cdot 10^3$ $\tau_{bbbb}=-57\,(6)$ $\tau_{cccc}\ =-25\,(4)$ $\tau'_{aabb}+\tau'_{aacc}=31\,(8)$ $\tau'_{bbcc}\ =-31\,(13)$	79SUN			
	CH_3NDOD	$v=0$	$A=32891.123\,(102)$ $B=\ 9282.157\,(7)$ $C=\ 8124.914\,(6)$	$\tau_{aaaa}=\ -0.60\,(35)\cdot 10^3$ $\tau_{bbbb}=-50\,(7)$ $\tau_{cccc}\ =-24\,(5)$ $\tau'_{aabb}+\tau'_{aacc}=31\,(10)$ $\tau'_{bbcc}\ =-17\,(16)$				
	CH_3NHOD	$v=0$	$A=37527.716\,(56)$ $B=\ 9363.483\,(13)$ $C=\ 8313.558\,(13)$					
	$^{13}CH_3NHOH$	$v=0$	$A=38721.804\,(89)$ $B=\ 9665.436\,(17)$ $C=\ 8470.287\,(17)$					
72 *6-51*	CH_3NO	$v=0$	$A=61284.01\,(5)$ $B=11452.77\,(3)$ $C=10254.54\,(3)$	$\delta_J=1.972\,(3)$	78TUR	78TUR	for structure, see also Vol. II/7 (3, 112) valence force constants calculated	2.6.3, 37 2.8, 112 3, 26
	$CH_3{}^{15}NO$	$v=0$	$A=59441.88\,(5)$ $B=11449.91\,(3)$ $C=10199.43\,(3)$	$\delta_J=1.85\,(1)$				
	$CH_3N^{18}O$	$v=0$	$A=60426.42\,(5)$ $B=10912.56\,(3)$ $C=\ 9797.26\,(3)$	$\delta_J=1.75\,(1)$				
	$^{13}CH_3NO$	$v=0$	$A=61054.3\,(10)$ $B=11104.83\,(3)$ $C=\ 9968.55\,(3)$	$(\delta_J=1.972)$ *)			*) fixed value	
	sym-CHD_2NO	$v=0$	$A=51006.404\,(140)$ $B=\ 9816.375\,(30)$ $C=\ 9182.894\,(30)$	$\Delta_J\ =\ 7.0756\pm0.8$ $\Delta_{JK}=\ 169.1122\pm11.2$ $\Delta_K\ =\ 298.0859\pm17.8$ $\delta_J\ =\ 0.6944\pm0.018$ $\delta_K\ =-503.3225\pm3.2$				
Cont.								

Nr.	Molecule	v	A, B, C MHz	D, H kHz	Ref.	Ref. struct.	Varia, remarks	See also Tab., Nr.
72 *6–51* Cont.	CD_3NO	$v=0$	$A=43927.865\,(140)$ $B=9638.078\,(40)$ $C=8776.835\,(30)$	$\Delta_J=9.7708\pm0.8$ $\Delta_{JK}=150.4854\pm10.7$ $\Delta_K=66.5474\pm15.2$ $\delta_J=0.9451\pm0.03$ $\delta_K=-202.7162\pm3.1$	78TUR		*) fixed value	
	$CD_3{}^{15}NO$	$v=0$	$A=42995.66\,(5)$ $B=9630.85\,(3)$ $C=8728.17\,(3)$	$\delta_J=(1.17)$ *)				
	$CD_3N^{18}O$	$v=0$	$A=43570.11\,(5)$ $B=9187.83\,(3)$ $C=8384.42\,(3)$	$\delta_J=1.05\,(1)$				
	$^{13}CD_3NO$	$v=0$	$A=43941.6\,(10)$ $B=9427.99\,(3)$ $C=8597.95\,(3)$	$\delta_J=(1.17)$ *)				
	sym-CH_2DNO	$v=0$	$A=51434.673\,(140)$ $B=11158.470\,(30)$ $C=9728.824\,(30)$	$\Delta_J=11.1740\pm0.4$ $\Delta_{JK}=51.5194\pm5.8$ $\Delta_K=385.4784\pm7.2$ $\delta_J=1.8317\pm0.04$ $\delta_K=-76.4010\pm1.0$				
73	CF_3NO	$v=0$	$A=5576.0$ $B=3200.701$ $C=3176.469$		76TUR2	76TUR2	for structure, see also Vol. II/7 (3, 26)	2.6.3, 38 2.8, 113
	$CF_3{}^{15}NO$	$v=0$	$A=5544.4$ $B=3180.160$ $C=3146.211$					
	$CF_3N^{18}O$	$v=0$	$A=5575.7\,(2)$ $B=3057.702\,(3)$ $C=3035.672\,(3)$					
74 *6–51a* *4–44*	CH_3NO_2	$v=0$	$A=13341.881\,(30)$ *) $B=10544.375\,(30)$ $C=5875.760\,(29)$	$D_J=6.14\,(62)$ *) $D_{JK}=17.75\,(37)$ $D_K=-7.54\,(40)$ $\delta_J=2.467\,(44)$ $\delta_K=15.75\,(19)$	75ROH		for structure, see also Vol. II/7 (3, 27) *) from $m=0$ spectrum, the coupling between internal and overall rotations has been studied see original literature [75ROH]	2.8, 114 2.9.3, 10

Nr.	Molecule	v	A, B, C MHz	D, H kHz	Ref.	Ref. struct.	Varia, remarks	See also Tab., Nr.
75	$CH_3O^{35}Cl$	$v=0$	$A=42064.286\,(11)$ $B=\;6296.861\,(6)$ $C=\;5670.610\,(5)$		76SUZ		for structure, see also Vol. II/7 (3, 90)	2.6.3, 39 2.7.3, 95 2.9.3, 11 3, 27
	$CD_3O^{35}Cl$	$v=0$	$A=32538.914\,(16)$ $B=\;5465.222\,(8)$ $C=\;4974.976\,(8)$					
76 *6–54*	$CD_3S^{35}Cl$	$v=0$	$A=14580.02\,(4)$ $B=\;4130.10\,(1)$ $C=\;3352.70\,(1)$		73GUA	73GUA	for structure, see also Vol. II/7 (3, 94)	2.6.3, 40 2.7.3, 96 3, 27a
	$CD_3S^{37}Cl$	$v=0$	$A=14531.13\,(4)$ $B=\;4021.25\,(2)$ $C=\;3278.09\,(2)$					
	$CH_3{}^{34}S^{35}Cl$	$v=0$	$A=16874.33\,(9)$ $B=\;4575.41\,(4)$ $C=\;3678.76\,(4)$					
77	CF_3OF	$v=0$	$A=5600\,(100)$ $B=3108.11\,(2)$ $C=3049.24\,(2)$		74BUC		for structure, see also Vol. II/7 (3, 34)	2.6.3, 41 2.8, 74
		$v_\tau=1$	$A=5530\,(100)$ $B=3100.49\,(3)$ $C=3048.87\,(3)$					
78	CF_3OOF	$v=0$	$A=4918.2\,(113)$ $B=1942.22\,(5)$ $C=1870.19\,(5)$		80DES	80DES		2.8, 75
		$v_\tau=1$	$A=4906.2\,(173)$ $B=1942.72\,(4)$ $C=1872.08\,(4)$				$\nu_\tau=70\,(5)\,cm^{-1}$, CF_3 torsion	
		$v_\tau=2$	$A=4895.3\,(250)$ $B=1943.11\,(6)$ $C=1873.66\,(6)$					

Nr.	Molecule	v	A, B, C MHz	D, H kHz	Ref.	Ref. struct.	Varia, remarks	See also Tab., Nr.
79	$CF_3OO^{35}Cl$	$v=0$	$A=4808\,(12)$ $B=1318.55\,(2)$ $C=1278.28\,(2)$		80DUR2	80DUR2		2.7.3, 97
	$CF_3OO^{37}Cl$	$v=0$	$A=4748\,(300)$ $B=1285.28\,(96)$ $C=1246.80\,(96)$					
80 *6-47* *4-41*	CH_3SCN						for structure, see also Vol. II/7 (3, 223)	2.8, 95
81	$CH_3{}^{80}SeCN$	$v=0$	$A=10143.867\,(31)$ $B=3483.6892\,(24)$ $C=2631.6454\,(24)$	$D_J=1.101\,(34)$ $D_{JK}=13.79\,(17)$ $D_K=223\,(16)$ $\delta_J=0.4026\,(47)$ $\delta_K=10.8\,(24)$	78LAN	78LAN, 78SAK		2.6.3, 42 2.8, 96 3, 28
		$v_{10}=1$	$A=10245.53\,(26)$ $B=3495.7129\,(49)$ $C=2631.8711\,(49)$				ν_{10}(C—Se—C bend) $=168\ cm^{-1}$	
		$v_{10}=2$	$A=10353.82\,(36)$ $B=3507.5426\,(73)$ $C=2632.0672\,(73)$					
		$v_\tau=1$	$A=10096.61\,(26)$ *) $B=3474.4203\,(51)$ $C=2629.9600\,(51)$				ν_τ torsion *) A state	
	$CH_3{}^{78}SeCN$	$v=0$	$A=10197.88\,(13)$ $B=3492.6571\,(25)$ $C=2640.4335\,(25)$					
		$v_{10}=1$	$A=10300.42\,(89)$ $B=3504.7249\,(17)$ $C=2640.6594\,(17)$					
Cont.	$CH_3{}^{77}SeCN$	$v=0$	$A=10226.57\,(20)$ $B=3497.3435\,(142)$ $C=2644.9260\,(142)$					

Nr.	Molecule	v	A, B, C MHz	D, H kHz	Ref.	Ref. struct.	Varia, remarks	See also Tab., Nr.
81 Cont.	$CH_3{}^{76}SeCN$	$v=0$	$A=10254.50\,(23)$ $B=\ 3501.9120\,(57)$ $C=\ 2649.5829\,(57)$		78LAN			
	$CH_3{}^{82}SeCN$	$v=0$	$A=10092.31\,(26)$ $B=\ 3475.0080\,(47)$ $C=\ 2623.1991\,(47)$					
	$^{13}CH_3{}^{80}SeCN$	$v=0$	$A=9725.64\,(16)$ $B=3454.9518\,(34)$ $C=2586.4540\,(34)$					
		$v_{10}=1$	$A=9826.70\,(31)$ $B=3466.8065\,(54)$ $C=2586.6358\,(54)$					
	$^{13}CH_3{}^{78}SeCN$	$v=0$	$A=9781.08\,(23)$ $B=3463.1436\,(45)$ $C=2594.9701\,(45)$					
	$CD_3{}^{80}SeCN$	$v=0$	$A=8322.31\,(81)$ $B=3308.06\,(2)$ $C=2436.92\,(1)$		78SAK			
		$v_\tau=1$	$A=8307.45\,(98)$ *) $B=3304.21\,(2)$ $C=2436.03\,(2)$				*) A species	
	$CD_3{}^{78}SeCN$	$v=0$	$A=8374.37\,(108)$ $B=3314.17\,(2)$ $C=2444.65\,(2)$					
		$v_\tau=1$	$A=8359.77\,(185)$ *) $B=3310.28\,(5)$ $C=2443.74\,(4)$					
82 *6–52* *4–44b* Cont.	CH_3ONO, cis	$v=0$	$A=20272.540\,(36)$ *) $B=\ 7437.914\,(4)$ $C=\ 5630.582\,(9)$	$\Delta_J=\ 6.74\,(20)$ *) $\Delta_{JK}=-12.90\,(130)$ $\Delta_K=\ 54.97\,(210)$ $\delta_J=\ 2.1\,(1)$ $\delta_K=\ 5.4\,(20)$	80GHO	79TUR1	for structure, see also Vol. II/7 (3, 113) *) for the A species	2.7.3, 21 2.8, 76 3, 29

Nr.	Molecule	v	A, B, C MHz	D, H kHz	Ref.	Ref. struct.	Varia, remarks	See also Tab., Nr.
82 *6-52* *4-446* Cont.	CH_3ONO, cis	$v_{10}=1$	$A=20315.989\ (205)$ $B=7500.508\ (85)$ $C=5639.314\ (127)$	**)	80GHO		$\nu_{10}=365\ cm^{-1}$, C—O—N deformation **) ground-state constants used	
		$v_{14}=1$	$A=20271.878\ (262)$ $B=7398.417\ (94)$ $C=5627.243\ (72)$	**)			$\nu_{14}=210\ (31)\ cm^{-1}$, N—O torsion	
		$v_{14}=2$	$A=20269.094\ (218)$ $B=7358.538\ (116)$ $C=5624.161\ (175)$	**)				
		$v_{15}=1$	$A=20371.541\ (395)$ $B=7353.297\ (152)$ $C=5594.568\ (98)$	**)			ν_{15}, methyl torsion	
		$v_{15}=2$	$A=19176.615\ (1412)$ $B=7341.923\ (386)$ $C=5592.613\ (259)$	**)				
		$v_{14}=1, v_{15}=1$	$A=20360.014\ (1100)$ $B=7325.444\ (380)$ $C=5588.846\ (190)$	**)				
	$CH_3O^{15}NO$, cis	$v=0$	$A=20054.83$ $B=7368.62$ $C=5574.13$		79TUR1			
	$CH_3ON^{18}O$, cis	$v=0$	$A=19761.81$ $B=7152.89$ $C=5427.79$					
	$CH_3{}^{18}ONO$, cis	$v=0$	$A=19559.16$ $B=7364.32$ $C=5532.57$					
	$^{13}CH_3ONO$, cis	$v=0$	$A=20051.83$ $B=7249.75$ $C=5505.66$					
	$^{13}CH_3O^{15}NO$ cis form	$v=0$	$A=19842.55$ $B=7179.22$ $C=5449.03$					
Cont.	sym-CH_2DONO cis form	$v=0$	$A=20216.25$ $B=6879.16$ $C=5302.21$					

Nr.	Molecule	v	A, B, C MHz	D, H kHz	Ref.	Ref. struct.	Varia, remarks	See also Tab., Nr.
82 *6-52* *4-446* Cont.	asym-CH_2DONO cis form	$v=0$	$A=18768.75$ $B=7234.08$ $C=5479.79$		79TUR1			
	CD_3ONO, cis	$v=0$	$A=17488.55$ $B=6534.51$ $C=5058.85$					
	CH_3ONO, trans	$v=0$	$B=4953.37$ $C=4627.32$				$E_{cis}-E_{trans}=314\,(22)\,cm^{-1}$ [80GHO]	
	$CH_3O^{15}NO$, trans	$v=0$	$B=4936.61$ $C=4604.82$					
	$CH_3ON^{18}O$, trans	$v=0$	$B=4713.70$ $C=4416.95$					
	$CH_3{}^{18}ONO$, trans	$v=0$	$B=4967.53$ $C=4589.26$					
	$^{13}CH_3ONO$, trans	$v=0$	$B=4846.28$ $C=4504.49$					
	CD_3ONO, trans	$v=0$	$B=4339.48$ $C=4105.50$					
84	CH_3ONH_2	$v=0$	$A=42488\,(150)$ $B=10049.59\,(3)$ $C=8962.85\,(3)$		74FON	74FON	the amine protons occupy a symmetrical trans position relative to the methyl group	2.7.3, 22 2.8, 77
		$v_x=1$	$A=42300$ $B=10092.1$ $C=8969.2$					
	CH_3ONHD	$v=0$	$A=40441\,(150)$ $B=9419.60\,(4)$ $C=8547.21\,(4)$					
		$v_x=1$	$A=40230\,(200)$ $B=9463.0\,(2)$ $C=8557.8\,(2)$					
	CH_3OND_2	$v=0$	$A=38513\,(150)$ $B=8877.75\,(5)$ $C=8194.86\,(5)$					
		$v_x=1$	$A=38300$ $B=8919.1$ $C=8207.5$					

Nr.	Molecule	v	A, B, C MHz	D, H kHz	Ref.	Ref. struct.	Varia, remarks	See also Tab., Nr.
85	CH_3OCH_2F *)	$v=0$	$A=17698.46\,(8)$ **) $B=5604.41\,(2)$ $C=4791.39\,(2)$	$d_J=7.36\,(148)$ $d_{JK}=-180\,(2)$ $d_K=-1.60\,(2)\cdot10^3$ $d_{EJ}=0.134\,(3)$ $d_{EK}=1.77\,(2)$	81NA1	81NA1	for structure, see also Vol. II/7 (3, 266) *) gauche form, skeletal dihedral angle = 69° 13′ **) for the A species, uncertainties are 2.5 σ $\nu_\tau=127\,(25)\ cm^{-1}$, methyl torsion $\nu_t=174\,(25)\ cm^{-1}$, skeletal torsion	2.6.3, 43 2.8, 78
		$v_\tau=1$	$A=17731.47\,(33)$ $B=5587.26\,(10)$ $C=4787.35\,(7)$	$d_J=6.1\,(27)$ $d_{JK}=-62\,(32)$				
		$v_t=1$	$A=17919.28\,(31)$ $B=5562.14\,(8)$ $C=4768.45\,(6)$	$d_J=6.1\,(23)$ $d_{JK}=-64\,(27)$				
	CH_3OCD_2F	$v=0$	$A=15384.46\,(10)$ **) $B=5407.14\,(3)$ $C=4624.83\,(4)$	$d_J=6.33\,(256)$ $d_{JK}=-112\,(3)$ $d_K=-5.00\,(2)\cdot10^3$ $d_{EJ}=0.073\,(4)$ $d_{EK}=5.11\,(2)$				
	CD_3OCH_2F	$v=0$	$A=15513.69\,(9)$ $B=4987.55\,(3)$ $C=4304.90\,(4)$	$d_J=5.38\,(207)$ $d_{JK}=-157\,(3)$ $d_K=1.90\,(2)\cdot10^3$ $d_{EJ}=0.129\,(4)$ $d_{EK}=-1.80\,(2)$				
	$CH_3O^{13}CH_2F$	$v=0$	$A=17443.16\,(46)$ **) $B=5580.22\,(13)$ $C=4762.71\,(10)$	$d_J=5.6\,(37)$ $d_{JK}=-46\,(42)$				
	$^{13}CH_3OCH_2F$	$v=0$	$A=17532.76\,(36)$ **) $B=5469.01\,(11)$ $C=4681.68\,(8)$	$d_J=5.9\,(29)$ $d_{JK}=-48\,(34)$				
	$CH_3{}^{18}OCH_2F$	$v=0$	$A=17228.46\,(45)$ **) $B=5539.76\,(12)$ $C=4726.05\,(9)$	$d_J=3.6\,(33)$ $d_{JK}=-38\,(39)$				
	CH_3OCHDF-1	$v=0$	$A=16460.26\,(33)$ $B=5476.26\,(9)$ $C=4769.29\,(7)$	$d_J=6.1\,(25)$ $d_{JK}=-36\,(29)$			for the numbering of the atoms, see Fig. 30	
	CH_3OCHDF-2	$v=0$	$A=16443.81\,(37)$ $B=5532.18\,(11)$ $C=4644.13\,(8)$	$d_J=7.5\,(27)$ $d_{JK}=-30\,(42)$				
Cont.								

Nr.	Molecule	v	A, B, C MHz	D, H kHz	Ref.	Ref. struct.	Varia, remarks	See also Tab., Nr.
85 Cont.	CH_2DOCH_2F-3	$v=0$	$A=17531.17\,(30)$	$d_J\ =\ 7.4\,(32)$	81NA1			
			$B=\ 5249.43\,(8)$	$d_{JK}=-46\,(27)$				
			$C=\ 4528.68\,(7)$					
	CH_2DOCH_2F-4	$v=0$	$A=16437.84\,(44)$	$d_J\ =\ 7.1\,(33)$				
			$B=\ 5535.73\,(14)$	$d_{JK}=-35\,(53)$				
			$C=\ 4647.98\,(9)$					
	CH_2DOCH_2F-5	$v=0$	$A=16750.50\,(36)$	$d_J\ =\ 7.8\,(29)$				
			$B=\ 5363.45\,(11)$	$d_{JK}=-45\,(34)$				
			$C=\ 4663.21\,(8)$					
86	$CH_3OCH_2{}^{35}Cl$ +)	$v=0$	$A=14233.873\,(13)$	$\Delta_J\ =\ 2.45\,(2)$	74IKE	80HAY3	for structure,	2.7.3, 98
			$B=\ 3481.650\,(2)$	$\Delta_{JK}=-15.04\,(13)$			see also	2.8, 79
			$C=\ 3032.799\,(3)$	$\Delta_K\ =\ 92.15\,(293)$			Vol. II/7 (3, 263)	3, 30
				$\delta_J\ =\ 0.5965\,(19)$			+) gauche form	
				$\delta_K\ =\ 8.83\,(36)$				
		$v_{\tau_1}=1$	$A=14275.775\,(21)$	$\Delta_J\ =\ 2.39\,(10)$			ν (CH_2Cl torsion)	
			$B=\ 3476.898\,(6)$	$\Delta_{JK}=-16.78\,(16)$			$=133\,(1)\,cm^{-1}$	
			$C=\ 3030.415\,(6)$	$\Delta_K\ =\ 103.3\,(37)$				
				$\delta_J\ =\ 0.6092\,(23)$				
				$\delta_K\ =\ 8.16\,(47)$				
		$v_{\tau_2}=1$	$A=14384.546\,(20)$	$\Delta_J\ =\ 2.36\,(10)$			ν (CH_3 torsion) $=179\,(1)\,cm^{-1}$	
			$B=\ 3457.859\,(6)$	$\Delta_{JK}=-17.09\,(16)$				
			$C=\ 3018.017\,(6)$	$\Delta_K\ =\ 108.6\,(31)$				
				$\delta_J\ =\ 0.5605\,(18)$				
				$\delta_K\ =\ 7.74\,(35)$				
	$CH_3OCH_2{}^{37}Cl$	$v=0$	$A=14186.229\,(20)$	$\Delta_J\ =\ 2.09\,(9)$				
			$B=\ 3398.509\,(5)$	$\Delta_{JK}=-14.85\,(17)$				
			$C=\ 2967.530\,(5)$	$\Delta_K\ =\ 100.7\,(37)$				
				$\delta_J\ =\ 0.5544\,(24)$				
				$\delta_K\ =\ 9.64\,(49)$				
	$CD_3OCH_2{}^{35}Cl$	$v=0$	$A=12484.28\,(50)$ *)		74HAY4		*) uncertainties are 2.5 σ	
			$B=\ 3174.79\,(12)$					
			$C=\ 2771.23\,(14)$					
	$CD_3OCH_2{}^{37}Cl$	$v=0$	$A=12440.27\,(46)$ *)					
			$B=\ 3099.28\,(11)$					
Cont.			$C=\ 2711.53\,(14)$					

Nr.	Molecule	v	A, B, C MHz	D, H kHz	Ref.	Ref. struct.	Varia, remarks	See also Tab., Nr.
86 Cont.	$CH_3OCD_2{}^{35}Cl$	$v=0$	A = 12214.08 (65) *) B = 3434.84 (16) C = 2960.05 (19)		74HAY4		*) uncertainties are 2.5 σ	
	$CH_3OCD_2{}^{37}Cl$	$v=0$	A = 12167.48 (54) *) B = 3353.97 (13) C = 2897.29 (16)					
	$CH_3O^{13}CH_2{}^{35}Cl$	$v=0$	A = 13919.88 (11) *) B = 3479.83 (3) C = 3019.98 (4)	d_J = 23 (7) *) d_{JK} = −1.7 (4)	79HAY2			
	$CH_3O^{13}CH_2{}^{37}Cl$	$v=0$	A = 13870.56 (22) *) B = 3396.69 (7) C = 2955.02 (7)	d_J = 25 (13) *) d_{JK} = −1.4 (8)				
	$CH_3OCHD^{35}Cl$-1	$v=0$	A = 13270.25 (7) *) B = 3437.41 (2) C = 3018.93 (2)	d_J = 19 (4) *) d_{JK} = −1.8 (2)			for the numbering of the atoms, see Fig. 3, 30	
	$CH_3OCHD^{37}Cl$-1	$v=0$	A = 13225.16 (16) *) B = 3356.05 (5) C = 2953.83 (6)	d_J = 24 (10) *) d_{JK} = −1.5 (6)				
	$CH_3OCHD^{35}Cl$-2	$v=0$	A = 13014.61 (12) *) B = 3479.81 (4) C = 2972.41 (4)	d_J = 22 (7) *) d_{JK} = −1.6 (4)				
	$CH_3OCHD^{37}Cl$-2	$v=0$	A = 12965.68 (19) *) B = 3397.22 (6) C = 2909.55 (6)	d_J = 21 (11) *) d_{JK} = −1.5 (7)				
	$^{13}CH_3OCH_2{}^{35}Cl$	$v=0$	A = 14026.22 (15) *) B = 3406.68 (4) C = 2967.31 (5)	d_J = 19 (9) *) d_{JK} = −1.7 (5)				
	$^{13}CH_3OCH_2{}^{37}Cl$	$v=0$	A = 13977.57 (11) *) B = 3324.82 (3) C = 2903.02 (4)	d_J = 24 (7) *) d_{JK} = −1.6 (4)				
	$CH_2DOCH_2{}^{35}Cl$-3	$v=0$	A = 13898.78 (20) *) B = 3310.42 (5) C = 2892.62 (6)	d_J = 20 (10) *) d_{JK} = −1.8 (6)				
Cont.	$CH_2DOCH_2{}^{37}Cl$-3	$v=0$	A = 13846.63 (7) *) B = 3232.03 (2) C = 2830.52 (2)	d_J = 18 (4) *) d_{JK} = −1.7 (3)				

Nr.	Molecule	v	A, B, C MHz	D, H kHz	Ref.	Ref. struct.	Varia, remarks	See also Tab., Nr.
86 Cont.	$CH_2DOCH_2{}^{35}Cl$-4	$v=0$	$A=13302.42(10)$ *) $B=3453.32(3)$ $C=2967.64(3)$	$d_J=23(6)$ *) $d_{JK}=-1.4(3)$	79HAY2		*) uncertainties are 2.5 σ	
	$CH_2DOCH_2{}^{37}Cl$-4	$v=0$	$A=13259.70(17)$ *) $B=3370.36(5)$ $C=2904.20(5)$	$d_J=23(9)$ *) $d_{JK}=-2.2(5)$				
	$CH_2DOCH_2{}^{35}Cl$-5	$v=0$	$A=13552.85(11)$ *) $B=3356.74(3)$ $C=2955.16(4)$	$d_J=24(7)$ *) $d_{JK}=-1.6(4)$				
	$CH_2DOCH_2{}^{37}Cl$-5	$v=0$	$A=13507.67(16)$ *) $B=3276.82(4)$ $C=2891.04(5)$	$d_J=22(8)$ *) $d_{JK}=-1.8(5)$				
	$CD_3OCD_2{}^{35}Cl$	$v=0$	$A=10874.68(17)$ *) $B=3134.82(5)$ $C=2707.35(5)$	$d_J=16(9)$ *) $d_{JK}=-1.3(6)$				
	$CD_3OCD_2{}^{37}Cl$	$v=0$	$A=10830.46(10)$ *) $B=3061.68(3)$ $C=2650.06(3)$	$d_J=18(6)$ *) $d_{JK}=-1.6(4)$				
	$CH_3{}^{18}OCH_2{}^{35}Cl$	$v=0$	$A=13968.12(12)$ *) $B=3409.25(4)$ $C=2974.86(4)$	$d_J=19.5(74)$ *) $d_{JK}=-1.72(43)$	80HAY3			
87	$CH_3OCH_2{}^{79}Br$	$v=0$	$A=13312.38(61)$ *) $B=2306.30(8)$ $C=2081.51(9)$		74HAY4	74HAY4	for structure, see also Vol. II/7 (3, 261) *) uncertainties are 2.5 σ	2.7.3, 148
	$CH_3OCH_2{}^{81}Br$	$v=0$	$A=13305.16(58)$ $B=2288.24(8)$ $C=2066.63(5)$					
	$CD_3OCH_2{}^{79}Br$	$v=0$	$A=11646.13(88)$ $B=2109.02(11)$ $C=1903.77(11)$					
Cont.	$CD_3OCH_2{}^{81}Br$	$v=0$	$A=11639.64(78)$ $B=2092.13(10)$ $C=1889.82(10)$					

Nr.	Molecule	v	A, B, C MHz	D, H kHz	Ref.	Ref. struct.	Varia, remarks	See also Tab., Nr.
87 Cont.	$CH_3OCD_2{}^{79}Br$	$v=0$	$A=11397.86(58)$ $B=2276.57(8)$ $C=2035.26(8)$		74HAY4			
	$CH_3OCD_2{}^{81}Br$	$v=0$	$A=11391.21(59)$ $B=2258.52(8)$ $C=2020.61(8)$					
88	CH_3OCH_2CN gauche form	$v=0$	$A=11893.02(21)$ $B=3423.22(7)$ $C=2871.61(7)$	$d_J=10.3(75)$ $d_{JK}=331(268)$ $d_K=874(650)$ $d_{EJ}=-0.00229(244)\cdot 10^{-3}$ $d_{EK}=-0.106(78)\cdot 10^{-3}$	74KEW	74KEW	$E_{trans}-E_{gauche}$ $=5.7(17)$ kJ/mol	2.6.3, 44 2.8, 80
		$v_\tau=1$	$A=11981.2(20)$ $B=3415.54(4)$ $C=2866.26(3)$	$D_J=3.37(42)$ $D_{JK}=-26.0(6)$			$v_\tau(O—CH_2)=93(25)$ cm^{-1} from relative intensity measurements	
		$v_\tau=2$	$A=12100.1(42)$ $B=3406.46(7)$ $C=2860.50(7)$	$D_J=3.39(97)$ $D_{JK}=-28.4(12)$				
	CH_3OCH_2CN trans form	$v=0$	$A=29610(90)$ $B=2470.96(2)$ $C=2348.07(2)$	$D_J=0.433(96)$ $D_{JK}=-17.9(2)$			$v_\tau(O—CH_2)=74(32)$ cm^{-1} from relative intensity measurements	
89	CH_3SCH_2CN gauche form	$v=0$	$A''=7715.324(175)$ $B''=2641.821(29)$ $C''=2174.500(32)$	$\tau_1=-9.121(2707)$ $\tau_2=-0.322(1393)$ $\tau_3=1056(162)$ $\tau_{aaaa}=-2888(84)$ $\tau_{bbbb}=-15.12(242)$ $\tau_{cccc}=-6.584(1448)$	78KEW	78KEW		2.6.3, 45
		$v_\tau=1$	$A=7751.14(220)$ $B=2631.93(4)$ $C=2169.03(4)$	$D_J=2.11(28)$ $D_{JK}=-2.93(68)$			$v_\tau(S—CH_2)=101(30)$ cm^{-1} from relative intensity measurements	
		$v_\tau=2$	$A=7795.11(246)$ $B=2622.83(4)$ $C=2164.18(7)$	$D_J=2.63(41)$ $D_{JK}=-3.93(74)$				

Nr.	Molecule	v	A, B, C MHz	D, H kHz	Ref.	Ref. struct.	Varia, remarks	See also Tab., Nr.
90	$CH_3N{=}CH_2$	$v=0$	$A=52527.265\pm0.12$*) $B=10666.1008\pm0.029$ $C=9377.3587\pm0.023$	$\Delta_J=6.566\pm0.75$*) $\Delta_{JK}=-18.327\pm11$ $\Delta_K=1041.818\pm17$ $\delta_J=1.3334\pm0.068$ $\delta_K=-45.788\pm1.1$ $H_K=3.82034\pm0.270$ $H_{KJ}=0.17672\pm0.025$	77BA2		for structure, see also Vol. II/7 (3, 268) *) for the A species	3, 31
	$CD_3N{=}CD_2$	$v=0$	$A=34577.274\pm0.11$ $B=8161.0836\pm0.027$ $C=7191.1612\pm0.029$	$\Delta_J=3.354\pm0.38$ $\Delta_{JK}=11.294\pm4.6$ $\Delta_K=125.885\pm4.1$ $\delta_J=0.8779\pm0.085$ $\delta_K=-59.730\pm3.7$				
91	$CH_3SO_2{}^{35}Cl$	$v=0$	$A=4636.421$ (5) $B=2787.938$ (5) $C=2682.889$ (5)	$\Delta_J=0.468$ (20) $\Delta_{JK}=1.894$ (2) $\Delta_K=-1.425$ (23) $\delta_J=13.6(2)\cdot10^{-3}$ $\delta_K=-1.041$ (8)	75VAN	75VAN	for structure, see also Vol. II/7 (3, 93)	2.7.3, 99
	$CH_3SO_2{}^{37}Cl$	$v=0$	$A=4636.319$ (8) $B=2714.847$ (6) $C=2615.081$ (11)	$\Delta_J=0.457$ (30) $\Delta_{JK}=1.836$ (8) $\Delta_K=-1.426$ (50) $\delta_J=13.6\cdot10^{-3}$ *) $\delta_K=-1.041$ *)			*) fixed value	
92	CH_3OSO_2F anti form	$v=0$	$A=4882.73$ (4) $B=2758.17$ (2) $C=2735.86$ (1)		77HAR1	77HAR1	for structure, see also Vol. II/7 (3, 101)	2.6.3, 46
		$v_x=1$	$A=4330$ (200) $B=2755.90$ (5) $C=2734.65$ (5)				$\nu\approx200\cdots250\ cm^{-1}$	

Nr.	Molecule	v	A, B, C MHz	D, H kHz	Ref.	Ref. struct.	Varia, remarks	See also Tab., Nr.
93 *6-56* *4-48*	CH_2F_2	$v=0$	$\tilde{A}=49142.818(18)$ $\tilde{B}=10604.7050(44)$ $\tilde{C}=9249.8437(33)$	$\tau_{aaaa}=-2300.61(77)$ $\tau_{bbbb}=-61.51(18)$ $\tau_{cccc}=-27.23(13)$ $T_1=157.5(23)$ *) $T_2=-37.56(19)$	78HIR2	78HIR4	for structure, see also Vol. II/7 (3, 74) cubic force constants determined, r_e(C—H) = 1.084 (3) Å r_e(C—F) = 1.3508 (5) Å $\measuredangle_e$ HCH = 112.8 (3)° $\measuredangle_e$ FCF = 108.49 (6)° [78HIR4] $\nu_4(a_1)=528.5\ cm^{-1}$ *) $T_1=\tau'_{aabb}+\frac{A-B}{A-C}\tau'_{ccaa}$ $T_2=\tau'_{bbcc}+\frac{B-C}{A-C}\tau'_{ccaa}$	2.6.3, 47 2.7.3, 5 2.9.3, 12
		$v_4=1$	$\tilde{A}=49480.557(19)$ $\tilde{B}=10582.8371(49)$ $\tilde{C}=9216.6002(36)$	$\tau_{aaaa}=-2386.10(87)$ $\tau_{bbbb}=-61.00(21)$ $\tau_{cccc}=-26.90(16)$ $T_1=156.8(28)$ *) $T_2=-36.82(23)$				
		$v_4=2$	$\tilde{A}=49815.676(49)$ $\tilde{B}=10560.939(17)$ $\tilde{C}=9181.533(15)$	$\tau_{aaaa}=-2489(14)$ $\tau_{bbbb}=-61.2(14)$ $\tau_{cccc}=-25.9(20)$ $T_1=125(17)$ *) $T_2=-35.8(15)$				
		$v_4=3$	$\tilde{A}=50146.827(90)$ $\tilde{B}=10539.192(87)$ $\tilde{C}=9143.79(10)$	$\tau_{aaaa}=-2.72(10)\cdot 10^3$ $\tau_{bbbb}=-60(13)$ $\tau_{cccc}=-27(11)$ $T_1=-20(140)$ *) $T_2=-56(13)$				
		$v_2=1$	$\tilde{A}=48950.175(48)$ $\tilde{B}=10607.912(22)$ $\tilde{C}=9344.645(16)$	$\tau_{aaaa}=-2247(16)$ $\tau_{bbbb}=-59.3(20)$ $\tau_{cccc}=-45.4(25)$ $T_1=377(24)$ *) $T_2=-35.9(22)$			$\nu_2(a_1)=1508.0\ cm^{-1}$	
		$v_5=1$	$\tilde{A}=48876.621(39)$ $\tilde{B}=10604.055(13)$ $\tilde{C}=9332.448(12)$	$\tau_{aaaa}=-2300(12)$ $\tau_{bbbb}=-64.4(13)$ $\tau_{cccc}=-43.0(16)$ $T_1=457(14)$ *) $T_2=-25.4(13)$			$\nu_5(a_2)=1262.0\ cm^{-1}$	
		$v_7=1$	$\tilde{A}=49893.532(58)$ $\tilde{B}=10612.860(19)$ $\tilde{C}=9153.129(17)$	$\tau_{aaaa}=-3792(15)$ $\tau_{bbbb}=-66.0(14)$ $\tau_{cccc}=-14.4(20)$ $T_1=-79(17)$ *) $T_2=-21.2(16)$			$\nu_7(b_1)=1177.9\ cm^{-1}$	
Cont.								

Nr.	Molecule	v	A, B, C MHz	D, H kHz	Ref.	Ref. struct.	Varia, remarks	See also Tab., Nr.
93 *6–56* *4–48* Cont.	CH_2F_2	$v_8=1$	$\tilde{A}=49023.273\,(42)$ $\tilde{B}=10608.340\,(17)$ $\tilde{C}=\ 9135.556\,(14)$	$\tau_{aaaa}=\ -2327\,(11)$ $\tau_{bbbb}=-65.2\,(15)$ $\tau_{cccc}=\ -9.7\,(17)$ $T_1=-46\,(16)$ *) $T_2=-41.1\,(15)$	78HIR2		$\nu_8(b_2)=1435.0\ cm^{-1}$ *) $T_1=\tau'_{aabb}+\frac{A-B}{A-C}\tau'_{ccaa}$ $T_2=\tau'_{bbcc}+\frac{B-C}{A-C}\tau'_{ccaa}$	
		$v_3=1$	$\tilde{A}=48295.44\,(51)$ $\tilde{B}=10544.10\,(19)$ $C^*=\ 9261.2\,(95)$	fixed at their ground state value			$\nu_3(a_1)=1113.2\ cm^{-1}$ Coriolis interaction between ν_3 and ν_9: $D=10189\,(336)$ MHz $F=\ -116\,(17)$ MHz $\nu_3-\nu_9=628.3\,(78)\cdot 10^3$ MHz C^* = unperturbed rotational constant $\nu_9(b_2)=32681\,172.3\,(30)$ MHz [77KAW1]	
		$v_9=1$	$\tilde{A}=48704.26\,(44)$ $\tilde{B}=10524.25\,(22)$ $C^*=\ 9147.3\,(96)$	fixed at their ground state value				
		$v_{3+4}=1$	$\tilde{A}=48645.900\,(57)$ $\tilde{B}=10530.623\,(68)$ $\tilde{C}=\ 9367.799\,(57)$	$\tau_{cccc}=-162.6\,(11)$ **)			**) other constants fixed on their ground state values	
		$v_{4+5}=1$	$\tilde{A}=49209.655\,(69)$ $\tilde{B}=10581.442\,(76)$ $\tilde{C}=\ 9293.765\,(69)$	**)				
		$v_{4+7}=1$	$\tilde{A}=50231.828\,(1)$ $\tilde{B}=10589.582\,(2)$ $\tilde{C}=\ 9124.164\,(1)$	$\tau_{cccc}=-52.1\,(2)$ **)				
		$v_{4+9}=1$	$\tilde{A}=49040.445\,(94)$ $\tilde{B}=10509.837$ ***) $\tilde{C}=\ 8983.087\,(26)$	$\tau_{cccc}=148.1\,(18)$ **)			***) fixed value	
	$^{13}CH_2F_2$	$v=0$	$\tilde{A}=47730.785\,(39)$ $\tilde{B}=10606.137\,(14)$ $\tilde{C}=\ 9199.117\,(11)$	$\tau_{aaaa}=-2241\,(13)$ $\tau_{bbbb}=\ -63.8\,(12)$ $\tau_{cccc}=\ -29.0\,(17)$ $T_1=\ 123\,(14)$ *) $T_2=-38.6\,(13)$				
Cont.		$v_4=1$	$\tilde{A}=48067.173\,(64)$ $\tilde{B}=10584.479\,(59)$ $\tilde{C}=\ 9165.404\,(37)$	$\tau_{aaaa}=-2300\,(45)$ $\tau_{bbbb}=\ -64.4\,(72)$ $\tau_{cccc}=\ -27.9\,(64)$ $T_1=\ 103\,(77)$ *) $T_2=-39.2\,(63)$				

Nr.	Molecule	v	A, B, C MHz	D, H kHz	Ref.	Ref. struct.	Varia, remarks	See also Tab., Nr.
93 *6-56* *4-48* Cont.	$CHDF_2$	$v=0$	$\tilde{A}=40682.063\,(15)$ $\tilde{B}=10454.4883\,(44)$ $\tilde{C}=9008.9616\,(40)$	$\tau_{aaaa}=-1349.8\,(17)$ $\tau_{bbbb}=-58.57\,(21)$ $\tau_{cccc}=-24.57\,(15)$ $T_1=78.2\,(20)$ *) $T_2=-35.14\,(21)$	75HIR1		*) $T_1=\tau'_{aabb}+\frac{A-B}{A-C}\tau'_{ccaa}$ $T_2=\tau'_{bbcc}+\frac{B-C}{A-C}\tau'_{ccaa}$	
	CD_2F_2	$v=0$	$\tilde{A}=34745.2599\,(77)$ $\tilde{B}=10241.4400\,(24)$ $\tilde{C}=8831.8185\,(26)$	$\tau_{aaaa}=-854.08\,(56)$ $\tau_{bbbb}=-54.251\,(68)$ $\tau_{cccc}=-21.98\,(11)$ $T_1=41.64\,(59)$ *) $T_2=-32.967\,(98)$				
		$v_4=1$	$\tilde{A}=34926.2948\,(64)$ $\tilde{B}=10218.9396\,(19)$ $\tilde{C}=8798.0526\,(16)$	$\tau_{aaaa}=-878.94\,(34)$ $\tau_{bbbb}=-53.718\,(40)$ $\tau_{cccc}=-21.872\,(27)$ $T_1=41.15\,(28)$ *) $T_2=-32.298\,(30)$			$\nu_4(a_1)=521.7\ cm^{-1}$	
		$v_4=2$	$\tilde{A}=35111.431\,(30)$ $\tilde{B}=10197.3207\,(86)$ $\tilde{C}=8765.0036\,(92)$	$\tau_{aaaa}=-917.7\,(35)$ $\tau_{bbbb}=-53.52\,(30)$ $\tau_{cccc}=-21.94\,(46)$ $T_1=35.8\,(26)$ *) $T_2=-33.05\,(48)$				
		$v_3=1$	$\tilde{A}=35227.665\,(82)$ $\tilde{B}=10230.237\,(23)$ $\tilde{C}=8944.370\,(25)$	$\tau_{aaaa}=-1476\,(11)$ $\tau_{bbbb}=-62.35\,(95)$ $\tau_{cccc}=-108.5\,(12)$ $T_1=940.5\,(64)$ *) $T_2=-47.1\,(13)$			$\nu_3(a_1)=1026.5\ cm^{-1}$	
		$v_5=1$	$\tilde{A}=34634.312\,(36)$ $\tilde{B}=10220.502\,(11)$ $\tilde{C}=8730.420\,(11)$	$\tau_{aaaa}=-863.4\,(40)$ $\tau_{bbbb}=-53.51\,(37)$ $\tau_{cccc}=2.52\,(56)$ $T_1=-30.1\,(31)$ *) $T_2=-33.97\,(59)$			$\nu_5(a_2)=907.3\ cm^{-1}$	
Cont.		$v_7=1$	$\tilde{A}=34129.131\,(17)$ $\tilde{B}=10192.1602\,(48)$ $\tilde{C}=8924.4522\,(49)$	$\tau_{aaaa}=-203.8\,(22)$ $\tau_{bbbb}=-52.86\,(15)$ $\tau_{cccc}=-44.95\,(22)$ $T_1=-48.5\,(16)$ *) $T_2=-59.54\,(35)$			$\nu_7(b_1)=962.1\ cm^{-1}$	

Nr.	Molecule	v	A, B, C MHz	D, H kHz	Ref.	Ref. struct.	Varia, remarks See also Tab., Nr.
93 *6-56* *4-48* Cont.	CD_2F_2	$v_9=1$	$\tilde{A}=34605.36\,(11)$	$\tau_{aaaa}=-988\,(14)$	75HIR1		$\nu_9(b_2)=1002.4\ cm^{-1}$
			$\tilde{B}=10259.664\,(28)$	$\tau_{bbbb}=\ -63.6\,(15)$			*) $T_1=\tau'_{aabb}+\frac{A-B}{A-C}\tau'_{ccaa}$
			$\tilde{C}=\ 8690.739\,(30)$	$\tau_{cccc}=\ 60.0\,(16)$			$T_2=\tau'_{bbcc}+\frac{B-C}{A-C}\tau'_{ccaa}$
				$T_1=-579\,(11)$ *)			
				$T_2=\ -4.9\,(17)$			
		$v_2=1$	$\tilde{A}=34662.001\,(210)$	fixed at their ground state value			$\nu_2(a_1)=1170.3\ cm^{-1}$
			$\tilde{B}=10220.5764\,(650)$				Coriolis interaction between
			$C^*=\ 8795.91\,(120)$				ν_2 and ν_8:
		$v_8=1$	$\tilde{A}=34655.450\,(220)$	fixed at their ground state value			$D\ =13912.7\,(210)$ MHz
			$\tilde{B}=10211.0689\,(680)$				$\Delta E=360147\,(36)$ MHz
			$C^*=\ 8800.12\,(120)$				$\nu_8(b_2)=1158.3\ cm^{-1}$
	$^{13}CD_2F_2$	$v=0$	$\tilde{A}=34138.387\,(19)$	$\tau_{aaaa}=-840.6\,(15)$			
			$\tilde{B}=10242.7414\,(59)$	$\tau_{bbbb}=\ -53.73\,(17)$			
			$\tilde{C}=\ 8792.5874\,(57)$	$\tau_{cccc}=\ -21.52\,(27)$			
				$T_1=\ 43.4\,(16)$ *)			
				$T_2=-32.22\,(29)$			
94 *4-49*	$CH_2{}^{35}Cl_2$	$v=0$	$A=32002.2653\,(83)$	$\Delta_J=\ 1.39208\,(75)$	81DAV	81DAV	for structure, see also Vol. II/7 (3, 70)
			$B=\ 3320.27890\,(94)$	$\Delta_{JK}=-26.176\,(18)$			
			$C=\ 3065.22685\,(95)$	$\Delta_K=\ 473.16\,(16)$			harmonic force field calculated [81DAV]
				$\delta_J=\ 0.176992\,(78)$			
				$\delta_K=\ 5.365\,(11)$			
				$H_{JK}=\ -0.0418\,(27)\cdot10^{-3}$			
				$H_{KJ}=\ -1.09\,(13)\cdot10^{-3}$			
				$H_K=\ 13.3\,(27)\cdot10^{-3}$			
				$h_J=\ 0.236\,(24)\cdot10^{-6}$			
	$CH_2{}^{37}Cl_2$	$v=0$	$A=31753.92\,(11)$ *)		79HAR		*) uncertainties are 2 σ
			$B=\ 3142.81\,(12)$				
			$C=\ 2911.25\,(11)$				
	$CH_2{}^{35}Cl^{37}Cl$	$v=0$	$A=31878.25\,(7)$ *)				
			$B=\ 3231.38\,(11)$				
Cont.			$C=\ 2988.09\,(7)$				

Nr.	Molecule	v	A, B, C MHz	D, H kHz	Ref.	Ref. struct.	Varia, remarks	See also Tab., Nr.
94 *4-49* Cont.	$CD_2{}^{35}Cl_2$	$v=0$	$A=23676.4143(63)$ $B=3285.16444(91)$ $C=2994.05616(94)$	$\Delta_J=1.2977(24)$ $\Delta_{JK}=-15.884(50)$ $\Delta_K=204.05(26)$ $\delta_J=0.19014(26)$ $\delta_K=4.079(24)$ $H_{KJ}=-0.956(41)\cdot 10^{-3}$ $H_K=8.92(76)\cdot 10^{-3}$ $h_J=0.859(49)\cdot 10^{-6}$	81DAV			
	$^{13}CH_2{}^{35}Cl_2$	$v=0$	$A=30746.544(52)$ $B=3320.656(53)$ $C=3053.487(52)$	$\Delta_J=1.3794$ **) $\Delta_{JK}=-25.0346$ $\Delta_K=446.7206$ $\delta_J=0.1834$ $\delta_K=5.3693$			**) constrained to the values predicted from the force field	
95 *6-57*	CH_2Br_2					75CH	for structure, see also Vol. II/7 (3, 68)	
96 *4-50*	$CH_2(CN)_2$	$v=0$	$\mathscr{A}=20882.7537(170)$ *) $\mathscr{B}=2942.3003(31)$ $\mathscr{C}=2616.7225(31)$	$\Delta_J=1.855455\pm0.014$ $\Delta_{JK}=-67.9218\pm0.27$ $\Delta_K=862.1628\pm1.3$ $\delta_J=0.4892607\pm0.0016$ $\delta_K=6.7501\pm0.29$ $H_{KJ}=(-7.7964\pm1.6)\cdot 10^{-3}$ $H_K=(1.2644\pm0.20)\cdot 10^{-1}$ $h_J=(5.3893\pm1.1)\cdot 10^{-6}$ $h_{JK}=(-4.9014\pm0.83)\cdot 10^{-4}$ $h_K=(1.5994\pm0.73)\cdot 10^{-2}$	74COO3		for structure, see also Vol. II/7 (3, 350) *) uncertainties are 2 σ	
97 *4-51* Cont.	$CH_2F^{35}Cl$	$v=0$	$\tilde{A}=41811.250(131)$ $\tilde{B}=5715.972(19)$ $\tilde{C}=5194.889(18)$	$d_J=-18.4011(2716)$ $d_{JK}=-734.316(15531)$ $d_K=-3170.95(11428)$ $d_{WJ}=4.1(1)\cdot 10^{-6}$ $d_{WK}=1.019(30)\cdot 10^{-4}$	75NAN	75NAN	for structure, see also Vol. II/7 (3, 71)	2.6.3, 48 2.7.3, 100

Nr.	Molecule	v	A, B, C MHz	D, H kHz	Ref.	Ref. struct.	Varia, remarks	See also Tab., Nr.
97 *4-51* Cont.	$CH_2F^{37}Cl$	$v=0$	$\tilde{A}=41738.954(132)$ $\tilde{B}=5580.782(18)$ $\tilde{C}=5081.904(17)$	$d_J=-17.2450(2666)$ $d_{JK}=-738.714(17051)$ $d_K=-3294.71(12588)$ $d_{WJ}=3.9(1)\cdot10^{-6}$ $d_{WK}=1.057(34)\cdot10^{-4}$	75NAN			
98 *6-58*	$CH_2{}^{79}BrCN$	$v=0$	$A=21874.07(75)$ $B=2194.82(10)$ $C=2023.04(8)$	$\Delta_J=-0.3(5)$ $\Delta_{JK}=-26(15)$ $\Delta_K=(-93.99\pm0.12)\cdot10^3$ $\delta_J=-0.13(25)$ $\delta_K=-1040.7(220)$	76GUM	76GUM		2.7.3, 149
	$CH_2{}^{81}BrCN$	$v=0$	$A=21874.28(84)$ $B=2179.95(10)$ $C=2008.45(9)$	$\Delta_J=-1.1(5)$ $\Delta_{JK}=-51(15)$ $\Delta_K=-67.42(12)\cdot10^3$ $\delta_J=0.02(25)$ $\delta_K=-0.57(22)\cdot10^3$				
99	CH_2ICN	$v=0$	$A=20037.32\pm3.00$ $B=1747.95(21)$ $C=1622.87(23)$		75AUL	75AUL		2.7.3, 158
100 *6-58a*	CH_2OHCN	$v=0$	$A=33605.57(22)$ $B=4840.44(4)$ $C=4377.60(3)$	$d_J=2.7(8)$ $d_{JK}=-81(5)$	73CAZ3		gauche form $\Delta E_{01}=110700$ $G_a=1000$ $G_b=289(3)$ MHz OH torsion potential function determined	2.7.3, 23 3, 32
		$v_\tau=1$	$A=33611.68(22)$ $B=4835.51(4)$ $C=4376.44(3)$					
	CH_2ODCN	$v=0$	$A=30651.04(22)$ $B=4726.75(2)$ $C=4262.36(2)$	$d_J=3.0(3)$ $d_{JK}=-53(1)$			$\Delta E_{01}=16749.65(29)$ $G_a=87(8)$ $G_b=94(1)$ MHz	
		$v_\tau=1$	$A=30657.23(22)$ $B=4724.51(2)$ $C=4262.13(2)$					

Nr.	Molecule	v	A, B, C MHz	D, H kHz	Ref.	Ref. struct.	Varia, remarks	See also Tab., Nr.
101 *6–58b*	NH_2CH_2CN	v=0	$A=30245.778\,(27)$ $B=4761.048\,(13)$ $C=4310.769\,(8)$	$D_J=3.29\,(38)$ $D_{JK}=-55.8\,(31)$	77BRO	73PIC2	for structure, see also Vol. II/7 (3, 241)	2.6.3, 49 2.7.3, 24
	$NHDCH_2CN$	v=0	$A=27585.42\,(70)$ $B=4631.179\,(60)$ $C=4188.763\,(60)$		73PIC2		N—H bond gauche to CN group	
	ND_2CH_2CN	v=0	$A=25380.65\,(70)$ $B=4504.939\,(60)$ $C=4082.280\,(60)$					
102	$CH_2{=}SO$	v=0	$A''=40404.513\,(30)$ $B''=9394.746\,(7)$ $C''=7607.111\,(6)$	$\tau_1=173.8\,(19)$ $\tau_2=13.89\,(34)$ $\tau_3=1117.9\,(48)$ $\tau_{aaaa}=-3201.8\,(79)$ $\tau_{bbbb}=-41.14\,(15)$ $\tau_{cccc}=-10.73\,(9)$ $H_J=-2.7\,(2)\cdot10^{-6}$ $H_{JK}=1.15\,(62)\cdot10^{-3}$ $H_{KJ}=-7.1\,(18)\cdot10^{-3}$ $H_K=50.6\,(51)\cdot10^{-3}$ $h_J=0.035\,(12)\cdot10^{-3}$ $h_{JK}=-1.18\,(76)\cdot10^{-3}$ $h_K=33\,(16)\cdot10^{-3}$	76PEN2	76PEN2	relative intensities of vibrational satellites measured	2.6.3, 50 3, 33
	CHD=SO, syn	v=0	$A=34644.5$ *) $B=9042.578\,(32)$ $C=7157.445\,(32)$				*) calculated using the inertial defect of the normal species	
	CHD=SO, anti	v=0	$A=40285.9$ *) $B=8584.340\,(20)$ $C=7063.348\,(19)$					
	$^{13}CH_2{=}SO$	v=0	$A=39919.7$ *) $B=9079.359\,(60)$ $C=7382.672\,(60)$					
Cont.	$CH_2{=}^{34}SO$	v=0	$A=39390.4$ *) $B=9394.777\,(28)$ $C=7570.531\,(28)$					

Nr.	Molecule	v	A, B, C MHz	D, H kHz	Ref.	Ref. struct.	Varia, remarks	See also Tab., Nr.
102 Cont.	$CH_2{=}S^{18}O$	$v=0$	$A=39316.7$ *) $B=$ 8911.215 (26) $C=$ 7250.855 (26)		76PEN2		*) calculated using the inertial defect of the normal species	
	$CD_2{=}SO$	$v=0$	$A=34401.2$ *) $B=$ 8309.722 (34) $C=$ 6681.283 (34)					
	$CD_2{=}^{34}SO$	$v=0$	$A=33599.5$ *) $B=$ 8307.632 (75) $C=$ 6649.124 (73)					
103 *6-58c*	CH_2NH	$v=0$	$A''=196211.046$ (45) $B''=$ 34642.3956 (72) $C''=$ 29352.2326 (69)	$\tau_1=-3017.4$ (82) $\tau_2=-481.0$ (13) $\tau_3=$ 14.66 (3) $\cdot 10^3$ $\tau_{aaaa}=-28152$ (13) $\tau_{bbbb}=-319.85$ (48) $\tau_{cccc}=-160.24$ (36) $H_J=(-0.1109\pm0.93)\cdot10^{-4}$ $H_{JK}=(0.41472\pm0.27)\cdot10^{-2}$ $H_{KJ}=(0.22216\pm0.36)\cdot10^{-1}$ $H_K=0.59874\pm0.30$ $h_J=(0.4996\pm0.21)\cdot10^{-4}$ $h_{JK}=(0.6205\pm0.19)\cdot10^{-2}$ $h_K=0.25551\pm0.053$	73KIR2	77PEA		2.6.3, 53 2.7.3, 25
	CH_2ND	$v=0$	$A''=157673.849$ (39) $B''=$ 32069.3621 (120) $C''=$ 26563.8470 (120)	$\tau_1=-1863.4$ (177) $\tau_2=-340.47$ (304) $\tau_3=$ 10.1 (1) $\cdot 10^3$ $\tau_{aaaa}=-21487$ (17) $\tau_{bbbb}=-303.59$ (134) $\tau_{cccc}=-129.30$ (80)	77PEA			
	$^{13}CH_2NH$	$v=0$	$A=196195.211$ (45) *) $B=$ 33747.872 (7) $C=$ 28707.045 (7)				*) the centrifugal distortion constants were used to calculate rigid rotor transition frequencies	
Cont.	$CH_2{}^{15}NH$	$v=0$	$A=195738.28$ (18) *) $B=$ 33736.130 (30) $C=$ 28688.616 (30)					

Nr.	Molecule	v	A, B, C MHz	D, H kHz	Ref.	Ref. struct.	Varia, remarks	See also Tab., Nr.
103 *6-58c* Cont.	CD_2NH	$v=0$	$A=117054.69(16)$ *) $B=29204.67(13)$ $C=23297.233(80)$		77PEA		*) the centrifugal distortion constants were used to calculate rigid rotor transition frequencies	
	CD_2ND	$v=0$	$A=102117.65(17)$ *) $B=27108.397(80)$ $C=21355.899(70)$					
104	$CH_2{=}N{-}CN$ *)	$v=0$	$A=63373.11(42)$ $B=5449.418(36)$ $C=5009.490(39)$	$\Delta_J=2.462(36)$ $\Delta_{JK}=-130.06(44)$ $\Delta_K=5794(83)$ $\delta_J=0.4761(77)$ $\delta_K=39(16)$ $H_{KJ}=-23.2(62)\cdot10^{-3}$	80BAK1	80BAK1	*) HCN dimer, $\Delta=0.170$ amu Å^2	3, 34
	$CD_2{=}N{-}CN$	$v=0$	$A=50640.377(99)$ $B=4993.206(14)$ $C=4538.181(11)$	$\Delta_J=2.066(92)$ $\Delta_{JK}=-77.2(23)$ $\Delta_K=2653(24)$ $\delta_J=0.4313(81)$ $\delta_K=24.9(20)$ $H_{KJ}=0$ **)			**) assumed	
105 *6-60*	$CH_3{-}N{=}ND$							2.6.3, 51
106	cis-$CH_3N{=}NCH_3$	$v=0$	$A=16285.20(44)$ *) $B=6739.10(22)$ $C=5067.39(20)$	$\Delta_J=11.1$ **) $\Delta_{JK}=-22.1$ $\Delta_K=5.6$ $\delta_J=-5.2$ $\delta_K=5.5$	79STE	79STE	for structure, see also Vol. II/7 (3, 292) *) uncertainties are 3 σ, AA species. **) the centrifugal distortion uncertainties are much larger than the constants	2.6.3, 52 2.8, 115
	cis-$CD_3N{=}NCD_3$	$v=0$	$A=12471.78(13)$ *) $B=5344.90(16)$ $C=4158.98(14)$	$\Delta_J=11.5$ **) $\Delta_{JK}=12.4$ $\Delta_K=-3.3$ $\delta_J=-0.2$ $\delta_K=-3.7$				
		$v_\tau=1$	$A-C=8615.98$ ***) $\kappa=-0.72934$				$v_\tau=120(30)$ cm^{-1} ***) AA species	

Nr.	Molecule	v	A, B, C MHz	D, H kHz	Ref.	Ref. struct.	Varia, remarks	See also Tab., Nr.
107	$CF_2{}^{35}Cl_2$	$v=0$	$A=4118.896\,(29)$	$\Delta_J = 0.480\,(56)$	82JON	77TA1		2.7.3, 101
6-61			$B=2638.704\,(26)$	$\Delta_{JK} = -0.409\,(28)$				
			$C=2233.691\,(26)$	$\Delta_K = 1.65\,(15)$				
				$\delta_J = 0.1192\,(18)$				
		$v_6=1$	$A=4115.545\,(36)$	$\Delta_J = 0.434\,(56)$				
			$B=2630.948\,(27)$	$\Delta_{JK} = -0.586\,(31)$			$\nu_6=923.23956\,(12)\ cm^{-1}$	
			$C=2225.758\,(28)$	$\Delta_K = 1.61\,(16)$				
				$\delta_J = 0.1289\,(33)$				
	$CF_2{}^{35}Cl^{37}Cl$	$v=0$	$A=4092.07\,(7)$		77TA1			
			$B=2582.30\,(5)$					
			$C=2185.54\,(6)$					
	${}^{13}CF_2{}^{35}Cl_2$	$v=0$	$A=4115.73\,(3)$					
			$B=2638.95\,(2)$					
			$C=2232.86\,(2)$					
108	HC(=O)H	$v=0$	$A=281970.5723\,(240)$ *)	$\Delta_J = 75.295319 \pm 0.021$	79WIN		for structure,	2.6.3, 54
6-62			$B= 38836.045576$	$\Delta_{JK} = 1290.5059 \pm 0.37$			see also	2.7.3, 6
4-58			± 0.0013	$\Delta_K = 19423.74 \pm 7$			Vol. II/7 (3, 79)	2.9.3, 14
			$C= 34002.203442$	$\delta_J = 10.456758 \pm 0.0009$			*) combination of MW, IR	
			± 0.0012	$\delta_K = 1026.0307 \pm 0.25$			and UV data; error limits	
				$H_J = (3.1435 \pm 2.1) \cdot 10^{-5}$			are 3σ	
				$H_{JK} = (2.90192 \pm 0.069) \cdot 10^{-2}$				
				$H_{KJ} = -0.11218 \pm 0.028$				
				$H_K = 4.499 \pm 0.2$				
				$h_J = (4.2399 \pm 0.17) \cdot 10^{-5}$				
				$h_{JK} = (1.56659 \pm 0.031) \cdot 10^{-2}$				
				$h_K = 1.37212 \pm 0.018$				
		$v_2=1$	$A=281800.95\,(87)$ **)	$\Delta_K = 19480\,(23)$ **)	81SWE		ν_2 (C—O stretch) =	
			$B= 38611.05\,(39)$	$\Delta_{JK} = 1299.6\,(36)$			$1746.0090\,(2)\ cm^{-1}$	
			$C= 33737.53\,(42)$	$\Delta_J = 71.7\,(11)$			**) diode laser results, all other	
Cont.				$H_K = 4.40\,(16)$			distortion terms held to	
							ground state values	

Nr.	Molecule	v	A, B, C MHz	D, H kHz	Ref.	Ref. struct.	Varia, remarks	See also Tab., Nr.
108 *6–62* *4–58* Cont.	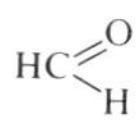	$v_3=1$	$A=9.46697\,(39)$ ***) $B=1.300078\,(7)$ $C=1.129318\,(10)$	$\Delta_J=2.825\,(120)\cdot10^{-6}$ ***) $\Delta_{JK}=2.421\,(193)\cdot10^{-5}$ $\Delta_K=7.586\,(78)\cdot10^{-4}$ $\delta_J=5.11\,(13)\cdot10^{-7}$ $\delta_K=5.00\,(11)\cdot10^{-5}$	81ALL, 78DAN		***) simultaneous analysis of ν_3, ν_5 and ν_6 all values in cm^{-1}, uncertainties are 3 σ $\nu_3=1500.176\ cm^{-1}$; $\nu_4=1167.258\ cm^{-1}$; $\nu_6=1249.091\ cm^{-1}$ $G^a_{46}=10.0728\,(535)\ cm^{-1}$ $F^{bc}_{46}=0.002498\,(119)\ cm^{-1}$ $G^b_{34}=1.28841\,(76)\ cm^{-1}$ $G^c_{36}=0.75368\,(154)\ cm^{-1}$ [81ALL]	
		$v_4=1$	$A=9.2890\,(129)$ ***) $B=1.287243\,(6)$ $C=1.135716\,(3)$	$\Delta_J=2.412\,(55)\cdot10^{-6}$ ***) $\Delta_{JK}=2.413\,(83)\cdot10^{-5}$ $\Delta_K=2.346\,(142)\cdot10^{-4}$ $\delta_J=2.71\,(8)\cdot10^{-7}$ $\delta_K=2.33\,(11)\cdot10^{-5}$				
		$v_6=1$	$A=9.4881\,(129)$ ***) $B=1.298244\,(5)$ $C=1.129886\,(10)$	$\Delta_J=2.539\,(68)\cdot10^{-6}$ ***) $\Delta_{JK}=8.025\,(131)\cdot10^{-5}$ $\Delta_K=11.861\,(148)\cdot10^{-4}$ $\delta_J=3.53\,(9)\cdot10^{-7}$ $\delta_K=4.64\,(9)\cdot10^{-5}$				
		$v_5=1$	$A=9.22787\,(34)$ **) $B=1.29349\,(49)$ $C=1.13061\,(56)$	$\Delta_K=6.58\,(36)\cdot10^{-4}$ **) $\Delta_{JK}=1.174\,(32)\cdot10^{-4}$	81ALL		$\nu_5=2843.321\,(4)\ cm^{-1}$ **) laser Stark results, in cm^{-1} uncertainties are 3 σ, Δ_K and Δ_{JK} are fixed values. Ground state values for Δ_J, δ_J, δ_K and H_K were assumed.	
	$H_2C^{17}O$	$v=0$	$A=281965.085\pm3.0$ +) $B=37812.28701\pm0.045$ $C=33214.523449\pm0.031$	$\Delta_J=71.07647\pm0.56$ $\Delta_{JK}=1239.44\pm18$ $\Delta_K=19412.5$ ++) $\delta_J=9.76778\pm0.087$ $\delta_K=987.28\pm19$ $H_J=2.03\cdot10^{-5}$ ++) $H_{JK}=2.8426\cdot10^{-2}$ ++) $H_{KJ}=1.607\pm3.9$ $H_K=3.37$ ++) $h_J=4.193\cdot10^{-5}$ ++) $h_{JK}=(3.195\pm3.4)\cdot10^{-2}$ $h_K=1.754\pm1.7$	80COR1, 80DAV1		+) uncertainties are 3 σ ++) constrained to H_2CO value	
Cont.								

Nr.	Molecule	v	A, B, C MHz	D, H kHz	Ref.	Ref. struct.	Varia, remarks	See also Tab., Nr.
108 *6-62* *4-58* Cont.	$H_2C^{18}O$	$v=0$	$A = 281961.9352 \pm 0.39$ ⁺) $B = 36904.16926 \pm 0.0066$ $C = 32511.53108 \pm 0.0063$	$\Delta_J = 68.455366 \pm 0.100$ ⁺⁺⁺) $\Delta_{JK} = 1177.4919 \pm 0.96$ $\Delta_K = 19412.5$ ⁺⁺) $\delta_J = 9.09612 \pm 0.016$ $\delta_K = 943.492 \pm 3.0$ $H_{KJ} = -0.06360 \pm 0.057$ $h_J = (4.7132 \pm 0.54) \cdot 10^{-5}$ $h_{JK} = (1.2720 \pm 0.58) \cdot 10^{-2}$ $h_K = 1.3547 \pm 0.30$	80COR2, 80DAV1		⁺) uncertainties are 3 σ ⁺⁺) constrained to H_2CO value ⁺⁺⁺) H_J, H_{JK}, H_K are constrained to the H_2CO values	
	$H_2{}^{13}CO$	$v=0$	$A = 281993.2578 \pm 0.135$ *) $B = 37811.088658 \pm 0.0025$ $C = 33213.978976 \pm 0.0025$	$\Delta_J = 71.1525480 \pm 0.0061$ $\Delta_{JK} = 1257.4888 \pm 0.64$ $\Delta_K = 19499.34 \pm 22$ $\delta_J = 9.631303 \pm 0.0022$ $\delta_K = 987.4468 \pm 1.00$ $H_J = (4.9849 \pm 3.9) \cdot 10^{-5}$ $H_{JK} = (2.74352 \pm 0.145) \cdot 10^{-2}$ $H_{KJ} = -0.06978 \pm 0.150$ $H_K = 5.052 \pm 0.9$ $h_J = (3.5772 \pm 0.46) \cdot 10^{-5}$ $h_{JK} = (1.45414 \pm 0.130) \cdot 10^{-2}$ $h_K = 1.38185 \pm 0.085$	79WIN, 80COR2		*) uncertainties are 3 σ, combination of MW, IR and UV data	
		$v_2=1$	$A = 281798.44\ (51)$ **) $B = 37602.55\ (48)$ $C = 32963.11\ (51)$	$\Delta_K = 19558\ (11)$ **) $\Delta_{JK} = 1253.3\ (19)$ $\Delta_J = 67.54\ (30)$ $H_K = 4.974\ (51)$	81SWE		$\nu_2 = 1707.9814\ (2)\ cm^{-1}$ **) diode laser results, all other distortion terms held to ground state values	
Cont.	$H_2{}^{13}C^{17}O$	$v=0$	$A = 281987.345 \pm 1.9$ ⁺) $B = 36776.79059 \pm 0.025$ $C = 32412.92097 \pm 0.019$	$\Delta_J = 67.97327 \pm 0.26$ ⁺⁺⁺) $\Delta_{JK} = 1203.80 \pm 11$ $\Delta_K = 19412.5$ ⁺⁺) $\delta_J = 8.96484 \pm 0.063$ $\delta_K = 947.11 \pm 11$ $H_{KJ} = 1.231 \pm 2.4$ $h_{JK} = (2.501 \pm 2.3) \cdot 10^{-2}$ $h_K = 1.7360 \pm 0.97$	80COR1, 80DAV1		⁺) uncertainties are 3 σ ⁺⁺) constrained to H_2CO value ⁺⁺⁺) H_J, H_{JK}, H_K and h_J are constrained to the H_2CO values	

Nr.	Molecule	v	A, B, C MHz	D, H kHz	Ref.	Ref. struct.	Varia, remarks	See also Tab., Nr.
108 *6-62* *4-58* Cont.	$H_2{}^{13}C^{18}O$	$v=0$	$A=281984.9973\pm0.93$ $^{+})$ $B=35859.25569\pm0.0100$ $C=31697.86825\pm0.0096$	$\Delta_J=64.83703\pm0.39$ $^{+++})$ $\Delta_{JK}=1142.373\pm2.6$ $\Delta_K=19412.5$ $^{++})$ $\delta_J=8.34127\pm0.022$ $\delta_K=900.876\pm3.9$ $H_{KJ}=-0.1597\pm0.53$ $h_J=(4.6949\pm0.58)\cdot10^{-5}$ $h_{JK}=(1.2957\pm0.87)\cdot10^{-2}$ $h_K=1.1532\pm0.39$	80COR2, 80DAV1		$^{+})$ uncertainties are 3 σ $^{++})$ constrained to H_2CO value $^{+++})$ H_J, H_{JK} and H_K are constrained to the H_2CO values	
	HDCO	$v=0$	$A=198111.99$ (6) $B=34910.84$ (2) $C=29561.07$ (2)	$\Delta_J=61.9$ (3) $\Delta_{JK}=711$ (2) $\Delta_K=602(3)\cdot10^1$ $\delta_J=11.68$ (2) $\delta_K=929$ (3) $H_J=-0.031$ (1) $H_{JK}=0.66$ (1) $H_{KJ}=-4.4$ (2) $H_K=991.1$ (78) $h_J=-0.108(5)\cdot10^{-2}$ $h_K=49.6$ (4)	78DAN			
		$v_5=1$	$A=203856$ $B=34945$ $C=29423$	$\Delta_J=68$ $\Delta_{JK}=-0.47\cdot10^3$ $\Delta_K=5.97\cdot10^3$ *) $\delta_J=9$ $\delta_K=928$ *)	79DAN		ν_5(CHD rock) $=1028.4$ cm^{-1} *) fixed on ground state value Coriolis interaction between ν_5 and ν_6 $\Delta=\nu_6-\nu_5=908161$ MHz $G_a=178240$ MHz $G_b=13969$ MHz $F_{bc}=44$ MHz [79DAN] ν_6(out-of-plane bend) $=1058.8$ cm^{-1}	
		$v_6=1$	$A=190729$ $B=34677$ $C=29612$	$\Delta_J=50$ $\Delta_{JK}=1.92\cdot10^3$ $\Delta_K=5.97\cdot10^3$ *) $\delta_J=-4$ $\delta_K=928$ *)				
Cont.	$HD^{13}CO$	$v=0$	$A=197569$ (12) $B=34193.06$ (3) $C=29034.02$ (3)	$\Delta_J=62$ (1) $\Delta_{JK}=723$ (3) $\Delta_K=6.02\cdot10^3$ **) $\delta_J=10$ (1) $\delta_K=929$ **)	78DAN		**) constrained to HDCO value	

Nr.	Molecule	v	A, B, C MHz	D, H kHz	Ref.	Ref. struct.	Varia, remarks	See also Tab., Nr.
108 *6-62* *4-58* Cont.	D_2CO	$v=0$	$A=141653.30(5)$ $B=32283.57(1)$ $C=26185.34(1)$	$\Delta_J=53.1(2)$ $\Delta_{JK}=623(2)$ $\Delta_K=4.40(3)\cdot10^3$ $\delta_J=11.43(2)$ $\delta_K=520(4)$ $H_J=0$ ***) $H_{JK}=-0.29(24)\cdot10^{-2}$ $H_{KJ}=0.3(1)$ $H_K=-6.2(11)$ $h_J=0$ ***) $h_{JK}=0$ ***) $h_K=0.20(8)$	78DAN		***) fixed value	
		$v_2=1$	$A=141696.6\pm7$ *) $B=32068.4\pm7$ $C=25998.6\pm10$	ground state values assumed	73JOH		$\nu_2=1701.620(1)\ cm^{-1}$, C—O stretch *) laser Stark	
		$v_3=1$	$A=142466(13)$ $B=32732.8(2)$ $C=26216.8(2)$	$\Delta_J=74(4)$ $\Delta_{JK}=-1.82(1)\cdot10^3$ $\Delta_K=4.40\cdot10^3$ *) $\delta_J=23(2)$ $\delta_K=520$ *)	78DAN		$\nu_3=1105.7\ cm^{-1}$ *) fixed on ground state value	
		$v_4=1$	$A=139612(23)$ $B=31872.3(7)$ $C=26239.2(7)$	$\Delta_J=37(3)$ $\Delta_{JK}=-0.60(1)\cdot10^3$ $\Delta_K=4.40\cdot10^3$ *) $\delta_J=0.4(20)$ $\delta_K=520$ *)			$\nu_4=938.0345(6)\ cm^{-1}$ [77COF1] *) fixed on ground state value Coriolis interaction between ν_4 and ν_6: $F_{bc}=7(4)$ MHz assuming $\Delta=\nu_6-\nu_4=1535459$ MHz $G_a=136178(770)$ MHz from [77COF1]	
		$v_6=1$	$A=143478(18)$ $B=32384.8(7)$ $C=25995.2(7)$	$\Delta_J=63(4)$ $\Delta_{JK}=3.61(1)\cdot10^3$ $\Delta_K=4.40\cdot10^3$ *) $\delta_J=13(2)$ $\delta_K=520$ *)			$\nu_6=989.2519(18)\ cm^{-1}$ [77COF1]	
Cont.	$D_2C^{18}O$	$v=0$	$A=141648(3)$ $B=30595.86(2)$ $C=25063.12(2)$	$\Delta_J=47.9(3)$ $\Delta_{JK}=563(3)$ $\Delta_K=4.40\cdot10^3$ **) $\delta_J=10.1(2)$ $\delta_K=520$ **)			**) constrained to D_2 CO value	

Nr.	Molecule	v	A, B, C MHz	D, H kHz	Ref.	Ref. struct.	Varia, remarks	See also Tab., Nr.
108 *6-62* *4-58* Cont.	$D_2{}^{13}CO$	$v=0$	$A=141661.3\,(8)$ $B=\ 31733.36\,(2)$ $C=\ 25822.50\,(2)$	$\Delta_J\ =\ 51.2\,(3)$ $\Delta_{JK}=632\,(1)$ $\Delta_K\ =\ 4.40\cdot 10^3$ **) $\delta_J\ =\ 10.50\,(8)$ $\delta_K\ =520$ **) $H_{JK}=\ 0.16\,(1)$ $H_{KJ}=\ 0.3$ **) $H_K\ =-248.3\,(10)$ $h_K\ =\ -7.9\,(6)$	78DAN		**) constrained to D_2CO value	
108a *6-62a* *4-59*	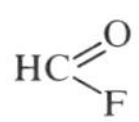	$v=0$	$A=91156.513\,(30)$ $B=11760.1507\,(45)$ $C=10396.7610\,(39)$	$D_J\ =\ 9.739\,(24)$ $D_{JK}=-105.92\,(49)$ $D_K\ =\ 3170.3\,(32)$ $d_1\ =\ -1.6773\,(15)$ $d_2\ =\ -0.10415\,(24)$ $H_J\ =\ -0.014\,(18)\cdot 10^{-3}$ $H_{JK}=\ -0.49\,(70)\cdot 10^{-3}$ $H_{KJ}=-43.0\,(160)\cdot 10^{-3}$ $H_K\ =0.7740\,(2700)$ $h_1\ =0.0029\,(13)\cdot 10^{-3}$ $h_2\ =0.00093\,(36)\cdot 10^{-3}$ $h_3\ =0.000163\,(60)\cdot 10^{-3}$ *)	81JON		*) calculated from planarity condition	
		$v_5=1$	$A=91848.213\,(93)$ $B=11746.915\,(12)$ $C=10365.715\,(11)$	$D_{JK}=-106.72\,(12)$ $D_K\ =\ 3316.8\,(23)$ $d_1\ =-1.69631\,(86)$ $d_2\ =-0.11951\,(40)$ **)			$v_5=660\ cm^{-1}$, FCO bend **) the other constants are transferred from ground state	
Cont.	DCOF	$v=0$	$A=65096.541\,(23)$ $B=11761.7594\,(50)$ $C=\ 9941.7422\,(38)$	$D_J\ =\ 9.050\,(28)$ $D_{JK}=-37.01\,(44)$ $D_K\ =1191.4\,(15)$ $d_1\ =-1.9738\,(31)$ $d_2=-0.18350\,(65)$ $H_J\ =-0.028\,(48)\cdot 10^{-3}$ $H_{JK}=-2.1\,(15)\cdot 10^{-3}$ $H_{KJ}=\ 16.0\,(150)\cdot 10^{-3}$ $H_K\ =\ 0.1000\,(1200)$				

Nr.	Molecule	v	A, B, C MHz	D, H kHz	Ref.	Ref. struct.	Varia, remarks	See also Tab., Nr.
108a *6–62a* *4–59* Cont.	DCOF	$v_5=1$	$A=65379.817\,(49)$ $B=11749.0522\,(87)$ $C=\ 9907.9537\,(82)$	$h_1 = 0.0009\,(70)\cdot 10^{-3}$ $h_2 = 0.0015\,(19)\cdot 10^{-3}$ $h_3 = 0.00057\,(29)\cdot 10^{-3}$ *) $D_{JK} = -38.324\,(82)$ $D_K = 1233.88\,(69)$ $d_1 = -1.9999\,(21)$ $d_2 = -0.200893\,(45)$ **)	81JON		*) calculated from planarity condition **) the other constants are transferred from ground state	
109	$HCO^{35}Cl$	$v=0$	$\tilde{A}=77971.445\,(8)$ $\tilde{B}=\ 6140.8058\,(6)$ $\tilde{C}=\ 5685.2362\,(6)$	$\Delta_J = 3.6694\,(7)$ $\Delta_{JK} = -59.382\,(16)$ $\Delta_K = 2788.5\,(7)$ $\delta_J = 0.38752\,(7)$ $\delta_K = 22.217\,(11)$ $H_J = 1.55\,(11)\cdot 10^{-6}$ $H_{KJ} = -11.23\,(22)\cdot 10^{-3}$ $H_K = 0.261\,(24)$ $h_J = 0.70\,(2)\cdot 10^{-6}$	81SUZ	76TAK	*) uncertainties are 3 σ harmonic force field calculated [81SUZ]	2.6.3, 55 2.7.3, 102 2.9.3, 17
	$HCO^{37}Cl$	$v=0$	$\tilde{A}=77858.545\,(14)$ $\tilde{B}=\ 6001.2164\,(8)$ $\tilde{C}=\ 5564.8049\,(8)$	$\Delta_J = 3.5190\,(8)$ $\Delta_{JK} = -58.150\,(19)$ $\Delta_K = 2779.1\,(14)$ $\delta_J = 0.36363\,(7)$ $\delta_K = 21.395\,(12)$ $H_J = 1.41\,(13)\cdot 10^{-6}$ $H_{KJ} = -11.20\,(78)\cdot 10^{-3}$ $H_K = 0.406\,(72)$ $h_J = 0.58\,(2)\cdot 10^{-6}$				
	$H^{13}CO^{35}Cl$	$v=0$	$A=75566.84\,(9)$ *) $B=\ 6106.35\,(8)$ $C=\ 5642.70\,(22)$		76TAK			
	$HC^{18}O^{35}Cl$	$v=0$	$A=76180.16\,(8)$ *) $B=\ 5786.71\,(6)$ $C=\ 5371.54\,(6)$					
Cont.	$DCO^{35}Cl$	$v=0$	$A=57079.12\,(8)$ *) $B=\ 6100.13\,(3)$ $C=\ 5503.71\,(3)$					

Nr.	Molecule	v	A, B, C MHz	D, H kHz	Ref.	Ref. struct.	Varia, remarks	See also Tab., Nr.
109 Cont.	$DCO^{37}Cl$	$v=0$	$A=57003.37\ (8)$ *) $B=\ 5958.79\ (3)$ $C=\ 5387.76\ (3)$		76TAK		*) uncertainties are 3 σ	
	$D^{13}CO^{35}Cl$	$v=0$	$A=55967.10\ (282)$ *) $B=\ 6066.52\ (19)$ $C=\ 5465.70\ (99)$					
110 *6–63* *4–60*	HC(=O)NH_2	$v=0$	$A''=72716.9496\pm0.022$ $B''=11373.4541\ (39)$ $C''=\ 9833.90416\pm0.0036$	$\tau_1=\ 188.353\pm3.1$ $\tau_2=\ \ 2.8776\pm0.44$ $\tau_3=1701\pm12$ $\tau_{aaaa}=-5373.00\pm15$ $\tau_{bbbb}=\ -44.0702\pm0.20$ $\tau_{cccc}=\ -18.9373\pm0.17$ $H_J\ =(-0.9815\pm2)\cdot10^{-4}$ $H_{JK}=\ (0.5787\pm6)\cdot10^{-3}$ $H_{KJ}=(-0.3260\pm0.5)\cdot10^{-1}$ $H_K\ =\ \ 0.2454\pm0.4$ $h_J\ =\ (0.899\pm4)\cdot10^{-5}$ *) $h_{JK}\ =(-0.1972\pm0.3)\cdot10^{-2}$ $h_K\ =\ (1.012\pm1.0)\cdot10^{-1}$	73KIR1	74HIR	for structure, see also Vol. II/7 (3, 111) planar potential function to the inversion: $V=V_2\,\tau^2+v_4\,\tau^4$ where τ is the inversion angle of the amino group, $V_2=156\ (34)\ cm^{-1}/rd^2$ $V_4=398\ (29)\ cm^{-1}/rd^4$ [74HIR] *) Watson uses $2\,h_J$ for this parameter	
		$v_\tau=1$	$A=71738.782\ (31)$ $B=11351.530\ (5)$ $C=\ 9839.326\ (5)$	$\tau_{aaaa}=-5029.0\ (40)$ $\tau_{bbbb}=\ -44.81\ (8)$ $\tau_{cccc}=\ -19.71\ (5)$ $\tau'_1=-81.6\ (5)$ $\tau'_2=\ -0.024\ (5)$	74HIR		ν_τ (amino inversion) $=295\ (10)\ cm^{-1}$ from relative intensity measurements relative intensities of other vibrational satellites also measured [74HIR]	
	$H^{13}CONH_2$	$v=0$	$B=11373.5$ $C=\ 9801.5$					
	$HC^{18}ONH_2$	$v=0$	$B=10774.9$ $C=\ 9370.3$					
	cis-HCONHD	$v_\tau=1$	$B=10992.7$ $C=\ 9348.5$					
Cont.	trans-HCONHD	$v_\tau=1$	$B=10467.6$ $C=\ 9154.0$					

Nr.	Molecule	v	A, B, C MHz	D, H kHz	Ref.	Ref. struct.	Varia, remarks	See also Tab., Nr.
110 *6-63* *4-60* Cont.	$HCOND_2$	$v=0$	$A=59716.223(14)$ $B=10193.511(2)$ $C=8710.415(2)$	$\tau_{aaaa}=-2613.6(18)$ $\tau_{bbbb}=-35.33(6)$ $\tau_{cccc}=-15.28(4)$ $\tau'_1=-74.9(4)$ $\tau'_2=0.140(6)$	74HIR			
		$v_\tau=1$	$A=58903.265(11)$ $B=10193.111(2)$ $C=8736.336(2)$	$\tau_{aaaa}=-2442.8(28)$ $\tau_{bbbb}=-35.30(9)$ $\tau_{cccc}=-15.54(7)$ $\tau'_1=-68.4(6)$ $\tau'_2=0.07(2)$				
	$H^{13}COND_2$	$v=0$	$B=10189.2$ $C=8680.7$					
	$HC^{18}OND_2$	$v=0$	$B=9655.6$ $C=8304.2$					
111 *6-64* *4-61*	$HC(=O)OH$ trans	$v=0$	$A=77512.2310(63)$ $B=12055.1045(8)$ $C=10416.1145(8)$	$\Delta_J=9.9894(15)$ $\Delta_{JK}=-86.250(620)$ $\Delta_K=1702.29(19)$ $\delta_J=1.94920(24)$ $\delta_K=42.600(29)$ $H_J=0.928(39)\cdot10^{-5}$ $H_{KJ}=-0.01026(14)$ $H_K=0.1195(18)$ $h_J=0.6035(48)\cdot10^{-5}$ $h_K=0.01255(48)$	80WIL	80DAV4	for structure, see also Vol. II/7 (3, 80) harmonic force field calculated [80DAV4]	2.6.3, 56 2.7.3, 7 2.9.3, 15 3, 35
		$v_{CH}=1$	$A=77221.2(84)$ *) $B=12051.0(14)$ $C=10402.4(12)$	ground state constants were used	74TAK		ν(C—H stretch) $=2942.06(1)$ cm^{-1} *) MW IR DR	
		$v_6=1$	$A=77853.19(3)$ $B=12003.043(8)$ $C=10352.251(8)$	$\Delta_J=10.26(9)$ $\Delta_{JK}=-92.2(2)$ $\Delta_K=1705(10)$ $\delta_J=2.07(1)$ $\delta_K=58.1(10)$	79WIL1		ν_6(C—O stretch) $=1104.851$ cm^{-1}	
Cont.								

Nr.	Molecule	v	A, B, C MHz	D, H kHz	Ref.	Ref. struct.	Varia, remarks	See also Tab., Nr.
111 *6-64* *4-61* Cont.	HC(=O)OH trans	$v_8=1$	$A=76724.20\,(6)$ $B=12001.46\,(1)$ $C=10419.90\,(1)$	$\Delta_J = 10.3\,(1)$ $\Delta_{JK}=-66.0\,(3)$ $\Delta_K=1289\,(29)$ $\delta_J = 1.814\,(5)$ $\delta_K=10\,(1)$	79WIL1		ν_8 (C—H wag.) $=1033.466\ cm^{-1}$	
		$v_7=1$	$A=75481.7\,(26)$ $B=11991.8\,(19)$ $C=10401.31\,(26)$	$\Delta_J = 17.6\,(8)$ $\Delta_{JK}=-248\,(3)$ $\Delta_K = -8.6\,(5)\cdot10^3$ $\delta_J=1.5\,(4)$ $\delta_K=3.55\,(8)\cdot10^3$	78WIL		ν_7 (O—C—O bend) $=625.4\ cm^{-1}$ Coriolis interaction between ν_7 and ν_9: $\Delta=\nu_9-\nu_7=436831\,(114)$ MHz $G_a = 700.0$ MHz (fixed value) $G_b = 6809\,(62)$ MHz $F_{ac}= 50.7\,(81)$ MHz $F_{bc}=-176.7\,(4)$ MHz [78WIL]	
		$v_9=1$	$A=79954.9\,(26)$ $B=12024.3\,(12)$ $C=10423.61\,(20)$	$\Delta_J = 7.8\,(1)$ $\Delta_{JK}=27\,(4)$ $\Delta_K=11.1\,(8)\cdot10^3$ $\delta_J=-9.9\,(4)$ $\delta_K=-1.77\,(6)\cdot10^3$				
	$H^{13}COOH$ trans	$v=0$	$A=75580.8474\,(57)$ $B=12053.56428\,(86)$ $C=10378.99597\,(74)$	$\Delta_J = 9.9134\,(22)$ $\Delta_{JK}=-85.013\,(33)$ $\Delta_K=1673.66\,(19)$ $\delta_J = 1.98138\,(22)$ $\delta_K=41.928\,(36)$ $H_{KJ}=-1.494\,(51)\cdot10^{-2}$ $H_K = 0.1656\,(65)$ $h_J = 4.82\,(17)\cdot10^{-6}$ $h_K = 1.84\,(12)\cdot10^{-2}$	80DAV4			
	$HC^{18}OOH$ trans	$v=0$	$A=76526.465\,(15)$ $B=11397.1065\,(25)$ $C= 9904.6859\,(20)$	$\Delta_J = 8.9757\,(338)$ $\Delta_{JK}=-82.560\,(687)$ $\Delta_K=1674.5\,(39)$ $\delta_J = 1.6934\,(21)$ $\delta_K=39.36\,(23)$ $H_{KJ}=-1.32\,(104)\cdot10^{-2}$ $H_K = 0.17\,(21)$ $h_J = 2.2\,(30)\cdot10^{-6}$ $h_K = 2.8\,(17)\cdot10^{-2}$	80DAV		^{18}O at the carbonyl group	
Cont.								

Nr.	Molecule	v	A, B, C MHz	D, H kHz	Ref.	Ref. struct.	Varia, remarks	See also Tab., Nr.
111 *6-64* *4-61* Cont.	$HCO^{18}OH$ trans	$v=0$	$A=77201.3076(88)$ $B=11383.7761(18)$ $C=9905.9491(12)$	$\Delta_J=8.9674(245)$ $\Delta_{JK}=-82.823(465)$ $\Delta_K=1701.5(25)$ $\delta_J=1.6723(12)$ $\delta_K=38.95(17)$ $H_{KJ}=-1.68(55)\cdot10^{-2}$ $H_K=0.25(10)$ $h_J=3.3(21)\cdot10^{-6}$ $h_K=2.3(14)\cdot10^{-2}$	80DAV		^{18}O at the hydroxyl group	
	HCOOD trans	$v=0$	$A=66099.557(50)$ $B=11762.5524(30)$ $C=9969.9561(33)$	$\Delta_J=10.1847(99)$ $\Delta_{JK}=-59.67(12)$ $\Delta_K=988(35)$ $\delta_J=2.1464(38)$ $\delta_K=44.38(84)$ $H_{KJ}=-0.0124(16)$ $H_K=1.87(45)$ $h_J=-0.80(18)\cdot10^{-5}$	80WIL			
		$v_7=1$	$A=65451.23(18)$ $B=11609.99(2)$ $C=9968.69(2)$	$\Delta_J=8.3(2)$ $\Delta_{JK}=84.2(8)$ $\Delta_K=988$ (fixed value) $\delta_J=-0.2(2)$ $\delta_K=44.7$ (fixed value)	78WIL		ν_7 (O—C—O bend) $=556.3\ cm^{-1}$	
		$v_9=1$	$A=67084.09(12)$ $B=11853.36(1)$ $C=9958.15(1)$	$\Delta_J=11.1(1)$ $\Delta_{JK}=-227.8(7)$ $\Delta_K=988$ (fixed) $\delta_J=4.5(1)$ $\delta_K=47.7$ (fixed)			ν_9 (O—D torsion) $=507.5\ cm^{-1}$	
Cont.	DCOOH trans	$v=0$	$A=57709.185(15)$ $B=12055.9737(27)$ $C=9955.6058(24)$	$\Delta_J=9.403(18)$ $\Delta_{JK}=-39.52(12)$ $\Delta_K=755.2(11)$ $\delta_J=2.2239(29)$ $\delta_K=37.73(22)$ $H_{KJ}=-0.31(22)\cdot10^{-2}$ $H_K=0.020(19)$ $h_J=0.59(26)\cdot10^{-5}$ $h_K=0.60(59)\cdot10^{-2}$	80WIL			

Nr.	Molecule	v	A, B, C MHz	D, H kHz	Ref.	Ref. struct.	Varia, remarks	See also Tab., Nr.
111 *6-64* *4-61* Cont.		$v_7=1$	$A=56746(3)$ $B=11941.5(5)$ $C=9949.3(4)$	$\Delta_J=5(4)$ $\Delta_{JK}=-0.45(4)\cdot10^3$ $\Delta_K=755$ (fixed) $\delta_J=-5(2)$ $\delta_K=38$ (fixed)	78WIL		Coriolis interaction between ν_7 and ν_9: $\nu_9-\nu_7=304531(401)$ MHz $G_a=176$ (fixed) MHz $G_b=6488.2(36)$ MHz $F_{ac}=34(20)$ MHz $F_{bc}=-180.6(19)$ MHz ν_9 (O—H torsion) $=629$ cm^{-1} ν_7 (O—C—O bend) $=620$ cm^{-1}	
		$v_9=1$	$A=58862.9(40)$ $B=12072.2(6)$ $C=9964.7(5)$	$\Delta_J=1.1(4)$ $\Delta_{JK}=1.10(9)\cdot10^3$ $\Delta_K=755$ (fixed) $\delta_J=-51(6)$ $\delta_K=38$ (fixed)				
	DCOOD trans	$v=0$	$A=50816.545(9)$ $B=11759.889(2)$ $C=9534.171(2)$	$\Delta_J=9.64(2)$ $\Delta_{JK}=-34.0(1)$ $\Delta_K=497(2)$ $\delta_J=2.420(1)$ $\delta_K=35.17(7)$ $H_{KJ}=-0.010(7)$ $h_J=7(1)\cdot10^{-6}$				
		$v_7=1$	$A=50596.08(6)$ $B=11616.446(7)$ $C=9531.999(7)$	$\Delta_J=7.98(8)$ $\Delta_{JK}=22.5(2)$ $\Delta_K=497$ (fixed) $\delta_J=0.94(4)$ $\delta_K=35$ (fixed)			ν_7 (O—C—O bend) $=558$ cm^{-1}	
		$v_9=1$	$A=51244.49(11)$ $B=11841.36(1)$ $C=9519.30(1)$	$\Delta_J=11.0(1)$ $\Delta_{JK}=-93.9(3)$ $\Delta_K=497$ (fixed) $\delta_J=3.9(1)$ $\delta_K=35$ (fixed)			ν_9 (O—D torsion) $=491$ cm^{-1}	
Cont.	HCOOH cis	$v=0$	$\tilde{A}=86461.565(22)$ $\tilde{B}=11689.1767(30)$ $\tilde{C}=10283.9868(30)$	$\Delta_J=8.353(35)$ $\Delta_{JK}=-71.07(77)$ $\Delta_K=2357.2(45)$ $\delta_J=1.4185(22)$ $\delta_K=41.102(82)$	76HO1	78BJ	the cis rotamer is found to lie higher energy than the trans rotamer by 1365 (30) cm^{-1} [76HO1]	2.6.3, 56 3, 35

Nr.	Molecule	v	A, B, C MHz	D, H kHz	Ref.	Ref. struct.	Varia, remarks	See also Tab., Nr.
111 *6-64* *4-61* Cont.	$HC^{18}OOH$ cis	$v=0$	$\tilde{A}=85388.388(35)$ $\tilde{B}=11058.5903(40)$ $\tilde{C}=9778.6232(40)$	$\Delta_J=7.428(52)$ $\Delta_{JK}=-72.7(12)$ $\Delta_K=2345.2(70)$ $\delta_J=1.2334(47)$ $\delta_K=37.55(20)$	78BJ		^{18}O at the carbonyl group	
	$HCO^{18}OH$ cis	$v=0$	$\tilde{A}=85175.320(26)$ $\tilde{B}=11093.7328(33)$ $\tilde{C}=9803.3164(33)$	$\Delta_J=7.682(28)$ $\Delta_{JK}=-68.72(67)$ $\Delta_K=2276.6(47)$ $\delta_J=1.2626(17)$ $\delta_K=37.936(91)$			^{18}O at the hydroxyl group	
	$H^{13}COOH$ cis	$v=0$	$\tilde{A}=84201.819(19)$ $\tilde{B}=11687.5239(28)$ $\tilde{C}=10249.6623(28)$	$\Delta_J=8.268(23)$ $\Delta_{JK}=-70.26(52)$ $\Delta_K=2315.0(32)$ $\delta_J=1.4433(15)$ $\delta_K=40.431(69)$				
	HCOOD cis	$v=0$	$\tilde{A}=83962.785(22)$ $\tilde{B}=10883.9413(29)$ $\tilde{C}=9624.9421(29)$	$\Delta_J=6.733(36)$ $\Delta_{JK}=-48.11(81)$ $\Delta_K=1984.4(48)$ $\delta_J=1.0628(21)$ $\delta_K=33.911(68)$				
	DCOOH cis	$v=0$	$\tilde{A}=62653.4395(87)$ $\tilde{B}=11690.1692(18)$ $\tilde{C}=9837.9145(18)$	$\Delta_J=7.899(29)$ $\Delta_{JK}=-22.94(48)$ $\Delta_K=954.4(11)$ $\delta_J=1.6833(34)$ $\delta_K=37.541(94)$				
	DCOOD cis	$v=0$	$\tilde{A}=61507.409(11)$ $\tilde{B}=10884.5356(22)$ $\tilde{C}=9237.0082(22)$	$\Delta_J=6.267(29)$ $\Delta_{JK}=-15.25(48)$ $\Delta_K=848.5(14)$ $\delta_J=1.2429(23)$ $\delta_K=31.315(78)$				

Nr.	Molecule	v	A, B, C MHz	D, H kHz	Ref.	Ref. struct.	Varia, remarks	See also Tab., Nr.
112	HC(=O)SH cis	$v=0$	$\tilde{A}=62927.7132\,(21)$ $\tilde{B}=6134.26108\,(19)$ $\tilde{C}=5584.75363\,(19)$	$\Delta_J=3.67858\,(36)$ $\Delta_{JK}=-47.9619\,(65)$ $\Delta_K=1305.68\,(18)$ $\delta_J=0.463122\,(37)$ $\delta_K=17.653\,(12)$ $H_J=2.79\,(30)\cdot10^{-6}$ $H_{JK}=-48\,(13)\cdot10^{-6}$ $H_{KJ}=-4.162\,(51)\cdot10^{-3}$ $H_K=80.6\,(47)\cdot10^{-3}$ $h_J=1.118\,(18)\cdot10^{-6}$ $h_{JK}=14\,(16)\cdot10^{-6}$ $h_K=7.6\,(19)\cdot10^{-3}$	76HO3	77HO	the cis rotamer is found to lie at higher energy than the trans rotamer by 2767 (7) J mol^{-1} [76HO5]	2.6.3, 57 2.9.3, 3, 36
	HC(:O)SH trans	$v=0$	$\tilde{A}=62036.0911\,(17)$ $\tilde{B}=6125.30490\,(14)$ $\tilde{C}=5569.64234\,(14)$	$\Delta_J=3.42871\,(27)$ $\Delta_{JK}=-43.1667\,(51)$ $\Delta_K=1250.74\,(15)$ $\delta_J=0.434730\,(32)$ $\delta_K=16.5158\,(84)$ $H_J=1.37\,(24)\cdot10^{-6}$ $H_{JK}=-2\,(16)\cdot10^{-6}$ $H_{KJ}=-4.737\,(53)\cdot10^{-3}$ $H_K=88.1\,(44)\cdot10^{-3}$ $h_J=0.737\,(37)\cdot10^{-6}$ $h_{JK}=19\,(18)\cdot10^{-6}$ $h_K=5.9\,(20)\cdot10^{-3}$				
	$H^{13}C$(:O)SH cis	$v=0$	$\tilde{A}=61295.6325\,(39)$ $\tilde{B}=6103.12450\,(39)$ $\tilde{C}=5545.74954\,(39)$	$\Delta_J=3.62686\,(81)$ $\Delta_{JK}=-46.448\,(14)$ $\Delta_K=1281.86\,(38)$ $\delta_J=0.466269\,(75)$ $\delta_K=17.485\,(28)$	77HO			
	$H^{13}C$(:O)SH trans	$v=0$	$\tilde{A}=60503.9010\,(37)$ $\tilde{B}=6094.40708\,(36)$ $\tilde{C}=5531.43327\,(36)$	$\Delta_J=3.38315\,(76)$ $\Delta_{JK}=-41.664\,(13)$ $\Delta_K=1232.70\,(35)$ $\delta_J=0.437076\,(72)$ $\delta_K=16.374\,(25)$				
Cont.								

Nr.	Molecule	v	A, B, C MHz	D, H kHz	Ref.	Ref. struct.	Varia, remarks	See also Tab., Nr.
112 Cont.	HC(:^{18}O)SH cis	$v=0$	$\tilde{A}$ = 61850.9798 (37) $\tilde{B}$ = 5785.66443 (36) $\tilde{C}$ = 5286.52128 (36)	Δ_J = 3.29749 (66) Δ_{JK} = −47.478 (13) Δ_K = 1296.41 (48) δ_J = 0.403526 (75) δ_K = 15.864 (25)	77HO			
	HC(:^{18}O)SH trans	$v=0$	$\tilde{A}$ = 60972.0319 (33) $\tilde{B}$ = 5775.67561 (31) $\tilde{C}$ = 5271.16418 (31)	Δ_J = 3.06850 (60) Δ_{JK} = −42.807 (12) Δ_K = 1240.91 (56) δ_J = 0.378236 (68) δ_K = 14.827 (20)				
	HC(:O)^{34}SH cis	$v=0$	$\tilde{A}$ = 62712.427 (15) $\tilde{B}$ = 5988.1552 (15) $\tilde{C}$ = 5461.7548 (15)	Δ_J = 3.534 (13) Δ_{JK} = −46.95 (36) Δ_K = 1297.3 (43) δ_J = 0.43568 (58) δ_K = 16.917 (51)	76HO4			
	HC(:O)^{34}SH trans	$v=0$	$\tilde{A}$ = 62013.8403 (92) $\tilde{B}$ = 5971.92667 (95) $\tilde{C}$ = 5442.36776 (95)	Δ_J = 3.2769 (70) Δ_{JK} = −41.66 (20) Δ_K = 1245.1 (23) δ_J = 0.40447 (28) δ_K = 15.823 (28)				
	HC(:O)SD cis	$v=0$	$\tilde{A}$ = 55038.2202 (35) $\tilde{B}$ = 5955.31696 (39) $\tilde{C}$ = 5369.84705 (39)	Δ_J = 3.1865 (10) Δ_{JK} = −23.553 (22) Δ_K = 746.23 (20) δ_J = 0.411391 (96) δ_K = 15.473 (21)				
	HC(:O)SD trans	$v=0$	$\tilde{A}$ = 50809.8095 (29) $\tilde{B}$ = 6086.24103 (31) $\tilde{C}$ = 5430.07687 (31)	Δ_J = 3.42625 (90) Δ_{JK} = −22.105 (18) Δ_K = 582.53 (13) δ_J = 0.486644 (83) δ_K = 16.642 (10)				
Cont.	DC(:O)SH cis	$v=0$	A = 48403.2600 (24) $\tilde{B}$ = 6091.93570 (31) $\tilde{C}$ = 5405.99264 (31)	Δ_J = 3.42549 (68) Δ_{JK} = −25.735 (11) Δ_K = 627.28 (11) δ_J = 0.523121 (76) δ_K = 16.072 (11)				

Nr.	Molecule	v	A, B, C MHz	D, H kHz	Ref.	Ref. struct.	Varia, remarks	See also Tab., Nr.
112 Cont.	DC(:O)SH trans	$v=0$	$\tilde{A}=47970.8473(19)$ $\tilde{B}=6083.92022(25)$ $\tilde{C}=5393.71191(25)$	$\Delta_J=3.20376(55)$ $\Delta_{JK}=-22.9348(95)$ $\Delta_K=609.78(11)$ $\delta_J=0.489187(62)$ $\delta_K=15.0373(88)$	76HO4			
113 *6-65* *4-63*	HC(=O)OCH$_3$	$v=0$	$A=19982.23889(499)$ *) $B=6914.01332(138)$ $C=5304.47963(129)$	$\Delta_J=6.0589(164)$ $\Delta_{JK}=-21.2897(423)$ $\Delta_K=79.133(154)$ $\delta_J=1.87101(188)$ $\delta_K=3.9508(619)$	79BAU1		for structure, see also Vol. II/7 (3, 245) *) for the A species	2.8, 81
114	HC(=O)SCH$_3$	$v=0$	$A=11035.13(19)$ $B=5099.99(13)$ $C=3561.57(3)$	$\Delta_J=3.1(5)$ $\Delta_{JK}=2.6(9)$ $\Delta_K=3.4(41)$ $\delta_J=1.07(5)$ $\delta_K=7.7(8)$	81CAM1, 76JON1		molecular conformation having a planar array of heavy atoms alkyl group cis to the carbonyl group	2.6.3, 58 2.8, 97 3, 37
115	HC(=O)NH(CH$_3$)	$v=0$	$A=19484.3$ $B=6330.1$ $C=4904.8$		73ELZ2		for structure, see also Vol. II/7 (3, 271) methyl group cis to the oxygen atom	2.6.3, 59 2.8, 116
116	HC(=O)N(CH$_3$)$_2$						for structure, see also Vol. II/7 (3, 425)	2.6.3, 60 2.8, 117
117 *4-64*	HC(=O)OCH=CH$_2$							2.9.3, 16
118	HC(=O)OCH$_2$C≡CH	$v=0$	$A=19205.81(20)$ $B=1750.304(5)$ $C=1622.039(5)$		75JON			3, 38

Nr.	Molecule	v	A, B, C MHz	D, H kHz	Ref.	Ref. struct.	Varia, remarks	See also Tab., Nr.
119 *6-72a*	$N_3C(=O)OCH_3$	$v=0$	$A=10702$		80KA1	80KA1	azido group and methyl group cis to the C=O group	2.8, 83
			$B=$ 1568.66					
			$C=$ 1380.22					
		$v_T=1$	$\Delta A=-200$ *)				v_T, torsion about the C—N bond	
			$\Delta B=$ 0.50					
			$\Delta C=$ 2.24				*) $\Delta A=A_v-A_{v-1}$	
		$v_T=2$	$\Delta A=-234$ *)					
			$\Delta B=$ 0.68					
			$\Delta C=$ 2.16					
		$v_a=1$	$\Delta A=-180$ *)					
			$\Delta B=$ 1.36					
			$\Delta C=$ −0.07					
		$v_b=1$	$\Delta A=-164$ *)				v_b, methyl torsion	
			$\Delta B=$ −3.19					
			$\Delta C=$ −1.51					
		$v_c=1$	$\Delta A=$ −48 *)					
			$\Delta B=$ 1.84					
			$\Delta C=$ 2.22					
		$v_d=1$	$\Delta A=$ 91 *)					
			$\Delta B=$ −2.12					
			$\Delta C=$ −3.21					
	$NN^{15}NC(=O)OCH_3$	$v=0$	$A=10548$					
			$B=$ 1566.20					
			$C=$ 1375.70					
		$v_T=1$	$\Delta A=-242$ *)					
			$\Delta B=$ 0.45					
			$\Delta C=$ 2.24					
		$v_T=2$	$\Delta A=-182$ *)					
			$\Delta B=$ 0.48					
			$\Delta C=$ 2.40					
		$v_a=1$	$\Delta A=$ 177 *)					
			$\Delta B=$ 1.35					
Cont.			$\Delta C=$ −0.08					

Nr.	Molecule	v	A, B, C MHz	D, H kHz	Ref.	Ref. struct.	Varia, remarks	See also Tab., Nr.
119	$^{15}NNNC(=O)OCH_3$	$v=0$	$A=10693$		80KA1			
6-72a			$B=1530.81$					
Cont.			$C=1350.65$					
		$v_T=1$	$\Delta A=-195$ *)				*) $\Delta A = A_v - A_{v-1}$	
			$\Delta B=0.51$					
			$\Delta C=2.21$					
		$v_T=2$	$\Delta A=-313$ *)					
			$\Delta B=0.64$					
			$\Delta C=2.36$					
		$v_a=1$	$\Delta A=210$ *)					
			$\Delta B=1.38$					
			$\Delta C=-0.02$					
	$N_3C(=O)OCD_3$	$v=0$	$A=10061$					
			$B=1441.87$					
			$C=1281.71$					
		$v_T=1$	$\Delta A=-167$ *)					
			$\Delta B=0.49$					
			$\Delta C=1.60$					
		$v_T=2$	$\Delta A=-143$ *)					
			$\Delta B=0.48$					
			$\Delta C=1.56$					
		$v_d=1$	$\Delta A=-62$ *)					
			$\Delta B=0.01$					
			$\Delta C=0.87$					
120	$HC(=O)OCH_2F$	$v=0$	$A=11195.15$	$\tau_{aaaa}=-335.84$	81LOP	81LOP	cis ester	2.6.3, 61
			$B=3359.40$	$\tau_{bbbb}=-55.05$			configuration	
			$C=3261.16$	$\tau_{cccc}=-12.12$			$\tau(OCO_eC)=1.5(10)°$	
				$\tau'_{aabb}=72.46$ *)			$\tau(FCO_eC)=84(1)°$	
				$\tau'_{bbcc}=-29.18$			*) $\tau'_{aabb}=\tau'_{aacc}$	
		$v_a=1$	$A=11442.4$				$v_a=96(6)$ cm^{-1}	
			$B=3803.6$					
Cont.			$C=3235.5$					

Nr.	Molecule	v	A, B, C MHz	D, H kHz	Ref.	Ref. struct.	Varia, remarks	See also Tab., Nr.
120 Cont.	$DC(O)OCH_2F$	$v=0$	$A=10854.59$ $B=3728.83$ $C=3164.46$	$\tau_{aaaa}=-287.97$ $\tau_{bbbb}=-47.02$ $\tau_{cccc}=-11.13$ $\tau'_{aabb}=56.53$ $\tau'_{bbcc}=-25.48$	81LOP			
	$HC(O)OCD_2F$	$v=0$	$A=10339.05$ $B=3704.13$ $C=3179.02$	$\tau_{aaaa}=-247.20$ $\tau_{bbbb}=-37.05$ $\tau_{cccc}=-2.67$ $\tau'_{aabb}=56.49$ $\tau'_{bbcc}=-16.35$				
	$DC(O)OCD_2F$	$v=0$	$A=10042.61$ $B=3580.70$ $C=3084.60$	$\tau_{aaaa}=-201.09$ $\tau_{bbbb}=-38.23$ $\tau_{cccc}=-9.04$ $\tau'_{aabb}=38.19$ $\tau'_{bbcc}=-21.20$				
	$HC(^{18}O)^{18}OCH_2F$	$v=0$	$A=10460.70$ $B=3729.97$ $C=3109.02$	$\tau_{aaaa}=-321.47$ $\tau_{bbbb}=-53.48$ $\tau_{cccc}=-11.66$ $\tau'_{aabb}=67.96$ $\tau'_{bbcc}=-28.27$				
	$HC(O)^{18}OCH_2F$	$v=0$	$A=10766.38$ $B=3858.67$ $C=3223.94$	$\tau_{aaaa}=-340.85$ $\tau_{bbbb}=-56.43$ $\tau_{cccc}=-12.92$ $\tau'_{aabb}=69.01$ $\tau'_{bbcc}=-30.00$				
	HC(O)OCFHD	$v=0$	$A=10736.89$ $B=3798.04$ $C=3253.07$	$\tau_{aaaa}=-311.44$ $\tau_{bbbb}=-48.58$ $\tau_{cccc}=-11.84$ $\tau'_{aabb}=63.98$ $\tau'_{bbcc}=-28.04$			$\tau=DO_eCF=121.5°$	
	HC(O)OCFHD	$v=0$	$A=10756.82$ $B=3762.19$ $C=3186.11$	$\tau_{aaaa}=-266.17$ $\tau_{bbbb}=-50.14$ $\tau_{cccc}=-10.34$ $\tau'_{aabb}=54.94$ $\tau'_{bbcc}=-24.98$			$\tau=FO_eCD=118.3°$	
Cont.								

Nr.	Molecule	v	A, B, C MHz	D, H kHz	Ref.	Ref. struct.	Varia, remarks	See also Tab., Nr.
120 Cont.	$HC(O)O^{13}CH_2F$	$v=0$	$A=11102.75$ $B=3826.02$ $C=3242.34$	$\tau_{aaaa}=-338.54$ $\tau_{bbbb}=-53.33$ $\tau_{cccc}=-11.88$ $\tau'_{aabb}=69.62$ $\tau'_{bbcc}=-28.24$	81LOP			
	$HC(^{18}O)OCH_2F$	$v=0$	$A=10885.14$ $B=3730.08$ $C=3145.00$	$\tau_{aaaa}=-335.74$ $\tau_{bbbb}=-54.06$ $\tau_{cccc}=-13.05$ $\tau'_{aabb}=69.80$ $\tau'_{bbcc}=-29.39$				
	$H^{13}C(O)OCH_2F$	$v=0$	$A=11161.13$ $B=3810.76$ $C=3226.14$	$\tau_{aaaa}=-385.18$ $\tau_{bbbb}=-51.91$ $\tau_{cccc}=-11.40$ $\tau'_{aabb}=73.99$ $\tau'_{bbcc}=-27.11$				
121 *4-62*	$H_2C{=}NOH$	$v=0$	$A''=67679.279\,(111)$ $B''=11859.136\,(17)$ $C''=10080.070\,(17)$	$\tau_1=129.80\,(126)$ $\tau_2=0$ (fixed) $\tau_3=1357.38\,(297)$ $\tau'_{aaaa}=-5683.17\,(4192)$ $\tau_{bbbb}=-45.15\,(19)$ $\tau_{cccc}=-17.06\,(26)$	78KAU2		for structure, see also Vol. II/7 (3, 110)	
122 Cont.	$HN(CHO)_2$ *)	$v=0$	$A=21490.671\,(18)$ $B=3029.943\,(2)$ $C=2655.748\,(2)$	$\tau_{aaaa}=-519.176$ $\tau_{bbbb}=-4.37662$ $\tau_{cccc}=-2.00539$ $\tau_{bbaa}=29.7826$ $\tau_{ccaa}=14.9522$ $\tau_{bbcc}=-2.90754$ $\tau_{baba}=-3.81457$ $\tau_{bcbc}=-6.95\cdot10^{-3}$	73STE1	73STE1	*) asymmetric cis-trans planar structure	2.7.3, 26

Nr.	Molecule	v	A, B, C MHz	D, H kHz	Ref.	Ref. struct.	Varia, remarks	See also Tab., Nr.
122 Cont.	$DN(CHO)_2$	$v=0$	$A=19396.690(30)$ $B=3025.547(4)$ $C=2617.488(4)$	$\tau_{bbbb}=-3.62099$ $\tau_{cccc}=-1.47052$ $\tau_{bbaa}=20.4654$ $\tau_{ccaa}=15.3173$ $\tau_{bbcc}=-2.33744$ $\tau_{baba}=-1.66630$ $\tau_{bcbc}=-0.875$	73STE1			
123	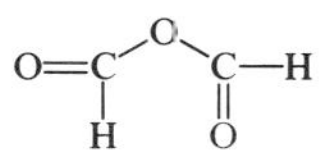	$v=0$	$A=21879.9192(41)$ *) $B=3245.71696(72)$ $C=2829.45712(56)$	$\Delta_J=0.8564(44)$ $\Delta_{JK}=-11.172(64)$ $\Delta_K=156.24(23)$ $\delta_J=0.1621(3)$ $\delta_K=3.760(53)$	77VAC	77VAC	for structure, see also Vol. II/7 (3, 199) *) $\Delta=-0.190859(50)$ amu Å^2	2.6.3, 62 3, 39
		$v_{15}=1$	$A=21436.406(36)$ $B=3263.092(3)$ $C=2847.731(3)$				$\nu_{15}(a'')=85(8)$ cm^{-1}	
		$v_{15}=2$	$A=21047.868(55)$ $B=3279.858(5)$ $C=2863.847(5)$					
		$v_{15}=3$	$A=20714.209(60)$ $B=3295.052(5)$ $C=2877.936(5)$					
		$v_{15}=4$	$A=20437.946(70)$ $B=3309.579(6)$ $C=2891.048(6)$					
		$v_{15}=5$	$A=20191.427(70)$ $B=3323.110(6)$ $C=2903.168(5)$					
		$v_{14}=1$	$A=21740.676(25)$ $B=3241.746(2)$ $C=2832.595(2)$				$\nu_{14}(a'')=211(10)$ cm^{-1}	
		$v_{11}=1$	$A=22294.871(35)$ $B=3236.887(3)$ $C=2821.649(3)$				$\nu_{11}(a')=236(20)$ cm^{-1}	
Cont.								

Nr.	Molecule	v	A, B, C MHz	D, H kHz	Ref.	Ref. struct.	Varia, remarks	See also Tab., Nr.
123 Cont.	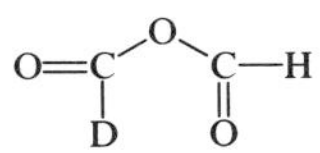	$v=0$	$A=20347.4274\ (65)$ $B=\ 3241.61332\ (66)$ $C=\ 2799.94590\ (63)$	$\Delta_J=\ 0.8510\ (41)$ $\Delta_{JK}=-9.409\ (43)$ $\Delta_K=110.11\ (26)$ $\delta_J=0.1629\ (9)$ $\delta_K=3.30\ (15)$	77VAC			
	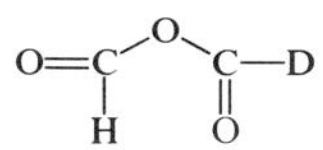	$v=0$	$A=20247.4041\ (75)$ $B=\ 3175.36413\ (92)$ $C=\ 2747.92994\ (86)$	$\Delta_J=\ 0.7591\ (49)$ $\Delta_{JK}=-7.255\ (56)$ $\Delta_K=105.32\ (40)$ $\delta_J=0.1472\ (14)$ $\delta_K=3.24\ (24)$				
	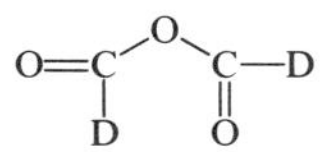	$v=0$	$A=18910.9780\ (53)$ $B=\ 3170.47837\ (65)$ $C=\ 2719.05272\ (59)$	$\Delta_J=\ 0.7499\ (45)$ $\Delta_{JK}=-6.238\ (38)$ $\Delta_K=\ 75.01\ (25)$ $\delta_J=0.1455\ (13)$ $\delta_K=3.17\ (17)$				
	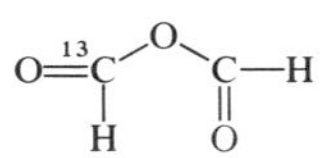	$v=0$	$A=21769.0975\ (62)$ $B=\ 3227.86738\ (64)$ $C=\ 2814.11174\ (57)$	$\Delta_J=\ 0.8480\ (36)$ $\Delta_{JK}=-11.083\ (40)$ $\Delta_K=\ 152.07\ (25)$ $\delta_J=0.1620\ (11)$ $\delta_K=3.46\ (19)$				
	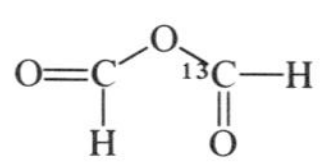	$v=0$	$A=21701.3148\ (134)$ $B=\ 3211.5775\ (14)$ $C=\ 2800.5280\ (9)$	$\Delta_J=\ 0.827\ (11)$ $\Delta_{JK}=-10.622\ (42)$ $\Delta_K=\ 150.86\ (70)$ $\delta_J=0.1568\ (25)$ $\delta_K=3.7599$ (fixed)				
	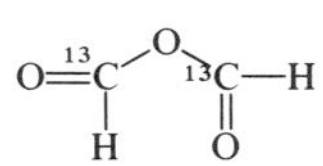	$v=0$	$A=21591.0663\ (150)$ $B=\ 3193.3913\ (10)$ $C=\ 2784.9328\ (9)$	centrifugal distortion constants fixed to the value of parent molecule.				
	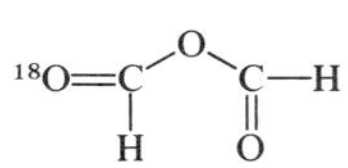	$v=0$	$A=21838.6649\ (110)$ $B=\ 3075.6893\ (14)$ $C=\ 2698.7721\ (13)$	$\Delta_J=\ 0.759\ (15)$ $\Delta_{JK}=-10.483\ (64)$ $\Delta_K=\ 157.97\ (237)$ $\delta_J=0.1402\ (20)$ $\delta_K=3.08\ (23)$				
Cont.								

Nr.	Molecule	v	A, B, C MHz	D, H kHz	Ref.	Ref. struct.	Varia, remarks	See also Tab., Nr.
123 Cont.	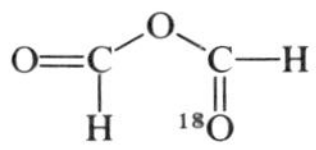	$v=0$	$A=21159.1834\,(86)$ $B=3121.0205\,(12)$ $C=2722.6067\,(10)$	$\Delta_J=0.8129\,(61)$ $\Delta_{JK}=-11.436\,(67)$ $\Delta_K=150.00\,(38)$ $\delta_J=0.1585\,(16)$ $\delta_K=3.16\,(27)$	77VAC			
	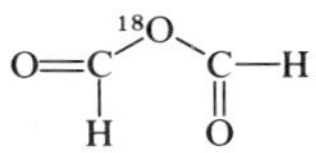	$v=0$	$A=20952.5609\,(68)$ $B=3245.9526\,(10)$ $C=2813.4711\,(8)$	$\Delta_J=0.8266\,(66)$ $\Delta_{JK}=-10.887\,(59)$ $\Delta_K=152.94\,(42)$ $\delta_J=0.1758\,(24)$ $\delta_K=2.37\,(37)$				
124	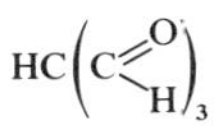	$v=0$	$A=4944.12\,(2)$ $B=2034.03\,(3)$ $C=1441.08\,(4)$		80TUR	80TUR	planar keto-enol structure with intramolecular hydrogen bond, C=O cis to C=C	3, 40
		$v_a=1$	$A=4918.38\,(1)$ $B=2035.59\,(2)$ $C=1443.56\,(5)$				$\nu_a=140\,(20)$ cm^{-1} out-of-plane	
		$v_b=1$	$A=4961.77\,(1)$ $B=2033.94\,(5)$ $C=1438.77\,(8)$				$\nu_b=220\,(20)$ cm^{-1} in-plane	
		$v_c=1$	$A=4914.17\,(1)$ $B=2035.36\,(7)$ $C=1440.96\,(14)$				$\nu_c=250\,(20)$ cm^{-1} out-of-plane	
	$(CHO)_2C{=}CH(OD)$	$v=0$	$A=4914.51\,(2)$ $B=2000.16\,(6)$ $C=1421.59\,(10)$					
		$v_a=1$	$A=4889.24\,(1)$ $B=2001.87\,(4)$ $C=1423.67\,(8)$					
		$v_b=1$	$A=4929.82\,(1)$ $B=1999.94\,(6)$ $C=1419.64\,(10)$					
Cont.		$v_c=1$	$A=4896.93\,(1)$ $B=2001.29\,(3)$ $C=1421.75\,(7)$					

Nr.	Molecule	v	A, B, C MHz	D, H kHz	Ref.	Ref. struct.	Varia, remarks	See also Tab., Nr.
124 Cont.	$(CHO)_2C{=}CH(OD)$	$v_d=1$	$A=4891.36$ (1) $B=2002.56$ (8) $C=1420.43$ (14)		80TUR		ν_d in plane cis vibration	
125 *6-67* *4-65*	OCF_2	$v=0$	$A=11813.5381$ (18) $B=11753.0547$ (18) $C=5880.8992$ (18)	$\Delta_J = 6.1339$ (147) *) $\Delta_{JK} = -3.2071$ (130) $\Delta_K = 13.3815$ (87) $\delta_J = 2.5555$ (22) $\delta_K = 4.3417$ (44) $H_{JK} = -0.1262 (8) \cdot 10^{-3}$ $H_K = 0.0952 (4) \cdot 10^{-3}$ $h_J = -0.0163 (4) \cdot 10^{-3}$ $h_K = 0.1353 (2) \cdot 10^{-3}$	74CAR, 76TY	80NA2	for structure, see also Vol. II/7 (3, 33) *) representation I^r used	2.9.3, 18
	$^{13}COF_2$	$v=0$	$A=11814.7989$ (15) $B=11747.3419$ (15) $C=5879.7866$ (15)	$\Delta_J = 6.0346$ (390)*) $\Delta_{JK} = -3.3711$ (100) $\Delta_K = 13.5137$ (68) $\delta_J = 2.5591$ (17) $\delta_K = 4.3281$ (35) $H_{JK} = -0.1251 (11) \cdot 10^{-3}$ $H_K = 0.0966 (6) \cdot 10^{-3}$ $h_J = -0.0149 (6) \cdot 10^{-3}$ $h_K = 0.1366 (3) \cdot 10^{-3}$	74CAR			
	$C^{18}OF_2$	$v=0$	$A=11813.4824$ (53) $B=10878.5953$ (53) $C=5653.2960$ (53)	$\Delta_J = 5.3709$ (2838)*) $\Delta_{JK} = -1.2742$ (922) $\Delta_K = 12.3908$ (697) $\delta_J = 2.1251$ (135) $\delta_K = 4.7867$ (368)				
	$^{13}C^{18}OF_2$	$v=0$	$A=11814.7576$ (50) $B=10876.3868$ (50) $C=5653.0047$ (50)	$\Delta_J = 5.5164$ (2667)*) $\Delta_{JK} = -1.2580$ (814) $\Delta_K = 12.3844$ (617) $\delta_J = 2.1448$ (119) $\delta_K = 4.7282$ (324)				

Nr.	Molecule	v	A, B, C MHz	D, H kHz	Ref.	Ref. struct.	Varia, remarks	See also Tab., Nr.
126	$OC^{35}Cl_2$	$v=0$	$A=7918.7942(36)$	$\Delta_J = 0.9712(28)$	78CAR	80NA1	for structure,	2.7.3, 103
6-67a			$B=3474.9594(17)$	$\Delta_{JK} = -1.5326(91)$			see also Vol. II/7 (3, 16)	
4-66			$C=2412.2064(14)$	$\Delta_K = 10.4143(89)$			harmonic force field	
				$\delta_J = 0.3481(9)$			calculated [78CAR]	
				$\delta_K = 1.7933(188)$			r_e(C=O) = 1.1756 (23) Å	
				$H_J = -0.0113(27) \cdot 10^{-3}$			r_e(C—Cl) = 1.7381 (19) Å	
				$H_{JK} = 0.1931(478) \cdot 10^{-3}$			$\measuredangle_e$ ClCCl = 111.79 (24)°	
				$H_{KJ} = -0.4871(1343) \cdot 10^{-3}$			[80NA1]	
				$H_K = 0.3650(860) \cdot 10^{-3}$				
				$h_J = 0.0071(13) \cdot 10^{-3}$				
				$h_{JK} = -0.1762(381) \cdot 10^{-3}$				
				$h_K = 0.7588(2296) \cdot 10^{-3}$				
	$OC^{35}Cl^{37}Cl$	$v=0$	$A=7867.8126(15)$	$\Delta_J = 0.9251(8)$				
			$B=3379.9604(5)$	$\Delta_{JK} = -1.5050(35)$				
			$C=2361.4535(5)$	$\Delta_K = 10.3128(24)$				
				$\delta_J = 0.3243(3)$				
				$\delta_K = 1.8093(48)$				
				$H_J = 0.0009(4) \cdot 10^{-3}$				
				$H_{JK} = -0.0034(32) \cdot 10^{-3}$				
				$H_{KJ} = -0.0038(95) \cdot 10^{-3}$				
				$H_K = 0.0490(61) \cdot 10^{-3}$				
				$h_J = 0.0002(2) \cdot 10^{-3}$				
				$h_{JK} = 0.0073(36) \cdot 10^{-3}$				
				$h_K = -0.0188(178) \cdot 10^{-3}$				
	$OC^{37}Cl_2$	$v=0$	$A=7814.5348(10)$	$\Delta_J = 0.8735(15)$				
			$B=3287.3556(4)$	$\Delta_{JK} = -1.4419(86)$				
			$C=2311.2182(4)$	$\Delta_K = 10.1623(55)$				
				$\delta_J = 0.3068(9)$				
				$\delta_K = 1.7421(151)$				
				$H_J = -0.0072(31) \cdot 10^{-3}$				
				$H_{JK} = 0.1833(831) \cdot 10^{-3}$				
				$H_{KJ} = -0.5679(2425) \cdot 10^{-3}$				
				$H_K = 0.3810(1601) \cdot 10^{-3}$				
				$h_J = 0.0043(20) \cdot 10^{-3}$				
				$h_{JK} = -0.0914(514) \cdot 10^{-3}$				
Cont.				$h_K = 1.0612(4831) \cdot 10^{-3}$				

Nr.	Molecule	v	A, B, C MHz	D, H kHz	Ref.	Ref. struct.	Varia, remarks	See also Tab., Nr.
126 *6–67a* *4–66* Cont.	$^{18}OC^{35}Cl_2$	$v=0$	$A_{cor}=7284.877(12)$ *) $B_{cor}=3474.968(9)$ $C_{cor}=2349.850(7)$	$\tau_{aaaa}=-31.6(15)$ *) $\tau_{bbbb}=-6.65(30)$ $\tau_{aabb}=7.6(13)$ $\tau_{abab}=-5.7(6)$	80NA1		*) uncertainties are 3σ, $A_{cor}=A+(1/4)(2\tau_{abab})$ $B_{cor}=B+(1/4)(2\tau_{abab})$ $C_{cor}=C-(1/4)(3\tau_{abab})$	
	$O^{13}C^{35}Cl_2$	$v=0$	$A_{cor}=7888.433(10)$ *) $B_{cor}=3475.294(5)$ $C_{cor}=2409.549(4)$	$\tau_{aaaa}=-39.6(11)$ $\tau_{bbbb}=-6.64(12)$ $\tau_{aabb}=8.4(5)$ $\tau_{abab}=-5.8(3)$				
	$^{18}OC^{37}Cl_2$	$v=0$	$A_{cor}=7183.561(15)$ *) $B_{cor}=3287.366(7)$ $C_{cor}=2252.634(4)$	$\tau_{aaaa}=-32.3(15)$ $\tau_{bbbb}=-5.91(26)$ $\tau_{aabb}=7.4(3)$ $\tau_{abab}=-5.2(2)$				
	$O^{13}C^{37}Cl_2$	$v=0$	$A_{cor}=7782.721(14)$ *) $B_{cor}=3287.674(6)$ $C_{cor}=2308.591(5)$	$\tau_{aaaa}=-38.2(20)$ $\tau_{bbbb}=-5.91(16)$ $\tau_{aabb}=7.5(7)$ $\tau_{abab}=-5.4(4)$				
	$^{18}O^{13}C^{35}Cl_2$	$v=0$	$A_{cor}=7262.628(9)$ *) $B_{cor}=3475.302(4)$ $C_{cor}=2347.681(4)$	$\tau_{aaaa}=-32.2(8)$ $\tau_{bbbb}=-6.59(8)$ $\tau_{aabb}=7.4(3)$ $\tau_{abab}=-5.4(2)$				
	$^{18}O^{13}C^{37}Cl_2$	$v=0$	$A_{cor}=7160.063(18)$ *) $B_{cor}=3287.687(11)$ $C_{cor}=2250.465(8)$	$\tau_{aaaa}=-33.2(21)$ $\tau_{bbbb}=-5.93(33)$ $\tau_{aabb}=7.3(14)$ $\tau_{abab}=-5.1(7)$				
127 Cont.	$OC^{79}Br_2$	$v=0$	$A=6224.8643(23)$ $B=1264.0118(6)$ $C=1049.9390(6)$	$\Delta_J=0.1739(2)$ $\Delta_{JK}=-0.838(3)$ $\Delta_K=8.881(16)$ $\delta_J=0.03916(4)$ $\delta_K=0.511(6)$ $H_J=0.096(17)\cdot10^{-6}$ $H_{JK}=0.4(9)\cdot10^{-6}$ $H_{KJ}=47(13)\cdot10^{-6}$ $H_K=0.10(12)\cdot10^{-3}$ $h_J=0.144(6)\cdot10^{-6}$ $h_{JK}=-0.0119(9)\cdot10^{-3}$ $h_K=6(33)\cdot10^{-6}$	77CAR	77CAR	harmonic force field calculated [77CAR]	2.7.3, 150

Nr.	Molecule	v	A, B, C MHz	D, H kHz	Ref.	Ref. struct.	Varia, remarks	See also Tab., Nr.
127 Cont.	$OC^{79}Br^{81}Br$	$v=0$	$A=6214.6356(21)$ $B=1248.3769(4)$ $C=1038.8430(4)$	$\Delta_J = 0.1697(2)$ $\Delta_{JK} = -0.854(2)$ $\Delta_K = 8.870(15)$ $\delta_J = 0.03760(4)$ $\delta_K = 0.545(4)$ $H_J = -0.065(14)\cdot10^{-6}$ $H_{JK} = 1.7(6)\cdot10^{-6}$ $H_{KJ} = -14(12)\cdot10^{-6}$ $H_K = 0.08(10)\cdot10^{-3}$ $h_J = -0.019(5)\cdot10^{-6}$ $h_{JK} = -0.6(7)\cdot10^{-6}$ $h_K = 0.034(21)\cdot10^{-3}$	77CAR			
	$OC^{81}Br_2$	$v=0$	$A=6204.3856(31)$ $B=1232.8231(6)$ $C=1027.7614(6)$	$\Delta_J = 0.1692(2)$ $\Delta_{JK} = -0.854(2)$ $\Delta_K = 8.982(54)$ $\delta_J = 0.03794(3)$ $\delta_K = 0.611(4)$ $H_J = 0.027(17)\cdot10^{-6}$ $H_{JK} = -1.0(7)\cdot10^{-6}$ $H_{KJ} = 25(17)\cdot10^{-6}$ $H_K = 0.62(30)\cdot10^{-3}$ $h_J = 0.016(5)\cdot10^{-6}$ $h_{JK} = -1.2(6)\cdot10^{-6}$ $h_K = -8(26)\cdot10^{-6}$				
128 *6-69* *4-67*	$OCF^{35}Cl$	$v=0$	$A=11830.35(3)$ $B=5286.93(3)$ $C=3648.69(3)$		76SCA1	80OBE	for structure, see also Vol. II/7 (3, 8)	2.7.3, 104 2.9.3, 19
129	$OC(NH_2)_2$	$v=0$	$A=11233.333(8)$ *) $B=10369.369(8)$ $C=5416.668(6)$	$\tau_{aaaa} = -44.23(13)$ $\tau_{bbbb} = -40.75(13)$ $\tau_{aabb} = 22.68(41)$ $\tau_{abab} = -28.06(13)$	75BRW1		*) $\Delta = -0.42619(16)$ amu Å^2	2.6.3, 63 2.7.3, 27

Nr.	Molecule	v	A, B, C MHz	D, H kHz	Ref.	Ref. struct.	Varia, remarks	See also Tab., Nr.
130	$F_2S{=}NC({=}O)F$	$v=0$	$A=4563.58\,(8)$ $B=1476.86\,(5)$ $C=1436.52\,(4)$		80BAI	80BAI	lone pairs of electrons on the S and N being cis to each other, the F on the COF group eclipsing the lone pair on the N	
	$F_2{}^{34}S{=}NC({=}O)F$	$v=0$	$A=4526.3\,(9)$ $B=1466.23\,(4)$ $C=1423.80\,(4)$					
131	$^{35}ClC({=}O)OCH_3$	$v=0$	$A=9766.09\,(8)$ *) $B=2692.21\,(1)$ $C=2138.04\,(1)$	$d_J = 0.3025\,(1199)$ $d_{JK} = -0.7935\,(13324)$ $d_K = 44.7171\,(202795)$ $d_{WJ} = 0.0274\,(258)$ $d_{WK} = 2.9107\,(10973)$	73LIS1	77DUR2	*) for the A species planar heavy atom skeleton with the CH_3 group cis to the C=O group	2.7.3, 105 2.8, 82
	$^{37}ClC({=}O)OCH_3$	$v=0$	$A=9739.25\,(24)$ *) $B=2623.03\,(3)$ $C=2092.94\,(3)$	$d_J = 0.2933\,(2377)$ $d_{JK} = -10.6278\,(40490)$ $d_K = 195.6926\,(605291)$ $d_{WJ} = -0.0295\,(629)$ $d_{WK} = 6.2178\,(31872)$				
	$^{35}ClC({=}O)OCD_3$	$v=0$	$A=9172.25\,(18)$ $B=2413.68\,(1)$ $C=1957.65\,(1)$		77DUR2			
	$^{35}ClC({=}O)O^{13}CH_3$	$v=0$	$A=9738.40\,(30)$ $B=2620.09\,(1)$ $C=2091.18\,(1)$					
	$^{35}ClC({=}O)OCH_2D$ out of plane	$v=0$	$A=9619.16\,(15)$ $B=2594.66\,(1)$ $C=2082.27\,(1)$					
	$^{35}ClC({=}O)OCH_2D$ in plane	$v=0$	$A=9465.36\,(30)$ $B=2582.49\,(1)$ $C=2054.86\,(1)$					
131a	$FC({=}O)SCH_3$	$v=0$	$A=7706.4\,(4)$ *) $B=3215.94\,(1)$ $C=2301.07\,(2)$	$\Delta_J = 0.5\,(2)$ $\Delta_{JK} = 0.2\,(7)$	81CAM2	81CAM2	syn conformation *) for the A species	2.6.3, 63a 2.8, 97a
Cont.		$v_t=1$	$A=7661.3\,(1)$ $B=3206.41\,(1)$ $C=2303.18\,(1)$	$\Delta_J = 0.42\,(7)$ $\Delta_{JK} = -2.3\,(3)$			$v_t=121\,(30)\ cm^{-1}$, skeletal torsion	

Nr.	Molecule	v	A, B, C MHz	D, H kHz	Ref.	Ref. struct.	Varia, remarks	See also Tab., Nr.
131a Cont.	FC(=O)SCH$_3$	$v_t=2$	$A=7610.5$ (3) $B=3195.29$ (1) $C=2305.46$ (3)	$\Delta_J=0.8$ (2) $\Delta_{JK}=1.7$ (7)	81CAM2			
		$v_\tau=1$	$A=7655.6$ (2) $B=3192.67$ (1) $C=2296.39$ (1)	$\Delta_J=0.65$ (12) $\Delta_{JK}=-13.0$ (6)			$v_\tau=109$ (30) cm^{-1}, methyl torsion	
		$v_\tau=2$	$A=8103.4$ (4) $B=3255.67$ (2) $C=2290.77$ (3)	$\Delta_J=0.2$ (3) $\Delta_{JK}=35$ (1)				
		$v_p=1$	$A=7800.6$ (4) $B=3216.73$ (2) $C=2294.44$ (2)	$\Delta_J=0.4$ (2) $\Delta_{JK}=-5.2$ (8)			$v_p=230$ (40) cm^{-1}, CSC bend	
		$v_x=1$	$A=7650$ (fixed) $B=3193$ (2) $C=2294$ (3)				$v_x=250$ (50) cm^{-1}	
		$v_y=1$	$A=7650$ (fixed) $B=3228$ (2) $C=2335$ (2)				$v_y=300$ (50) cm^{-1}	
132	^{35}ClC(=O)SCH$_3$	$v=0$	$A=6986.1$ (3) $B=2033.35$ (1) $C=1590.28$ (1)	$D_J=0.20$ (3)	80CAM3	80CAM3	syn conformation	2.7.3, 106 2.8, 98
	^{37}ClC(=O)SCH$_3$	$v=0$	$A=6971.6$ (5) $B=1979.57$ (1) $C=1556.45$ (1)	$D_J=0.21$ (9)				
132a	NCC(=O)SCH$_3$	$v=0$	$A=6774.8$ (2) *) $B=1961.851$ (6) $C=1535.030$ (7)	$\Delta_J=0.23$ (2)	81CAM3		syn conformation *) for the A species	2.8, 98a
		$v_\tau=1$	$A=6612$ (9) $B=1945.3$ (2) $C=1534.7$ (3)	$\Delta_J=10$ (1) $\Delta_{JK}=-16$ (4)			$v_\tau=110$ (20) cm^{-1}, methyl torsion	
Cont.		$v_\tau=2$	$A=7394.9$ (5) $B=1976.65$ (1) $C=1530.42$ (2)	$\Delta_J=0.29$ (6) $\Delta_{JK}=4.8$ (5)				

Nr.	Molecule	v	A, B, C MHz	D, H kHz	Ref.	Ref. struct.	Varia, remarks	See also Tab., Nr.
132a Cont.	$NCC(=O)SCH_3$	$v_t=1$	$A=6722\ (13)$ $B=1957.2\ (3)$ $C=1537.1\ (4)$	$\Delta_J\ =0.23$ (fixed)	81CAM3		$\nu_t=104\ (20)\ cm^{-1}$, skeletal torsion	
		$v_p=1$	$A=6939.4\ (4)$ $B=1968.72\ (1)$ $C=1535.18\ (1)$	$\Delta_J\ =0.24\ (7)$ $\Delta_{JK}=6.6\ (5)$			$\nu_p=171\ (30)\ cm^{-1}$, CSC in plane bend	
133 *6-66*	$HC(=O)OCH_2CH_3$	$v=0$	$A=9985.580\ (10)$ $B=3839.596\ (5)$ $C=3212.869\ (4)$	$d_J\ =\ -38.676\ (391)$ $d_{JK}\ =-260.78\ (1029)$ $d_K\ =-190.72\ (1661)$ $d_{WJ}=0.12620\ (111)\cdot 10^{-4}$ $d_{WK}=0.41726\ (2769)\cdot 10^{-4}$	80KU1			
		$v_\tau=1$	$A=10245.938\ (49)$ $B=\ 3755.641\ (11)$ $C=\ 3179.266\ (10)$	$d_J\ =\ -51.971\ (1369)$ $d_{JK}\ =-313.65\ (4664)$ $d_K\ =-169.93\ (8937)$ $d_{WJ}=0.17127\ (404)\cdot 10^{-4}$ $d_{WK}=0.43575\ (12904)\cdot 10^{-4}$			ν_τ, C—O torsion	
134	$FC(=O)OCH_2CH_3$ extended form *)	$v=0$	$A=9191\ (1)$ $B=2112.61\ (1)$ $C=1756.73\ (1)$		78TR1		*) syn-anti planar heavy atom structure: τ_1 (O=COC)$=0°$, τ_2 (COCC)$=180°$ $\nu_a=70\ (10)\ cm^{-1}$, torsion around the O-ethyl bond	
		$v_a=1$	$A=9134\ (1)$ $B=2113.08\ (1)$ $C=1761.02\ (1)$					
		$v_a=2$	$A=9071\ (2)$ $B=2114.09\ (2)$ $C=1766.28\ (2)$					
		$v_a=3$	$A=9013\ (1)$ $B=2115.70\ (1)$ $C=1772.87\ (1)$					
		$v_a=4$	$A=8950\ (30)$ $B=2118.5\ (3)$ $C=1781.5\ (3)$					
Cont.		$v_b=1$	$A=9135\ (2)$ $B=2111.94\ (2)$ $C=1758.29\ (2)$				$\nu_b=110\ (15)\ cm^{-1}$, torsion around the O-ester bond	

Nr.	Molecule	v	A, B, C MHz	D, H kHz	Ref.	Ref. struct.	Varia, remarks	See also Tab., Nr.
134 Cont.	FC(=O)OCH$_2$CH$_3$ compact form**)	$v=0$	$A=7760\,(3)$ $B=2388.34\,(4)$ $C=2102.47\,(3)$		78TR1		**) syn-gauche structure: τ_1 (O=COC) = 0°, τ_2 (COCC) ≃ 90° The 2 conformers have approximately equal energy: (0 ± 40) cm^{-1} $v_a=90\,(10)$ cm^{-1}, O-ethyl torsion	
		$v_a=1$	$A=7761\,(4)$ $B=2368.61\,(3)$ $C=2090.05\,(2)$					
		$v_a=2$	$A=7798\,(4)$ $B=2344.62\,(4)$ $C=2073.57\,(4)$					
		$v_a=3$	$A=8295\,(7)$ $B=2331.52\,(7)$ $C=2060.91\,(5)$					
		$v_b=1$	$A=7720\,(10)$ $B=2388.89\,(7)$ $C=2104.47\,(8)$				$v_b=120\,(20)$ cm^{-1}, O-ester torsion	
135	^{35}ClC(=O)OCH$_2$CH$_3$ extended form**)	$v=0$	$B+C=2819\,(3)$*)		76TR1		*) LRMW **) cis-trans heavy-atom planar structure: τ (O=COC) = 0°, τ (COCC) = 180° ***) cis-gauche structure: τ (O=COC) = 0°, τ (COCC) ≃ 90° $^+$) gauche-gauche structure: τ (O=COC) ≃ 45°, τ (COCC) ≃ 240° the extended and compact forms have nearly equal energy and the intermediate species have 1⋯2 kcal/mol higher energy	
	^{37}ClC(=O)OCH$_2$CH$_3$ extended form**)	$v=0$	$B+C=2762\,(3)$*)					
	^{35}ClC(=O)OCH$_2$CH$_3$ compact form***)	$v=0$	$B+C=3061\,(3)$*)					
	^{37}ClC(=O)OCH$_2$CH$_3$ compact form***)	$v=0$	$B+C=2989\,(3)$*)					
	^{35}ClC(=O)OCH$_2$CH$_3$ intermediate form$^+$)	$v=0$	$B+C=2954\,(5)$*)					
	^{37}ClC(=O)OCH$_2$CH$_3$ intermediate form$^+$)	$v=0$	$B+C=2885\,(17)$*)					

Nr.	Molecule	v	A, B, C MHz	D, H kHz	Ref.	Ref. struct.	Varia, remarks	See also Tab., Nr.
136	$CNC(=O)OCH_2CH_3$ extended form *)	$v=0$	$A''=6453.32\,(36)$ $B''=1500.474\,(6)$ $C''=1236.362\,(6)$	$\tau_1=6.67\,(59)$ $\tau_2=0.655\,(109)$ $\tau_3=16\,(17)$ $\tau_{bbbb}=-0.956\,(95)$ $\tau_{cccc}=-0.17\,(8)$	78SU1		*) syn-anti structure: $\tau_1\,(O{=}COC)=0°$, $\tau_2\,(COCC)=180°$ the extended form is more stable than the compact form: $\Delta E=(55\pm27)\,cm^{-1}$	2.6.3, 64
		$v_x=1$	$A=6428.3\,(29)$ $B=1501.25\,(3)$ $C=1238.83\,(3)$					
	compact form **)	$v=0$	$A''=6787.82\,(71)$ $B''=1549.375\,(8)$ $C''=1406.801\,(8)$	$\tau_1=-1.8\,(10)$ $\tau_2=-1.0\,(2)$ $\tau_3=21\,(28)$ $\tau_{bbbb}=-1.080\,(117)$ $\tau_{cccc}=-0.944\,(122)$			**) syn-gauche structure: $\tau_1\,(O{=}COC)=0°$, $\tau_2\,(COCC)\simeq90°$ ***) gauche-gauche structure: $\tau_1\,(O{=}COC)\simeq45°$, $\tau_2\,(COCC)\simeq240°$ relative energy: $(600\pm140)\,cm^{-1}$	
		$v_x=1$	$A=6898\,(16)$ $B=1549.71\,(12)$ $C=1401.04\,(17)$					
	intermediate form ***)	$v=0$	$B+C=2870\,(2)$ +)		76TR1		+) LRMW	
137	$^{35}ClC(=O)SCH_2CH_3$ extended form +)	$v=0$	$B+C=2171\,(3)$*)		77SIL		*) LRMW +) syn-anti structure: $\tau_1\,(OCSC)=0°$, $\tau_2\,(CSCC)=180°$ ++) syn-gauche structure: $\tau_1\,(OCSC)=0°$, $\tau_2\,(CSCC)\simeq90°$ +++) gauche-gauche structure: $\tau_1\,(OCSC)\simeq60°$, $\tau_2\,(CSCC)\simeq240°$ the relative energies of the extended, compact and intermediate forms are 0, 0.5 and ≈2.0 kcal/mol, respectively	
	$^{35}ClC(=O)SCH_2CH_3$ compact form ++)	$v=0$ $v_x=1$	$B+C=2486.2\,(6)$*) $B+C=2472.8\,(6)$*)					
	$^{37}ClC(=O)SCH_2CH_3$ compact form ++)	$v=0$ $v_x=1$	$B+C=2428\,(1)$*) $B+C=2413\,(1)$*)					
	$^{35}ClC(=O)SCH_2CH_3$ intermediate form +++)	$v=0$	$B+C=2368\,(3)$*)					
	$^{37}ClC(=O)SCH_2CH_3$ intermediate form +++)	$v=0$	$B+C=2314\,(4)$*)					

Nr.	Molecule		v	A, B, C MHz	D, H kHz	Ref.	Ref. struct.	Varia, remarks	See also Tab., Nr.
138	$HC(=O)OCH_2CH_2CH_3$,	EA	$v=0$	$B+C=3063\,(2)$*)		78TR5		*) LRMW E: $\tau_1\,(O{=}COC)=0°$, $\tau_2\,(COCC)=180°$ C: $\tau_1=0°$, $\tau_2\simeq 90°$ I: $\tau_1\simeq 50°$, $\tau_2\simeq 230°$ G: $\tau_3\,(OCCC)=60°$ A: $\tau_3=180°$	
		EG	$v=0$	$B+C=3454\,(1)$*)					
			$v_x=1$	$B+C=3466\,(1)$*)					
			$v_x=2$	$B+C=3482\,(1)$*)					
			$v_x=3$	$B+C=3500\,(1)$*)					
			$v_x=4$	$B+C=3521\,(1)$*)					
			$v_y=1$	$B+C=3488\,(1)$*)					
139	$FC(=O)OCH_2CH_2CH_3$,	EA	$v=0$	$B+C=2198\,(1)$*)		78TR5		*) LRMW see Nr. 138 for definitions	
		EG	$v=0$	$B+C=2595\,(5)$*)					
		CA	$v=0$	$B+C=2512\,(1)$*)					
			$v_x=1$	$B+C=2501\,(1)$*)					
		CG	$v=0$	$B+C=2989\,(1)$*)					
140	$^{35}ClC(=O)OCH_2CH_2CH_3$,	EA	$v=0$	$B+C=1668\,(2)$*)		78TR5		*) LRMW see Nr. 138 for definitions	
		EG	$v=0$	$B+C=1999\,(3)$*)					
		CA	$v=0$	$B+C=1830\,(2)$*)					
141	$NC{-}C(=O)OCH_2CH_2CH_3$,	EA	$v=0$	$B+C=1676\,(1)$*)		78TR5		*) LRMW see Nr. 138 for definitions the I species have ~2 kcal/mol higher energy	
		EG	$v=0$	$B+C=2004\,(2)$*)					
		CA	$v=0$	$B+C=1825\,(1)$*)					
			$v_x=1$	$B+C=1820\,(1)$*)					
		IA	$v=0$	$B+C=1761\,(1)$*)					
		IG	$v=0$	$B+C=2101\,(2)$*)					
142	$FC(=O)SCH_2CH_2CH_3$,	CA	$v=0$	$B+C=2156.2\,(5)$*)		78SIL		*) LRMW see Nr. 138 for definitions	
			$v_x=1$	$B+C=2140\,(1)$					
143	$^{35}ClC(=O)SCH_2CH_2CH_3$,	CA	$v=0$	$B+C=1579.0\,(5)$*)		78SIL		*) LRMW see Nr. 138 for definitions The CA conformer is the most stable	
			$v_x=1$	$B+C=1571\,(1)$*)					
		IA	$v=0$	$B+C=1533\,(2)$*)					
	$^{37}ClC(=O)SCH_2CH_2CH_3$,	CA	$v=0$	$B+C=1545.7\,(5)$*)					
			$v_x=1$	$B+C=1533\,(2)$*)					

Nr.	Molecule	v	A, B, C MHz	D, H kHz	Ref.	Ref. struct.	Varia, remarks / See also Tab., Nr.
144	NC—C(=O)SCH$_2$CH$_2$CH$_3$, CA IA IG	$v=0$ $v_x=1$ $v=0$ $v=0$	$B+C=1600.3\,(5)$*) $B+C=1593.0\,(5)$*) $B+C=1556\,(1)$*) $B+C=1760\,(1)$*)		78SIL		*) LRMW see Nr. 138 for definitions The CA conformer is the most stable
145 *6-73*	HC(=S)H	$v=0$ $v_4=1$ $v_6=1$ $v_3=1$	$A=9.726991\,(216)$ *) $B=0.5903919\,(33)$ $C=0.5554514\,(33)$ $A=9.65447\,(182)$ *) $B=0.58521\,(30)$ $C=0.55651\,(40)$ $A=9.8052\,(185)$ *) $B=0.58758$ $C=0.56075$ $A=9.716001\,(717)$ *) $B=0.58759\,(17)$ $C=0.55204\,(17)$	$\Delta_J = 0.659\,(13)\cdot 10^{-6}$ *) $\Delta_{JK} = 0.1784\,(64)\cdot 10^{-4}$ $\Delta_K = 0.7622\,(207)\cdot 10^{-3}$ $\delta_J = 0.406\,(83)\cdot 10^{-7}$ $\delta_K = 0.1109\,(89)\cdot 10^{-4}$ $\Delta_J = -0.38\,(69)\cdot 10^{-6}$ $\Delta_{JK} = 0.975\,(161)\cdot 10^{-4}$ $\Delta_K = -0.777\,(113)\cdot 10^{-3}$ **) $\Delta_J = 0.32\,(69)\cdot 10^{-6}$ $\Delta_{JK} = 0.873\,(150)\cdot 10^{-4}$ $\Delta_K = -0.613\,(118)\cdot 10^{-3}$ **) $\Delta_K = 0.764\,(52)\cdot 10^{-3}$ **)	80BED		for structure, 2.6.3, 65 see also 2.9.3, 20 Vol. II/7 (3, 81) *) Laser Stark data, all values in cm^{-1}, uncertainties are 3σ **) The other constants are fixed at the ground state values $\nu_4=990.1866\,(33)$ cm^{-1}, out of plane bend $\nu_6=991.0149\,(35)$ cm^{-1}, HCH wag $\nu_3=1059.2037\,(33)$ cm^{-1}, CS stretch $\xi^a_{64}=10.004199\,(789)$ cm^{-1} $\simeq 2A\,\zeta^a_{64}$ $\eta^{bc}_{64}=0.001623\,(144)$ cm^{-1} $\xi^b_{34}=0.0547\,(101)$ cm^{-1} $\simeq 2B\,\zeta^b_{34}$ $\xi^c_{36}=0.3085\,(116)$ cm^{-1} $\simeq 2C\,\zeta^c_{36}$
146	F_2CS $F_2C^{34}S$	$v=0$ $v=0$	$A=11892.57\,(50)$ *) $B=5133.034\,(30)$ $C=3580.316\,(30)$ $A=11891.94\,(100)$ $B=4959.011\,(50)$ $C=3494.899\,(50)$		73CAR	73CAR, 80CHR2	for structure, 2.6.3, 66 see also Vol. II/7 (3, 35) *) $\Delta=0.203\,(4)$ amu Å^2

Nr.	Molecule	v	A, B, C MHz	D, H kHz	Ref.	Ref. struct.	Varia, remarks	See also Tab., Nr.
147	$SC^{35}Cl_2$	$v=0$	$A=3557.3089(3)$ $B=3464.1108(3)$ $C=1753.1000(2)$	$\Delta_J = 0.6508(6)$ *) $\Delta_{JK} = -0.2680(11)$ $\Delta_K = 1.2997(7)$ $\delta_J = 0.2782(2)$ $\delta_K = 0.5395(6)$	75CAR	75CAR	for structure, see also Vol. II/7 (3, 18) harmonic force field calculated [75CAR] *) representation I^r	2.7.3, 107
	$SC^{35}Cl^{37}Cl$	$v=0$	$A=3503.3107(4)$ $B=3391.2987(3)$ $C=1721.3039(2)$	$\Delta_J = 0.6652(4)$ *) $\Delta_{JK} = -0.1910(6)$ $\Delta_K = 0.9504(4)$ $\delta_J = 0.2865(1)$ $\delta_K = 0.6164(2)$				
	$SC^{37}Cl_2$	$v=0$	$A=3402.751(6)$ $B=3365.254(5)$ $C=1690.109(3)$	$\Delta_J = 0.863(16)$ *) $\Delta_{JK} = -1.429(47)$ $\Delta_K = 1.911(56)$ $\delta_J = 0.380(3)$ $\delta_K = 0.287(17)$				
	$^{34}SC^{35}Cl_2$	$v=0$	$A=3557.279(71)$ $B=3327.156(46)$ $C=1717.308(5)$	$\Delta_J = 1.170(397)$ *) $\Delta_{JK} = -7.120(4939)$ $\Delta_K = 10.119(6426)$ $\delta_J = 0.534(201)$ $\delta_K = -1.898(1821)$				
148 Cont.	$SCF^{35}Cl$	$v=0$	$A=7984.462(1)$ $B=3437.360(1)$ $C=2400.045(1)$	$\Delta_J = 0.6930(7)$ $\Delta_{JK} = -0.6026(47)$ $\Delta_K = 11.0249(32)$ $\delta_J = 0.2452(4)$ $\delta_K = 1.7165(91)$ $H_J = -3.98(86) \cdot 10^{-6}$ $H_{JK} = 52.92(1780) \cdot 10^{-6}$ $H_{KJ} = -196.6(476) \cdot 10^{-6}$ $H_K = 185.03(3170) \cdot 10^{-6}$ $h_J = 1.306(465) \cdot 10^{-6}$ $h_{JK} = -30.45(1350) \cdot 10^{-6}$ $h_K = 238.1(851) \cdot 10^{-6}$	77HAM	78HAM	harmonic force field calculated [78HAM]	2.7.3, 108

Nr.	Molecule	v	A, B, C MHz	D, H kHz	Ref.	Ref. struct.	Varia, remarks	See also Tab., Nr.
148 Cont.	$SCF^{35}Cl$	$v_3=1$	$A=7989.756(9)$ $B=3428.370(4)$ $C=2394.754(4)$	$\Delta_J = 0.572(52)$ $\Delta_{JK}=-0.751(189)$ $\Delta_K = 11.65(100)$ $\delta_J = 0.243(7)$ $\delta_K=1.79(12)$	78HAM		ν(C—Cl)$=611.9$ cm^{-1}	
		$v_4=1$	$A=7973.018(4)$ $B=3441.661(3)$ $C=2403.216(3)$	$\Delta_J = 0.753(31)$ $\Delta_{JK}=-0.520(74)$ $\Delta_K = 10.83(38)$ $\delta_J = 0.253(3)$ $\delta_K = 1.51(6)$			ν_4(SCF)$=427.1$ cm^{-1}	
		$v_5=1$	$A=8001.594(10)$ $B=3438.451(5)$ $C=2394.201(4)$	$\Delta_J = 0.474(51)$ $\Delta_{JK}=-0.482(155)$ $\Delta_K= 12.12(85)$ $\delta_J = 0.262(7)$ $\delta_K = 1.60(12)$			ν_5(SCCl)$=322.6$ cm^{-1}	
		$v_5=2$	$A=8018.777(9)$ $B=3439.496(4)$ $C=2388.408(4)$	$\Delta_J = 0.516(46)$ $\Delta_{JK}=-0.345(122)$ $\Delta_K = 10.52(60)$ $\delta_J = 0.258(6)$ $\delta_K = 1.71(10)$				
		$v_6=1$	$A=7973.878(12)$ $B=3436.888(7)$ $C=2402.108(7)$	$\Delta_J = 0.984(92)$ $\Delta_{JK}=-0.395(142)$ $\Delta_K = 10.04(73)$ $\delta_J = 0.250(7)$ $\delta_K=1.61(12)$			ν_6(out-of-plane)$=539.8$ cm^{-1}	
	$SCF^{37}Cl$	$v=0$	$A=7928.589(3)$ $B=3345.859(1)$ $C=2350.175(1)$	$\Delta_J = 0.6572(18)$ $\Delta_{JK}=-0.5790(88)$ $\Delta_K = 10.8202(120)$ $\delta_J = 0.23217(36)$ $\delta_K = 1.6362(63)$	75KOH			
	$^{34}SCF^{35}Cl$	$v=0$	$A=7929.024(3)$ $B=3332.375(1)$ $C=2343.547(1)$	$\Delta_J = 0.645(15)$ $\Delta_{JK}=-0.584(38)$ $\Delta_K = 10.859(269)$ $\delta_J = 0.2294(15)$ $\delta_K = 1.650(28)$				
Cont.								

Nr.	Molecule	v	A, B, C MHz	D, H kHz	Ref.	Ref. struct.	Varia, remarks	See also Tab., Nr.
148 Cont.	$^{34}SCF^{37}Cl$	$v=0$	$A=7870.566\,(6)$ $B=3243.625\,(4)$ $C=2294.345\,(4)$	$\Delta_J=0.699\,(27)$ $\Delta_{JK}=-0.424\,(82)$ $\Delta_K=10.18\,(46)$ $\delta_J=0.2254\,(35)$ $\delta_K=1.462\,(68)$	76HAM2			
	$S^{13}CF^{35}Cl$	$v=0$	$A=7977.512$ $B=3437.395$ $C=2399.434$		75KOH			
149	HC(=S)NH_2	$v=0$	$A=61753.80$ *) $B=6101.74$ $C=5549.96$		74SUG	74SUG	for structure, see also Vol. II/7 (3, 117)	2.6.3, 67 2.7.3, 28
		$v_I=1$	$B=6097.50$ $C=5552.70$				*) $\Delta=0.0508$ amu Å^2 v_I (NH_2 wag) = 393 (40) cm^{-1}	
		$v_{II}=1$	$B=6101.18$ $C=5544.68$				v_{II} (N—C—S bend) = 457 (50) cm^{-1}	
	HC(=S)$^{15}NH_2$	$v=0$	$A=61297.1$ **) $B=5919.61$ $C=5395.43$				**) Calculated using estimated inertial defects	
	$H^{13}C$(=S)NH_2	$v=0$	$A=59980.3$ **) $B=6085.09$ $C=5521.35$					
	HC(=^{34}S)NH_2	$v=0$	$A=61630.62$ $B=5952.24$ $C=5425.12$					
	HC(=S)NHD, cis ***)	$v=0$	$A=52178.52$ $B=5935.81$ $C=5326.07$				***) D cis to C=S	
		$v_I=1$	$B=5930.24$ $C=5327.04$					
Cont.		$v_{II}=1$	$B=5934.61$ $C=5321.50$					

Nr.	Molecule	v	A, B, C MHz	D, H kHz	Ref.	Ref. struct.	Varia, remarks	See also Tab., Nr.
149 Cont.	HC(=S)NHD, trans [+])	$v=0$	$A=61506.00$ $B=5672.84$ $C=5192.28$		74SUG		[+]) D trans to C=S	
		$v_I=1$	$B=5674.38$ $C=5200.88$					
		$v_{II}=1$	$B=5673.15$ $C=5188.15$					
	HC(=S)ND_2	$v=0$	$A=51827.45$ $B=5536.72$ $C=5000.90$					
		$v_I=1$	$B=5539.67$ $C=5010.77$				$\nu_I=293\,(30)\ cm^{-1}$	
		$v_{II}=1$	$B=5533.44$ $C=4997.64$				$\nu_{II}=393\,(40)\ cm^{-1}$	
150	HC(=S)SH trans	$v=0$	$A=49226.804\pm86$ $B=3447.5312\,(89)$ $C=3219.4954\,(96)$	$\Delta_J=1.100\,(100)$ $\Delta_{JK}=-39.049\,(240)$	78BAK1	79BA1		2.6.3, 68 3, 41
		$v_x=1$	$A=49464.119\pm110$ $B=3445.7699\,(110)$ $C=3216.7914\,(120)$	$\Delta_J=0.988\,(120)$ $\Delta_{JK}=-37.182\,(300)$				
		$v_y=1$	$A=49008.192\pm110$ $B=3441.1572\,(110)$ $C=3217.4529\,(120)$	$\Delta_J=1.037\,(120)$ $\Delta_{JK}=-39.052\,(290)$				
	HC(=^{34}S)SH trans	$v=0$	$A=49130.328\pm110$ $B=3346.2303\,(100)$ $C=3130.2542\,(100)$	$\Delta_J=1.131\pm0.12$ $\Delta_{JK}=-38.362\pm0.28$	79BA1			
	HC(=S)^{34}SH trans	$v=0$	$A=49572.301\pm170$ $B=3346.9168\pm0.018$ $C=3131.2455\pm0.016$	$\Delta_J=1.365\pm0.22$ $\Delta_{JK}=-46.488\pm0.44$				
Cont.	HC(=S)SD trans	$v=0$	$A=41649.500\pm84$ $B=3422.1357\pm0.013$ $C=3160.0450\pm0.012$	$\Delta_J=0.901\pm0.20$ $\Delta_{JK}=-24.919\pm0.50$				

Nr.	Molecule	v	A, B, C MHz	D, H kHz	Ref.	Ref. struct.	Varia, remarks	See also Tab., Nr.
150 Cont.	HC(=S)SH cis	$v=0$	$A=48947.359 \pm 380$ $B=3498.7192 \pm 0.024$ $C=3261.4335 \pm 0.030$	$\Delta_J=1.022 \pm 0.42$ $\Delta_{JK}=-48.380 \pm 1.2$	78BAK1		the trans conformer is more stable by $\approx 350\ cm^{-1}$/mol	
		$v_x=1$	$A=48897.542 \pm 440$ $B=3491.2760 \pm 0.027$ $C=3258.0100 \pm 0.034$	$\Delta_J=1.643 \pm 0.47$ $\Delta_{JK}=-50.513 \pm 1.4$				
		$v_y=1$	$A=48489.065 \pm 200$ $B=3501.4692 \pm 0.017$ $C=3262.5140 \pm 0.017$	$\Delta_J=0.960 \pm 0.31$ $\Delta_{JK}=-40.082 \pm 0.72$				
	HC(=S)SD cis	$v=0$	$A=45263.816 \pm 310$ $B=3382.9189 \pm 0.030$ $C=3144.9742 \pm 0.030$	$\Delta_J=2.817 \pm 0.51$ $\Delta_{JK}=-29.913 \pm 1.3$	79BA1			
151 *6-74* *4-68*	CH_3CH_2F	$v=0$	$A=36070.12\,(66)$ $B=9364.53\,(4)$ $C=8199.63\,(4)$	$d_{JK}=-0.264\,(64) \cdot 10^3$ $d_{EJ}=169\,(19)$ *)	80SH2	80ZEI	for structure, see also Vol. II/7 (3, 265)	2.8, 4
		$v_\tau=1$	$A=35961.47\,(79)$ $B=9340.29\,(4)$ $C=8189.96\,(4)$	$d_{JK}=-0.211\,(71) \cdot 10^3$ $d_{EJ}=170\,(21)$ *)			v_τ, methyl torsion	
		$v_\tau=2$	$A=35886.26\,(57)$ $B=9315.83\,(3)$ $C=8180.16\,(3)$	$d_{JK}=-0.095\,(50) \cdot 10^3$ $d_{EJ}=175\,(15)$ *)				
		$v_\delta=1$	$A=36347.78\,(37)$ $B=9367.05\,(2)$ $C=8182.00\,(2)$	$d_{JK}=-0.218\,(33) \cdot 10^3$ $d_{EJ}=173\,(10)$ *)			v_δ, skeletal bend Fermi resonance between $v_\delta=1$ and $v_\tau=2$ analyzed *) coefficient of $-E_rJ(J+1)$ where E_r is the reduced energy	
	CD_3CH_2F	$v=0$	$A=29258.66\,(39)$ $B=2972.49\,(2)$ $C=7105.33\,(2)$	$d_{JK}=-0.129\,(20) \cdot 10^3$ $d_{EJ}=115\,(7)$ *)				
		$v_\tau=1$	$A=29177.29\,(57)$ $B=7953.69\,(3)$ $C=7102.95\,(3)$	$d_{JK}=-0.125\,(29) \cdot 10^3$ $d_{EJ}=116\,(11)$ *)				
		$v_\tau=2$	$A=29194.64\,(12)$ $B=7949.96\,(4)$ $C=7100.33\,(4)$	$d_{JK}=-0.085\,(20) \cdot 10^3$ $d_{EJ}=96\,(13)$ *)				
		$v_\delta=1$	$A=29364.19\,(60)$ $B=7963.75\,(4)$ $C=7094.39\,(4)$	$d_{JK}=-0.124\,(30) \cdot 10^3$ $d_{EJ}=83\,(16)$ *)				

Nr.	Molecule	v	A, B, C MHz	D, H kHz	Ref.	Ref. struct.	Varia, remarks	See also Tab., Nr.
152 *4–69*	$CH_2DCH_2{}^{35}Cl$, symmetric	$v=0$	$I_a = 16.14961$ *) $I_b = 99.11836$ $I_c = 108.97414$		80ZEI	80ZEI	for structure, see also Vol. II/7 (3, 262) *) all values in amu Å^2	2.7.3, 109 2.9, 13
	$CH_2DCH_2{}^{35}Cl$, asymmetric	$v=0$	$I_a = 17.85325$ *) $I_b = 95.39749$ $I_c = 105.61887$					
	$CD_2HCD_2{}^{35}Cl$, symmetric	$v=0$	$I_a = 24.65870$ *) $I_b = 101.12846$ $I_c = 113.19159$					
	$CD_2HCD_2{}^{35}Cl$, asymmetric	$v=0$	$I_a = 22.83574$ *) $I_b = 104.59011$ $I_c = 116.59560$					
	$CD_3CH_2{}^{35}Cl$	$v=0$	$I_a = 19.63298$ *) $I_b = 105.78617$ $I_c = 115.96633$					
	$^{13}CH_3CH_2{}^{35}Cl$	$v=0$	$I_a = 16.25982$ *) $I_b = 94.89482$ $I_c = 104.88162$					
	$CH_3{}^{13}CH_2{}^{35}Cl$	$v=0$	$I_a = 16.55705$ *) $I_b = 92.34902$ $I_c = 102.63603$					
	$CD_3CD_2{}^{35}Cl$	$v=0$	$I_a = 24.69586$ *) $I_b = 107.89548$ $I_c = 119.97494$					
	$CH_3CDH^{35}Cl$	$v=0$	$I_a = 18.54812$ *) $I_b = 93.06741$ $I_c = 104.02575$					
153 *4–70*	$CH_2DCH_2{}^{79}Br$ symmetric	$v=0$	$I_a = 16.88010$ *) $I_b = 143.08166$ $I_c = 153.66795$		80ZEI	80ZEI	for structure, see also Vol. II/7 (3, 260) *) all values in amu Å^2	
	$CH_2DCH_2{}^{79}Br$ asymmetric	$v=0$	$I_a = 18.67586$ *) $I_b = 137.99162$ $I_c = 149.03416$					
Cont.	$CH_2DCH_2{}^{81}Br$ symmetric	$v=0$	$I_a = 16.88469$ *) $I_b = 143.96003$ $I_c = 154.55162$					

Nr.	Molecule	v	A, B, C MHz	D, H kHz	Ref.	Ref. struct.	Varia, remarks	See also Tab., Nr.
153 *4–70* Cont.	$CH_2DCH_2{}^{81}Br$ asymmetric	$v=0$	$I_a = 16.68126$ *) $I_b = 138.84161$ $I_c = 149.88983$		80ZEI		*) all values in amu $Å^2$	
154 *4–71*	CH_3CH_2I	$v=0$	$A' = 29116.32103\,(788)$ $B' = 2979.56389\,(77)$ $C' = 2796.45201\,(90)$	$D'_J = 1.1655\,(24)$ $D'_{JK} = -11.9521\,(84)$ $D'_K = 260.287\,(745)$ $\delta_J = 0.113329\,(156)$ $R'_6 = -0.004071\,(205)$	80BOU2		for structure, see also Vol. II/7 (3, 267)	2.7.3, 159 2.8, 5
		$v_\tau=1$	$A = 29515\,(41)$ $B = 2976.74\,(7)$ $C = 2791.59\,(6)$	ground state constants used	80BOU3		ν_{18}(a″) = 259 cm^{-1}, CH_3 torsion	
		$v_{11}=1$	$A = 28596\,(40)$ $B = 2969.52\,(4)$ $C = 2790.81\,(5)$				ν_{11}(a′) = 258 cm^{-1}, in-plane CCI bend	
155 *6–75* *4–72*	CH_3CH_2CN	$v=0$	$A' = 27663.68273\,(221)$ $B' = 4714.18782\,(36)$ $C' = 4235.08521\,(36)$	$D'_J = 3.008281\,(445)$ $D'_{JK} = -47.26068\,(375)$ $D'_K = 547.7751\,(585)$ $\delta_J = 0.6857343\,(992)$ $R'_6 = -0.0327961\,(363)$ $H'_J = 0.009340\,(163)\cdot 10^{-3}$ $H'_{JK} = -0.11811\,(214)\cdot 10^{-3}$ $H'_{KJ} = -1.54516\,(525)\cdot 10^{-3}$ $H'_K = 30.823\,(285)\cdot 10^{-3}$ $H'_5 = 0.0035876\,(461)\cdot 10^{-3}$ $1/2\,H'_6 = 0.002201\,(109)\cdot 10^{-3}$ $H'_{10} = 0.0004149\,(516)\cdot 10^{-3}$	78BUR	74HEI	for structure, see also Vol. II/7 (3, 393)	2.6.3, 69 2.7.3, 29 2.8, 6 3, 42
		$v_{23}=1$	$A = 29794.20\,(148)$ $B = 4713.81\,(38)$ $C = 4239.66\,(27)$	$D_J = 29\,(8)$ $D_{JK} = -160\,(31)$	74MAD		$\nu_{23} = 210.5$ cm^{-1}, CH_3 torsion	
Cont.		$v_{13}=1$	$A = 25656.24\,(33)$ $B = 4717.47\,(9)$ $C = 4296.95\,(6)$	$D_J = 1\,(2)$ $D_{JK} = 120\,(8)$			$\nu_{13} = 206.9\,(5)$ cm^{-1}, CCN—bend in-plane	

Nr.	Molecule	v	A, B, C MHz	D, H kHz	Ref.	Ref. struct.	Varia, remarks	See also Tab., Nr.
155	$^{13}CH_3CH_2CN$	$v=0$	$A=27342.259\,(50)$		74HEI			
6-75			$B=4598.057\,(27)$					
4-72			$C=4133.767\,(20)$					
Cont.	$CH_3{}^{13}CH_2CN$	$v=0$	$A=27045.499\,(51)$		74HEI			
			$B=4697.959\,(28)$					
			$C=4207.070\,(20)$					
	$CH_3CH_2{}^{13}CN$	$v=0$	$A=27635.046\,(57)$		74HEI			
			$B=4689.914\,(31)$					
			$C=4214.797\,(23)$					
	$CH_3CH_2C^{15}N$	$v=0$	$A=27541.965\,(67)$	$D_J=2.8\,(3)$	76HEI			
			$B=4574.777\,(15)$	$D_{JK}=-41\,(2)$				
			$C=4119.516\,(11)$	$D_K=549\,(14)$				
		$v_\tau=1$	$A=29161.55\,(48)$	$D_J=6\,(3)$				
			$B=4573.59\,(12)$	$D_{JK}=-49\,(21)$				
			$C=4121.18\,(8)$					
		$v_{13}=1$	$A=26044.19\,(11)$	$D_J=4.3\,(7)$	76HEI		$\nu_{13}=203.4\,(8)\ cm^{-1}$	
			$B=4582.83\,(3)$	$D_{JK}=32\,(5)$				
			$C=4128.53\,(2)$					
	CH_2DCH_2CN symmetric	$v=0$	$A=27650.897\,(49)$		74MAD			
			$B=4425.142\,(27)$					
			$C=4000.821\,(19)$					
	CH_2DCH_2CN asymmetric	$v=0$	$A=25022.652\,(41)$		74MAD			
			$B=4583.476\,(22)$					
			$C=4110.264\,(16)$					
	CH_3CD_2CN	$v=0$	$A=21943.195\,(70)$	$D_J=2.6\,(3)$	76HEI			
			$B=4600.206\,(15)$	$D_{JK}=-20\,(2)$				
			$C=4087.873\,(11)$	$D_K=252\,(15)$				
		$v_\tau=1$	$A=23021.49\,(7)$	$D_J=0.8\,(4)$	76HEI			
			$B=4599.55\,(2)$	$D_{JK}=-71\,(3)$				
			$C=4088.84\,(1)$					
		$v_{13}=1$	$A=20296.07\,(10)$	$D_J=4.3\,(6)$	76HEI		$\nu_{13}=204.7\,(5)\ cm^{-1}$	
			$B=4607.78\,(3)$	$D_{JK}=24\,(4)$				
			$C=4095.17\,(2)$					
	CD_3CH_2CN	$v=0$	$A=22865.104\,(53)$		74HEI			
			$B=4204.788\,(29)$					
Cont.			$C=3800.247\,(21)$					

Nr.	Molecule	v	A, B, C MHz	D, H kHz	Ref.	Ref. struct.	Varia, remarks	See also Tab., Nr.
155	CD_3CHDCN	$v=0$	$B=4169.499(62)$		74MAD			
6-75			$C=3736.824(62)$					
4-72	CD_3CD_2CN	$v=0$	$A=18584.64(53)$		74LI			
Cont.			$B=\ 4126.94(8)$					
			$C=\ 3682.72(7)$					
		$v_\tau=1$	$B=4121.73(21)$					
			$C=3682.11(19)$					
		$v_\tau=2$	$B=4117.03(10)$					
			$C=3681.22(9)$					
		$v_\tau=3$	$B=4112.33(16)$					
			$C=3680.13(14)$					
		$v_{13}=1$	$B=4142.12(15)$					
			$C=3686.94(14)$					
		$v_{13}=2$	$B=4157.07(13)$					
			$C=3691.00(11)$					
		$v_\tau=1;\ v_{13}=1$	$B=4135.69(21)$					
			$C=3686.52(19)$					
156	CH_3CH_2OH	$v=0$	$A_+=34015.81$ *)		80KA2		for structure, see also	2.6.3, 70
6-77	gauche		$B_+=\ 9188.637$				Vol. II/7 (3, 296)	2.8, 7
4-73			$C_+=\ 8099.983$				*) effective	
			$A_-=34331.86$				energy operator:	
			$B_-=\ 9194.421$				$H_{++}=A_+P_z^2+B_+P_y^2+C_+P_x^2$	
			$C_-=\ 8100.850$				$H_{--}=A_-P_z^2+B_-P_y^2+C_-P_x^2$	
	CH_3CH_2OD	$v=0$	$A_+=31765.0$ *)				$+\Delta$	
	gauche		$B_+=\ 8897.06$				$H_{+-}=D(P_xP_y+P_yP_x)$	
			$C_+=\ 7833.64$				$+E(P_xP_z+P_zP_x)$	
			$A_-=31798.1$				$D=25.24$ MHz; $E=159.00$ MHz	
			$B_-=\ 8894.40$				$\Delta=96739.27$ MHz (for OH)	
			$C_-=\ 7833.96$				$D=20.6$ MHz; $E=235.2$ MHz	
							$\Delta=17096.8$ MHz (for OD)	
	CH_2DCH_2OH	$v=0$	$A_+=33466.3$ *)				$\Delta=97273.0$ MHz	
	symmetric, gauche		$B_+=\ 8483.52$				D and E undetermined.	
			$C_+=\ 7532.63$					
			$A_-=33515.4$					
			$B_-=\ 8490.00$					
Cont.			$C_-=\ 7535.79$					

Nr.	Molecule	v	A, B, C MHz	D, H kHz	Ref.	Ref. struct.	Varia, remarks	See also Tab., Nr.
156 *6-77* *4-73* Cont.	CH_2DCH_2OH asymmetric, gauche	$v=0$	$A_I=31145.58$ **) $B_I=8807.33$ $C_I=7768.23$ $A_{II}=30951.84$ $B_{II}=8795.96$ $C_{II}=7783.41$		80KA2		**) with the hydroxyl hydrogen (deuterium) on the same side (form I) or on the opposite side (form II) of the molecular plane with respect to the methyl deuterium	
	CH_2DCH_2OD asymmetric, gauche	$v=0$	$A_I=28989.79$ **) $B_I=8550.29$ $C_I=7507.13$ $A_{II}=29017.79$ $B_{II}=8503.69$ $C_{II}=7542.14$				$E_{gauche}-E_{trans}=41.2\,(50)\ cm^{-1}$ potential of the OH internal rotation: $V_1=57.0\ cm^{-1}$ $V_2=0.8\ cm^{-1}$ $V_3=395.0\ cm^{-1}$	
	CD_3CD_2OH	$v_\tau=1$	$A-C=15778.5$ $\kappa=-0.87551$		74KAD		ν_τ (C—C torsion) $=186\ cm^{-1}$	
		$v_\delta=1$	$A-C=14980.0$ $\kappa=-0.88545$				ν_δ (C—C—O deformation) $=265\ cm^{-1}$	
157 *6-78*	CH_3CH_2SH gauche	$v=0$	$A=28747.69\,(50)$ *) $B=5294.78\,(12)$ $C=4846.18\,(12)$		76NAK2	76NAK2	*) uncertainties are 2.5 σ for the definitions of Δ and D see CH_3CH_2OH, Nr. 156, $\Delta=1753.84\,(29)$ MHz, $D=12.43\,(37)$ MHz	2.6.3, 71 2.8, 8 3, 43
		$v_{\tau_1}=1$	$B=5292.39\,(228)$ *) $C=4839.44\,(225)$		76NAK2		ν_{τ_1} = S-H torsion	
		$v_{\tau_2}=1$	$B=5278.99\,(17)$ *) $C=4837.01\,(16)$		76NAK2		$\nu_{\tau_2}=CH_3$ torsion	
	CH_3CH_2SD gauche	$v=0$	$A=26063.97\,(21)$*) $B=5190.04\,(5)$ $C=4768.64\,(5)$		76NAK2		$\Delta=70.33\,(14)$ MHz (tunneling frequency)	
		$v_{\tau_1}=1$	$B=5186.32$ $C=4763.73$		75SC			
Cont.	$CH_3CH_2{}^{34}SH$ gauche	$v=0$	$A=28709.50\,(62)$ *) $B=5175.20\,(15)$ $C=4744.98\,(15)$		76NAK2		$\Delta=1730.53\,(35)$ MHz $D=12.07\,(46)$ MHz	

Nr.	Molecule	v	A, B, C MHz	D, H kHz	Ref.	Ref. struct.	Varia, remarks	See also Tab., Nr.
157 6–78 Cont.	$^{13}CH_3CH_2SH$ gauche	$v=0$	$A=28540.59$ (36) *) $B=$ 5136.85 (9) $C=$ 4708.01 (9)		76NAK2		*) uncertainties are 2.5 σ $\Delta=1742.79$ (21) MHz $D=11.93$ (28) MHz	
	$CH_3{}^{13}CH_2SH$ gauche	$v=0$	$A=28056.43$ (36) *) $B=$ 5276.21 (9) $C=$ 4811.70 (9)		76NAK2		$\Delta=1723.91$ (21) MHz $D=13.32$ (24) MHz	
	s-CH_2DCH_2SH gauche	$v=0$	$A=28704.62$ (38)*) $B=$ 4925.35 (9) $C=$ 4534.49 (9)		76NAK2		$\Delta=1789.73$ (23) MHz $D=10.76$ (27) MHz	
	a-CH_2DCH_2SH-1 ⁺) gauche	$v=0$	$A=26169.06$ (72) *) $B=$ 5107.21 (13) $C=$ 4680.65 (14)	$D_J=3.7$ (65)	76NAK2		⁺) see Fig. 43 potential of the SH internal rotation: $V_1=$ −207 (67) cal/mol $V_2=$ −386 (27) cal/mol $V_3=1305$ (20) cal/mol	
	a-CH_2DCH_2SH-2 ⁺) gauche	$v=0$	$A=26208.79$ (48) *) $B=$ 5112.49 (9) $C=$ 4666.76 (10)	$D_J=3.0$ (43)	76NAK2			
	CH_3CHDSH-1 ⁺) gauche	$v=0$	$A=25313.90$ (54) *) $B=$ 5245.62 (8) $C=$ 4742.81 (9)	$D_J=3.6$ (68)	76NAK2			
	CH_3CHDSH-2 gauche	$v=0$	$A=25243.72$ (60) *) $B=$ 5229.55 (9) $C=$ 4749.93 (10)	$D_J=4.8$ (68)	76NAK2			
	CH_3CH_2SH trans	$v=0$	$A=28416.59$ $B=$ 5485.76 $C=$ 4881.81	$\tau_{aaaa}=$ −2.676 (21) $\tau_{bbbb}=$ −1.408 (46) $\tau_{cccc}=-92.6$ (17) $T_4=$ 2.77 (35) **) $T_5=-74.52$ (66)	75SC, 74HAY1	75SC	$E_{trans}-E_{gauche}=142$ (15) cm^{-1} **) $T_4=\tau_{aabb}+\tau_{bbcc}+\tau_{ccaa}$ $T_5=\frac{A-C}{B-C}\tau_{bbcc}+\tau_{ccaa}$	
		$v_{\tau_1}=1$	$A=27995.30$ $B=$ 5427.58 $C=$ 4870.60					
		$v_{\tau_2}=1$	$A=27928.77$ (300) $B=$ 5466.55 (36) $C=$ 4873.64 (37)	$D_J=2.28$ (162)	76NAK2			
Cont.	CH_3CH_2SD trans	$v=0$	$A=27155.91$ $B=$ 5304.36 $C=$ 4702.60		75SC, 74HAY1		$E_{trans}-E_{gauche}=138$ (30) cm^{-1}	

Nr.	Molecule	v	A, B, C MHz	D, H kHz	Ref.	Ref. struct.	Varia, remarks	See also Tab., Nr.
157 *6–78* Cont.	CH_2DCH_2SH symmetric, trans	$v=0$	$A=38387.09$ $B=\ 5095.35$ $C=\ 4570.29$		75SC, 74HAY1			
	CH_2DCH_2SH asymmetric, trans	$v=0$	$A=25913.77$ $B=\ 5294.04$ $C=\ 4714.44$	$\tau_{aaaa}=\ -2.25\,(45)$ $\tau_{bbbb}=\ -1.02\,(56)$ $\tau_{cccc}=-71.8\,(58)$ $T_4=\ 7.7\,(61)$ **) $T_5=-5.4\,(22)$	75SC, 74HAY1		**) $T_4=\tau_{aabb}+\tau_{bbcc}+\tau_{ccaa}$ $T_5=\frac{A-C}{B-C}\tau_{bbcc}+\tau_{ccaa}$	
	$CH_3CH_2{}^{34}SH$ trans	$v=0$	$A=28330.19\,(58)$ *) $B=\ 5367.00\,(7)$ $C=\ 4785.13\,(7)$	$D_J=5.06\,(333)$	74HAY1		*) uncertainties are 2.5 σ	
	$^{13}CH_3CH_2SH$ trans	$v=0$	$A=28212.58\,(56)$ *) $B=\ 5322.41\,(5)$ $C=\ 4746.32\,(5)$	$D_J=10.05\,(325)$	74HAY1			
	$CH_3{}^{13}CH_2SH$ trans	$v=0$	$A=27706.34\,(49)$ *) $B=\ 5466.02\,(6)$ $C=\ 4844.68\,(6)$	$D_J=0.29\,(569)$	74HAY1			
	CH_3CHDSH trans	$v=0$	$A=24950.46\,(65)$ *) $B=\ 5420.40\,(8)$ $C=\ 4778.38\,(8)$	$D_J=4.26\,(340)$				
	CH_3CD_2SH trans	$v=0$	$A=22275.58\,(81)$ *) $B=\ 5342.61\,(9)$ $C=\ 4691.45\,(9)$	$D_J=2.56\,(673)$				
158	$CH_3CH_2{}^{80}SeH$ gauche	$v=0$	$A=26827.11\pm145.68$ $B=\ 3623.84\,(2)$ $C=\ 3398.47\,(2)$		77DUR3	77DUR3	potential of the SeH internal rotation: $V_2=-96.4\,(10)$ $V_3=\ 432\,(4)$ $V_6=-20\,(2)\ cm^{-1}$	2.6.3, 73 2.8, 9
	$CH_3CH_2{}^{78}SeH$ gauche	$v=0$	$A=26904.35\pm183.12$ $B=\ 3645.61\,(2)$ $C=\ 3417.61\,(2)$					
	$CH_3CH_2{}^{80}SeD$ gauche	$v=0$	$A=24197.05\pm140$ $B=\ 3584.16\,(1)$ $C=\ 3375.76\,(1)$					
Cont.	$CH_3CH_2{}^{78}SeD$ gauche	$v=0$	$A=24151.77\pm158$ $B=\ 3605.16\,(2)$ $C=\ 3394.36\,(2)$					

Nr.	Molecule	v	A, B, C MHz	D, H kHz	Ref.	Ref. struct.	Varia, remarks	See also Tab., Nr.
158 Cont.	$CH_3CH_2{}^{80}SeH$ trans	$v=0$	$A=26431.28\,(7)$ $B=3781.24\,(2)$ $C=3453.18\,(2)$		77DUR3		$E_{trans}-E_{gauche}=66\ cm^{-1}$	
	$CH_3CH_2{}^{78}SeH$ trans	$v=0$	$A=26444.32\,(5)$ $B=3803.40\,(1)$ $C=3471.88\,(1)$					
	$CH_3CH_2{}^{80}SeD$ trans	$v=0$	$A=24550.13\,(15)$ $B=3735.19\,(3)$ $C=3382.00\,(3)$					
	$CH_3CH_2{}^{78}SeD$ trans	$v=0$	$A=24562.49\,(9)$ $B=3755.94\,(2)$ $C=3399.07\,(2)$					
159 *6-165b* *4-128a*	$CH_3CH_2OCH_3$ trans	$v=0$	$A=27991.97\,(31)$ *) $B=4159.48\,(4)$ $C=3891.19\,(4)$	$D_J=-1.26\,(56)$	75HAY	75HAY	*) uncertainties are 2.5 σ	2.6.3, 74 2.8, 10 3, 44
	$^{13}CH_3CH_2OCH_3$ trans	$v=0$	$A=27954.74\,(32)$ *) $B=4045.73\,(4)$ $C=3790.79\,(4)$	$D_J=-1.23\,(53)$				
	$CH_3{}^{13}CH_2OCH_3$ trans	$v=0$	$A=27576.63\,(33)$ *) $B=4150.96\,(4)$ $C=3875.58\,(4)$	$D_J=-1.26\,(59)$				
	$CH_3CH_2O^{13}CH_3$ trans	$v=0$	$A=27980.74\,(30)$ *) $B=4053.86\,(4)$ $C=3798.41\,(4)$	$D_J=-1.20\,(49)$				
	$CD_3CH_2OCH_3$ trans	$v=0$	$A=23846.68\,(16)$ *) $B=3681.05\,(2)$ $C=3470.23\,(2)$	$D_J=-0.86\,(32)$				
		$v_{\tau_1}=1$	$A=23816.17\,(31)$ *) $B=3674.61\,(2)$ $C=3467.29\,(2)$	$D_J=-0.63\,(71)$			v_{τ_1} = O—CH_3 torsion	
	$CH_3CH_2OCD_3$ trans	$v=0$	$A=23838.25\,(20)$ *) $B=3704.04\,(2)$ $C=3491.59\,(3)$	$D_J=-0.90\,(34)$				
Cont.		$v_{\tau_2}=1$	$A=23793.62\,(26)$ *) $B=3699.04\,(2)$ $C=3489.13\,(2)$	$D_J=-0.76\,(56)$			v_{τ_2} = C—CH_3 torsion	

Nr.	Molecule	v	A, B, C MHz	D, H kHz	Ref.	Ref. struct.	Varia, remarks	See also Tab., Nr.
159 Cont.	$CH_3CHDOCH_3$ trans	$v=0$	$A=25040.09(28)$ *) $B=$ 4131.91 (3) $C=$ 3843.28 (4)	$D_J=-1.24(50)$	75HAY		*) uncertainties are 2.5 σ	
	$CH_3CD_2OCH_3$ trans	$v=0$	$A=22669.15(25)$ *) $B=$ 4045.92 (3) $C=$ 3804.69 (3)	$D_J=-0.98(35)$				
	$CH_2DCH_2OCH_3$ symmetric, trans	$v=0$	$A=27505.27(31)$ *) $B=$ 3931.93 (4) $C=$ 3683.26 (4)	$D_J=-1.11(50)$				
	$CH_2DCH_2OCH_3$ asymmetric, trans	$v=0$	$A=26057.36(28)$ *) $B=$ 4008.91 (3) $C=$ 3760.58 (4)	$D_J=-1.15(43)$				
	$CH_3CH_2OCH_2D$ symmetric, trans	$v=0$	$A=27218.87(45)$ *) $B=$ 3956.91 (5) $C=$ 3699.99 (6)	$D_J=-1.05(74)$				
	$CH_3CH_2OCH_2D$ asymmetric, trans	$v=0$	$A=26194.74(27)$ *) $B=$ 4009.67 (3) $C=$ 3765.13 (4)	$D_J=-1.23(48)$				
160	$CH_3CH_2SCH_3$ trans	$v=0$	$A=16017.73(16)$ *) $B=$ 3186.68 (2) $C=$ 2799.36 (2)	$d_J=0.976(349)$ *)	81HAY	81HAY	*) uncertainties are 2.5 σ	2.6.3, 75 2.8, 100
		$v_T=1$	$A=15813.44$ $B=$ 3183.64 $C=$ 2804.43	$D_J=1.333$	74IMA		v_T=S—CH_3 torsion	
		$v_T=2$	$A=15608.41$ $B=$ 3181.09 $C=$ 2810.27	$D_J=0.007$				
	$^{13}CH_3CH_2SCH_3$	$v=0$	$A=16017.08(19)$ *) $B=$ 3094.30 (2) $C=$ 2727.82 (3)	$d_J=0.637(425)$	81HAY			
	$CH_3{}^{13}CH_2SCH_3$	$v=0$	$A=15794.94(20)$ *) $B=$ 3173.24 (2) $C=$ 2782.17 (3)	$d_J=0.800(417)$				
Cont.								

Nr.	Molecule	v	A, B, C MHz	D, H kHz	Ref.	Ref. struct.	Varia, remarks	See also Tab., Nr.
160 Cont.	$CH_3CH_2S^{13}CH_3$	$v=0$	$A=15886.69\,(16)$ *) $B=\ 3115.01\,(2)$ $C=\ 2740.01\,(2)$	$d_J=0.812\,(338)$	81HAY		*) uncertainties are 2.5 σ	
	$CH_3CH_2{}^{34}SCH_3$	$v=0$	$A=15701.52\,(33)$ *) $B=\ 3177.57\,(4)$ $C=\ 2782.51\,(4)$	$d_J=0.378\,(703)$				
	$CH_3CH_2SCD_3$	$v=0$	$A=14165.37\,(20)$ *) $B=\ 2886.58\,(2)$ $C=\ 2553.42\,(3)$	$d_J=0.546\,(443)$				
	$CH_3CHDSCH_3$	$v=0$	$A=14857.71\,(39)$ *) $B=\ 3162.12\,(4)$ $C=\ 2764.98\,(5)$	$d_J=0.567\,(785)$				
	$CH_2DCH_2SCH_3$ symmetric	$v=0$	$A=15776.55\,(27)$ *) $B=3017.25\,(3)$ $C=\ 2661.25\,(4)$	$d_J=0.767\,(587)$				
	$CH_2DCH_2SCH_3$ asymmetric	$v=0$	$A=15452.98\,(21)$ *) $B=\ 3069.89\,(2)$ $C=\ 2713.60\,(3)$	$d_J=0.585\,(438)$				
	$CH_3CH_2SCH_2D$ symmetric	$v=0$	$A=15999.48\,(24)$ *) $B=\ 3037.38\,(3)$ $C=\ 2683.50\,(3)$	$d_J=0.667\,(527)$				
	$CH_3CH_2SCH_2D$ asymmetric	$v=0$	$A=15043.08\,(20)$ *) $B=\ 3099.51\,(2)$ $C=\ 2723.08\,(3)$	$d_J=0.798\,(428)$				
161	CH_3CH_2NCS *)	$v=0$	$A=14194\,(10)$ $B=\ 1779.27\,(2)$ $C=\ 1612.14\,(2)$	$D_J=\ 1.58\,(2)$ $D_{JK}=-36.9\,(2)$	76SAK	76SAK	*) cis form: the methyl group eclipses the isothiocyanato group ν_τ(C—N torsion) $=58\,(30)\ cm^{-1}$ ν_δ(C—N—C bend) $=107\,(30)\ cm^{-1}$	2.6.3, 76 3, 45
		$v_\tau=1$	$A=14599\,(10)$ $B=\ 1748.42\,(2)$ $C=\ 1598.18\,(2)$	$D_J=\ 1.70\,(2)$ $D_{JK}=-48.6\,(2)$				
		$v_\delta=1$	$A=15215\,(15)$ $B=\ 1739.03\,(2)$ $C=\ 1590.22\,(2)$	$D_J=\ 2.31\,(2)$ $D_{JK}=-61.6\,(2)$				

Nr.	Molecule	v	A, B, C MHz	D, H kHz	Ref.	Ref. struct.	Varia, remarks	See also Tab., Nr.
162	CH_3CH_2NCO *)	$v=0$	$A=14110(3)$ $B=3067.01(2)$ $C=2605.01(2)$	$D_J=4.17(22)$ $D_{JK}=-39.1(5)$	76SA	76SA	*) cis form: the methyl group eclipses the isocyanato group ν_τ(C—N torsion)$=100\ cm^{-1}$	2.6.3, 77 3, 46
		$v_\tau=1$	$A=14657(20)$ $B=2982.36(9)$ $C=2578.47(9)$	$D_J=5.69(104)$ $D_{JK}=-69.3(29)$				
		$v_\tau=2$	$A=15215(25)$ $B=2922.95(8)$ $C=2556.90(8)$	$D_J=7.15(99)$ $D_{JK}=-95.0(28)$				
		$v_\tau=3$	$A=15919(56)$ $B=2861.01(12)$ $C=2533.53(14)$	$D_J=9.38(144)$ $D_{JK}=-145.5(62)$				
		$v_\tau=4$	$A=16853(110)$ $B=2792.80(17)$ $C=2505.87(13)$	$D_J=11.48(173)$ $D_{JK}=-428.4(51)$				
163	CH_3CH_2ONO cis, trans *)	$v=0$	$A=18021.0(1)$ $B=3019.936(5)$ $C=2674.593(5)$		79TUR3	80END1	*) τ_1(ON—OC)$=0°$ τ_2(CC—ON) $=180°$ ν_τ(C—O torsion)$=111(23)\ cm^{-1}$	2.6.3, 78 3, 47
		$v_\tau=1$	$A=17798.0(2)$ $B=3016.566(6)$ $C=2678.047(6)$					
		$v_\tau=2$	$A=17564(6)$ $B=3013.95(1)$ $C=2682.59(1)$					
		$v_\tau=3$	$A=17342(24)$ $B=3012.25(4)$ $C=2688.55(4)$					
		$v_\tau=4$	$A=17085(4)$ $B=3011.674(7)$ $C=2696.296(7)$					
		$v_a=1$	$A=18181(65)$ $B=3015.0(1)$ $C=2672.8(1)$				$\nu_a=236(20)\ cm^{-1}$, CH_3 torsion or O—N torsion	
Cont.								

Nr.	Molecule	v	A, B, C MHz	D, H kHz	Ref.	Ref. struct.	Varia, remarks	See also Tab., Nr.
163 Cont.	$CH_3CH_2O^{15}NO$ cis, trans *)	$v=0$	$A=17893\,(52)$ $B=\ 2987.17\,(5)$ $C=\ 2645.35\,(5)$		80END1		*) $\tau_1(ON—OC)=0°$ $\tau_2(CC—ON)=180°$	
	CD_3CD_2ONO cis, trans *)	$v=0$	$A=14470\,(26)$ $B=\ 2646.35\,(4)$ $C=\ 2370.70\,(4)$					
	CH_3CH_2ONO cis, gauche **)	$v=0$	$A=10185\,(4)$ $B=\ 3979.09\,(7)$ $C=\ 3325.06\,(7)$		79TUR3		**) $\tau_1=0°$, $\tau_2=80°$	
		$v_\tau=1$	$A=10386\,(7)$ $B=\ 3906.0\,(1)$ $C=\ 3293.8\,(1)$				$E_{cis\text{-}gauche}-E_{cis\text{-}trans}$ $=238\,(50)\ cm^{-1}$ ν_τ(CO torsion)$=103\,(18)\ cm^{-1}$	
	$CH_3CH_2O^{15}NO$ cis, gauche **)	$v=0$	$A=10157\,(5)$ $B=\ 3924.98\,(5)$ $C=\ 3287.94\,(5)$		80END1			
	CD_3CD_2ONO cis, gauche **)	$v=0$	$A=8592.5\,(3)$ $B=\ 3451.67\,(5)$ $C=\ 2953.47\,(5)$					
	CH_3CH_2ONO trans, gauche ***)	$v=0$	$A=17639.3\,(2)$ $B=\ 2746.44\,(2)$ $C=\ 2619.46\,(2)$	$\Delta_J=\ 4.2\,(1)$ $\Delta_{JK}=-106.6\,(2)$ $\Delta_K=\ 1243\,(22)$ $\delta_J=\ 1.28\,(1)$ $\delta_K=25\,(6)$	79TUR3		***) $\tau_1=180°$, $\tau_2=90°$ $E_{trans\text{-}gauche}-E_{cis\text{-}trans}$ $=81\,(20)\ cm^{-1}$	
		$v_\tau=1$	$A=18786\,(95)$ $B=\ 2687.16\,(2)$ $C=\ 2586.41\,(2)$	$\Delta_J=\ 4.3\,(1)$ $\Delta_{JK}=-142.1\,(4)$ $\Delta_K=\ 1243$ (fixed) $\delta_J=\ 0.67\,(15)$ $\delta_K=25$ (fixed)			ν_τ(C—O torsion)$=52\,(16)\ cm^{-1}$ potential function for the C—O torsion: $V_1=461$, $V_2=447$ and $V_3=154\ cm^{-1}$	
	$CH_3CH_2O^{15}NO$ trans, gauche ***)	$v=0$	$A=17469.8\,(5)$ $B=2727.91\,(6)$ $C=\ 2604.98\,(6)$		80END1		barrier height separating the two equivalent gauche forms: $137\,(7)\ cm^{-1}$	
	CD_3CD_2ONO trans, gauche ***)	$v=0$	$A=13362\,(4)$ $B=\ 2445.04\,(2)$ $C=\ 2315.98\,(2)$	$\Delta_J=\ 3.36\,(3)$ $\Delta_{JK}=-57.0\,(1)$ $\delta_J=0.90\,(7)$	79TUR3			
		$v_\tau=1$	$A=14058\,(8)$ $B=\ 2398.64\,(1)$ $C=\ 2287.40\,(1)$	$\Delta_J=\ 3.36\,(4)$ $\Delta_{JK}=-72.4\,(2)$ $\delta_J=0.77\,(6)$				

Nr.	Molecule	v	A, B, C MHz	D, H kHz	Ref.	Ref. struct.	Varia, remarks	See also Tab., Nr.
164 *6–79b*	$CH_3CH_2ONO_2$ trans	$v=0$	$A=9974.95\,(2)$ $B=2208.08\,(1)$ $C=1850.94\,(1)$		74SCR	74SCR		2.6.3, 79
		$v_a=1$	$A=10053.27$ $B=2208.33$ $C=1851.68$				ν_a (C—O torsion) $=100\,(15)\ cm^{-1}$	
		$v_a=2$	$A=10119.61$ $B=2209.14$ $C=1852.64$					
		$v_b=1$	$A=10000.61$ $B=2208.55$ $C=1852.53$				$\nu_b=114\,(20)\ cm^{-1}$, NO_2 torsion or in-plane skeletal bend	
		$v_b=2$	$A=10176.70$ $B=2209.58$ $C=1852.68$					
		$v_c=1$	$A=9998.18$ $B=2207.97$ $C=1850.60$				$\nu_c=600\,(50)\ cm^{-1}$	
	$CH_3CH_2ONO_2$ gauche	$v=0$	$A=7945.62\,((2)$ $B=2557.14\,(1)$ $C=2256.88\,(1)$				the gauche form is less stable by $143\,(70)\ cm^{-1}$	
		$v_a=1$	$A=7920.03$ $B=2557.20$ $C=2261.74$				ν_a (C—O torsion)$=90\,(20)\ cm^{-1}$	
		$v_a=2$	$A=7883.15$ $B=2557.34$ $C=2266.65$				potential function for internal rotation about the C—O bond: $V_1=6.099$ kcal/mol $V_2=-4.050$ kcal/mol $V_3=2.974$ kcal/mol $V_4=0.830$ kcal/mol gauche-gauche barrier$=8.65\pm6.0$ kcal/mol above gauche minimum trans-gauche$=2.30\,(60)$ kcal/mol above trans minimum	
		$v_a=3$	$A=7818.31$ $B=2557.57$ $C=2271.65$					
		$v_a=4$	$A=7757.46$ $B=2557.86$ $C=2276.71$					
		$v_a=5$	$A=8000$ $B=2557.82$ $C=2281.66$					

Nr.	Molecule	v	A, B, C MHz	D, H kHz	Ref.	Ref. struct.	Varia, remarks	See also Tab., Nr.
165 *6–79c*	$CH_3CH_2NHCH_3$ *)	$v=0$	$A=25935.85\,(27)$ $B=\ 3919.76\,(6)$ $C=\ 3668.24\,(6)$	$d_J\ =-4\,(2)$ $d_{JK}=-5\,(43)$ $d_K\ =\ 31\,(14)$ $d_{WJ}=\ 1\,(1)\cdot 10^{-6}$ $d_{WK}=-6\,(10)\cdot 10^{-6}$	73PEN1	73PEN1	*) trans form inversion splitting: $\Delta v=1981.39$ MHz Amino inversion barrier ≈ 5.2 kcal/mol	2.6.3, 80 2.7.3, 30
		$v_T=1$	$A=25518.41\,(16)$ $B=\ 3919.68\,(9)$ $C=\ 3675.36\,(3)$				v_T = heavy-atom torsion $\Delta v=2028.57$ MHz	
	$CH_3CH_2NDCH_3$	$v=0$	$A=23768.12\,(17)$ $B=\ 3904.29\,(5)$ $C=\ 3627.54\,(3)$	$d_J\ =\ -1\,(3)$ $d_{JK}=\ 28\,(27)$ $d_K\ =\ 6\,(9)$ $d_{WJ}=\ 0\,(1)\cdot 10^{-6}$ $d_{WK}=-11\,(2)\cdot 10^{-6}$			$\Delta v=64.71$ MHz	
		$v_T=1$	$A=23423.78\,(35)$ $B=\ 3903.94\,(11)$ $C=\ 3634.31\,(5)$				$\Delta v=42.68$ MHz	
		$v_T=2$	$(A-C)/2=9721.26\,(17)$ $\kappa=-0.97303\,(3)$				$\Delta v=32.0$ MHz	
166	$CH_3CH_2O^{35}Cl$	$v=0$	$A''=31582.01\,(7)$ $B''=\ 2691.084\,(10)$ $C''=2560.916\,(8)$	$\tau_1\ =\ 1.3\,(46)$ $\tau_2\ =-1.13\,(48)$ $\tau_3\ =\ 450\,(80)$ $\tau_{aaaa}=-577\,(90)$ $\tau_{bbbb}=\ -1.77\,(20)$ $\tau_{cccc}=\ -1.37\,(19)$	78SU2	78SU2		2.6.3, 81 2.7.3, 110
		$v_\tau=1$	$A=30750.34\,(7)$ $B=\ 2691.943\,(5)$ $C=\ 2566.633\,(5)$				$v_\tau=125\,(23)\ cm^{-1}$, most likely torsional motion about the C—O bond	
	$CH_3CH_2O^{37}Cl$	$v=0$	$A=31568.64\,(8)$ $B=\ 2629.909\,(4)$ $C=\ 2505.372\,(4)$					
		$v_\tau=1$	$A=30734.25\,(19)$ $B=\ 2630.813\,(6)$ $C=\ 2510.923\,(6)$					

Nr.	Molecule	v	A, B, C MHz	D, H kHz	Ref.	Ref. struct.	Varia, remarks	See also Tab., Nr.
167	$CH_3CH_2OCH_2F$ ⁺)	$v=0$	$A=14958.51(38)$ *) $B=2554.72(3)$ $C=2385.86(2)$	$d_J=0.65(28)$ *) $d_{JK}=-20(8)$	80HAY1	80HAY1	⁺) gauche-trans isomer *) uncertainties are 2.5 σ v_τ, methyl torsion	2.6.3, 82 2.8, 11 3, 48
		$v_\tau=1$	$A=14991.18(48)$ *) $B=2550.43(3)$ $C=2382.10(2)$	$d_J=0.58(31)$ *) $d_{JK}=-22(8)$				
	$CH_3CH_2O^{13}CH_2F$	$v=0$	$A=14792.11(42)$ *) $B=2536.68(4)$ $C=2367.05(3)$	$d_J=0.61(31)$ *) $d_{JK}=-20(8)$				
	$CH_3{}^{13}CH_2OCH_2F$	$v=0$	$A=14844.38(37)$ *) $B=2540.87(3)$ $C=2373.96(2)$	$d_J=0.72(27)$ *) $d_{JK}=-19(7)$				
	$^{13}CH_3CH_2OCH_2F$	$v=0$	$A=14938.88(35)$ *) $B=2484.42(3)$ $C=2324.54(2)$	$d_J=0.61(25)$ *) $d_{JK}=-20(7)$				
	$CH_3CH_2{}^{18}OCH_2F$	$v=0$	$A=14599.38(44)$ *) $B=2553.23(4)$ $C=2377.66(3)$	$d_J=0.55(31)$ *) $d_{JK}=-22(9)$				
	CH_3CH_2OCHDF-1	$v=0$	$A=14096.21(40)$ *) $B=2514.47(4)$ $C=2365.41(2)$	$d_J=0.57(29)$ *) $d_{JK}=-17(9)$				
	CH_3CH_3OCHDF-2	$v=0$	$A=14114.07(39)$ *) $B=2516.89(4)$ $C=2331.61(2)$	$d_J=0.66(28)$ *) $d_{JK}=-13(9)$			see Fig. 48 for identification numbers of atoms	
	$CH_3CHDOCH_2F$-3	$v=0$	$A=14152.24(36)$ *) $B=2551.33(3)$ $C=2362.55(2)$	$d_J=0.63(26)$ *) $d_{JK}=-18(8)$				
	$CH_3CHDOCH_2F$-4	$v=0$	$A=14070.80(37)$ *) $B=2515.76(3)$ $C=2374.82(2)$	$d_J=0.57(27)$ *) $d_{JK}=-18(7)$				
	$CH_2DCH_2OCH_2F$-5	$v=0$	$A=14638.27(31)$ *) $B=2462.98(3)$ $C=2298.34(2)$	$d_J=0.50(22)$ *) $d_{JK}=-20(6)$				
Cont.								

Nr.	Molecule	v	A, B, C MHz	D, H kHz	Ref.	Ref. struct.	Varia, remarks	See also Tab., Nr.
167 Cont.	$CH_2DCH_2OCH_2F$-6	$v=0$	$A=14290.95\,(31)$ *) $B=2465.38\,(3)$ $C=2322.80\,(2)$	$d_J=0.74\,(22)$ *) $d_{JK}=-19\,(6)$	80HAY1		*) uncertainties are 2.5 σ	
	$CH_2DCH_2OCH_2F$-7	$v=0$	$A=14633.59\,(39)$ *) $B=2438.48\,(3)$ $C=2277.93\,(2)$	$d_J=0.54\,(26)$ *) $d_{JK}=-19\,(7)$				
	$CH_3CH_2OCD_2F$	$v=0$	$A=13352.68\,(35)$ *) $B=2477.66\,(4)$ $C=2313.24\,(2)$	$d_J=0.61\,(25)$ *) $d_{JK}=-11\,(9)$				
	$CH_3CD_2OCH_2F$	$v=0$	$A=13359.60\,(32)$ *) $B=2512.71\,(3)$ $C=2351.82\,(2)$	$d_J=0.71\,(22)$ *) $d_{JK}=-12\,(8)$				
168	$CH_3CH_2OCH_2{}^{35}Cl$ *)	$v=0$	$A=10949.33\,(11)$ **) $B=1805.65\,(2)$ $C=1657.13\,(2)$	$d_J=0.81\,(52)$ **) $d_{JK}=-17\,(2)$	79HAY1	80HAY3	*) gauche-trans form; see Fig. 49 for identification numbers of atoms **) uncertainties are 2.5 σ	2.7.3, 111 3, 49
	$CH_3CH_2OCH_2{}^{37}Cl$	$v=0$	$A=10871.80\,(19)$ **) $B=1766.29\,(3)$ $C=1622.23\,(4)$	$d_J=0.41\,(84)$ **) $d_{JK}=-19\,(3)$				
	$CH_3CH_2O^{13}CH_2{}^{35}Cl$	$v=0$	$A=10732.79\,(18)$ **) $B=1803.17\,(3)$ $C=1650.38\,(3)$	$d_J=1.09\,(79)$ **) $d_{JK}=-16\,(3)$				
	$CH_3CH_2OCHD^{35}Cl$-1	$v=0$	$A=10317.61\,(20)$ **) $B=1793.19\,(4)$ $C=1648.14\,(4)$	$d_J=1.05\,(96)$ **) $d_{JK}=-13\,(3)$				
	$CH_3CH_2OCHD^{35}Cl$-2	$v=0$	$A=10214.31\,(13)$ **) $B=1798.09\,(2)$ $C=1634.07\,(3)$	$d_J=0.46\,(59)$ **) $d_{JK}=-14\,(2)$				
	$CH_3{}^{13}CH_2OCH_2{}^{35}Cl$	$v=0$	$A=10885.69\,(12)$ **) $B=1792.52\,(2)$ $C=1646.59\,(2)$	$d_J=0.41\,(52)$ **) $d_{JK}=-16\,(2)$				
	$CH_3CHDOCH_2{}^{35}Cl$-3	$v=0$	$A=10536.29\,(19)$ **) $B=1801.86\,(3)$ $C=1646.05\,(4)$	$d_J=1.10\,(84)$ **) $d_{JK}=-16\,(3)$				
Cont.	$CH_3CHDOCH_2{}^{35}Cl$-4	$v=0$	$A=10466.52\,(16)$ **) $B=1778.27\,(3)$ $C=1645.81\,(3)$	$d_J=1.06\,(70)$ **) $d_{JK}=-12\,(2)$				

Nr.	Molecule	v	A, B, C MHz	D, H kHz	Ref.	Ref. struct.	Varia, remarks	See also Tab., Nr.
168 Cont.	$^{13}CH_3CH_2OCH_2{}^{35}Cl$	$v=0$	$A=10905.28(12)$ **) $B=1758.26(2)$ $C=1616.48(2)$	$d_J=0.66(58)$ **) $d_{JK}=-17(2)$	79HAY1		**) uncertainties are 2.5 σ	
	$CH_2DCH_2OCH_2{}^{35}Cl$-5	$v=0$	$A=10875.07(22)$ **) $B=1739.33(4)$ $C=1599.87(4)$	$d_J=0.51(107)$ **) $d_{JK}=-17(4)$				
	$CH_2DCH_2OCH_2{}^{35}Cl$-6	$v=0$	$A=10507.48(25)$ **) $B=1751.87(4)$ $C=1616.38(5)$	$d_J=0.83(110)$ **) $d_{JK}=-18(4)$				
	$CH_2DCH_2OCH_2{}^{35}Cl$-7	$v=0$	$A=10635.84(19)$ **) $B=1735.69(4)$ $C=1592.25(4)$	$d_J=0.61(85)$ **) $d_{JK}=-18(4)$				
	$CH_3CH_2OCD_2{}^{35}Cl$	$v=0$	$A=9669.53(17)$ **) $B=1785.29(3)$ $C=1625.92(3)$	$d_J=0.65(81)$ **) $d_{JK}=-16(3)$				
	$CH_3CD_2OCH_2{}^{35}Cl$	$v=0$	$A=10091.26(16)$ **) $B=1774.78(3)$ $C=1634.92(3)$	$d_J=1.08(70)$ **) $d_{JK}=-13(2)$				
	$CH_3CH_2{}^{18}OCH_2{}^{35}Cl$	$v=0$	$A=10694.92(20)$ **) $B=1799.47(3)$ $C=1647.90(4)$	$d_J=0.68(94)$ **) $d_{JK}=15.4(32)$	80HAY3			
169 *4-74*	CH_3CHF_2	$v=0$	$A=9491.98(1)$ $B=8962.74(1)$ $C=5170.46(1)$		74AGE		for structure, see also Vol. II/7 (3, 237)	2.8, 12
		$v_\tau=1$	$A=9491.34(1)$ $B=8934.30(1)$ $C=5163.87(1)$				$\nu_\tau=222\ cm^{-1}$, CH_3 torsion	
		$v_\tau=2$	$A=9497.66$ *) $B=8910.04$ $C=5158.72$				*) A state	
		$v_x=1$	$A=9532.74$ $B=8992.70$ $C=5168.50$				$\nu_x=392\ cm^{-1}$	

Nr.	Molecule	v	A, B, C MHz	D, H kHz	Ref.	Ref. struct.	Varia, remarks	See also Tab., Nr.
170	$CH_3CHF^{35}Cl$	$v=0$	$A=9016.25$ $B=4666.68$ $C=3329.35$	$\Delta_J=0$ (constrained) $\Delta_{JK}=2.84\,(28)$ $\Delta_K=6.57\,(98)$ $\delta_J=0.424\,(17)$ $\delta_K=4.08\,(32)$	74THO	74THO		2.6.3, 83 2.7.3, 112 2.8, 13
	$CH_3CHF^{37}Cl$	$v=0$	$A=9014.84$ $B=4537.01$ $C=3263.16$					
171	$CH_3CH{=}NH$ cis	$v=0$	$A=53120.528\,(80)$ *) $B=9778.4950\,(106)$ $C=8701.3167\,(166)$	$\Delta_J=6.617\,(100)$ $\Delta_{JK}=-15.19\,(91)$ $\Delta_K=570.7\,(58)$ $\delta_J=1.102\,(16)$ $\delta_K=-72.5\,(35)$	80LOV	80LOV	*) uncertainties are 3 σ	2.6.3, 84 2.7.3, 31 2.8, 14
		$v_\tau=1$	$A=52985.60$ $B=9770.85$ $C=8713.57$	$\Delta_J=5.89$ $\Delta_{JK}=177.1$ $\Delta_K=570.7$ (fixed) $\delta_J=0.789$ $\delta_K=-262.3$			ν_τ CH_3 torsion	
	$CH_3CH{=}NH$ trans	$v=0$	$A=49960.6\,(117)$ *) $B=9828.21\,(15)$ $C=8650.29\,(15)$	$\Delta_J=6.91\,(63)$ $\Delta_{JK}=-12.4\,(30)$ $\Delta_K=570.7$ (fixed) $\delta_J=1.13\,(60)$ $\delta_K=-53\,(69)$				
172 *6–80a*	$CH_3CH{=}NCH_3$	$v=0$	$I_a=13.209340\,(33)$ *) $I_b=123.89070\,(61)$ $I_c=130.87669\,(13)$		79BOS		for structure, see also Vol. II/7 (3, 419) *) in amu Å², conversion factor of 505376 amu Å² MHz	2.8, 15
	$CH_3CD{=}NCH_3$	$v=0$	$I_a=15.3223900\,(70)$ *) $I_b=123.9384\,(14)$ $I_c=133.0431\,(12)$					
	$CH_3CH{=}NCD_3$	$v=0$	$A=30918.491\,(51)$ $B=3621.2874\,(53)$ $C=3450.6303\,(68)$					
	$CD_3CD{=}NCH_3$	$v=0$	$A=27366.246\,(35)$ $B=3601.3035\,(54)$ $C=3383.4353\,(59)$					

Nr.	Molecule	v	A, B, C MHz	D, H kHz	Ref.	Ref. struct.	Varia, remarks	See also Tab., Nr.
173	syn—CH_3CH=SO	$v=0$	$A=$ 14.2 · 10³ *) $B=$4678.53 $C=$3594.63		78PEN3		*) for the A species	2.6.3, 85 2.8, 16
173a	$HOCHCH_3CN$	$v_I=0$	$A=$8790.9 (5) $B=$4005.77 (2) $C=$2975 · 77 (2)		81COR		I and II are probably the 2 non-equivalent rotamers in which OH is gauche to CN	
		$v_{II}=0$	$A=$8585.5 (9) $B=$4028.62 (3) $C=$2987.79 (3)					
174	CH_2ODCH_2OD	$v=0$	$A=$14394.63 (17) $B=$ 5276.27 (24) $C=$ 4323.53 (20)		80WAL	81CAM4	a potential function for the internal rotation of the OD groups has been derived	2.6.3, 86
	$CH_2OH'CH_2OD$ *)	$v=0$	$A=$15126.955 (34) $B=$ 5311.085 (12) $C=$ 4412.157 (12)	Δ_J = 6.75 (29) Δ_{JK} = −32.4 (5) Δ_K = 93 (6) δ_J = 1.909 (15) δ_K = 16.4 (7)	81CAM4		*) primed Hydrogen participates in the hydrogen bond, the other is trans with respect to the C—C bond; one O is gauche with respect to the other	
	$CH_2OD'CH_2OH$ *)	$v=0$	$A=$14620.287 (22) $B=$ 5548.481 (13) $C=$ 4517.854 (13)	Δ_J = 8.27 (31) Δ_{JK} = −30.4 (4) Δ_K = 51 (3) δ_J = 2.48 (1) δ_K = 12.5 (4)				
175	$HSCH_2CH_2OH$*)	$v=0$	$A=$12152.641 (12) $B=$ 3467.078 (10) $C=$ 2965.554 (9)		77SUN	77SUN	*) all gauche (GGG) conformation, see figure intramolecular hydrogen bond OH···S $v_\tau=180\,(20)\ cm^{-1}$, C—C torsion	2.6.3, 87 3, 50
Cont.		$v_\tau=1$	$A=$12249.98 ± 0.012 $B=$ 3441.74 (1) $C=$ 2949.85 (1)					

Nr.	Molecule	v	A, B, C MHz	D, H kHz	Ref.	Ref. struct.	Varia, remarks	See also Tab., Nr.
175 Cont.	$HSCH_2CH_2OD$	$v=0$	$A=11632.710(11)$ $B=3446.367(28)$ $C=2918.793(8)$		77SUN			
		$v_\tau=1$	$A=11714.38$ $B=3420.44$ $C=2902.72$					
	$DSCH_2CH_2OH$	$v=0$	$A=11665.347(37)$ $B=3376.599(142)$ $C=2928.546(32)$					
	$DSCH_2CH_2OD$	$v=0$	$A=11185.594(16)$ $B=3356.121(12)$ $C=2882.437(9)$					
	$H^{34}SCH_2CH_2OH$	$v=0$	$A=11111.855(90)$ $B=3380.006(14)$ $C=2899.306(14)$					
175a *6-82*	CH_2FCH_2F gauche	$v=0$	$A=17322.36(1)$ $B=5013.148(4)$ $C=4382.700(5)$	$\Delta_J=7.34(4)$ $\Delta_{JK}=-52.5(5)$ $\Delta_K=155.1(2)$ $\delta_J=2.08(1)$ $\delta_K=24.1(7)$	81DRE		for structure, see also Vol. II/7 (3, 238)	
		$v_{10}=1$	$A=17466.832(5)$ $B=5001.151(9)$ $C=4376.638(9)$	$\Delta_J=7.67(6)$ $\Delta_{JK}=-52.7(6)$ $\Delta_K=167(1)$ $\delta_J=2.08(3)$ $\delta_K=28(4)$			$\nu_{10}=148$ cm^{-1}, torsion	
		$v_9=1$	$A=17276.647(6)$ $B=5020.60(1)$ $C=4381.80(1)$	$\Delta_J=7.67(7)$ $\Delta_{JK}=-51.6(7)$ $\Delta_K=155(1)$ $\delta_J=2.09(3)$ $\delta_K=27(4)$			$\nu_9=327$ cm^{-1}, CCF bend	
Cont.		$v_9=2$	$A=17228.73(8)$ $B=5030.0(3)$ $C=4379.7(3)$	$\Delta_J=15(1)$ $\Delta_{JK}=-140(20)$ $\Delta_K=160(10)$ $\delta_J=3.1(4)$ $\delta_K=700(100)$				

Nr.	Molecule	v	A, B, C MHz	D, H kHz	Ref.	Ref. struct.	Varia, remarks	See also Tab., Nr.
175a *6–82* Cont.	CH_2FCH_2F gauche	$v_x=1$	$A=17436.56(2)$ $B=4981.15(2)$ $C=4364.75(4)$	$\Delta_J=7.3(3)$ $\Delta_{JK}=-45(3)$ $\Delta_K=154(5)$ $\delta_J=1.9(1)$ $\delta_K=-20(20)$	81DRE		$\nu_x=479\ cm^{-1}$	
		$v_J=1$	$A=17406.792(9)$ $B=5013.98(2)$ $C=4377.34(2)$	$\Delta_J=7.7(1)$ $\Delta_{JK}=-55(1)$ $\Delta_K=171(2)$ $\delta_J=2.14(5)$ $\delta_K=36(7)$			$\nu_J=441\ cm^{-1}$	
176	$CH_2NH_2CH_2NH_2$ gauche I *), state (+)	$v=0$	$A=14472.133(12)$ $B=5229.2072(50)$ $C=4405.5612(51)$	$\Delta_J=4.85(11)$ $\Delta_{JK}=-22.52(14)$ $\Delta_K=80.1(16)$ $\delta_J=1.3666(30)$ $\delta_K=11.42(21)$ $H_{JK}=0.317(17)\cdot10^{-3}$ $H_K=-11.9(62)\cdot10^{-3}$	78MAR	78MAR	for structure, see also Vol. II/7 (3, 315) *) see figure separation between the (+) and (−) states = 86.356 (21) MHz	2.6.3, 88 3, 51
	$CH_2NH_2CH_2NH_2$ gauche I *), state (−)	$v=0$	$A=14471.900(12)$ $B=5229.2043(50)$ $C=4405.4596(51)$	$\Delta_J=4.80(11)$ $\Delta_{JK}=-20.13(13)$ $\Delta_K=48.2(16)$ $\delta_J=1.3903(30)$ $\delta_K=10.54(22)$ $H_{JK}=0.383(19)\cdot10^{-3}$ $H_K=-44.0(68)\cdot10^{-3}$			conformer I is favoured by 0.3 kcal/mol relative to II.	
	$CH_2NH_2CH_2NH_2$ gauche II *)	$v=0$	$A=14355.536(20)$ $B=5125.3014(90)$ $C=4356.2924(85)$	$\Delta_J=3.85(32)$ **) $\Delta_{JK}=-19.73(41)$ $\Delta_K=51.0(51)$ $\delta_J=1.3271(95)$ $\delta_K=9.14(45)$			**) average constants of the (+) and (−) states separation between the (+) and (−) states = 1.332 (18) MHz	
		$v_\tau=1$	$A=14451.195(93)$ **) $B=5090.458(36)$ $C=4338.028(35)$	$\Delta_J=8.0(12)$ **) $\Delta_{JK}=-37.6(20)$ $\Delta_K=152(25)$ $\delta_J=1.185(43)$ $\delta_K=57.1(22)$			$\nu_\tau=181(20)\ cm^{-1}$, C—C torsion separation between the (+) and (−) states = 1.564 (66) MHz	

Nr.	Molecule	v	A, B, C MHz	D, H kHz	Ref.	Ref. struct.	Varia, remarks	See also Tab., Nr.
177	$CH_2NH_2CH_2F$ conformer I *)	$v=0$	$A=15437.6657(82)$ $B=5302.3944(28)$ $C=4439.5230(21)$	$\Delta_J=7.305(15)$ $\Delta_{JK}=-37.544(77)$ $\Delta_K=91.494(38)$ $\delta_J=2.1469(31)$ $\delta_K=15.54(31)$ $H_J=0.0176(68)\cdot 10^{-3}$ $H_{JK}=0.60(14)\cdot 10^{-3}$ $H_{KJ}=-2.40(30)\cdot 10^{-3}$ $H_K=4.44(27)\cdot 10^{-3}$ $h_J=0.0006(22)\cdot 10^{-3}$ $h_{JK}=0.42(14)\cdot 10^{-3}$ $h_K=3.4(14)\cdot 10^{-3}$	80MAR1	80MAR1	*) see figure the conformers are stabilized by intramolecular hydrogen bond	2.6.3, 89 2.7.3, 32 3, 52
		$v_\tau=1$	$A=15579.291(18)$ $B=5263.6823(63)$ $C=4421.1920(59)$	$\Delta_J=6.990(40)$ $\Delta_{JK}=-35.74(26)$ $\Delta_K=97.26(17)$ $\delta_J=2.106(13)$ $\delta_K=10.4(11)$ $H_J=0.09(14)\cdot 10^{-3}$ $H_{JK}=4.2(20)\cdot 10^{-3}$ $H_{KJ}=13.5(13)\cdot 10^{-3}$ $H_K=-1.7(24)\cdot 10^{-3}$ $h_J=0.246(15)\cdot 10^{-3}$ $h_{JK}=-8.9(24)\cdot 10^{-3}$ $h_K=-69.6(116)\cdot 10^{-3}$			$\nu_\tau=195(15)$ cm^{-1}, C—C torsion	
		$v_\tau=2$	$A=15716.82(29)$ $B=5225.778(75)$ $C=4403.780(74)$					
		$v_\delta=1$	$A=15470.01(21)$ $B=5273.061(91)$ $C=4425.575(65)$				$\nu_\delta=227(25)$ cm^{-1}, CCN bend	
		$v_T=1$	$A=15397.21(14)$ $B=5305.652(47)$ $C=4436.164(45)$				$\nu_T=350(30)$ cm^{-1}, C—N torsion	
Cont.		$v_\tau=1, v_\delta=1$	$A=15609.10(33)$ $B=5235.10(22)$ $C=4408.01(22)$					

Nr.	Molecule	v	A, B, C MHz	D, H kHz	Ref.	Ref. struct.	Varia, remarks	See also Tab., Nr.
177 Cont.	$CH_2NH_2CH_2F$ conformer II *)	$v=0$	$A=15186.4940(77)$ $B=5222.0616(28)$ $C=4400.7600(23)$	$\Delta_J=6.943(13)$ $\Delta_{JK}=-34.775(68)$ $\Delta_K=86.028(36)$ $\delta_J=1.9911(27)$ $\delta_K=13.17(27)$ $H_J=-0.0008(65)\cdot10^{-3}$ $H_{JK}=-1.57(17)\cdot10^{-3}$ $H_{KJ}=2.91(33)\cdot10^{-3}$ $H_K=-0.68(34)\cdot10^{-3}$ $h_J=-0.0118(21)\cdot10^{-3}$ $h_{JK}=0.481(83)\cdot10^{-3}$ $h_K=-22.0(15)\cdot10^{-3}$	80MAR1		the enthalpy difference is 0.1 (3) kcal/mol with conformer I as the more stable *) see figure	
		$v_\tau=1$	$A=15303\ 333(13)$ $B=5192.3159(45)$ $C=4385.4414(40)$	$\Delta_J=6.716(17)$ $\Delta_{JK}=-34.75(11)$ $\Delta_K=91.965(39)$ $\delta_J=1.9061(32)$ $\delta_K=14.72(21)$ $H_J=-0.052(11)\cdot10^{-3}$ $H_{JK}=0.192(22)\cdot10^{-3}$ $H_{KJ}=-4.10(50)\cdot10^{-3}$ $H_K=6.69(65)\cdot10^{-3}$			$\nu_\tau=162(15)\ cm^{-1}$, C—C torsion	
		$v_\tau=2$	$A=15403.98(79)$ $B=5163.74(14)$ $C=4369.41(13)$					
		$v_T=1$	$A=15176.01(19)$ $B=5230.49(11)$ $C=4404.34(12)$				$\nu_T=324(30)\ cm^{-1}$, C—N torsion	
178 *6-86a*	$CH_2ND_2CH_2OD$	$v=0$	$A=13116.26(7)$ $B=5024.67(2)$ $C=4176.92(2)$		76PEN4	76PEN4	for structure, see also Vol. II/7 (3, 311)	3, 53
	$CH_2ND_2CH_2{}^{18}OD$	$v=0$	$A=12917.22(7)$ $B=4829.19(2)$ $C=4023.88(2)$					
Cont.	$CH_2{}^{15}ND_2CH_2OD$	$v=0$	$A=13032.07(6)$ $B=4935.03(2)$ $C=4106.96(2)$					

Nr.	Molecule	v	A, B, C MHz	D, H kHz	Ref.	Ref. struct.	Varia, remarks	See also Tab., Nr.
178 *6–86a* Cont.	CH_2NDHCH_2OD *)	$v=0$	$A=13354\,(6)$ $B=\ 5288.97\,(6)$ $C=\ 4365.11\,(6)$		76PEN4		*) the first and second amino H are labelled (2) and (3) in the figure	
	CH_2NHDCH_2OD *)	$v=0$	$A=13555\,(6)$ $B=\ 5232.80\,(6)$ $C=\ 4283.12\,(6)$					
	$CH_2ND_2CH_2OH$	$v=0$	$A=13685\,(6)$ $B=\ 5045.55\,(5)$ $C=\ 4249.14\,(6)$					
179	$CH_2NH(CH_3)CH_2OH$ GT form	$v=0$	$A=12123.35\,(31)$ $B=\ 2653.774\,(15)$ $C=\ 2400.756\,(13)$		75PEN2	75PEN2	$E_{GG}-E_{GT}=0.46\,(7)$ kcal/mol	2.6.3, 90
	$CH_2NH(CH_3)CH_2OH$ GG form	$v=0$	$A=8985.12\,(31)$ $B=3076.388\,(14)$ $C=2868.664\,(15)$					
180	$CH_2N(CH_3)_2CH_2OH$ *)	$v=0$	$A=5814.0\,(2)$ $B=2214.54\,(2)$ $C=2037.96\,(2)$		76PEN3	76PEN3	*) gauche O—C—C—N configuration	2.6.3, 91 3, 54
	$CH_2N(CH_3)_2CH_2OD$	$v=0$	$A=5688.95\,(38)$ $B=2190.17\,(4)$ $C=2028.00\,(4)$					
181 *6–86b*	$CH_3OCH_2CH_2OH$							2.8, 84
182	$CH_3OCH_2CH_2NH_2$*)	$v=0$	$A=12710.34\,(1)$ $B=\ 2653.355\,(4)$ $C=\ 2407.078\,(2)$	$\Delta_J=\ 1.13\,(2)$ $\Delta_{JK}=-10.22\,(17)$ $\Delta_K=\ 82.60\,(28)$ $\delta_J=0.22\,(1)$ $\delta_K=2.4\,(7)$	80CAM1	80CAM1	*) gauche form with an intramolecular hydrogen bond of the N—H···O type	2.6.3, 92 2.7.3, 33 2.8, 85 3, 55
		$v_{t_2}=1$, A species	$A=12817.68\,(5)$ $B=\ 2645.063\,(4)$ $C=\ 2400.783\,(4)$	$\Delta_J=\ 1.11\,(2)$ $\Delta_{JK}=-10.8\,(2)$ $\Delta_K=\ 86\,(1)$ $\delta_J=0.22\,(1)$ $\delta_K=2.42$ (fixed)			$v_{t_2}=CH_3$ torsion	
Cont.								

Nr.	Molecule	v	A, B, C MHz	D, H kHz	Ref.	Ref. struct.	Varia, remarks	See also Tab., Nr.
182 Cont.	$CH_3OCH_2CH_2NH_2$	$v_{t_2}=1$, E species	$A=12818.13(4)$ $B=2645.062(3)$ $C=2400.784(4)$	$\Delta_J=1.11(2)$ $\Delta_{JK}=-10.8(2)$ $\Delta_K=83(1)$ $\delta_J=0.22(1)$ $\delta_K=2.42$ (fixed) constants fixed at the ground state values	80CAM1			
		$v_{t_2}=2$, A species	$A=12925(3)$ $B=2637.10(1)$ $C=2394.73(1)$					
		$v_{t_1}=1$	$A=12592.06(7)$ **) $B=2660.494(2)$ $C=2412.506(2)$				$v_{t_1}=65(15)\ cm^{-1}$, O—C_{ethyl} torsional mode	
		$v_{t_1}=2$	$A=12475.8(1)$ **) $B=2667.952(4)$ $C=2418.014(3)$				**) calculated using the ground state centrifugal distortion constants for the C—O torsional barrier $V_1=9(3)$ kcal/mol	
		$v_{t_1}=3$	$A=12374.6(2)$ **) $B=2675.870(3)$ $C=2423.561(3)$					
		$v_{t_1}=4$	$A=12285.(3)$ **) $B=2683.88(1)$ $C=2428.95(1)$					
		$v_{t_1}=6$	$B=2695.4(5)$ **) $C=2445.9(5)$					
		$v_{t_1}=7$	$B=2704.8(5)$ **) $C=2450.8(5)$					
		$v_{t_1}=8$	$B=2714.7(5)$ **) $C=2456.3(5)$					
		$v_{t_1}=9$	$B=2726.7(5)$ **) $C=2461.1(5)$					
		$v'_{t_3}=1$	$A=12738(7)$ **) $B=2649.15(2)$ $C=2403.32(2)$				$v_{t_3}=150(30)\ cm^{-1}$, C—C torsion, this state is split into (+) and (−) levels	
Cont.		$v''_{t_3}=1$	$B=2648.4(5)$ $C=2403.0(6)$					

Nr.	Molecule	v	A, B, C MHz	D, H kHz	Ref.	Ref. struct.	Varia, remarks	See also Tab., Nr.
182 Cont.	$CH_3OCH_2CH_2NHD'$ +)	$v=0$	$A=12054.0(2)$ $B=2628.05(1)$ $C=2362.56(2)$	$\Delta_J=1.01(12)$ $\Delta_{JK}=-8.5(9)$ $\Delta_K=-49(40)$ $\delta_J=0.28(2)$ $\delta_K=-4(6)$	80CAM1		+) primed hydrogen participates in the hydrogen bond	
	$CH_3OCH_2CH_2NDH'$ +)	$v=0$	$A=12158.2(3)$ $B=2579.63(1)$ $C=2352.55(2)$	$\Delta_J=1.13(8)$ $\Delta_{JK}=-10.4(9)$ $\Delta_K=84(44)$ $\delta_J=0.23(3)$ $\delta_K=-2(7)$				
	$CH_3OCH_2CH_2ND_2$	$v=0$	$A=11571.8(2)$ $B=2555.952(7)$ $C=2311.67(1)$	$\Delta_J=1.25(6)$ $\Delta_{JK}=-9.06(54)$ $\Delta_K=102(22)$ $\delta_J=0.21(1)$ $\delta_K=7(4)$				
183	$CH_3OCH_2CH_2CN$ *)	$v=0$	$A=17821(14)$ $B=1425.526(5)$ $C=1353.909(5)$	$D_J=0.162(10)$ $D_{JK}=-10.28(3)$	76LO1	76LO1	*) fully trans form	2.6.3, 93 3, 56
		$v_T=1$	$A=17214(18)$ $B=1426.71(1)$ $C=1356.36(1)$	$D_J=0.192(20)$ $D_{JK}=-10.27(4)$			$\nu_T=70(10)\ cm^{-1}$, CH_2—CH_2 torsional mode	
		$v_T=2$	$A=16833(42)$ $B=1427.87(1)$ $C=1358.81(1)$	$D_J=0.179(40)$ $D_{JK}=-9.98(15)$				
		$v_T=3$	$A=16524(65)$ $B=1428.98(2)$ $C=1361.31(2)$	$D_J=0.329(49)$ $D_{JK}=-9.13(101)$				
		$v_T=4$	$A=15497(60)$ $B=1429.99(2)$ $C=1364.10(4)$	$D_J=0.434(51)$ $D_{JK}=-26.2(43)$				
		$v_Y=1$	$A=16925(59)$ $B=1426.64(1)$ $C=1357.25(1)$	$D_J=0.077(35)$ $D_{JK}=-14.18(35)$			$\nu_Y=90(17)\ cm^{-1}$, CH_2—O torsion	
Cont.		$v_Y=2$	$A=16034(121)$ $B=1427.01(2)$ $C=1360.86(2)$	$D_J=0.123(30)$ $D_{JK}=-25.8(3)$				

Nr.	Molecule	v	A, B, C MHz	D, H kHz	Ref.	Ref. struct.	Varia, remarks	See also Tab., Nr.
183 Cont.	$CH_3OCH_2CH_2CN$	$v_B=1$	$A=17489\,(28)$ $B=\ 1425.86\,(2)$ $C=\ 1354.77\,(2)$	$D_J=\ 0.091\,(30)$ $D_{JK}=-11.87\,(56)$	76LO1		ν_B, C—C—N bend	
184	CHF_2CH_2OH	$v=0$	$A=9069.926\,(16)$ $B=3745.2704\,(54)$ $C=2857.0128\,(52)$	$\Delta_J=\ 0.945\,(95)$ $\Delta_{JK}=\ 5.84\,(43)$ $\Delta_K=50.0\,(21)$ $\delta_J=\ 0.116\,(22)$ $\delta_K=\ 7.51\,(56)$	80MAR2	80MAR2	conformation with intramolecular hydrogen bond	2.6.3, 94
		$v_\tau=1$	$A=9030.87\,(24)$ $B=3739.0625\,(70)$ $C=2854.8745\,(95)$	$\Delta_J=\ 0.95\,(12)$ $\Delta_{JK}=\ 7.57\,(81)$			$\nu_\tau=124\,(10)\,cm^{-1}$, C—C torsion	
		$v_\tau=2$	$A=9004.54\,(50)$ $B=3733.969\,(20)$ $C=2852.196\,(23)$	$\Delta_J=1.0\,(3)$ $\Delta_{JK}=4.5\,(16)$				
		$v_\tau=3$	$A=8983.8\,(45)$ $B=3729.18\,(14)$ $C=2849.74\,(30)$	$\Delta_J=-5.2\,(36)$				
		$v_a=1$	$A=9083.10\,(28)$ $B=3746.7274\,(94)$ $C=2854.770\,(12)$	$\Delta_J=0.66\,(16)$ $\Delta_{JK}=8.6\,(85)$			$\nu_a=243\,(20)\,cm^{-1}$, bending mode	
		$v_b=1$	$A=9028.24\,(48)$ $B=3732.143\,(15)$ $C=2853.045\,(22)$	$\Delta_J=0.0\,(28)$			$\nu_b=309\,(40)\,cm^{-1}$, bending mode	
	CHF_2CH_2OD	$v=0$	$A=8953.16\,(24)$ $B=3640.0796\,(79)$ $C=2785.9475\,(104)$	$\Delta_J=0.46\,(14)$ $\Delta_{JK}=9.21\,(78)$				
185 6-90 4-78 Cont.	$CH_3C(=O)H$	$v=0$	$A=56609.474\,(537)$ $B=10162.7664\,(875)$ $C=\ 9100.4121\,(864)$	$\Delta_J=\ 7.6986\,(7293)$ $\Delta_{JK}=\ 0.9764\,(79215)$ $\Delta_K=678.5932\,(512701)$ $\delta_J=\ 1.2216\,(1345)$ $\delta_K=-135.3705\,(122858)$	76BAU	73NOS1	for structure, see also Vol. II/7 (3, 243)	2.6.3, 95 2.8, 17 3, 57

Nr.	Molecule	v	A, B, C MHz	D, H kHz	Ref.	Ref. struct.	Varia, remarks	See also Tab., Nr.
185 6–90 4–78 Cont.	$CH_2DC(=O)H$ symmetric	$v=0$	$A=48508.862\,(60)$ $B=9864.667\,(10)$ $C=8637.098\,(10)$	$\Delta_J=7.6822\pm0.04$ $\Delta_{JK}=-3.6110\pm0.4$ $\Delta_K=397.0430\pm0.6$ $\delta_J=1.4003\,(50)$ $\delta_K=-54.6695\pm0.1$	76TUR1			
	$CD_2HC(=O)H$ symmetric	$v=0$	$A=47940.735\,(100)$ $B=8778.724\,(20)$ $C=8175.086\,(20)$	$\Delta_J=5.4025\pm0.4$ $\Delta_{JK}=30.8607\pm6$ $\Delta_K=306.6783\pm11$ $\delta_J=0.6049\,(90)$ $\delta_K=-350.1144\pm2$				
	$CH_2DC(=O)H$ asymmetric	$v=0^+$	$A=52330.14\,(6)$ $B=9314.08\,(3)$ $C=8687.68\,(3)$	$\Delta_J=337.8\,(9)$ $\Delta_{JK}=-972\,(3)$ $\Delta_K=636\,(2)$ $\delta_J=161.6\,(4)$ $\delta_K=-321\,(1)$	81TUR		interaction term: $H_{+-}=(\mu_{bc}+\varepsilon_1 P^2)(P_bP_c$ $+P_cP_b)+\varepsilon_2[P_c^2(P_bP_c+P_cP_b)$ $+(P_bP_c+P_cP_b)P_c^2]$ $+$ = symmetric level; $-$ = antisymmetric level $\Delta E_{+-}=804.5\,(1)$ MHz $\mu_{bc}=266.92\,(3)$ MHz $\varepsilon_1=9.5\,(4)$ kHz $\varepsilon_2=-4.02\,(2)$ kHz	
		$v=0^-$	$A=52328.81\,(6)$ $B=9312.93\,(3)$ $C=8687.85\,(3)$	$\Delta_J=338.8\,(8)$ $\Delta_{JK}=-975\,(3)$ $\Delta_K=638\,(2)$ $\delta_J=162.3\,(5)$ $\delta_K=-322\,(1)$				
	$CHD_2C(=O)H$ asymmetric	$v=0^+$	$A=45141.52\,(5)$ $B=8941.6\,(2)$ $C=8422.0\,(2)$	$\Delta_J=193\,(2)$ $\Delta_{JK}=-564\,(12)$ $\Delta_K=382\,(11)$ $\delta_J=95\,(1)$ $\delta_K=-190\,(5)$			$\Delta E_{+-}=183.6\,(1)$ MHz $\mu_{bc}=421.3\,(1)$ MHz $\varepsilon_1=-2\,(2)$ kHz $\varepsilon_2=-2.7\,(4)$ kHz	
		$v=0^-$	$A=45141.72\,(5)$ $B=8940.9\,(2)$ $C=8422.0\,(2)$	$\Delta_J=195\,(2)$ $\Delta_{JK}=-567\,(12)$ $\Delta_K=384\,(11)$ $\delta_J=96\,(1)$ $\delta_K=-191\,(5)$				

Nr.	Molecule	v	A, B, C MHz	D, H kHz	Ref.	Ref. struct.	Varia, remarks	See also Tab., Nr.
186 *4-82*	$CH_3C(=O)CN$	$v=0$	$A=10185.67(3)$ *) $B=4157.57(3)$ $C=3002.77(2)$		76SCA2	77PAN1	for structure, see also Vol. II/7 (3, 365) *) for the A species $\nu_\tau \simeq 128\ cm^{-1}$, CH_3 torsion	2.7.3, 34 2.8, 18
		$v_\tau=1$	$A=10174.91(3)$ *) $B=4141.75(3)$ $C=3000.75(2)$					
		$v_\delta=1$	$A=10158.02(3)$ *) $B=4176.19(3)$ $C=3005.56(2)$				$\nu_\delta=176.0(5)\ cm^{-1}$, C—C—N in-plane bend	
	$CH_3C(=O)C^{15}N$	$v=0$	$A=10183.659(8)$ $B=4011.073(2)$ $C=2925.282(2)$	$d_J=-2.145(300)$ $d_{JK}=-127.1(60)$ $d_K=-274.9(100)$ $d_{WJ}=0.8450(800)\cdot10^{-6}$ $d_{WK}=40.03(100)\cdot10^{-4}$	77PAN3			
		$v_\tau=1$	$A=10172.244$ *) $B=3996.111$ $C=2923.288$		78PAN2			
		$v_\delta=1$	$A=10156.983$ *) $B=4028.932$ $C=2928.086$					
	$CH_2DC(=O)CN$ symmetric	$v=0$	$A=9987.22(8)$ $B=3982.71(3)$ $C=2894.76(2)$		77PAN1			
	$CH_2DC(=O)CN$ asymmetric	$v=0$	$A=9948.54(8)$ $B=4103.74(3)$ $C=2931.69(2)$					
187	$CH_3C(=O)NCO$	$v=0$	$A=10773.160(25)$ *) $B=2222.693(5)$ $C=1862.934(5)$	$\Delta_J=0.4882(414)$ *) $\Delta_{JK}=8.6492(2429)$ $\Delta_K=10.5489(55544)$ $\delta_J=0.0728(28)$ $\delta_K=5.0906(2451)$	80LAD	80LAD	*) for the A species	2.6.3, 96 2.8, 19

Nr.	Molecule	v	A, B, C MHz	D, H kHz	Ref.	Ref. struct.	Varia, remarks	See also Tab., Nr.
188 *6–91a* *4–83*	$CH_3C(=O)OH$	$v=0$	$A=11335.580\,(13)$ *) $B=\ 9478.721\,(11)$ $C=\ 5324.990\,(7)$	$\Delta_J\ =\ 4.359\,(10)$ *) $\Delta_{JK}=\ 3.98\,(8)$ $\Delta_K\ =\ 3.05\,(16)$ $\delta_J\ =\ 1.768\,(4)$ $\delta_K\ =\ 6.54\,(3)$	82DEM	81VAN1	for structure, see also Vol. II/7 (3, 246) *) representation I^r used	2.8, 19a 3, 58
	CD_3COOH	$v=0$	$A=10576.022\,(12)$ $B=\ 7851.826\,(5)$ $C=\ 4765.942\,(5)$	$\Delta_{JK}=\ 4.0\,(2)$ $\Delta_K\ =\ 2.0\,(2)$ $\delta_J\ =\ 1.08\,(3)$ $\delta_K\ =\ 4.4\,(2)$	81VAN1			
	CH_3COOD	$v=0$	$A=11190.770\,(29)$ $B=\ 9004.472\,(27)$ $C=\ 5141.843\,(12)$	$\Delta_{JK}=\ -1.8\,(11)$ $\Delta_K\ =\ 10.9\,(10)$ $\delta_J\ =\ 1.32\,(17)$ $\delta_K\ =\ 5.9\,(6)$				
	$CH_3CO^{18}OH$	$v=0$	$A=10839.930\,(42)$ $B=\ 9274.700\,(28)$ $C=\ 5150.520\,(11)$	$\Delta_{JK}=\ -3.0\,(14)$ $\Delta_K\ =\ 9.7\,(10)$ $\delta_J\ =\ 1.75\,(22)$ $\delta_K\ =\ 5.0\,(6)$				
	$CH_3C^{18}OOH$	$v=0$	$A=10778.356\,(15)$ $B=\ 9299.845\,(13)$ $C=\ 5144.229\,(6)$	$\Delta_{JK}=\ 5.4\,(9)$ $\Delta_K\ =\ 0.2\,(6)$ $\delta_J\ =\ 1.61\,(14)$ $\delta_K\ =\ 7.1\,(4)$				
	$^{13}CH_3COOH$	$v=0$	$A=11334.313\,(51)$ $B=\ 9121.769\,(16)$ $C=\ 5210.065\,(5)$	$\Delta_{JK}=\ 3.7\,(5)$ $\Delta_K\ =\ 3.4\,(4)$ $\delta_J\ =\ 1.60\,(7)$ $\delta_K\ =\ 6.7\,(3)$				
	$^{13}CH_3COOD$	$v=0$	$A=11177.738\,(38)$ $B=\ 8677.802\,(17)$ $C=\ 5030.875\,(6)$	$\Delta_{JK}=\ 1.5\,(5)$ $\Delta_K\ =\ 8.0\,(5)$ $\delta_J\ =\ 1.50\,(7)$ $\delta_K\ =\ 5.0\,(3)$				
	CH_2DCOOH	$v=0$	$A=11192.410\,(7)$ $B=\ 8821.641\,(7)$ $C=\ 5087.298\,(5)$	$\Delta_{JK}=\ 7.3\,(6)$ $\Delta_K\ =\ -0.5\,(5)$ $\delta_J\ =\ 1.56\,(8)$ $\delta_K\ =\ 1.4\,(3)$				
Cont.								

Nr.	Molecule	v	A, B, C MHz	D, H kHz	Ref.	Ref. struct.	Varia, remarks	See also Tab., Nr.
188 *6-91a* *4-83* Cont.	CHD_2COOH	$v=0$	$A=10841.226(4)$ $B=8259.696(4)$ $C=4969.904(3)$	$\Delta_{JK}=20.4(2)$ $\Delta_K=-14.2(2)$ $\delta_J=1.19(3)$ $\delta_K=-1.20(1)$	81VAN1			
	CH_2DCOOD	$v=0$	$A=11110.850(16)$ **) $B=8357.437(16)$ $C=4913.759(16)$	$\Delta_J=1.8(11)$ $\Delta_{JK}=5.84(88)$ $\Delta_K=1.6(12)$ $\delta_J=1.32(11)$ $\delta_K=1.32(47)$	79CAM		**) $\Delta=-3.1060(4)$ amu Å^2	
188a *6-92*	$CH_3C(=O)OCH_3$	$v=0$	$A=10246.60(10)$ $B=4170.278(59)$ $C=3076.527(52)$		80SHE			2.8, 19b
189	$CH_3C(=O)OOH$	$v=0$	$I_a=46.7304(74)$ *) $I_b=118.2175(17)$ $I_c=161.8460(23)$		76CUG	76CUG	*) in amu Å^2 planar structure for the peroxy acid group	2.6.3, 97 2.8, 20 3, 59
	$CH_3C(=O)OOD$	$v=0$	$I_a=47.292(40)$ *) $I_b=120.520(7)$ $I_c=164.729(12)$					
190	$CH_3C(=O)SH$	$v=0$	$A=9913.29(56)$ $B=4923.11(23)$ $C=3354.60(24)$		77NAI	77NAI		2.6.3, 98 2.8, 21 3, 60
	$CH_3C(=O)SD$	$v=0$	$A=9662.80(78)$ $B=4810.74(26)$ $C=3273.92(18)$					
191	$CF_3C(=O)OCH_3$	$v=0$	$A=3615.10(40)$ $B=1521.14(1)$ $C=1332.23(1)$		74JON	74JON		3, 61
	$CF_3C(=O)OCD_3$	$v=0$	$A=3529.88(50)$ $B=1393.16(1)$ $C=1232.50(1)$					

Nr.	Molecule	v	A, B, C MHz	D, H kHz	Ref.	Ref. struct.	Varia, remarks	See also Tab., Nr.
192	$CF_3C(=O)OCH_2CH_3$ extended **)	$v=0$	$B+C=1842\,(2)$ *)		76TR1		*) LRMW **) $\tau_1(O=COC)=0°$; $\tau_2(COCC)=180°$ ***) $\tau_1=0°$; $\tau_2\simeq 90°$ +) $\tau_1\simeq 45°$; $\tau_2\simeq 240°$	
	$CF_3C(=O)OCH_2CH_3$ compact ***)	$v=0$	$B+C=1947\,(2)$ *)					
	$CF_3C(=O)OCH_2CH_3$ intermediate +)	$v=0$	$B+C=1904\,(5)$ *)					
193	$CF_3C(=O)SCH_2CH_3$ compact **)	$v=0$ $v=1$ $v=2$	$B+C=1599.2\,(2)$ *) $B+C=1593.2\,(3)$ $B+C=1584.5\,(6)$		77SIL		*) LRMW **) $\tau_1(OCSC)=0°$; $\tau_2(CSCC)\simeq 90°$ ***) $\tau_1\simeq 60°$; $\tau_2\simeq 240°$ the intermediate form is ≈2.5 kcal/mol higher in energy than the compact form	
	$CF_3C(=O)SCH_2CH_3$ intermediate ***)	$v=0$	$B+C=1554\,(1)$ *)					
194	$CF_3C(=O)OCH_2CH_2CH_3$ EA EG CA CG IA	$v=0$ $v=0$ $v=0$ $v=0$ $v=0$	$B+C=1165\,(1)$ *) $B+C=1378\,(1)$ *) $B+C=1245\,(1)$ *) $B+C=1667\,(2)$ *) $B+C=1213\,(1)$ *)		78TR5		*) LRMW E: $\tau_1(O{=}COC)=0°$; $\tau_2(COCC)=180°$ C: $\tau_1\simeq 0°$; $\tau_2\simeq 90°$ I: $\tau_1\simeq 50°$; $\tau_2=230°$ G: $\tau_3(OCCC)\simeq 60°$; A: $\tau_3=180°$	
195	$CF_3C(=O)SCH_2CH_2CH_3$ CA IA	$v=0$ $v=0$	$B+C=1088.8\,(5)$ *) $B+C=1064\,(11)$ *)		78SIL		*) LRMW see Nr. 194 for definitions of CA and IA	
196	$CF_3C(=O)OCH(CH_3)_2$	$v=0$	$B+C=1509\,(4)$ *)		77TR2		*) LRMW $\tau_1(O{=}COC)=0°$; $\tau_2(COCH)\simeq 40°$	

Nr.	Molecule	v	A, B, C MHz	D, H kHz	Ref.	Ref. struct.	Varia, remarks	See also Tab., Nr.
197	$CF_3C(=O)SCH(CH_3)_2$	$v=0$	$B+C=1211.7(6)$ *)		79SIL		*) LRMW τ_1 (OCSC) = 0°; τ_2 (CSCH) = 0° or 60°	
	$CF_3C(=O)SCH(CH_3)_2$ syn-anti	$v=0$	$B+C=1310.2(5)$ *)				$\tau_1=0°$; $\tau_2=180°$	
198	$CF_3C(=O)OCH_2CH(CH_3)_2$ IG	$v=0$	$B+C=1042(3)$ *)		78TR3		*) LRMW see Nr. 194 for definition of IG	
199	$CH_2FC(=O)OH$ trans	$v=0$	$A=10681.156(5)$ $B=3908.222(2)$ $C=2914.371(2)$	$\tau_{aaaa}=-39.3(2)$ $\tau_{bbbb}=-4.93(2)$ $\tau_{cccc}=-1.68(2)$ $\tau_{abab}=-11.7(2)$	78VAN2	78VAN2	for structure, see also Vol. II/7 (3, 216)	3, 62
	$CH_2FC(=O)OD$ trans	$v=0$	$A=10677.301(4)$ $B=3742.082(2)$ $C=2820.850(2)$	$\tau_{aaaa}=-40.5(2)$ $\tau_{bbbb}=-4.27(2)$ $\tau_{cccc}=-1.54(2)$ $\tau_{abab}=-10.6(2)$				
	$CH_2FC(={}^{18}O)OH$ trans	$v=0$	$A=9995.443(18)$ $B=3886.078(6)$ $C=2848.826(5)$	$\tau_{aaaa}=-38(6)$ $\tau_{bbbb}=-5.23(24)$ $\tau_{cccc}=-1.71(15)$ $\tau_{abab}=-11.5(6)$				
	$CH_2FC(=O){}^{18}OH$ trans	$v=0$	$A=10471.131(19)$ $B=3756.760(7)$ $C=2814.315(5)$	$\tau_{aaaa}=-56(8)$ $\tau_{bbbb}=-3.58(28)$ $\tau_{cccc}=-0.94(19)$ $\tau_{abab}=-12.8(7)$				
	$CH_2FC(={}^{18}O){}^{18}OH$ trans	$v=0$	$A=9778.056(16)$ $B=3741.104(4)$ $C=2753.128(4)$	$\tau_{aaaa}=-38(5)$ $\tau_{bbbb}=-4.40(10)$ $\tau_{cccc}=-1.51(5)$ $\tau_{abab}=-10.7(2)$				
	${}^{13}CH_2FC(=O)OH$ trans	$v=0$	$A=10562.381(18)$ $B=3893.038(10)$ $C=2897.095(6)$	$\tau_{aaaa}=-28(7)$ $\tau_{bbbb}=-5.3(6)$ $\tau_{cccc}=-1.9(4)$ $\tau_{abab}=-10.6(12)$				
Cont.								

Nr.	Molecule	v	A, B, C MHz	D, H kHz	Ref.	Ref. struct.	Varia, remarks	See also Tab., Nr.
199 Cont.	$CH_2F^{13}C(=O)OH$ trans	$v=0$	$A=10680.074(13)$ $B=3899.608(6)$ $C=2909.501(5)$	$\tau_{aaaa}=-47(6)$ $\tau_{bbbb}=-5.0(4)$ $\tau_{cccc}=-1.7(3)$ $\tau_{abab}=-11.6(9)$	78VAN2			
	$CD_2FC(=O)OH$ trans	$v=0$	$A=9615.701(9)$ $B=3828.248(4)$ $C=2836.967(3)$	$\tau_{aaaa}=-24(2)$ $\tau_{bbbb}=-4.53(13)$ $\tau_{cccc}=-1.61(8)$ $\tau_{abab}=-7.1(3)$				
	$CD_2FC(=O)OD$ trans	$v=0$	$A=9611.225(13)$ $B=3666.720(3)$ $C=2747.045(3)$	$\tau_{aaaa}=-24(4)$ $\tau_{bbbb}=-3.78(10)$ $\tau_{cccc}=-1.36(7)$ $\tau_{abab}=-6.8(3)$				
	$CHDFC(=O)OH$ trans	$v=0$	$A=10115.974(10)$ $B=3869.565(3)$ $C=2873.970(2)$	$\tau_{aaaa}=-38(3)$ $\tau_{bbbb}=-4.95(8)$ $\tau_{cccc}=-1.68(5)$ $\tau_{abab}=-9.5(2)$				
	$CH_2FC(=O)OH$ cis	$v=0$	$A=10287.17(7)$ $B=4128.442(11)$ $C=3002.659(9)$	$\tau_{aaaa}=-53(7)$ $\tau_{bbbb}=-3.9(8)$ $\tau_{cccc}=-0.6(6)$ $\tau_{abab}=-15(2)$			$E_{cis}-E_{trans}=590(28)$ cal/mol	
	$CH_2FC(=O)OD$ cis	$v=0$	$A=9827.4(6)$ $B=4123.74(3)$ $C=2959.78(2)$	*)			*) fixed on value of parent molecule	
	$CH_2FC(={}^{18}O)OH$ cis	$v=0$	$A=10140.7(8)$ $B=3946.91(3)$ $C=2893.58(2)$	*)				
	$CH_2FC(=O){}^{18}OH$ cis	$v=0$	$A=9656.9(7)$ $B=4108.49(3)$ $C=2936.12(2)$	*)				
Cont.	${}^{13}CH_2FC(=O)OH$ cis	$v=0$	$A=10153.3(5)$ $B=4113.35(3)$ $C=2983.19(2)$	*)				

Nr.	Molecule	v	A, B, C MHz	D, H kHz	Ref.	Ref. struct.	Varia, remarks	See also Tab., Nr.
199 Cont.	$CH_2F^{13}C(=O)OH$ cis	$v=0$	$A=10287.3(6)$ $B=4115.08(3)$ $C=2995.64(2)$	*)	78VAN2		*) fixed on value of parent molecule	
	$CD_2FC(=O)OH$ cis	$v=0$	$A=9233.4(6)$ $B=4039.09(3)$ $C=2914.74(2)$	*)				
	$CD_2FC(=O)OD$ cis	$v=0$	$A=8838.2(6)$ $B=4036.18(3)$ $C=2872.71(2)$	*)				
	$CHDFC(=O)OH$	$v=0$	$A=9726.8(8)$ $B=4085.23(4)$ $C=2956.77(3)$	*)				
200	$CH_2{}^{35}ClC(=O)OH$	$v=0$	$A=10514.872(17)$ $B=2351.358(4)$ $C=1946.028(4)$	$\tau_{aaaa}=-38.0(1)$ $\tau_{bbbb}=-2.07(3)$ $\tau_{cccc}=-0.99(2)$ $\tau_{abab}=-6.1(1)$	74VAN2			2.6.3, 99 2.7.3, 113
		$v_\tau=1$	$A=10422.687(20)$ $B=2351.876(4)$ $C=1951.356(5)$	$\tau_{aaaa}=-40.4(2)$ $\tau_{bbbb}=-2.18(3)$ $\tau_{cccc}=-1.14(4)$ $\tau_{abab}=-5.4(3)$			$\nu_\tau=62\ cm^{-1}$, torsion	
		$v_\tau=2$	$A=10329.059(20)$ $B=2352.652(5)$ $C=1957.094(5)$					
		$v_\tau=3$	$A=10232.716(16)$ $B=2353.772(4)$ $C=1963.451(4)$					
		$v_\tau=4$	$A=10131.443(24)$ $B=2355.287(8)$ $C=1970.644(8)$					
Cont.	$CH_2{}^{37}ClC(=O)OH$	$v=0$	$A=10514.166(21)$ $B=2287.986(6)$ $C=1902.400(6)$	$\tau_{aaaa}=-39$ (fixed) $\tau_{bbbb}=-1.5(2)$ $\tau_{cccc}=-0.6(1)$ $\tau_{abab}=-7.7(6)$				

Nr.	Molecule	v	A, B, C MHz	D, H kHz	Ref.	Ref. struct.	Varia, remarks	See also Tab., Nr.
200 Cont.	$CH_2{}^{35}ClC(=O)OD$	$v=0$	$A=10514.942\,(45)$ $B=2266.218\,(10)$ $C=1887.445\,(9)$	$\tau_{aaaa}=-39.4$ (fixed) $\tau_{bbbb}=-1.9\,(3)$ $\tau_{cccc}=-1.0\,(2)$ $\tau_{abab}=-4.9\,(11)$	74VAN2			
201	$CH_2{}^{79}BrC(=O)OH$	$v=0$	$A=10460.377\,(19)$ $B=1514.793\,(5)$ $C=1335.161\,(4)$	$\tau_{aaaa}=-40$ (fixed) $\tau_{bbbb}=-0.98\,(4)$ $\tau_{cccc}=-0.60\,(4)$ $\tau_{abab}=-3.2\,(6)$	78VAN1	78VAN1	conformation found to be trans with respect to the atoms Br CCOH	2.7.3, 151
		$v_\tau=1$	$A=10338.070\,(34)$ $B=1516.178\,(7)$ $C=1339.256\,(7)$	$\tau_{aaaa}=-40$ (fixed) $\tau_{bbbb}=-1.04\,(5)$ $\tau_{cccc}=-0.70\,(6)$ $\tau_{abab}=-1.6\,(10)$				
	$CH_2{}^{79}BrC(=O)OD$	$v=0$	$A=10459.644\,(57)$ $B=1461.719\,(8)$ $C=1293.810\,(6)$	$\tau_{aaaa}=-40$ (fixed) $\tau_{bbbb}=-0.73\,(9)$ $\tau_{cccc}=-0.45\,(8)$ $\tau_{abab}=-3$ (fixed)				
	$CH_2{}^{81}BrC(=O)OH$	$v=0$	$A=10460.249\,(23)$ $B=1500.928\,(6)$ $C=1324.378\,(4)$	$\tau_{aaaa}=-40$ (fixed) $\tau_{bbbb}=-0.89\,(6)$ $\tau_{cccc}=-0.56\,(4)$ $\tau_{abab}=-3$ (fixed)				
201 *6-95* *4-85*	$CH_2FC(=O)F$ trans	$v=0$	$A=10508.903\,(6)$ $B=3855.058\,(3)$ $C=2870.712\,(3)$	$\tau_{aaaa}=-41.4\,(6)$ $\tau_{bbbb}=-4.95\,(4)$ $\tau_{cccc}=-1.66\,(3)$ $\tau_{abab}=-12.6\,(2)$	78VAN2	78VAN2	for structure, see also Vol. II/7 (3, 187)	3, 63
		$v_\tau=1$	$A=10457.030\,(13)$ $B=3850.867\,(9)$ $C=2875.593\,(6)$	$\tau_{aaaa}=-34\,(5)$ $\tau_{bbbb}=-4.5\,(6)$ $\tau_{cccc}=-1.5\,(4)$ $\tau_{abab}=-12.3\,(13)$				
Cont.		$v_\tau=2$	$A=10410.743\,(13)$ $B=3847.494\,(6)$ $C=2880.303\,(5)$	$\tau_{aaaa}=-46\,(6)$ $\tau_{bbbb}=-4.7\,(3)$ $\tau_{cccc}=-1.6\,(2)$ $\tau_{abab}=-12.0\,(7)$				

Nr.	Molecule	v	A, B, C MHz	D, H kHz	Ref.	Ref. struct.	Varia, remarks	See also Tab., Nr.
201 *6-95* *4-85* Cont.	$CH_2FC(=^{18}O)F$ trans	$v=0$	$A=9817.685(7)$ $B=3834.776(3)$ $C=2805.589(3)$	$\tau_{aaaa}=-35.2(7)$ $\tau_{bbbb}=-5.04(6)$ $\tau_{cccc}=-1.62(4)$ $\tau_{abab}=-12.4(2)$	78VAN2			
	$^{13}CH_2FC(=O)F$ trans	$v=0$	$A=10397.782(4)$ $B=3839.622(2)$ $C=2853.860(2)$	$\tau_{aaaa}=-38.7(4)$ $\tau_{bbbb}=-4.90(3)$ $\tau_{cccc}=-1.62(2)$ $\tau_{abab}=-12.1(1)$				
	$CH_2F^{13}C(=O)F$ trans	$v=0$	$A=10505.374(4)$ $B=3848.169(2)$ $C=2866.630(2)$	$\tau_{aaaa}=-41.7(4)$ $\tau_{bbbb}=-4.94(3)$ $\tau_{cccc}=-1.68(2)$ $\tau_{abab}=-12.5(1)$				
	$CD_2FC(=O)F$ trans	$v=0$	$A=9486.888(5)$ $B=3778.036(3)$ $C=2796.988(3)$	$\tau_{aaaa}=-26.4(5)$ $\tau_{bbbb}=-4.63(4)$ $\tau_{cccc}=-1.56(3)$ $\tau_{abab}=-8.3(2)$				
		$v_\tau=1$	$A=9451.018(11)$ $B=3772.235(7)$ $C=2801.237(5)$	$\tau_{aaaa}=-31(4)$ $\tau_{bbbb}=-4.8(4)$ $\tau_{cccc}=-1.7(3)$ $\tau_{abab}=-7.2(6)$				
		$v_\tau=2$	$A=9415.489(11)$ $B=3766.808(7)$ $C=2805.746(5)$	$\tau_{aaaa}=-16(4)$ $\tau_{bbbb}=-5.5(4)$ $\tau_{cccc}=-2.4(3)$ $\tau_{abab}=-4.9(7)$				
	$CHDFC(=O)F$ trans	$v=0$	$A=9967.699(5)$ $B=3817.729(2)$ $C=2832.294(2)$	$\tau_{aaaa}=-31.8(5)$ $\tau_{bbbb}=-4.78(4)$ $\tau_{cccc}=-1.56(3)$ $\tau_{abab}=-10.7(2)$				
		$v_\tau=1$	$A=9924.943(17)$ $B=3812.589(9)$ $C=2836.825(6)$	$\tau_{aaaa}=-36(6)$ $\tau_{bbbb}=-4.7(5)$ $\tau_{cccc}=-1.5(3)$ $\tau_{abab}=-9.9(10)$				
Cont.		$v_\tau=2$	$A=9883.377(60)$ $B=3807.939(27)$ $C=2841.575(18)$	$\tau_{aaaa}=-52(18)$ $\tau_{bbbb}=-5.7(13)$ $\tau_{cccc}=-2.1(9)$ $\tau_{abab}=-7.8(27)$				

Nr.	Molecule	v	A, B, C MHz	D, H kHz	Ref.	Ref. struct.	Varia, remarks	See also Tab., Nr.
201	$CH_2FC(=O)F$	$v=0$	$A=10223.839\,(7)$	$\tau_{aaaa}=-51\,(2)$	78VAN2			
6–95			$B=4110.527\,(2)$	$\tau_{bbbb}=-6.16\,(9)$				
4–85	cis		$C=2991.998\,(2)$	$\tau_{cccc}=-2.45\,(6)$				
Cont.				$\tau_{abab}=-10.0\,(2)$				
		$v_\tau=1$	$A=10155.070\,(48)$	$\tau_{aaaa}=-61\,(14)$				
			$B=4105.222\,(12)$	$\tau_{bbbb}=-6.4\,(5)$				
			$C=3003.716\,(8)$	$\tau_{cccc}=-2.6\,(3)$				
				$\tau_{abab}=-10.7\,(9)$				
		$v_\tau=2$	$A=10090.310\,(71)$	fixed at ground state values				
			$B=4100.767\,(38)$					
			$C=3014.426\,(7)$					
	$CH_2FC(={}^{18}O)F$	$v=0$	$A=10054.381\,(18)$	$\tau_{aaaa}=-48\,(5)$				
			$B=3936.102\,(5)$	$\tau_{bbbb}=-5.31\,(20)$				
	cis		$C=2884.718\,(4)$	$\tau_{cccc}=-2.17\,(12)$				
				$\tau_{abab}=-9.4\,(4)$				
	$^{13}CH_2FC(=O)F$	$v=0$	$A=10094.640\,(19)$	$\tau_{aaaa}=-34\,(5)$				
			$B=4092.930\,(5)$	$\tau_{bbbb}=-6.06\,(19)$				
	cis		$C=2971.706\,(4)$	$\tau_{cccc}=-2.31\,(11)$				
				$\tau_{abab}=-10.0\,(4)$				
	$CH_2F^{13}C(=O)F$	$v=0$	$A=10222.163\,(21)$	$\tau_{aaaa}=-54\,(5)$				
			$B=4098.168\,(5)$	$\tau_{bbbb}=-5.91\,(18)$				
	cis		$C=2985.300\,(3)$	$\tau_{cccc}=-2.31\,(11)$				
				$\tau_{abab}=-10.5\,(4)$				
	$CD_2FC(=O)F$	$v=0$	$A=9188.805\,(16)$	$\tau_{aaaa}=-32\,(4)$				
			$B=4015.423\,(9)$	$\tau_{bbbb}=-6.6\,(5)$				
	cis		$C=2901.971\,(6)$	$\tau_{cccc}=-3.6\,(3)$				
				$\tau_{abab}=0.2\,(8)$				
		$v_\tau=1$	$A=9143.776\,(19)$	$\tau_{aaaa}=-47\,(6)$			$\nu_\tau=49.5\ cm^{-1}$, torsion	
			$B=4007.164\,(10)$	$\tau_{bbbb}=-5.4\,(5)$				
			$C=2913.439\,(6)$	$\tau_{cccc}=-2.7\,(3)$				
				$\tau_{abab}=-2.9\,(8)$				
		$v_\tau=2$	$A=9100.988\,(30)$	$\tau_{aaaa}=-44\,(7)$				
			$B=3999.846\,(19)$	$\tau_{bbbb}=-4.9\,(12)$				
			$C=2923.968\,(13)$	$\tau_{cccc}=-2.4\,(8)$				
Cont.				$\tau_{abab}=-4.4\,(19)$				

Nr.	Molecule	v	A, B, C MHz	D, H kHz	Ref.	Ref. struct.	Varia, remarks	See also Tab., Nr.
201	CHDFC(=O)F	$v=0$	$A=9672.061\ (22)$	$\tau_{aaaa}=-27\ (7)$	78VAN2			
6-95			$B=4065.365\ (11)$	$\tau_{bbbb}=-7.1\ (6)$				
4-85	cis		$C=2944.528\ (7)$	$\tau_{cccc}=-3.1\ (4)$				
Cont.				$\tau_{abab}=-5.3\ (11)$				
		$v_\tau=1$	$A=9617.709\ (23)$	$\tau_{aaaa}=-40\ (8)$				
			$B=4057.870\ (11)$	$\tau_{bbbb}=-5.6\ (5)$				
			$C=2956.349\ (7)$	$\tau_{cccc}=-2.3\ (4)$				
				$\tau_{abab}=-8.6\ (9)$				
		$v_\tau=2$	$A=9565.884\ (76)$	$\tau_{aaaa}=-58\ (21)$				
			$B=4051.500\ (42)$	$\tau_{bbbb}=-4.6\ (24)$				
			$C=2967.109\ (27)$	$\tau_{cccc}=-1.7\ (17)$				
				$\tau_{abab}=-10.1\ (42)$				
202	$CH_2FC(=O)^{35}Cl$ *)	$v=0$	$A=9025.82\ (4)$		74SZA	74SZA	*) F and Cl trans to each other	2.7.3, 114
			$B=2403.92\ (4)$					
			$C=1920.70\ (2)$					
		$v_\tau=1$	$A=8964.59\ (8)$					
			$B=2404.06\ (8)$					
			$C=1923.37\ (2)$					
	$CH_2FC(=O)^{37}Cl$	$v=0$	$A=8994.95\ (8)$					
			$B=2342.24\ (8)$					
			$C=1879.75\ (2)$					
203	$CH_2{}^{35}ClC(=O)H$ cis *)	$v=0$	$A=15189.90\ (60)$		76FOR	76FOR	*) Cl and O cis	2.7.3, 115
			$B=3858.50\ (20)$					
			$C=3135.55\ (20)$					
	$CH_2{}^{35}ClC(=O)H$ trans	$v=0$	$A=33609.44\ (30)$					
			$B=2693.64\ (5)$					
			$C=2549.46\ (5)$					
		$v_\tau=1$	$A=31694.40\ (30)$				torsional potential function, see [77MAL2]	
			$B=2720.43\ (5)$					
Cont.			$C=2583.94\ (5)$					

Nr.	Molecule	v	A, B, C MHz	D, H kHz	Ref.	Ref. struct.	Varia, remarks	See also Tab., Nr.
203 Cont.	$CH_2{}^{35}ClC(=O)H$ trans	$v_\tau=2$	$A=30976.98$ (30) $B=2734.71$ (5) $C=2601.19$ (5)		76FOR			
		$v_\tau=3$	$A=29981.77$ (30) $B=2753.66$ (5) $C=2624.40$ (5)					
		$v_\tau=4$	$A=29215.91$ (30) $B=2770.37$ (5) $C=2644.65$ (5)					
		$v_\tau=5$	$A=28534.96$ (30) $B=2786.51$ (5) $C=2664.16$ (5)					
		$v_\tau=6$	$A=27935.76$ (30) $B=2801.81$ (20) $C=2682.86$ (20)					
	$CH_2{}^{37}ClC(=O)H$ trans	$v=0$	$A=33540.95$ (30) $B=2630.21$ (20) $C=2492.09$ (20)					
204	$CH_2{}^{35}ClC(=O){}^{35}Cl$	$v=0$	$A=9030.255$ (7) $B=1503.976$ (1) $C=1299.588$ (1)	$\tau_{aaaa}=-72.9$ (2) $\tau_{bbbb}=-0.629$ (3) $\tau_{cccc}=-0.365$ (2) $\tau_{abab}=-2.51$ (3)	76VAN	76VAN	planar conformation with both Cl trans $\Delta=-3.1174$ amu Å^2 *) fixed on ground state values of parent molecule	2.7.3., 116
		$v_\tau=1$	$A=8944.60$ (4) $B=1505.48$ (1) $C=1302.19$ (1)	*)				
		$v_\tau=2$	$A=8861.89$ (6) $B=1507.16$ (2) $C=1304.91$ (2)	*)				
	$CH_2{}^{37}ClC(=O){}^{35}Cl$	$v=0$	$A=9024.64$ (10) $B=1462.53$ (2) $C=1268.41$ (2)	*)				
	$CH_2{}^{35}ClC(=O){}^{37}Cl$	$v=0$	$A=8993.09$ (5) $B=1465.67$ (1) $C=1270.14$ (1)	*)				

Nr.	Molecule	v	A, B, C MHz	D, H kHz	Ref.	Ref. struct.	Varia, remarks	See also Tab., Nr.
205	$CH_2FC(=O)NH_2$	$v=0$	$A=9884.358(4)$ $B=4059.671(2)$ $C=2932.474(1)$	$d_J=-2.06(5)$ $d_{JK}=-31.2(13)$ $d_K=-50.7(24)$ $d_{WJ}=8.01(14)\cdot10^{-7}$ $d_{WK}=8.41(38)\cdot10^{-6}$	74MAR2	74MAR2	planar skeleton, CF and CO trans to each other, hydrogen bond between F and the nearest H amide	2.6.3, 100 2.7.3, 35 3, 64
		$v_\tau=1$	$A=9839.787(7)$ $B=4056.206(3)$ $C=2938.174(2)$	$d_J=-2.38(11)$ $d_{JK}=-28.8(22)$ $d_K=-46.3(42)$ $d_{WJ}=9.09(30)\cdot10^{-7}$ $d_{WK}=7.73(66)\cdot10^{-6}$			$\nu_\tau=114(14)\ cm^{-1}$, C—C torsion	
		$v_\tau=2$	$A=9795.62(4)$ $B=4052.86(2)$ $C=2943.81(2)$	$d_J=-0.16(125)$ $d_{JK}=-63.4(267)$ $d_K=-120.0(443)$ $d_{WJ}=2.9(33)\cdot10^{-7}$ $d_{WK}=18.4(77)\cdot10^{-6}$				
		$v_\tau=3$	$A=9752.10(7)$ $B=4049.69(4)$ $C=2949.29(4)$	$d_J=6.1(34)$ $d_{JK}=-305.8(661)$ $d_K=-533.5(1115)$ $d_{WJ}=-19.7(82)\cdot10^{-7}$ $d_{WK}=88.5(192)\cdot10^{-6}$				
		$v_\tau=4$	$A=9709.14(11)$ $B=4046.61(7)$ $C=2954.93(6)$	$d_J=-3.1(54)$ $d_{JK}=20.2(853)$ $d_K=-9.4(1438)$ $d_{WJ}=4.3(136)\cdot10^{-7}$ $d_{WK}=-3.2(251)\cdot10^{-6}$				
		$v_\tau=5$	$A=9666.18(8)$ $B=4043.57(5)$ $C=2960.42(5)$					
		$v_\tau=6$	$A=9624.44(10)$ $B=4040.69(7)$ $C=2965.92(7)$					
		$v_\delta=1$	$A=9891.93(7)$ $B=4054.75(4)$ $C=2922.70(4)$	$d_J=5.0(23)$ $d_{JK}=-168.5(353)$ $d_K=-294.4(552)$ $d_{WJ}=59.1(467)\cdot10^{-7}$ $d_{WK}=48.5(101)\cdot10^{-6}$			$\nu_\delta=239(20)\ cm^{-1}$, in-plane bend	
Cont.								

Nr.	Molecule	v	A, B, C MHz	D, H kHz	Ref.	Ref. struct.	Varia, remarks	See also Tab., Nr.
205 Cont.	$CH_2FC(=O)NH_2$	$v_\tau=1, v_\delta=1$	$A=9847.3\,(3)$ $B=4051.0\,(1)$ $C=2928.9\,(1)$		74MAR2		—NH_2 out-of plane deformation mode tentatively assigned	
	$CH_2FC(=O)ND(1)\,H(2)$	$v=0$	$A=9329.76\,(3)$ $B=4055.98\,(3)$ $C=2879.88\,(3)$	$d_J = -2.4\,(11)$ $d_{JK} = -45.5\,(118)$ $d_K = -82.5\,(179)$ $d_{WJ} = 6.9\,(19)\cdot 10^{-7}$ $d_{WK} = 13.3\,(34)\cdot 10^{-6}$				
	$CH_2FC(=O)NH(1)\,D(2)$	$v=0$	$A=9293.07\,(4)$ $B=3994.87\,(2)$ $C=2845.52\,(2)$	$d_J = -1.0\,(10)$ $d_{JK} = -54.4\,(191)$ $d_K = -96.1\,(286)$ $d_{WJ} = 7.0\,(29)\cdot 10^{-7}$ $d_{WK} = 15.7\,(56)\cdot 10^{-6}$				
	$CH_2FC(=O)ND_2$	$v=0$	$A=8804.18\,(4)$ $B=3992.53\,(2)$ $C=2796.93\,(2)$	$d_J = -1.2\,(9)$ $d_{JK} = -57.4\,(113)$ $d_K = -48.5\,(172)$ $d_{WJ} = 5.8\,(20)\cdot 10^{-7}$ $d_{WK} = 15.7\,(34)\cdot 10^{-6}$				
206	$NH_2CH_2C(=O)OH$ conformer I	$v=0$	$A''=10341.76\,(17)$ *) $B''=3876.195\,(9)$ $C''=2912.361\,(10)$	$\tau_1=-24.84\,(13)$ $\tau_2=-6.003\,(99)$ $\tau_3=75\,(10)$ $\tau_{aaaa}=-36\,(30)$ $\tau_{bbbb}=-4.604\,(90)$ $\tau_{cccc}=-1.61\,(11)$	80SUE		*) $\Delta=-5.719$ amu Å^2	2.6.3, 101 3, 65
		$v_\tau=1$	$A''=10294.06\,(63)$ $B''=3874.984\,(51)$ $C''=2918.745\,(69)$	$\tau_1=-24.45\,(78)$ $\tau_2=-5.78\,(43)$ $\tau_3=109\,(42)$ $\tau_{bbbb}=-4.25\,(54)$ $\tau_{cccc}=-2.8\,(10)$			$v_\tau=62\,(20)$ cm^{-1}, out of plane torsion, probably around the C—C bond	
	$NH_2CH_2C(=O)OH$ conformer II	$v=0$	$A''=10130.47\,(62)$ *) $B''=4071.473\,(36)$ $C''=3007.538\,(44)$	$\tau_1=-28.52\,(64)$ $\tau_2=-6.89\,(44)$ $\tau_3=33\,(36)$ $\tau_{aaaa}=-94\,(100)$ $\tau_{bbbb}=-2.98\,(52)$ $\tau_{cccc}=-1.75\,(51)$	78SU3		*) $\Delta=-5.98$ amu Å^2 $E_{II}-E_I=490\,(150)$ cm^{-1}	
Cont.								

Nr.	Molecule	v	A, B, C MHz	D, H kHz	Ref.	Ref. struct.	Varia, remarks	See also Tab., Nr.
206 Cont.	$NH_2CH_2C(=O)OH$ conformer II	$v=1$	$A''=10079.79\,(62)$ $B''=4077.206\,(74)$ $C''=3017.844\,(94)$	$\tau_1=-21.4\,(11)$ $\tau_2=-5.54\,(63)$ $\tau_3=41\,(43)$ $\tau_{bbbb}=-1.7\,(12)$ $\tau_{cccc}=-3.7\,(11)$	78SU3			
207 6-97	$HO^{13}CH_2C(=O)H$	$v=0$	$A=18126.88\,(10)$ $B=6487.47\,(7)$ $C=4923.03\,(5)$		73MAS2	73MAS2	for structure, see also Vol. II/7 (3, 247)	2.6.3, 102 3, 66
	$HOCH_2{}^{13}C(=O)H$	$v=0$	$A=18259.524\,(51)$ $B=6472.328\,(17)$ $C=4924.556\,(20)$	$d_J=-10.0\,(18)$ $d_{JK}=-313\,(53)$ $d_K=-433\,(113)$ $d_{WJ}=3.45\,(30)\cdot10^{-6}$ $d_{WK}=43\,(9)\cdot10^{-6}$			parent species: relative intensity measurements yielded 195 (30) cm^{-1} for the C—C torsional fundamental and 260 (40) cm^{-1} for the lowest in-plane skeletal bending mode	
	$HOCH_2C(={}^{18}O)H$	$v=0$	$A=18087.032\,(26)$ $B=6242.805\,(9)$ $C=4778.491\,(12)$	$d_J=-17.1\,(16)$ $d_{JK}=-268\,(26)$ $d_K=-396\,(64)$ $d_{WJ}=4.15\,(27)\cdot10^{-6}$ $d_{WK}=35.2\,(52)\cdot10^{-6}$				
207a 6-97a	$HOCH_2C(=O)OH$	$v=0$	$A=10696.0950\,(24)$ *) $B=4051.0323\,(8)$ $C=2994.6632\,(8)$	$\Delta_J=0.8162\,(57)$ $\Delta_{JK}=3.276\,(29)$ $\Delta_K=5.602\,(96)$ $\delta_J=0.2119\,(15)$ $\delta_K=2.581\,(40)$	81BLO		*) $\Delta=-3.2422$ amu $Å^2$	2.6.3, 102a
		$v_{21}=1$	$A=10636.1914\,(41)$ $B=4045.1649\,(26)$ $C=2998.7414\,(26)$	$\Delta_J=0.8629\,(537)$ $\Delta_{JK}=3.352\,(69)$ $\Delta_K=4.988\,(454)$ $\delta_J=0.2119\,(23)$ $\delta_K=2.435\,(62)$				
		$v_{21}=2$	$A=10578.2469\,(50)$ $B=4039.8048\,(29)$ $C=3002.8843\,(29)$	$\Delta_J=0.7256\,(631)$ $\Delta_{JK}=3.464\,(82)$ $\Delta_K=4.474\,(552)$ $\delta_J=0.2132\,(28)$ $\delta_K=2.256\,(76)$				

Nr.	Molecule	v	A, B, C MHz	D, H kHz	Ref.	Ref. struct.	Varia, remarks	See also Tab., Nr.
208	$HOCH_2C(=O)OCH_3$	$v=0$	$A=10002.7\pm0.02$ *) $B=2204.0\pm0.05$ $C=1849.25\pm0.05$		73BRO1	73BRO1	*) $\Delta=-6.54$ amu Å^2 CO(H) and CO(CH_3) groups trans to each other and the hydroxyl H directed towards the nearest lone pair of the O carbonyl	3, 67
		$v_\tau=1$	$A=9920$ (10) $B=2203.38$ (5) $C=1851.87$ (5)				v_τ torsion	
		$v_\tau=2$	$A=9825$–9860 $B=2202.82\pm0.1$ $C=1854.5\pm0.1$					
		$v_\delta=1$	$A=10325$ (30) $B=2204.9$ (1) $C=1844.1$ (1)				v_δ in-plane vibration	
	$DOCH_2C(=O)OCH_3$	$v=0$	$A=9831.5$ (10) $B=2165.64$ (5) $C=1865.52$ (5)					
209	$CH_3OCH_2C(=O)OH$	$v=0$	$A=9409.08$ (79) *) $B=2088.817$ (6) $C=1748.884$ (6)		73MAS3	73MAS3	*) $\Delta=-6.6845$ (47) amu Å^2 planar HCOCCOOH skeleton. Hydrogen bond between hydroxyl H and ether O	2.6.3, 103 3, 68
		$v_{T_1}=1$	$A=9384.9$ (12) $B=2090.167$ (8) $C=1753.505$ (8)				v_{T_1}: C—C torsion	
		$v_{T_1}=2$	$A=9355.2$ (16) $B=2092.124$ (10) $C=1758.097$ (10)					
		$v_{T_1}=3$	$A=9331.5$ (25) $B=2094.735$ (17) $C=1763.044$ (17)					
		$v_{T_2}=1$	$A=9337.0$ (20) $B=2090.145$ (15) $C=1752.039$ (14)				v_{T_2}: C—O torsion	
		$v_{T_2}=2$	$A=9173.4$ (50) $B=2091.309$ (37) $C=1754.582$ (37)					
Cont.								

Nr.	Molecule	v	A, B, C MHz	D, H kHz	Ref.	Ref. struct.	Varia, remarks	See also Tab., Nr.
209 Cont.	$CH_3OCH_2C{\overset{O}{\lessgtr}}_{OH}$	$v_{T_1}=1$, $v_{T_2}=1$	$A=9309.8\,(46)$ $B=2091.096\,(24)$ $C=1756.712\,(24)$		73MAS3			
		$v_{T_3}=1$	$A=9325.7\,(51)$ $B=2088.348\,(31)$ $C=1748.055\,(30)$				ν_{T_3}: CH_3 torsion	
	$CH_3OCH_2C{\overset{O}{\lessgtr}}_{OD}$	$v=1$	$A=8999.9\,(12)$ $B=2088.627\,(10)$ $C=1734.114\,(12)$					
210	$CHF_2C{\overset{O}{\lessgtr}}_{OH}$ gauche	$v=0$	$A=5537.412\,(25)$ $B=3245.155\,(28)$ $C=2289.681\,(29)$	$\Delta_J = 5.2\,(12)$ $\Delta_{JK} = -10.60\,(16)$ $\Delta_K = 8.1\,(5)$ $\delta_J = 1.16\,(1)$ $\delta_K = 16.3\,(1)$ $H_J = 11\,(17)\cdot10^{-3}$ $H_{JK} = 3\,(3)\cdot10^{-3}$ $H_{KJ} = -10\,(9)\cdot10^{-3}$ $H_K = 14\,(10)\cdot10^{-3}$ $h_J = 0.07\,(12)\cdot10^{-3}$ $h_{JK} = -2.0\,(20)\cdot10^{-3}$ $h_K = 4\,(7)\cdot10^{-3}$	78MAA	78MAA		
		$v_\tau=1$	$A=5536.463\,(15)$ $B=3237.981\,(20)$ $C=2294.400\,(21)$	$\Delta_J = 6.8\,(6)$ $\Delta_{JK} = -11.82\,(13)$ $\Delta_K = 9.6\,(4)$ $\delta_J = 1.25\,(1)$ $\delta_K = 18.8\,(1)$ $H_J = 20\,(5)\cdot10^{-3}$ $H_{JK} = 1\,(2)\cdot10^{-3}$ $H_{KJ} = -4\,(6)\cdot10^{-3}$ $H_K = 10\,(8)\cdot10^{-3}$ $h_J = -0.02\,(8)\cdot10^{-3}$ $h_{JK} = -1.0\,(14)\cdot10^{-3}$ $h_K = 1\,(5)\cdot10^{-3}$				
Cont.								

Nr.	Molecule	v	A, B, C MHz	D, H kHz	Ref.	Ref. struct.	Varia, remarks	See also Tab., Nr.
210 Cont.	$CHF_2C(=O)OH$ gauche	$v_\tau=2$	$A=5535.634\,(35)$ *) $B=3228.605\,(42)$ $C=2300.461\,(46)$	$\Delta_J = 5.9\,(6)$ *) $\Delta_{JK} = -14.36\,(21)$ $\Delta_K = 13.2\,(8)$ $\delta_J = 1.38\,(1)$ $\delta_K = 23.6\,(1)$ $H_J = 17.6$ (fixed) $\cdot 10^{-3}$ $H_{JK} = -3\,(4) \cdot 10^{-3}$ $H_{KJ} = 5\,(9) \cdot 10^{-3}$ $H_K = 11.8$ (fixed) $\cdot 10^{-3}$ $h_J = -0.2\,(1) \cdot 10^{-3}$ $h_{JK} = 1.2\,(27) \cdot 10^{-3}$ $h_K = -3\,(10) \cdot 10^{-3}$	78MAA		*) calculated from mean values of the observed doublets	
	$CDF_2C(=O)OH$ gauche	$v_\tau=0$	$A=5405.518\,(10)$ $B=3186.914\,(21)$ $C=2283.275\,(26)$	$\Delta_J = 3.9\,(3)$ **) $\Delta_{JK} = -9.54\,(19)$ $\Delta_K = 8.7\,(7)$ $\delta_J = 1.07\,(1)$ $\delta_K = 14.7\,(1)$			**) some sextic constants determined	
		$v_\tau=1$	$A=5404.796\,(20)$ $B=3179.612\,(35)$ $C=2288.074\,(44)$	$\Delta_J = 5.5\,(6)$ **) $\Delta_{JK} = -10.44\,(30)$ $\Delta_K = 8.0\,(11)$ $\delta_J = 1.15\,(2)$ $\delta_K = 17.2\,(2)$				
		$v_\tau=2$	$A=5404.41\,(5)$ *) $B=3170.19\,(1)$ $C=2294.58\,(7)$	$\Delta_J = 5.0$ (fixed) *) ***) $\Delta_{JK} = -13.15\,(28)$ $\Delta_K = 12.8\,(5)$ $\delta_J = 1.26\,(5)$ $\delta_K = 21.0\,(5)$			***) sextic constants assumed	
	$CHF_2C(=O)OD$ gauche	$v=0$	$A=5512.767\,(11)$ $B=3121.552\,(13)$ $C=2223.667\,(18)$	$\Delta_J = 5.6\,(2)$ **) $\Delta_{JK} = -9.81\,(20)$ $\Delta_K = 7.9\,(6)$ $\delta_J = 1.10\,(2)$ $\delta_K = 15.6\,(2)$				
Cont.								

Nr.	Molecule	v	A, B, C MHz	D, H kHz	Ref.	Ref. struct.	Varia, remarks	See also Tab., Nr.
210 Cont.	$CHF_2C(=O)OD$ gauche	$v_\tau=1$	$A=5510.91$ (8) $B=3115.17$ (7) $C=2228.06$ (18)	$\Delta_J=$ 7.1 (fixed) ***) $\Delta_{JK}=-10.7$ (20) $\Delta_K=$ 9 (5) $\delta_J=$ 1.21 (11) $\delta_K=18.7$ (20)	78MAA		**) some sextic constants determined ***) sextic constants assumed	
	$CDF_2C(=O)OD$ gauche	$v=0$	$A=5381.534$ (13) $B=3067.092$ (19) $C=2217.369$ (24)	$\Delta_J=$ 4.9 (2) **) $\Delta_{JK}=-9.17$ (20) $\Delta_K=$ 8.4 (9) $\delta_J=$ 0.97 (1) $\delta_K=14.7$ (1)				
	$CHF_2C(=O){}^{18}OH$ gauche	$v=0$	$A=5438.034$ (12) $B=3141.274$ (15) $C=2224.107$ (16)	$\Delta_J=$ 3.3 (3) **) $\Delta_{JK}=-11.02$ (14) $\Delta_K=$ 10.1 (5) $\delta_J=$ 1.20 (1) $\delta_K=15.5$ (1)				
	$CHF_2C(={}^{18}O)OH$ gauche	$v=0$	$A=5358.393$ (17) $B=3179.775$ (22) $C=2232.742$ (24)	$\Delta_J=$ 3.9 (4) **) $\Delta_{JK}=-8.53$ (31) $\Delta_K=$ 6.7 (7) $\delta_J=$ 1.03 (2) $\delta_K=15.1$ (2)				
	$CHF_2C(=O)OH$ trans	$v=0$ $v_\tau=1$ $v_\tau=2$	$B=2732.224$ (17)$^+$) $C=2721.682$ (17) $B=2739.306$ (29) $C=2724.161$ (29) $B=2743.765$ (42) $C=2728.562$ (64)	$D'_J=4.11$ (16) $R'_6=1.73$ (6) $D'_J=4.28$ (30) $R'_6=1.94$ (14) $D'_J=4.20$ (46) $R'_6=3.79$ (16)			$^+$) with $A=5550$ MHz assumed	
211 Cont.	$CHF_2C(=O)F$ trans	$v=0$ $v_\tau=1$	$A=5538.94$ (2) $B=2737.89$ (2) $C=2737.73$ (3) $A=5529.95$ (2) $B=2741.42$ (5) $C=2738.10$ (5)	$D_J=2.71$ (4) $R_6=1.00$ (14) *)	77VAN1	77VAN1	*) ground state values assumed	2.6.3, 104 3, 69

Nr.	Molecule	v	A, B, C MHz	D, H kHz	Ref.	Ref. struct.	Varia, remarks	See also Tab., Nr.
211 Cont.	$CHF_2C(=O)F$ trans	$v_\tau=2$	$A=5521.24(2)$ $B=2745.16(5)$ $C=2738.47(5)$	*)	77VAN1		*) ground state values assumed	
	$CDF_2C(=O)F$ trans	$v=0$	$A=5391.71(2)$ $B=2727.46(2)$ $C=2692.00(2)$	$D_J=2.52(6)$ $R_6=0.95(1)$				
	$CHF_2C(=^{18}O)F$ trans	$v=0$	$A=5448.73(2)$ $B=2653.24(1)$ $C=2632.34(1)$	$D_J=2.38(3)$ $R_6=0.95(1)$			$E_G-E_T=264(55)$ cal/mol	
	$CHF_2C(=O)F$ gauche	$v=0$	$A=5482.287(15)$ $B=3271.434(17)$ $C=2243.186(18)$	$\Delta_J=2.7(5)$ $\Delta_{JK}=-4.14(8)$ $\Delta_K=3.72(22)$ $\delta_J=0.671(7)$ $\delta_K=6.27(8)$				
		$v_\tau=1$	$A=5480.700(15)$ $B=3263.024(17)$ $C=2246.672(18)$	$\Delta_J=1.6(5)$ $\Delta_{JK}=-4.53(8)$ $\Delta_K=4.10(22)$ $\delta_J=0.713(7)$ $\delta_K=7.04(8)$				
		$v_\tau=2$	$A=5479.101(15)$ $B=3254.886(17)$ $C=2250.520(18)$	$\Delta_J=3.2(5)$ $\Delta_{JK}=-4.94(8)$ $\Delta_K=4.34(22)$ $\delta_J=0.771(7)$ $\delta_K=7.98(8)$				
	$CDF_2C(=O)F$ gauche	$v=0$	$A=5353.863(15)$ $B=3214.046(17)$ $C=2236.163(18)$	$\Delta_J=2.0(5)$ $\Delta_{JK}=-3.71(8)$ $\Delta_K=3.36(22)$ $\delta_J=0.597(7)$ $\delta_K=5.61(8)$				
Cont.		$v_\tau=1$	$A=5352.461(15)$ $B=3205.809(17)$ $C=2239.633(18)$	$\Delta_J=2.3(5)$ $\Delta_{JK}=-4.02(8)$ $\Delta_K=3.47(22)$ $\delta_J=0.638(7)$ $\delta_K=6.30(8)$				

Nr.	Molecule	v	A, B, C MHz	D, H kHz	Ref.	Ref. struct.	Varia, remarks	See also Tab., Nr.
211 Cont.	$^{13}CHF_2C(=O)F$ gauche	$v=0$	$A=5472.949\ (15)$ $B=3258.744\ (17)$ $C=2238.617\ (18)$	$\Delta_J = 2$ (fixed) $\Delta_{JK} = -4.03\ (8)$ $\Delta_K = 3.81\ (22)$ $\delta_J = 0.658\ (7)$ $\delta_K = 6.16\ (8)$	77VAN1			
	$CHF_2{}^{13}C(=O)F$ gauche	$v=0$	$A=5479.755\ (15)$ $B=3259.425\ (17)$ $C=2237.127\ (18)$	$\Delta_J = 2$ (fixed) $\Delta_{JK} = -4.06\ (8)$ $\Delta_K = 3.74\ (22)$ $\delta_J = 0.659\ (7)$ $\delta_K = 6.26\ (8)$				
	$CHF_2C(={}^{18}O)F$ gauche	$v=0$	$A=5299.133\ (15)$ $B=3206.847\ (17)$ $C=2186.537\ (18)$	$\Delta_J = 1.6\ (5)$ $\Delta_{JK} = -3.35\ (8)$ $\Delta_K = 2.88\ (22)$ $\delta_J = 0.594\ (7)$ $\delta_K = 5.99\ (8)$				
212 *6-97b*	$HC(=O)-C(=O)-H$ cis	$v=0$	$A=26713.466\ (51)$ $B=6190.831\ (33)$ $C=5032.482\ (16)$	$\Delta_J = 6.35\ (64)$ $\Delta_{JK} = -46.2\ (58)$ $\Delta_K = 188.0\ (36)$ $\delta_J = 1.733\ (71)$ $\delta_K = 13.0\ (48)$	80MAM	76COL	for structure, see also Vol. II/7 (3, 198)	
		$v_\tau=1$	$A=27005.245\ (39)$ $B=6121.0190\ (80)$ $C=5033.7365\ (85)$				$\nu_\tau=108\ (10)\ cm^{-1}$, torsion	
		$v_\delta=1$	$A-C=21863.23\ (15)$ $\kappa=-0.893734\ (13)$				$\nu_\delta=288\ (20)\ cm^{-1}$, C—C=O bend	
	$HC(=O)-C(=O)-D$ cis	$v=0$	$A=23857.40\ (36)$ $B=6075.89\ (9)$ $C=4849.65\ (8)$		76COL			
	$DC(=O)-C(=O)-D$ cis	$v=0$	$A=21521.82\ (21)$ $B=5968.99\ (4)$ $C=4680.61\ (5)$					

Nr.	Molecule	v	A, B, C MHz	D, H kHz	Ref.	Ref. struct.	Varia, remarks	See also Tab., Nr.
213 6–97 c	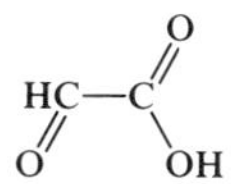 trans 1	$v=0$	$A=10966.813\,(15)$ *) $B=4605.988\,(3)$ $C=3242.092\,(4)$	$\tau_{aaaa}=-39.7\,(58)$ $\tau_{bbbb}=-3.31\,(16)$ $\tau_{cccc}=1.8\,(8)$ $\tau'_{aabb}=-36.79\,(76)$	73MAS1	77VAN2	*) $\Delta=0.07578\,(22)$ amu Å^2 CO groups trans to each other $\nu_\tau=167\,(12)\ \mathrm{cm}^{-1}$, C—C torsion	2.6.3, 105 3, 70
		$v_\tau=1$	$A=10893.489\,(52)$ $B=4601.746\,(3)$ $C=3246.763\,(7)$	$\tau_{aaaa}=-62.0\,(121)$ $\tau_{bbbb}=-4.00\,(37)$ $\tau_{cccc}=0.06\,(80)$ $\tau'_{aabb}=-33.5\,(12)$				
		$v_\tau=2$	$A=10830.244\,(209)$ $B=4597.379\,(8)$ $C=3250.386\,(29)$	$\tau_{aaaa}=-131.8\,(395)$ $\tau_{bbbb}=-6.1\,(13)$ $\tau_{cccc}=-5.1\,(28)$ $\tau'_{aabb}=-24.0\,(40)$				
		$v_\tau=3$	$A=10773.86\,(57)$ $B=4594.44\,(13)$ $C=3253.18\,(8)$					
		$v_\delta=1$	$A=10976.945\,(57)$ $B=4606.017\,(3)$ $C=3233.132\,(7)$	$\tau_{aaaa}=-66.3\,(116)$ $\tau_{bbbb}=-4.17\,(37)$ $\tau_{cccc}=-0.1\,(8)$ $\tau'_{aabb}=-33.7\,(13)$			$\nu_\delta=288\,(26)\ \mathrm{cm}^{-1}$, in-plane bend	
	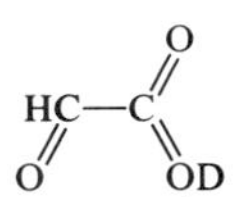 trans 1	$v=0$	$A=10422.262\,(136)$ $B=4600.668\,(4)$ $C=3190.305\,(10)$	$\tau_{aaaa}=-138.6\,(122)$ $\tau_{bbbb}=-6.45\,(55)$ $\tau_{cccc}=-6.5\,(10)$ $\tau'_{aabb}=-23.3\,(26)$				
	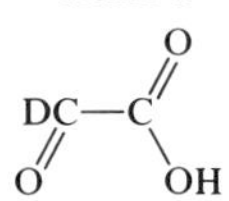 trans 1	$v=0$	$A=10276.7\,(8)$ $B=4573.66\,(6)$ $C=3163.85\,(6)$		77VAN2			
	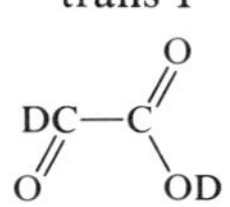 trans 1	$v=0$	$A=9777.0\,(10)$ $B=4569.88\,(7)$ $C=3113.17\,(7)$					
Cont.	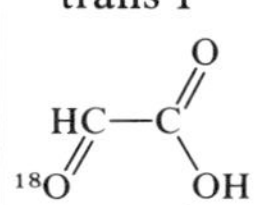 trans 1	$v=0$	$A=10965.74\,(27)$ $B=4349.089\,(10)$ $C=3112.754\,(25)$	$\tau_{aaaa}=-70.5\,(480)$ $\tau_{bbbb}=-6.2\,(13)$ $\tau_{cccc}=-5.0\,(26)$ $\tau_{aabb}=-47.2\,(48)$	76CHR			

Nr.	Molecule	v	A, B, C MHz	D, H kHz	Ref.	Ref. struct.	Varia, remarks	See also Tab., Nr.
213 6–97 *c* Cont.	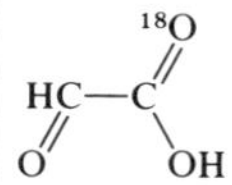 trans 1	$v=0$	$A=10737.525\ (92)$ $B=4414.702\ (5)$ $C=3126.948\ (10)$	$\tau_{aaaa}=-47.2\ (190)$ $\tau_{bbbb}=-2.69\ (54)$ $\tau_{cccc}=1.3\ (11)$ $\tau'_{aabb}=-29.5\ (23)$	76CHR			
	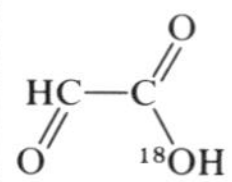 trans 1	$v=0$	$A=10264.66\ (13)$ $B=4579.313\ (7)$ $C=3165.071\ (18)$	$\tau_{aaaa}=-87.8\ (230)$ $\tau_{bbbb}=-5.11\ (93)$ $\tau_{cccc}=-2.8\ (21)$ $\tau'_{aabb}=-24.9\ (25)$				
	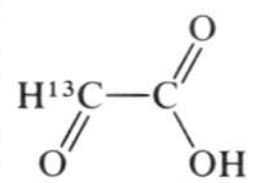 trans 1	$v=0$	$A=10861.3\ (7)$ $B=4577.03\ (6)$ $C=3218.60\ (6)$		77VAN2			
	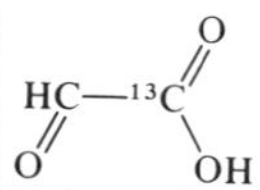 trans 1	$v=0$	$A=10975.1\ (95)$ $B=4591.17\ (26)$ $C=3236.77\ (24)$		76CHR			
	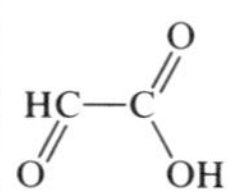 trans 2	$v=0$	$A=10933.4\ (8)$ $B=4468.40\ (6)$ $C=3173.25\ (6)$		77VAN2		$E_{trans\,2}-E_{trans\,1}=1.2\ (5)$ kcal/mol	
		$v_\tau=1$	$A=10866.1\ (6)$ $B=4465.63\ (5)$ $C=3182.72\ (5)$					
		$v_\tau=2$	$A=10804.8\ (8)$ $B=4463.04\ (6)$ $C=3191.77\ (6)$					
	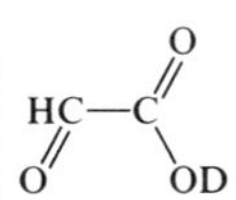 trans 2	$v=0$	$A=10452.2\ (8)$ $B=4374.17\ (6)$ $C=3084.89\ (6)$					
214 Cont.	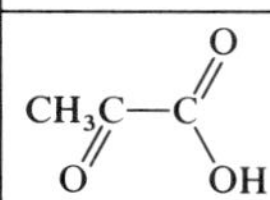	$v=0$	$A=5536.0436\ (38)$ *) $B=3583.7074\ (12)$ $C=2204.8571\ (8)$	$\Delta_J=0.705\ (10)$ *) $\Delta_{JK}=-0.613\ (69)$ $\Delta_K=1.382\ (111)$ $\delta_J=0.274\ (6)$ $\delta_K=0.606\ (52)$	73KAL	77DYL	planar HCCOCOOH skeleton with the two C=O groups trans to each other *) for the A species	2.6.3, 106 2.8, 22 3, 71

Nr.	Molecule	v	A, B, C MHz	D, H kHz	Ref.	Ref. struct.	Varia, remarks	See also Tab., Nr.
214 Cont.	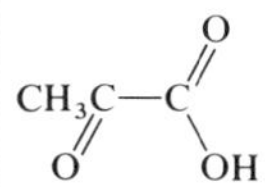	$v_{T_1}=1$	$A=5533.204\,(555)$ *) $B=3570.623\,(241)$ $C=2207.114\,(143)$	$\Delta_J = -1.27\,(118)$ *) $\Delta_{JK} = -28.89\,(856)$ $\Delta_K = 54.28\,(1200)$ $\delta_J = -2.04\,(68)$ $\delta_K = 28.46\,(241)$	77DYL		$\nu_{T_1}=101\,(9)\,cm^{-1}$, C—C torsion *) for the A species	
		$v_{T_1}=2$	$A=5537.931\,(319)$ *) $B=3561.577\,(71)$ $C=2209.268\,(33)$	$\Delta_J = 0.86\,(42)$ *) $\Delta_{JK} = -0.39\,(391)$ $\Delta_K = 3.69\,(633)$ $\delta_J = 0.33\,(20)$ $\delta_K = 1.11\,(455)$				
		$v_{T_1}=3$	$A=5535.209\,(122)$ *) $B=3549.057\,(21)$ $C=2211.603\,(6)$					
		$v_{T_2}=1$	$A=5524.469\,(534)$ *) $B=3565.093\,(140)$ $C=2203.701\,(38)$	$\Delta_J = 0.95\,(73)$ *) $\Delta_{JK} = 6.67\,(970)$ $\Delta_K = -13.91\,(1553)$ $\delta_J = 0.83\,(81)$ $\delta_K = -3.24\,(690)$			$\nu_{T_2}=134\,(7)\,cm^{-1}$, CH_3 torsion	
	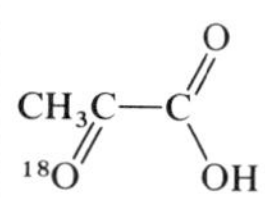	$v=0$	$A=5340.633\,(86)$ *) $B=3524.446\,(8)$ $C=2151.239\,(8)$					
	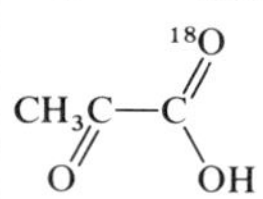	$v=0$	$A=5345.111\,(115)$ *) $B=3515.273\,(12)$ $C=2148.577\,(14)$					
	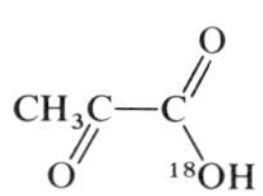	$v=0$	$A=5452.908\,(117)$ *) $B=3467.121\,(12)$ $C=2147.418\,(12)$					
	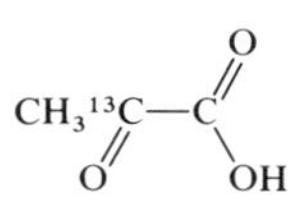	$v=0$	$A=5534.792\,(45)$ *) $B=3569.485\,(4)$ $C=2199.269\,(3)$					
Cont.	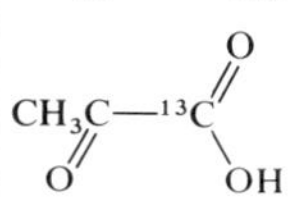	$v=0$	$A=5534.105\,(40)$ *) $B=3570.081\,(4)$ $C=2199.386\,(3)$					

Nr.	Molecule	v	A, B, C MHz	D, H kHz	Ref.	Ref. struct.	Varia, remarks	See also Tab., Nr.
214 Cont.	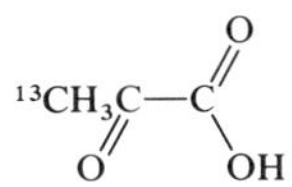	$v=0$	$A=5482.437\ (36)$ *) $B=3508.864\ (3)$ $C=2167.946\ (3)$		77DYL			
	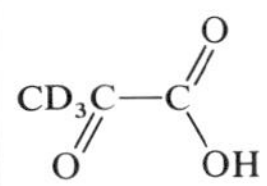	$v=0$	$A=5181.898\ (55)$ $B=3255.022\ (5)$ $C=2049.608\ (4)$					
	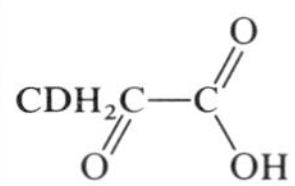 symmetric	$v=0$	$A=5515.109\ (62)$ $B=3402.324\ (5)$ $C=2132.279\ (4)$					
	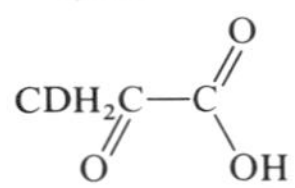 asymmetric	$v=0$	$A=5347.164\ (64)$ $B=3505.045\ (6)$ $C=2157.937\ (5)$					
	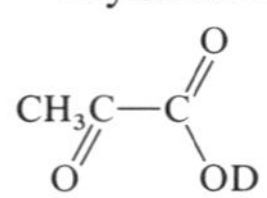	$v=0$	$A=5377.903\ (181)$ *) $B=3560.768\ (28)$ $C=2170.810\ (39)$		77DYL, 74MAR1		*) for the A species	
215	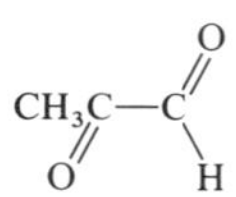	$v=0$	$A=9102.4332\ (31)$ *) +) $B=4439.8832\ (27)$ $C=3038.9404\ (22)$	$\Delta_J=1.327\ (39)$ *) $\Delta_{JK}=7.18\ (22)$ $\Delta_K=-1.41\ (70)$ $\delta_J=0.464\ (12)$ $\delta_K=5.68\ (18)$	78DYL	78DYL	for structure, see also Vol. II/7 (3, 377) +) trans conformer: $\Delta=-3.04733\ (14)$ amu Å^2 *) for the A species	2.6.3, 107 2.8, 23 3, 72
		$v_{T_1}=1$	$A=9077.988\ (25)$ *) $B=4396.249\ (17)$ $C=3038.609\ (14)$	$\Delta_J=-3.46\ (25)$ *) $\Delta_{JK}=20.0\ (17)$ $\Delta_K=-11.9\ (48)$ $\delta_J=-1.87\ (10)$ $\delta_K=-16.3\ (16)$			$\nu_{T_1}=90.2\ (41)\ cm^{-1}$, C–C torsion	
		$v_{T_2}=2$	$A=9069.908\ (17)$ *) $B=4409.380\ (13)$ $C=3043.276\ (10)$	$\Delta_J=-4.09\ (17)$ $\Delta_{JK}=17.6\ (13)$ $\Delta_K=-3.5\ (36)$ $\delta_J=-2.33\ (8)$ $\delta_K=-12.2\ (13)$			$\nu_{T_2}=122.7\ (49)\ cm^{-1}$, CH_3 torsion	

Nr.	Molecule	v	A, B, C MHz	D, H kHz	Ref.	Ref. struct.	Varia, remarks	See also Tab., Nr.
216 *6-94*	$CH_3C(=NOD)H$ syn	$v=0$	$A=45682\,(500)$ $B=4057.20\,(4)$ $C=3806.13\,(4)$		76OH	76OH		2.6.3, 108 2.8, 24 3, 73
	$CH_3C(=NOD)H$ anti	$v=0$	$A=17250$ *) $B=6284.33$ $C=4704.96$		77OH1	77OH1	*) for the A species	
217	$CH_3C(=S)H$	$v=0$	$A=48574.210\,(57)$ *) $B=5750.977\,(5)$ $C=5307.571\,(5)$	$\Delta_J=2.72\,(3)$ *) $\Delta_{JK}=-13.74\,(85)$ $\Delta_K=1462.3\,(57)$ $\delta_J=0.372\,(2)$ $\delta_K=-10.35\,(18)$ $H_{KJ}=0.2492\,(211)$ $h_{JK}=1.77\,(18)\cdot 10^{-3}$	76KRO	76KRO	*) for the A species	2.6.3, 109 2.8, 25
		$v_\tau=1$	$B=5748.72\,(3)$ *) $C=5314.76\,(3)$	$\Delta_{JK}=-350.6\,(45)$ *)			$v_\tau \simeq 170\ cm^{-1}$, CH_3 torsion	
	$CH_3C(=S)D$	$v=0$	$A=39020.981\,(37)$ *) $B=5729.923\,(8)$ $C=5151.290\,(6)$	$\Delta_J=2.40\,(10)$ *) $\Delta_{JK}=-14.89\,(177)$ $\Delta_K=643.3\,(65)$ $\delta_J=0.416\,(3)$ $\delta_K=-4.95\,(57)$				
		$v_\tau=1$	$B=5729.36\,(3)$ *) $C=5156.68\,(3)$	$\Delta_{JK}=-170.4\,(51)$ *)				
	$CD_3C(=S)H$	$v=0$	$A=36946.87\,(7)$ *) $B=4947.587\,(5)$ $C=4611.369\,(4)$	$\Delta_J=1.83\,(4)$ *) $\Delta_{JK}=3.09\,(67)$ $\Delta_K=419.8\,(150)$ $\delta_J=0.230\,(2)$ $\delta_K=-32.40\,(18)$				
		$v_\tau=1$	$B=4948.43\,(3)$ *) $C=4620.48\,(3)$	$\Delta_{JK}=-81.3\,(33)$ *)				
Cont.	$CH_2DC(=S)H$ asymmetric	$v=0$	$A=45465.6\,(86)$ $B=5374.64\,(2)$ $C=5021.48\,(2)$	$\Delta_J=2.6\,(8)$ $\delta_J=0.260\,(5)$ $\delta_K=-19.5\,(91)$				

Nr.	Molecule	v	A, B, C MHz	D, H kHz	Ref.	Ref. struct.	Varia, remarks	See also Tab., Nr.
217 Cont.	$CH_2DC(=S)H$ symmetric	$v=0$	$A=41723.5\ (51)$ $B=5598.82\ (2)$ $C=5089.77\ (2)$	$\Delta_{JK}=-19.8\ (43)$ $\delta_J=0.403\ (7)$ $\delta_K=-6.7\ (93)$	76KRO			
	$CHD_2C(=S)H$ asymmetric	$v=0$	$B=5248.54\ (1)$ $C=4831.48\ (1)$					
	$CHD_2C(=S)H$ symmetric	$v=0$	$B=5053.85\ (1)$ $C=4779.45\ (1)$	$\Delta_{JK}=5.7\ (22)$				
	$CH_3C(=^{34}S)H$	$v=0$	$B=5618.98\ (3)$ $C=5618.98\ (3)^{+})$				$^{+})$ mistake in [76KRO]? the true value should be 5193.6	
	$^{13}CH_3C(=S)H$	$v=0$	$B=5563.75\ (3)$ $C=5144.36\ (3)$					
	$CH_3{}^{13}C(=S)H$	$v=0$	$B=5738.14\ (3)$ $C=5281.62\ (3)$					
218	$CH_3C(=^{80}Se)H$	$v=0$	$B=3916.372\ (12)$ $C=3690.207\ (12)$	$D_J=1.25\ (18)$ $D_{JK}=-20.57\ (65)$	78HUT	78HUT		2.8, 26
		$v_\tau=1$	$B=3916.64\ (4)$ $C=3694.28\ (4)$	$D_{JK}=-116\ (2)$				
	$CH_3C(=^{78}Se)H$	$v=0$	$B=3939.63\ (6)$ $C=3710.95\ (6)$	$D_J=1.34\ (21)$ $D_{JK}=-20.00\ (79)$				
		$v_\tau=1$	$B=3939.97\ (5)$ $C=3715.01\ (5)$	$D_{JK}=-123\ (2)$				
	$CH_3C(=^{76}Se)H$	$v=0$	$B=3964.068\ (9)$ $C=3732.731\ (9)$	$D_J=1.25\ (22)$ $D_{JK}=-18.79\ (82)$				
219 *6-98* *4-86*	$CH_3CH_2CH_3$	$\tilde{v}_n=1_1\ ^{+})$	$A=29166.55$ *) $B=8432.45$ $C=7449.47$		73TRI		for structure, see also Vol. II/7 (3, 428) *) for the $A_1 A_2$ species $^{+})$ CH_3 torsion **) for the $A_2 A_1$ species	2.9.3, 22
		$\tilde{v}_n=1_2\ ^{+})$	$A=29088.18$ **) $B=8415.31$ $C=7445.81$					

Nr.	Molecule	v	A, B, C MHz	D, H kHz	Ref.	Ref. struct.	Varia, remarks	See also Tab., Nr.
220 *4-87*	$CH_3CH_2CH_2F$ gauche	$v=0$	$A=14503.658$	$d_J = -32.102$	79KAU			
			$B=5085.828$	$d_{JK} = -369.75$				
			$C=4295.284$	$d_K = -478.13$				
				$d_{WJ} = 9.2057 \cdot 10^{-6}$				
				$d_{WK} = 53.899 \cdot 10^{-6}$				
221 *4-88*	$CH_3CH_2CH_2{}^{35}Cl$ gauche	$v=0$	$A=11829.927$	$d_J = -13.669$	77KAU2			
			$B=3322.581$	$d_{JK} = -226.18$				
			$C=2853.014$	$d_K = -413.14$				
				$d_{WJ} = 5.1097 \cdot 10^{-6}$				
				$d_{WK} = 54.810 \cdot 10^{-6}$				
	$CH_3CH_2CH_2{}^{37}Cl$ gauche	$v=0$	$A=11788.265$	$d_J = -11.134$				
			$B=3249.078$	$d_{JK} = -273.63$				
			$C=2796.472$	$d_K = -501.47$				
				$d_{WJ} = 4.8044 \cdot 10^{-6}$				
				$d_{WK} = 71.405 \cdot 10^{-6}$				
221 a *6-99*	$CH_3CH_2CH_2{}^{79}Br$ gauche	$v=0$	$A=11034.346(67)$		81NII	81NII		2.7.3, 151 a
			$B=2277.725(14)$					
			$C=2024.525(14)$					
		$v_T=1$	$A=11086.627(76)$					
			$B=2272.450(15)$				$v_T=127$ cm^{-1}, C—C torsion	
			$C=2020.464(15)$					
		$v_T=2$	$B=2266.160(45)$					
			$C=2015.840(40)$					
	$CH_3CH_2CH_2{}^{81}Br$ gauche	$v=0$	$A=11027.924(63)$					
			$B=2261.019(13)$					
			$C=2011.115(13)$					
		$v_T=1$	$A=11080.304(93)$					
			$B=2255.769(18)$					
			$G=2007.079(17)$					
		$v_T=2$	$B=2249.521(42)$					
			$C=2002.495(46)$					

Nr.	Molecule	v	A, B, C MHz	D, H kHz	Ref.	Ref. struct.	Varia, remarks	See also Tab., Nr.
222 *6-100*	$CH_3CH_2CH_2I$ trans	$v=0$	$B+C=2573.7\,(14)$ *)		77ST		*) LRMW	
	$CH_3CH_2CH_2I$ gauche	$v=0$	$B+C=3387\,(3)$ *)					
223 *4-85*	$CH_3CH_2CH_2CN$ gauche	$v=0$	$A=10060.307$	$d_J = -17.788$	79KAU			
			$B=3267.702$	$d_{JK}=-261.98$				
			$C=2705.472$	$d_K=-404.87$				
				$d_{WJ}=7.3021\cdot10^{-6}$				
				$d_{WK}=63.690\cdot10^{-6}$				
224	$CH_3CH_2CH_2NC$ trans	$v=0$	$A=23700\,(100)$		75FUL	75FUL		2.6.3, 110
			$B=2407.632\,(20)$					
			$C=2278.853\,(30)$					
		$v_\tau=1$	$B=2408.31$				$\nu_\tau=114\,(20)\,cm^{-1}$, torsion around the central C—C bond	
			$C=2283.96$					
		$v_\tau=2$	$B=2409.19$					
			$C=2289.19$					
		$v_\tau=3$	$B=2410.28$					
			$C=2294.60$					
		$v_\tau=4$	$B=2411.60$					
			$C=2300.27$					
		$v_\delta=1$	$B=2409.02$				ν_δ (CNC bend) $=168\,(20)\,cm^{-1}$	
			$C=2282.98$					
		$v_T=1$	$B=2405.33$				ν_T (methyl torsion) $=243\,(25)\,cm^{-1}$	
			$C=2277.49$					
		$v_\tau=1,\ v_\delta=1$	$B=2409.80$					
			$C=2288.00$					
		$v_\tau=1,\ v_T=1$	$B=2406.06$					
			$C=2282.32$					
	$CH_3CH_2CH_2NC$ gauche	$v=0$	$A=10208.983\,(30)$	$\tau_{aaaa}=-172.734$			$E_{trans}-E_{gauche}=99\,(45)\,cm^{-1}$	
			$B=3479.219\,(15)$	$\tau_{bbbb}=-24.571$				
			$C=2859.981\,(15)$	$\tau_{cccc}=-5.649$				
				$\tau'_{aabb}=27.304$				
				$\tau'_{aacc}=27.304$				
Cont.				$\tau'_{bbcc}=-11.999$				

Nr.	Molecule	v	A, B, C MHz	D, H kHz	Ref.	Ref. struct.	Varia, remarks	See also Tab., Nr.
224 Cont.	$CH_3CH_2CH_2NC$ gauche	$v_\tau=1$	$A=10254.79$ $B=3472.75$ $C=2855.92$		75FUL		$\nu_\tau=123\,(20)\,cm^{-1}$, torsion around the central C—C bond potential function for this torsion: $V_1=0.186\,(53)$ $V_2=-0.478\,(58)$ $V_3=3.588\,(69)$ $V_4=-0.114\,(28)$ kcal/mol	
		$v_\tau=2$	$A=10303.25$ $B=3465.53$ $C=2851.75$					
		$v_\tau=3$	$B=3457.07$ $C=2848.13$					
		$v_\delta=1$	$A=10233.82$ $B=3482.91$ $C=2861.57$				$\nu_\delta=184\,(20)\,cm^{-1}$, C—N—C bend	
		$v_T=1$	$A=10169.07$ $B=3480.83$ $C=2860.42$				$\nu_T=235\,(25)\,cm^{-1}$, methyl torsion	
		$v_\tau=1, v_\delta=1$	$A=10309.38$ $B=3475.76$ $C=2857.51$					
225 *6-101*	$CH_3CH_2CH_2OH$ gauche	$v=0$	$A''=14348.54\,(21)$ $B''=5177.044\,(67)$ $C''=4434.398\,(84)$	$\tau_{aaaa}=-1.08\,(19)\cdot10^3$ $\tau_{bbbb}=-135.1\,(71)$ $\tau_{cccc}=-124.4\,(83)$ $\tau_1=-685\,(28)$ $\tau_2=-204.8\,(77)$ $\tau_3=7\,(73)\cdot10^3$	79AB, 81AB	81AB	for structure, see also Vol. II/7 (3, 429)	2.6.3, 110a 2.8, 27
		$v_\tau=1$	$A=14351.1$ *) $B=5176.4$ $C=4434.3$		81AB		ν_τ, methyl torsion *) for the A species	
	$CH_3CH_2CH_2OH$ trans	$v=0$	$A=26401.671\,(50)$ $B=3802.154\,(11)$ $C=3549.543\,(20)$	$\Delta_J=1.672\,(73)$ $\Delta_{JK}=9.57\,(99)$ $\Delta_K=141.5\,(47)$ $\delta_J=0.0943\,(45)$ $\delta_K=-29.6\,(69)$	81DRE1			
		$v_\tau=1$	$A=26404$ *) $B=3804$ $C=3551$		72AB			

Nr.	Molecule	v	A, B, C MHz	D, H kHz	Ref.	Ref. struct.	Varia, remarks	See also Tab., Nr.
226	$CH_3CH_2CH_2OCH_3$ trans-trans	$v=0$	$A=19755.36\,(14)$ $B=\ 2125.45\,(1)$ $C=\ 2017.60\,(1)$	$d_J=0.35\,(10)$	80KAT	80KAT	uncertainties are 3 σ	2.6.3, 111 2.8, 28 3, 74
		$v_\tau=1$	$A-(B+C)/2$ $=17569.60\,(29)$ $(B-C)/2=52.08\,(1)$				v_τ CH_3C excited torsional state	
		$v_x=1$	$A-(B+C)/2$ $=18048.24\,(33)$ $(B-C)/2=54.83\,(2)$				v_x skeletal torsion	
	$^{13}CH_3CH_2CH_2OCH_3$ trans-trans	$v=0$	$A=19687.10\,(12)$ $B=\ 2071.67\,(1)$ $C=\ 1968.40\,(1)$	$d_J=0.29\,(9)$				
	$CH_3{}^{13}CH_2CH_2OCH_3$ trans-trans	$v=0$	$A=19547.65\,(14)$ $B=\ 2112.54\,(1)$ $C=\ 2003.80\,(1)$	$d_J=0.29\,(11)$				
	$CH_3CH_2{}^{13}CH_2OCH_3$ trans-trans	$v=0$	$A=19660.44\,(13)$ $B=\ 2125.57\,(1)$ $C=\ 2016.74\,(1)$	$d_J=0.29\,(10)$				
	$CH_3CH_2CH_2O^{13}CH_3$ trans-trans	$v=0$	$A=19697.35\,(15)$ $B=\ 2077.20\,(1)$ $C=\ 1973.49\,(1)$	$d_J=0.29\,(11)$				
	$CD_3CH_2CH_2OCH_3$ trans-trans	$v=0$	$A=17431.37\,(14)$ $B=\ 1921.60\,(1)$ $C=\ 1831.32\,(1)$	$d_J=0.29\,(11)$				
	$CH_3CD_2CH_2OCH_3$ trans-trans	$v=0$	$A=16953.14\,(14)$ $B=\ 2087.55\,(1)$ $C=\ 1974.79\,(1)$	$d_J=0.22\,(10)$				
	$CH_3CH_2CD_2OCH_3$ trans-trans	$v=0$	$A=17324.33\,(19)$ $B=\ 2111.88\,(1)$ $C=\ 2002.22\,(2)$	$d_J=0.31\,(14)$				
	$CH_3CH_2CH_2OCD_3$ trans-trans	$v=0$	$A=17418.82\,(24)$ $B=\ 1939.01\,(2)$ $C=\ 1847.38\,(2)$	$d_J=0.29\,(18)$				
Cont.	s-$CH_2DCH_2CH_2OCH_3$ trans-trans	$v=0$	$A=19668.08\,(17)$ $B=\ 2028.67\,(1)$ $C=\ 1929.56\,(1)$	$d_J=0.32\,(12)$				

Nr.	Molecule	v	A, B, C MHz	D, H kHz	Ref.	Ref. struct.	Varia, remarks	See also Tab., Nr.
226 Cont.	a-$CH_2DCH_2CH_2OCH_3$ trans-trans	$v=0$	$A=18581.64(12)$ $B=2063.05(1)$ $C=1959.65(1)$	$d_J=0.29(9)$	80KAT			
	$CH_3CH_2CH_2OCH_2D$-s trans-trans	$v=0$	$A=19615.59(11)$ $B=2038.76(1)$ $C=1938.19(1)$	$d_J=0.26(9)$				
	$CH_3CH_2CH_2OCH_2D$-a trans-trans	$v=0$	$A=18601.17(8)$ $B=2068.10(1)$ $C=1964.65(1)$	$d_J=0.29(6)$				
227	$CH_3CH_2CH_2SH$ gauche	$v=0$	$A=23907.47(9)$ $B=2345.597(6)$ $C=2250.338(9)$	$D_J=0.22(8)$ $D_{JK}=-5.7(15)$	79NAK	81NA	$D=1.32(2)$ MHz, $E=58$ MHz, $\Delta=1613.01(4)$ MHz, for definitions, see CH_3CH_2OH Nr. 156	2.6.3, 112
		$v_\tau=1$	$A=24483(934)$ $B=2346.314(43)$ $C=2253.911(42)$		77OH2		v_τ, $-CH_2-CH_2$-torsional motion	
	$CH_3CH_2CH_2SD$ gauche	$v=0$	$A=22082.94(10)$ $B=2306.716(7)$ $C=2217.804(11)$	$D_J=0.21(10)$ $D_{JK}=-6.8(19)$	79NAK		$\Delta=58.30(5)$ MHz	
	$CH_3CH_2CD_2SH$ gauche	$v=0$	$A=19740.00(7)$ $B=2328.654(6)$ $C=2221.559(9)$	$d_J=0.22(9)$ $d_{JK}=-4.9(14)$	81NA		$\Delta=1628.47(4)$ MHz, $D=1.193(7)$ MHz, $E=43.0$ MHz	
	$CH_3CD_2CH_2SH$ gauche	$v=0$	$A=20007.63(7)$ $B=2309.584(6)$ $C=2207.319(9)$	$d_J=0.23(7)$ $d_{JK}=-3.5(14)$			$\Delta=1566.09(4)$; $D=0.682(11)$; E=47.0 MHz	
	$CD_3CH_2CH_2SH$ gauche	$v=0$	$A=20884.85(6)$ $B=2094.557(4)$ $C=2019.009(6)$	$d_J=0.21(4)$ $d_{JK}=-2.3(9)$			$\Delta=1604.02(3)$; $D=0.241(6)$; E=45.5 MHz	
	CH_3CH_2CHDSH-1 [+]) gauche	$v=0$	$A=21587.27(29)$ $B=2336.990(11)$ $C=2235.399(10)$	$d_J=0.19(12)$ $d_{JK}=-6.5(24)$			[+]) 1 and 2 denote the D atoms which are situated in the same and opposite direction to the H atom of the thiol group with respect to the CCCS plane	
Cont.	CH_3CH_2CHDSH-2 [+]) gauche	$v=0$	$A=21647.72(30)$ $B=2339.929(12)$ $C=2233.443(10)$	$d_J=0.21(13)$ $d_{JK}=-5.9(25)$				

Nr.	Molecule	v	A, B, C MHz	D, H kHz	Ref.	Ref. struct.	Varia, remarks	See also Tab., Nr.
227 Cont.	CH_3CHDCH_2SH-1 ⁺) gauche	$v=0$	$A=21769.30\,(27)$ $B=2329.602\,(11)$ $C=2225.685\,(9)$	$d_J=0.24\,(10)$ $d_{JK}=-5.0\,(24)$	81NA		potential function for the SH group: $V_1=-207$; $V_2=-353\,(6)$ $V_3=1310\,(9)$ cal/mol [81NA] ⁺) 1 and 2 denote the D atoms which are situated in the same and opposite direction to the H atom of the thiol group with respect to the CCCS plane	
	CH_3CHDCH_2SH-2 ⁺) gauche	$v=0$	$A=21784.11\,(30)$ $B=2327.638\,(11)$ $C=2228.657\,(10)$	$d_J=0.21\,(12)$ $d_{JK}=-5.6\,(24)$				
	$CH_3CH_2CH_2SH$ trans	$v=0$	$A=23845.24\,(22)$ $B=2393.469\,(12)$ $C=2269.621\,(16)$	$d_J=0.51\,(14)$				
	$CH_3CH_2CH_2SD$ trans	$v=0$	$A=22769.40\,(15)$ $B=2335.348\,(8)$ $C=2207.905\,(10)$	$d_J=0.47\,(9)$				
228	$CH_3CH_2CH_2NCO$, I *)	$v=0$	$A=6263.1\,(10)$ $B=2372.064\,(30)$ $C=1864.604\,(33)$	$\Delta_J=7.95\,(32)$ $\Delta_{JK}=-14.6\,(21)$	80ELL	80ELL	*) CH_3 gauche to C—N bond	3, 75
		$v_{T_1}=1$	$A=6530.9\,(18)$ $B=2303.860\,(47)$ $C=1803.665\,(56)$	$\Delta_J=4.60\,(43)$ $\Delta_{JK}=-40.1\,(43)$			$\nu_{T_1}=38\,(20)\,cm^{-1}$, C_b—N torsion	
		$v_{T_2}=1$	$A=6550.9\,(27)$ $B=2279.384\,(75)$ $C=1820.107\,(72)$	$\Delta_J=3.23\,(60)$ $\Delta_{JK}=-68\,(14)$			$\nu_{T_2}=74\,(34)\,cm^{-1}$, C_a—N=C_b bend	
		$v_{T_3}=1$	$A=6622.6\,(36)$ $B=2246.756\,(45)$ $C=1814.926\,(68)$	$\Delta_J=5.91\,(62)$ $\Delta_{JK}=30.3\,(94)$			ν_{T_3} central C_b—C_c torsion	
	$CH_3CH_2CH_2NCO$, II **)	$v=0$	$A=11116.3\,(65)$ $B=1635.104\,(15)$ $C=1466.650\,(16)$	$\Delta_J=0.922\,(83)$ $\Delta_{JK}=-23.7\,(15)$			**) CH_3 anti to NCO group $\Delta G^0=-0.23\,(15)$ kcal/mol at 195 °C	
229	$CH_3CH_2CH_2ONO$ syn-anti-gauche ⁺)	$v=0$	$B+C=3537\,(3)$ *)		79TR		*) LRMW ⁺) conformers about τ_1 (N—O), τ_2 (O—C) and τ_3 (C—C)	
	$CH_3CH_2CH_2ONO$ syn-anti-anti ⁺)	$v=0$	$B+C=3138\,(1)$ *)					
	$CH_3CH_2CH_2ONO$ anti-gauche-anti ⁺)	$v=0$ $v_\tau=1$	$B+C=2936\,(1)$ *) $B+C=2899\,(1)$ *)				potential function for rotation about the O—C (τ_2) bond determined	

Nr.	Molecule	v	A, B, C MHz	D, H kHz	Ref.	Ref. struct.	Varia, remarks	See also Tab., Nr.
230	$CH_3CH_2CH_2O$ Cl trans	$v=0$	$B+C=2829$ (2) *)		78SU2		*) LRMW	
231	$H_2NCH_2CH_2CH_2OH$	$v=0$	$A=7405.168$ (459) *) $B=3868.258$ (165) $C=2829.432$ (165)	$\Delta_J=$ 3.47 (177) *) $\Delta_{JK}=$ 6.11 (204) $\Delta_K=-19.26$ (966) $\delta_J=$ 0.76 (30) $\delta_K=-5.98$ (354)	79Mc	79Mc	*) incertainties are 3σ	2.6.3, 113 3, 76
		$v_\tau=1$	$A=7414.582$ (375) *)$^+$) $B=3861.782$ (120) $C=2832.210$ (72)				$\nu_\tau=122$ (25) cm^{-1}, probably heavy-atom torsion $^+$) centrifugal distortion corrections of the ground state used	
		$v_\tau=2$	$A=7414.219$ (2766) *)$^+$) $B=3855.302$ (123) $C=2834.760$ (114)					
		$v_x=1$	$A=7436.598$ (1428) *)$^+$) $B=3835.271$ (72) $C=2810.578$ (189)					
	$H_2NCH_2CH_2CH_2OD$	$v=0$	$A=7218.074$ (1065) *) $B=3862.646$ (51) $C=2799.007$ (45)					
	$HDNCH_2CH_2CH_2OH$ D_{II} **)	$v=0$	$A=7214.675$ (744) *) $B=3724.066$ (33) $C=2730.691$ (30)				**) see Fig. 76	
	$DHNCH_2CH_2CH_2OH$ D_{III} **)	$v=0$	$A=7180.556$ (1938) *) $B=3767.535$ (87) $C=2773.094$ (78)					
232	$CH_3CH(HO)CH_2(OH)$, T *)	$v=0$	$A=8393.2$ (4) $B=3648.53$ (1) $C=2778.31$ (2)	$\Delta_J=0.6$ (2) **) $\Delta_{JK}=4.4$ (4)	81CAM	81CAM	*) CH_3 trans with respect to the $C_1—O_1$ bond **) other constants kept fixed: $\Delta_K=2.95$, $\delta_J=0.13$, $\delta_K=2.41$ kHz ***) fixed on the ground state values $\nu_{t_1}=105$ (20) cm^{-1}, $C_1—C_2$ torsion?	2.6.3, 114
		$v_{t_1}=1$	$A=8327.5$ (6) $B=3641.97$ (2) $C=2776.89$ (3)	***)				
Cont.		$v_{t_1}=2$	$A=8287.1$ (1.4) $B=3636.10$ (4) $C=2774.80$ (3)	***)				

Nr.	Molecule	v	A, B, C MHz	D, H kHz	Ref.	Ref. struct.	Varia, remarks	See also Tab., Nr.
232 Cont.	CH_3CHCH_2 T *) HO OH	$v_{t_2}=1$	$A=8450$ (2) $B=3648.69$ (5) $C=2773.12$ (5)	***)	81CAM		$\nu_{t_2}=180$ (30) cm^{-1}, CH_3 torsion? *) CH_3 trans with respect to the C_1—O_1 bond **) other constants kept fixed: $\Delta_K=2.95$, $\delta_J=0.13$, $\delta_K=2.41$ kHz ***) fixed on the ground state values	
	CH_3CHCH_2, T *) HO OD	$v=0$	$A=8301.0$ (4) $B=3565.38$ (2) $C=2720.26$ (2)	$\Delta_J=0.5$ (3) **) $\Delta_{JK}=5.2$ (5)				
	CH_3CHCH_2, T *) DO OH	$v=0$	$A=7941.1$ (5) $B=3614.88$ (4) $C=2710.41$ (4)	$\Delta_J=1.1$ (6) **) $\Delta_{JK}=2.8$ (12)				
	CH_3CHCH_2, T *) DO OD	$v=0$	$A=7852.7$ (3) $B=3534.11$ (2) $C=2655.12$ (2)	$\Delta_J=1.1$ (2) **) $\Delta_{JK}=4.5$ (5)				
	CH_3CHCH_2, G $^+$) HO OH	$v=0$	$A=6642.7$ (2) $B=4163.59$ (1) $C=3365.35$ (1)	$\Delta_J=1.6$ (2) $^{++}$) $\Delta_{JK}=6.6$ (4)			$\Delta E_{T-G}=-0.58$ kcal/mol $^+$) CH_3 gauche with respect to the C_1—O_1 bond $^{++}$) other constants kept fixed: $\Delta_K=-0.02$, $\delta_J=0.34$, $\delta_K=-0.03$ kHz $^{+++}$) fixed on the value of the parent species	
	CH_3CHCH_2, G $^+$) HO OD	$v=0$	$A=6525$ (14) $B=4072$ (1) $C=3318$ (1)	$^{+++}$)				
	CH_3CHCH_2, G $^+$) DO OH	$v=0$	$A=6505$ (25) $B=4037$ (2) $C=3300$ (2)	$^{+++}$)				
	CH_3CHCH_2, G $^+$) DO OD	$v=0$	$A=6368.0$ (6) $B=3973.82$ (5) $C=3225.68$ (5)	$\Delta_J=1.6$ (9) $^{++}$) $\Delta_{JK}=6$ (4)				
233	$CH_3CH(NH_2)CH_2OH$ form I *)	$v=0$	$A=7948.403$ (77) $B=3640.570$ (11) $C=2747.727$ (11)	$\Delta_J=0.51$ (18) $\Delta_{JK}=3.84$ (63) $\Delta_K=1.5$ (32) $\delta_J=0.163$ (24) $\delta_K=3.23$ (56)	78ELL	78ELL	*) CH_3 anti to C—O bond	2.6.3, 115
		$v_{T_1}=1$	$A=7904.88$ (103) $B=3636.404$ (86) $C=2745.944$ (87)				$\nu_{T_1}=126$ (10) cm^{-1}, heavy-atom torsion	
Cont.		$v_{T_1}=2$	$A=7873.80$ (42) $B=3632.502$ (16) $C=2743.808$ (14)					

Nr.	Molecule	v	A, B, C MHz	D, H kHz	Ref.	Ref. struct.	Varia, remarks	See also Tab., Nr.
233 Cont.	$CH_3CH(NH_2)CH_2OH$	$v_{T_2}=1$	$A=7930.10\ (82)$ $B=3639.024\ (34)$ $C=2745.458\ (40)$		78ELL		$\nu_{T_2}=181\ (30)\ cm^{-1}$, methyl torsion	
		$v_\delta=1$	$A=7957.3\ (22)$ $B=3633.753\ (51)$ $C=2743.646\ (48)$				$\nu_\delta=199\ (49)\ cm^{-1}$, bending mode, or C—N or C—O torsion	
	$CH_3CH(NH_2)CH_2OH$ form II **)	$v=0$	$A=6461.620\ (32)$ $B=4144.442\ (30)$ $C=3336.163\ (30)$	$\Delta_J = 1.65\ (69)$ $\Delta_{JK} = 6.35\ (15)$ $\Delta_K = -7.02\ (40)$ $\delta_J = 0.3420\ (94)$ $\delta_K = -0.034\ (117)$			**) CH_3 gauche to C—O bond, *I* is energetically favoured with $\Delta G^0 = -0.58\ (17)$ kcal/mol $\Delta H^0 = -0.50\ (19)$ kcal/mol $\Delta S^0 = 0.3\ (1)$ e.u.	
		$v_{T_1}=1$	$A=6471.64\ (113)$ $B=4138.216\ (73)$ $C=3323.088\ (65)$				$\nu_{T_1}=144\ (26)\ cm^{-1}$, heavy-atom torsion	
		$v_x=1$	$A=6454.80\ (72)$ $B=4141.01\ (33)$ $C=3332.80\ (25)$				$\nu_x=225\ (26)\ cm^{-1}$, perhaps methyl torsion	
234	$CH_3CH(OH)CH_2NH_2$	$v=0$	$A=8487.1037\ (37)$ $B=3569.3804\ (15)$ $C=2767.2212\ (12)$	$\Delta_J = 0.6577\ (22)$ $\Delta_{JK} = 4.237\ (31)$ $\Delta_K = 2.947\ (15)$ $\delta_J = 0.1311\ (26)$ $\delta_K = 2.411\ (57)$	76MAR1	76MAR1	CH_3 anti to NH_2	2.8, 29 3, 77
		$v_{T_1}=1$	$A=8439.409\ (24)$ $B=3566.021\ (19)$ $C=2765.596\ (19)$	$\Delta_J = 1.03\ (29)$ $\Delta_{JK} = 4.23\ (13)$ $\Delta_K = 4.19\ (97)$ $\delta_J = 0.1391\ (67)$ $\delta_K = 2.30\ (15)$			$\nu_{T_1}=139\ (15)\ cm^{-1}$, heavy-atom torsion	
Cont.		$v_{T_1}=2$	$A=8400.356\ (51)$ $B=3562.207\ (37)$ $C=2763.095\ (37)$	$\Delta_J = -0.45\ (56)$ $\Delta_{JK} = 3.82\ (57)$ $\Delta_K = 6.4\ (34)$ $\delta_J = 0.128\ (22)$ $\delta_K = 2.54\ (54)$				

Nr.	Molecule	v	A, B, C MHz	D, H kHz	Ref.	Ref. struct.	Varia, remarks	See also Tab., Nr.
234 Cont.	$CH_3CH(OH)CH_2NH_2$	$v_{T_2}=1$	$A=8487.565\ (28)$ $B=3566.995\ (24)$ $C=2764.017\ (24)$	$\Delta_J=8.55\ (35)$ $\Delta_{JK}=4.37\ (12)$ $\Delta_K=1.48\ (85)$ $\delta_J=0.1529\ (64)$ $\delta_K=2.07\ (14)$	76MAR1		$\nu_{T_2}=255\ (25)\ cm^{-1}$, high-frequency heavy-atom torsion	
		$v_\tau=1$	$A=8475.87\ (5)$ $B=3566.03\ (4)$ $C=2764.81\ (4)$				ν_τ, methyl torsion	
235	$CH_3CH(OH)CH_2F$	$v=0$	$A=8570.5576\ (43)$ $B=3566.4735\ (17)$ $C=2742.4504\ (21)$	$\Delta_J=0.7646\ (61)$ $\Delta_{JK}=6.416\ (58)$ $\Delta_K=1.640\ (20)$ $\delta_J=0.1747\ (38)$ $\delta_K=3.56\ (11)$ $H_J=0.001079\ (51)\cdot 10^{-3}$ $H_{JK}=-0.0209\ (11)\cdot 10^{-3}$	77MAR	77MAR	C–F bond anti to CH_3 group	2.6.3, 116 2.8, 30 3, 78
		$v_\delta=1$	$A=8592.314\ (51)$ $B=3561.633\ (20)$ $C=2733.601\ (19)$	$\Delta_{JK}=8.08\ (88)$ $\Delta_K=-11.5\ (60)$ $\delta_J=0.228\ (52)$ $\delta_K=2.96\ (105)$			$\nu_\delta=151\ (15)\ cm^{-1}$, lowest bending mode	
		$v_b=1$	$A=8519.227\ (40)$ $B=3560.406\ (18)$ $C=2737.034\ (18)$	$\Delta_{JK}=7.17\ (47)$ $\Delta_K=0.8\ (35)$ $\delta_J=0.264\ (30)$ $\delta_K=2.82\ (59)$			ν_b, second lowest bending mode	
		$v_T=1$	$A=8525.0492\ (56)$ $B=3561.8778\ (20)$ $C=2739.9096\ (24)$	$\Delta_J=0.7825\ (72)$ $\Delta_{JK}=6.326\ (64)$ $\Delta_K=1.158\ (25)$ $\delta_J=0.1714\ (59)$ $\delta_K=3.51\ (12)$ $H_J=1.47\ (19)\cdot 10^{-6}$ $H_{JK}=-14.4\ (37)\cdot 10^{-6}$			$\nu_T=109\ (10)\ cm^{-1}$, heavy-atom torsion	
Cont.		$v_T=2$	$A=8485.663\ (29)$ $B=3557.479\ (11)$ $C=2737.203\ (12)$	$\Delta_J=0.85\ (11)$ $\Delta_{JK}=6.25\ (39)$ $\Delta_K=1.3\ (26)$ $\delta_J=0.197\ (24)$ $\delta_K=2.80\ (48)$				

Nr.	Molecule	v	A, B, C MHz	D, H kHz	Ref.	Ref. struct.	Varia, remarks	See also Tab., Nr.
235 Cont.	$CH_3CH(OH)CH_2F$	$v_\tau=1$	$A=8555.31$ (7) $B=3563.57$ (7) $C=2740.37$ (6)		77MAR		$\nu_\tau=210$ (15) cm^{-1}, methyl torsion	
	$CH_3CH(OD)CH_2F$	$v=0$	$A=8158.248$ (16) $B=3566.051$ (13) $C=2698.411$ (12)	$\Delta_J=0.59$ (25) $\Delta_{JK}=7.32$ (42) $\Delta_K=0.15$ (224) $\delta_J=0.182$ (20) $\delta_K=3.38$ (42)				
236	CH_3CHFCH_3	$v=0$	$A=8693.82$ (2) $B=8102.51$ (2) $C=4788.83$ (1)		73GRI	73GRI		2.6.3, 117 2.8, 31
		$v_\tau=1^+$	$A=8681.87$ (4) $B=8095.78$ (5) $C=4784.12$ (2)				$\nu_\tau^+=230$ (20) cm^{-1}, symmetric methyl torsion	
		$v_\tau=1^-$	$A=8686.29$ (4) $B=8079.28$ (3) $C=4782.70$ (3)				$\nu_\tau^-=272$ (20) cm^{-1}, anti-symmetric methyl torsion	
		$v_\delta=1$	$A=8686.89$ (3) $B=8114.33$ (2) $C=4801.32$ (1)				$\nu_\delta=390$ (30) cm^{-1}, probably wagging of the CF group	
237	$(CH_3)_2CHI$	$v=0$	$A=8022.052$ (165) $B=2187.147$ (6) $C=1817.649$ (5)		78GRO	78GRO		2.7.3, 160
238 *6–103*	$(CH_3)_2CHCN$	$v=0$	$A=7941.13$ (22) $B=3968.00$ (2) $C=2901.02$ (2)		74DUR8	74DUR8		2.6.3, 118 2.8, 32
		$v_\alpha=1$	$A=7933.22$ (41) $B=3964.91$ (2) $C=2898.17$ (2)				$\nu_\alpha=200$ (20) cm^{-1}, CH_3 torsion	
		$v_\beta=1$	$A=7995.15$ (105) $B=3971.74$ (4) $C=2904.66$ (4)				$\nu_\beta=180$ (20) cm^{-1}, C—C—N in-plane bend	
		$v_\gamma=1$	$A=7922.81$ (101) $B=3970.26$ (4) $C=2902.79$ (4)				$\nu_\gamma=249$ (20) cm^{-1}, CH_3 torsion	

Nr.	Molecule	v	A, B, C MHz	D, H kHz	Ref.	Ref. struct.	Varia, remarks	See also Tab., Nr.
239	$(CH_3)_2CHNH_2$	$v=0$	$A=8331.86\,(5)$ $B=7977.25\,(5)$ $C=4637.1\,(62)$		77MEH	77MEH	trans form	2.6.3, 119 2.7.3, 36
	$(CH_3)_2CHNHD$	$v=0$	$A=8075.82\,(5)$ $B=7706.77\,(6)$ $C=4498.3\,(4.4)$					
	$(CH_3)_2CHND_2$	$v=0$	$A=7806.12\,(5)$ $B=7490.49\,(5)$ $C=4307\,(10)$					
240 *6-102*	$(CH_3)_2CHOH$ trans	$v=0$	$A=8488.996\,(26)$ $B=8041.911\,(33)$ $C=4765.210\,(29)$	$\tau_{aaaa}=-27.2\,(88)$ $\tau_{bbbb}=-24.1\,(88)$ $\tau_{cccc}=\ 0.7\,(89)$ $\tau_1=-22.2\,(91)$ *) $\tau_2=\ -6\,(17)$	79HIR3		for structure, see also Vol. II/7 (3, 430) *) $\tau_1=\tau'_{aabb}+\frac{A-B}{A-C}\tau'_{ccaa}$ $\tau_2=\tau'_{bbcc}+\frac{B-C}{A-C}\tau'_{ccaa}$ $E_{trans}-E_{gauche}=450\,(210)$ cal/mol **) symmetric, with $Q_b=-153.0$ MHz; $Q_c=-279.663$ MHz $\Delta=46798.50\,(11)$ MHz (tunneling splitting) $R_{ab}=-58.215\,(12)$ MHz; $R_{ca}=\ 89.5559\,(92)$ MHz ***) antisymmetric	2.6.3, 120
	$(CH_3)_2CHOH$ gauche	$v=0$	$A_s=8638.994\,(15)$ **) $B_s=8065.126\,(14)$ $C_s=4769.546\,(10)$ $A_a=8641.350\,(14)$ ***) $B_a=8060.500\,(14)$ $C_a=4766.142\,(11)$					
	$(CH_3)_2CHOD$ trans	$v=0$	$A=8099.065\,(19)$ $B=7918.010\,(24)$ $C=4680.729\,(22)$	$\tau_{aaaa}=-34.2\,(64)$ $\tau_{bbbb}=-30.9\,(64)$ $\tau_{cccc}=\ -9.7\,(65)$ $\tau_1=-39.0\,(68)$ *) $\tau_2=-28\,(13)$				
241	$(CH_3)_2C(OH)CN$	$v=0$	$A=4748.4$ $B=2830.83$ $C=2762.99$		77LIS	77LIS		
242	$(CH_3)_2CHSH$ trans	$v=0$	$A=7892.65\,(30)$ $B=4414.42\,(2)$ $C=3158.03\,(1)$		75GRI	75GRI	the trans form is more stable by 57 cal/mol	2.6.3, 121
Cont.		$v_\tau=1$	$A=7890.17\,(26)$ $B=4418.48\,(1)$ $C=3152.33\,(1)$				$\nu_\tau=208\,(30)\ cm^{-1}$, S—H torsion V_3 (SH) $=1.98$ kcal/mol	

Nr.	Molecule	v	A, B, C MHz	D, H kHz	Ref.	Ref. struct.	Varia, remarks	See also Tab., Nr.
242 Cont.	$(CH_3)_2CHSD$ trans	$v=0$	$A=7680.75(19)$ $B=4300.03(1)$ $C=3132.67(1)$		75GRI			
		$v_\tau=1$	$A=7679.87(30)$ $B=4305.11(1)$ $C=3125.69(1)$				$\nu_\tau=140(20)\ cm^{-1}$	
	$(CH_3)_2CH^{34}SH$ trans	$v=0$	$A=7893.82(22)$ $B=4291.51(1)$ $C=3094.56(1)$					
	$(CH_3)_2CH^{34}SD$ trans	$v=0$	$A=7680.17(44)$ $B=4185.95(2)$ $C=3071.66(1)$					
	$(CH_3)_2CHSH$ gauche	$v=0$	$A=7880.13(103)$ *) $B=4524.60(2)$ *) $C=3169.36(47)$ *)				*) mean of the $v=0$, $v=1$ inversion states $G_a=24.97(7)$; $G_b=15.98(2)$ MHz $\Delta E=562.38(10)$ MHz, inversion-level splitting $\Delta E=10.06(20)$ MHz	
	$(CH_3)_2CHSD$ gauche	$v=0$	$A=7700.00(3)$ $B=4416.58(1)$ $C=3094.81(1)$					
242a	$CH_3C(=O)-CH_2OH$	$v=0$	$A=10342.914(6)$ *) $B=3819.074(3)$ $C=2865.100(3)$	$\tau_{aaaa}=-822.5(20)$ *) $\tau_{bbbb}=-4.9(2)$ $\tau_{cccc}=-1.5(2)$ $\tau_{bbcc}=2.2(2)$ $\tau_{aabb}+\tau_{aacc}=-80.4(4)$	80KAT1	80KAT1	*) for the A species	2.6.3, 121b 2.8, 32a
	$CH_3C(=O)-CH_2OD$	$v=0$	$A=9926.237(99)$ $B=3738.834(8)$ $C=2813.128(4)$					
243	$(CH_3)_2CHO^{35}Cl$-gauche $(CH_3)_2CHO^{35}Cl$-trans		$B+C=4392(5)$ *) $B+C=2829(2)$		78SUE2		*) LRMW	

Nr.	Molecule	v	A, B, C MHz	D, H kHz	Ref.	Ref. struct.	Varia, remarks	See also Tab., Nr.
244	$(CH_3)_2CHC^{79}Br(CH_3)_2$ gauche $(CH_3)_2CHC^{79}Br(CH_3)_2$ trans $(CH_3)_2CHC^{81}Br(CH_3)_2$ gauche $(CH_3)_2CHC^{81}Br(CH_3)_2$ trans		$B+C=2232$ *) $B+C=2555$ *) $B+C=2189$ *) $B+C=2540$ *)		79AND		*) LRMW $\Delta E_{\text{gauche-trans}}=500\,(300)$ cal/mol	
245	$(CH_3)_2CHCI(CH_3)_2$ gauche $(CH_3)_2CHCI(CH_3)_2$ trans		$B+C=1719$ *) $B+C=2027$ *)		79AND		*) LRMW	
246	$(CH_3)_2CHOC(=O)H$	$v=0$	$B+C=4641\,(5)$ *)		77TR2		*) LRMW, a single rotamer found with τ_1 (O=COC) = 0°, τ_2 (COCH) ≃ 40°	
247	$(CH_3)_2CHOC(=O)F$	$v=0$ $v_{T_1}=1$ $v_{T_2}=1$	$A=4967.0\,(8)$ $B=1704.69\,(2)$ $C=1468.86\,(1)$ $A=4963\,(5)$ $B=1694.11\,(7)$ $C=1471.43\,(4)$ $A=4998\,(6)$ $B=1705.21\,(7)$ $C=1471.10\,(4)$		78TR2		τ_1 (O=COC) = 0°, τ_2 (COCH) ≃ 35° $\nu_{T_1}=70\,(20)\ cm^{-1}$, O-isopropyl torsion $\nu_{T_2}=120\,(20)\ cm^{-1}$, C—O torsion	3, 79
248	$(CH_3)_2CHOC(=O)^{35}Cl$ $(CH_3)_2CHOC(=O)^{37}Cl$	$v=0$ $v=0$	$B+C=2275\,(1)$ *) $B+C=2226\,(1)$ *)		77TR2		*) LRMW, a single rotamer found with: τ_1 (O=COC) = 0°, τ_2 (COCH) ≃ 40°	
249	$(CH_3)_2CHOC(=O)CN$	$v=0$	$B+C=2257\,(1)$ *)		77TR2		*) LRMW, rotamer with τ_1 (O=COC) = 0°, τ_2 (COCH) ≃ 40°	

Nr.	Molecule	v	A, B, C MHz	D, H kHz	Ref.	Ref. struct.	Varia, remarks	See also Tab., Nr.
250	$(CH_3)_2CHSC(=O)F$ syn-gauche *)	$v=0$	$A=4196\,(1)$ $B=1374.23\,(3)$ $C=1159.80\,(4)$		79SIL		*) $\tau_1\,(OCSC)=0°$, $\tau_2\,(CSCH)\simeq 40°$	
251	$(CH_3)_2CHSC(=O){}^{35}Cl$ $(CH_3)_2CHSC(=O){}^{37}Cl$	$v=0$ $v=0$	$B+C=1814.6\,(5)$ *) $B+C=1774.4\,(5)$ *)		79SIL		*) LRMW; either syn-syn $\tau_1\,(OCSC)=0°$, $\tau_2\,(CSCH)=0°$) or syn-gauche ($\tau_1=0°$, $\tau_2\simeq 60°$)	
252	$(CH_3)_2CHSC(=O)CN$	$v=0$	$B+C=1839.6\,(7)$ *)		79SIL		*) LRMW	
252 a	$CH_3CF_2CH_3$	$v=0$ $v_\tau=1_1$ $v_\tau=1_2$	$A=5149.79\,(4)$ $B=4840.11\,(2)$ $C=4805.62\,(2)$ $A=5146.3\,(4)$ $B=4834.4\,(2)$ $C=4797.7\,(2)$ $A=5146.3\,(4)$ $B=4840.3\,(2)$ $C=4799.0\,(2)$		81DUR5	81DUR5	 $\nu_\tau(A_2)=217\,(15)\,cm^{-1}$ methyl torsion $\nu_\tau(B_2)=250\,(30)\,cm^{-1}$	2.6.3, 121 a
253	$(CH_3)_2C{=}S$	$v=0$	$A=8728.77\,(1)$ $B=4390.550\,(1)$ $C=3274.951\,(1)$		74KRO			2.8, 33
254	$(CH_2F)_2C{=}O$ *) cis-trans Cont.	$v=0$ $v_1=1$	$A=6024.843\,(6)$ **) $B=2454.414\,(1)$ $C=1783.897\,(1)$ $A=6033.966\,(5)$ **) $B=2446.628\,(1)$ $C=1785.541\,(1)$	$\Delta_J=0.532\,(10)$ **) $\Delta_{JK}=-1.28\,(3)$ $\Delta_K=7.7\,(2)$ $\delta_J=0.172\,(1)$ $\delta_K=0.78\,(3)$ $\Delta_J=0.532\,(8)$ **) $\Delta_{JK}=-1.32\,(4)$ $\Delta_K=7.8\,(2)$ $\delta_J=0.173\,(2)$ $\delta_K=0.80\,(4)$	75FIN	75FIN	*) all heavy atoms lie in a plane, one F lies cis and the other trans to O **) uncertainties are 2 σ $\nu_1=82\,(12)\,cm^{-1}$, trans-CH_2F torsion	2.6.3, 122 3, 80

Nr.	Molecule	v	A, B, C MHz	D, H kHz	Ref.	Ref. struct.	Varia, remarks	See also Tab., Nr.
254 Cont.	$(CH_2F)_2C{=}O$ *) cis-trans	$v_1=2$	$A=6044.946$ (7) **) $B=2438.939$ (2) $C=1786.883$ (2)	$\Delta_J = 0.54$ (1) **) $\Delta_{JK}=-1.26$ (5) $\Delta_K= 7.5$ (3) $\delta_J=0.178$ (2) $\delta_K=0.76$ (5)	75FIN		*) all heavy atoms lie in a plane, F lies cis and the other trans to O **) uncertainties are 2σ	
		$v_1=3$	$A=6057.61$ (1) **) $B=2431.309$ (3) $C=1787.968$ (2)	$\Delta_J = 0.54$ (1) **) $\Delta_{JK}=-1.5$ (1) $\Delta_K= 9.8$ (8) $\delta_J=0.176$ (4) $\delta_K=0.9$ (1)				
		$v_1=4$	$A=6071.59$ (25) **) $B=2423.693$ (10) $C=1788.881$ (8)	$\Delta_J = 0.54$ (3) **) $\Delta_{JK}=-1.1$ (4) $\Delta_K= 22$ (36) $\delta_J=0.15$ (2) $\delta_K=1.7$ (18)				
		$v_2=1$	$A=5998.130$ (8) **) $B=2451.297$ (2) $C=1784.460$ (2)	$\Delta_J = 0.52$ (1) **) $\Delta_{JK}=-1.13$ (6) $\Delta_K= 6.7$ (3) $\delta_J=0.171$ (2) $\delta_K=0.65$ (5)			$\nu_2=127$ (15) cm^{-1}, cis-CH_2F torsion	
		$v_2=2$	$A=5971.852$ (7) **) $B=2449.113$ (2) $C=1785.352$ (2)	$\Delta_J = 0.53$ (1) **) $\Delta_{JK}=-1.06$ (9) $\Delta_K= 6.1$ (5) $\delta_J=0.168$ (3) $\delta_K=0.62$ (7)				
		$v_1=1, v_2=1$	$A=6010.46$ (13) **) $B=2443.720$ (6) $C=1785.859$ (6)	$\Delta_J = 0.51$ (3) **) $\Delta_{JK}=-2.1$ (2) $\Delta_K= 0.8$ (20) $\delta_J=0.21$ (2) $\delta_K=-0.1$ (12)				
		$v_3=1$	$A=6019.84$ (13) **) $B=2455.657$ (8) $C=1782.612$ (6)	$\Delta_J = 0.56$ (3) **) $\Delta_{JK}=-1.4$ (2) $\Delta_K=-6.6$ (17) $\delta_J=0.17$ (2) $\delta_K=0.5$ (8)			$\nu_3=285$ (25) cm^{-1}, in-plane bending motion	

Nr.	Molecule	v	A, B, C MHz	D, H kHz	Ref.	Ref. struct.	Varia, remarks	See also Tab., Nr.
255	$(CH_2{}^{35}Cl)_2C{=}O$ $^{+}$)	$v=0$	$B+C=2038.2\,(5)$ *)		76KAR3		*) LRMW	
	$(CH_2{}^{37}Cl)_2C{=}O$	$v=0$	$B+C=1990.7\,(5)$ *)				$^{+}$) either O-cis, trans,	
	$(CD_2{}^{35}Cl)_2C{=}O$	$v=0$	$B+C=1992.2\,(5)$ *)				O-cis, skew,	
	$(CD_2{}^{37}Cl)_2C{=}O$	$v=0$	$B+C=1947.0\,(5)$ *)				or O-gauche, skew	
	$(CH_2{}^{35}Cl)_2C{=}{}^{18}O$	$v=0$	$B+C=2024.4\,(5)$ *)					
	$(CH_2{}^{37}Cl)_2C{=}{}^{18}O$	$v=0$	$B+C=1977.6\,(6)$ *)					
256	HOCH=CHC(=O)H *)	$v=0$	$A=9833.857\,(2)$	$\Delta_J = 3.83\,(4)$	78BAU,	81BAU	*) enolic hydrogen	2.6.3, 123
			$B=5212.535\,(2)$	$\Delta_{JK}=-0.30\,(4)$	76ROW		bonded form	2.7.3, 79
			$C=3404.362\,(2)$	$\Delta_K = 1.32\,(11)$				3, 81
				$\delta_J = 1.42\,(1)$				
				$\delta_K = 7.84\,(5)$				
		$v=1$	$A=9846.052\,(3)$	$\Delta_J = 3.11\,(5)$			$\nu=26\,(10)\,cm^{-1}$,	
			$B=5158.738\,(3)$	$\Delta_{JK}=-10.96\,(6)$			symmetric double minimum	
			$C=3383.236\,(2)$	$\Delta_K = 12.17\,(15)$			potential	
				$\delta_J = 1.10\,(1)$				
				$\delta_K = 0.30\,(7)$				
	DOCH=CH—C(=O)H	$v=0$	$A=9579.83\,(3)$		81BAU,		Coriolis interaction between the	
			$B=5154.05\,(2)$		78BAU		ground state and the low	
			$C=3348.62\,(2)$				lying state:	
		$v=1$	$A=9581.16\,(3)$				$\Delta E_{01}=87402\,(131)$ MHz	
			$B=5144.50\,(2)$				$F=100.52\,(22)$ MHz	
			$C=3344.85\,(2)$					
	HOCH=CD—C(=O)H	$v=0$	$A=8980.74\,(4)$					
			$B=5210.88\,(3)$					
			$C=3295.39\,(3)$					
		$v=1$	$A=8992.88\,(4)$					
			$B=5157.41\,(2)$					
			$C=3275.95\,(2)$					
	HOCD=CH—C(=O)H	$v=0$	$A=9675.30\,(2)$					
			$B=4961.55\,(2)$					
			$C=3277.83\,(1)$					
		$v=1$	$A=9682.34\,(2)$					
			$B=4931.92\,(2)$					
Cont.			$C=3265.96\,(1)$					

Nr.	Molecule	v	A, B, C MHz	D, H kHz	Ref.	Ref. struct.	Varia, remarks	See also Tab., Nr.
256 Cont.	$HC(=^{18}O)-CH_2-C(=O)-H$	$v=0$	$A=9590.59\ (1)$		81BAU, 78BAU			
			$B=5042.13\ (1)$					
			$C=3302.46\ (1)$					
		$v=1$	$A=9603.09\ (5)$					
			$B=4990.36\ (40)$					
			$C=3282.02\ (4)$					
	$HC(=^{17}O)-CH_2-C(=O)-H$	$v=0$	$A=9704.46\ (9)$					
			$B=5125.77\ (8)$					
			$C=3351.82\ (2)$					
		$v=1$	$A=9716.88\ (4)$					
			$B=5072.79\ (2)$					
			$C=3330.98\ (1)$					
	$HC(=^{18}O)-CH_2-C(=^{18}O)-H$	$v=0$	$A=9340.02\ (2)$					
			$B=4883.68\ (2)$					
			$C=3204.70\ (2)$					
		$v=1$	$A=9351.70\ (14)$					
			$B=4832.37\ (13)$					
			$C=3184.26\ (12)$					
	$HC(=^{18}O)-CH_2-C(=^{17}O)-H$	$v=0$	$A=9457.95$					
			$B=4961.23$					
			$C=3252.04$					
	$DOCH{=}CD-C(=O)H$	$v=0$	$A=8758.98\ (1)$				Coriolis interaction between $v=0$ and $v=1$:	
			$B=5152.68\ (1)$				$\Delta E_{01}=86448.12\ (59)$ MHz	
			$C=3241.92\ (1)$				$F=87.90\ (12)$ MHz	
		$v=1$	$A=8760.41\ (1)$					
			$B=5143.21\ (1)$					
			$C=3238.45\ (1)$					
	$DOCH{=}CD-C(=O)H$ mono-^{18}O	$v=0$	$A=8567.17\ (2)$					
			$B=4977.51\ (2)$					
			$C=3146.440\ (2)$					
		$v=1$	$A=8560.86\ (5)$					
			$B=4976.75\ (5)$					
Cont.			$C=3145.49\ (4)$					

Nr.	Molecule	v	A, B, C MHz	D, H kHz	Ref.	Ref. struct.	Varia, remarks	See also Tab., Nr.
256 Cont.	D^{18}OCH=CDC(=^{18}O)H	$v=0$	$A=8353.10(5)$		81BAU, 78BAU		Coriolis interaction between $v=0$ and $v=1$:	
			$B=4826.32(6)$				$\Delta E_{01}=81405(58)$ MHz	
			$C=3056.70(4)$				$F=82.00(27)$ MHz	
		$v=1$	$A=8354.38(5)$					
			$B=4817.40(6)$					
			$C=3053.48(4)$					
	DOCD=CDC(=O)H	$v=0$	$A=8630.732(1)$	$\Delta_J=2.64(2)$				
			$B=4914.726(2)$	$\Delta_{JK}=-3.53(2)$				
			$C=3130.010(1)$	$\Delta_K=4.09(6)$				
				$\delta_J=0.986(4)$				
				$\delta_K=2.96(2)$				
	DOCH=CDC(=O)D	$v=0$	$A=8645.182(1)$	$\Delta_J=2.73(4)$				
			$B=4906.101(2)$	$\Delta_{JK}=-4.13(2)$				
			$C=3128.514(2)$	$\Delta_K=4.66(7)$				
				$\delta_J=0.992(3)$				
				$\delta_K=2.68(2)$				
	HO^{13}CH=CH—C(=O)H	$v=0$	$A=9800.15(3)$					
			$B=5135.52(7)$					
			$C=3367.38(2)$					
		$v=1$	$A=9812.02(1)$					
			$B=5083.47(4)$					
			$C=3346.73(1)$					
	HOCD=CH—C(=O)D	$v=0$	$A=9492.27(4)$					
			$B=4753.73(4)$					
			$C=3165.79(3)$					
		$v=1$	$A=9500.37(2)$					
			$B=4708.68(2)$					
			$C=3146.98(2)$					
	HO^{13}CH=^{13}CH—C(=O)H	$v=0$	$A=9563.78(1)$					
			$B=5135.55(1)$					
			$C=3339.00(1)$					
		$v=1$	$A=9576.12(1)$					
			$B=5083.63(1)$					
Cont.			$C=3318.91(1)$					

Nr.	Molecule	v	A, B, C MHz	D, H kHz	Ref.	Ref. struct.	Varia, remarks	See also Tab., Nr.
256 Cont.	$HO^{13}CH{=}CH{-}^{13}C(=O)H$	$v=0$	$A=9765.57(2)$ $B=5059.39(2)$ $C=3330.47(2)$		81BAU, 78BAU			
		$v=1$	$A=9777.11(1)$ $B=5088.88(2)$ $C=3310.26(1)$					
	$DOCD{=}CH{-}C(=O)D$	$v=0$	$A=9241.62(2)$ $B=4705.06(2)$ $C=3115.85(1)$					
		$v=1$	$A=9242.26(2)$ $B=4697.07(2)$ $C=3112.49(1)$					
	$DO^{13}CH{=}CD{-}C(=O)H$	$v=0$	$A=8733.80(3)$ $B=5077.36(2)$ $C=3208.57(1)$					
		$v=1$	$A=8735.41(3)$ $B=5068.04(2)$ $C=3205.16(1)$					
	$HO^{13}CH{=}^{13}CH{-}^{13}C(=O)H$	$v=0$	$A=9533.102(3)$ $B=5059.399(3)$ $C=3302.982(3)$					
		$v=1$	$A=9545.173(6)$ $B=5009.032(6)$ $C=3283.306(6)$					
	$DOCD{=}CD{-}C(=O)D$	$v=0$	$A=8491.53(2)$ $B=4703.86(2)$ $C=3025.32(1)$					
		$v=1$	$A=8492.41(2)$ $B=4695.93(2)$ $C=3022.22(1)$					
	$DO^{13}CH{=}^{13}CD{-}C(=O)H$	$v=0$	$A=8550.73(2)$ $B=5077.53(2)$ $C=3180.60(1)$					
Cont.		$v=1$	$A=8552.66(2)$ $B=5068.16(2)$ $C=3180.27(1)$					

Nr.	Molecule	v	A, B, C MHz	D, H kHz	Ref.	Ref. struct.	Varia, remarks	See also Tab., Nr.
256 Cont.	$DO^{13}CH{=}CD^{13}C(=O)H$	$v=0$	$A=8707.81\,(2)$		81BAU, 78BAU			
			$B=5002.77\,(2)$					
			$C=3175.18\,(2)$					
		$v=1$	$A=8709.05\,(2)$					
			$B=4993.48\,(2)$					
			$C=3172.54\,(2)$					
	$D^{18}OCD{=}CDC(=O)H$	$v=0$	$A=8407.743\,(4)$					
			$B=4773.292\,(4)$					
			$C=3043.252\,(4)$					
	$DOCH{=}CDC(={}^{18}O)D$	$v=0$	$A=8414.625\,(6)$					
			$B=4769.812\,(5)$					
			$C=3042.840\,(5)$					
	$DOCD{=}CDC(={}^{18}O)H$	$v=0$	$A=8456.581\,(10)$					
			$B=4747.028\,(9)$					
			$C=3038.884\,(9)$					
	$D^{18}OCH{=}CDC(=O)D$	$v=0$	$A=8477.341\,(7)$					
			$B=4734.716\,(7)$					
			$C=3036.603\,(6)$					
	$DOCD{=}CH^{13}C(=O)D$	$v=0$	$A=8606.823\,(3)$					
			$B=4841.483\,(3)$					
			$C=3097.043\,(3)$					
	$DO^{13}CH{=}CDC(=O)D$	$v=0$	$A=8609.011\,(3)$					
			$B=4840.393\,(3)$					
			$C=3096.980\,(3)$					
	$DO^{13}CH{=}{}^{13}CD^{13}C(=O)H$	$v=0$	$A=8527.750\,(10)$					
			$B=5002.816\,(8)$					
			$C=3150.963\,(7)$					
		$v=1$	$A=8529.162\,(10)$					
			$B=4993.944\,(8)$					
			$C=3147.732\,(7)$					

Nr.	Molecule	v	A, B, C MHz	D, H kHz	Ref.	Ref. struct.	Varia, remarks	See also Tab., Nr.
257	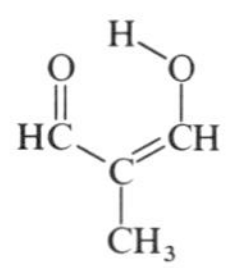	$v=0$	$A_+=5064.803\,(26)$ *) $B_+=3507.710\,(6)$ $C_+=2094.590\,(4)$ $A_-=5011.108\,(26)$ $B_-=3508.572\,(6)$ $C_-=2093.695\,(4)$		81SAN	81SAN	Coriolis interaction between the 2 lowest-lying states: $\Delta_\pm=84012.0\,(40)$ MHz C-type $=31.49\,(5)$ MHz *) for the A species $\nu\simeq100\,(20)\ \mathrm{cm}^{-1}$	2.6.3, 124 2.8, 34
		$v=1$	$A_+=4857.47\,(21)$ *) $B_+=3505.780\,(36)$ $C_+=2093.315\,(19)$ $A_-=5109.30\,(15)$ $B_-=3502.532\,(22)$ $C_-=2096.894\,(19)$					
	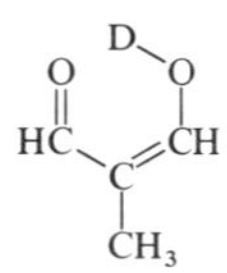	$v=0$	$A_+=5014.56\,(4)$ *) $B_+=3444.954\,(9)$ $C_+=2066.647\,(5)$ $A_-=5007.46\,(4)$ $B_-=3445.064\,(9)$ $C_-=2066.502\,(5)$				Coriolis interaction: $\Delta_\pm=10777\,(1)$ MHz C-type $=48.33\,(2)$ MHz	
	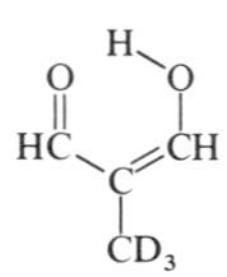	$v=0$	$A_+=4912.4\,(1)$ *) $B_+=3055.973\,(15)$ $C_+=1924.236\,(15)$ $A_-=4853.3\,(1)$ $B_-=3056.254\,(15)$ $C_-=1923.974\,(15)$				Coriolis interaction: $\Delta_\pm=27777\,(2)$ MHz C-type $=28.14\,(3)$ MHz	
258 *4-92*	$CH_3CH_2C(=O)H$ gauche	$v=0$	$A=23035\,(42)$ $B=4314.969\,(4)$ $C=4147.878\,(4)$		74PIC		Coriolis interaction between the 2 lowest-lying states: $\Delta=471.80\,(7)$ MHz $F=23.77\,(1)$ MHz (coefficient of the $P_bP_c+P_cP_b$ term)	
		$v=1$	$A=23035\,(42)$ $B=4315.017\,(4)$ $C=4148.102\,(4)$					

Nr.	Molecule	v	A, B, C MHz	D, H kHz	Ref.	Ref. struct.	Varia, remarks	See also Tab., Nr.
258a *6-105*	$CH_3CH_2C(=O)F$ cis	$v=0$	$A=10042.5425\,(56)$ $B=3762.2088\,(21)$ $C=2832.6919\,(18)$	$\Delta_J=0.776\,(18)$ $\Delta_{JK}=3.474\,(66)$ $\Delta_K=4.20\,(36)$ $\delta_J=0.1832\,(29)$ $\delta_K=1.774\,(89)$	81SCA1			2.8, 34a
259	$CH_3CH_2C(=O)^{35}Cl$ cis *)	$v=0$	$A=8854.770\,(4)$ $B=2377.940\,(1)$ $C=1918.681\,(1)$	$\Delta_J=0.291\,(6)$ $\Delta_{JK}=1.14\,(2)$ $\Delta_K=13.0\,(3)$ $\delta_J=0.0584\,(5)$ $\delta_K=0.73\,(2)$	76KAR1, 79MAT	76KAR1	*) heavy-atom planar configuration, CH_3 and C=O cis to each other	2.7.3, 117 2.8, 35
		$v_\tau=1$	$A=8787.555\,(7)$ $B=2379.498\,(1)$ $C=1922.147\,(1)$	$\Delta_J=0.262\,(9)$ $\Delta_{JK}=1.11\,(5)$ $\Delta_K=12.9\,(7)$ $\delta_J=0.0578\,(9)$ $\delta_K=0.54\,(5)$			$v_\tau=86\,(10)\ cm^{-1}$, C—C torsion	
		$v_\tau=2$	$A=8719.657\,(8)$ $B=2381.493\,(2)$ $C=1926.000\,(2)$	$\Delta_J=0.30\,(1)$ $\Delta_{JK}=1.06\,(8)$ $\Delta_K=12\,(1)$ $\delta_J=0.0589\,(14)$ $\delta_K=0.25\,(8)$				
		$v_\tau=3$	$A=8647.6\,(3)$ $B=2383.83\,(1)$ $C=1930.295\,(6)$					
		$v_\tau=4$	$A=8577.33\,(3)$ $B=2387.026\,(6)$ $C=1935.133\,(7)$	$\Delta_J=0.36\,(5)$ $\Delta_{JK}=1.9\,(5)$ $\Delta_K=-1\,(7)$ $\delta_J=0.071\,(7)$ $\delta_K=-1.3\,(4)$				
		$v_\alpha=1$	$A=8822.73\,(2)$ $B=2375.925\,(3)$ $C=1917.372\,(3)$	$\Delta_J=0.24\,(2)$ **) $\Delta_{JK}=0.9\,(2)$ $\Delta_K=14\,(3)$ $\delta_J=0.052\,(3)$ $\delta_K=0.8\,(2)$			$v_\alpha=197\ cm^{-1}$, CH_3 torsion **) for the A species	
Cont.								

Nr.	Molecule	v	A, B, C MHz	D, H kHz	Ref.	Ref. struct.	Varia, remarks	See also Tab., Nr.
259 Cont.	$CH_3CH_2C(=O)^{35}Cl$ cis	$v=0$	$A=8828.082(9)$ $B=2320.275(2)$ $C=1879.747(2)$	$\Delta_J=0.27(1)$ $\Delta_{JK}=1.04(3)$ $\Delta_K=14.0(6)$ $\delta_J=0.0548(7)$ $\delta_K=0.73(4)$	76KAR1, 79MAT			
260 *6-105 a*	$CH_3CH_2C(=O)OH$ cis	$v=0$	$A=10165.36(2)$ *) $B=3817.89(1)$ $C=2875.17(1)$	$D_J=0.756(30)$ *) $D_{JK}=3.423(250)$ $D_K=3.709(250)$ $\delta_J=0.180(20)$ $\delta_K=0.417(70)$	75ST1	75ST2	for structure, see also Vol. II/7 (3, 408) *) uncertainties are 3σ	2.6.3, 125 2.8, 36 3, 82
		$v_{T_1}=1$	$A=10090.01(2)$ *) $B=3815.21(1)$ $C=2883.29(1)$	$D_J=0.758(20)$ *) $D_{JK}=3.627(240)$ $D_K=3.814(230)$ $\delta_J=0.185(10)$ $\delta_K=0.012(60)$			$\nu_{T_1}=64(3)$ cm^{-1}, C—C torsion	
		$v_{T_2}=1$	$A=10133.86(5)$ *) $B=3811.06(5)$ $C=2871.95(5)$	$D_J=0.320\pm0.6$ *) $D_{JK}=3.093\pm0.3$ $D_K=3.709$ (assumed) $\delta_J=0.173\pm0.02$ $\delta_K=0.417$ (assumed)			$\nu_{T_2}=190.4$ cm^{-1}, CH_3 torsion	
		$v_D=1$	$A=10194.79(5)$ *) $B=3817.13(5)$ $C=2869.42(5)$	$D_J=0.507\pm0.9$ *) $D_{JK}=3.784\pm0.25$ $D_K=3.709$ (assumed) $\delta_J=0.206\pm0.02$ $\delta_K=0.417$ (assumed)			$\nu_D=270(30)$ cm^{-1}, CCC deformation	
		$v_\delta=1$	$A=10155.70(11)$ *) $B=3814.77(6)$ $C=2870.73(6)$	**)			**) assumed as in ground state $\nu_\delta=530(30)$ cm^{-1}, CCO deformation	
Cont.		$v_\tau=1$	$A=10106.56(16)$ *) $B=3812.34(9)$ $C=2878.10(9)$	**)			$\nu_\tau=547(20)$ cm^{-1}, O—H torsion	

Nr.	Molecule	v	A, B, C MHz	D, H kHz	Ref.	Ref. struct.	Varia, remarks	See also Tab., Nr.
260 *6-105 a* Cont.	$^{13}CH_3CH_2C(=O)OH$ cis	$v=0$	$A=10154.57$ (6) *) $B=3702.87$ (2) $C=2809.40$ (2)		75ST2		*) uncertainties are 3σ	
		$v_{T_1}=1$	$A=10088.18$ (8) $B=3700.48$ (3) $C=2817.27$ (3)					
	$CH_3{}^{13}CH_2C(=O)OH$ cis	$v=0$	$A=10038.44$ (5) *) $B=3801.46$ (2) $C=2856.51$ (2)					
		$v_{T_1}=1$	$A=9975.56$ (6) $B=3798.34$ (3) $C=2864.59$ (3)					
	$CH_3CH_2{}^{13}C(=O)OH$ cis	$v=0$	$A=10154.98$ (5) *) $B=3811.51$ (2) $C=2871.53$ (2)					
		$v_{T_1}=1$	$A=10085.50$ (5) $B=3808.83$ (2) $C=2879.62$ (2)					
	$CH_3CH_2C(=O){}^{18}OH$ cis	$v=0$	$A=9984.50$ (5) *) $B=3676.93$ (3) $C=2781.37$ (3)					
		$v_{T_1}=1$	$A=9916.31$ (5) $B=3675.60$ (2) $C=2789.33$ (2)					
	$CH_3CH_2C(={}^{18}O)OH$ cis	$v=0$	$A=9541.25$ (4) *) $B=3802.24$ (2) $C=2815.02$ (2)					
		$v_{T_1}=1$	$A=9484.55$ (4) $B=3798.99$ (2) $C=2822.51$ (2)					
Cont.	$CH_3CH_2C(=O)OD$ cis	$v=0$	$A=10155.13$ (3) *) $B=3666.38$ (2) $C=2788.54$ (2)					

Nr.	Molecule	v	A, B, C MHz	D, H kHz	Ref.	Ref. struct.	Varia, remarks	See also Tab., Nr.
260 *6–105 a* Cont.	$CH_3CH_2C(=O)OD$ cis	$v_{T_1}=1$	$A=10087.48\,(3)$ $B=3664.25\,(2)$ $C=2796.42\,(2)$		75ST2			
		$v_{T_2}=1$	$A=10132.40\,(6)$ $B=3660.12\,(3)$ $C=2785.52\,(3)$					
		$v_D=1$	$A=10195.49\,(6)$ $B=3665.75\,(3)$ $C=2783.25\,(3)$					
	$CH_3CH_2C(=O)^{18}OD$ cis	$v=0$	$A=9985.03\,(5)$ *) $B=3541.01\,(3)$ $C=2703.08\,(3)$				*) uncertainties are 3 σ	
		$v_{T_1}=1$	$A=9915.10\,(6)$ $B=3540.02\,(3)$ $C=2710.81\,(3)$					
	$CH_3CH_2C(={}^{18}O)OD$ cis	$v=0$	$A=9538.79\,(4)$ *) $B=3653.78\,(2)$ $C=2732.76\,(2)$					
		$v_{T_1}=1$	$A=9479.66\,(3)$ $B=3651.14\,(2)$ $C=2740.05\,(2)$					
	$CH_3CD_2C(=O)OH$ cis	$v=0$	$A=9147.21\,(4)$ *) $B=3745.20\,(3)$ $C=2796.45\,(3)$					
		$v_{T_1}=1$	$A=9100.64\,(4)$ $B=3740.20\,(2)$ $C=2804.57\,(2)$					
		$v_{T_2}=1$	$A=9133.86\,(9)$ $B=3738.54\,(5)$ $C=2793.22\,(5)$					
Cont.		$v_D=1$	$A=9172.13\,(7)$ $B=3743.89\,(4)$ $C=2792.15\,(4)$					

Nr.	Molecule	v	A, B, C MHz	D, H kHz	Ref.	Ref. struct.	Varia, remarks	See also Tab., Nr.
260	$CD_2HCH_2C(=O)OH$	$v=0$	$A=9840.97\ (5)$ *)		75ST2		*) uncertainties are 3σ	
6-105 a			$B=3469.60\ (2)$					
Cont.	asymmetric, cis		$C=2671.61\ (2)$					
		$v_{T_1}=1$	$A=9781.73\ (4)$					
			$B=3467.00\ (2)$					
			$C=2678.66\ (2)$					
	$CD_2HCH_2C(=O)OH$	$v=0$	$A=9683.94\ (5)$ *)					
			$B=3547.00\ (2)$					
	symmetric, cis		$C=2729.03\ (2)$					
		$v_{T_1}=1$	$A=9621.31\ (5)$					
			$B=3545.06\ (2)$					
			$C=2736.35\ (2)$					
	$CH_3CD_2C(=O)OD$	$v=0$	$A=9146.67\ (3)$ *)					
			$B=3597.15\ (1)$					
			$C=2713.17\ (1)$					
		$v_{T_1}=1$	$A=9098.22\ (3)$					
			$B=3592.81\ (1)$					
			$C=2721.01\ (1)$					
		$v_{T_2}=1$	$A=9132.28\ (8)$					
			$B=3591.05\ (4)$					
			$C=2710.17\ (4)$					
		$v_D=1$	$A=9172.89\ (6)$					
			$B=3596.14\ (3)$					
			$C=2709.08\ (3)$					
	$CD_3CH_2C(=O)OH$	$v=0$	$A=9600.70\ (6)$ *)					
			$B=3357.97\ (3)$					
			$C=2609.94\ (3)$					
		$v_{T_1}=1$	$A=9543.16\ (5)$					
			$B=3355.69\ (2)$					
			$C=2616.64\ (2)$					
		$v_{T_2}=1$	$A=9578.74\ (6)$					
			$B=3353.62\ (3)$					
			$C=2608.95\ (3)$					
		$v_D=1$	$A=9634.42\ (6)$					
			$B=3357.34\ (4)$					
			$C=2604.99\ (4)$					

Nr.	Molecule	v	A, B, C MHz	D, H kHz	Ref.	Ref. struct.	Varia, remarks	See also Tab., Nr.
260a	$CH_3CH_2C(=O)SH$ cis	$v=0$	$A=8729.128\,(30)$ $B=2389.545\,(8)$ $C=1922.584\,(9)$	$d_J=-0.93\,(15)$ $d_{JK}=8.3\,(64)$ $d_K=48\,(20)$ $d_{WJ}=0.58\,(6)\cdot10^{-3}$ $d_{WK}=-5.2\,(31)\cdot10^{-3}$	81TAK1			2.6.3, 125a
261	$CH_3CH_2CH_2CH_2{}^{35}Cl$ all trans	$v=0$	$B+C=2586.6\,(32)$ *)		77ST		for structure, see also Vol. II/7 (3, 542) *) LRMW	
	$CH_3CH_2CH_2CH_2{}^{37}Cl$ all trans	$v=0$	$B+C=2528.0\,(29)$ *)					
	$CH_3CH_2CH_2CH_2{}^{35}Cl$ trans-trans-gauche	$v=0$	$B+C=2920.5\,(24)$ *)					
262	$CH_3CH_2CH_2CH_2{}^{79}Br$ all trans	$v=0$	$B+C=1799.8\,(9)$ *)		77ST		for structure, see also Vol. II/7 (3, 541) *) LRMW	
	$CH_3CH_2CH_2CH_2{}^{81}Br$ all trans	$v=0$	$B+C=1785.7\,(13)$ *)					
	$CH_3CH_2CH_2CH_2{}^{81}Br$ gauche-trans-trans	$v=0$	$B+C=2307\,(24)$ *)					
	$CH_3CH_2CH_2CH_2{}^{79}Br$ trans-trans-gauche	$v=0$	$B+C=1994.8\,(12)$ *)					
	$CH_3CH_2CH_2CH_2{}^{81}Br$ trans-trans-gauche	$v=0$	$B+C=1978.4\,(17)$ *)					
263	$CH_3CH_2CH_2CH_2I$ all trans	$v=0$	$B+C=1445.3\,(4)$ *)		77ST		*) LRMW	
	$CH_3CH_2CH_2CH_2I$ gauche-trans-trans	$v=0$	$B+C=1868.4\,(30)$ *)					
	$CH_3CH_2CH_2CH_2I$ trans-trans-gauche	$v=0$	$B+C=1583.5\,(4)$ *)					
264	$CH_3CH_2CH_2C(=O)F$ syn-anti	$v=0$	$B+C=3359\,(5)$ *)		78THO		*) LRMW syn-anti: τ_1 (O=CCC) $=0°$, τ_2 (CCCC) $=180°$ syn-gauche: $\tau_1=0°$, $\tau_2\simeq70°$ skew-anti: $\tau_1=120°$, $\tau_2=180°$ this conformer is less stable	
Cont.	$CH_3CH_2CH_2C(=O)F$ syn-gauche	$v=0$	$B+C=4114\,(7)$ *)					

Nr.	Molecule	v	A, B, C MHz	D, H kHz	Ref.	Ref. struct.	Varia, remarks	See also Tab., Nr.
264 Cont.	$CH_3CH_2CH_2C(=O)F$ skew-anti	$v=0$	$B+C=3523$ (6) *)		78THO			
265	$CH_3CH_2CH_2C(=O){}^{35}Cl$ syn-anti	$v=0$	$B+C=2446$ (3) *)		78THO		*) LRMW for definitions see Nr. 264	
	$CH_3CH_2CH_2C(=O){}^{37}Cl$ syn-anti	$v=0$	$B+C=2399$ (3) *)					
	$CH_3CH_2CH_2C(=O){}^{35}Cl$ syn-gauche	$v=0$	$B+C=2804$ (3) *)					
	$CH_3CH_2CH_2C(=O){}^{37}Cl$ syn-gauche	$v=0$	$B+C=2746$ (5) *)					
266	$CH_3CH_2CH_2C(=O){}^{79}Br$ syn-anti	$v=0$	$B+C=1720$ (3) *)		78THO		*) LRMW for definitions see Nr. 264	
	$CH_3CH_2CH_2C(=O){}^{81}Br$ syn-anti	$v=0$	$B+C=1704$ (3) *)					
	$CH_3CH_2CH_2C(=O){}^{79}Br$ syn-gauche	$v=0$	$B+C=1899$ (3) *)					
	$CH_3CH_2CH_2C(=O){}^{81}Br$ syn-gauche	$v=0$	$B+C=1883$ (3) *)					
267	${}^{35}ClCH_2CH(CH_3)—CH_3$*)	$v=0$	$A=7527.05$ (100) $B=2146.21$ (2) $C=1793.59$ (2)	$\Delta_{JK}=$ 4.15 (30) $\delta_J=-0.080$ (5)	81BRO	81BRO	*) asymmetrical form	3, 83
	${}^{37}ClCH_2CH(CH_3)—CH_3$	$v=0$	$A=7524.40$ (200) $B=2091.73$ (5) $C=1755.54$ (5)	$\Delta_{JK}=2.5$ (10) $\delta_J=0.20$ (15)				

Nr.	Molecule	v	A, B, C MHz	D, H kHz	Ref.	Ref. struct.	Varia, remarks / See also Tab., Nr.
268	$(CH_3)_2CHCH_2CH_2{}^{35}Cl$ $(CH_3)_2CHCH_2CH_2{}^{37}Cl$	$v=0$ $v=0$	$B+C=2133.1$ (8) *) $B+C=2081.0$ (6) *)		77ST		*) LRMW
269	$(CH_3)_2CHCH_2CH_2{}^{79}Br$ $(CH_3)_2CHCH_2CH_2{}^{81}Br$	$v=0$ $v=0$	$B+C=1438.4$ (1) *) $B+C=1425.4$ (1) *)		77ST		*) LRMW
270	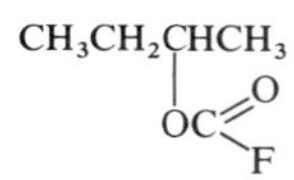	$v=0$	 $B+C=2530$ (4) *) $B+C=2399$ (3) *) $B+C=2118$ (2) *) $B+C=2026$ (4) *)		76TR2		*) LRMW τ_1 $(O_1{=}C_2O_3C_4)=0°$; τ_2 $(C_2O_3C_4H_8)=+50°$; τ_3 $(O_3C_4C_5C_6)=180°$. $\tau_1=0°$; $\tau_2\simeq-50°$; $\tau_3=180°$ $\tau_1=0°$; $\tau_2\simeq-50°$; $\tau_3\simeq60°$ $\tau_1=0°$; $\tau_2\simeq-50°$; $\tau_3\simeq60°$
271	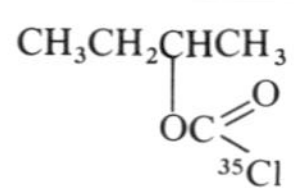	$v=0$	 $B+C=1867$ (2) *) $B+C=1807$ (2) *) $B+C=1571$ (3) *)		76TR2		*) LRMW for definitions, see Nr. 270 $\tau_1=0°$; $\tau_2\simeq50°$; $\tau_3=180°$ $\tau_1=0°$; $\tau_2\simeq-50°$; $\tau_3=180°$ $\tau_1=0°$; $\tau_2\simeq50°$ or $-50°$; $\tau_3\simeq60°$
272	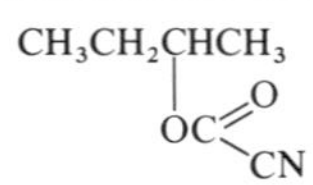		 $B+C=1877$ (4) *) $B+C=1815$ (2) *) $B+C=1587$ (2) *)		76TR2		*) LRMW for definitions, see Nr. 270 $\tau_1=0°$; $\tau_2\simeq50°$; $\tau_3=180°$ $\tau_1=0°$; $\tau_2\simeq-50°$; $\tau_3=180°$ $\tau_1=0°$; $\tau_2\simeq50°$ or $-50°$; $\tau_3\simeq60°$
273	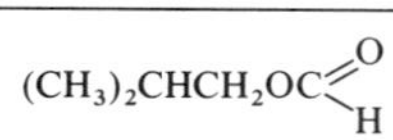 EG	$v=0$	$B+C=2534$ (5) *)		78TR3		*) LRMW for definitions, see Nr. 274
274	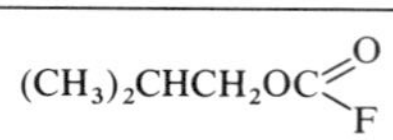 EG EA CG IG IA		$B+C=1852$ (2) *) $B+C=2060$ (5) *) $B+C=2113$ (4) *) $B+C=1979$ (2) *) $B+C=2191$ (5) *)		78TR3		*) LRMW, $v=0$. E: τ_1 (O=COC)$=0°$; τ_2 (COCC)$=180°$ C: $\tau_1\simeq0°$; $\tau_2\simeq$ 90° I: $\tau_1\neq0°$; $\tau_2\simeq240°$ G: τ_3 (COCH)$\approx120°$ A: $\tau_3=180°$

Nr.	Molecule	v	A, B, C MHz	D, H kHz	Ref.	Ref. struct.	Varia, remarks	See also Tab., Nr.
275	$(CH_3)_2CHCH_2OC(=O)^{35}Cl$ IG		$B+C=1497(2)$ *)		78TR3		*) LRMW, $v=0$, for definitions, see Nr. 274	
	IA		$B+C=1687(5)$ *)					
	$(CH_3)_2CHCH_2OC(=O)^{37}Cl$ IG		$B+C=1467(2)$ *)					
	IA		$B+C=1654(3)$ *)					
276	$(CH_3)_2CHCH_2OC(=O)CN$ IG		$B+C=1516(1)$ *)		78TR3		*) LRMW, $v=0$, for definitions, see Nr. 274	
	IA		$B+C=1708(1)$ *)					
277	$CH_3(CH_2)_4{}^{79}Br$ all trans	$v=0$	$B+C=1091(2)$ *)		77ST		*) LRMW	
	$CH_3(CH_2)_4{}^{81}Br$ all trans	$v=0$	$B+C=1082(2)$ *)					
	$CH_3(CH_2)_4{}^{79}Br$ trans···trans-gauche	$v=0$	$B+C=1254.6$ *)					
	$CH_3(CH_2)_4{}^{81}Br$ trans···trans-gauche	$v=0$	$B+C=1243.1$ *)					
278	$CH_3(CH_2)_4I$ all trans	$v=0$	$B+C=879.3(3)$ *)		77ST		*) LRMW	
	$CH_3(CH_2)_4I$ gauche-trans···trans	$v=0$	$B+C=1189.0(16)$ *)					
	$CH_3(CH_2)_4I$ trans···trans-gauche	$v=0$	$B+C=1004.56(14)$ *)					
	$CH_3(CH_2)_4I$ TGT	$v=0$	$B+C=971.5(23)$ *)					
279	$CH_3(CH_2)_5F$, TTGT	$v=0$	$B+C=1630.1(9)$ *)		77ST		*) LRMW	
280	$CH_3(CH_2)_5I$, all trans	$v=0$	$B+C=586.9(2)$ *)		77ST		*) LRMW	
	$CH_3(CH_2)_5I$, GT···T	$v=0$	$B+C=766.3(8)$ *)					
	$CH_3(CH_2)_5I$, T···TG	$v=0$	$B+C=628.6(4)$ *)					
	$CH_3(CH_2)_5I$, TTGT	$v=0$	$B+C=686.7(4)$ *)					

Nr.	Molecule	v	A, B, C MHz	D, H kHz	Ref.	Ref. struct.	Varia, remarks	See also Tab., Nr.
2.5.2.3 Aliphatic molecules with double and triple carbon-bonds								
280a	$CH_2{=}CD_2$	$v=0$	$A=97496.7\,(15)$ *) $B=25675.260\,(18)$ $C=20268.791\,(15)$	$\Delta_J = 32.73\,(56)$ *) $\Delta_{JK} = 191.9\,(28)$ $\Delta_K = 1224$ **) $\delta_J = 7.88\,(5)$ $\delta_K = 206\,(6)$	81HIR2	81HIR2	for structure, see also Vol. II/7 (3, 225) *) uncertainties are 2.5 σ **) fixed to the value calculated from the force field	2.6.3, 125b
	$CH_2{=}CHD$	$v=0$	$A=120093.521\,(70)$ *) $B=27470.737\,(26)$ $C=22297.745\,(14)$	$\Delta_J = 39.03\,(53)$ *) $\Delta_{JK} = 130.4\,(45)$ $\Delta_K = 2116.4\,(34)$ $\delta_J = 8.405\,(63)$ $\delta_K = 243.8\,(43)$				
	$CHD{=}CHD$ cis	$v=0$	$A=99667.262\,(89)$ *) $B=25417.208\,(41)$ $C=20199.073\,(24)$	$\Delta_J = 35.9\,(12)$ *) $\Delta_{JK} = 117\,(11)$ $\Delta_K = 1458.9\,(63)$ $\delta_J = 8.648\,(98)$ $\delta_K = 196.6\,(66)$				
	$CD_2{=}CD_2$	$v=0$	$A=2.4419\,(2)$ ***) $B=0.734923\,(8)$ $C=0.563531\,(7)$	$\Delta_J = 7.8\,(2)\cdot 10^{-7}$ ***) $\Delta_{JK} = 2.75\,(9)\cdot 10^{-6}$ $\Delta_K = 2\,(2)\cdot 10^{-5}$ $\delta_J = 1.93\,(7)\cdot 10^{-7}$ $\delta_K = 3.8\,(9)\cdot 10^{-6}$	81OHA		***) in cm^{-1} (diode laser results)	
		$v_{11}=1$	$A'-A'' = -0.01484\,(3)$ ***) $B'-B'' = -0.00153\,(2)$ $C'-C'' = -0.00143\,(2)$	$\Delta'_J-\Delta''_J = -4\,(2)\cdot 10^{-8}$ ***) $\Delta'_{JK}-\Delta''_{JK} = 4.4\,(1)\cdot 10^{-6}$ $\Delta'_K-\Delta''_K = 1.5\,(4)\cdot 10^{-6}$ $\delta'_J-\delta''_J = -3\,(2)\cdot 10^{-8}$ $\delta'_K-\delta''_K = -4.3\,(5)\cdot 10^{-6}$			$\nu_{11} = 2200.9783\,(5)\ cm^{-1}$	
281 *4–93* Cont.	$CH_2{=}CHF$	$v=0$	$\tilde{A}=64584.693\,(15)$ *) $\tilde{B}=10636.8837\,(26)$ $\tilde{C}=9118.0357\,(21)$	$\Delta_J = 8.4352\,(96)$ *) $\Delta_{JK} = -75.93\,(15)$ $\Delta_K = 1329.39\,(60)$ $\delta_J = 1.76591\,(85)$ $\delta_K = 35.654\,(56)$ $H_J = -0.9\,(17)\cdot 10^{-5}$ $H_{JK} = -0.7\,(28)\cdot 10^{-4}$ $H_{KJ} = -4.2\,(35)\cdot 10^{-3}$ $H_K = 3.5\,(36)\cdot 10^{-2}$	73GER1		for structure, see also Vol. II/7 (3, 212) *) uncertainties are 2 σ	

Nr.	Molecule	v	A, B, C MHz	D, H kHz	Ref.	Ref. struct.	Varia, remarks	See also Tab., Nr.
281 *4–93* Cont.	$CH_2{=}CHF$			$h_J = 6.3(13)\cdot10^{-6}$ $h_{JK} = -6.5(39)\cdot10^{-4}$ $h_K = 3.9(13)\cdot10^{-2}$	73GER1			
282 *6–109* *4–94*	$CH_2{=}CH^{35}Cl$	$v=0$	$A = 56839.704(144)$ *) $B = 6029.954(16)$ $C = 5445.235(16)$	$\tau_{aaaa} = -4868(43)$ *) $\tau_{bbbb} = -16.49(22)$ $\tau_{aabb} = 147.4(30)$ $\tau_{abab} = -48.40(43)$	71GER		for structure, see also Vol. II/7 (3, 205) *) confidence limits of 95 %	2.7.3, 118
	$CH_2{=}CH^{37}Cl$	$v=0$	$A = 56756.744(308)$ *) $B = 5903.562(34)$ $C = 5341.209(34)$	$\tau_{aaaa} = -4871(89)$ *) $\tau_{bbbb} = -15.72(43)$ $\tau_{aabb} = 145.4(55)$ $\tau_{abab} = -47.10(86)$				
283 *6–110* *4–95*	$CH_2{=}CH^{79}Br$	$v_{10}=1$	$A = 54588.29 \pm 150$ $B = 4160.01(5)$ $C = 3857.27(5)$	$T_1 = -7.1$ *) $T_2 = -8.5$ $T_3 = -4900$ $T_4 = 105$ $T_5 = -1150$	72MAR		for structure, see also Vol. II/7 (3, 201) $\nu_{10} = 348\ cm^{-1}$, in-plane vibration due to the C—Br deformation *) $T_1 = \tau_{xxxx}$ $T_2 = \tau_{yyyy}$ $T_3 = \tau_{zzzz}$ $T_4 = \tau_1$ $T_5 = \tau'_2$	2.7.3, 152
	$CH_2{=}CH^{81}Br$	$v_{10}=1$	$A = 54339.56$ $B = 4135.02$ $C = 3836.92$	$T_1 = -5.3$ *) $T_2 = -2$ $T_3 = -4830$ $T_4 = 94$ $T_5 = -1870$				
	$CD_2{=}CD^{79}Br$	$v=0$	$A = 36452.07$ $B = 3676.999$ $C = 3337.155$	$T_1 = -5.82$ *) $T_2 = -3.47$ $T_3 = -1391$ $T_4 = 28.1$ $T_5 = -409$	73MAR			
	$CD_2{=}CD^{81}Br$	$v=0$	$A = 36443.511$ $B = 3653.993$ $C = 3318.154$	$T_1 = -5.88$ *) $T_2 = -3.75$ $T_3 = -1358$ $T_4 = 27.5$ $T_5 = -457$				

Nr.	Molecule	v	A, B, C MHz	D, H kHz	Ref.	Ref. struct.	Varia, remarks	See also Tab., Nr.
284 *4-97*	$CH_2{=}CHCN$	$v=0$	$\tilde{A}=49850.700\,(13)$ $\tilde{B}=4971.2141\,(8)$ $\tilde{C}=4513.8289\,(8)$	$\Delta_J = 2.2448\,(21)$ $\Delta_{JK} = -85.442\,(18)$ $\Delta_K = 2718.3\,(21)$ $\delta_J = 0.45716\,(37)$ $\delta_K = 24.575\,(97)$ $H_J = (4.73 \pm 2.0) \cdot 10^{-6}$ $H_{JK} = (2.30 \pm 1.7) \cdot 10^{-4}$ $H_{KJ} = (-9.03 \pm 0.60) \cdot 10^{-3}$ $H_K = (4.51 \pm 0.80) \cdot 10^{-1}$ $h_J = (3.38 \pm 0.83) \cdot 10^{-6}$ $h_{JK} = (-3.25 \pm 3.25) \cdot 10^{-4}$ $h_K = (7.49 \pm 2.2) \cdot 10^{-2}$	73GER2		for structure, see also Vol. II/7 (3, 363)	
285 *6-112*	$CH_2{=}CHNC$	$v=0$	$\tilde{A}=51479.4578\,(30)$ $\tilde{B}=5386.64856\,(25)$ $\tilde{C}=4868.94021\,(25)$	$\Delta_J = 2.47615\,(63)$ $\Delta_{JK} = -100.0082\,(69)$ $\Delta_K = 3421.08\,(40)$ $\delta_J = 0.549200\,(89)$ $\delta_K = 27.203\,(20)$ $H_J = 0.00860\,(55) \cdot 10^{-3}$ $H_{JK} = -0.444\,(68) \cdot 10^{-3}$ $H_{KJ} = 4.08\,(22) \cdot 10^{-3}$ $H_K = 328.6\,(118) \cdot 10^{-3}$ $h_J = 0.00329\,(23) \cdot 10^{-3}$ $h_{JK} = 0.326\,(98) \cdot 10^{-3}$ $h_K = 50.6\,(75) \cdot 10^{-3}$	75YAM2		for structure, see also Vol. II/7 (3, 363)	
286 *6-113* *4-98*	$CH_2{=}CHNO_2$	$v=0$	$A=11811.792\,(23)$ $B=4680.825\,(16)$ $C=3353.790\,(13)$	$\tau_{aaaa} = -28.6\,(164)$ $\tau_{bbbb} = -7.8\,(19)$ $\tau_{aabb} = -9.4\,(104)$ $\tau_{abab} = -6.9\,(38)$	73NOS2	75NOS	for structure, see also Vol. II/7 (3, 222)	3, 84
		$v_{18}=1$	$A=11737.286\,(67)$ $B=4679.236\,(46)$ $C=3363.387\,(37)$	$\tau_{aaaa} = -61.7\,(451)$ $\tau_{bbbb} = -5.3\,(52)$ $\tau_{aabb} = -8.7\,(284)$ $\tau_{abab} = -14.5\,(101)$			v_{18}, NO_2 torsion	
Cont.								

Nr.	Molecule	v	A, B, C MHz	D, H kHz	Ref.	Ref. struct.	Varia, remarks	See also Tab., Nr.
286	$CH_2{=}CHNO_2$	$v_{18}=2$	$A=11667.197(55)$	$\tau_{aaaa}=-48.1(349)$	73NOS2			
6-113			$B=4677.517(36)$	$\tau_{bbbb}=-7.3(40)$				
4-98			$C=3372.706(29)$	$\tau_{aabb}=-5.1(216)$				
Cont.				$\tau_{abab}=-6.6(75)$				
		$v_{18}=3$	$A=11606.001(313)$	$\tau_{aaaa}=-46.1$ *)			*) assumed values	
			$B=4675.503(115)$	$\tau_{bbbb}=-6.8$ *)				
			$C=3381.608(94)$	$\tau_{aabb}=-2.0$ *)				
				$\tau_{abab}=-5.0$ *)				
		$v_{18}=4$	$A=11540.949(287)$	$\tau_{aaaa}=-46.1$ *)				
			$B=4673.265(34)$	$\tau_{bbbb}=-6.8$ *)				
			$C=3390.262(33)$	$\tau_{aabb}=-2.0$ *)			**) values of $v_{18}=3$ assumed	
				$\tau_{abab}=-5.0$ *)				
		$v_{13}=1$	$A=11835.188(160)$	**)			ν_{13}, in-plane skeletal	
			$B=4576.972(34)$				bending mode	
			$C=3346.052(33)$					
		$v_{13}=1$, $v_{18}=1$	$A=11758.814(409)$	**)				
			$B=4675.330(12)$					
			$C=3356.393(15)$					
		$v_{13}=1$, $v_{18}=2$	$A=11693.580(280)$	**)				
			$B=4673.790(8)$					
			$C=3365.276(10)$					
	$CH_2{=}CDNO_2$	$v=0$	$A=11042.801(19)$	$\tau_{aaaa}=-37.3(116)$				
			$B=4658.508(13)$	$\tau_{bbbb}=-4.8(18)$				
			$C=3278.062(10)$	$\tau_{aabb}=-9.1(99)$				
				$\tau_{abab}=-12.8(35)$				
		$v_{18}=1$	$A=10984.825(110)$	$\tau_{aaaa}=-32.3$ *)			*) assumed values	
			$B=4653.926(30)$	$\tau_{bbbb}=-6.3$ *)				
			$C=3287.422(30)$	$\tau_{aabb}=-3.9$ *)				
				$\tau_{abab}=-6.8$ *)				
		$v_{18}=2$	$A=10929.200(124)$	***)			***) values of $v_{18}=1$ assumed	
			$B=4649.097(34)$					
			$C=3296.536(34)$					
		$v_{18}=3$	$A=10874.712(180)$	***)				
			$B=4643.994(18)$					
Cont.			$C=3305.482(18)$					

Nr.	Molecule	v	A, B, C MHz	D, H kHz	Ref.	Ref. struct.	Varia, remarks	See also Tab., Nr.
286	$CH_2{=}CDNO_2$	$v_{18}=4$	$A=10819.28\,(122)$	***)	73NOS2		***) values of $v_{18}=1$ assumed	
6-113			$B=4638.594\,(122)$					
4-98			$C=3314.458\,(123)$					
Cont.		$v_{13}=1$	$A=11066.004\,(175)$	***)				
			$B=4654.889\,(26)$					
			$C=3270.840\,(25)$					
		$v_{13}=1,\ v_{18}=1$	$A=11005.907\,(346)$	***)				
			$B=4650.294\,(51)$					
			$C=3280.559\,(57)$					
		$v_{13}=1,\ v_{18}=2$	$A=10951.702\,(427)$	***)				
			$B=4645.416\,(63)$					
			$C=3290.039\,(70)$					
	$CD_2{=}CHNO_2$	$v=0$	$A=11420.128\,(10)$	$\tau_{aaaa}=-21.6\,(79)$				
			$B=4221.516\,(6)$	$\tau_{bbbb}=-5.7\,(7)$				
			$C=3083.694\,(4)$	$\tau_{aabb}=-4.9\,(44)$				
				$\tau_{abab}=-7.5\,(16)$				
		$v_{18}=1$	$A=11350.500\,(20)$	$\tau_{aaaa}=-31.9\,(165)$				
			$B=4219.279\,(12)$	$\tau_{bbbb}=-5.6\,(15)$				
			$C=3092.336\,(9)$	$\tau_{aabb}=-3.0\,(91)$				
				$\tau_{abab}=-8.0\,(32)$				
		$v_{18}=2$	$A=11284.495\,(7)$	$\tau_{aaaa}=-29.4\,(52)$				
			$B=4216.860\,(4)$	$\tau_{bbbb}=-6.3\,(5)$				
			$C=3100.690\,(3)$	$\tau_{aabb}=-6.0\,(29)$				
				$\tau_{abab}=-7.8\,(10)$				
		$v_{18}=3$	$A=11221.791\,(32)$	$\tau_{aaaa}=-21.9\,(277)$				
			$B=4214.221\,(16)$	$\tau_{bbbb}=-7.1\,(20)$				
			$C=3108.781\,(12)$	$\tau_{aabb}=-12.2\,(119)$				
				$\tau_{abab}=-3.2\,(42)$				
		$v_{18}=4$	$A=11162.204\,(212)$	$\tau_{aaaa}=-20.7$ *)			*) assumed values	
			$B=4211.372\,(61)$	$\tau_{bbbb}=-56.0\,(1)$				
			$C=3116.613\,(61)$	$\tau_{aabb}=-4.0$ *)				
				$\tau_{abab}=-6.6$ *)				
		$v_{13}=1$	$A=11428.413\,(147)$	$\tau_{aaaa}=-20.7$ *)				
			$B=4217.739\,(21)$	$\tau_{bbbb}=-4.2\,(2)$				
			$C=3077.153\,(21)$	$\tau_{aabb}=-4.0$ *)				
Cont.				$\tau_{abab}=-6.6$ *)				

Nr.	Molecule	v	A, B, C MHz	D, H kHz	Ref.	Ref. struct.	Varia, remarks	See also Tab., Nr.
286	$CD_2{=}CHNO_2$	$v_{13}=1, v_{18}=1$	$A=11538.704\,(122)$	$\tau_{aaaa}= -20.7$ *)	73NOS2		*) assumed values	
6–113			$B=\ 4215.318\,(38)$	$\tau_{bbbb}= -6.2$ *)				
4–98			$C=\ 3086.383\,(38)$	$\tau_{aabb}= -4.0$ *)				
Cont.				$\tau_{abab}= -6.6$ *)				
	$CHD{=}CHNO_2$	$v=0$	$A=11502.276\,(377)$	$\Delta_K=20.24$ *)	75NOS			
	cis		$B=\ 4526.237\,(23)$	$\delta_J=\ 0.16$ *)				
			$C=\ 3249.682\,(20)$	$\delta_K=\ 4.42$ *)				
		$v_{18}=1$	$A=11429.282\,(7)$	$\tau_{aaaa}= -32.3$ *)	73NOS2			
			$B=\ 4523.798\,(2)$	$\tau_{bbbb}= -6.3$ *)				
			$C=\ 3259.196\,(2)$	$\tau_{aabb}= -3.9$ *)				
				$\tau_{abab}= -6.8$ *)				
		$v_{18}=2$	$A=11360.229\,(7)$	$^{+}$)			$^{+}$) values of $v_{18}=1$ assumed	
			$B=\ 4521.093\,(2)$					
			$C=\ 3268.356\,(2)$					
		$v_{18}=3$	$A=11294.946\,(796)$	$^{+}$)				
			$B=\ 4518.174\,(25)$					
			$C=\ 3277.252\,(20)$					
		$v_{18}=4$	$A=11231.89\,(102)$	$^{+}$)				
			$B=\ 4515.003\,(33)$					
			$C=\ 3285.946\,(26)$					
		$v_{13}=1$	$A=11523.023\,(513)$	$^{+}$)				
			$B=\ 4521.817\,(12)$					
			$C=\ 3242.027\,(12)$					
		$v_{13}=1, v_{18}=1$	$A=11446.261$					
			$B=\ 4519.262$					
			$C=\ 3252.171$					
	$CHD{=}CHNO_2$	$v=0$	$A=11752.809\,(211)$	$\Delta_K=20.24$ *)	75NOS			
	trans		$B=\ 4347.079\,(14)$	$\delta_J=\ 0.16$ *)				
			$C=\ 3174.631\,(11)$	$\delta_K=\ 4.42$ *)				
		$v_{18}=1$	$A=11680.380\,(22)$	$\tau_{aaaa}= -32.3$ *)	73NOS2			
			$B=\ 4345.808\,(4)$	$\tau_{bbbb}= -6.3$ *)				
			$C=\ 3183.471\,(4)$	$\tau_{aabb}= -3.9$ *)				
				$\tau_{abab}= -6.8$ *)				
		$v_{18}=2$	$A=11612.181\,(33)$	$^{+}$)				
			$B=\ 4344.331\,(7)$					
Cont.			$C=\ 3191.880\,(7)$					

Nr.	Molecule	v	A, B, C MHz	D, H kHz	Ref.	Ref. struct.	Varia, remarks	See also Tab., Nr.
286	$CHD{=}CHNO_2$	$v_{18}=3$	$A=11549.120(475)$	+)	73NOS2		+) values of $v_{18}=1$ assumed	
6-113	trans		$B=4342.623(9)$					
4-98			$C=3199.970(9)$					
Cont.		$v_{18}=4$	$A=11489.029(735)$	+)				
			$B=4340.713(14)$					
			$C=3207.766(14)$					
		$v_{13}=1$	$A=11781.89(586)$	+)				
			$B=4343.998(81)$					
			$C=3168.052(81)$					
	$CH_2{=}^{13}CHNO_2$	$v=0$	$A=11692.884(244)$	++)	75NOS		++) constants of $CHD{=}CHNO_2$ used	
			$B=4652.789(18)$					
			$C=3329.825(12)$					
	$^{13}CH_2{=}CHNO_2$	$v=0$	$A=11811.806(277)$	++)				
			$B=4524.804(18)$					
			$C=3272.860(14)$					
	$CH_2{=}CH^{15}NO_2$	$v=0$	$A=11812.880(342)$	++)				
			$B=4670.939(24)$					
			$C=3348.779(17)$					
	$CH_2{=}CHN^{18}O^{16}O$	$v=0$	$A=11039.855(247)$	++)				
	cis		$B=4657.864(17)$					
			$C=3277.041(12)$					
	$CH_2{=}CHN^{18}O^{16}O$	$v=0$	$A=11562.617(293)$	++)				
	trans		$B=4501.852(20)$					
			$C=3241.597(14)$					
	$CH_2{=}^{13}CDNO_2$	$v=0$	$A=10950.390(229)$	++)				
			$B=4630.685(17)$					
			$C=3256.143(9)$					
	$^{13}CH_2{=}CDNO_2$	$v=0$	$A=11041.280(239)$	++)				
			$B=4505.508(17)$					
			$C=3201.390(12)$					
	$CD_2{=}^{13}CHNO_2$	$v=0$	$A=11303.165(277)$	++)				
			$B=4202.558(17)$					
			$C=3065.057(16)$					
	$^{13}CD_2{=}CHNO_2$	$v=0$	$A=11420.845(185)$	++)				
			$B=4102.284(14)$					
			$C=3019.621(9)$					

Nr.	Molecule	v	A, B, C MHz	D, H kHz	Ref.	Ref. struct.	Varia, remarks	See also Tab., Nr.
287	$CH_2{=}CHNH_2$	$v=0$	$A''=56312.9\,(190)$ *) $B''=10034.756\,(58)$ $C''=8564.930\,(60)$	$\tau_1=142\,(7)$ *) $\tau_2=13.3\,(50)$ $\tau_3=4.75\,(106)\cdot 10^3$ $\tau_{bbbb}=-50.2\,(82)$ $\tau_{cccc}=-21.3\,(81)$	75LOV	75LOV	*) uncertainties are 2 σ	2.6.3, 126 2.7.3, 37
		$v_i=1$	$A''=55877.8\,(373)$ *) $B''=10029.65\,(17)$ $C''=8570.03\,(16)$	$\tau_1=119\,(26)$ *) $\tau_2=4\,(18)$ $\tau_3=1.6\,(18)\cdot 10^3$ $\tau_{bbbb}=-29\,(22)$ $\tau_{cccc}=-22\,(18)$			$\nu_i=65\,(25)$ cm^{-1}, inversion state	
288	$CH_2{=}CHN_3$ cis	$v=0$	$A=14698.23\,(60)$ $B=4088.44\,(10)$ $C=3195.64\,(10)$		77FOR2	77FOR2	planar	2.6,3, 127 3, 85
	$CH_2{=}CHN_3$ trans	$v=0$	$A=46761.19\,(90)$ $B=2751.40\,(10)$ $C=2598.83\,(10)$				$E_{trans}-E_{cis}=460\,(100)$ cal/mol	
289	$CH_2{=}CHNCO$ trans	$v=0$	$A=58900\,(2900)$ $B=2437.75\,(13)$ $C=2346.52\,(13)$	$\Delta_J<0.2$ $\Delta_{JK}=-5.2\,(8)$ $\delta_J<0.4$	77BOU2	78KIR, 77BOU3		2.6.3, 128 2.7.3, 38 3, 86
		$v_{18}=1$	$A=54601\,(85)$ $B=2442.552\,(2)$ $C=2357.718\,(2)$	$\Delta_J=0.299\,(15)$ $\Delta_{JK}=2.72\,(13)$	78KIR		$\nu_{18}=77\,(10)$ cm^{-1}, out-of-plane bend	
		$v_{18}=2$	$A=50023\,(163)$ $B=2447.978\,(4)$ $C=2368.054\,(4)$	$\Delta_J=0.29\,(3)$ $\Delta_{JK}=43.29\,(35)$				
		$v_{17}=1$	$A=70532\,(133)$ $B=2442.393\,(1)$ $C=2350.196\,(1)$	$\Delta_J=0.270\,(10)$ $\Delta_{JK}=-76.83\,(28)$			$\nu_{17}=200\,(20)$ cm^{-1}, in-plane bending	
		$v_{17}=1, v_{18}=1$	$A=55772\,(311)$ $B=2446.852\,(5)$ $C=2366.140\,(5)$	$\Delta_J=0.35\,(4)$ $\Delta_{JK}=-71.7\,(8)$				
	$CH_2{=}CHNCO$ cis	$v=0$	$A=19878\,(80)$ $B=3107.43\,(9)$ $C=2689.52\,(9)$		77BOU3			

Nr.	Molecule	v	A, B, C MHz	D, H kHz	Ref.	Ref. struct.	Varia, remarks	See also Tab., Nr.
290	$CH_2{=}CHOH$	$v=0$	$A=59660.513$ $B=10561.877$ $C=\ 8965.590$	$d_J\ =\ -34.848$ $d_{JK}=\ -3367.8$ $d_K\ =-15093$ $d_{WJ}=0.44681\cdot10^{-5}$ $d_{WK}=0.31554\cdot10^{-3}$	77KAU1, 76SAI	76SAI		2.6.3, 129 3, 87
	$CH_2{=}CHOD$	$v=0$	$A=52588.624$ $B=10320.813$ $C=\ 8620.886$	$d_J\ =\ -35.315$ $d_{JK}=\ -4167.9$ $d_K\ =-14050$ $d_{WJ}=0.46607\cdot10^{-5}$ $d_{WK}=0.41291\cdot10^{-3}$				
291 *6-114* *4-99a*	$CH_2{=}CHOCH_3$	$v=0$	$\tilde{A}=18225.008$ $\tilde{B}=\ 6388.988$ $\tilde{C}=\ 4875.886$	$d_J\ =-13.505$ $d_{JK}=-65.163$ $d_K\ =\ 4.761$ $d_{WJ}=\ 3.076\cdot10^{-6}$ $d_{WK}=\ 3.325\cdot10^{-6}$	78KAU1		for structure, see also Vol. II/7 (3, 405)	2.9.3, 24
292	$CH_2{=}CHOCH_2CH_3$	$v=0$	$A=16307.085\,(7)$ $B=\ 2816.9430\,(13)$ $C=\ 2478.0885\,(12)$	$\Delta_J\ =\ 0.4790\,(30)$ $\Delta_{JK}=-2.208\,(52)$ $\Delta_K\ =\ 37.03\,(14)$ $\delta_J\ =0.08218\,(38)$ $\delta_K\ =0.515\,(76)$	79OW		all heavy atoms coplanar	2.6.3, 130 2.8, 37 3, 88
		$v_{T_1}=1$	$A=16115.581\,(13)$ $B=\ 2814.995\,(2)$ $C=\ 2481.020\,(3)$	$\Delta_J\ =\ 0.445\,(12)$ $\Delta_{JK}=-2.30\,(10)$ $\Delta_K\ =\ 25.92\,(35)$ $\delta_J\ =0.0871\,(22)$ $\delta_K\ =0.4632$ (fixed)			$\nu_{T_1}=115\,(10)\ cm^{-1}$, $O{-}C_2H_5$ torsion	
		$v_{T_1}=2$	$A=15946.862\,(16)$ $B=\ 2813.614\,(4)$ $C=\ 2484.179\,(4)$	$\Delta_J\ =\ 0.499\,(13)$ $\Delta_{JK}=-2.44\,(11)$ $\Delta_K\ =\ 34.50\,(38)$ $\delta_J\ =0.0787\,(22)$ $\delta_K\ =0.4632$ (fixed)				
		$v_{T_1}=3$	$A=15772.706\,(58)$ $B=\ 2812.619\,(9)$ $C=\ 2487.935\,(8)$	$\Delta_J\ =\ 0.697\,(140)$ $\Delta_{JK}=-6.22\,(34)$ $\Delta_K\ =\ 36.87$ (fixed) $\delta_J\ =0.0770\,(41)$ $\delta_K\ =0.4632$ (fixed)				
Cont.								

Nr.	Molecule	v	A, B, C MHz	D, H kHz	Ref.	Ref. struct.	Varia, remarks	See also Tab., Nr.
292 Cont.	$CH_2{=}CHOCH_2CH_3$	$v_{T_2}=1$	$A=16313.757(5)$ $B=2811.5780(10)$ $C=2476.1551(11)$	$\Delta_J=0.4757(43)$ $\Delta_{JK}=-2.903(36)$ $\Delta_K=36.58(12)$ $\delta_J=0.0824(6)$ $\delta_K=0.4407$ (fixed)	79OW		$\nu_{T_2}=210(10)$ cm^{-1}, C—CH_3 torsion	
		$v_{T_3}=1$	$A=16292.923(14)$ $B=2810.200(3)$ $C=2477.211(3)$	$\Delta_J=0.428(11)$ $\Delta_{JK}=-1.62(10)$ $\Delta_K=28.55(26)$ $\delta_J=0.0687(20)$ $\delta_K=0.4407$ (fixed)			$\nu_{T_3}=263(15)$ cm^{-1}, C—O torsion	
		$v_\delta=1$	$A=16318.040(3)$ $B=2812.6116(7)$ $C=2476.9370(9)$	$\Delta_J=0.4863(30)$ $\Delta_{JK}=-2.387(25)$ $\Delta_K=27.39(10)$ $\delta_J=0.0851(5)$ $\delta_K=0.4632$ (fixed)			$\nu_\delta=348(20)$ cm^{-1}, COC bend	
293 *6-116*	$CH_2{=}CHOCH{=}CH_2$ cis, trans	$v=0$	$A=15669.73(14)$ *) $B=3220.98(3)$ $C=2790.20(2)$	$\tau_{aaaa}=-1338.9(16)$ *) $\tau_{bbbb}=-10.6(2)$ $\tau_{cccc}=-5.9(2)$ $\tau_{aabb}=164.6(22)$ $\tau_{bbcc}=-5.7(1)$	78HIR1		*) uncertainties are 2.5 σ analysis of the doublet splittings: $H_{\text{V-rot}}=G_a\hat{p}_v\hat{J}_a+(\alpha_v\hat{J}_a^2+\beta_v\hat{J}_a^4+\gamma_v\hat{J}_a^2\hat{J}^2)$ where the term in parentheses apply only for upper levels $G_a=2.764(37)$; $\alpha_v=-0.391(39)$ $\beta_v=-0.0067(11)$; $\gamma_v=0.00065(7)$ MHz tunneling doubling: $\Delta_v=27.28(14)$ MHz	
294	$CH_2{=}CHSH$ syn	$v=0$	$A=49815.28(6)$ *) $B=5835.716(14)$ $C=5222.081(11)$	$D_J=2.85(17)$ *) $D_{JK}=-33.22(208)$ $\delta_J=0.425(65)$	79TAN2	79TAN2	planar *) uncertainties are 2.5 σ	2.6.3, 131
		$v_\tau=1$	$A=49394(900)$ *) $B=5822.41(10)$ $C=5228.21(10)$				$\nu_\tau=155(20)$ cm^{-1}, SH torsion	
Cont.		$v_\tau=2$	$A=48898(820)$ *) $B=5811.40(8)$ $C=5235.96(8)$					

Nr.	Molecule	v	A, B, C MHz	D, H kHz	Ref.	Ref. struct.	Varia, remarks	See also Tab., Nr.
294 Cont.	$CH_2{=}CHSH$ syn	$v_\delta=1$	$A=49261.2(36)$ *) $B=5832.21(22)$ $C=5213.15(45)$		79TAN2		$\nu_\delta=334(20)$ cm^{-1}, CCS bend *) uncertainties are 2.5 σ	
	$CH_2{=}CHSD$ syn	$v=0$	$A=42210.90(4)$ *) $B=5797.486(10)$ $C=5096.362(7)$	$D_J=2.77(16)$ *) $D_{JK}=-23.05(143)$ $\delta_J=0.464(38)$				
	$CH_2{=}CHSH$ anti	$v=0$	$A=49422.75(5)$ *) $B=5897.215(9)$ $C=5279.436(9)$	$D_J=3.12(11)$ *) $D_{JK}=-38.50(171)$ $\delta_J=0.498(51)$	79TAN3	79TAN3	$E_{anti}-E_{syn}=50(25)$ cm^{-1} the ground state is planar	
		$v_\tau=1$	$A=48889.25(6)$ *) $B=5891.532(15)$ $C=5298.692(13)$	$D_J=3.07(17)$ *) $D_{JK}=-36.28(213)$ $\delta_J=0.450(74)$			$\nu_\tau=74(30)$ cm^{-1}, SH torsion	
		$v_\tau=2$	$A=48592.28(13)$ *) $B=5880.469(43)$ $C=5305.592(45)$	$D_J=3.20(46)$ *) $D_{JK}=-33.65(538)$ $\delta_J=0.521(54)$			potential barrier of 19 cm^{-1} at the planar conformation	
		$v_\tau=3$	$A=48245.44(8)$ *) $B=5868.029(21)$ $C=5313.306(16)$	$D_J=3.13(22)$ *) $D_{JK}=-33.36(345)$ $\delta_J=0.472(90)$				
	$CH_2{=}CHSD$ anti	$v=0$	$A=44785.79(4)$ *) $B=5718.114(11)$ $C=5087.542(7)$	$D_J=2.78(11)$ *) $D_{JK}=-23.51(116)$ $\delta_J=0.415(43)$				
		$v_\tau=1$	$A=44207.27(5)$ *) $B=5709.692(14)$ $C=5113.101(10)$	$D_J=2.69(15)$ *) $D_{JK}=-22.37(170)$ $\delta_J=0.349(58)$				
295 *6-117* *4-100* Cont.	$CH_2{=}CF_2$	$v=0$	$A=0.367005867(33)$ *) $B=0.347873598(16)$ $C=0.178302671(14)$	$\Delta_J=1.6035(26)\cdot10^{-7}$ *) $\Delta_{JK}=5.099(91)\cdot10^{-8}$ $\Delta_K=2.5859(75)\cdot10^{-7}$ $\delta_J=6.687(10)\cdot10^{-8}$ $\delta_K=1.8898(45)\cdot10^{-7}$ $H_K=4.81(24)\cdot10^{-13}$	81LAF, 76BAS		for structure, see also Vol. II/7 (3, 185) *) diode laser results, all values in cm^{-1}, representation I^r used	2.6.3, 132 2.9.3, 23

Nr.	Molecule	v	A, B, C MHz	D, H kHz	Ref.	Ref. struct.	Varia, remarks	See also Tab., Nr.
295 *6-117* *4-100* Cont.	$CH_2{=}CF_2$	$v_4=1$	$A=0.36598861\ (27)$ *) $B=0.34740751\ (24)$ $C=0.17773941\ (22)$	$\Delta_J=1.5970\ (23)\cdot 10^{-7}$ *) $\Delta_{JK}=5.52\ (20)\cdot 10^{-8}$ $\Delta_K=2.596\ (32)\cdot 10^{-7}$ $\delta_J=6.8028\ (69)\cdot 10^{-8}$ $\delta_K=1.9879\ (66)\cdot 10^{-7}$ $H_J=-1.98\ (85)\cdot 10^{-13}$ $H_{JK}=-2.65\ (89)\cdot 10^{-12}$ $H_{KJ}=-6.4\ (23)\cdot 10^{-12}$ $H_K=164\ (29)\cdot 10^{-13}$	81LAF, 78DUX		$\nu_4=925.771992\ (44)\ cm^{-1}$ *) all values in cm^{-1}	
		$v_9=1$	$A=0.36782001\ (23)$ *) $B=0.34812118\ (21)$ $C=0.17832848\ (20)$	$\Delta_J=1.6640\ (24)\cdot 10^{-7}$ *) $\Delta_{JK}=3.029\ (86)\cdot 10^{-8}$ $\Delta_K=2.8663\ (87)\cdot 10^{-7}$ $\delta_J=6.7707\ (61)\cdot 10^{-8}$ $\delta_K=1.8370\ (28)\cdot 10^{-7}$ $H_J=12.43\ (75)\cdot 10^{-13}$ $H_K=4.81.10^{-13}$ (fixed)			$\nu_9=953.804922\ (39)\ cm^{-1}$	
296	$CH_2{=}C(CN)_2$	$v=0$	$A=6579.266\ (5)$ $B=2882.032\ (2)$ $C=2000.925\ (2)$	$\Delta_J=1.2506\ (47)$ $\Delta_{JK}=-8.5636\ (252)$ $\Delta_K=32.2825\ (490)$ $\delta_J=0.5568\ (21)$ $\delta_K=1.3311\ (335)$	78TAN1		harmonic force field calculated [79TAN1]	
		$v_7=1$	$A=6586.931\ (13)$ $B=2893.867\ (7)$ $C=2002.350\ (7)$	*)			*) centrifugal constants of the ground state used	
		$v_7=2$	$A=6595.035\ (13)$ $B=2905.625\ (10)$ $C=2003.745\ (10)$	*)			$\nu_7=157\ cm^{-1}$, in-plane C—$(CN)_2$ bend, sym. a_1	
		$v_7=3$	$A=6603.541\ (21)$ $B=2917.310\ (16)$ $C=2005.120\ (15)$	*)				
		$v_7=4$	$A=6612.401\ (31)$ $B=2928.906\ (18)$ $C=2006.459\ (16)$	*)				
Cont.		$v_7=5$	$A=6621.586\ (101)$ $B=2940.377\ (73)$ $C=2007.711\ (68)$	*)				

Nr.	Molecule	v	A, B, C MHz	D, H kHz	Ref.	Ref. struct.	Varia, remarks	See also Tab., Nr.
296 Cont.	CH_2=C(CN)$_2$	$v_{10}=1$	$A=6598.018(158)$ $B=2867.508(102)$ $C=2005.042(101)$	*)	78TAN1		$\nu_{10}=239\ cm^{-1}$, out-of-plane C—(CN)$_2$ bend, sym. b_1 *) centrifugal constants of the ground state used	
		$v_{10}=2$	$A=6618.225(24)$ $B=2856.458(10)$ $C=2009.329(7)$	*)				
		$v_{10}=1, v_7=1$	$A=6604.819(125)$ $B=2879.321(101)$ $C=2006.626(100)$	*)				
		$v_{12}=1$	$A=6569.007(6)$ $B=2889.897(4)$ $C=2004.944(4)$	*)			$\nu_{12}(a_2)=368\ cm^{-1}$, out-of-plane C—(CN)$_2$ bend *) centrifugal constants of the ground state used	
		$v_{18}=1$	$A=6579.179(115)$ $B=2904.766(78)$ $C=2003.396(74)$	*)			$\nu_{18}(b_2)=258\ cm^{-1}$, in-plane C—(CN)$_2$ bend	
		$v_{18}=1, v_7=1$	$A=6586.648(125)$ $B=2916.621(87)$ $C=2004.810(82)$	*)				
	CD_2=C(CN)$_2$	$v=0$	$A=5768.503(3)$ $B=2853.059(2)$ $C=1905.970(2)$	$\Delta_J=1.1947(42)$ $\Delta_{JK}=-6.8234(266)$ $\Delta_K=22.8766(262)$ $\delta_J=0.5424(26)$ $\delta_K=1.0501(364)$				
		$v_7=1$	$A=5773.940(6)$ $B=2864.383(3)$ $C=1906.941(3)$	$\Delta_J=1.2689(246)$ $\Delta_{JK}=-6.9210(644)$ $\Delta_K=23.0140(1950)$ $\delta_J=0.5473(45)$ $\delta_K=1.2642(601)$			$\nu_7=156\ cm^{-1}$.	
		$v_7=2$	$A=5779.664(17)$ $B=2875.606(10)$ $C=1907.889(10)$	**)			**) centrifugal constants of CD_2=C(CN)$_2$, $v=0$ used	
		$v_{10}=1$	$A=5783.721(72)$ $B=2812.548(43)$ $C=1910.672(32)$	**)			$\nu_{10}=239\ cm^{-1}$ Coriolis interaction between ν_{10} and ν_{18}: $\Delta E_{10,18}=5.9(1)\ cm^{-1}$ ${}^b\zeta_{10,18}=0.40(5)$	
Cont.								

Nr.	Molecule	v	A, B, C MHz	D, H kHz	Ref.	Ref. struct.	Varia, remarks	See also Tab., Nr.
296 Cont.	$CD_2{=}C(CN)_2$	$v_{18}=1$	$A=5767.635\,(84)$ $B=2901.800\,(44)$ $C=1909.071\,(33)$	**)	78TAN1			
		$v_{12}=1$	$A=5764.898\,(13)$ $B=2860.017\,(9)$ $C=1909.249\,(7)$	**)			$\nu_{12}=341\ cm^{-1}$	
297	$CH_2{=}C^{35}ClCN$	$v=0$	$A=6973.273\,(8)$ *) $B=3148.155\,(3)$ $C=2165.951\,(3)$	$\Delta_J = 1.07656\pm0.012$ *) $\Delta_{JK}=-3.48734\pm0.078$ $\Delta_K = 16.4726\pm0.13$ $\delta_J = 0.449964\pm0.0060$ $\delta_K = 2.26745\pm0.094$	75AVI3	75AVI3	*) uncertainties are 2 σ	2.7.3, 119
	$CH_2{=}C^{37}ClCN$	$v=0$	$A=6909.35\,(2)$ *) $B=3081.17\,(1)$ $C=2127.98\,(1)$	$\Delta_J = 0.9090\pm0.044$ *) $\Delta_{JK}=-4.039\pm0.19$ $\Delta_K = 18.54\pm0.65$ $\delta_J = 0.4359\pm0.015$ $\delta_K = 2.216\pm0.23$				
298 *4-135*	$CH_2{=}C(CH_3)_2$	$v=0$	$(A-C)/2$ $=2258.7216\,(30)$ *) $\kappa=0.6673037\,(8)$	$\Delta_{JK}=-16.70\,(7)$ *) $\Delta_K = 7.88\,(46)$ $\delta_J = -0.0653\,(30)$ $\delta_K = -19.385\,(48)$	75DEM		for structure, see also Vol. II/7 (3, 527) *) uncertainties are 2 σ, representation III^r used, for the AA species $\tilde{v}=1_1, 1_2$ first two excited torsional state	2.8, 41
		$\tilde{v}=1_1$	$A=9132.01\,(3)$ *) $B=8369.09\,(3)$ $C=4613.46\,(2)$					
		$\tilde{v}=1_2$	$A=9140.61\,(3)$ *) $B=8353.22\,(3)$ $C=4613.41\,(2)$					
299 *6-119* *4-119* Cont.	$CD_2{=}C{=}O$	$v=0$	$\tilde{A}=141503.7\,(34)$ $\tilde{B}= 9121.2370\,(37)$ $\tilde{C}= 8552.2946\,(37)$	$\Delta_J = 2.7085\,(30)$ $\Delta_{JK} = 321.389\,(38)$ $\Delta_K = 4322.8$ (fixed) $\delta_J = 0.21972\,(50)$ $\delta_K = 203.0\,(19)$ $H_{JK} = 2.97\,(15)\cdot10^{-3}$ $H_{KJ} = -0.14188\,(49)$	76NEM2	76NEM1	for structure, see also Vol. II/7 (3, 196)	2.6.3, 133 2.7.3, 8 2.9.3, 25

Nr.	Molecule	v	A, B, C MHz	D, H kHz	Ref.	Ref. struct.	Varia, remarks	See also Tab., Nr.
299 *6-119* *4-119* Cont.	CHD=C=O	$v=0$	$\tilde{A}=194352.2\,(229)$ $\tilde{B}=9647.543\,(15)$ $\tilde{C}=9174.166\,(15)$	$\Delta_J=3.1637\,(67)$ $\Delta_{JK}=326.970\,(77)$ $\Delta_K=13670.9$ (fixed) $\delta_J=0.22525\,(92)$ $\delta_K=239.2\,(74)$ $H_{JK}=2.79\,(40)\cdot 10^{-3}$ $H_{KJ}=-0.2779\,(14)$	76NEM2			
300 *6-119a* *4-120a*	$(CH_3)_2C=C=O$	$v=0$	$\tilde{A}=8267.3709\,(76)$ *) $\tilde{B}=3884.1616\,(32)$ $\tilde{C}=2728.8372\,(25)$	$\Delta_J=0.678\,(23)$ *) $\Delta_{JK}=12.91\,(23)$ $\Delta_K=-2.63\,(74)$ $\delta_J=0.2362\,(76)$ $\delta_K=7.967\,(94)$	76DEM3		for structure, see also Vol. II/7 (3, 512) *) uncertainties are 2 σ ν_r, in-plane vibration, probably the skeletal rock: $\nu_{21}\,(b_1)=178.2\ cm^{-1}$ Coriolis interaction between ν_r and $\nu_\tau\,(b_2)$ (methyl torsion): $\nu_r-\nu_\tau=5.8\ cm^{-1}$, $^a\zeta_{r\tau}=0.24$	
		$v_r=1$	$\tilde{A}=8372.858\,(30)$ *) $\tilde{B}=3898.7644\,(85)$ $\tilde{C}=2729.4949\,(77)$	$\Delta_J=0.447\,(72)$ *) $\Delta_{JK}=9.95\,(35)$ $\Delta_K=46.3\,(28)$ $\delta_J=0.350\,(12)$ $\delta_K=26.56\,(37)$				
		$v_r=2$	$A=8457.846\,(484)$ *) $B=3913.051\,(19)$ $C=2729.943\,(15)$					
		$v_\tau=1$	$A=8151.135\,(361)$ *) $B=3886.457\,(12)$ $C=2729.632\,(12)$					
301	$CH_2=C=S$	$v=0$	$A=286655\,(82)$ $B=5659.47596\,(72)$ $C=5544.51269\,(75)$	$D_J=1.08569\,(4)$ $D_{JK}=168.269\,(77)$ $D_K=23.5\cdot 10^3$ (fixed) $d_1=-25.46\,(68)\cdot 10^{-3}$ $d_2=-5.21\,(35)\cdot 10^{-3}$ $H_{JK}=0.716\,(20)\cdot 10^{-3}$ $H_{KJ}=-0.4087\,(73)$ $L_{KJ}=0.65\,(24)\cdot 10^{-3}$ $S_{KJ}=-0.0533\,(24)\cdot 10^{-3}$	80WIN2	79GEO		2.6.3, 134
Cont.		$v_9=1$	$B=5670.84\,(2)$ $C=5559.47\,(2)$	$D_{JK}=117\,(4)$	79GEO		$\nu_9\,(b_2)=320\,(50)\ cm^{-1}$	

Nr.	Molecule	v	A, B, C MHz	D, H kHz	Ref.	Ref. struct.	Varia, remarks	See also Tab., Nr.
301 Cont.	CH_2=C=S	$v_6=1$	$B=5659.93\,(3)$ $C=5565.15\,(3)$	$D_{JK}=202\,(4)$	79GEO		$\nu_6\,(b_1)=410\,(50)\,cm^{-1}$	
		$v_5=1$	$B+C=11216.70$				$\nu_5\,(b_1)=680\,(100)\,cm^{-1}$	
		$v_8=1$	$B+C=11207.44$				$\nu_8\,(b_2)=650\,(100)\,cm^{-1}$	
	CHD=C=S	$v=0$	$B=5323.082\,(45)$ $C=5176.999\,(45)$	$D_{JK}=109\,(8)$	79GEO, 79BA2			
	CD_2=C=S	$v=0$	$B=5039.7599\,(27)$ $C=4863.1018\,(27)$	$D_{JK}=120.5\,(7)$	79GEO			
	CH_2=C=^{34}S	$v=0$	$B=5531.555\,(18)$ $C=5421.675\,(18)$	$D_{JK}=159\,(3)$	79GEO, 79BA2			
	CH_2=^{13}C=S	$v=0$	$B=5642.620\,(19)$ $C=5528.357\,(19)$	$D_{JK}=171\,(3)$				
	$^{13}CH_2$=C=S	$v=0$	$B=5456.715\,(14)$ $C=5349.778\,(14)$	$D_{JK}=170$ (fixed)				
302	CH_3CH=C=S	$v=0$	$A=35528\,(7)$ *) $B=2708.362\,(21)$ $C=2552.969\,(22)$	$\Delta_J=0.39\,(14)$ $\Delta_{JK}=-46.55\,(30)$	80BAK3	80BAK3	*) $\Delta=$ -2.867 amu Å^2 assumed	2.6.3, 135 2.8, 42
		$v_\tau=1$	$A=36303\,(11)$ *) $B=2705.158\,(35)$ $C=2554.038\,(36)$	$\Delta_J=0.91\,(29)$ $\Delta_{JK}=-134.7\,(19)$			$\nu_\tau=140\,cm^{-1}$, CH_3 torsion	
		$v_\tau=2$	$A=31932\,(8)$ *) $B=2732.985\,(44)$ $C=2553.994\,(43)$	$\Delta_J=-0.86\,(41)$ $\Delta_{JK}=340.0\,(9)$				
		$v_b=1$	$A=34819\,(7)$ *) $B=2717.566\,(24)$ $C=2557.390\,(24)$	$\Delta_J=0.24\,(20)$ $\Delta_{JK}=-46.79\,(42)$			$\nu_b=157\,(5)\,cm^{-1}$, skeletal in-plane deformation	
		$v_b=2$	$A=34158\,(3)$ *) $B=2726.599\,(14)$ $C=2561.738\,(16)$	$\Delta_J=0.22\,(13)$ $\Delta_{JK}=-340.9\,(30)$				
		$v_\tau=1$, $v_b=1$	$A=35736.\,(7)$ *) $B=2713.904\,(28)$ $C=2558.964\,(31)$	$\Delta_J=0.82\,(28)$ $\Delta_{JK}=-134.6\,(14)$				
	CH_3CH=C=^{34}S	$v=0$	$A=35449\,(4)$ *) $B=2639.524\,(20)$ $C=2491.327\,(20)$	$\Delta_J=1.08\,(18)$ $\Delta_{JK}=-45.79\,(21)$				

Nr.	Molecule	v	A, B, C MHz	D, H kHz	Ref.	Ref. struct.	Varia, remarks	See also Tab., Nr.
303	$CH_2{=}C{=}^{80}Se$	$v=0$	$B=3867.911\,(12)$	$\Delta_J = 0.47\,(20)$	78BAK3	81BAK	$\Delta=0.05$ amu Å^2 assumed	
			$C=3813.849\,(12)$	$\Delta_{JK}=91.27\,(45)$				
	$CH_2{=}C{=}^{78}Se$	$v=0$	$B=3889.837\,(12)$	$\Delta_J = 0.46\,(20)$				
			$C=3835.162\,(12)$	$\Delta_{JK}=92.15\,(45)$				
	$^{13}CH_2{=}C{=}^{80}Se$	$v=0$	$B=3704.453\,(15)$	$\Delta_J = 0.64\,(27)$	81BAK			
			$C=3654.827\,(15)$	$\Delta_{JK}=84.36\,(43)$				
	$^{13}CH_2{=}C{=}^{78}Se$	$v=0$	$B=3726.097\,(14)$	$\Delta_J = 0.64$ (fixed)				
			$C=3675.904\,(14)$	$\Delta_{JK}=85.1\,(6)$				
	$CH_2{=}^{13}C{=}^{80}Se$	$v=0$	$B=3832.833\,(17)$	$\Delta_J = 0.73\,(30)$				
			$C=3779.735\,(17)$	$\Delta_{JK}=90.32\,(48)$				
	$CH_2{=}^{13}C{=}^{78}Se$	$v=0$	$B=3855.101\,(17)$	$\Delta_J = 0.60\,(31)$				
			$C=3801.388\,(17)$	$\Delta_{JK}=91.20\,(49)$				
304	$CF_2{=}CHF$	$v=0$	$A=10665.31$		73BHA	73BHA	for structure,	2.6.3, 136
6-180			$B=3872.36$				see also	3, 89
4-103			$C=2837.97$				Vol. II/7 (3, 174)	
		$v_9=1$	$A=10624.19$				$\nu_9=232$ cm^{-1}	
			$B=3870.65$					
			$C=2835.71$					
		$v_9=2$	$A=10582.23$					
			$B=3868.71$					
			$C=2833.42$					
		$v_{12}=1$	$A=10697.95$				$\nu_{12}=305$ cm^{-1}	
			$B=3879.12$					
			$C=2841.55$					
		$v_8=1$	$A=10668.49$				$\nu_8=485$ cm^{-1}	
			$B=3876.61$					
			$C=2837.10$					
		$v_{11}=1$	$A=10635.74$				$\nu_{11}=555$ cm^{-1}	
			$B=3866.40$					
			$C=2837.38$					
305	$CH_3CH{=}CHNO_2$	$v=0$	$A=10650\,(20)$		74FOR1			2.6.3, 137
	trans		$B=2028.56\,(3)$					2.8, 43
Cont.			$C=1722.66\,(3)$					

Nr.	Molecule	v	A, B, C MHz	D, H kHz	Ref.	Ref. struct.	Varia, remarks	See also Tab., Nr.
305 Cont.	$CH_3CH{=}CHNO_3$ trans	$v_\tau=1$	$A=10615$ $B=\ 2028.47$ $C=\ 1725.11$		74FOR1		v_τ, NO_2 torsion	
		$v_\tau=2$	$A=10570$ $B=\ 2028.31$ $C=\ 1727.32$					
		$v_\tau=3$	$A=10512$ $B=\ 2028.11$ $C=\ 1729.37$					
305a *6-126a* *4-112a*	$CH_3CCl{=}CH_2$						for structure, see also Vol. II/7 (3, 385)	2.6.3, 138
306	$CH_3CI{=}CH_2$	$v=0$	$A=9285.153\,(20)$ $B=2337.2198\,(14)$ $C=1887.5871\,(14)$	$\Delta_J\ =\ 0.300\,(52)$ $\Delta_{JK}=\ 2.22\,(11)$ $\Delta_K\ =11.2\,(21)$ $\delta_J\ =0.0800\,(15)$ $\delta_K\ =2.65\,(14)$	79GRO			2.7.3, 161 2.8, 44
		$v_{21}=1$	$A=9286.072\,(460)$ $B=2333.6572\,(33)$ $C=1885.6525\,(23)$				v_{21}, torsion	
		$v_{14}=1$	$A=9306.602\,(323)$ $B=2336.3726\,(20)$ $C=1882.4257\,(19)$				v_{14}, in-plane excited state	
307 *6-129c*	$CF_3CH{=}CH_2$	$v=0$	$A=5442.67\,(10)$ $B=2943.47\,(3)$ $C=2873.14\,(3)$		74SAI	74SAI	for structure, see also Vol. II/7 (3, 360) $v_\tau=88\,(25)\ cm^{-1}$, CF_3 torsion	2.6.3, 139 2.8, 45 3, 90
		$v_\tau=1$	$A=5436.56\,(10)$ $B=2942.68\,(3)$ $C=2881.26\,(3)$					
Cont.		$v_\tau=2$	$A=5430.25\,(10)$ $B=2941.77\,(3)$ $C=2889.53\,(3)$					

Nr.	Molecule	v	A, B, C MHz	D, H kHz	Ref.	Ref. struct.	Varia, remarks	See also Tab., Nr.
307 *6-129c* Cont.	$CF_3{}^{13}CH{=}CH_2$	$v=0$	$A=5122(170)$ $B=2924.11(5)$ $C=2847.82(5)$		74SAI			
	$CF_3CH{=}CHD$ cis	$v=0$	$A=5520(70)$ $B=2862.92(5)$ $C=2779.54(5)$					
	$CF_3CH{=}CHD$ trans	$v=0$	$A=5500(140)$ $B=2782.74(5)$ $C=2716.74(5)$					
	$CF_3CD{=}CH_2$	$v=0$	$A=5213(50)$ $B=2924.28(5)$ $C=2806.77(5)$					
308 *6-129a*	$CF_3CF{=}CF_2$	$v=0$	$A=2557.88(12)$ $B=1255.033(6)$ $C=987.082(7)$		73JAC	73JAC, 80CHR1		2.8, 46
		$v_\tau=1$	$A=2555.99(16)$ $B=1254.945(7)$ $C=988.377(8)$				v_τ, CF_3 torsion	
		$v_\tau=2$	$A=2554.02(17)$ $B=1254.852(8)$ $C=989.754(10)$					
		$v_\tau=3$	$A=2552.64(45)$ $B=1254.772(27)$ $C=991.154(35)$					
		$v_\tau=4$	$A=2552.27\ 42)$ $B=1254.794(23)$ $C=992.541(31)$					
		$v_x=1$	$A=2553.60(20)$ $B=1254.942(11)$ $C=988.067(14)$					
309	$CH_2{=}CHCH_2{}^{79}Br$ skew	$v=0$	$A=19247.56(27)$ $B=1975.339(1)$ $C=1914.761(1)$		76NI	76NI		2.7.3, 153 3, 91
Cont.		$v_\tau=1$	$A=19380.12(61)$ $B=1980.202(43)$ $C=1917.520(43)$		79SAS		$v_\tau=109\ cm^{-1}$, C—C torsion	

Nr.	Molecule	v	A, B, C MHz	D, H kHz	Ref.	Ref. struct.	Varia, remarks	See also Tab., Nr.
309 Cont.	$CH_2{=}CHCH_2{}^{79}Br$ skew	$v_\tau=2$	$A=19505.72\,(31)$ $B=1984.967\,(21)$ $C=1920.120\,(21)$		79SAS			
	$CH_2{=}CHCH_2{}^{81}Br$ skew	$v=0$	$A=19234.26\,(41)$ $B=1961.417\,(2)$ $C=1901.563\,(2)$		76NI			
		$v_\tau=1$	$A=19366.35\,(50)$ $B=1966.298\,(36)$ $C=1904.352\,(36)$		79SAS			
		$v_\tau=2$	$A=19491.94\,(58)$ $B=1971.055\,(34)$ $C=1906.940\,(34)$					
310	$CH_2{=}CHCH_2I$ skew	$v=0$	$A=17644.34\,(25)$ $B=1588.12\,(1)$ $C=1538.64\,(1)$		77SAS	77SAS		2.7.3, 162
		$v_\tau=1$	$A=17783.84\,(77)$ $B=1591.26\,(2)$ $C=1540.24\,(2)$		79TAK2		$v_\tau=114\,(34)\ cm^{-1}$, CH_2I torsion	
		$v_\tau=2$	$A=17915.85\,(138)$ $B=1594.49\,(3)$ $C=1541.85\,(3)$					
311 *6–135*	$CH_2{=}CHCH_2NH_2$ N cis, lone pair gauche	$v=0$	$A^+=16308.87\,(2)$ *) $B^+=5799.50\,(1)$ $C^+=4458.10\,(1)$ $A^-=16304.17\,(2)$ $B^-=5796.20\,(1)$ $C^-=4458.20\,(1)$		78BOT		*) uncertainties are 2σ the low barrier restricting torsion about CN bond removes the degeneracy. Effective Hamiltonian: $\langle +\|H\|+\rangle=A^+P_a^2+B^+P_b^2+C^+P_c^2$ $\langle -\|H\|-\rangle=A^-P_a^2+B^-P_b^2+C^-P_c^2+\Delta$ $\langle +\|H\|-\rangle=\mathrm{i}\,Q_aP_a+Q_bP_b$ where $\Delta=4574.98\,(9)$ MHz $Q_a=108.98\,(13)$; $Q_b=94.9\,(8)$ MHz (for N cis) [78BOT]	2.6.3, 140 2.7.3, 39 3, 92
	$CH_2{=}CHCH_2NH_2$ N gauche, lone pair trans	$v=0$	$A''=23957.304\,(56)$ $B''=4229.8602\,(115)$ $C''=4154.7833\,(111)$	$\tau_1=249.10\,(252)$ $\tau_2=26.48\,(92)$ $\tau_3=858.84\,(26675)$ $\tau_{aaaa}=-1952.57\,(19817)$ $\tau_{bbbb}=-7.12\,(103)$ $\tau_{cccc}=-14.31\,(103)$	78KAU3			
Cont.								

Nr.	Molecule	v	A, B, C MHz	D, H kHz	Ref.	Ref. struct.	Varia, remarks	See also Tab., Nr.
311 *6-135* Cont.	CH_2=$CHCH_2NH_2$ N gauche, lone pair trans	$v_{20}=1$, $v_{27}=2^+$)	$A^- = 24119.0 \pm 75.0$ $B^- = 4218.2 \pm 0.6$ $C^- = 4152.3 \pm 1.5$		80BOT		$^+$) IRMWDR, ν_{27}, C—C torsion; ν_{20}, out-of-plane vinyl motion. $\nu_0 = 919.531\ (3)\ cm^{-1}$ $^{++}$) IRMWDR, $\nu_{20} = 920.908\ (1)\ cm^{-1}$, out-of-plane vinyl motion	
	CH_2=$CHCH_2NH_2$ N gauche, lone pair gauche 1	$v_{20}=1$	$A = 24268\ (120)$ $^{++}$) $B = 4219\ (33)$ $C = 4106\ (33)$					
	CH_2=$CHCH_2ND_2$ N cis, lone pair trans	$v=0$	$A = 14433.8$ $B = 5414.8$ $C = 4213.3$		80FIS			
312	H_2C=$CHCH_2OC(=O)H$ EG $^+$)		$B+C = 3298\ (3)$		77TR3		$^+$) LRMW, for definitions see Nr. 312a	
312a	H_2C=$CHCH_2OC(=O)F$ EG ES CG IG		$B+C = 2406\ (1)$ *) $B+C = 2764\ (1)$ $B+C = 3075\ (5)$ $B+C = 2815\ (5)$		77TR3		*) LRMW, E: τ_1(O=COC)$=0°$; τ_2(COCC) $=180°$ C: $\tau_1 = 0°$, $\tau_2 \simeq 90°$ I: $\tau_1 \simeq 90°$, $\tau_2 \simeq 240°$ S: τ_3(OCCC)$=0°$; G: $\tau_3 = 120°$	
313	H_2C=$CHCH_2OC(=O)^{35}Cl$ EG CG CG′ IG H_2C=$CHCH_2OC(=O)^{37}Cl$ EG CG		$B+C = 1834.5\ (3)$ *) $B+C = 2015\ (1)$ $B+C = 2213\ (2)$ $B+C = 2081\ (2)$ $B+C = 1799.5\ (5)$ *) $B+C = 1988\ (1)$		77TR3		*) LRMW, for definitions see Nr. 312a	
314	H_2C=$CHCH_2OC(=O)CN$ EG CG CG′ IG		$B+C = 1838.1\ (3)$ *) $B+C = 2019.8\ (3)$ $B+C = 2188.0\ (7)$ $B+C = 2069\ (2)$		77TR3		*) LRMW, for definitions see Nr. 312a	
315	H_2C=$CHCH_2OC(=O)CF_3$		$B+C = 1423\ (5)$ *)		78STE		*) LRMW	

Nr.	Molecule	v	A, B, C MHz	D, H kHz	Ref.	Ref. struct.	Varia, remarks	See also Tab., Nr.
316	$CH_2{=}CFCH_2F$ gauche	$v=0$	$A=8594.89\,(4)$*) $B=3896.65\,(5)$ $C=2998.39\,(4)$		76ENG	76ENG	*) 95 % confidence interval $E_{gauche}-E_{cis}=145\,(60)\,cm^{-1}$ Internal rotation potential function: $V_1=208\,(23)$; $V_2=143\,(396)$; $V_3=982\,(6)$; $V_4=-76\,(16)$; $V_5=-98\,(6)$; $V_6=14\,(2)$, all in cm^{-1}	2.6.3, 141
		$v_\tau=1$	$A=8584.61\,(4)$ $B=3900.74\,(5)$ $C=2998.42\,(4)$					
		$v_\tau=2$	$A=8574.85\,(23)$ $B=3905.00\,(21)$ $C=2998.29\,(14)$					
		$v_\tau=3$	$A=8565.21\,(65)$ $B=3909.09\,(25)$ $C=2998.22\,(25)$					
		$v_\tau=4$	$A=8555.76\,(12)$ $B=3913.18\,(12)$ $C=2998.02\,(12)$					
317	$CH_2{=}CHCHF_2$ cis	$v=0$	$A=9126.08\,(16)$*) $B=3721.120\,(75)$ $C=2946.598\,(32)$	$\tau_{aaaa}=-29\,(24)$*) $\tau_{bbbb}=0.1\,(35)$ $\tau_{cccc}=1.8\,(77)$ $T_1=-104.6\,(62)$**) $T_2=1.8\,(95)$	76BOT		*) uncertainties are 2.5 σ **) $T_1=\tau'_{aabb}+\frac{A-B}{A-C}\tau'_{ccaa}$ $T_2=\tau'_{bbcc}+\frac{B-C}{A-C}\tau'_{ccaa}$ $v_\tau=63\,(8)\,cm^{-1}$, C—C torsion	2.6.3, 142
		$v_\tau=1$	$A=9124.73\,(24)$*) $B=3739.92\,(13)$ $C=2948.05\,(14)$	$\tau_{aaaa}=1\,(44)$*) $\tau_{bbbb}=-4.2\,(67)$ $\tau_{cccc}=-4\,(16)$ $T_1=-104\,(12)$**) $T_2=17\,(19)$				
		$v_\tau=2$	$A=9122.42\,(51)$*) $B=3758.24\,(37)$ $C=2949.27\,(39)$	$\tau_{aaaa}=151\,(86)$*) $\tau_{bbbb}=-46\,(20)$ $\tau_{cccc}=56\,(41)$ $T_1=-66\,(23)$**) $T_2=-39\,(26)$				
Cont.		$v_\tau=3$	$A=9121.66\,(60)$*) $B=3776.68\,(20)$ $C=2950.79\,(23)$	$\tau_{bbbb}=-2\,(15)$*) $\tau_{cccc}=-1\,(23)$***)			***) τ_{aaaa}, T_1, T_2 were fixed to the ground state values	

Nr.	Molecule	v	A, B, C MHz	D, H kHz	Ref.	Ref. struct.	Varia, remarks	See also Tab., Nr.
317 Cont.	$CH_2{=}CHCHF_2$ cis	$v_\tau=4$	$A=9117\,(11)$ *) $B=3795.98\,(24)$ $C=2952.14\,(20)$	+)	76BOT		+) centrifugal distortion constants of ground state used	
		$v_\tau=5$	$A=9123\,(12)$ *) $B=3815.32\,(26)$ $C=2953.33\,(28)$	+)				
	$CH_2{=}CHCHF_2$ gauche	$v=0$	$A=8901.64\,(15)$ *) $B=4192.759\,(83)$ $C=3107.718\,(78)$	$\tau_{aaaa}=-45\,(19)$ *) $\tau_{bbbb}=-9.3\,(31)$ $\tau_{cccc}=-2.0\,(34)$ $T_1=-22.6\,(65)$ **) $T_2=-7.6\,(37)$			*) uncertainties are 2.5 σ $E_{gauche}-E_{cis}=260\,(80)\ cm^{-1}$	
		$v_\tau=1$	$A=8877.75\,(37)$ *) $B=4190.40\,(22)$ $C=3114.04\,(22)$	$\tau_{aaaa}=-59\,(42)$ *) $\tau_{bbbb}=-8.9\,(88)$ $\tau_{cccc}=7.5\,(96)$ $T_1=-12\,(18)$ **) $T_2=-7\,(10)$			$\nu_\tau=85\,(10)\ cm^{-1}$, C—C torsion potential function restricting the torsion around the C—C bond: $V_1=266\,(40)\ cm^{-1}$ $V_3=508\,(200)\ cm^{-1}$	
		$v_\tau=2$	$A=8858.79\,(18)$ *) $B=4186.343\,(70)$ $C=3119.151\,(83)$	$\tau_{aaaa}=-54\,(20)$ *) $\tau_{bbbb}=-5.5\,(26)$ $\tau_{cccc}=-2.8\,(31)$ $T_1=-10.0\,(55)$ **) $T_2=-1.5\,(32)$			**) $T_1=\tau'_{aabb}+\frac{A-B}{A-C}\tau'_{ccaa}$ $T_2=\tau'_{bbcc}+\frac{B-C}{A-C}\tau'_{ccaa}$	
		$v_\tau=3$	$A=8844.79\,(67)$ *) $B=4179.40\,(47)$ $C=3121.93\,(74)$	+)				
		$v_\delta=1$	$A=8926.04\,(50)$ *) $B=4185.03\,(3)$ $C=3096.06\,(3)$	+)			ν_δ, bending state	
318 *6–136* *4–99*	$H_2C{=}CHC(=O)H$ trans	$v=0$	$A=47353.7106\,(44)$ $B=4659.49924\,(19)$ $\tilde{C}=4242.68964\,(19)$	$\Delta_J=1.04153\,(33)$ $\Delta_{JK}=-8.78252\,(28)$ $\Delta_K=359.91\,(46)$ $\delta_J=0.119849\,(84)$ $\delta_K=5.7986\,(22)$ $H_{JK}=-0.0123\,(53)\cdot10^{-3}$ $H_{KJ}=-0.4756\,(93)\cdot10^{-3}$	75WIN2		for structure, see also Vol. II/7 (3, 375)	2.9.3, 26

Nr.	Molecule	v	A, B, C MHz	D, H kHz	Ref.	Ref. struct.	Varia, remarks	See also Tab., Nr.
319	$H_2C{=}CHC(=S)H$ trans	$v=0$	$A=45777.041(29)$ $B=2795.9768(14)$ $C=2635.2206(15)$	$\Delta_J=0.397(5)$ $\Delta_{JK}=-5.9(2)$ $\Delta_K=368(6)$ $\delta_J=0.02827(41)$ $\delta_K=2.47(24)$ $H_{JK}=0.090(50)\cdot10^{-3}$ $h_{JK}=-0.52(22)\cdot10^{-3}$	80GEO3	80GEO3		2.6.3, 143
		$v_a=1$	$A=45947(63)$ $B=2794.225(4)$ $C=2633.374(4)$	$\Delta_J=0.375(39)$ $\Delta_{JK}=-7.37(26)$			$\nu_a=465(30)\ \mathrm{cm}^{-1}$	
		$v_{18}=1$	$A=43812(48)$ $B=2801.013(4)$ $C=2644.276(4)$	$\Delta_J=0.402(40)$ $\Delta_{JK}=-4.51(26)$			$\nu_{18}=140(30)\ \mathrm{cm}^{-1}$	
		$v_{18}=2$	$A=44358(263)$ $B=2803.137(19)$ $C=2647.103(19)$	$\Delta_J=-0.05(18)$ $\Delta_{JK}=345(3)$				
		$v_{18}=3$	$A=41557(87)$ $B=2809.766(7)$ $C=2658.914(7)$	$\Delta_J=0.550(65)$ $\Delta_{JK}=-86.9(11)$				
		$v_{18}=4$	$A=39792(253)$ $B=2815.224(21)$ $C=2668.749(21)$	$\Delta_J=0.55(20)$ $\Delta_{JK}=-88.4(34)$				
		$v_{18}=5$	$A=38505(58)$ $B=2820.932(7)$ $C=2679.171(9)$	$\Delta_J=0.505(55)$ $\Delta_{JK}=-19.5(4)$				
		$v_{18}=6$	$A=37329(74)$ $B=2826.232(9)$ $C=2688.799(11)$	$\Delta_J=0.524(70)$ $\Delta_{JK}=-9.6(5)$				
		$v_{13}=1$	$A=46808(241)$ $B=2800.336(19)$ $C=2640.499(19)$	$\Delta_J=0.80(18)$ $\Delta_{JK}=-387.5(12)$			$\nu_{13}=280(30)\ \mathrm{cm}^{-1}$	
Cont.		$v_{13}=1,\ v_{18}=1$	$A=44785(298)$ $B=2803.484(22)$ $C=2647.078(22)$	$\Delta_J=0.29(20)$ $\Delta_{JK}=36.7(36)$				

Nr.	Molecule	v	A, B, C MHz	D, H kHz	Ref.	Ref. struct.	Varia, remarks	See also Tab., Nr.
319 Cont.	$H_2C{=}CHC({=}^{34}S)H$	$v=0$	$A=45747.051$ *) $B=$ 2725.4929 (36) $C=$ 2752.4231 (42)	Δ_J = 0.347 (38) $\Delta_{JK}=-6.6$ (2)	80GEO3		*) calculated assuming $\Delta=-0.01359$ amu Å^2	
	$H_2{}^{13}C{=}CHC({=}S)H$	$v=0$	$A=45756.908$ *) $B=$ 2710.3301 (37) $C=$ 2558.9422 (44)	Δ_J = 0.323 (39) $\Delta_{JK}=-6.0$ (2)				
	$H_2C{=}{}^{13}CHC({=}S)H$	$v=0$	$A=45101.396$ *) $B=$ 2775.8061 (66) $C=$ 2615.0553 (71)	Δ_J = 0.371 (56) $\Delta_{JK}=-6.3$ (3)				
	$H_2C{=}CH{}^{13}C({=}S)H$	$v=0$	$A=45006.897$ *) $B=$ 2796.0991 (49) $C=$ 2632.6153 (64)	Δ_J = 0.397 **) $\Delta_{JK}=-5.9$ **)			**) constrained values	
320 *6–139*	$H_2C{=}CHC({=}O)OH$ s-cis	$v=0$	$A=11078.80$ $B=$ 4251.94 $C=$ 3073.39		74BOL2	74BOL2		2.6.3, 144 3, 93
		$v_\tau=1$	$A=11007.38$ $B=$ 4249.62 $C=$ 3080.97				$\nu_\tau=105$ (20) cm^{-1}, torsion	
		$v_\tau=2$	$A=10942.95$ $B=$ 4247.43 $C=$ 3087.91					
		$v_\tau=3$	$A=10883.71$ $B=$ 4245.31 $C=$ 3094.42					
		$v_\tau=4$	$A=10828.57$ $B=$ 4243.20 $C=$ 3100.64					
	$H_2C{=}CHC({=}O)OH$ s-trans	$v=0$	$A=10716.11$ $B=$ 4388.28 $C=$ 3114.31				$E_{s\text{-}trans}-E_{s\text{-}cis}=58$ (20) cm^{-1}	
Cont.		$v_\tau=1$	$A=10657.37$ $B=$ 4384.99 $C=$ 3122.12				$\nu_\tau=95$ (20) cm^{-1}, torsion	

Nr.	Molecule	v	A, B, C MHz	D, H kHz	Ref.	Ref. struct.	Varia, remarks	See also Tab., Nr.
320 *6–139* Cont.	$H_2C{=}CHC(=O)OH$ s-trans	$v_\tau=2$	$A=10602.72$		74BOL2		potential function hindering the internal rotation:	
			$B=4381.60$				$V_2=1340\,(500)\ cm^{-1}$	
			$C=3129.45$					
		$v_\tau=3$	$A=10552.26$					
			$B=4378.10$					
			$C=3136.34$					
		$v_\tau=4$	$A=10504.09$					
			$B=4374.45$					
			$C=3142.92$					
	$H_2C{=}CHC(=O)OD$ s-cis	$v=0$	$A=11078.81$					
			$B=4251.96$					
			$C=3073.40$					
		$v_\tau=1$	$A=10992.52$					
			$B=4074.05$					
			$C=2986.90$					
	$H_2C{=}CHC(=O)OD$ s-trans	$v=0$	$A=10242.21$					
			$B=4307.84$					
			$C=3033.39$					
		$v_\tau=1$	$A=10191.35$					
			$B=4303.86$					
			$C=3040.76$					
321	$H_2C{=}C(CH_3)C(=O)F$ s-trans	$v=0$	$A=5324.714\,(4)$		80LAT	80LAT		2.6.3, 145
			$B=3475.505\,(4)$					2.8, 47
			$C=2131.559\,(8)$					
		$v_\tau=1$	$A=5322.64\,(74)$				$\nu_\tau=76\,(8)\ cm^{-1}$, C—COF torsion potential function:	
			$B=3463.35\,(9)$				$V_1=140$; $V_2=1931$	
			$C=2135.63\,(10)$				$V_3=230$; $V_4=-33\ cm^{-1}$	
		$v_\tau=2$	$A=5318.28\,(50)$					
			$B=3450.99\,(6)$					
			$C=2139.91\,(7)$					
		$v_\tau=3$	$A=5315.75\,(178)$					
			$B=3438.89\,(22)$					
Cont.			$C=2144.02\,(23)$					

Nr.	Molecule	v	A, B, C MHz	D, H kHz	Ref.	Ref. struct.	Varia, remarks	See also Tab., Nr.
321 Cont.	$H_2C{=}C(CH_3)C({=}O)F$ s-trans	$v_\tau=4$	$A=5312.16\,(86)$ $B=3426.67\,(10)$ $C=2148.42\,(11)$		80LAT			
		$v_\tau=5$	$A=5308.77\,(60)$ $B=3414.57\,(7)$ $C=2152.66\,(8)$					
	$H_2C{=}C(CH_3)C({=}O)F$ s-cis	$v=0$	$A=5224.089\,(30)$ $B=3535.906\,(4)$ $C=2138.638\,(4)$				$E_{cis}-E_{trans}=359\,(11)\ cm^{-1}$	
		$v_\tau=1$	$A=5222.08\,(104)$ $B=3520.92\,(14)$ $C=2144.51\,(16)$				$\nu_\tau=52\,(11)\ cm^{-1}$	
322	$CH_2{=}CHCH{=}NH$ anti	$v=0$	$A''=45773.628\,(18)$ $B''=\ 4560.916\,(4)$ $C''=\ 4148.242\,(3)$	$\tau_1=\ 26.4\,(53)$ $\tau_2=-0.3\,(5)$ $\tau_3=432\,(24)$ $\tau_{aaaa}=-1182\,(55)$ $\tau_{bbbb}=\ -4.2\,(2)$ $\tau_{cccc}=\ -2.6\,(1)$	81BRO2	78PEN1	torsional excited states measured: $\nu_{21}=143\,(10)\ cm^{-1}$	2.6.3, 146 2.7.3, 40 3, 94
	$CH_2{=}CHCH{=}ND$ anti	$v=0$	$B=4320.80\,(2)$ $C=3940.95\,(2)$		78PEN1		both forms are planar	
	$CH_2{=}CH^{13}CH{=}NH$ anti	$v=0$	$B=4546.39\,(2)$ $C=4131.47\,(2)$					
	$CH_2{=}^{13}CHCH{=}NH$ anti	$v=0$	$B=4545.43\,(2)$ $C=4129.76\,(2)$					
	$^{13}CH_2{=}CHCH{=}NH$ anti	$v=0$	$B=4427.88\,(3)$ $C=4037.34\,(3)$					
	$CH_2{=}CHCH{=}NH$ s-trans	$v=0$	$A=43759.52\,(18)$ $B=\ 4564.581\,(39)$ $C=\ 4134.423\,(30)$	$D_J\ =0\,(1)$ $D_{JK}=1\,(6)$	81BRO2		$E_{s\text{-}trans}-E_{anti}=0.9\,(1)$ kcal/mol	
	$CH_2{=}CHCH{=}ND$ s-trans	$v=0$	$B=4435.61\,(2)$ $C=3992.71\,(2)$		78PEN1			

Nr.	Molecule	v	A, B, C MHz	D, H kHz	Ref.	Ref. struct.	Varia, remarks	See also Tab., Nr.
323 *6-141*	$CH_2{=}CHCH_2CH_3$ cis	$v=0$	$A=15302.566\,(141)$ $B=5574.977\,(80)$ $C=4303.124\,(80)$	$d_J=-12.34\,(419)$ $d_{JK}=-117.5\,(520)$ $d_K=-121.6\,(935)$ $d_{WJ}=3.233\,(296)\cdot10^{-6}$ $d_{WK}=0.150\,(107)\cdot10^{-4}$	77KAU5	81BOU	for structure, see also Vol. II/7 (3, 525)	2.9.3, 28
	$CD_2{=}CHCH_2CH_3$ cis	$v=0$	$A=14187.125\,(11)$ $B=5148.806\,(4)$ $C=3962.463\,(4)$	$\Delta_J=2.959\,(56)$ $\Delta_{JK}=-7.94\,(32)$ $\Delta_K=24.8\,(12)$ $\delta_J=0.888\,(25)$ $\delta_K=3.67\,(70)$	81BOU			
	$^{13}CH_2{=}CHCH_2CH_3$ cis		$A=15148.537\,(16)$ $B=5442.077\,(3)$ $C=4211.638\,(3)$	$\Delta_J=3.370\,(60)$ $\Delta_{JK}=-9.899\,(270)$ $\Delta_K=31.69\,(340)$ $\delta_J=0.991\,(14)$ $\delta_K=4.15\,(36)$				
	$CD_2{=}CDCD_2CD_3$ cis	$v=0$	$A=10597.801\,(8)$ $B=4412.849\,(6)$ $C=3374.997\,(6)$	$\Delta_J=2.009\,(120)$ $\Delta_{JK}=-4.157\,(150)$ $\Delta_K=11.56\,(100)$ $\delta_J=0.590\,(10)$ $\delta_K=1.06\,(22)$				
	$^{13}CH_2{=}CHCH_2CH_3$ skew	$v=0$	$A=18597\,(1000)$ $B=4044.66\,(30)$ $C=3952.64\,(30)$	$\Delta_{JK}=-41\,(7)$	79BOU1	79BOU1		
	$CH_2{=}CHCH_2{}^{13}CH_3$ skew	$v=0$	$A=21618\,(200)$ $B=4053.71\,(2)$ $C=3956.64\,(2)$	$\Delta_J=3.64\,(40)$ $\Delta_{JK}=-66.2\,(8)$				
	$CD_2{=}CDCD_2CD_3$ skew	$v=0$	$A=14365\,(20)$ $B=3313.19\,(1)$ $C=3321.19\,(1)$	$\Delta_J=2.13\,(6)$ $\Delta_{JK}=-22.8\,(1)$ $\delta_J=-0.52\,(10)$				
	$CD_2{=}CHCH_2CH_3$ skew	$v=0$	$A=20858\,(200)$ $B=3795.009\,(34)$ $C=3710.571\,(34)$	$\Delta_J=3.13\,(24)$ $\Delta_{JK}=-53.89\,(72)$ $\delta_J=-0.82\,(40)$	80BOU5			

Nr.	Molecule	v	A, B, C MHz	D, H kHz	Ref.	Ref. struct.	Varia, remarks	See also Tab., Nr.
323a	$H_2C{=}CHCH_2CH_2OH$	$v=0$	$A=9185.7124\,(74)$	$\Delta_J = 5.3802\,(58)$	81MAR	81MAR	hydrogen-bonded conformation	2.6.3, 146a
			$B=3330.0652\,(26)$	$\Delta_{JK} = -31.77\,(11)$				
			$C=2742.3434\,(24)$	$\Delta_K = 73.35\,(11)$				
				$\delta_J = 1.7362\,(63)$				
				$\delta_K = 8.23\,(21)$				
				$H_J = -0.0137\,(15)\cdot 10^{-3}$				
				$H_{JK} = -0.918\,(82)\cdot 10^{-3}$				
		$v_{T_1}=1$	$A=9275.001\,(20)$	$\Delta_J = 4.70\,(20)$			$\nu_{T_1}=84\,(20)\ cm^{-1}$, C_2—C_3 torsion	
			$B=3313.1092\,(98)$	$\Delta_{JK} = -29.73\,(17)$				
			$C=2731.9126\,(95)$	$\Delta_K = 80.4\,(21)$				
				$\delta_J = 1.5547\,(76)$				
				$\delta_K = 8.32\,(26)$				
		$v_{T_1}=2$	$A=9362.265\,(27)$	$\Delta_J = 4.75\,(23)$				
			$B=3295.982\,(12)$	$\Delta_{JK} = -27.53\,(33)$				
			$C=2721.573\,(11)$	$\Delta_K = 69.0\,(43)$				
				$\delta_J = 1.440\,(10)$				
				$\delta_K = 7.68\,(32)$				
		$v_{T_2}=1$	$A=9217.074\,(24)$	$\Delta_J = 5.36\,(31)$			$\nu_{T_2}=172\,(15)\ cm^{-1}$, C_3—C_4 torsion	
			$B=3318.202\,(14)$	$\Delta_{JK} = -31.60\,(24)$				
			$C=2734.290\,(14)$	$\Delta_K = 76.0\,(28)$				
				$\delta_J = 1.7498\,(92)$				
				$\delta_K = 7.80\,(28)$				
	$H_2C{=}CHCH_2CH_2OD$	$v=0$	$A=8852.609\,(33)$	$\Delta_J = 4.75\,(30)$				
			$B=3324.908\,(14)$	$\Delta_{JK} = -29.41\,(22)$				
			$C=2710.228\,(14)$	$\Delta_K = 67.1\,(29)$				
				$\delta_J = 1.771\,(11)$				
				$\delta_K = 7.96\,(30)$				
324 *6-140*	$CD_3CH{=}CHCD_3$ cis	$v=0$	$A=12684.2\,(2)$		79GUY	79GUY	for structure, see also Vol. II/7 (3, 526)	2.8, 48; 2.9.3, 28
			$B=4008.11\,(4)$					
			$C=3346.45\,(4)$					
	$CD_3CD{=}CDCD_3$ cis	$v=0$	$A=11136.1\,(3)$					
			$B=3997.05\,(14)$					
			$C=3170.96\,(11)$					

Nr.	Molecule	v	A, B, C MHz	D, H kHz	Ref.	Ref. struct.	Varia, remarks	See also Tab., Nr.
324a	$(CH_3)_2C{=}CH^{35}Cl$	$v=0$	$A=8400.77\,(118)$ $B=2258.18\,(2)$ $C=1818.72\,(2)$		81GUI	81GUI		2.7.3, 119a 2.8, 48a
	$(CH_3)_2C{=}CH^{37}Cl$	$v=0$	$A=8399.31\,(66)$ $B=2199.93\,(2)$ $C=1780.73\,(2)$					
325	$CH_2{=}CHCH(CH_3)_2$ trans	$v=0$	$A=7536.335\,(350)$ *) $B=3550.987\,(24)$ $C=2741.614\,(26)$	$\Delta_J=0.6\,(2)$ *) $\Delta_{JK}=20.4\,(6)$ $\Delta_K=-26.8\,(944)$ $\delta_J=0.2\,(2)$ $\delta_K=8.5\,(54)$	79CRE		*) uncertainties are 2 σ	2.6.3, 147
		$v_\tau=1$	$A=7536.561\,(368)$ *) $B=3567.316\,(30)$ $C=2742.475\,(30)$	$\Delta_J=0.5\,(2)$ *) $\Delta_{JK}=21.4\,(8)$ $\Delta_K=-43.2\,(1030)$ $\delta_J=0.2\,(2)$ $\delta_K=6.4\,(66)$			$v_\tau=90\,(10)\ cm^{-1}$, torsional motion of the vinyl group torsional potential constants: $V_1=300\,(100)\ cm^{-1}$, $V_2=-50\,(100)\ cm^{-1}$ $V_3=750\,(75)\ cm^{-1}$, $V_4=-100\,(50)\ cm^{-1}$	
		$v_\tau=2$	$A=7536.938\,(256)$ *) $B=3583.788\,(21)$ $C=2743.243\,(22)$	$\Delta_J=0.6\,(1)$ *) $\Delta_{JK}=22.2\,(6)$ $\Delta_K=-13.5\,(683)$ $\delta_J=0.2\,(1)$ $\delta_K=\ (46)$				
		$v_\tau=3$	$A=7536.950\,(250)$ *) $B=3600.419\,(22)$ $C=2743.938\,(22)$	$\Delta_J=0.8\,(2)$ *) $\Delta_{JK}=22.9\,(7)$ $\Delta_K=5.1\,(638)$ $\delta_J=0.1\,(1)$ $\delta_K=12.3\,(47)$			*) uncertainties are 2 σ	
	$CH_2{=}CHCH(CH_3)_2$ gauche	$v=0$	$A=7294.39\,(48)$ *) $B=3916.06\,(2)$ $C=2879.23\,(2)$				$E_{gauche}-E_{trans}=130\,(20)\ cm^{-1}$	
		$v_\tau=1$	$A=7280.47\,(50)$ *) $B=3914.90\,(2)$ $C=2883.36\,(2)$				$v_\tau=104\,(10)\ cm^{-1}$	
		$v_\tau=2$	$A=7268.91\,(56)$ *) $B=3913.03\,(3)$ $C=2887.03\,(2)$					

Nr.	Molecule	v	A, B, C MHz	D, H kHz	Ref.	Ref. struct.	Varia, remarks	See also Tab., Nr.
326	$CH_3CH{=}CHCH_2CH_3$ cis $^{+}$)	$v=0$	$A=10987.805(41)$ *) $B=2601.395(5)$ $C=2371.405(5)$	$\Delta_J=4.72(4)$ *) $\Delta_{JK}=-81.2(2)$ $\Delta_K=478(13)$ $\delta_J=0.569(2)$ $\delta_K=-41.1(3)$ $H_J=-0.6(1)\cdot10^{-3}$ $H_{JK}=-5.0(9)\cdot10^{-3}$ $H_{KJ}=122(42)\cdot10^{-3}$ $H_K=-1800(1200)\cdot10^{-3}$ $h_J=0.036(5)\cdot10^{-3}$ $h_{JK}=-0.5(9)\cdot10^{-3}$ $h_K=-43(33)\cdot10^{-3}$	81VAN	81VAN	*) for the A-species $^{+}$) skew conformation	2.6.3, 148 2.8, 49 2.9.3, 31
		$v_\tau=1$	$A=10868.956(53)$ $B=2637.681(10)$ $C=2396.975(10)$	$\Delta_J=6.74(3)$ $\Delta_{JK}=-107.2(2)$ $\Delta_K=567(7)$ $\delta_J=1.063(2)$ $\delta_K=-55.1(2)$ $H_{JK}=-6.1(7)\cdot10^{-3}$ $h_J=0.085(8)\cdot10^{-3}$ $h_{JK}=0.2(11)\cdot10^{-3}$			$v_\tau\simeq34\ cm^{-1}$, C—C torsion Barrier: $V_3=210(20)\ cm^{-1}$ (tunneling between the 2 equivalent skew conformations)	
		$v_\tau=2$	$A=10811.562(38)$ $B=2670.390(8)$ $C=2417.859(8)$	$\Delta_J=7.74(3)$ $\Delta_{JK}=-115.0(1)$ $\Delta_K=581(5)$ $\delta_J=1.453(1)$ $\delta_K=-56.5(1)$ $H_{JK}=-3.3(4)\cdot10^{-3}$ $h_J=0.081(6)\cdot10^{-3}$ $h_{JK}=5.5(7)\cdot10^{-3}$				
327	$CH_3HC{=}CHC({=}O)^{35}Cl$ s-cis		$B+C=2554.47(229)$ *)		78LUM		*) LRMW	
Cont.	$CH_3HC{=}CHC({=}O)^{37}Cl$ s-cis		$B+C=2505.59(403)$ *)					

Nr.	Molecule v	A, B, C MHz	D, H kHz	Ref.	Ref. struct.	Varia, remarks	See also Tab., Nr.
327 Cont.	$CH_3HC{=}CHC(=O)^{35}Cl$ s-trans $CH_3HC{=}CHC(=O)^{37}Cl$ s-trans	$B+C=2979.92\,(146)$ *) $B+C=2943.34\,(253)$ *)		78LUM			
328	$CH_2{=}CHOCH_2CF_3$ s-cis	$B+C=2181.9\,(3)$ *) $A=4424\,(7)$		75TRU		*) LRMW	
329	$CH_3CH{=}CHCH_2{}^{35}Cl$ trans $CH_3CH{=}CHCH_2{}^{37}Cl$ trans	$B+C=2905.4\,(14)$ *) $B+C=2843.2\,(15)$ *)		78STE		*) LRMW	
330	$CH_3CH{=}CHCH_2{}^{79}Br$ trans $CH_3CH{=}CHCH_2{}^{81}Br$ trans	$B+C=2073\,(1)$ *) $B+C=2057\,(1)$ *)		78STE		*) LRMW	
331	$CH_2{=}CHCH_2CH_2{}^{79}Br$ $CH_2{=}CHCH_2CH_2{}^{81}Br$	$B+C=1975.3\,(4)$ *) $B+C=1960.1\,(5)$ *)		78STE		*) LRMW	
332	$CH_3CH{=}CHCH_2OH$ $CH_3CH{=}CHCH_2OD$	$B+C=4184.19\,(61)$ *) $B+C=4112.16\,(107)$ *)		78LUM		*) LRMW	
333	$CH_3HC{=}CHCH_2OC(=O)CF_3$	$B+C=1234\,(11)$ *)		78STE		*) LRMW	
334	$H_2C{=}CH(CH_2)_2OC(=O)CF_3$	$B+C=882\,(22)$ *)		78STE		*) LRMW	
335	$CH_2{=}CH(CH_2)_2CH_2{}^{79}Br$ $CH_2{=}CH(CH_2)_2CH_2{}^{81}Br$	$B+C=1453.2\,(8)$ *) $^{+}$) $B+C=1209\,(3)$ *) $^{++}$) $B+C=1440.7\,(3)$ *) $^{+}$) $B+C=1198\,(2)$ *) $^{++}$)		78STE		*) LRMW $^{+}$) s-trans-gauche-skew $^{++}$) trans-trans-skew	
336	$CH_3CH{=}CHCH_2CH_2{}^{79}Br$ $CH_3CH{=}CHCH_2CH_2{}^{81}Br$	$B+C=1182.7\,(2)$ *) $B+C=1172.9\,(2)$ *)		78STE		*) LRMW, trans-skew	

Nr.	Molecule	v	A, B, C MHz	D, H kHz	Ref.	Ref. struct.	Varia, remarks	See also Tab., Nr.
337	$CH_3HC{=}CH(CH_2)_2OC({=}O)CF_3$		$B+C=606\,(5)$ *) +) $B+C=690\,(10)$ *) ++)		78STE		*) LRMW +) trans ++) cis	
338	$H_2C{=}CH(CH_2)_4OC({=}O)H$		$B+C=1077.9\,(7)$ *)		78STE		*) LRMW	
339	$CH_3CH_2HC{=}CH(CH_2)_2OC({=}O)CF_3$		$B+C=456\,(2)$ *)		78STE		*) LRMW	
340	$CH_2{=}CH{-}CH{=}CHF$ trans	$v=0$	$A=34371\,(18)$ $B=\ 2170.669\,(1)$ $C=\ 2041.682\,(1)$	$\Delta_J=\ 0.16\,(2)$ $\Delta_{JK}=-5.8\,(2)$	80CED1	80CED1		2.6.3, 149 3, 95
		$v_\tau=1$	$A=32507\,(42)$ $B=\ 2172.664\,(4)$ $C=\ 2046.624\,(4)$	$\Delta_J=\ 0.23\,(5)$ $\Delta_{JK}=-4.4\,(5)$			$\nu_\tau=138\,(11)\ cm^{-1}$, torsion	
		$v_\tau=2$	$A=31191\,(87)$ $B=\ 2174.666\,(8)$ $C=\ 2051.534\,(8)$	$\Delta_J=\ 0.25\,(11)$ $\Delta_{JK}=-15.9\,(11)$				
341	$CH_2{=}CH{-}CH{=}CH^{35}Cl$ cis	$v=0$	$A=12621.88\,(1)$ $B=\ 1953.906\,(1)$ $C=\ 1691.702\,(1)$	$\Delta_J=\ 0.457\,(1)$ $\Delta_{JK}=-7.16\,(1)$ $\Delta_K=\ 65.98\,(8)$ $\delta_J=\ 0.0983\,(4)$ $\delta_K=\ 1.32\,(6)$	80KAR	80KAR		2.7.3, 120 3, 96
		$v_\tau=1$	$A=12403.20\,(1)$ $B=\ 1959.889\,(1)$ $C=\ 1697.354\,(1)$	$\Delta_J=\ 0.482\,(3)$ $\Delta_{JK}=-6.74\,(1)$ $\Delta_K=\ 44.5\,(2)$ $\delta_J=\ 0.1033\,(8)$ $\delta_K=\ 0.7\,(2)$			$\nu_\tau=115\,(13)\ cm^{-1}$, torsion	
		$v_\tau=2$	$A=12211.17\,(1)$ $B=\ 1965.364\,(3)$ $C=\ 1702.581\,(3)$	$\Delta_J=\ 0.509\,(6)$ $\Delta_{JK}=-6.26\,(3)$ $\delta_J=\ 0.106\,(1)$				
		$v_\tau=3$	$A=12041.3\,(1)$ $B=\ 1970.39\,(1)$ $C=\ 1707.46\,(1)$	$\Delta_J=\ 0.55\,(2)$ $\Delta_{JK}=-5.29\,(5)$ $\delta_J=\ 0.10\,(2)$				
Cont.								

Nr.	Molecule	v	A, B, C MHz	D, H kHz	Ref.	Ref. struct.	Varia, remarks	See also Tab., Nr.
341 Cont.	CH_2=CH—CH=$CH^{35}Cl$ cis	$v_\delta=1$	$A=12797\,(1)$ $B=1950.54\,(1)$ $C=1688.56\,(1)$	$\Delta_J=0.46\,(2)$ $\Delta_{JK}=-8.03\,(5)$ $\delta_J=0.10\,(2)$	80KAR		$\nu_\delta=187\,(25)\ cm^{-1}$, skeletal in-plane bending	
	CH_2=CH—CH=$CH^{37}Cl$ cis	$v=0$	$A=12524.15\,(1)$ $B=1912.017\,(2)$ $C=1658.501\,(2)$	$\Delta_J=0.444\,(6)$ $\Delta_{JK}=-7.11\,(4)$ $\Delta_K=65.6\,(7)$ $\delta_J=0.094\,(1)$ $\delta_K=1.4\,(3)$				
		$v_\tau=1$	$A=12306\,(1)$ $B=1917.90\,(2)$ $C=1664.04\,(2)$	$\Delta_J=0.45\,(5)$ $\Delta_{JK}=-6.6\,(1)$ $\delta_J=0.09\,(2)$				
		$v_\delta=1$	$A=12696\,(2)$ $B=1908.75\,(3)$ $C=1655.41\,(3)$	$\Delta_J=0.46\,(9)$ $\Delta_{JK}=-8.0\,(2)$ $\delta_J=0.09\,(3)$				
	CH_2=CH—CD=$CH^{35}Cl$ cis	$v=0$	$A=11703.03\,(1)$ $B=1941.584\,(2)$ $C=1665.013\,(2)$	$\Delta_J=0.43\,(1)$ $\Delta_{JK}=-6.32\,(2)$ $\delta_J=0.100\,(1)$ $\Delta_K=53.29$ *) $\delta_K=1.20$ *)			*) calculated from the force field	
	CH_2=CD—CH=$CH^{35}Cl$ cis	$v=0$	$A=12100.04\,(2)$ $B=1948.571\,(3)$ $C=1678.125\,(3)$	$\Delta_J=0.47\,(1)$ $\Delta_{JK}=-6.40\,(2)$ $\delta_J=0.101\,(2)$ $\Delta_K=50.86$ *) $\delta_K=1.27$ *)				
	CH_2=CH—CH=$CD^{35}Cl$ cis	$v=0$	$A=11590.07\,(1)$ $B=1942.908\,(2)$ $C=1663.693\,(2)$	$\Delta_J=0.41\,(1)$ $\Delta_{JK}=-5.82\,(2)$ $\delta_J=0.100\,(2)$ $\Delta_K=48.84$ *) $\delta_K=1.23$ *)			*) calculated from the force field	
	CHD=CH—CH=$CH^{35}Cl$ cis	$v=0$	$A=12114.86\,(2)$ $B=1872.549\,(3)$ $C=1621.593\,(3)$	$\Delta_J=0.43\,(1)$ $\Delta_{JK}=-6.76\,(3)$ $\delta_J=0.092\,(3)$ $\Delta_K=63.01$ *) $\delta_K=1.21$ *)			D in position H_8 in Fig. 96	
Cont.								

Nr.	Molecule	v	A, B, C MHz	D, H kHz	Ref.	Ref. struct.	Varia, remarks	See also Tab., Nr.
341 Cont.	$CHD{=}CH{-}CH{=}CH^{35}Cl$ cis	$v=0$	$A=12544.49\,(2)$ $B=\ 1871.670\,(3)$ $C=\ 1628.500\,(3)$	$\Delta_J\ =\ 0.42\,(1)$ $\Delta_{JK}=-6.38\,(3)$ $\delta_J\ =\ 0.083\,(2)$ $\Delta_K\ =\ 60.67$ *) $\delta_K\ =\ 1.16$ *)	80KAR		D in position H_9 in Fig. 96 *) calculated from the force field	
	$CH_2{=}CH{-}CH{=}{}^{13}CH^{35}Cl$ cis	$v=0$	$A=12342\,(1)$ $B=\ 1950.88\,(1)$ $C=\ 1684.31\,(1)$	$\Delta_J\ =\ 0.46\,(2)$ $\Delta_{JK}=-6.87\,(5)$ $\delta_J\ =\ 0.07\,(2)$ $\Delta_K\ =\ 62.64$ *) $\delta_K\ =\ 1.30$ *)				
	$CH_2{=}CH{-}{}^{13}CH{=}CH^{35}Cl$ cis	$v=0$	$A=12397\,(1)$ $B=\ 1950.35\,(1)$ $C=\ 1684.94\,(1)$	$\Delta_J\ =\ 0.42\,(2)$ $\Delta_{JK}=-6.98\,(6)$ $\delta_J\ =\ 0.08\,(2)$ $\Delta_K\ =\ 65.24$ *) $\delta_K\ =\ 1.29$ *)				
	$CH_2{=}{}^{13}CH{-}CH{=}CH^{35}Cl$ cis	$v=0$	$A=12578\,(1)$ $B=\ 1938.07\,(1)$ $C=\ 1679.07\,(1)$	$\Delta_J\ =\ 0.46\,(2)$ $\Delta_{JK}=-6.96\,(5)$ $\delta_J\ =\ 0.09\,(2)$ $\Delta_K\ =\ 64.12$ *) $\delta_K\ =\ 1.31$ *)				
	${}^{13}CH_2{=}CH{-}CH{=}CH^{35}Cl$ cis	$v=0$	$A=12576\,(1)$ $B=\ 1896.77\,(1)$ $C=\ 1647.91\,(1)$	$\Delta_J\ =\ 0.45\,(2)$ $\Delta_{JK}=-6.91\,(8)$ $\delta_J\ =\ 0.05\,(3)$ $\Delta_K\ =\ 66.19$ *) $\delta_K\ =\ 1.26$ *)				
	$CH_2{=}CH{-}CH{=}CH^{35}Cl$ trans	$v=0$	$A=30388\,(38)$ $B=\ 1453.572\,(1)$ $C=\ 1387.426\,(1)$	$\Delta_J\ =\ 0.081\,(6)$ $\Delta_{JK}=-6.15\,(24)$	80CED2	80CED2		
	$CH_2{=}CH{-}CH{=}CH^{37}Cl$ trans	$v=0$	$A=30503\,(37)$ $B=\ 1419.811\,(2)$ $C=\ 1356.494\,(2)$	$\Delta_J\ =\ 0.065\,(11)$ $\Delta_{JK}=-7.26\,(27)$				
Cont.	${}^{13}CH_2{=}CH{-}CH{=}CH^{35}Cl$ trans	$v=0$	$A=30296\,(142)$ $B=\ 1414.931\,(8)$ $C=\ 1351.777\,(7)$	$\Delta_J\ =\ 0.093\,(44)$ $\Delta_{JK}=-5.93\,(109)$				

Nr.	Molecule	v	A, B, C MHz	D, H kHz	Ref.	Ref. struct.	Varia, remarks	See also Tab., Nr.
341 Cont.	$CH_2{=}^{13}CH{-}CH{=}CH^{35}Cl$ trans	$v=0$	$A=29964\,(102)$	$\Delta_J = 0.158\,(60)$	80CED2			
			$B=1437.775\,(7)$	$\Delta_{JK}=-4.76\,(69)$				
			$C=1372.491\,(10)$					
	$CH_2{=}CH{-}^{13}CH{=}CH^{35}Cl$ trans	$v=0$	$A=30718\,(74)$	$\Delta_J = 0.085\,(20)$				
			$B=1451.967\,(3)$	$\Delta_{JK}=-6.86\,(36)$				
			$C=1385.780\,(4)$					
	$CH_2{=}CH{-}CH{=}^{13}CH^{35}Cl$ trans	$v=0$	$A=29884\,(84)$	$\Delta_J = 0.081$ *)			*) calculated from the force field	
			$B=1452.702\,(4)$	$\Delta_{JK}=-6.08$ *)				
			$C=1385.528\,(7)$					
	$CH_2{=}CH{-}CH{=}CD^{35}Cl$ trans	$v=0$	$A=26379\,(35)$	$\Delta_J = 0.084\,(20)$				
			$B=1452.469\,(3)$	$\Delta_{JK}=-5.01\,(52)$				
			$C=1376.631\,(3)$					
	$CH_2{=}CH{-}CD{=}CH^{35}Cl$ trans	$v=0$	$A=27687\,(51)$	$\Delta_J = 0.101\,(24)$				
			$B=1452.681\,(4)$	$\Delta_{JK}=-4.75\,(61)$				
			$C=1380.290\,(4)$					
	$CH_2{=}CD{-}CH{=}CH^{35}Cl$ trans	$v=0$	$A=27053\,(22)$	$\Delta_J = 0.063\,(11)$				
			$B=1436.597\,(2)$	$\Delta_{JK}=-5.20\,(29)$				
			$C=1364.102\,(2)$					
	$CHD{=}CH{-}CH{=}CH^{35}Cl$ trans	$v=0$	$A=30447\,(105)$	$\Delta_J = 0.089\,(31)$			D in position H(8) in Fig. 96	
			$B=1387.901\,(5)$	$\Delta_{JK}=-5.87\,(73)$				
			$C=1327.439\,(5)$					
	$CHD{=}CH{-}CH{=}CH^{35}Cl$ trans	$v=0$	$A=27339\,(70)$	$\Delta_J = 0.088\,(31)$			D in position H(9) in Fig. 96	
			$B=1415.706\,(5)$	$\Delta_{JK}=-6.89\,(82)$				
			$C=1345.752\,(6)$					
342	$CH_2{=}CH{-}CH{=}CHOCH_3$ s-trans	$v=0$	$A=24524\,(16)$	$D_J = 0.131\,(20)$	77KEW	77KEW		3, 97
			$B=1336.51\,(1)$	$D_{JK}=-1.70\,(5)$				
			$C=1277.76\,(1)$					
		$v_{T_1}=1$	$B^*=2620.56$				$B^*=B+C$	
		$v_{T_1}=2$	$B^*=2626.04$				$\nu_{T_1}=60\,(20)\,cm^{-1}$, O—CH torsion	
		$v_{T_1}=3$	$B^*=2630.77$					
		$v_{T_1}=4$	$B^*=2634.86$					
		$v_{T_2}=1$	$B^*=2615.63$				$\nu_{T_2}=80\,(30)\,cm^{-1}$, =C—C= torsion	
		$v_{T_2}=2$	$B^*=2616.71$					
		$v_{T_1}=1, v_{T_2}=1$	$B^*=2621.75$					
Cont.		$v_{T_1}=2, v_{T_2}=1$	$B^*=2627.08$					

Nr.	Molecule	v	A, B, C MHz	D, H kHz	Ref.	Ref. struct.	Varia, remarks	See also Tab., Nr.
342 Cont.	CH_2=CH—CH=CHOCH_3 s-cis	$v=0$	$A=12642.73\,(23)$ $B=\ 1607.61\,(1)$ $C=\ 1439.64\,(1)$	$D_J=\ 0.147\,(22)$ $D_{JK}=-1.75\,(4)$	77KEW		one series of satellites has been observed and assigned to the =C—C= torsion: $\nu_T=103\,(30)\ cm^{-1}$	
343	CH_2=CH—CH=C=O trans	$v=0$	$A''=39557\,(20)$ $B''=\ 2392.9024\,(54)$ $C''=\ 2255.9539\,(49)$	$\tau_{bbbb}=-3.96\,(33)$ $\tau_{cccc}=-2.38\,(33)$ $\tau_1=129.77\,(85)$ $\tau_2=\ 4.24\,(33)$ $H_J=0.00202\,(51)$	79BR, 79BJ	79BJ		2.6.3, 150 3, 98
		$v_{21}=1$	$A=36580$ $B=\ 2395.1$ $C=\ 2262.8$		79BJ		$\nu_{21}=105\,(30)\ cm^{-1}$, vinyl torsion	
		$v_{15}=1$	$A=42686$ $B=\ 2400.0$ $C=\ 2259.4$				$\nu_{15}=155\,(30)\ cm^{-1}$, in-plane skeletal bend	
344	CH_2=CH—C^{35}Cl=CH_2	$v=0$	$A=5438.865\,(4)$ $B=3743.080\,(4)$ $C=2216.321\,(4)$	$\Delta_J=\ 0.83\,(8)$ $\Delta_{JK}=-1.80\,(9)$ $\Delta_K=\ 3.3\,(1)$ $\delta_J=0.50\,(1)$ $\delta_K=0.96\,(6)$	79KAR, 81CED	81CED		2.7.3, 121
	CH_2=CH—C^{37}Cl=CH_2	$v=0$	$A=5359.405\,(8)$ $B=3685.570\,(8)$ $C=2182.939\,(7)$	$\Delta_J=\ 0.54\,(16)$ $\Delta_{JK}=-2.72\,(55)$ $\Delta_K=\ 4.3\,(7)$ $\delta_J=0.44\,(5)$ $\delta_K=1.26\,(35)$	81CED			
	CH_2=CH—C^{35}Cl=$^{13}CH_2$	$v=0$	$A=5262.03\,(2)$ $B=3727.32\,(1)$ $C=2180.97\,(1)$	$\Delta_{JK}=-1.46\,(16)$ $\delta_J=0.49\,(3)$ $\delta_K=0.93\,(14)$				
	CH_2=CH—$^{13}C^{35}$Cl=CH_2	$v=0$	$A=5417.71\,(2)$ $B=3743.02\,(2)$ $C=2212.78\,(1)$	$\Delta_{JK}=-1.79\,(3)$ $\Delta_K=\ 2.8\,(2)$ $\delta_J=0.465\,(9)$				
Cont.	CH_2=^{13}CH—C^{35}Cl=CH_2	$v=0$	$A=5426.30\,(1)$ $B=3692.52\,(1)$ $C=2196.45\,(1)$	$\Delta_K=3.1\,(1)$				

Nr.	Molecule	v	A, B, C MHz	D, H kHz	Ref.	Ref. struct.	Varia, remarks	See also Tab., Nr.
344 Cont.	$^{13}CH_2=CH-C^{35}Cl=CH_2$	$v=0$	$A=5414.95(2)$ $B=3630.24(1)$ $C=2172.42(1)$	$\Delta_{JK}=-1.34(13)$ $\Delta_K=2.5(4)$ $\delta_K=1.03(4)$	81CED			
	$CH_2=CH-C^{35}Cl=CHD$ (D cis to Cl)	$v=0$	$A=5262.809(4)$ $B=3647.444(3)$ $C=2153.546(3)$	$\Delta_J=0.82(6)$ $\Delta_{JK}=-0.77(22)$ $\Delta_K=2.4(3)$ $\delta_J=0.42(2)$ $\delta_K=0.96(14)$				
	$CH_2=CH-C^{35}Cl=CHD$ (D trans to Cl)	$v=0$	$A=5045.631(6)$ $B=3742.129(5)$ $C=2147.757(4)$	$\Delta_J=0.68(10)$ $\Delta_{JK}=-1.75(5)$ $\delta_J=0.49(1)$ $\delta_K=0.94(4)$				
	$CHD=CH-C^{35}Cl=CH_2$ (D trans to Cl)	$v=0$	$A=5417.81(1)$ $B=3489.54(1)$ $C=2121.69(1)$	$\Delta_{JK}=-1.27(6)$ $\Delta_K=3.4(2)$ $\delta_K=1.12(4)$				
	$CHD=CH-C^{35}Cl=CH_2$ (D cis to Cl)	$v=0$	$A=5295.70(3)$ $B=3676.95(2)$ $C=2169.45(2)$	$\delta_K=0.92(7)$				
345	$CH_2=C=CHF$	$v=0$	$A=41089(2000)$ $B=4300.11(4)$ $C=4039.26(4)$		75ELL			2.6.3, 151
346 *6-146*	$CH_2=C=CH^{79}Br$	$v=0$	$B=2009.89(1)$ $C=1897.43(1)$	$\Delta_J=0.7(3)$	76DEM2			2.7.3, 154
	$CH_2=C=CH^{81}Br$	$v=0$	$B=1996.06(1)$ $C=1884.96(1)$	$\Delta_J=1.0(3)$				
347 *6-147*	$CH_2=C=CHCN$	$v=0$	$A=25981.00(8)$ *) $B=2689.29(3)$ $C=2474.81(2)$	$\Delta_J=1.28(4)$ *) $\Delta_{JK}=-85.5(4)$ $\Delta_K=1.94(2)\cdot 10^3$ $\delta_J=0.310(4)$ $\delta_K=6(10)$	73BOU		*) uncertainties are 2 σ	2.6.3, 152 2.7.3, 41 3, 99
Cont.		$v_1=1$	$A=26318.4(3)$ *) $B=2701.01(3)$ $C=2479.96(3)$					

Nr.	Molecule	v	A, B, C MHz	D, H kHz	Ref.	Ref. struct.	Varia, remarks	See also Tab., Nr.
347 Cont.	$CH_2{=}C{=}CHCN$	$v_I=2$	$A=26622.1\,(10)$ *)		73BOU		*) uncertainties are 2 σ	
			$B=\ 2712.57\,(2)$					
			$C=\ 2485.15\,(2)$					
		$v_{II}=1$	$B=\ 2680.80\,(8)$ *)					
			$C=\ 2472.20\,(8)$					
		$v_{I+II}=1$	$B=\ 2692.5\,(2)$ *)					
			$C=\ 2477.8\,(2)$					
348	$CH_2{=}C{=}CHCH_2{}^{35}Cl$ skew	$v=0$	$A=15484\,(12)$	$\Delta_J=\ 0.94\,(1)$	77KAR	77KAR		2.7.3, 122
			$B=\ 1606.07\,(1)$	$\Delta_{JK}=-56.38\,(3)$				3, 100
			$C=\ 1557.55\,(1)$	$\delta_J=\ 0.19\,(2)$			$\nu_\tau=60-70\ cm^{-1}$, torsion	
		$v_\tau=1$	$A=15606\,(17)$	$\Delta_J=\ 1.04\,(2)$				
			$B=\ 1612.84\,(2)$	$\Delta_{JK}=-58.40\,(5)$				
			$C=\ 1560.80\,(2)$	$\delta_J=\ 0.23\,(4)$				
		$v_\tau=2$	$A=15722\,(18)$	$\Delta_J=\ 1.04\,(3)$				
			$B=\ 1622.59\,(2)$	$\Delta_{JK}=-60.62\,(7)$				
			$C=\ 1564.01\,(2)$	$\delta_J=\ 0.22\,(5)$				
		$v_\delta=1$	$A=15668\,(18)$	$\Delta_J=\ 0.95\,(2)$			$\nu_\delta=170-190\ cm^{-1}$, skeletal bending	
			$B=\ 1601.19\,(2)$	$\Delta_{JK}=-58.87\,(4)$				
			$C=\ 1555.65\,(2)$	$\delta_J=\ 0.17\,(4)$				
	$CH_2{=}C{=}CHCH_2{}^{37}Cl$ skew	$v=0$	$A=15396\,(13)$	$\Delta_J=\ 0.91\,(1)$				
			$B=\ 1571.26\,(1)$	$\Delta_{JK}=-55.54\,(3)$				
			$C=\ 1523.94\,(1)$	$\delta_J=\ 0.19\,(2)$				
	$CHD{=}C{=}CHCH_2{}^{35}Cl$ *) skew	$v=0$	$A=15296\,(16)$	$\Delta_J=\ 0.84\,(3)$			*) D in position H_2 of Fig. 100	
			$B=\ 1535.80\,(1)$	$\Delta_{JK}=-52.24\,(4)$				
			$C=\ 1492.75\,(1)$	$\delta_J=\ 0.14\,(3)$				
	$CHD{=}C{=}CHCH_2{}^{35}Cl$ +) skew	$v=0$	$A=14470\,(12)$	$\Delta_J=\ 0.89\,(2)$			+) D in position H_1 of Fig. 100	
			$B=\ 1556.25\,(2)$	$\Delta_{JK}=-51.71\,(6)$				
			$C=\ 1500.67\,(2)$	$\delta_J=\ 0.17\,(4)$				
349	$CH_2{=}C{=}CHCH_2CH_3$ cis	$v=0$	$A=10101$		79FOR			
			$B=\ 3237.21$					
			$C=\ 2571.20$					
	$CH_2{=}C{=}CHCH_2CH_3$ skew	$v=0$	$A=17900$					
			$B=\ 2279.38$					
			$C=\ 2240.72$					

Nr.	Molecule	v	A, B, C MHz	D, H kHz	Ref.	Ref. struct.	Varia, remarks	See also Tab., Nr.
350	$CH_2{=}C{=}CF_2$	$v=0$	$A=10881.9\,(28)$		74DUR10	74DUR10	for structure, see also Vol. II/7 (3, 349)	2.6.3, 153
			$B=\ 3667.34\,(3)$					
			$C=\ 2791.41\,(3)$					
		$v_I=1$	$A=10758.36\,(84)$					
			$B=\ 3683.68\,(2)$					
			$C=\ 2795.19\,(2)$					
		$v_{II}=1$	$A=11017.02\,(90)$					
			$B=\ 3676.42\,(2)$					
			$C=\ 2799.40\,(2)$					
	$^{13}CH_2{=}C{=}CF_2$	$v=0$	$A=10879.3\,(45)$					
			$B=\ 3534.65\,(6)$					
			$C=\ 2713.90\,(5)$					
	$CH_2{=}^{13}C{=}CF_2$	$v=0$	$A=10877.9\,(27)$					
			$B=\ 3642.16\,(5)$					
			$C=\ 2776.89\,(4)$					
	$CHD{=}C{=}CF_2$	$v=0$	$A=10695.1\,(57)$					
			$B=\ 3444.71\,(5)$					
			$C=\ 2671.91\,(4)$					
		$v_I=1$	$A=10557.08\,(58)$					
			$B=\ 3459.53\,(1)$					
			$C=\ 2675.52\,(1)$					
		$v_{II}=1$	$A=10845.42\,(66)$					
			$B=\ 3453.13\,(1)$					
			$C=\ 2679.41\,(1)$					
	$CD_2{=}C{=}CF_2$	$v=0$	$A=10489.0\,(24)$					
			$B=\ 3251.49\,(3)$					
			$C=\ 2566.26\,(3)$					
		$v_I=1$	$A=10337.86\,(79)$					
			$B=\ 3265.03\,(1)$					
			$C=\ 2569.83\,(1)$					
		$v_{II}=1$	$A=10654.86\,(89)$					
			$B=\ 3259.27\,(1)$					
			$C=\ 2573.09\,(1)$					
351	$CHF{=}C{=}CHF$	$v=0$	$A=25101.15\,(12)$		75ELL			2.6.3, 154
			$B=\ 2170.02\,(7)$					
			$C=\ 2117.40\,(4)$					

Nr.	Molecule	v	A, B, C MHz	D, H kHz	Ref.	Ref. struct.	Varia, remarks	See also Tab., Nr.
352 *6-148*	$CH_2{=}C{=}C(CH_3)_2$	$v=0$	$A=8263.99082\,(298)$ *) $B=3614.19398\,(136)$ $C=2639.48826\,(89)$	$\Delta_J = 0.4970\,(95)$ *) $\Delta_{JK}=10.490\,(60)$ $\Delta_K = 1.985\,(204)$ $\delta_J = 0.15723\,(176)$ $\delta_K = 6.108\,(40)$	76DEM1		*) for the EE species, uncertainties are 2σ	
		$v_{I}=1$	$A=8259.16\,(20)$ *) $B=3617.640\,(20)$ $C=2640.214\,(20)$		77DEM2		Coriolis interaction between ν_{I} and ν_{II}: $^{(a)}\zeta_{I,II}=0.36\,(2A^{(a)}\zeta_{I,II} = 5989.4\,(4)$ MHz) $E_{II}-E_{I}=48761.2\,(10)$ MHz $\nu_{I}(b_1)=191\ cm^{-1}$, skeletal deformation	
		$v_{II}=1$	$A=8248.25\,(10)$ *) $B=3626.382\,(20)$ $C=2640.565\,(20)$				$\nu_{II}(b_2)$, out-of-phase torsion	
		$v_{III}=1$	$A=8251.23\,(2)$ *) $B=3613.797\,(3)$ $C=2638.384\,(3)$				$\nu_{III}(a_2)=236\ cm^{-1}$, in-phase torsion	
		$v_{IV}=1$	$A=8270.73\,(4)$ *) $B=3618.352\,(5)$ $C=2643.832\,(5)$				$\nu_{IV}(b_2)=241\ cm^{-1}$, skeletal deformation (wag)	
		$v_{IV}=2$	$A=8274.1\,(25)$ *) $B=3622.60\,(9)$ $C=2648.28\,(5)$					
		$v_{V}=1$	$A=8206.9\,(5)$ *) $B=3621.87\,(2)$ $C=2639.99\,(1)$				$\nu_{V}(a_1)=355\ cm^{-1}$, $C{-}C_2$ scissoring	
		$v_{VI}=1$	$A=8343$ $B=3601.5$ $C=2730.5$					
		$v_{VII}=1$	$A=8577$ $B=3602.6$ $C=2678.4$					
		$v_{VIII}=1$	$A=8486$ $B=3628.9$ $C=2646.4$					

Nr.	Molecule	v	A, B, C MHz	D, H kHz	Ref.	Ref. struct.	Varia, remarks	See also Tab., Nr.
353	CH_2=C=C=O [++])	$v=0$	$A-D_K$ $=149830.319\,(24)$ [+]) $B=4387.04599\,(46)$ $C=4258.11229\,(46)$	$D_J = 1.5558\,(58)$ $D_{JK} = -789.67\,(42)$ $d_1 = -0.228662\,(97)$ $d_2 = -0.0340\,(24)$ $H_J = -0.0533\,(33)\cdot 10^{-3}$ $H_{KJ} = -3.872\,(84)$ $L_{JKKK} = 0.0320\,(39)$	81BRO3	81BRO3	[+]) $D_K \sim 10$ MHz $\Delta = 0.1151\,(2)$ amuÅ^2 [++]) C_s symmetry heavy chain of atoms bent at the middle C by $\sim 26°$	2.6.3, 155
	CHD=C=C=O cis	$v=0$	$A-D_K=108238\,(1580)$ $B=4227.9482\,(46)$ $C=4059.2171\,(44)$	$D_J = 1.98\,(16)$ $D_{JK} = -464.6\,(18)$ $H_{KJ} = -1.48\,(18)$				
	CHD=C=C=O trans	$v=0$	$A-D_K=140756\,(2891)$ $B=4124.7758\,(44)$ $C=4001.0764\,(38)$	$D_J = 1.43\,(11)$ $D_{JK} = -428.3\,(12)$ $H_{KJ} = -0.53\,(16)$				
	$^{13}CH_2$=C=C=O	$v=0$	$A-D_K=148421\,(3482)$ $B=4255.7767\,(46)$ $C=4133.5165\,(60)$	$D_J = 1.02\,(18)$ $D_{JK} = -769.0\,(16)$ $H_{KJ} = -1.93\,(16)$				
	CH_2=C=C=^{18}O	$v=0$	$A-D_K=165076\,(5024)$ $B=4167.8057\,(41)$ $C=4050.9399\,(34)$	$D_J = 1.457\,(78)$ $D_{JK} = -746.4\,(11)$ $H_{KJ} = -3.163\,(59)$				
	CD_2=C=C=O	$v=0$	$B=3992.03\,(20)$ *) $C=3830.19\,(20)$	$D_J = 2.8 \pm 5.6$ *) $D_{JK} = -257\,(20)$	77BLA2		*) uncertainties are 2σ	
353a	CH_2=C=C=C=O	$v=0$	$b_p = -(2.22 \pm 0.36)\cdot 10^{-5}$ $B=2160.70\,(4)$ $C=2146.82\,(4)$	$D_J = 0.241\,(10)$ $D_{JK} = -427\,(40)$ $H_{JJK} = -(3.32 \pm 20)\cdot 10^{-3}$ $H_{JKK} = 10.9\,(6)$ $L_{JJJK} = -(1.7 \pm 2.6)\cdot 10^{-4}$ $L_{JJKK} = (1.31 \pm 1.64)\cdot 10^{-3}$ $L_{JKKK} = 2.56\,(22)$	79BR1	79BR1	C_{2v} symmetry uncertainties are 2σ	2.6.3, 155a
	CHD=C=C=C=O	$v=0$	$b_p = -(5.91 \pm 0.70)\cdot 10^{-5}$ $B=2073.61\,(4)$ $C=2053.85\,(4)$	$D_J = -0.277\,(72)$ $D_{JK} = -243\,(20)$ $H_{JJK} = -(6.35 \pm 20)\cdot 10^{-2}$ $H_{JKK} = 6.79 \pm 2.2$ $L_{JJJK} = (5.1 \pm 10)\cdot 10^{-4}$ $L_{JJKK} = -(3.29 \pm 6)\cdot 10^{-3}$ $L_{JKKK} = 0.511\,(78)$				
Cont.								

Nr.	Molecule	v	A, B, C MHz	D, H kHz	Ref.	Ref. struct.	Varia, remarks	See also Tab., Nr.
353a Cont.	$CD_2{=}C{=}C{=}C{=}O$	$v=0$	$b_p = -(9.53 \pm 0.20) \cdot 10^{-5}$ $B = 1996.12\,(2)$ $C = 1970.75\,(2)$	$D_J = -2.68\,(26) \cdot 10^{-3}$ $D_{JK} = -135\,(5)$ $H_{JJK} = (5.9 \pm 20) \cdot 10^{-3}$ $H_{JKK} = 3.73\,(76)$ $L_{JJJK} = (1.2 \pm 3) \cdot 10^{-4}$ $L_{JJKK} = -(1.2 \pm 2) \cdot 10^{-3}$ $L_{JKKK} = 0.133\,(28)$	79BR1			
354	$CH_2{=}C{=}C{=}CH^{35}Cl$	$v=0$	$A = 25181\,(50)$ $B = 1551.10\,(2)$ $C = 1459.80\,(2)$	$D_J = 0.34\,(5)$ $D_{JK} = -45.4\,(1)$	74KAR2	74KAR2	diagonal force constants estimated	2.7.3, 123
		$v_1=1$	$A = 25304\,(50)$ $B = 1556.82\,(2)$ $C = 1463.05\,(2)$	$D_J = 0.34\,(5)$ $D_{JK} = -44.7\,(1)$			$\nu_1 = 100\,(50)\ cm^{-1}$, in-plane skeletal bending	
		$v_1=2$	$A = 25475\,(50)$ $B = 1562.52\,(2)$ $C = 1466.30\,(2)$	$D_J = 0.35\,(5)$ $D_{JK} = -43.7\,(1)$				
		$v_2=1$	$A = 25200\,(50)$ $B = 1551.22\,(2)$ $C = 1461.38\,(2)$	$D_J = 0.35\,(5)$ $D_{JK} = -47.4\,(1)$			$\nu_2 = 200\,(50)\ cm^{-1}$, out-of-plane skeletal bending	
355	$HC{\equiv}CCH(CH_3)_2$	$v=0$	$A = 7969.48\,(2)$ $B = 3828.801\,(2)$ $C = 2833.216\,(2)$	$d_J = -2.29\,(17)$ $d_{JK} = -71.0\,(35)$ $d_K = -112.8\,(38)$ $d_{WJ} = 0.89\,(4) \cdot 10^{-6}$ $d_{WK} = 23.2\,(10) \cdot 10^{-6}$	73MOC2	73MOC2	for structure, see also Vol. II/7 (3, 604)	2.6.3, 156 2.8, 38
		$v_1=1$	$A = 8035.7\,(5)$ $B = 3833.19\,(2)$ $C = 2837.93\,(2)$				$\nu_1 \simeq 192\ cm^{-1}$, C—C≡C bending	
		$v_2=1$	$A = 7963.4\,(11)$ $B = 3825.45\,(4)$ $C = 2830.30\,(3)$				$\nu_2 \simeq 222\ cm^{-1}$, symmetric CH_3 torsion	
Cont.		$v_3=1$	$A = 7916.7\,(4)$ $B = 3840.43\,(2)$ $C = 2832.66\,(2)$				$\nu_3 \simeq 253\ cm^{-1}$, asymmetric CH_3 torsion	

Nr.	Molecule	v	A, B, C MHz	D, H kHz	Ref.	Ref. struct.	Varia, remarks	See also Tab., Nr.
355 Cont.	$DC{\equiv}CCH(CH_3)_2$	$v=0$	$A=7957.2\,(5)$ $B=3553.60\,(1)$ $C=2681.19\,(1)$		73MOC2			
	$HC{\equiv}CCH(CH_3)^{13}CH_3$	$v=0$	$A=7776.4\,(12)$ $B=3791.51\,(4)$ $C=2788.91\,(4)$					
	$DC{\equiv}CCH(CH_3)^{13}CH_3$	$v=0$	$A=7763.0\,(7)$ $B=3519.01\,(2)$ $C=2640.07\,(2)$					
356 *6-151*	$HC{\equiv}CCH_2NH_2$ trans	$v=0$	$A=29619.92\,(6)$ $B=4595.172\,(14)$ $C=4160.570\,(9)$	$d_J=-22\,(1)$ $d_{JK}=-0.75\,(9)\cdot10^3$ $d_K=-2.5\,(6)\cdot10^3$ $d_{WJ}=5.6\,(2)\cdot10^{-6}$ $d_{WK}=1.3\,(2)\cdot10^{-4}$	77CER	77CER		2.6.3, 157 3, 101
	$HC{\equiv}CCH_2NHD$ trans	$v=0$	$A=27008.29\,(24)$ $B=4478.533\,(23)$ $C=4048.826\,(19)$					
	$DC{\equiv}CCH_2NH_2$ trans	$v=0$	$A=29019.91\,(10)$ $B=4272.227\,(10)$ $C=3882.219\,(8)$					
	$HC{\equiv}{}^{13}CCH_2NH_2$ trans	$v=0$	$A=29577.31\,(14)$ $B=4570.848\,(6)$ $C=4139.779\,(6)$					
	$H^{13}C{\equiv}CCH_2NH_2$ trans	$v=0$	$B=4450.186\,(8)$ $C=4039.980\,(7)$					
	$HC{\equiv}C^{13}CH_2NH_2$ trans	$v=0$	$B=4579.933\,(28)$ $C=4134.373\,(32)$					
	$HC{\equiv}CCH_2{}^{15}NH_2$ trans	$v=0$	$B=4488.360\,(9)$ $C=4066.566\,(8)$					
	$HC{\equiv}CCH_2ND_2$ trans	$v=0$	$A=24853.13\,(16)$ $B=4364.445\,(21)$ $C=3951.046\,(21)$					
Cont.	$DC{\equiv}CCH_2NHD$ trans	$v=0$	$A=26459.67\,(5)$ $B=4167.719\,(12)$ $C=3781.874\,(12)$					

Nr.	Molecule	v	A, B, C MHz	D, H kHz	Ref.	Ref. struct.	Varia, remarks	See also Tab., Nr.
356 *6-151* Cont.	$DC{\equiv}CCH_2ND_2$ trans	$v=0$	$A=24349.69\,(10)$ $B=\;4065.451\,(30)$ $C=\;3692.896\,(28)$		77CER			
357 *6-153*	$HC{\equiv}CCH_2OH$	$v=0$			73KAD	73KAD	for structure, see also Vol. II/7 (3, 373)	
358 *6-152*	$HC{\equiv}CCH_2SH$	$v=0$	$A^+=22350.46\,(33)$ $B^+=\;3045.98\,(4)$ $C^+=\;2757.32\,(4)$ $A^-=22342.15\,(30)$ $B^-=\;3045.43\,(5)$ $C^-=\;2757.44\,(4)$	$D_J\;=\;2$ (fixed) $D_{JK}=-50$ (fixed)	76MIR	76MIR	interaction constants (for their definition see Nr. 156 and [76MIR]): $Q_a\;=186.1\pm2$; $Q_b\;=-48.6\pm5$; $\Delta\;=6891.76$; $R_{ac}=-12.2$; $R_{bc}=0.4$ MHz	2.6.3, 158
	$HC{\equiv}CCH_2SD$	$v=0$	$A^+=20534.28\,(11)$ $B^+=\;3005.67\,(2)$ $C^+=\;2725.02\,(5)$ $A^-=20533.28\,(10)$ $B^-=\;3005.64\,(3)$ $C^-=\;2725.05\,(3)$	$D_J\;=\;2$ (fixed) $D_{JK}=-30$ (fixed)			$Q_a\;=14.7\,(5)$; $Q_b\;=-8\pm26$; $\Delta\;=388.5\,(1)$; $R_{ac}=-13.9$; $R_{bc}=1.9$ MHz barrier determined	
359	$HC{\equiv}CCH_2OCH_3$ gauche	$v=0$	$A=11756.7861\,(70)$ $B=\;3323.7926\,(20)$ $C=\;2795.3321\,(16)$	$\Delta_J\;=\;2.8783\,(60)$ $\Delta_{JK}=-20.712\,(45)$ $\Delta_K\;=\;98.475\,(20)$ $\delta_J\;=0.90293\,(94)$ $\delta_K\;=8.98\,(11)$ $H_J\;=\;0.0018\,(15)\cdot10^{-3}$ $H_{JK}=\;0.206\,(28)\cdot10^{-3}$ $H_{KJ}=-1.263\,(63)\cdot10^{-3}$ $H_K\;=\;4.15\,(11)\cdot10^{-3}$ $h_J\;=\;0.00381\,(38)\cdot10^{-3}$ $h_{JK}=-0.245\,(35)\cdot10^{-3}$ $h_K\;=\;4.43\,(32)\cdot10^{-3}$	76MAR2	80HAY4		2.6.3, 159 2.8, 86 3, 102
Cont.		$v_{T_1}=1$	$A=11830.487\,(14)$ $B=\;3317.5594\,(42)$ $C=\;2790.8917\,(36)$	$\Delta_J\;=\;3.002\,(14)$ $\Delta_{JK}=-22.11\,(15)$ $\Delta_K\;=\;110.433\,(86)$ $\delta_J\;=\;0.9257\,(36)$ $\delta_K\;=10.96\,(38)$			v_{T_1}, C—O torsion	

Nr.	Molecule	v	A, B, C MHz	D, H kHz	Ref.	Ref. struct.	Varia, remarks	See also Tab., Nr.
359 Cont.	$HC{\equiv}CCH_2OCH_3$ gauche			$H_J = 0.018(15)\cdot 10^{-3}$ $H_{JK} = 0.11(40)\cdot 10^{-3}$ $H_{KJ} = -0.80(76)\cdot 10^{-3}$ $H_K = 5.27(94)\cdot 10^{-3}$ $h_J = 0.0008(35)\cdot 10^{-3}$ $h_{JK} = 0.32(24)\cdot 10^{-3}$ $h_K = 0.1(47)\cdot 10^{-3}$	76MAR2			
		$v_{T_2}=1$	$A = 11798.74(7)$ *) $B = 3323.75(3)$ $C = 2794.83(3)$	$\Delta_J = 4.82(29)$ *) $\Delta_{JK} = -32.49(77)$ $\Delta_K = 244.2(65)$ $\delta_J = 0.7185(77)$ $\delta_K = 18.5(18)$			ν_{T_2}, CH_3 torsion *) for the A species	
	$HC{\equiv}CCH_2O^{13}CH_3$	$v=0$	$A = 11547.30(16)$ **) $B = 3271.11(8)$ $C = 2747.12(7)$	$d_J = 2.6(28)$ **) $d_{JK} = -19(8)$ $d_K = 100(30)$	80HAY4		**) uncertainties are 2.5 σ	
	$HC{\equiv}CCH_2{}^{18}OCH_3$	$v=0$	$A = 11610.71(22)$ **) $B = 3259.27(10)$ $C = 2750.56(9)$	$d_J = 4.0(36)$ **) $d_{JK} = -19(12)$ $d_K = 90(50)$				
	$HC{\equiv}CCH_2OCH_2D$-1	$v=0$	$A = 11359.17(17)$ **) $B = 3194.95(8)$ $C = 2685.60(8)$	$d_J = 1.1(32)$ **) $d_{JK} = -21(9)$ $d_K = 120(40)$			see Fig. 102	
	$HC{\equiv}CCH_2OCH_2D$-2	$v=0$	$A = 11067.11(15)$ **) $B = 3313.52(7)$ $C = 2747.77(6)$	$d_J = 2.3(27)$ **) $d_{JK} = -15(8)$ $d_K = 80(30)$				
	$HC{\equiv}CCH_2OCH_2D$-3	$v=0$	$A = 11228.58(16)$ **) $B = 3223.62(8)$ $C = 2737.04(7)$	$d_J = 2.9(33)$ **) $d_{JK} = -17(9)$ $d_K = 100(30)$				
	$HC{\equiv}CCHDOCH_3$-4	$v=0$	$A = 11040.87(18)$ **) $B = 3285.56(8)$ $C = 2777.12(7)$	$d_J = 2.8(18)$ **) $d_{JK} = -17(9)$ $d_K = 90(40)$				
	$HC{\equiv}CCHDOCH_3$-5	$v=0$	$A = 10829.09(17)$ **) $B = 3323.00(8)$ $C = 2740.03(7)$	$d_J = 1.9(31)$ **) $d_{JK} = -16(9)$ $d_K = 80(40)$				
Cont.	$HC{\equiv}CCH_2OCD_3$	$v=0$	$A = 10316.96(16)$ **) $B = 3093.99(8)$ $C = 2594.43(7)$	$d_J = 2.0(29)$ **) $d_{JK} = -16(9)$ $d_K = 80(30)$				

Nr.	Molecule	v	A, B, C MHz	D, H kHz	Ref.	Ref. struct.	Varia, remarks / See also Tab., Nr.
359 Cont.	$DC{\equiv}CCH_2OCH_3$	$v=0$	$A=11462.29\,(15)$ **) $B=\ 3129.10\,(7)$ $C=\ 2642.28\,(7)$	$d_J\ =2.0\,(28)$ **) $d_{JK}=-14\,(8)$ $d_K\ =90\,(30)$	76MAR2		**) uncertainties are 2.5 σ
360	$HC{\equiv}CCH_2OC({=}O)CF_3$ E C I		$B+C=1342\,(2)$ *) $B+C=1507\,(1)$ *) $B+C=1467\,(3)$ *)		77TR1		E: τ_1 (O=COC) = 0°; τ_2 (COCC) = 180° C: $\tau_1 = 0°$; $\tau_2 \simeq 90°$ I: $\tau_1 \simeq 60°$; $\tau_2 \simeq 270°$ *) LRMW
361	$HC{\equiv}CCH_2OC({=}O)F$ E C		$B+C=2515\,(1)$ *) $B+C=3175\,(1)$ *)		77TR1		*) LRMW, for the definitions see Nr. 360; excited state also measured
362	$HC{\equiv}CCH_2OC({=}O){}^{35}Cl$ E C $HC{\equiv}CCH_2OC({=}O){}^{37}Cl$ C		$B+C=1938\,(2)$ *) $B+C=2280\,(2)$ *) $B+C=2235\,(2)$ *)		77TR1		*) LRMW, for the definitions, see Nr. 360
363	$HC{\equiv}CCH_2OC({=}O)CN$ E C I		$B+C=1935\,(3)$ *) $B+C=2253\,(1)$ *) $B+C=2142\,(1)$ *)		77TR1		*) LRMW, for the definitions, see Nr. 360; excited states also measured
364 *6-154* *4-124*	$HC{\equiv}C{-}C({=}O)H$	$v=0$	$A=68035.299\,(43)$ $B=\ 4826.3014\,(73)$ $C=\ 4499.5107\,(69)$	$D_J\ =1.917\,(12)$ $D_{JK}=-148.102\,(47)$ $D_K\ =\ 8991.240\,(9730)$ $D_1(\delta_J)=\ 0.3458\,(14)$ $D_2(R_5)=-19.500\,(1590)$ $H_{KJ}=-8.31\,(13)\cdot10^{-3}$ $H_{JK}=-0.741\,(59)\cdot10^{-3}$	73WIN		for structure, see also Vol. II/7 (3, 351)
		$v_2=1$	$A=67953.26\,(28)$ *) $B=\ 4826.008\,(41)$ $C=\ 4497.528\,(41)$	$\Delta_J\ =\ 1.96\,(7)$ *) $\Delta_{JK}=\ 561\,(18)$ $\Delta_K\ =8991$ **) $\delta_J\ =\ 0.345$ **) $\delta_K\ =39$ **)	78TAK		*) IRMWDR **) fixed on the value of the ground state $\nu_2=2858.231\,(2)\ cm^{-1}$ [76TA]
Cont.							

Nr.	Molecule	v	A, B, C MHz	D, H kHz	Ref.	Ref. struct.	Varia, remarks	See also Tab., Nr.
364	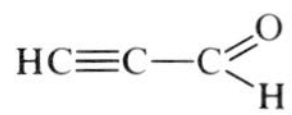	$v_9=1$	$A=63454.6\,(20)$*)		80JON		$\nu_9=$ 205.3 cm^{-1}	
6-154			$B=$ 4848.90 (10)					
4-124			$C=$ 4512.60 (10)					
Cont.		$v_9=2$	$B=4871.04\,(10)$*)				*) IRMWDR	
			$C=4524.55\,(10)$					
		$v_9=3$	$B\simeq4881$ *) $^{+}$)				$^{+}$) strongly perturbed	
			$C\simeq4529$					
		$v_{12}=1$	$B=4834.40\,(10)$*)				$\nu_{12}=260.6$ cm^{-1}	
			$C=4515.23\,(10)$					
		$v_{12}=2$	$B=4842.18\,(10)$*)					
			$C=4531.35\,(10)$					
		$v_{12}=3$	$B=4849.23\,(10)$*)					
			$C=4548.08\,(10)$					
		$v_9=1;\ v_{12}=1$	$B=4854.39\,(10)$*)					
			$C=4530.42\,(10)$					
		$v_9=2;\ v_{12}=1$	$B=4874.69\,(10)$*)					
			$C=4544.45\,(10)$					
		$v_9=1;\ v_{12}=2$	$B=4858.77\,(10)$*)					
			$C=4549.74\,(10)$					
		$v_9=2;\ v_{12}=2$	$B=4893.83\,(20)$*)					
			$C=4557.20\,(20)$					
		$v_{11}=1$?	$B=4828.50\,(10)$*)				$\nu_{11}\approx600$ cm^{-1}	
			$C=4500.48\,(10)$					
		$v_8=1$?	$B=4831.08\,(10)$*)					
			$C=4501.45\,(10)$				$\nu_8\approx600$ cm^{-1}	
		$v_7=1$?	$B=4827.93\,(10)$*)					
			$C=4488.22\,(10)$				$\nu_7\approx750$ cm^{-1}	
		$v_9=1;\ v_{11}=1$	$B=4849.88\,(10)$*)					
			$C=4516.64\,(10)$					
		$v=$?	$B=4848.47\,(10)$*)				$\nu\approx950$ cm^{-1}	
Cont.			$C=4538.07\,(10)$					

Nr.	Molecule	v	A, B, C MHz	D, H kHz	Ref.	Ref. struct.	Varia, remarks	See also Tab., Nr.
364 *6-154* *4-124* Cont.	DC≡C—C(=O)H	$v=0$	$A=66778.016(12)$ $B=4463.8489(7)$ $C=4177.7950(7)$	$\Delta_J=1.5919(5)$ $\Delta_{JK}=-139.214(13)$ $\Delta_K=9432.8(18)$ $\delta_J=0.2885(4)$ $\delta_K=30.69(4)$ $H_{JK}=-0.817(13)\cdot10^{-3}$ $H_{KJ}=-9.62(4)\cdot10^{-3}$ $H_K=2.55(8)$ $h_J=0.0047(3)\cdot10^{-3}$	80TA2			
		$v_2=1$	$A=66811(302)$ *) $B=4466(26)$ $C=4178(26)$	fixed at the ground state constants	80TA3		$\nu_2=2857.888(1)\ \mathrm{cm}^{-1}$ *) IRMWDR	
365	HC≡C—C(=O)F	$v=0$	$A=11826.610(6)$ $B=4169.236(4)$ $C=3077.434(4)$	$\Delta_J=0.549(73)$ $\Delta_{JK}=22.43(14)$ $\Delta_K=-10.52(87)$ $\delta_J=0.1798(35)$ $\delta_K=12.44(9)$	75DAV	75DAV	planar	2.6.3, 160
		$v_9=1$	$A=11696.203(20)$ $B=4188.930(6)$ $C=3081.448(6)$	$\Delta_J=0.74(12)$ $\Delta_{JK}=21.5(5)$ $\Delta_K=-25(4)$ $\delta_J=0.210(11)$ $\delta_K=11.1(3)$			$\nu_9=218(25)\ \mathrm{cm}^{-1}$	
		$v_{12}=1$	$A=11963.822(20)$ $B=4182.257(6)$ $C=3087.993(6)$	$\Delta_J=0.68(11)$ $\Delta_{JK}=24.4(5)$ $\Delta_K=-5.6(38)$ $\delta_J=0.193(9)$ $\delta_K=13.5(3)$			$\nu_{12}=188(25)\ \mathrm{cm}^{-1}$	
	DC≡C—C(=O)F	$v=0$	$A=11826.173(5)$ $B=3835.805(2)$ $C=2891.717(2)$	$\Delta_J=0.436(13)$ $\Delta_{JK}=21.49(8)$ $\Delta_K=-9.90(62)$ $\delta_J=0.1367(20)$ $\delta_K=11.99(6)$				

Nr.	Molecule	v	A, B, C MHz	D, H kHz	Ref.	Ref. struct.	Varia, remarks	See also Tab., Nr.
366	HC≡C—C(=O)^{35}Cl	$v=0$	$A=7190.008\,(6)$ $B=3149.241\,(2)$ $C=2186.745\,(2)$		74DAV3	74DAV3		2.7.3, 124
	HC≡C—C(=O)^{37}Cl	$v=0$	$A=7111.709\,(12)$ $B=3084.763\,(4)$ $C=2148.352\,(2)$					
367 *6-154a*	HC≡C—C(=O)OH	$v=0$	$A=12110.0172\,(46)$ $B=4146.9388\,(14)$ $C=3084.4861\,(12)$	$\Delta_J=0.5376\,(146)$ $\Delta_{JK}=21.32\,(10)$ $\Delta_K=-7.715\,(45)$ $\delta_J=0.1605\,(43)$ $\delta_K=12.19\,(13)$ $H_J=-0.220\,(63)\cdot10^{-3}$ $H_{JK}=6.88\,(158)\cdot10^{-3}$ $H_{KJ}=-19.9\,(41)\cdot10^{-3}$ $H_K=13.7\,(29)\cdot10^{-3}$ $h_J=0.117\,(31)\cdot10^{-3}$ $h_{JK}=-5.90\,(136)\cdot10^{-3}$ $h_K=58.6\,(119)\cdot10^{-3}$	76DAV1			
	HC≡C—C(=O)OD	$v=0$	$A=11858.4445\,(50)$ $B=4015.7137\,(15)$ $C=2995.5965\,(11)$	$\Delta_J=0.5205\,(140)$ $\Delta_{JK}=18.91\,(15)$ $\Delta_K=-4.026\,(70)$ $\delta_J=0.1593\,(70)$ $\delta_K=11.06\,(16)$ $H_J=-0.073\,(81)\cdot10^{-3}$ $H_{JK}=4.99\,(264)\cdot10^{-3}$ $H_{KJ}=-12.7\,(65)\cdot10^{-3}$ $H_K=9.4\,(49)\cdot10^{-3}$ $h_J=0.076\,(56)\cdot10^{-3}$ $h_{JK}=-2.91\,(221)\cdot10^{-3}$ $h_K=39.5\,(193)\cdot10^{-3}$				

Nr.	Molecule	v	A, B, C MHz	D, H kHz	Ref.	Ref. struct.	Varia, remarks	See also Tab., Nr.
368	$HC{\equiv}C{-}C({=}O)NH_2$	$v=0$	$A=11417.938\,(5)$ $B=4135.483\,(2)$ $C=3032.595\,(2)$	$\Delta_J=0.580\,(8)$ $\Delta_{JK}=21.09\,(6)$ $\Delta_K=-10.03\,(3)$ $\delta_J=0.194\,(6)$ $\delta_K=11.76\,(1)$	78LIT	78LIT	planar	2.6.3, 161 2.7.3, 42 3, 103
		$v_{in}=1$	$A=11306.856\,(9)$ $B=4153.947\,(3)$ $C=3036.029\,(4)$	$\Delta_J=0.586\,(21)$ $\Delta_{JK}=19.33\,(16)$ $\Delta_K=-16.98\,(46)$ $\delta_J=0.187\,(12)$ $\delta_K=11.08\,(23)$			$v_{in}=273\,(75)\,cm^{-1}$, C≡C—C in-plane bend	
		$v_{out}=1$	$A=11537.970\,(16)$ $B=4146.317\,(5)$ $C=3041.814\,(5)$	$\Delta_J=0.836\,(36)$ $\Delta_{JK}=22.78\,(24)$ $\Delta_K=-10.82\,(154)$ $\delta_J=0.232\,(15)$ $\delta_K=12.32\,(35)$			$v_{out}=275\,cm^{-1}$	
		$v_{inv}=1$	$A=11397.294\,(20)$ $B=4131.114\,(13)$ $C=3032.370\,(10)$	$\Delta_J=0.639\,(164)$ $\Delta_{JK}=20.66\,(43)$ $\Delta_K=-10.09\,(263)$ $\delta_J=0.175\,(28)$ $\delta_K=12.13\,(56)$			$v_{inv}=333\,(75)\,cm^{-1}$, NH_2 out-of-plane wagging	
	$HC{\equiv}C{-}C({=}O)ND_2$	$v=0$	$A=10267.745\,(9)$ $B=3979.454\,(4)$ $C=2865.486\,(3)$	$\Delta_J=0.463\,(14)$ $\Delta_{JK}=17.80\,(13)$ $\Delta_K=-8.74\,(22)$ $\delta_J=0.256\,(5)$ $\delta_K=8.87\,(15)$				
	$DC{\equiv}C{-}C({=}O)ND_2$	$v=0$	$A=10264.225\,(14)$ $B=3672.302\,(5)$ $C=2702.349\,(5)$	$\Delta_J=0.484\,(31)$ $\Delta_{JK}=17.19\,(14)$ $\Delta_K=-12.94\,(120)$ $\delta_J=0.159\,(8)$ $\delta_K=9.47\,(19)$				
369 Cont.	$HC{\equiv}C{-}OCH_2CH_3$ anti	$v=0$	$A=29500\,(200)$ $B=2502.178\,(7)$ $C=2380.034\,(8)$		74BJ1	74BJ1		2.6.3, 162 3, 104

Nr.	Molecule	v	A, B, C MHz	D, H kHz	Ref.	Ref. struct.	Varia, remarks	See also Tab., Nr.
369 Cont.	$HC{\equiv}C{-}OCH_2CH_3$ anti	$v_T=1$	$A=31000\,(300)$ $B=2507.968\,(8)$ $C=2387.767\,(8)$		74BJ1		$\nu_T=100\,(7)\,cm^{-1}$, internal torsion	
		$v_T=2$	$A=31100\,(900)$ $B=2515.728\,(7)$ $C=2395.685\,(7)$					
		$v_T=3$	$A=33000\,(2000)$ $B=2520.503\,(12)$ $C=2405.776\,(7)$					
	$HC{\equiv}C{-}OCH_2CH_3$ gauche	$v=0$	$A=11799.84\,(2)$ $B=3422.405\,(7)$ $C=2887.691\,(7)$	$d_J=-29.3\,(3)$ $d_{JK}=-473.8\,(96)$ $d_K=-796.1\,(239)$ $d_{WJ}=10.58\,(5)\cdot10^{-6}$ $d_{WK}=110.7\,(31)\cdot10^{-6}$			$E_{gauche}-E_{anti}=138\,(90)$ cal/mol	
		$v_T=1$	$A=11958.3\,(78)$ $B=3396.06\,(5)$ $C=2875.97\,(4)$				$\nu_T=78\,(15)\,cm^{-1}$, internal torsion potential function for rotation about the O—C_{ethyl} bond derived	
		$v_T=2$	$A=12255.9\,(160)$ $B=3364.72\,(9)$ $C=2862.03\,(8)$					
		$v_M=1$	$A=11807.5\,(48)$ $B=3421.54\,(3)$ $C=2885.80\,(4)$				$\nu_M=270\,(30)\,cm^{-1}$, CH_3 torsion	
370	$HC{\equiv}C{-}SCH_2CH_3$ gauche*)	$v=0$	$A=6345.494\,(7)$ $B=2960.155\,(4)$ $C=2189.444\,(4)$	$d_J=-14.16\,(10)$ $d_{JK}=-76.6\,(15)$ $d_K=-11.5\,(18)$ $d_{WJ}=6.679\,(30)\cdot10^{-6}$ $d_{WK}=13.42\,(60)\cdot10^{-6}$	74BJ2	74BJ2	*) CH_3 and $C{\equiv}CH$ groups gauche to each other	2.6.3, 163
Cont.		$v_\alpha=1$	$A=6403.407\,(14)$ $B=2942.734\,(8)$ $C=2183.360\,(8)$	$d_J=-14.58\,(20)$ $d_{JK}=-84.6\,(32)$ $d_K=-17.2\,(37)$ $d_{WJ}=6.945\,(70)\cdot10^{-6}$ $d_{WK}=15.64\,(120)\cdot10^{-6}$			$\nu_\alpha=90\,(20)\,cm^{-1}$, presumably C—S torsion	

Nr.	Molecule	v	A, B, C MHz	D, H kHz	Ref.	Ref. struct.	Varia, remarks	See also Tab., Nr.
370 Cont.	HC≡C—SCH_2CH_3 gauche *)	$v_\alpha=2$	$A=6465.066(30)$ $B=2925.559(20)$ $C=2177.421(20)$	$d_J = -14.71(90)$ $d_{JK} = -108.4(131)$ $d_K = 6.6(148)$ $d_{WJ} = 7.02(30)\cdot10^{-6}$ $d_{WK} = 23.6(51)\cdot10^{-6}$	74BJ2		*) CH_3 and C≡CH groups gauche to each other	
		$v_\beta=1$	$A=6359.64(9)$ $B=2965.89(2)$ $C=2189.82(2)$	$d_J = -12.1(14)$ $d_{JK} = -113.6(201)$ $d_K = -5.4(180)$ $d_{WJ} = 5.89(40)\cdot10^{-6}$ $d_{WK} = 27.3(76)\cdot10^{-6}$			$v_\beta=157(30)\ cm^{-1}$, presumably C—S—C bending	
		$v_\beta=2$	$A=6375.26(18)$ $B=2971.61(13)$ $C=2189.98(13)$	**)			**) assumed equal to $v_\beta=1$	
371	$^{35}ClC≡CCH_2{}^{35}Cl$	$v=0$	$A=16401.03(1)$ $B=9638.43(1)$ $C=9148.78(1)$	$\Delta_J = 0.1635(4)$ $\Delta_{JK} = -17.26(2)$ $\Delta_K = 767.5(6)$ $\delta_J = 0.0239(1)$ $\delta_K = 1.86(4)$	79GUN	75GUN		2.7.3, 125
		$v=1$	$A=16487.84(2)$ $B=9669.48(2)$ $C=9167.09(2)$	$\Delta_J = 0.1676(7)$ $\Delta_{JK} = -17.21(3)$ $\Delta_K = 786.0(1)$ $\delta_J = 0.0246(1)$ $\delta_K = 2.28(6)$			$v=85\ cm^{-1}$	
	$^{37}ClC≡CCH_2{}^{35}Cl$	$v=0$	$A=16348.76(1)$ $B=9380.71(1)$ $C=8914.69(1)$	$\Delta_J = 0.1536(4)$ $\Delta_{JK} = -16.80(2)$ $\Delta_K = 764.5(5)$ $\delta_J = 0.0219(1)$ $\delta_K = 1.82(4)$				
	$^{35}ClC≡CCH_2{}^{37}Cl$	$v=0$	$A=16192.34(1)$ $B=9415.66(1)$ $C=8941.51(1)$	$\Delta_J = 0.1590(5)$ $\Delta_{JK} = -17.05(2)$ $\Delta_K = 752.5(7)$ $\delta_J = 0.0225(1)$ $\delta_K = 1.88(4)$				

Nr.	Molecule	v	A, B, C MHz	D, H kHz	Ref.	Ref. struct.	Varia, remarks	See also Tab., Nr.
372	HC≡CCH$_2$CH$_2$OH gauche	$v=0$	$A=10438.35$ (15) $B=3385.87$ (5) $C=2760.54$ (5)		75SZA	75SZA		2.6.3, 164 3, 105
		$v_\tau=1$	$A=10520.77$ (15) $B=3373.54$ (5) $C=2752.17$ (5)				v_τ, probably C—C torsion	
		$v_\tau=2$	$A=10604.36$ (15) $B=3361.56$ (5) $C=2744.08$ (5)					
	HC≡CCH$_2$CH$_2$OD gauche	$v=0$	$A=9998.35$ (15) $B=3378.14$ (5) $C=2723.79$ (5)					
373 *6-158* *4-126*	HC≡C—CH=CH$_2$	$v=0$	$A=50308$ (55) $B=4744.9317$ (77) $C=4329.7899$ (77)	$D_J=1.93$ (22) $D_{JK}=-83.22$ (67)	80TON		for structure, see also Vol. II/7 (3, 482) harmonic force field calculated ν_{13} (a′) = 217 cm^{-1}	
		$v_{13}=1$	$A=49274$ (59) $B=4763.276$ (11) $C=4337.5385$ (99)	$D_J=2.17$ (28) $D_{JK}=-70.90$ (86)				
		$v_{18}=1$	$A=51676$ (53) $B=4747.7146$ (81) $C=4338.6433$ (83)	$D_J=1.86$ (21) $D_{JK}=-93.19$ (64)			ν_{18} (a″) = 304 cm^{-1} Coriolis constants: $^a\zeta_{13,18}=0.58$ (5); $^b\zeta_{13,18}=0.28$ (5)	
	DC≡C—CH=CH$_2$	$v=0$	$A=49393$ (34) $B=4403.9538$ (40) $C=4037.8007$ (40)	$D_J=1.56$ (11) $D_{JK}=-77.35$ (34)				
		$v_{13}=1$	$A=48337$ (36) $B=4420.6331$ (43) $C=4044.8607$ (43)	$D_J=1.60$ (12) $D_{JK}=-66.43$ (46)				
		$v_{18}=1$	$A=50866$ (49) $B=4406.1168$ (51) $C=4045.6294$ (51)	$D_J=1.60$ (14) $D_{JK}=-88.66$ (44)			Coriolis constants: $^a\zeta_{13,18}=0.61$ (5); $^b\zeta_{13,18}=0.26$ (5)	

Nr.	Molecule	v	A, B, C MHz	D, H kHz	Ref.	Ref. struct.	Varia, remarks	See also Tab., Nr.
374	$HC{\equiv}C{-}C^{35}Cl{=}CH_2$	$v=0$	$A=6875.239(4)$ $B=3096.353(2)$ $C=2131.950(2)$	$\Delta_J=1.05(2)$ $\Delta_{JK}=-3.08(5)$ $\Delta_K=16.0(2)$ $\delta_J=0.427(3)$ $\delta_K=2.39(5)$	74KAR1	74KAR1	harmonic force field calculated	2.7.3, 126 3, 106
	$HC{\equiv}C{-}C^{37}Cl{=}CH_2$	$v=0$	$A=6804.974(6)$ $B=3034.415(3)$ $C=2095.763(3)$	$\Delta_J=1.00(2)$ $\Delta_{JK}=-3.46(10)$ $\Delta_K=16.6(3)$ $\delta_J=0.425(6)$ $\delta_K=2.17(10)$				
375	$HC{\equiv}C{-}CH{=}CH^{35}Cl$ trans	$v=0$	$A=48604(41)$ $B=1509.612(2)$ $C=1463.459(2)$	$D_J=0.110(3)$ $D_{JK}=-16.19(2)$	73KAR2	73KAR2		2.7.3, 127 3, 107
	$HC{\equiv}C{-}CH{=}CH^{37}Cl$ trans	$v=0$	$A=48500(62)$ $B=1474.986(2)$ $C=1430.895(2)$	$D_J=0.109(4)$ $D_{JK}=-15.72(2)$				
	$DC{\equiv}C{-}CH{=}CH^{35}Cl$ trans	$v=0$	$A=47730(49)$ $B=1434.222(2)$ $C=1391.832(2)$	$D_J=0.087(3)$ $D_{JK}=-17.06(2)$				
	$HC{\equiv}C{-}CH{=}CH^{35}Cl$ cis	$v=0$	$A=8942.40(1)$ $B=2572.271(4)$ $C=1994.994(3)$	$\Delta_J=2.20(2)$ $\Delta_{JK}=-17.8(1)$ $\Delta_K=53(1)$ $\delta_J=0.700(2)$ $\delta_K=3.3(1)$	75KA2	75KA2		
	$HC{\equiv}C{-}CH{=}CH^{37}Cl$ cis	$v=0$	$A=8857.50(2)$ $B=2519.494(5)$ $C=1958.967(5)$	$\Delta_J=2.16(5)$ $\Delta_{JK}=-17.8(2)$ $\Delta_K=56(2)$ $\delta_J=0.661(5)$ $\delta_K=3.7(2)$				
	$DC{\equiv}C{-}CH{=}CH^{35}Cl$	$v=0$	$A=8569.43(2)$ $B=2450.872(10)$ $C=1903.284(9)$	$\Delta_J=2.07(5)$ $\Delta_{JK}=-16.3(5)$ $\Delta_K=45(5)$ $\delta_J=0.672(9)$ $\delta_K=2.3(4)$				

Nr.	Molecule	v	A, B, C MHz	D, H kHz	Ref.	Ref. struct.	Varia, remarks	See also Tab., Nr.
376	^{35}ClC≡C—CH=CH$_2$	$v=0$	$A=41927\,(50)$ $B=1410.388\,(4)$ $C=1363.563\,(4)$	$D_J=0.12\,(1)$ $D_{JK}=-28.55\,(2)$	75KA1	75KA1		2.7.3, 128 3, 108
	^{37}ClC≡C—CH=CH$_2$	$v=0$	$A=41907\,(50)$ $B=1377.027\,(4)$ $C=1332.289\,(4)$	$D_J=0.11\,(1)$ $D_{JK}=-27.94\,(2)$				
	^{35}ClC≡C—CH=CHD	$v=0$	$A=41705\,(50)$ $B=1347.939\,(4)$ $C=1304.846\,(4)$	$D_J=0.13\,(1)$ $D_{JK}=-29.65\,(2)$			D trans to the chloroacetylene group	
	^{35}ClC≡C—CH=CHD	$v=0$	$A=35145\,(50)$ $B=1382.398\,(6)$ $C=1328.929\,(6)$	$D_J=0.13\,(2)$ $D_{JK}=-21.05\,(3)$			D cis to the chloroacetylene group	
	^{35}ClC≡C—CD=CH$_2$	$v=0$	$A=35341\,(60)$ $B=1387.239\,(8)$ $C=1333.992\,(8)$	$D_J=0.13\,(2)$ $D_{JK}=-14.85\,(5)$				
377	HC≡CCH$_2$C≡CH	$v=0$	$A=19076.77\,(2)$ $B=2859.224\,(3)$ $C=2520.801\,(3)$	$\tau_{aaaa}=-2963\,(3)$ $\tau_{bbbb}=-11.38\,(1)$ $\tau_{cccc}=-3.165\,(11)$ $\tau_1=239.2\,(1)$ $\tau_2=21.82\,(1)$ $\tau_3=606.8\,(11)$	81KUC	81KUC	C$_{2v}$ symmetry	2.6.3, 165
		$v_x=1$	$A=19300.82\,(112)$ $B=2870.209\,(2)$ $C=2524.656\,(5)$	$\tau_{aaaa}=-10231\,(4466)$ $\tau_{bbbb}=-12.96\,(146)$ $\tau_{cccc}=-3.94\,(83)$ $\tau_1=237.9\,(90)$ $\tau_2=24.91\,(38)$ $\tau_3=1579\,(900)$			$\nu_x=140\,(20)\,cm^{-1}$, symmetric vibrational state	
	DC≡CCH$_2$C≡CD	$v=0$	$A=17408.28\,(112)$ $B=2574.372\,(13)$ $C=2270.134\,(13)$	$\tau_{aaaa}=-17460\,(45000)$ $\tau_{bbbb}=-10.59\,(171)$ $\tau_{cccc}=-3.16\,(57)$ $\tau_1=201.2\,(63)$ $\tau_2=20.7\,(89)$ $\tau_3=1252\,(2195)$				

2.5.2.4 Elemento-organic compounds

Nr.	Molecule	v	A, B, C MHz	D, H kHz	Ref.	Ref. struct.	Varia, remarks	See also Tab., Nr.
378	$^{28}SiH_2F_2$	$v=0$	$A = 24690.7782\ (27)$	$\Delta_J = 7.6734\ (55)$	80DAV3	80DAV3	for structure, see also	
4-137			$B = 7802.06045\ (84)$	$\Delta_{JK} = -58.501\ (42)$			Vol. II/7 (2, 188)	
			$C = 6377.12825\ (69)$	$\Delta_K = 307.6450\ (92)$			harmonic force field calculated	
				$\delta_J = 2.40688\ (41)$				
				$\delta_K = 12.864\ (28)$				
				$H_J = 1.969\ (78) \cdot 10^{-5}$				
				$H_{JK} = -2.116\ (88) \cdot 10^{-4}$				
				$H_{KJ} = -2.050\ (34) \cdot 10^{-3}$				
				$H_K = 1.2339\ (58) \cdot 10^{-2}$				
				$h_J = 9.72\ (14) \cdot 10^{-6}$			*) octic constant: $L_K P_a^8$	
				$L_K = -2.25\ (13) \cdot 10^{-7}$ *)				
		$v_4=1$	$A = 24933.6399\ (53)$	$\Delta_J = 7.632\ (22)$			$\nu_4 = 322\ cm^{-1}$	
			$B = 7799.9658\ (16)$	$\Delta_{JK} = -58.48\ (20)$				
			$C = 6364.2383\ (18)$	$\Delta_K = 320.906\ (45)$				
				$\delta_J = 2.4027\ (37)$				
				$\delta_K = 14.154\ (96)$				
				$H_{KJ} = -2.86\ (10) \cdot 10^{-3}$				
				$H_K = 1.446\ (23) \cdot 10^{-2}$				
				$h_J = 53.1\ (36) \cdot 10^{-6}$				
	$^{29}SiH_2F_2$	$v=0$	$A = 24402.7975\ (56)$	$\Delta_J = 7.656\ (32)$				
			$B = 7802.2340\ (17)$	$\Delta_{JK} = -57.80\ (30)$				
			$C = 6357.7169\ (22)$	$\Delta_K = 303.243\ (74)$				
				$\delta_J = 2.4185\ (87)$			**) H_J, H_{JK}, H_{KJ}, h_J are	
				$\delta_K = 12.63\ (14)$			constrained to the values	
				$H_K = 1.2137\ (27) \cdot 10^{-2}$ **)			obtained for $^{28}SiH_2F_2$	
	$^{30}SiH_2F_2$	$v=0$	$A = 24129.9862\ (96)$	$\Delta_J = 7.626\ (58)$				
			$B = 7802.3893\ (29)$	$\Delta_{JK} = -57.19\ (31)$				
			$C = 6339.0057\ (57)$	$\Delta_K = 298.980\ (90)$				
				$\delta_J = 2.436\ (14)$				
				$\delta_K = 12.38\ (53)$				
Cont.				$H_K = 1.1980\ (29) \cdot 10^{-2}$ **)				

Nr.	Molecule	v	A, B, C MHz	D, H kHz	Ref.	Ref. struct.	Varia, remarks	See also Tab., Nr.
378 *4-137* Cont.	$^{28}SiD_2F_2$	$v=0$	$A=18884.769\,(86)$ $B=7447.513\,(19)$ $C=6126.770\,(86)$	$\Delta_J=6.4866$ ***) $\Delta_{JK}=-33.4\,(20)$ $\Delta_K=128\,(22)$ $\delta_J=2.14\,(14)$ $\delta_K=7.4765$ ***)	80DAV3		***) constrained to the values obtained in the derived force field	
379	$^{28}SiH_2{}^{35}Cl_2$	$v=0$	$A=14135.006\,(8)$ *) $B=2573.581\,(2)$ $C=2232.613\,(2)$	$\Delta_J=1.022\,(4)$ *) $\Delta_{JK}=-15.46\,(5)$ $\Delta_K=142.1\,(1)$ $\delta_J=0.2410\,(4)$ $\delta_K=3.117\,(32)$	76DAV2	76DAV2	*) uncertainties are 2σ	2.6.3, 176 2.7.3, 129
	$^{28}SiH_2{}^{35}Cl^{37}Cl$	$v=0$	$A=14034.340\,(14)$ *) $B=2504.480\,(3)$ $C=2177.997\,(3)$	$\Delta_J=0.969\,(6)$ *) $\Delta_{JK}=-14.91\,(8)$ $\Delta_K=139.5\,(2)$ $\delta_J=0.2334\,(7)$ $\delta_K=2.490\,(118)$				
	$^{29}SiH_2{}^{35}Cl_2$	$v=0$	$A=13887.804\,(36)$ *) $B=2573.609\,(19)$ $C=2226.358\,(18)$					
380	$SiH_3CH_2{}^{79}Br$	$v=0$	$A=20075.59\,(49)$ *) $B=2151.50\,(2)$ $C=2012.49\,(3)$	$D_J=1.7\,(3)$ *)	74HAY2	74HAY2	*) uncertainties are 2.5σ	2.7.3, 155
	$SiH_3CH_2{}^{81}Br$	$v=0$	$A=20064.50\,(43)$ *) $B=2134.48\,(2)$ $C=1997.49\,(2)$	$D_J=1.7\,(3)$ *)				
	$SiD_3CH_2{}^{79}Br$	$v=0$	$A=16440.09\,(45)$ *) $B=1963.01\,(2)$ $C=1849.16\,(2)$	$D_J=1.2\,(4)$ *)				
	$SiD_3CH_2{}^{81}Br$	$v=0$	$A=16432.87\,(34)$ *) $B=1947.07\,(2)$ $C=1834.91\,(2)$	$D_J=1.0\,(3)$ *)				
381 *4-141*	$^{28}SiH_3CH_2CH_3$	$v=0$	$A=23348.40$ $B=4941.49$ $C=4515.28$		77DUR1		for structure, see also Vol. II/7 (3, 317)	2.8, 60

Nr.	Molecule	v	A, B, C MHz	D, H kHz	Ref.	Ref. struct.	Varia, remarks	See also Tab., Nr.
382	$CH_3CH_2CH_2SiH_3$	$v=0$	$A=20065.10\,(23)$ *) $B=2254.478\,(13)$ $C=2159.131\,(16)$	$d_J=0.57\,(14)$ *)	80HAY2	80HAY2	*) uncertainties are 2.5 σ	2.6.3, 177
		$v_\tau=1$	$A=19700$ $B=2251$ $C=2158$				$v_\tau=100\ cm^{-1}$, skeletal torsion	
	$CH_3CH_2CH_2SiD_3$	$v=0$	$A=16399.77\,(20)$ *) $B=2093.062\,(11)$ $C=2012.176\,(12)$	$d_J=0.40\,(12)$ *)				
	$CD_3CH_2CH_2SiH_3$	$v=0$	$A=17892.13\,(18)$ *) $B=2021.186\,(9)$ $C=1944.914\,(12)$	$d_J=0.39\,(9)$ *)				
	$CH_3CD_2CH_2SiH_3$	$v=0$	$A=17268.03\,(20)$ *) $B=2221.042\,(11)$ $C=2119.647\,(14)$	$d_J=0.49\,(11)$ *)				
	$CH_3CH_2CD_2SiH_3$	$v=0$	$A=16976.86\,(20)$ *) $B=2238.955\,(12)$ $C=2131.382\,(13)$	$d_J=0.45\,(19)$ *)				
383 *6-167c*	$CH_3SiH_2{}^{35}Cl$	$v=0$	$A=16110.945\,(121)$ $B=3904.428\,(4)$ $C=3325.555\,(3)$	$\Delta_J=2.1831\,(868)$ $\Delta_{JK}=-15.5892\,(4540)$ $\Delta_K=163.8733\pm9.07$ $\delta_J=0.5640\,(27)$ $\delta_K=7.4589\pm0.710$	75ZEI	80ZEI		2.7.3, 130
	$CH_3SiH_2{}^{37}Cl$	$v=0$	$A=16060.830\,(166)$ $B=3806.205\,(4)$ $C=3251.962\,(3)$	$\Delta_J=2.2474\,(1060)$ $\Delta_{JK}=-14.5190\pm0.457$ $\Delta_K=176.8706\pm15.8$ $\delta_J=0.5228\,(42)$ $\delta_K=8.5665\pm1.01$				
	$CH_3SiD_2{}^{35}Cl$	$v=0$	$A=13173.762\,(108)$ $B=3790.510\,(5)$ $C=3219.300\,(5)$	$\Delta_J=1.6682\pm0.108$ $\Delta_{JK}=-8.2091\pm0.243$ $\Delta_K=83.4795\pm6.08$ $\delta_J=0.5039\,(35)$ $\delta_K=6.8032\pm0.603$				
Cont.								

Nr.	Molecule	v	A, B, C MHz	D, H kHz	Ref.	Ref. struct.	Varia, remarks	See also Tab., Nr.
383 *6-167c* Cont.	$CH_3SiD_2{}^{37}Cl$	$v=0$	$A=13137.899\,(168)$ $B=3694.791\,(7)$ $C=3147.911\,(6)$	$\Delta_J=1.9825\,(1620)$ $\Delta_{JK}=-9.2778\pm0.425$ $\Delta_K=104.0023\pm11.1$ $\delta_J=0.4627\,(53)$ $\delta_K=6.5225\pm0.867$	75ZEI			
	$CD_3SiD_2{}^{35}Cl$	$v=0$	$A=11523.804\,(85)$ $B=3403.434\,(5)$ $C=2896.453\,(4)$	$\Delta_J=1.6293\,(1310)$ $\Delta_{JK}=-9.2621\pm0.456$ $\Delta_K=61.5918\pm3.92$ $\delta_J=0.4402\,(24)$ $\delta_K=3.0165\pm0.566$				
	$CD_3SiD_2{}^{37}Cl$	$v=0$	$A=11486.886\,(173)$ $B=3318.997\,(5)$ $C=2832.784\,(5)$	$\Delta_J=1.5447\,(1520)$ $\Delta_{JK}=-9.0378\pm0.293$ $\Delta_K=51.4730\pm9.83$ $\delta_J=0.4227\pm0.006$ $\delta_K=1.1926\pm1.04$				
	$CD_3SiH_2{}^{35}Cl$	$v=0$	$I_a=36.46352$ *) $I_b=144.96476$ $I_c=169.42291$		80ZEI		*) in amu A^2	
	$CD_3SiH_2{}^{37}Cl$	$v=0$	$I_a=36.59237$ *) $I_b=148.64985$ $I_c=173.23738$					
	$CH_2DSiH_2{}^{35}Cl$ symmetric	$v=0$	$I_a=31.49611$ *) $I_b=137.42655$ $I_c=160.05477$					
	$CH_2DSiH_2{}^{35}Cl$ asymmetric	$v=0$	$I_a=33.93097$ *) $I_b=133.21628$ $I_c=156.91893$					
	$CD_2HSiH_2{}^{35}Cl$ symmetric	$v=0$	$I_a=36.40420$ *) $I_b=137.15615$ $I_c=161.58985$					
	$CD_2HSiH_2{}^{35}Cl$ asymmetric	$v=0$	$I_a=34.01774$ *) $I_b=141.12201$ $I_c=164.87349$					
Cont.	$^{13}CH_3SiH_2{}^{35}Cl$	$v=0$	$I_a=31.89296$ *) $I_b=132.80829$ $I_c=155.86746$					

Nr.	Molecule	v	A, B, C MHz	D, H kHz	Ref.	Ref. struct.	Varia, remarks	See also Tab., Nr.
383 *6-167c* Cont.	$CH_3{}^{29}SiH_2{}^{35}Cl$	$v=0$	$I_a=31.73663$ *) $I_b=129.75491$ $I_c=152.65125$		80ZEI		*) in amu Å^2	
384	$CH_3{}^{28}SiH^{35}Cl_2$	$v=0$	$A=4342.232(8)$ $B=2433.310(5)$ $C=1706.065(5)$	$\Delta_J=67(6)$ $\Delta_{JK}=-72(21)$ $\Delta_K=493(64)$ $\delta_J=24(1)$ $\delta_K=109(14)$	79KRA, 77EN	79KRA		2.7.3, 131, 2.8, 61
	$CH_3{}^{28}SiH^{35}Cl^{37}Cl$	$v=0$	$A=4313.709(6)$ $B=2370.967(4)$ $C=1671.243(3)$	$\Delta_J=64(3)$ $\Delta_{JK}=-79(17)$ $\Delta_K=468(48)$ $\delta_J=21(1)$ $\delta_K=126(13)$				
	$CH_3{}^{28}SiH^{37}Cl_2$	$v=0$	$A=4283.332(8)$ $B=2310.723(5)$ $C=1636.966(3)$					
	$CH_3{}^{29}SiH^{35}Cl_2$	$v=0$	$A=4328.515(8)$ $B=2430.977(7)$ $C=1705.133(3)$		79KRA			
	$CH_3{}^{30}SiH^{35}Cl_2$	$v=0$	$A=4315.206(8)$ $B=2428.711(7)$ $C=1704.206(3)$					
	$^{13}CH_3{}^{28}SiH^{35}Cl_2$	$v=0$	$A=4184.484(9)$ $B=2432.011(7)$ $C=1681.789(4)$					
385	CH_3OSiH_3	$v=0$	$I_a=15.059$ *) $I_b=88.513$ *) $C=5353.28$		74LEC		for structure, see also Vol. II/7 (3, 135) *) in amu A^2	2.6.3, 178 2.8,8 7
	CD_3OSiH_3	$v=0$	$I_a=18.297$ *) $I_b=101.329$ *) $C=4710.52$					
Cont.	CH_3OSiD_3	$v=0$	$I_a=20.643$ *) $I_b=98.091$ *) $C=4872.56$					

Nr.	Molecule	v	A, B, C MHz	D, H kHz	Ref.	Ref. struct.	Varia, remarks	See also Tab., Nr.
385 Cont.	CD_3OSiD_3	$v=0$	$I_a = 23.856$ *) $I_b = 111.676$ $C = 4306.73$		74LEC			
386	CH_3OSiF_3	$v=0$	$A = 4092$ *) $B = 2265.68$ (7) $C = 2248.71$ (7)		78OD2	78OD2	*) calculated from the structure	
	CD_3OSiF_3	$v=0$	$A = 3996$ *) $B = 2051.17$ (6) $C = 2034.61$ (6)					
387	$(SiH_3)_2O$						for structure, see also Vol. II/7 (2, 292) quasi-linear double rotor molecule [78DOS2]	
388	$(SiH_3)_2S$	$v=0$	$A = 9543.346$ (9) *) $B = 2852.534$ (2) $C = 2308.506$ (2)	$D'_J = 2.116$ (25) *) $D'_{JK} = -13.39$ (24) $D'_K = 60.7$ (17) $\delta_J = 0.651$ (8) $R'_6 = 0.012$ (9)	78DÖS2	78DÖS2	for structure, see also Vol. II/7 (2, 302) *) for the AA species	2.6.3, 179 2.8, 106
389	CH_3SSiH_3	$v=0$	$A = 13820.98$ (45) $B = 4316.28$ (5) $C = 3489.16$ (4)	$d_J = 15$ (12) $d_{JK} = -0.16 (2) \cdot 10^3$	80SH1	80SH1		2.6.3, 180 2.8, 101
	CH_3SSiD_3	$v=0$	$A = 11876.41$ (44) $B = 3898.61$ (5) $C = 3207.64$ (4)	$d_J = 10$ (5) $d_{JK} = -0.12 (2) \cdot 10^3$				
390	$SiH_3CH_2OCH_3$ trans	$v=0$	$A = 22992.79$ (27) *) $B = 2500.67$ (2) $C = 2387.26$ (2)	$d_J = 0.68$ (18) *) $d_{JK} = -6.0$ (48)	81SHI	81SHI	*) uncertainties are 2.5 σ	2.6.3, 181 2.8, 88 3, 109
	$^{29}SiH_3CH_2OCH_3$ trans	$v=0$	$A = 22983.35$ (29) $B = 2466.98$ (1) $C = 2356.44$ (2)	$d_J = 0.69$ (15) $d_{JK} = -5.6$ (44)				
Cont.	$SiH_3CH_2{}^{18}OCH_3$ trans	$v=0$	$A = 22602.58$ (27) $B = 2477.00$ (2) $C = 2361.51$ (2)	$d_J = 0.70$ (18) $d_{JK} = -5.3$ (50)				

Nr.	Molecule	v	A, B, C MHz	D, H kHz	Ref.	Ref. struct.	Varia, remarks	See also Tab., Nr.
390 Cont.	$SiH_3CH_2O^{13}CH_3$ trans	$v=0$	$A=22992.73$ (21) $B=2435.71$ (1) $C=2328.00$ (2)	$d_J=0.61$ (14) $d_{JK}=-5.9$ (39)	81SHI			
	$SiH_2DCH_2OCH_3$ symmetric, trans	$v=0$	$A=21986.87$ (25) $B=2417.37$ (2) $C=2300.92$ (2)	$d_J=0.58$ (17) $d_{JK}=-3.4$ (45)				
	$SiH_2DCH_2OCH_3$ asymmetric, trans	$v=0$	$A=20860.78$ (23) $B=2442.60$ (1) $C=2337.96$ (2)	$d_J=0.67$ (16) $d_{JK}=-4.7$ (43)				
	$SiH_3CD_2OCH_3$ trans	$v=0$	$A=18904.54$ (26) $B=2480.78$ (2) $C=2351.32$ (2)	$d_J=0.60$ (20) $d_{JK}=-6.0$ (fixed)				
	$SiH_3CH_2OCH_2D$ symmetric, trans	$v=0$	$A=22264.38$ (23) $B=2396.10$ (1) $C=2284.50$ (2)	$d_J=0.63$ (15) $d_{JK}=-6.0$ (42)				
	$SiH_3CH_2OCH_2D$ asymmetric, trans	$v=0$	$A=21874.07$ (23) $B=2414.82$ (1) $C=2312.52$ (2)	$d_J=0.59$ (16) $d_{JK}=-5.2$ (43)				
	$SiD_3CH_2OCH_3$ trans	$v=0$	$A=18267.47$ (24) $B=2312.19$ (2) $C=2216.95$ (2)	$d_J=0.56$ (17) $d_{JK}=-0.7$ (46)				
	$SiH_3CH_2OCD_3$ trans	$v=0$	$A=20135.74$ (26) $B=2248.36$ (2) $C=2157.20$ (2)	$d_J=0.49$ (17) $d_{JK}=-4.2$ (47)				
		$v_{T_1}=1$	$A=19766.40$ (20) $B=2252.04$ (1) $C=2163.09$ (1)	$d_J=0.55$ (11) $d_{JK}=-3.3$ (28)			ν_{T_1}, skeletal torsion	
		$v_{T_2}=1$	$A=19981.26$ (23) $B=2245.94$ (1) $C=2156.48$ (1)	$d_J=0.52$ (12) $d_{JK}=-1.3$ (30)			ν_{T_2}, silyl torsion	
		$v_{T_3}=1$	$A=20123.60$ (21) $B=2246.77$ (1) $C=2156.33$ (1)	$d_J=0.51$ (10) $d_{JK}=-20.4$ (24)			ν_{T_3}, methyl torsion	

Nr.	Molecule	v	A, B, C MHz	D, H kHz	Ref.	Ref. struct.	Varia, remarks	See also Tab., Nr.
391 *4-134*	$(CH_3)_2SiH_2$	$v_n=1_1$	$A=16696.408$ *) $B=5934.794$ $C=4891.172$		73TRI		for structure, see also Vol. II/7 (3, 317) *) for the AA species	2.6.3, 183 2.8, 62
		$v_n=1_2$	$A=16621.371$ *) $B=5930.181$ $C=4892.781$					
392	$(SiH_3)_2CH_2$	$v=0$	$A=18449.77\,(24)$ $B=4941.88\,(3)$ $C=2739.60\,(4)$	$D_J=2.67\,(51)$	78SHI	78SHI	for structure, see also Vol. II/7 (3, 140) uncertainties are 2.5 σ	2.6.3, 182 2.8, 63
	$^{29}SiH_3CH_2SiH_3$	$v=0$	$A=18428.16\,(22)$ $B=4900.69\,(2)$ $C=2703.38\,(3)$	$D_J=2.66\,(47)$				
	$(SiH_3)_2CHD$	$v=0$	$A=16649.79\,(23)$ $B=2931.13\,(3)$ $C=2705.85\,(3)$	$D_J=2.51\,(49)$				
	$SiH_3CH_2SiH_2D$, symmetric	$v=0$	$A=18145.00\,(24)$ $B=2816.37\,(3)$ $C=2624.96\,(4)$	$D_J=2.24\,(51)$				
	$SiH_3CH_2SiH_2D$, asymmetric	$v=0$	$A=16926.19\,(27)$ $B=2876.29\,(3)$ $C=2684.14\,(4)$	$D_J=2.56\,(56)$				
	$(SiD_3)_2CH_2$	$v=0$	$A=13144.63\,(11)$ $B=2482.94\,(1)$ $C=2344.03\,(1)$	$D_J=1.23\,(22)$				
393	$(CH_3)_2Si^{35}Cl_2$	$v=0$	$A=2584.140\,(12)$ *) $B=1842.514\,(19)$ $C=1628.643\,(22)$	$d_J=-2.52\,(18)$ *) $d_{JK}=-1.54\,(5)$ $d_K=0.561\,(14)$ $d_{WJ}=1.67\,(7)\cdot10^{-6}$	80NA3	80NA3	*) uncertainties are 3 σ	2.7.3, 132
	$(CH_3)_2Si^{35}Cl^{37}Cl$	$v=0$	$A=2568.830\,(19)$ *) $B=1806.538\,(21)$ $C=1594.588\,(22)$	$d_J=-2.24\,(31)$ *) $d_{JK}=-1.39\,(14)$ $d_K=0.612\,(25)$ $d_{WJ}=1.43\,(17)\cdot10^{-6}$				

Nr.	Molecule	v	A, B, C MHz	D, H kHz	Ref.	Ref. struct.	Varia, remarks	See also Tab., Nr.
394	$(CH_3)_2SiHCN$	$v=0$	$A=5311.22$ (23) $B=2845.97$ (2) $C=2086.17$ (1)		75DUR1	75DUR1		2.6.3, 184
		$v_\alpha=1$	$A=5291.62$ (25) $B=2855.28$ (2) $C=2085.66$ (2)				$\nu_\alpha=145$ (50) cm^{-1}	
		$v_\beta=1$	$A=5312.26$ (22) $B=2856.17$ (2) $C=2087.86$ (2)				$\nu_\beta=273$ (30) cm^{-1}	
		$v_\gamma=1$	$A=5244.22$ (37) $B=2862.93$ (2) $C=2089.84$ (2)				$\nu_\gamma=303$ (15) cm^{-1}	
395	$CH_3CH_2SiH_2F$ trans	$v=0$	$B=2871.63$ (4) *) $C=2619.48$ (4)	$d_J=0.72$ (38) *)	81HAY1	81HAY1	*) uncertainties are 2.5σ	2.6.3, 185 2.8, 49a 3, 109a
		$v_{T_1}=1$	$B=2869.85$ (8) $C=2625.31$ (7)	$d_J=1.01$ (73)			$\nu_{T_1}=90\ cm^{-1}$, skeletal torsion	
		$v_{T_2}=1$	$B=2873.07$ (4) $C=2619.00$ (6)	$d_J=2.64$ (44)			$\nu_{T_2}=170\ cm^{-1}$, CH_3 torsion	
		$v_\delta=1$	$B=2876.27$ (11) $C=2619.98$ (10)	$d_J=0.77$ (105)			$\nu_\delta=250\ cm^{-1}$, C—Si—F bend [77HAY1]	
	$^{13}CH_3CH_2SiH_2F$	$v=0$	$B=2788.06$ (4) $C=2549.62$ (4)	$d_J=0.54$ (38)				
	$CH_3{}^{13}CH_2SiH_2F$	$v=0$	$B=2856.26$ (2) $C=2600.71$ (2)	$d_J=0.38$ (23)				
	$CH_3CH_2{}^{29}SiH_2F$	$v=0$	$B=2867.97$ (6) $C=2613.40$ (4)	$d_J=0.65$ (45)				
	s-$CH_2DCH_2SiH_2F$	$v=0$	$B=2717.51$ (7) $C=2487.17$ (6)	$d_J=0.04$ (60)				
	a-$CH_2DCH_2SiH_2F$	$v=0$	$B=2769.86$ (5) $C=2536.77$ (4)	$d_J=0.85$ (43)				
	$CH_3CHDSiH_2F$	$v=0$	$B=2846.57$ (7) $C=2583.85$ (6)	$d_J=1.02$ (61)				
Cont.	CH_3CH_2SiHDF	$v=0$	$B=2846.68$ (5) $C=2591.20$ (4)	$d_J=0.51$ (44)				

Nr.	Molecule	v	A, B, C MHz	D, H kHz	Ref.	Ref. struct.	Varia, remarks	See also Tab., Nr.
395 Cont.	$CH_3CH_2SiD_2F$ trans	$v=0$	$B=2816.97(7)$ $C=2568.00(6)$	$d_J=0.64(64)$	81HAY1			
	$CH_3CH_2SiH_2F$ gauche	$v=0$	$A=10253.32(15)$ $B=3615.08(2)$ $C=3031.08(2)$	$d_J=5.2(13)$ d_{JK} $-98(3)$ $d_K=-592(30)$ $d_{EJ}=75(2)$ **) $d_{EK}=652(28)$			$E_{trans}-E_{gauche}=55(50)$ cal/mol [81HAY1]	
		$v_{T_1}=1$	$A=10315.44(65)$ *) $B=3603.69(8)$ $C=3027.49(7)$	$d_J=4.4(18)$ $d_{JK}=-15(32)$			$\nu_{T_1}=75\ cm^{-1}$, skeletal torsion *) uncertainties are 2.5 σ	
		$v_{T_2}=1$	$A=10287.49(100)$ $B=3612.65(12)$ $C=3028.30(10)$	$d_J=4.7(28)$ $d_{JK}=-13(50)$			$\nu_{T_2}=140\ cm^{-1}$, methyl torsion [77HAY1] **) d_{EJ} is the coefficient of $-E_r J(J+1)$ and d_{EK} that of $-E_r \langle P_z^2 \rangle$ where E_r is the reduced energy	
	$^{13}CH_3CH_2SiH_2F$ gauche	$v=0$	$A=10178.53(43)$ $B=3517.82(9)$ $C=2957.74(7)$	$d_J=3.8(19)$ $d_{JK}=-18(31)$				
	$CH_3{}^{13}CH_2SiH_2F$ gauche	$v=0$	$A=10162.01(52)$ $B=3579.67(11)$ $C=3004.84(8)$	$d_J=3.3(22)$ $d_{JK}=-17(36)$				
	$CH_3CH_2{}^{29}SiH_2F$ gauche	$v=0$	$A=10186.64(22)$ $B=3603.47(3)$ $C=3018.17(2)$	constants of parent species assumed				
	$CH_2DCH_2SiH_2F$-1 gauche	$v=0$	$A=10189.86(44)$ $B=3398.77(9)$ $C=2873.36(7)$	$d_J=3.3(19)$ $d_{JK}=-15(40)$			for the numbering of the atoms, see Fig. 109a	
	$CH_2DCH_2SiH_2F$-2 gauche	$v=0$	$A=9773.18(38)$ $B=3555.79(9)$ $C=2946.97(6)$	$d_J=3.4(18)$ $d_{JK}=-13(40)$				
	$CH_2DCH_2SiH_2F$-3 gauche	$v=0$	$A=9890.79(47)$ $B=3473.49(9)$ $C=2955.65(7)$	$d_J=5.3(20)$ $d_{JK}=-22(37)$				
Cont.	$CH_3CHDSiH_2F$-4 gauche	$v=0$	$A=9800.56(51)$ $B=3553.80(10)$ $C=2950.80(8)$	$d_J=4.5(22)$ $d_{JK}=-18(36)$				

Nr.	Molecule	v	A, B, C MHz	D, H kHz	Ref.	Ref. struct.	Varia, remarks	See also Tab., Nr.
395 Cont.	$CH_3CHDSiH_2F$-5 gauche	$v=0$	$A=9766.26(52)$ $B=3529.73(11)$ $C=3006.77(8)$	$d_J=3.9(24)$ $d_{JK}=-27(28)$	81HAY1			
	CH_3CH_2SiHDF-6 gauche	$v=0$	$A=9718.35(56)$ $B=3554.24(11)$ $C=2948.50(9)$	$d_J=3.5(24)$ $d_{JK}=-10(34)$				
	CH_3CH_2SiHDF-7 gauche	$v=0$	$A=9631.72(51)$ $B=3536.53(10)$ $C=3012.62(9)$	$d_J=4.3(23)$ $d_{JK}=-17(32)$				
	$CH_3CH_2SiD_2F$ gauche	$v=0$	$A=9159.59(33)$ $B=3476.08(1)$ $C=2933.28(1)$	$d_J=3.1(7)$ $d_{JK}=-77(5)$ $d_K=1804(87)$ $d_{EJ}=49(1)$ *) $d_{EK}=-3206(44)$			*) uncertainties are 2.5 σ	
396 *6-167d*	$CH_3CH_2SiH_2{}^{35}Cl$ gauche	$v=0$	$A=7191.8752(59)$ $B=2585.1897(33)$ $C=2083.6754(33)$	$\Delta_J=2.46(12)$ $\Delta_{JK}=-14.382(55)$ $\Delta_K=40.91(56)$ $\delta_J=0.8317(18)$ $\delta_K=4.175(63)$	74TYP	74TYP	for structure, see also Vol. II/7 (3, 309)	2.7.3, 133 3, 110
	$CH_3CH_2SiH_2{}^{37}Cl$ gauche	$v=0$	$A=7140.041(14)$ $B=2528.0910(24)$ $C=2042.2961(22)$	$\Delta_J=2.233(41)$ $\Delta_{JK}=-13.28(20)$ $\Delta_K=31.9(21)$ $\delta_J=0.8051(42)$ $\delta_K=3.23(16)$				
	$CH_3CH_2SiD_2{}^{35}Cl$ gauche	$v=0$	$A=6392.1844(92)$ $B=2543.2853(41)$ $C=2031.9035(40)$	$\Delta_J=2.354(71)$ $\Delta_{JK}=-12.59(22)$ $\Delta_K=31.7(16)$ $\delta_J=0.7835(61)$ $\delta_K=3.14(19)$				
	$CH_3CD_2SiH_2{}^{35}Cl$ gauche	$v=0$	$A=6836.3687(50)$ $B=2461.0140(19)$ $C=2013.1686(18)$	$\Delta_J=2.034(24)$ $\Delta_{JK}=-10.163(73)$ $\Delta_K=29.46(68)$ $\delta_J=0.6427(21)$ $\delta_K=3.386(78)$				
Cont.								

Nr.	Molecule	v	A, B, C MHz	D, H kHz	Ref.	Ref. struct.	Varia, remarks	See also Tab., Nr.
396 *6–167 d* Cont.	$CD_3CH_2SiH_2{}^{35}Cl$ gauche	$v=0$	$A=6558.1270\,(76)$ $B=2372.0360\,(45)$ $C=1918.0068\,(43)$	$\Delta_J=1.923\,(95)$ $\Delta_{JK}=-13.22\,(32)$ $\Delta_K=33.5\,(24)$ $\delta_J=0.7400\,(101)$ $\delta_K=3.40\,(38)$	74TYP			
	$CH_3CH_2SiH_2{}^{35}Cl$ trans	$v=0$	$A=14742.5\,(21)$ $B=1876.7444\,(37)$ $C=1733.8894\,(33)$	$\Delta_J=0.3240\,(93)$ $\Delta_{JK}=-5.14\,(39)$ $\Delta_K=97.84$ (fixed) $\delta_J=0.051\,(12)$ $\delta_K=1.045$ (fixed)				
	$CH_3CH_2SiH_2{}^{37}Cl$ trans	$v=0$	$A=14693.9\,(34)$ $B=1833.0635\,(77)$ $C=1695.8528\,(62)$	$\Delta_J=0.349\,(22)$ $\Delta_{JK}=-5.77\,(26)$ $\Delta_K=98.34$ (fixed) $\delta_J=0.056\,(31)$ $\delta_K=1.025$ (fixed)				
	$CH_3CH_2SiD_2{}^{35}Cl$ trans	$v=0$	$A=12192.6\,(15)$ $B=1856.1307\,(42)$ $C=1708.6652\,(42)$	$\Delta_J=0.318\,(13)$ $\Delta_{JK}=-4.10\,(35)$ $\Delta_K=57.62$ (fixed) $\delta_J=0.095\,(19)$ $\delta_K=0.973$ (fixed)				
	$CH_3CD_2SiH_2{}^{35}Cl$ trans	$v=0$	$A=12874.5\,(15)$ $B=1845.5939\,(41)$ $C=1696.6074\,(52)$	$\Delta_J=0.307\,(12)$ $\Delta_{JK}=-4.59\,(27)$ $\Delta_K=56.74$ (fixed) $\delta_J=0.036\,(20)$ $\delta_K=0.858$ (fixed)				
	$CD_3CH_2SiH_2{}^{35}Cl$ trans	$v=0$	$A=13592.61\,(21)$ $B=1684.5735\,(57)$ $C=1569.7507\,(36)$	$\Delta_J=0.264\,(10)$ $\Delta_{JK}=-3.19\,(34)$ $\Delta_K=72.17$ (fixed) $\delta_J=0.130\,(17)$ $\delta_K=0.449$ (fixed)				
397 *6–167 a*	$CH_3CH_2SiH_2CH_3$ trans	$v=0$	$A=15114.85\,(20)$ *) $B=2703.36\,(10)$ $C=2463.25\,(10)$		72HAY	72HAY	for structure, see also Vol. II/7 (3, 456) *) for the A species	2.6.3, 186
	$CH_3CH_2SiD_2CH_3$ trans	$v=0$	$A=12645.98\,(20)$ *) $B=2656.80\,(10)$ $C=2415.73\,(10)$					

Nr.	Molecule	v	A, B, C MHz	D, H kHz	Ref.	Ref. struct.	Varia, remarks	See also Tab., Nr.
398	$CH_2{=}CHSiH_2CH_3$ cis	$v=0$	$A=9599.30\,(45)$ $B=3984.78\,(6)$ $C=2962.59\,(5)$	$d_J=2.4\,(26)$ $d_{JK}=-29\,(15)$ $d_{EJ}=26\,(3)$ **)	81IMA	81IMA	**) for definition of d_{EJ}, see Nr. 395	2.6.3, 187 2.8, 64 3, 110a
	$CH_2{=}CHSiH_2{}^{13}CH_3$ cis	$v=0$	$A=9409.44\,(30)$ $B=3918.34\,(5)$ $C=2907.78\,(5)$	$d_J=2.9\,(37)$ $d_{JK}=-33\,(10)$ $d_{EJ}=25\,(3)$				
	$CH_2{=}CHSiD_2CH_3$ cis	$v=0$	$A=8533.99\,(40)$ $B=3825.72\,(5)$ $C=2860.35\,(4)$	$d_J=2.2\,(34)$ $d_{JK}=-22\,(14)$ $d_{EJ}=18\,(3)$				
	$CH_2{=}CHSiH_2CD_3$ cis	$v=0$	$A=8452.21\,(31)$ $B=3668.77\,(4)$ $C=2724.47\,(3)$	$d_J=2.5\,(18)$ $d_{JK}=-29\,(11)$ $d_{EJ}=21\,(2)$				
	$CH_2{=}CHSiHDCH_3$ cis	$v=0$	$A=9031.96\,(41)$ $B=3906.53\,(5)$ $C=2908.61\,(5)$	$d_J=2.5\,(35)$ $d_{JK}=-25\,(14)$ $d_{EJ}=21\,(3)$				
	$CH_2{=}CHSiH_2CH_2D$ cis, asymmetric	$v=0$	$A=9025.32\,(43)$ $B=3919.18\,(5)$ $C=2894.11\,(5)$	$d_J=2.9\,(36)$ $d_{JK}=-32\,(15)$ $d_{EJ}=24\,(3)$				
	$CH_2{=}CHSiH_2CH_2D$ cis, symmetric	$v=0$	$A=9446.09\,(34)$ $B=3791.82\,(4)$ $C=2841.30\,(4)$	$d_J=2.7\,(30)$ $d_{JK}=-29\,(11)$ $d_{EJ}=24\,(3)$				
	$CH_2{=}CHSiH_2CH_3$ skew	$v=0$	$A=13728.12\,(10)$ $B=3048.64\,(1)$ $C=2825.24\,(1)$	$d_J=1.9\,(2)$ $d_{JK}=-52\,(2)$ $d_K=270\,(20)$ $d_{EJ}=28\,(1)$				
	$CH_2{=}CHSiH_2{}^{13}CH_3$ skew	$v=0$	$A=13550.08\,(6)$ $B=2983.45\,(1)$ $C=2762.37\,(1)$	$d_J=1.7\,(1)$ $d_{JK}=-61\,(1)$ $d_{EJ}=34\,(1)$				
	$CH_2{=}CHSiD_2CH_3$ skew	$v=0$	$A=11626.37\,(9)$ $B=2983.80\,(1)$ $C=2752.54\,(1)$	$d_J=1.7\,(2)$ $d_{JK}=-38\,(2)$ $d_{EJ}=25\,(1)$				
	$CH_2{=}CHSiH_2CD_3$ skew	$v=0$	$A=12094.61\,(8)$ $B=2773.14\,(1)$ $C=2564.66\,(1)$	$d_J=1.8\,(2)$ $d_{JK}=-59\,(2)$ $d_{EJ}=36\,(1)$				
Cont.								

Nr.	Molecule	v	A, B, C MHz	D, H kHz	Ref.	Ref. struct.	Varia, remarks	See also Tab., Nr.
398 Cont.	$CH_2{=}CHSiHDCH_3$-4 skew	$v=0$	$A=12542.62\,(13)$ $B=3016.46\,(1)$ $C=2774.75\,(1)$	$d_J=1.7\,(3)$ $d_{JK}=-61\,(3)$ $d_{EJ}=41\,(2)$	81IMA		for the numbering of the atoms, see Fig. 110a	
	$CH_2{=}CHSiHDCH_3$-5 skew	$v=0$	$A=12631.62\,(25)$ $B=3022.93\,(2)$ $C=2795.47\,(2)$	$d_J=2.0\,(6)$ $d_{JK}=-41\,(6)$ $d_{EJ}=19\,(4)$				
	$CH_2{=}CHSiH_2CH_2D$-6 skew	$v=0$	$A=12760.71\,(9)$ $B=2983.45\,(1)$ $C=2749.63\,(1)$	$d_J=2.1\,(2)$ $d_{JK}=-70\,(2)$ $d_{EJ}=47\,(1)$				
	$CH_2{=}CHSiH_2CH_2D$-7 skew	$v=0$	$A=12991.82\,(17)$ $B=2963.96\,(1)$ $C=2736.31\,(1)$	$d_J=1.7\,(4)$ $d_{JK}=-55\,(4)$ $d_{EJ}=29\,(2)$				
	$CH_2{=}CHSiH_2CH_2D$-8 skew	$v=0$	$A=13652.59\,(27)$ $B=2905.21\,(2)$ $C=2703.14\,(2)$	$d_J=1.9\,(6)$ $d_{JK}=-49\,(6)$ $d_{EJ}=29\,(4)$				
399	$CH_2{=}CHCH_2SiH_3$	$v=0$	$A=15744.81\,(26)$ *) $B=2773.12\,(13)$ $C=2655.81\,(7)$		77HAY2	77HAY2	*) uncertainties are 2.5σ	2.6.3, 188 2.8, 65
		$v_{T_1}=1$	$A=15729.64\,(53)$ *) $B=2791.97\,(26)$ $C=2663.99\,(15)$				v_{T_1}, skeletal torsion	
		$v_{T_2}=1$	$A=15860.65\,(23)$ *) $B=2759.87\,(11)$ $C=2646.74\,(6)$				v_{T_2}, SiH_3 torsion	
	$CH_2{=}CHCH_2SiD_3$	$v=0$	$A=13365.33\,(17)$ *) $B=2568.72\,(8)$ $C=2468.69\,(5)$					
400	SiF_3BF_2	$v=0$	$A(BF_2)=10373\,(30)$ $B=1504.90\,(5)$ $C=1314.79\,(3)$		74OGA	74OGA	for structure, see also Vol. II/7 (2, 42)	2.6.3, 189 2.8, 127
401 *6–166a* Cont.	$^{28}SiH_3PH_2$	$v=0$	$A=51675.9\,(1)$ $B=5581.48\,(5)$ $C=5556.50\,(5)$		75VÁR	76DUR2	for structure, see also Vol. II/7 (2, 297)	2.6.3, 190 2.8, 131

Nr.	Molecule	v	A, B, C MHz	D, H kHz	Ref.	Ref. struct.	Varia, remarks	See also Tab., Nr.
401 *6-166a* Cont.	SiH_3PD_2	$v=0$	$A=37589.06$ (11) $B=\ 5315.70$ (2) $C=\ 5258.70$ (2)		76DUR2			
402	$CH_3SiH_2PH_2$ trans	$v=0$	$A=16094$ (25) $B=\ 3904.30$ (12) $C=\ 3325.58$ (12)		79DU1	79DU1		
	$CH_3SiH_2PH_2$ gauche	$v=0$	$A=16048$ (20) $B=\ 3806.07$ (8) $C=\ 3251.97$ (8)					
403		$v=0$	$A=5473.39$ (4) $B=4137.01$ (2) $C=2661.55$ (2)		76DUR7	76DUR7	twisted C_2 conformation	2.6.3, 191
		$v_{39}=1$	$A=5467.92$ (48) $B=4140.86$ (7) $C=2664.52$ (8)				ring-puckering mode	
		$v_{39}=2$	$A=5462.45$ (27) $B=4143.89$ (4) $C=2667.36$ (5)					
		$v_{39}=3$	$A=5457.43$ (42) $B=4146.34$ (6) $C=2669.89$ (6)					
		$v_{39}=4$	$A=5452.68$ (34) $B=4148.22$ (5) $C=2672.21$ (5)					
		$v_{39}=5$	$A=5449.28$ (24) $B=4149.69$ (4) $C=2674.26$ (3)					
		$v=0$	$A=5473.42$ (20) $B=4085.26$ (4) $C=2639.99$ (2)					
		$v=0$	$A=5308.25$ (16) $B=3774.14$ (3) $C=2542.43$ (3)					

Nr.	Molecule	v	A, B, C MHz	D, H kHz	Ref.	Ref. struct.	Varia, remarks	See also Tab., Nr.
404	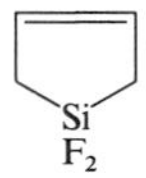	$v=0$	$A=3546.36$ (140) $B=1798.07$ (2) $C=1706.25$ (2)		74DUR1	74DUR1	C_{2v} symmetry, five-membered ring structure planar $v=38$ (7) cm^{-1}, ring-puckering vibration; nearly harmonic vibration	2.6.3, 192
		$v=1$	$A=3543.54 \pm 2.0$ $B=1801.89$ (4) $C=1714.02$ (2)					
		$v=2$	$A=3542.17 \pm 2.4$ $B=1805.19$ (19) $C=1720.47$ (4)					
		$v=3$	$A=3539 \pm 3.0$ $B=1808.01$ (20) $C=1726.22$ (4)					
405		$v=0$	$A=2744.775$ (2) $B=2662.017$ (2) $C=2145.994$ (2)	$\tau_{aaaa}= -1.16$ (4) $\tau_{bbbb}= -0.75$ (3) $\tau_{cccc}= -0.75$ (2) $\tau_{bbcc}= 14.1$ (3) $\tau_{aacc}= -17.8$ (4)	77IK	77IK	C_s symmetry	2.6.3, 193 3, 111
		$v_\tau=1$	$A=2745.94$ (2) $B=2662.79$ (2) $C=2148.779$ (3)	$\tau_{aaaa}= -0.8$ (4) $\tau_{bbbb}= -0.7$ (2) $\tau_{cccc}= -0.78$ (4) $\tau_{bbcc}= 16$ (4) $\tau_{aacc}= -20$ (5)			$v_\tau=70$ (25) cm^{-1}, torsional mode of the skeleton	
		$v_\tau=2$	$A=2746.88$ (5) $B=2663.59$ (6) $C=2150.931$ (6)	$\tau_{aaaa}= -3.2$ (7) $\tau_{bbbb}= -2.0$ (4) $\tau_{cccc}= -0.60$ (6) $\tau_{bbcc}= -14$ (9) $\tau_{aacc}= 20$ (10)			$2v_\tau=160$ (60) cm^{-1}	
406	$^{74}GeH_3CH_2F$	$v=0$	$A=26166.35$ $B=3515.274$ $C=3288.729$		74KR2	74KR2		2.6.3, 194 2.8, 68 3, 112
	$^{72}GeH_3CH_2F$	$v=0$	$A=26170.64$ $B=3539.347$ $C=3310.488$					
Cont.								

Nr.	Molecule	v	A, B, C MHz	D, H kHz	Ref.	Ref. struct.	Varia, remarks	See also Tab., Nr.
406 Cont.	$^{70}GeH_3CH_2F$	$v=0$	$A=26175.86$ $B=3565.257$ $C=3332.738$		74KR2			
407	$^{74}GeH_3CH_2{}^{35}Cl$	$v=0$	$A=20655.77$ (46) *) $B=2086.54$ (4) $C=1964.04$ (4)	$D_J=1.0$ (11)	76NAK1	76NAK1	* uncertainties are 2.5 σ	2.7.3, 134 2.8, 69
		$v_\tau=1$	$A=20634.71$ (46) *) $B=2078.25$ (2) $C=1958.63$ (2)	$D_J=0.3$ (12)			v_τ, GeH_3 torsion	
	$^{72}GeH_3CH_2{}^{35}Cl$	$v=0$	$A=20665.75$ (52) *) $B=2105.33$ (6) $C=1980.75$ (5)	$D_J=1.0$ (10)				
		$v_\tau=1$	$A=20646.26$ (45) *) $B=2096.90$ (1) $C=1975.31$ (1)	$D_J=-0.2$ (13)				
	$^{70}GeH_3CH_2{}^{35}Cl$	$v=0$	$A=20676.58$ (47) *) $B=2125.04$ (5) $C=1998.29$ (4)	$D_J=0.9$ (9)				
	$^{74}GeH_3CH_2{}^{37}Cl$	$v=0$	$A=20581.30$ (53) *) $B=2018.46$ (3) $C=1902.96$ (3)	$D_J=1.3$ (10)				
	$^{74}GeD_3CH_2{}^{35}Cl$	$v=0$	$A=16705.07$ (41) *) $B=2004.01$ (3) $C=1894.59$ (3)	$D_J=0.7$ (8)				
408	$^{74}GeH_3CH_2CH_3$	$v=0$	$A=22254.227$ (378) $B=3582.885$ (17) $C=3339.232$ (16)		77DUR4	77DUR4	rotational transitions in torsional excited states measured	2.6.3, 195 2.8, 40
	$^{72}GeH_3CH_2CH_3$	$v=0$	$A=22258.586$ (346) $B=3604.455$ (16) $C=3358.060$ (15)					
Cont.	$^{70}GeH_3CH_2CH_3$	$v=0$	$A=22263.027$ (566) $B=3627.177$ (25) $C=3377.829$ (25)					

Nr.	Molecule	v	A, B, C MHz	D, H kHz	Ref.	Ref. struct.	Varia, remarks	See also Tab., Nr.
408 Cont.	$^{74}GeD_3CH_2CH_3$	$v=0$	$A=17658.508$ (495) $B=$ 3407.891 (31) $C=$ 3192.965 (33)		77DUR4			
	$^{72}GeD_3CH_2CH_3$	$v=0$	$A=17660.733$ (503) $B=$ 3425.539 (32) $C=$ 3208.509 (34)					
	$^{70}GeD_3CH_2CH_3$	$v=0$	$A=17662.691$ (528) $B=$ 3444.012 (34) $C=$ 3224.796 (35)					
	$^{74}GeH_3CH_2CD_3$	$v=0$	$A=19235.172$ (355) $B=$ 3117.600 (12) $C=$ 2924.293 (12)					
	$^{72}GeH_3CH_2CD_3$	$v=0$	$A=19237.902$ (658) $B=$ 3137.601 (16) $C=$ 2941.953 (16)					
	$^{70}GeH_3CH_2CD_3$	$v=0$	$A=19240.480$ (1497) $B=$ 3158.628 (21) $C=$ 2960.499 (21)					
409	$^{70}GeH_3CH{=}CH_2$	$v=0$	$A=32731.06$ $B=$ 3885.84 $C=$ 3627.26		74DUR7	74DUR7		2.6.3, 196 2.7.3, 146a 2.8, 70 3, 113
	$^{72}GeH_3CH{=}CH_2$	$v=0$	$A=32726.53$ $B=$ 3862.15 $C=$ 3606.59					
	$^{73}GeH_3CH{=}CH_2$	$v=0$	$A=32724.49$ $B=$ 3850.73 $C=$ 3596.61					
	$^{74}GeH_3CH{=}CH_2$	$v=0$	$A=32721.94$ $B=$ 3839.67 $C=$ 3586.89					
	$^{76}GeH_3CH{=}CH_2$	$v=0$	$A=32718.37$ $B=$ 3818.19 $C=$ 3568.14					
Cont.	$^{70}GeD_3CH{=}CH_2$	$v=0$	$B=3684.29$ $C=3456.31$					

Nr.	Molecule	v	A, B, C MHz	D, H kHz	Ref.	Ref. struct.	Varia, remarks	See also Tab., Nr.
409 Cont.	$^{72}GeD_3CH{=}CH_2$	$v=0$	$B=3665.14$ $C=3439.42$		74DUR7			
	$^{73}GeD_3CH{=}CH_2$	$v=0$	$B=3655.90$ $C=3431.24$					
	$^{74}GeD_3CH{=}CH_2$	$v=0$	$B=3646.90$ $C=3423.29$					
	$^{76}GeD_3CH{=}CH_2$	$v=0$	$B=3629.45$ $C=3407.88$					
410	$CH_3{}^{74}GeH_2F$	$v=0$	$A=13360.28$ $B=6251.57$ $C=4625.07$		76ROB1, 76KRI	76ROB1	rotational transitions in excited torsional states measured	2.6.3, 197 2.8, 71 3, 114
	$CH_3{}^{72}GeH_2F$	$v=0$	$A=13439.42$ $B=6251.54$ $C=4634.50$					
	$CH_3{}^{70}GeH_2F$	$v=0$	$A=13521.26$ $B=6251.56$ $C=4644.31$					
	$^{13}CH_3{}^{74}GeH_2F$	$v=0$	$A=13171.59$ $B=6059.28$ $C=4496.95$		76ROB1			
	$^{13}CH_3{}^{72}GeH_2F$	$v=0$	$A=13250.98$ $B=6059.25$ $C=4506.21$					
	$^{13}CH_3{}^{70}GeH_2F$	$v=0$	$A=13334.39$ $B=6059.26$ $C=4515.88$					
	$CD_3{}^{74}GeH_2F$	$v=0$	$A=11797.53$ $B=5375.76$ $C=4067.09$					
	$CD_3{}^{72}GeH_2F$	$v=0$	$A=11866.78$ $B=5376.39$ $C=4075.69$					
	$CD_3{}^{70}GeH_2F$	$v=0$	$A=11939.23$ $B=5377.00$ $C=4084.62$					

Nr.	Molecule	v	A, B, C MHz	D, H kHz	Ref.	Ref. struct.	Varia, remarks	See also Tab., Nr.
411	$CH_3{}^{74}GeH_2{}^{35}Cl$	$v=0$	$I_a=$ 45.85621 *) $I_b=$150.03044 $I_c=$186.64447		80ZEI	80ZEI	*) in amu A^2	
	$CH_3{}^{72}GeH_2{}^{35}Cl$	$v=0$	$I_a=$ 45.58635 *) $I_b=$149.65640 $I_c=$185.99712					
	$CH_3{}^{74}GeH_2{}^{37}Cl$	$v=0$	$I_a=$ 46.02616 *) $I_b=$155.21374 $I_c=$191.99296					
	$CH_2D^{74}GeH_2{}^{35}Cl$ symmetric	$v=0$	$I_a=$ 46.73726 *) $I_b=$156.71174 $I_c=$194.15836					
	$CH_2D^{74}GeH_2{}^{35}Cl$ asymmetric	$v=0$	$I_a=$ 49.83738 *) $I_b=$152.56447 $I_c=$191.76333					
	${}^{13}CH_3{}^{74}GeH_2{}^{35}Cl$	$v=0$	$I_a=$ 47.29855 *) $I_b=$152.40896 $I_c=$190.46849					
	${}^{13}CH_3{}^{74}GeH_2{}^{37}Cl$	$v=0$	$I_a=$ 47.47941 *) $I_b=$157.63821 $I_c=$195.88368					
	$CH_3{}^{74}GeHD^{35}Cl$	$v=0$	$I_a=$ 48.85784 *) $I_b=$151.91618 $I_c=$188.51002					
412 *6-171* *4-146*	$CH_3{}^{11}BF_2$	$v=0$	$A(BF_2)=10587.2538$ ± 0.007 $B=8329.0795\pm 0.007$ $C=4650.4572\pm 0.007$	$\Delta_J = 4.03\pm 0.2$ $\Delta_{JK}= 18.09\pm 0.4$ $\Delta_K = -6.99\pm 0.5$ $\delta_J = 1.685\pm 0.04$ $\delta_K = 14.05\pm 0.2$	79LAG	79LAG		2.6.3, 198 2.7.3, 10
	$CH_3{}^{10}BF_2$	$v=0$	$A(BF_2)=10585.9080$ ± 0.007 $B=8328.1610\pm 0.006$ $C=4649.8893\pm 0.006$	$\Delta_J = 2.93\pm 0.3$ $\Delta_{JK}= 19.04\pm 0.5$ $\Delta_K = -8.79\pm 0.6$ $\delta_J = 1.741\pm 0.05$ $\delta_K = 13.38\pm 0.2$				
Cont.								

Nr.	Molecule	v	A, B, C MHz	D, H kHz	Ref.	Ref. struct.	Varia, remarks	See also Tab., Nr.
412 *6-171* *4-146* Cont.	$^{13}CH_3{}^{11}BF_2$	$v=0$	$A(BF_2)=10587.161\pm0.02$ $B=8000.586\pm0.01$ $C=4546.123\pm0.01$	**)	79LAG		**) centrifugal distortion constants of parent species used	
	$CD_3{}^{11}BF_2$	$v=0$	$A(BF_2)=10582.081\pm0.08$ $B=6876.177\pm0.02$ $C=4157.164\pm0.02$	$\Delta_J = 3.18\pm0.8$ $\Delta_{JK} = 21.67\pm0.9$ $\Delta_K = -6.99$ (fixed) $\delta_J = 1.017\pm0.07$ $\delta_K = 14.05$ (fixed)				
413	$CH_3CH_2{}^{11}BF_2$	$v=0$	$A=8882.84$ (9) $B=3449.29$ (3) $C=2564.13$ (3)		78OD1	78OD1	all heavy atoms are coplanar	2.6.3, 199 2.8, 128
		$v_\tau=1$	$A=8828.49$ (9) $B=3445.38$ (3) $C=2572.32$ (3)				$\nu_\tau=44\ cm^{-1}$, BF_2 torsion	
		$v_\tau=2$	$A=8771.54$ (78) $B=3442.39$ (2) $C=2581.35$ (2)					
		$v_\tau=3$	$A=8711.45$ (87) $B=3440.26$ (2) $C=2591.44$ (2)					
		$v_\tau=4$	$A=8647.49$ (86) $B=3438.98$ (2) $C=2602.83$ (2)					
	$CH_3CH_2{}^{10}BF_2$	$v=0$	$A=8881.86$ (104) $B=3454.00$ (3) $C=2566.70$ (3)					
414	$CH_2{=}CH^{11}BF_2$	$v=0$	$A=9632.65$ (5) $B=3836.03$ (1) $C=2743.00$ (1)		74DUR3	76DUR6	planar molecule	2.6.3, 200 2.8, 130
		$v_{18}=1$	$A=9568.6$ (10) $B=3832.88$ (4) $C=2749.59$ (3)				harmonic force field calculated [76DUR6] $\nu_{18}=103\ cm^{-1}$, $B{-}F_2$ torsion	
Cont.		$v_{17}=1$	$A=9573.3$ (10) $B=3833.80$ (3) $C=2748.44$ (2)				$\nu_{17}=198\ cm^{-1}$, CBF_2 bend	

Nr.	Molecule	v	A, B, C MHz	D, H kHz	Ref.	Ref. struct.	Varia, remarks	See also Tab., Nr.
414 Cont.	$CH_2{=}CH^{11}BF_2$	$v_{13}=1$	$A=9603.6$ (18) $B=3835.09$ (6) $C=2745.23$ (5)		76DUR3		$\nu_{13}=215\ cm^{-1}$, BCC in-plane bend	
	$CH_2{=}CH^{10}BF_2$	$v=0$	$A=9632.8$ (8) $B=3840.47$ (3) $C=2745.22$ (2)					
	$CH_2{=}^{13}CH^{11}BF_2$	$v=0$	$A=9541.92$ (7) $B=3808.86$ (2) $C=2721.89$ (2)		76DUR6			
	$CD_2{=}CD^{11}BF_2$	$v=0$	$A=8853.74$ (50) $B=3452.87$ (1) $C=2484.08$ (1)					
		$v_{18}=1$	$A=8778.9$ (11) $B=3444.29$ (3) $C=2492.62$ (3)				$\nu_{18}=95\ cm^{-1}$, BF_2 torsion	
		$v_{17}=1$	$A=8803.2$ (7) $B=3447.48$ (2) $C=2489.74$ (2)				$\nu_{17}=185\ cm^{-1}$, BF_2 wagging	
		$v_{13}=1$	$A=8859.9$ (2) $B=3453.14$ (1) $C=2482.38$ (1)				$\nu_{13}=198\ cm^{-1}$, C—C—B bend	
	$CD_2{=}CD^{10}BF_2$	$v=0$	$A=8853.59$ (60) $B=3458.16$ (2) $C=2486.74$ (2)					
	$CD_2{=}^{13}CD^{11}BF_2$	$v=0$	$A=8791.52$ (280) $B=3434.21$ (4) $C=2468.17$ (4)					
	$CH_2{=}CD^{11}BF_2$	$v=0$	$A=9090.27$ (40) $B=3801.38$ (2) $C=2680.22$ (1)					
	$CHD{=}CH^{11}BF_2$ cis *)	$v=0$	$A=9425.91$ (50) $B=3715.63$ (1) $C=2664.78$ (1)				*) cis and trans refer to the relative positions of the D and BF_2 group with respect to the double bond	
	$CHD{=}CH^{11}BF_2$ trans *)	$v=0$	$A=9604.38$ (80) $B=3576.35$ (2) $C=2605.51$ (2)					
Cont.								

Nr.	Molecule	v	A, B, C MHz	D, H kHz	Ref.	Ref. struct.	Varia, remarks	See also Tab., Nr.
414 Cont.	$CHD{=}CH^{10}BF_2$ cis *)	$v=0$	$A=9425.91$ (90) $B=3720.44$ (3) $C=2667.21$ (2)		76DUR6		*) cis and trans refer to the relative positions of the D and BF_2 group with respect to the double bond	
	$CHD{=}CH^{10}BF_2$ trans *)	$v=0$	$A=9603.91$ (110) $B=3581.12$ (3) $C=2607.96$ (2)					
	$CHD{=}CD^{11}BF_2$ cis *)	$v=0$	$A=9074.90$ (3) $B=3547.19$ (1) $C=2550.10$ (1)					
	$CHD{=}CD^{11}BF_2$ trans *)	$v=0$	$A=8880.84$ (9) $B=3687.47$ (1) $C=2605.36$ (1)					
	$CHD{=}CD^{10}BF_2$ cis *)	$v=0$	$A=9074.86$ (7) $B=3552.14$ (3) $C=2552.57$ (3)					
	$CHD{=}CD^{10}BF_2$ trans *)	$v=0$	$A=8881.43$ (140) $B=3692.55$ (6) $C=2607.79$ (5)					
415	$H_2N^{11}B_2H_5$	$v=0$	$A=19448.87$ *) $B=14228.18$ $C=10098.81$		74LAU	74LAU	*) estimated errors are 0.05 MHz	2.6.3, 201 2.7.3, 44 3, 115
	$H_2N^{10}B^{11}BH_5$	$v=0$	$A=19617.02$ *) $B=14604.47$ $C=10334.28$					
	$H_2N^{10}B_2H_5$	$v=0$	$A=19762.49$ *) $B=15001.03$ $C=10573.53$					
	$H_2{}^{15}N^{11}B_2H_5$	$v=0$	$A=19011.60$ *) $B=14228.69$ $C=9979.44$					
	$H_2{}^{15}N^{10}B^{11}BH_5$	$v=0$	$A=19186.54$ *) $B=14603.39$ $C=10212.62$					
	$^{11}B^{11}BD_t$		$A=18550.72$ *) $B=12988.15$ $C=9569.89$				see Fig. 115	
Cont.								

Nr.	Molecule	v	A, B, C MHz	D, H kHz	Ref.	Ref. struct.	Varia, remarks	See also Tab., Nr.
415 Cont.	$^{11}B^{10}BD_t$		$A = 18649.80$ *) $B = 13303.78$ $C = 9764.05$		74LAU			
	$^{10}B^{11}BD_t$		$A = 18750.05$ *) $B = 13297.33$ $C = 9791.81$					
	$^{11}B^{11}BD_b$		$A = 18096.28$ *) $B = 14261.12$ $C = 9738.27$				see Fig. 115	
	$^{11}B^{10}BD_b$		$A = 18225.19$ *) $B = 14636.43$ $C = 9950.81$					
416	$(CH_3)_2N^{10}B^{11}BH_5$	$v=0$	$A = 5813.08$ *) $B = 4993.011$ $C = 4073.837$		73COH	73COH	for structure, see also Vol. II/7 (3, 324) *) uncertainties are 0.02 MHz for B and C, and 0.2 MHz for A	2.6.3, 202 3, 116
	$(CH_3)_2N^{11}B_2H_5$	$v=0$	$A = 5749.66$ *) $B = 4888.543$ $C = 4034.675$					
	$(CH_3)_2N^{11}B_2D_5$	$v=0$	$A = 4977.96$ *) $B = 4112.250$ $C = 3615.493$					
	$(CH_3)_2N^{10}B^{11}BD_5$	$v=0$	$A = 5025.09$ *) $B = 4178.799$ $C = 3641.608$					
	$(CH_3)_2N^{11}B_2D_4H_t$	$v=0$	$A = 5162.15$ *) $B = 4246.965$ $C = 3675.136$				see Fig. 116	
	$(CH_3)_2{}^{15}N^{10}B^{11}BH_5$	$v=0$	$A = 5812.72$ *) $B = 4992.952$ $C = 4073.886$					
	$(CH_3)_2{}^{15}N^{11}B_2H_5$	$v=0$	$A = 5749.87$ *) $B = 4888.287$ $C = 4034.642$					

Nr.	Molecule	v	A, B, C MHz	D, H kHz	Ref.	Ref. struct.	Varia, remarks	See also Tab., Nr.
417	$CH_3{}^{11}B_2H_5$	$v=0$	$A=31781$ *) $B=$ 6940.505 $C=$ 6194.995		77PEN1	77PEN1	*) estimated errors: in $A \approx 100$ MHz, in B and $C \approx 0.02$ MHz	2.6.3, 203 2.8, 129 3, 117
		$v=1$	$A=32371$ *) $B=$ 6930.945 $C=$ 6183.465					
		$v=2$	$A=30957$ *) $B=$ 6924.638 $C=$ 6172.233					
	$CH_3{}^{10}B_c{}^{11}B_tH_5$	$v=0$	$A=32737$ *) $B=$ 6930.850 $C=$ 6218.920					
	$CH_3{}^{11}B_c{}^{10}B_tH_5$	$v=0$	$A=32093$ *) $B=$ 7176.743 $C=$ 6388.623					
	$CH_3{}^{10}B_2H_5$	$v=0$	$A=32512$ *) $B=$ 7175.925 $C=$ 6413.485					
	$CH_3{}^{11}B_2D_5$	$v=0$	$A=21935$ **) $B=$ 6015.409 $C=$ 5252.874				**) estimated errors: in $A \approx 50$ MHz in B and $C \approx 0.02$ MHz see Fig. 117	
	$CH_3{}^{10}B_c{}^{11}B_tD_5$	$v=0$	$A=22332$ **) $B=$ 6016.811 $C=$ 5270.016					
	$CH_3{}^{11}B_c{}^{10}B_tD_5$	$v=0$	$A=21990$ **) $B=$ 6165.011 $C=$ 5370.516					
	$CH_3{}^{10}B_2D_5$	$v=0$	$A=22414$ **) $B=$ 6165.735 $C=$ 5387.525					
418	B_4H_{10}	$v=0$	$A=11013.388$ (19) $B=$ 6198.643 (23) $C=$ 5592.817 (21)	$\Delta_{JK}= -4.77$ (59) $\Delta_K =$ 14.89 (91) $\delta_K =$ 4.3 (27)	81SIM	81SIM	C_{2v} symmetry for the numbering of the atoms, see Fig. 118	2.6.3, 204 3, 118
	1-^{10}B	$v=0$	$A=11248.386$ (15) $B=$ 6215.416 (20) $C=$ 5638.440 (20)	$\Delta_{JK}= -5.36$ (54) $\Delta_K =$ 14.77 (72) $\delta_K =$ 5.8 (27)				
Cont.								

Nr.	Molecule	v	A, B, C MHz	D, H kHz	Ref.	Ref. struct.	Varia, remarks	See also Tab., Nr.
418 Cont.	2-^{10}B	$v=0$	$A=11055.969\,(17)$ $B=6368.152\,(20)$ $C=5718.786\,(18)$	$\Delta_{JK}=-5.24\,(48)$ $\Delta_K=14.9\,(10)$ $\delta_K=3.9\,(20)$	81SIM			
	μ-B_4H_9D *)	$v=0$	$A=10621.95\,(19)$ $B=6078.70\,(17)$ $C=5390.91\,(20)$	$\Delta_{JK}=-2.6\,(10)$ $\Delta_K=76\,(14)$ $\delta_K=4.0$ (fixed)			*) bridge substituted	
419	$^{11}B_6H_{10}$	$v=0$	$A=5026.03\,(2)$ $B=4984.13\,(2)$ $C=3037.46\,(4)$		79SCH	79SCH	for the numbering of the atoms, see Fig. 119	2.6.3, 205 3, 119
	3-^{10}B	$v=0$	$A=5135.38\,(2)$ $B=4987.04\,(2)$ $C=3077.74\,(1)$					
	4-^{10}B	$v=0$	$A=5115.15\,(10)$ $B=5010.39\,(7)$ $C=3078.83\,(4)$					
	1-^{10}B	$v=0$	$A=5061.13\,(5)$ $B=5018.62\,(5)$ $C=3037.47\,(6)$					
	2-^{10}B	$v=0$	$A=5099.62\,(2)$ $B=5026.49\,(2)$ $C=3079.57\,(1)$					
	$^{10}B_6H_{10}$	$v=0$	$A=5352.17\,(3)$ $B=5321.23\,(3)$ $C=3252.02\,(4)$					
420	CB_5H_7	$v=0$	$A=5714.38$ *) $B=5642.20$ $C=5324.22$		76MCK	76MCK	*) estimated error: 0.1 MHz for the numbering of the atoms, see Fig. 120	2.6.3, 206 3, 120
	2-^{10}B	$v=0$	$A=5788.30$ *) $B=5687.01$ $C=5402.62$					
Cont.	4-^{10}B	$v=0$	$A=5783.89$ *) $B=5687.82$ $C=5392.05$					

Nr.	Molecule	v	A, B, C MHz	D, H kHz	Ref.	Ref. struct.	Varia, remarks	See also Tab., Nr.
420 Cont.	6-^{10}B	$v=0$	$A=5821.45$ *) $B=5710.90$ $C=5354.98$		76MCK		*) estimated error: 0.1 MHz	
	2,6-^{10}B	$v=0$	$A=5891.66$ *) $B=5756.81$ $C=5439.37$					
	4,6-^{10}B	$v=0$	$A=5889.46$ *) $B=5753.66$ $C=5431.14$					
	$^{13}CB_5H_7$	$v=0$	$A=5641.51$ *) $B=5598.66$ $C=5299.67$					
	^{13}C, 2-^{10}B	$v=0$	$A=5721.12$ *) $B=5641.92$ $C=5371.11$					
	^{13}C, 4-^{10}B	$v=0$	$A=5716.12$ *) $B=5642.13$ $C=5362.19$					
	^{13}C, 6-^{10}B	$v=0$	$A=5748.09$ *) $B=5658.09$ $C=5339.56$					
421	1,7-$C_2B_6H_8$	$v=0$	$A=3885.03$ *) $B=3124.79$ $C=3091.11$		76ROG	76ROG	*) estimated error: 0.04 MHz or less for the numbering of the atoms, see Fig. 121	2.6.3, 207 3, 121
	2-^{10}B	$v=0$	$A=3902.60$ *) $B=3168.02$ $C=3129.77$					
	3-^{10}B	$v=0$	$A=3932.73$ *) $B=3148.87$ $C=3102.46$					
	5-^{10}B	$v=0$	$A=3936.42$ *) $B=3148.20$ $C=3102.41$					
Cont.								

Nr.	Molecule	v	A, B, C MHz	D, H kHz	Ref.	Ref. struct.	Varia, remarks	See also Tab., Nr.
421 Cont.	2,4-^{10}B	$v=0$	$A=3950.88$ *) $B=3195.05$ $C=3137.75$		76ROG			
	2,5-^{10}B	$v=0$	$A=3954.51$ *) $B=3187.74$ $C=3145.86$					
	2,6-^{10}B	$v=0$	$A=3954.40$ *) $B=3195.58$ $C=3138.62$					
	3,5-^{10}B	$v=0$	$A=3984.03$ *) $B=3178.81$ $C=3107.45$					
	3,6-^{10}B	$v=0$	$A=3985.41$ *) $B=3159.25$ $C=3126.70$					
422	1,6 $C_2B_7H_9$		$A=2867.79$ *) $B=2806.37$ $C=2366.36$		76LAU	76LAU	*) estimated errors: in A and $B\simeq 0.03$ MHz, in $C=0.01$ MHz for the numbering of the atoms, see Fig. 122	2.6.3, 208 3, 122
	4-^{10}B		$A=2898.63$ *) $B=2823.30$ $C=2380.76$					
	8-^{10}B		$A=2867.70$ *) $B=2848.90$ $C=2396.49$					
	4,8-^{10}B		$A=2900.17$ *) $B=2865.04$ $C=2411.49$					
	3,4-^{10}B		$A=2913.05$ *) $B=2856.81$ $C=2395.34$					
	3-^{10}B	$v=0$	$A=2881.58$ $B=2839.91$ $C=2380.58$					

Nr.	Molecule	v	A, B, C MHz	D, H kHz	Ref.	Ref. struct.	Varia, remarks	See also Tab., Nr.
423	2,3,4,5-$C_4B_2H_6$	$v=0$	$A=6506.48$ *) $B=6382.17$ $C=4203.83$		74PAS	74PAS	for structure, see also Vol. II/7 (3, 505) *) estimated error: 0.02 MHz for the numbering of the atoms, see Fig. 123	2.6.3, 209 3, 123
	1-^{10}B	$v=0$	$A=6589.36$ *) $B=6462.02$ $C=4203.64$					
	6-^{10}B	$v=0$	$A=6538.67$ *) $B=6511.46$ $C=4268.67$					
	1,6-^{10}B	$v=0$	$A=6624.93$ *) $B=6591.38$ $C=4268.55$					
	2-^{13}C	$v=0$	$A=6416.87$ *) $B=6344.70$ $C=4150.64$					
	3-^{13}C	$v=0$	$A=6478.39$ *) $B=6284.38$ $C=4153.62$					
	2,3-^{13}C	$v=0$	$A=6352.10$ *) $B=6282.36$ $C=4102.25$					
	2-^{13}C, 1-^{10}B	$v=0$	$A=6498.01$ *) $B=6425.47$ $C=4150.42$					
	3-^{13}C, 1^{10}-B	$v=0$	$A=6560.88$ *) $B=6362.95$ $C=4152.88$					
	2,3-^{13}C, 1-^{10}B	$v=0$	$A=6431.35$ *) $B=6360.55$ $C=4102.10$					
424 *6–176a*	CH_3PF_2	$v=0$	$A=7155.34$ *) $B=6815.25$ $C=4364.54$		74CO2	74CO2	*) estimated error: 0.05 MHz	2.6.3, 210 2.8, 124 3, 124
Cont.		$v_\tau=1$	$A=7158.23$ *)**) $B=6792.16$ $C=4358.40$				**) for the A species v_τ, CH_3 torsion	

Nr.	Molecule	v	A, B, C MHz	D, H kHz	Ref.	Ref. struct.	Varia, remarks	See also Tab., Nr.
424 *6–176a* Cont.		$v_a=1$	$A=7168.00$ *) $B=6818.97$ $C=4340.26$		74CO2		$v_a=325\ cm^{-1}$	
		$v_b=1$	$A=7154.43$ *) $B=6814.23$ $C=4372.80$					
425	$CH_3CH_2PH_2$ trans	$v=0$	$A=25691\ (46)$ $B=5016.48\ (3)$ $C=4600.28\ (3)$		76DUR5	76DUR5		2.6.3, 211
	CH_3CH_2PHD trans	$v=0$	$A=23278\ (30)$ $B=4933.27\ (2)$ $C=4521.69\ (2)$					
	$CH_3CH_2PD_2$ trans	$v=0$	$A=21254\ (36)$ $B=4840.61\ (3)$ $C=4457.67\ (3)$					
	$CH_3CH_2PH_2$ gauche	$v=0$	$A=25572.20\ (5)$ $B=5126.24\ (3)$ $C=4658.06\ (2)$				$E_{gauche}-E_{trans}=200\ (100)\ cm^{-1}$	
	$CH_3CH_2PD_2$ gauche	$v=0$	$A=22212.45\ (2)$ $B=4830.565\ (8)$ $C=4419.881\ (6)$					
426 *6–177a*	CH_3OPF_2	$v=0$	$A=5980.13\pm2.0$ $B=3641.82\ (6)$ $C=3123.46\ (6)$		74CO1	74CO1	for structure, see also Vol. II/7 (3, 100) methyl group cis to the fluorine atoms	2.8, 90 3, 125
	$^{13}CH_3OPF_2$	$v=0$	$A=5962.10\pm2.0$ $B=3542.68\ (6)$ $C=3054.87\ (6)$					
	$CH_3{}^{18}OPF_2$	$v=0$	$A=5924.79\pm2.0$ $B=3566.41\ (6)$ $C=3082.56\ (6)$					
	CD_3OPF_2	$v=0$	$A=5715.34\pm2.0$ $B=3252.64\ (6)$ $C=2843.72\ (6)$					

Nr.	Molecule	v	A, B, C MHz	D, H kHz	Ref.	Ref. struct.	Varia, remarks	See also Tab., Nr.
427	CH_3POF_2	$v=0$	$A=4495.91$ (4) $B=4271.86$ (3) $C=4125.90$ (3)		82DUR	82DUR		2.6.3, 212 2.8, 125
		$v_\tau=1$	$A=4505.11$ (253) $B=4270.10$ (20) $C=4124.06$ (17)		76DUR1		$\nu_\tau=230\ cm^{-1}$, CH_3 torsion	
		$v_\tau=2$	$A=4512.86$ (241) $B=4267.80$ (15) $C=4122.55$ (13)					
428	$CH_2{=}CHPF_2$	$v=0$	$A=7158.3$ (10) $B=3041.28$ (2) $C=2486.57$ (1)		81FON	81FON	fluorines trans to the vinyl group	2.6.3, 213
429	$CH_2{=}PH$	$v=0$	$A=138503.2892$ (394) $B=16417.7945$ (41) $C=14649.4073$ (42)	$D_J=16.419$ (53) $D_{JK}=216.07$ (83) $D_K=2441.0$ (124) $d_1=-1.8876$ (97) $d_2=-0.28404$ (252)	81BRO1, 81KRO1	81BRO1, 81KRO1		2.6.3, 214
	$^{13}CH_2{=}PH$	$v=0$	$A=138496.2925$ (451) $B=15764.5552$ (46) $C=14126.9406$ (48)	$D_J=15.277$ (122) $D_{JK}=204.11$ (179) $D_K=2437.7$ (225) $d_1=-1.6945$ (234) $d_2=-0.24433$ (500)				
	$CD_2{=}PH$	$v=0$	$A=94409.6638$ (299) $B=13875.9901$ (32) $C=12074.2573$ (31)	$D_J=11.329$ (46) $D_{JK}=127.65$ (82) $D_K=1246.3$ (37) $d_1=-1.6801$ (64) $d_2=-0.29624$ (233)				
	$CD_2{=}PD$	$v=0$	$A=71043.3933$ (170) $B=13605.3405$ (27) $C=11395.0659$ (27)	$D_J=10.076$ (76) $D_{JK}=106.91$ (113) $D_K=589.7$ (37) $d_1=-1.8474$ (111) $d_2=-0.41920$ (410)				
Cont.								

Nr.	Molecule	v	A, B, C MHz	D, H kHz	Ref.	Ref. struct.	Varia, remarks	See also Tab., Nr.
429	$CDH{=}PH_{(Z)}$	$v=0$	$A=114588.7475\,(272)$	$D_J=13.641\,(47)$	81BRO1,		Z=D and H cis	
Cont.			$B=14960.4801\,(33)$	$D_{JK}=131.50\,(98)$	81KRO1			
			$C=13207.0388\,(37)$	$D_K=2091.1\,(56)$				
				$d_1=-1.8563\,(37)$				
				$d_2=-0.27900\,(254)$				
	$CDH{=}PH_{(E)}$	$v=0$	$A=115090.6240\,(302)$	$D_J=13.792\,(49)$			E=D and H trans	
			$B=14993.9840\,(41)$	$D_{JK}=136.89\,(102)$				
			$C=13240.0623\,(38)$	$D_K=1990.8\,(70)$				
				$d_1=-1.8402\,(49)$				
				$d_2=-0.27166\,(322)$				
	$CH_2{=}PD$	$v=0$	$A=93513.75\,(13)$	$\Delta_J=15.39\,(36)$	81KRO1			
			$B=16098.885\,(20)$	$\Delta_{JK}=220.7\,(64)$				
			$C=13701.898\,(24)$	$\Delta_K=1137\,(18)$				
				$\delta_J=2.376\,(27)$				
				$\delta_K=152.96\,(70)$				
430	$CH_2{=}P^{35}Cl$	$v=0$	$A=22711.17$		76HOP	76HOP		
			$B=4667.23$					
			$C=3865.45$					
431	$CF_2{=}PH$	$v=0$	$A=11107.33$		76HOP	76HOP		
			$B=4766.37$					
			$C=3330.79$					
432	$CH_3PH_2BH_3$	$v_{16}=1$	$A=16390.56\,(16.84)$		75DUR4		for structure,	2.8, 122
5-176			$B=5939.14\,(7)$				see also Vol. II/7 (3, 139)	
			$C=4973.77\,(7)$				harmonic force field calculated	
		$v_{26}=1$	$A=20018.36\,(24.26)$				$\nu_{16}=214\ cm^{-1}$, skeletal	
			$B=5970.07\,(6)$				C—P—B bending	
			$C=4973.41\,(6)$				$\nu_{26}=200\ cm^{-1}$, methyl torsion	
		$v_{27}=1$	$A=17484.86\,(15.33)$				$\nu_{27}=138\ cm^{-1}$, BH_3 torsion	
			$B=5959.23\,(5)$					
Cont.			$C=4973.13\,(5)$					

Nr.	Molecule	v	A, B, C MHz	D, H kHz	Ref.	Ref. struct.	Varia, remarks	See also Tab., Nr.
432 *5-176* Cont.	$CH_3PD_2BH_3$	$v_{16}=1$	$A=14031.81$ (12.86) $B=5781.00$ (8) $C=4802.53$ (8)		75DUR4			
		$v_{26}=1$	$A=15415.90$ (9.00) $B=5807.09$ (4) $C=4806.24$ (4)					
		$v_{27}=1$	$A=14567.06$ (5.48) $B=5794.82$ (5) $C=4804.82$ (5)					
433	$CH_3PF_2{}^{11}BH_3$	$v=0$	$A=4193.95$ *) $B=3659.55$ $C=3507.26$		75CR2	75CR2	*) estimated uncertainty: 0.05 MHz	2.6.3, 215 2.8, 123 3, 126
	$CH_3PF_2{}^{10}BH_3$	$v=0$	$A=4198.83$ *) $B=3762.36$ $C=3605.20$					
	$CH_3PF_2{}^{11}BD_3$	$v=0$	$A=4046.13$ *) $B=3262.63$ $C=3135.43$					
	$CH_3PF_2{}^{10}BD_3$	$v=0$	$A=4047.16$ *) $B=3340.56$ $C=3207.98$					
433a	$CH_2{=}CHC{\equiv}P$	$v=0$	$A=46694$ (24) $B=2807.7100$ (21) $C=2645.8356$ (21)	$\Delta_J=0.589$ (18) $\Delta_{JK}=-48.279$ (32)	81OHN	81OHN		2.6.3, 215a
		$v_{11}=1$	$A=46027$ (26) $B=2817.2040$ (28) $C=2650.6227$ (28)	$\Delta_J=0.621$ (26) $\Delta_{JK}=-43.137$ (37)			ν_{11} (a')$=184$ (30) cm^{-1}	
		$v_{15}=1$	$A=47614$ (35) $B=2809.7734$ (32) $C=2650.5402$ (32)	$\Delta_J=0.628$ (30) $\Delta_{JK}=-53.197$ (42)			ν_{15} (a'')$=263$ (30) cm^{-1}	
433b Cont.	$CH_3CH_2PH_2{}^{11}BH_3$ trans	$v=0$	$A=15850$ (17) $B=2729.35$ (2) $C=2502.85$ (2)		81DUR6	81DUR6		2.6.3, 215b

Nr.	Molecule	v	A, B, C MHz	D, H kHz	Ref.	Ref. struct.	Varia, remarks	See also Tab., Nr.
433b Cont.	$CH_3CH_2PD_2{}^{11}BH_3$ trans	$v=0$	$A=13496\,(7)$ $B=\;2694.39\,(2)$ $C=\;2461.93\,(2)$		81DUR6			
	$CH_3CH_2PH_2{}^{11}BD_3$ trans	$v=0$	$A=13848\,(12)$ $B=\;2458.63\,(2)$ $C=\;2268.29\,(2)$					
	$CH_3CH_2PD_2{}^{11}BD_3$ trans	$v=0$	$A=11966\,(5)$ $B=\;2431.24\,(1)$ $C=\;2234.09\,(1)$					
	$CH_3CH_2PH_2{}^{10}BH_3$ trans	$v=0$	$A=15953\,(12)$ $B=\;2804.27\,(2)$ $C=\;2568.41\,(2)$					
	2.5.2.5 Elements with two or three internal rotors							
434 *6-161* *4-128*	$(CH_3)_2O$	$v=0$	$A''=38789.3953\,(109)$ *) $B''=10056.5381\,(29)$ $C''=\;8886.80566\,(278)$	$\tau_1=\;-5.843\,(633)$ *) $\tau_2=-20.397\,(134)$ $\tau_3=\;1.62\,(2)\,10^3$ $\tau_{aaaa}=-1357.4\,(10)$ $\tau_{bbbb}=\;-50.720\,(51)$ $\tau_{cccc}=\;-22.237\,(38)$	79LOV		for structure, see also Vol. II/7 (3, 297) *) for the AA species	2.8, 89
		$v_n=1_1$	$A=38797.34\,(30)$ *) $B=10024.95\,(5)$ $C=\;8867.34\,(10)$		78LUT			
		$v_n=1_2$	$A=38770.43\,(30)$ *) $B=\;9993.07\,(5)$ $C=\;8870.13\,(10)$					
		$v_n=2_1$	$A=39394.67\,(44)$ *) $B=10003.35\,(9)$ $C=\;8851.08\,(13)$					
		$v_n=2_2$	$A=39571.26\,(17)$ *) $B=\;9984.53\,(4)$ $C=\;8879.56\,(5)$					
Cont.		$v_n=2_3$	$A=39131.42\,(32)$ *) $B=\;9946.58\,(6)$ $C=\;8852.25\,(9)$					

Nr.	Molecule	v	A, B, C MHz	D, H kHz	Ref.	Ref. struct.	Varia, remarks	See also Tab., Nr.
434 *6-161* *4-128* Cont.	CD_3OCH_3	$v=0$	$A=30912.15$ (50) **) $B=$ 8635.50 (15) $C=$ 7747.45 (20)		79LOV, 76DUR3		**) for the A^- torsional sublevels	
		$v_n=1_1$	$A=30911.09$ (7) **) $B=$ 8609.50 (2) $C=$ 7736.91 (2)					
		$v_n=1_2$	$A=30904.07$ (10) **) $B=$ 8590.73 (3) $C=$ 7733.92 (4)					
	$(CD_3)_2O$	$v=0$	$A=25696.17$ (11) $B=$ 7483.79 (3) $C=$ 6798.01 (6)		78LUT			
		$v_n=1_1$	$A=25701.73$ (9) *) $B=$ 7466.97 (3) $C=$ 6785.67 (5)				*) for the AA species	
		$v_n=1_2$	$A=25674.80$ (9) *) $B=$ 7448.76 (3) $C=$ 6793.73 (5)					
		$v_n=2_1$	$A=25723.78$ (8) *) $B=$ 7448.79 (2) $C=$ 6772.76 (3)					
		$v_n=2_2$	$A=25697.70$ (8) *) $B=$ 7433.62 (2) $C=$ 6776.71 (3)					
		$v_n=2_3$	$A=25728.74$ (6) *) $B=$ 7427.18 (2) $C=$ 6783.64 (2)					
435 *6-162* *4-129*	$(CH_3)_2S$	$v=0$	$A=17809.7298$ (40) *) $B=$ 7621.0976 (17) $C=$ 5717.7694 (14)	Δ_J = 8.0585 (94) Δ_{JK} = −35.509 (59) Δ_K = 139.29 (42) δ_J = 2.8268 (20) δ_K = 3.580 (50)	80DEM		for structure, see also Vol. II/7 (3, 300) *) uncertainties are 2σ	2.6.3, 166 2.8, 102
Cont.		$v_n=1_1$	$A=17817.77$ **) $B=$ 7590.16 $C=$ 5700.81		73TRI		**) for the AA species	

Nr.	Molecule	v	A, B, C MHz	D, H kHz	Ref.	Ref. struct.	Varia, remarks	See also Tab., Nr.
435 *6-162* *4-129*	$(CH_3)_2S$	$v_n=1_2$	$A=17831.76$ **) $B=7574.20$ $C=5710.71$		73TR1		**) for the AA species	
Cont.	$(CD_3)_2S$	$v=0$	$A=13590.4972(33)$ $B=5792.6391(12)$ $C=4525.4179(12)$	$\Delta_J=4.4625(53)$ $\Delta_{JK}=-15.146(35)$ $\Delta_K=58.337(15)$ $\delta_J=1.44150(69)$ $\delta_K=-1.250(32)$	81DEM1			
		$v_n=1_1$	$A=13594.562(43)$ $B=5776.429(22)$ $C=4512.941(21)$	$\Delta_J=4.12(58)$ $\Delta_{JK}=-4.2(23)$ $\Delta_K=68(12)$ $\delta_J=1.568(86)$ $\delta_K=5.9(36)$				
		$v_n=1_2$	$A=13593.913(48)$ $B=5767.670(25)$ $C=4526.327(23)$	$\Delta_J=-0.66(67)$ $\Delta_{JK}=-31.9(25)$ $\Delta_K=91(14)$ $\delta_J=1.421(90)$ $\delta_K=-6.9(26)$				
		$v_n=2_1$	$A=13606.730(167)$ **) $B=5758.810(100)$ $C=4494.010(62)$					
		$v_n=2_2$	$A=13602.077(436)$ **) $B=5754.070(81)$ $C=4524.574(109)$					
		$v_n=2_3$	$A=13605.573(116)$ **) $B=5743.677(64)$ $C=4522.375(65)$					
		$v_\delta=1$	$A=13713.680(42)$ $B=5789.791(16)$ $C=4515.223(20)$	$\Delta_J=2.33(41)$ $\Delta_{JK}=-15.3(20)$ $\Delta_K=76.1(97)$ $\delta_J=1.22(18)$ $\delta_K=-2.7(45)$			$\nu_\delta=\nu_7(a_1)=242\ cm^{-1}$	
Cont.		$v_\delta=2$	$A=13836.153(644)$ $B=5786.878(343)$ $C=4505.231(381)$					

Nr.	Molecule	v	A, B, C MHz	D, H kHz	Ref.	Ref. struct.	Varia, remarks	See also Tab., Nr.
435 *6-162* *4-129* Cont.	$(CD_3)_2{}^{34}S$	$v=0$	$A=13287.743(37)$ $B=5792.828(21)$ $C=4491.255(18)$	$\Delta_J=4.59(66)$ $\Delta_{JK}=-15.4(18)$ $\Delta_K=72.4(86)$ $\delta_J=1.64(18)$ $\delta_K=-2.0(36)$	81DEM1			
	$^{13}CD_3SCD_3$	$v=0$	$A=13500.789(91)$ $B=5672.065(73)$ $C=4441.646(80)$	$\Delta_J=10.2(23)$ $\Delta_{JK}=-25.6(92)$ $\Delta_K=91(27)$ $\delta_J=2.52(94)$ $\delta_K=14(15)$				
436 *6-163*	$(CH_3)_2{}^{76}Se$	$v=0$	$A=11442.978(7)$ $B=6912.763(7)$ $C=4551.511(6)$	$\Delta_J=4.92(20)$ $\Delta_{JK}=-33.3(6)$ $\Delta_K=86.9(10)$ $\delta_J=3.39(6)$ $\delta_K=0.27(48)$	80DR	77PAN2	for structure, see also Vol. II/7 (3, 303)	2.8, 103
	$(CH_3)_2{}^{77}Se$	$v=0$	$A=11405.953(8)$ $B=6912.807(8)$ $C=4545.652(7)$	$\Delta_J=6.11(23)$ $\Delta_{JK}=-32.3(7)$ $\Delta_K=84.7(12)$ $\delta_J=3.48(6)$ $\delta_K=-0.63(55)$				
	$(CH_3)_2{}^{78}Se$	$v=0$	$A=11369.972(16)$ $B=6912.844(17)$ $C=4539.929(14)$	$\Delta_J=9.32(45)$ $\Delta_{JK}=-32.5(13)$ $\Delta_K=84.7(23)$ $\delta_J=3.45(13)$ $\delta_K=-0.35(107)$				
	$(CH_3)_2{}^{80}Se$	$v=0$	$A=11300.409(19)$ $B=6912.862(19)$ $C=4528.768(17)$	$\Delta_J=8.08(52)$ $\Delta_{JK}=-33.0(15)$ $\Delta_K=85.2(27)$ $\delta_J=3.41(14)$ $\delta_K=-0.01(121)$				
Cont.	$(CH_3)_2{}^{82}Se$	$v=0$	$A=11234.103(12)$ $B=6912.885(13)$ $C=4518.053(11)$	$\Delta_J=7.22(34)$ $\Delta_{JK}=-32.7(10)$ $\Delta_K=84.0(17)$ $\delta_J=3.45(9)$ $\delta_K=-0.43(78)$				

Nr.	Molecule	v	A, B, C MHz	D, H kHz	Ref.	Ref. struct.	Varia, remarks	See also Tab., Nr.
436 *6-163* Cont.	$(CD_3)_2{}^{78}Se$	$v=0$	$A=8839.401$ $B=5296.497$ $C=3612.906$	$d_J=-14.600$ $d_{JK}=-19.581$ $d_K=59.199$ $d_{WJ}=4.367\cdot10^{-6}$ $d_{WK}=-3.695\cdot10^{-6}$	76PAN1			
	$(CD_3)_2{}^{80}Se$	$v=0$	$A=8782.406$ $B=5296.524$ $C=3603.330$	$d_J=-13.994$ $d_{JK}=-20.536$ $d_K=55.828$ $d_{WJ}=4.288\cdot10^{-6}$ $d_{WK}=-3.326\cdot10^{-6}$				
437 *4-132*	$(CH_3)_2PH$	$v_\tau=(1,0)$	$A=16050.23$ *) $B=6999.63$ $C=5394.83$		77DUR5	81DUR1	for structure, see also Vol. II/7 (3, 313) *) for the AA species v_τ, torsional state	2.8, 126
		$v_\tau=(0,1)$	$A=16054.38$ *) $B=6996.08$ $C=5396.34$					
	CH_3CD_3PH	$v=0$	$A=14129.75$ $B=6102.60$ $C=4791.65$		81DUR1			
		$v_\tau=(0,1)$	$A=14113.68$ $B=6087.57$ $C=4787.02$				$0,1=CD_3$ torsion	
		$v_\tau=(1,0)$	$A=14112.70$ $B=6086.96$ $C=4784.92$				$1,0=CH_3$ torsion	
	$(CD_3)_2PH$	$v=0$	$A=12514.04$ $B=5379.36$ $C=4281.14$					
		$v_\tau=(0,1)$	$A=12502.37$ $B=5369.22$ $C=4275.01$					
		$v_\tau=(1,0)$	$A=12495.21$ $B=5366.18$ $C=4279.57$					

Nr.	Molecule	v	A, B, C MHz	D, H kHz	Ref.	Ref. struct.	Varia, remarks	See also Tab., Nr.
438	$(CH_3)_2NCN$	$v=0$	$A=8885.6\,(17)$ $B=4260.41\,(8)$ $C=3037.30\,(8)$		73LI2	73LI2		2.6.3, 167 3, 127
		$v_\delta=1$	$A=8936.5\,(21)$ $B=4274.34\,(8)$ $C=3037.99\,(8)$				$v_\delta=220\ cm^{-1}$, N—C≡N bending	
439	$(CH_3)_2NPF_2$	$v=0$	$A=4315.6\,(27)$ $B=2164.75\,(1)$ $C=2090.95\,(1)$	$D_J=0.9019\,(770)$ $D_{JK}=0.6928\,(3480)$	73FOR	73FOR		3, 128
	$(CD_3)_2NPF_2$	$v=0$	$A=3748.8\,(10)$ $B=1940.08\,(2)$ $C=1803.41\,(2)$	$D_J=0.6047\,(1444)$ $D_{JK}=0.2452\,(3766)$				
440 *6-164a*	$^{13}CH_3CH_3NNO$, syn	$v=0$	$A=8724.20\,(13)$ $B=4609.73\,(6)$ $C=3127.32\,(5)$		75GUA	79GUA	for structure, see also Vol. II/7 (3, 292)	2.7.3, 43 3, 129
	$^{13}CH_3CH_3NNO$, anti	$v=0$	$A=8987.54\,(4)$ $B=4495.49\,(3)$ $C=3106.33\,(3)$					
	$(CH_3)_2{}^{15}N^{14}NO$	$v=0$	$A=9053.47\,(3)$ $B=4606.48\,(1)$ $C=3167.15\,(1)$					
	$(CH_3)_2{}^{14}N^{15}NO$	$v=0$	$A=8962.71\,(7)$ $B=4595.68\,(3)$ $C=3150.85\,(3)$					
	CH_3CD_3NNO, d_3-anti	$v=0$	$A=8371.488\,(93)$ $B=4086.571\,(70)$ $C=2890.257\,(63)$		79GUA			
	CH_3CD_3NNO, d_3-syn	$v=0$	$A=7551.093\,(62)$ $B=4464.099\,(42)$ $C=2960.588\,(41)$					
	$(CD_3)_2NNO$	$v=0$	$A=7022.607\,(44)$ $B=3976.069\,(55)$ $C=2710.547\,(49)$					
	$(CD_3)_2{}^{15}N^{15}NO$	$v=0$	$A=6957.073\,(22)$ $B=3957.482\,(22)$ $C=2692.243\,(51)$					

Nr.	Molecule	v	A, B, C MHz	D, H kHz	Ref.	Ref. struct.	Varia, remarks	See also Tab., Nr.
441	$(CH_3)_2PCN$	$v=0$	$A=5763.1$ (3) $B=3116.55$ (2) $C=2418.34$ (1)		74DUR4	74DUR4		2.6.3, 168
		$v_\alpha=1$	$A=5750.6$ (2) $B=3114.57$ (1) $C=2415.29$ (1)				$\nu_\alpha=252$ (20) cm^{-1}	
		$v_\beta=1$	$A=5797.0$ (3) $B=3117.95$ (3) $C=2422.26$ (1)				$\nu_\beta=201$ (10) cm^{-1}	
		$v_\gamma=1$	$A=5759.2$ (4) $B=3116.78$ (3) $C=2417.18$ (1)				$\nu_\gamma=178$ (15) cm^{-1}	
442	$(CH_3)_2CHPH_2$ gauche	$v=0$	$A=7633.34$ (9) $B=4243.36$ (2) $C=3045.84$ (2)		78DUR	78DUR		2.6.3, 169
	$(CH_3)_2CHPD_2$ gauche	$v=0$	$A=7226.47$ (5) $B=4041.06$ (2) $C=2946.85$ (2)					
443	$(CH_3)_2HP^{11}BH_3$	$v=0$	$A=5925.81$ (13) $B=5275.85$ (3) $C=3291.54$ (3)		74DUR5	74DUR5		2.6.3, 170
	$(CH_3)_2HP^{10}BH_3$	$v=0$	$A=5930.72$ (8) $B=5479.72$ (3) $C=3368.26$ (2)					
	$(CH_3)_2DP^{11}BH_3$	$v=0$	$A=5732.80$ (8) $B=5120.10$ (2) $C=3290.42$ (2)					
	$(CH_3)_2DP^{10}BH_3$	$v=0$	$A=5737.75$ (20) $B=5313.00$ (13) $C=3367.35$ (2)					
	$(CH_3)_2HP^{11}BD_3$	$v=0$	$A=5642.00$ (13) $B=4526.64$ (2) $C=2985.37$ (2)					
Cont.								

Nr.	Molecule	v	A, B, C MHz	D, H kHz	Ref.	Ref. struct.	Varia, remarks	See also Tab., Nr.
443 Cont.	$(CH_3)_2HP^{11}BD_2H$, C_s symmetry	$v=0$	$A=5778.78$ (8) $B=4739.23$ (2) $C=3037.24$ (1)		74DUR5			
	$(CH_3)_2HP^{11}BD_2H$, C_1 symmetry	$v=0$	$A=5733.43$ (11) $B=4737.56$ (3) $C=3099.58$ (1)					
443a	$CH_3NHNHCH_3$ inner-outer	$v=0$	$A=17308.31$ (13) $B=4946.56$ (3) $C=4505.97$ (3)		81NAK1	81NAK1	uncertainties are 3 σ	3, 129a
	$CH_3NHNHCH_3$ outer-outer	$v=0$	$A=27735.55$ (26) $B=4173.21$ (5) $C=3937.41$ (4)					
444 *4-133a*	CH_3SSCD_3	$v=0$	$A=7385.070$ (4) $B=2597.454$ (3) $C=2363.105$ (3)	$d_J=-16.74$ (40) $d_{JK}=-523.2$ (20) $d_K=-927.9$ (30) $d_{WJ}=7.405$ (10) $\cdot 10^{-3}$ $d_{WK}=0.1940$ (10)	74KUH	74KUH	for structure, see also Vol. II/7 (3, 301)	2.8, 104
		$v_9=1$	$A=7397.367$ (12) $B=2592.371$ (11) $C=2361.842$ (11)	$d_J=-16.71$ (30) $d_{JK}=-528.2$ (120) $d_K=-944.6$ (230) $d_{WJ}=7.393$ (30) $\cdot 10^{-3}$ $d_{WK}=0.1967$ (50)			ν_9, S—S torsion	
		$v_9=2$	$A=7407.236$ (20) $B=2587.912$ (10) $C=2360.676$ (10)	$d_J=-17.16$ (30) $d_{JK}=-515.4$ (220) $d_K=-923.7$ (440) $d_{WJ}=7.608$ (120) $\cdot 10^{-3}$ $d_{WK}=0.1918$ (10)				
		$v_x=1$	$A=7407.128$ (30) $B=2594.271$ (25) $C=2360.222$ (24)	$d_J=-16.46$ (60) $d_{JK}=-502.8$ (250) $d_K=-902.6$ (500) $d_{WJ}=7.262$ (140) $\cdot 10^{-3}$ $d_{WK}=0.1861$ (100)				
Cont.								

Nr.	Molecule	v	A, B, C MHz	D, H kHz	Ref.	Ref. struct.	Varia, remarks	See also Tab., Nr.
445 *6-165c*	$(CH_3CH_2)_2O$ $^+$)	$v=0$	$A=17956.15\,(29)$ *) $B=2244.19\,(2)$ $C=2101.77\,(2)$	$D_J=-0.158$	74HAY3	74HAY3	*) uncertainties are 2.5 σ $^+$) trans-trans isomer	2.6.3, 171
	$CH_3CHDOCH_2CH_3$	$v=0$	$A=16684.45\,(30)$ *) $B=2223.20\,(2)$ $C=2077.15\,(2)$	$D_J=-0.356$				
	$CH_2DCH_2OCH_2CH_3$, symmetric	$v=0$	$A=17931.67\,(43)$ *) $B=2141.47\,(3)$ $C=2011.30\,(3)$	$D_J=-0.300$				
	$CH_2DCH_2OCH_2CH_3$, antisymmetric	$v=0$	$A=16881.35\,(30)$ *) $B=2183.50\,(2)$ $C=2045.26\,(2)$	$D_J=-0.455$				
	$^{13}CH_3CH_2OCH_2CH_3$,	$v=0$	$A=17859.99\,(27)$ *) $B=2190.00\,(2)$ $C=2052.89\,(2)$	$D_J=-0.406$				
	$CH_3{}^{13}CH_2OCH_2CH_3$	$v=0$	$A=17782.77\,(15)$ *) $B=2230.74\,(1)$ $C=2087.59\,(1)$	$D_J=-0.382$				
446	$(CH_3)_3CSH$	$v=0$	$A=4478.1537\pm0.05$ *) $B=2998.9655\pm0.1$ $C=2970.0228\pm0.1$		74VAL		*) 95% confidence level estimates	2.6.3, 172 2.8, 105 3, 130
447	$(CH_3)_3COC(=O)H$		$B+C=3950\,(1)$ *)		78TR4		*) LRMW	
448	$(CH_3)_3COC(=O)CF_3$		$B+C=1301.1\,(5)$ *)		78TR4		*) LRMW	
449	$(CH_3)_3CCH_2OC(=O)H$ E I		$B+C=2195.1\,(5)$ *) $B+C=2327.4\,(5)$ *)		77TR4		*) LRMW. E: $\tau_1(O{=}COC)=0°$, $\tau_2(COCC)=180°$ I: $\tau_1\simeq80°$, $\tau_2\sim210°$	
450	$(CH_3)_3CCH_2OC(=O)F$ E I		$B+C=1625.0\,(5)$ *) $B+C=1703.0\,(6)$ *)		77TR4		*) LRMW; for definitions, see Nr. 449	

Nr.	Molecule	v	A, B, C MHz	D, H kHz	Ref.	Ref. struct.	Varia, remarks	See also Tab., Nr.
451	$(CH_3)_3CCH_2OC(=O){}^{35}Cl$ E		$B+C=1250\ (2)$ *)		77TR4		*) LRMW; for definitions, see Nr. 449	
	I		$B+C=1208.2\ (2)$ *)					
	$(CH_3)_3CCH_2OC(=O){}^{37}Cl$ I		$B+C=1282.2\ (3)$ *)					
452	$(CH_3)_3CCH_2OC(=O)CN$ E		$B+C=1292.4\ (3)$ *)		77TR4		*) LRMW; for definitions, see Nr. 449	
	I		$B+C=1337.1\ (2)$ *)					
453	$(CH_3)_3CCH_2OC(=O)CF_3$ I		$B+C=884\ (2)$ *)		77TR4		*) LRMW; for definitions, see Nr. 449	
454	$(CH_3)_3CNO$							2.6.3, 173 2.8, 118
455	$(CH_3)_3CNO_2$	$v=0$	$A(CMe_3)=4533.15\pm0.8$ $A(NO_2)=13341.8$ (fixed) $B=2506.275\pm0.14$ $C=2112.963\pm0.01$		80LAN	80LAN		2.6.3, 174 2.8, 119
	$(CH_3)_3C^{15}NO_2$	$v=0$	$A(CMe_3)=4534.15\pm0.5$ $A(NO_2)=13343.0$ (fixed) $B=2496.636\pm0.07$ $C=2106.341\pm0.01$					
455a	$(CH_3)_3CONO$	$v=0$	$A=4492.3\ (2)$ $B=1723.06\ (8)$ $C=1719.19\ (8)$		81END	81END		
	$(CH_3)_3CO^{15}NO$	$v=0$	$A=4492.9\ (2)$ $B=1709.98\ (5)$ $C=1705.98\ (5)$					
	$(CD_3)_3CONO$	$v=0$	$A=3470\ (50)$ $B=1519.21\ (7)$ $C=1517.89\ (7)$					
456 Cont.	$(CH_3)_3N\cdot BH_2F$	$v=0$	$A=4620.91\ (2)$ *) $B=2585.20\ (1)$ $C=2562.28\ (1)$		78CAX1	78CAX1	*) uncertainties are 2σ	

Nr.	Molecule	v	A, B, C MHz	D, H kHz	Ref.	Ref. struct.	Varia, remarks	See also Tab., Nr.
456 Cont.	$(CH_3)_3{}^{15}NBH_2F$	v=0	A = 4621.24 (10) *) B = 2582.25 (1) C = 2559.38 (1)		78CAX1			
	$(CH_3)_3N^{10}BH_2F$	v=0	A = 4655.95 (10) *) B = 2595.79 (1) C = 2583.49 (1)					
	$(CH_3)_3{}^{15}N^{10}BH_2F$	v=0	A = 4656.05 (10) *) B = 2592.95 (2) C = 2580.64 (2)					
457	$(CH_3)_3CPH_2$	v=0	A = 4397.63 (4) B = 2878.88 (2) C = 2870.86 (2)		78LI	78LI		2.6.3, 175
	$(CH_3)_3CPHD$	v=0	B = 2831.45 (2) C = 2801.50 (2)					
	$(CH_3)_3CPD_2$	v=0	A = 4261.98 (4) B = 2769.82 (2) C = 2752.71 (2)					

2.5.2.6 Saturated cyclic compounds

Nr.	Molecule	v	A, B, C MHz	D, H kHz	Ref.	Ref. struct.	Varia, remarks	See also Tab., Nr.
458 *4-150*	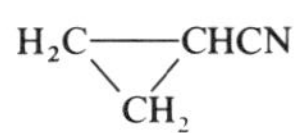	v=0	A = 15786.068 (22) B = 3465.112 (7) C = 3286.252 (10)	D_N = 1.18 (30)	80BRO1	75PE2	for structure, see also Vol. II/7 (3, 497)	2.7.3, 45
	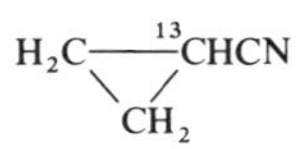	v=0	A = 15615.04 (5) B = 3454.818 (4) C = 3284.411 (3)		75PE2			
	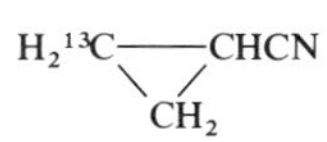	v=0	A = 15496.75 (1) B = 3421.920 (2) C = 3236.900 (2)					
459	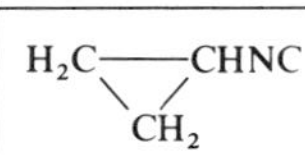	v=0	A = 16006.42 (10) B = 3733.82 (3) C = 3515.92 (1)	D_J = −3.35 (12) D_{JK} = 14.08 (90) D_K = 2.37 (90) δ_J = −0.53 (1)	81PER			2.6.3, 216

Nr.	Molecule	v	A, B, C MHz	D, H kHz	Ref.	Ref. struct.	Varia, remarks	See also Tab., Nr.
460 *6-182*	H_2C—$^{13}CHNH_2$ (ring, CH_2)	$v=0$	$A=16144.44\,(5)$ $B=\;6698.45\,(2)$ $C=\;5793.12\,(1)$		78MAT2	78MAT2	for structure, see also Vol. II/7 (3, 420)	
	$H_2{}^{13}C$—$CHNH_2$ (ring, CH_2)	$v=0$	$A=15976.83\,(3)$ $B=\;6645.05\,(1)$ $C=\;5702.48\,(1)$		75HA1			
461	H_2C—$CHOH$, gauche (ring, CH_2)	$v=0$	$A^+=16668.59\,(45)$ *) $B^+=\;6911.28\,(18)$ $C^+=\;5955.48\,(9)$ $A^-=16660.96\,(45)$ $B^-=\;6911.36\,(9)$ $C^-=\;5954.60\,(18)$		78MCD1	78MCD1	*) uncertainties are 3 σ $\Delta E(O^-—O^+)=4115.26\,(42)$ MHz, $G_a=120.61\,(1.26)$ MHz $G_c=\;42.84\,(18)$ MHz. $\Delta E(O^-—O^+)=163.74$. (1.80) MHz potential function determined [78MCD1]	2.6.3, 217 3, 131
	H_2C—$CHOD$, gauche (ring, CH_2)	$v=0$	$A=16343.79\,(45)$ *) $B=\;6602.71\,(18)$ $C=\;5688.65\,(18)$					
462 *6-181*b	H_2C—$CHNO_2$ (ring, CH_2)	$v=0$	$A=7953.49\,(2)$ $B=2937.40\,(1)$ $C=2583.60\,(1)$	$d_J\;=-1.08\,(2)$ $d_{JK}=361\,(4)$ $d_K\;=679\,(7)$ $d_{WJ}=\;4.49\,(29)\cdot 10^{-7}$ $d_{WK}=-1.30\,(1)\cdot 10^{-4}$	73MOC1	73MOC1	for structure, see also Vol. II/7 (3, 394)	2.6.3, 218
		$v_\tau=1$	$A=7927.69\,(7)$ $B=2937.56\,(2)$ $C=2586.57\,(2)$	$d_J\;=-1.09\,(65)$ $d_{JK}=442\,(26)$ $d_K\;=829\,(48)$ $d_{WJ}=\;6.73\,(1.88)\cdot 10^{-7}$ $d_{WK}=-1.60\,(10)\cdot 10^{-4}$			$v_\tau=70\,(20)\;cm^{-1}$, torsional frequency of the NO_2 group, $V=1100\,(500)\;cm^{-1}$	
		$v_\tau=2$	$A=7904.58\,(203)$ $B=2937.58\,(1)$ $C=2589.36\,(2)$					
		$v_\tau=3$	$A=7884.34\,(355)$ $B=2937.54\,(2)$ $C=2592.17\,(3)$					
Cont.		$v_\tau=4$	$A=7855.62\,(287)$ $B=2937.42\,(2)$ $C=2594.99\,(2)$					

Nr.	Molecule	v	A, B, C MHz	D, H kHz	Ref.	Ref. struct.	Varia, remarks	See also Tab., Nr.
462 *6-181b* Cont.	H_2C—$CHNO_2$ (ring with CH_2)	$v_\tau=5$	$A=7837.10\ (366)$ $B=2937.24\ (2)$ $C=2597.76\ (3)$		73MOC1			
	H_2C—$CDNO_2$ (ring with CH_2)	$v=0$	$A=7597.98\ (2)$ $B=2934.19\ (1)$ $C=2542.77\ (1)$					
463	H_2C—$CHNO$ (ring with CH_2), trans	$v=0$	$A=16406.47\ (5)$ $B=\ \ 3453.34\ (2)$ $C=\ \ 3253.84\ (2)$		78COR			2.6.3, 219 2.7.3, 46
		$v_\tau=1$	$A=16297.00\ (5)$ $B=\ \ 3467.04\ (2)$ $C=\ \ 3259.13\ (2)$				$\nu_\tau=140\ (20)\ \mathrm{cm}^{-1}$, nitroso torsion	
		$v_\tau=2$	$A=16192.62\ (5)$ $B=\ \ 3479.97\ (2)$ $C=\ \ 3264.21\ (2)$					
		$v_\tau=3$	$A=16091\ (1)$ $B=\ \ 3492.42\ (2)$ $C=\ \ 3268.76\ (2)$					
464	H_2C—$CHC(=O){}^{35}Cl$ (ring with CH_2)	$v=0$	$A=6541.77\ (5)$ $B=1796.675\ (7)$ $C=1583.898\ (7)$	$d_J\ =-0.694\ (147)$ $d_{JK}=202.0\ (50)$ $d_K\ =586\ (16)$ $d_{WJ}=\ \ 4.95\ (87)\cdot 10^{-7}$ $d_{WK}=-1.199\ (30)\cdot 10^{-4}$	76NAI	76NAI	for structure, see also Vol. II/7 (3, 494) cis conformer	
	H_2C—$CHC(=O){}^{37}Cl$ (ring with CH_2)	$v=0$	$A=6512.78\ (1.00)$ $B=1755.147\ (8)$ $C=1549.737\ (8)$					
465 Cont.	H_2C—CF_2 (ring with CH_2)	$v=0$	$A=7183.38\ (7)$ $B=5564.80\ (1)$ $C=4243.35\ (1)$		75PER	75PER		2.6.3, 220

Nr.	Molecule	v	A, B, C MHz	D, H kHz	Ref.	Ref. struct.	Varia, remarks	See also Tab., Nr.
465 Cont	cyclo-H_2C–CF_2–CH_2	$v_a=1$	$A=7175.29$ (21) $B=5479.22$ (1) $C=4239.27$ (1)		75PER		ν_a(a)$\simeq 327$ cm^{-1}, probably CF_2 torsion	
		$v_b=1$	$A=7190.94$ (31) $B=5648.03$ (2) $C=4234.59$ (2)				ν_b(b)$\simeq 351$ cm^{-1}	
	cyclo-D_2C–CF_2–CD_2	$v=0$	$A=6296.34$ (15) $B=4799.25$ (1) $C=3871.85$ (1)					
	cyclo-$H_2{}^{13}C$–CF_2–CH_2	$v=0$	$A=7124.51$ (56) $B=5447.91$ (5) $C=4195.22$ (5)					
	cyclo-D_2C–${}^{13}CF_2$–CD_2	$v=0$	$A=6296.68$ (36) $B=4798.71$ (6) $C=3871.60$ (5)					
	cyclo-$D_2{}^{13}C$–CF_2–CD_2	$v=0$	$A=6250.90$ (14) $B=1719.10$ (1) $C=3836.45$ (1)					
466	cyclo-H_2C–$C(CN)_2$–CH_2	$v=0$	$A''=3779.6481$ (17) $B''=2549.2078$ (15) $C''=1726.7509$ (14)	$\tau_{aaaa}=-19.996$ (43) $\tau_{bbbb}=-5.663$ (26) $\tau_{cccc}=-0.273$ (29) $\tau_1=2.200$ (78) $\tau_2=0.205$ (26) $\tau_3=45.50$ (42)	75PE1	75PE1		
467	cyclo-F_2C–CF_2–CH_2	$v=0$	$A=3841.19$ $B=2518.11$ $C=1969.71$		75LAU			
	cyclo-F_2C–CF_2–CD_2	$v=0$	$A=3588.40$ $B=2427.43$ $C=1956.60$					
	cyclo-$F_2{}^{13}C$–CF_2–CD_2	$v=0$	$A=3588.60$ $B=2421.20$ $C=1952.44$					

Nr.	Molecule	v	A, B, C MHz	D, H kHz	Ref.	Ref. struct.	Varia, remarks	See also Tab., Nr.
468	FHC——CHF, trans \ / CH_2	$v=0$	$A=11546.50$ $B=\ 3559.93$ $C=\ 3071.37$		76SEN			
	FDC——CDF, trans \ / CH_2	$v=0$	$A=10438.53$ $B=\ 3442.69$ $C=\ 3056.94$					
	FHC——CHF, trans \ / CD_2	$v=0$	$A=9838.18$ $B=3517.52$ $C=2944.82$					
	FHC——CHF, cis \ / CH_2	$v=0$	$A=7433.41$ $B=4678.50$ $C=3560.89$		77SEN			
469	H_2C——$C(CH_3)_2$ \ / CH_2	$v=0$	$A=6135.257\,(16)$ $B=5203.342\,(15)$ $C=3810.509\,(19)$	$\tau_{aaaa}=-15.1\,(24)$ $\tau_{bbbb}=-10.5\,(22)$ $\tau_{cccc}=-14.5\,(24)$ $\tau'_{bbcc}+\tau'_{aacc}=-11.6\,(44)$ $\tau'_{aabb}=-8.4\,(22)$	79MTH	79MTH		2.6.3, 221
470 *6-184a*	H_2C——$CHOCH_3$, gauche \ / CH_2	$v=0$	$A=12901.29\,(5)$ $B=\ 3536.18\,(1)$ $C=\ 3150.96\,(1)$	$d_J=-6.6\,(10)$ $d_{JK}=-98\,(21)$ $d_K=-180\,(62)$ $d_{WJ}=2.3\,(3)\cdot 10^{-3}$ $d_{WK}=2.1\,(6)\cdot 10^{-2}$	73PEN2	73PEN2		2.6.3, 222 2.8, 91 3, 132
		$v_T=1$	$A=12903.11\,(8)$ $B=\ 3524.15\,(8)$ $C=\ 3146.06\,(8)$				$\nu_T\simeq 173\ cm^{-1}$, heavy-atom torsion	
		$v_T=2$	$A=12904.89\,(6)$ $B=\ 3515.91\,(6)$ $C=\ 3143.36\,(6)$					
		$v_\tau=1$	$(A-C)/2=4872.47\,(33)$ $\kappa=-0.921287$				ν_τ, methyl torsion	
471	H_2C——$CHSCH_3$, gauche \ / CH_2	$v=0$	$A=8777.38\,(9)$ $B=2681.43\,(2)$ $C=2280.94\,(2)$		75PEN1	75PEN1		2.6.3, 223 2.8, 107 3, 133

Nr.	Molecule	v	A, B, C MHz	D, H kHz	Ref.	Ref. struct.	Varia, remarks	See also Tab., Nr.
472 *6-186*	H_2C—CH_2—CH—CH=CH_2 (cyclopropyl)	$v=0$	$A=15050(15)$ *) $B=3061.41(5)$ $C=2941.34(5)$		74CO3	74CO3	*) from Stark effect	2.6.3, 224 3, 134
		$v_T=1$	$A=15092(150)$ $B=3071.43(5)$ $C=2944.02(5)$				$\nu_T=125\ cm^{-1}$, torsion	
		$v_T=2$	$A=14794(150)$ $B=3081.38(5)$ $C=2946.70(5)$					
		$v_T=3$	$A=14875(150)$ $B=3091.22(5)$ $C=2949.27(5)$					
	$H_2{}^{13}C$—CH_2—CH—CH=CH_2 (cyclopropyl)	$v=0$	$A=15067$ (assumed) $B=3022.66(10)$ $C=2896.45(10)$					
473 *6-181a*	H_2C—CH_2—$CHCH_2{}^{35}Cl$, cis (cyclopropyl)	$v=0$	$A=7799.5$ $B=2858.9$ $C=2550.0$		77FU2			
	H_2C—CH_2—$CHCH_2{}^{35}Cl$, gauche (cyclopropyl)	$v=0$	$A=11745.65(4)$ $B=2047.274(6)$ $C=1886.622(6)$	$\Delta_J=0.85(5)$ $\Delta_{JK}=-0.9(3)$ $\Delta_K=44(12)$ $\delta_J=-0.099(2)$ $\delta_K=19.1(6)$	78MO1	78MO1		
	H_2C—CH_2—$CHCH_2{}^{37}Cl$, gauche (cyclopropyl)	$v=0$	$A=11691.61(48)$ $B=1997.664(10)$ $C=1842.823(10)$	$\Delta_J=0.7(1)$ $\Delta_{JK}=-64.6(9)$ $\Delta_K=2400(600)$ $\delta_J=0.19(4)$ $\delta_K=-67(13)$				
	H_2C—CH_2—$CHCD_2{}^{35}Cl$, gauche (cyclopropyl)	$v=0$	$A=10389.56(10)$ $B=2031.354(6)$ $C=1858.039(6)$	$\Delta_J=-0.10(4)$ $\Delta_K=-344(18)$ $\delta_K=-2.2(2)$	78MO3			

Nr.	Molecule	v	A, B, C MHz	D, H kHz	Ref.	Ref. struct.	Varia, remarks	See also Tab., Nr.
474	H_2C——$CHCH_2{}^{79}Br$, cis \ / CH_2	$v=0$	$A=8759.918\ (45)$ $B=1597.413\ (7)$ $C=1522.141\ (7)$	$\Delta_J = -0.07\ (3)$ $\delta_J = -0.0152\ (8)$ $\delta_K = 24.6\ (7)$ $H_K = 8.9\ (5)$	79MO1	79MO1		3, 135
	H_2C——$CHCH_2{}^{81}Br$, cis \ / CH_2	$v=0$	$A=8716.552\ (40)$ $B=1583.761\ (6)$ $C=1509.017\ (6)$	$\Delta_J = -0.43\ (4)$ $\Delta_{JK} = 2.9\ (2)$ $\Delta_K = -602\ (37)$ $\delta_J = -0.2340\ (7)$ $\delta_K = 168.1\ (7)$ $H_K = -7\ (3)$ $h_K = 0.18\ (4)$				
	H_2C——$CHCH_2{}^{79}Br$, gauche \ / CH_2	$v=0$	$A=11469.285\ (67)$ $B=1374.777\ (4)$ $C=1295.394\ (4)$	$\Delta_{JK} = 1.2\ (1)$ $\Delta_K = 3959\ (16)$ $\delta_K = -11.7\ (2)$				
	H_2C——$CHCH_2{}^{81}Br$, gauche \ / CH_2	$v=0$	$A=11400.100\ (59)$ $B=1364.088\ (3)$ $C=1283.952\ (3)$	$\Delta_{JK} = 0.9\ (2)$ $\Delta_K = 3520\ (15)$ $\delta_K = -9.9\ (2)$				
475 *6-187*	H_2C——$CHCH_2OH$ \ / CH_2	$v=0$	$A=12484.99\ (8)$ $B=3236.46\ (1)$ $C=2894.38\ (1)$		78BRO			2.6.3, 225 3, 136
	H_2C——$CHCH_2OD$ \ / CH_2	$v=0$	$A=12063.95\ (10)$ $B=3177.23\ (2)$ $C=2826.18\ (2)$					
	H_2C——$CHCH_aD_bOH$ \ / CH_2	$v=0$	$A=11742.86\ (10)$ $B=3205.17\ (2)$ $C=2830.41\ (2)$				for D_a and D_b see Fig. 136	
	H_2C——$CHCD_aH_bOH$ \ / CH_2	$v=0$	$A=11853.44\ (11)$ $B=3188.10\ (3)$ $C=2878.28\ (3)$					
	H_2C——$CHCH_aD_bOD$ \ / CH_2	$v=0$	$A=11321.30\ (10)$ $B=3145.87\ (2)$ $C=2762.23\ (2)$					
	H_2C——$CHCD_aH_bOD$ \ / CH_2	$v=0$	$A=11433.85\ (6)$ $B=3129.30\ (2)$ $C=2809.81\ (2)$					

Nr.	Molecule	v	A, B, C MHz	D, H kHz	Ref.	Ref. struct.	Varia, remarks	See also Tab., Nr.
476	H_2C—$CHSiH_3$ (ring closed by CH_2)	$v=0$	$A=13996.4091\ (35)$ $B=\ 3591.2915\ (10)$ $C=\ 3360.7719\ (9)$	$D'_J\ =\ 1.3976\ (28)$ $D'_{JK}=\ 1.047\ (30)$ $D'_K\ =\ 11.403\ (15)$ $\delta_J\ =\ 0.03283\ (38)$ $R'_6\ =-0.00328\ (15)$	79TYP	79TYP		2.6.3, 226 3, 137
477	H_2C—$CH^{11}BF_2$ *) (ring closed by CH_2)	$v=0$	$A=6929.79\ (103)$ $B=2467.69\ (1)$ $C=2119.84\ (1)$		80ODO	80ODO	*) bisected conformation: BF_2 group lying in the plane of symmetry $\nu_\tau=72\ (15)\ cm^{-1}$, BF_2 torsion $V_2=\ 4.23$ kcal/mol	2.6.3, 227 2.8, 138
		$v_\tau=1$	$A=6904.65\ (89)$ $B=2468.96\ (1)$ $C=2122.86\ (1)$					
		$v_\tau=2$	$A=6883.47\ (140)$ $B=2469.93\ (2)$ $C=2125.57\ (2)$					
		$v_\tau=3$	$A=6857.92\ (85)$ $B=2470.67\ (1)$ $C=2128.04\ (1)$					
		$v_\tau=4$	$A=6839.77\ (161)$ $B=2471.19\ (2)$ $C=2130.19\ (2)$					
		$v_\tau=5$	$A=6820.23\ (270)$ $B=2471.40\ (2)$ $C=2132.17\ (2)$					
		$v_w=1$	$A=6934.61\ (85)$ $B=2464.97\ (1)$ $C=2116.39\ (1)$				$\nu_w=206\ cm^{-1}$, BF_2 wag	
		$v_w=2$	$A=6904.68\ (122)$ $B=2465.83\ (2)$ $C=2119.79\ (2)$					
	H_2C—$CH^{10}BF_2$ (ring closed by CH_2)	$v=0$	$A=6931.80\ (95)$ $B=2474.00\ (1)$ $C=2124.43\ (1)$					

Nr.	Molecule	v	A, B, C MHz	D, H kHz	Ref.	Ref. struct.	Varia, remarks	See also Tab., Nr.
478 *4-152*	H_2C—CH_2 (O)	$v=0$	$A'=25483.79\,(3)$ $B'=22120.76\,(3)$ $C'=14097.71\,(3)$	$\tau_{aaaa}=-218.6\,(20)$ $\tau_{bbbb}=-73.0\,(20)$ $\tau'_{cccc}=29.4\,(80)$ $\tau_{aabb}=-153.4\,(20)$	74HI1, 74CR2	74HI1, 74HI2	for structure, see also Vol. II/7 (3, 244)	2.6.3, 228 2.7.3, 80 2.9.3, 32
		$v_a=1$	$A=25453.45$ $B=22003.60$ $C=14010.75$	$\tau_{aaaa}=-290\,(270)$ $\tau_{bbbb}=-10\,(280)$ $\tau_{cccc}=-60\,(300)$ $t_1=-590\,(370)$ *) $t_2=-910\,(490)$ *)	75YOS		$\nu_a=803\,(18)\,cm^{-1}$, probably $\nu_{15}\,(b_2)$, CH_2 rock *) $t_1=\tau'_{aabb}+\tau'_{ccaa}(A-B)/(A-C)$ $t_2=\tau'_{bbcc}+\tau'_{ccaa}(B-C)/(A-C)$	
		$v_b=1$	$A=25222.09$ $B=22250.62$ $C=13626.91$	$\tau_{aaaa}=-300\,(200)$ $\tau_{bbbb}=-290\,(210)$ $\tau_{cccc}=460\,(230)$ $t_1=110\,(260)$ *) $t_2=650\,(370)$ *)			$\nu_b=814\,(23)\,cm^{-1}$, probably $\nu_{12}\,(b_1)$, ring deform	
		$v_c=1$	$A=25521.92$ $B=22041.25$ $C=14417.47$	$\tau_{aaaa}=-360\,(170)$ $\tau_{bbbb}=-210\,(180)$ $\tau_{cccc}=-430\,(190)$ $t_1=-350\,(230)$ *) $t_2=-290\,(310)$ *)			$\nu_c=872\,(34)\,cm^{-1}$, probably $\nu_5\,(a_1)$, ring deform	
		$v_d=1$	$2B-A-C=4799.59$ $A-C=11206.23$				$\nu_d(a)=1020\,(33)\,cm^{-1}$	
		$v_e=1$	$2B-A-C=5404.57$ $A-C=10749.00$				$\nu_e=1109\,(42)\,cm^{-1}$	
	$H_2{}^{13}C$—CH_2 (O)	$v=0$	$A'=25291.89\,(5)$ $B'=21597.77\,(6)$ $C'=13825.75\,(6)$	$\tau_{aaaa}=-230.1\,(130)$ $\tau_{bbbb}=-75.0\,(60)$ $\tau_{cccc}=-48.3\,(320)$ $\tau'_{aabb}=-152.0\,(40)$	74HI1, 74CR2			
	H_2C—CH_2 (^{18}O)	$v=0$	$A'=23992.83\,(11)$ $B'=22121.58\,(12)$ $C'=13628.56\,(13)$	$\tau_{aaaa}=-184.3\,(90)$ $\tau_{bbbb}=-70.0\,(50)$ $\tau_{cccc}=95.2\,(64)$ $\tau'_{aabb}=-140.7\,(40)$				
	DHC—CHD (O), trans	$v=0$	$A'=22943.19\,(3)$ $B'=18198.47\,(3)$ $C'=12585.27$	$\tau_{aaaa}=-166.7\,(30)$ $\tau_{bbbb}=-35.8\,(10)$ $\tau_{cccc}=25.1\,(70)$ $\tau'_{aabb}=-81.5\,(10)$	74HI1			
Cont.								

Nr.	Molecule	v	A, B, C MHz	D, H kHz	Ref.	Ref. struct.	Varia, remarks	See also Tab., Nr.
478 *4-152* Cont.	$DH^{13}C$—CHD, trans O	$v=0$	$A'=22786.84\,(6)$ $B'=17852.32\,(7)$ $C'=12377.50\,(6)$	$\tau_{aaaa}=-161.7\,(150)$ $\tau_{bbbb}=-37.9\,(40)$ $\tau_{cccc}=20.1\,(17)$ $\tau'_{aabb}=-84.2\,(30)$	74HI1			
	DHC—CHD, trans ^{18}O	$v=0$	$A'=21646.74\,(9)$ $B'=18202.46\,(10)$ $C'=12186.66\,(9)$	$\tau_{aaaa}=-137.7\,(90)$ $\tau_{bbbb}=-38.5\,(30)$ $\tau_{cccc}=21.4\,(220)$ $\tau'_{aabb}=-81.7\,(20)$				
	DHC—CHD, cis O	$v=0$	$A'=22700.41\,(5)$ $B'=18318.39\,(5)$ $C'=12650.08\,(5)$	$\tau_{aaaa}=-175.4\,(60)$ $\tau_{bbbb}=-40.1\,(20)$ $\tau_{cccc}=6.1\,(130)$ $\tau'_{aabb}=-76.8\,(20)$				
	$DH^{13}C$—CHD, cis O	$v=0$	$A'=22555.61\,(4)$ $B'=17963.42\,(4)$ $C'=12438.85\,(4)$	$\tau_{aaaa}=-162.7\,(40)$ $\tau_{bbbb}=-43.2\,(20)$ $\tau_{cccc}=-6.7\,(100)$ $\tau'_{aabb}=-81.7\,(20)$				
	DHC—CHD, cis ^{18}O	$v=0$	$A'=21424.03\,(5)$ $B'=18317.20\,(5)$ $C'=12243.45\,(5)$	$\tau_{aaaa}=-133.6\,(50)$ $\tau_{bbbb}=-41.3\,(20)$ $\tau_{cccc}=33.4\,(130)$ $\tau'_{aabb}=-77.6\,(20)$				
	H_2C—CHD O	$v=0$	$A'=24252.47\,(6)$ $B'=19905.34\,(6)$ $C'=13327.40\,(6)$	$\tau_{aaaa}=-179.1\,(80)$ $\tau_{bbbb}=-48.6\,(20)$ $\tau_{cccc}=39.0\,(100)$ $\tau'_{aabb}=-109.5\,(20)$				
479 *4-153*	H_2C—$CHCH_3$ O	$v=0$	$A=18023.874\,(5)$ $B=6682.139\,(2)$ $C=5951.389\,(2)$	$\Delta_J=2.896\,(6)$ $\Delta_{JK}=3.300\,(80)$ $\Delta_K=19.820\,(12)$ $\delta_J=0.186\,(3)$ $\delta_K=2.926\,(161)$	77CR1	77CR1	for structure, see also Vol. II/7 (3, 406)	
	$H_2{}^{13}C$—$CHCH_3$ O	$v=0$	$A=17723.991\,(9)$ $B=6601.264\,(6)$ $C=5856.374\,(7)$					
Cont.	H_2C—$^{13}CHCH_3$ O	$v=0$	$A=17871.199\,(7)$ $B=6657.968\,(5)$ $C=5948.682\,(5)$					

Nr.	Molecule	v	A, B, C MHz	D, H kHz	Ref.	Ref. struct.	Varia, remarks	See also Tab., Nr.
479 Cont.	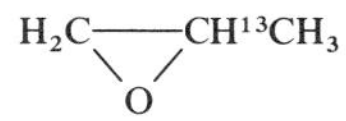	$v=0$	$A=18002.316(9)$ $B=6475.750(6)$ $C=5789.034(7)$		77CR1			
480	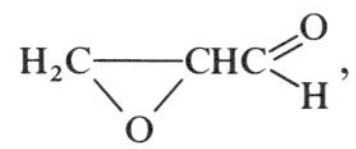 , trans *)	$v=0$	$A=18241.0997(53)$ $B=3272.9349(7)$ $C=3137.7074(9)$	$\Delta_J=0.57250(179)$ $\Delta_{JK}=12.38995(2621)$ $\Delta_K=17.29416(9140)$ $\delta_J=0.02388(39)$ $\delta_K=3.11717(6544)$	77CR2	77CR2	*) O of CHO trans to the oxirane ring	2.6.3, 229
		$v_\tau=1$	$A=18102.2423(71)$ $B=3285.8290(10)$ $C=3142.8648(12)$	$\Delta_J=0.59244(231)$ $\Delta_{JK}=11.64298(3299)$ $\Delta_K=14.21107(12720)$ $\delta_J=0.02882(44)$ $\delta_K=3.75123(7434)$			$v_\tau=140(10)\ \mathrm{cm}^{-1}$, torsion	
		$v_\tau=2$	$A=17967.2093(37)$ $B=3298.5789(8)$ $C=3147.9752(9)$	$\Delta_J=0.61994(188)$ $\Delta_{JK}=10.94006(2757)$ $\Delta_K=12.11776(9302)$ $\delta_J=0.03515(36)$ $\delta_K=4.33203(5938)$				
		$v_\tau=3$	$A=17848.14(33)$ $B=3309.87(3)$ $C=3152.25(4)$					
		$v_\tau=4$	$A=17738.68(40)$ $B=3320.43(4)$ $C=3156.27(5)$					
		$v_\tau=5$	$A=17637.52(41)$ $B=3330.30(4)$ $C=3159.89(5)$					
		$v_\tau=6$	$A=17552.91(53)$ $B=3338.65(5)$ $C=3162.71(6)$					
Cont.		$v_\tau=7$	$A=17370.00$ (fixed) $B=3345.11(3)$ $C=3165.10(3)$					

Nr.	Molecule	v	A, B, C MHz	D, H kHz	Ref.	Ref. struct.	Varia, remarks	See also Tab., Nr.
480 Cont.	$H_2{}^{13}C$—CHC(=O)H (epoxide, O), trans	$v=0$	$A=17907.681\,(161)$ $B=3230.515\,(14)$ $C=3089.287\,(19)$	**)	77CR2		**) centrifugal distortion constants of the parent species assumed	
	H_2C—^{13}CHC(=O)H (epoxide, O), trans	$v=0$	$A=18099.170\,(90)$ $B=3267.170\,(8)$ $C=3136.587\,(10)$	**)				
	H_2C—$CH^{13}C$(=O)H (epoxide, O), trans	$v=0$	$A=18156.788\,(166)$ $B=3250.550\,(16)$ $C=3119.416\,(20)$	**)				
481	H_2C—CHCH=CH_2 (epoxide, O), s-trans	$v=0$	$A=17367.285\,(11)$ $B=3138.219\,(3)$ $C=3043.686\,(2)$	$\tau_{aaaa}=-83.6\,(9)$ $\tau_{bbbb}=-1.94\,(2)$ $\tau_{cccc}=-1.77\,(3)$ $\tau_1=-45.8\,(3)$ $\tau_2=-7.31\,(6)$ $\tau_3=316.1\,(20)$ $H_J=-0.30\,(40)\cdot10^{-5}$ $H_{JK}=0.26\,(100)\cdot10^{-3}$ $H_{KJ}=-0.17\,(30)\cdot10^{-2}$ $H_K=0.10\,(70)\cdot10^{-2}$ $h_J=0.18\,(16)\cdot10^{-5}$ $h_{JK}=-0.41\,(31)\cdot10^{-3}$ $h_K=0.50\,(200)\cdot10^{-1}$	75IKE	75IKE		2.6.3, 230 3, 138
		$v_\tau=1$	$A=17224.63\,(20)$ $B=3148.95\,(1)$ $C=3047.04\,(1)$				$v_\tau=142\,(20)\ cm^{-1}$, torsion	
482	FHC—CHF (epoxide, O), cis	$v=0$	$A=9063.03\,(2)$ $B=5184.51\,(2)$ $C=3857.90\,(1)$		78GIL	78GIL		2.6.3, 231
	FDC—CHF (epoxide, O), cis	$v=0$	$A=8761.41\,(2)$ $B=5021.81\,(2)$ $C=3781.79\,(1)$					
Cont.	FDC—CDF (epoxide, O), cis	$v=0$	$A=8468.35\,(2)$ $B=4873.67\,(1)$ $C=3707.97\,(1)$					

Nr.	Molecule	v	A, B, C MHz	D, H kHz	Ref.	Ref. struct.	Varia, remarks	See also Tab., Nr.
482 Cont.	FHC—CHF (ring O: ^{18}O), cis	$v=0$	$A=8610.56$ (2) $B=5162.04$ (1) $C=3785.34$ (1)		78GIL			
	$FH^{13}C$—CHF (ring O), cis	$v=0$	$A=9018.02$ (6) $B=5147.61$ (5) $C=3839.51$ (3)					
483 *6–195*	H_2C—$CHCH_2F$ (ring O), gauche III	$v=0$	$A=14833.1$ $B=3210.2$ $C=2933.9$		77FU1			2.6.3, 232 3, 139
		$v_\tau=1$	$A=14788.1$ $B=3211.1$ $C=2937.1$				v_τ, torsion	
		$v_\tau=2$	$A=14758.9$ $B=3210.2$ $C=2939.9$					
		$v_\tau=3$	$A=14737$ $B=3210$ $C=2942$					
	H_2C—$CHCH_2F$ (ring O), gauche I	$v=0$	$A=14662.8$ $B=3129.4$ $C=2863.3$					
	H_2C—$CHCH_2F$ (ring O), cis II	$v=0$	$A=10645.8$ $B=4050.2$ $C=3679.1$					
484 *6–195a*	H_2C—$CHCH_2{}^{35}Cl$ (ring O), gauche 1 *)	$v=0$	$A=13373.02$ (4) $B=2080.353$ (7) $C=1932.469$ (7)	$\Delta_{JK}=-6$ (1) $\Delta_K=2400$ (300) $\delta_J=-0.43$ (7) $\delta_K=17$ (1) $H_{KJ}=-0.13$ (3) $H_K=570$ (70) $h_{JK}=0.061$ (5) $h_K=-5.1$ (9)	78MO2, 77FU2	78MO2	*) Cl twisted toward the O side of the ring	3, 140
Cont.								

Nr.	Molecule	v	A, B, C MHz	D, H kHz	Ref.	Ref. struct.	Varia, remarks	See also Tab., Nr.
484 *6–195 a* Cont.	H_2C—$CHCH_2{}^{37}Cl$, (ring O) gauche 1	$v=0$	$A=13361.24(3)$ $B=2028.853(1)$ $C=1887.990(1)$	$\Delta_{JK}=0.31(6)$ $\Delta_K=1660(160)$ $\delta_J=-0.16(1)$ $\delta_K=54.1(5)$	78MO2, 77FU2			
	H_2C—$CHCH_2{}^{35}Cl$, (ring O) gauche 2 **)	$v=0$	$A=12739.351(87)$ $B=2066.829(9)$ $C=1881.486(9)$	$\Delta_{JK}=5.2(4)$ $\Delta_K=4772(25)$ $\delta_K=-0.8(3)$	79MO3, 77FU2	79MO3	**) Cl trans to O	
	H_2C—$CHCH_2{}^{35}Cl$, cis ***) (ring O)	$v=0$	$A=8378.665(23)$ $B=2840.666(9)$ $C=2510.554(9)$	$\Delta_{JK}=-0.9(1)$ $\Delta_K=191(2)$ $\delta_K=3.98(7)$			***) Cl cis to the ring	
485	H_2C—$CHCH_2{}^{79}Br$, (ring O) gauche 1 *)	$v=0$	$A=12296.050(44)$ $B=1391.677(3)$ $C=1317.360(3)$	$\Delta_K=-726(10)$ $\delta_J=0.024(3)$ $\delta_K=-8.5(3)$	79MO2	79MO2	*) Br near the O atom	3, 141
	H_2C—$CHCH_2{}^{81}Br$, (ring O) gauche 1	$v=0$	$A=12199.162(63)$ $B=1378.321(4)$ $C=1309.142(4)$	$\Delta_K=-3410(14)$ $\delta_J=0.063(1)$ $\delta_K=-28.10(5)$				
	H_2C—$CHCH_2{}^{79}Br$, (ring O) gauche 2 +)	$v=0$	$A=12278.436(53)$ $B=1378.830(4)$ $C=1304.852(4)$	$\Delta_K=842(12)$ $\delta_J=0.020(5)$ $\delta_K=-4.7(4)$			+) Br near the CH_2 of the ring	
	H_2C—$CHCH_2{}^{81}Br$, (ring O) gauche 2	$v=0$	$A=12189.869(60)$ $B=1369.696(5)$ $C=1301.584(5)$	$\Delta_K=767(13)$ $\delta_J=0.022(6)$ $\delta_K=-5.9(5)$				
	H_2C—$CHCH_2{}^{79}Br$, cis (ring O)	$v=0$	$A=7733.314(74)$ $B=1808.087(7)$ $C=1737.340(7)$	$\Delta_{JK}=-0.9(3)$ $\Delta_K=-387(23)$ $\delta_K=-1.5(5)$				
	H_2C—$CHCH_2{}^{81}Br$, cis (ring O)	$v=0$	$A=7726.16(10)$ $B=1801.159(6)$ $C=1730.135(6)$	$\Delta_{JK}=0.5(3)$ $\Delta_K=-75(27)$ $\delta_K=-5.6(6)$				
486 Cont.	H_2C—$CHCH_2OH$ (ring O)	$v=0$	$A=10347.871(5)$ $B=4102.360(2)$ $C=3781.950(2)$	$\Delta_J=2.38(4)$ $\Delta_{JK}=-311(4)$ $\Delta_K=5.2(6)$ $\delta_J=0.3159(4)$ $\delta_K=-9.76(5)$	75BRO	75BRO		2.6.3, 233

Nr.	Molecule	v	A, B, C MHz	D, H kHz	Ref.	Ref. struct.	Varia, remarks	See also Tab., Nr.
486 Cont.	H_2C—$CHCH_2OD$ (O bridge)	$v=0$	$A=10010.314(8)$ $B=4056.734(4)$ $C=3717.023(3)$	Δ_J 2.53 (5) $\Delta_{JK}=-197(4)$ $\Delta_K=7.7(4)$ $\delta_J=0.3532(5)$ $\delta_K=-7.19(4)$	75BRO			
487	H_2C—$CHCH_2CH_3$ (O bridge)	$v=0$	$A=13063.3939(32)$ $B=3282.54701(80)$ $C=2959.03650(85)$	$\Delta_J=1.6594(34)$ $\Delta_{JK}=-10.400(35)$ $\Delta_K=42.319(19)$ $\delta_J=0.29824(45)$ $\delta_K=3.106(48)$ $H_J=-3.9(38)\cdot10^{-6}$ $H_{JK}=-0.168(81)\cdot10^{-3}$ $H_{KJ}=0.063(21)\cdot10^{-3}$ $H_K=-0.29(13)\cdot10^{-3}$	78LJ	78LJ		2.6.3, 234 2.8, 50
		$v_T=1$	$A=13051.2757(40)$ $B=3276.4977(10)$ $C=2957.2385(10)$	$\Delta_J=1.6377(38)$ $\Delta_{JK}=-10.217(40)$ $\Delta_K=42.279(20)$ $\delta_J=0.29422(45)$ $\delta_K=3.121(49)$ $H_J=-1.8(39)\cdot10^{-6}$ $H_{JK}=-0.122(83)\cdot10^{-3}$ $H_{KJ}=0.056(32)\cdot10^{-3}$ $H_K=-0.16(14)\cdot10^{-3}$			v_T, heavy-atom torsion	
		$v_T=2$	$A-C=10077.410(16)$ $\kappa=-0.9371815(64)$	$\Delta_{JK}=-10.09(12)$ $\Delta_K=54.4(45)$ $\delta_J=0.31484(95)$ $\delta_K=4.676(98)$ *)			*) the other constants were held fixed at their ground state value	
		$v_\tau=1$A	$A-C=10098.515(22)$ $\kappa=-0.9361546(89)$	$\Delta_{JK}=-11.12(24)$ $\delta_J=0.2769(18)$ $\delta_K=2.49(26)$ *)			v_τ, methyl torsion, A species	
		$v_\tau=1$E	$A-C=10098.809(12)$ $\kappa=-0.9361546(48)$	$\Delta_{JK}=-10.88(13)$ $\delta_J=0.27268(97)$ $\delta_K=2.51(14)$ *)			v_τ, methyl torsion, E species	

Nr.	Molecule	v	A, B, C MHz	D, H kHz	Ref.	Ref. struct.	Varia, remarks	See also Tab., Nr.
488	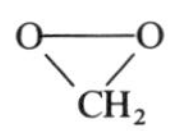	$v=0$	$A''=28976.762\ (78)$ $B''=25056.382\ (86)$ $C''=14779.889\ (65)$	$\tau_{aaaa}=-325\ (18)$ $\tau_{bbbb}=-398\ (19)$ $\tau_{cccc}=-43\ (11)$ $\tau_1=-371\ (60)$ $\tau_2=-106\ (20)$ $\tau_3=6.95\ (16)\cdot 10^3$	78SU4	78SU4		2.6.3, 235
	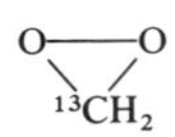	$v=0$	$A''=28012.641\ (68)$ $B''=25056.772\ (86)$ $C''=14524.118\ (75)$	$\tau_{aaaa}=-323\ (16)$ $\tau_{bbbb}=-418\ (19)$ $\tau_{cccc}=-58.2\ (80)$ $\tau_1=-422\ (64)$ $\tau_2=-123\ (22)$ $\tau_3=8.41\ (26)\cdot 10^3$				
	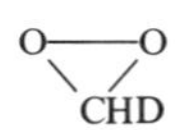	$v=0$	$A''=25157.48\ (11)$ $B''=24154.97\ (10)$ $C''=14003.461\ (72)$	$\tau_{aaaa}=-267\ (23)$ $\tau_{bbbb}=-398\ (23)$ $\tau_{cccc}=-41\ (10)$ $\tau_1=-482\ (65)$ $\tau_2=-147\ (22)$ $\tau_3=17.1\ (13)\cdot 10^3$				
	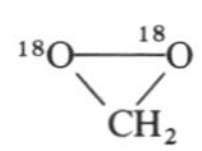	$v=0$	$A=28120.96\ (21)$ $B=22491.83\ (22)$ $C=13648.58\ (23)$					
489 *6-198* *4-154*	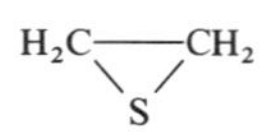	$v=0$	$A=21973.86\ (27)$ *) $B=10825.11\ (13)$ $C=8026.29\ (10)$	$\tau_{aaaa}=-182\ (8)$ *) $\tau_{bbbb}=-67\ (3)$ $\tau_{cccc}=-23\ (2)$ $\tau_{aabb}=-179\ (10)$ $\tau_{bbcc}=-48\ (3)$	76HIR1	74OKI	for structure, see also Vol. II/7 (3, 255) *) uncertainties are 2.5 σ	
		$v_a=1$	$\alpha^A=24.69\ (93)$ *) $\alpha^B=42.95\ (43)$ $\alpha^C=143.81\ (29)$	$\tau_{aaaa}=-509\ (270)$ *) $\tau_{bbbb}=-27\ (120)$ $\tau_{cccc}=31\ (120)$ $\tau_{aabb}=-54\ (220)$ $\tau_{bbcc}=23\ (150)$			$v_a(A)\simeq 680\ cm^{-1}=v_5$ ring deformation (624 cm^{-1})	
		$v_b=1$	$\alpha^A=-70.48\ (110)$ *) $\alpha^B=79.84\ (57)$ $\alpha^C=54.57\ (38)$	$\tau_{aaaa}=300\ (330)$ *) $\tau_{bbbb}=-87\ (180)$ $\tau_{cccc}=-69\ (180)$ $\tau_{aabb}=-191\ (330)$ $\tau_{bbcc}=-97\ (210)$			$v_b(B)\simeq 710\ cm^{-1}=v_{12}$ ring deformation (660 cm^{-1})	
Cont.								

Nr.	Molecule	v	A, B, C MHz	D, H kHz	Ref.	Ref. struct.	Varia, remarks	See also Tab., Nr.
489 *6–198* *4–154* Cont.	H_2C——CH_2 S	$v_c=1$	$\alpha^A = -16.12\,(180)$ $\alpha^B = 34.50\,(88)$ $\alpha^C = -84.73\,(58)$	$\tau_{aaaa} = -407\,(570)$ *) $\tau_{bbbb} = -71\,(310)$ $\tau_{cccc} = -63\,(310)$ $\tau_{aabb} = -100\,(560)$ $\tau_{bbcc} = -46\,(370)$	76HIR1		$\nu_c(A) \simeq 920\ cm^{-1} = \nu_7$ or ν_8, CH_2 rock or CH_2 twist ($970\ cm^{-1}$) *) uncertainties are 2.5 σ	
		$v_d=1$	$\alpha^A = -19.73\,(83)$ *) $\alpha^B = 45.87\,(36)$ $\alpha^C = 36.57\,(25)$	$\tau_{aaaa} = -123\,(210)$ *) $\tau_{bbbb} = -85\,(90)$ $\tau_{cccc} = -49\,(90)$ $\tau_{aabb} = -172\,(170)$ $\tau_{bbcc} = -76\,(110)$			$\nu_d(B) \simeq 950\ cm^{-1} = \nu_{15}$ CH_2 twist, ($824\ cm^{-1}$)	
		$v_e=1$	$\alpha^A = 67.50\,(82)$ *) $\alpha^B = -0.20\,(39)$ $\alpha^C = 99.92\,(27)$	$\tau_{aaaa} = -211\,(230)$ *) $\tau_{bbbb} = -36\,(110)$ $\tau_{cccc} = 16\,(110)$ $\tau_{aabb} = -105\,(190)$ $\tau_{bbcc} = -6\,(130)$			$\nu_e(B) \simeq 800\ cm^{-1} = \nu_{14}$ CH_2 rock, ($945\ cm^{-1}$)	
	H_2C——CH_2 ^{34}S	$v=0$	$A = 21972.97\,(67)$ *) $B = 10551.03\,(34)$ $C = 7874.58\,(22)$	$\tau_{aaaa} = -253\,(260)$ *) $\tau_{bbbb} = -16\,(140)$ $\tau_{cccc} = 14\,(140)$ $\tau_{aabb} = -42\,(250)$ $\tau_{bbcc} = 18\,(170)$				
	$H_2{}^{13}C$——CH_2 S	$v=0$	$A = 21481.71\,(53)$ *) $B = 10653.40\,(29)$ $C = 7866.33\,(20)$	$\tau_{aaaa} = -417\,(200)$ *) $\tau_{bbbb} = -82\,(120)$ $\tau_{cccc} = -55\,(120)$ $\tau_{aabb} = -153\,(220)$ $\tau_{bbcc} = -54\,(150)$				
490	H_2C——$C{=}CH_2$ S	$v=0$	$A = 13783.57\,(3)$ $B = 5276.87\,(1)$ $C = 3913.46\,(1)$		77PEN2			2.6.3, 236
491 *6–202*	H_2C——$CHCH_3$, trans NH	$v=0$	$A = 16892.437$ $B = 6533.607$ $C = 5761.299$	$d_J = -4.57\,(17)$ $d_{JK} = 20\,(7)$ $d_K = 72\,(13)$ $d_{WJ} = 1.228\,(31) \cdot 10^{-6}$ $d_{WK} = -5.0\,(12) \cdot 10^{-6}$	73SU		for structure, Vol. II/7 (3, 421)	2.7.3, 47

Nr.	Molecule	v	A, B, C MHz	D, H kHz	Ref.	Ref. struct.	Varia, remarks	See also Tab., Nr.
492	H_2C—CHCN, cis (ring with NH)	$v=0$	$A=16877.718\ (32)$ $B=\ \ 3528.931\ (4)$ $C=\ \ 3373.065\ (4)$	$D_N=1.10\ (4)$	80BRO2			2.7.3, 48 3, 142
493	H_2C—CH_2 *) (ring with N) N—CH_2CH_2OH	$v=0$	$A=8528.87\ (25)$ $B=2069.74\ (2)$ $C=2020.41\ (2)$		78PEN2	78PEN2	*) gauche O—C—C—N configuration, maintained by an O—H···N-type hydrogen bond	2.6.3, 237
	H_2C—CH_2 (ring with N) N—CH_2CH_2OD	$v=0$	$A=8275.07\ (19)$ $B=2051.16\ (2)$ $C=1995.43\ (2)$					
494	H_2C—CHCN *) H_2C—CH_2	$v=0$	$A=9761.607\ (35)$ $B=2383.359\ (8)$ $C=2105.109\ (9)$		73FON	73DUR1	*) bent equilibrium ground state for the ring with the CN in the equatorial position ν_R, ring-puckering vibration	2.6.3, 238 2.7.3, 49
		$v_R=1$	$A=9610.195\ (65)$ $B=2389.091\ (7)$ $C=2111.556\ (7)$					
		$v_R=2$	$A=9565.661\ (147)$ $B=2394.458\ (22)$ $C=2117.754\ (26)$					
		$v_R=3$	$A\simeq 9720.78$ $B=2400.06\ (16)$ $C=2122.78\ (9)$		73DUR1			
		$v_\beta=1$	$A=9867.902\ (49)$ $B=2387.670\ (9)$ $C=2106.324\ (10)$		73FON		ν_β, out-of-plane CN vibration	
		$v_\beta=2$	$A\simeq 9973.95$ $B=2391.83\ (15)$ $C=2107.59\ (9)$		73DUR1			
		$v_\gamma=1$	$A=9958.524\ (518)$ $B=2390.422\ (37)$ $C=2106.983\ (42)$		73FON		ν_γ, in-plane CN vibration	

Nr.	Molecule	v	A, B, C MHz	D, H kHz	Ref.	Ref. struct.	Varia, remarks	See also Tab., Nr.
495	H_2C—CHOH *) H_2C—CH_2	$v=0$	$A=10119.67$ (10) $B=$ 4282.93 (4) $C=$ 3421.04 (4)		78MCD2	78MCD2	*) trans equatorial with respect to the ring H atom	2.6.3, 239
		$v_R=1$	$A=10050.76$ (840) $B=$ 4287.83 (24) $C=$ 3424.88 (20)				$\nu_R=165$ (15) cm^{-1}, ring-puckering vibration	
	H_2C—CHOD H_2C—CH_2	$v=0$	$A=9911.77$ (12) $B=4132.74$ (6) $C=3347.25$ (6)					
496 *6-209* *4-159*	H_2C—C=^{18}O H_2C—CH_2	$v=0$	$A=10786.2$ (8) $B=$ 4536.17 (10) $C=$ 3408.03 (10)		76STI2	76STI2	for structure, see also Vol. II/7 (3, 516)	
	D_2C—C=O H_2C—CD_2	$v=0$	$A=8110.43$ (80) $B=4636.68$ (10) $C=3263.24$ (10)					
497 *6-240*	H_2C=C—C=CH_2 *) H_2C—CH_2	$v=0$	$A=4925.217$ $B=4089.880$ $C=2301.665$		75AVI1	75AVI1	*) ring skeleton planar	2.6.3, 240
		$v_R=1$	$A=4930.092$ $B=4080.071$ $C=2308.365$				ν_R, ring-puckering vibration	
		$v_R=2$	$A=4934.096$ $B=4071.403$ $C=2313.958$					
		$v_R=3$	$A=4937.205$ $B=4064.896$ $C=2318.945$					
		$v_R=4$	$A=4939.677$ $B=4059.131$ $C=2323.442$					
498 Cont.	H_2C—CHCN *) H_2C=C—CH_2	$v=0$	$A=9207.15$ (25) $B=1521.72$ (2) $C=1395.33$ (2)		74DUR9	74DUR9	*) non-planar heavy-atom skeleton with the CN group in the equatorial position	2.6.3, 241 2.7.3, 50

Nr.	Molecule	v	A, B, C MHz	D, H kHz	Ref.	Ref. struct.	Varia, remarks	See also Tab., Nr.
498 Cont.	$H_2C—CHCN$ *) $H_2C{=}C—CH_2$	$v_\alpha=1$ $v_\beta=1$ $v_\gamma=1$	$A=8997.4\,(120)$ $B=1527.52\,(2)$ $C=1404.41\,(2)$ $A=9113.5\,(54)$ $B=1528.24\,(2)$ $C=1403.11\,(2)$ $A=9142\,(22)$ $B=1524.44\,(4)$ $C=1398.99\,(4)$		74DUR9		*) non-planar heavy-atom skeleton with the CN group in the equatorial position	
499	$H_2C—C{=}O$ *) $H_2C—C{=}O$	$v=0$ $v_R=1$ $v_R=2$ $v_R=3$	$A=5329.07\,(1)$ $B=4050.486\,(2)$ $C=2371.062\,(2)$ $A=5333.54\,(1)$ $B=4042.836\,(3)$ $C=2376.392\,(3)$ $A=5337.66\,(1)$ $B=4035.605\,(3)$ $C=2381.415\,(3)$ $A=5341.53\,(3)$ $B=4028.73\,(1)$ $C=2386.21\,(1)$		79LE1	79LE1	*) planar equilibrium conformation of the heavy atoms $\nu_R=68\,(30)\,cm^{-1}$, ring-puckering vibration	2.6.3, 242
500	$H_2C—CF_2$ $H_2C—CF_2$	$v_R=0,1$ $v_R=2,3$ $v_\alpha=1$ $v_\beta=1$ $v_\gamma=1$	$A=2665.93$ $B=1938.35$ $C=1909.90$ $A=2663.76$ $B=1936.78$ $C=1906.77$ $A=2661.34$ $B=1935.18$ $C=1904.09$ $A=2658.89$ $B=1933.49$ $C=1901.10$ $A=2669.53$ $B=1932.22$ $C=1912.11$		76DUR4		$\nu_R=80\,(15)\,cm^{-1}$, ring-puckering mode	2.6.3, 243

Nr.	Molecule	v	A, B, C MHz	D, H kHz	Ref.	Ref. struct.	Varia, remarks	See also Tab., Nr.
501 *4–158*	H_2C—CH_2 H_2C—O	$v=0$	$Z=11734.0807\ (81)$ $X=6730.6894\ (81)$ $Y=12045.4908\ (81)$	$\Delta_J=8.15147\ (107025)$ $\Delta_{JK}=-1.29785\ (7091)$ $\Delta_K=4.62165\ (4781)$ $\delta_J=-1.59951\ (1149)$ $\delta_K=0.891412\ (22321)$ $h_{JK}=9.5439\ (939)\cdot 10^{-6}$	74CR1	75CR3	for structure, see also Vol. II/7 (3, 407)	
		$v_R=1$	$Z=11726.5819\ (96)$ $X=6772.2732\ (96)$ $Y=12058.0086\ (96)$	$\Delta_J=12.10885\ (126829)$ $\Delta_{JK}=-7.18386\ (9439)$ $\Delta_K=10.16362\ (6360)$ $\delta_J=-1.47727\ (1535)$ $\delta_K=-1.59584\ (2969)$ $h_{JK}=26.5605\ (1618)\cdot 10^{-6}$			ν_R, ring-puckering vibration	
		$v_R=2$	$Z=11718.8730\ (63)$ $X=6789.0443\ (63)$ $Y=12058.9670\ (63)$	$\Delta_J=4.23903\ (82564)$ $\Delta_{JK}=-4.98927\ (6302)$ $\Delta_K=8.17707\ (4253)$ $\delta_J=-1.58038\ (1016)$ $\delta_K=-0.803786\ (19700)$ $h_{JK}=19.4628\ (1406)\cdot 10^{-6}$				
		$v_R=3$	$Z=11709.9570\ (92)$ $X=6809.7182\ (92)$ $Y=12060.3357\ (92)$	$\Delta_J=3.74359\ (120702)$ $\Delta_{JK}=-4.93263\ (9498)$ $\Delta_K=8.10082\ (6403)$ $\delta_J=-1.59551\ (1541)$ $\delta_K=-0.835808\ (29716)$ $h_{JK}=19.9964\ (2055)\cdot 10^{-6}$				
		$v_R=4$	$Z=11700.5422\ (94)$ $X=6827.1698\ (94)$ $Y=12059.5098\ (94)$	$\Delta_J=8.05044\ (124538)$ $\Delta_{JK}=-5.15417\ (9306)$ $\Delta_K=8.17820\ (6281)$ $\delta_J=-1.54192\ (1504)$ $\delta_K=-0.951816\ (29025)$ $h_{JK}=19.6151\ (2097)\cdot 10^{-6}$				
		$v_R=5$	$Z=11690.6230\ (88)$ $X=6842.9721\ (88)$ $Y=12057.1909\ (88)$	$\Delta_J=11.19695\ (115931)$ $\Delta_{JK}=-4.71939\ (9399)$ $\Delta_K=7.81342\ (6339)$ $\delta_J=-1.59184\ (1527)$ $\delta_K=-0.872737\ (29342)$ $h_{JK}=19.5839\ (2203)\cdot 10^{-6}$				
Cont.								

Nr.	Molecule	v	A, B, C MHz	D, H kHz	Ref.	Ref. struct.	Varia, remarks	See also Tab., Nr.
501	D_2C—CH_2	$v=0$	$Z=$ 9983.1178 (96)	Δ_J = 4.75546 (124621)	74CR1			
4-158	H_2C—O		$X=$ 6237.2536 (96)	Δ_{JK} = 3.71455 (8363)				
Cont.			$Y=$11603.3169 (96)	Δ_K = −2.54140 (6081)				
				δ_J = −1.45246 (1376)				
				δ_K = 1.65166 (2262)				
				h_{JK} = −136.726 (1061) · 10^{-6}				
		$v_R=1$	$Z=$ 9983.1310 (87)	Δ_J = 10.69633 (112054)				
			$X=$ 6274.9981 (87)	Δ_{JK} = −7.87604 (7804)				
			$Y=$11610.5960 (87)	Δ_K = 8.87827 (5650)				
				δ_J = −1.40137 (1311)				
				δ_K = −0.603181 (2126)				
				h_{JK} = 187.515 (2133) · 10^{-6}				
		$v_R=2$	$Z=$9980.3296 (87)	Δ_J = 10.06267 (112407)				
			$X=$ 6290.2021 (87)	Δ_{JK} = −4.01449 (7850)				
			$Y=$11609.5607 (87)	Δ_K = 5.22799 (5684)				
				δ_J = −1.45902 (1320)				
				δ_K = 0.0500895 (215673)				
				h_{JK} = 26.4108 (33924) · 10^{-6}				
		$v_R=3$	$Z=$ 9977.1301 (84)	Δ_J = 4.98793 (109071)				
			$X=$ 6308.8209 (84)	Δ_{JK} = −4.30882 (7612)				
			$Y=$11608.4259 (84)	Δ_K = 5.49132 (5511)				
				δ_J = −1.47962 (1286)				
				δ_K = 0.0127188 (209659)				
				h_{JK} = 41.9960 (33561) · 10^{-6}				
		$v_R=4$	$Z=$ 9973.2778 (96)	Δ_J = 4.71312 (124189)				
			$X=$ 6324.5121 (96)	Δ_{JK} = −4.29462 (8998)				
			$Y=$11605.6068 (96)	Δ_K = 5.44814 (6494)				
				δ_J = −1.45469 (1528)				
				δ_K = −0.0261221 (246976)				
				h_{JK} = 36.5475 (64687) · 10^{-6}				
		$v_R=5$	$Z=$ 9968.9778 (164)	Δ_J = 10.13030 (212110)				
			$X=$ 6338.7534 (164)	Δ_{JK} = −4.50919 (16107)				
			$Y=$11601.6572 (164)	Δ_K = 5.58292 (11624)				
				δ_J = −1.43201 (2743)				
				δ_K = −0.0770304 (429277)				
Cont.				h_{JK} = 18.5307 (114930) · 10^{-6}				

Nr.	Molecule	v	A, B, C MHz	D, H kHz	Ref.	Ref. struct.	Varia, remarks	See also Tab., Nr.
501 *4-158* Cont.	D_2C-CH_2 H_2C-O (ring)	$v_R=6$	$Z=9964.2239\ (145)$ $X=6351.7207\ (145)$ $Y=11596.6691\ (145)$	$\Delta_J=13.46507\ (190813)$ $\Delta_{JK}=-4.46038\ (15741)$ $\Delta_K=5.55451\ (11388)$ $\delta_J=-1.51508\ (2414)$ $\delta_K=0.0637459\ (393894)$ $h_{JK}=67.4521\ (374597)\cdot 10^{-6}$	74CR1			
	H_2C-CD_2 D_2C-O (ring)	$v=0$	$Z=8976.2414\ (25)$ $X=5860.9278\ (25)$ $Y=10910.7844\ (25)$	$\Delta_J=3.84124\ (15992)$ $\Delta_{JK}=-3.42066\ (2865)$ $\Delta_K=3.49812\ (2039)$ $\delta_J=-1.28010\ (384)$ $\delta_K=1.18775\ (566)$				
		$v_R=1$	$Z=8991.5765\ (34)$ $X=5894.4805\ (34)$ $Y=10899.2725\ (33)$	$\Delta_J=4.30881\ (21339)$ $\Delta_{JK}=-5.45628\ (3765)$ $\Delta_K=5.56903\ (2687)$ $\delta_J=-1.25527\ (492)$ $\delta_K=0.114014\ (7366)$				
		$v_R=2$	$Z=8996.4802\ (21)$ $X=5908.0524\ (21)$ $Y=10890.6578\ (21)$	$\Delta_J=3.84171\ (13189)$ $\Delta_{JK}=-4.41200\ (2326)$ $\Delta_K=4.62191\ (1660)$ $\delta_J=-1.29812\ (304)$ $\delta_K=0.393728\ (4554)$				
		$v_R=3$	$Z=9002.5032\ (21)$ $X=5924.7283\ (21)$ $Y=10880.4981\ (21)$	$\Delta_J=3.45112\ (13359)$ $\Delta_{JK}=-4.41693\ (2464)$ $\Delta_K=4.64041\ (1759)$ $\delta_J=-1.27037\ (324)$ $\delta_K=0.333315\ (4877)$				
		$v_R=4$	$Z=9006.8800\ (35)$ $X=5938.7486\ (35)$ $Y=10870.3277\ (35)$	$\Delta_J=3.59330\ (22098)$ $\Delta_{JK}=-4.37261\ (3978)$ $\Delta_K=4.60022\ (2838)$ $\delta_J=-1.25346\ (511)$ $\delta_K=0.293164\ (7678)$				
Cont.		$v_R=5$	$Z=9010.2780\ (50)$ $X=5951.4228\ (50)$ $Y=10859.9606\ (50)$	$\Delta_J=3.77085\ (29970)$ $\Delta_{JK}=-4.14447\ (4162)$ $\Delta_K=4.42618\ (2972)$ $\delta_J=-1.24903\ (536)$ $\delta_K=0.276499\ (8033)$				

Nr.	Molecule	v	A, B, C MHz	D, H kHz	Ref.	Ref. struct.	Varia, remarks	See also Tab., Nr.
501 *4-158* Cont.	D_2C—CD_2 | | D_2C—O	$v=0$	$Z=8728.4536\,(22)$	$\Delta_J\ =2.68081\,(14425)$	74CR1			
			$X=5479.8980\,(23)$	$\Delta_{JK}=0.436901\,(47071)$				
			$Y=9371.8390\,(22)$	$\Delta_K\ =0.354084\,(33280)$				
				$\delta_J\ =-0.822210\,(6409)$				
				$\delta_K\ =\ 1.16837\,(1160)$				
				$H_{JK}=\ 1.1924\,(11775)\cdot10^{-3}$				
				$H_{KJ}=-1.6297\,(15543)\cdot10^{-3}$				
				$H_K\ =\ 0.6551\,(6068)\cdot10^{-3}$				
				$h_J\ =-0.1116\,(1134)\cdot10^{-3}$				
				$h_{JK}=\ 0.4655\,(4310)\cdot10^{-3}$				
				$h_K\ =-0.3051\,(2802)\cdot10^{-3}$				
		$v_R=1$	$Z=8739.9922\,(23)$	$\Delta_J\ =\ 2.55650\,(14838)$				
			$X=5510.3818\,(24)$	$\Delta_{JK}=-3.39848\,(4476)$				
			$Y=9366.9684\,(23)$	$\Delta_K\ =\ 4.18993\,(3168)$				
				$\delta_J\ =-0.798225\,(6074)$				
				$\delta_K\ =-0.773557\,(10996)$				
				$H_{JK}=-0.8447\,(11310)\cdot10^{-3}$				
				$H_{KJ}=\ 1.0973\,(14979)\cdot10^{-3}$				
				$H_K\ =-0.4096\,(5870)\cdot10^{-3}$				
				$h_J\ =\ 0.07620\,(10851)\cdot10^{-3}$				
				$h_{JK}=-0.3016\,(4147)\cdot10^{-3}$				
				$h_K\ =\ 0.1775\,(2708)\cdot10^{-3}$				
		$v_R=2$	$Z=8743.2489\,(26)$	$\Delta_J\ =\ 2.80822\,(17350)$				
			$X=5522.6936\,(28)$	$\Delta_{JK}=-2.04131\,(5091)$				
			$Y=9362.3528\,(26)$	$\Delta_K\ =\ 2.90440\,(3594)$				
				$\delta_J\ =-0.839340\,(6987)$				
				$\delta_K\ =-0.203717\,(12620)$				
				$H_{JK}=-0.2020\,(12779)\cdot10^{-3}$				
				$H_{KJ}=\ 0.2096\,(16918)\cdot10^{-3}$				
				$H_K\ =-0.05467\,(66289)\cdot10^{-3}$				
				$h_J\ =\ 0.02207\,(12308)\cdot10^{-3}$				
				$h_{JK}=-0.05834\,(47005)\cdot10^{-3}$				
Cont.				$h_K\ =\ 0.02019\,(30700)\cdot10^{-3}$				

Nr.	Molecule	v	A, B, C MHz	D, H kHz	Ref.	Ref. struct.	Varia, remarks	See also Tab., Nr.
501 *4-158* Cont.	$D_2C—CD_2$ / $D_2C—O$ (ring)	$v_R=3$	$Z=8747.2636(28)$	$\Delta_J = 2.89592(18711)$	74CR1			
			$X=5537.7775(30)$	$\Delta_{JK} = -2.10778(5327)$				
			$Y=9356.9591(28)$	$\Delta_K = 2.97514(3761)$				
				$\delta_J = -0.826252(7306)$				
				$\delta_K = -0.268371(13214)$				
				$H_{JK} = -0.1860(13446)\cdot10^{-3}$				
				$H_{KJ} = 0.2141(17834)\cdot10^{-3}$				
				$H_K = -0.06781(70030)\cdot10^{-3}$				
				$h_J = 0.02054(12931)\cdot10^{-3}$				
				$h_{JK} = -0.06145(49550)\cdot10^{-3}$				
				$h_K = 0.02887(32455)\cdot10^{-3}$				
		$v_R=4$	$Z=8749.9044(32)$	$\Delta_J = 1.84142(21364)$				
			$X=5550.4294(34)$	$\Delta_{JK} = -2.09506(6322)$				
			$Y=9351.2962(32)$	$\Delta_K = 2.96046(4453)$				
				$\delta_J = -0.816569(8805)$				
				$\delta_K = -0.293175(15813)$				
				$H_{JK} = -1.8241(15720)\cdot10^{-3}$				
				$H_{KJ} = 2.3890(20809)\cdot10^{-3}$				
				$H_K = -0.9214(8155)\cdot10^{-3}$				
				$h_J = 0.1766(1518)\cdot10^{-3}$				
				$h_{JK} = -0.6646(5796)\cdot10^{-3}$				
				$h_K = 0.4245(3784)\cdot10^{-3}$				
		$v_R=5$	$Z=8751.7470(60)$	$\Delta_J = 2.89708(36008)$				
			$X=5561.8820(60)$	$\Delta_{JK} = -2.00264(8269)$				
			$Y=9345.4000(60)$	$\Delta_K = 2.88509(5834)$				
				$\delta_J = -0.815098(11371)$				
				$\delta_K = -0.301750(20588)$				
				$H_{JK} = 0.5521(21271)\cdot10^{-3}$				
				$H_{KJ} = -0.7694(28255)\cdot10^{-3}$				
				$H_K = 0.3180(11117)\cdot10^{-3}$				
				$h_J = -0.04925(20463)\cdot10^{-3}$				
				$h_{JK} = 0.2144(7862)\cdot10^{-3}$				
Cont.				$h_K = -0.1529(5163)\cdot10^{-3}$				

Nr.	Molecule	v	A, B, C MHz	D, H kHz	Ref.	Ref. struct.	Varia, remarks	See also Tab., Nr.
501 *4-158* Cont.	$H_2C—CD_2$ $H_2C—O$	$v=0$	$Y=11311.4956\ (11)$ $Z=10322.2421\ (11)$ $X=6280.3971\ (11)$	$\Delta_{JK}=-1.3277\ (282)$ $\Delta_K=2.4951\ (193)$ $\delta_J=-1.4399\ (43)$ $\delta_K=1.3692\ (76)$	74MAL			
		$v_R=1$	$Y=11301.6514\ (12)$ $Z=10336.4766\ (12)$ $X=6317.8322\ (12)$	$\Delta_{JK}=-6.1966\ (252)$ $\Delta_K=7.3954\ (173)$ $\delta_J=-1.4081\ (39)$ $\delta_K=-0.75095\ (689)$				
		$v_R=2$	$Y=11293.3817\ (10)$ $Z=10339.9137\ (10)$ $X=6332.9898\ (10)$	$\Delta_{JK}=-4.3640\ (192)$ $\Delta_K=5.6668\ (131)$ $\delta_J=-1.4571\ (30)$ $\delta_K=-0.13897\ (529)$				
		$v_R=3$	$Y=11283.6887\ (10)$ $Z=10344.1624\ (10)$ $X=6351.5819\ (9)$	$\Delta_{JK}=-4.3885\ (168)$ $\Delta_K=5.7115\ (115)$ $\delta_J=-1.4391\ (27)$ $\delta_K=-0.21537\ (466)$				
		$v_R=4$	$Y=11273.7570\ (11)$ $Z=10346.5429\ (11)$ $X=6367.2133\ (10)$	$\Delta_{JK}=-4.2282\ (192)$ $\Delta_K=5.5831\ (131)$ $\delta_J=-1.4324\ (31)$ $\delta_K=-0.23483\ (535)$				
		$v_R=5$	$Y=11263.4966\ (12)$ $Z=10347.7428\ (12)$ $X=6381.3419\ (11)$	$\Delta_{JK}=-4.1406\ (219)$ $\Delta_K=5.5040\ (150)$ $\delta_J=-1.4167\ (34)$ $\delta_K=-0.27425\ (604)$				
	$D_2{}^{13}C—CH_2$ $H_2C—O$	$v=0$	$A=11603.55\ (3)$ $B=9767.57\ (2)$ $C=6152.46\ (2)$	$\Delta_{JK}=2.4\ (19)$ $\Delta_K=-1.8\ (12)$ $\delta_J=-1.9\ (3)$ $\delta_K=1.7\ (3)$	75CR3			
	$D_2C—{}^{13}CH_2$ $H_2C—O$	$v=0$	$A=11328.06\ (1)$ $B=9982.05\ (1)$ $C=6156.59\ (1)$	$\Delta_{JK}=3.38\ (12)$ $\Delta_K=-2.09\ (8)$ $\delta_J=-1.38\ (2)$ $\delta_K=1.71\ (3)$				

Nr.	Molecule	v	A, B, C MHz	D, H kHz	Ref.	Ref. struct.	Varia, remarks	See also Tab., Nr.
502 *6-210a* *4-160*	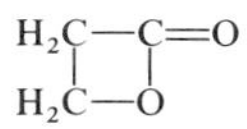	$v=0$	$A'=12406.012\,(5)$ $B'=\ 5244.456\,(2)$ $C'=\ 3869.191\,(2)$	$d_J\ =\ -2.08\,(5)$ $d_{JK}=-23.63\,(95)$ $d_K\ =-33.43\,(169)$ $d_{WJ}=\ 0.56\,(1)\cdot10^{-6}$ $d_{WK}=\ 5.08\,(22)\cdot10^{-6}$	74BO1	74BO2		
503 *6-211a*	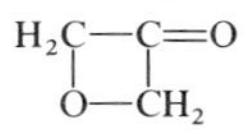						for structure, see also Vol. II/7 (3, 378)	2.9.3, 33
504	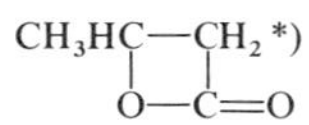	$v=0$	$A=9267.912\,(16)$ $B=2698.357\,(2)$ $C=2380.008\,(3)$	$\tau_{aaaa}=-15\,(100)$ $\tau_{bbbb}=\ -3.48\,(36)$ $\tau_{cccc}=\ -4.32\,(30)$ $\tau_{aabb}=\ 86\,(20)$ $\tau_{aacc}=-74.7\,(200)$	76COF		*) ring planar potential is single minima	2.6.3, 244
		$v_R=1$	$A=9238.513\,(18)$ $B=2701.660\,(3)$ $C=2386.327\,(3)$	*)			$\nu_R=111\ cm^{-1}$, ring puckering vibration	
		$v_R=2$	$A=9208.283\,(6)$ $B=2704.834\,(1)$ $C=2392.448\,(1)$	*)			*) ground state values assumed	
		$v_R=3$	$A=9177.401\,(25)$ $B=2707.854\,(2)$ $C=2398.327\,(3)$	*)				
505 *6-213*	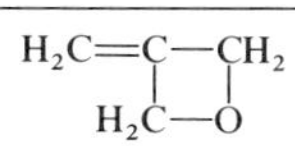						for structure, see also Vol. II/7 (3, 515)	2.9.3, 34
505a *6-215* *4-158a* Cont.	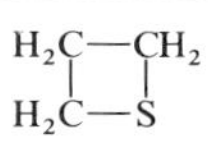	$v=0$	$A=10107.2018\,(16)$ $B=\ 6670.1426\,(11)$ $C=\ 4443.6175\,(11)$	$\Delta_J\ =\ 2.386\,(19)$ $\Delta_{JK}=-2.171\,(60)$ $\Delta_K\ =\ 6.69\,(10)$ $\delta_J\ =\ 0.6119\,(44)$ $\delta_K\ =\ 2.085\,(33)$ $h_{JK}=\ 0.0943\,(37)\cdot10^{-3}$	77NIE		for structure, see also Vol. II/7 (3, 414) effective two-level vibration-rotation Hamiltonian used: Energy difference: $W_{01}=8231.961\,(15)$ MHz, coupling constant: $\mu_{yz}=121.3323\,(40)$ MHz	

Nr.	Molecule	v	A, B, C MHz	D, H kHz	Ref.	Ref. struct.	Varia, remarks	See also Tab., Nr.
505a *6-215* *4-158a* Cont.	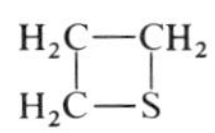	$v_R=1$	$A=10107.5431\ (13)$ $B=\ 6670.0768\ (11)$ $C=\ 4444.2911\ (11)$	$\Delta_J\ =\ \ 2.388\ (20)$ $\Delta_{JK}=-2.113\ (49)$ $\Delta_K\ =\ \ 6.769\ (90)$ $\delta_J\ =\ \ 0.599\ (30)$ $\delta_K\ =\ \ 1.954\ (31)$ $h_{JK}=\ \ 0.0757\ (19)\cdot 10^{-3}$	77NIE			
	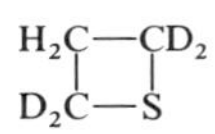	$v_R=0$	$X=7677.1777\ (68)$ $Z=6329.9639\ (57)$ $Y=3980.8752\ (71)$	$\Delta_J\ =\ \ 2.304\ (170)$ $\Delta_{JK}=-2.925\ (103)$ $\Delta_K\ =\ \ 3.726\ (77)$ $\delta_J\ =\ \ 0.6743\ (108)$ $\delta_K\ =-0.1781\ (193)$	80SON		$W_{01}=3314.475\ (897)$ MHz, $(\mu_{xy})_{01}=96.9182\ (934)$ MHz	
		$v_R=1$	$X=7677.3770\ (69)$ $Z=6329.8881\ (66)$ $Y=3981.1632\ (71)$	$\Delta_J\ =\ \ 2.358\ (170)$ $\Delta_{JK}=-3.028\ (92)$ $\Delta_K\ =\ \ 3.762\ (71)$ $\delta_J\ =\ \ 0.6717\ (98)$ $\delta_K\ =-0.1890\ (191)$				
		$v_R=2$	$X=7655.6833\ (39)$ $Z=6332.2845\ (36)$ $Y=3953.4104\ (37)$	$\Delta_J\ =\ \ 2.386\ (137)$ $\Delta_{JK}=-3.157\ (75)$ $\Delta_K\ =\ \ 3.788\ (59)$ $\delta_J\ =\ \ 0.6411\ (79)$ $\delta_K\ =-0.2266\ (155)$			$W_{23}=174678.162\ (720)$ MHz, $(\mu_{xy})_{23}=78.8633\ (61)$ MHz, quartic coupling constant $^{+}$): $(T_{xyJJ})_{23}=-0.5437\ (105)$ kHz $^{+}$) The coupling term is: $(\mu_{xy}+T_{xyJJ}J^2)[J_x,J_y]_+$	
		$v_R=3$	$X=7663.2883\ (39)$ $Z=6329.1379\ (36)$ $Y=3964.4940\ (36)$	$\Delta_J\ =\ \ 2.162\ (137)$ $\Delta_{JK}=-2.511\ (87)$ $\Delta_K\ =\ \ 3.334\ (68)$ $\delta_J\ =\ \ 0.7021\ (64)$ $\delta_K\ =-0.3016\ (137)$				
		$v_R=4$	$X=7640.7186\ (52)$ $Z=6333.8049\ (52)$ $Y=3933.9371\ (52)$	$\Delta_J\ =\ \ 1.643\ (222)$ $\Delta_{JK}=-3.173\ (76)$ $\Delta_K\ =\ \ 4.374\ (57)$ $\delta_J\ =\ \ 0.7228\ (62)$ $\delta_K\ =-0.0889\ (119)$			$W_{45}=1307.1006\ (807)$ MHz, $(\mu_{xy})_{45}=59.9369\ (18)$ MHz, $(T_{xyJJ})_{45}=-0.3764\ (353)$ kHz	
		$v_R=5$	$X=7659.4071\ (54)$ $Z=6324.7923\ (52)$ $Y=3962.2423\ (52)$	$\Delta_J\ =\ \ 1.912\ (204)$ $\Delta_{JK}=\ \ 0.7968\ (1090)$ $\Delta_K\ =\ \ 0.1951\ (819)$ $\delta_J\ =\ \ 0.6999\ (57)$ $\delta_K\ =-0.6845\ (130)$				

Nr.	Molecule	v	A, B, C MHz	D, H kHz	Ref.	Ref. struct.	Varia, remarks	See also Tab., Nr.
506	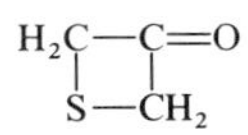	$v=0$	$A=10205.06\,(59)$ *) $B=3266.63\,(1)$ $C=2559.70\,(1)$		75AVI2	75AVI2	*) uncertainties are 2σ	2.6.3, 245
		$v_R=1$	$A=10117.74\,(58)$ $B=3277.54\,(1)$ $C=2574.95\,(1)$				$\nu_R=53.64\ \mathrm{cm^{-1}}$, ring-puckering vibration, single minimum anharmonic potential function determined: $V=9.81\,(Z^4+6.41\,Z^2)$ where Z is a reduced ring-puckering coordinate	
		$v_R=2$	$A=10041.06\,(75)$ $B=3287.38\,(1)$ $C=2588.56\,(1)$					
		$v_R=3$	$A=9970.36\,(78)$ $B=3296.49\,(1)$ $C=2601.17\,(1)$					
		$v_R=4$	$A=9905.80\,(86)$ $B=3305.13\,(2)$ $C=2613.03\,(1)$					
		$v_R=5$	$A=9844.70\,(69)$ $B=3313.33\,(1)$ $C=2624.32\,(1)$					
		$v_R=6$	$A=9786.86\,(42)$ $B=3321.16\,(1)$ $C=2635.11\,(1)$					
		$v_R=7$	$A=9732.27\,(21)$ $B=3328.67\,(1)$ $C=2645.48\,(1)$					
		$v_R=8$	$A=9681.29\,(149)$ $B=3335.89\,(3)$ $C=2655.47\,(2)$					
		$v_R=9$	$A=9632.24\,(106)$ $B=3342.80\,(2)$ $C=2665.11\,(2)$					
		$v_R=10$	$A=9586.31\,(263)$ $B=3349.41\,(5)$ $C=2674.45\,(4)$					

Nr.	Molecule	v	A, B, C MHz	D, H kHz	Ref.	Ref. struct.	Varia, remarks	See also Tab., Nr.
507	$H_2C—CH_2$ / $H_2C—SO$	$v=0$	$A=8851.17$ (4) $B=3587.792$ (4) $C=2990.659$ (4)		77BEV	77BEV	for structure, see also Vol. II/7 (3, 414)	3, 143
		$v_R=1$	$A=8790.80$ (4) $B=3589.776$ (4) $C=2992.533$ (4)					
		$v_R=2$	$A=8740.64$ (4) $B=3592.033$ (4) $C=2995.228$ (4)					
		$v_R=3$	$A=8697.96$ (57) $B=3594.891$ (9) $C=2999.090$ (9)					
		$v_R=4$	$A=8680.1$ (8) $B=3598.850$ (12) $C=3005.694$ (12)					
		$v_x=1$	$A=8830.2$ $B=3584.0$ $C=2989.2$					
		$v_y=1$	$A=8876.6$ $B=3585.6$ $C=2985.8$					
	$H_2C—{}^{13}CH_2$ / $H_2C—SO$	$v=0$	$A=8655.69$ (49) $B=3572.828$ (11) $C=2959.054$ (11)					
	$H_2{}^{13}C—CH_2$ / $H_2C—SO$	$v=0$	$A=8851.15$ (163) $B=3509.601$ (6) $C=2936.214$ (6)					
	$H_2C—CH_2$ / $H_2C—{}^{34}SO$	$v=0$	$A=8799.27$ (38) $B=3563.035$ (6) $C=2979.428$ (6)					
	$H_2C—CH_2$ / $H_2C—S^{18}O$	$v=0$	$A=8800.55$ (70) $B=3427.488$ (14) $C=2883.901$ (14)					
	$H_2C—CHD_\alpha$ / $H_2C—SO$	$v=0$	$A=8374.28$ (98) $B=3543.425$ (12) $C=2954.984$ (12)					
Cont.								

Nr.	Molecule	v	A, B, C MHz	D, H kHz	Ref.	Ref. struct.	Varia, remarks	See also Tab., Nr.
507 Cont.	H_2C—CHD_β H_2C—SO	$v=0$	$A=8232.98$ (71) $B=3565.991$ (11) $C=2905.177$ (11)		77BEV			
	$D_\alpha HC$—CH_2 H_2C—SO	$v=0$	$A=8704.81$ (64) $B=3448.029$ (11) $C=2909.030$ (11)					
	$D_\beta HC$—CH_2 H_2C—SO	$v=0$	$A=8777.84$ (45) $B=3411.089$ (6) $C=2874.812$ (6)					
508	H_2C—SO S—CH_2							2.6.3, 246
509	H_2C—NCH_3 H_2C—NCH_3	$v=0$	$A=5053.003$ (3) $B=3801.114$ (3) $C=2436.404$ (3)	$\Delta_J=0.907$ (34) $\Delta_{JK}=-2.249$ (30) $\Delta_K=3.842$ (34) $\delta_J=0.383$ (4) $\delta_K=0.168$ (20)	80GEB	80GEB	CH_3 groups in quasi-equatorial position	2.7.3, 51
510 *6-818* *4-161*	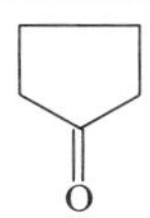	$v=0$	$\tilde{A}=6620.0611$ (47) $\tilde{B}=3351.5334$ (20) $\tilde{C}=2410.4190$ (16)	$\Delta_J=0.291$ (12) $\Delta_{JK}=1.171$ (37) $\Delta_K=0.533$ (57) $\delta_J=0.0540$ (25) $\delta_K=0.494$ (54)	78POZ1	78MAM	for structure, see also Vol. II/7 (3, 613)	
	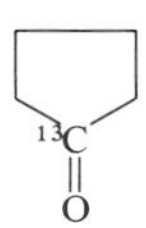	$v=0$	$A=6620.09$ (12) $B=3335.9948$ (64) $C=2402.4469$ (63)		78MAM			
	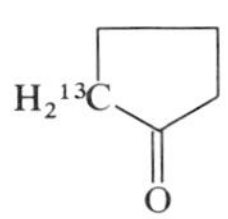	$v=0$	$A=6490.48$ (17) $B=3351.2092$ (85) $C=2393.2834$ (80)					
	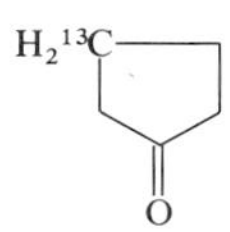	$v=0$	$A=6570.95$ (12) $B=3304.4324$ (59) $C=2380.7603$ (53)					

Nr.	Molecule	v	A, B, C MHz	D, H kHz	Ref.	Ref. struct.	Varia, remarks	See also Tab., Nr.
511	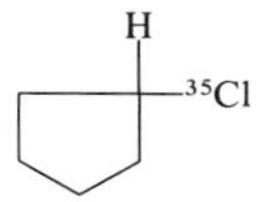 *)	$v=0$	$A=4547.77$ (1) $B=2290.22$ (1) $C=2073.34$ (1)		78LOY	78LOY	*) bent axial form	2.7.3, 135 3, 144
		$v_R=1$	$A=4600.64$ (4) $B=2269.84$ (8) $C=2046.50$ (4)				$\nu_R=52$ (5) cm^{-1}, pseudo-rotational ring mode	
		$v_R=2$	$A=4648.53$ (2) $B=2255.39$ (3) $C=2027.60$ (1)					
		$v_R=3$	$A=4715.52$ (303) $B=2233.21$ (6) $C=1998.41$ (6)					
		$v_R=4$	$A=4756.17$ (394) $B=2221.30$ (7) $C=1983.08$ (7)					
	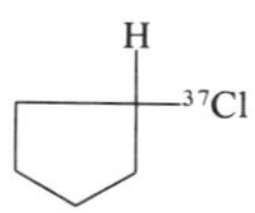	$v=0$	$A=4539.20$ (2) $B=2235.08$ (2) $C=2029.81$ (1)					
	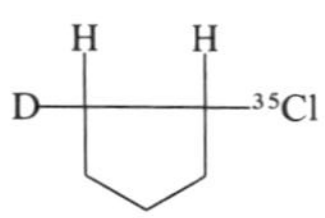	$v=0$	$A=4416.38$ (2) $B=2249.33$ (2) $C=2067.57$ (1)					
512	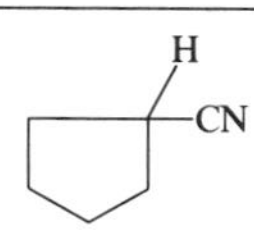 equatorial	$v=0$	$A=6324.905$ (170) $B=1790.937$ (4) $C=1497.792$ (3)	$\tau_{aaaa}=750$ (160) $\tau_{bbbb}=-0.68$ (8) $\tau_{cccc}=-0.47$ (6) $\tau'_{bbcc}=-1.2$ (2) $\tau'_{aabb}+\tau'_{aacc}=-4.0$ (2)	80CHO	80CHO		3, 145
		$v_R=1$	$A=6182.46$ (72) $B=1807.61$ (2) $C=1518.29$ (1)					
Cont.		$v_R=2$	$A=6078.27$ (271) $B=1821.05$ (8) $C=1534.90$ (4)					

Nr.	Molecule	v	A, B, C MHz	D, H kHz	Ref.	Ref. struct.	Varia, remarks	See also Tab., Nr.
512 Cont.	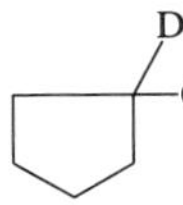 CN, equatorial	$v=0$	$A=6143.293\,(365)$ $B=1775.865\,(10)$ $C=1497.696\,(8)$		80CHO			
	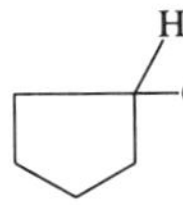 CN, axial	$v=0$	$A=4297.196\,(12)$ $B=2210.245\,(9)$ $C=2057.205\,(10)$	$\tau_{aaaa}=-12.7\,(4)$ $\tau_{bbbb}=-3.7\,(3)$ $\tau_{cccc}=-5.9\,(3)$ $\tau'_{bbcc}=-4.4\,(3)$ $\tau'_{aabb}+\tau'_{aacc}=-13.4\,(2)$			energy difference of the two conformers: 0 ± 200 cal/mol	
	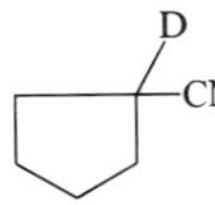 CN, axial	$v=0$	$A=4165.757\,(41)$ $B=2169.846\,(12)$ $C=2052.813\,(13)$					
513	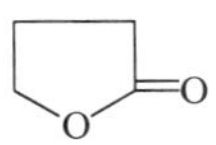	$v=0$	$A=7280.09\,(32)$ $B=3585.37\,(1)$ $C=2562.54\,(1)$		73DUR7			2.6.3, 247
		$v_R=1$	$A=7261.60\,(44)$ $B=3586.98\,(2)$ $C=2563.84\,(2)$				$\nu_R=148\ cm^{-1}$, ring-puckering vibration	
		$v_R=2$	$A=7252.91\,(93)$ $B=3588.11\,(5)$ $C=2564.50\,(5)$					
		$v_t=1$	$A=7266.11\,(46)$ $B=3581.85\,(2)$ $C=2556.84\,(1)$				$\nu_t=225\ cm^{-1}$, ring-twisting mode	
514	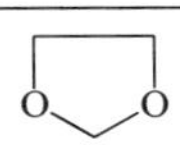	$v=0$	$A=7827.07$ $B=7534.36$ $C=4294.37$		74BAR		$v=0$ and $v_R=1$ are in interaction: $\Delta E_{01}=64840.65$ MHz, $F=435\,(30)$ MHz (coupling constant). ν_R pseudorotation mode. potential function for pseudorotation derived	2.6.3, 248
		$v_R=1$	$A=7834.07$ $B=7525.26$ $C=4290.47$					
Cont.		$v_R=2$	$A=7813.09$ $B=7553.26$ $C=4301.36$					

Nr.	Molecule	v	A, B, C MHz	D, H kHz	Ref.	Ref. struct.	Varia, remarks	See also Tab., Nr.
514 Cont.	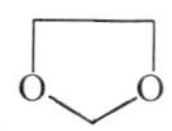	$v_R=3$	$A=7821.44$ $B=7543.99$ $C=4296.18$		74BAR			
		$v_R=4$	$A=7817.42$ $B=7540.98$ $C=4298.80$					
		$v_R=5$	$A=7832.65$ $B=7524.84$ $C=4280.89$					
		$v_R=6$	$A=7799.84$ $B=7562.40$ $C=4317.82$					
		$v_R=7$	$A=7820.11$ $B=7540.95$ $C=4250.76$					
		$v_R=8$	$A=7809.06$ $B=7551.05$ $C=4354.32$					
514a *4-161a*	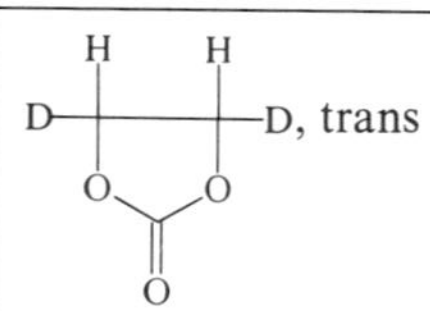	$v=\mathrm{I}$	$A=7544.20\,(151)$ $B=3616.57\,(6)$ $C=2566.47\,(6)$		80BAC		I and II denote different vibrational states of the pseudorotation.	
		$v=\mathrm{II}$	$A=7490.00\,(123)$ $B=3628.64\,(5)$ $C=2611.89\,(5)$					
	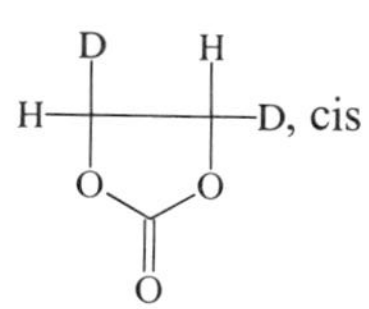	$v=\mathrm{I}$	$A=7536.14\,(409)$ $B=3620.79\,(16)$ $C=2588.54\,(18)$					
		$v=\mathrm{II}$	$A=7542.51\,(740)$ $B=3623.07\,(28)$ $C=2589.71\,(32)$					

Nr.	Molecule	v	A, B, C MHz	D, H kHz	Ref.	Ref. struct.	Varia, remarks	See also Tab., Nr.
515	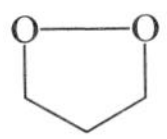	$v=0$	$A=7500.0$ $B=7328.4$ $C=4209.8$		80KON	80KON		
515a	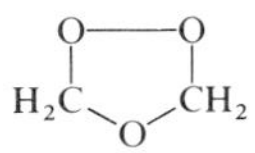	$v=0$	$A=8243.84$ (4) *) $B=8093.80$ (4) $C=4584.74$ (2)		72GIL	76KUC, 77MAZ	for structure, see also Vol. II/7 (3, 248) *) uncertainties are 2 σ $\nu_R=200$ (40) cm^{-1}, ring-bending	2.6.3, 249 3, 146
		$v_R=1$	$A=8236.06$ (4) *) $B=8092.18$ (4) $C=4579.94$(2)					
		$v_R=2$	$A=8228.96$ (8) *) $B=8089.94$ (12) $C=4575.13$ (6)					
		$v_R=3$	$A=8221.25$ (2) *) $B=8089.35$ (24) $C=4570.14$ (2)					
	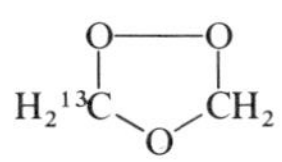	$v=0$	$A=8233.28$ (4) *) $B=7928.25$ (4) $C=4530.18$ (2)					
	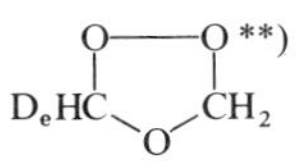	$v=0$	$A=8203.59$ (2) *) $B=7576.80$ (1) $C=4418.74$ (1)				**) e = equatorial a = axial	
	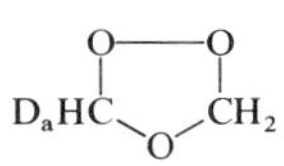	$v=0$	$A=8050.30$ (4) *) $B=7668.24$ (2) $C=4500.84$ (2)					
	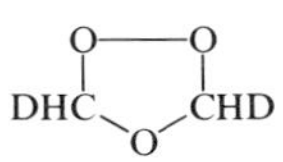 , cis	$v=0$	$A=7971.71$ (2) *) $B=7225.89$ (2) $C=4340.32$ (2)		76KUC			
	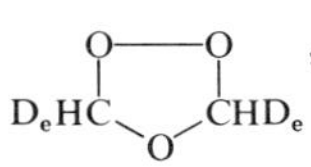 , trans**)	$v=0$	$A=8084.21$ (1) *) $B=7175.83$ (1) $C=4260.35$ (1)					
	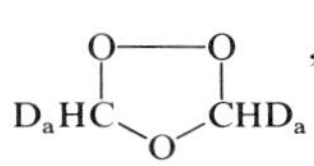 , trans**)	$v=0$	$A=7861.33$ (3) *) $B=7287.46$ (1) $C=4413.55$ (1)					
Cont.	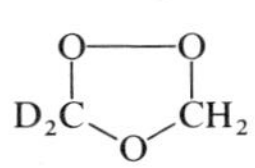	$v=0$	$A=8005.35$ (3) *) $B=7205.29$ (3) $C=4345.87$ (1)					

Nr.	Molecule	v	A, B, C MHz	D, H kHz	Ref.	Ref. struct.	Varia, remarks	See also Tab., Nr.
515a Cont.	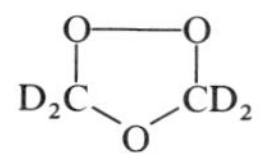	$v=0$	$A=7689.05$ (2) *) $B=6534.36$ (2) $C=4120.83$ (1)		76KUC		*) uncertainties are 2 σ	
	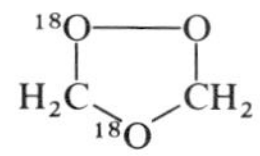	$v=0$	$A=8026.57$ (2) *) $B=7588.67$ (2) $C=4366.09$ (1)					
	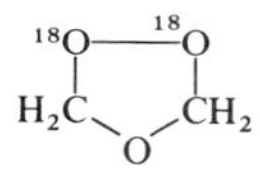	$v=0$	$A=7825.73$ (16) *) $B=7793.92$ (6) $C=4383.33$ (5)					
	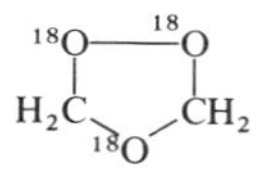	$v=0$	$A=7825.94$ (9) *) $B=7440.34$ (5) $C=4269.37$ (3)					
	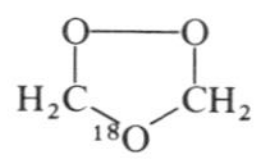	$v=0$	$A=8094.14$ (16) *) $B=7877.01$ (3) $C=4469.11$ (2)		74GIL			
	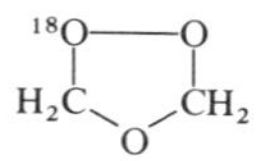	$v=0$	$A=8143.49$ (2) *) $B=7831.46$ (2) $C=4480.96$ (1)					
	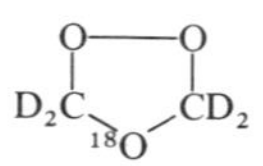	$v=0$	$A=7380.51$ (2) *) $B=6534.59$ (2) $C=4030.64$ (1)		77MAZ			
	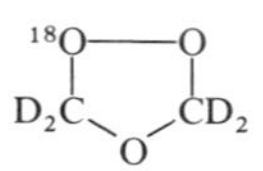	$v=0$	$A=7486.41$ (1) *) $B=6436.31$ (2) $C=4034.60$ (1)					
	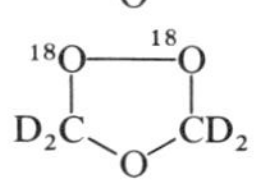	$v=0$	$A=7272.88$ (2) *) $B=6358.60$ (2) $C=3953.37$ (1)					
	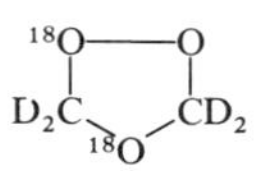	$v=0$	$A=7188.61$ (2) $B=6431.49$ (2) $C=3944.70$ (1)					
	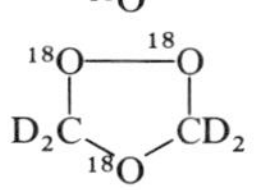	$v=0$	$A=6975.42$ (3) $B=6358.86$ (2) $C=3863.88$ (1)					
	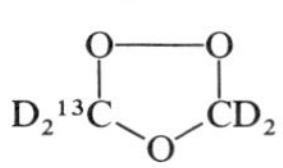	$v=0$	$A=7676.72$ (3) $B=6430.32$ (3) $C=4076.90$ (2)					

Nr.	Molecule	v	A, B, C MHz	D, H kHz	Ref.	Ref. struct.	Varia, remarks	See also Tab., Nr.
516	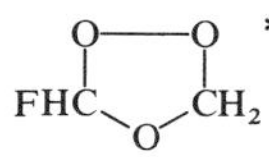 *)	$v=0$	$A=6774.04\,(2)$ $B=3916.32\,(1)$ $C=3122.66\,(1)$		74LA1	76LA	*) twisted half-chair conformation. F in an axial position	2.6.3, 250
	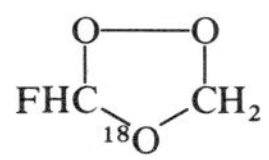	$v=0$	$A=6534.6$ $B=3912.9$ $C=3071.7$		76LA			
	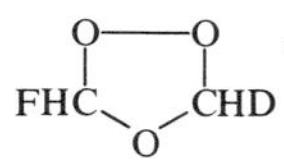 , cis	$v=0$	$A=6582$ $B=3791$ $C=3040$					
517	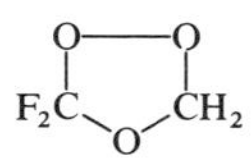	$v=0$	$A=4966.52\,(1)$ $B=2570.64\,(1)$ $C=2456.79\,(1)$		79GIL			2.6.3, 251
		$v_1=1$	$A=4963.28\,(1)$ $B=2573.16\,(1)$ $C=2459.33\,(1)$				$\nu_1=103\,(21)\ \mathrm{cm}^{-1}$	
		$v_2=1$	$A=4959.94\,(2)$ $B=2575.69\,(2)$ $C=2461.86\,(2)$				$\nu_2=195\,(22)\ \mathrm{cm}^{-1}$	
518	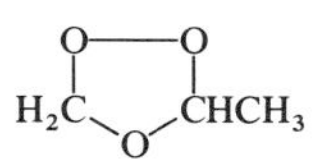	$v=0$	$A=7574.36\,(3)$ $B=3476.33\,(2)$ $C=2644.79\,(2)$		74LA2	74LA2	O—O half-chair conformation with equatorial CH_3 groups	2.6.3, 252 3, 147
		$v_x=1$	$A=7538.01\,(3)$ $B=3481.22\,(2)$ $C=2650.43\,(2)$				$\nu_x=100\,(15)\ \mathrm{cm}^{-1}$, probably bending mode	
		$v_x=2$	$A=7505.13\,(6)$ $B=3486.02\,(3)$ $C=2655.69\,(2)$					
	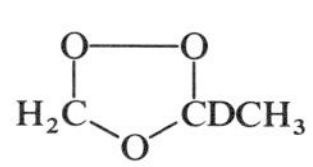	$v=0$	$A=7326.45\,(3)$ $B=3410.82\,(2)$ $C=2636.66\,(2)$					
Cont.	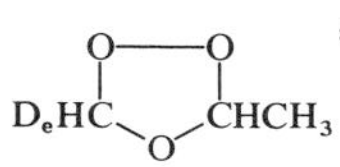 *)	$v=0$	$A=7410.52\,(4)$ $B=3372.04\,(2)$ $C=2566.58\,(2)$				*) e = equatorial a = axial	

Nr.	Molecule	v	A, B, C MHz	D, H kHz	Ref.	Ref. struct.	Varia, remarks	See also Tab., Nr.
518 Cont.	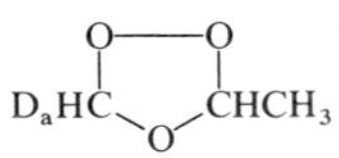 *)	$v=0$	$A=7316.41$ (3) $B=3387.18$ (2) $C=2609.76$ (2)		74LA2		*) e = equatorial a = axial	
	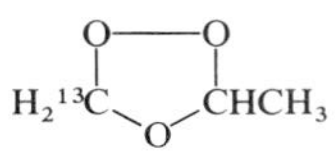	$v=0$	$A=7515.66$ (4) $B=3434.42$ (2) $C=2615.48$ (2)					
	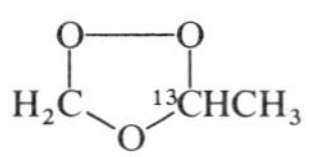	$v=0$	$A=7555.92$ (5) $B=3460.27$ (3) $C=2637.80$ (1)					
	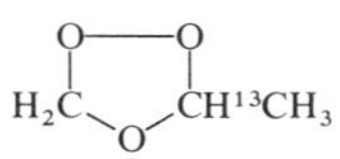	$v=0$	$A=7570.19$ (5) $B=3373.21$ (4) $C=2585.72$ (2)					
	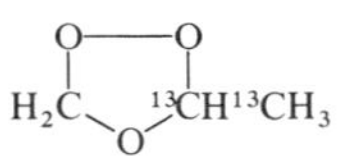	$v=0$	$A=7551.24$ (5) $B=3359.94$ (2) $C=2579.64$ (2)					
	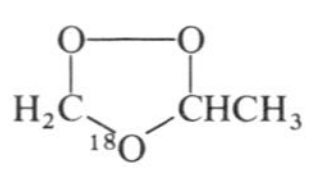	$v=0$	$A=7292.36$ (4) $B=3475.90$ (2) $C=2609.88$ (2)					
	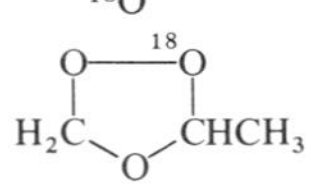	$v=0$	$A=7313.40$ (5) $B=3475.09$ (2) $C=2613.08$ (2)					
	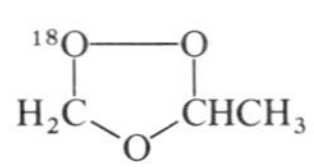	$v=0$	$A=7476.71$ (6) $B=3388.34$ (6) $C=2583.17$ (2)					
519	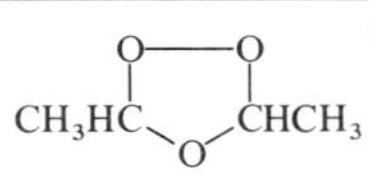 *)	$v=0$	$A=6022.0553$ (95) $B=2103.1359$ (68) $C=1695.6843$ (58)	$\Delta_J = 0.086$ (66) $\Delta_{JK} = -0.014$ (131) $\Delta_K = 2.44$ (134) $\delta_J = 0.0245$ (42) $\delta_K = 0.301$ (142)	80KÜH 74LA2		*) trans form	2.6.3, 253
520 Cont.	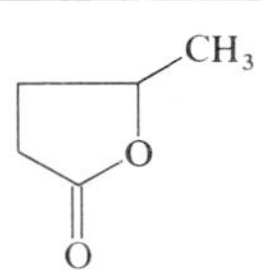 equatorial	$v=0$	$A''=5337.160$ (75) $B''=2193.593$ (17) $C''=1668.033$ (15)	$\tau_{aaaa} = -421.0$ (169) $\tau_{bbbb} = -1.283$ (710) $\tau_{cccc} = -0.4681$ (5383) $\tau_1 = -1.572$ (1080) $\tau_2 = 0.3892$ (4936) $\tau_3 = 114.7$ (417)	81KEW			2.8, 50a

Nr.	Molecule	v	A, B, C MHz	D, H kHz	Ref.	Ref. struct.	Varia, remarks	See also Tab., Nr.
520 Cont.	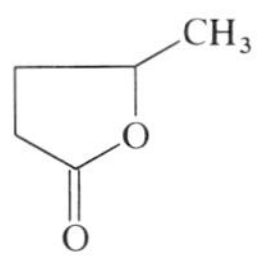	$v_R=1$	$A''=4980.69\,(1323)$ $B''=2173.40\,(134)$ $C''=1691.61\,(158)$	$\tau_{aaaa}=-7.424\,(56)\cdot 10^4$ $\tau_{bbbb}=-213.5\,(228)$ $\tau_{cccc}=105.0\,(321)$ $\tau_1=318.6\,(96)$ $\tau_2=347.5\,(56)$ $\tau_3=327.2\,(57)\cdot 10^2$	81KEW		$\nu_R=149\,(60)\ \mathrm{cm}^{-1}$, ring-puckering	
		$v_t=1$	$A''=4958.69\,(247)$ $B''=2173.97\,(15)$ $C''=1684.89\,(14)$	$\tau_{aaaa}=-0.88\,(10)\cdot 10^4$ $\tau_{bbbb}=-46.3\,(36)$ $\tau_{cccc}=-9.0\,(24)$ $\tau_1=106\,(5)$ $\tau_2=834.3\,(34)$ $\tau_2=78.3\,(34)\cdot 10^2$			$\nu_t=131\,(67)\ \mathrm{cm}^{-1}$, ring-twisting	
521 *6-220*	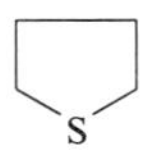	$v=0$	$A=6020.5098\,(47)$ $B=4373.9342\,(39)$ $C=2801.0312\,(40)$	$\Delta_J=0.807\,(73)$ $\Delta_K=1.203\,(44)$ $\delta_J=0.2416\,(28)$ $\delta_K=0.299\,(35)$	76POZ3	79MAM	for structure, see also Vol. II/7 (3, 539)	2.6.3, 254
		$v_\varphi=1$	$A=6018.373$ $B=4370.436$ $C=2801.980$		73MAM		$\nu_\varphi=115\ \mathrm{cm}^{-1}$, pseudorotation	
		$v_\varphi=2$	$A=6016.554$ $B=4367.276$ $C=2802.730$					
		$v_\varphi=3$	$A=6014.633$ $B=4364.018$ $C=2803.687$					
		$v_\varphi=4$	$A=6012.728$ $B=4360.002$ $C=2804.685$					
		$v_R=1$	$A=6010.909$ $B=4368.870$ $C=2796.550$				$\nu_R=311\ \mathrm{cm}^{-1}$	
Cont.		$v_\varphi=1$, $v_R=1$	$A=6009.090$ $B=4364.970$ $C=2793.070$					

Nr.	Molecule	v	A, B, C MHz	D, H kHz	Ref.	Ref. struct.	Varia, remarks	See also Tab., Nr.
521 Cont.	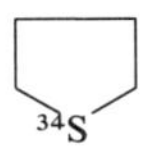	$v=0$	$A=6020.56(18)$ $B=4264.552(15)$ $C=2755.785(26)$		79MAM			
	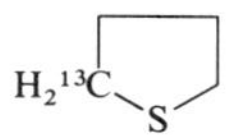	$v=0$	$A=5895.65(10)$ $B=4373.344(11)$ $C=2774.042(12)$					
	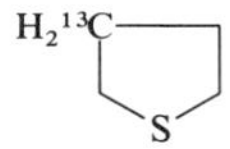	$v=0$	$A=5981.33(13)$ $B=4303.207(13)$ $C=2765.871(15)$					
521 a		$v=0$	$A=3780.814(326)$ $B=2047.243(6)$ $C=1899.572(6)$	$\Delta_J = 0.14(5)$ $\Delta_{JK} = 4.36(19)$ $\Delta_K = -113(72)$ $\delta_K = -4.50(148)$	81ALO		for structure, see also Vol. II/7 (3, 538)	
		$v_R=1$	$A=3784.885(411)$ $B=2049.909(7)$ $C=1902.996(8)$	$\Delta_J = 0.14(3)$ $\Delta_{JK} = 3.71(15)$ $\Delta_K = 140(93)$			$\nu_R = 24(25)\ \mathrm{cm}^{-1}$	
		$v_R=2$	$A=3788.047(517)$ $B=2052.213(11)$ $C=1906.203(12)$	$\Delta_J = 0.15(5)$ $\Delta_{JK} = 3.60(9)$ $\Delta_K = 377(138)$				
		$v_R=3$	$A=3791.807(579)$ $B=2054.255(5)$ $C=1909.302(5)$	$\Delta_J = 0.14$ *) $\Delta_{JK} = 4.36$ *) $\Delta_K = 414(40)$			*) fixed value.	
522 *6-221*	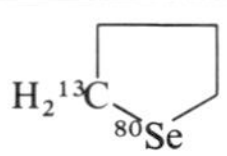	$v=0$	$A=5608.98(18)$ $B=2819.532(8)$ $C=2022.624(8)$		76MAM	76MAM	for structure, see also Vol. II/7 (3, 540)	
	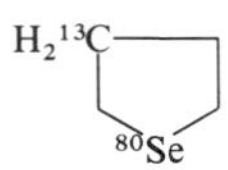	$v=0$	$A=5695.94(10)$ $B=2770.714(4)$ $C=2009.166(4)$					
523	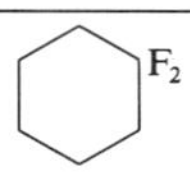	$v=0$	$A=3191.781(13)$ $B=1759.768(5)$ $C=1512.293(3)$	$D_J = 0.124(10)$ $D_{JK} = 0.292(92)$ $D_K = -8.20(34)$	74DAM	74DAM	for structure, see also Vol. II/7 (3, 693)	2.6.3, 255

Nr.	Molecule	v	A, B, C MHz	D, H kHz	Ref.	Ref. struct.	Varia, remarks	See also Tab., Nr.
524	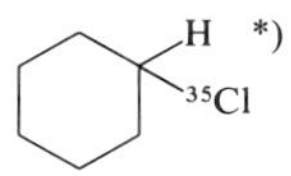	$v=0$	$A=4291.64\,(26)$ $B=1396.96\,(5)$ $C=1127.34\,(5)$		73DAM	73DAM	for structure, see also Vol. II/7 (3, 698) *) equatorial	2.7.3, 136
	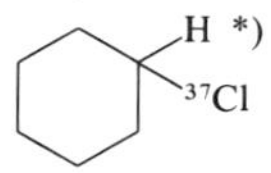	$v=0$	$A=4289.77\,(55)$ $B=1359.67\,(5)$ $C=1102.96\,(5)$					
525	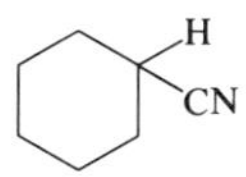	$v=0$	$A=4238.766\,(134)$ *) $B=1399.172\,(3)$ $C=1128.845\,(2)$		77DAM	77DAM	*) equatorial isomer	
		$v=0$	$A=3005.584\,(219)$ **) $B=1763.483\,(8)$ $C=1558.615\,(7)$				**) axial isomer	
526 *6-226*	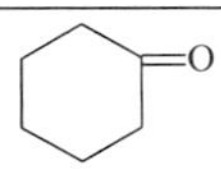	$v=0$	$A=4195.316\,(59)$ $B=2502.627\,(5)$ $C=1754.443\,(5)$	$D_J=0.42\,(3)$ $D_{JK}=-1.22\,(14)$	81ALO1	80DIL		
		$v_R=1$	$A=4188.1\,(2)$ $B=2506.23\,(1)$ $C=1756.847\,(8)$				$\nu_R=100\,(20)\ \mathrm{cm}^{-1}$, ring-bending	
		$v_R=2$	$A=4188.7\,(16)$ $B=2509.89\,(8)$ $C=1759.01\,(4)$					
		$v_R=3$	$A=4180.0\,(25)$ $B=2512.70\,(13)$ $C=1761.274\,(49)$					
		$v_t=1$	$A=4198.70\,(80)$ $B=2500.787\,(33)$ $C=1753.399\,(18)$				$\nu_t=180\,(18)\ \mathrm{cm}^{-1}$, ring-twisting	
		$v_a=1$	$A=4222.1\,(16)$ $B=2498.555\,(69)$ $C=1755.202\,(30)$					
		$v_b=1$	$A=4217.3\,(25)$ $B=2495.94\,(12)$ $C=1752.254\,(48)$					

Nr.	Molecule	v	A, B, C MHz	D, H kHz	Ref.	Ref. struct.	Varia, remarks	See also Tab., Nr.
527	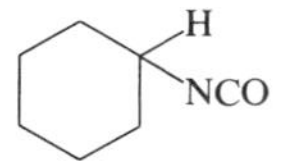		$B+C=1702.9\,(2)$ *)		80KIT		*) LRMW	
528	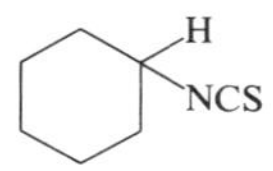		$B+C=1155.9\,(1)$ *)		80KIT		*) LRMW	
529	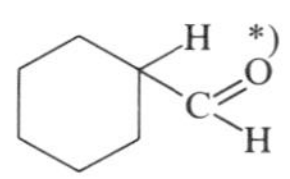	$v=0$ $v=0$ $v_T=1$ $v_T=2$	$A=4131.52\,(9)$ **) $B=1327.701\,(5)$ $C=1090.131\,(6)$ $A=3951.70\,(9)$ ***) $B=1424.402\,(2)$ $C=1137.299\,(1)$ $A=3944.5\,(2)$ $B=1424.705\,(6)$ $C=1138.627\,(2)$ $A=3938.4\,(2)$ $B=1425.178\,(6)$ $C=1139.885\,(3)$	$\Delta_J=0.03\,(1)$ $\Delta_{JK}=3.16\,(2)$	79KAO	79KAO	*) chair ring with equatorial aldehyde **) cis form (C=O eclipsing axial C_1—H) ***) gauche form (C=O eclipsing ring C_1—C_2) $E_{cis}-E_{gauche}=720\,(200)$ cal/mol v_T, C—CHO torsion	
530	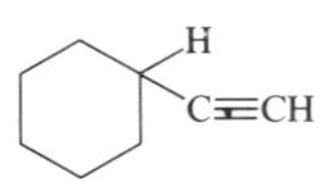	$v=0$ $v_x=1$ $v=0$	$A=4246.1\,(1)$ *) $B=1386.451\,(1)$ $C=1121.707\,(1)$ $A=4194.6\,(1)$ $B=1387.459\,(2)$ $C=1122.864\,(2)$ $A=2994.441\,(6)$ **) $B=1730.834\,(7)$ $C=1540.004\,(6)$		80DAM	80DAM	*) equatorial chair form $v_x=130\,(30)\ cm^{-1}$ **) axial chair form $E_{ax}-E_{eq}=600\,(200)$ cal/mol	
531 *6-227*	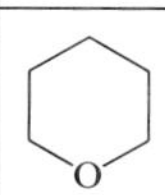	$v=0$	$\tilde{A}=4673.502\,(9)$ $\tilde{B}=4495.074\,(9)$ $\tilde{C}=2601.281\,(9)$	$d_J=2.516\,(227)$ $d_{JK}=171.2\,(34)$ $d_K=-74.13\,(145)$ $d_{WJ}=-3.242\,(201)\cdot10^{-7}$ $d_{WK}=-3.787\,(73)\cdot10^{-5}$	76LO2			2.6.3, 256

Nr.	Molecule	v	A, B, C MHz	D, H kHz	Ref.	Ref. struct.	Varia, remarks	See also Tab., Nr.
532	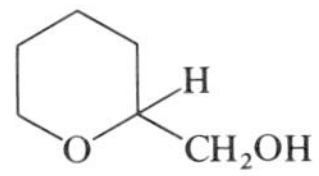	$v=0$	$A=4064.0$ (3) $B=1412.043$ (5) $C=1124.141$ (3)		79TUR2	79TUR2		3, 148
	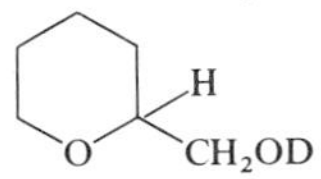	$v=0$	$A=4017.8$ (2) $B=1390.046$ (4) $C=1106.652$ (2)					
533	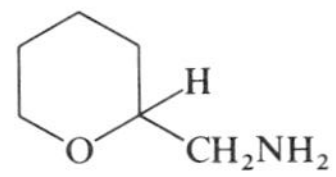	$v=0$	$A=4041.0$ (2) $B=1391.137$ (4) $C=1114.462$ (3)		79TUR2	79TUR2		3, 148
	—ND_2 species	$v=0$	$A=3947.1$ (3) $B=1327.79$ (1) $C=1072.55$ (2)					
	—NDH″ species	$v=0$	$A=3989.6$ (4) $B=1365.384$ (9) $C=1094.15$ (1)					
	—NH′D species	$v=0$	$B+C=2443.3$					
534	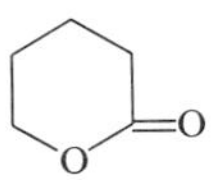	$v=0$	$A=4661.21$ *) $B=2558.57$ $C=1771.16$		81PHI		*) half-chair form	2.6.3, 256a
		$v=0$	$A=4547.06$ **) $B=2595.87$ $C=1867.86$				**) half-boat form (less abundant)	
535	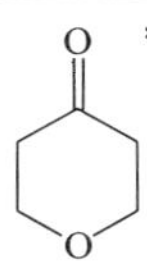 *)	$v=0$	$A=4566.882$ (33) $B=2538.316$ (3) $C=1805.878$ (4)	$\Delta_J=0.37$ (2) $\Delta_{JK}=-0.98$ (5) $\delta_J=0.07$ (2)	80ALO2	80ALO2	*) chair conformation $v_R \simeq 100\ cm^{-1}$, ring-bending	2.6.3, 257
		$v_R=1$	$A=4557.423$ (40) $B=2542.386$ (3) $C=1808.805$ (4)	$\Delta_J=0.35$ (3) $\Delta_{JK}=-0.88$ (6) $\delta_J=0.06$ (2)				
		$v_R=2$	$A=4548.751$ (40) $B=2546.130$ (4) $C=1811.461$ (5)	$\Delta_J=0.35$ (3) $\Delta_{JK}=-0.88$ (6) $\delta_J=0.05$ (2)				
		$v_R=3$	$A=4540.581$ (43) $B=2549.571$ (4) $C=1813.872$ (6)	$\Delta_J=0.42$ (4) $\Delta_{JK}=-0.93$ (7)				
Cont.								

Nr.	Molecule	v	A, B, C MHz	D, H kHz	Ref.	Ref. struct.	Varia, remarks	See also Tab., Nr.
535 Cont.	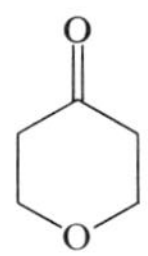	$v_R=4$	$A=4533.030\,(51)$ $B=2552.722\,(5)$ $C=1816.040\,(7)$	$\Delta_J = 0.43\,(4)$ $\Delta_{JK} = -0.91\,(7)$	80ALO2			
		$v_t=1$	$A=4571.688\,(42)$ $B=2536.298\,(4)$ $C=1804.592\,(6)$	$\Delta_J = 0.40\,(3)$ $\Delta_{JK} = -0.95\,(7)$ $\delta_J = 0.06\,(2)$			$\nu_t \simeq 185\ \mathrm{cm}^{-1}$, ring-twisting	
		$v_a=1$	$A=4578.308\,(184)$ $B=2535.068\,(14)$ $C=1801.943\,(22)$					
		$v_b=1$	$A=4586.652\,(241)$ $B=2535.518\,(24)$ $C=1802.807\,(27)$					
		$v_c=1$	$A=4565.827\,(77)$ $B=2540.393\,(7)$ $C=1806.812\,(9)$					
		$v_d=1$	$A=4568.370\,(112)$ $B=2540.821\,(9)$ $C=1807.041\,(11)$					
536	 *)	$v=0$	$A=3992.719\,(12)$ $B=3005.812\,(12)$ $C=1914.683\,(14)$		75KI1	75KI1	*) chair conformation	2.6.3, 258
536a	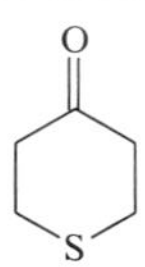	$v=0$	$A=3935.149\,(31)$ $B=1829.444\,(1)$ $C=1364.609\,(1)$	$\Delta_J = 0.22\,(2)$ $\Delta_{JK} = -0.70\,(2)$ $\Delta_K = -3.22\,(216)$ $\delta_J = 0.04\,(1)$ $\delta_K = -0.68\,(11)$	81ALO2	81ALO2	chair form	2.6.3, 258a
		$v_R=1$	$A=3940.184\,(225)$ $B=1832.940\,(11)$ $C=1366.745\,(7)$	$\Delta_J = 0.06\,(2)$ $\Delta_{JK} = 0.50\,(15)$ $\Delta_K = 359\,(12)$ $\delta_K = 15.73\,(157)$			$\nu_R = 77\,(22)\ \mathrm{cm}^{-1}$, ring-bending	
Cont.		$v_R=2$	$A=3923.456\,(220)$ $B=1835.620\,(10)$ $C=1369.136\,(2)$	$\Delta_J = 0.14\,(5)$ $\Delta_{JK} = 0.13\,(9)$ $\Delta_K = 255\,(19)$ $\delta_K = 11.48\,(165)$				

Nr.	Molecule	v	A, B, C MHz	D, H kHz	Ref.	Ref. struct.	Varia, remarks	See also Tab., Nr.
536a Cont.	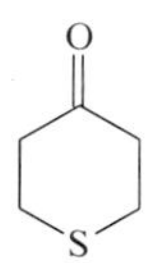	$v_R=3$	$A=3914.149\,(200)$	$\Delta_J = 0.20\,(8)$	81ALO2			
			$B=1838.314\,(10)$	$\Delta_{JK} = -0.44\,(10)$				
			$C=1371.253\,(3)$	$\Delta_K = 336\,(20)$				
				$\delta_K = 16.95\,(196)$				
		$v_R=4$	$A=3896.586\,(881)$	$\Delta_J = 0.20$ (fixed)				
			$B=1840.621\,(22)$	$\Delta_{JK} = -0.20\,(15)$				
			$C=1373.473\,(10)$	$\Delta_K = 47\,(25)$				
		$v_R=5$	$A=3885.497\,(742)$	$\Delta_J = 0.20$ (fixed)				
			$B=1842.969\,(19)$	$\Delta_{JK} = -0.39\,(22)$				
			$C=1375.407\,(8)$	$\Delta_K = 50\,(18)$				
		$v_t=1$	$A=3937.888\,(85)$	$\Delta_J = 0.20\,(3)$			$v_t=204\,(27)\ cm^{-1}$, ring-twisting	
			$B=1828.369\,(3)$	$\Delta_{JK} = -0.76\,(5)$				
			$C=1363.946\,(3)$	$\Delta_K = 3.6\,(29)$				
		$v_a=1$	$A=3931.446\,(32)$	$\Delta_J = -0.26\,(16)$				
			$B=1831.549\,(25)$	$\Delta_{JK} = 2.43\,(98)$				
			$C=1365.961\,(19)$	$\Delta_K = -190\,(79)$				
				$\delta_K = -20.8\,(41)$				
		$v_b=1$	$A=3923.534\,(846)$	$\Delta_J = 0.52\,(15)$				
			$B=1831.522\,(18)$	$\Delta_{JK} = -2.40\,(92)$				
			$C=1366.227\,(13)$	$\Delta_K = 363\,(88)$				
				$\delta_K = 23.9\,(47)$				
537 *6-228* *4-162a*	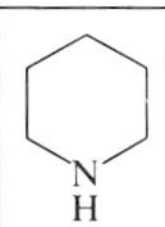 axial	$v=0$	$A=4494.368\,(10)$	$D_J = 1.277\,(170)$	81PAR	81PAR		2.6.3, 258b 2.7.3, 51a
			$B=4395.271\,(10)$	$D_{JK} = -1.418\,(6)$				
			$C=2535.611\,(10)$	$D_K = 0.4644\,(360)$				
	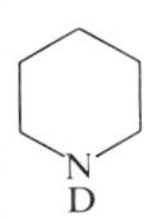 axial	$v=0$	$A=4440.515\,(19)$	$D_J = 0.868\,(144)$				
			$B=4254.109\,(15)$	$D_{JK} = -1.330\,(24)$				
			$C=2505.661\,(10)$	$D_K = 0.5964\,(405)$				
	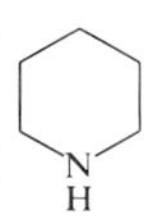 equatorial	$v=0$	$A=4517.227\,(8)$	$D_J = 0.964\,(69)$			$E_{axial} - E_{eq} = 3.1\,(3)$ kJ/mol	
			$B=4437.239\,(8)$	$D_{JK} = -1.472\,(7)$				
			$C=2542.989\,(5)$	$D_K = 0.6637\,(273)$				
	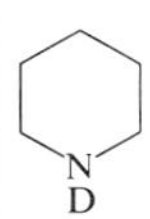 equatorial	$v=0$	$A=4527.629\,(32)$	$D_J = 0.824\,(327)$				
			$B=4232.970\,(41)$	$D_{JK} = -1.326\,(58)$				
			$C=2474.904\,(27)$	$D_K = 0.4067\,(947)$				

Nr.	Molecule	v	A, B, C MHz	D, H kHz	Ref.	Ref. struct.	Varia, remarks	See also Tab., Nr.
538	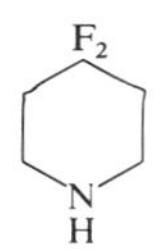 *)	$v=0$	$A=3308.072$ (923) $B=1770.409$ (22) $C=1535.052$ (18)		77BEL	77BEL	*) N—H equatorial conformer	
	N—D species	$v=0$	$A=3308.979$ (1156) $B=1716.309$ (31) $C=1494.376$ (29)					
539 *6-229*	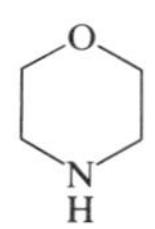	$v=0$	$A=4925.006$ (13) $B=4625.140$ (14) $C=2684.244$ (15)	d_J = 0.73 (54) d_{JK} = 110.08 (758) d_K = −48.03 (324) d_{WJ} = −0.01 (3) $\cdot 10^{-6}$ d_{WK} = −23.52 (158) $\cdot 10^{-6}$	79FIL			2.6.3, 259
540	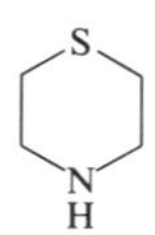 *)	$v=0$	$A=4198.443$ (7) $B=3061.117$ (3) $C=1961.006$ (2)		80FAN	80FAN	*) chair equatorial conformer	2.6.3, 260 2.7.3, 52
		$v_I=1$	$A=4191.173$ (84) $B=3061.875$ (13) $C=1960.102$ (6)				$\nu_I \simeq 180\ cm^{-1}$	
		$v_{II}=1$	$A=4197.460$ (109) $B=3057.918$ (17) $C=1960.013$ (8)				$\nu_{II} \simeq 230\ cm^{-1}$	
	N—D species	$v=0$	$A=4198.843$ (29) $B=2935.651$ (5) $C=1908.937$ (3)					
		$v_I=1$	$A=4191.266$ (75) $B=2936.492$ (13) $C=1908.332$ (4)					
		$v_{II}=1$	$A=4195.221$ (199) $B=2933.729$ (37) $C=1907.936$ (11)					

Nr.	Molecule	v	A, B, C MHz	D, H kHz	Ref.	Ref. struct.	Varia, remarks	See also Tab., Nr.
541	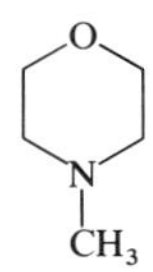 *)	$v=0$	$A=4821.235\,(77)$		74DAS		*) chair form	2.6.3, 261
			$B=2342.579\,(4)$				with equatorial	2.7.3, 53
			$C=1719.452\,(3)$				CH_3	2.8, 121
		$v_R=1$	$A=4804.77$				$\nu_R=180\,(10)\ cm^{-1}$,	
			$B=2343.38$				ring deformation	
			$C=1720.15$					
		$v_2=1$	$A=4820.30$				$\nu_2=226\,(10)\ cm^{-1}$	
			$B=2340.21$					
			$C=1718.56$					
		$v_3=1$	$A=4820.96$				$\nu_3=267\,(10)\ cm^{-1}$	
			$B=2340.90$					
			$C=1718.06$					
		$v_4=1$	$A=4792.75$				$\nu_4=270\,(15)\ cm^{-1}$	
			$B=2343.91$					
			$C=1720.33$					
		$v_5=1$	$A=4806.66$				$\nu_5=357\,(25)\ cm^{-1}$	
			$B=2341.10$					
			$C=1719.12$					
		$v_6=1$	$A=4804.60$				$\nu_6=361\,(25)\ cm^{-1}$	
			$B=2341.66$					
			$C=1718.68$					
542 *6–227a*	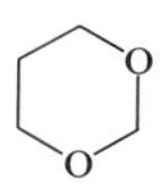	$v=0$	$\tilde{A}=4999.927\,(10)$	$d_J = 1.977\,(301)$	76LO2			
			$\tilde{B}=4807.613\,(10)$	$d_{JK} = -208.8\,(28)$				
			$\tilde{C}=2757.111\,(11)$	$d_K = 91.18\,(123)$				
				$d_{WJ} = -1.839\,(358)\cdot10^{-7}$				
				$d_{WK} = 4.221\,(57)\cdot10^{-5}$				
543	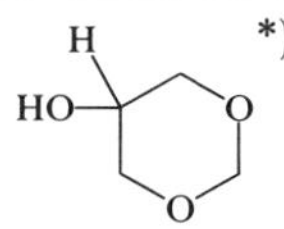 *)	$v=0$	$A=3913.5525\,(22)$	$\Delta_J = 0.637\,(18)$	81ROM	80ALO1	*) chair	2.6.3, 262
			$B=2922.7841\,(22)$	$\Delta_{JK} = -0.958\,(41)$			conformation with axial OH	
			$C=2193.4469\,(30)$	$\Delta_K = 1.128\,(21)$				
				$\delta_J = 0.097\,(9)$				
				$\delta_K = 0.16\,(6)$				
		$v_a=1$	$A=3916.577\,(12)$	$\Delta_J = 3.19\,(28)$			$\nu_a\simeq160\ cm^{-1}$	
			$B=2920.343\,(12)$	$\Delta_{JK} = -0.85\,(11)$				
			$C=2194.674\,(12)$	$\Delta_K = 1.07\,(10)$				
Cont.				$\delta_J = 0.11\,(6)$				

Nr.	Molecule	v	A, B, C MHz	D, H kHz	Ref.	Ref. struct.	Varia, remarks	See also Tab., Nr.
543 Cont.	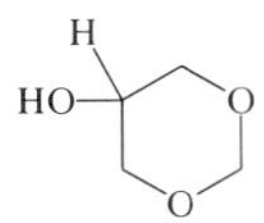	$v_b=1$	$A=3918.2612\,(93)$ $B=2914.8667\,(93)$ $C=2184.5284\,(98)$	$\Delta_J = 2.26\,(26)$ $\Delta_{JK} = -0.82\,(10)$ $\Delta_K = 1.03\,(10)$ $\delta_J = 0.10\,(4)$	81ROM		$\nu_b \simeq 225\ cm^{-1}$	
	hydroxy—d_1	$v=0$	$A=3851.201\,(16)$ $B=2867.789\,(23)$ $C=2183.161\,(27)$		80ALO1			
544	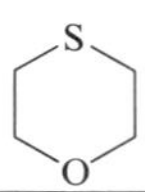 *)	$v=0$	$A=4333.85\,(5)$ $B=3076.144\,(7)$ $C=1991.956\,(7)$		75KI2	75KI2	for structure, see also Vol. II/7 (3, 537) *) chair conformation	2.6.3, 263
545	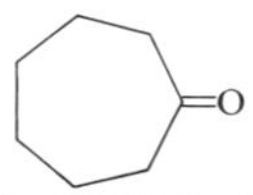	$v=0$	$A=2821.67\,(6)$ $B=1941.46\,(1)$ $C=1342.52\,(1)$		79AKA			
546	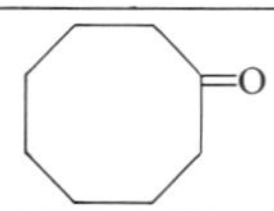	$v=0$	$A=1910\,(5)$ $B=1680\,(30)$ $C=1100\,(100)$		78ROU			
547	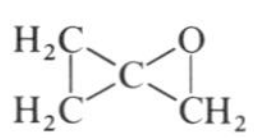	$v=0$	$A=11616.911\,(2)$ $B=4394.777\,(1)$ $C=4321.157\,(1)$	$\tau_{aaaa} = -31.04\,(8)$ $\tau_{bbbb} = -3.05\,(1)$ $\tau_{cccc} = -2.78\,(1)$ $\tau_{aabb} = -17.99\,(1)$ $\tau_{bbcc} = -4.26\,(1)$	75SLA	75SLA		2.6.3, 264 3, 149
	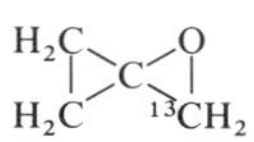	$v=0$	$A=11497.045\,(5)$ $B=4314.287\,(2)$ $C=4259.635\,(2)$	$\tau_{aaaa} = -30.66\,(11)$ $\tau_{bbbb} = -2.88\,(1)$ $\tau_{cccc} = -2.69\,(1)$ $\tau_{aabb} = -17.17\,(10)$ $\tau_{bbcc} = -4.04\,(1)$				
	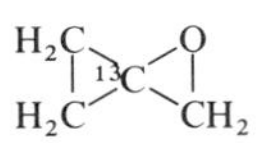	$v=0$	$A=11614.440\,(4)$ $B=4394.981\,(1)$ $C=4321.662\,(1)$	$\tau_{aaaa} = -30.94\,(9)$ $\tau_{bbbb} = -3.05\,(1)$ $\tau_{cccc} = -2.80\,(1)$ $\tau_{aabb} = -18.00\,(9)$ $\tau_{bbcc} = -4.23\,(1)$				
Cont.								

Nr.	Molecule	v	A, B, C MHz	D, H kHz	Ref.	Ref. struct.	Varia, remarks	See also Tab., Nr.
547 Cont.	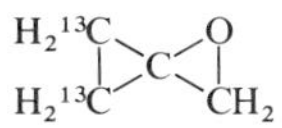	$v=0$	$A=11464.034\,(5)$ $B=4334.695\,(2)$ $C=4242.230\,(2)$	$\tau_{aaaa}=-29.78\,(16)$ $\tau_{bbbb}=-2.92\,(2)$ $\tau_{cccc}=-2.60\,(2)$ $\tau_{aabb}=-16.96\,(16)$ $\tau_{bbcc}=-4.10\,(1)$	75SLA			
	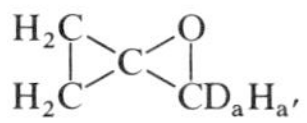	$v=0$	$A=11124.967\,(2)$ $B=4258.849\,(1)$ $C=4169.958\,(1)$	$\tau_{aaaa}=-28.51\,(8)$ $\tau_{bbbb}=-3.24\,(1)$ $\tau_{cccc}=-2.83\,(1)$ $\tau_{aabb}=-15.47\,(7)$ $\tau_{bbcc}=-2.82\,(1)$			see Fig. 149 for atom labeling	
	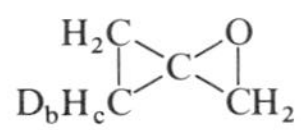	$v=0$	$A=11015.749\,(2)$ $B=4280.715\,(1)$ $C=4155.955\,(1)$	$\tau_{aaaa}=-27.54\,(11)$ $\tau_{bbbb}=-3.13\,(1)$ $\tau_{cccc}=-2.68\,(2)$ $\tau_{aabb}=-15.61\,(11)$ $\tau_{bbcc}=-3.35\,(1)$				
	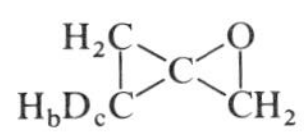	$v=0$	$A=11030.490\,(3)$ $B=4293.265\,(1)$ $C=4170.073\,(1)$	$\tau_{aaaa}=-26.65\,(23)$ $\tau_{bbbb}=-3.20\,(3)$ $\tau_{cccc}=-2.74\,(4)$ $\tau_{aabb}=-16.16\,(23)$ $\tau_{bbcc}=-3.43\,(1)$				
548 *6-231a*	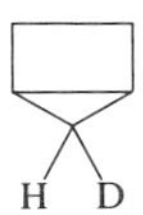, exo	$v=0$	$A=9093.40\,(2)$ $B=5730.10\,(2)$ $C=4545.03\,(3)$		76MAT	76MAT	for structure, see also Vol. II/7 (3, 610)	3, 150
	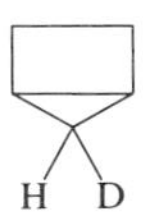, endo	$v=0$	$A=8725.62\,(1)$ $B=5856.68\,(2)$ $C=4726.13\,(2)$					
	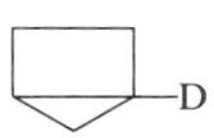	$v=0$	$A=8583.42\,(1)$ $B=5975.33\,(2)$ $C=4657.24\,(6)$					
Cont.	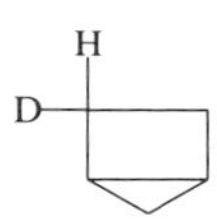, endo	$v=0$	$A=8673.28\,(1)$ $B=5934.14\,(4)$ $C=4650.18\,(4)$					

Nr.	Molecule	v	A, B, C MHz	D, H kHz	Ref.	Ref. struct.	Varia, remarks	See also Tab., Nr.
548 *6-231 a* Cont.	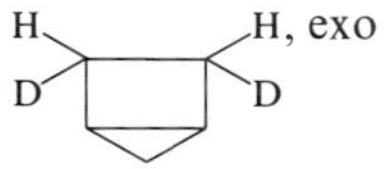	$v=0$	$A=8581.75\,(1)$ $B=5661.04\,(3)$ $C=4400.20\,(2)$		76MAT			
	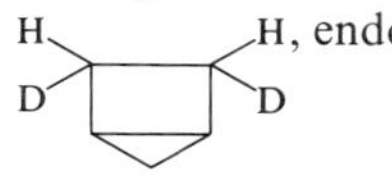	$v=0$	$A=8268.93\,(1)$ $B=5769.08\,(2)$ $C=4529.48\,(1)$					
549	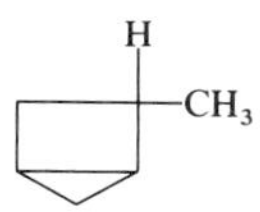, endo	$v=0$	$A=6257.36\,(1)$ $B=3412.19\,(1)$ $C=2843.34\,(1)$		75HA2	75HA2		2.6.3, 265 3, 151
	exo	$v=0$	$A=7418.75\,(1)$ $B=3101.66\,(1)$ $C=2652.57\,(1)$					
550	 *)	$v=0$	$A=5026.243\,(3)$ $B=2833.813\,(3)$ $C=2411.679\,(3)$	$\Delta_J = 0.465\,(36)$ $\Delta_{JK} = -0.363\,(6)$ $\Delta_K = 1.081\,(38)$ $\delta_J = 26.8\,(3)\cdot 10^{-3}$ $\delta_K = 0.260\,(9)$	76MJÖ	76MJÖ	*) boat form	2.6.3, 266 3, 152
		$v_R=1$	$A=5028.609\,(5)$ $B=2826.677\,(6)$ $C=2402.598\,(8)$	$\Delta_J = 0.442\,(83)$ $\Delta_{JK} = -0.384\,(66)$ $\Delta_K = 0.960\,(230)$ $\delta_J = 27\,(4)\cdot 10^{-3}$ $\delta_K = 0.288\,(108)$			v_R, ring puckering mode	
		$v_R=2$	$A=5029.13\,(109)$ $B=2818.981\,(76)$ $C=2394.374\,(75)$					
551	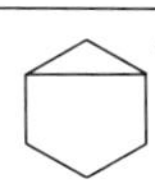 *)	$v=0$	$A=5542.96$ $B=4236.82$ $C=3127.04$		74COO2	74COO2	*) boat form	2.6.3, 267 3, 153
552	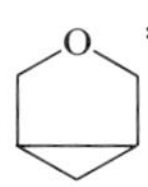 *)	$v=0$	$A=6038.06$ $B=4432.47$ $C=3303.43$		74MAO	74MAO	*) boat form	2.6.3, 268
Cont.		$v_R=1$	$A=6042.12$ $B=4222.46$ $C=3290.52$				$v_R=239.1\ \mathrm{cm}^{-1}$, ring-puckering mode	

Nr.	Molecule	v	A, B, C MHz	D, H kHz	Ref.	Ref. struct.	Varia, remarks	See also Tab., Nr.
552 Cont.	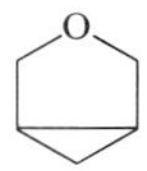	$v_R=2$	$A=6048.16$ $B=4411.69$ $C=3276.38$		74MAO			
553	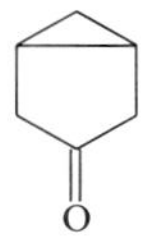	$v=0$	$A=5356.835\,(1)$ $B=2461.202\,(1)$ $C=2037.243\,(1)$	$\Delta_J = 0.545\,(8)$ $\Delta_{JK} = -2.200\,(5)$ $\Delta_K = 5.80\,(4)$ $\delta_J = -0.0296\,(2)$ $\delta_K = 0.154\,(6)$	78BEV	78BEV		2.6.3, 269 3, 154
		$v_B=1$	$A=5370.7$ $B=2458.384$ $C=2034.577$				$v_B=77.56$ cm^{-1}, bending mode potential function determined	
		$v_B=2$	$A=5383.6$ $B=2455.578$ $C=2031.991$					
		$v_B=3$	$A=5397.3$ $B=2452.835$ $C=2029.560$					
		$v_B=4$	$A=5410.6$ $B=2450.242$ $C=2027.387$					
		$v_B=5$	$A=5422.7$ $B=2447.863$ $C=2025.572$					
	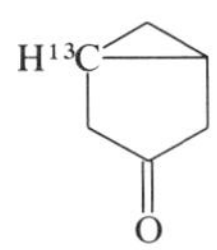	$v=0$	$A=5318.89\,(5)$ $B=2443.072\,(6)$ $C=2021.074\,(6)$					
	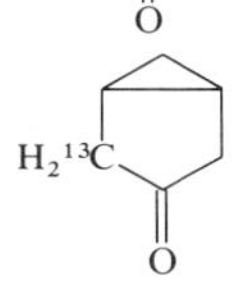	$v=0$	$A=5263.48\,(4)$ $B=2459.186\,(8)$ $C=2023.861\,(9)$					
Cont.	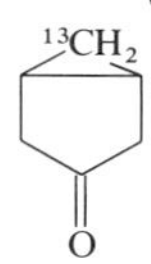	$v=0$	$A=5315.48\,(4)$ $B=2418.246\,(2)$ $C=2013.600\,(2)$					

Nr.	Molecule	v	A, B, C MHz	D, H kHz	Ref.	Ref. struct.	Varia, remarks	See also Tab., Nr.
553 Cont.	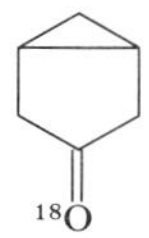	$v=0$	$A=5344.44$ (5) $B=2347.101$ (6) $C=1960.129$ (7)		78BEV			
554	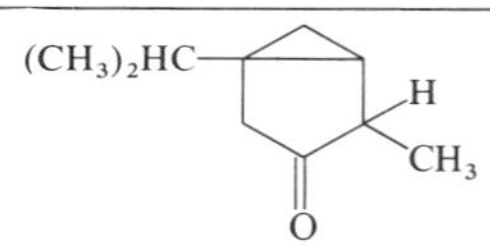		$B+C=1522.1$ (5) *) $^{+}$) $B+C=1416.8$ (18) *) $^{++}$)		78KIE		*) LRMW $^{+}$) thujone ($\varphi=15°$), see Fig. 155 $^{++}$) isothujone ($\varphi=25°$)	3, 155
555	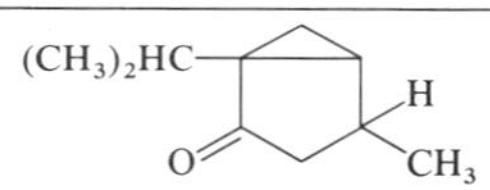		$B+C=1540.2$ (12) *)		78KIE		*) LRMW	3, 155
556	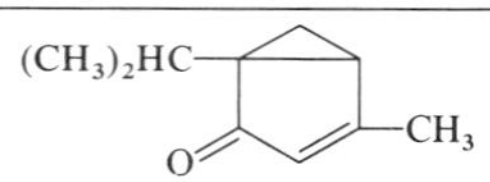		$B+C=1585$ (2) *)		78KIE		*) LRMW	3, 155
557	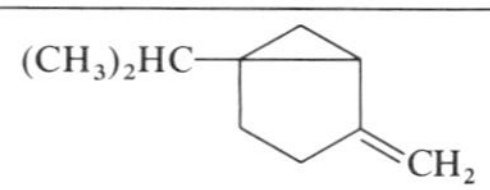		$B+C=1772.0$ (6) *)		78KIE		*) LRMW	3, 155
558	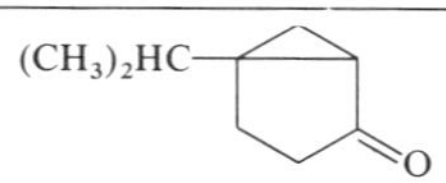		$B+C=1763.5$ (7) *)		78KIE		*) LRMW	3, 155
559	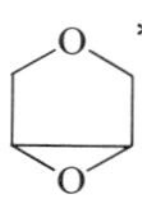 *)	$v=0$ $v_R=1$	$A=6287.302$ (11) $B=4683.546$ (8) $C=3358.517$ (89) $A=6297.262$ (14) $B=4671.466$ (11) $C=3344.545$ (110)		73CRE	73CRE	*) boat form $\nu_R=200$ (12) cm^{-1}, ring puckering mode	2.6.3, 270 3, 156
559a Cont.	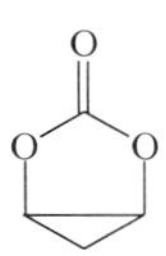	$v=0$ $v_B=1$	$A=6730.57$ (17) $B=2674.733$ (23) $C=2207.211$ (23) $A=6718.31$ (45) $B=2676.550$ (10) $C=2210.382$ (10)		81BOM	81BOM	$\nu_B=124$ (12) cm^{-1}, five membered ring bending	

Nr.	Molecule	v	A, B, C MHz	D, H kHz	Ref.	Ref. struct.	Varia, remarks	See also Tab., Nr.
559 a Cont.	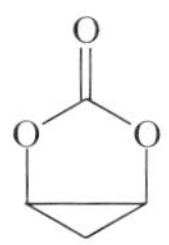	$v_B=2$	$A=6704.83\,(49)$ $B=2678.373\,(10)$ $C=2213.564\,(11)$		81BOM			
		$v_B=3$	$A=6694.28\,(65)$ $B=2680.221\,(15)$ $C=2216.684\,(16)$					
		$v_B=4$	$A=6678.95\,(45)$ $B=2682.038\,(10)$ $C=2219.855\,(11)$					
		$v_B=5$	$A=6664.54\,(72)$ $B=2683.844\,(17)$ $C=2222.977\,(17)$					
	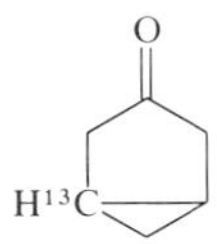	$v=0$	$A=6668.88\,(37)$ $B=2656.078\,(71)$ $C=2190.578\,(74)$					
	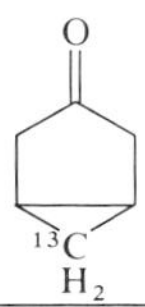	$v=0$	$A=6680.50\,(129)$ $B=2623.835\,(25)$ $C=2177.670\,(20)$					
560	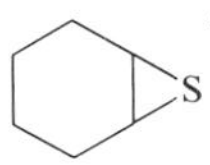 *)	$v=0$	$A=3512.086\,(10)$ $B=2057.969\,(3)$ $C=1623.023\,(3)$		73KEW	73KEW	*) twisted half-chair structure	2.6.3, 271 3, 157
		$v_b=0$	$A=3512.416\,(30)$ $B=2055.729\,(5)$ $C=1622.196\,(4)$				$v_b=115\,(20)\ \mathrm{cm^{-1}}$, bending mode	
		$v_t=1$	$A=3512.232\,(41)$ $B=2054.758\,(5)$ $C=1622.139\,(4)$				$v_t=200\,(60)\ \mathrm{cm^{-1}}$, twisting mode	
561	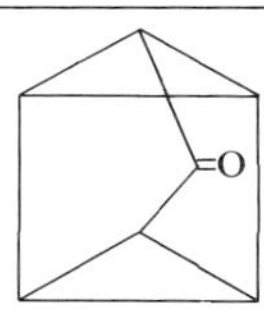	$v=0$	$A=4361.92\,(7)$ $B=2469.960\,(3)$ $C=2028.872\,(3)$		77COF2	77COF2		2.6.3, 272 3, 158

Nr.	Molecule	v	A, B, C MHz	D, H kHz	Ref.	Ref. struct.	Varia, remarks	See also Tab., Nr.
562		$v=0$	$A=3694.246\,(5)$ $B=3212.566\,(3)$ $C=2774.813\,(2)$		80CHP	80CHP		2.6.3, 273 3, 159
562a	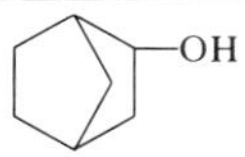 exo	$v=0$	$A=3605.9374\,(9)$ $B=1935.4207\,(8)$ $C=1752.3947\,(8)$	$\Delta_J=0.099\,(7)$ $\Delta_{JK}=0.246\,(1)$ $\Delta_K=0.115\,(9)$ $\delta_J=0.0067\,(8)$ $\delta_K=0.013\,(2)$	81TYB	81TYB		2.6.3, 273a
	endo	$v=0$	$A=3151.4865\,(15)$ $B=2095.2483\,(24)$ $C=1914.7057\,(25)$	$\Delta_J=0.376\,(8)$ $\Delta_{JK}=0.614\,(7)$ $\Delta_K=0.964\,(8)$				
		$v=1$	$A-C=1230.66$ $\kappa=-0.7068$					
	exo-OD	$v=0$	$A=3583.437\,(7)$ $B=1881.556\,(10)$ $C=1703.179\,(4)$	*)			*) constants of parent species used	
	endo-OD	$v=0$	$A=3138.761\,(17)$ $B=2034.469\,(27)$ $C=1859.925\,(56)$	*)				
563	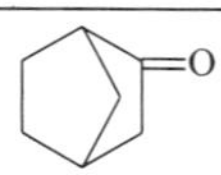	$v=0$	$A=3433.04\,(59)$ $B=2105.575\,(15)$ $C=1887.634\,(16)$		74COF			2.6.3, 274
		$v_a=1$	$A=3434.37\,(72)$ $B=2106.877\,(19)$ $C=1889.990\,(20)$					
		$v_a=2$	$A=3433.34\,(19)$ $B=2107.967\,(5)$ $C=1891.999\,(5)$					
		$v_a=3$	$A=3432.56\,(86)$ $B=2108.844\,(24)$ $C=1893.641\,(26)$					
564	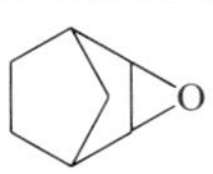	$v=0$	$A=3759.41\,(7)$ $B=2324.893\,(2)$ $C=2125.750\,(2)$		74COF			

Nr.	Molecule	v	A, B, C MHz	D, H kHz	Ref.	Ref. struct.	Varia, remarks	See also Tab., Nr.
565	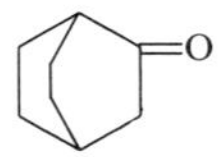	$v=0$	$A=2600\,(160)$ $B=1635.8\,(10)$ $C=1609.7\,(7)$		74COF			
566	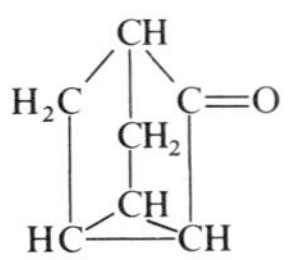	$v=0$	$A=3397.63\,(16)$ $B=2261.260\,(4)$ $C=2103.531\,(4)$		74COF			
567	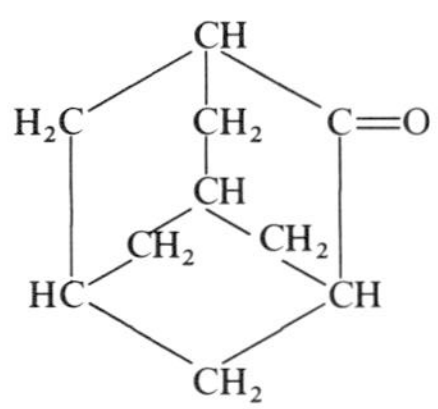	$v=0$	$B+C=2444.94\,(5)$ $B-C<0.4$		74COF			
568	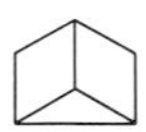	$v=0$	$A=5954.071\,(3)$ $B=5256.393\,(3)$ $C=4174.399\,(3)$		75SUE	75SUE		2.6.3, 275 3, 160
	1-^{13}C	$v=0$	$A=5867.626\,(3)$ $B=5210.559\,(4)$ $C=4160.403\,(2)$				see Fig. 160 for atom numbering	
	2-^{13}C	$v=0$	$A=5901.561\,(2)$ $B=5216.169\,(3)$ $C=4131.711\,(1)$					
	3-^{13}C	$v=0$	$A=5920.845\,(2)$ $B=5167.746\,(3)$ $C=4110.496\,(2)$					
	4-^{13}C	$v=0$	$A=5876.278\,(10)$ $B=5240.566\,(9)$ $C=4146.176\,(7)$					
Cont.	1-D	$v=0$	$A=5634.473\,(4)$ $B=5085.973\,(4)$ $C=4119.973\,(5)$					

Nr.	Molecule	v	A, B, C MHz	D, H kHz	Ref.	Ref. struct.	Varia, remarks	See also Tab., Nr.
568 Cont.	2-D	$v=0$	$A=5763.411$ (3)		75SUE			
			$B=5117.623$ (3)					
			$C=4020.351$ (4)					
	3-endo-D	$v=0$	$A=5762.491$ (2)					
			$B=5077.017$ (2)					
			$C=4069.970$ (2)					
	3-exo-D	$v=0$	$A=5924.913$ (2)					
			$B=5004.566$ (2)					
			$C=4002.300$ (2)					
	4-D	$v=0$	$A=5649.614$ (4)					
			$B=5182.197$ (5)					
			$C=4067.030$ (5)					
	3 endo-, 5-endo D	$v=0$	$A=5553.196$ (11)					
			$B=4938.978$ (14)					
			$C=3968.920$ (7)					
569	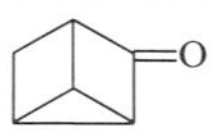	$v=0$	$A=5748.2$ (4)		75COF			2.6.3, 276
			$B=2904.950$ (8)					
			$C=2507.114$ (7)					
570	 *)	$v=0$	$A=3925.787$ (5)		75CR1	75CR1	for structure,	2.6.3, 277
			$B=3305.386$ (6)				see also	3, 161
			$C=2847.545$ (34)				Vol. II/7 (3, 695)	
		$v_a=1$	$A=3926.431$ (3)				*) C_{2v} symmetry	
			$B=3306.055$ (4)				$\nu_a=120$ (30) cm^{-1}	
			$C=2852.233$ (17)					
		$v_a=2$	$A=3927.097$ (5)					
			$B=3306.522$ (5)					
			$C=2856.354$ (26)					
		$v_a=3$	$A=3927.771$ (3)					
			$B=3306.803$ (3)					
			$C=2860.051$ (15)					
		$v_b=1$	$A=3927.235$ (3)				$\nu_b=330$ (30) cm^{-1}	
			$B=3301.755$ (3)					
Cont.			$C=2844.074$ (16)					

Nr.	Molecule	v	A, B, C MHz	D, H kHz	Ref.	Ref. struct.	Varia, remarks	See also Tab., Nr.
570 Cont.	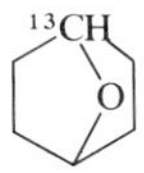	$v=0$	$A=3888.006\,(6)$ $B=3303.341\,(8)$ $C=2829.324\,(45)$		75CR1			
	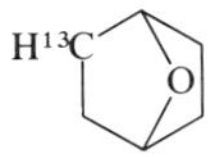	$v=0$	$A=3901.424\,(3)$ $B=3265.715\,(4)$ $C=2812.935\,(22)$					
571	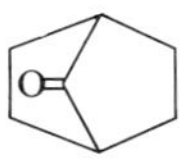	$v=0$	$A=2773.55\,(9)$ $B=2301.74\,(1)$ $C=2133.93\,(1)$		75DUR5			2.6.3, 278 3, 162
572	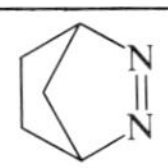	$v=0$	$A=4088.065\,(6)$ $B=3565.843\,(6)$ $C=3219.200\,(5)$		76SUE	76SUE		2.6.3, 279 3, 163
	1-^{13}C	$v=0$	$A=4045.482\,(6)$ $B=3563.353\,(6)$ $C=3194.706\,(4)$				see Fig. 163 for atom numbering	
	5-^{13}C	$v=0$	$A=4063.883\,(5)$ $B=3522.784\,(7)$ $C=3176.491\,(4)$					
	7-^{13}C	$v=0$	$A=4025.042\,(7)$ $B=3517.452\,(6)$ $C=3219.023\,(4)$					
	2-^{15}N	$v=0$	$A=4065.189\,(46)$ $B=3523.856\,(172)$ $C=3184.280\,(14)$					
	1-D	$v=0$	$A=3932.481\,(11)$ $B=3556.913\,(15)$ $C=3126.144\,(10)$					
	5-endo-D	$v=0$	$A=3981.455\,(9)$ $B=3481.121\,(13)$ $C=3156.189\,(6)$					
	5-exo-, 6-exo-D	$v=0$	$A=3996.645\,(12)$ $B=3361.061\,(11)$ $C=3000.270\,(9)$					
Cont.								

Nr.	Molecule	v	A, B, C MHz	D, H kHz	Ref.	Ref. struct.	Varia, remarks	See also Tab., Nr.
572 Cont.	7-endo-D	$v=0$	$A=3966.228(7)$ $B=3444.596(8)$ $C=3195.113(4)$		76SUE			
	7-exo-D	$v=0$	$A=3958.173(14)$ $B=3450.200(16)$ $C=3205.167(7)$					
573 *4-164*	$CH_3COOH \cdot CF_3COOH$							2.6.3, 280
574	$CF_3COOH \cdot CH_3CH_2COOH$		$B+C=636.9(30)$ *)		75BEL		*) LRMW	
575	$CF_3COOH \cdot (CH_3)_2CHCOOH$		$B+C=533.2(20)$ *)		75BEL		*) LRMW	
576	$CF_3COOH \cdot (CH_3)_3CCOOH$		$B+C=466.0(20)$ *)		75BEL		*) LRMW	2.6.3, 281
577	$CF_3COOH \cdot CH_2{=}CHCOOH$		$B+C=656.0(40)$ *)		75BEL		*) LRMW	2.6.3, 282
578	$CF_3COOH \cdot (CH_3)_2C{=}CHCOOH$		$B+C=420.5(30)$ *)		75BEL		*) LRMW	2.6.3, 283
579	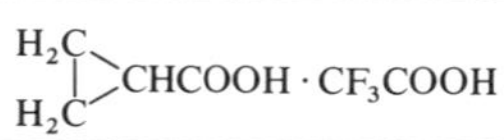		$B+C=539.6(20)$ *)		75BEL		*) LRMW	2.6.3, 284
580			$B+C=439(10)$ *)		75BEL		*) LRMW	
581	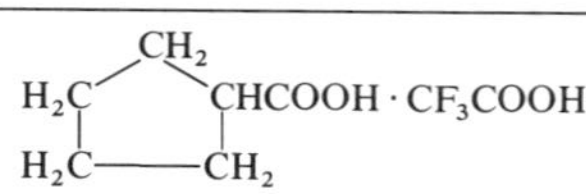		$B+C=382(13)$ *)		75BEL		*) LRMW	
582	$CF_3COOH \cdot CH_3CONH_2$		$B+C=817.9(20)$ *)		75BEL		*) LRMW	2.6.3, 284a
583	$CF_3COOH \cdot CH_3CH_2CONH_2$		$B+C=624.4(20)$ *)		75BEL		*) LRMW	2.6.3, 285
584	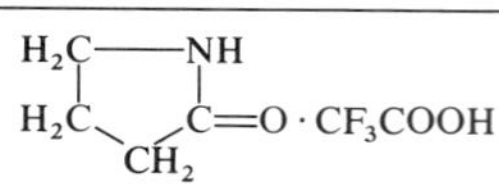		$B+C=579.1(20)$ *)		75BEL		*) LRMW	2.6.3, 286

Nr.	Molecule	v	A, B, C MHz	D, H kHz	Ref.	Ref. struct.	Varia, remarks	See also Tab., Nr.
	2.5.2.7 Unsaturated cyclic and aromatic compounds							
585 *6-233* *4-166*	HC=CH, CH_2 (ring)	$v=0$	$A=30061.00(13)$ $B=21825.84(6)$ $C=13795.83(7)$	$\tau_{aaaa}=-269.2(826)$ $\tau_{bbbb}=-220.8(374)$ $\tau_{cccc}=-66.8(713)$ $\tau_1=-708.6(962)$ $\tau_2=-191.0(318)$ $\tau_3=4192.5(2924)$	75STI	75STI	for structure, see also Vol. II/7 (3, 369)	
	HC=CH, $^{13}CH_3$ (ring)	$v=0$	$A=30062.00(31)$ $B=21129.22(22)$ $C=13513.62(22)$					
	$H^{13}C$=CH, CH_2 (ring)	$v=0$	$A=29369.26(24)$ $B=21601.94(18)$ $C=13560.06(18)$					
	DC=CD, CH_2 (ring)	$v=0$	$A=23178.01(27)$ $B=20102.21(14)$ $C=11585.57(14)$					
	HC=CH, CHD (ring)	$v=0$	$A=28792.96(21)$ $B=19356.66(19)$ $C=13011.74(19)$					
586	HC=CH, CF_2 (ring)	$v=0$	$A=8379.008(6)$ $B=5786.382(5)$ $C=4329.415(5)$	$\tau_{aaaa}=-23.5(15)$ $\tau_{bbbb}=-9.653(876)$ $\tau_{cccc}=-4.40(90)$ $\tau_1=-33.92(264)$ $\tau_2=-10.38(88)$ $\tau_3=146.672(15049)$	76RAM	76RAM		2.6.3, 287
	$H^{13}C$=CH, CF_2 (ring)	$v=0$	$A=8320.89(2)$ $B=5659.875(9)$ $C=4273.371(8)$					
	DC=CD, CF_2 (ring)	$v=0$	$A=7737.97(2)$ $B=5119.340(12)$ $C=4105.390(12)$	$\tau_{aaaa}=-41.26(777)$ $\tau_{bbbb}=-8.57(146)$ $\tau_{cccc}=-6.081(1468)$ $\tau_1=-34.18(442)$ $\tau_2=-10.94(145)$ $\tau_3=91.181(28000)$				
Cont.								

Nr.	Molecule	v	A, B, C MHz	D, H kHz	Ref.	Ref. struct.	Varia, remarks	See also Tab., Nr.
586 Cont.	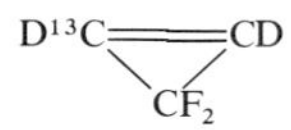	$v=0$	$A=7688.38(5)$ $B=5025.643(7)$ $C=4058.637(6)$		76RAM			
587		$v=0$	$A=6872.96(21)$ *) $B=5353.25(3)$ $C=3846.52(3)$		78WIB	78WIB	*) uncertainties are 3σ	2.6.3, 288
588	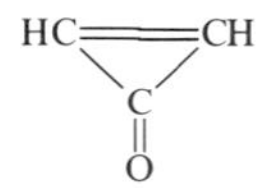	$v=0$	$A=32046.00(50)$ $B=\ 7824.95(6)$ $C=\ 6280.72(6)$		73BEN1	73BEN1		2.6.3, 289 2.9.3, 36
	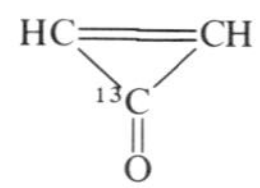	$v=0$	$A=32049.3(50)$ $B=\ 7816.52(5)$ $C=\ 6275.25(5)$					
	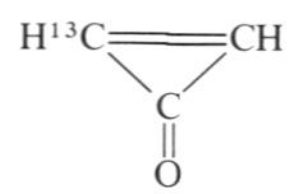	$v=0$	$A=31235.4(50)$ $B=\ 7709.36(5)$ $C=\ 6172.72(5)$					
	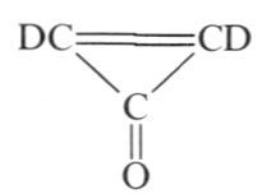	$v=0$	$A=24538.85(90)$ $B=\ 7254.85(3)$ $C=\ 5591.75(3)$					
	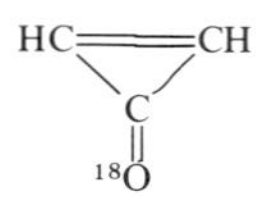	$v=0$	$A=32046.82(60)$ $B=\ 7344.82(4)$ $C=\ 5967.51(4)$					
589	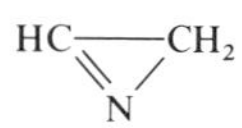	$v=0$	$A=35615.43$ $B=22223.72$ $C=15064.81$		77FOR1			2.6.3, 290 2.7.3, 54
590 *4-167*	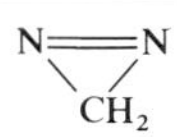							2.7.3, 55
591	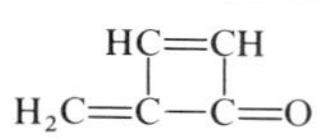	$v=0$	$A=5775.664(9)$ $B=4312.314(7)$ $C=2467.814(8)$		74CZ1	74CZ1		2.6.3, 291 2.9.3, 38 3, 164

Nr.	Molecule	v	A, B, C MHz	D, H kHz	Ref.	Ref. struct.	Varia, remarks	See also Tab., Nr.
592 *6-240*	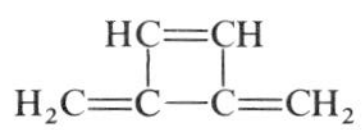							2.9.3, 37
593	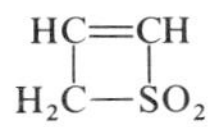	$v=0$	$A=5463.98\,(59)$		73RAL2	73RAL2	for structure, see also Vol. II/7 (3, 665)	
			$B=3052.679\,(6)$					
			$C=2894.365\,(6)$					
		$v_x=1$	$A=5465.6$				$\nu_x=184\ \mathrm{cm}^{-1}$	
			$B=3051.76$					
			$C=2894.24$					
		$v_x=2$	$A=5457.7$					
			$B=3050.67$					
			$C=2893.54$					
		$v_y=1$	$A=5460.4$				$\nu_y=303\ \mathrm{cm}^{-1}$	
			$B=3055.42$					
			$C=2891.35$					
		$v_z=1$	$A=5465.0$				$\nu_z=392\ \mathrm{cm}^{-1}$	
			$B=3021.02$					
			$C=2891.98$					
593a	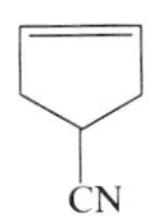 equatorial	$v=0$	$A=6762.657\,(26)$	$\tau_{aaaa}=5.4\,(43)$	81CHO	81CHO		2.6.3, 291a
			$B=1916.722\,(2)$	$\tau_{bbbb}=-0.78\,(40)$				
			$C=1590.595\,(2)$	$\tau_{cccc}=-0.76\,(40)$				
				$\tau'_{bbcc}=-0.76\,(60)$				
				$\tau'_{aabb}+\tau'_{aacc}=-1.9\,(1)$				
		$v_\alpha=1$	$\Delta A=-97.95\,(139)$ *)				$\nu_\alpha=95\,(20)\ \mathrm{cm}^{-1}$	
			$\Delta B=7.27\,(2)$				*) change from ground state values, uncertainties are 3σ	
			$\Delta C=7.88\,(2)$					
		$v_\alpha=2$	$\Delta A=-198.68\,(105)$ *)					
			$\Delta B=17.96\,(2)$					
			$\Delta C=19.68\,(2)$					
		$v_\beta=1$	$\Delta A=30.93\,(128)$ *)				$\nu_\beta=160\,(30)\ \mathrm{cm}^{-1}$	
			$\Delta B=2.74\,(3)$					
			$\Delta C=0.27\,(2)$					
		$v_\beta=2$	$\Delta A=67.27\,(277)$ *)					
			$\Delta B=5.48\,(8)$					
Cont.			$\Delta C=0.56\,(8)$					

Nr.	Molecule	v	A, B, C MHz	D, H kHz	Ref.	Ref. struct.	Varia, remarks	See also Tab., Nr.
593a Cont.	 equatorial	$v_\beta=3$	$\Delta A=$ 104.94 (623) *) $\Delta B=$ 8.29 (25) $\Delta C=$ 0.86 (14)		81CHO		*) change from ground state values, uncertainties are 3 σ	
	$2-{}^{13}C$	$v=0$	$A=6614.76$ (1513) $B=1914.64$ (32) $C=1582.27$ (34)				the double bond is C_3C_4	
	$3-{}^{13}C$	$v=0$	$A=6738.03$ (1291) $B=1891.54$ (25) $C=1570.65$ (26)					
	$6-{}^{13}C$ $^{+})$	$v=0$	$A=6758.66$ (2215) $B=1895.83$ (42) $C=1576.47$ (45)				$^{+})$ cyano-C	
	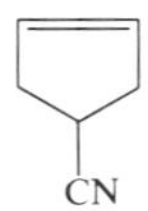 axial	$v=0$	$A=5095.774$ (27) $B=2185.811$ (14) $C=1936.496$ (11)	$\tau_{aaaa}=-40$ (2) $\tau_{bbbb}=-5.3$ (4) $\tau_{cccc}=-7.0$ (3) $\tau'_{bbcc}=-4.4$ (5) $\tau'_{aabb}+\tau'_{aacc}=1.7$ (4)			$E_{ax}-E_{eq}=400$ (60) cal/mol	
		$v_\gamma=1$	$\Delta A=$ 9.71 (835) *) $\Delta B=$ 2.44 (39) $\Delta C=-0.25$ (51)				$v_\gamma=230$ (40) cm^{-1}	
594 *6-242a*	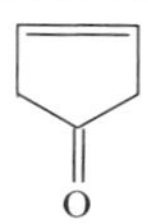	$v=0$	$A=7378.33$ (1) $B=3615.209$ (1) $C=2503.587$ (1)		73BEV1	73BEV1	for structure, see also Vol. II/7 (3, 592)	2.6.3, 292
		$v_B=1$	$A=7352.03$ (1) $B=3620.805$ (1) $C=2511.911$ (1)				$v_B=82.7\ cm^{-1}$, ring-bending vibration potential function determined [73BEV2]	
		$v_B=2$	$A=7326.42$ (1) $B=3626.199$ (1) $C=2519.813$ (1)					
Cont.		$v_B=3$	$A=7301.33$ (1) $B=3631.476$ (1) $C=2527.432$ (1)					

Nr.	Molecule	v	A, B, C MHz	D, H kHz	Ref.	Ref. struct.	Varia, remarks	See also Tab., Nr.
594 *6–242 a* Cont.	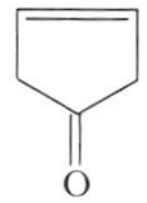	$v_B=4$	$A=7276.69(1)$ $B=3636.685(1)$ $C=2534.846(1)$		73BEV1			
		$v_B=5$	$A=7252.39(1)$ $B=3641.845(1)$ $C=2542.103(1)$					
		$v_B=6$	$A=7229.54(20)$ $B=3646.553(5)$ $C=2548.719(5)$					
		$v_B=7$	$A=7204.72(20)$ $B=3652.055(5)$ $C=2556.249(5)$					
		$v_T=1$	$A=7379.54(20)$ $B=3612.222(5)$ $C=2504.172(5)$				$v_T=379.0\ cm^{-1}$, ring twisting mode	
	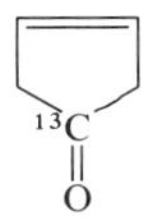	$v=0$	$A=7378.68(20)$ $B=3598.79(1)$ $C=2495.77(1)$					
	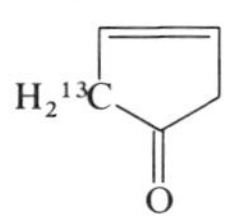	$v=0$	$A=7217.16(20)$ $B=3615.07(1)$ $C=2484.67(1)$					
	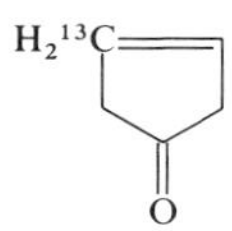	$v=0$	$A=7332.10(20)$ $B=3559.75(1)$ $C=2471.66(1)$					
	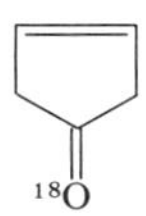	$v=0$	$A=7378.18(1)$ $B=3421.510(1)$ $C=2409.136(1)$					
Cont.	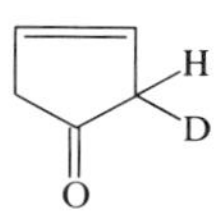	$v=0$	$A=6951.27(1)$ $B=3596.396(1)$ $C=2460.956(1)$					

Nr.	Molecule	v	A, B, C MHz	D, H kHz	Ref.	Ref. struct.	Varia, remarks	See also Tab., Nr.
594 *6–242 a* Cont.	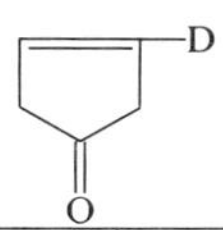	$v=0$	$A=7217.99(20)$ $B=3473.83(1)$ $C=2417.32(1)$		73BEV1			
595 *6–243*	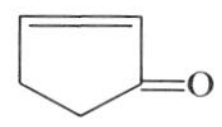	$v=0$	$A=7410.0(1)$ $B=3586.366(3)$ $C=2492.588(3)$		79CHA	79CHA		2.6.3, 293
		$v_B=1$	$A=7392.6(3)$ $B=3590.206(4)$ $C=2498.936(4)$				$\nu_B=94.7\ cm^{-1}$, ring-bending vibration	
		$v_B=2$	$A=7375.3(1)$ $B=3593.992(8)$ $C=2504.810(8)$					
		$v_B=3$	$A=7359.0(3)$ $B=3597.37(1)$ $C=2509.75(1)$					
		$v_B=4$	$A=7340.8(4)$ $B=3600.99(1)$ $C=2515.28(1)$					
		$v_B=5$	$A=7320.9(6)$ $B=3604.59(2)$ $C=2520.56(2)$					
		$v_T=1$	$A=7390.3(7)$ $B=3589.70(2)$ $C=2496.29(2)$				ν_T, ring-twisting mode	
	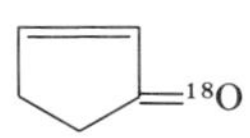	$v=0$	$A=7410.0(4)$ $B=3392.70(1)$ $C=2397.46(1)$					
596 Cont.	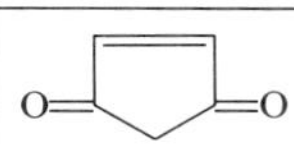	$v=0$	$A=6668.970(3)$ $B=2202.614(2)$ $C=1673.893(2)$	$\Delta_J=0.098(9)$ $\Delta_{JK}=0.152(13)$ $\Delta_K=2.304(156)$ $\delta_J=0.0247(6)$ $\delta_K=0.310(17)$	78MJÖ	78MJÖ		2.6.3, 294

Nr.	Molecule	v	A, B, C MHz	D, H kHz	Ref.	Ref. struct.	Varia, remarks	See also Tab., Nr.
596 Cont.	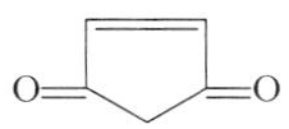	$v_B=1$	$A=6634.868(4)$ $B=2206.204(3)$ $C=1677.982(3)$	$\Delta_J=0.095(19)$ $\Delta_{JK}=0.166(18)$ $\Delta_K=2.140(173)$ $\delta_J=0.0270(7)$ $\delta_K=0.235(19)$	78MJÖ		$\nu_B \simeq 90\ \mathrm{cm}^{-1}$, ring-bending vibration	
		$v_B=2$	$A=6601.753(6)$ $B=2209.754(5)$ $C=1682.000(5)$	$\Delta_J=0.144(25)$ $\Delta_{JK}=0.082(36)$ $\Delta_K=2.802(299)$ $\delta_J=0.0260(12)$ $\delta_K=0.261(39)$				
		$v_B=3$	$A=6569.470(13)$ $B=2213.248(11)$ $C=1685.932(11)$	$\Delta_J=0.020(71)$ $\Delta_{JK}=0.154(50)$ $\Delta_K=1.537(433)$ $\delta_J=0.0296(17)$ $\delta_K=0.173(52)$				
		$v_B=4$	$A=6538.082(25)$ $B=2216.740(5)$ $C=1689.853(4)$	$\Delta_J=0.098$ $\Delta_{JK}=0.152$ $\Delta_K=1.830(539)$ $\delta_J=0.0247$ $\delta_K=0.140(69)$				
	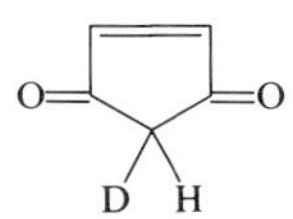	$v=0$	$A=6354.794(7)$ $B=2195.611(4)$ $C=1657.454(4)$	$\Delta_J=0.070(21)$ $\Delta_{JK}=0.167(23)$ $\Delta_K=2.331(217)$ $\delta_J=0.0265(12)$ $\delta_K=0.236(31)$				
597	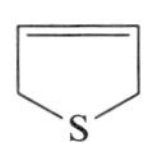	$v=0$	$A=6621.59(12)$ $B=4798.42(3)$ $C=2885.27(1)$		81DUR2	81DUR2	for structure, see also Vol. II (3, 491) $\nu_R=87\ \mathrm{cm}^{-1}$, ring-puckering vibration	2.6.3, 295
		$v_R=1$	$A=6628.19(7)$ $B=4796.56(2)$ $C=2895.74(1)$					
Cont.		$v_R=2$	$A=6633.35(6)$ $B=4794.19(1)$ $C=2904.35(1)$					

Nr.	Molecule	v	A, B, C MHz	D, H kHz	Ref.	Ref. struct.	Varia, remarks	See also Tab., Nr.
597 Cont.	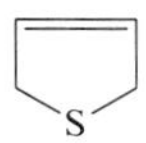	$v_R=3$	$A=6637.63(7)$ $B=4791.51(2)$ $C=2912.02(1)$		81DUR2			
		$v_R=4$	$A=6641.26(5)$ $B=4788.62(1)$ $C=2919.07(1)$					
		$v_R=5$	$A=6650.57(15)$ $B=4786.44(3)$ $C=2924.69(1)$					
597a	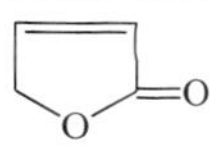	$v=0$	$A=8221.354(5)$ $B=3883.734(4)$ $C=2682.493(4)$	$\Delta_J=0.226(67)$ $\Delta_{JK}=1.232(86)$ $\Delta_K=0.92(33)$ $\delta_J=0.091(4)$ $\delta_K=0.713(50)$	81ALO3		heavy atoms coplanar	2.6.3, 295a
		$v_B=1$	$A=8190.193(5)$ $B=3887.283(4)$ $C=2686.706(4)$	$\Delta_J=0.263(72)$ $\Delta_{JK}=1.183(93)$ $\Delta_K=1.18(35)$ $\delta_J=0.090(4)$ $\delta_K=0.774(62)$			$\nu_B=199(7)$ cm^{-1}, ring-bending mode	
		$v_B=2$	$A=8159.670(6)$ $B=3890.869(4)$ $C=2690.938(4)$	$\Delta_J=0.396(85)$ $\Delta_{JK}=1.24(14)$ $\Delta_K=0.71(51)$ $\delta_J=0.096(7)$ $\delta_K=0.630(110)$				
		$v_B=3$	$A=8129.802(19)$ $B=3894.484(6)$ $C=2695.173(6)$	$\Delta_J=0.450(99)$ $\Delta_{JK}=0.94(47)$ $\Delta_K=2.1(17)$ $\delta_J=0.088(20)$ $\delta_K=0.78(28)$				
		$v_T=1$	$A=8205.334(25)$ $B=3884.575(8)$ $C=2684.623(6)$	$\Delta_J=0.29(14)$ $\Delta_{JK}=0.89(76)$ $\Delta_K=2.2(27)$ $\delta_J=0.076(30)$ $\delta_K=0.95(41)$			$\nu_T=349(12)$ cm^{-1}, ring-twisting mode	

Nr.	Molecule	v	A, B, C MHz	D, H kHz	Ref.	Ref. struct.	Varia, remarks	See also Tab., Nr.
598 *6-245*	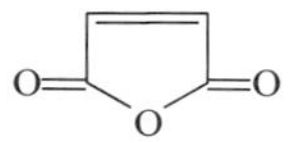	$v=0$	$A=6842.486\,(13)$ *) $B=2467.614\,(8)$ $C=1813.627\,(8)$	$\tau_{aaaa}=-11.25\,(42)$ *) $\tau_{bbbb}=-0.78\,(5)$ $\tau_{abab}=-1.22\,(17)$ $\tau_{aabb}=0.710\,(35)$	77STI	77STI	for structure, see also Vol. II/7 (3, 475) *) uncertainties are 3σ	2.9.3, 39
	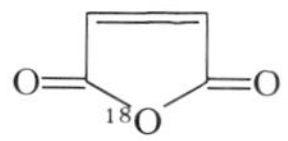	$v=0$	$A=6688.643\,(24)$ *) $B=2467.808\,(8)$ $C=1802.755\,(5)$	$\tau_{aaaa}=-12.2\,(36)$ *) $\tau_{bbbb}=-0.75\,(15)$ $\tau_{abab}=-0.78\,(30)$ **)			**) with $\tau_{aabb}=0.710$ kHz held fixed	
	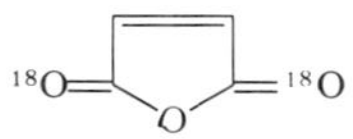	$v=0$	$A=6790.337\,(44)$ *) $B=2354.397\,(11)$ $C=1748.276\,(8)$	$\tau_{aaaa}=-9.3\,(69)$ *) $\tau_{bbbb}=-0.75\,(13)$ $\tau_{abab}=-1.24\,(45)$ **)				
	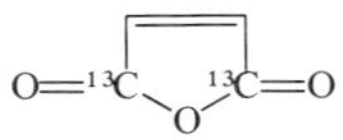	$v=0$	$A=6841.501\,(33)$ *) $B=2452.607\,(9)$ $C=1805.442\,(6)$	$\tau_{aaaa}=-10.8\,(48)$ *) $\tau_{bbbb}=-0.71\,(13)$ $\tau_{abab}=-1.14\,(39)$ **)				
	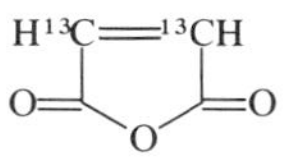	$v=0$	$A=6691.282\,(23)$ *) $B=2462.240\,(6)$ $C=1799.954\,(5)$	$\tau_{aaaa}=-11.5\,(33)$ *) $\tau_{bbbb}=-0.80\,(9)$ $\tau_{abab}=-1.15\,(27)$ **)				
	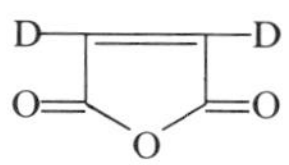	$v=0$	$A=6106.159\,(7)$ *) $B=2423.551\,(3)$ $C=1734.998\,(3)$	$\tau_{aaaa}=-8.69\,(12)$ *) $\tau_{bbbb}=-0.71\,(2)$ $\tau_{abab}=-1.09\,(6)$ $\tau_{aabb}=0.61\,(11)$				
	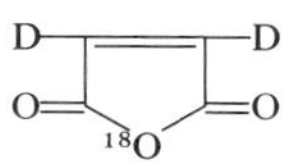	$v=0$	$A=5971.996\,(16)$ *) $B=2423.742\,(6)$ $C=1724.104\,(4)$	$\tau_{aaaa}=-7.6\,(13)$ *) $\tau_{bbbb}=-0.69\,(7)$ $\tau_{abab}=-1.04\,(16)$ ***)			***) with $\tau_{aabb}=0.61$ kHz held fixed	
	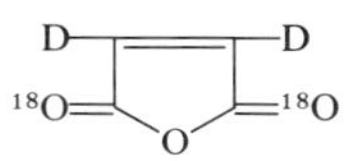	$v=0$	$A=6058.195\,(14)$ *) $B=2314.121\,(5)$ $C=1674.540\,(3)$	$\tau_{aaaa}=-8.9\,(14)$ *) $\tau_{bbbb}=-0.66\,(5)$ $\tau_{abab}=-1.01\,(12)$ ***)				
	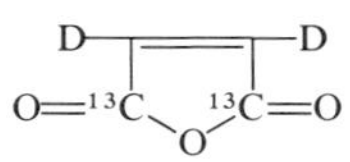	$v=0$	$A=6104.567\,(25)$ *) $B=2409.078\,(8)$ $C=1727.440\,(5)$	$\tau_{aaaa}=-10.1\,(30)$ *) $\tau_{bbbb}=-0.77\,(9)$ $\tau_{abab}=-0.96\,(27)$ ***)				
	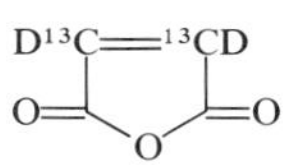	$v=0$	$A=5993.203\,(28)$ *) $B=2418.373\,(7)$ $C=1723.126\,(5)$	$\tau_{aaaa}=-8.3\,(30)$ *) $\tau_{bbbb}=-0.72\,(11)$ $\tau_{cccc}=-0.98\,(27)$ ***)				
599 *6-246* *4-177*	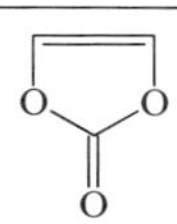						for structure, see also Vol. II/7 (3, 353)	2.9.3, 40

Nr.	Molecule	v	A, B, C MHz	D, H kHz	Ref.	Ref. struct.	Varia, remarks	See also Tab., Nr.
600	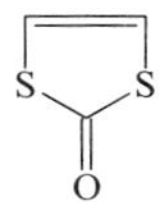	$v=0$	$A=3279.666(14)$ *) $B=3202.398(10)$ $C=1619.978(6)$	$\tau_{aaaa}=-5.18(9)$ *) $\tau_{bbbb}=-2.92(5)$ $\tau_{abab}=0.28(5)$	80KUM		*) uncertainties are 2σ force field calculated	2.6.3, 296
601	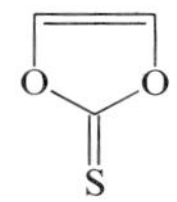	$v=0$	$A=9453.22(15)$ *) $B=2313.040(5)$ $C=1857.950(3)$	$\tau_{aaaa}=-33.6(170)$ *) $\tau_{bbbb}=-0.78(11)$ $\tau_{abab}=-1.98(40)$	80IQB		*) uncertainties are 2σ	2.6.3, 297
		$v_{12}=1$	$A=9388.23(200)$ *) $B=2315.27(2)$ $C=1860.82(2)$				$\nu_{12}=221\ cm^{-1}$, ring-bending "envelope" mode	
		$v_{12}=2$	$A=9324.18(200)$ *) $B=2317.52(2)$ $C=1863.72(2)$				force field calculated	
		$v_{12}=3$	$A=9624.96(360)$ *) $B=2319.74(4)$ $C=1866.59(4)$					
602	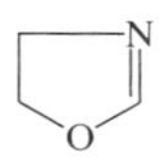	$v=0$	$A=8473.39(4)$ $B=8122.16(4)$ $C=4384.12(2)$		74DUR12			2.6.3, 298 2.7.3, 56
		$v_R=1$	$A=8476.83(17)$ $B=8134.68(17)$ $C=4406.47(8)$				$\nu_R=79(8)\ cm^{-1}$, ring-puckering vibration potential function determined	
		$v_R=2$	$A=8476.49(2)$ $B=8139.55(2)$ $C=4418.83(2)$					
		$v_R=3$	$A=8475.75(8)$ $B=8143.48(7)$ $C=4430.72(7)$					
		$v_R=4$	$A=8474.47(3)$ $B=8146.05(2)$ $C=4441.22(2)$					
		$v_R=5$	$A=8472.57(2)$ $B=8147.73(2)$ $C=4450.88(2)$					

Nr.	Molecule	v	A, B, C MHz	D, H kHz	Ref.	Ref. struct.	Varia, remarks	See also Tab., Nr.
603		$v=0$	$A=7993.8121\,(75)$ $B=7620.5743\,(75)$ $C=4249.7905\,(75)$	$\Delta_J = 1.718\,(300)$ $\Delta_{JK} = 10.002\,(284)$ $\Delta_K = -9.339\,(208)$ $\delta_J = 0.513\,(48)$ $\delta_K = -5.139\,(106)$ $\Phi_J = 0.228\,(19) \cdot 10^{-3}$ $\Phi_{JK} = 1.104\,(174) \cdot 10^{-3}$	80HAL	80HAL	the $v_R=0$ and 1 states have also been analyzed using a coupled Hamiltonian: $E_1 - E_0 = 231.6361\,(102)$ GHz, $\mu_{ac} = 35.853\,(34)$ MHz, $\mu' = -0.241\,(14)$ kHz with $H_{01} = h(\mu_{ac} + \mu' \hat{J}^2) \cdot (J_a J_c + J_c J_a)$ representation I^r used v_R, ring-puckering vibration potential function determined	2.7.3, 57
		$v_R=1$	$A=8000.8397\,(106)$ $B=7623.3602\,(108)$ $C=4262.0743\,(103)$	$\Delta_J = 2.193\,(420)$ $\Delta_{JK} = -9.710\,(358)$ $\Delta_K = 11.524\,(271)$ $\delta_J = 0.993\,(57)$ $\delta_K = 4.920\,(137)$ $\Phi_J = -0.079\,(19) \cdot 10^{-3}$ $\Phi_{JK} = -0.142\,(55) \cdot 10^{-3}$				
		$v_R=2$	$A=7984.6906\,(48)$ $B=7614.8875\,(47)$ $C=4238.1841\,(48)$	$\Delta_J = 1.587\,(202)$ $\Delta_{JK} = -0.450\,(122)$ $\Delta_K = 1.433\,(86)$ $\delta_J = 0.712\,(19)$ $\delta_K = 0.164\,(45)$ $\Phi_{JK} = 0.00065\,(16) \cdot 10^{-3}$				
		$v_R=3$	$A=7998.4193\,(77)$ $B=7619.4257\,(76)$ $C=4264.2926\,(75)$	$\Delta_J = 1.855\,(309)$ $\Delta_{JK} = -1.671\,(177)$ $\Delta_K = 2.541\,(125)$ $\delta_J = 0.619\,(27)$ $\delta_K = 0.708\,(67)$ $\Phi_{JK} = -0.0069\,(3) \cdot 10^{-3}$				
		$v_R=4$	$A=8000.6682\,(51)$ $B=7618.7441\,(50)$ $C=4271.9204\,(49)$	$\Delta_J = 2.249\,(200)$ $\Delta_{JK} = -1.645\,(85)$ $\Delta_K = 2.464\,(61)$ $\delta_J = 0.599\,(13)$ $\delta_K = 0.675\,(32)$ $\Phi_{JK} = -0.0049\,(6) \cdot 10^{-3}$				
Cont.								

Nr.	Molecule	v	A, B, C MHz	D, H kHz	Ref.	Ref. struct.	Varia, remarks	See also Tab., Nr.
603 Cont.	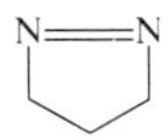	$v_R=5$	$A=8004.2515(81)$ $B=7618.5163(80)$ $C=4282.3520(75)$	$\Delta_J = 2.341(291)$ $\Delta_{JK} = -1.346(183)$ $\Delta_K = 2.261(129)$ $\delta_J = 0.641(29)$ $\delta_K = 0.569(70)$ $\Phi_{JK} = -0.0059(26) \cdot 10^{-3}$	80HAL			
604		$v=0$	$A=6378.64(7)$ $B=4524.72(2)$ $C=2828.27(1)$		80DUR3	80DUR3		2.6.3, 299
		$v_R=1$	$A=6373.73(17)$ $B=4520.96(2)$ $C=2826.84(1)$				$\nu_R=160(15)\ cm^{-1}$, ring puckering vibration	
		$v_R=2$	$A=6367.83(4)$ $B=4517.97(5)$ $C=2824.53(0)$					
		$v=0$	$A=6217.59(8)$ $B=4385.57(2)$ $C=2806.17(1)$					
605 *4-169*		$v=0$	$A=8129.894(6)$ $B=7859.476(5)$ $C=4145.094(4)$		76DAM	76DAM	for structure, see also Vol. II/7 (3, 590)	2.6.3, 300
	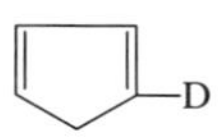	$v=0$	$A=8414.008(6)$ $B=7591.871(6)$ $C=4091.147(3)$					
	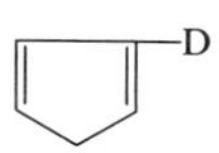	$v=0$	$A=8306.997(6)$ $B=7678.059(4)$ $C=4090.256(3)$					
	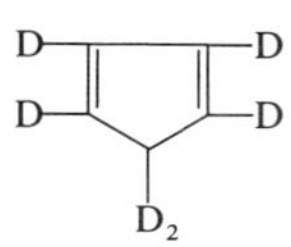	$v=0$	$A=6608.380(9)$ $B=6607.320(5)$ $C=3444.119(2)$					
	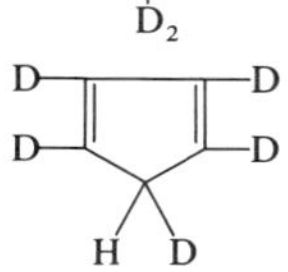	$v=0$	$A=7707.857(5)$ $B=6681.870(5)$ $C=3529.283(4)$					

Nr.	Molecule	v	A, B, C MHz	D, H kHz	Ref.	Ref. struct.	Varia, remarks	See also Tab., Nr.
606		$v=0$	$A=8356(5)$ $B=1904.24(2)$ $C=1565.36(2)$		78FOR			2.6.3, 301
607	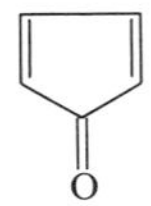	$v=0$	$A=8152.104(65)$ $B=3939.526(13)$ $C=2655.786(12)$		73BAR			2.6.3, 302
		$v_x=1$	$A=8113.391(53)$ $B=3946.508(9)$ $C=2661.374(10)$					
608 *6-248*		$v=0$	$A=8084.044(33)$ $B=3607.605(10)$ $C=2494.354(10)$		73SUE2	73SUE2		2.9.3, 41
609	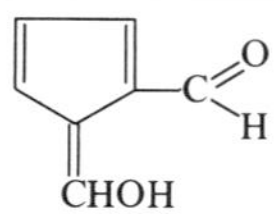	$v=0$	$A=2816.401(20)$ **) $B=1613.434(3)$ $C=1025.852(3)$		73PIC1	73PIC1	C_{2v} symmetry with the proton centered in the hydrogen bond **) uncertainties are 95% confidence limits	
		$v_x=1$	$A=2812.340(60)$ $B=1614.719(7)$ $C=1026.681(4)$					
	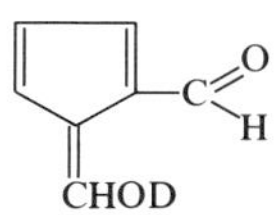	$v=0$	$A=2786.574(100)$ **) $B=1591.607(10)$ $C=1013.075(3)$					
610 *6-249* *4-170*	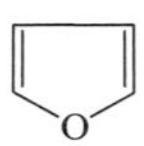	$v=0$	$A=9447.1210(10)$ $B=9246.7419(10)$ $C=4670.8234(10)$	$D_1=-10.864(6)$ *) $D_2=1.361(3)$ $D_3=-0.8184(10)$	78MAT	78MAT	for structure, see also Vol. II/7 (3, 488) *) with $D_4=-1.5$ kHz held fixed $D_1=\tau_{abab}+2\tau_{cccc}$ $D_2=\tau_{aaaa}+\tau_{bbbb}-2\tau_{abab}-8\tau_{cccc}$ $D_3=\tau_{aaaa}-\tau_{bbbb}$ $D_4=\tau_{cccc}$	2.9.3, 42
	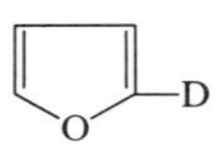	$v=0$	$A=9280.3433(14)$ $B=8638.6241(14)$ $C=4472.1266(13)$	$D_1=-9.322(6)$ *) $D_2=-0.597(6)$ $D_3=-2.144(5)$				
	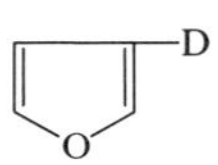	$v=0$	$A=9383.709(2)$ $B=8490.524(2)$ $C=4455.483(2)$	$D_1=-9.368(16)$ *) $D_2=-0.64(2)$ $D_3=-2.988(18)$				
Cont.	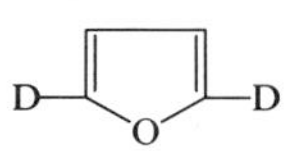	$v=0$	$A=9033.6024(7)$ $B=8160.7819(6)$ $C=4285.8610(6)$	$D_1=-8.652(6)$ *) $D_2=-0.682(9)$ $D_3=-2.322(11)$				

Nr.	Molecule	v	A, B, C MHz	D, H kHz	Ref.	Ref. struct.	Varia, remarks	See also Tab., Nr.
610 *6–249* *4–170* Cont.	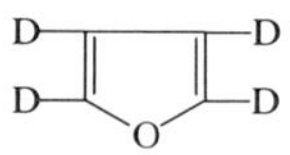	$v=0$	$A=8056.5022\,(9)$ $B=7688.0556\,(9)$ $C=3932.5863\,(9)$	$D_1=-7.022\,(7)$ *) $D_2=\ \ 0.288\,(4)$ $D_3=-0.750\,(6)$	78MAT		*) see preceding page	
611 *6–251*	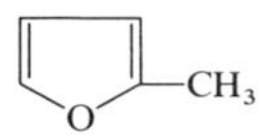						for structure, see also Vol. II/7 (3, 593)	2.9.3, 43
612 *6–251 a*	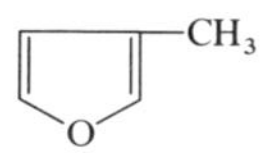						for structure, see also Vol. II/7 (3, 594)	2.9.3, 44
612a	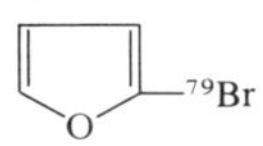 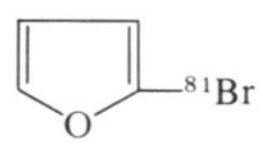	$v=0$ $v=0$	$A=9231.722\,(12)$ $B=1345.671\,(1)$ $C=1174.243\,(1)$ $A=9231.755\,(11)$ $B=1332.366\,(1)$ $C=1164.095\,(1)$		81STO			2.7.3, 155a
612b	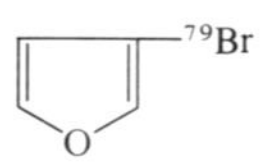 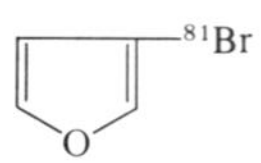	$v=0$ $v=0$	$A=9297.815\,(22)$ $B=1268.089\,(1)$ $C=1115.641\,(1)$ $A=9297.792\,(20)$ $B=1255.437\,(1)$ $C=1105.837\,(1)$		81STO			2.7.3, 155b
613	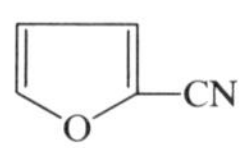	$v=0$	$A=9220.106\,(20)$ $B=2029.262\,(11)$ $C=1662.640\,(9)$		76EN, 76KOJ	76EN, 76KOJ		2.6.3, 303 2.7.3, 58
613a Cont.	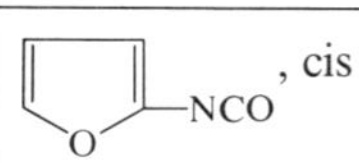	$v=0$ $v_i=1$ $v_i=2$	$A=7787.5\,(6)$ $B=1284.48\,(1)$ $C=1102.65\,(1)$ $A=7648.3\,(6)$ $B=1283.01\,(1)$ $C=1104.09\,(1)$ $A=7516.1\,(6)$ $B=1281.65\,(1)$ $C=1105.58\,(1)$		81MAS	81MAS	 v_i, C—N torsion	

Nr.	Molecule	v	A, B, C MHz	D, H kHz	Ref.	Ref. struct.	Varia, remarks	See also Tab., Nr.
613a Cont.	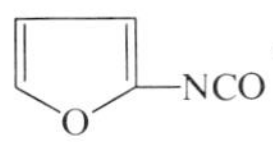 , cis	$v_j=1$	$A=7959.8\,(5)$ $B=1285.51\,(1)$ $C=1102.22\,(1)$		81MAS		v_j, CNC bending	
		$v_j=2$	$A=8139.5\,(9)$ $B=1286.33\,(2)$ $C=1101.65\,(2)$					
		$v_x=1$	$A=7808.0\,(5)$ $B=1283.51\,(1)$ $C=1103.66\,(1)$					
	trans form	$v=0$	$A=8242.4\,(8)$ $B=1213.14\,(2)$ $C=1057.61\,(1)$				$E_{trans}-E_{cis}\simeq 3$ kcal/mol	
		$v_i=1$	$A=8097.5\,(5)$ $B=1213.29\,(1)$ $C=1059.90\,(1)$					
		$v_i=2$	$A=7962.4\,(17)$ $B=1213.70\,(3)$ $C=1062.15\,(2)$					
		$v_j=1$	$A=8414.5\,(6)$ $B=1214.23\,(2)$ $C=1057.71\,(1)$					
614	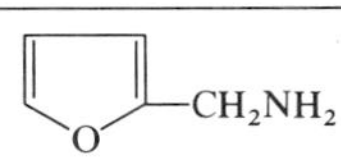	$v=0$	$A=6830.66\,(18)$ $B=1943.9753\,(25)$ $C=1653.8378\,(23)$	$D_4=1.783\,(58)$ $D_5=-0.2925\,(81)$	80PED		for definition of D-constants, see Nr. 610	
615	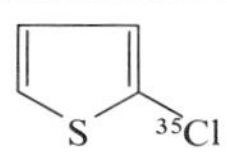	$v=0$	$A=5396.039\,(25)$ $B=1875.224\,(2)$ $C=1391.162\,(2)$	$\Delta_J=79.5\,(30)\cdot10^{-3}$ $\Delta_{JK}=0.843\,(14)$ $\Delta_K=5.324\,(2894)$ $\delta_J=10.5\,(22)\cdot10^{-3}$ $\delta_K=0.496\,(127)$	73MJÖ	73MJÖ	for structure, see also Vol. II/7 (3, 477)	2.7.3, 137
Cont.		$v_x=1$	$A=5346.45\,(153)$ $B=1876.33\,(4)$ $C=1392.65\,(3)$					

Nr.	Molecule	v	A, B, C MHz	D, H kHz	Ref.	Ref. struct.	Varia, remarks	See also Tab., Nr.
615 Cont.	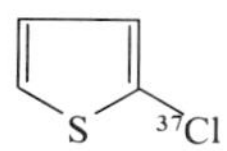	$v=0$	$A=5395.819(37)$ $B=1818.821(2)$ $C=1359.835(2)$	$\Delta_J = 80.1(49)\cdot10^{-3}$ $\Delta_{JK} = 0.794(19)$ $\Delta_K = -7.924(3918)$ $\delta_J = 40.0(31)\cdot10^{-3}$ $\delta_K = 0.167(157)$	73MJÖ			
		$v_x=1$	$A=5343.75(534)$ $B=1819.92(16)$ $C=1361.33(17)$					
616	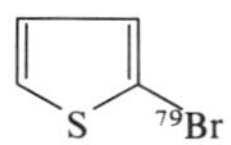	$v=0$	$A=5403.432(111)$ $B=1139.0689(10)$ $C=940.5142(18)$	$\Delta_J = 52.0(51)\cdot10^{-3}$ $\Delta_{JK} = 0.538(28)$ $\Delta_K = -1.305(13004)$ $\delta_J = 3.9(26)\cdot10^{-3}$ $\delta_K = 0.865(428)$	75MJ1	75MJ1	for structure, see also Vol. II/7 (3, 476)	2.7.3, 156
		$v_x=1$	$A=5330.55(64)$ $B=1139.961(6)$ $C=941.578(6)$				$v_x=180\ cm^{-1}$, probably out-of-plane C—Br bending	
	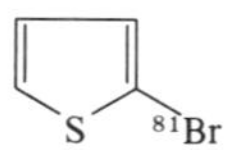	$v=0$	$A=5403.563(94)$ $B=1126.5173(11)$ $C=931.9315(9)$	$\Delta_J = 47.9(20)\cdot10^{-3}$ $\Delta_{JK} = 0.434(52)$ $\Delta_K = 23.833(15257)$ $\delta_J = 10.0(9)\cdot10^{-3}$ $\delta_K = 0.741(241)$				
		$v_x=1$	$A=5329.52(80)$ $B=1127.401(8)$ $C=932.998(7)$					
617	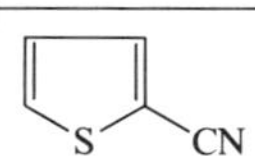	$v=0$	$A=5402.387(24)$ $B=1816.253(5)$ $C=1358.614(5)$	$\Delta_J = -0.092(44)$ $\Delta_{JK} = 2.889(80)$ $\Delta_K = -6.683(740)$ $\delta_J = 0.098(4)$ $\delta_K = -4.852(172)$	77WIE1, 75AVI4	75AVI4		2.6.3, 304 2.7.3, 59
618 Cont.	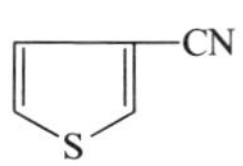	$v=0$	$A=7115.180(13)$ $B=1491.299(2)$ $C=1232.535(2)$	$\Delta_J = 0.043(7)$ $\Delta_{JK} = 1.028(15)$ $\Delta_K = 1.97(72)$ $\delta_J = 0.0089(9)$ $\delta_K = 0.670(76)$	77WIE2	80WI1		2.6.3, 305 2.7.3, 60

Nr.	Molecule	v	A, B, C MHz	D, H kHz	Ref.	Ref. struct.	Varia, remarks	See also Tab., Nr.
618 Cont.	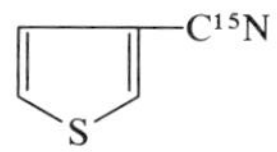	$v=0$	$A=7112.55\,(308)$ $B=1446.842\,(3)$ $C=1201.918\,(3)$		80WI1			
	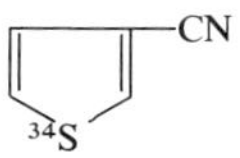	$v=0$	$A=7076.131\,(936)$ $B=1463.308\,(7)$ $C=1211.308\,(7)$					
619 *6–253*	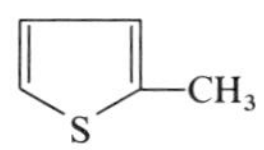	$v_\tau=1$	$A=5096.233\,(10)$ *) $B=3113.139\,(5)$ $C=1986.156\,(5)$		76POZ1		v_τ, methyl torsion *) for the A species	2.8, 51
620 *6–253a*	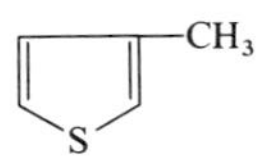	$v=0$	$A=6982.945$ *) $B=2455.694$ $C=1836.824$		73GUN		for structure, see also Vol. II/7 (3, 595) *) for the A species v_τ, methyl torsion	2.6.3, 306 2.8, 52
		$v_\tau=1$	$A=6811.44$ *) $B=2454.72$ $C=1836.98$					
		$v_\tau=2$	$A=7811.12$ *) $B=2456.24$ $C=1836.74$					
621	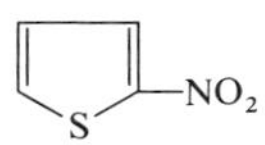	$v=0$	$A=3818.49\,(8)$ $B=1518.602\,(3)$ $C=1086.887\,(3)$		77RAL			3, 165
		$v_\tau=1$	$A=3817.72\,(9)$ $B=1515.521\,(3)$ $C=1087.777\,(3)$				$v_\tau \simeq 62\ cm^{-1}$, NO_2 torsion	
		$v_\tau=2$	$A=3816.81\,(13)$ $B=1512.447\,(5)$ $C=1088.672\,(4)$					
		$v_\tau=3$	$A=3816.21\,(14)$ $B=1509.399\,(5)$ $C=1089.566\,(4)$					
		$v_\tau=4$	$A=3815.49\,(11)$ $B=1506.351\,(4)$ $C=1090.473\,(3)$					

Nr.	Molecule	v	A, B, C MHz	D, H kHz	Ref.	Ref. struct.	Varia, remarks	See also Tab., Nr.
622	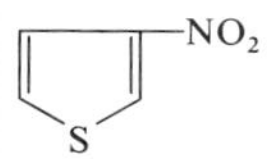	$v=0$	$A=4622.61\,(7)$ $B=1231.751\,(1)$ $C=973.062\,(1)$		77ISV			
		$v_\tau=1$	$A=4621.03\,(10)$ $B=1230.088\,(2)$ $C=974.051\,(2)$				$v_\tau=60\ cm^{-1}$, NO_2 torsion, $V_2=1340\ cm^{-1}$	
		$v_\tau=2$	$A=4619.57\,(9)$ $B=1228.411\,(2)$ $C=975.027\,(2)$					
		$v_\tau=3$	$A=4618.15\,(9)$ $B=1226.716\,(2)$ $C=975.996\,(1)$					
		$v_\tau=4$	$A=4616.65\,(9)$ $B=1225.008\,(2)$ $C=976.957\,(1)$					
		$v_\tau=5$	$A=4615.27\,(10)$ $B=1223.280\,(2)$ $C=977.917\,(2)$					
		$v_\tau=6$	$A=4614.07\,(10)$ $B=1221.540\,(2)$ $C=978.874\,(2)$					
623 *6-254*		$v=0$	$\tilde{A}=7597.610\,(14)$ $\tilde{B}=3383.4592\,(35)$ $\tilde{C}=2340.0666\,(36)$	$\Delta_J=0.434\,(45)$ $\Delta_{JK}=-0.15$ (fixed) $\Delta_K=2.13\,(31)$ $\delta_J=0.1434\,(17)$ $\delta_K=0.13$ (fixed)	76GUN		for structure, see also Vol. II/7 (3, 492)	
	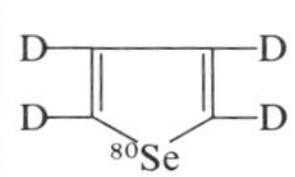	$v=0$	$\tilde{A}=6248.464\,(28)$ $\tilde{B}=3080.9082\,(44)$ $\tilde{C}=2062.8197\,(47)$	$\Delta_J=0.293\,(44)$ $\Delta_{JK}=-0.17$ (fixed) $\Delta_K=1.37\,(55)$ $\delta_J=0.1178\,(30)$ $\delta_K=0.042$ (fixed)				

Nr.	Molecule	v	A, B, C MHz	D, H kHz	Ref.	Ref. struct.	Varia, remarks	See also Tab., Nr.
624	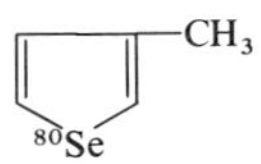	$v=0$	$A=6270.70$ *) $B=1679.77$ $C=1335.61$		73SHA		*) for the A species	2.6.3, 307 2.8, 53 3, 166
		$v_\tau=1$	$A=6202.85$ *) $B=1679.61$ $C=1335.76$				v_τ, methyl torsion	
	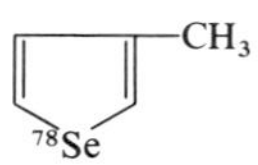	$v=0$	$A=6276.40$ *) $B=1693.26$ $C=1344.36$					
		$v_\tau=1$	$A=6210.30$ *) $B=1693.07$ $C=1344.49$					
625	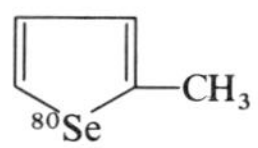	$v=0$	$A=3373.469(20)$ *) $B=2859.347(5)$ $C=1561.948(5)$		78POZ2		*) for the A species	2.6.3, 308 2.8, 54
		$v_\tau=1$	$A=3353.465(23)$ *) $B=2854.335(5)$ $C=1562.500(4)$				v_τ, methyl torsion	
626	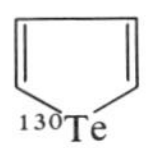	$v=0$	$A=7082.15(48)$ *) $B=2519.31(1)$ $C=1857.64(2)$		73BRO2	73BRO2	*) uncertainties are 95% confidence limits	2.6.3, 309
		$v=0$	$A=7082.20(43)$ *) $B=2529.26(1)$ $C=1863.04(1)$					
		$v=0$	$A=7082.42(64)$ *) $B=2539.52(2)$ $C=1868.59(2)$					
627 *6-255* *4-176*		$v=0$	$A=9130.637(10)$ $B=9001.360(10)$ $C=4532.120(10)$		74BOL1		for structure, see also Vol. II/7 (3, 498)	2.7.3, 61

Nr.	Molecule	v	A, B, C MHz	D, H kHz	Ref.	Ref. struct.	Varia, remarks	See also Tab., Nr.
628	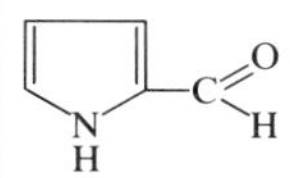	$v=0$	$A=7890.35(44)$ $B=2085.04(5)$ $C=1649.363(5)$		74MAR3	74MAR3		2.6.3, 310 3, 167
		$v_T=1$	$A=7805.51(44)$ $B=2085.368(5)$ $C=1652.165(5)$				$\nu_T=151(11)\ cm^{-1}$, C—C torsion	
		$v_T=2$	$A=7718.60(90)$ $B=2085.556(11)$ $C=1655.005(10)$					
		$v_T=3$	$A=7621.75(81)$ $B=2085.512(11)$ $C=1657.941(9)$					
		$v_S=1$	$A=7973.07(74)$ $B=2085.209(10)$ $C=1647.960(8)$				$v_S=210(17)\ cm^{-1}$, symmetrical aldehyde group deformation	
		$v_A=1$	$A=7893.57(76)$ $B=2084.869(10)$ $C=1650.049(7)$				$v_A=270(38)\ cm^{-1}$, antisymmetrical aldehyde group deformation	
		$v_T=1, v_S=1$	$A=7882.28(115)$ $B=2085.445(14)$ $C=1650.829(11)$					
	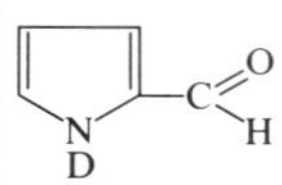	$v=0$	$A=7521.55(84)$ $B=2084.379(12)$ $C=1632.015(12)$					
	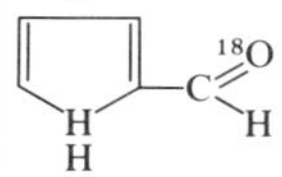	$v=0$	$A=7861.22(102)$ $B=1986.329(12)$ $C=1585.807(12)$					
629	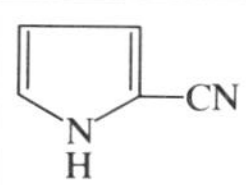	$v=0$	$A=8946.6297(88)$ $B=1988.5936(21)$ $C=1626.5440(16)$	$\Delta_J=0.0690(26)$ $\Delta_{JK}=2.812(11)$ $\delta_J=0.0169(26)$ $\delta_K=1.26(10)$	80DAV2	80DAV2		3, 168
	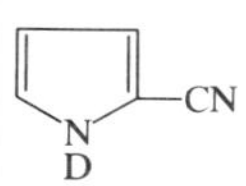	$v=0$	$A=8367.099(11)$ $B=1988.4938(19)$ $C=1606.2658(16)$	$\Delta_J=0.0839(33)$ $\Delta_{JK}=2.734(13)$ $\delta_J=0.0150(12)$ $\delta_K=1.461(81)$				

Nr.	Molecule	v	A, B, C MHz	D, H kHz	Ref.	Ref. struct.	Varia, remarks	See also Tab., Nr.
630 *6–257* *4–176 a*	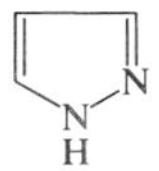	$v=0$	$A=9618.770(13)$ $B=9412.535(13)$ $C=4755.853(13)$	$D_1=-10.5890(140)$ $D_2=1.1881(43)$ $D_3=0.9062(43)$	74NY2	74NY2	for structure, see also Vol. II/7 (3, 371) for definition of *D*-constants, see Nr. 610	2.6.3, 311 2.7.3, 62 3, 169
	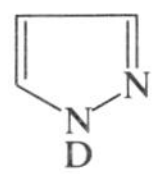	$v=0$	$A=9455.230(5)$ $B=8859.733(5)$ $C=4572.847(4)$	$D_1=-9.3640(280)$ $D_2=-0.5944(250)$ $D_3=-3.4083(200)$				
	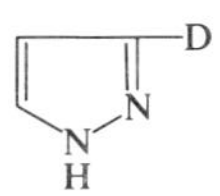	$v=0$	$A=9435.783(4)$ $B=8774.190(4)$ $C=4545.189(3)$	$D_1=-9.3640(410)$ $D_2=-0.4028(950)$ $D_3=-2.5176(170)$				
	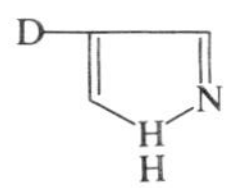	$v=0$	$A=9566.176(10)$ $B=8617.847(9)$ $C=4532.324(9)$	$D_1=-9.3256(860)$ $D_2=-0.2588(220)$ $D_3=-2.7291(890)$				
	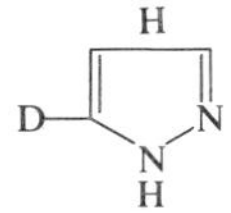	$v=0$	$A=9537.304(4)$ $B=8677.868(3)$ $C=4542.381(3)$	$D_1=-9.2527(260)$ $D_2=-0.3901(220)$ $D_3=-2.4582(120)$				
	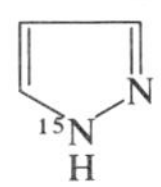	$v=0$	$A=9488.641(4)$ $B=9339.851(4)$ $C=4705.424(4)$	$D_1=-9.8907(88)$ $D_2=-1.0558(22)$ $D_3=-1.7526(27)$				
	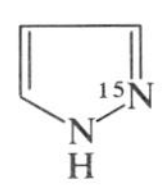	$v=0$	$A=9618.304(4)$ $B=9180.010(4)$ $C=4695.610(3)$	$D_1=-10.3101(350)$ $D_2=1.0964(590)$ $D_3=-0.4326(180)$				
	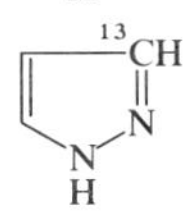	$v=0$	$A=9457.566(6)$ $B=9340.094(6)$ $C=4697.835(6)$	$D_1=-9.8854(69)$ $D_2=-1.0266(34)$ $D_3=0.7680(20)$				
	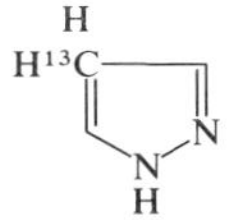	$v=0$	$A=9582.358(9)$ $B=9193.394(8)$ $C=4690.511(8)$	$D_1=-10.0738(160)$ $D_2=-0.1870(96)$ $D_3=-0.3679(120)$				
Cont.	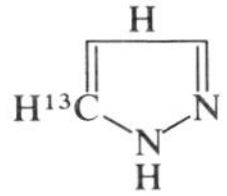	$v=0$	$A=9571.674(5)$ $B=9223.004(5)$ $C=4695.634(4)$	$D_1=-10.3213(150)$ $D_2=0.5582(100)$ $D_3=0.7919(83)$				

Nr.	Molecule	v	A, B, C MHz	D, H kHz	Ref.	Ref. struct.	Varia, remarks	See also Tab., Nr.
630 *6-257* *4-176a* Cont.	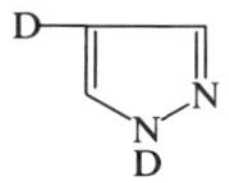	$v=0$	$A=9376.687$ $B=8157.643$ $C=4361.464$		73NAS			
	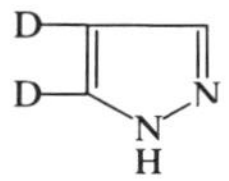	$v=0$	$A=8970.521$ $B=8410.042$ $C=4339.398$					
	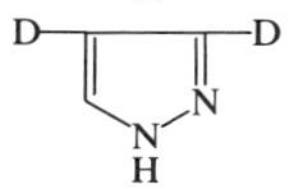	$v=0$	$A=8833.560$ $B=8545.057$ $C=4342.613$					
	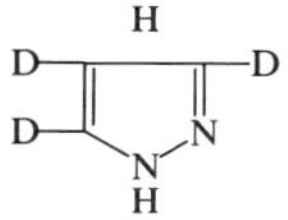	$v=0$	$A=8550.630$ $B=8106.798$ $C=4160.365$					
	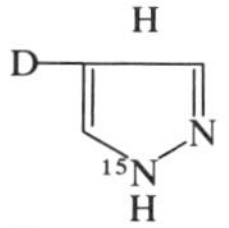	$v=0$	$A=9488.372$ $B=8507.051$ $C=4484.413$					
	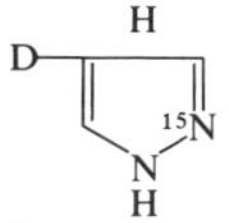	$v=0$	$A=9507.335$ $B=8459.330$ $C=4475.110$					
	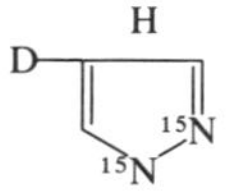	$v=0$	$A=9407.433$ $B=8372.314$ $C=4428.627$					
631	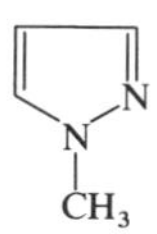	$v=0$	$A=8970.107(104)$ *) $B=3745.544(92)$ $C=2682.667(86)$	$D_J = 1.08(170)$ $D_{JK} = -4.76(150)$ $D_K = -30.94(144)$ $D_{DK} = -1.35(90)$	77KRI	77KRI	*) for the A species	2.8, 120
632 *6-258*	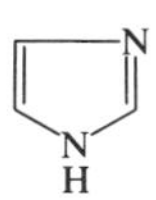	$v=0$	$A=9725.27$ $B=9373.97$ $C=4771.93$		76BLA1			2.7.3, 63 3, 170
	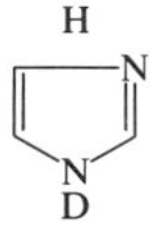	$v=0$	$A=9668.87$ $B=8699.52$ $C=4578.38$					

Nr.	Molecule	v	A, B, C MHz	D, H kHz	Ref.	Ref. struct.	Varia, remarks	See also Tab., Nr.
633	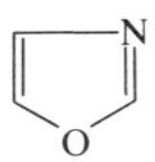	$v=0$	$A=10050.99(2)$ *)	$\tau_{aaaa}=-14.92(5)$	78KUM1	78KUM1	*) uncertainties	2.6.3, 312
6-259			$B=9645.73(2)$	$\tau_{bbbb}=-15.73(5)$			are 3σ	2.7.3, 64
			$C=4919.41(2)$	$\tau_{abab}=-7.85(2)$ **)			**) with $\tau_{aabb}=0$	2.9.3, 45
	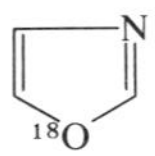	$v=0$	$A=10014.29(3)$ *)	$\tau_{aaaa}=-14.89(40)$				3, 171
			$B=9232.01(3)$	$\tau_{bbbb}=-14.33(27)$				
			$C=4801.01(2)$	$\tau_{abab}=-7.68(10)$ **)				
	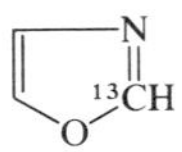	$v=0$	$A=9835.02(3)$ *)	$\tau_{aaaa}=-14.52(5)$				
			$B=9636.68(3)$	$\tau_{bbbb}=-16.28(5)$				
			$C=4864.75(3)$	$\tau_{abab}=-7.34(3)$ **)				
	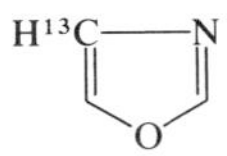	$v=0$	$A=9899.08(2)$ *)	$\tau_{aaaa}=-14.77(8)$				
			$B=9539.28(2)$	$\tau_{bbbb}=-14.80(8)$				
			$C=4855.27(2)$	$\tau_{abab}=-7.91(3)$ **)				
	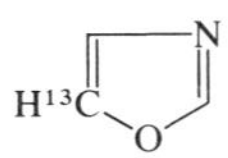	$v=0$	$A=9938.09(3)$ *)	$\tau_{aaaa}=-14.99(16)$				
			$B=9512.53(3)$	$\tau_{bbbb}=-16.10(15)$				
			$C=4857.67(3)$	$\tau_{abab}=-7.07(5)$ **)				
	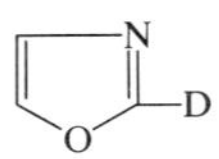	$v=0$	$A=9649.62(2)$ *)	$\tau_{aaaa}=-14.26(7)$				
			$B=9207.25(2)$	$\tau_{bbbb}=-12.98(6)$				
			$C=4709.21(2)$	$\tau_{abab}=-7.09(3)$ **)				
	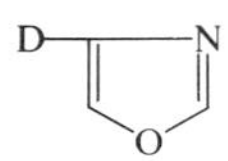	$v=0$	$A=9801.49(2)$ *)	$\tau_{aaaa}=-15.79(28)$				
			$B=9005.67(2)$	$\tau_{bbbb}=-12.96(19)$				
			$C=4690.92(2)$	$\tau_{abab}=-6.47(9)$ **)				
	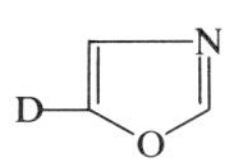	$v=0$	$A=9884.48(3)$ *)	$\tau_{aaaa}=-14.68(73)$				
			$B=8959.59(3)$	$\tau_{bbbb}=-13.58(46)$				
			$C=4697.26(3)$	$\tau_{abab}=-6.22(12)$ **)				
	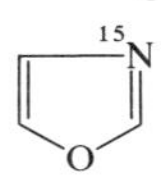	$v=0$	$A=10042.256(7)$	$\tau_{aaaa}=-15.081(69)$	78STI1			
			$B=9395.346(7)$	$\tau_{bbbb}=-15.300(51)$				
			$C=4851.376(6)$	$\tau_{abab}=-7.400(20)$ **)				
634	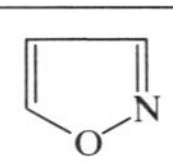	$v=0$	$A=9786.297(6)$	$\tau_{aaaa}=-11.7(11)$	75ST5	75ST5		2.6.3, 313
6-260			$B=9575.096(6)$	$\tau_{bbbb}=-14.7(10)$				2.7.3, 65
			$C=4836.927(5)$	$\tau_{aabb}=6.9(31)$				2.9.3, 46
Cont.				$\tau_{abab}=-10.2(10)$				3, 172

Nr.	Molecule	v	A, B, C MHz	D, H kHz	Ref.	Ref. struct.	Varia, remarks	See also Tab., Nr.
634 *6-260* Cont.	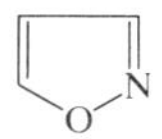	$v_{18}=1$	$A=9759.75(3)$ $B=9549.86(2)$ $C=4838.36(2)$	$\tau_{aaaa}=-12.3(4)$ $\tau_{bbbb}=-14.7(3)$ $\tau_{abab}=-10.6(2)$ *)	75ST3		$\nu_{18}=595\ \mathrm{cm}^{-1}$, out-of-plane bending mode *) with $\tau_{aabb}=6.9$ kHz fixed.	
		$v_{17}=1$	$A=9764.29(2)$ $B=9536.05(2)$ $C=4837.90(1)$	$\tau_{aaaa}=-12.1(4)$ $\tau_{bbbb}=-14.2(4)$ $\tau_{abab}=-9.6(2)$ *)			$\nu_{17}=632\ \mathrm{cm}^{-1}$, out-of-plane bending mode	
		$v_{17}=2$	$A=9733.76(8)$ $B=9525.22(8)$ $C=4839.66(3)$	$\tau_{aaaa}=-11.04(147)$ $\tau_{bbbb}=-14.54(130)$ $\tau_{abab}=-10.32(62)$ *)	75ST5			
		$v_{16}=1$	$A=9768.99(3)$ $B=9543.68(3)$ $C=4838.68(2)$	$\tau_{aaaa}=-10.7(4)$ $\tau_{bbbb}=-12.4(4)$ $\tau_{abab}=-9.7(2)$ *)	75ST3		$\nu_{16}=764\ \mathrm{cm}^{-1}$, out-of-plane bending mode	
		$v_{13}=1$	$A=9838.19(3)$ $B=9620.96(3)$ $C=4829.68(2)$	$\tau_{aaaa}=-95.5(8)$ $\tau_{bbbb}=-97.8(7)$ $\tau_{abab}=-60.1(4)$ *)			$\nu_{13}=844\ \mathrm{cm}^{-1}$, in-plane ring-bending mode	
		$v_{12}=1$	$A=9802.70(2)$ $B=9585.95(2)$ $C=4836.05(1)$	$\tau_{aaaa}=-14.4(4)$ $\tau_{bbbb}=-17.3(3)$ $\tau_{abab}=-11.4(2)$ *)			$\nu_{12}=856\ \mathrm{cm}^{-1}$, in-plane ring-bending mode	
		$v_{15}=1$	$A=9748.82(3)$ $B=9520.95(3)$ $C=4838.33(2)$	$\tau_{aaaa}=67.2(8)$ $\tau_{bbbb}=94.7(6)$ $\tau_{abab}=44.4(3)$ *)			$\nu_{15}=889\ \mathrm{cm}^{-1}$, out-of-plane bending mode	
		$v_{11}=1$	$A=9782.03(3)$ $B=9566.36(3)$ $C=4831.09(2)$	$\tau_{aaaa}=-13.9(6)$ $\tau_{bbbb}=-16.3(4)$ $\tau_{abab}=-11.8(3)$ *)			$\nu_{11}=1021\ \mathrm{cm}^{-1}$, in-plane CH bending mode	
		$v_{14}=1$	$A=9793.89(2)$ $B=9582.95(2)$ $C=4839.50(2)$	$\tau_{aaaa}=-11.5(3)$ $\tau_{bbbb}=-14.3(3)$ $\tau_{abab}=-10.4(1)$ *)			$\nu_{14}=1033\ \mathrm{cm}^{-1}$, out-of-plane bending mode	
		$v_{10}=1$	$A=9795.33(3)$ $B=9566.42(2)$ $C=4833.12(2)$	$\tau_{aaaa}=-11.5(4)$ $\tau_{bbbb}=-14.3(4)$ $\tau_{abab}=-10.0(2)$ *)			$\nu_{10}=1089\ \mathrm{cm}^{-1}$, in-plane CH bending mode	
		$v_9=1$	$A=9774.31(2)$ $B=9558.60(2)$ $C=4833.68(2)$	$\tau_{aaaa}=-30.9(5)$ $\tau_{bbbb}=-46.5(4)$ $\tau_{abab}=-22.7(2)$ *)			$\nu_9=1128\ \mathrm{cm}^{-1}$, in-plane ring-bending mode	
Cont.		$v_8=1$	$A=9745.22(9)$ $B=9522.14(7)$ $C=4831.90(7)$	$\tau_{aaaa}=11.6(9)$ $\tau_{bbbb}=23.3(9)$ $\tau_{abab}=4.3(5)$ *)			$\nu_8=1217\ \mathrm{cm}^{-1}$, in-plane CH bending mode	

Nr.	Molecule	v	A, B, C MHz	D, H kHz	Ref.	Ref. struct.	Varia, remarks	See also Tab., Nr.
634 *6-260* Cont.	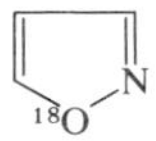	$v=0$	$A=9578.854(13)$ $B=9348.834(14)$ $C=4728.436(12)$	$\tau_{aaaa}=-12.9(2)$ $\tau_{bbbb}=-12.3(2)$ $\tau_{abab}=-10.2(1)$ *)	75ST3			
	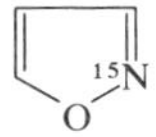	$v=0$	$A=9773.551(28)$ $B=9348.057(27)$ $C=4775.229(25)$	$\tau_{aaaa}=-11.9(4)$ $\tau_{bbbb}=-13.5(3)$ $\tau_{abab}=-10.2(2)$ *)				
	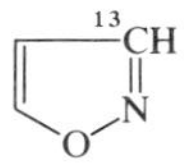	$v=0$	$A=9707.406(27)$ $B=9414.313(27)$ $C=4776.512(22)$	$\tau_{aaaa}=-8.5(6)$ $\tau_{bbbb}=-15.1(5)$ $\tau_{abab}=-8.5(2)$ *)				
	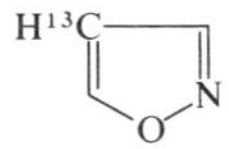	$v=0$	$A=9674.010(28)$ $B=9414.434(24)$ $C=4768.436(15)$	$\tau_{aaaa}=-10.3(8)$ $\tau_{bbbb}=-12.8(5)$ $\tau_{abab}=-9.5(1)$ *)				
	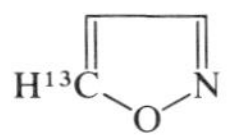	$v=0$	$A=9781.994(13)$ $B=9351.523(12)$ $C=4778.143(11)$	$\tau_{aaaa}=-12.0(2)$ $\tau_{bbbb}=-14.5(2)$ $\tau_{abab}=-9.9(5)$ *)				
	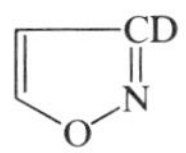	$v=0$	$A=9685.71(9)$ $B=8837.85(7)$ $C=4618.61(6)$	$\tau_{aaaa}=-12.8(9)$ $\tau_{bbbb}=-11.6(3)$ $\tau_{abab}=-8.7(2)$ *)				
	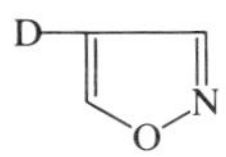	$v=0$	$A=9643.227(19)$ $B=8821.695(13)$ $C=4604.486(8)$	**)	78LOW		**) the distortion constants were taken to be the same as those of parent species	
	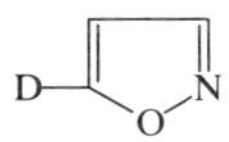	$v=0$	$A=9770.34(12)$ $B=8786.69(8)$ $C=4623.65(5)$	$\tau_{aaaa}=-16.8(44)$ $\tau_{bbbb}=-12.2(18)$ $\tau_{abab}=-8.9(6)$ *)	75ST3			
635	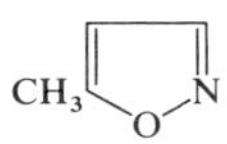	$v=0$	$A=9241.360(31)$ *) $B=3559.377(11)$ $C=2610.221(11)$	$D_K=21.4(16)$ $DD_K=2.1(2)$	77SRI, 75MJ2		*) for the A species	2.8, 55 3, 173
636 *5-263* *4-172*	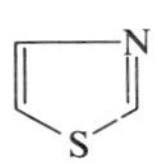	$v=0$	$A=8529.425(4)$ $B=5505.755(3)$ $C=3344.298(3)$		80WI2		for structure, see also Vol. II/7 (3, 366)	2.7.3, 66 2.9.3, 47 3, 174
637 *6-264*	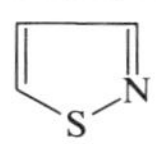	$v=0$	$A=8275.575(5)$ $B=5846.228(3)$ $C=3424.193(3)$		80WI2			2.7.3, 67 2.9.3, 48 3, 175

Nr.	Molecule	v	A, B, C MHz	D, H kHz	Ref.	Ref. struct.	Varia, remarks	See also Tab., Nr.
638 *4-174*	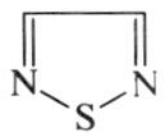	$v=0$	$A=8538.436(13)$ *) $B=6333.038(8)$ $C=3633.926(7)$	$\tau_{aaaa}=-15.74(30)$ $\tau_{bbbb}=-9.65(12)$ $\tau_{abab}=-1.62(36)$ **)	78STI3	78STI3	for structure, see also Vol. II/7 (3, 193) *) uncertainties are 3 σ **) $\tau_{aabb}=-7.7$ kHz assumed.	2.6.3, 314 2.7.3, 68
	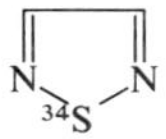	$v=0$	$A=8538.670(12)$ $B=6166.960(9)$ $C=3578.661(8)$	$\tau_{aaaa}=-15.54(30)$ $\tau_{bbbb}=-9.34(15)$ $\tau_{abab}=-1.54(30)$ **)				
	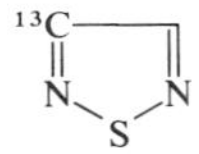	$v=0$	$A=8472.029(15)$ $B=6215.713(11)$ $C=3583.153(10)$	$\tau_{aaba}=-15.47(37)$ $\tau_{bbbb}=-9.22(17)$ $\tau_{abab}=-1.30(30)$ **)				
	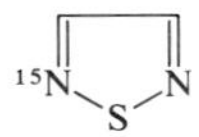	$v=0$	$A=8324.235(17)$ $B=6333.303(14)$ $C=3594.629(10)$	$\tau_{aaaa}=-15.43(45)$ $\tau_{bbbb}=-9.62(30)$ $\tau_{cccc}=-1.33(49)$ **)				
	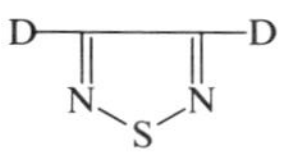	$v=0$	$A=8041.747(10)$ $B=5717.785(8)$ $C=3339.958(6)$	$\tau_{aaaa}=-13.15(27)$ $\tau_{bbbb}=-7.72(14)$ $\tau_{abab}=-0.81(25)$ **)				
	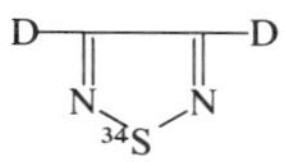	$v=0$	$A=8041.972(13)$ $B=5569.697(9)$ $C=3288.903(8)$	$\tau_{aaaa}=-13.21(39)$ $\tau_{bbbb}=-7.43(17)$ $\tau_{abab}=-0.55(42)$ **)				
	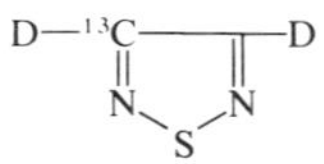	$v=0$	$A=7981.850(14)$ $B=5629.614(11)$ $C=3299.474(10)$	$\tau_{aaaa}=-13.11(54)$ $\tau_{bbbb}=-7.57(13)$ $\tau_{cccc}=-0.79(33)$ **)				
	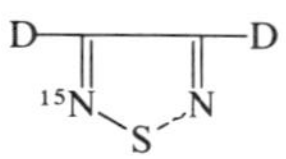	$v=0$	$A=7851.445(16)$ $B=5717.895(13)$ $C=3306.690(12)$	$\tau_{aaaa}=-13.31(54)$ $\tau_{bbbb}=-7.72(16)$ $\tau_{cccc}=-0.49(42)$ **)				
639	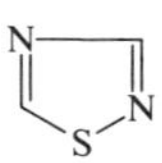	$v=0$	$A=8811.08(2)$ *) $B=5938.20(1)$ $C=3545.34(1)$	$\tau_{aaaa}=-15.2(17)$ $\tau_{bbbb}=-7.87(31)$ $\tau_{aabb}=-0.94(50)$ $\tau_{abab}=-3.47(20)$	76ST3	76ST3	*) uncertainties are 3 σ	3, 176
	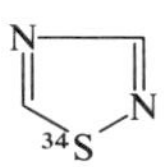	$v=0$	$A=8809.75(3)$ *) $B=5774.93(2)$ $C=3486.27(1)$	$\tau_{aaaa}=-13.3(27)$ $\tau_{bbbb}=-7.17(50)$ $\tau_{abab}=-3.34(27)$ **)			**) with $\tau_{aabb}=-0.94$ kHz assumed	
Cont.	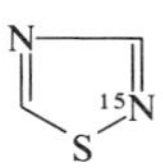	$v=0$	$A=8591.19(3)$ *) $B=5937.86(1)$ $C=3509.06(1)$	$\tau_{aaaa}=-13.8(26)$ $\tau_{bbbb}=-7.84(42)$ $\tau_{abab}=-3.33(15)$ **)				

Nr.	Molecule	v	A, B, C MHz	D, H kHz	Ref.	Ref. struct.	Varia, remarks	See also Tab., Nr.
639 Cont.	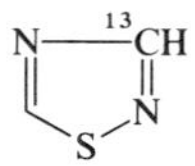	$v=0$	$A=8731.40(3)$ *) $B=5818.03(2)$ $C=3500.25(1)$	$\tau_{aaaa}=-13.4(30)$ $\tau_{bbbb}=-7.48(50)$ $\tau_{abab}=-3.48(21)$ **)	76ST3			
	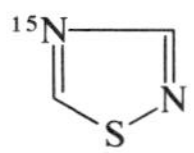	$v=0$	$A=8759.14(4)$ *) $B=5813.92(2)$ $C=3492.42(1)$	$\tau_{aaaa}=-15.0(38)$ $\tau_{bbbb}=-7.30(57)$ $\tau_{abab}=-3.23(26)$ **)				
	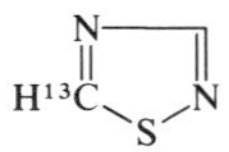	$v=0$	$A=8598.97(3)$ *) $B=5936.25(1)$ $C=3509.80(1)$	$\tau_{aaaa}=-13.0(25)$ $\tau_{bbbb}=-7.63(40)$ $\tau_{abab}=-3.43(15)$ **)				
	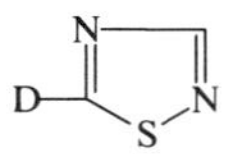	$v=0$	$A=8091.37(1)$ *) $B=5937.91(1)$ $C=3422.73(1)$	$\tau_{aaaa}=-11.6(9)$ $\tau_{bbbb}=-7.76(16)$ $\tau_{abab}=-3.27(12)$ **)				
	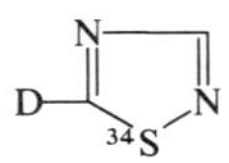	$v=0$	$A=8090.91(4)$ *) $B=5774.57(2)$ $C=3367.73(1)$	$\tau_{aaaa}=-11.4(30)$ $\tau_{bbbb}=-7.11(50)$ $\tau_{abab}=-2.74(30)$ **)				
	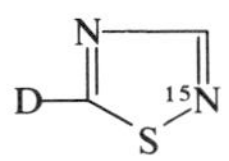	$v=0$	$A=7897.87(5)$ *) $B=5937.52(2)$ $C=3387.47(1)$	$\tau_{aaaa}=-13.0(30)$ $\tau_{bbbb}=-8.14(60)$ $\tau_{abab}=-3.40(55)$ **)				
	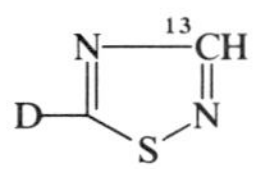	$v=0$	$A=8019.29(3)$ *) $B=5847.74(2)$ $C=3379.82(1)$	$\tau_{aaaa}=-12.9(23)$ $\tau_{bbbb}=-7.51(60)$ $\tau_{abab}=-2.79(45)$ **)				
	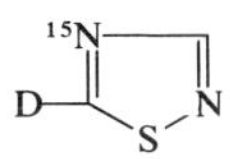	$v=0$	$A=8052.52(6)$ *) $B=5813.19(3)$ $C=3374.11(1)$	$\tau_{aaaa}=-13.5(36)$ $\tau_{bbbb}=-7.20(90)$ $\tau_{abab}=-2.50(70)$ **)				
	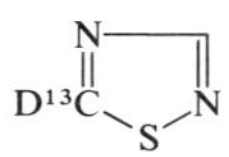	$v=0$	$A=7920.20(3)$ *) $B=5935.86(2)$ $C=3391.03(1)$	$\tau_{aaaa}=-12.0(15)$ $\tau_{bbbb}=-8.05(40)$ $\tau_{abab}=-3.13(12)$ **)				
640	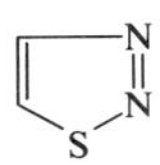	$v=0$	$A=8711.38(3)$ *) $B=5847.09(2)$ $C=3496.59(2)$	$\tau_{aaaa}=-12.0(30)$ $\tau_{bbbb}=-8.3(4)$ $\tau_{aabb}=-1.6(8)$ $\tau_{abab}=-4.1(2)$	76ST1	76ST1	*) uncertainties are 3 σ	2.6.3, 315 3, 177
Cont.	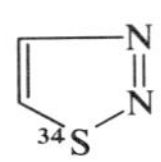	$v=0$	$A=8711.41(4)$ *) $B=5686.12(2)$ $C=3438.37(2)$	$\tau_{aaaa}=-12.3(45)$ $\tau_{bbbb}=-7.9(8)$ $\tau_{abab}=-4.1(4)$ **)			**) $\tau_{aabb}=-1.6$ kHz assumed	

Nr.	Molecule	v	A, B, C MHz	D, H kHz	Ref.	Ref. struct.	Varia, remarks	See also Tab., Nr.
640 Cont.	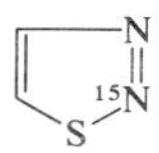	$v=0$	$A=8502.61\,(4)$ *) $B=5846.91\,(3)$ $C=3462.38\,(2)$	$\tau_{aaaa}=-11.8\,(42)$ $\tau_{bbbb}=-8.4\,(9)$ $\tau_{abab}=-4.0\,(5)$ **)	76ST1		*) uncertainties are 3 σ **) $\tau_{aabb}=-1.6$ kHz assumed	
	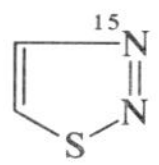	$v=0$	$A=8650.57\,(5)$ *) $B=5740.43\,(3)$ $C=3448.54\,(2)$	$\tau_{aaaa}=-11.1\,(48)$ $\tau_{bbbb}=-8.1\,(8)$ $\tau_{abab}=-3.9\,(5)$ **)				
	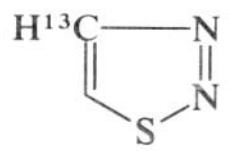	$v=0$	$A=8641.88\,(5)$ *) $B=5742.95\,(2)$ $C=3448.07\,(2)$	$\tau_{aaaa}=-11.7\,(40)$ $\tau_{bbbb}=-7.6\,(8)$ $\tau_{abab}=-3.8\,(5)$ **)				
	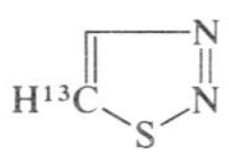	$v=0$	$A=8486.04\,(4)$ *) $B=5847.30\,(2)$ $C=3459.77\,(2)$	$\tau_{aaaa}=-11.2\,(38)$ $\tau_{bbbb}=-8.1\,(8)$ $\tau_{abab}=-3.7\,(4)$ **)				
	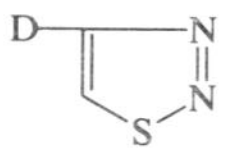	$v=0$	$A=8518.49\,(5)$ *) $B=5530.15\,(2)$ $C=3351.31\,(2)$	$\tau_{aaaa}=-13.3\,(50)$ $\tau_{bbbb}=-7.1\,(8)$ $\tau_{abab}=-3.4\,(4)$ **)				
	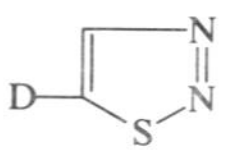	$v=0$	$A=7998.72\,(2)$ *) $B=5839.27\,(2)$ $C=3373.24\,(1)$	$\tau_{aaaa}=-9.6\,(15)$ $\tau_{bbbb}=-8.4\,(3)$ $\tau_{aabb}=-1.4\,(7)$ $\tau_{abab}=-3.9\,(3)$				
	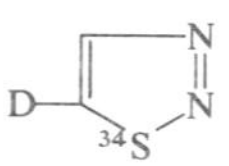	$v=0$	$A=7997.22\,(3)$ *) $B=5680.08\,(2)$ $C=3319.24\,(2)$	$\tau_{aaaa}=-8.8\,(28)$ $\tau_{bbbb}=-7.7\,(6)$ $\tau_{abab}=-3.4\,(3)$ ***)			***) with $\tau_{aabb}=-1.4$ kHz assumed.	
	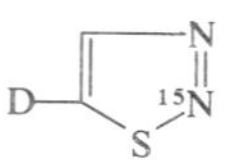	$v=0$	$A=7815.75\,(4)$ *) $B=5838.42\,(2)$ $C=3339.97\,(2)$	$\tau_{aaaa}=-12.0\,(22)$ $\tau_{bbbb}=-8.4\,(5)$ $\tau_{abab}=-3.4\,(2)$ ***)				
	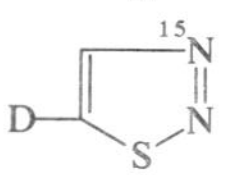	$v=0$	$A=7949.74\,(4)$ *) $B=5728.83\,(2)$ $C=3327.53\,(2)$	$\tau_{aaaa}=-10.9\,(30)$ $\tau_{bbbb}=-7.9\,(6)$ $\tau_{abab}=-3.6\,(3)$ ***)				
	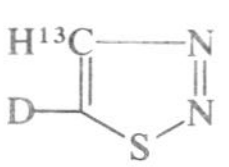	$v=0$	$A=7939.11\,(3)$ *) $B=5737.70\,(2)$ $C=3328.66\,(1)$	$\tau_{aaaa}=-10.0\,(21)$ $\tau_{bbbb}=-7.9\,(5)$ $\tau_{abab}=-3.6\,(3)$ ***)				
	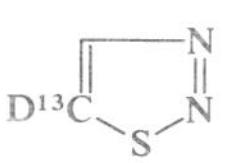	$v=0$	$A=7816.59\,(3)$ *) $B=5839.34\,(2)$ $C=3340.43\,(2)$	$\tau_{aaaa}=-10.4\,(16)$ $\tau_{bbbb}=-8.2\,(4)$ $\tau_{abab}=-3.5\,(2)$ ***)				
	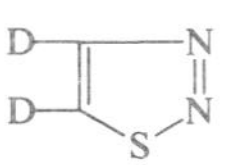	$v=0$	$A=7831.12\,(2)$ *) $B=5528.41\,(2)$ $C=3238.85\,(2)$	$\tau_{aaaa}=-9.2\,(19)$ $\tau_{bbbb}=-6.7\,(3)$ $\tau_{abab}=-3.3\,(2)$ **)				

Nr.	Molecule	v	A, B, C MHz	D, H kHz	Ref.	Ref. struct.	Varia, remarks	See also Tab., Nr.
641	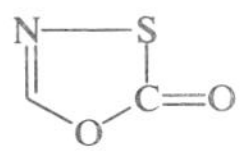	$v=0$	$A=5587.411\,(11)$ $B=3645.8735\,(89)$ $C=2205.4635\,(89)$	$T_1=-0.061\,(20)$ $T_2=0.099\,(17)$ $T_3=-0.749\,(95)$ $T_4=1.186\,(22)$ $T_5=-0.05\,(10)$	78BAK2	78BAK2	for definition of *T*-constants, see [71CAS]	
642 *6-268*	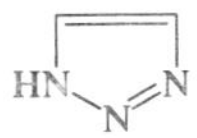	$v=0$	$A=10030.71$ $B=9870.56$ $C=4972.96$		77BLA1			2.7.3, 69 3, 178
	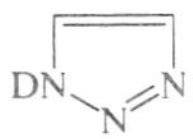	$v=0$	$A=9967.73$ $B=9160.17$ $C=4771.78$					
643 *6-267a*	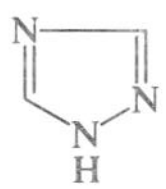	$v=0$	$A=10245.058$ $B=9832.056$ $C=5015.161$		75BL2	75BOL		2.6.3, 316 2.7.3, 70 3,179
	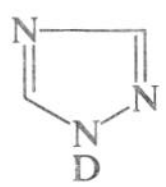	$v=0$	$A=10055.167$ $B=9226.569$ $C=4809.797$					
	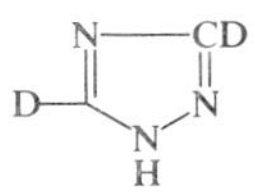	$v=0$	$A=9700.31$ $B=8679.69$ $C=4579.20$		75BOL			
644	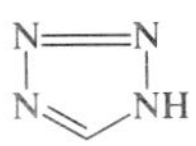	$v=0$	$A=10667.32\,(18)$ *) $B=10310.91\,(18)$ $C=5240.37\,(15)$		74KRU		*) uncertainties are 3 σ	2.6.3, 317
	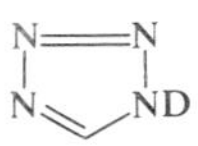	$v=0$	$A=10378.40\,(12)$ *) $B=9752.13\,(12)$ $C=5025.45\,(9)$					
	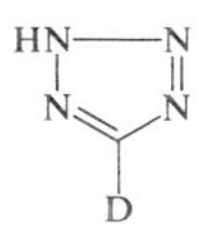	$v=0$	$A=10406.37\,(12)$ *) $B=9456.92\,(9)$ $C=4952.08\,(6)$				C-deuteriotetrazole is 1 H-tetrazole	

Nr.	Molecule	v	A, B, C MHz	D, H kHz	Ref.	Ref. struct.	Varia, remarks	See also Tab., Nr.
645	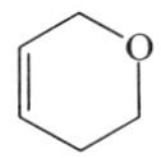	$v=0$ $v_R=1$	$A=5085.152(12)$ $B=4849.365(13)$ $C=2712.487(13)$ $A=5078.965(14)$ $B=4849.044(15)$ $C=2712.187(16)$		74WE1	74WE1	*) twisted ring skeleton ν_R, ring-bending mode	2.6.3, 318
646	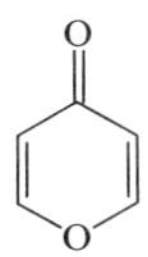	$v=0$	$A=5859.17$ $B=2715.09$ $C=1855.55$		73NOR1	73NOR1		2.9.3, 53 3,180
647	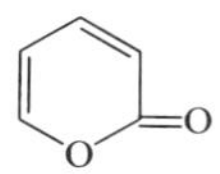	$v=0$	$A=5677.643(20)$ $B=2882.240(5)$ $C=1912.133(3)$		73NOR1	73NOR1		3, 181
648	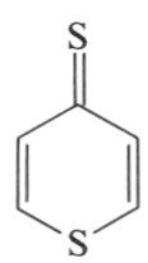	$v=0$	$A=5078.1(2)$ $B=1180.288(5)$ $C=\ \ 957.892(5)$	$D_J\ \ =0.017(8)$ $D_{JK}=0.050(9)$	76COR	76COR		2.6.3, 319
649	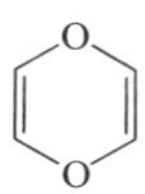	$v=0$ $v_R=1$ $v_R=2$ $v_T=1$ $v_R=1, v_T=1$	$A=5697.70$ $B=4968.41$ $C=2865.18$ $A=5685.58$ $B=4974.22$ $C=2866.23$ $A=5672.82$ $B=4979.36$ $C=2867.26$ $A=5691.31$ $B=4963.92$ $C=2862.81$ $A=5681.24$ $B=4969.91$ $C=2863.46$		74WE2	74WE2	C_2 symmetry ν_R, ring-bending mode ν_T=ring-twisting mode	2.6.3, 320

Nr.	Molecule	v	A, B, C MHz	D, H kHz	Ref.	Ref. struct.	Varia, remarks	See also Tab., Nr.
650	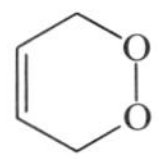	$v=0$	$A=5287.43(6)$ $B=5110.21(6)$ $C=2857.85(3)$		78KON	78KON		2.6.3, 321
651	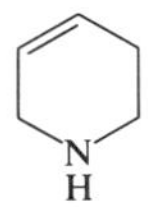 axial NH equatorial NH 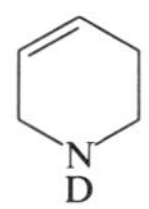 axial ND equatorial ND	$v=0$ $v=0$ $v=0$ $v=0$	$A=4897.47(1)$ $B=4709.16(1)$ $C=2641.48(1)$ $A=4950.50(1)$ $B=4743.05(1)$ $C=2647.82(1)$ $A=4785.44(1)$ $B=4594.74(1)$ $C=2616.00(1)$ $A=4882.41(1)$ $B=4575.05(1)$ $C=2577.26(2)$		76CH	76CH	*) half-chair conformer $E_{ax.}-E_{eq.}=50(30)\ cm^{-1}$	2.6.3, 322 2.7.3, 71
652	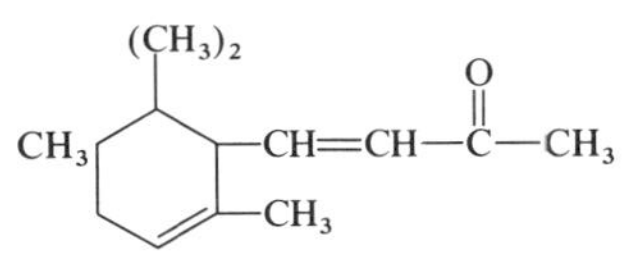		$B+C=642.0(1)$ *) **) $B+C=687.0(2)$ *) $^{+}$) $^{++}$) $B+C=694.6(10)$ *) $^{+}$)		74STE		*) LRMW **) cis form $^{+}$) trans form $^{++}$) more stable conformer	
653	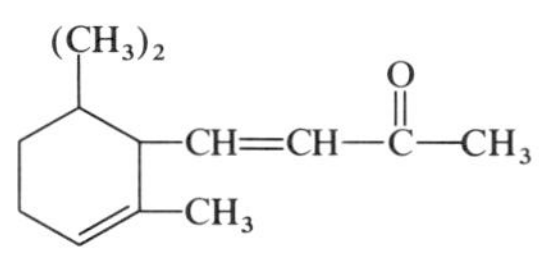		$B+C=775.1(1)$ *) $^{+}$) $B+C=749.2(1)$ *)		74STE		*) LRMW $^{+}$) more stable conformer	
654	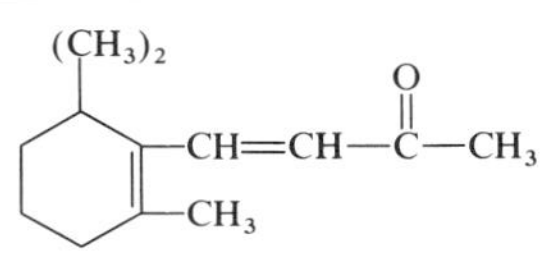		$B+C=752.5(2)$ *)		74STE		*) LRMW	
655	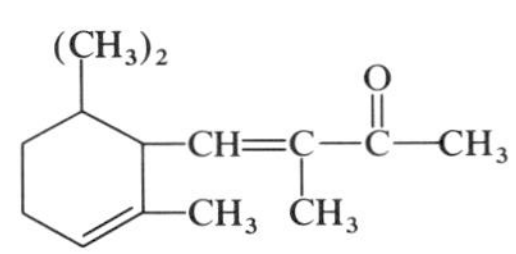		$B+C=726.8(1)$ *) $^{+}$) $B+C=702.9(1)$ *)		74STE		*) LRMW $^{+}$) more stable conformer	

Nr.	Molecule	v	A, B, C MHz	D, H kHz	Ref.	Ref. struct.	Varia, remarks	See also Tab., Nr.
656	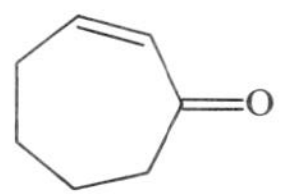	$v=0$	$A=2997.267(38)$ $B=2049.235(9)$ $C=1399.760(9)$	$D_J = 0.2588(332)$ $D_{JK} = -0.2925(1561)$	79KIT	79KIT		2.6.3, 323
		$v_x=1$	$A=3000.510(46)$ $B=2046.652(7)$ $C=1398.875(7)$	$D_J = 0.1417(299)$				
657	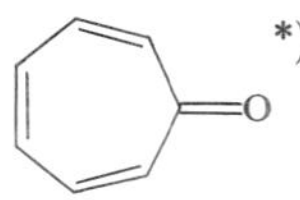 *)	$v=0$	$A=3712.34(5)$ $B=2027.64(1)$ $C=1313.14(1)$		74CR3		for structure, see also Vol. II/7 (3, 726) *) C_{2v} symmetry	2.6.3, 324 2.9.3, 54
		$v_x=1$	$A=3699.19(23)$ $B=2033.53(2)$ $C=1318.55(1)$					
		$v_x=2$	$A=3687.40(21)$ $B=2038.62(2)$ $C=1323.31(1)$					
		$v_x=3$	$A=3676.62(14)$ $B=2043.27(1)$ $C=1327.27(1)$					
		$v_x=4$	$A=3666.49(16)$ $B=2047.63(1)$ $C=1331.83(1)$					
658		$v=0$	$A=3419.424(2)$ $B=3297.290(2)$ $C=1799.961(3)$	$\Delta_J = 0.6948928 \pm 0.018$ $\Delta_{JK} = 4.385128 \pm 0.025$ $\Delta_K = -4.947697 \pm 0.036$ $\delta_J = -0.02920363 \pm 0.0056$ $\delta_K = -59.66475 \pm 0.16$ $H_{JK} = (-3.5355 \pm 0.031) \cdot 10^{-3}$ $H_{KJ} = (1.1603 \pm 0.011) \cdot 10^{-2}$ $H_K = (-8.2678 \pm 0.14) \cdot 10^{-3}$ $h_J = (-1.6553 \pm 0.15) \cdot 10^{-5}$	79AVI1	79AVI1	for structure, see also Vol. II/7 (3, 738)	2.6.3, 325
Cont.		$v_\alpha=1$	$A=3417.646(3)$ $B=3309.165(3)$ $C=1809.368(3)$	$\Delta_J = 0.70554 \pm 0.022$ $\Delta_{JK} = -1.6432(99)$ $\Delta_K = 1.0895(110)$ $\delta_J = -0.010476(840)$ $\delta_K = 7.5148(840)$				

Nr.	Molecule	v	A, B, C MHz	D, H kHz	Ref.	Ref. struct.	Varia, remarks	See also Tab., Nr.
658 Cont.		$v_\beta=1$	$A=3416.607(4)$ $B=3314.519(4)$ $C=1813.785(4)$	$\Delta_J = 0.88277 \pm 0.032$ $\Delta_{JK} = -1.5257(920)$ $\Delta_K = 0.82470 \pm 0.077$ $\delta_J = -0.052891 \pm 0.0049$ $\delta_K = 4.0077 \pm 0.92$	79AVI1			
		$v_\gamma=1$	$A=3415.379(5)$ $B=3319.787(5)$ $C=1818.154(3)$	$\Delta_J = 0.68068 \pm 0.034$ $\Delta_{JK} = -1.0558 \pm 0.16$ $\Delta_K = 0.47646 \pm 0.14$ $\delta_J = -0.027142 \pm 0.0045$ $\delta_K = -0.42821 \pm 1.7$				
		$v_\delta=1$	$A=3413.71(10)$ $B=3324.53(10)$ $C=1821.94(3)$					
		$v_\varepsilon=1$	$A=3412.77(8)$ $B=3328.31(8)$ $C=1825.36(3)$					
659	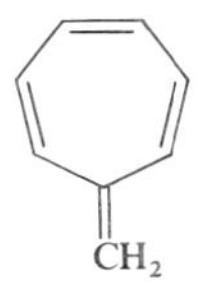	$v=0$	$A=3665.972(66)$ $B=2004.3165(59)$ $C=1297.5029(90)$	$\Delta_J = 0.182(35)$ $\Delta_{JK} = -0.47(22)$ $\delta_K = 0.78(38)$	76BAU1	76BAU1		2.6.3, 326
		$v_{42}=1$	$A=3654.53(59)$ $B=2008.489(56)$ $C=1302.525(99)$	$\Delta_J = -0.40(49)$ $\Delta_{JK} = 2.7(28)$ $\delta_K = -4.7(47)$			$\nu_{42}(b_2)=70(11)\,cm^{-1}$	
		$v_{42}=2$	$A=3640.46(95)$ $B=2012.436(96)$ $C=1307.64(10)$					
		$v_{42}=3$	$A=3620.9(23)$ $B=2015.72(25)$ $C=1313.47(27)$					
		$v_{42}=4$	$A=3617.9(24)$ $B=2020.74(26)$ $C=1317.34(29)$					
660	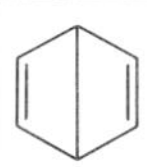	$v=0$	$A=7770.391(12)$ $B=4479.298(8)$ $C=3488.945(32)$		74GRI	74GRI	for structure, see also Vol. II/7 (3, 668)	2.6.3, 327

Nr.	Molecule	v	A, B, C MHz	D, H kHz	Ref.	Ref. struct.	Varia, remarks	See also Tab., Nr.
661	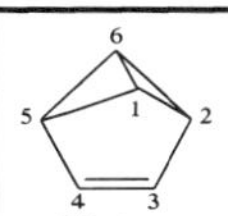	$v=0$	$A=7389.233(7)$ $B=5275.919(4)$ $C=3889.779(3)$		73SUE1	73SUE1	for structure, see also Vol. II/7 (3, 669)	
	1-^{13}C	$v=0$	$A=7334.510(47)$ $B=5192.586(4)$ $C=3859.480(3)$					
	2-^{13}C	$v=0$	$A=7267.656(92)$ $B=5273.698(7)$ $C=3854.647(6)$					
	3-^{13}C	$v=0$	$A=7344.398(162)$ $B=5191.750(11)$ $C=3831.499(10)$					
	1-D	$v=0$	$A=7175.631(4)$ $B=4997.498(3)$ $C=3793.592(3)$					
	2-D	$v=0$	$A=6948.922(7)$ $B=5256.549(3)$ $C=3754.365(2)$					
	3-D	$v=0$	$A=7226.229(7)$ $B=5035.768(3)$ $C=3714.995(3)$					
662	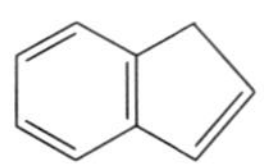	$v=0$	$A=3775.01(13)$ $B=1580.86(1)$ $C=1122.24(1)$		79LI		for structure, see also Vol. II/7 (3, 780)	
663		$v=0$	$A=3923.772(66)$ *) $B=3432.367(9)$ $C=3014.925(9)$		77CHI	77CHI	*) uncertainties are 3 σ	2.6.3, 328 3, 182
664	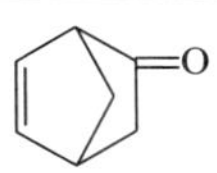	$v=0$	$A=3648.74(22)$ $B=2217.753(5)$ $C=2004.478(6)$		74COF			2.6.3, 329
		$v_x=1$	$A=3648.62(35)$ $B=2218.971(8)$ $C=2006.373(6)$					
Cont.								

Nr.	Molecule	v	A, B, C MHz	D, H kHz	Ref.	Ref. struct.	Varia, remarks	See also Tab., Nr.
664 Cont.	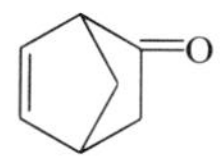	$v_x=2$	$A=3648.36\,(42)$ $B=2220.183\,(9)$ $C=2008.242\,(7)$		74COF			
		$v_x=3$	$A=3747.81\,(44)$ $B=2221.301\,(9)$ $C=2010.094\,(6)$					
665	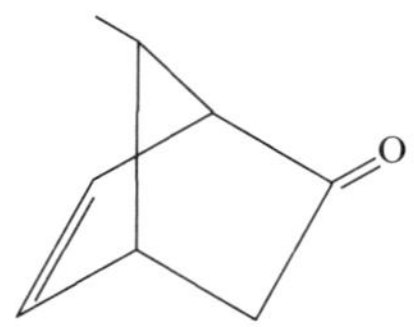		$B+C=2985\,(10)$ *)		76GIL2		*) LRMW exo isomer	
666	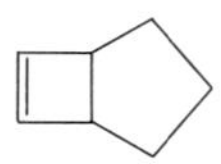	$v=0$	$A=4419.20$ $B=3080.16$ $C=2375.33$		79AVI2	79AVI2	for structure, see also Vol. II/7 (3, 736)	2.6.3, 330
		$v_a=1$	$A=4420.06$ $B=3078.47$ $C=2374.86$				v_a = twisting vibration of the five-membered ring	
		$v_b=1$	$A=4424.16$ $B=3073.22$ $C=2369.70$				v_b = puckering vibration of the five-membered ring	
		$v_b=2$	$A=4429.44$ $B=3066.49$ $C=2363.85$					
		$v_b=1,\ v_c=1$	$A=4427.19$ $B=3071.57$ $C=2366.83$				v_c = rocking of the four-membered ring	
	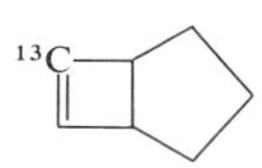	$v=0$	$A=4389.85$ $B=3033.45$ $C=2346.73$					

Nr.	Molecule	v	A, B, C MHz	D, H kHz	Ref.	Ref. struct.	Varia, remarks	See also Tab., Nr.
667	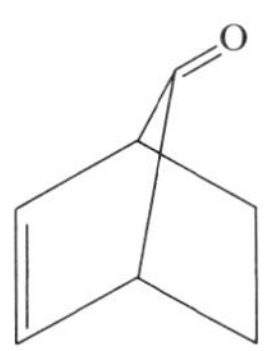	$v=0$	$A=2979.22(8)$ $B=2418.60(452)$ $C=2235.51(1)$		75DUR5			2.6.3, 331
		$v_\alpha=1$	$A=2979.48(15)$ $B=2423.12(435)$ $C=2236.70(1)$					
		$v_\alpha=2$	$A=2979.74(10)$ $B=2427.47(417)$ $C=2237.79(1)$					
		$v_\alpha=3$	$A=2979.71(65)$ $B=2431.64(5)$ $C=2238.86(5)$					
		$v_\beta=1$	$A=2979.88(19)$ $B=2419.99(1)$ $C=2234.96(1)$					
	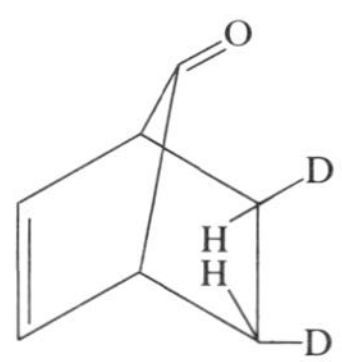	$v=0$	$A=2789.67(6)$ $B=2385.24(1)$ $C=2150.60(1)$					
668	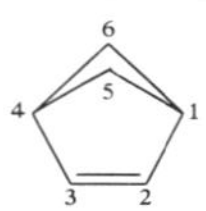	$v=0$	$A=5820.653(121)$ $B=4689.605(7)$ $C=4289.591(7)$		76WAN	76WAN	for structure, see also Vol. II/7 (3, 677) C_{2v} symmetry	2.6.3, 332
	1-^{13}C	$v=0$	$A=5752.524(103)$ $B=4685.847(6)$ $C=4249.150(7)$					
	2-^{13}C	$v=0$	$A=5792.950(58)$ $B=4627.840(3)$ $C=4223.050(3)$					
	5-^{13}C	$v=0$	$A=5748.483(79)$ $B=4612.760(4)$ $C=4264.360(4)$					

Nr.	Molecule	v	A, B, C MHz	D, H kHz	Ref.	Ref. struct.	Varia, remarks	See also Tab., Nr.
669	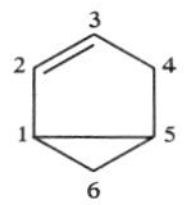	$v=0$	$A=6306.121\,(6)$ $B=4516.667\,(4)$ $C=3208.823\,(2)$		78MAT1	78MAT1		2.6.3, 333
	1-^{13}C	$v=0$	$A=6249.016\,(4)$ $B=4489.502\,(4)$ $C=3187.593\,(1)$					
	2-^{13}C	$v=0$	$A=6200.995\,(8)$ $B=4498.162\,(36)$ $C=3172.108\,(4)$					
	5-^{13}C	$v=0$	$A=6249.495\,(26)$ $B=4492.845\,(10)$ $C=3188.046\,(4)$					
	6-^{13}C	$v=0$	$A=6276.204\,(55)$ $B=4418.036\,(24)$ $C=3166.229\,(5)$					
670	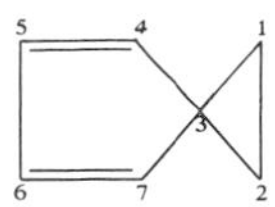	$v=0$	$A=6102.149\,(14)$ $B=2454.223\,(1)$ $C=2028.890\,(1)$		81HAR	81HAR	for structure, see also Vol. II/7 (3, 729)	2.6.3, 334
	1-^{13}C	$v=0$	$A=6062.345\,(114)$ $B=2410.269\,(5)$ $C=2003.061\,(3)$					
	3-^{13}C	$v=0$	$A=6102.025\,(205)$ $B=2451.877\,(8)$ $C=2027.310\,(5)$					
	4-^{13}C	$v=0$	$A=6003.000\,(94)$ $B=2452.051\,(5)$ $C=2016.335\,(3)$					
	5-^{13}C	$v=0$	$A=6063.945\,(136)$ $B=2419.796\,(8)$ $C=2001.166\,(4)$					
Cont.	1-D	$v=0$	$A=5934.441\,(19)$ $B=2387.133\,(1)$ $C=1988.101\,(1)$					

Nr.	Molecule	v	A, B, C MHz	D, H kHz	Ref.	Ref. struct.	Varia, remarks	See also Tab., Nr.
670 Cont.	4-D 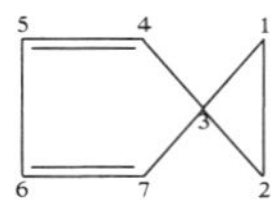	$v=0$	$A=5766.973\,(71)$ $B=2454.162\,(2)$ $C=1990.343\,(3)$		81HAR			
	5-D	$v=0$	$A=5977.960\,(57)$ $B=2375.489\,(2)$ $C=1961.592\,(2)$					
671	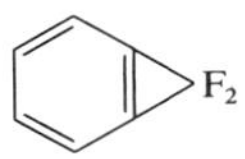	$v=0$	$A=3730.203$ $B=1209.099$ $C=1088.952$		75POZ	75POZ		2.6.3, 335
		$v_x=1$	$A=3726.854$ $B=1210.738$ $C=1090.829$				$v_x \simeq 116\ cm^{-1}$ probably out-of-plane bending	
		$v_x=2$	$A=3723.051$ $B=1212.354$ $C=1092.681$					
		$v_x=3$	$A=3720.55$ $B=1213.966$ $C=1094.501$					
672	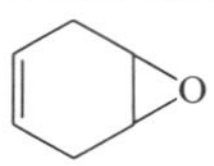	$v=0$	$A=4073.154\,(7)$ $B=3394.537\,(7)$ $C=2174.066\,(12)$		78CHA	78CHA	boat conformation	2.6.3, 336
673	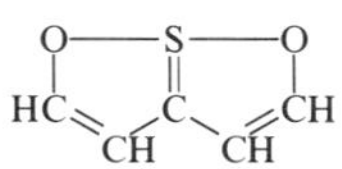	$v=0$	$A=4234.7556\,(13)$ $B=1677.60185\,(87)$ $C=1201.58046\,(81)$	$\Delta_J=0.0618\,(25)$ $\Delta_{JK}=0.1225\,(41)$ $\Delta_K=0.319\,(26)$ $\delta_J=0.01784\,(32)$ $\delta_K=0.1600\,(62)$	77PED		C_{2v} symmetry	2.6.3, 337
		$v_a=1$	$A=4223.6115\,(15)$ $B=1676.84772\,(89)$ $C=1202.50119\,(84)$					
		$v_a=2$	$A=4212.7454\,(24)$ $B=1676.0595\,(13)$ $C=1203.41205\,(56)$					
Cont.		$v_s=1$	$A=4232.4713\,(19)$ $B=1675.8359\,(11)$ $C=1202.03472\,(92)$					

Nr.	Molecule	v	A, B, C MHz	D, H kHz	Ref.	Ref. struct.	Varia, remarks	See also Tab., Nr.
673 Cont.	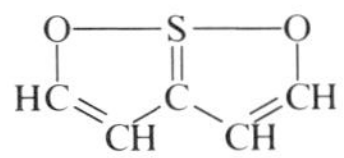	$v_s=2$	$A=4230.1801\,(11)$		77PED			
			$B=1674.09777\,(69)$					
			$C=1202.44316\,(66)$					
		$v_{a'}=1$	$A=4233.7674\,(20)$					
			$B=1668.5049\,(10)$					
			$C=1195.4475\,(10)$					
	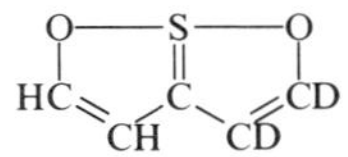	$v=0$	$A=3860.2278\,(22)$					
			$B=1653.90721\,(92)$					
			$C=1157.83237\,(89)$					
674	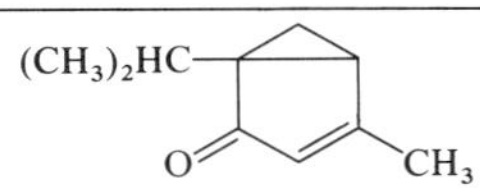		see Nr. 556					
675 *6-274a* *4-180*	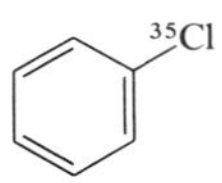	$v=0$	$A=5672.87\,(26)$		76ROU	76MIC	for structure, see also Vol. II/7 (3, 656)	2.7.3, 138
			$B=1576.775\,(4)$					
			$C=1233.661\,(4)$					
	1-^{13}C	$v=0$	$A=5672.68\,(77)$					
			$B=1575.827\,(15)$					
			$C=1233.108\,(19)$					
	2-^{13}C	$v=0$	$A=5580.91\,(34)$					
			$B=1576.510\,(6)$					
			$C=1229.096\,(5)$					
	3-^{13}C	$v=0$	$A=5582.51\,(25)$					
			$B=1563.588\,(4)$					
			$C=1221.325\,(3)$					
	4-^{13}C	$v=0$	$A=5672.33\,(38)$					
			$B=1550.319\,(7)$					
			$C=1217.423\,(6)$					
	2-D	$v=0$	$A=5396.43\,(29)$					
			$B=1576.371\,(5)$					
			$C=1219.782\,(4)$					
	3-D	$v=0$	$A=5400.34\,(38)$					
			$B=1553.311\,(6)$					
			$C=1206.151\,(6)$					
	4-D	$v=0$	$A=5672.8\,(12)$					
			$B=1521.199\,(15)$					
Cont.			$C=1199.405\,(15)$					

Nr.	Molecule	v	A, B, C MHz	D, H kHz	Ref.	Ref. struct.	Varia, remarks	See also Tab., Nr.
675 *6–274 a* *4–180* Cont.	2,6-D_2	$v=0$	$A=5139.71(10)$ $B=1575.918(3)$ $C=1205.982(2)$		76ROU			
	3,5-D_2	$v=0$	$A=5137.73(70)$ $B=1531.815(11)$ $C=1179.862(10)$		76MIC			
	3,4,5-D_3	$v=0$	$A=5139.29(117)$ $B=1480.444(17)$ $C=1149.169(14)$					
	2,4,6-D_3	$v=0$	$A=5141.06(85)$ $B=1520.298(14)$ $C=1173.119(13)$					
	2,3,5,6-D_4	$v=0$	$A=4696.95(19)$ $B=1530.759(6)$ $C=1154.370(6)$					
	D_5	$v=0$	$A=4696.47(52)$ $B=1479.310(9)$ $C=1124.892(7)$					
	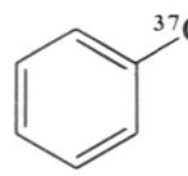 ^{37}Cl	$v=0$	$A=5672.72(65)$ $B=1532.781(10)$ $C=1206.562(7)$		76ROU			
	2,6-D_2	$v=0$	$A=5139.71(9)$ $B=1532.165(2)$ $C=1180.191(2)$					
	2-D	$v=0$	$A=5396.35(43)$ $B=1532.475(6)$ $C=1193.357(7)$		76MIC			
	3-D	$v=0$	$A=5399.60(56)$ $B=1509.831(8)$ $C=1179.787(7)$					
	3,5-D_2	$v=0$	$A=5139.23(51)$ $B=1488.829(8)$ $C=1154.167(7)$					
	D_5	$v=0$	$A=4697.35(37)$ $B=1438.310(8)$ $C=1101.022(8)$					

Nr.	Molecule	v	A, B, C MHz	D, H kHz	Ref.	Ref. struct.	Varia, remarks	See also Tab., Nr.
675a *6–276* *4–182*	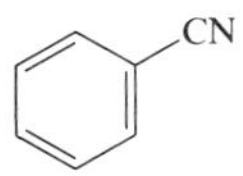	$v=0$	$A=5655.154\,(50)$ $B=1546.877\,(1)$ $C=1214.404\,(1)$	$\Delta_J=0.0535\,(80)$ $\Delta_{JK}=0.772\,(30)$ $\Delta_K=6.1\,(12)$ $\delta_J=0.0043\,(20)$ $\delta_K=1.33\,(20)$	81FLI1		for structure, see also Vol. II/7 (3, 721)	2.7.3, 71a
676	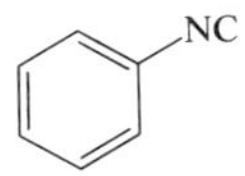	$v=0$	$A=5659.5190\,(56)$ $B=1639.7757\,(12)$ $C=1271.1538\,(12)$	$T_4=0.8526\,(90)$ $T_5=-0.0473\,(38)$	73BAK	73BAK	for definition of *T*-constants, see [71CAS]	
	2-D	$v=0$	$A=5384.321\,(14)$ $B=1638.7005\,(23)$ $C=1256.1018\,(22)$	$T_4=0.775\,(17)$ $T_5=-0.027\,(11)$				
	2,4-D_2	$v=0$	$A=5384.564\,(30)$ $B=1583.367\,(2)$ $C=1223.344\,(1)$					
	2,4,6-D_3	$v=0$	$A=5129.733\,(54)$ $B=1582.291\,(3)$ $C=1209.064\,(2)$					
677 *6–277* *4–184*	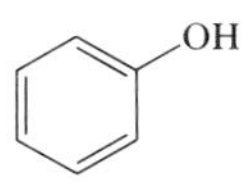	$v=0$	$A=5650.5154\,(11)$ $B=2619.2360\,(5)$ $C=1789.8520\,(3)$	$\tau_{aaaa}=-4.7679$ $\tau_{bbbb}=-0.8898$ $\tau_{cccc}=-0.1821$ $\tau_{abab}+\alpha\tau_{caca}=-1.2346$ *)	79LAR	79LAR	for structure, see also Vol. II/7 (3, 671) *) $\alpha=(A-B)/(A-C)$ **) the constants were fixed on the values for phenol	2.6.3, 338 2.8, 92
	1-^{13}C	$v=0$	$A=5650.7939\,(11)$ $B=2608.1275\,(8)$ $C=1784.6853\,(7)$	**)				
	2-^{13}C	$v=0$	$A=5560.0490\,(17)$ $B=2618.6547\,(10)$ $C=1780.4015\,(10)$	**)				
	6-^{13}C	$v=0$	$A=5560.3673\,(21)$ $B=2618.6162\,(13)$ $C=1780.4155\,(13)$	**)				
	3-^{13}C	$v=0$	$A=5562.2801\,(18)$ $B=2600.5200\,(10)$ $C=1772.2219\,(10)$	**)				
Cont.								

Nr.	Molecule	v	A, B, C MHz	D, H kHz	Ref.	Ref. struct.	Varia, remarks	See also Tab., Nr.
677 *6–277* *4–184* Cont.	5-^{13}C	$v=0$	$A=5562.1737(15)$ $B=2600.8023(8)$ $C=1772.3423(8)$	**)	79LAR		**) the constants were fixed on the values for phenol	
	4-^{13}C	$v=0$	$A=5650.7203(14)$ $B=2572.6609(9)$ $C=1767.9981(9)$	**)				
678	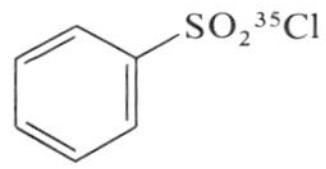	$v=0$	$A=2077.4$ $B=734.7$ $C=614.2$		78BOG	78BOG		3, 183
	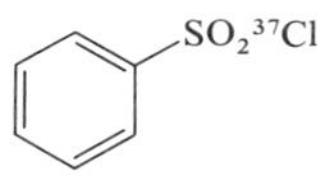	$v=0$	$A=2053.7$ $B=725.8$ $C=605.9$					
679 *6–279*	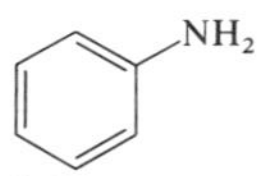	$v=0$	$A=5615.57$ $B=2592.24$ $C=1776.73$		74LIS, 73HAT	74LIS	for structure, see also Vol. II/7 (3, 673)	2.6.3, 339 2.7.3, 72
	2-D	$v=0$	$A=5346.09$ $B=2583.96$ $C=1744.99$		74LIS			
	4-D	$v=0$	$A=5615.91$ $B=2481.85$ $C=1724.17$					
	3,5-D_2	$v=0$	$A=5092.47$ $B=2518.99$ $C=1688.15$					
	2,4,6-D_3	$v=0$	$A=5092.70$ $B=2466.40$ $C=1664.38$					
	2,3,4,5,6-D_5	$v=0$	$A=4657.89$ $B=2401.07$ $C=1586.93$					
Cont.	^{15}N	$v=0$	$A=5615.24$ $B=2522.21$ $C=1743.56$					

Nr.	Molecule	v	A, B, C MHz	D, H kHz	Ref.	Ref. struct.	Varia, remarks	See also Tab., Nr.
679	1-^{13}C	$v=0$	$A=5615.63$		74LIS			
6–279			$B=2580.70$					
Cont.			$C=1771.33$					
	2-^{13}C	$v=0$	$A=5526.84$					
			$B=2591.58$					
			$C=1767.46$					
	3-^{13}C	$v=0$	$A=5528.46$					
			$B=2574.30$					
			$C=1759.57$					
	4-^{13}C	$v=0$	$A=5615.70$					
			$B=2546.83$					
			$C=1755.30$					
	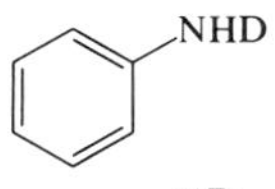	$v=0$	$A=5569.16$					
			$B=2492.90$					
			$C=1726.28$					
	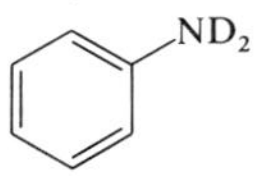	$v=0$	$A=5519.74$		74LIS,			
			$B=2403.75$		73HAT			
			$C=1679.04$					
680	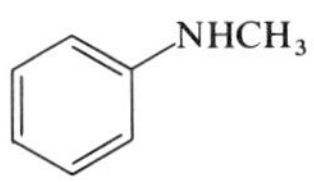	$v=0$	$A=4969.64(5)$ *)		81CER	81CER	*) state a	
			$B=1520.999(1)$				**) state b, analogous to the	
			$C=1175.670(3)$				O^+, O^- inversion doubling,	
			$A=4958.14(10)$ **)				$\Delta E(O^+ - O^-) \simeq 50$ cm^{-1}	
			$B=1522.312(3)$					
			$C=1177.659(2)$					
		$v_c=1$	$A=4947.11(12)$				$\nu_c=100(30)$ cm^{-1}, $HNCH_3$	
			$B=1521.942(3)$				torsion	
			$C=1178.728(2)$					
		$v_d=1$	$A=4940.15(11)$				ν_d, $HNCH_3$ torsional state of	
			$B=1522.792(3)$				the O^- inversion state	
			$C=1179.556(3)$					
		$v_e=1$	$A=4869(17)$					
			$B=1523.10(4)$					
			$C=1180.32(58)$					
		$v_f=1$	$A=4915.76(12)$					
			$B=1522.078(3)$					
Cont.			$C=1182.289(3)$					

Nr.	Molecule	v	A, B, C MHz	D, H kHz	Ref.	Ref. struct.	Varia, remarks	See also Tab., Nr.
680 Cont.	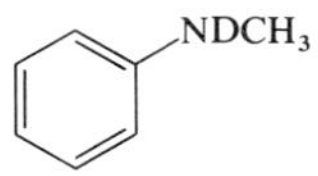	$v=0$	$A=4857.63(5)$ *) $B=1502.006(1)$ $C=1158.964(2)$ $A=4849.82(6)$ **) $B=1503.112(1)$ $C=1160.462(2)$		81CER		*) state a **) state b, analogous to the O^+, O^- inversion doubling, $\Delta E(O^+-O^-)\simeq 50\ cm^{-1}$	
		$v_c=1$	$A=4839.88(15)$ $B=1502.358(1)$ $C=1161.193(3)$					
		$v_d=1$	$A=4837.46(8)$ $B=1503.030(1)$ $C=1161.857(2)$					
680a *6-280* *4-183a*	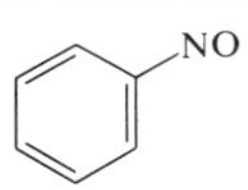	$v=0$	$A=5248.87(50)$ $B=1643.261(20)$ $C=1252.002(10)$	$\Delta_J=0.007(100)$ $\Delta_{JK}=0.36(50)$ $\Delta_K=-262(200)$ $\delta_J=0.03(5)$ $\delta_K=-9.4(70)$	81FLI2		for structure, see also Vol. II/7 (3, 662)	2.7.3, 72a
681	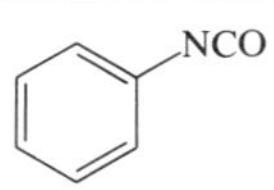	$v=0$	$A=5202.3(4)$ $B=972.685(7)$ $C=819.623(6)$	$\Delta_J=0.068(4)$ $\Delta_{JK}=-0.208(30)$ $\delta_J=0.015(5)$	77BOU4	79MAS		2.6.3, 340 3, 184
		$v_\tau=1$	$A=5144.6(3)$ $B=972.788(6)$ $C=821.139(6)$	$\Delta_J=0.071(4)$ $\Delta_{JK}=0.157(56)$ $\delta_J=0.018(4)$			$v_\tau=57(16)\ cm^{-1}$, torsion $V_2=3.4(1.0)$ kcal/mol	
		$v_\tau=2$	$A=5096.5(6)$ $B=973.03(1)$ $C=822.53(2)$	$\Delta_J=0.088(19)$				
		$v_\tau=3$	$A=5048(2)$ $B=973.34(2)$ $C=823.87(2)$	$\Delta_J=0.069(15)$ $\delta_J=0.029(10)$				
Cont.		$v_\tau=4$	$A=5002(2)$ $B=973.65(3)$ $C=825.26(4)$	$\Delta_J=0.084(32)$				

Nr.	Molecule	v	A, B, C MHz	D, H kHz	Ref.	Ref. struct.	Varia, remarks	See also Tab., Nr.
681 Cont.	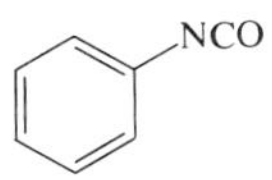	$v_i=1$	$A=5263.4(5)$ $B=972.955(9)$ $C=817.495(9)$	$\Delta_J=0.0716(70)$ $\delta_J=0.015(6)$	77BOU4		ν_i, in-plane wagging of NCO.	
		$v_x=1$	$A=5198(6)$ $B=973.1(1)$ $C=819.92(5)$	$\Delta_J=0.248(50)$ $\delta_J=0.106(40)$			$\nu_x=166(13)$ cm^{-1}, in-plane vibration	
	2-D trans	$v=0$	$A=4929.3(3)$ $B=971.932(7)$ $C=812.005(6)$	$\Delta_J=0.068(5)$ $\delta_J=0.014(5)$	79MAS			
	2-D cis	$v=0$	$A=5027.9(4)$ $B=972.359(9)$ $C=814.968(7)$	$\Delta_J=0.059(6)$ $\Delta_{JK}=0.217(113)$ $\delta_J=0.006(5)$				
682	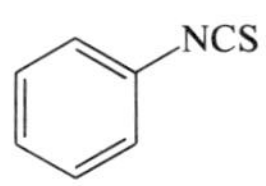	$v=0$	$A=5217.45(500)$ *) $B=634.447(10)$ $C=565.627(10)$		76OND3	76OND3	*) uncertainties are 2.5σ	
		$v_x=1$	$A=5184.15(15.00)$ *) $B=633.519(20)$ $C=565.584(20)$					
683	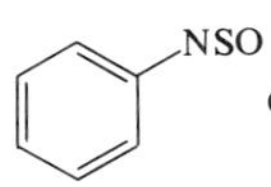 cis	$v=0$	$A=4026.669(6)$ $B=860.636(1)$ $C=709.531(1)$	$D_J=0.0371(12)$ $D_{JK}=-0.160(20)$	77CAM1			2.6.3, 341
		$v_\tau=1$	$A=4013.42(2)$ $B=861.359(4)$ $C=711.246(4)$	$D_J=0.0295(49)$ $D_{JK}=-0.236(51)$			$\nu_\tau=41.1$ cm^{-1}, torsion, $V_2=2.3$ kcal/mol	
		$v_\tau=2$	$A=4000.0(6)$ $B=862.106(13)$ $C=713.013(17)$	$D_J=0.035(18)$				
		$v_\tau=3$	$A=3987.6(8)$ $B=862.873(17)$ $C=714.815(22)$	$D_J=0.045(21)$				
684 *6–283* Cont.	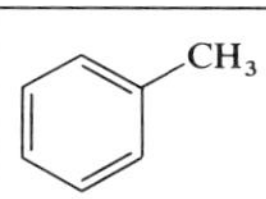	$v=0$	$A'=5729.325(217)$ *) $B=2517.447(9)$ $C=1748.872(20)$		81AMI	81AMI	for structure, see also Vol. II/7 (3, 728) *) rotational constant of the framework part only	

Nr.	Molecule	v	A, B, C MHz	D, H kHz	Ref.	Ref. struct.	Varia, remarks	See also Tab., Nr.
684 *6–283* Cont.	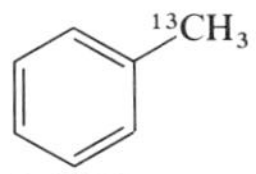	$v=0$	$A'=5729.49(15)$ *) $B=2444.590(10)$ $C=1713.364(10)$		81AMI		*) rotational constant of the framework part only	
	1-^{13}C	$v=0$	$A'=5730.215(141)$ *) $B=2506.420(6)$ $C=1743.528(6)$					
	2-^{13}C	$v=0$	$A'=5638.549(277)$ *) $B=2516.859(16)$ $C=1739.904(15)$					
	3-^{13}C	$v=0$	$A'=5638.572(100)$ *) $B=2500.599(4)$ $C=1732.217(43)$					
	4-^{13}C	$v=0$	$A'=5729.63(12)$ *) $B=2474.926(5)$ $C=1728.233(5)$					
685	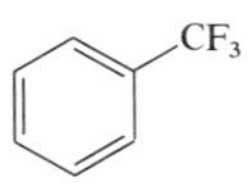	$v=0$	$B=946.31(8)$ $C=810.63(4)$		76OGA			2.8, 56
686	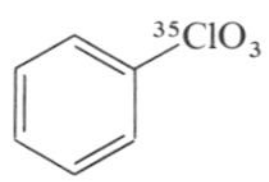		$B+C=1608.6(2)$		77CAM2	77CAM2		
	$C_6H_5{}^{37}ClO_3$		$B+C=1598.6(2)$					
	$C_6D_5{}^{35}ClO_3$		$B+C=1505.6(2)$					
	$C_6D_5{}^{37}ClO_3$		$B+C=1436.2(2)$					
687	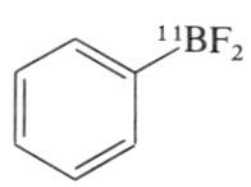	$v=0$	$A=3686.14(7)$ $B=1099.431(2)$ $C=847.199(1)$		74CHR	74CHR	planar molecule	2.8, 139
		$v_\tau=1$	$A=3685.40(10)$ $B=1097.401(2)$ $C=847.924(2)$				$\nu_\tau=49\ cm^{-1}$, torsion	
Cont.		$v_\tau=2$	$A=3684.95(10)$ $B=1095.342(2)$ $C=848.642(2)$					

Nr.	Molecule	v	A, B, C MHz	D, H kHz	Ref.	Ref. struct.	Varia, remarks	See also Tab., Nr.
687 Cont.	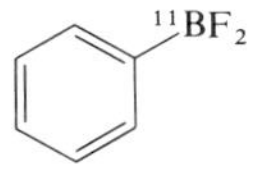	$v_\tau=3$	$A=3684.15(10)$ $B=1093.266(2)$ $C=849.362(2)$		74CHR			
	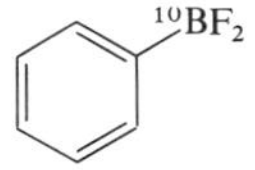	$v=0$	$A=3685.80(11)$ $B=1105.685(3)$ $C=850.900(2)$					
		$v_\tau=1$	$A=3685.51(15)$ $B=1103.629(3)$ $C=851.626(4)$					
		$v_\tau=2$	$A=3684.31(18)$ $B=1101.549(4)$ $C=852.361(4)$					
688	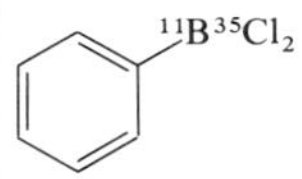	$v=0$	$A=2060.74(7)$ $B=769.375(3)$ $C=560.497(4)$		78CAM	78CAM		2.8, 140
		$v_\tau=1$	$A=2059.86(14)$ $B=768.549(6)$ $C=561.131(10)$				$v_\tau=37(2)\ cm^{-1}$, torsion	
		$v_\tau=2$	$A=2058.89(13)$ $B=767.714(6)$ $C=561.758(10)$					
		$v_\tau=3$	$A=2058.7(3)$ $B=768.89(1)$ $C=562.33(2)$					
		$v_\tau=4$	$A=2057.9(4)$ $B=766.05(2)$ $C=562.94(3)$					
	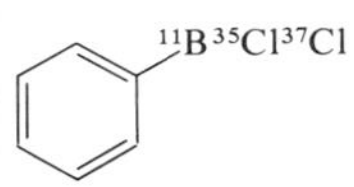	$v=0$	$A=2024.9(2)$ $B=760.417(9)$ $C=553.095(9)$					
		$v_\tau=1$	$A=2024.4(7)$ $B=759.62(3)$ $C=553.72(2)$					
Cont.		$v_\tau=2$	$A=2023.0(6)$ $B=758.78(2)$ $C=554.31(2)$					

Nr.	Molecule	v	A, B, C MHz	D, H kHz	Ref.	Ref. struct.	Varia, remarks	See also Tab., Nr.
688 Cont.	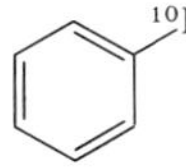 $^{10}B^{35}Cl_2$	$v=0$	$A=2060.74$ (fixed) $B=$ 770.64 (1) $C=$ 561.15 (3)		78CAM			
689	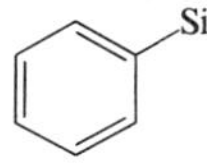 SiH_3	$v=0$	$A'=5702.96$ (27) *) $B=1499.574$ (3) $C=1187.118$ (3)		75CA1	76CA2	for structure, see also Vol. II/7 (3, 682) *) rotational constant of the framework part only	2.6.3, 342 2.8, 72
	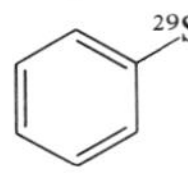 $^{29}SiH_3$	$v=0$	$A'=5703$ (assumed) *) $B=1475.637$ (10) $C=1172.056$ (10)		76CA2			
	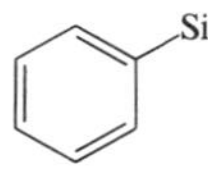 SiD_3	$v=0$	$A'=5700.2$ (5) *) $B=1388.993$ (5) $C=1116.559$ (5)					
690 *4-186*	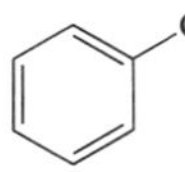 C≡CH	$v=0$	$A=5680.3$ (3) $B=1529.742$ (3) $C=1204.952$ (3)	$D_J=0.0298$ (30) $D_{JK}=1.047$ (4)	75COX	75COX	C_{2v} symmetry	2.6.3, 343
	2-^{13}C	$v=0$	$B+C=2730.442$ (5)					
	3-^{13}C	$v=0$	$A=5589.3$ (20) $B=1518.985$ (15) $C=1194.185$ (15)					
	4-^{13}C	$v=0$	$B=1507.181$ (10) *) $C=1190.910$ (10)				*) $A=5680.3$ MHz, assumed	
	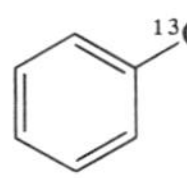 ^{13}C≡CH	$v=0$	$B=1511.566$ (10) *) $C=1193.646$ (10)					
	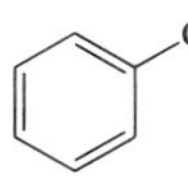 C≡^{13}CH	$v=0$	$B=1483.818$ (10) *) $C=1176.276$ (10)					
	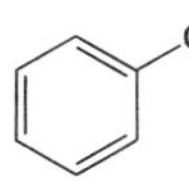 C≡CD	$v=0$	$A=5680.8$ (6) $B=1450.244$ (4) $C=1155.061$ (4)	$D_J=0.0234$ (30) $D_{JK}=1.076$ (10)				

Nr.	Molecule	v	A, B, C MHz	D, H kHz	Ref.	Ref. struct.	Varia, remarks	See also Tab., Nr.
691	 CH=CH^{79}Br trans		$B+C=731.9$ *)		73RAL3		*) LRMW	
	 CH=CH^{81}Br trans		$B+C=724.3$ *)					
692	 CH=CHCHO s-trans trans s-cis **)		$B+C=1106.3\,(1)$ * $B+C=1179.1\,(1)$ *)		74STE		*) LRMW **) less stable conformer	
693 *6-284*	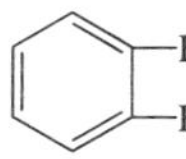 F, F						for structure, see also Vol. II/7 (3, 650)	2.9.3, 49
694 *6-288*	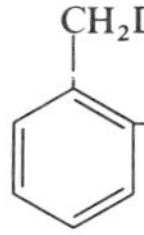 CH_2D, F, symmetric *)	$v=0$	$A=3121.553\,(10)$ **) $B=2150.948\,(6)$ $C=1283.713\,(6)$		75SCH		*) D in plane (or H). **) uncertainties are 2 σ. ***) D out-of-plane (or H)	2.8, 57
	asymmetric ***)	$v=0^+$	$A^+=3203.18\,(2)$ **) $B^+=2118.645\,(2)$ $C^+=1290.07\,(2)$				$\Delta\nu=1018.8\,(20)$ MHz, $Q_a=$ 15.05 (10) MHz, $Q_b=$ 10.6 (12) MHz, $R_{bc}=$ 5.68 MHz [1]) $R_{ac}=$ 16.74 MHz [1]) [1]) calculated from structure	
		$v=0^-$	$A^-=3202.72\,(2)$ $B^-=2118.31\,(3)$ $C^-=1290.08\,(3)$					
	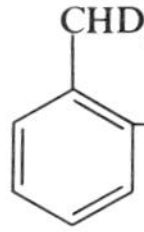 CHD_2, F, symmetric *)	$v=0$	$A=3164.644\,(8)$ **) $B=2059.581\,(10)$ $C=1266.899\,(10)$					
	asymmetric ***)	$v=0^+$	$A^+=3090.48\,(2)$ **) $B^+=2088.775\,(5)$ $C^+=1260.81\,(7)$				$\Delta\nu=111.8\,(16)$ MHz, $Q_a=$ 3.98 (6) MHz, $Q_b=$ 1.5 (7) MHz, $R_{bc}=$ 5.00 MHz [1]) $R_{ac}=$ 16.18 MHz [1]) [1]) calculated from structure	
		$v=0^-$	$A^-=3090.17\,(2)$ $B^-=2088.808\,(8)$ $C^-=1260.81\,(2)$					

Nr.	Molecule	v	A, B, C MHz	D, H kHz	Ref.	Ref. struct.	Varia, remarks	See also Tab., Nr.
695	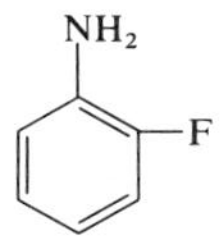	$v=0^+$	$A=3310.30$ $B=2218.72$ $C=1329.82$		77CHR			
		$v=0^-$	$A=3307.21$ $B=2218.25$ $C=1329.45$					
		$v_x=1$	$A=3302.43$ $B=2217.48$ $C=1330.68$				v_x, out-of-plane vibration, probably torsion around C—N bond	
696	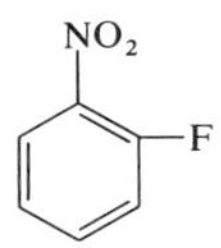	$v=0$	$A=2319.098(15)$ $B=1253.0196(62)$ $C=837.5323(34)$	$D_2=-0.038(6)$ $D_3=-0.011(3)$ $D_5=0.077(67)$	80COR3	80COR3	for definition of D-constants, see Nr. 610	
697	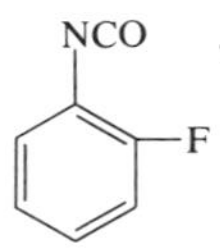 , cis	$v=0$	$A=3211.4(2)$ $B=965.669(3)$ $C=742.543(2)$		80MAS	80MAS		3, 185
		$v_i=1$	$A=3183.5(2)$ $B=966.221(4)$ $C=743.522(3)$				v_i, NCO group torsion	
		$v_i=2$	$A=3157.2(2)$ $B=966.730(4)$ $C=744.425(4)$					
		$v_i=3$	$A=3131.1(3)$ $B=967.239(6)$ $C=745.321(4)$					
		$v_i=4$	$A=3104.7(3)$ $B=967.758(7)$ $C=746.211(5)$					
		$v_j=1$	$A=3225.5(2)$ $B=965.042(4)$ $C=741.307(4)$				v_j, NCO in-plane vibration	
		$v_i=1, v_j=1$	$A=3194.9(5)$ $B=965.641(20)$ $C=742.393(4)$					

Nr.	Molecule	v	A, B, C MHz	D, H kHz	Ref.	Ref. struct.	Varia, remarks	See also Tab., Nr.
698	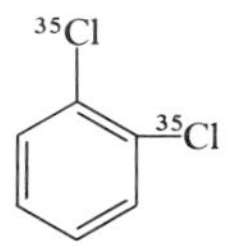	$v=0$	$A=1930.02(13)$ *) $B=1431.16(7)$ $C=821.670(4)$		76OND1	76OND1	*) uncertainties are 2.5 σ	2.7.3, 139
		$v_a=1$	$A=1927.4(12)$ $B=1430.21(73)$ $C=822.125(11)$				$\nu_a=152\ cm^{-1}$, out-of-plane C—Cl bending	
		$v_a=2$	$A=1921.9(14)$ $B=1431.03(93)$ $C=822.564(13)$					
		$v_a=3$	$A=1913(45)$ $B=1434(26)$ $C=823.013(19)$					
	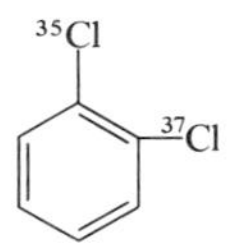	$v=0$	$A=1895.87(18)$ $B=1414.10(14)$ $C=809.846(8)$					
699	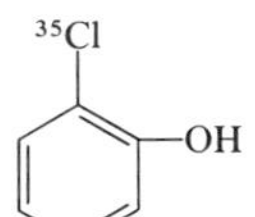	$v=0$	$A=2985.454(8)$ $B=1549.861(3)$ $C=1020.195(2)$	$\tau_{aaaa}=-1.9(3)$ $\tau_{bbbb}=-0.36(4)$ $\tau_{aabb}=-0.1(2)$ $\tau_{abab}=-0.48(7)$	78OND1	78OND2	s-cis-form: the hydroxyl hydrogen pointed toward the chlorine	2.7.3, 140
	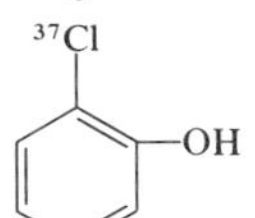	$v=0$	$A=2976.812(11)$ $B=1510.810(4)$ $C=1002.147(1)$	$\tau_{aaaa}=-1.7(3)$ $\tau_{bbbb}=-0.29(3)$ $\tau_{abab}=-0.50(2)$ $\tau_{aabb}=0$ (fixed)				
	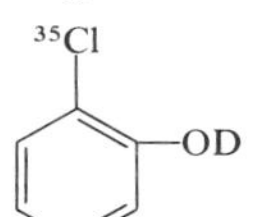	$v=0$	$A=2930.937(9)$ $B=1539.217(4)$ $C=1009.203(2)$	$\tau_{aaaa}=-2.0(3)$ $\tau_{bbbb}=-0.36(3)$ $\tau_{aabb}=-0.1(2)$ $\tau_{abab}=-0.46(7)$	78OND2			
700	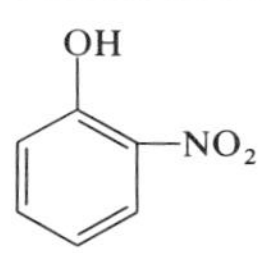	$v=0$	$A=2368.738(50)$ $B=1276.110(6)$ $C=829.812(7)$		73LEA	73LEA	$\Delta=-0.360(6)$ amu Å^2	3, 186
Cont.		$v_\tau=1$	$A=2368.98(10)$ $B=1275.031(12)$ $C=830.760(18)$				ν_τ, NO_2 torsion	

Nr.	Molecule	v	A, B, C MHz	D, H kHz	Ref.	Ref. struct.	Varia, remarks	See also Tab., Nr.
700 Cont.	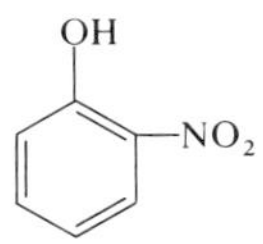	$v_\tau=2$	$A=2368.90(13)$ $B=1273.873(16)$ $C=\ 831.730(23)$		73LEA			
		$v_\tau=3$	$A=2368.93(12)$ $B=1272.705(15)$ $C=\ 832.660(25)$					
		$v_\tau=4$	$A=2369.6(4)$ $B=1271.535(46)$ $C=\ 833.485(65)$					
	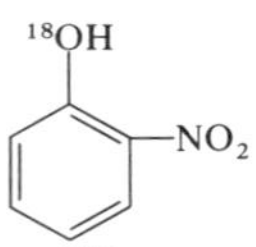	$v=0$	$A=2269.967(36)$ $B=1276.040(5)$ $C=\ 817.329(5)$					
	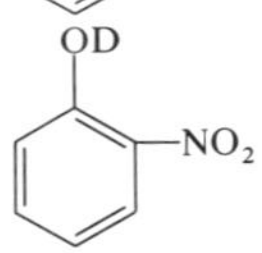	$v=0$	$A=2325.514(49)$ $B=1272.217(6)$ $C=\ 822.816(5)$					
	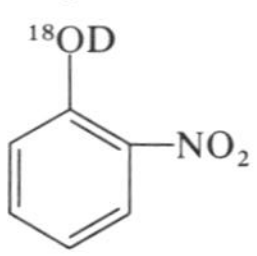	$v=0$	$A=2231.529(42)$ $B=1272.113(5)$ $C=\ 810.706(5)$					
701	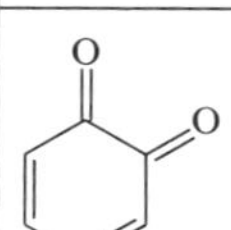	$v=0$	$A=3309.90$ $B=2495.41$ $C=1421.195$		75BL1			2.6.3, 344
702 *6–285*	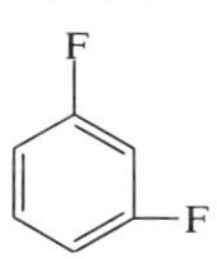						for structure, see also Vol. II/7 (3, 651)	2.9.3, 50
703	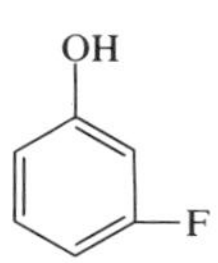	$v=0$	$\tilde{A}=3748.487(34)$ $\tilde{B}=1797.713(17)$ $\tilde{C}=1215.048(17)$	$d_J\ =\ 0.2257(2116)$ $d_{JK}=-10.977(2059)$ $d_K\ =-15.005(3281)$ $d_{WJ}=-2.0(10)\cdot 10^{-7}$ $d_{WK}=7.2(14)\cdot 10^{-6}$	81JAM	81JAM		3, 187

Nr.	Molecule	v	A, B, C MHz	D, H kHz	Ref.	Ref. struct.	Varia, remarks	See also Tab., Nr.
704	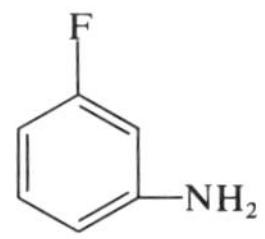	$v=0^+$	$A=3723.242\,(21)$ $B=1791.924\,(1)$ $C=1210.736\,(1)$		73CAZ1	73CAZ1		3, 188
		$v=0^-$	$A=3721.648\,(21)$ $B=1790.957\,(2)$ $C=1210.493\,(2)$					
	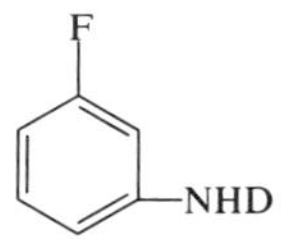 , cis *)	$v=0^+$	$A=3622.927\,(18)$ $B=1754.221\,(1)$ $C=1183.341\,(1)$				*) cis and trans refer to the position of D relative to F	
		$v=0^-$	$A=3622.527\,(37)$ $B=1753.324\,(2)$ $C=1183.378\,(2)$					
	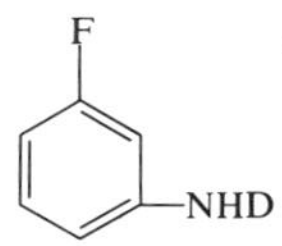 , trans *)	$v=0^+$	$A=3714.738\,(31)$ $B=1726.069\,(2)$ $C=1179.884\,(2)$					
		$v=0^-$	$A=3712.914\,(60)$ $B=1725.778\,(4)$ $C=1179.927\,(4)$					
	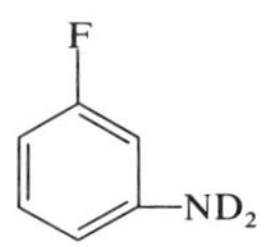	$v=0^+$	$A=3617.612\,(32)$ $B=1691.373\,(2)$ $C=1154.349\,(2)$					
		$v=0^-$	$A=3616.770\,(35)$ $B=1691.214\,(2)$ $C=1154.516\,(2)$					
705	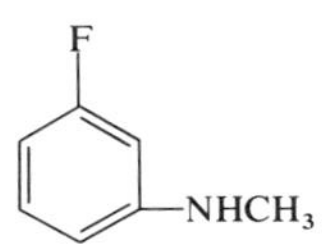 , cis *)	$v=0$	$B+C=1931.6\,(3)$ **)		80CER		*) with respect to the amino hydrogen	
	trans *)	$v=0$	$B+C=2158.2\,(6)$ **)				**) LRMW	
706	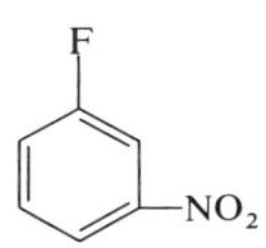	$v=0$	$A=2718.169\,(29)$ $B=968.2496\,(11)$ $C=714.4387\,(7)$		80COR3			
Cont.		$v_\tau=1$	$A=2717.086\,(45)$ $B=967.3942\,(18)$ $C=715.3107\,(23)$				v_τ, NO_2 torsion	

Nr.	Molecule	v	A, B, C MHz	D, H kHz	Ref.	Ref. struct.	Varia, remarks	See also Tab., Nr.
706 Cont.	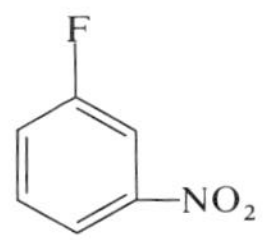	$v_\tau=2$	$A=2715.249(35)$ $B=\ 966.4912(13)$ $C=\ 716.1763(7)$		80COR3		$V_2=1428\ cm^{-1}$	
707	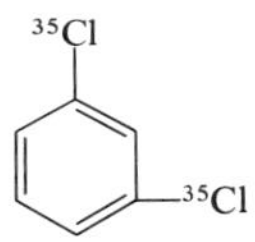	$v=0$	$A=2832.337(10)$ *) $B=\ 862.812(3)$ $C=\ 661.262(3)$		76OND2	76OND2	*) uncertainties are 2.5 σ	2.7.3, 141
	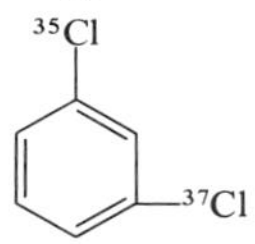	$v=0$	$A=2811.610(15)$ *) $B=\ 842.222(6)$ $C=\ 648.010(4)$					
708 *6–286a* *4–187*	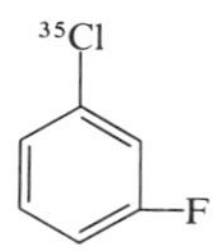						for structure, see also Vol. II/7 (3, 648)	2.7.3, 142
709	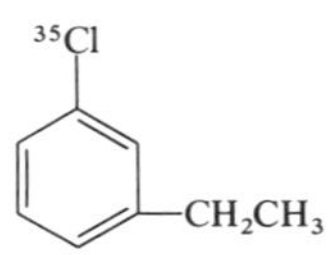		$B+C=1505(7)$ *)		74FAR		*) LRMW	
710	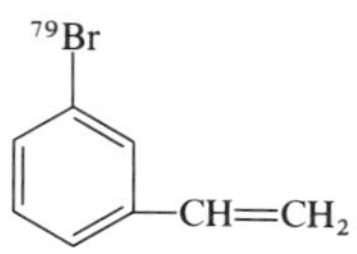		$B+C=1141.0$ *)		73RAL3		*) LRMW	
	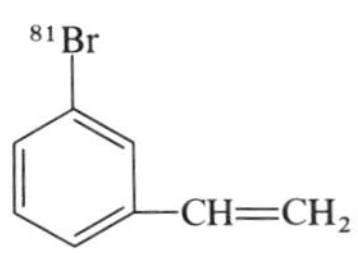		$B+C=1130.0$ *)					

Nr.	Molecule	v	A, B, C MHz	D, H kHz	Ref.	Ref. struct.	Varia, remarks	See also Tab., Nr.
711	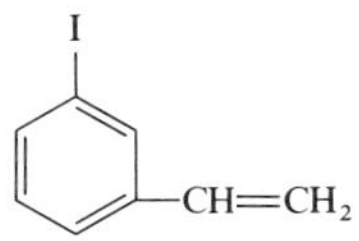		$B+C=887$ *) $B+C=801$ **)		75RAL		*) cis form, LRMW **) trans form, LRMW	
712	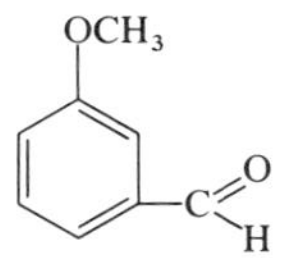		$B+C=1416(1)$ *) $^{+}$) $B+C=1570(2)$ *) $B+C=1623(4)$ *)		74STE		*) LRMW. $^{+}$) principal conformer	
713	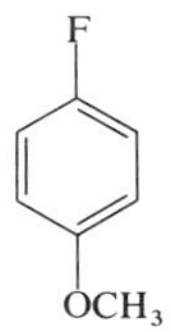	$v=0$	$A=4883.87(2)$ $B=\ 972.475(2)$ $C=\ 815.540(1)$		73LIS2			
		$v_\tau=1$	$A=4868.3(3)$ $B=\ 972.661(1)$ $C=\ 816.490(3)$		80LIS		$v_\tau=70\ cm^{-1}$, torsion of the methoxy group	
		$v_\tau=2$	$A=4850.8(8)$ $B=\ 972.774(2)$ $C=\ 817.440(80)$					
		$v_\tau=3$	$A=4842.5(6)$ $B=\ 972.963(2)$ $C=\ 818.482(7)$					
		$v_\tau=4$	$A=4835.9(3)$ $B=\ 973.088$ $C=\ 819.378(4)$					
		$v_\tau=5$	$A=4830.9(18)$ $B=\ 973.090(4)$ $C=\ 820.230(19)$					
714	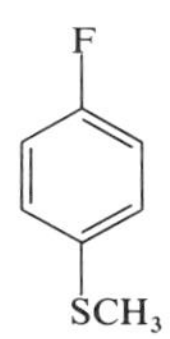		$B+C=1390.7(2)$ *) $B+C=1383.9(3)$ *) $B+C=1368.9(6)$ *)		76LIS1		*) LRMW relative intensity: 0.3 relative intensity: 0.05	

Nr.	Molecule	v	A, B, C MHz	D, H kHz	Ref.	Ref. struct.	Varia, remarks	See also Tab., Nr.
715	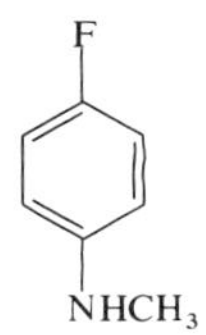	$v=0^+$	$A=4804.32(12)$ $B=953.0525(4)$ $C=800.746(1)$		81CER1	81CER1	inversion splitting of the $HNCH_3$ group: $0^+-0^- \sim 50\ cm^{-1}$	
		$v=0^-$	$A=4795.16(17)$ $B=953.6042(5)$ $C=801.713(2)$					
		$v_\tau=1^+$	$A=4789.65(17)$ $B=952.877(1)$ $C=801.646(1)$				$v_\tau \sim 100\ cm^{-1}$, torsion of the $HNCH_3$ group about the N-aryl bond	
		$v_\tau=1^-$, state d	$A=4781.62(15)$ $B=953.5386(5)$ $C=802.365(1)$					
		$v_\tau=1^-$, state e	$A=4778.06(22)$ $B=953.910(1)$ $C=802.417(2)$					
	$DNCH_3-C_6H_4-F$	$v=0^+$	$A=4713.60(16)$ $B=940.422(1)$ $C=789.875(1)$					
		$v=0^-$	$A=4708.48(16)$ $B=940.835(1)$ $C=790.490(1)$					
		$v_\tau=1^+$	$A=4695.98(28)$ $B=940.674(1)$ $C=791.030(2)$					
		$v_\tau=1^-$	$A=4693.19(24)$ $B=941.010(1)$ $C=791.283(2)$					
	$DNCD_3-C_6H_4-F$	$v=0$	$B+C=1620.9(5)$ *)		79CER		*) LRMW	
	$HNCD_3-C_6H_4-F$	$v=0$	$B+C=1640.4(2)$ *)					
716	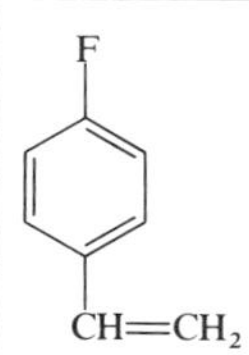	$v=0$	$A=5036.69(11)$ $B=971.2642(8)$ $C=815.2652(8)$		76RAL2			
Cont.		$v_\tau=1$	$A=5013.30(13)$ $B=972.3811(9)$ $C=817.5351(9)$				$v_\tau=30(15)\ cm^{-1}$, torsion	

Nr.	Molecule	v	A, B, C MHz	D, H kHz	Ref.	Ref. struct.	Varia, remarks	See also Tab., Nr.
716 Cont.	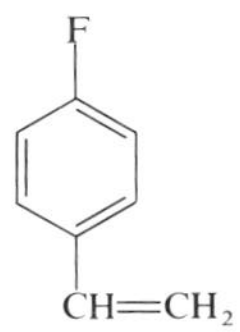	$v_\tau=2$	$A=4997.76(14)$ $B=973.0779(10)$ $C=819.1290(9)$		76RAL2			
		$v_\tau=3$	$A=4980.63(14)$ $B=973.6712(10)$ $C=820.5439(9)$					
		$v_i=1$	$A=5060.17(18)$ $B=970.9280(11)$ $C=814.8871(11)$				$\nu_i=205\ cm^{-1}$, in-plane deformation	
		$v_a=1$	$A=5017.69(17)$ $B=971.3729(12)$ $C=815.8596(12)$				$\nu_a=125\ cm^{-1}$, F sensitive out-of-plane vibration	
		$v_b=1$	$A=4983.84(18)$ $B=973.1836(13)$ $C=819.7580(12)$				$\nu_b=205\ cm^{-1}$	
		$v_c=1$	$A=4982.16(21)$ $B=973.3070(15)$ $C=819.7910(15)$				$\nu_c=210\ cm^{-1}$	
		$v_d=1$	$A=4981.78(9)$ $B=973.2460(7)$ $C=820.1092(6)$				$\nu_d=170\ cm^{-1}$	
717	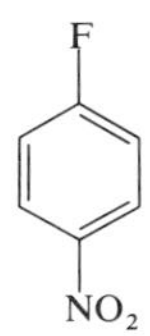	$v=0$	$A=3952.5070(85)$ $B=803.2069(10)$ $C=667.9809(10)$	$T_5=-0.0102(15)$	80COR3		for definition of T_5, see [71CAS]	
		$v_\tau=1$	$A=3951.348(86)$ $B=802.6597(10)$ $C=668.7156(9)$				ν_τ, NO_2 torsion	
		$v_\tau=2$	$A=3950.09(11)$ $B=802.0772(12)$ $C=669.4124(13)$				$V_2=1213\ cm^{-1}$	

Nr.	Molecule	v	A, B, C MHz	D, H kHz	Ref.	Ref. struct.	Varia, remarks	See also Tab., Nr.
718	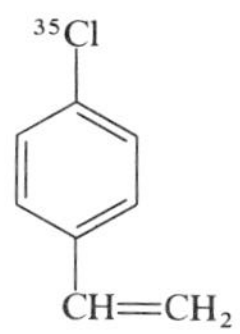	$v=0$	$A=4996.03(17)$ $B=691.9402(7)$ $C=608.3518(7)$		76RAL3		planar	
		$v_\tau=1$	$A=4972.20(23)$ $B=692.6817(9)$ $C=609.7744(9)$				$\nu_\tau=35(15)\ cm^{-1}$, torsion	
		$v_\tau=2$	$A=4951.42(26)$ $B=693.1893(9)$ $C=610.8070(10)$					
		$v_a=1$	$A=4965.73(27)$ $B=692.4153(11)$ $C=609.3341(10)$				$\nu_a=115\ cm^{-1}$	
		$v_b=1$	$A=4951.99(27)$ $B=692.8491(11)$ $C=610.2563(11)$				$\nu_b=110\ cm^{-1}$	
719	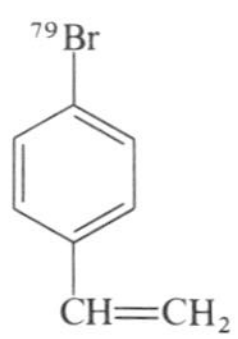		$B+C=897.1$ *)		73RAL3	73RAL3, 74NY1	*) LRMW	3, 189
	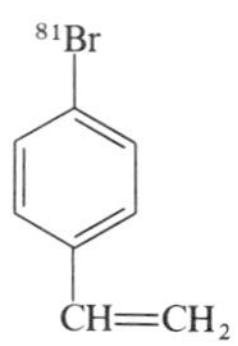		$B+C=888.4$ *)					
720	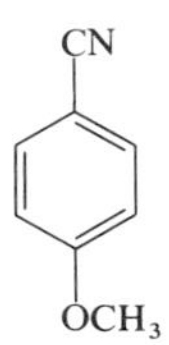		$B+C=1333.5(1)$ *)		74STE		*) LRMW	

Nr.	Molecule	v	A, B, C MHz	D, H kHz	Ref.	Ref. struct.	Varia, remarks	See also Tab., Nr.
721	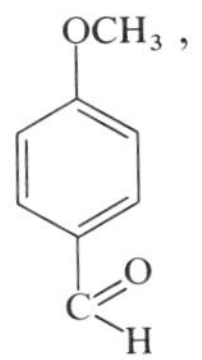 s-trans non planar s-cis		$B+C=1311.2(1)$ *) $B+C=1323.5(1)$ *) $B+C=1334.9(1)$ *)		74STE		*) LRMW	
	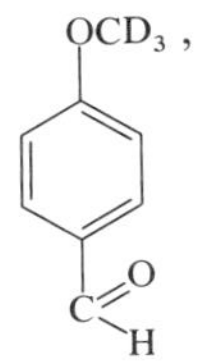 s-trans non-planar s-cis		$B+C=1235.0(4)$ *) $B+C=1248.1(5)$ *) $B+C=1260.6(9)$ *)					
722	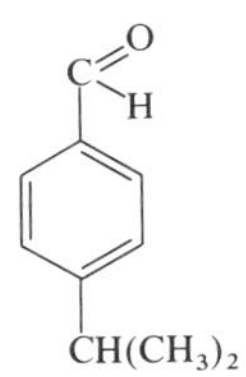		$B+C=1058.8(3)$ *)		74FAR		*) LRMW	
723	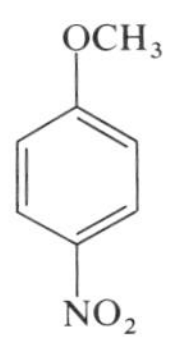		$B+C=1094.5(1)$ *)		74STE		*) LRMW	
724	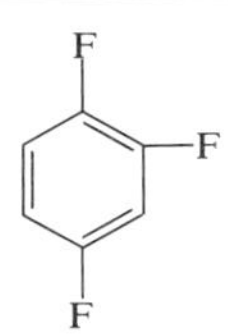	$v=0$	$\tilde{A}=3084.0215(62)$ $\tilde{B}=1278.3595(27)$ $\tilde{C}=903.6835(27)$	$\Delta_J=-0.0674(54)$ $\Delta_{JK}=0.728(52)$ $\Delta_K=-0.657(16)$ $\delta_J=0.0528(32)$ $\delta_K=-0.547(59)$	78DOR, 76NAN	78DOR		2.6.3, 345 3, 190

Nr.	Molecule	v	A, B, C MHz	D, H kHz	Ref.	Ref. struct.	Varia, remarks	See also Tab., Nr.
725	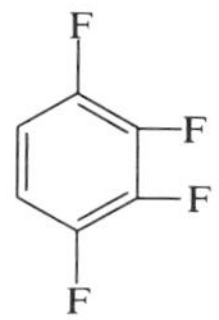	$v=0$	$A=2090.711\,(12)$ *) $B=1196.257\,(14)$ $C=\ \ 760.880\,(5)$		74HAR3	74HAR3	*) uncertainties are 2 σ	2.6.3, 346 3, 191
726	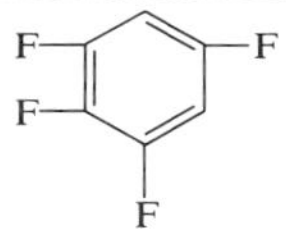	$v=0$	$\mathscr{A}=1757.070\,(12)$ $\mathscr{B}=1243.1151\,(87)$ $\mathscr{C}=\ \ 727.9903\,(79)$	$\tau_{aaaa}=-1.30\,(126)$ $\tau_{bbbb}=-4.05\,(40)$ $\tau_{cccc}=-0.75\,(40)$ $\tau_1=-15.21\,(134)$ $\tau_2=\ \ -5.09\,(37)$	77SHA	77SHA		2.6.3, 347
727	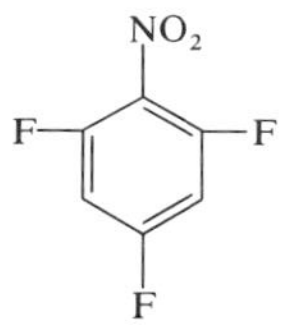	$v=0$	$A=1544.921\,(26)$ $B=\ \ 755.8967\,(25)$ $C=\ \ 534.6446\,(16)$	$D_2=-0.052\,(42)$ $D_3=-0.035\,(16)$ $D_5=-0.054\,(49)$	80COR3		for definition of D-constants, see Nr. 610	
728 *6-292*	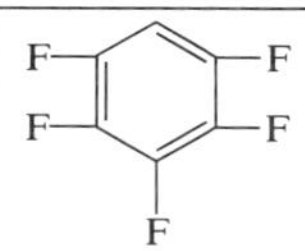	$v=0$	$A=1480.856\,(3)$ $B=1030.066\,(3)$ $C=\ \ 607.496\,(2)$		74DO2	74DO2		2.6.3, 348
729	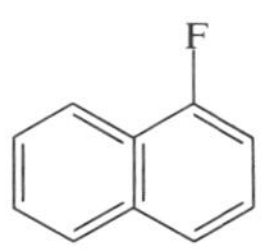	$v=0$	$A=1920.5617\,(8)$ $B=1122.2318\,(10)$ $C=\ \ 708.4778\,(4)$		73KAR1			
730 *6-294* *4-190* Cont.	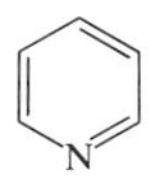 2-D	$v=0$ $v=0$	$A=6039.2516\,(6)$ $B=5804.9116\,(6)$ $C=2959.2117\,(6)$ $A=5900.8828\,(5)$ $B=5558.5214\,(5)$ $C=2861.7137\,(5)$	$D_1=-4.351\,(3)$ $D_2=-0.642\,(3)$ $D_3=\ \ 0.108\,(2)$ *) $D_1=-4.1952\,(14)$ $D_2=\ \ 0.4021\,(12)$ $D_3=-0.5932\,(13)$ *)	77MAT	77MAT	for structure, see also Vol. II/7 (3, 583) *) with $D_4=-0.5$ kHz, assumed for definition of D-constants, see Nr. 610	2.6.3, 349 2.7, 3.73 2.9.3, 51

Nr.	Molecule	v	A, B, C MHz	D, H kHz	Ref.	Ref. struct.	Varia, remarks	See also Tab., Nr.
730 *6–294* *4–190* Cont.	3-D	$v=0$	$A=5889.1923(10)$ $B=5555.0518(10)$ $C=2858.0310(10)$	$D_1=-4.212(3)$ $D_2=\ \ 0.446(3)$ $D_3=-0.628(3)$ *)	77MAT			
	4-D	$v=0$	$A=6039.9967(10)$ $B=5420.0697(9)$ $C=2855.8194(8)$	$D_1=-3.870(18)$ $D_2=-0.87(5)$ $D_3=-0.70(4)$ *)				
	2-^{13}C	$v=0$	$A=5963.13(4)$ **) $B=5758.90(3)$ $C=2928.96(1)$		74SOR		**) uncertainties are 3 σ	
	3-^{13}C	$v=0$	$A=5956.57(5)$ **) $B=5756.00(4)$ $C=2926.63(1)$					
	4-^{13}C	$v=0$	$A=6039.46(3)$ **) $B=5676.03(2)$ $C=2925.40(1)$					
	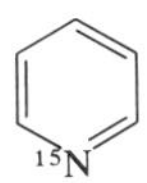	$v=0$	$A=6039.45(8)$ **) $B=5680.37(6)$ $C=2926.54(2)$					
731 *6–297*	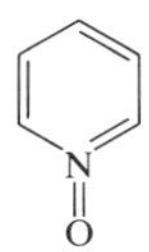	$v=0$	$A=5899.3555(23)$ $B=2794.6037(12)$ $C=1896.2563(9)$	$\tau_{aaaa}=-4.95(31)$ $\tau_{bbbb}=-1.05(3)$ $\tau_{cccc}=-0.24(2)$ $\tau_{abab}=-1.23(6)$	75SNE	75SNE	for structure, see also Vol. II/7 (3, 585)	
	^{15}N	$v=0$	$A=5899.632(14)$ $B=2781.4091(23)$ $C=1890.2013(6)$	*)			*) $\tau_{aaaa}=-5.00$; $\tau_{bbbb}=-1.05$; $\tau_{cccc}=-0.24$; $\tau_{abab}=-1.23$ kHz, assumed	
	2-^{13}C	$v=0$	$A=5805.942(33)$ $B=2793.781(58)$ $C=1886.123(15)$	*)				
	3-^{13}C	$v=0$	$A=5804.820(17)$ $B=2774.501(3)$ $C=1877.1931(8)$	*)				
	4-^{13}C	$v=0$	$A=5899.617(18)$ $B=2742.362(3)$ $C=1872.0856(9)$	*)				

Nr.	Molecule	v	A, B, C MHz	D, H kHz	Ref.	Ref. struct.	Varia, remarks	See also Tab., Nr.
732 *6-294a*	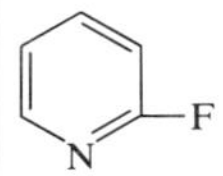	$v=0$	$A=5870.856(11)$ $B=2699.971(5)$ $C=1849.244(4)$	$\tau_{aaaa}=-2.465(16)$ $\tau_{bbbb}=-0.422(3)$ $\tau_{abab}=-0.618(15)$ *)	72SHA		*) with $\tau_{cccc}=0$ fixed	
733 *6-294d*	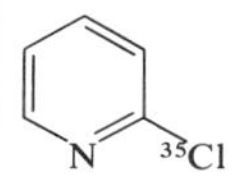	$v=0$	$A=5872.52(12)$ *) $B=1637.83(3)$ $C=1280.48(3)$	$D_{JK}=0.32(4)$	74WAL	74WAL	for structure, see also Vol. II/7 (3, 576) *) uncertainties are for 95% confidence $\nu_x=188\ cm^{-1}$, C—Cl out-of-plane vibration	2.7.3, 143
		$v_x=1$	$A=5843.61(30)$ $B=1638.97(6)$ $C=1282.02(6)$	$D_{JK}=0.70(15)$				
	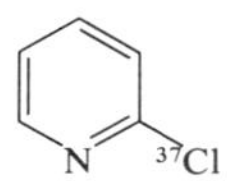	$v=0$	$A=5872.16(28)$ *) $B=1591.76(5)$ $C=1252.17(5)$	$D_{JK}=0.29(4)$				
734	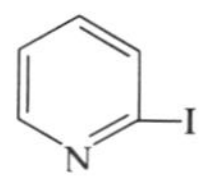	$v=0$	$A=5872.0(30)$ $B=\ 768.919(9)$ $C=\ 679.788(7)$		74CA1	74CA1		2.7.3, 163
735 *6-294b*	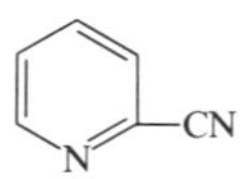							2.6.3, 351
736	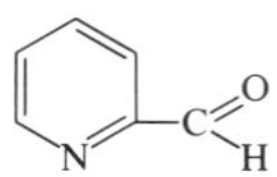	$v=0$	$A=5362.52(219)$ $B=1620.07(3)$ $C=1244.42(3)$		75KAW	75KAW	s-trans form	2.6.3, 350 3, 192
		$v_1=1$	$A=5342.21(271)$ $B=1620.65(4)$ $C=1246.31(4)$				$\nu_1=127(20)\ cm^{-1}$	
		$v_2=1$	$A=5256.60(422)$ $B=1615.70(7)$ $C=1245.49(5)$				$\nu_2=187(15)\ cm^{-1}$	
		$v_3=1$	$A=5432.22(466)$ $B=1621.67(6)$ $C=1246.67(4)$				$\nu_3=219(19)\ cm^{-1}$	
Cont.		$v_4=1$	$A=5374.64(294)$ $B=1622.03(3)$ $C=1245.65(3)$				$\nu_4=239(13)\ cm^{-1}$	

Nr.	Molecule	v	A, B, C MHz	D, H kHz	Ref.	Ref. struct.	Varia, remarks	See also Tab., Nr.
736 Cont.	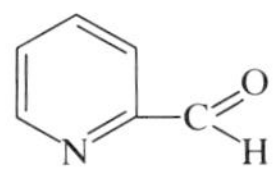	$v_5=1$	$A=5342.23\,(292)$ $B=1622.79\,(4)$ $C=1247.91\,(4)$		75KAW		$\nu_5=312\,(35)\ \text{cm}^{-1}$	
737	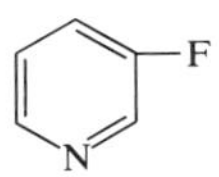	$v=0$	$\tilde{A}=5829.714\,(6)$ $\tilde{B}=2637.484\,(4)$ $\tilde{C}=1815.684\,(1)$	$\Delta_J=0.82\,(2)$ $\Delta_{JK}=3.07\,(13)$ $\Delta_K=1.81\,(5)$ $\delta_J=-0.036\,(9)$ $\delta_K=1.93\,(15)$	76SHA	76SHA		2.6.3, 352 2.7.3, 74 3, 193
738	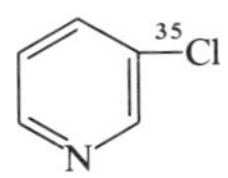	$v=0$	$\tilde{A}=5839.521\,(11)$ $\tilde{B}=1604.181\,(3)$ $\tilde{C}=1258.322\,(3)$	$\Delta_J=0.10\,(1)$ $\Delta_{JK}=0.36\,(9)$ $\Delta_K=1.18\,(7)$ $\delta_J=-0.008\,(5)$ $\delta_K=0.88\,(20)$	75SHA			2.7.3, 75
	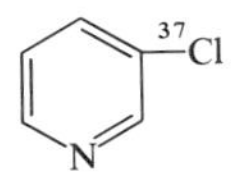	$v=0$	$A=5839.575\,(24)$ $B=1558.825\,(5)$ $C=1230.299\,(5)$		75CHA			
738a	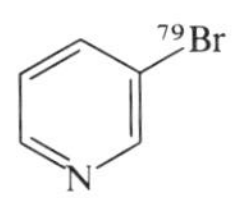	$v=0$	$A=5835.3954\,(78)$ $B=1005.0052\,(35)$ $C=857.2564\,(35)$	$\tau_{aaaa}=4.665\,(1908)$ $\tau_{bbbb}=-0.1795\,(164)$ $\tau_{cccc}=-0.101\,(16)$ $\tau_1=-1.524\,(65)$ $\tau_2=-0.257\,(17)$ $\tau_3=9.998\,(857)$	81DOR	81DOR		2.7.3, 156a
	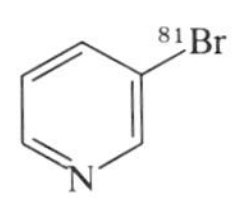	$v=0$	$A=5835.3969\,(82)$ $B=994.5550\,(35)$ $C=849.6406\,(34)$	$\tau_{aaaa}=5.036\,(2085)$ $\tau_{bbbb}=-0.1613\,(163)$ $\tau_{cccc}=-0.08798\,(1620)$ $\tau_1=-1.524\,(68)$ $\tau_2=-0.2524\,(175)$ $\tau_3=8.344\,(863)$				
739	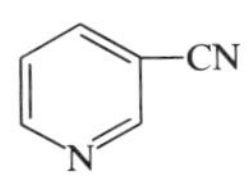	$v=0$	$A=5823.01\,(5)$ $B=1571.34\,(3)$ $C=1237.17\,(3)$		75FOR			2.6.3, 353

Nr.	Molecule	v	A, B, C MHz	D, H kHz	Ref.	Ref. struct.	Varia, remarks	See also Tab., Nr.
740	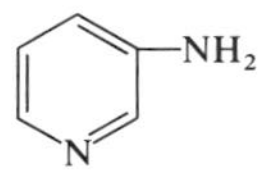	$v=0^+$	$A=5781.94(1)$ $B=2663.374(5)$ $C=1825.996(3)$		75CHR	75CHR		
		$v=0^-$	$A=5779.68(2)$ $B=2661.506(8)$ $C=1825.651(4)$				$E(0^-)-E(0^+)\simeq 50\ \mathrm{cm}^{-1}$	
	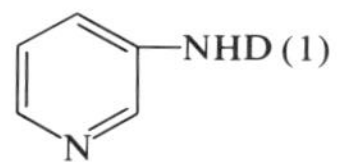	$v=0^+$	$A=5739.02(8)$ $B=2555.690(11)$ $C=1771.823(5)$					
		$v=0^-$	$A=5736.45(2)$ $B=2554.852(6)$ $C=1772.019(3)$					
	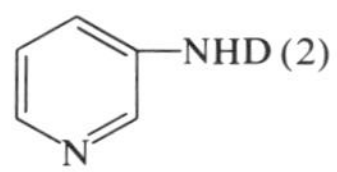	$v=0^+$	$A=5728.70(3)$ $B=2560.335(10)$ $C=1772.832(5)$					
		$v=0^-$	$A=5726.38(2)$ $B=2559.533(11)$ $C=1772.937(5)$					
	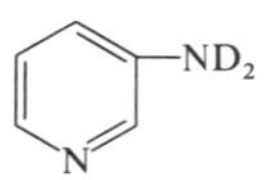	$v=0^+$	$A=5679.43(3)$ $B=2463.729(11)$ $C=1722.783(5)$					
		$v=0^-$	$A=5677.50(2)$ $B=2463.548(9)$ $C=1723.106(4)$					
741	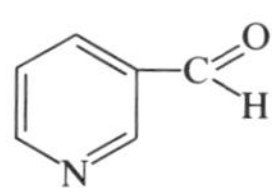	$v=0$	$A=5417.8(4)$ $B=1583.299(9)$ $C=1225.389(9)$		77HAE	77HAE	s-trans form	2.6.3, 354
		$v_\tau=1$	$A=5390.7(46)$ $B=1584.03(8)$ $C=1227.30(7)$				$v_\tau=135(30)\ \mathrm{cm}^{-1}$, probably C_3C torsion	

Nr.	Molecule	v	A, B, C MHz	D, H kHz	Ref.	Ref. struct.	Varia, remarks	See also Tab., Nr.
742	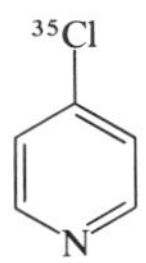	$v=0$	$A=6018.9(6)$ $B=1572.621(8)$ $C=1246.73(1)$		76CA3			2.6.3, 355 2.7.3, 144
	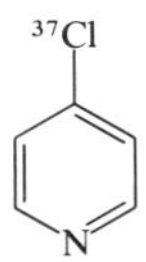	$v=0$	$A=6022.3(17)$ $B=1528.24(2)$ $C=1218.61(2)$					
743	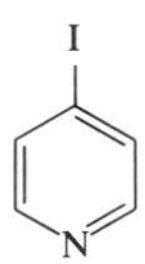	$v=0$	$A=6015.0(50)$ $B=\ \ 741.91(1)$ $C=\ \ 660.45(1)$		74CA1			2.7.3, 164
744	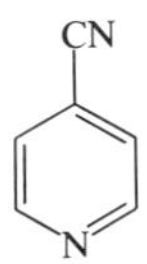	$v=0$	$A=6000(5)$ $B=1541.19(3)$ $C=1226.00(3)$		75FOR			2.6.3, 356
745	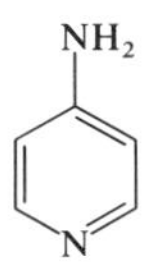	$v=0^+$	$A=5955.19(17)$ $B=2609.099(7)$ $C=1815.875(6)$		75CHR	75CHR		
		$v=0^-$	$A=5951.85(21)$ $B=2606.384(8)$ $C=1815.334(7)$					
	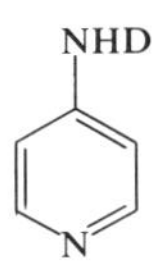	$v=0^+$	$A=5905.43(30)$ $B=2506.008(11)$ $C=1761.492(12)$					
		$v=0^-$	$A=5901.52(32)$ $B=2504.608\ 12)$ $C=1761.770(12)$					
	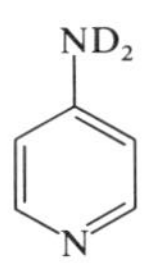	$v=0$	$A=5849.02(28)$ $B=2413.512(11)$ $C=1711.287(11)$					

Nr.	Molecule	v	A, B, C MHz	D, H kHz	Ref.	Ref. struct.	Varia, remarks	See also Tab., Nr.
746	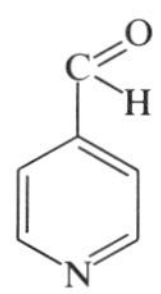	$v=0$	$A''=5519.12(16)$ *) $B''=1559.194(21)$ $C''=1216.136(30)$	$\Delta_J=0.0586(78)$ *) $\Delta_{JK}=0.140(10)$ $\Delta_K=0.91(40)$	80GEO4	80GEO4	*) uncertainties are 2 σ	2.6.3, 357 2.7.3, 76 3, 194
		$v_\tau=1$	$A=5497.19(12)$ $B=1559.647(44)$ $C=1217.995(28)$				$\nu_\tau=124(30)$ cm^{-1}, CHO torsion	
		$v_\tau=2$	$A-C=4255.46(40)$ $\varkappa=-0.8399$					
747	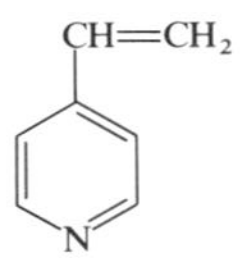	$v=0$	$A=5436.82(11)$ $B=1544.3972(16)$ $C=1204.6997(14)$		76RAL1			
		$v_\tau=1$	$A=5415.71(10)$ $B=1546.1491(16)$ $C=1208.8379(15)$				$\nu_\tau=35(15)$ cm^{-1}, torsion	
		$v_\tau=2$	$A=5399.67(16)$ $B=1547.0950(24)$ $C=1211.6658(23)$					
		$v_\tau=3$	$A=5383.71(13)$ $B=1547.8404(18)$ $C=1214.2809(19)$					
		$v_\tau=4$	$A=5366.72(31)$ $B=1548.3733(44)$ $C=1216.6872(42)$					
		$v_\tau=5$	$A=5356.08(16)$ $B=1548.7684(25)$ $C=1218.7912(23)$					
		$v_\tau=6$	$A=5339.09(18)$ $B=1549.1834(31)$ $C=1220.8475(27)$					
748 Cont.	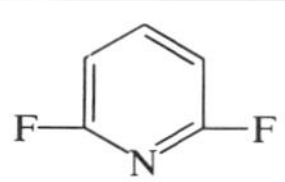	$v=0$	$\tilde{A}=3747.673(6)$ $\tilde{B}=1905.817(3)$ $\tilde{C}=1263.248(1)$	$\Delta_J=0.046(14)$ $\Delta_{JK}=0.230(40)$ $\Delta_K=0.431(29)$ $\delta_J=0.038(1)$ $\delta_K=-0.211(52)$	76DOR	75ST4		2.6.3, 358 2.7.3, 77

Nr.	Molecule	v	A, B, C MHz	D, H kHz	Ref.	Ref. struct.	Varia, remarks	See also Tab., Nr.
748 Cont.	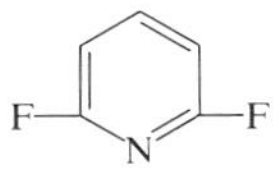	$v_{13}=1$	$A=3745.70(3)$ *) $B=1905.09(1)$ $C=1264.20(1)$	$\tau_{aaaa}=-2.73(75)$ *) $\tau_{bbbb}=-0.48(9)$	76ST2		$\nu_{13}=219\ cm^{-1}$ *) uncertainties are 3 σ; $\tau_{aabb}=0.35$ kHz; $\tau_{abab}=-0.47$ kHz	
		$v_{18}=1$	$A=3735.89(2)$ *) $B=1907.10(1)$ $C=1264.43(1)$	$\tau_{aaaa}=-2.40(63)$ *) $\tau_{bbbb}=-0.49(6)$			$\nu_{18}=247\ cm^{-1}$	
		$v_{10}=1$	$A=3757.18(3)$ *) $B=1906.79(1)$ $C=1262.56(1)$	$\tau_{aaaa}=-2.60(67)$ *) $\tau_{bbbb}=-0.48(5)$			$\nu_{10}=347\ cm^{-1}$	
		$v_{17}=1$	$A=3747.56(3)$ *) $B=1905.07(1)$ $C=1263.86(1)$	$\tau_{aaaa}=-2.26(63)$ *) $\tau_{bbbb}=-0.45(5)$			$\nu_{17}=460\ cm^{-1}$	
		$v_{27}=1$	$A=3748.21(3)$ *) $B=1906.71(2)$ $C=1261.95(1)$	$\tau_{aaaa}=-3.01(78)$ *) $\tau_{bbbb}=-0.55(6)$			$\nu_{27}=500\ cm^{-1}$	
		$v_9=1?$	$A=3748.36(3)$ *) $B=1904.93(2)$ $C=1263.78(2)$	$\tau_{aaaa}=-2.23(90)$ *) $\tau_{bbbb}=-0.44(8)$			$\nu_9\simeq 550\ cm^{-1}$	
		$v_{26}=1?$	$A=3741.52(3)$ *) $B=1905.42(1)$ $C=1263.39(1)$	$\tau_{aaaa}=-2.22(75)$ *) $\tau_{bbbb}=-0.44(6)$			$\nu_{26}\simeq 600\ cm^{-1}$	
		$v_{13}=2$	$A=3743.69(4)$ *) $B=1904.35(2)$ $C=1265.14(1)$	$\tau_{aaaa}=-3.09(99)$ *) $\tau_{bbbb}=-0.49(9)$				
		$v_{13}=1, v_{18}=1$	$A=3734.16(2)$ *) $B=1906.36(2)$ $C=1265.37(2)$	$\tau_{aaaa}=-2.3(13)$ *) $\tau_{bbbb}=-0.48(12)$				
		$v_{18}=2$	$A=3724.87(4)$ *) $B=1908.32(1)$ $C=1265.54(1)$	$\tau_{aaaa}=-2.3(11)$ *) $\tau_{bbbb}=-0.48(9)$				
Cont.		$v_{13}=1, v_{10}=1$	$A=3754.71(3)$ *) $B=1906.03(1)$ $C=1263.53(1)$	$\tau_{aaaa}=-2.62(90)$ *) $\tau_{bbbb}=-0.45(7)$				

Nr.	Molecule	v	A, B, C MHz	D, H kHz	Ref.	Ref. struct.	Varia, remarks	See also Tab., Nr.
748 Cont.	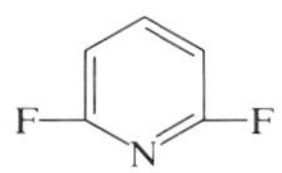	$v_{18}=1, v_{10}=1$	$A=3745.31(4)$ *) $B=1908.10(2)$ $C=1263.81(1)$	$\tau_{aaaa}=-4.3(62)$ *) $\tau_{bbbb}=-0.6(7)$	76ST2		*) uncertainties are 3 σ; $\tau_{aabb}=0.35$ kHz; $\tau_{abab}=-0.47$ kHz	
		$v_{13}=3$	$A=3741.8(6)$ *) $B=1903.8(4)$ $C=1266.1(3)$	$\tau_{aaaa}=-2.56$ (fixed) $\tau_{bbbb}=-0.49$ (fixed)				
		$v_{13}=1, v_{17}=1$	$A=3745.62(9)$ *) $B=1904.35(7)$ $C=1264.76(5)$	$\tau_{aaaa}=-3.3(37)$ *) $\tau_{bbbb}=-0.8(4)$				
		$v_{13}=2, v_{18}=1$	$A=3732.41(4)$ *) $B=1905.64(2)$ $C=1266.30(2)$	$\tau_{aaaa}=-2.3(11)$ *) $\tau_{bbbb}=-0.4(1)$				
		$v_{10}=2$	$A=3765.6(2)$ *) $B=1907.73(9)$ $C=1261.99(6)$	$\tau_{aaaa}=-4.5(55)$ *) $\tau_{bbbb}=-0.61(48)$				
		$v_{13}=1, v_{18}=2$	$A=3723.33(3)$ *) $B=1907.58(1)$ $C=1266.47(1)$	$\tau_{aaaa}=-1.78(84)$ *) $\tau_{bbbb}=-0.41(8)$				
		$v_{18}=3$	$A=3714.49(7)$ *) $B=1909.51(3)$ $C=1266.60(2)$	$\tau_{aaaa}=-1.1(17)$ *) $\tau_{bbbb}=-0.43(14)$				
		$v_{13}=4$	$A=3739.54(9)$ *) $B=1902.90(2)$ $C=1266.99(1)$	$\tau_{aaaa}=-0.6(23)$ *) $\tau_{bbbb}=-0.33(15)$				
		$v_{13}=5$	$A=3737.6(2)$ *) $B=1902.30(6)$ $C=1267.94(5)$	$\tau_{aaaa}=-2.9(42)$ *) $\tau_{bbbb}=-0.73(30)$				
	1-^{15}N	$v=0$	$A=3723.250(32)$ +) $B=1905.911(9)$ $C=1260.498(8)$	$\tau_{aaaa}=-2.6(12)$ +) $\tau_{abab}=-0.47(14)$	75ST4		+) uncertainties are 3 σ; $\tau_{bbbb}=-0.49$ kHz; $\tau_{aabb}=0.35$ kHz, assumed if not otherwise stated	
	2-^{13}C	$v=0$	$A=3746.188(33)$ +) $B=1897.056(9)$ $C=1259.222(6)$	$\tau_{aaaa}=-3.1(15)$ +) $\tau_{bbbb}=-0.48(6)$ $\tau_{abab}=-0.36(12)$				
Cont.	3-^{13}C	$v=0$	$A=3712.639(38)$ +) $B=1895.389(11)$ $C=1254.677(9)$	$\tau_{aaaa}=-2.0(15)$ +) $\tau_{bbbb}=-0.45(6)$ $\tau_{abab}=-0.52(14)$				

Nr.	Molecule	v	A, B, C MHz	D, H kHz	Ref.	Ref. struct.	Varia, remarks	See also Tab., Nr.
748 Cont.	4-^{13}C	$v=0$	$A=3656.646(33)$ $^{+}$) $B=1905.848(11)$ $C=1252.740(7)$	$\tau_{aaaa}=-1.6(13)$ $^{+}$) $\tau_{bbbb}=-0.45(6)$ $\tau_{abab}=-0.54(12)$	76ST2			
	4-D	$v=0$	$A=3525.486(17)$ $^{+}$) $B=1905.810(9)$ $C=1236.965(7)$	$\tau_{aaaa}=-1.99(23)$ $^{+}$) $\tau_{bbbb}=-0.45(5)$ $\tau_{aabb}=\ 0.37(15)$ $\tau_{abab}=-0.44(10)$				
	1-^{15}N, 4-D	$v=0$	$A=3502.659(33)$ $^{+}$) $B=1905.883(9)$ $C=1234.176(5)$	$\tau_{aaaa}=-2.2(12)$ $^{+}$) $\tau_{bbbb}=-0.47(7)$ $\tau_{abab}=-0.42(12)$				
	2-^{13}C, 4-D	$v=0$	$A=3523.834(39)$ $^{+}$) $B=1897.041(12)$ $C=1233.062(8)$	$\tau_{aaaa}=-1.7(12)$ $^{+}$) $\tau_{bbbb}=-0.47(7)$ $\tau_{abab}=-0.51(11)$				
	3-^{13}C, 4-D	$v=0$	$A=3495.843(28)$ $^{+}$) $B=1895.376(9)$ $C=1228.916(6)$	$\tau_{aaaa}=-2.6(9)$ $^{+}$) $\tau_{bbbb}=-0.49(6)$ $\tau_{abab}=-0.39(9)$				
	4-^{13}C, 4-D	$v=0$	$A=3447.011(22)$ $^{+}$) $B=1905.836(7)$ $C=1227.167(5)$	$\tau_{aaaa}=-2.3(6)$ $^{+}$) $\tau_{bbbb}=-0.49(4)$ $\tau_{abab}=-0.44(6)$				
	3-D, 4-D, 5-D	$v=0$	$A=3405.138(9)$ $^{+}$) $B=1840.087(5)$ $C=1194.489(5)$	$\tau_{aaaa}=-1.97(16)$ $^{+}$) $\tau_{bbbb}=-0.43(3)$ $\tau_{aabb}=\ 0.27(8)$ $\tau_{abab}=-0.39(5)$				
749	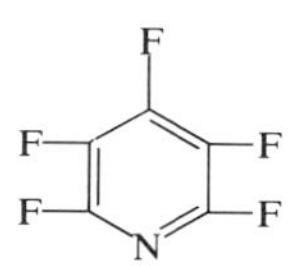	$v=0$	$A=1481.539(3)$ $B=1075.348(4)$ $C=\ 623.101(1)$		74DO1	74DO1		2.6.3, 359 2.7.3, 78
750		$v=0$	$A=4871.03(18)$ $B=2295.87(1)$ $C=1560.10(1)$		75LAT		for structure, see also Vol. II/7 (3, 577)	2.6.3, 360 2.7.3, 146
751 Cont.	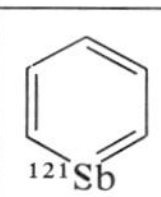	$v=0$	$A=4512.69(42)$ $B=1738.00(1)$ $C=1254.51(1)$		78FO1	78FO1		2.7.3, 157

Nr.	Molecule	v	A, B, C MHz	D, H kHz	Ref.	Ref. struct.	Varia, remarks	See also Tab., Nr.
751 Cont.	$^{123}SbC_5H_5$	$v=0$	$A=4512.84(30)$		78FO1			
			$B=1729.80(1)$					
			$C=1250.22(1)$					
	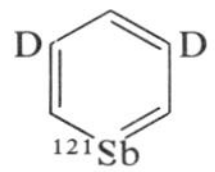	$v=0$	$A=4176.18(33)$					
			$B=1660.94(1)$					
			$C=1188.15(1)$					
	$^{123}SbC_5H_3D_2$	$v=0$	$A=4176.60(61)$					
			$B=1652.94(3)$					
			$C=1184.03(3)$					
752	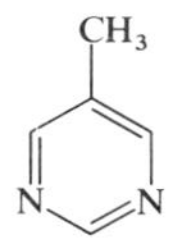	$v=0$	$A'=6108.41(24)$ *)		75CA2		*) rotational constant of the framework part only	2.6.3, 361; 2.8, 58
			$B=2642.198(8)$					
			$C=1844.196(8)$					
753	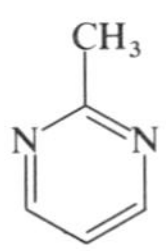	$v=0$	$A'=6096.9(1)$ *)		76CA4		*) rotational constant of the framework part only	2.6.3, 362; 2.8, 59
			$B=2751.687(5)$					
			$C=1895.583(5)$					
754	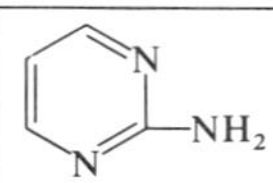	$v=0$	$A=5948.07(31)$		76LIS2	76LIS2		
			$B=2881.03(2)$					
			$C=1942.03(1)$					
		$v_i=1$	$A=5939.86(31)$				v_i, inversion vibration	
			$B=2878.69(2)$					
			$C=1942.52(1)$					
		$v_i=2$	$A=5937.63(38)$					
			$B=2880.43(2)$					
			$C=1943.67(2)$					
		$v_i=3$	$A=5920.66(50)$					
			$B=2880.59(3)$					
Cont.			$C=1945.39(2)$					

Nr.	Molecule	v	A, B, C MHz	D, H kHz	Ref.	Ref. struct.	Varia, remarks	See also Tab., Nr.
754 Cont.	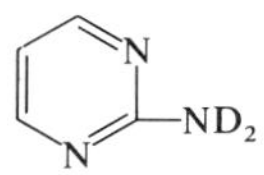	$v=0$ $v_i=1$	$A=5838.56(27)$ $B=2663.57(1)$ $C=1832.56(1)$ $A=5822.23(55)$ $B=2663.93(2)$ $C=1833.23(2)$		76LIS2			
755	$Ar \cdot CO_2$	$v=0$	$A=11914(80)$ $B=\ 1979.87(25)$ $C=\ 1682.75(70)$	$\tau_{aaaa}=\ -15.97$ *) $\tau_{bbbb}=\ -98.3(20)$ $\tau_{aabb}=\ 0$ *) $\tau_{abab}=-907(120)$	79ST1	79ST1	C_{2v} symmetry *) fixed value	2.6.3, 363
756	$Ar \cdot OCS$	$v=0$	$A=6786.014(20)$ $B=1509.909(6)$ $C=1227.220(6)$		76HAR	76HAR	T-shaped structure	2.6.3, 364
756a	$Ar^{35}ClCN$	$v=0$	$A''=6152.5411(21)$ $B''=1577.0362(8)$ $C''=1246.7514(6)$	$\tau_{aaaa}=-234.8(17)$ $\tau_{bbbb}=\ -55.97(10)$ $\tau_{cccc}=\ -21.60(7)$ $\tau_1=-576.0(2)$ $\tau_2=\ -97.28(8)$ $\tau_3=\ 2059(3)$	81KEE1	81KEE1	T-shaped structure	2.7.3, 144a
757	$Ar \cdot N_2O$							2.6.3, 365
758	$Ar \cdot H_2S$							2.6.3, 366
759	$Ar \cdot SO_2$	$v=0$	$A=9146.4(7)$ $B=1485.2(2)$ $C=1318.6(2)$		80DEL3	80DEL3	C_s symmetry inversion frequency: $\Delta=975.1(9)$ MHz	2.6.3, 367
760	$Ar \cdot O_3$	$v=0$	$A=12686$ $B=\ 1947$ $C=\ 1716$		80DEL1	80DEL1		2.6.3, 368
761 Cont.	$Ar \cdot CH_3{}^{35}Cl$		$A=13393(20)$ $B=\ 1650(20)$ $C=\ 1507(20)$		81DEL	81DEL	Ar to CH_3Cl center of mass direction nearly perpendicular to the CH_3Cl threefold axis	

Nr.	Molecule	v	A, B, C MHz	D, H kHz	Ref.	Ref. struct.	Varia, remarks	See also Tab., Nr.
761 Cont.	$Ar \cdot CH_3{}^{37}Cl$	$v=0$	$A=13160\,(40)$ $B=1620\,(40)$ $C=1490\,(40)$		81DEL			
762	$Ar \cdot HC{\equiv}CH$ $Ar \cdot DC{\equiv}CD$	$v=0$ $v=0$	$B-C=218.0497\,(7)$ $B=3170\,(100)$ $C=2960\,(100)$ $B-C=262.7112\,(9)$ $B=2870\,(20)$ $C=2610\,(20)$		80DEL2	80DEL2	T-shaped structure	2.6.3, 369 2.7.3, 9
762a	$HC{\equiv}CH \cdot H^{35}Cl$ $HC{\equiv}CH \cdot H^{37}Cl$	$v=0$ $v=0$	$A=36084\,(838)$ $B=2481.065\,(6)$ $C=2308.602\,(6)$ $A=36065\,(857)$ $B=2424.423\,(6)$ $C=2259.470\,(6)$	$D_J=7.9\,(4)$ $D_{JK}=497\,(8)$ $D_J=7.6\,(4)$ $D_{JK}=476\,(8)$	81LEG	81LEG	T-shaped geometry of C_{2v} symmetry	2.7.3, 144b
763	$(CO)_2$	$v=0$	$A=58 \cdot 10^3$ $B=2.95 \cdot 10^3$ $C=2.81 \cdot 10^3$		79VAN			
763a	$N_2O \cdot HF$ $^{15}N_2O \cdot HF$ $N_2O \cdot DF$ $^{15}N_2O \cdot DF$	$v=0$ $v=0$ $v=0$ $v=0$	$A=26328.5\,(3)$ $B=2718.6\,(10)$ $C=2441.0\,(10)$ $A=25773.02\,(5)$ $B=2646.57\,(4)$ $C=2377.91\,(4)$ $A=26004.1\,(2)$ $B=2699.7\,(1)$ $C=2423.5\,(1)$ $A=25456.46\,(8)$ $B=2626.68\,(6)$ $C=2360.47\,(6)$		81JOY	81JOY		2.6.3, 369a
763b Cont.	$HF \cdot C_3H_6$	$v=0$	$A=19948\,(60)$ $B=2753.706\,(2)$ $C=2539.115\,(2)$	$D_J=5.08\,(27)$ $D_{JK}=43.5\,(36)$	81BUX3	81BUX3	HF perpendicular to the cyclopropane edge and in the plane of the ring	2.7.3, 9a

Nr.	Molecule	v	A, B, C MHz	D, H kHz	Ref.	Ref. struct.	Varia, remarks	See also Tab., Nr.
763 b Cont.	$DF \cdot C_3H_6$	$v=0$	$B=$ 2729.22 (70) $C=$ 2517.96 (70)		81BUX3			
	D_2◁ · HF	$v=0$	$B=$ 2560.3 $C=$ 2388.5					
	◁ · HF D_2	$v=0$	$B=$ 2727.2 $C=$ 2498.6					
764	$H^{35}Cl \cdot C_3H_6$	$v=0$	$A=$ 20004 (10) $B=$ 1522.3313 (9) $C=$ 1454.7435 (9)	D_J = 1.91 (3) D_{JK} = 29.96 (17)	80LG3	80LG3		2.7.3, 145 3, 195
764 a	$(CH_2)_2O \cdots HF$	$v=0$	$A=$ 16099 (67) $B=$ 3638.75 (4) $C=$ 3462.31 (4)	Δ_J = 15.3 (6) Δ_{JK} = 25.8 (17)	81GEO	81GEO	C_s symmetry	2.6.3, 369 b
	$(CH_2)_2O \cdots DF$	$v=0$	$A=$ 15984 (40) $B=$ 3605.66 (2) $C=$ 3434.85 (2)	Δ_J = 14.1 (4) Δ_{JK} = 20.9 (9)				
	$(CH_2)_2{}^{18}O \cdots HF$	$v=0$	$B+C=$ 7059.7 (1) $B-C=$ 135.0 (15)					
765	$(CH_2)_3O \cdots HF$	$v=0$	$A=$ 9217 (138) $B=$ 2575.1 (11) $C=$ 2350.6 (11)	Δ_J = 11.3 (2) Δ_{JK} = −57.0 (4)	80GEO2	80GEO2		3, 196
	$(CH_2)_3O \cdots DF$	$v=0$	$A=$ 9157 (ass.) $B=$ 2544.7 (14) $C=$ 2329.3 (15)	Δ_J = 9.9 (10) Δ_{JK} = −56 (2)				
766	$CH_2{=}CH_2 \cdot H^{35}Cl$	$v=0$	$A=$ 25457 (349) $B=$ 2308.143 (3) $C=$ 2167.970 (3)	D_J = 7.2 (2) D_{JK} = 282 (5)	81ALD	81ALD	C_{2v} symmetry with the HCl on the C_2 symmetry axis ⊥ C=C	2.7.3, 145 a
	$CH_2{=}CH_2 \cdot H^{37}Cl$	$v=0$	$A=$ 25618 (335) $B=$ 2254.141 (3) $C=$ 2120.248 (3)	D_J = 6.7 (2) D_{JK} = 268 (4)				
	$CH_2{=}CH_2 \cdot D^{35}Cl$	$v=0$	$A=$ 25312 (415) $B=$ 2312.457 (4) $C=$ 2171.997 (4)	D_J = 6.7 (3) D_{JK} = 272 (6)				

2.5.3 References for 2.3, 2.4, 2.5, 2.6.2, 2.6.3, 2.7.2, 2.7.3 and 2.8

36WIL Wilson, E.B., Howard, J.B.: J. Chem. Phys. **4** (1936) 230.
52KIV Kivelson, D., Wilson, E.B.: J. Chem. Phys. **20** (1952) 1575.
53RAM Ramsey, N.F.: Nuclear Moments, John Wiley & Sons, New York, **1953**.
55TOW Townes, C.H., Schawlow, A.L.: Microwave Spectroscopy; McGraw Hill Book Co, New York, **1955**.
57EDM Edmonds, A.R.: Angular Momentum in Quantum Mechanics; Princeton Univ. Press, Princeton, **1957**.
59LIN Lin, C.C., Swalen, J.D.: Rev. Mod. Phys. **31** (1959) 841.
61HAY Hayashi, M., Pierce, L.: Tables for the Internal Rotation Problem; Department of Chemistry, Notre Dame University, Ind., **1961**.
64THA Thaddeus, P., Krisher, L.C., Loubser, J.H.N.: J. Chem. Phys. **40** (1964) 257.
64TIN Tinkham, M.: Group Theory and Quantum Mechanics; McGraw Hill Book Co, New York, **1964**.
66ROS Rose, M.E.: Elementary Theory of Angular Momentum; John Wiley & Sons, New York, **1966**.
66WOO Woods, R.C.: J. Mol. Spectrosc. **21** (1966) 4.
67WAT Watson, J.K.G.: J. Chem. Phys. **46** (1967) 1935.
67WOO Woods, R.C.: J. Mol. Spectrosc. **22** (1967) 49.
68DRE Dreizler, H.: Fortschr. Chem. Forsch. **10** (1968) 59.
68KRA Krakow, B., Lord, R.L.: J. Mol. Spectrosc. **27** (1968) 148.
68LAF Lafferty, W.J., Sams, R.L.: Symp. Mol. Struct. Spectr. Columbus Ohio, **1968**, paper 09.
68LEE Lees, R.M., Baker, J.G.: J. Chem. Phys. **48** (1968) 5299.
68MUE Muenter, J.S.: J. Chem. Phys. **48** (1968) 4544.
69MCK McKnight, J.S., Gordy, W.: Bull. Am. Phys. Soc. **14** (1969) 621.
70GOR Gordy, W., Cook, R.L.: Microwave Molecular Spectroscopy; Wiley-Interscience, New York, **1970**.
70LEE Leeuw, F.H. de, Dymanus, A.: Chem. Phys. Lett. **7** (1970) 288.
70TOM Tomasevich, G.: Ph.D. Thesis, Harvard Univ. **1970**.
71AMA Amat, G., Nielsen, H.H., Tarrago, G.: Rotation-vibration of Polyatomic Molecules; Marcel Dekker, Inc., New York, **1971**.
71CAS Casado, J., Nygaard, L., Sørensen, G.O.: J. Mol. Struct **8** (1971) 211.
71DAV Davies, P.B., Neumann, R.M., Wofsy, S.C., Klemperer, W.: J. Chem. Phys. **55** (1971) 3564.
71GER Gerry, M.C.L.: Can. J. Chem. **49** (1971) 255.
71KUK Kukolich, S.G., Nelson, A.C., Yamanashi, B.S.: J. Amer. Chem. Soc. **93** (1971) 6769.
71TUC Tucker, K.D., Tomasevich, G.R., Thaddeus, P.: Astrophys. J. **169** (1971) 429.
72AB Abdurakhmanov, A.A., Ragimova, R.A., Imanov, L.M.: Izv. Akad. Nauk Azerb. S.S.R. **2** (1972) 3.
72DRE Dreizler, H.: Molecular Spectroscopy: Modern Research, **1972**, 59; 2.4. Rotational Spectra of molecules with two internal degrees of freedom, Academic Press, Inc., New York and London.
72DYK Dyke, T.R., Howard, B.J., Klemperer, W.: J. Chem. Phys. **56** (1972) 2442.
72GIL Gillies, C.W., Kuczkowski, R.L.: J. Amer. Chem. Soc. **94** (1972) 6337.
72HAY Hayashi, M., Matsumura, C.: Bull. Chem. Soc. Jpn. **45** (1972) 732.
72HOU Hougen, J.T.: J. Chem. Phys. **57** (1972) 4207.
72KIR Kirchhoff, W.H.: J. Mol. Spectrosc. **41** (1972) 333.
72LE Lees, R.M.: J. Chem. Phys. **57** (1972) 2249.
72MAR Maroor, J., Hemptinne, M. de: Bull. Acad. R. Belg. **58** (1972) 956.
72MIL Mills, I.M.: Molecular Spectroscopy: Modern Research, **1972**, 115; 3.2 Vibration-rotation structure in asymmetric- and symmetric-top molecules, Academic Press, Inc., New York and London.
72PEA Pearson, E.F., Kim, H.: J. Chem. Phys. **57** (1972) 4230.
72POC Pochan, J.M., Flygare, W.H.: J. Phys. Chem. **76** (1972) 2249.
72RAO Rao, V.M., Kewley, R.: Can. J. Chem. **50** (1972) 955.
72SHA Sharma, S.D., Doraiswamy, S.: Proc. Indian Acad. Sci. **76A** (1972) 221.
72SLO Sloan, J.J., Kewley, R.: Can. J. Chem. **50** (1972) 955.

72WIN Winnewisser, M., Winnewisser, B.P.: J. Mol. Spectrosc. **41** (1972) 143.

73BAK Bak, B., Van Eijck, B.P., Kierkegaard, C.: J. Mol. Struct. **18** (1973) 429.
73BAR Baron, P.A., Brown, R.D.: Chem. Phys. **1** (1973) 444.
73BEL Bellet, J., Lafferty, W.J., Steenbeckeliers, G.: J. Mol. Spectrosc. **47** (1973) 388.
73BEN1 Benson, R.C., Flygare, W.H., Oda, M., Breslow, R.: J. Amer. Chem. Soc. **95** (1973) 2772.
73BEN2 Benson, R.C., Flygare, W.H.: J. Chem. Phys. **58** (1973) 2366.
73BEN3 Benson, R.C., Flygare, W.H.: J. Chem. Phys. **58** (1973) 2651.
73BEV1 Bevan, J.W., Legon, A.C.: J. Chem. Soc. Faraday Trans. 2 **69** (1973) 902.
73BEV2 Bevan, J.W., Legon, A.C.: J. Chem. Soc. Faraday Trans. 2 **69** (1973) 916.
73BHA Bhaumik, A., Brooks, W.V.F., Dass, S.C.: J. Mol. Struct. **16** (1973) 29.
73BLA Blackman, G.L., Bolton, K., Brown, R.D., Burden, F.R., Mishra, A.: J. Mol. Spectrosc. **47** (1973) 457.
73BOU Bouchy, A., Demaison, J., Roussy, G., Barriol, J.: J. Mol. Struct. **18** (1973) 211.
73BRO1 Brochu, M., Buckley, P.: Can. J. Spectrosc. **18** (1973) 165.
73BRO2 Brown, R.D., Crofts, J.G.: Chem. Phys. **1** (1973) 217.
73CAR Careless, A.J., Kroto, H.W., Landsberg, B.M.: Chem. Phys. **1** (1973) 371.
73CAS Casleton, K.H., Kukolich, S.G.: Chem. Phys. Lett. **22** (1973) 331.
73CAZ1 Cazzoli, G., Damiani, D., Lister, D.G.: J. Chem. Soc. Faraday Trans. 2 **69** (1973) 119.
73CAZ2 Cazzoli, G., Lister, D.G.: J. Mol. Spectrosc. **45** (1973) 467.
73CAZ3 Cazzoli, G., Lister, D.G., Mirri, A.M.: J. Chem. Soc. Faraday Trans. 2 **69** (1973) 569.
73CLO Clough, S.A., Beers, Y., Klein, G.P., Rothman, L.S.: J. Chem. Phys. **59** (1973) 2254.
73COH Cohen, E.A., Beaudet, R.A.: Inorg. Chem. **12** (1973) 1570.
73COO Cook, R.L., Winnewisser, G., Lindsey, D.C.: J. Mol. Spectrosc. **46** (1973) 276.
73CRE Creswell, R.A., Lafferty, W.J.: J. Mol. Spectrosc. **46** (1973) 371.
73CZI Czieslik, W., Andresen, U., Dreizler, H.: Z. Naturforsch. **28a** (1973) 1906.
73DAM Damiani, D., Ferretti, L.: Chem. Phys. Lett. **21** (1973) 592.
73DUR1 Durig, J.R., Carreira, L.A., Lafferty, W.J.: J. Mol. Spectrosc. **46** (1973) 187.
73DUR2 Durig, J.R., Chen, M.M., Li, Y.S.: J. Mol. Struct. **15** (1973) 37.
73DUR3 Durig, J.R., George, W.O., Li, Y.S., Carter, R.O.: J. Mol. Struct. **16** (1973) 47.
73DUR4 Durig, J.R., Chen, M.M., Li, Y.S., Turner, J.B.: J. Phys. Chem. **77** (1973) 227.
73DUR5 Durig, J.R., Li, Y.S., Carreira, L.A., Odom, J.D.: J. Amer. Chem. Soc. **95** (1973) 2491.
73DUR6 Durig, J.R., Li, Y.S., Odom, J.D.: J. Mol. Struct. **16** (1973) 443.
73DUR7 Durig, J.R., Li, Y.S., Tong, C.C.: J. Mol. Struct. **18** (1973) 269.
73DYK Dyke, T.R., Muenter, J.S.: J. Chem. Phys. **59** (1973) 3125.
73ELZ1 Elzaro, R.A., Brittain, A.H., Thompson, H.B., Schwendeman, R.H.: Symp. Mol. Struct. Spectr. Columbus, Ohio, **1973**, paper CC7.
73ELZ2 Elzaro, R.A., Schwendeman, R.H.: Symp. Mol. Struct. Spectr. Columbus, Ohio, **1973**, paper CC6.
73ENG Engelbrecht, L., Sutter, D., Dreizler, H.: Z. Naturforsch. **28a** (1973) 709.
73FON Fong, M.Y., Harmony, M.D.: J. Chem. Phys. **58** (1973) 4260.
73FOR Forti, P., Damiani, D., Favero, P.G.: J. Amer. Chem. Soc. **95** (1973) 756.
73GAI Gaines, L., Tomasevich, G.R.: Symp. Mol. Struct. Spectr. Columbus, Ohio, **1973**, pap. F5.
73GER1 Gerry, M.C.L.: J. Mol. Spectrosc. **45** (1973) 71.
73GER2 Gerry, M.C.L., Winnewisser, G.: J. Mol. Spectrosc. **48** (1973) 1.
73GRI Griffiths, J.H., Owen, N.L., Sheridan, J.: J. Chem. Soc. Faraday Trans. 2 **69** (1973) 1359.
73GUA Guarnieri, A., Charpentier, L., Kück, B.: Z. Naturforsch. **28a** (1973) 1721.
73GUN Gunderova, L.N., Shapkin, A.A., Pozdeev, N.M.: Opt. Spektrosk. **34** (1973) 1211; Opt. Spectry (English translation) **34** (1973) 703.
73HAR Hartford, S.L., Allen, W.C., Norris, C.L., Pearson, E.F., Flygare, W.H.: Chem. Phys. Lett. **18** (1973) 153.
73HAT Hatta, A., Suzuki, M., Kozima, K.: Bull. Chem. Soc. Jpn **46** (1973) 2321.
73HAY Hayashi, M., Kuwada, K.: Bull. Chem. Soc. Jpn **46** (1973) 2691.
73HIL Hilderbrandt, R.L., Wieser, J.D.: J. Mol. Struct. **15** (1973) 27.
73HIR Hirota, E., Matsumura, C.: J. Chem. Phys. **59** (1973) 3038.
73HOY Hoy, A.R., Bertram, M., Mills, I.M.: J. Mol. Spectrosc. **46** (1973) 429.
73JAC Jacob, E.J., Lide, D.R.: J. Chem. Phys. **59** (1973) 5877.
73JOH Johns, J.W.C., McKellar, A.R.W.: J. Mol. Spectrosc. **48** (1973) 354.

73KAD	Kadzhar, C.O., Abdullaev, G.A., Imanov, L.M.: Izv. Akad. Nauk. Az. SSR, Fiz.-Tekh. Mat. **1** (1973) 38.
73KAL	Kaluza, C.E., Bauder, A., Günthard, H.H.: Chem. Phys. Lett. **22** (1973) 454.
73KAR1	Karlsson, H.: Acta Chem. Scand. **27** (1973) 1435.
73KAR2	Karlsson, F., Vestin, R.: Acta Chem. Scand. **27** (1973) 3033.
73KAW	Kawashima, Y., Hirose, C.: Bull. Chem. Soc. Jpn **46** (1973) 2969.
73KEW	Kewley, R.: Can. J. Chem. **51** (1973) 529.
73KIR1	Kirchhoff, W.H., Johnson, D.R.: J. Mol. Spectrosc. **45** (1973) 159.
73KIR2	Kirchhoff, W.H., Johnson, D.R., Lovas, F.J.: J. Phys. Chem. Ref. Data **2** (1973) 1.
73KIR3	Kirchhoff, W.H., Johnson, D.R., Powell, F.X.: J. Mol. Spectrosc. **48** (1973) 157.
73KIR4	Kirchhoff, W.H., Lide, D.R., Powell, F.X.: J. Mol. Spectrosc. **47** (1973) 491.
73KIR5	Kirty, K.K., Srivastava, S.L.: Indian J. Pure Appl. Phys. **11** (1973) 293.
73KIRS	Kirschner, S.M., Watson, J.K.G.: J. Mol. Spectrosc. **47** (1973) 347.
73KRI	Krisher, L.C., Wolf, S.N.: J. Chem. Phys. **58** (1973) 396.
73KUC	Kuczkowski, R.L.: J. Mol. Spectrosc. **45** (1973) 261.
73KUK1	Kukolich, S.G., Casleton, K.H.: Chem. Phys. Lett. **18** (1973) 408.
73KUK2	Kukolich, S.G., Nelson, A.C.: J. Amer. Chem. Soc. **95** (1973) 680.
73KUK3	Kukolich, S.G., Ruben, D.J., Wang, J.H.S., Williams, J.R.: J. Chem. Phys. **58** (1973) 3155.
73KUK4	Kukolich, S.G., Wang, J.H.S., Ruben, D.J.: J. Chem. Phys. **58** (1973) 5474.
73LAR	Larsen, N.W., Mathier, E., Bauder, A., Günthard, H.H.: J. Mol. Spectrosc. **47** (1973) 183.
73LAU	Laurie, V.W., Stigliani, W.M.: J. Amer. Chem. Soc. **95** (1973) 4154.
73LEA	Leavell, S., Curl, R.F.: J. Mol. Spectrosc. **45** (1973) 428.
73LEE	Lees, R.M., Lovas, F.J., Kirchhoff, W.H., Johnson, D.R.: J. Phys. Chem. Ref. Data **2** (1973) 205.
73LEG	Legon, A.C.: J. Chem. Soc. Faraday Trans. 2 **69** (1973) 29.
73LI1	Li, Y.S., Durig, J.R.: Inorg. Chem. **12** (1973) 306.
73LI2	Li, Y.S., Durig, J.R.: J. Mol. Struct. **16** (1973) 433.
73LI3	Li, Y.S., Durig, J.R.: J. Mol. Spectrosc. **47** (1973) 179.
73LIS1	Lister, D.G., Owen, N.L.: J. Chem. Soc. Faraday Trans. 2 **69** (1973) 1036.
73LIS2	Lister, D.G., Owen, N.L.: J. Chem. Soc. Faraday Trans. 2 **69** (1973) 1304.
73LOV	Lovas, F.J., Johnson, D.R.: J. Chem. Phys. **59** (1973) 2347.
73LUC	Lucia, F.C. de, Helminger, P., Gordy, W., Morgan, H.W., Staats, P.A.: Phys. Rev. A **8** (1973) 2785.
73LUCK	Lucken, E.A.C., Pozzi, R., Ramaprasad, K.R.: J. Mol. Struct. **18** (1973) 377.
73MAM	Mamleev, A.K., Akulin, O.B., Pozdeev, N.M.: Deponirovannye Stat'i viniti (1973) 7359 DEP.
73MAR	Maroor, J., Hemptinne, M. de: Bull. Acad. R. Belg. **58** (1973) 969.
73MAS1	Marstokk, K.M., Møllendal, H.: J. Mol. Struct. **15** (1973) 137.
73MAS2	Marstokk, K.M., Møllendal, H.: J. Mol. Struct. **16** (1973) 259.
73MAS3	Marstokk, K.M., Møllendal, H.: J. Mol. Struct. **18** (1973) 247.
73MIR	Mirri, A.M., Caminati, W.: J. Mol. Spectrosc. **47** (1973) 204.
73MJO	Mjöberg, P.J., Ljunggren, S.: Z. Naturforsch. **28a** (1973) 729.
73MOC1	Mochel, A.R., Britt, C.O., Boggs, J.E.: J. Chem. Phys. **58** (1973) 3221.
73MOC2	Mochel, A.R., Bjørseth, A., Britt, C.O., Boggs, J.E.: J. Mol. Spectrosc. **48** (1973) 107.
73NAS	Nasibullin, R.S., Latypova, R.G., Troitskaya, V.S., Vinokurov, V.G., Pozdeev, N.M.: Opt. Spektrosk. **35** (1973) 92; Opt. Spectry (english translation) **35** (1973) 52.
73NAV	Nave, C.R., Sheridan, J.: J. Mol. Struct. **15** (1973) 391.
73NOR1	Norris, C.L., Benson, R.C., Beak, P., Flygare, W.H.: J. Amer. Chem. Soc. **95** (1973) 2766.
73NOR2	Norris, C.L., Tigelaar, H.L., Flygare, W.H.: Chem. Phys. **1** (1973) 1.
73NOS1	Nösberger, P., Bauder, A., Günthard, H.H.: Chem. Phys. **1** (1973) 418.
73NOS2	Nösberger, P., Bauder, A., Günthard, H.H.: Chem. Phys. **1** (1973) 426.
73NOV	Novick, S.E., Davies, P., Harris, S.J., Klemperer, W.: J. Chem. Phys. **59** (1973) 2273.
73PEA1	Pearson, E.F., McCormick, R.V.: J. Chem. Phys. **58** (1973) 1619.
73PEA2	Pearson, E.F., Trueblood, M.B.: Astrophys. J. **179** (1973) L145.
73PEA3	Pearson, E.F., Trueblood, M.B.: J. Chem. Phys. **58** (1973) 826.
73PEN1	Penn, R.E., Boggs, J.E.: J. Mol. Spectrosc. **47** (1973) 340.
73PEN2	Penn, R.E., Boggs, J.E.: J. Chem. Phys. **59** (1973) 4208.

73PIC1	Pickett, H.M.: J. Amer. Chem. Soc. **95** (1973) 1770.
73PIC2	Pickett, H.M.: J. Mol. Spectrosc. **46** (1973) 335.
73RAL1	Raley, J.M., Wollrab, J.E., Lovejoy, R.W.: J. Mol. Spectrosc. **48** (1973) 100.
73RAL2	Ralowski, W., Ljunggren, S., Mjöberg, J.: Acta Chem. Scand. **27** (1973) 3128.
73RAL3	Ralowski, W., Wettermark, G., Ljunggren, S.: Acta Chem. Scand. **27** (1973) 1565.
73RIM	Rimmer, D.F., Smith, J.G., Whiffen, D.H.: J. Mol. Spectrosc. **45** (1973) 114.
73RIN	Rinehart, E.A., Reinhart, P.B., Wollrab, J.E.: J. Mol. Spectrosc. **47** (1973) 556.
73ROC1	Rock, S.L., McGurk, J.C., Flygare, W.H.: Chem. Phys. Lett. **19** (1973) 153.
73ROC2	Rock, S.L., Pearson, E.F., Appleman, E.H., Norris, C.L., Flygare, W.H.: J. Chem. Phys. **59** (1973) 3940.
73SAI	Saito, S., Takagi, K.: J. Mol. Spectrosc. **47** (1973) 99.
73SCH	Schmidt, R., Beeson, E.L.: J. Chem. Phys. **59** (1973) 3070.
73SHA	Shapkin, A.A., Gunderova, L.N., Magdesieva, N.N., Pozdeev, N.M.: Zh. Strukt. Khim. **14** (1973) 1037; J. Struct. Chem. (english translation) **14** (1973) 974.
73SHO	Shoji, H., Tanaka, T., Hirota, E.: J. Mol. Spectrosc. **47** (1973) 268.
73SLA	Slafer, W.D., Harris, D.O.: J. Mol. Spectrosc. **45** (1973) 412.
73STE	Steenbeckeliers, G., Bellet, J.: J. Mol. Spectrosc. **45** (1973) 10.
73STE1	Steinmetz, W.E.: J. Amer. Chem. Soc. **95** (1973) 2777.
73STE2	Steinmetz, W.E.: J. Chem. Phys. **59** (1973) 3872 Err.
73SU	Su, C.F., Beeson, E.L.: J. Chem. Phys. **59** (1973) 759; errata: J. Chem. Phys. **60** (1974) 4638.
73SUE1	Suenram, R.D., Harmony, M.D.: J. Amer. Chem. Soc. **95** (1973) 4506.
73SUE2	Suenram, R.D., Harmony, M.D.: J. Chem. Phys. **58** (1973) 5842.
73SUZ	Suzeau, P., Jurek, R., Chanussot, J.: C.R. Acad. Sci. **276B** (1973) 777.
73THO	Thomas, C.H.: J. Chem. Phys. **59** (1973) 70.
73TOM	Tomasevich, G.R., Tucker, K.D., Thaddeus, P.: J. Chem. Phys. **59** (1973) 131.
73TRI	Trinkaus, A., Dreizler, H., Rudolph, H.D.: Z. Naturforsch. **28a** (1973) 750.
73TUC	Tucker, K.D., Tomasevich, G.R.: J. Mol. Spectrosc. **48** (1973) 475.
73WAN	Wang, J.H.S., Kukolich, S.G.: J. Amer. Chem. Soc. **95** (1973) 4138.
73WEN	Wenger, H.U., Bauder, A., Günthard, H.H.: Chem. Phys. **1** (1973) 441.
73WIN	Winnewisser, G.: J. Mol. Spectrosc. **46** (1973) 16.
73YAM	Yamaguchi, I., Ohashi, O., Takahashi, H., Sakaizumi, T., Onda, M.: Bull. Chem. Soc. Jpn **46** (1973) 1560.
73ZEI	Zeil, W., Haas, B.: Z. Naturforsch. **28a** (1973) 1230.
74AGE	Ageno, H.Y., Raley, J.M., Wollrab, J.E.: J. Mol. Spectrosc. **52** (1974) 163.
74AND	Andresen, U., Dreizler, H.: Z. Naturforsch. **29a** (1974) 797.
74ATT	Attanasio, A., Bauder, A., Günthard, H.H.: Chem. Phys. **6** (1974) 373.
74BAR	Baron, P.A., Harris, D.O.: J. Mol. Spectrosc. **49** (1974) 70.
74BJ1	Bjørseth, A.: J. Mol. Struct. **20** (1974) 61.
74BJ2	Bjørseth, A.: J. Mol. Struct. **23** (1974) 1.
74BJO	Bjorvatten, T.: J. Mol. Struct. **20** (1974) 75.
74BO1	Boggia, L.M., Filgueira, R.R., Sorarrain, O.M., Damiani, D.: Z. Naturf. **29a** (1974) 95.
74BO2	Boggia, L.M., Sorarrain, O.M., Fornes, J.A., Villani, M.C.: Z. Phys. Chem. Leipzig **255** (1974) 44.
74BOL1	Bolton, K., Brown, R.D.: Aust. J. Phys. **27** (1974) 143.
74BOL2	Bolton, K., Lister, D.G., Sheridan, J.: J. Chem. Soc. Faraday Trans. 2 **70** (1974) 113.
74BRA	Braun, W., Günther, H., Umbrecht, H., Zeil, W.: Z. Phys. Chem. Frankfurt **93** (1974) 247.
74BRE	Brewer, R.G., Shoemaker, R.L., Stenhom, S.: Phys. Rev. Lett. **33** (1974) 63.
74BUC	Buckley, P., Weber, J.P.: Can. J. Chem. **52** (1974) 942.
74CA1	Caminati, W., Forti, P.: Chem. Phys. Lett. **29** (1974) 239.
74CA2	Caminati, W., Cervellati, R., Mirri, A.M.: J. Mol. Spectrosc. **51** (1974) 288.
74CAR	Carpenter, J.H.: J. Mol. Spectrosc. **50** (1974) 182.
74CAZ	Cazzoli, G., Favero, P.G., Dal Borgo, A.: J. Mol. Spectrosc. **50** (1974) 82.
74CHR	Christen, D., Lister, D.G., Sheridan, J.: J. Chem. Soc. Faraday Trans. 2 **70** (1974) 1953.
74CO1	Codding, E.G., Jones, C.E., Schwendeman, R.H.: Inorg. Chem. **13** (1974) 178.
74CO2	Codding, E.G., Creswell, R.A., Schwendeman, R.H.: Inorg. Chem. **13** (1974) 856.

74CO3	Codding, E.G., Schwendeman, R.H.: J. Mol. Spectrosc. **49** (1974) 226.
74COF	Coffey, D.: Austin Symp. Gas Phase Mol. Struct. **1974**, paper M. 18.
74COH	Cohen, E.A., Poynter, R.L.: J. Mol. Spectrosc. **53** (1974) 131.
74COO1	Cook, R.L., Lucia, F.C. de, Helminger, P.: J. Mol. Spectrosc. **53** (1974) 62.
74COO2	Cook, R.L., Malloy, T.B.: J. Amer. Chem. Soc. **96** (1974) 1703.
74COO3	Cook, R.L., Walden, R.T., Jones, G.E.: J. Mol. Spectrosc. **53** (1974) 370.
74COX	Cox, A.P., Finnigan, D.J., Whittle, M.J.: J. Mol. Spectrosc. **49** (1974) 124.
74CR1	Creswell, R.A., Mills, I.M.: J. Mol. Spectrosc. **52** (1974) 392.
74CR2	Creswell, R.A., Schwendeman, R.H.: Chem. Phys. Lett. **27** (1974) 521.
74CR3	Creswell, R.A.: J. Mol. Spectrosc. **51** (1974) 111.
74CZ1	Czieslik, W., Carpentier, L., Sutter, D.H.: Z. Naturforsch. **29a** (1974) 1498.
74CZ2	Czieslik, W., Sutter, D.H.: Z. Naturforsch. **29a** (1974) 1820.
74DAM	Damiani, D., Ferretti, L.: Chem. Phys. Lett. **24** (1974) 357.
74DAS	Dass, S.C., Kewley, R.: Can. J. Chem. **52** (1974) 434.
74DAV1	Davidson, J.R., Burnham, A.K., Siegel, B., Beak, P., Flygare, W.H.: J. Amer. Chem. Soc. **96** (1974) 7394.
74DAV2	Davis, R.E., Muenter, J.S.: Chem. Phys. Lett. **24** (1974) 343.
74DAV3	Davis, R.W., Gerry, M.C.L., Visaisouk, S., Balfour, W.J.: Chem. Phys. Lett. **26** (1974) 561.
74DO1	Doraiswamy, S., Sharma, S.D.: Chem. Phys. **6** (1974) 76.
74DO2	Doraiswamy, S., Sharma, S.D.: Pramana **2** (1974) 219.
74DUB	Dubrulle, A., Boucher, D.: C.R. Acad. Sci. **278B** (1974) 211.
74DUR1	Durig, J.R., Carreira, L.A., Laane, J.: J. Mol. Struct. **21** (1974) 281.
74DUR2	Durig, J.R., Carreira, L.A., Odom, J.D.: J. Amer. Chem. Soc. **96** (1974) 2688.
74DUR3	Durig, J.R., Carter, R.O., Odom, J.D.: Inorg. Chem. **13** (1974) 701.
74DUR4	Durig, J.R., Cox, A.W., Li, Y.S.: Inorg. Chem. **13** (1974) 2302.
74DUR5	Durig, J.R., Hudgens, B.A., Li, Y.S., Odom, J.D.: J. Chem. Phys. **61** (1974) 4890.
74DUR6	Durig, J.R., Hudgens, B.A., Odom, J.D.: Inorg. Chem. **13** (1974) 2306.
74DUR7	Durig, J.R., Kizer, K.L., Li, Y.S.: J. Amer. Chem. Soc. **96** (1974) 7400.
74DUR8	Durig, J.R., Li, Y.S.: J. Mol. Struct. **21** (1974) 289.
74DUR9	Durig, J.R., Li, Y.S., Harmony, M.D., Fong, M.Y.: J. Mol. Struct. **23** (1974) 377.
74DUR10	Durig, J.R., Li, Y.S., Tong, C.C., Zens, A.P., Ellis, P.D.: J. Amer. Chem. Soc. **96** (1974) 3805.
74DUR11	Durig, J.R., Li, Y.S., Turner, J.B.: Inorg. Chem. **13** (1974) 1495.
74DUR12	Durig, J.R., Riethmiller, S., Li, Y.S.: J. Chem. Phys. **61** (1974) 253.
74FAB	Fabricant, B., Muenter, J.S.: J. Mol. Spectrosc. **53** (1974) 57.
74FAR	Farag, M., Bohn, R.K.: Austin Symp. Gas Phase Mol. Struct. **1974** paper W.15.
74FON	Fong, M.Y., Johnson, L.J., Harmony, M.D.: J. Mol. Spectrosc. **53** (1974) 45.
74FOR1	Ford, R.G.: J. Mol. Spectrosc. **49** (1974) 117.
74FOR2	Foreman, P.B., Chien, K.R., Williams, J.R., Kukolich, S.G.: J. Mol. Spectrosc. **52** (1974) 251.
74FRE	Freund, S.M., Duxbury, G., Römheld, M., Tiedje, J.T., Oka, T.: J. Mol. Spectrosc. **52** (1974) 38.
74GAI	Gaines, L., Casleton, K.H., Kukolich, S.G.: Astrophys. J. **191** (1974) L99.
74GAR	Garvey, R.M., Lucia, F.C. de: J. Mol. Spectrosc. **50** (1974) 38.
74GIL	Gillies, C.W., Lattimer, R.P., Kuczkowski, R.L.: J. Amer. Chem. Soc. **96** (1974) 1536.
74GOD	Godon, M., Bauer, A.: C.R. Acad. Sci. **278B** (1974) 113.
74GRI	Griffith, D.W.T., Kent, J.E.: Chem. Phys. Lett. **25** (1974) 290.
74HAQ	Haque, S.S., Lees, R.M., Saint Clair, J.M., Beers, Y., Johnson, D.R.: Astrophys. J. **187** (1974) L15.
74HAR1	Harris, S.J., Novick, S.E., Klemperer, W., Falconer, W.E.: J. Chem. Phys. **61** (1974) 193.
74HAR2	Harris, S.J., Novick, S.E., Klemperer, W.: J. Chem. Phys. **60** (1974) 3208.
74HAR3	Hartmann, T., Botskor, I.: J. Mol. Struct. **22** (1974) 337.
74HAY1	Hayashi, M., Imaishi, H., Kuwada, K.: Bull. Chem. Soc. Jpn **47** (1974) 2382.
74HAY2	Hayashi, M., Kuwada, K.: Bull. Chem. Soc. Jpn **46** (1974) 2691.
74HAY3	Hayashi, M., Kuwada, K.: Bull. Chem. Soc. Jpn **47** (1974) 3006.
74HAY4	Hayashi, M., Kuwada, K., Imaishi, H.: Chem. Lett. **(1974)** 913.
74HEI	Heise, H.M., Lutz, H., Dreizler, H.: Z. Naturforsch. **29a** (1974) 1345.

74HEL1	Helminger, P., Lucia, F.C. de, Gordy, W., Morgan, H.W., Staats, P.A.: Phys. Rev. A **9** (1974) 12.
74HEL2	Helminger, P., Lucia, F.C. de, Gordy, W., Staats, P.A., Morgan, H.W.: Phys. Rev. A **10** (1974) 1072.
74HI1	Hirose, C.: Bull. Chem. Soc. Jpn **47** (1974) 1311.
74HI2	Hirose, C.: Bull. Chem. Soc. Jpn **47** (1974) 976.
74HIR	Hirota, E., Sugisaki, R., Nielsen, C.J., Sørensen, G.O.: J. Mol. Spectrosc. **49** (1974) 251.
74HOC	Hocking, W.H., Gerry, M.C.L., Winnewisser, G.: Astrophys. J. **187** (1974) L89.
74IKE	Ikeda, T., Curl, R.F., Karlsson, H.: J. Mol. Spectrosc. **53** (1974) 101.
74IMA	Imaishi, H., Hayashi, M.: J. Sci. Hiroshima Univ., Ser. A **38** (1974) 21.
74JON	Jones, G.I.L., Summers, T.D., Owen, N.L.: J. Chem. Soc. Faraday Trans 2 **70** (1974) 100.
74KAD	Kadzhar, C.O., Askerov, A.B., Abbasov, A.A., Imanov, L.M.: Izv. Akad. Nauk Az. SSR Fiz. Techn. Mat. Nauk **4** (1974) 58.
74KAR1	Karlsson, F., Granberg, M., Vestin, R.: Acta Chem. Scand. **28A** (1974) 206.
74KAR2	Karlsson, F., Granberg, M., Vestin, R.: Acta Chem. Scand. **28A** (1974) 201.
74KEW	Kewley, R.: Can. J. Chem. **52** (1974) 509.
74KR1	Krisher, L.C., Watson, W.A., Morrison, J.A.: J. Chem. Phys. **61** (1974) 3429.
74KR2	Krisher, L.C., Watson, W.A., Morrison, J.A.: J. Chem. Phys. **60** (1974) 3417.
74KRO	Kroto, H.W., Landsberg, B.M., Suffolk, R.J., Vodden, A.: Chem. Phys. Lett. **29** (1974) 265.
74KRU	Krugh, W.D., Gold, L.P.: J. Mol. Spectrosc. **49** (1974) 423.
74KUH	Kuhler, M., Charpentier, L., Sutter, D., Dreizler, H.: Z. Naturforsch. **29a** (1974) 1335.
74LA1	Lattimer, R.P., Kuczkowski, R.L.: Symp. Mol. Struct. Spectr. Columbus Ohio, **1974**, paper TB3.
74LA2	Lattimer, R.P., Kuczkowski, R.L., Gillies, C.W.: J. Amer. Chem. Soc. **96** (1974) 348.
74LAU	Lau, K.K., Burg, A.B., Beaudet, R.A.: Inorg. Chem. **13** (1974) 2787.
74LEC	LeCroix, C.D., Curl, R.F., McKinney, P.M., Myers, R.J.: J. Mol. Spectrosc. **53** (1974) 250.
74LI	Li, Y.S., Durig, J.R.: J. Mol. Spectrosc. **54** (1974) 296.
74LIS	Lister, D.G., Tyler, J.K., Høg, J.H., Larsen, N.W.: J. Mol. Struct. **23** (1974) 253.
74LOV	Lovas, F.J., Tiemann, E., Johnson, D.R.: J. Chem. Phys. **60** (1974) 5005.
74MAD	Mäder, H., Heise, H.M., Dreizler, H.: Z. Naturforsch. **29a** (1974) 164.
74MAK	Maki, A.G.: J. Phys. Chem. Ref. Data **3** (1974) 221.
74MAL	Mallinson, P.D., Robiette, A.G.: J. Mol. Spectrosc. **52** (1974) 413.
74MAM	Mamleev, A.K., Akulinin, O.B., Pozdeev, N.M.: Zh. Prikl. Spektrosk. **20** (1974) 546; J. Applied Spectry (english translation) **20** (1974) 420.
74MAO	Malloy, T.B.: J. Mol. Spectrosc. **49** (1974) 432.
74MAR1	Marstokk, K.M., Møllendal, H.: J. Mol. Struct. **20** (1974) 257.
74MAR2	Marstokk, K.M., Møllendal, H.: J. Mol. Struct. **22** (1974) 287.
74MAR3	Marstokk, K.M., Møllendal, H.: J. Mol. Struct. **23** (1974) 93.
74MAT	Matsuura, H., Kawashima, Y., Hirose, C.: J. Mol. Struct. **20** (1974) 205.
74NAS	Nasibullin, R.S., Latypova, R.G., Troitskaya, V.S., Vinokurov, V.G., Pozdeev, N.M.: Zh. Strukt. Khim. **15** (1974) 47; J. Struct. Chem. (english translation) **15** (1974) 41.
74NEL	Nelson, A.C., Kukolich, S.G., Ruben, D.J.: J. Mol. Spectrosc. **51** (1974) 107.
74NOR	Norris, C.L., Pearson, E.F., Flygare, W.H.: J. Chem. Phys. **60** (1974) 1758.
74NY1	Nygaard, L.: Acta Chem. Scand. **28A** (1974) 118.
74NY2	Nygaard, L., Christen, D., Nielsen, J.T., Pedersen, E.J., Snerling, O., Vestergaard, E., Sørensen, G.O.: J. Mol. Struct. **22** (1974) 401.
74OGA	Ogata, T., Cox, A.P., Smith, D.L., Timms, P.L.: Chem. Phys. Lett. **26** (1974) 186.
74OKI	Okiye, K., Hirose, C., Lister, D.G., Sheridan, J.: Chem. Phys. Lett. **24** (1974) 111.
74PAR	Parent, C.R., Gerry, M.C.L.: J. Mol. Spectrosc. **49** (1974) 343.
74PAS	Pasinski, J.P., Beaudet, R.A.: J. Chem. Phys. **61** (1974) 683.
74PEA	Pearson, E.F., Norris, C.L., Flygare, W.H.: J. Chem. Phys. **60** (1974) 1761.
74PIC	Pickett, H.M., Scroggin, D.G.: J. Chem. Phys. **61** (1974) 3954.
74REI	Reinartz, J.M.L.J., Dymanus, A.: Chem. Phys. Lett. **24** (1974) 346.
74REJ	Reijnders, J.M.H., Ellenbroek, A.W., Dymanus, A.: Chem. Phys. Lett. **26** (1974) 470.
74RU1	Ruben, D.J., Kukolich, S.G.: J. Chem. Phys. **61** (1974) 3780.

74RU2	Ruben, D.J., Kukolich, S.G.: J. Chem. Phys. **60** (1974) 100.
74SAI	Saito, S., Makino, F.: Bull. Chem. Soc. Jpn **47** (1974) 1863.
74SAS	Sastry, K.V.L.N., Guarnieri, A.: Z. Naturforsch. **29a** (1974) 1495.
74SCR	Scroggin, D.G., Riveros, J.M., Wilson, E.B.: J. Chem. Phys. **60** (1974) 1376.
74SLA	Slafer, W.D., Harris, D.O.: J. Mol. Spectrosc. **50** (1974) 435.
74SOR	Sørensen, G.O., Mahler, L., Rastrup-Andersen, N.: J. Mol. Struct. **20** (1974) 119.
74STE	Steinmetz, W.E.: J. Amer. Chem. Soc. **96** (1974) 685.
74SUG	Sugisaki, R., Tanaka, T., Hirota, E.: J. Mol. Spectrosc. **49** (1974) 241.
74SUT	Sutter, D.H.: Z. Naturforsch. **29a** (1974) 786.
74SZA	Szalanski, L.B., Ford, R.G.: J. Mol. Spectrosc. **53** (1974) 428.
74TAK	Takami, M., Shimoda, K.: Jap. J. Appl. Phys. **13** (1974) 1699; errata: Jap. J. Appl. Phys. **14** (1975) 171.
74TAM1	Tamagake, K., Tsuboi, M.: Bull. Chem. Soc. Jpn **47** (1974) 73.
74TAM2	Tamagake, K., Tsuboi, M.: J. Mol. Spectrosc. **53** (1974) 189.
74TAM3	Tamagake, K., Tsuboi, M.: J. Mol. Spectrosc. **53** (1974) 204.
74THO	Thomas, C.H., Nisbet, K.D., Graner, G.: J. Chem. Phys. **61** (1974) 5072.
74TIE	Tiemann, E., Hoeft, J., Lovas, F.J., Johnson, D.R.: J. Chem. Phys. **60** (1974) 5000.
74TYP	Typke, V., Dakkouri, M., Zeil, W.: Z. Naturforsch. **29a** (1974) 1081.
74VAL	Valenzuela, E.A., Woods, R.C.: J. Chem. Phys. **61** (1974) 4119.
74VAN1	Van Eijck, B.P.: J. Mol. Spectrosc. **53** (1974) 246.
74VAN2	Van Eijck, B.P., Maagdenberg, A.A.J., Wanrooy, J.: J. Mol. Struct. **22** (1974) 61.
74WAL	Walden, R.T., Cook, R.L.: J. Mol. Spectrosc. **52** (1974) 244.
74WE1	Wells, J.A., Malloy, T.B.: J. Chem. Phys. **60** (1974) 3987.
74WE2	Wells, J.A., Malloy, T.B.: J. Chem. Phys. **60** (1974) 2132.
74WIN	Winnewisser, M., Winnewisser, B.P.: Z. Naturforsch. **29a** (1974) 633.
75AUL	Ault, G.M., Graybeal, J.D.: Symp. Mol. Struct. Spectr. Columbus, Ohio, **1975**, paper RG14.
75AVI1	Avirah, T.K., Cook, R.L., Malloy, T.B.: J. Mol. Spectrosc. **54** (1975) 231.
75AVI2	Avirah, T.K., Cook, R.L., Malloy, T.B.: J. Mol. Spectrosc. **55** (1975) 464.
75AVI3	Avirah, T.K., Malloy, T.B., Cook, R.L.: J. Mol. Struct. **26** (1975) 267.
75AVI4	Avirah, T.K., Malloy, T.B., Cook, R.L.: J. Mol. Struct. **29** (1975) 47.
75BAR	Bartke, T.C., Bjørseth, A., Haaland, A., Marstokk, K.M., Møllendal, H.: J. Organometal. Chem. **85** (1975) 271.
75BAU1	Bauer, A., Godon, M.: Can. J. Phys. **53** (1975) 1154.
75BAU2	Bauer, A., Tarrago, G., Remy, A.: J. Mol. Spectrosc. **58** (1975) 111.
75BEL	Bellott, E.M., Wilson, E.B.: Tetrahedron **31** (1975) 2896.
75BEN	Bendtsen, J., Winnewisser, M.: Chem. Phys. Lett. **33** (1975) 141.
75BL1	Blackman, G.L., Brown, R.D., Porter, A.P.: J. Chem. Soc. Chem. Commun. **1975**, 499.
75BL2	Blackman, G.L., Brown, R.D., Burden, F.R., Mishra, A.: J. Mol. Spectrosc. **57** (1975) 294.
75BOG	Bogey, M.: J. Phys. B: Atom. Mol. Phys. **8** (1975) 1934.
75BOL	Bolton, K., Brown, R.D., Burden, F.R., Mishra, A.: J. Mol. Struct. **27** (1975) 261.
75BRO	Brooks, W.V.F., Sastry, K.V.L.N.: Can. J. Chem. **53** (1975) 2247.
75BRW1	Brown, R.D., Godfrey, P.D., Storey, J.: J. Mol. Spectrosc. **58** (1975) 445.
75BRW2	Brown, R.D., Matouskova, J.: J. Mol. Struct. **29** (1975) 33.
75CA1	Caminati, W., Cazzoli, G., Mirri, A.M.: Chem. Phys. Lett. **35** (1975) 475.
75CA2	Caminati, W., Cazzoli, G., Mirri, A.M.: Chem. Phys. Lett. **31** (1975) 104.
75CAE	Careless, A.J., Kroto, H.W.: J. Mol. Spectrosc. **57** (1975) 198.
75CAE2	Careless, A.J., Kroto, H.W.: J. Mol. Spectrosc. **57** (1975) 189.
75CAR	Carpenter, J.H., Rimmer, D.F., Smith, J.G., Whiffen, D.H.: J. Chem. Soc. Faraday Trans. 2 **71** (1975) 1752.
75CAS	Casleton, K.H., Kukolich, S.G.: J. Chem. Phys. **62** (1975) 2696.
75CAX	Cassoux, P., Kuczkowski, R.L., Bryan, P.S., Taylor, R.C.: Inorg. Chem. **14** (1975) 126.
75CAZ	Cazzoli, G., Cervellati, R., Mirri, A.M.: J. Mol. Spectrosc. **56** (1975) 422.
75CH	Chadwick, D., Millen, D.J.: J. Mol. Struct. **25** (1975) 216.
75CHA	Chatterjee, A., Ghosh, D.K.: Indian J. Phys. **49** (1975) 717.
75CHR	Christen, D., Norbury, D., Lister, D.G., Palmieri, P.: J. Chem. Soc. Faraday Trans. 2 **71** (1975) 438.
75COF	Coffey, D., Ho, C.Y.: J. Mol. Spectrosc. **55** (1975) 487.

75COL	Colmont, J.M.: J. Mol. Spectrosc. **58** (1975) 220.
75COO	Cook, R.L., Lucia, F.C. de, Helminger, P.: J. Mol. Struct. **28** (1975) 237.
75COX	Cox, A.P., Ewart, I.C., Stigliani, W.M.: J. Chem. Soc. Faraday Trans. 2 **71** (1975) 504.
75CR1	Creswell, R.A.: J. Mol. Spectrosc. **56** (1975) 133.
75CR2	Creswell, R.A., Elzaro, R.A., Schwendeman, R.H.: Inorg. Chem. **14** (1975) 2256.
75CR3	Creswell, R.A.: Mol. Phys. **30** (1975) 217.
75DAG	Dagg, I.R., Reesor, G.E., Wong, A.M.C.: Can. J. Phys. **53** (1975) 861.
75DAV	Davis, R.W., Gerry, M.C.L.: J. Mol. Spectrosc. **57** (1975) 118.
75DEM	Demaison, J., Rudolph, H.D.: J. Mol. Struct. **24** (1975) 325.
75DUB	Dubrulle, A., Boucher, D., Burie, J.: C.R. Acad. Sci. **281B** (1975) 631.
75DUR1	Durig, J.R., Cooper, P.J., Li, Y.S.: Inorg. Chem. **14** (1975) 2845.
75DUR2	Durig, J.R., Cooper, P.J., Li, Y.S.: J. Mol. Spectrosc. **57** (1975) 169.
75DUR3	Durig, J.R., Hellams, K.L.: J. Mol. Struct. **29** (1975) 349.
75DUR4	Durig, J.R., Kalasinsky, V.F., Li, Y.S., Odom, J.D.: J. Phys. Chem. **79** (1975) 468.
75DUR5	Durig, J.R., Lopata, A.D., Li, Y.S.: J. Mol. Struct. **26** (1975) 85.
75ELL	Ellis, P.D., Li, Y.S., Tong, C.C., Zens, A.P., Durig, J.R.: J. Chem. Phys. **62** (1975) 1311.
75ENG	Engelbrecht, L., Sutter, D.H.: Z. Naturforsch. **30a** (1975) 1265.
75FIL	Filgueira, R.R., Forti, P., Corbelli, G.: J. Mol. Spectrosc. **57** (1975) 97.
75FIN	Finnigan, D.J., Gillies, C.W., Suenram, R.D., Wilson, E.B., Karlsson, H.: J. Mol. Spectrosc. **57** (1975) 363.
75FOR	Ford, R.G.: J. Mol. Spectrosc. **58** (1975) 178.
75FUL	Fuller, M.J., Wilson, E.B.: J. Mol. Spectrosc. **58** (1975) 414.
75GRI	Griffiths, J.H., Boggs, J.E.: J. Mol. Spectrosc. **56** (1975) 257.
75GUA	Guarnieri, A., Rohwer, F., Scappini, F.: Z. Naturforsch. **30a** (1975) 904.
75GUN	Günther, H.: Dissertation Tübingen, **1975**.
75HA1	Harmony, M.D., Bostrom, R.E., Hendricksen, D.K.: J. Chem. Phys. **62** (1975) 1599.
75HA2	Harmony, M.D., Wang, C.S., Wiberg, K.B., Bishop, K.C.: J. Chem. Phys. **63** (1975) 3312.
75HAY	Hayashi, M., Kuwada, K.: J. Mol. Struct. **28** (1975) 147.
75HEL	Helminger, P., Lucia, F.C. de: J. Mol. Spectrosc. **58** (1975) 375.
75HIR1	Hirota, E., Sahara, M.: J. Mol. Spectrosc. **56** (1975) 21.
75HIR2	Hirota, E., Imachi, M.: Can. J. Phys. **53** (1975) 2023.
75HO1	Hocking, W.H., Gerry, M.C.L., Winnewisser, G.: Can. J. Phys. **53** (1975) 1869.
75HO2	Hocking, W.H., Williams, M.L., Gerry, M.C.L.: J. Mol. Spectrosc. **58** (1975) 250.
75HOL	Holt, C.W., Gerry, M.C.L., Ozier, I.: Can. J. Phys. **53** (1975) 1791.
75IKE	Ikeda, T., Sastry, K.V.L.N., Curl, R.F.: J. Mol. Spectrosc. **56** (1975) 411.
75JOH	Johns, J.W.C., McKellar, A.R.W.: J. Chem. Phys. **63** (1975) 1682.
75JON	Jones, G.I.L., Lister, D.G., Owen, N.L.: J. Chem. Soc. Faraday. Trans. 2 **71** (1975) 1330.
75KA1	Karlsson, F., Granberg, M., Vestin, R.: Acta Chem. Scand. **29A** (1975) 111.
75KA2	Karlsson, F., Granberg, M., Vestin, R.: Acta Chem. Scand. **29A** (1975) 855.
75KAW	Kawashima, Y., Suzuki, M., Kozima, K.: Bull. Chem. Soc. Jpn **48** (1975) 2009.
75KI1	Kitchin, R.W., Malloy, T.B., Cook, R.L.: J. Mol. Spectrosc. **57** (1975) 179.
75KI2	Kitchin, R.W., Avirah, T.K., Malloy, T.B., Cook, R.L.: J. Mol. Struct. **24** (1975) 337.
75KOH	Kohrmann, H.J., Zeil, W.: Z. Naturforsch. **30a** (1975) 183.
75KUI	Kuijpers, P., Törring, T., Dymanus, A.: Z. Naturforsch. **30a** (1975) 1256.
75KUK	Kukolich, S.G.: J. Amer. Chem. Soc. **97** (1975) 5704.
75LAT	Lattimer, R.P., Kuczkowski, R.L., Ashe, A.J., Meinzer, A.L.: J. Mol. Spectrosc. **57** (1975) 428.
75LAU	Laurie, V.W., Stigliani, W.M.: Symp. Mol. Struct. Spectr. Columbus, Ohio, **1975** paper TF5.
75LEG	Legon, A.C., Millen, D.J., Rogers, S.C.: J. Chem. Soc. Chem. Commun. (**1975**) 580.
75LI1	Li, Y.S., Durig, J.R.: J. Chem. Phys. **63** (1975) 2263 L.
75LI2	Li, Y.S., Durig, J.R.: J. Mol. Spectrosc. **54** (1975) 296.
75LOV	Lovas, F.J., Clark, F.O., Tiemann, E.: J. Chem. Phys. **62** (1975) 1925.
75LUC1	Lucia, F.C. de: J. Mol. Spectrosc. **55** (1975) 271.
75LUC2	Lucia, F.C. de, Helminger, P.: J. Mol. Spectrosc. **56** (1975) 138.
75LUC3	Lucia, F.C. de, Helminger, P.: J. Mol. Spectrosc. **54** (1975) 200.
75LUK	Lucken, E.A.C., Pozzi, R., Ramaprasad, K.R.: J. Mol. Struct. **26** (1975) 259.
75MAL	Mallinson, P.D.: J. Mol. Spectrosc. **55** (1975) 94.

75MIR Mirri, A.M., Scappini, F., Mäder, H.: J. Mol. Spectrosc. **57** (1975) 264.
75MJ1 Mjöberg, P.J., Ralowski, W.M., Ljunggren, S.O.: Z. Naturforsch. **30a** (1975) 541.
75MJ2 Mjöberg, P.J., Ralowski, W.M., Ljunggren, S.O.: Z. Naturforsch. **30a** (1975) 1279.
75NAN Nandi, R.N., Chatterji, A.: Spectrochim. Acta **31a** (1975) 603.
75NOS Nösberger, P., Bauder, A., Günthard, H.H.: Chem. Phys. **8** (1975) 245.
75NOV Novick, S.E., Harris, S.J., Janda, K.C., Klemperer, W.: Can. J. Phys. **53** (1975) 2007.
75ODO Odom, J.D., Kalasinsky, V.F., Durig, J.R.: Inorg. Chem. **14** (1975) 2837.
75PAS Pasinski, J., McMahon, S.A., Beaudet, R.A.: J. Mol. Spectrosc. **55** (1975) 88.
75PE1 Pearson, R., Choplin, A., Laurie, V.W., Schwartz, J.: J. Chem. Phys. **62** (1975) 2949.
75PE2 Pearson, R., Choplin, A., Laurie, V.W.: J. Chem. Phys. **62** (1975) 4859.
75PEN1 Penn, R.E., Grossman, R.W.: J. Mol. Spectrosc. **58** (1975) 333.
75PEN2 Penn, R.E., Buxton, L.W.: J. Mol. Spectrosc. **56** (1975) 229.
75PER Perretta, A.T., Laurie, V.W.: J. Chem. Phys. **62** (1975) 2469.
75POZ Pozzi, R., Ramaprasad, K.R., Lucken, E.A.C.: J. Mol. Struct. **28** (1975) 111.
75RAL Ralowski, W.M., Mjöberg, P.J., Almhöf, J.E.: J. Chem. Soc. Faraday. Trans. 2 **71** (1975) 1109.
75ROH Rohart, F.: J. Mol. Spectrosc. **57** (1975) 301.
75SAM Sams, R.L., Maki, A.G.: J. Mol. Struct. **26** (1975) 107.
75SC Schmidt, R.E., Quade, C.R.: J. Chem. Phys. **62** (1975) 3864.
75SCH Schwoch, D., Rudolph, H.D.: J. Mol. Spectrosc. **57** (1975) 47.
75SHA Sharma, S.D., Doraiswamy, S.: J. Mol. Spectrosc. **57** (1975) 377.
75SHI Shimizu, F.: J. Phys. Soc. Jpn **38** (1975) 1106.
75SLA Slafer, W.D., English, A.D., Harris, D.O., Aue, D.H., Shellhamer, D.F., Meshishnek, M.J.: J. Amer. Chem. Soc. **97** (1975) 6638.
75SNE Snerling, O., Nielsen, C.J., Nygaard, L. Pedersen, E.J., Sørensen, G.O.: J. Mol. Struct. **27** (1975) 205.
75ST1 Stiefvater, O.L.: J. Chem. Phys. **62** (1975) 233.
75ST2 Stiefvater, O.L.: J. Chem. Phys. **62** (1975) 244.
75ST3 Stiefvater, O.L.: J. Chem. Phys. **63** (1975) 2560.
75ST4 Stiefvater, O.L.: Z. Naturforsch. **30a** (1975) 1765.
75ST5 Stiefvater, O.L., Nösberger, P., Sheridan, J.: Chem. Phys. **9** (1975) 435.
75STI Stigliani, W.M., Laurie, V.W., Li, J.C.: J. Chem. Phys. **62** (1975) 1890.
75SUE Suenram, R.D.: J. Amer. Chem. Soc. **97** (1975) 4869.
75SUZ Suzuki, M., Guarnieri, A.: Z. Naturforsch. **30a** (1975) 497.
75SZA Szalanski, L.B., Ford, R.G.: J. Mol. Spectrosc. **54** (1975) 148.
75TAN Tanaka, T., Hirota, E.: J. Mol. Spectrosc. **54** (1975) 437.
75TRU True, N.S., Bohn, R.K.: J. Chem. Phys. **62** (1975) 3951.
75VAC Vaccani, S., Bauder, A., Günthard, H.H.: Chem. Phys. Lett. **35** (1975) 457.
75VAN Van Eijck, B.P., Korthof, A.J., Mijlhoff, F.C.: J. Mol. Struct. **24** (1975) 222.
75VAR Varma, R., Ramaprasad, K.R., Nelson, J.F.: J. Chem. Phys. **63** (1975) 915.
75WAL Walls, C., Lister, D.G., Sheridan, J.: J. Chem. Soc. Faraday Trans. 2 **71** (1975) 1091.
75WIN1 Winnewisser, B.P., Winnewisser, M.: J. Mol. Spectrosc. **56** (1975) 471.
75WIN2 Winnewisser, M., Winnewisser, G., Honda, T., Hirota, E.: Z. Naturforsch. **30a** (1975) 1001.
75YAM1 Yamada, K., Winnewisser, B.P., Winnewisser, M.: J. Mol. Spectrosc. **56** (1975) 449.
75YAM2 Yamada, K., Winnewisser, M.: Z. Naturforsch. **30a** (1975) 672.
75YOS Yoshimizu, N., Hirose, C., Maeda, S.: Bull. Chem. Soc. Jpn **48** (1975) 3529.
75ZEI Zeil, W., Braun, W., Haas, B., Knehr, H., Rückert, F., Dakkouri, M.: Z. Naturforsch. **30a** (1975) 1441.

76ALE Alexander, A.J., Kroto, H.W., Walton, D.R.M.: J. Mol. Spectrosc. **62** (1976) 175.
76AND Andreev, B.A., Burenin, A.V., Karyakin, E.N., Krupnov, A.F., Shapin, S.M.: J. Mol. Spectrosc. **62** (1976) 125.
76BAL Baldacci, A., Ghersetti, S., Hurlock, S.C., Rao, K.N.: J. Mol. Spectrosc. **59** (1976) 116.
76BAS Baskakov, O.I., Moskienko, M.V., Dyubko, S.F.: Zh. Prikl. Spektrosk. **24** (1976) 1083; J. Appl. Spectry. (english translation) **24** (1976) 772.
76BAU Bauder, A., Günthard, H.H.: J. Mol. Spectrosc. **60** (1976) 290.
76BAU1 Bauder, A., Keller, C., Neuenschwander, M.: J. Mol. Spectrosc. **63** (1976) 281.
76BAU2 Bauer, A., Godon, M., Maes, S.: J. Mol. Spectrosc. **59** (1976) 421.

76BLA1	Blackman, G.L., Brown, R.D., Burden, F.R., Elsum, I.R.: J. Mol. Spectrosc. **60** (1976) 63.
76BLA2	Blackman, G.L., Brown, R.D., Godfrey, P.D., Gunn, H.I.: Nature **261** (1976) 395.
76BOD	Bodenseh, H.K., Hüttner, W., Nowicki, P.: Z. Naturforsch. **31a** (1976) 1638.
76BOT	Botskor, I., Hirota, E.: J. Mol. Spectrosc. **61** (1976) 79.
76BRI	Brier, P.N., Jones, S.R., Baker, J.G.: J. Mol. Spectrosc. **60** (1976) 18.
76CA2	Caminati, W., Cazzoli, G.: Chem. Phys. Lett. **38** (1976) 218.
76CA3	Caminati, W., Forti, P.: Chem. Phys. Lett. **38** (1976) 222.
76CA4	Caminati, W., Cazzoli, G., Troiano, D.: Chem. Phys. Lett. **43** (1976) 65.
76CAS	Casleton, K.H., Gierke, T.D., Wang, J.H.S., Kukolich, S.G.: J. Chem. Phys. **64** (1976) 471.
76CAZ	Cazzoli, G., Caminati, W.: J. Mol. Spectrosc. **62** (1976) 1.
76CH	Chao, S., Avirah, T.K., Cook, R.L., Malloy, T.B.: J. Phys. Chem. **80** (1976) 1141.
76CHR	Christiansen, I., Marstokk, K.M., Møllendal, H.: J. Mol. Struct. **30** (1976) 137.
76CLA	Clark, W.W., Lucia, F.C. de: J. Mol. Struct. **32** (1976) 29.
76COF	Coffey, D., Hershberger, M.V.: J. Mol. Spectrosc. **59** (1976) 28.
76COL	Cole, A.R.H., Li, Y.S., Durig, J.R.: J. Mol. Spectrosc. **61** (1976) 346.
76COR	Corkill, M.J., Cox, A.P., Ewart, I.C.: J. Chem. Soc. Chem. Commun. **1976**, 546.
76CRE	Creswell, R.A., Pearson, E.F., Winnewisser, M., Winnewisser, G.: Z. Naturforsch. **31a** (1976) 221.
76CUG	Cugley, J.A., Bossert, W., Bauder, A., Günthard, H.H.: Chem. Phys. **16** (1976) 229.
76CZI	Czieslik, W., Wiese, J., Sutter, D.H.: Z. Naturforsch. **31a** (1976) 1210.
76DAM	Damiani, D., Ferretti, L., Gallinella, E.: Chem. Phys. Lett. **37** (1976) 265.
76DAV1	Davis, R.W., Gerry, M.C.L.: J. Mol. Spectrosc. **59** (1976) 407.
76DAV2	Davis, R.W., Gerry, M.C.L.: J. Mol. Spectrosc. **60** (1976) 117.
76DEM1	Demaison, J., Schwoch, D., Tan, B.T., Rudolph, H.D.: J. Mol. Spectrosc. **60** (1976) 324.
76DEM2	Demaison, J.: C.R. Acad. Sci. **283B** (1976) 249.
76DEM3	Demaison, J., Schwoch, D., Tan, B.T., Rudolph, H.D.: J. Mol. Spectrosc. **59** (1976) 226.
76DOR	Doraiswamy, S., Sharma, S.D.: Chem. Phys. Lett. **37** (1976) 527.
76DUC	Duckett, J.A., Robiette, A.G., Mills, I.M.: J. Mol. Spectrosc. **62** (1976) 34.
76DUD	Duda, L.E., Flygare, W.H.: J. Chem. Phys. **65** (1976) 2841.
76DUR1	Durig, J.R., Kalasinsky, K.S., Kalasinsky, V.F.: J. Mol. Struct. **34** (1976) 9.
76DUR2	Durig, J.R., Li, Y.S., Chen, M.M., Odom, J.D.: J. Mol. Spectrosc. **59** (1976) 74.
76DUR3	Durig, J.R., Li, Y.S., Groner, P.: J. Mol. Spectrosc. **62** (1976) 159.
76DUR4	Durig, J.R., Li, Y.S., Hudgens, B.A., Cohen, E.A.: J. Mol. Spectrosc. **63** (1976) 459.
76DUR5	Durig, J.R., Cox, A.W.: J. Chem. Phys. **64** (1976) 1930.
76DUR6	Durig, J.R., Hall, L.W., Carter, R.O., Wurrey, C.J., Kalasinsky, V.F., Odom, J.D.: J. Phys. Chem. **80** (1976) 1188.
76DUR7	Durig, J.R., Lafferty, W.J., Kalasinsky, V.F.: J. Phys. Chem. **80** (1976) 1199.
76EN	Engelbrecht, L., Sutter, D.H.: Z. Naturforsch. **31a** (1976) 670.
76ENG	English, A.D., Scharpen, Le R.H., Ewool, K.M., Strauss, H.L., Harris, D.O.: J. Mol. Spectrosc. **60** (1976) 210.
76FOR	Ford, R.G.: J. Chem. Phys. **65** (1976) 354.
76GAR	Garvey, R.M., Lucia, F.C. de, Cederberg, J.W.: Mol. Phys. **31** (1976) 265.
76GEO	Georghiou, C., Baker, J.G., Jones, S.R.: J. Mol. Spectrosc. **63** (1976) 89.
76GER	Gerry, M.C.L., Lees, R.M., Winnewisser, G.: J. Mol. Spectrosc. **61** (1976) 231.
76GIL1	Gillies, C.W.: J. Mol. Spectrosc. **59** (1976) 482.
76GIL2	Gillies, C.W., Heck, J.: Austin Symp. Gas Phase Mol. Struct. **1976**, paper MA8 B.
76GOU	Goulet, P., Jurek, R., Chanussot, J.: J. Phys. Paris **37** (1976) 495.
76GRA	Graybeal, J.D.: Austin Symp. Gas Phase Mol. Struct. **1976**, paper TA7.
76GUM	Gum, M.L., Graybeal, J.D.: J. Mol. Spectrosc. **62** (1976) 364.
76GUN	Gunderova, L.N., Kostyuchenko, L.S., Latypova, R.G., Pozdeev, N.M., Sverdlov, L.M.: Zh. Strukt. Khim. **17** (1976) 726; J. Struct. Chem. (english translation) **17** (1976) 625.
76HAL	Hall, J.L., Magyar, J.A., in "High resolution laser spectroscopy", K. Shimoda ed., **1976**, 174, Springer Verlag, Berlin.
76HAM1	Hamer, E., Sutter, D.H.: Z. Naturforsch. **31a** (1976) 265.
76HAM2	Hamm, R., Kohrmann, H.J., Günther, H., Zeil, W.: Z. Naturforsch. **31a** (1976) 594.
76HAR	Harris, S.J., Janda, K.C., Novick, S.E., Klemperer, W.: J. Chem. Phys. **63** (1976) 881.

76HAY	Hayashi, M., Imachi, M.: Chem. Lett. **1976**, 121.
76HEI	Heise, H.M., Mäder, H., Dreizler, H.: Z. Naturforsch. **31a** (1976) 1228.
76HER	Herlemont, F., Thibault, J., Lemaire, J.: J. Mol. Spectrosc. **61** (1976) 138.
76HIR1	Hirose, C., Okiye, K., Maeda, S.: Bull. Chem. Soc. Jpn **49** (1976) 916.
76HIR2	Hirota, E., Tanaka, T., Saito, S.: J. Mol. Spectrosc. **63** (1976) 478.
76HO1	Hocking, W.H.: Z. Naturforsch. **31a** (1976) 1113.
76HO2	Hocking, W.H., Gerry, M.C.L.: J. Mol. Spectrosc. **59** (1976) 338.
76HO3	Hocking, W.H., Winnewisser, G.: Z. Naturforsch. **31a** (1976) 422.
76HO4	Hocking, W.H., Winnewisser, G.: Z. Naturforsch. **31a** (1976) 438.
76HO5	Hocking, W.H., Winnewisser, G.: Z. Naturforsch. **31a** (1976) 995.
76HOG	Høg, J., Pedersen, T.: J. Mol. Spectrosc. **61** (1976) 243.
76HOP	Hopkinson, M.J., Kroto, H.W., Nixon, J.F., Simmons, N.P.C.: J. Chem. Soc. Chem. Commun. (1976) 513.
76IM	Imachi, M., Tanaka, T., Hirota, E.: J. Mol. Spectrosc. **63** (1976) 265.
76INO	Inoue, H., Naruse, N., Hirota, E.: Bull. Chem. Soc. Jpn **49** (1976) 1260.
76JOH	Johnson, D.R., Suenram, R.D., Lafferty, W.J.: Astrophys. J. **208** (1976) 245.
76JON1	Jones, G.I.L., Lister, D.G., Owen, N.L., Gerry, M.C.L., Palmieri, P.: J. Mol. Spectrosc. **60** (1976) 348.
76JON2	Jones, H., Kohler, F., Rudolph, H.D.: J. Mol. Spectrosc. **63** (1976) 205.
76KAG	Kagann, R.H., Ozier, I., Gerry, M.C.L.: J. Chem. Phys. **64** (1976) 3487.
76KAR1	Karlsson, H.: J. Mol. Struct. **33** (1976) 227.
76KAR2	Karlsson, H.: J. Mol. Struct. **33** (1976) 319.
76KAR3	Karlsson, H., Nielsen, J.T., Nygaard, L., Sørensen, K.: Acta Chem. Scand. **30A** (1976) 820.
76KAW	Kawashima, Y., Cox, A.P.: J. Mol. Spectrosc. **61** (1976) 435.
76KOJ	Kojima, T., Ogata, T., Maeda, S.: Chem. Lett. **1976**, 607.
76KRI	Krisher, L.C., Morrison, J.A.: J. Chem. Phys. **64** (1976) 3556.
76KRO	Kroto, H.W., Landsberg, B.M.: J. Mol. Spectrosc. **62** (1976) 346.
76KUC	Kuczkowski, R.L., Gillies, C.W., Gallaher, K.L.: J. Mol. Spectrosc. **60** (1976) 361.
76KUI	Kuijpers, P., Törring, T., Dymanus, A.: Chem. Phys. **15** (1976) 457.
76LA	Lattimer, R.P., Mazur, U., Kuczkowski, R.L.: J. Amer. Chem. Soc. **98** (1976) 4012.
76LAT	Latypova, R.G., Mamleev, A.K., Gunderova, L.N., Pozdeev, N.M.: Zh. Strukt. Khim **17** (1976) 849; J. Struct. Chem. (english translation) **17** (1976) 728.
76LAU	Lau, K.K., Beaudet, R.A.: Inorg. Chem. **15** (1976) 1059.
76LIS1	Lister, D.G., Palmieri, P., Zauli, C.: J. Mol. Struct. **35** (1976) 299.
76LIS2	Lister, D.G., Lowe, S.E., Palmieri, P.: J. Chem. Soc. Faraday Trans. 2 **72** (1976) 920.
76LO1	Lowe, R.S., Kewley, R.: J. Mol. Spectrosc. **63** (1976) 216.
76LO2	Lowe, R.S., Kewley, R.: J. Mol. Spectrosc. **60** (1976) 312.
76MAM	Mamleev, A.H., Pozdeev, N.M., Magdesieva, N.N.: J. Mol. Struct. **33** (1976) 211.
76MAR1	Marstokk, K.M., Møllendal, H.: J. Mol. Struct. **35** (1976) 57.
76MAR2	Marstokk, K.M., Møllendal, H.: J. Mol. Struct. **32** (1976) 191.
76MAT	Mathur, S.N., Harmony, M.D., Suenram, R.D.: J. Chem. Phys. **64** (1976) 4340.
76MCK	McKown, G.L., Don, B.P., Beaudet, R.A., Vergamini, P.J., Jones, L.H.: J. Amer. Chem. Soc. **98** (1976) 6909.
76MIC	Michel, F., Nery, H., Nösberger, P., Roussy, G.: J. Mol. Struct. **30** (1976) 409.
76MIR	Mirri, A.M., Scappini, F., Cervellati, R., Favero, P.G.: J. Mol. Spectrosc. **63** (1976) 509.
76MJO	Mjöberg, P.J., Ralowski, W.M., Ljunggren, S.O., Bäckvall, J.E.: J. Mol. Spectrosc. **60** (1976) 179.
76MUR	Murray, J.T., Little, W.A., Williams, Q., Weatherly, T.L.: J. Chem. Phys. **65** (1976) 985.
76NAI	Nair, K.P.R., Boggs, J.E.: J. Mol. Struct. **33** (1976) 45.
76NAK1	Nakagawa, J., Hayashi, M.: Bull. Chem. Soc. Jpn **49** (1976) 3441.
76NAK2	Nakagawa, J., Kuwada, K., Hayashi, M.: Bull. Chem. Soc. Jpn **49** (1976) 3420.
76NAK3	Nakagawa, J., Hayashi, M.: Chem. Lett. **1976**, 349.
76NAN	Nandi, R.N., Chatterjee, A., Ghosh, D.K.: J. Mol. Struct. **33** (1976) 63.
76NEM1	Nemes, L.: Kém. Közl. **46** (1976) 434.
76NEM2	Nemes, L., Winnewisser, M.: Z. Naturforsch. **31a** (1976) 272.
76NI	Niide, Y., Takano, M., Satoh, T., Sasada, Y.: J. Mol. Spectrosc. **63** (1976) 108.
76NOV1	Novick, S.E., Janda, K.C., Klemperer, W.: J. Chem. Phys. **65** (1976) 5115.

76NOV2	Novick, S.E., Janda, K.C., Holmgren, S.L., Waldman, M., Klemperer, W.: J. Chem. Phys. **65** (1976) 1114.
76OGA	Ogata, T., Cox, A.P.: J. Mol. Spectrosc. **61** (1976) 265.
76OH	Ohashi, O., Ishihara, R., Murakami, K., Sakaizumi, T., Onda, M., Yamaguchi, I.: Bull. Chem. Soc. Jpn **49** (1976) 891.
76OKA	Oka, T.: "Forbidden rotational transitions" in "Molecular Spectroscopy: modern research", volume II, K.N. RAO ed., **1976**, Academic Press, New York.
76OND1	Onda, M, Yamaguchi, I.: J. Mol. Struct. **34** (1976) 1.
76OND2	Onda, M., Ohashi, O., Yamaguchi, I.: J. Mol. Struct. **31** (1976) 203.
76OND3	Onda, M., Kambayashi, S., Sakaizumi, T., Yamaguchi, I.: J. Mol. Struct. **34** (1976) 299.
76OZ1	Ozier, I., Lees, R.M., Gerry, M.C.L.: Can. J. Phys. **54** (1976) 1094.
76OZ2	Ozier, I., Lees, R.M., Gerry, M.C.L.: J. Chem. Phys. **65** (1976) 1795.
76OZ3	Ozier, I., Lee, S.S., Ramsey, N.F.: J. Chem. Phys. **65** (1976) 3985.
76PAN1	Pandey, G.K., Dreizler, H.: Z. Naturforsch. **31a** (1976) 357.
76PAN2	Pandey, G.K., Lutz, H., Dreizler, H.: Z. Naturforsch. **31a** (1976) 1413.
76PEA1	Pearson, E.F., Creswell, R.A., Winnewisser, M., Winnewisser, G.: Z. Naturforsch. **31a** (1976) 1394.
76PEA2	Pearson, E.F., Winnewisser, B.P., Trueblood, M.B.: Z. Naturforsch. **31a** (1976) 1259.
76PEN1	Penn, R.A.: Austin Symp. Gas Phase Mol. Struct. **1976**, paper WM2.
76PEN2	Penn, R.E., Olsen, R.J.: J. Mol. Spectrosc. **61** (1976) 21.
76PEN3	Penn, R.E., Birkenmeier, J.A.: J. Mol. Spectrosc. **62** (1976) 416.
76PEN4	Penn, R.E., Olsen, R.J.: J. Mol. Spectrosc. **62** (1976) 423.
76POZ1	Pozdeev, N.M., Latypova, R.G., Gunderova, L.N.: Zh. Strukt. Khim **17** (1976) 362; J. Struct. Chem. (english translation) **17** (1976) 313.
76POZ3	Pozdeev, N.M., Gunderova, L.N., Mamleev, A.K.: Opt. Spectrosk. **40** (1976) 773; Opt. Spectry (english translation) **40** (1976) 444.
76RAL1	Ralowski, W.M., Mjöberg, P.J., Ljunggren, S.O.: J. Amer. Chem. Soc. **98** (1976) 6897.
76RAL2	Ralowski, W.M., Mjöberg, P.J., Ljunggren, S.O.: J. Mol. Struct. **30** (1976) 1.
76RAL3	Ralowski, W.M., Mjöberg, P.J., Ljunggren, S.O.: J. Mol. Struct. **31** (1976) 169.
76RAM	Ramaprasad, K.R., Laurie, V.W., Craig, N.C.: J. Chem. Phys. **64** (1976) 4832.
76ROB1	Roberts, R.F., Varma, R., Nelson, J.F.: J. Chem. Phys. **64** (1976) 5035.
76ROB2	Robiette, A.G.: J. Mol. Struct. **35** (1976) 81.
76ROB3	Robiette, A.G., Georghiou, C., Baker, J.G.: J. Mol. Spectrosc. **63** (1976) 391.
76ROG	Rogers, H.N., Lau, K.K., Beaudet, R.A.: Inorg. Chem. **15** (1976) 1775.
76ROU	Roussy, G., Michel, F.: J. Mol. Struct. **30** (1976) 399.
76ROW	Rowe, W.F., Duerst, R.W., Wilson, E.B.: J. Amer. Chem. Soc. **98** (1976) 4021.
76SA	Sakaizumi, T., Yamada, O., Ushida, K., Ohashi, O., Yamaguchi, I.: Bull. Chem. Soc. Jpn. **49** (1976) 2908.
76SAI	Saito, S.: Chem. Phys. Lett. **42** (1976) 399.
76SAK	Sakami, T., Ohashi, O., Yamaguchi, I.: Bull. Chem. Soc. Jpn. **49** (1976) 948.
76SAY	Saykally, R.J., Szanto, P.G., Anderson, T.G., Woods, R.C.: Symp. Mol. Struct. Spectr. Columbus, Ohio **1976**, paper TS1.
76SCA1	Scappini, F., Guarnieri, A.: Z. Naturforsch. **31a** (1976) 369.
76SCA2	Scappini, F., Dreizler, H.: Z. Naturforsch. **31a** (1976) 840.
76SCA3	Scappini, F., Mäder, H., Dreizler, H.: Z. Naturforsch. **31a** (1976) 1398.
76SEN	Sengupta, S.K., Laurie, V.W.: Symp. Mol. Struct. Spectr. Columbus, Ohio, **1976**, paper MS4.
76SHA	Sharma, S.D., Doraiswamy, S.: J. Mol. Spectrosc. **59** (1976) 216.
76SM1	Smith, J.G.: Mol. Phys. **32** (1976) 621.
76SM2	Smith, J.G., Thompson, I.: Mol. Phys. **32** (1976) 1247.
76ST1	Stiefvater, O.L.: Chem. Phys. **13** (1976) 73.
76ST2	Stiefvater, O.L., Lui, S., Ladd, J.A.: Z. Naturforsch. **31a** (1976) 53.
76ST3	Stiefvater, O.L.: Z. Naturforsch. **31a** (1976) 1681.
76STI1	Stigliani, W.M., Laurie, V.W.: J. Mol. Spectrosc. **60** (1976) 188.
76STI2	Stigliani, W.M., Laurie, V.W., Scharpen, L.H.: J. Mol. Spectrosc. **62** (1976) 85.
76SUE	Suenram, R.D.: J. Mol. Struct. **33** (1976) 1.
76SUZ	Suzuki, M., Guarnieri, A.: Z. Naturforsch. **31a** (1976) 1242.
76TA	Takami, M., Shimoda, K.: J. Mol. Spectrosc. **59** (1976) 35.

76TAK	Takeo, H., Matsumura, C.: J. Chem. Phys. **64** (1976) 4536.
76TAN	Tanaka, T., Yamada, C., Hirota, E.: J. Mol. Spectrosc. **63** (1976) 142.
76TR1	True, N.S., Bohn, R.K.: J. Amer. Chem. Soc. **98** (1976) 1188.
76TR2	True, N.S., Bohn, R.K.: Symp. Mol. Struct. Spectr. Columbus, Ohio, **1976**, paper WB4.
76TSU	Tsunekawa, S.: J. Phys. Soc. Jpn. **41** (1976) 2077.
76TUR1	Turner, P.H., Cox, A.P.: Chem. Phys. Lett. **42** (1976) 84.
76TUR2	Turner, P.H., Cox, A.P.: Chem. Phys. Lett. **39** (1976) 585.
76TY	Typke, V.: J. Mol. Spectrosc. **63** (1976) 170.
76VAN	Van Eijck, B.P., Kaleveld, E.W.: J. Mol. Struct. **34** (1976) 161.
76VAR	Varma, R., Curl, R.F.: J. Phys. Chem. **80** (1976) 402.
76WAN	Wang, C.S., Harmony, M.D.: J. Amer. Chem. Soc. **98** (1976) 1108.
76WIN	Winnewisser, M., Christiansen, J.J.: Chem. Phys. Lett. **37** (1976) 270.
77BA1	Bak, B., Svanholt, H.: J. Mol. Struct. **37** (1977) 153.
77BA2	Bak, B., Larsen, N.W., Svanholt, H.: Acta Chem. Scand. **31 A** (1977) 755.
77BAR	Barbe, A., Secroun, C., Jouve, P., Monnanteuil, N., Depannemaecker, J.C., Duterage, B., Bellet, J., Pinson, P.: J. Mol. Spectrosc. **64** (1977) 343.
77BAU	Bauer, A., Godon, M.: J. Mol. Spectrosc. **66** (1977) 237.
77BEL	Bellott, E.M.: Tetrahedron **33** (1977) 1707.
77BEN	Bendtsen, J.: J. Raman Spectrosc. **6** (1977) 306.
77BEV	Bevan, J.W., Legon, A.C., Millen, D.J.: Proc. Roy. Soc. London **354a** (1977) 491.
77BLA1	Blackman, G.L., Brown, R.D., Burden, F.R., Garland, W.: J. Mol. Spectrosc. **65** (1977) 313.
77BLA2	Blackman, G.L., Brown, R.D., Brown, R.F.C., Eastwood, F.W., McMullen, G.L.: J. Mol. Spectrosc. **68** (1977) 488.
77BOU1	Boucher, D., Burie, J., Demaison, J., Dubrulle, A., Legrand, J., Segard, B.: J. Mol. Spectrosc. **64** (1977) 290.
77BOU2	Bouchy, A., Roussy, G.: C.R. Acad. Sci. **284c** (1977) 411.
77BOU3	Bouchy, A., Roussy, G.: J. Mol. Spectrosc. **68** (1977) 156.
77BOU4	Bouchy, A., Roussy, G.: J. Mol. Spectrosc. **65** (1977) 395.
77BRI	Brier, P.N., Jones, S.R., Baker, J.G., Georghiou, C.: J. Mol. Spectrosc. **64** (1977) 415.
77BRO	Brown, R.D., Godfrey, P.D., Ottrey, A.L., Storey, J.W.V.: J. Mol. Spectrosc. **68** (1977) 359.
77BUR	Burie, J., Demaison, J., Dubrulle, A., Boucher, D.: J. Mol. Spectrosc. **68** (1977) 495.
77CAM1	Caminati, W., Mirri, A.M., Maccagnani, G.: J. Mol. Spectrosc. **66** (1977) 368.
77CAM2	Caminati, W., Cervellati, R., Lunelli, B.: J. Phys. Chem. **81** (1977) 1494.
77CAR	Carpenter, J.H., Smith, J.G., Thompson, I., Whiffen, D.H.: J. Chem. Soc. Faraday Trans. 2 **73** (1977) 384.
77CAX1	Cassoux, P., Kuczkowski, R.L., Creswell, R.A.: Inorg. Chem. **16** (1977) 2959.
77CAX2	Cassoux, P., Kuczkowski, R.L., Serafini, A.: Inorg. Chem. **16** (1977) 3005.
77CAZ	Cazzoli, G., Favero, P.G., Esposti, C.D.: Chem. Phys. Lett. **50** (1977) 336.
77CER	Cervellati, R., Caminati, W., Esposti, C.D., Mirri, A.M.: J. Mol. Spectrosc. **66** (1977) 389.
77CHA	Chatterjee, A., Nandi, R.N., Ghosh, D.K.: Indian J. Phys. **51b** (1977) 17.
77CHI	Chiang, J.F., Chiang, R., Lu, K.C., Sung, E.M., Harmony, M.D.: J. Mol. Struct. **41** (1977) 67.
77CHR	Christen, D., Damiani, D., Lister, D.G.: J. Mol. Struct. **41** (1977) 315.
77COF1	Coffey, D., Yamada, C., Hirota, E.: J. Mol. Spectrosc. **64** (1977) 98.
77COF2	Coffey, D.: J. Mol. Spectrosc. **68** (1977) 21.
77CR1	Creswell, R.A., Schwendeman, R.H.: J. Mol. Spectrosc. **64** (1977) 295.
77CR2	Creswell, R.A., Manor, P.J., Assink, R.A., Schwendeman, R.H.: J. Mol. Spectrosc. **64** (1977) 365.
77DAM	Damiani, D., Ferretti, L., Corbelli, G.: Chem. Phys. Lett. **47** (1977) 335.
77DAV	Davis, R.W., Gerry, M.C.L.: J. Mol. Spectrosc. **65** (1977) 455.
77DEM1	Demaison, J., Dubrulle, A., Boucher, D., Burie, J.: J. Chem. Phys. **67** (1977) 254.
77DEM2	Demaison, J., Schwoch, D., Tan, B.T., Rudolph, H.D.: J. Mol. Spectrosc. **68** (1977) 97.
77DEP	Depannemaecker, J.C., Bellet, J.: J. Mol. Spectrosc. **66** (1977) 106.
77DER	Deroche, J.C.: Ph.D. Thesis, Paris-Sud **1977**.
77DO1	Dössel, K.F., Robiette, A.G.: Z. Naturforsch. **32a** (1977) 462.
77DO2	Dössel, K.F., Sutter, D.H.: Z. Naturforsch. **32a** (1977) 473.
77DO3	Dössel, K.F., Sutter, D.H.: Z. Naturforsch. **32a** (1977) 1444.

77DUB	Dubrulle, A., Boucher, D., Burie, J., Demaison, J.: Chem. Phys. Lett. **45** (1977) 559.
77DUR1	Durig, J.R., Groner, P., Lopata, A.D.: Chem. Phys. **21** (1977) 401.
77DUR2	Durig, J.R., Griffin, M.G.: J. Mol. Spectrosc. **64** (1977) 252.
77DUR3	Durig, J.R., Bucy, W.E.: J. Mol. Spectrosc. **64** (1977) 474.
77DUR4	Durig, J.R., Lopata, A.D., Groner, P.: J. Chem. Phys. **66** (1977) 1888.
77DUR5	Durig, J.R., Groner, P., Li, Y.S.: J. Chem. Phys. **67** (1977) 2216.
77DUX	Duxbury, G., Freund, S.M.: J. Mol. Spectrosc. **67** (1977) 219.
77DYK	Dyke, T.R., Mack, K.M., Muenter, J.S.: J. Chem. Phys. **66** (1977) 498.
77DYL	Dyllick-Brenzinger, C.E., Bauder, A., Günthard, H.H.: Chem. Phys. **23** (1977) 195.
77EN	Endo, K., Takeo, H., Matsumura, C.: Bull. Chem. Soc. Jpn. **50** (1977) 626.
77FAB	Fabricant, B., Krieger, D., Muenter, J.S.: J. Chem. Phys. **67** (1977) 1576.
77FOR1	Ford, R.G.: J. Amer. Chem. Soc. **99** (1977) 2389.
77FOR2	Ford, R.G.: J. Mol. Spectrosc. **65** (1977) 273.
77FU1	Fujiwara, F.G., Painter, J.L., Kim, H.: J. Mol. Struct. **41** (1977) 169.
77FU2	Fujiwara, F.G., Chang, J.C., Kim, H.: J. Mol. Struct. **41** (1977) 177.
77GAR	Garvey, R.M., Lucia, F.C. de: Can. J. Phys. **55** (1977) 1115.
77GOD	Godon, M., Bauer, A.: J. Mol. Struct. **38** (1977) 9.
77HAE	Haeck, L., Bouchy, A., Roussy, G.: Chem. Phys. Lett. **52** (1977) 512.
77HAM	Hamm, R., Günther, H., Zeil, W.: Z. Naturforsch. **32a** (1977) 754.
77HAR1	Hargittai, I., Seip, R., Nair, K.P.R., Britt, C.O., Boggs, J.E., Cyvin, B.N.: J. Mol. Struct. **39** (1977) 1.
77HAR2	Harmony, M.D., Baron, P.A.: J. Mol. Struct. **38** (1977) 1.
77HAY1	Hayashi, M., Hikino, H., Imachi, M.: Chem. Lett. **1977**, 1.
77HAY2	Hayashi, M., Imachi, M., Saito, M.: Chem. Lett. **1977**, 221.
77HEL	Helms, D.A., Gordy, W.: J. Mol. Spectrosc. **66** (1977) 206.
77HIR	Hirota, E., Matsumura, K., Imachi, M., Fujio, M., Tsuno, Y., Matsumura, C.: J. Chem. Phys. **66** (1977) 2660.
77HO	Hocking, W.H., Winnewisser, G.: Z. Naturforsch. **32a** (1977) 1108.
77IK	Ikeura, T., Tanaka, K., Hirota, E., Inaba, S., Nagai, Y.: Bull. Chem. Soc. Jpn. **50** (1977) 1309.
77ISV	Isvén, M.K., Ralowski, W.M., Ljunggren, S.O.: Chem. Phys. Lett. **47** (1977) 294.
77JAN	Janda, K.C., Steed, J.M., Novick, S.E., Klemperer, W.: J. Chem. Phys. **67** (1977) 5162.
77JO1	Johns, J.W.C., McKellar, A.R.W.: J. Chem. Phys. **66** (1977) 1217.
77JO2	Johns, J.W.C., McKellar, A.R.W.: J. Mol. Spectrosc. **64** (1977) 327.
77JON	Jones, H., Sheridan, J., Stiefvater, O.L.: Z. Naturforsch. **32a** (1977) 866.
77KAR	Karlsson, F., Granberg, M., Vestin, R.: Acta Chem. Scand. **31A** (1977) 379.
77KAU1	Kaushik, V.K.: Chem. Phys. Lett. **49** (1977) 89.
77KAU2	Kaushik, V.K.: Spectrochim. Acta **33a** (1977) 463.
77KAU3	Kaushik, V.K., Venkateswarlu, P.: Chem. Phys. Lett. **46** (1977) 426.
77KAU4	Kaushik, V.K., Venkateswarlu, P.: Chem. Phys. Lett. **48** (1977) 115.
77KAU5	Kaushik, V.K., Venkateswarlu, P.: Chem. Phys. Lett. **51** (1977) 516.
77KAW1	Kawaguchi, K., Tanaka, T.: J. Mol. Spectrosc. **68** (1977) 125.
77KAW2	Kawaguchi, K., Yamada, C., Tanaka, T., Hirota, E.: J. Mol. Spectrosc. **64** (1977) 125.
77KAW	Kawashima, Y., Cox, A.P.: J. Mol. Spectrosc. **65** (1977) 319.
77KEW	Kewley, R., Dass, S.C.: Can. J. Chem. **55** (1977) 3480.
77KRE	Kreiner, W.A., Andresen, U., Oka, T.: J. Chem. Phys. **66** (1977) 4662.
77KRE1	Kreiner, W.A., Orr, B.J., Andresen, U., Oka, T.: Phys. Rev. A **15** (1977) 2298.
77KRI	Krishnaji, Srivastava, S.L., Narain, N.K.: Indian J. Phys. **51b** (1977) 8.
77KRO	Kroto, H.W., Maier, M.: J. Mol. Spectrosc. **65** (1977) 280.
77KUI	Kuijpers, P., Törring, T., Dymanus, A.: Z. Naturforsch. **32a** (1977) 930.
77KUK	Kukolich, S.G.: J. Chem. Phys. **66** (1977) 4345.
77LAF	Lafferty, W.J., Suenram, R.D., Johnson, D.R.: J. Mol. Spectrosc. **64** (1977) 147.
77LEG	Legon, A.C., Millen, D.J., Mjöberg, P.J.: Chem. Phys. Lett. **47** (1977) 589.
77LIS	Lister, D.G., Lowe, S.E.: J. Mol. Struct. **41** (1977) 318.
77LUC	Lucia, F.C. de, Helminger, P.A.: J. Chem. Phys. **67** (1977) 4262.
77MAK1	Maki, A.G., Sams, R.L., Pearson, R.: J. Mol. Spectrosc. **64** (1977) 452.
77MAK2	Maki, A.G., Freund, S.M.: J. Mol. Spectrosc. **66** (1977) 493.
77MAL1	Mallinson, P.D.: J. Mol. Spectrosc. **68** (1977) 68.

77MAL2	Malloy, T.B., Carreira, L.A.: J. Chem. Phys. **66** (1977) 4246.
77MAR	Marstokk, K.M., Møllendal, H.M.: J. Mol. Struct. **40** (1977) 1.
77MAT	Mata, F., Quintana, M.J., Sørensen, G.O.: J. Mol. Struct. **42** (1977) 1.
77MAZ	Mazur, U., Kuczkowski, R.L.: J. Mol. Spectrosc. **65** (1977) 84.
77MEE	Meerts, W.L., Stolte, S., Dymanus, A.: Chem. Phys. **19** (1977) 467.
77MEH	Mehrotra, S.C., Griffin, L.L., Britt, C.O., Boggs, J.E.: J. Mol. Spectrosc. **64** (1977) 244.
77MOS1	Moskienko, M.V., Dyubko, S.F.: Opt. Spektrosk. **43** (1977) 850; Opt. Spectry (English translation) **43** (1977) 503.
77MOS2	Moskienko, M.V., Dyubko, S.F.: Ukr. Phys. J. **22** (1977) 235.
77NAI	Naito, T., Ohashi, O., Yamaguchi, I.: J. Mol. Spectrosc. **68** (1977) 32.
77NIE	Nielsen, C.J.: Acta Chem. Scand. **31A** (1977) 791.
77OH1	Ohashi, O., Hara, H., Noji, K., Sakaizumi, T., Onda, M., Yamaguchi, I.: Bull. Chem. Soc. Jpn. **50** (1977) 834.
77OH2	Ohashi, O., Ohnishi, M., Tagui, A., Sakaizumi, T., Yamaguchi, I.: Bull. Chem. Soc. Jpn. **50** (1977) 1749.
77OND	Onda, M., Yamaguchi, I.: J. Mol. Struct. **36** (1977) 158.
77PAN1	Pandey, G.K., Andolfatto, M., Dreizler, H.: Z. Naturforsch. **32a** (1977) 1301.
77PAN2	Pandey, G.K., Dreizler, H.: Z. Naturforsch. **32a** (1977) 482.
77PAN3	Pandey, G.K., Dreizler, H.: Z. Naturforsch. **32a** (1977) 761.
77PEA	Pearson, R., Lovas, F.J.: J. Chem. Phys. **66** (1977) 4149.
77PED	Pedersen, T., Skaarup, S.V., Pedersen, C.T.: Acta Chem. Scand. **31B** (1977) 711.
77PEN1	Penn, R.E., Buxton, L.W.: J. Chem. Phys. **67** (1977) 831.
77PEN2	Penn, R.E., Owens, T.A.: Symp. Mol. Struct. Spectr. Columbus, Ohio **1977**, paper TG4.
77POR	Porter, A.P., Godfrey, P.D.: J. Mol. Spectrosc. **68** (1977) 492.
77RAL	Ralowski, W., Ljunggren, S.: Acta Chem. Scand. **31A** (1977) 492.
77SAS	Sasada, Y., Niide, Y., Takano, M., Satoh, T.: J. Mol. Spectrosc. **66** (1977) 421.
77SCH	Schwoch, D., Burg, A.B., Beaudet, R.A.: Inorg. Chem. **16** (1977) 3219.
77SEN	Sengupta, S.K., Laurie, V.W.: Symp. Mol. Struct. Spectr. Columbus, Ohio **1977**, paper TG5.
77SHA	Sharma, S.D., Doraiswamy, S.: Can. J. Phys. **55** (1977) 1211.
77SIL	Silvia, C.J., True, N.S., Bohn, R.K.: Symp. Mol. Struct. Spectr. Columbus, Ohio **1977**, paper RE4.
77SNY	Snyder, L.E., Hollis, J.M., Buhl, D.: Astrophys. J. **215** (1977) 87.
77SRI	Srivastava, S.L.: Indian J. Phys. **51b** (1977) 1.
77ST	Steinmetz, W.E., Hickernell, F., Mum, I.K., Scharpen, L.H.: J. Mol. Spectrosc. **68** (1977) 173.
77STE	Stevens, J.F., Bevan, J.W., Curl, R.F., Geanangel, R.A., Hu, M.G.: J. Amer. Chem. Soc. **99** (1977) 1442.
77STI	Stiefvater, O.L.: Z. Naturforsch. **32a** (1977) 1480.
77SU	Su, C.F., Beeson, E.L.: J. Chem. Phys. **66** (1977) 330.
77SUE	Suenram, R.D., Johnson, D.R.: J. Mol. Spectrosc. **65** (1977) 239.
77SUN	Sung, E.M., Harmony, M.D.: J. Amer. Chem. Soc. **99** (1977) 5603.
77TA1	Takeo, H., Matsumura, C.: Bull. Chem. Soc. Jpn. **50** (1977) 636.
77TA2	Takeo, H., Matsumura, C.: Bull. Chem. Soc. Jpn. **50** (1977) 1633.
77TR1	True, N.S., Bohn, R.K.: J. Amer. Chem. Soc. **99** (1977) 3575.
77TR2	True, N.S., Bohn, R.K.: J. Mol. Struct. **36** (1977) 173.
77TR3	True, N.S., Bohn, R.K.: J. Phys. Chem. **81** (1977) 1671.
77TR4	True, N.S., Bohn, R.K.: J. Phys. Chem. **81** (1977) 1667.
77VAC	Vaccani, S., Roos, U., Bauder, A., Günthard, H.H.: Chem. Phys. **19** (1977) 51.
77VAN1	Van Eijck, B.P.: J. Mol. Struct. **37** (1977) 1.
77VAN2	Van Eijck, B.P., Van Duijneveldt, F.B.: J. Mol. Struct. **39** (1977) 157.
77VEN	Venkatachar, A.C., Taylor, R.C., Kuczkowski, R.L.: J. Mol. Struct. **38** (1977) 17.
77WAT	Watson, J.K.G.: Amsterdam Elsevier **1977** "Aspects of quartic and sextic centrifugal effects of rotational energy levels" in "Vibrational Spectra and Structure" A serie of Advances, vol. 6, J.R. Durig, ed.
77WIE1	Wiese, J., Engelbrecht, L., Dreizler, H.: Z. Naturforsch. **32a** (1977) 152.
77WIE2	Wiese, J., Sutter, D.H.: Z. Naturforsch. **32a** (1977) 890.
77YAM1	Yamada, C., Hirota, E.: J. Mol. Spectrosc. **64** (1977) 31.

77YAM2	Yamada, K., Winnewisser, M., Winnewisser, G., Szalanski, L.B., Gerry, M.C.L.: J. Mol. Spectrosc. **64** (1977) 401.
77YAM3	Yamada, K., Winnewisser, M.: J. Mol. Spectrosc. **68** (1977) 307.
77YAM4	Yamada, K.: J. Mol. Spectrosc. **68** (1977) 423.
78ALE	Alexander, A.J., Kroto, H.W., Maier, M., Walton, D.R.M.: J. Mol. Spectrosc. **70** (1978) 84.
78ALL	Allegrini, M., Johns, J.W.C., McKellar, A.R.W.: J. Mol. Spectrosc. **73** (1978) 168.
78AMA	Amano, T., Schwendeman, R.H.: J. Chem. Phys. **68** (1978) 530.
78ARI	Arimondo, E., Glorieux, P., Oka, T.: Phys. Rev. A **17** (1978) 1375.
78BJ	Bjarnov, E., Hocking, W.H.: Z. Naturforsch. **33a** (1978) 610.
78BAK1	Bak, B., Nielsen, O.J., Svanholt, H.: J. Mol. Spectrosc. **69** (1978) 401.
78BAK2	Bak, B., Nielsen, O.J., Svanholt, H., Almenningen, A., Bastiensen, O.B., Fernholt, L., Gundersen, G., Nielsen, C.J., Cyvin, B.N., Cyvin, S.J.: Acta Chem. Scand. **32A** (1978) 1005.
78BAK3	Bak, B., Nielsen, O.J., Svanholt, H., Holm, A.: Chem. Phys. Lett. **53** (1978) 374.
78BAU	Baughcum, S.L.: Ph. D. Thesis, Harvard University, 1978.
78BEV	Bevan, J.W., Legon, A.C., Ljunggren, S.O., Mjöberg, P.J.: J. Amer. Chem. Soc. **100** (1978) 8161.
78BOG	Boggia, L.M., Filgueira, R.R., Marañon, J., Sorarrain, O.M.: Spectr. Lett. **11** (1978) 143.
78BOS	Bossert, W., Ekkers, J., Bauder, A., Günthard, H.H.: Chem. Phys. **27** (1978) 433.
78BOT	Botskor, I., Rudolph, H.D.: J. Mol. Spectrosc. **71** (1978) 430.
78BOU	Boucher, D., Burie, J., Dangoisse, D., Demaison, J., Dubrulle, A.: Chem. Phys. **29** (1978) 323.
78BRA	Braun, W.: Z. Naturforsch. **33a** (1978) 455.
78BRI	Briggs, T.S., Gwinn, W.D., Jolly, W.L., Thorne, L.R.: J. Amer. Chem. Soc. **100** (1978) 7762.
78BRO	Brooks, W.V.F., Sastri, C.K.: Can. J. Chem. **56** (1978) 530.
78BUR	Burie, J., Demaison, J., Dubrulle, A., Boucher, D.: J. Mol. Spectrosc. **72** (1978) 275.
78CAL	Caldow, G.L., Duxbury, G., Evans, L.A.: J. Mol. Spectrosc. **69** (1978) 239.
78CAM	Caminati, W., Lister, D.G.: J. Chem. Soc. Faraday Trans. 2 **74** (1978) 896.
78CAR	Carpenter, J.H., Rimmer, D.F.: J. Chem. Soc. Faraday Trans. 2 **74** (1978) 466.
78CAX1	Cassoux, P., Kuczkowski, R.L., Fong, G.D., Geanangel, R.A.: J. Mol. Struct. **48** (1978) 25.
78CAX2	Cassoux, P., Serafini, A., Fong, G., Kuczkowski, R.L.: Inorg. Chem. **17** (1978) 1204.
78CAZ1	Cazzoli, G., Forti, P., Lunelli, B.: J. Mol. Spectrosc. **69** (1978) 71.
78CAZ2	Cazzoli, G., Esposti, C.D., Favero, P.G.: J. Mol. Struct. **48** (1978) 1.
78CHA	Chao, S., Cook, R.L., Malloy, T.B.: J. Chem. Phys. **68** (1978) 4027.
78COR	Corkill, M.J., Cox, A.P., Norris, J.A.: J. Chem. Soc. Chem. Commun. **1978**, 388.
78COX	Cox, A.P., Corkill, M.J., Stevens, R.: Austin Symp. Gas Phase Mol. Struct. **1978**, paper WA4.
78DAN	Dangoisse, D., Willemot, E., Bellet, J.: J. Mol. Spectrosc. **71** (1978) 414.
78DER	Deroche, J.C.: J. Mol. Spectrosc. **69** (1978) 19.
78DIC	Dickinson, R., Kirby, G.W., Sweeny, J.G., Tyler, J.K.: J. Chem. Soc. Faraday Trans. 2 **74** (1978) 1393.
78DOR	Doraiswamy, S., Sharma, S.D.: J. Phys. Soc. Jpn. **44** (1978) 598.
78DOS1	Dössel, K.F., Wiese, J., Sutter, D.H.: Z. Naturforsch. **33a** (1978) 21.
78DOS2	Dössel, K.F., Sutter, D.H.: Z. Naturforsch. **33a** (1978) 1190.
78DUB	Dubrulle, A., Boucher, D., Burie, J., Demaison, J.: J. Mol. Spectrosc. **72** (1978) 158.
78DUR	Durig, J.R., Li, Y.S.: J. Mol. Spectrosc. **70** (1978) 27.
78DUX	Duxbury, G., Herman, H.: J. Mol. Spectrosc. **73** (1978) 444.
78DYL	Dyllick-Brenzinger, C.E., Bauder, A.: Chem. Phys. **30** (1978) 147.
78EL1	Ellenbroek, A.W., Dymanus, A.: Chem. Phys. **35** (1978) 227.
78EL2	Ellenbroek, A.W., Dymanus, A.: Chem. Phys. **33** (1978) 107.
78ELL	Ellingsen, B.H., Marstokk, K.M., Møllendal. H.: J. Mol. Struct. **48** (1978) 9.
78FO1	Fong, G.D., Kuczkowski, R.L., Ashe, A.J.: J. Mol. Spectrosc. **70** (1978) 197.
78FO2	Fong, G.D., Kuczkowski, R.L.: Symp. Mol. Struct. Spectr. Columbus, Ohio, **1978**, paper TI7.
78FOR	Ford, R.G., Seitzman, H.A.: J. Mol. Spectrosc. **69** (1978) 326.

78GHE	Gheorghiou, C., Brier, P.N., Baker, J.G., Jones, S.R.: J. Mol. Spectrosc. **72** (1978) 282.
78GIL	Gillies, C.W.: J. Mol. Spectrosc. **71** (1978) 85.
78GRO	Groner, P., Li, Y.S., Durig, J.R.: J. Mol. Spectrosc. **72** (1978) 20.
78HAL	Halonen, L., Mills, I.M.: J. Mol. Spectrosc. **73** (1978) 494.
78HAM	Hamm, R.: Z. Naturforsch. **33a** (1978) 571.
78HEL1	Helminger, P., Lucia, F.C. de: J. Mol. Spectrosc. **70** (1978) 263.
78HEL2	Helms, D.A., Gordy, W.: J. Mol. Spectrosc. **69** (1978) 473.
78HIR1	Hirose, C., Maeda, S.: J. Mol. Spectrosc. **72** (1978) 62.
78HIR2	Hirota, E.: J. Mol. Spectrosc. **69** (1978) 409.
78HIR3	Hirota, E., Saito, S., Endo, Y.: J. Mol. Spectrosc. **70** (1978) 469.
78HIR4	Hirota, E.: J. Mol. Spectrosc. **71** (1978) 145.
78HOK	Hoke, W.E., Voss, H.L., Campbell, E.J., Flygare, W.H.: Chem. Phys. Lett. **58** (1978) 441.
78HUT	Hutchinson, M., Kroto, H.W.: J. Mol. Spectrosc. **70** (1978) 347.
78HUT1	Hüttner, W., Bodenseh, H.K., Nowicki, P., Morgenstern, K.: J. Mol. Spectrosc. **71** (1978) 246.
78HUT2	Hüttner, W., Bodenseh, H.K., Nowicki, P.: Mol. Phys. **35** (1978) 729.
78IEK	Ieki, M., Kumamoto, E., Kawaguchi, K., Yamada, C., Tanaka, T., Hirota, E.: J. Mol. Spectrosc. **71** (1978) 229.
78JAN	Janda, K.C., Bernstein, L.S., Steed, J.M., Novick, S.E., Klemperer, W.: J. Amer. Chem. Soc. **100** (1978) 8074.
78JOH	Johns, J.W.C., McKellar, A.R.W.: Can. J. Phys. **56** (1978) 737.
78JON	Jones, H., Takami, M., Sheridan, J.: Z. Naturforsch. **33a** (1978) 156.
78KAG	Kagann, R.H., Ozier, I., Gerry, M.C.L.: J. Mol. Spectrosc. **71** (1978) 281.
78KAU1	Kaushik, V.K.: Spectrochim. Acta **34A** (1978) 1105.
78KAU2	Kaushik, V.K., Takagi, K.: J. Phys. Soc. Jpn. **45** (1978) 1975.
78KAU3	Kaushik, V.K., Venkateswarlu, P.: J. Mol. Spectrosc. **70** (1978) 341.
78KAU4	Kaushik, V.K., Venkateswarlu, P.: J. Mol. Spectrosc. **70** (1978) 1.
78KAW1	Kawashima, Y., Cox, A.P.: J. Mol. Spectrosc. **72** (1978) 423.
78KAW2	Kawashima, Y., Takeo, H., Matsumura, C.: Chem. Phys. Lett. **57** (1978) 145.
78KEW	Kewley, R.: Austin Symp. Gas Phase Mol. Struct. **1978**, paper A11.
78KEW1	Kewley, R.: Can. J. Chem. **56** (1978) 772.
78KIE	Kiesel, Z., Legon, A.C.: J. Amer. Chem. Soc. **100** (1978) 8166.
78KIR	Kirby, C., Kroto, H.W.: J. Mol. Spectrosc. **70** (1978) 216.
78KON	Kondo, T., Matsumoto, M., Tanimoto, M.: Tetrahedron Lett. (**1978**) 3819.
78KRE	Kreiner, W.A., Oka, T., Robiette, A.G.: J. Chem. Phys. **68** (1978) 3236.
78KUM1	Kumar, A., Sheridan, J., Stiefvater, O.: Z. Naturforsch. **33a** (1978) 145.
78KUM2	Kumar, A., Sheridan, J.; Stiefvater, O.L.: Z. Naturforsch. **33a** (1978) 549.
78KUZ	Kuznesof, P.M., Kuczkowski, R.L.: Inorg. Chem. **17** (1978) 2308.
78LAF	Lafferty, W.J., Lovas, F.J.: J. Phys. Chem. Ref. Data **7** (1978) 441.
78LAN	Landsberg, B.M.: Z. Naturforsch. **33a** (1978) 1333.
78LI	Li, Y.S., Cox, A.W., Durig, J.R.: J. Mol. Spectrosc. **70** (1978) 34.
78LIT	Little, G.B., Gerry, M.C.L.: J. Mol. Spectrosc. **71** (1978) 321.
78LJ	Ljunggren, S.O., Mjöberg, P.J., Bäckvall, J.E.: Z. Naturforsch. **33a** (1978) 1312.
78LOP	Lopata, A.D., Kuczkowski, R.L.: Symp. Mol. Struct. Spectr. Columbus, Ohio, **1978**, paper TI6.
78LOV	Lovas, F.J.: J. Phys. Chem. Ref. Data **7** (1978) 1445.
78LOW	Lowe, S.E., Sheridan, J.: Chem. Phys. Lett. **58** (1978) 79.
78LOY	Loyd, R.C., Mathur, S.N., Harmony, M.D.: J. Mol. Spectrosc. **72** (1978) 359.
78LUM	Lum, D.K., Bauman, L.E., Malloy, T.B., Cook, R.L.: J. Mol. Struct. **50** (1978) 1.
78LUT	Lutz, H., Dreizler, H.: Z. Naturforsch. **33a** (1978) 1498.
78MAA	Maagdenberg, A.A.J.: Dissertation, Utrecht University, **1978**.
78MAL	Mallinson, P.D., Zafra, R.L. de: Mol. Phys. **36** (1978) 827.
78MAM	Mamleev, A.K., Gunderova, L.N., Pozdeev, N.M.: Zh. Strukt. Khim. **19** (1978) 744; J. Struct. Chem. (english translation) **19** (1978) 642.
78MAR	Marstokk, K.M., Møllendal, H.: J. Mol. Struct. **49** (1978) 221.
78MAT	Mata, F., Martin, M.C., Sørensen, G.O.: J. Mol. Struct. **48** (1978) 157.
78MAT1	Mathur, S.N., Harmony, M.D.: J. Mol. Spectrosc. **69** (1978) 37.
78MAT2	Mathur, S.N., Harmony, M.D.: J. Chem. Phys. **69** (1978) 4316.

78MCD1	MacDonald, J.N., Norbury, D., Sheridan, J.: J. Chem. Soc. Faraday Trans. 2 **74** (1978) 1365.
78MCD2	MacDonald, J.N., Norbury, D., Sheridan, J.: Spectrochim. Acta **34A** (1978) 815.
78MEE	Meerts, W.L., Ozier, I.: Phys. Rev. Lett. **41** (1978) 1109.
78MIR	Mirri, A.M., Cervellati, R., Cazzoli, G.: J. Mol. Spectrosc. **71** (1978) 386.
78MJO	Mjöberg, P.J., Ljunggren, S.O., Bäckvall, J.E.: Z. Naturforsch. **33a** (1978) 581.
78MO1	Mohammadi, M.A., Brooks, W.V.F.: J. Mol. Spectrosc. **73** (1978) 347.
78MO2	Mohammadi, M.A., Brooks, W.V.F.: J. Mol. Spectrosc. **73** (1978) 353.
78MO3	Mohammadi, M.A., Brooks, W.V.F.: Can. J. Spectrosc. **23** (1978) 181.
78MON	Monnanteuil, N., Depannemaecker, J.C., Bellet, J., Barbe, A., Secroun, C., Jouve, P., Giorcianni, S., Hoh, Y.S., Rao, K.N.: J. Mol. Spectrosc. **71** (1978) 399.
78MOS1	Moskienko, M.V., Dyubko, S.F.: Zh. Prikl. Spektrosk. **28** (1978) 558; J. Applied Spectry (english translation) **28** (1978) 387.
78MOS2	Moskienko, M.V., Dyubko, S.F.: Radiophys. Quantum Electron. **21** (1978) 668.
78OD1	Odom, J.D., Li, Y.S., Stampf, F.J., Durig, J.R.: J. Amer. Chem. Soc. **100** (1978) 3751.
78OD2	Odom, J.D., Stampf, E.J., Li, Y.S., Durig, J.R.: J. Mol. Struct. **49** (1978) 1.
78OND1	Onda, M., Oshima, Y., Yamaguchi, I.: Bull. Chem. Soc. Jpn. **51** (1978) 65.
78OND2	Onda, M., Yamaguchi, I.: Bull. Chem. Soc. Jpn. **51** (1978) 3121.
78OZI	Ozier, I., Meerts, W.L.: Phys. Rev. Lett. **40** (1978) 226.
78PAN1	Pandey, G.K., Sutter, D.H.: Z. Naturforsch. **33a** (1978) 29.
78PAN2	Pandey, G.K., Dreizler, H.: Z. Naturforsch. **33a** (1978) 204.
78PEN1	Penn, R.E.: J. Mol. Spectrosc. **69** (1978) 373.
78PEN2	Penn, R.E., Olsen, R.J.: J. Mol. Spectrosc. **70** (1978) 229.
78PEN3	Penn, R.E., Linneman, M.: Symp. Mol. Struct. Spectr. Columbus, Ohio, **1978**, paper TI1.
78PEN4	Penn, R.E.: Austin Symp. Gas Phase Mol. Struct. **1978,** paper TA4.
78POZ1	Pozdeev, N.M., Mamleev, A.K., Gunderova, L.N.: Opt. Spektrosk. **44** (1978) 600; Opt. Spectry (english translation) **44** (1978) 348.
78POZ2	Pozdeev, N.M., Gunderova, L.N., Latypova, R.G., Magdesieva, N.M.: Zh. Strukt. Khim. **19** (1978) 747; J. Struct. Chem. (english translation) **19** (1978) 645.
78REI	Reinartz, J.M.L.J., Meerts, W.L., Dymanus, A.: Chem. Phys. **31** (1978) 19.
78ROM	Römheld, M.: Dissertation, Ulm, **1978.**
78ROU	Rounds, T.C., Strauss, H.L.: J. Chem. Phys. **69** (1978) 268.
78SAK	Sakaizumi, T., Kohri, Y., Ohashi, O., Yamaguchi, I.: Bull. Chem. Soc. Jpn. **51** (1978) 3411.
78SHI	Shiki, Y., Kuginuki, Y., Hasegawa, A., Hayashi, M.: J. Mol. Spectrosc. **73** (1978) 9.
78SIL	Silvia, C.J., True, N.S., Bohn, R.K.: J. Phys. Chem. **82** (1978) 483.
78SMA	Small, C.E., Smith, J.G.: J. Mol. Spectrosc. **73** (1978) 215.
78SMI	Smith, J.G.: Mol. Phys. **35** (1978) 461.
78STE	Steinmetz, W.E., Hollenberg, J., Hickernell, F., Orr, C.E., Scharpen, L.H.: J. Phys. Chem. **82** (1978) 940.
78STI1	Stiefvater, O.: Z. Naturforsch. **33a** (1978) 153.
78STI2	Stiefvater, O.: Z. Naturforsch. **33a** (1978) 1518.
78STI3	Stiefvater, O.: Z. Naturforsch. **33a** (1978) 1511.
78SU1	Suenram, R.D., True, N.S., Bohn, R.K.: J. Mol. Spectrosc. **69** (1978) 435.
78SU2	Suenram, R.D., Lovas, F.J., Johnson, D.R.: J. Mol. Spectrosc. **69** (1978) 458.
78SU3	Suenram, R.D., Lovas, F.J.: J. Mol. Spectrosc. **72** (1978) 372.
78SU4	Suenram, R.D., Lovas, F.J.: J. Amer. Chem. Soc **100** (1978) 5117.
78SUG	Sugie, M., Takeo, H., Matsumura, C.: Bull. Chem. Soc. Jpn. **51** (1978) 3065.
78SZA	Szalanski, L.B., Gerry, M.C.L., Winnewisser, G., Yamada, K., Winnewisser, M.: Can. J. Phys. **56** (1978) 1297.
78TAK	Takami, M., Suzuki, M.: J. Mol. Spectrosc. **73** (1978) 144.
78TAN1	Tan, B.T., Demaison, J., Rudolph, H.D.: J. Mol. Spectrosc. **71** (1978) 471.
78TAN2	Tanaka, K., Tanaka, T.: J. Mol. Spectrosc. **69** (1978) 335.
78THO	Thomas, L.P., True, N.S., Bohn, R.K.: J. Phys. Chem. **82** (1978) 480.
78TR	Truchetet, F., Jurek, R., Chanussot, J.: Can. J. Phys. **56** (1978) 601.
78TR1	True, N.S., Bohn, R.K.: J. Mol. Struct. **50** (1978) 205.
78TR2	True, N.S., Bohn, R.K.: J. Mol. Struct. **50** (1978) 217.
78TR3	True, N.S., Bohn, R.K.: J. Phys. Chem. **82** (1978) 474.

78TR4	True, N.S., Bohn, R.K.: J. Phys. Chem. **82** (1978) 478.
78TR5	True, N.S., Bohn, R.K.: J. Phys. Chem. **82** (1978) 466.
78TSU	Tsunekawa, S., Kojima, T.: J. Phys. Soc. Jpn. **44** (1978) 1925.
78TUR	Turner, P.H., Cox, A.P.: J. Chem. Soc. Faraday Trans. 2 **74** (1978) 533.
78VAN1	Van Eijck, B.P., Dijkerman, H.A., Smits, J.: J. Mol. Spectrosc. **73** (1978) 305.
78VAN2	Van Eijck, B.P., Brandts, P., Maas, J.P.M.: J. Mol. Struct. **44** (1978) 1.
78VOS	Voss, H.L., Krajnovich, D., Hoke, W.E., Flygare, W.H.: J. Chem. Phys. **68** (1978) 1439.
78WIB	Wiberg, K.B., Ellison, G.B., Wendoloski, J.J., Pratt, W.E., Harmony, M.D.: J. Amer. Chem. Soc. **100** (1978) 7837.
78WIL	Willemot, E., Dangoisse, D., Bellet, J.: J. Mol. Spectrosc. **73** (1978) 96.
78WIN	Winnewisser, M., Vogt, J.: Z. Naturforsch. **33a** (1978) 1323.
78YAM	Yamada, K., Winnewisser, M.: J. Mol. Spectrosc. **72** (1978) 484.
78YOU	Young, L.D.G., Young, A.T.: J. Quant. Spectrosc. Radiat. Transfer **20** (1978) 533.
79AB	Abdurakhmanov, A.A., Burenin, A.V., Veliyulin, E.I., Imanova, A.L., Ragimova, R.A., Shapin, S.M.: Izv. Akad. Nauk Az. SSR, Ser. Fiz. Tekh. Mat. Nauk 1 (1979) 96.
79AKA	Akavipat, S., Malloy, T.B., Cook, R.L.: Conf. MW Spectr., Duke University, Durham, **1979,** paper F10.
79AND	Anderson, J.E., Doecke, C.W.: Nouv. J. Chim. **3** (1979) 81.
79ANT	Anttila, R., Hietanen, J., Kauppinen, J.: Mol. Phys. **37** (1978) 925.
79ARI	Arimondo, E., Glorieux, P.: Phys. Rev. A **19** (1979) 1067.
79AVI1	Avirah, T.K., Malloy, T.B., Cook, R.L.: J. Chem. Phys. **71** (1979) 2194.
79AVI2	Avirah, T.K., Cook, R.L., Malloy, T.B.: J. Chem. Phys. **71** (1979) 3478.
79BA1	Bak, B., Nielsen, O., Svanholt, H., Christiansen, J.J.: J. Mol. Spectrosc. **75** (1979) 134.
79BA2	Bak, B., Nielsen, O., Svanholt, H., Holm, A., Toubro, N.H., Krantz, A., Laureni, J.: Acta Chem. Scand. **33A** (1979) 161.
79BA3	Bak, B., Svanholt, H., Holm, A.: Acta Chem. Scand. **33A** (1979) 597.
79BAU1	Bauder, A.: J. Phys. Chem. Ref. Data **8** (1979) 583.
79BAU2	Bauer, A., Boucher, D., Burie, J., Demaison, J., Dubrulle, A.: J. Phys. Chem. Ref. Data **8** (1979) 537.
79BEL	Bellet, J., Jurek, R., Chanussot, J.: J. Mol. Spectrosc. **78** (1979) 16.
79BEN	Bendtsen, J., Winnewisser, M.: Chem. Phys. **40** (1979) 359.
79BEV	Bevan, J.W., Stevens, J.F., Curl, R.F.: J. Mol. Spectrosc. **78** (1979) 514.
79BJ	Bjarnov, E.: Z. Naturforsch. **34a** (1979) 1269.
79BOS	Bossert, W., Bauder, A., Günthard, H.H.: Chem. Phys. **39** (1979) 367.
79BOU1	Bouchy, A., Roussy, G., Ledoux, M.J., Gault, F.G.: J. Chim. Phys. **76** (1979) 357.
79BOU2	Bouchy, A., Roussy, G.: J. Mol. Spectrosc. **77** (1979) 145.
79BR	Brown, R.D., Godfrey, P.D., Woodruff, M.: Austr. J. Chem. **32** (1979) 2103.
79BR1	Brown, R.D., Brown, R.F.C., Eastwood, F.W., Godfrey, P.D., McNaughton, D.: J. Amer. Chem. Soc. **101**, (1979) 4705.
79BUR1	Burenin, A.V., Krupnov, A.F., Mart'yanov, S.V., Mel'nikov, A.A., Nikolayev, L.I.: J. Mol. Spectrosc. **75** (1979) 333.
79BUR2	Burenin, A.V., Karyakin, E.N., Krupnov, A.F., Shapin, S.M.: J. Mol. Spectrosc. **78** (1979) 181.
79CAM	Caminati, W., Scappini, F., Corbelli, G.: J. Mol. Spectrosc. **75** (1979) 327.
79CAZ	Cazzoli, G., Lucia, F.C. de: J. Mol. Spectrosc. **76** (1979) 131.
79CED	Cederberg, J.: J. Mol. Spectrosc. **77** (1979) 102.
79CER	Cervellati, R., Dal Borgo, A., Scappini, F.: J. Mol. Struct. **56** (1979) 69.
79CHA	Chadwick, D., Legon, A.C., Millen, D.J.: J. Chem. Soc. Faraday Trans. 2 **75** (1979) 302.
79CHN	Chance, K.V., Bowen, K.H., Winn, J.S., Klemperer, W.: J. Chem. Phys. **70** (1979) 5157.
79COG	Cogley, C.D., Fry, H.A., Kukolich, S.G.: J. Mol. Spectrosc. **75** (1979) 447.
79CR	Cremer, D., Christen, D.: J. Mol. Spectrosc. **74** (1979) 480.
79CRE	Creswell, R.A., Pagitsas, M., Shoja-Chaghervand, P., Schwendeman, R.H.: J. Phys. Chem. **83** (1979) 1427.
79DAL	Dal Borgo, A., Di Lonardo, G., Scappini, F., Trombetti, A.: Chem. Phys. Lett. **63** (1979) 115.
79DAN	Dangoisse, D., Willemot, E., Bellet, J.: J. Mol. Spectrosc. **75** (1979) 474.
79DEM	Demaison, J., Dubrulle, A., Boucher, D., Burie, J., Typke, V.: J. Mol. Spectrosc. **76** (1979) 1.

79DIX	Dixon, T.A., Baiocchi, F.A., Joyner, C.H., Steed, J.M., Klemperer, W.: Symp. Mol. Struct. Spectr. Columbus, Ohio, **1979,** paper TG4.
79DO1	Dössel, K.F., Sutter, D.H.: Z. Naturforsch. **34a** (1979) 469.
79DO2	Dössel, K.F., Sutter, D.H.: Z. Naturforsch. **34a** (1979) 482.
79DU1	Durig, J.R., Jalilian, M., Li, Y.S., Carter, R.O.: J. Mol. Struct. **55** (1979) 177.
79DU2	Durig, J.R., Li, Y.S., Sullivan, J.F.: J. Chem. Phys. **71** (1979) 1041.
79EN1	Endo, K.: Nippon Kagaku Kaishi (J. Chem. Soc. Jpn.) **1979**, 1129.
79EN2	Endo, Y., Saito, S., Hirota, E., Chikaraishi, T.: J. Mol. Spectrosc. **77** (1979) 222.
79EN3	Endo, K., Shimada, S., Morita, S., Kamura, Y.: Nippon Kagaku Kaishi (J. Chem. Soc. Jpn.) **1979**, 1.
79FER	Ferguson, A.C., Flygare, W.H.: J. Phys. Chem. **83** (1979) 3161.
79FIL	Filgueira, R.R., Fantoni, A.C., Boggia, L.M.: J. Mol. Spectrosc. **78** (1979) 175.
79FOR	Ford, R.G.: Conf. MW Spectr., Duke University, Durham, **1979,** paper D10.
79GEO	Georgiou, K., Kroto, H.W., Landsberg, B.M.: J. Mol. Spectrosc. **77** (1979) 365.
79GIL	Gillies, C.W., Sponseller, S.P., Kuczkowski, R.L.: J. Phys. Chem. **83** (1979) 1545.
79GRO	Groner, P., Bauder, A.: J. Mol. Spectrosc. **74** (1979) 259.
79GUA	Guarnieri, A., Nicolaisen, R.: Z. Naturforsch. **34a** (1979) 620.
79GUN	Günther, H., Zeil, W.: Z. Naturforsch. **34a** (1979) 1243.
79GUY	Guyon, P., Bouchy, A., Roussy, G.: J. Mol. Struct. **57** (1979) 53.
79HAR	Harmony, M.D., Mathur, S.N., Merdian, S.J.: J. Mol. Spectrosc. **75** (1979) 144.
79HAY1	Hayashi, M., Kato, H.: J. Mol. Struct. **53** (1979) 179.
79HAY2	Hayashi, M., Kato, H.: J. Mol. Spectrosc. **76** (1979) 412.
79HEN	Henningsen, J.O., Petersen, J.C., Petersen, F.R., Jennings, D.A., Evenson, K.M.: J. Mol. Spectrosc. **77** (1979) 298.
79HIR1	Hirota, E., Saito, S., Endo, Y.: J. Chem. Phys. **71** (1979) 1183.
79HIR2	Hirota, E.: J. Mol. Spectrosc. **77** (1979) 213.
79HIR3	Hirota, E.: J. Phys. Chem. **83** (1979) 1457.
79JON1	Jones, H.: J. Mol. Spectrosc. **78** (1979) 452.
79JON2	Jones, H., Rudolph, H.D.: Z. Naturforsch. **34a** (1979) 340.
79KAG	Kagann, R.H., Ozier, I., McRae, G.A., Gerry, M.C.L.: Can. J. Phys. **57** (1979) 593.
79KAO	Kao, P.N., Turner, P.H.: J. Amer. Chem. Soc. **101** (1979) 4497.
79KAR	Karlsson, F., Smith, Z.: Acta Chem. Scand. **33A** (1979) 483.
79KAU	Kaushik, V.K.: Spectrochim. Acta **35A** (1979) 851.
79KAW1	Kawashima, Y., Takeo, H., Matsumura, C.: Chem. Phys. Lett. **63** (1979) 119.
79KAW2	Kawashima, Y., Takeo, H., Matsumura, C.: J. Mol. Spectrosc. **78** (1979) 493.
79KIT	Kitchin, R.W., Malloy, T.B., Cook, R.L.: J. Mol. Spectrosc. **75** (1979) 429.
79KRA	Kraft, H.G., Baas, B., Zeil, W.: Z. Naturforsch. **34a** (1979) 1458.
79KRO	Kroto, H.W., Nixon, J.F., Simmons, N.P.C.: J. Mol. Spectrosc. **77** (1979) 270.
79KRU	Krupnov, A.F.: Modern aspects of microwave spectroscopy, Chantry G.W. ed., Academic Press, London **1979**, 217.
79LAD	Landsberg, B.M.: Chem. Phys. Lett. **60** (1979) 265.
79LAG	Langridge-Smith, P.R.R., Stevens, R., Cox, A.P.: J. Chem. Soc. Faraday Trans. 2 **75** (1979) 1620.
79LAR	Larsen, N.W.: J. Mol. Struct. **51** (1979) 175.
79LE1	Legon, A.C.: J. Chem. Soc. Faraday Trans. 2 **75** (1979) 651.
79LE2	Legon, A.C., Millen, D.J., Samsam-Baktiari, A.: J. Mol. Struct. **52** (1979) 71.
79LI	Li, Y.S., Jalilian, M.R., Durig, J.R.: J. Mol. Struct. **51** (1979) 171.
79LOV	Lovas, F.J., Lutz, H., Dreizler, H.: J. Phys. Chem. Ref. Data **8** (1979) 1051.
79MC	McMahan, M.A., Sharma, S.D., Curl, R.F.: J. Mol. Spectrosc. **75** (1979) 220.
79MAM	Mamleev, A.K., Pozdeev, N.M.: Zh. Strukt. Khim. **20** (1979) 1114; J. Struct. Chem. (english translation) **20** (1979) 949.
79MAR	Marsden, C.J., Brown, R.D., Godfrey, P.D.: J. Chem. Soc. Chem. Commun. **1979,** 399.
79MAS	Masson, M., Bouchy, A., Dormoy, J.R., Roussy, G.: J. Chim. Phys. **76** (1979) 262.
79MAT	Mata, F., Alonso, J.L.: J. Mol. Struct. **56** (1979) 199.
79MTH	Mathur, S.N., Harmony, M.D.: J. Mol Struct. **57** (1979) 63.
79ME1	Meerts, W.L., Ozier, I.: Symp. Mol. Struct. Spectr. Columbus, Ohio, **1979,** paper TG9.
79ME2	Meerts, W.L., Ozier, I., Dymanus, A.: Can. J. Phys. **57** (1979) 1163.
79MO1	Mohammadi, M.A., Brooks, W.V.F.: J. Mol. Spectrosc. **77** (1979) 42.

79MO2	Mohammadi, M.A., Brooks, W.V.F.: J. Mol. Spectrosc. **77** (1979) 85.
79MO3	Mohammadi, M.A., Brooks, W.V.F.: J. Mol. Spectrosc. **78** (1979) 89.
79NAK	Nakagawa, J., Hayashi, M.: Chem. Lett. **1979**, 1321.
79OW	Owen, N.L., Sørensen, G.O.: J. Phys. Chem. **83** (1979) 1483.
79PAT	Patel, D., Margolese, D., Dyke, T.R.: J. Chem. Phys. **70** (1979) 2740.
79PHI	Philip, T., Cook, R.L., Malloy, T.B.: Symp. Mol. Struct. Spectr. Columbus, Ohio, **1979**, paper RF4.
79SAN	Sanders, N.D.: Ph. D. Thesis, Harvard University, **1979.**
79SAS	Sasada, Y., Niide, Y., Takano, M., Satoh, T.: J. Mol. Spectrosc. **75** (1979) 87.
79SCH	Schwoch, D., Don, B., Burg, A.B., Beaudet, R.A.: J. Phys. Chem. **83** (1979) 1465.
79SHI	Shiki, Y., Oyamada, M., Hayashi, M.: Chem. Lett. **1979,** 775.
79SIL	Silvia, C.J., True, N.S., Bohn, R.K.: J. Mol. Struct. **51** (1979) 163.
79SMA	Small, C.E., Smith, J.G.: Mol. Phys. **37** (1979) 665.
79SMI	Smith, J.G.: J. Mol. Spectrosc. **77** (1979) 169.
79ST1	Steed, J.M., Dixon, T.A., Klemperer, W.: J. Chem. Phys. **70** (1979) 4095; errata: J. Chem. Phys. **75** (1981) 5977.
79ST2	Steed, J.M., Bernstein, L.S., Dixon, T.A., Janda, K.C., Klemperer, W.: J. Chem. Phys. **71** (1979) 4189.
79STE	Stevens, J.F., Curl, R.F., Engel, P.S.: J. Phys. Chem. **83** (1979) 1432.
79SUG	Sugie, M., Takeo, H., Matsumura, C.: Chem. Phys. Lett. **64** (1979) 573.
79SUN	Sung, E.M., Harmony, M.D.: J. Mol. Spectrosc. **74** (1979) 228.
79TA	Tagaki, K.: J. Mol. Spectrosc. **75** (1979) 97.
79TAK1	Takami, M.: J. Chem. Phys. **71** (1979) 4164.
79TAK2	Takano, M.: J. Mol. Spectrosc. **75** (1979) 41.
79TAN1	Tan, B.T., Demaison, J., Rudolph, H.D.: J. Mol. Spectrosc. **76** (1979) 104.
79TAN2	Tanimoto, M., Almond, V., Charles, S.W., McDonald, J.N., Owen, N.L.: J. Mol. Spectrosc. **78** (1979) 95.
79TAN3	Tanimoto, M., McDonald, J.N.: J. Mol. Spectrosc. **78** (1979) 106.
79TR	True, N.S., Bohn, R.K.: Chem. Phys. Lett. **60** (1979) 332.
79TUR1	Turner, P.H., Corkill, M.J., Cox, A.P.: J. Phys. Chem. **83** (1979) 1473.
79TUR2	Turner, P.H.: J. Amer. Chem. Soc. **101** (1979) 4499.
79TUR3	Turner, P.H.: J. Chem. Soc. Faraday Trans. 2 **75** (1979) 317.
79TYP	Typke, V.: J. Mol. Spectrosc. **77** (1979) 117.
79VAN	Van den Bout, P.A., Steed, J.M., Bernstein, L.S., Klemperer, W.: Astrophys. J. **234** (1979) 503.
79VIS	Viswanathan, R., Dyke, T.R.: Symp. Mol. Struct. Spectr. Columbus, Ohio, **1979**, pap. TG6.
79WAT	Watson, J.K.G., Takami, M., Oka, T.: J. Chem. Phys. **70** (1979) 5376.
79WIL1	Willemot, E., Dangoisse, D., Bellet, J.: J. Mol. Spectrosc. **77** (1979) 161.
79WIL2	Williams, J.R., Kukolich, S.G.: J. Mol. Spectrosc. **74** (1979) 242.
79WIN	Winnewisser, G., Cornet, R.A., Birss, F.W., Gordon, R.M., Ramsay, D.A., Till, S.M.: J. Mol. Spectrosc. **74** (1979) 327.
79YA	Yamada, K., Winnewisser, M., Winnewisser, G., Szalanski, L.B., Gerry, M.C.L.: J. Mol. Spectrosc. **78** (1979) 189.
79YAM	Yamazaki, A., Mogi, K., Koyama, M., Yamaguchi, I.: J. Mol. Struct. **55** (1979) 185.
80ALL	Allegrini, M., Johns, J.W.C., McKellar, A.R.W., Pinson, P.: J. Mol. Spectrosc. **79** (1980) 446.
80ALO1	Alonso, J.L., Wilson, E.B.: J. Amer. Chem. Soc. **102** (1980) 1248.
80ALO2	Alonso, J.L.: J. Mol. Spectrosc. **84** (1980) 520.
80AR1	Arimondo, E., Baker, J.G., Glorieux, P., Oka, T., Sakai, J.: J. Mol. Spectrosc. **82** (1980) 54.
80AR2	Arimondo, E., Glorieux, P.: J. Mol. Spectrosc. **84** (1980) 559.
80BAC	Bäckvall, J.E., Karlsson, O., Ljunggren, S.O.: Tetrahedron Lett. **21** (1980) 4985.
80BAI	Bailey, S.R., Graybeal, J.D.: Austin symp. gas phase mol. struct., **1980**, paper MA4.
80BAK1	Bak, B., Svanholt, H.: Chem. Phys. Lett. **75** (1980) 528.
80BAK2	Bak, B., Svanholt, H.: Acta chem. Scand. **34A** (1980) 57.
80BAK3	Bak, B., Svanholt, H., Holm, A.: Acta chem. Scand. **34A** (1980) 625.
80BAL	Balle, T.J., Campbell, E.J., Keenan, M.R., Flygare, W.H.: J. Chem. Phys. **72** (1980) 922.

80BAR1	Barton, A.E., Henderson, T.J., Langridge-Smith, P.R.R., Howard, B.J.: Chem. Phys. **45** (1980) 429.
80BAR2	Barton, A.E., Howlett, D.J.B., Howard, B.J.: Mol. Phys. **41** (1980) 619.
80BED	Bedwell, D.J., Duxbury, G.: J. Mol. Spectrosc. **84** (1980) 531.
80BEL	Belov, S.P., Gershstein, L.I., Krupnov, A.F., Maslovskij, A.V., Urban, Š., Špirko, V., Papoušek, D.: J. Mol. Spectrosc. **84** (1980) 288.
80BES	Bestmann, G., Dreizler, H., Mäder, H., Andresen, U.: Z. Naturforsch. **35a** (1980) 392.
80BEV1	Bevan, J.W., Legon, A.C., Millen, D.J., Rogers, S.C.: Proc. Roy. Soc. London **370A** (1980) 239.
80BEV2	Bevan, J.W., Kisiel, Z., Legon, A.C., Millen, D.J., Rogers, S.C.: Proc. Roy. Soc. London **372A** (1980) 441.
80BOG	Bogey, M., Bauer, A.: J. Mol. Spectrosc. **84** (1980) 170.
80BOT	Botskor, I., Jones, H.: J. Mol. Spectrosc. **81** (1980) 1.
80BOU1	Boucher, D., Burie, J., Bauer, A., Dubrulle, A., Demaison, J.: J. Phys. Chem. Ref. Data **9** (1980) 659.
80BOU2	Boucher, D., Dubrulle, A., Demaison, J.: J. Mol. Spectrosc. **84** (1980) 375.
80BOU3	Boucher, D., Dubrulle, A., Demaison, J.: Z. Naturforsch. **35a** (1980) 442.
80BOU4	Boucher, D., Dubrulle, A., Demaison, J., Dreizler, H.: Z. Naturforsch. **35a** (1980) 1136.
80BOU5	Bouchy, A., Ledoux, M.J.: J. Mol. Spectrosc. **80** (1980) 453.
80BOW	Bowen, K.H., Leopold, K.R., Chance, K.V., Klemperer, W.: J. Chem. Phys. **73** (1980) 137.
80BRO1	Brown, R.D., Godfrey, P.D., Ottrey, A.L.: J. Mol. Spectrosc. **81** (1980) 303.
80BRO2	Brown, R.D., Godfrey, P.D., Ottrey, A.L.: J. Mol. Spectrosc. **82** (1980) 73.
80BRO3	Brown, R.D., Godfrey, P.D., Woodruff, M.: Austin Symp. Gas Phase Mol. Struct., **1980**, paper TA7.
80BUR1	Burenin, A.V., Kazakov, V.P., Krupnov, A.F., Mel'nikov, A.A., Shapin, S.M.: All-union symposium on high resolution molecular spectroscopy. Tomsk, **1980,** p. 75.
80BUR2	Burrell, P.M., Bjarnov, E., Schwendeman, R.H.: J. Mol. Spectrosc. **82** (1980) 193.
80CAM1	Caminati, W., Wilson, E.B.: J. Mol. Spectrosc. **81** (1980) 356.
80CAM2	Caminati, W., Wilson, E.B.: J. Mol. Spectrosc. **81** (1980) 507.
80CAM3	Caminati, W., Bohn, R.K., True, N.S.: J. Mol. Spectrosc. **84** (1980) 355.
80CAR	Carlier, J., Bauer, A.: Mol. Phys. **40** (1980) 231.
80CAR1	Carlotti, M., Trombetti, A., Velino, B., Vrbancich, J.: J. Mol. Spectrosc. **83** (1980) 401.
80CAZ	Cazzoli, G., Degli Esposti, C., Favero, P.G.: J. Phys. Chem. **84** (1980) 1756.
80CED1	Cederbalk, P.: Acta chem. Scand. **34A** (1980) 409.
80CED2	Cederbalk, P., Karlsson, F.: Acta Chem. Scand. **34A** (1980) 541.
80CER	Cervellati, R., Scappini, F.: J. Mol. Struct. **62** (1980) 81.
80CHO	Choe, J.I., Harmony, M.D.: J. Mol. Spectrosc. **81** (1980) 480.
80CHP	Choplin, A.: Chem. Phys. Lett. **71** (1980) 503.
80CHR1	Christen, D., Hoffmann, V., Klaboe, P.: J. Mol. Struct. **60** (1980) 151.
80CHR2	Christen, D., Oberhammer, H., Zeil, W., Haas, A., Darmadi, A.: J. Mol. Struct. **66** (1980) 203.
80COH	Cohen, E.A.: J. Mol. Spectrosc. **79** (1980) 496.
80COL1	Colmont, J.M.: J. Mol. Spectrosc. **80** (1980) 166.
80COL2	Colmont, J.M.: J. Mol. Struct. **62** (1980) 85.
80COO	Cooper, T.A., Kroto, H.W., Nixon, J.F., Ohashi, O.: J. Chem. Soc. Chem. Commun. **1980,** 333.
80COR1	Cornet, R., Landsberg, B.M., Winnewisser, G.: J. Mol. Spectrosc. **82** (1980) 253.
80COR2	Cornet, R., Winnewisser, G.: J. Mol. Spectrosc. **80** (1980) 438.
80COR3	Correll, T., Larsen, N.W., Pedersen, T.: J. Mol. Struct. **65** (1980) 43.
80COX	Cox, A.P., Duxbury, G., Hardy, J.A., Kawashima, Y.: J. Chem. Soc. Faraday Trans. 2 **76** (1980) 339.
80DAM	Damiani, D., Corbelli, G., Scappini, F.: J. Mol. Struct. **63** (1980) 221.
80DAS1	Das, P.P., Devi, V.M., Rao, K.N.: J. Mol. Spectrosc. **84** (1980) 305.
80DAS2	Das, P.P., Devi, V.M., Rao, K.N.: J. Mol. Spectrosc. **84** (1980) 313.
80DAV1	Davies, D.T., Richards, R.J., Gerry, M.C.L.: J. Mol. Spectrosc. **80** (1980) 307.
80DAV2	Davis, R.W., Gerry, M.C.L.: J. Phys. Chem. **84** (1980) 1767.
80DAV3	Davis, R.W., Robiette, A.G., Gerry, M.C.L.: J. Mol. Spectrosc. **83** (1980) 185.

80DAV4	Davis, R.W., Robiette, A.G., Gerry, M.C.L., Bjarnov, E., Winnewisser, G.: J. Mol. Spectrosc. **81** (1980) 93.
80DEL1	DeLeon, R.L., Mack, K.M., Muenter, J.S.: J. Chem. Phys. **71** (1980) 4487.
80DEL2	DeLeon, R.L., Muenter, J.S.: J. Chem. Phys. **72** (1980) 6020.
80DEL3	DeLeon, R.L., Yokozeki, A., Muenter, J.S.: J. Chem. Phys. **73** (1980) 2044.
80DEL4	DeLeon, R.L., Muenter, J.S.: J. Mol. Spectrosc. **83** (1980) 469.
80DEM	Demaison, J., Schwoch, D., Tan, B.T., Rudolph, H.D.: J. Mol. Spectrosc. **83** (1980) 391.
80DES	DesMarteau, D.D., Li, Y.S., Durig, J.R.: Inorg. Chem. **19** (1980) 1699.
80DIL	Dillen, J., Geise, H.J.: J. Mol. Struct. **69** (1980) 137.
80DLO	Di Lonardo, G., Trombetti, A.: Chem. Phys. Lett. **76** (1980) 307.
80DRE	Dreizler, H., Fliege, E., Pandey, G.K.: Z. Naturforsch. **35a** (1980) 1223.
80DUB	Dubrulle, A., Demaison, J., Burie, J., Boucher, D.: Z. Naturforsch. **35a** (1980) 471.
80DUR1	Durig, J.R., Chatterjee, K.K., Li, Y.S., Jalilian, M., Zozulin, A.J., Odom, J.D.: J. Chem. Phys. **73** (1980) 21.
80DUR2	Durig, JR., Li, Y.S., DesMarteau, D.D.: J. Mol. Struct. **68** (1980) 81.
80DUR3	Durig, J.R., Streusand, B.J., Li, Y.S., Richardson, L., Laane, J.: J. Chem. Phys. **73** (1980) 5564.
80ELL	Ellingsen, B.H.: Acta Chem. Scand. **34A** (1980) 115.
80END1	Endo, K., Koshiba, T., Saito, H., Kamura, Y.: Nippon Kagaku Kaishi **1980**, 1661.
80END2	Endo, Y., Yoshida, K., Saito, S., Hirota, E.: J. Chem. Phys. **73** (1980) 3511.
80FAN	Fantoni, A.C., Filgueira, R.R., Boggia, L.M., Caminati, W.: J. Mol. Spectrosc. **84** (1980) 493.
80FIS	Fischer, E., Botskor, I.: Z. Naturforsch. **35a** (1980) 415.
80GEB	Gebhardt, K.F., Oberhammer, H., Zeil, W.: J. Chem. Soc. Faraday. Trans. 2 **76** (1980) 1293.
80GEO1	Georgiou, A.S., Legon, A.C., Millen, D.J.: Proc. Roy. Soc. London **370A** (1980) 257.
80GEO2	Georgiou, A.S., Legon, A.C., Millen, D.J.: J. Mol. Struct. **69** (1980) 69.
80GEO3	Georgiou, K., Kroto, H.W.: J. Mol. Spectrosc. **83** (1980) 94.
80GEO4	Georgiou, K., Roussy, G.: J. Mol. Spectrosc. **82** (1980) 176.
80GHO	Ghosh, P.N., Bauder, A., Günthard, H.H.: Chem. Phys. **53** (1980) 39.
80GIL	Gillis, J.R., Edwards, T.H.: J. Mol. Spectrosc. **81** (1980) 373.
80GUN	Günther, H., Oberhammer, H., Eujen, R.: J. Mol. Struct. **64** (1980) 249.
80HAL	Halonen, L., Friz, E., Robiette, A.G., Mills, I.M.: J. Mol. Spectrosc. **79** (1980) 432.
80HAR	Harmony, M.D., Strand, M.R.: J. Mol. Spectrosc. **81** (1980) 308.
80HAY1	Hayashi, M., Kato, H., Oyamada, M.: J. Mol. Spectrosc. **83** (1980) 408.
80HAY2	Hayashi, M., Nakagawa, J., Aguni, Y.: Bull. Chem. Soc. Jpn. **53** (1980) 2468.
80HAY3	Hayashi, M., Kato, H.: Bull. Chem. Soc. Jpn. **53** (1980) 2701.
80HAY4	Hayashi, M., Nakagawa, J., Kato, H.: J. Mol. Spectrosc. **84** (1980) 362.
80HUT	Hutchinson, M., Kroto, H.W., Walton, D.R.M.: J. Mol. Spectrosc. **82** (1980) 394.
80IBI	Ibisch, E., Andresen, U.: Z. Naturforsch. **35a** (1980) 1020.
80IQB	Iqbal, K., Owen, N.L., Sheridan, J.: J. Mol. Struct. **64** (1980) 137.
80JAC	Jackson, K.C., Langridge-Smith, P.R.R., Howard, B.J.: Mol. Phys. **39** (1980) 817.
80JOH	Johnston, L.H., Srivastava, R.P., Lees, R.M.: J. Mol. Spectrosc. **84** (1980) 1.
80JON	Jones, H.: J. Mol. Spectrosc. **81** (1980) 21.
80KA1	Kakar, R.K., Quade, C.R., Lwowski, W., Wilde, R.E.: J. Chem. Phys. **72** (1980) 4123.
80KA2	Kakar, R.K., Quade, C.R.: J. Chem. Phys. **72** (1980) 4300.
80KAR	Karlsson, F., Smith, Z.: J. Mol. Spectrosc. **81** (1980) 327.
80KAT	Kato, H., Nakagawa, J., Hayashi, M.: J. Mol. Spectrosc. **80** (1980) 272.
80KAT1	Kattiyà-Ari, M., Harmony, M.D.: Int. J. Quant. Chem., Quantum Chemistry Symposium N°. 14 (1980) 443.
80KAU	Kaushik, V.K., Takagi, K., Matsumura, C.: J. Mol. Spectrosc. **82** (1980) 418.
80KAU1	Kaushik, V.K.: Chem. Phys. Lett. **70** (1980) 317.
80KAW	Kawaguchi, A., Tanaka, T., Hirota, E., Mochida, K., Sakurai, H.: J. Mol. Spectrosc. **79** (1980) 168.
80KE1	Keenan, M.R., Campbell, E.J., Balle, T.J., Buxton, L.W., Minton, T.K., Soper, P.D., Flygare, W.H.: J. Chem. Phys. **72** (1980) 3070.
80KE2	Keenan, M.R., Buxton, L.W., Campbell, E.J., Balle, T.J., Flygare, W.H.: J. Chem. Phys. **73** (1980) 3523.

80KE3	Keenan, M.R., Minton, T.K., Legon, A.C., Balle, T.J., Flygare, W.H.: Proc. Nat. Acad. Sci. USA **77** (1980) 5583.
80KIN	King, M.C., Kroto, H.W.: J. Chem. Soc. Chem. Commun. **1980**, 915.
80KIR1	Kirby, C., Kroto, H.W.: J. Mol. Spectrosc. **83** (1980) 1.
80KIR2	Kirby, C., Kroto, H.W.: J. Mol. Spectrosc. **83** (1980) 130.
80KIR3	Kirby, C., Kroto, H.W., Walton, D.R.M.: J. Mol. Spectrosc. **83** (1980) 261.
80KIT	Kitchin, R.W., Pittman, K.W., Combs, L.L., Cook, R.L.: J. Mol. Struct. **63** (1980) 303.
80KOH	Kohler, F., Jones, H., Rudolph, H.D.: J. Mol. Spectrosc. **80** (1980) 56.
80KOJ	Kojima, T.: J. Phys. Soc. Jpn. **49** (1980) 1197.
80KON	Kondo, T., Tanimoto, M., Matsumoto, M., Nomoto, K., Achiba, Y., Kimura, K.: Tetrahedron Lett. **21** (1980) 1649.
80KRO1	Kroto, H.W., Nixon, J.F., Ohno, K., Simmons, N.P.C.: J. Chem. Soc. Chem. Commun. **1980**, 709.
80KRO2	Kroto, H.W., Nixon, J.F., Simmons, N.P.C.: J. Mol. Spectrosc. **82** (1980) 185; errata: J. Mol. Spectrosc. **85** (1981) 496.
80KÜH	Kühne, H., Forster, M., Hulliger, J., Ruprecht, H., Bauder, A., Günthard, H.H.: Helv. Chem. Acta **63** (1980) 1971.
80KUM	Kumar, A., Iqbal, K., Owen, N.L.: Sheridan, J.: J. Mol. Struct. **64** (1980) 149.
80LAD	Landsberg, B.M., Iqbal, K.: J. Chem. Soc. Faraday Trans. 2 **76** (1980) 1208.
80LAN	Langridge-Smith, P.R.R., Stevens, R., Cox, A.P.: J. Chem. Soc. Faraday Trans. 2 **76** (1980) 330.
80LAT	Latypova, R.G., Gunderova, L.N., Mamleev, A.K., Pozddeev, N.M.: Zh. Strukt. Khim. **21** (1980) 78; J. Struct. Chem. (english transl.) **21** (1980) 315.
80LEE	Lees, R.M., Mohammadi, M.A.: Can. J. Phys. **58** (1980) 1640.
80LG1	Legon, A.C., Millen, D.J., Rogers, S.C.: Proc. Roy. Soc. London **370A** (1980) 213.
80LG2	Legon, A.C., Soper, P.D., Keenan, M.R., Minton, T.K., Balle, T.J., Flygare, W.H.: J. Chem. Phys. **73** (1980) 583.
80LG3	Legon, A.C., Aldrich, P.D., Flygare, W.H.: J. Amer. Chem. Soc. **102** (1980) 7584.
80LG4	Legon, A.C., Millen, D.J., Rogers, S.C.: J. Mol. Struct. **67** (1980) 29.
80LIS	Lister, D.G.: J. Mol. Struct. **68** (1980) 33.
80LOV	Lovas, F.J., Suenram, R.D., Johnson, D.R., Clark, F.O., Tiemann, E.: J. Chem. Phys. **72** (1980) 4964.
80MA	Marshall, M.D., Muenter, J.S.: J. Mol. Spectrosc. **83** (1980) 279.
80MAM	Mamleev, A.K., Latypova, R.G., Gunderova, L.N., Tyolin, V.I., Pozdeev, N.M.: Zh. Strukt. Khim. **21** (1980) 46; J. Struct. Chem. (engl. translation) **21** (1980) 600.
80MAR1	Marstokk, K.M., Møllendal, H.: Acta Chem. Scand. **34A** (1980) 15.
80MAR2	Marstokk, K.M., Møllendal, H.: Acta Chem. Scand. **34A** (1980) 765.
80MAS	Masson, M.A., Bouchy, A., Roussy, G., Serratrice, G., Delpuech, J.J.: J. Mol. Struct. **68** (1980) 307.
80MAT1	Matsumura, K., Tanaka, T., Endo, Y., Saito, S., Hirota, E.: J. Phys. Chem. **84** (1980) 1793.
80MAT2	Matsumura, K., Tanaka, K., Yamada, C., Tanaka, T.: J. Mol. Spectrosc. **80** (1980) 209.
80MEY	Meyer, C., Sergent-Rozey, M.: J. Mol. Spectrosc. **83** (1980) 343.
80MIR	Mirri, A.M., Cervellati, R., Cazzoli, G.: J. Mol. Spectrosc. **83** (1980) 202.
80NA1	Nakata, M., Fukuyama, T., Kuchitsu, K., Takeo, H., Matsumura, C.: J. Mol. Spectrosc. **83** (1980) 118.
80NA2	Nakata, M., Kohata, K., Fukuyama, T., Kuchitsu, K., Wilkins, C.J.: J. Mol. Struct. **68** (1980) 271.
80NA3	Nakata, M., Takeo, H., Matsumura, C.: J. Mol. Spectrosc. **82** (1980) 117.
80OBE	Oberhammer, H.: J. Chem. Phys. **73** (1980) 4310.
80ODO	Odom, J.D., Szafran, Z., Johnston, S.A., Li, Y.S., Durig, J.R.: J. Amer. Chem. Soc. **102** (1980) 7173.
80ODU	Odutola, J.A., Dyke, T.R.: J. Chem. Phys. **72** (1980) 5062.
80OHK	Ohkoshi, I., Niide, Y., Takano, M., Sasada, Y.: J. Mol. Spectrosc. **82** (1980) 27.
80ORR	Orr, J., Oka, T.: Appl. Phys. **21** (1980) 293
80PED	Pedersen, T.: J. Mol. Struct. **64** (1980) 277.
80QUA	Quade, C.R., Suenram, R.D.: J. Chem. Phys. **73** (1980) 1127.
80REI	Reinartz, J.M.L.J., Meerts, W.L., Dymanus, A.: Chem. Phys. **45** (1980) 387.

80RIC	Richards, R.J., Davis, R.W., Gerry, M.C.L.: J. Chem. Soc. Chem. Commun. **1980**, 915.
80ROH	Rohwer, F., Guarnieri, A.: Z. Naturforsch. **35a** (1980) 336.
80SAS	Sasada, H.: J. Mol. Spectrosc. **83** (1980) 15.
80SHE	Sheridan, J., Bossert, W., Bauder, A.: J. Mol. Spectrosc. **80** (1980) 1.
80SH1	Shiki, Y., Ohta, T., Nakagawa, J., Hayashi, M.: Chem. Lett. **1980**, 447.
80SH2	Shiki, Y., Yamaoka, F., Hayashi, M.: Chem. Lett. **1980**, 699.
80SHI	Shimoda, K., Ueda, Y., Iwahori, J.: Appl. Phys. **21** (1980) 181.
80SON	Sont, W.N., Wieser, H., Mills, I.M.: Mol. Phys. **41** (1980) 1081.
80STA	Staley, S.W., Howard, A.E., Harmony, M.D., Mathur, S.N., Kattija-Ari, M., Choe, J.I., Lind, G.: J. Amer. Chem. Soc. **102** (1980) 3639.
80SUE	Suenram, R.D., Lovas, F.J.: J. Amer. Chem. Soc. **102** (1980) 7180.
80TA1	Takami, M.: J. Chem. Phys. **73** (1980) 2665.
80TA2	Takami, M.: J. Mol. Spectrosc. **80** (1980) 301.
80TA3	Takami, M., Suzuki, M.: J. Chem. Phys. **72** (1980) 4089.
80TON	Tørneng, E., Nielsen, C.J., Klaeboe, P., Hopf, H., Priebe, H.: Spectrochim. Acta **36A** (1980) 975.
80TOR	Törring, T., Bekooy, J.P., Meerts, W.L., Hoeft, J., Tiemann, E., Dymanus, A.: J. Chem. Phys. **73** (1980) 4875.
80TSU	Tsunekawa, S., Kojima, T.: J. Phys. Soc. Jpn. **49** (1980) 1957.
80TUR	Turner, P.H.: J. Chem. Soc. Faraday Trans. 2 **76** (1980) 383.
80URB	Urban, Š., Špirko, V., Papoušek, D., McDowell, R.S., Nereson, N.G., Belov, S.P., Gershstein, L.I., Maslovskij, A.V., Krupnov, A.F., Curtis, J., Rao, K.N.: J. Mol. Spectrosc. **79** (1980) 455.
80WAL	Walder, E., Bauder, A., Günthard, H.H.: Chem. Phys. **51** (1980) 223.
80WI1	Wiese, J., Schwarz, R., Sutter, D.H.: Z. Naturforsch. **35a** (1980) 770.
80WI2	Wiese, J., Sutter, D.H.: Z. Naturforsch. **35a** (1980) 712.
80WIL	Willemot, E., Dangoisse, D., Monnanteuil, N., Bellet, J.: J. Phys. Chem. Ref. Data **9** (1980) 59.
80WIN1	Winnewisser, B.P.: J. Mol. Spectrosc. **82** (1980) 220.
80WIN2	Winnewisser, M., Schäfer, E.: Z. Naturforsch. **35a** (1980) 483.
80YA1	Yamada, K.: J. Mol. Spectrosc. **79** (1980) 323.
80YA2	Yamada, K., Schieder, R., Winnewisser, G., Mantz, A.W.: Z. Naturforsch. **35a** (1980) 690.
80YA3	Yamada, K., Winnewisser, M., Winnewisser, G., Szalanski, L.B., Gerry, M.C.L.: J. Mol. Spectrosc. **79** (1980) 295.
80ZEI	Zeil, W., Christen, D.: J. Phys. Chem. **84** (1980) 1790.
81AB	Abdurakhmanov, A.A., Veliyulin, E.I., Ragimova, R.A., Imanov, L.M.: Zh. Strukt. Khim. **22** (1981) 39; J. Struct. Chem. (english translation) **22** (1981) 28.
81ALD	Aldrich, P.D., Legon, A.C., Flygare, W.H.: J. Chem. Phys. **75** (1981) 2126.
81ALL	Allegrini, M., Johns, J.W.C., McKellar, A.R.W.: to be published.
81ALO	Alonso, J.L.: J. Mol. Struct. **73** (1981) 71.
81ALO1	Alonso, J.L.: J. Mol. Struct. **73** (1981) 63.
81ALO2	Alonso, J.L.: J. Mol. Spectrosc. **87** (1981) 110.
81ALO3	Alonso, J.L., Legon, A.C.: J. Chem. Soc. Faraday Trans 2 **77** (1981) 2191.
81AMA	Amano, T.: J. Mol. Spectrosc. **88** (1981) 194.
81AMI	Amir-Ebrahimi, V., Choplin, A., Demaison, J., Roussy, G.: J. Mol. Spectrosc. **89** (1981) 42.
81BAI	Baiocchi, F.A., Dixon, T.A., Joyner, C.H., Klemperer, W.: J. Chem. Phys. **74** (1981) 6544.
81BAI1	Baiocchi, F.A., Dixon, T.A., Joyner, C.H., Klemperer, W.: J. Chem. Phys. **75** (1981) 2041.
81BAK	Bak, B., Kristiansen, N.A., Svanholt, H., Holm, A., Rosenkilde, S.: Chem. Phys. Lett. **78** (1981) 301.
81BAL	Balikci, B., Brier, P.N.: J. Mol. Spectrosc. **89** (1981) 254.
81BAL1	Balikci, B., Brier, P.N.: J. Mol. Spectrosc. **85** (1981) 109.
81BAR	Bartell, L.S., Doun, S., Marsden, C.J.: J. Mol. Struct. **75** (1981) 271.
81BAU	Baughcum, S.L., Duerst, R.W., Rowe, W.F., Smith, Z., Wilson, E.B.: J. Amer. Chem. Soc. **103** (1981) 6296.
81BEL	Belov, S.P., Burenin, A.V., Gershtein, L.I., Krupnov, A.F., Markov, V.N., Maslovsky, A.V., Shapin, S.M.: J. Mol. Spectrosc. **86** (1981) 184.

81BEL1	Belov, S.P., Burenin, A.V., Polyansky, O.L., Shapin, S.M.: J. Mol. Spectrosc. **90** (1981) 579.
81BLO	Blom, C.E., Bauder, A.: Chem. Phys. Lett. **82** (1981) 492.
81BOM	Bomar, S., Bay, E., Kraft, H.G., Bevan, J.W.: J. Mol. Spectrosc. **87** (1981) 482.
81BOU	Bouchy, A., Roussy, G.: J. Chim. Phys. **78** (1981) 669.
81BOW1	Bowman, W.C., Lucia, F.C. de, Helminger, P.: J. Mol. Spectrosc. **88** (1981) 431.
81BOW2	Bowman, W.C., Lucia, F.C. de, Helminger, P.: J. Mol. Spectrosc. **87** (1981) 571.
81BRO	Brooks, W.V.F., Gosselin, J.A., Mohammadi, M.A., Thibault, J.D.: J. Mol. Struct. **72** (1981) 17.
81BRO1	Brown, R.D., Godfrey, P.D., McNaughton, D.: Aust. J. Chem. **34** (1981) 465.
81BRO2	Brown, R.D., Godfrey, P.D., Winkler, D.A.: Chem. Phys. **59** (1981) 243.
81BRO3	Brown, R.D., Godfrey, P.D., Champion, R., McNaughton, D.: J. Amer. Chem. Soc. **103** (1981) 5711.
81BRO4	Brown, R.D., Godfrey, P.D., Winkler, D.A.: J. Mol. Spectrosc. **89** (1981) 352.
81BUR	Burenin, A.V., Karyakin, E.N., Krupnov, A.F., Shapin, S.M., Val'dov, S.M.: J. Mol. Spectrosc. **85** (1981) 1.
81BUX	Buxton, L.W., Campbell, E.J., Keenan, M.R., Balle, T.J., Flygare, W.H.: Chem. Phys. **54** (1981) 173.
81BUX1	Buxton, L.W., Campbell, E.J., Flygare, W.H.: Chem. Phys. **56** (1981) 399.
81BUX2	Buxton, L.W., Campbell, E.J., Flygare, W.H.: Chem. Phys. **59** (1981) 55.
81BUX3	Buxton, L.W., Aldrich, P.D., Shea, J.A., Legon, A.C., Flygare, W.H.: J. Chem. Phys. **75** (1981) 2681.
81CAL	Caldow, G.L., Duxbury, G.: J. Mol. Spectrosc. **89** (1981) 93. S.
81CAM	Caminati, W.: J. Mol. Spectrosc. **86** (1981) 193.
81CAM1	Caminati, W., Van Eijck, B.P., Lister, D.G.: J. Mol. Spectrosc. **90** (1981) 15.
81CAM2	Caminati, W., Meyer, R.: J. Mol. Spectrosc. **90** (1981) 303.
81CAM3	Caminati, W.: J. Mol. Spectrosc. **90** (1981) 315.
81CAM4	Caminati, W., Corbelli, G.: J. Mol. Spectrosc. **90** (1981) 572.
81CAM5	Campbell, E.J., Buxton, L.W., Keenan, M.R., Flygare, W.H.: Phys. Rev. A **24** (1981) 812.
81CAR	Carlier, J., Bauer, A.: J. Mol. Spectrosc. **88** (1981) 219.
81CAZ	Cazzoli, G., Degli Esposti, C., Damiani, D., Mirri, A.M.: Chem. Phys. Lett. **80** (1981) 365.
81CED	Cederbalk, P., Karlsson, F.: Acta Chem. Scand. **35A** (1981) 7.
81CER	Cervellati, R., Corbelli, G., Dal Borgo, A., Lister, D.G.: J. Mol. Struct. **73** (1981) 31.
81CER1	Cervellati, R., Dal Borgo, A., Lister, D.G., Scappini, F.: J. Mol. Struct. **77** (1981) 75.
81CHO	Choe, J.I., Harmony, M.D.: J. Mol. Spectrosc. **89** (1981) 333.
81COH	Cohen, E.A., Pickett, H.M.: J. Mol. Spectrosc. **87** (1981) 582.
81CRA	Cradock, S., McKean, D.C., McKenzie, M.W.: J. Mol. Struct. **74** (1981) 265.
81COR	Corbelli, G., Lister, D.G.: J. Mol. Struct. **74** (1981) 39.
81DAS	Das, P.P., Devi, V.M., Rao, K.N.: J. Mol. Spectrosc. **86** (1981) 202
81DAV	Davis, R.W., Robiette, A.G., Gerry, M.C.L.: J. Mol. Spectrosc. **85** (1981) 399.
81DEL	DeLeon, R.L., Muenter, J.S.: J. Chem. Phys. **75** (1981) 1113.
81DEM1	Demaison, J., Tan, B.T., Typke, V., Rudolph, H.D.: J. Mol. Spectrosc. **86** (1981) 406.
81DEM2	Demaison, J.: to be published.
81DEV	Devi, V.M., Das, P.P., Rao, K.N., Urban, Š., Papoušek, D., Špirko, V.: J. Mol. Spectrosc. **88** (1981) 293.
81DI1	Di Lonardo, G., Trombetti, A., Velino, B.: Chem. Phys. Lett. **80** (1981) 352.
81DI2	Di Lonardo, G., Trombetti, A.: Chem. Phys. Lett. **84** (1981) 327.
81DIX	Dixon, T.A., Baiocchi, F.A., Klemperer, W.: Symp. Mol. Struct. Spectr. Columbus, Ohio, **1981**, paper RC 2.
81DIX1	Dixon, T.A., Joyner, C.H., Baiocchi, F.A., Klemperer, W.: J. Chem. Phys. **74** (1981) 6539
81DOR	Doraiswamy, S., Sharma, S.D.: J. Mol. Spectrosc. **88** (1981) 95.
81DRE	Dreizler, H., Steffek, D.: Z. Naturforsch. **36a** (1981) 1239.
81DRE1	Dreizler, H., Scappini, F.: Naturforsch. **36a** (1981) 1187.
81DUB	Dubrulle, A., Burie, J., Boucher, D., Herlemont, F., Demaison, J.: J. Mol. Spectrosc. **88** (1981) 394.
81DUC	Duckett, J.A., Robiette, A.G., Gerry, M.C.L.: J. Mol. Spectrosc. **90** (1981) 374.
81DUR1	Durig, J.R., Hudson, S.D., Jalilian, M.R., Li, Y.S.: J. Chem. Phys. **74** (1981) 772.

81DUR2	Durig, J.R., Li, Y.S., Durig, D.T.: J. Chem. Phys. **74** (1981) 1564.
81DUR3	Durig, J.R., Mohamad, A.B., Trowell, P.L., Li, Y.S.: J. Chem. Phys. **75** (1981) 2147.
81DUR4	Durig, J.R., Bradley, C.B., Li, Y.S., Odom, J.D.: J. Mol. Struct. **74** (1981) 205.
81DUR5	Durig, J.R., Guirgis, G.A., Li, Y.S.: J. Chem. Phys. **74** (1981) 5946.
81DUR6	Durig, J.R., Brletic, P.A., Li, Y.S., Johnston, S.A., Odom, J.D.: J. Chem. Phys. **75** (1981) 1644.
81END	Endo, K., Okamatsu, H., Amada, H., Kamura, Y.: Nippon Kagaku Kaishi (1981) 317
81ETO	Etoh, T., Matsumura, K., Tanaka, T.: J. Mol. Spectrosc. **89** (1981) 511.
81FAY	Fayt, A.: Private communication.
81FLI	Fliege, E., Bestmann, G., Andolfatto, M., Dreizler, H.: Z. Naturforsch. **36a** (1981) 1126.
81FLI1	Fliege, E., Bestmann, G., Schwarz, R., Dreizler, H.: Z. Naturforsch. **36a** (1981) 1124
81FLI2	Fliege, E., Dreizler, H., Andresen, U.: Z. Naturforsch. **36a** (1981) 1122.
81FON	Fong, G.D., Kuczkowski, R.L.: Inorg. Chem. **20** (1981) 2342.
81FUS	Fusina, L., Mills, I.M.: J. Mol. Spectrosc. **86** (1981) 488.
81GEO	Georgiou, A.S., Legon, A.C., Millen, D.J.: Proc. Roy. Soc. London **373A** (1981) 511.
81GI1	Gillis, J.R., Edwards, T.H.: J. Mol. Spectrosc. **85** (1981) 55.
81GI2	Gillis, J.R., Edwards, T.H.: J. Mol. Spectrosc. **85** (1981) 74.
81GUA	Guarnieri, A., Scappini, F., Di Lonardo, G.: Chem. Phys. Lett. **82** (1981) 321.
81GUA1	Guarnieri, A., Andolfatto, M.: Z. Naturforsch. **36a** (1981) 899.
81GUI	Guirgis, G.A., Li, Y.S., Durig, J.R.: J. Chem. Phys. **75** (1981) 3274.
81HAR	Harmony, M.D., Mathur, S.N., Choe, J.I., Kattija-Ari, M., Howard, A.E., Staley, S.W.: J. Amer. Chem. Soc. **103** (1981) 2961.
81HAY	Hayashi, M., Adachi, M., Nakagawa, J.: J. Mol. Spectrosc. **86** (1981) 129.
81HAY1	Hayashi, M., Imachi, M., Oyamada, M.: J. Mol. Struct. **74** (1981) 97.
81HEL	Helminger, P., Bowman, W.C., Lucia, F.C. de: J. Mol. Spectrosc. **85** (1981) 120.
81HEN	Henningsen, J.O.: J. Mol. Spectrosc. **85** (1981) 282.
81HER	Herlemont, F., Lyszyk, M., Lemaire, J., Demaison, J.: Z. Naturforsch. **36a** (1981) 944.
81HER1	Herlemont, F., Fleury, J., Lemaire, J., Demaison, J.: J. Chem. Phys. **76** (1982) 4705.
81HIR	Hirota, E.: J. Mol. Spectrosc. to be published.
81HIR1	Hirota, E., Endo, Y., Saito, S., Duncan, J.L.: J. Mol. Spectrosc. **89** (1981) 285.
81HIR2	Hirota, E., Endo, Y., Saito, S., Yoshida, K., Yamaguchi, I., Machida, K.: J. Mol. Spectrosc. **89** (1981) 223.
81HUT	Hutson, J.M., Howard, B.J.: J. Chem. Phys. **74** (1981) 6520.
81IMA	Imachi, M., Nagayama, A., Nakagawa, J., Hayashi, M.: J. Mol. Struct. **77** (1981) 81.
81JAM	Jaman, A.I., Nandi, R.N., Ghosh, D.K.: J. Mol. Spectrosc. **86** (1981) 269.
81JON	Jones, H., Typke, V.: Z. Naturforsch. **36a** (1981) 1057.
81JOY	Joyner, C.H., Dixon, T.A., Baiocchi, F.A., Klemperer, W.: J. Chem. Phys. **74** (1981) 6550.
81KAW	Kawashima, Y., Takeo, H., Matsumura, C.: J. Chem Phys. **74** (1981) 5430.
81KEE	Keenan, M.R., Buxton, L.W., Campbell, E.J., Legon, A.C., Flygare, W.H.: J. Chem. Phys. **74** (1981) 2133.
81KEE1	Keenan, M.R., Wozniak, D.B., Flygare, W.H.: J. Chem. Phys. **75** (1981) 631.
81KEW	Kewley, R.: J. Mol. Spectrosc. **89** (1981) 548.
81KRE	Kreiner, W.A., Robiette, A.G.: J. Chem. Phys. **74** (1981) 3713.
81KRO	Kroto, H.W., Nixon, J.F., Simmons, N.P.C.: J. Mol. Spectrosc. **85** (1981) 496.
81KRO1	Kroto, H.W., Nixon, J.F., Ohno, K.: J. Mol. Spectrosc. **90** (1981) 367.
81KRO2	Kroto, H.W., Nixon, J.F., Ohno, K.: J. Mol. Spectrosc. **90** (1981) 512.
81KUC	Kuczkowski, R.L., Lovas, F.J., Suenram, R.D., Lattimer, R.P., Hillig, K.W., Ashe, A.J.: J. Mol. Struct. **72** (1981) 143.
81KUC1	Kuczkowski, R.L., Suenram, R.D., Lovas, F.J.: J. Amer. Chem. Soc. **103** (1981) 2561.
81LAF	Lafferty, W.J., Sattler, J.P., Worchesky, T.L., Ritter, K.J.: J. Mol. Spectrosc. **87** (1981) 416.
81LEG	Legon, A.C., Aldrich, P.D., Flygare, W.H.: J. Chem. Phys. **75** (1981) 625.
81LEG1	Legon, A.C., Soper, P.D., Flygare, W.H.: J. Chem. Phys. **74** (1981) 4944.
81LEG2	Legon, A.C., Soper, P.D., Flygare, W.H.: J. Chem. Phys. **74** (1981) 4936.
81LEO	Leopold, K.R., Bowen, K.H., Klemperer, W.: J. Chem. Phys. **74** (1981) 4211.
81LOP	Lopata, A.D., Kuczkowski, R.L.: J. Amer. Chem. Soc. **103** (1981) 3304.
81MAR	Marstokk, K.M., Møllendal, H.: Act. Chem. Scand. **35A** (1981) 395.
81MAS	Masson, M.A., Bouchy, A., Dormoy, J.R., Roussy, G.: J. Mol. Struct. **74** (1981) 11.

81MAT	Matsumura, K., Etoh, T., Tanaka, T.: J. Mol. Spectrosc. **90** (1981) 106.
81MEE	Meerts, W.L., Ozier, I.: J. Chem. Phys. **75** (1981) 596.
81NA	Nakagawa, J., Hayashi, M.: J. Mol. Spectrosc. **85** (1981) 327.
81NA1	Nakagawa, J., Kato, H., Hayashi, M.: J. Mol. Spectrosc. **90** (1981) 467.
81NAK	Nakata, M., Sugie, M., Takeo, H., Matsumura, C., Fukuyama, T., Kuchitsu, K.: J. Mol. Spectrosc. **86** (1981) 241.
81NAK1	Nakata, M., Takeo, H., Matsumura, C., Yamanouchi, K., Kuchitsu, K., Fukuyama, T.: Chem. Phys. Lett. **83** (1981) 246.
81NII	Niide, Y., Ohkoshi, I., Takano, M.: J. Mol. Spectrosc. **89** (1981) 387.
81OHA	Ohashi, N., Kawaguchi, K., Hirota, E.: J. Mol. Spectrosc. **85** (1981) 427.
81OHN	Ohno, K., Kroto, H.W., Nixon, J.F.: J. Mol. Spectrosc. **90** (1981) 507.
81PAR	Parkin, J.E., Buckley, P.J., Costain, C.C.: J. Mol. Spectrosc. **89** (1981) 465.
81PER	Permanand, R.R., Sheridan, J.: J. Mol. Struct. **70** (1981) 297.
81PHI	Philip, T., Cook, R.L., Malloy, T.B., Allinger, N.L., Chang, S., Yuh, Y.: J. Amer. Chem. Soc. **103** (1981) 2151.
81ROB	Robiette, A.G., Parent, C.R., Gerry, M.C.L.: J. Mol. Spectrosc. **86** (1981) 455.
81ROM	Roman, A., Alonso, J.L., Mata, F.: Z. Naturforsch. **36a** (1981) 622.
81SAN	Sanders, N.G.: J. Mol. Spectrosc. **86** (1981) 27.
81SCA	Scappini, F., Oka, T.: J. Mol. Spectrosc. **85** (1981) 390.
81SAT	Sattler, J.P., Worchesky, T.L., Lafferty, W.J.: J. Mol. Spectrosc. **88** (1981) 364.
81SCA1	Scappini, F., Dreizler, H.: Z. Naturforsch. **36a** (1981) 1327.
81SCH	Schäfer, E., Winnewisser, M., Christiansen, J.J.: Chem. Phys. Lett. **81** (1981) 380.
81SHI	Shiki, Y., Ibushi, N., Oyamada, M., Nakagawa, J., Hayashi, M.: J. Mol. Spectrosc. **87** (1981) 357.
81SIM	Simmons, N.P.C., Burg, A.B., Beaudet, R.A.: Inorg. Chem. **20** (1981) 533.
81SMI	Smith, J.G.: J. Mol. Spectrosc. **88** (1981) 126.
81SOP	Soper, P.D., Legon, A.C., Flygare, W.H.: J. Chem. Phys. **74** (1981) 2138.
81SOP1	Soper, P.D., Legon, A.C., Read, W.G., Flygare, W.H.: J. Phys. Chem. **85** (1981) 3440.
81STO	Stolze, M., Hübner, D. Sutter, D.H.: Z. Naturforsch. **36a** (1981) 886.
81SUE	Suenram, R.D., Thorne, L.R.: Chem. Phys. Lett. **78** (1981) 157.
81SUZ	Suzuki, M., Yamada, K., Takami, M.: J. Mol. Spectrosc. **88** (1981) 207.
81SUZ1	Suzuki, S., Yamaguchi, M., Onda, M., Sakaizumi, T., Ohashi, O., Yamaguchi, I.: J. Mol. Struct. **73** (1981) 41.
81SWE	Sweger, D.M., Sams, R.L.: J. Mol. Spectrosc. **87** (1981) 18.
81TAK	Takami, M.: J. Chem. Phys. **74** (1981) 4276.
81TAK1	Takahashi, Y., Yamaguchi, I.: J. Mol. Struct. **74** (1981) 43.
81TUR	Turner, P.H., Cox, A.P., Hardy, J.A.: J. Chem. Soc. Faraday Trans. 2 **77** (1981) 1217.
81TYB	Tyblewski, M., Vaccani, S., Bauder, A.: J. Mol. Struct. **74** (1981) 277.
81URB	Urban, Š., Špirko, V., Papoušek, D., Kauppinen, J., Belov, S.P., Gershtein, L.I., Krupnov, A.F.: J. Mol. Spectrosc. **88** (1981) 274.
81VAN	Van Eijck, B.P.: J. Mol. Spectrosc. **85** (1981) 189.
81VAN1	Van Eijck, B.P., Van Opheusden, J., Van Schaik, M.M.M., Van Zoeren, E.: J. Mol. Spectrosc. **86** (1981) 465.
81VER	Verry, C., Jurek, R., Goulet, P., Lamalle, B.: J. Phys. Paris **42** (1981) 223
82DEM	Demaison, J., Dubrulle, A., Boucher, D., Burie, J., Van Eijck, B.P.: J. Mol. Spectrosc. **94** (1982) 211.
82DUR	Durig, J.R., Stanley, A.E., Li, Y.S.: J. Mol. Struct. **78** (1982) 247.
82FRY	Fry, H.A., Kukolich, S.G.: J. Chem. Phys. **76** (1982) 4387.
82JON	Jones, H., Morillon-Chapey, M.: J. Mol. Spectrosc. **91** (1982) 87.

2.6 Dipole moments

2.6.1 Diatomic molecules

2.6.1.1 Preliminary remarks

1. Introduction

The Stark effect of a rotational transition is determined by the electric dipole moment μ and the anisotropy of the electric polarizability $(\alpha_{\parallel}-\alpha_{\perp})$ of the molecule. The Hamiltonian of these interactions is given in Vol. II/6 Landolt-Börnstein, New Series, including energy expressions up to the second order.

Because the electric dipole moment μ and the polarizability α depend on the internuclear distance, the measurements of these constants will result in a function on the vibrational quantum number v and in principal on the rotational quantum number J. The typical expansion is used:

$$\mu(v)=\mu_e+\mu_I(v+1/2)+\mu_{II}(v+1/2)^2+\cdots. \qquad (1)$$

The polarizability $(\alpha_{\parallel}-\alpha_{\perp})$ normally contributes little to the whole Stark effect, therefore the parameter is not determinable or only with low accuracy in most MW experiments.

Signs of the electric dipole moments are only known in few cases from the isotope dependence of the molecular rotational g_J-factor (compare the definition in part 2.9).

The following tables contain the Stark-effect parameters of $^1\Sigma$ and non-$^1\Sigma$ states.

2. Explanations of the columns in the tabulation

Column	Parameters	Definitions
1	Nr. *Nr.*	serial number of the molecule of table 2.2. The following number represents the serial number of the molecule in 2.6.1 of Vol. II/6 Landolt-Börnstein, New Series
2	v	vibrational quantum number of the state with the measured value of μ in column 3
3	μ [D]	measured absolute value of the electric dipole moment
4	μ_e [D]	equilibrium value of the electric dipole moment according to eq. (1)
5	μ_I [D]	compare eq. (1)
6	μ_{II} [D]	compare eq. (1)
7	Ref.	references for columns 3···6
8	Meth. LC QB OS ED MW MB EPR DR	method of measurement for the values in this line of the table level crossing spectroscopy quantum beat spectroscopy optical spectroscopy electric deflection method microwave spectroscopy molecular beam resonance experiment-electric or magnetic resonance electron paramagnetic resonance double resonance experiments (MODR, RF/DR)
9	State gr ex	electronic state ground state excited state

Column	Parameters	Definitions
10		Varia, remarks
	J	rotational quantum number of the states with measured values μ in columns 3···6, compare introduction
	$\alpha_{\parallel} - \alpha_{\perp}$ [Å^3]	anisotropy of the electric polarizability
	μ_m/μ_n	ratio of electric dipole moments in different vibrational states m and n
11		reference to tables 2.2.2, 2.7.1, 2.9.1, and, in subvolume b, 4.1

Signs of electric dipole moments, which must be obtained by techniques other than Stark effect measurements, are listed in the column "Varia, remarks". There exist some molecular beam deflection data of electric dipole moments which have reached the accuracy of dipole moments determined by microwave spectroscopy or may even be better. These values have been included in columns 3···6. If there are experimental dipole moments from other spectroscopic regions they are listed too. The methods are given with abbreviations as explained in part 1.

3. Conversion factors

Dipole moments μ and polarizabilities α are given in units Debye ($1\,\mathrm{D}=10^{-18}\,\mathrm{esu}=10^{-18}\,\mathrm{g}^{1/2}\,\mathrm{cm}^{5/2}\,\mathrm{s}^{-1}$) and Å^3, respectively, which are related to the cgs-system. But the applied electric field strength E is measured in the unit $\mathrm{V\,cm^{-1}}$ related to the SI-system.

For the calculation of the frequency shift $\Delta\nu$ from the numerical values of the field strength ($\mathrm{E/V\,cm^{-1}}$) and the dipole moment (μ/D) or the polarizability (α/Å^3), respectively, the two conversion factors should be used:

Dipole-conversion factor: $h\Delta\nu=\mu E$

$$\frac{(\Delta\nu/\mathrm{MHz})}{(\mu/\mathrm{D})(E/\mathrm{V\,cm^{-1}})}=\frac{10^{-25}}{(h/\mathrm{Js})(\mathrm{c/ms^{-1}})}=0.5034076\,(12).$$

Polarizability-conversion factor: $h\Delta\nu=\alpha E^2$

$$\frac{(\Delta\nu/\mathrm{MHz})}{(\alpha/\text{Å}^3)(E^2/\mathrm{V^2\,cm^{-2}})}=\frac{10^{-25}}{(h/\mathrm{Js})(\mathrm{c^2/m^2\,s^{-2}})}=1.679186\,(4)\cdot 10^{-9}$$

where $(h/\mathrm{Js})=6.626124(13)\cdot 10^{-34}$ and $(\mathrm{c/ms^{-1}})=299792458.0(12)$.

Sometimes slightly different values are found in the literature, which will not alter the values of the molecular parameters within their experimental accuracy in most cases. For the comparison of highly accurate values the reader should refer to the original literature.

4. Arrangement of substances

Compare 2.2.1.

2.6.1.2 Data

Nr.	Molecule	v	μ D	μ_e D	μ_I D	μ_{II} D	Ref.	Meth.	State		Varia, remarks	See Tab.
2	$^7Li^1H$	$v=1$	1.88 (19)				80DAG	LC	$A^1\Sigma$	ex		2.9.1
6–2		$v=2$	1.49 (18)				80DAG	LC	$A^1\Sigma$	ex		
		$v=5$	0.29 (5)				80DAG	LC	$A^1\Sigma$	ex		
	$^7Li^1H$	$v=2$	1.54 (7)				80BRI	QB	$A^1\Sigma$	ex	quantum beat spectroscopy	
		$v=4$	0.59 (4)				80BRI	QB	$A^1\Sigma$	ex		
		$v=5$	<0.15				80BRI	QB	$A^1\Sigma$	ex		
3	$^{23}Na^1H$	$v=0$	6.4 (7)				79DAG	ED	$X^1\Sigma^+$	gr		
	$^{23}Na^1H$	$v=12$	6.58 (13)				81BRI	QB	$A^1\Sigma^+$	ex	quantum beat spectroscopy	
		$v=13$	7.01 (14)				81BRI	QB	$A^1\Sigma^+$	ex		
		$v=14$	7.47 (15)				81BRI	QB	$A^1\Sigma^+$	ex		
		$v=15$	7.92 (16)				81BRI	QB	$A^1\Sigma^+$	ex		
5 *6–1c*	$^{23}Na^{39}K$	$v=0$	$\mu_0=2.73$ (9)				81WOR	MODR	$X^1\Sigma^+$	gr	recalculated with data of [72DAG]	2.2.2
8 *6–2a*	$^{12}C^1H$	$v=0$	0.887 (45)				74SCA	OS	$A^2\Delta$	ex	optical spectroscopy	4.1.5
9 *6–2b*	$^{14}N^1H$	$v=0$	1.389 (75)				74SCA	OS	$X^3\Sigma^-$	gr	optical spectroscopy	4.1.3
11	$^{16}O^1H$	$v=0$	1.6676 (9)				73MEE	MB	$X^2\Pi_{1/2}$	gr		2.7.1
6–3	$^{16}O^2D$	$v=0$	1.65312 (14)				73MEE	MB	$X^2\Pi_{1/2}$	gr		4.1.2
	$^{16}O^2D$	$v=0$	1.72 (10)				73WEI	LC	$A^2\Sigma^+$	ex		4.1.5
12 *6–4*	$^{32}S^1H$	$v=0$ $J=5/2$	0.7580 (1)				74MEE	MB	$^2\Pi_{3/2}$	gr		2.7.1
	$^{32}S^2D$	$v=0$	0.7571 (1)				75MEE	MB	$^2\Pi_{3/2}$	gr		4.1.5
16 *6–6*	$^1H^{35}Cl$	$J=1$ $v=0$	1.1086 (3)				73LEE	MB	$^1\Sigma^+$	gr		2.2.2 2.7.1 2.9.1
17	$^1H^{79}Br$	$v=0$	0.8271 (3)				74VAN	MW/MB	$X^1\Sigma^+$	gr	$\alpha_\parallel-\alpha_\perp=0.58$ (14) Å^3	2.2.2
6–7	$^1H^{81}Br$	$v=0$	0.8274 (3)				74VAN	MW/MB	$X^1\Sigma^+$	gr	[73DAB]	2.7.1
	$^2D^{79}Br$	$v=0$	0.8235 (3)				74VAN	MW/MB	$X^1\Sigma^+$	gr		2.9.1
	$^2D^{81}Br$	$v=0$	0.8233 (3)				74VAN	MW/MB	$X^1\Sigma^+$	gr		

Nr.	Molecule	v	μ D	μ_e D	μ_I D	μ_{II} D	Ref.	Meth.	State		Varia, remarks	See Tab.
19 *6-9*	$^{6}Li^{19}F$	$v=0$	6.32736 (20)	6.28409 (25)	0.08627 (5)	0.00054 (2)	76HEB	MB	$^{1}\Sigma^{+}$	gr	same data as given	2.2.2
		$v=1$	6.41472 (20)								in 2.6.1 of Vol. II/6	2.7.1
		$v=2$	6.50317 (20)								Landolt-Börnstein,	2.9.1
		$v=3$	6.59326 (100)								New Series, from [H15]	
	$^{7}Li^{19}F$	$v=0$	6.3248 (10)	6.2839 (12)	0.08153 (30)	0.00044 (12)	76HEB	MB	$^{1}\Sigma^{+}$	gr		
		$v=1$	6.4072 (10)									
		$v=2$	6.4905 (10)									
23 *6-19a*	KI			10.82 (20)			76STO1	ED	$^{1}\Sigma^{+}$	gr		2.7.1
24 *6-21a*	Rb Br			10.86 (20)			76STO1	ED	$^{1}\Sigma^{+}$	gr		2.2.2 2.7.1
25 *6-21b*	Rb I			11.48 (20)			76STO1	ED	$^{1}\Sigma^{+}$	gr		2.7.1
26 *6-22*	$^{133}Cs^{19}F$	$v=0$, $v=1$					78CHA	MB	$X^{1}\Sigma^{+}$	gr	effective polarizability $(\gamma_{\parallel}-\gamma_{\perp})'=2.09(4)$ Å^3 $(\alpha_{\parallel}-\alpha_{\perp})^{der}=0.90(36)$ Å^3 using: $\mu_1=19.8(10)$ D $\mu_2=-20.7$ D as expansion coefficients for a power series of μ versus $\xi=\frac{r-r_e}{r_e}$	2.2.2
28 *6-23a*	Cs Br			10.82 (20)			76STO1	ED	$^{1}\Sigma^{+}$	gr		2.2.2
29 *6-23b*	Cs I			11.69 (20)			76STO1	ED	$^{1}\Sigma^{+}$	gr		2.2.2
36 *6-28*	$^{138}Ba^{16}O$	$v=1$	2.63 (13)				79DOH	LC	$A^{1}\Sigma^{+}$	ex		2.2.2
		$v=3$	2.41 (15)				79DOH	LC	$A^{1}\Sigma^{+}$	ex		
		$v=7$	2.20 (11)				77WOR	MODR	$A^{1}\Sigma^{+}$	ex		
37 *6-29*	$^{138}Ba^{32}S$			10.853 (14)	0.022 (5)		76TIE3/ 69MEC	MB	$X^{1}\Sigma^{+}$	gr		2.2.2

Nr.	Molecule	v	μ D	μ_e D	μ_I D	μ_{II} D	Ref.	Meth.	State		Varia, remarks	See Tab.
46	$^{12}C^{14}N$	v=10	0.56(3)				73COO		$A^2\Pi_{3/2}$	ex	level-anticrossing spectroscopy with $B^2\Sigma^+$	2.7.1 4.1.2 4.1.5
59 *6–34*	$^{12}C^{16}O$	v=0	0.10980(3)	$\mu_e^{ev}=0.1222$			75MUE	MB	$X^1\Sigma^+$	gr	$^-CO^+$	2.2.2 2.9.1
	$^{12}C^{18}O$	v=0					77MEE	MB	$X^1\Sigma^+$	gr	$\mu/\mu(^{12}C^{16}O)=1.0027(4)$	
	$^{13}C^{16}O$	v=0					77MEE	MB	$X^1\Sigma^+$	gr	$\mu/\mu(^{12}C^{16}O)=1.0027(4)$ $^-CO^+$	
	$^{12}C^{16}O$	v=3	0.48(1)	0.72	0.067(5)		79LIS2	MB	$d^3\Delta_2$	ex	laser spectroscopy $^-CO^+$ (v=3, 4, 6)	
		v=4	0.42(2)				77HEM/ 79LIS2	MB	$d^3\Delta_2$	ex		
		v=6	0.28(2)				79LIS2	MB	$d^3\Delta_2$	ex		
	$^{12}C^{16}O$	v=6					75WIC1	MB	$a^3\Pi$	ex	Stark coefficient measured (v=6, 7)	4.1.6
		v=7					75WIC1	MB	$a^3\Pi$	ex		
									$a^3\Pi$	ex	dipole moment function in [75WIC2, 75WIC3]	
60 *6–35*	$^{12}C^{32}S$	v=0					73McG1	MW	$^1\Sigma^+$	gr	$^-CS^+$	
61	$^{12}C^{80}Se$	v=0	1.99(4)				73McG1	MW	$^1\Sigma^+$	gr		2.2.2 2.9.1
64 *6–39*	GeO						74DAV	MB	$^1\Sigma^+$	gr	$^+GeO^-$	2.2.2 2.9.1
78 *6–51*	$^{12}C^{19}F$	v=0	0.645(5)				80SAI	MW	$X^2\Pi_{1/2}$	gr		4.1.5
80 *6–52*	$^{14}N^{16}O$	v=1	0.1416(4)				75HOY		$X^2\Pi_{1/2}$	gr	Laser Stark-spectroscopy of the vibrational transition $v'=1-v''=0$ (75HOY rows)	2.7.1 2.9.1 4.1.2 4.1.5
					−0.01703(40)		75HOY		$X^2\Pi_{1/2}$	gr		
					−0.01735(8)		75HOY		$X^2\Pi_{3/2}$	gr		
	$^{14}N^{16}O$	v=0	1.10(3)				74BER	RF/DR	$A^2\Sigma^+$	ex		
	$^{14}N^{16}O$		0.20(2)				79LIS1		$a^4\Pi$	ex	molecular beam focused by an electrostatic quadrupole lens	
		v=3	0.20(4)				79LIS1		$a^4\Pi$	ex	from the analysis of the vibrational population of the excited molecular beam	

Nr.	Molecule	v	μ D	μ_e D	μ_I D	μ_{II} D	Ref.	Meth.	State		Varia, remarks	See Tab.
82 *6–53a*	$^{14}N^{19}F$	$v=0$	0.37(6)				73CUR	EPR	$a^1\,\Delta$	ex		2.7.1 4.1.7
95 *6–58*	$^{32}S^{19}F$	$v=0$	0.794(12)				73AMO	MW	$X^2\,\Pi_{3/2}$	gr		4.1.5
97	$^{79}Br^{19}F$	$v=0$	1.422(16)				79NAI2	MW	$^1\Sigma^+$	gr		2.2.2 2.7.1 2.9.1
98	$^{127}I^{19}F$	$v=0$	1.948(20)				79NAI1	MW	$^1\Sigma^+$	gr		2.2.2 2.9.1
99	$^{79}Br^{35}Cl$	$v=0$	0.519(4)				78NAI	MW	$^1\Sigma^+$	gr		2.2.2 2.9.1
100 *6–59*	I Cl	$v=7$	2.00(15)				74CUM	OS	$A^3\,\Pi_1$	ex	sign opposite	2.2.2
		$v=9$	1.95(5)				74CUM	OS	$A^3\,\Pi_1$	ex	to that of the	2.7.1
		$v=10$	1.90(10)				74CUM	OS	$A^3\,\Pi_1$	ex	ground state $X^1\,\Sigma^+$	4.1.6
		$v=11$	1.85(10)				74CUM	OS	$A^3\,\Pi_1$	ex		
		$v=12$	1.80(10)				74CUM	OS	$A^3\,\Pi_1$	ex		
		$v=13$	1.60(15)				74CUM	OS	$A^3\,\Pi_1$	ex		
		$v=14$	1.75(5)				74CUM	OS	$A^3\,\Pi_1$	ex		
		$v=17$	1.50(15)				74CUM	OS	$A^3\,\Pi_1$	ex		
		$v=18$	1.25(10)				74CUM	OS	$A^3\,\Pi_1$	ex		
		$v=21$	1.15(15)				74CUM	OS	$A^3\,\Pi_1$	ex		
		$v=25$	0.60(15)				74CUM	OS	$A^3\,\Pi_1$	ex		
		$v=27$	0.60(15)				74CUM	OS	$A^3\,\Pi_1$	ex		
102	I Br	$v=0$	0.737(10)				77TIE3	MW	$^1\Sigma^+$	gr		2.2.2 2.7.1
103	$^{127}I_2$	$v=0$					80CAL	MB/LIF	$X^1\,\Sigma^+$	gr	$\alpha_{\parallel}-\alpha_{\perp}=6.69(19)\,Å^3$	2.7.1 2.9.1 4.1.6

2.6.2 Linear and symmetric top molecules

2.6.2.1 Preliminary remarks

1. Introduction

In general the molecular dipole moment in vibrational state $v_1, \dots v_j, \dots v_{3N-6}$ is determined by an analysis of the Stark effect of the rotational lines. If two or three components along the directions of the principal inertia axes exist, then at least two or three Stark components of different projection quantum number M of one line or at least two or three Stark components of different lines are needed for analysis.

All values listed in this table are the values given in the original papers. Most dipole moments are determined using the value μ(OCS)=0.71521 (20) D, [68MUE], although the most accurate value at present is μ(OCS)=0.71519 (3) D, [74REI]. Note that $1\,D = 3.33564 \cdot 10^{-30}$ Cm.

2. Explanation of the columns in the tabulation

Column	Parameters		Definitions
1	Nr. *Nr.*		serial number of the molecule in this table the following number represents the serial number of the molecule in II/6 Landolt-Börnstein, New Series
2	v	v_j	vibrational quantum number of the j-th normal vibration: $j=1, \dots 3N-6$ (for linear molecules $j=1, \dots 3N-5$) vibrational ground state: $v_1=v_2=\cdots v_j \cdots = v_{3N-6}=0;$ in short $v=0$ vibrational state: $v_1=\cdots v_{j-1}=0; \quad v_j=v; \quad v_{j+1}=\cdots v_{3N-6}=0;$ in short $v_j=v$
3	μ [D]	or μ_v μ_0	molecular dipole moment in vibrational state $v_j=v$ molecular dipole moment in the vibrational ground state
4	Ref.		references for column 3
5	Meth.	 MW MBER LaSt IRIRDR IRMWDR	method of measurement for μ microwave spectroscopy molecular beam electric resonance Laser Stark spectroscopy infrared-infrared double resonance infrared-microwave double resonance
6	Varia, remarks		remarks concerning the previous columns
7			reference to Tables 2.3 and 2.4

3. Arrangement of substances

The ordering of the molecules in the tables 2.3 and 2.4 is retained.

2.6.2.2 Data

Nr.	Molecule	v	μ D	Ref.	Meth.	Varia, remarks	See Tab., Nr.
1	$^{16}OC^{18}O$	$v=0$	$7.0(1.5)\cdot 10^{-4}$	80END2	MW		2.3, 1
2 *6-1*	OCS	$v=0$	0.71519 (3)	74REI	MBER		2.3, 2
		$v_2=1^1$	0.70433 (3)				
		$v_2=2^2$	0.6936 (3)	74FAB	MBER		
		$v_2=3^1$	0.68173 (20)	78TAN2	LaSt		
	$OC^{33}S$	$v=0$	$\mu_{rel.}=1.00024$ (25)	74REI	MBER	$\mu_{rel.}=\frac{\mu(\text{isotopic species})}{\mu(\text{parent}, v=0)}$	
		$v_2=1^1$	$\mu_{rel.}=1.00012$ (9)				
	$OC^{34}S$	$v=0$	$\mu_{rel.}=1.00031$ (2)	70LEE	MBER		
		$v_2=1^1$	$\mu_{rel.}=1.00033$ (2)	74REI			
	^{18}OCS	$v=0$	$\mu_{rel.}=0.99904$ (3)	74REI			
		$v_2=1^1$	$\mu_{rel.}=0.99907$ (4)				
	$O^{13}CS$	$v=0$	$\mu_{rel.}=1.00017$ (2)	70LEE			
		$v_2=1^1$	$\mu_{rel.}=1.00067$ (2)	74REI			
3 *6-2b*	HCN	$v=0$	2.984594 (1500)	70TOM			2.3, 4
		$v_2=1^1$	2.941642 (1500)				
		$v_2=2^0$	2.89865 (150)				
		$v_2=2^2$	2.89813 (150)				
	DCN	$v=0$	2.990198 (1500)				
4	FCN	$v=0$	2.1203 (10)	77MAK2	LaSt		2.3, 5
		$v_2=1^1$	2.09563 (45)				
		$v_2=2^0$	2.0728 (28)				
		$v_3=1$	$\Delta\mu=-0.02852$ (71) *)			*) $\Delta\mu=\mu'-\mu''$	
5 *6-2a*	$^{35}ClCN$	$v=0$	2.83312 (15)	80REI	MBER		2.3, 6
		$v_2=1^1$	2.804 (3)				
6	HNC	$v=0$	3.05 (10)	76BLA2	MW		2.3, 9
7	FCP	$v=0$	0.279 (1)	80KRO2	MW		2.3, 11
8	HBS	$v=0$	1.298 (5)	74PEA	MW	from Zeeman studies: $\mu(+HBS-)$	2.3, 12
9	ClBS	$v_2=1^1$	1.45 (8)	80KIR2	MW		2.3, 13

Nr.	Molecule	v	μ D	Ref.	Meth.	Varia, remarks	See Tab., Nr.
10	N≡C—C≡P	$v=0$	3.44 (3)	80COO	MW		2.3, 15
11	O=C=C=C=S	$v=0$	0.634 (30)	76WIN	MW		2.3, 16
12 6–5	N_2O	$v=0$	0.160880 (23)	78REI	MBER		2.3, 18
		$v_2=1^1$	0.173484 (9)				
13	HC≡CD	$v=0$	0.01001 (15) *)	80MAT1	MW	uncertainties are 2.5 σ *) extrapolated value, C—D moiety more positive	2.3, 23
		$v_4=1$	0.02359 (5)				
		$v_4=3$	0.09077 (26)				
		$v_5=1$	0.05601 (9)				
		$v_5=3$	0.1472 (21)				
	DC≡CD		0.0358 (20) **)	77LAF	MW	**) vibrational transition dipole moment from the $\nu_5^1-\nu_4^1$ difference band	
14	HC≡CF	$v=0$	0.7207 (3)	76TAN	LaSt		2.3, 24
		$v_3=1$	0.7447 (3)				
		$v_4=1$	0.6557 (3)				
		$v_5=1$	0.7441 (3)				
		$v_3=1, v_4=1$	0.6769 (3)				
		$v_3=1, v_5=1$	0.7689 (3)				
	DC≡CF	$v=0$	0.73292 (22)	80MAT2	LaSt		
		$v_3=1$	0.75772 (30)				
		$v_4=1$	0.68412 (24)				
		$v_5=1$	0.75656 (17)				
		$v_5=2, \Sigma^+$	0.78063 (21)				
		$v_3=1, v_4=1$	0.70822 (17)				
		$v_3=1, v_5=1$	0.78270 (18)				
		$v_3=1, v_5=2, \Sigma^+$	0.80808 (25)				
		$v_4=1, v_5=1$	0.70698 (19)				
		$v_3=1, v_4=1, v_5=1$	0.73329 (17)				
15 6–6*b*	HC≡CBr	$v=0$	0.2340 (12)	77POR, 77JON	MW		2.3, 26
16	HC≡CCN	$v=0$	3.724 (30)	78LAF	MW		2.3, 27
16a	^{35}Cl—C≡C—F	$v=0$	0.25 (10)	81GUA1	MW		2.3, 27a

Nr.	Molecule	v	μ D	Ref.	Meth.	Varia, remarks	See Tab., Nr.
17	ClC≡CCN	$v=0$	3.38 (5)	74BJO	MW		2.3, 28
17a	HC≡C—C≡P	$v_7=1$	0.745 (5)	81KRO2	MW		2.3, 30a
17b	HC≡C—C≡CH		0.0787 (10) *)	81MAT	MW	*) transition dipole moment of the $\nu_8-\nu_6$ band	2.3, 30b
	HC≡C—C≡CD	$v_6=1$	0.0907 (6)	81ETO			
		$v_6=2$	0.1681 (14)				
		$v_6=1$, $v_9=1$	0.0900 (4)				
18	HC≡C—C≡CCN	$v=0$	4.33 (3)	76ALE	MW		2.3, 31
19	HCN···HF	$v=0$	5.612 (10)	80LG1	MW		2.3, 33
19a	HCN···HCN	$v=0$	6.0 (15)	81BRO4	MW		2.3, 34
19b	CO_2 · HF	$v=0$	2.2465 (4)	81BAI	MBER		2.3, 37a
	CO_2 · DF	$v=0$	2.3024 (1)				
19c	SCO · HF	$v=0$	3.2085 (20)	81BAI	MBER		2.3, 37b
20	Ne · $D^{35}Cl$	$v=0$	0.47507 (13)	80BAR2	MBER	$\mu_J=-0.00253$ (6), centrifugal distortion term	2.3, 38
21	Ar · HF	$v=0$	1.3353 (4)	81DIX1	MBER		2.3, 39
	Ar · DF	$v=0$	1.6768 (6)				
22	Ar · $H^{35}Cl$	$v=0$	0.811714 (16)	81HUT	MBER	$\mu_J=-0.21\,(5)\cdot10^{-4}$	2.3, 40
	Ar · $D^{35}Cl$	$v=0$	1.00955 (12)			$\mu_J=-1.13\,(18)\cdot10^{-4}$	
23	Ar · ^{35}ClF	$v=0$	1.053 (3)	74HAR1	MBER		2.3, 42
24	Ar · $H^{79}Br$	$v=0$	0.57417 (40)	80JAC	MBER	$\mu_J=-0.43\,(19)\cdot10^{-3}$	2.3, 41
	Ar · $H^{81}Br$	$v=0$	0.57241 (34)			$\mu_J=-0.23\,(16)\cdot10^{-3}$	
	Ar · $D^{79}Br$	$v=0$	0.75633 (32)			$\mu_J=-0.21\,(12)\cdot10^{-3}$	
	Ar · $D^{81}Br$	$v=0$	0.75630 (29)			$\mu_J=-0.14\,(10)\cdot10^{-3}$	
25	^{84}Kr · $H^{35}Cl$	$v=0$	0.956692 (11) *)	80BAR1	MBER	$\mu_J=0.294\,(25)\cdot10^{-4}$ distortion term *) uncertainties are 2 σ	2.3, 44
	^{84}Kr · $H^{37}Cl$	$v=0$	0.957922 (74) *)			$\mu_J=0.16\,(11)\cdot10^{-4}$	
	^{86}Kr · $H^{35}Cl$	$v=0$	0.956958 (84) *)			$\mu_J=0.19\,(12)\cdot10^{-4}$	
	^{84}Kr · $D^{35}Cl$	$v=0$	1.106371 (95) *)			$\mu_J=0.43\,(13)\cdot10^{-4}$	

Nr.	Molecule	v	μ D	Ref.	Meth.	Varia, remarks	See Tab., Nr.
26	$^{84}Kr \cdot {}^{35}ClF$	$v=0$	1.136 (3)	75NOV	MBER		2.3, 46
26a	$^{129}Xe \cdot HF$	$v=0$	1.7359 (4)	81BAI1	MBER		2.3, 46a
	$^{129}Xe \cdot DF$	$v=0$	1.9537 (7)				
27	$^{132}Xe \cdot H^{35}Cl$	$v=0$	1.0901 (30)	79CHN	MBER		2.3, 47
28	$^{14}NH_3$	$v=0$	1.47147 (6) $+1.809(22)\cdot 10^{-4}J(J+1)$ $-3.467(25)\cdot 10^{-4}K^2$	80SHI	LaSt		2.4, 1
		$v_2=1$	1.2480 (14) $+4.15(59)\cdot 10^{-4}J(J+1)$ $-7.05(69)\cdot 10^{-4}K^2$				
		$v_4=1^1$	1.459 (2)	74COH	MW		
	$^{15}NH_3$	$v=0$	1.47066 (11) $+1.938(26)\cdot 10^{-4}J(J+1)$ $-3.575(29)\cdot 10^{-4}K^2$	80SHI	LaSt		
		$v=0, (J, K)=(5,3)$	1.4760 (9)	81DI1	LaSt		
		$v_2=1$	1.2439 (24) $+5.75(60)\cdot 10^{-4}J(J+1)$ $-8.79(70)\cdot 10^{-4}K^2$	80SHI	LaSt		
		$v_2=1, (J, K)=(4,4)$	1.253 (6)	81DI1	LaSt		
		$v_2=1, (J, K)=(7,7)$	1.241 (10)				
		$v_2=1, (J, K)=(8,8)$	1.230 (5)				
		$v_2=1, (J, K)=(11,9)$	1.271 (6)				
	NH_2D	$v=0$	$\mu_c=1.45674(26)$ $+0.93(100)\cdot 10^{-5}P^2$ $+1.11(14)\cdot 10^{-4}(P_a^2+P_b^2-P_c^2)$	80SHI	LaSt	P_a, P_b, P_c = angular momentum components	
		$v_2=1$	$\mu_a=0.1567(37)$ $\mu_c=1.2663(17)$				
	ND_3	$v=0, (J, K)=(13,13)$	1.497 (11)	81DI2	LaSt		
		$v=0, (J, K)=(13,11)$	1.492 (8)				
		$v_2=1, (J, K)=(14,13)$	1.355 (5)				
		$v_2=1, (J, K)=(14,11)$	1.352 (3)				
29	$N^{35}Cl_3$	$v=0$	0.39 (1)	74CAZ	MW		2.4, 2

Nr.	Molecule	v	μ D	Ref.	Meth.	Varia, remarks	See Tab., Nr.
30	PH_3	$v=0$	0.57397 (30)	71DAV	MBER		2.4, 3
		$v_2=1$	0.5740 (2)	80DLO	LaSt		
		$v_4=1$	0.5784 (1)				
31	AsH_3	$v=0$	$\hbar^4 \tau_{xxxz} \mu_z / 2h(B-C)$ $= 3.938\ (17) \cdot 10^{-5}$	78HEL2	MW		2.4, 6
32	CH_3D	$v=0$	$\mp\{(5.657 \pm 0.004)$ $-(0.0427 \pm 0.0009) J(J+1)$ $+(0.0696 \pm 0.0026) K^2\} \cdot 10^{-3}$ *)	79WAT	IRIRDR	*) upper sign consistent with ab initio calculations	2.4, 11
		$v_6=1$	$3.75\ (5) \cdot 10^{-3}$	80ORR	IRIRDR		
	CH_2D_2	$v=0$	0.014 (5)	75HIR2	MW		
33	CH_3F	$v=0$, $J=1, K=1$	1.85840 (2) *)	80MA	MBER	*) $\mu(CH_3F)/\mu_{v_2=1}(OCS) = 2.638905\ (23)$	2.4, 12
		$v=0$, $J=2, K=2$	1.85842 (6) $^+$)			$^+$) $\mu(CH_3F)/\mu_{v_2=1}(OCS) = 2.63894\ (10)$	
		$v_3=1$	$\mu_v - \mu_0 = 0.04686\ (26)$	74FRE	LaSt		
		$v_3=2$	$\mu_v - \mu_0 = 0.0934\ (17)$				
	$^{13}CH_3F$	$v=0$	1.85794 (63)	74FRE,	LaSt		
		$v_3=1$	$\mu_v - \mu_0 = 0.04594\ (39)$	74BRE			
	CD_3F	$v_5=1$	1.8771 (7)	77DUX	LaSt		
			$\mu_v - \mu_0 = 0.0069\ (7)$ **)			**) $\mu_0 = 1.8702$	
		$v_5=1$	1.8751 (25)	81CAL	LaSt		
34	$CD_3{}^{35}Cl$	$v=0$	1.90597 (33)	77YAM1	LaSt		2.4, 13
		$v_2=1$	1.90741 (33)				
		$v_5=1$	1.90607 (36)				
35	$CH_3{}^{79}Br$	$v=0$	1.82167 (83)	78IEK	LaSt		2.4, 14
		$v_6=1$	1.81868 (90)				
	$CH_3{}^{81}Br$	$v=0$	1.82179 (75)				
		$v_6=1$	1.81820 (99)				
36	CD_3I	$v=0$	1.6504 (4)	77KAW2	LaSt		2.4, 15
		$v_2=1$	1.6479 (5)				
37	CH_3CN	$v=0$	3.92455 (14)	78ROM	LaSt		2.4, 16
		$v_4=1$	$\mu_v - \mu_0 = 0.01044\ (8)$				

Nr.	Molecule	v	μ D	Ref.	Meth.	Varia, remarks	See Tab., Nr.
38	CH_3NC	$v=0$ $v_4=1$	3.88674 (7) $\mu_v-\mu_0=0.030272$ (31)	78ROM	LaSt		2.4, 17
39	CF_3H CF_3D	$v=0$ $v=0$ $v_5=1$	1.65150 (10) 1.6403 (26) *) 1.6499 (32)	81MEE 80IBI	MBER LaSt	*) [81MEE] gives 1.65350 (10)	2.4, 18
40	CF_4	$v_3=1$	0.036 (4)	79TAK1	IRMWDR		2.4, 19
41	$CF_3{}^{79}Br$	$v=0$	0.639 (2)	80COX	MW		2.4, 20
42	CF_3I	$v=0$	1.048 (3)	80COX	MW		2.4, 21
43	SiH_4	$v=0$	$3.34 (4) \cdot 10^{-5}$	76KAG, 78KRE	MW LaSt		2.4, 22
44	SiH_3F SiD_3F	$v=0$ $v=0$	1.298 (4) 1.300 (4)	76GEO	MW		2.4, 23
45	$SiH_3{}^{81}Br$	$v=0$	1.319 (8)	77DO3	MW	from Zeeman studies $\mu(+H_3SiBr-)$ [79DO1]	2.4, 24
46	GeH_4	$v=0$	$3.33 (5) \cdot 10^{-5}$	77KRE1	Laser		2.4, 28
47	OPF_3	$v=0$ $v_4=1$ $v_5=1$ $v_6=1$	1.86847 (10) $\mu_J=3.28 (13) \cdot 10^{-6}$ $\mu_D=5.856 (20) \cdot 10^{-6}$ $(\mu_4-\mu_0)/\mu_0=-16.1 (3) \cdot 10^{-3}$ *) $(\mu_5-\mu_0)/\mu_0=-3.49 (4) \cdot 10^{-3}$ $(\mu_6-\mu_0)/\mu_0=-0.65 (4) \cdot 10^{-3}$	79ME2 78OZI 78AMA 79ME2	MBER MBER LaSt MBER MBER	μ_D=distortion dipole moment *) $\mu_4=1.83943$ (40) D	2.4, 31
48	$^{35}ClO_3F$	$v=0$	0.0228 (4)	80DEL4	MBER		2.4, 35
49	$CH_3{}^{11}BS$	$v=0$	2.573 (5)	80KIR1	MW		2.4, 40
50	$CH_3C{\equiv}P$	$v=0$	1.499 (1)	79KRO	MW		2.4, 41
51	CF_3CN	$v=0$	1.262 (10)	74FOR2	MW		2.4, 42
52	SiH_3NCS	$v=0$ $v_{10}=1^1$	2.38 (2) 2.36 (2)	77DO2	MW	from Zeeman studies: $\mu(+H_3SiNCS-)$ [79DO2]	2.4, 45
53	$CH_3C{\equiv}CH$	$v=0$ $v_5=1$	0.7839 (10) 0.7954 (10)	80BUR2	MW		2.4, 47

Nr.	Molecule	v	μ D	Ref.	Meth.	Varia, remarks	See Tab., Nr.
54	BH_3CO	$v=0$	1.698 (20)	77VEN	MW		2.4, 51
55	BF_3CO	$v=0$	0.592 (1)	78JAN	MBER		2.4, 52
56	$(CH_3)_3N \cdot BH_3$	$v=0$	4.84 (10)	75CAX	MW		2.4, 68
57	CH_3CD_3 $^{13}CH_3CD_3$ $CH_3{}^{13}CD_3$	$v=0$ $v=0$ $v=0$	0.01078 (9) $\mu(^{13}C)/\mu(^{12}C)=0.990$ (3) $\mu(^{13}C)/\mu(^{12}C)=1.015$ (5)	77HIR	MW		2.4, 72
58	CH_3CF_3	$v=0$	2.3470 (5)	78MEE	MBER		2.4, 73
59	SiH_3SiF_3	$v=0$	2.03 (7)	75PAS	MW		2.4, 78
60	$^{11}BH_3 \cdot NH_3$	$v=0$	5.216 (17)	81SUE	MW		2.4, 79
61	$BF_3 \cdot PH_3$	$v=0$	3.73 (30)	75ODO	MW		2.4, 80
62	$BH_3 \cdot PH_3$	$v=0$	4.00 (3)	73DUR5	MW		2.4, 81
63	$CH_2{=}C{=}CD_2$	$v=0$	0.0031 (3)	73HIR	MW		2.4, 82
64	$^{35}ClF_5$	$v=0$	0.536 (10)	76BOD	MW		2.4, 88
65		$v=0$	3.89 (2)	76GIL1	MW		2.4, 91
66	$C_7H_{10}O_3$	$v=0$	3.01 (3)	73SLA	MW		2.4, 94
67	C_5H_5BeH	$v=0$	2.08 (1)	75BAR	MW		2.4, 96
68	$Ar \cdot {}^{11}BF_3$ $Ar \cdot {}^{10}BF_3$	$v=0$ $v=0$	0.176 (2) 0.176 (2)	78JAN	MBER		2.4, 102
69	$Ar \cdot SO_3$ $Ar \cdot {}^{34}SO_3$	$v=0$ $v=0$	0.2676 (3) 0.2676 (3)	80BOW	MBER		2.4, 103
69a	$^{84}KrSO_3$	$v=0$	0.369 (2)	81LEO	MBER		2.4, 103a
70	$N_2 \cdot {}^{10}BF_3$	$v=0$	0.35 (2)	78JAN	MBER		2.4, 104
71	$N_2 \cdot SO_3$	$v=0$	0.46 (1)	80BOW	MBER		2.4, 105
72	$CH_3NC \cdot BH_3$	$v=0$	6.73 (7)	77STE	MW		2.4, 107

2.6.3 Asymmetric top molecules

2.6.3.1 Preliminary remarks

1. Introduction

See 2.6.2.1, Preliminary remarks.

2. Explanations of the columns in the tabulation

Column	Parameters		Definitions
1	Nr.		serial number of the molecule in this table
2	v	v_j	vibrational quantum number of the j-th normal vibration: $j=1, \dots 3N-6$, vibrational ground state: $v_1=v_2=\cdots v_j\cdots=v_{3N-6}=0$; in short $v=0$ vibrational state: $v_1=\cdots v_{j-1}=0;\quad v_j=v;\quad v_{j+1}=\cdots v_{3N-6}=0$; in short $v_j=v$
3	μ [D]		molecular dipole moment in vibrational state $v_j=v$
4	components of μ	μ_a, μ_b, μ_c	dipole moment components in direction of the ordered principal inertia axes a, b, and c or in relation to other specified coordinate systems
5	Ref.		references for columns 3 and 4
6	Varia, remarks	 $\nu_j, \tilde{\nu}_j$ $\alpha_{gg}-\alpha_{g'g'}$ $(g, g'=a, b, c)$	remarks concerning the previous columns type, frequency, or wave number of the j-th normal vibration polarizabilities references to figures in Chapter 3, subvolume b
7			references to tables 2.5

2.6.3.2 Data

Nr.	Molecule	v	μ D	Components of μ D	Ref.	Varia, remarks Fig. in Chap. 3, subvol. b	See Tab., Nr.
1	$HN{=}SOF_2$	$v=0$	2.43 (1)	$\mu_a=0.65$ (1) $\mu_b=2.343$ (5)	77CAX1	Fig. 9	2.5, 4
	$DN{=}SOF_2$	$v=0$	2.48 (3)	$\mu_a=0.45$ (5) $\mu_b=2.44$ (1)			
2	HNCO	$v=0$		μ_a *) $=1.575$ (5) $\mu_b=1.35$ (10)	75HO1	Fig. 10 *) rotational dependence observed	2.5, 5
	DNCO	$v=0$	2.10 (15)	$\mu_b=1.35$ (10)	74HOC		
3	DNCS	$v=0$	1.99 (15)	$\mu_a=1.67$ (3) $\mu_b=1.08$ (15)	78SZA	Fig. 11	2.5, 6
4	HNCSe	$v=0$		$\mu_a=1.98$	79LAD		2.5, 7
5	NCNCO	$v=0$	2.533 (11)	$\mu_a=2.488$ (10) $\mu_b=0.476$ (20)	76HO2	Fig. 12	2.5, 9
6	HNO *)	$v=0$	1.62 (2) *)	$\mu_a=0.996$ (6) *) $\mu_b=1.28$ (2)	77JO1	Fig. 13 *) Laser Stark data **) fixed at the ground state values	2.5, 12
		$v_2=1$	1.62	$\mu_a=1.000$ (8) $\mu_b=1.28$ **)		$v_2=1565.345$ (2) cm^{-1} N—O stretch	
		$v_3=1$	1.62	$\mu_a=0.986$ (9) $\mu_b=1.28$ **)		$v_3=1500.818$ (3) cm^{-1} bend	
	DNO *)	$v=0$	1.65 (2) *)	$\mu_a=1.102$ (5) $\mu_b=1.23$ (2)			
		$v_2=1$	1.65	$\mu_a=1.102$ (4) $\mu_b=1.23$ **)		$v_2=1546.8801$ (7) cm^{-1}	
7	NOF *)	$v=0$	1.730 (3) *)	$\mu_a=1.6899$ (10) *) $\mu_b=0.3702$ (72)	78ALL	*) Laser Stark data, the uncertainties are 3σ	2.5, 13
		$v_1=1$		$\mu_a=1.6018$ (10) $\mu_b=0.3625$ (55)		$v_1=1844.099$ (1) cm^{-1}	
8	NOCN	$v=0$	1.365 (3)	$\mu_a=1.241$ (3) $\mu_b=0.57$ (8)	78DIC	Fig. 14	2.5, 16

Nr.	Molecule	v	μ D	Components of μ D	Ref.	Varia, remarks Fig. in Chap. 3, subvol. b	See Tab., Nr.
9	HNO_2 trans *)	$v=0$	1.930 (17)	$\mu_a=1.3829$ (72)	80ALL	*) Laser Stark data	2.5, 19
				$\mu_b=1.3468$ (170)			
		$v_2=1$	1.852	$\mu_a=1.4436$ (59)		$\nu_2=1699.760$ (1) cm^{-1}	
				$\mu_b=1.1607$ (99)		N—O stretch	
	HNO_2 cis *)	$v=0$	1.428 (7)	$\mu_a=0.3071$ (39)			
				$\mu_b=1.3943$ (60)			
		$v_2=1$	1.441	$\mu_a=0.3472$ (35)		$\nu_2=1640.517$ (1) cm^{-1}	
				$\mu_b=1.3989$ (44)			
10	HN_3	$v=0$	1.70 (5)	$\mu_a=0.837$	75BEN	Fig. 15	2.5, 21
				$\mu_b=1.48$ (5)			
	DN_3	$v=0$	1.76 (5)	$\mu_a=0.894$ (10)	79BEN		
				$\mu_b=1.51$ (5)			
11	$ClNO_3$	$v=0$		$\mu_a=0.72$ (7)	77SUE		2.5, 24
				$\mu_b=0.28$ (2)			
12	NHDNHD species 4	$v=0$		$\mu_a=\mu_b=0$	76TSU	See Fig. 17 for numbering	2.5, 25
				$\mu_c=1.86$			
	NHDNHD species 5	$v=0$		$\mu_a=0.14$			
				$\mu_b=1.29$			
				$\mu_c=1.43$			
	NHDNHD species 6	$v=0$		$\mu_a=\mu_c=0$			
				$\mu_b=1.92$			
	N_2D_4	$v=0$	1.88		78TSU		
13	SO_2	$v=0$	1.63305 (4) *)		79PAT	*) MBER	2.5, 28
		$v_1=1$	1.63322 (4)			$\alpha_{aa}-\alpha_{bb}=1.62$ (40) $Å^3$	
		$v_1=2$	1.63331 (4)			$\alpha_{aa}-\alpha_{cc}=0.1$ (7) $Å^3$	
		$v_2=1$	1.62614 (4)				
		$v_2=2$	1.61915 (4)				
		$v_1=1, v_2=1$	1.62630 (4)				
		$v_3=1$	1.65246 (4)				
		$v_2=1, v_3=1$	1.64539 (4)				
14	$(SO)_2$	$v=0$	3.17 (10)	$\mu_b=3.17$ (10)	74LOV	Fig. 18	2.5, 29
15	$S^{35}Cl_2$	$v=0$	0.36 (1)	$\mu_b=0.36$ (1)	76MUR		2.5, 37

Nr.	Molecule	v	μ D	Components of μ D	Ref.	Varia, remarks Fig. in Chap. 3, subvol. b	See Tab., Nr.
15a	H_2SO_4	$v=0$	2.725 (15)	$\mu_c=2.725$ (15)	81KUC1		2.5, 37a
16	$(HF)_2$ $(DF)_2$ HFDF	$v=0$ $v=0$ $v=0$		$\mu_a=2.987$ (3) *) $\mu_a=2.9919$ (6) *) $\mu_a=3.029$ (3) *)	72DYK	*) MBER	2.5, 38
17	$HF^{35}ClF$ $HF^{37}ClF$ $DF^{35}ClF$	$v=0$ $v=0$ $v=0$		$\mu_a=2.313$ (1) *) $\mu_a=2.314$ (1) *) $\mu_a=2.295$ (1) *)	76NOV1	Fig. 19 *) MBER	2.5, 39
18	$HFH^{35}Cl$ $HFH^{37}Cl$ $HFD^{35}Cl$	$v=0$ $v=0$ $v=0$		$\mu_a=2.4095$ (5) *) $\mu_a=2.410$ (1) *) $\mu_a=2.483$ (1) *)	77JAN	Fig. 20 *) MBER	2.5, 40
19	HOF DOF	$v=0$ $v=0$	2.23 (10)	$\mu_a=0.37$ (10) $\mu_b=2.20$ (10) $\mu_a=0.28$ (10) $\mu_b=2.22$ (10)	73ROC2		2.5, 41
20	$FClO_2$	$v=0$	1.722 (30)	$\mu_a=0.551$ (20) $\mu_c=1.632$ (20)	74PAR		2.5, 42
21	$SPHF_2$ $SPDF_2$	$v=0$ $v=0$	1.87 (3) 1.86 (3)	$\mu_a=1.08$ $\mu_c=1.53$ $\mu_a=1.09$ $\mu_c=1.51$	73NAV	Fig. 21	2.5, 43
22	P_2D_3H(B) P_2D_3H(C) P_2D_4	$v=0$ $v=0$ $v=0$	0.92 (1) 0.92 (1)	$\mu_a=0.008$ (7) $\mu_b=0.92$ (1) $\mu_c=0.04$ (2) $\mu_a=0.020$ (7) $\mu_b=0.15$ (2) $\mu_c=0.91$ (1) $\mu_a=\mu_b=0$ $\mu_c=0.928$ (5)	74DUR2		2.5, 44
23	BF_2NH_2	$v=0$	2.595 (30)	$\mu_a=2.595$ (30)	73LOV	Fig. 22	2.5, 46
24	BH_2NH_2	$v=0$	1.844 (15)	$\mu_a=1.844$ (15)	79SUG		2.5, 47

Nr.	Molecule	v	μ D	Components of μ D	Ref.	Varia, remarks Fig. in Chap. 3, subvol. b	See Tab., Nr.
24a	$^{11}BH_2OH$	$v=0$	1.506 (15)	$\mu_a=0.495$ (7) $\mu_b=1.423$ (15)	81KAW	uncertainties are 3σ	2.5 47a
	$^{11}BD_2OD$	$v=0$	1.518 (47)	$\mu_a=0.554$ (12) $\mu_b=1.414$ (45)			
25	$^{11}BH(OH)_2$	$v=0$	1.47 (2)	$\mu_a=1.46$ (2) $\mu_b=0.15$ (12)	78KAW2		2.5, 48
26	$^{11}BFHOH$	$v=0$	2.06 (8)	$\mu_a=1.85$ (3) $\mu_b=0.91$ (20)	78KAW2		2.5, 49
27	$^{11}BF(OH)_2$	$v=0$	1.818 (7) *)	$\mu_a=0.195$ (45) *) $\mu_b=1.807$ (7) *)	79KAW2	Fig. 23 *) error limits are 3σ	2.5, 50
	$^{11}BF(OD)_2$	$v=0$	1.811 (24) *)	$\mu_a=0.529$ (73) *) $\mu_b=1.732$ (24) *)			
28	$HB(NH_2)_2$	$v=0$	1.25 (1)		78BRI		2.5, 51
29	O_3	$v=0$	0.53373 (7) *)		77MEE	$\alpha_{aa}-\alpha=2.04$ (40) $Å^3$ *) $\alpha_{bb}-\alpha=-0.98$ (13) $Å^3$ $\alpha_{cc}-\alpha=-1.03$ (42) $Å^3$ *) MBER	2.5, 52
30	H_2O	$v=0$		$\mu_b=1.8546$ (4) *)	73DYK	*) MBER, rotationless moment see also 5, appendix (subvolume b)	2.5, 53
	D_2O	$v=0$		$\mu_b=1.8545$ (4) *)	73DYK, 75DAG		
	HDO	$v=0$		$\mu_a=0.6567$ (4) $\mu_b=1.7318$ (9)	73CLO		
	$H_2{}^{17}O$	$v_2=1$		$\mu_b(v_2=1)-\mu_b(v=0)$ $=-0.0310$ (42) **)	78JOH	**) Laser Stark	
31	$(H_2O)_2$	$v=0$		$\mu_a=2.6429$ (2) *) $\mu_\perp=0.38$ (10)	80ODU	Fig. 24 *) MBER $\mu_\perp=\sqrt{\mu_b^2+\mu_c^2}$	2.5, 60
	$(D_2O)_2$	$v=0$; E_1		$\mu_a=2.6061$ (3) $\mu_\perp=0.52$ (10)			
		$v=0$; E_2		$\mu_a=2.6206$ (4) $\mu_\perp=0.43$ (10)			

Nr.	Molecule	v	μ D	Components of μ D	Ref.	Varia, remarks Fig. in Chap. 3, subvol. b	See Tab., Nr.
32	$H_2O\cdots HF$	$v=0$ $v_{OB}=1$	4.07 (1) 3.80 (1)	$\mu_a=4.07$ (1) $\mu_a=3.80$ (1)	80BEV2	$\nu_{OB}=(94\pm20)\ cm^{-1}$ out-of-plane bend	2.5, 61
32a	H_2O_2		1.5728 (6) *)		81COH	*) from the $0_{00}\leftarrow1_{10}$, $\tau=4\leftarrow2$, $n=0$ transition	2.5, 62
33	CH_3OH	$v=0$		$\mu_a=0.8957$ (5) $\mu_b=1.40$ (2)	81AMA		2.5, 65
		$v_\tau=1$		$\mu_a=0.930$ (5) **) $\mu_b=1.43$ (1)	80JOH	$\nu_\tau=890.761$ GHz, torsion **) Laser Stark	
		$v_{CO}=1$		$\mu_a=$ 0.9213 (8) $\mu_b=-1.39$ (9)	81AMA	ν_{CO}, C—O stretch	
	CH_3OD	$v=0$	1.705 (17)	$\mu_a=0.833$ (8) $\mu_b=1.488$ (15)	80KAU		
33a	CF_2	$v=0$	0.469 (26)	$\mu_b=0.469$ (26)	73KIR4		2.5, 64
34	CH_3SeH	$v=0$	1.291 (10)	$\mu_a=1.253$ (10) $\mu_b=0.314$ (10)	73THO		2.5, 67
35	CH_3NHCN	$v=0$		$\mu_a=4.72$ $\mu_b=1.30$	80BAK2		2.5, 70
36	CH_3NHOH	$v=0$	0.712 (49) *)	$\mu_a=0.611$ (8) *) $\mu_b=0.366$ (37) $\mu_c=0$	79SUN	*) uncertainties are 3σ	2.5, 71
37	$CH_3{}^{15}NO$	$v=0$	2.320 (4)	$\mu_a=2.262$ (4) $\mu_b=0.516$ (2)	78TUR	orientation of μ determined, see Fig. 26	2.5, 72
38	CF_3NO	$v=0$	0.18 (1)	$\mu_a=0.145$ (1) $\mu_b=0.10$ (1)	76TUR2		2.5, 73
39	$CH_3O^{35}Cl$	$v=0$	1.788 (8)	$\mu_a=1.373$ (4) $\mu_b=1.146$ (7)	76SUZ	for orientation and sign of μ, see Fig. 27	2.5, 75
	$CD_3O^{35}Cl$	$v=0$	1.791 (8)	$\mu_a=1.409$ (4) $\mu_b=1.106$ (8)			

Nr.	Molecule	v	μ D	Components of μ D	Ref.	Varia, remarks Fig. in Chap. 3, subvol. b	See Tab., Nr.
40	$CH_3S^{35}Cl$	$v=0$	2.01 (4)	$\mu_a=1.78$ (2) $\mu_b=0.95$ (5)	73GUA	for orientation and sign of μ, see Fig. 27a	2.5, 76
	$CD_3S^{35}Cl$	$v=0$	2.01 (4)	$\mu_a=1.85$ (2) $\mu_b=0.80$ (5)			
41	CF_3OF	$v=0$		$\mu_a=0.30$ (2) $\mu_b=0.10$ (6) $\mu_c=0$ *)	74BUC	*) assumed	2.5, 77
42	$CH_3{}^{80}SeCN$	$v=0$	4.42 (5)	$\mu_a=4.35$ (4) $\mu_b=0.76$ (10)	78SAK	for orientation of μ, see Fig. 28	2.5, 81
	$CD_3{}^{80}SeCN$	$v=0$	4.36 (4)	$\mu_a=4.31$ (4) $\mu_b=0.67$ (7)			
43	CH_3OCH_2F	$v=0$	1.744 (24)	$\mu_a=1.055$ (13) $\mu_b=0.334$ (99) $\mu_c=1.348$ (24)	81NA1		2.5, 85
	CD_3OCH_2F	$v=0$	1.756 (17)	$\mu_a=1.063$ (7) $\mu_b=0.394$ (62) $\mu_c=1.340$ (9)			
	CH_3OCD_2F	$v=0$	1.743 (26)	$\mu_a=1.025$ (13) $\mu_b=0.408$ (83) $\mu_c=1.350$ (17)			
44	CH_3OCH_2CN gauche form	$v=0$	2.98 (7)	$\mu_a=2.40$ (3) $\mu_b=1.37$ (10) $\mu_c=1.11$ (11)	74KEW		2.5, 88
45	CH_3SCH_2CN gauche form	$v=0$	3.02 (3)	$\mu_a=2.37$ (2) $\mu_b=1.63$ (2) $\mu_c=0.93$ (3)	78KEW		2.5, 89
46	CH_3OSO_2F	$v=0$	3.66 (7)	$\mu_a=3.47$ (5) $\mu_b=1.17$ (6) $\mu_c=0.02$ (1)	77HAR1		2.5, 92
47	CH_2F_2	$v=0$ $v_9=1$	1.9785 (21) 2.0100 (14)		77KAW1	Laser Stark ν_9 (antisymmetric C—F stretching) = 32681172.3 (30) MHz	2.5, 93

Nr.	Molecule	v	μ D	Components of μ D	Ref.	Varia, remarks Fig. in Chap. 3, subvol. b	See Tab., Nr.
48	CD_2ClF	$v=0$	1.834 (10)	$\mu_a=0.302$ (3) $\mu_b=1.810$ (10)	78SUG		2.5, 97
	CHDClF	$v=0$	1.819 (10)	$\mu_a=0.279$ (2) $\mu_b=1.787$ (10) $\mu_c=0.198$ (13)			
49	NH_2CH_2CN	$v=0$	2.640 (7)	$\mu_a=2.577$ (7) $\mu_b=0.5754$ (10)	73PIC2		2.5, 101
50	CH_2=SO	$v=0$	2.994 (5)	$\mu_a=2.952$ (5) $\mu_b=0.500$ (2)	76PEN2	Fig. 33 direction of μ: 25.50° relative to the S=O bond and 9.61° relative to the "a" axis of the normal species	2.5, 102
	CD_2=SO	$v=0$	2.995 (5)	$\mu_a=2.945$ (4) $\mu_b=0.550$ (18)			
50a	CH_3SOH	$v=0$	1.87 (7)	$\mu_a=1.08$ (2) $\mu_b=0.74$ (12) $\mu_c=1.34$ (6)	78PEN4		2.5, 102a
51	CH_3N=ND, trans	$v=0$	0.67 (10)	$\mu_a=0.45$ (2) $\mu_b=0.50$ (10)	73STE2		2.5, 105
52	cis-CH_3N=NCH_3	$v=0$	3.27 (10)		79STE		2.5, 106
53	CH_2NH	$v=0$		$\mu_a=1.325$ (20) $\mu_b=1.53$ (4)	73KIR2		2.5, 103
54	H_2CO	$v=0$	$\mu(1_{01})=2.33168$ (1) *) $\mu(1_{10})=2.33148$ (2) $\mu(2_{02})=2.33167$ (1) $\mu(3_{21})=2.33145$ (1)		77FAB	*) MBER **) Laser Stark	2.5, 108
		$v_2=1$	2.3469 (15) **)		81ALL		
		$v_2=2$	2.3605 (20) **)		75JOH		
		$v_3=1$	2.3250 (25) **)		81ALL		
		$v_4=1$	2.3086 (3) *)		77FAB		
		$v_5=1$	2.2844 (47) **)		81ALL		
		$v_6=1$	2.3285 (3) *)		77FAB		
	$H_2{}^{13}CO$	$v=0$	2.33191 (2) *)				
	HDCO	$v=0$	2.3383 (23) **)		77JO2		
Cont.		$v_3=1$	2.3563 (31) **)				

Nr.	Molecule	v	μ D	Components of μ D	Ref.	Varia, remarks Fig. in Chap. 3, subvol. b	See Tab., Nr.
54 Cont.	D_2CO	$v=0$ $v_2=1$ $v_4=1$ $v_6=1$	2.347122 (2) *) 2.3672 (15) **) 2.319 (10) **) 2.347 (4) **)		77FAB 81ALL 77COF1	*) MBER **) Laser Stark	
55	HCOCl	$v=0$	1.6 (2)	$\mu_a=0.3$ (2) $\mu_b=1.6$ (2)	76TAK		2.5, 109
	DCOCl	$v=0$	1.809 (15)	$\mu_a=0.322$ (3) $\mu_b=1.780$ (15)	78SUG		
56	$HC(=O)OH$, cis	$v=0$	3.787 (10)	$\mu_a=2.647$ (11) $\mu_b=2.708$ (9)	76HO1	Fig. 35	2.5, 111
57	$HC(=O)SH$, trans	$v=0$	1.536 (1)	$\mu_a=1.366$ (1) $\mu_b=0.702$ (1)	76HO5	Fig. 36	2.5, 112
	HC(:O)SD, trans	$v=0$	1.542 (2)	$\mu_a=1.361$ (1) $\mu_b=0.726$ (4)			
	DC(:O)SH, trans	$v=0$	1.536 (2)	$\mu_a=1.353$ (1) $\mu_b=0.727$ (3)			
	HC(:O)SH, cis	$v=0$	2.868 (1)	$\mu_a=1.805$ (1) $\mu_b=2.228$ (1)			
	HC(:O)SD, cis	$v=0$	2.870 (3)	$\mu_a=1.862$ (3) $\mu_b=2.185$ (3)			
	DC(:O)SH, cis	$v=0$	2.873 (2)	$\mu_a=1.768$ (2) $\mu_b=2.264$ (2)			
58	$HC(=O)SCH_3$	$v=0$	1.58 (5) *)	$\mu_a=1.52$ (3) $\mu_b=0.43$ (14) $\mu_c=0$ (assumed)	76JON1	*) from capacitance measurements $\mu=1.6$ (1) D Fig. 37	2.5, 114
59	$HC(=O)NH(CH_3)$	$v=0$	3.86 (2)		73ELZ2		2.5, 115
60	$HC(=O)N(CH_3)_2$	$v=0$	3.85 (2)		73ELZ1		2.5, 116

Nr.	Molecule	v	μ D	Components of μ D	Ref.	Varia, remarks Fig. in Chap. 3, subvol. b	See Tab., Nr.
61	$HC(=O)OCH_2F$	$v=0$ $v_\tau=1$	2.24 (2) 2.16 (2)	$\mu_a=0.10$ (2) $\mu_b=2.09$ (1) $\mu_c=0.80$ (1) $\mu_a=0.13$ (2) $\mu_b=1.99$ (1) $\mu_c=0.83$ (1)	81LOP	 $v_\tau=96$ (6) cm^{-1}, O—CH_2F torsion	2.5, 120
62	O=C(H)—O—C(=O)—H	$v=0$	1.813 (3)	$\mu_a=1.704$ (2) $\mu_b=0.620$ (4) $\mu_c=0.0$ (0.044)	75VAC	Fig. 39	2.5, 123
63	$OC(NH_2)_2$	$v=0$	3.83 (4)	$\mu_b=3.83$ (4)	75BRW1		2.5, 129
63 a	$FC(=O)SCH_3$		2.91 (2)	$\mu_a=2.89$ (2) $\mu_b=0.30$ (8)	81CAM2		2.5, 131 a
64	CN—C(=O)OCH_2CH_3 extended form *) CN—C(=O)OCH_2CH_3 compact form **)	$v=0$ $v=0$		$\mu_a=4.44$ (7) $\mu_b\sim 0$ $\mu_a=4.25$ (7) $\mu_b=0$ $\mu_c=1.08$ (23)	78SU1	*) τ_1(O=COC)$=0°$, τ_2(COCC)$=180°$ **) τ_1(O=COC)$=0°$, τ_2(COCC)$\simeq 90°$.	2.5, 136
65	H_2CS	$v=0$ $v_3=1$ $v_4=1$ $v_6=1$	1.6491 (4) *) 1.6576 (33) **) 1.635 (11) **) 1.6420 (48) **)		77FAB 80BED	*) MBER **) Laser Stark	2.5, 145
66	F_2CS	$v=0$		$\mu_a=0.080$ (1)	73CAR		2.5, 146
67	$HC(=S)NH_2$	$v=0$	4.01 (3)	$\mu_a=3.99$ (2) $\mu_b=0.13$ (25) $\mu_c=0$ (assumed)	74SUG		2.5, 149
68 Cont.	$HC(=S)SH$, trans	$v=0$		$\mu_a=1.53$	78BAK1	Fig. 41	2.5, 150

Nr.	Molecule	v	μ D	Components of μ D	Ref.	Varia, remarks Fig. in Chap. 3, subvol. b	See Tab., Nr.
68 Cont.	$HC(=S)SH$, cis	$v=0$		$\mu_a=2.10$	78BAK1		
69	CH_3CH_2CN	$v=0$	4.05 (3) *)	$\mu_a=3.85$ (2) $\mu_b=1.23$ (2)	74HEI	Direction of μ: see Fig. 42 *) $\measuredangle(\vec{\mu}, a)=19.6\,(2)°$	2.5, 155
	CD_3CH_2CN	$v=0$	4.07 (2)	$\mu_a=3.84$ (1) $\mu_b=1.37$ (1)			
	CH_3CD_2CN	$v=0$	4.10 (3)	$\mu_a=3.92$ (2) $\mu_b=1.19$ (1)			
70	CH_3CH_2OH gauche	$v=0$		$\mu_a=1.264$ (10) $\mu_b=0.104$ (8) $\mu_c=1.101$ (16)	80KA2		2.5, 156
71	CH_3CH_2SH gauche	$v=0$	1.61 (5)	$\mu_a=1.49$ (2) $\mu_b=0.19$ (10) $\mu_c=0.59$ (2)	75SC	Fig. 43	2.5, 157
	CH_3CH_2SD gauche	$v=0$	1.62 (5)	$\mu_a=1.48$ (2) $\mu_b=0.1$ (1) $\mu_c=0.64$ (3)			
	CH_2DCH_2SH symmetric, gauche	$v=0$	1.69 (7)	$\mu_a=1.49$ (3) $\mu_b=0.29$ (10) $\mu_c=0.74$ (10)			
	CH_2DCH_2SH asymmetric, gauche	$v=0$	1.68 (7)	$\mu_a=1.49$ (3) $\mu_b=0.28$ (10) $\mu_c=0.72$ (10)			
	CH_3CH_2SH trans	$v=0$	1.58 (4)	$\mu_a=1.06$ (3) $\mu_b=1.17$ (3)			
		$v_{\tau_1}=1$	1.55 (4)	$\mu_a=1.14$ (3) $\mu_b=1.05$ (3)		S—H torsion	
	CH_3CH_2SD trans	$v=0$	1.58 (6)	$\mu_a=1.03$ (4) $\mu_b=1.12$ (5)			
	CH_2DCH_2SH symmetric, trans	$v=0$	1.58 (5)	$\mu_a=1.06$ (3) $\mu_b=1.16$ (5)			
	CH_2DCH_2SH asymmetric, trans	$v=0$	1.58 (5)	$\mu_a=1.03$ (2) $\mu_b=1.19$ (5)			

Nr.	Molecule	v	μ D	Components of μ D	Ref.	Varia, remarks Fig. in Chap. 3, subvol. b	See Tab., Nr.
73	$CH_3CH_2{}^{80}SeH$ gauche	$v=0$	1.465 (12)	$\mu_a=1.417$ (5) $\mu_c=0.373$ (29)	77DUR3		2.5, 158
	$CH_3CH_2{}^{80}SeH$ trans	$v=0$	1.485 (2)	$\mu_a=1.217$ (2) $\mu_b=0.850$ (1)			
74	$CH_3CH_2OCH_3$ trans	$v=0$	1.174 (22)	$\mu_a=0.146$ (10) $\mu_b=1.165$ (20)	75HAY	$\sphericalangle(\vec{\mu}, b)=7°\,5'\,(32')$ Fig. 44	2.5, 159
	$CD_3CH_2OCH_3$ trans	$v=0$	1.187 (27)	$\mu_a=0.149$ (15) $\mu_b=1.178$ (23)			
	$CH_3CH_2OCD_3$ trans	$v=0$	1.207 (27)	$\mu_a=0.167$ (15) $\mu_b=1.195$ (22)			
75	$CH_3CH_2SCH_3$ trans	$v=0$	1.560 (22)	$\mu_a=0.050$ (52) $\mu_b=1.559$ (22)	81HAY	$\sphericalangle(\vec{\mu}, b)=1°\,49'$	2.5, 160
	$CH_3CH_2SCD_3$ trans	$v=0$	1.550 (24)	$\mu_a=0.042$ (61) $\mu_b=1.549$ (24)			
76	CH_3CH_2NCS *)	$v=0$	3.33 (3)	$\mu_a=3.33$ (3) $\mu_b=0.07$ (2)	76SAK	*) cis form Fig. 45	2.5, 161
		$v_\tau=1$	3.38 (3)	$\mu_a=3.38$ (3) $\mu_b=0.05$ (2)		v_τ=C—N torsion	
		$v_\delta=1$	3.50 (5)	$\mu_a=3.50$ (5) $\mu_b=0.04$ (2)		v_δ=C—N—C bend	
77	CH_3CH_2NCO*)	$v=0$		$\mu_a=2.81$ (2) $\mu_b=0.03$	76SA	*) cis form Fig. 46 v_τ=C—N torsion	2.5, 162
		$v_\tau=1$		$\mu_a=2.99$ (4) $\mu_b=0.13$			
		$v_\tau=2$		$\mu_a=3.01$ (4) $\mu_b=0.11$			
		$v_\tau=3$		$\mu_a=3.02$ (2) $\mu_b=0.03$			
		$v_\tau=4$		$\mu_a=3.05$ (5) $\mu_b=0.262$			

Nr.	Molecule	v	μ D	Components of μ D	Ref.	Varia, remarks Fig. in Chap. 3, subvol. b	See Tab., Nr.
78	CH_3CH_2ONO cis-trans*)	$v=0$	2.306 (3)	$\mu_a=2.024$ (2) $\mu_b=1.105$ (3)	79TUR3	Fig. 47 *) cis-trans: τ_1(ON—OC)=0°, τ_2(CC—ON)=180° cis-gauche: $\tau_1=0°$, $\tau_2=80°$ trans-gauche: $\tau_1=180°$, $\tau_2=90°$	2.5, 163
	CH_3CH_2ONO cis-gauche*)	$v=0$	2.221 (6)	$\mu_a=2.029$ (3) $\mu_b=0.49$ (2) $\mu_c=0.76$ (1)			
	CH_3CH_2ONO trans-gauche*)	$v=0$	2.50 (2)	$\mu_a=2.37$ (1) $\mu_b=0.16$ (14) $\mu_c=0.78$ (3)			
79	$CH_3CH_2ONO_2$ trans	$v=0$	3.39 (2)	$\mu_a=3.32$ (1) $\mu_b=0.739$ (29)	74SCR		2.5, 164
	$CH_3CH_2ONO_2$ gauche	$v=0$	3.23 (4)	$\mu_a=3.17$ (2) $\mu_b=0.45$ (16) $\mu_c=0.34$ (22)			
80	$CH_3CH_2NHCH_3$ *)	$v=0$	0.88 (2)	$\mu_a=0.00$ (3) $\mu_b=0.25$ (3) $\mu_c=0.84$ (1)	73PEN1	*) trans form	2.5, 165
81	CH_3CH_2OCl	$v=0$	1.97 (1)	$\mu_a=1.623$ (10) $\mu_b=1.097$ (5)	78SU2		2.5, 166
82	$CH_3CH_2OCH_2F$	$v=0$	1.806 (12)	$\mu_a=1.073$ (6) $\mu_b=0.490$ (35) $\mu_c=1.367$ (7)	80HAY1	∢(μ, CF)=136° 50′ ∢(μ, FCH_2—O)=107° 40′ Fig. 48	2.5, 167
	$CH_3CH_2OCD_2F$	$v=0$	1.818 (24)	$\mu_a=1.043$ (10) $\mu_b=0.570$ (58) $\mu_c=1.376$ (18)			
	$CH_3CD_2OCH_2F$	$v=0$	1.805 (12)	$\mu_a=1.074$ (7) $\mu_b=0.443$ (39) $\mu_c=1.382$ (8)			
83	CH_3CHFCl	$v=0$	2.068 (14)	$\mu_a=0.953$ (8) $\mu_b=1.836$ (15) $\mu_c=0.013$ (24)	78SUG		2.5, 170

Nr.	Molecule	v	μ D	Components of μ D	Ref.	Varia, remarks Fig. in Chap. 3, subvol. b	See Tab., Nr.
84	$CH_3CH{=}NH$, cis	$v=0$	2.058	$\mu_a=0.834\ (23)$ *) $\mu_b=1.882\ (5)$	80LOV	*) uncertainties are 3σ	2.5, 171
	$CH_3CH{=}NH$, trans	$v=0$	2.56	$\mu_a=2.408\ (14)$ *) $\mu_b=0.88\ (16)$			
85	syn-$CH_3CH{=}SO$	$v=0$		$\mu_a=2.7$ $\mu_b=1.5$	78PEN3		2.5, 173
86	CH_2ODCH_2OD	$v=0$	2.41 (15)	$\mu_a=2.166\ (37)$ $\mu_b=0.74\ (18)$ $\mu_c=0.75\ (21)$	80WAL		2.5, 174
	CH_2ODCH_2OH	$v=0$	2.28 (12)	$\mu_a=2.00\ (4)$ $\mu_b=1.00\ (17)$ $\mu_c=0.45\ (34)$	81CAM4	the H participates in the hydrogen bond	
87	CH_2OHCH_2SH	$v=0$	1.871 (19) *)	$\mu_a=0.619\ (15)$ *) $\mu_b=1.478\ (12)$ $\mu_c=0.965\ (9)$	77SUN	*) uncertainties are 3σ Fig. 50	2.5, 175
88	$CH_2NH_2CH_2NH_2$ gauche I	$v=0$	1.770 (33)	$\mu_a=1.059\ (7)$ $\mu_b=0.787\ (32)$ $\mu_c=1.179\ (23)$	78MAR	Fig. 51	2.5, 176
	$CH_2NH_2CH_2NH_2$ gauche II	$v=0$	2.203 (6)	$\mu_a=1.952\ (2)$ $\mu_b=0.867\ (6)$ $\mu_c=0.538\ (6)$			
89	$CH_2NH_2CH_2F$ conformer I *)	$v=0$	1.826 (12)	$\mu_a=1.148\ (3)$ $\mu_b=1.380\ (8)$ $\mu_c=0.332\ (21)$	80MAR1	*) see Fig. 52	2.5, 177
	$CH_2NH_2CH_2F$ conformer II *)	$v=0$	1.671 (10)	$\mu_a<0.03$ $\mu_b=1.337\ (8)$ $\mu_c=1.002\ (5)$			
90	$CH_2NH(CH_3)CH_2OH$ GT form	$v=0$	2.756 (13)	$\mu_a=2.272\ (8)$ $\mu_b=1.336\ (20)$ $\mu_c=0.808\ (18)$	75PEN2		2.5, 179
	$CH_2NH(CH_3)CH_2OH$ GG form	$v=0$	2.864 (11)	$\mu_a=2.610\ (11)$ $\mu_b=1.132\ (11)$ $\mu_c=0.331\ (34)$			

Nr.	Molecule	v	μ D	Components of μ D	Ref.	Varia, remarks Fig. in Chap. 3, subvol. b	See Tab., Nr.
91	$CH_2N(CH_3)_2CH_2OH$	$v=0$	2.56 (2)	$\mu_a=2.27$ (2) $\mu_b=0.3$ (1) $\mu_c=1.16$ (5)	76PEN3	Fig. 54	2.5, 180
92	$CH_3OCH_2CH_2NH_2$	$v=0$	2.21 (2)	$\mu_a=1.63$ (1) $\mu_b=0.77$ (5) $\mu_c=1.28$ (2)	80CAM1	Fig. 55	2.5, 182
93	$CH_3OCH_2CH_2CN$	$v=0$	3.47 (20)	$\mu_a=3.43$ (10) $\mu_b=0.56$ (85)	76LO1	Fig. 56	2.5, 183
94	CHF_2CH_2OH	$v=0$	0.690 (20)	$\mu_a=0.608$ (7) $\mu_b=0.282$ (12) $\mu_c=0.163$ (40)	80MAR2		2.5, 184
95	$CH_3C(=O)H$	$v=0$	2.750 (6)	$\mu_a=2.537$ (4) $\mu_b=1.062$ (5)	78TUR, 78BOS	for orientation of μ, see Fig. 57 [78TUR], [78BOS]	2.5, 185
	$CH_2DC(=O)H$ symmetric	$v=0$	2.72 (2)	$\mu_a=2.46$ (2) $\mu_b=1.156$ (4)	78TUR		
96	$CH_3C(=O)NCO$	$v=0$	1.76 (4)	$\mu_a=0.95$ (2) $\mu_b=1.48$ (3)	80LAD		2.5, 187
97	$CH_3C(=O)OOH$	$v=0$	2.384 (5)	$\mu_a=2.294$ (3) $\mu_b=0.650$ (4)	76CUG	Fig. 59	2.5, 189
98	$CH_3C(=O)SH$	$v=0$	1.821 (13)	$\mu_a=0.191$ (10) $\mu_b=1.811$ (13)	77NAI	Fig. 60	2.5, 190
99	$CH_2{}^{35}ClC(=O)OH$	$v=0$	2.685 (6)	$\mu_a=1.658$ (8) $\mu_b=2.112$ (66)	74VAN2		2.5, 200
100	$CH_2FC(=O)NH_2$	$v=0$	2.10 (2)	$\mu_a=1.27$ (1) $\mu_b=1.67$ (2)	74MAR2	Fig. 64	2.5, 205

Nr.	Molecule	v	μ D	Components of μ D	Ref.	Varia, remarks Fig. in Chap. 3, subvol. b	See Tab., Nr.
101	$NH_2CH_2C(=O)OH$ conformer I **)	$v=0$		$\mu_a=1.0\pm0.15$ *) $\mu_b<\mu_a$	80SUE	*) rate of growth technique	2.5, 206
	$NH_2CH_2C(=O)OH$ conformer II **)	$v=0$		$\mu_a>3$	78SU3	**) see Fig. 65	
102	$HOCH_2C(=O)H$	$v=0$	2.34 (1)	$\mu_a=0.262$ (2) $\mu_b=2.33$ (1)	73MAS2	Fig. 66	2.5, 207
102a	$CH_2OHC(=O)OH$	$v=0$	2.156 (9)	$\mu_a=1.913$ (5) $\mu_b=0.995$ (14)	81BLO		2.5, 207a
103	$CH_3OCH_2C(=O)OH$	$v=0$	4.72 (4)	$\mu_a=4.72$ (4) $\mu_b=0.15$ (2)	73MAS3	Fig. 68	2.5, 209
104	$CHF_2C(=O)F$ trans	$v=0$		$\mu_a=0.85$ (1) $\mu_b=0$ $\mu_c=1.19$ (4)	77VAN1	Fig. 69	2.5, 211
	$CHF_2C(=O)F$ gauche	$v=0$		$\mu_a=0.55$ (3) $\mu_b=0.91$ (2) $\mu_c=1.76$ (3)			
105	$HC(=O)-C(=O)OH$ trans 1	$v=0$		$\mu_a=1.85$ (3) $\mu_b=0.20$ (10) $\mu_c=1.86$ (4)	73MAS1	Fig. 70	2.5, 213
106	$CH_3C(=O)-C(=O)OH$, trans	$v=0$	2.30 (3)	$\mu_a=2.27$ (2) $\mu_b=0.35$ (2)	74MAR1	Fig. 71	2.5, 214
107	$CH_3C(=O)-C(=O)H$, trans	$v=0$	0.9751 (7)	$\mu_a=0.1597$ (11) $\mu_b=0.9620$ (7)	78DYL	Fig. 72	2.5, 215

Nr.	Molecule	v	μ D	Components of μ D	Ref.	Varia, remarks Fig. in Chap. 3, subvol. b	See Tab., Nr.
108	$CH_3C(=NOD)H$, syn	$v=0$	0.938 (9)	$\mu_a=0.865$ (4) $\mu_b=0.363$ (15)	76OH		2.5, 216
	$CH_3C(=NOD)H$, anti	$v=0$	0.828 (13)	$\mu_a=0.753$ (4) $\mu_b=0.34$ (3)	77OH1		
109	$CH_3C(=S)H$	$v=0$	2.33 (2)	$\mu_a=2.26$ (2) $\mu_b=0.56$ (1)	76KRO		2.5, 217
110	$CH_3CH_2CH_2NC$ trans	$v=0$	4.16 (2)	$\mu_a=4.01$ (2) $\mu_b=1.13$ (1) $\mu_c=0.078$ (80)	75FUL		2.5, 224
	$CH_3CH_2CH_2NC$ gauche	$v=0$	4.10 (9)	$\mu_a=3.25$ (4) $\mu_b=2.40$ (8) $\mu_c=0.68$ (41)			
110a	$CH_3CH_2CH_2OH$ gauche-gauche	$v=0$	1.55 (3)	$\mu_a=0.58$ (4) $\mu_b=1.18$ (2) $\mu_c=0.81$ (3)	81AB		2.5, 225
111	$CH_3CH_2CH_2OCH_3$ trans-trans	$v=0$	1.107 (13)	$\mu_a=0.082$ (10) $\mu_b=1.104$ (13)	80KAT	Fig. 74	2.5, 226
	$CH_3CH_2CH_2OCD_3$ trans-trans	$v=0$	1.113 (15)	$\mu_a=0.052$ (8) $\mu_b=1.112$ (15)			
	$CD_3CH_2CH_2OCH_3$ trans-trans	$v=0$	1.116 (18)	$\mu_a=0.109$ (23) $\mu_b=1.110$ (18)			
112	$CH_3CH_2CH_2SH$, gauche	$v=0$	1.683 (10)	$\mu_a=1.575$ (7) $\mu_b=0.115$ (82) $\mu_c=0.580$ (9)	79NAK, 81NA		2.5, 227
	$CH_3CH_2CH_2SD$, gauche	$v=0$	1.677 (11)	$\mu_a=1.566$ (11) $\mu_b=0.067$ (40) $\mu_c=0.594$ (13)			
	$CH_3CH_2CH_2SH$, trans	$v=0$	1.598 (54)	$\mu_a=1.179$ (18) $\mu_b=1.078$ (75) $\mu_c=0$	81NA		

Nr.	Molecule	v	μ D	Components of μ D	Ref.	Varia, remarks Fig. in Chap. 3, subvol. b	See Tab., Nr.
113	$H_2NCH_2CH_2CH_2OH$	$v=0$	3.13 (4) *)	$\mu_a=2.88$ (3) *) $\mu_b=1.23$ (4) $\mu_c=0.06$ (1)	79MC	*) uncertainties are 3σ Fig. 76	2.5, 231
114	$H_3C-CH(OH)-CH_2(OH)$, T	$v=0$	2.568 (15)	$\mu_a=2.524$ (15) $\mu_b=0.0$ (1) $\mu_c=0.471$ (24)	81CAM	CH_3 trans with respect to the C(1)—O(1) bond	2.5, 232
	$H_3C-CH(OH)-CH_2(OH)$, G	$v=0$	2.320 (2)	$\mu_a=2.247$ (2) $\mu_b=0.258$ (20) $\mu_c=0.518$ (1)		CH_3 gauche with respect to the C(1)—O(1) bond	
115	$CH_3CH(NH_2)CH_2OH$, I *)	$v=0$	3.13 (5)	$\mu_a=3.03$ (1) $\mu_b=0.54$ (3) $\mu_c=0.57$ (19)	78ELL	*) CH_3 anti to C—O bond	2.5, 233
	$CH_3CH(NH_2)CH_2OH$, II **)	$v=0$	2.88 (7)	$\mu_a=2.19$ (3) $\mu_b=1.46$ (8) $\mu_c=1.172$ (3)		**) CH_3 gauche to C—O bond	
116	$CH_3CH(OH)CH_2F$	$v=0$	1.608 (30)	$\mu_a=0.510$ (9) $\mu_b=1.496$ (26) $\mu_c=0.298$ (14)	77MAR	Fig. 78	2.5, 235
117	CH_3CHFCH_3	$v=0$	1.96 (3)	$\mu_a=0$ (assumed) $\mu_b=1.88$ (2) $\mu_c=0.54$ (2)	73GRI		2.5, 236
118	$(CH_3)_2CHCN$	$v=0$	4.29 (10)	$\mu_a=4.05$ (2) $\mu_c=1.4$ (2)	74DUR8		2.5, 238
119	$(CH_3)_2CHNH_2$	$v=0$	1.19 (3)	$\mu_b=0.10$ (4) $\mu_c=1.19$ (3)	77MEH	trans form	2.5, 239
120	$(CH_3)_2CHOH$ gauche	$v=0$	1.564 (26)	$\mu_a=1.114$ (15) $\mu_b=0.737$ (25) $\mu_c=0.8129$ (49)	79HIR3		2.5, 240
121 Cont.	$(CH_3)_2CHSH$ trans	$v=0$	1.612 (20)	$\mu_a=1.593$ (4) $\mu_c=0.239$ (9)	75GRI		2.5, 242

Nr.	Molecule	v	μ D	Components of μ D	Ref.	Varia, remarks Fig. in Chap. 3, subvol. b	See Tab., Nr.
121 Cont.	$(CH_3)_2CHSH$ gauche	$v=0$	1.532 (20)	$\mu_a=1.376$ (3) $\mu_b=0.490$ (5) $\mu_c=0.463$ (9)	75GRI		
121 a	$CH_3CF_2CH_3$	$v=0$	2.40 (2)	$\mu_b=2.40$ (2)	81DUR5		2.5, 252 a
121 b	$CH_3C(=O)CH_2OH$	$v=0$	3.10 (6)	$\mu_a=2.22$ (4) $\mu_b=2.17$ (2)	80KAT1		2.5, 242 a
122	$(CH_2F)_2C{=}O$	$v=0$	2.54 (3)	$\mu_a=2.38$ (3) $\mu_b=0.89$ (3)	75FIN	Fig. 80	2.5, 254
123	HOCH=CHC(=O)H	$v=0$ $v=1$		$\mu_b=2.59$ (2) $\mu_b=2.58$ (1)	78BAU, 81BAU	Fig. 81	2.5, 256
	DOCH=CDC(=O)H	$v=0$ $v=1$		$\mu_b=2.58$ (2) $\mu_b=2.57$ (3)			
	HOCH=CDC(=O)H	$v=0$		$\mu_b=2.58$ (1)			
	HOCH=CHC(=O)H mono—^{18}O	$v=0$		$\mu_a=0.076$ (1) $\mu_b=2.570$ (3)			
	DOCH=CDC(=O)H mono—^{18}O	$v=1$		$\mu_a=0.23$ (6) $\mu_b=2.59$ (1)			
	HOCD=CHC(=O)H	$v=0$		$\mu_a=0.371$ (1) $\mu_b=2.561$ (4)			
		$v=1$		$\mu_a=0.140$ (7) $\mu_b=2.586$ (13)			
	DOCD=CDC(=O)H	state a		$\mu_a=0.443$ (2) $\mu_b=2.565$ (3)			
	DOCD=CDC(=O)H	state b		$\mu_a=0.192$ (1) $\mu_b=2.584$ (2)			

Nr.	Molecule	v	μ D	Components of μ D	Ref.	Varia, remarks Fig. in Chap. 3, subvol. b	See Tab., Nr.
124	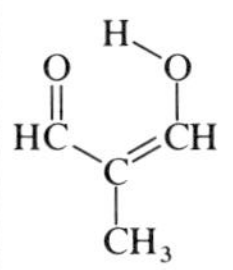	$v=0$		$\mu_a=2.9$ (1)	81SAN, 79SAN		2.5, 257
125	$CH_3CH_2C(=O)OH$ cis	$v=0$	1.55 (3)	$\mu_a=0.19$ (2) $\mu_b=1.54$ (2)	75ST1	for orientation of μ, see Fig. 82	2.5, 260
	$CH_3CD_2C(=O)OH$ cis	$v=0$	1.56 (3)	$\mu_a=0.17$ (2) $\mu_b=1.55$ (2)			
125a	$CH_3CH_2C(=O)SH$ cis	$v=0$	1.61 (20)	$\mu_a=0.40$ (2) $\mu_b=1.56$ (19)	81TAK1		2.5, 260a
125b	$CH_2{=}CD_2$	$v=0$	0.0091 (4)	$\mu_a=0.0091$ (4)	81HIR2		2.5, 280a
126	$CH_2{=}CHNH_2$	$v=0$ $v_i=1$		$\mu_a=1.083$ (8) $\mu_b=0.37$ $\mu_a=0.953$ (20)	75LOV		2.5, 287
127	$CH_2{=}CHN_3$ cis	$v=0$	1.376 (14)	$\mu_a=1.337$ (14) $\mu_b=0.326$ (3)	77FOR2	for orientation of μ, see Fig. 85	2.5, 288
	$CH_2{=}CHN_3$ trans	$v=0$	1.465 (17)	$\mu_a=1.156$ (14) $\mu_b=0.900$ (10)			
128	$CH_2{=}CHNCO$ trans	$v=0$	2.12 (3)	$\mu_a=2.00$ (2) $\mu_b=0.70$ (2)	77BOU3	for orientation of μ, see Fig. 86	2.5, 289
	$CH_2{=}CHNCO$ cis	$v=0$	2.14 (2)	$\mu_a=2.14$ (2) $\mu_b=0.09$ (2)	77BOU3, 79BOU2		
129	$CH_2{=}CHOH$	$v=0$	1.016 (9) *)	$\mu_a=0.616$ (7) $\mu_b=0.807$ (6)	76SAI	*) uncertainties are 2.5σ Fig. 87	2.5, 290
130	$CH_2{=}CHOCH_2CH_3$	$v=0$	0.98 (3)	$\mu_a=0.51$ (2) $\mu_b=0.84$ (2)	79OW	Fig. 88	2.5, 292

Nr.	Molecule	v	μ D	Components of μ D	Ref.	Varia, remarks Fig. in Chap. 3, subvol. b	See Tab., Nr.
131	$CH_2{=}CHSH$, syn	$v=0$	0.896 (3) *)	$\mu_a=0.813$ (1) $\mu_b=0.376$ (4)	79TAN2	*) uncertainties are 2.5σ	2.5, 294
	$CH_2{=}CHSH$, anti	$v=0$	1.117 (14) *)	$\mu_a=0.425$ (10) $\mu_b=1.033$ (10)	79TAN3		
		$v_\tau=1$	1.097 (14) *)	$\mu_a=0.523$ (10) $\mu_b=0.964$ (10)			
		$v_\tau=2$	1.063 (18) *)	$\mu_a=0.561$ (10) $\mu_b=0.903$ (15)			
		$v_\tau=3$	1.061 (42) *)	$\mu_a=0.630$ (30) $\mu_b=0.854$ (30)			
132	$CH_2{=}CF_2$	$v=0$ $v_4=1$ $v_9=1$	1.368 (1) 1.382 (1) 1.372 (1)		78DUX	Laser-Stark results	2.5, 295
133	$H_2C{=}C{=}O$ $D_2C{=}C{=}O$	$v=0$ $v=0$	1.42215 *) 1.42460 (3) *)		77FAB	*) MBER, sign of μ from Zeeman studies: [74SAS]	2.5, 299
134	$CH_2{=}C{=}S$	$v=0$	1.01 (3)		80WIN2		2.5, 301
135	$CH_3CH{=}C{=}S$	$v_\tau=2$		$\mu_a=1.54$	80BAK3	v_τ, CH_3 torsion	2.5, 302
136	$CF_2{=}CHF$	$v=0$	1.30 (6)	$\mu_a=0.075$ (15) $\mu_b=1.30$ (6)	73BHA	Fig. 89	2.5, 304
137	$CH_3CH{=}CHNO_2$ trans	$v=0$	4.54 (8)	$\mu_a=4.52$ (5) $\mu_b=0.42$ (15)	74FOR1		2.5, 305
138	$CH_3CCl{=}CH_2$	$v=0$	1.647 (10)	$\mu_a=1.633$ (10) $\mu_b=0.211$ (14)	78SUG		2.5, 305 a
139	$CF_3CH{=}CH_2$	$v=0$	2.433 (20)	$\mu_a=2.331$ (14) $\mu_b=0.698$ (25)	74SAI	Fig. 90	2.5, 307
140	$CH_2{=}CHCH_2NH_2$ N cis, lone pair gauche	$v=0$		$\mu_a=0.804$ (3) *) $\mu_b=0.78$ (1) $\mu_c=0.78$ (1)	78BOT	*) uncertainties are 2σ	2.5, 311
	$CH_2{=}CHCH_2ND_2$ N cis lone pair trans	$v=0$		$\mu_a=0.921$ (5) *) $\mu_b=0.12$ (2) $\mu_c=0.928$ (6)	80FIS	for orientation of μ, see Fig. 92	

Nr.	Molecule	v	μ D	Components of μ D	Ref.	Varia, remarks Fig. in Chap. 3, subvol. b	See Tab., Nr.
141	$CH_2{=}CFCH_2F$ gauche	$v=0$	2.67 (1)	$\mu_a=0.950\ (10)$ $\mu_b=1.982\ (80)$ $\mu_c=1.135\ (50)$	76ENG		2.5, 316
142	$CH_2{=}CHCHF_2$ cis $CH_2{=}CHCHF_2$ gauche	$v=0$ $v=0$	2.47 (3) 2.12 (5)	$\mu_a=2.369\ (15)$ $\mu_c=0.70\ (3)$ $\mu_a=1.535\ (15)$ $\mu_b=0.53\ (4)$ $\mu_c=1.36\ (3)$	76BOT		2.5, 317
143	$H_2C{=}CHC(=S)H$	$v=0$	2.61 (2)	$\mu_a=2.58\ (2)$ $\mu_b=0.36\ (4)$	80GEO3		2.5, 319
144	$H_2C{=}CHC(=O)OH$ s-cis $H_2C{=}CHC(=O)OH$ s-trans	$v=0$ $v=0$	1.46 (10) 2.02 (5)	$\mu_a=0.56\ (10)$ $\mu_b=1.35\ (5)$ $\mu_a=1.70\ (4)$ $\mu_b=1.10\ (5)$	74BOL2	Fig. 93	2.5, 320
145	$H_2C{=}C(CH_3)C(=O)F$ s-trans	$v=0$	3.23 (4)	$\mu_a=2.95\ (3)$ $\mu_b=1.31\ (2)$	76LAT		2.5, 321
146	$CH_2{=}CHCH{=}NH$ anti $CH_2{=}CHCH{=}NH$ s-trans	$v=0$ $v=0$	2.01 (1) 2.51 (1)	$\mu_a=1.13\ (1)$ $\mu_b=1.66\ (1)$ $\mu_a=2.39\ (1)$ $\mu_b=0.77\ (2)$	78PEN1	Fig. 94	2.5, 322
146a	$CH_2{=}CHCH_2CH_2OH$	$v=0$	1.65 (5)	$\mu_a=1.23\ (2)$ $\mu_b=1.02\ (2)$ $\mu_c=0.44\ (7)$	81MAR		2.5, 323a
147	$CH_2{=}CHCH(CH_3)_2$ trans $CH_2{=}CHCH(CH_3)_2$ gauche	$v=0$ $v=0$	0.320 (10) 0.398 (4)	$\mu_a=0.312\ (3)$ $\mu_c=0.071\ (42)$ $\mu_a=0.367\ (4)$ $\mu_b\simeq 0$ $\mu_c=0.154\ (6)$	79CRE		2.5, 325

Nr.	Molecule	v	μ D	Components of μ D	Ref.	Varia, remarks Fig. in Chap. 3, subvol. b	See Tab., Nr.
148	$CH_3CH=CHCH_2CH_3$, cis	$v=0$		$\mu_a=0.002$ (2) $\mu_b=0.274$ (3) $\mu_c=0.070$ (12)	81VAN		2.5, 326
149	$CH_2=CH-CH=CHF$, trans	$v=0$	1.380 (17)	$\mu_a=1.193$ (17) $\mu_b=0.694$ (6)	80CED1	Fig. 95	2.5, 340
150	$CH_2=CH-CH=C=O$ trans	$v=0$	0.97	$\mu_a=0.872$ (1) $\mu_b=0.416$ (57)	79BR, 79BJ	Fig. 98	2.5, 343
151	$CH_2=C=CHF$	$v=0$	1.97 (10)	$\mu_a=1.44$ (3) $\mu_b=1.34$ (10)	75ELL		2.5, 345
152	$CH_2=C=CHCN$	$v=0$	4.28	$\mu_a=4.07$ $\mu_b=1.33$	73BOU	Fig. 99	2.5, 347
153	$CH_2=C=CF_2$	$v=0$	2.07 (3)	$\mu_a=2.07$ (3)	74DUR10		2.5, 350
154	$CHF=C=CHF$	$v=0$	1.774 (10)	$\mu_b=1.774$ (10)	75ELL		2.5, 351
155	$CH_2=C=C=O$	$v=0$	2.297 (3)	$\mu_a=2.156$ (3) $\mu_b=0.7914$ (6) $\mu_c<0.05$	81BRO3		2.5, 353
155a	$CH_2=C=C=C=O$ $CHD=C=C=C=O$ $CD_2=C=C=C=O$	$v=0$ $v=0$	1.967 (19) 1.966 (17)	$\mu_a=1.967$ (19) $\mu_a=1.953$ (26) $\mu_a=1.966$ (17)	79BR1		2.5, 353a
156	$HC\equiv CCH(CH_3)_2$	$v=0$	0.721 (9)	$\mu_a=0.684$ (8) $\mu_c=0.227$ (4)	73MOC2		2.5, 355
157	$HC\equiv CCH_2NH_2$ trans	$v=0$		$\mu_a=0.586$ (2) $\mu_b=0.445$ (1)	77CER	Fig. 101	2.5, 356
158	$HC\equiv CCH_2SH$	$v=0$		$\mu_a=0.718$ (3) $\mu_b=0.495$ (10) $\mu_c=0.809$ (15)	75MIR		2.5, 358
159 Cont.	$HC\equiv CCH_2OCH_3$ gauche	$v=0$	1.170 (18)	$\mu_a=0.289$ (5) $\mu_b=0.502$ (28) $\mu_c=1.017$ (14)	80HAY4	direction of μ determined Fig. 102	2.5, 359

Nr.	Molecule	v	μ D	Components of μ D	Ref.	Varia, remarks Fig. in Chap. 3, subvol. b	See Tab., Nr.
159 Cont.	DC≡CCH$_2$OCH$_3$	$v=0$	1.174 (13)	$\mu_a=0.305$ (3) $\mu_b=0.505$ (21) $\mu_c=1.015$ (11)			
	HC≡CCH$_2$OCD$_3$	$v=0$	1.181 (11)	$\mu_a=0.247$ (3) $\mu_b=0.544$ (16) $\mu_c=1.019$ (10)			
159a	HC≡CC(=O)H	$v=0$	2.74	μ_a=2.35 $\mu_b=1.41$	80BRO3		2.5, 364
160	HC≡C—C(=O)F	$v=0$	2.98 (2)	$\mu_a=2.915$ (20) $\mu_b=0.630$ (80)	75DAV		2.5, 365
161	HC≡C—C(=O)NH$_2$	$v=0$		$\mu_a=1.08$ (2) $\mu_b=3.51$ (2)	78LIT	Fig. 103	2.5, 368
162	HC≡C—OCH$_2$CH$_3$ anti	$v=0$	1.91 (5)	$\mu_a=1.62$ (3) $\mu_b=1.01$ (3)	74BJ1	Fig. 104	2.5, 369
	HC≡C—OCH$_2$CH$_3$ gauche	$v=0$	2.02 (7)	$\mu_a=1.72$ (4) $\mu_b=0.64$ (7) $\mu_c=0.85$ (2)			
163	HC≡C—SCH$_2$CH$_3$ gauche	$v=0$	1.65 (3)	$\mu_a=1.15$ (1) $\mu_b=1.09$ (2) $\mu_c=0.46$ (4)	74BJ2		2.5, 370
164	HC≡CCH$_2$CH$_2$OH gauche	$v=0$	1.38 (6)	$\mu_a=0.91$ (3) $\mu_b=0.85$ (4) $\mu_c=0.60$ (2)	75SZA	Fig. 105	2.5, 372
165	HC≡CCH$_2$C≡CH	$v=0$	0.516 (5)	$\mu_b=0.516$ (5)	81KUC		2.5, 377
166	$(CH_3)_2S$ $(CD_3)_2S$	$v=0$ $v=0$	1.554 (4) 1.551 (10)	$\mu_b=1.554$ (4) $\mu_b=1.551$ (10)	81HAY 81DEM1		2.5, 435
167	$(CH_3)_2$NCN	$v=0$		$\mu_a=4.7$ $\mu_c=0.8$	73LI2	Fig. 127	2.5, 438

Nr.	Molecule	v	μ D	Components of μ D	Ref.	Varia, remarks Fig. in Chap. 3, subvol. b	See Tab., Nr.
168	$(CH_3)_2PCN$	$v=0$	4.11 (8)	$\mu_a=3.83$ (5) $\mu_c=1.5$ (1)	74DUR4		2.5, 441
169	$(CH_3)_2CHPH_2$ gauche	$v=0$	1.23 (1)	$\mu_a=1.15$ (1) $\mu_b=0.43$ (1) $\mu_c=0.03$ (2)	78DUR		2.5, 442
170	$(CH_3)_2DP^{11}BH_3$	$v=0$	4.85 (6)	$\mu_a=4.77$ (2) $\mu_c=0.90$ (5)	74DUR5		2.5, 443
171	$(CH_3CH_2)_2O$	$v=0$	1.061 (18)		74HAY3	trans-trans isomer.	2.5, 445
172	$(CH_3)_3CSH$	$v=0$	1.66 (4)	$\mu_{\parallel}=1.51$ (3) *) $\mu_{\perp}=0.68$ (4)	74VAL	*) $\mu_{\parallel}$ and $\mu_{\perp}$ are the components of μ along the ϱ axis and the ξ axis, respectively, see Fig. 130	2.5, 446
173	$(CH_3)_3CNO$	$v=0$		$\mu_a=2.03$ $\mu_b=1.54$	78COX		2.5, 454
174	$(CH_3)_3CNO_2$	$v=0$	3.74 (4)		80LAN		2.5, 455
175	$(CH_3)_3CPH_2$	$v=0$	1.17 (2)	$\mu_a=1.06$ (2) $\mu_c=0.49$ (2)	78LI		2.5, 457
176	$^{28}SiH_2{}^{35}Cl_2$	$v=0$	1.13 (2)	$\mu_b=1.13$ (2)	76DAV2		2.5, 379
177	$CH_3CH_2CH_2SiH_3$ $CH_3CH_2CH_2SiD_3$ $CH_3CD_2CH_2SiH_3$	$v=0$ $v=0$ $v=0$	0.811 (8) 0.796 (10) 0.809 (6)	$\mu_a=0.783$ (7) $\mu_b=0.212$ (8) $\mu_a=0.766$ (8) $\mu_b=0.215$ (22) $\mu_a=0.776$ (6) $\mu_b=0.227$ (4)	80HAY2		2.5, 382
178	CH_3OSiH_3	$v=0$	1.15 (2)	$\mu_a=0.647$ (6) $\mu_b=0.95$ (2)	74LEC		2.5, 385
179	$(SiH_3)_2S$	$v=0$		$\mu_b=0.896$ (8)	78DOS2		2.5, 388
180 Cont.	CH_3SSiH_3	$v=0$	1.380 (8)	$\mu_a=0.330$ (9) $\mu_b=1.340$ (7)	80SH1		2.5, 389

Nr.	Molecule	v	μ D	Components of μ D	Ref.	Varia, remarks Fig. in Chap. 3, subvol. b	See Tab., Nr.
180 Cont.	CH_3SSiD_3	$v=0$	1.383 (12)	$\mu_a=0.310$ (16) $\mu_b=1.348$ (11)			
181	$SiH_3CH_2OCH_3$ trans $SiD_3CH_2OCH_3$ trans $SiH_3CH_2OCD_3$ trans $SiH_3CD_2OCH_3$ trans	$v=0$ $v=0$ $v=0$ $v=0$	1.521 (31) 1.499 (13) 1.526 (16) 1.517 (12)	$\mu_a=0.871$ (19) $\mu_b=1.247$ (34) $\mu_a=0.851$ (14) $\mu_b=1.234$ (12) $\mu_a=0.895$ (17) $\mu_b=1.236$ (16) $\mu_a=0.860$ (13) $\mu_b=1.249$ (12)	81SHI	for orientation of μ, see Fig. 109	2.5, 390
182	$(SiH_3)_2CH_2$ $(SiD_3)_2CH_2$	$v=0$ $v=0$	0.819 (2) 0.811 (2)		78SHI		2.5, 392
183	$(CH_3)_2SiH_2$ $(CH_3)_2SiD_2$	$v=0$ $v=0$	0.713 (1) 0.703 (2)		78SHI		2.5, 391
184	$(CH_3)_2SiHCN$	$v=0$	3.8 (1)	$\mu_a=3.65$ (10) $\mu_c=1.04$ (10)	75DUR1		2.5, 394
185	$CH_3CH_2SiH_2F$ gauche $CH_3CH_2SiD_2F$ gauche $CH_3CH_2SiH_2F$ trans $CH_3CH_2SiD_2F$ trans	$v=0$ $v=0$ $v=0$ $v=0$	1.711 (51) 1.710 (31) 1.787 (11) 1.783 (18)	$\mu_a=1.368$ (22) $\mu_b=1.025$ (55) $\mu_c=0.066$ (281) $\mu_a=1.332$ (12) $\mu_b=1.056$ (32) $\mu_c=0.182$ (146) $\mu_a=1.785$ (9) $\mu_b=0.070$ (94) $\mu_a=1.776$ (18) $\mu_b=0.150$ (45)	81HAY1		2.5, 395
186	$CH_3CH_2SiH_2CH_3$ trans $CH_3CH_2SiD_2CH_3$ trans	$v=0$ $v=0$	0.758 (5) 0.758 (5)	$\mu_a=0.036$ (30) $\mu_b=0.757$ (3) $\mu_a=0.024$ (30) $\mu_b=0.758$ (3)	72HAY		2.5, 397
187 Cont.	$CH_2{=}CHSiH_2CH_3$ cis	$v=0$	0.729 (2)	$\mu_a=0.222$ (1) $\mu_b=0.694$ (2)	81IMA		2.5, 398

Nr.	Molecule	v	μ D	Components of μ D	Ref.	Varia, remarks Fig. in Chap. 3, subvol. b	See Tab., Nr.
187 Cont.	CH_2=$CHSiH_2CD_3$ cis	$v=0$	0.731 (7)	$\mu_a=0.154$ (25) $\mu_b=0.715$ (4)	81IMA		
	CH_2=$CHSiD_2CH_3$ cis	$v=0$	0.718 (3)	$\mu_a=0.249$ (2) $\mu_b=0.674$ (2)			
	CH_2=$CHSiH_2CH_3$ skew	$v=0$	0.670 (8)	$\mu_a=0.008$ (16) $\mu_b=0.665$ (5) $\mu_c=0.060$ (47)			
	CH_2=$CHSiH_2CD_3$ skew	$v=0$	0.670 (13)	$\mu_a=0.057$ (40) $\mu_b=0.664$ (8) $\mu_c=0.062$ (62)			
	CH_2=$CHSiD_2CH_3$ skew	$v=0$	0.664 (21)	$\mu_a=0.015$ (23) $\mu_b=0.656$ (13) $\mu_c=0.104$ (77)			
188	CH_2=$CHCH_2SiH_3$	$v=0$	0.526 (22)	$\mu_a=0.186$ (1) $\mu_b=0.489$ (6) $\mu_c=0.006$ (144)	77HAY2		2.5, 399
	CH_2=$CHCH_2SiD_3$	$v=0$	0.512 (15)	$\mu_a=0.183$ (9) $\mu_b=0.477$ (5) $\mu_c=0.030$ (60)			
189	SiF_3BF_2	$v=0$	1.09 (5)		74OGA		2.5, 400
190	SiH_3PH_2	$v=0$	0.59 (5)	$\mu_a=0.236$ (5) $\mu_c=0.54$ (4)	75VAR		2.5, 401
	SiH_3PD_2	$v=0$	0.60 (1)	$\mu_a=0.22$ (1) $\mu_c=0.56$ (1)	76DUR2		
191	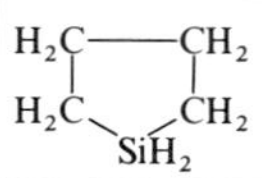	$v=0$	0.726 (6)	$\mu_a=726$ (5) $\mu_c<0.01$	76DUR7		2.5, 403
192	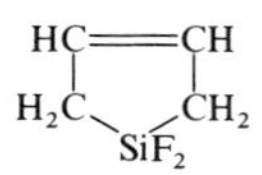	$v=0$		$\mu_a=2.02$ (6) $\mu_b=0$ $\mu_c=0.0$ (9)	74DUR1		2.5, 404
193	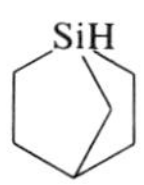	$v=0$	0.5 (1)	$\mu_b=0.5$ (1) $\mu_c=0.05$ (1)	77IK	Fig. 111	2.5, 405

Nr.	Molecule	v	μ D	Components of μ D	Ref.	Varia, remarks Fig. in Chap. 3, subvol. b	See Tab., Nr.
194	GeH_3CH_2F	$v=0$	1.64 (3)	$\mu_a=1.00$ (3) $\mu_b=1.30$ (3)	74KR2	Fig. 112	2.5, 406
195	$^{74}GeH_3CH_2CH_3$	$v=0$	0.76 (2)	$\mu_a=0.71$ (1) $\mu_b=0.27$ (1)	77DUR4		2.5, 408
196	$^{74}GeH_3CH{=}CH_2$	$v=0$	0.50 (3)	$\mu_a=0.49$ (2) $\mu_b=0.12$ (2)	74DUR7	Fig. 113	2.5, 409
197	CH_3GeH_2F	$v=0$	2.53 (15)	$\mu_a=2.32$ (10) $\mu_b=1.00$ (10)	76KRI	for orientation of μ, see Fig. 114	2.5, 410
	$CD_3{}^{74}GeH_2F$	$v=0$	2.59 (16)	$\mu_a=2.42$ (12) $\mu_b=0.93$ (14)	76ROB1		
198	CH_3BF_2	$v=0$	1.668 (3)	$\mu_a=1.668$ (3)	79LAG		2.5, 412
199	$CH_3CH_2BF_2$	$v=0$	1.69 (2)	$\mu_a=1.51$ (2) $\mu_b=0.75$ (2)	78OD1		2.5, 413
200	$CH_2{=}CHBF_2$	$v=0$	1.74 (4)	$\mu_a=1.69$ (4) $\mu_b=0.37$ (2)	74DUR3		2.5, 414
201	$H_2NB_2H_5$	$v=0$	2.67 (3)	$\mu_b=2.67$ (3)	74LAU	Fig. 115	2.5, 415
202	$(CH_3)_2NB_2H_5$	$v=0$	2.78 (2)	$\mu_a=2.78$ (2)	73COH	Fig. 116	2.5, 416
203	$CH_3B_2H_5$	$v=0$	0.566 (6) *)	$\mu_a=0.562$ (5) $\mu_b=0.06$ (3)	77PEN1	*) uncertainties are 2σ Fig. 117	2.5, 417
204	B_4H_{10}	$v=0$	0.486 (2)	$\mu_c=0.486$ (2)	81SIM	Fig. 118	2.5, 418
205	B_6H_{10}	$v=0$	2.50 (4)	$\mu_a=1.68$ (2) $\mu_c=1.85$ (4)	79SCH	Fig. 119	2.5, 419
206	CB_5H_7	$v=0$	1.43 (4)	$\mu_b=1.43$ (4) $\mu_c=0.06$ (4)	76MCK	Fig. 120	2.5, 420
207	1,7-$C_2B_6H_8$	$v=0$	1.40 (1)	$\mu_c=1.40$ (1)	76ROG	Fig. 121	2.5, 421
208	1, 6-$C_2B_7H_9$	$v=0$	2.14 (17)	$\mu_a=2.14$ (17)	76LAU	Fig. 122	2.5, 422

Nr.	Molecule	v	μ D	Components of μ D	Ref.	Varia, remarks Fig. in Chap. 3, subvol. b	See Tab., Nr.
209	2, 3, 4, 5-$C_4B_2H_6$	$v=0$	2.26 (6)	$\mu_a=2.24$ (2) $\mu_c=0.32$ (1)	74PAS	Fig. 123	2.5, 423
210	CH_3PF_2		2.056 (6)	$\mu_a=2.047$ (2) $\mu_c=0.195$ (62)	74CO2	for orientation of μ, see Fig. 124	2.5, 424
211	$CH_3CH_2PH_2$, gauche	$v=0$	1.22 (7)	$\mu_a=0.86$ (1) $\mu_b=0.73$ (7) $\mu_c=0.45$ (5)	76DUR5		2.5, 425
	$CH_3CH_2PH_2$, trans	$v=0$	1.226 (5)	$\mu_a=1.226$ (5)			
212	CH_3POF_2		3.69 (26)	$\mu_a=3.50$ (25) $\mu_b=1.15$ (7)	82DUR		2.5, 427
213	$CH_2{=}CHPF_2$	$v=0$		$\mu_a=2.42$ (2) $\mu_c=0.08$ (5)	78FO2, 81FON		2.5, 428
214	$CH_2{=}PH$			$\mu_a=0.7233$ (7) $\mu_b=0.4657$ (11)	81BRO1		2.5, 429
215	$CH_3PF_2BH_3$	$v=0$	3.94 (5)	$\mu_a=3.52$ (5) $\mu_b=1.76$ (5)	75CR2	for orientation of μ, see Fig. 126	2.5, 433
215a	$CH_2{=}CHC{\equiv}P$	$v=0$	1.183 (2)	$\mu_a=1.181$ (2) $\mu_b=0.074$ (1)	81OHN		2.5, 433a
215b	$CH_3CH_2PH_2{}^{11}BH_3$ trans	$v=0$	4.85 (2)	$\mu_a=4.66$ (1) $\mu_b=1.34$ (3)	81DUR6		2.5, 433b
216	H_2C——$CHNC$ \ / CH_2	$v=0$	4.03 (10)	$\mu_a=3.91$ (3) $\mu_c=0.95$ (30)	81PER		2.5, 459
217	H_2C——$CHOH$ \ / CH_2 gauche	$v=0$	1.46 (20)	$\mu_a=0.48$ (3) $\mu_b=1.12$ (15) $\mu_c=0.81$ (17)	78MCD1	Fig. 131	2.5, 461
218	H_2C——$CHNO_2$ \ / CH_2	$v=0$	3.95 (3)	$\mu_a=3.78$ (3) $\mu_b=1.14$ (1)	73MOC1		2.5, 462

Nr.	Molecule	v	μ D	Components of μ D	Ref.	Varia, remarks Fig. in Chap. 3, subvol. b	See Tab., Nr.
219	H_2C——$CHNO$ \ / CH_2	$v=0$	2.970	$\mu_a=2.846$ $\mu_c=0.852$	78COR		2.5, 463
220	H_2C——CF_2 \ / CH_2	$v=0$	2.32 (2)	$\mu_a=2.32\,(2)$	75PER		2.5, 465
221	H_2C——$C(CH_3)_2$ \ / CH_2	$v=0$	0.142 (1)	$\mu_a=0.142\,(1)$	79MTH		2.5, 469
222	H_2C——$CHOCH_3$, gauche \ / CH_2	$v=0$	1.20 (5)	$\mu_a=0.02\,(2)$ $\mu_b=1.01\,(4)$ $\mu_c=0.65\,(4)$	73PEN2	μ approximately bisects the (COC) angle. Fig. 132	2.5, 470
223	H_2C——$CHSCH_3$, gauche \ / CH_2	$v=0$	1.495 (8)	$\mu_a=0.1\,(1)$ $\mu_b=1.418\,(4)$ $\mu_c=0.457\,(1)$	75PEN1	Fig. 133	2.5, 471
224	H_2C——$CH-CH=CH_2$ \ / CH_2	$v=0$	0.498 (7)	$\mu_a=0.486\,(7)$ $\mu_c=0.110\,(10)$	74CO3	Fig. 134	2.5, 472
225	H_2C——$CHCH_2OH$ \ / CH_2	$v=0$	1.87 (2)	$\mu_a=1.81\,(1)$ $\mu_b=0.36\,(7)$ $\mu_c=0.31\,(4)$	78BRO		2.5, 475
	H_2C——$CHCH_2OD$ \ / CH_2	$v=0$	1.84 (2)	$\mu_a=1.78\,(2)$ $\mu_b=0.32\,(1)$ $\mu_c=0.30\,(2)$			
	H_2C——$CHCH_aD_bOH$ *) \ / CH_2	$v=0$	1.91 (3)	$\mu_a=1.816\,(5)$ $\mu_b=0.30\,(6)$ $\mu_c=0.51\,(12)$		*) see Fig. 136	
	H_2C——$CHCH_aD_bOD$ *) \ / CH_2	$v=0$	1.873 (1)	$\mu_a=1.810\,(1)$ $\mu_b=0.364\,(1)$ $\mu_c=0.317\,(3)$			
226	H_2C——$CHSiH_3$ \ / CH_2	$v=0$	0.890 (18)	$\mu_a=0.847\,(17)$ $\mu_c=0.273\,(10)$	79TYP	Fig. 137	2.5, 476
227	H_2C——$CH^{11}BF_2$ \ / CH_2	$v=0$	2.18 (6)	$\mu_a=2.17\,(5)$ $\mu_b=0.18\,(14)$	80ODO		2.5, 477

Nr.	Molecule	v	μ D	Components of μ D	Ref.	Varia, remarks Fig. in Chap. 3, subvol. b	See Tab., Nr.
228	H_2C—CH_2 (O ring)					sign of μ: + H_2C—H_2C >O − [76HAM1]	2.5, 478
229	H_2C—CHC(=O)H (O ring), trans	$v=0$	2.469 (31)	$\mu_a=1.932$ (5) $\mu_b=1.511$ (17) $\mu_c=0.277$ (156)	77CR2		2.5, 480
230	H_2C—CHCH=CH_2 (O ring)	$v=0$	1.875 (5)	$\mu_a=0.72$ (1) $\mu_b=1.688$ (3) $\mu_c=0.39$ (2)	75IKE	Fig. 138	2.5, 481
231	FHC—CHF (O ring), cis	$v=0$	2.89 (2)	$\mu_b=1.36$ (2) $\mu_c=2.55$ (1)	78GIL		2.5, 482
232	H_2C—$CHCH_2F$ (O ring), gauche III	$v=0$	3.08 (2)	$\mu_a=0.99$ (2) $\mu_b=2.90$ (1) $\mu_c=0.00$ (30)	77FU1	see Fig. 139	2.5, 483
233	H_2C—$CHCH_2OH$ (O ring)	$v=0$	1.44 (8)	$\mu_a=0.61$ (2) $\mu_b=1.20$ (9) $\mu_c=0.51$ (12)	75BRO		2.5, 486
234	H_2C—$CHCH_2CH_3$ (O ring)	$v=0$	1.891 (11)	$\mu_a=0.352$ (4) $\mu_b=1.806$ (3) $\mu_c=0.437$ (10)	78LJ		2.5, 487
235	O—O (CH_2 ring)	$v=0$	2.48 (7)	$\mu_b=2.48$ (7)	78SU4		2.5, 488
236	H_2C—C=CH_2 (S ring)	$v=0$	1.36 (1)	$\mu_a=0.68$ (1) $\mu_b=1.17$ (1)	77PEN2		2.5, 490
237	H_2C—CH_2 (N ring), N—CH_2CH_2OH	$v=0$	2.77 (5)	$\mu_a=2.34$ (3) $\mu_b=1.45$ (6) $\mu_c=0.3$ (3)	78PEN2		2.5, 493
238	H_2C—CHCN / H_2C—CH_2 (four-membered ring)	$v=0$	4.11 (3)	$\mu_a=4.00$ (3) $\mu_c=0.95$ (2)	73FON		2.5, 494

Nr.	Molecule	v	μ D	Components of μ D	Ref.	Varia, remarks Fig. in Chap. 3, subvol. b	See Tab., Nr.
239	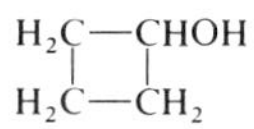	$v=0$	1.62 (5)	$\mu_a=1.39$ (5) $\mu_c=0.81$ (5)	78MCD2		2.5, 495
240	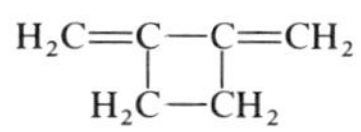	$v=0$	0.457 (2)	$\mu_b=0.457$ (2)	75AVI1		2.5, 497
241	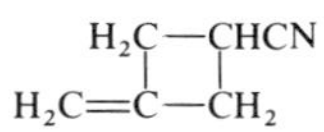	$v=0$	3.88 (3)	$\mu_a=3.66$ (3) $\mu_c=1.27$ (2)	74DUR9		2.5, 498
242	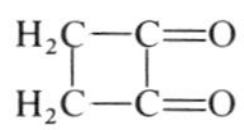	$v=0$	3.831 (5)	$\mu_b=3.831$ (5)	79LE1		2.5, 499
243	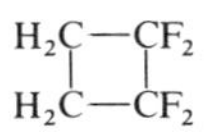	$v=0$	3.17 (3)	$\mu_b=3.17$ (3)	76DUR4		2.5, 500
244	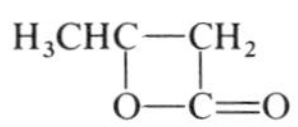	$v=0$	4.40 (6)	$\mu_a=3.81$ (6) $\mu_b^2+\mu_c^2=4.80$ (21)	76COF		2.5, 504
245	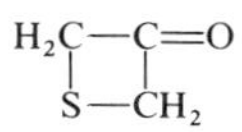	$v=0$ $v_R=1$ $v_R=2$ $v_R=3$ $v_R=4$ $v_R=5$		$\mu_a=0.999$ (3) $\mu_a=1.019$ (4) $\mu_a=1.041$ (2) $\mu_a=1.060$ (3) $\mu_a=1.080$ (5) $\mu_a=1.108$ (18)	75AVI2		2.5, 506
246	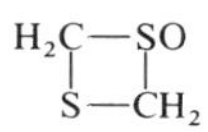	$v=0$	2.35	$\mu_a=2.34$ $\mu_b=0.15$	76PEN1		2.5, 508
247	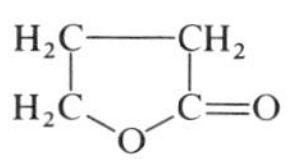	$v=0$	4.27 (3)	$\mu_a=4.04$ (3) $\mu_b=1.42$ (3) $\mu_c=0.33$ (2)	73DUR7		2.5, 513
248	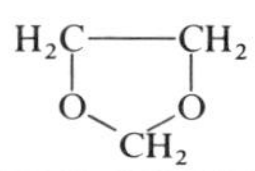	$v=0$	1.19 (3)	$\mu_b=1.19$ (3)	74BAR		2.5, 514
249	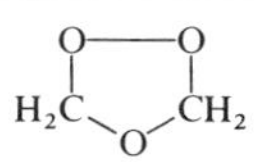	$v=0$	1.09 (1)	$\mu_b=1.09$ (1)	72GIL	Fig. 146	2.5, 515 a

Nr.	Molecule	v	μ D	Components of μ D	Ref.	Varia, remarks Fig. in Chap. 3, subvol. b	See Tab., Nr.
250	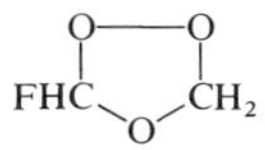	$v=0$	2.32 (2)	$\mu_a=1.46\ (1)$ $\mu_b=1.35\ (3)$ $\mu_c=1.20\ (1)$	74LA1		2.5, 516
251	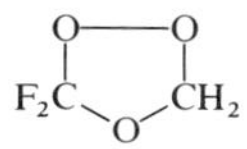	$v=0$	2.74 (3)	$\mu_a=2.20\ (1)$ $\mu_b=1.50\ (9)$ $\mu_c=0.65\ (18)$	79GIL		2.5, 517
252	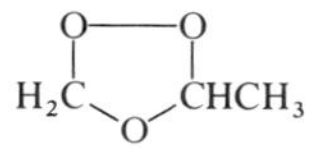	$v=0$	1.29 (2)	$\mu_a=0.74\ (1)$ $\mu_b=1.06\ (2)$ $\mu_c=0.10\ (8)$	74LA2	Fig. 147	2.5, 518
253	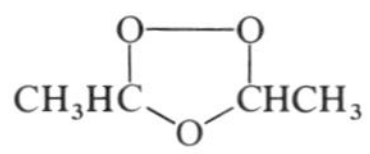	$v=0$	1.34 (3)	$\mu_b=1.34\ (3)$	74LA2		2.5, 519
254	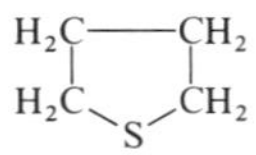	$v=0$		$\mu_b=2.01$	73MAM		2.5, 521
255	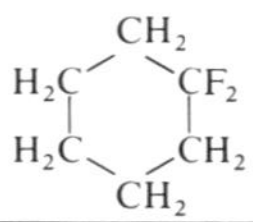	$v=0$	2.556 (10)	$\mu_a=2.415\ (7)$ $\mu_c=0.837\ (10)$	74DAM		2.5, 523
256	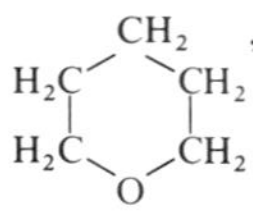 , chair	$v=0$	1.58 (3)	$\mu_a=1.39\ (2)$ $\mu_c=0.74\ (2)$	72RAO	erratum	2.5, 531
256a	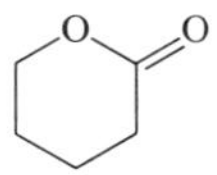 half-chair	$v=0$	4.71	$\mu_a=4.53\ (2)$ $\mu_b=1.30\ (3)$ $\mu_c<0.1$	81PHI		2.5, 534
257	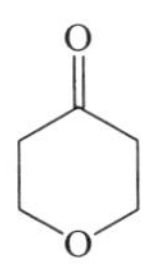 chair	$v=0$	1.720 (3)	$\mu_a=1.693\ (1)$ $\mu_c=0.300\ (13)$	80ALO2		2.5, 535

Nr.	Molecule	v	μ D	Components of μ D	Ref.	Varia, remarks Fig. in Chap. 3, subvol. b	See Tab., Nr.
258		$v=0$	1.781 (10)	$\mu_a=1.684$ (9) $\mu_c=0.578$ (2)	75KI1		2.5, 536
258a	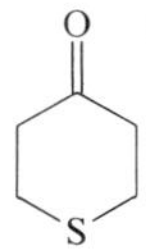 chair form	$v=0$	1.462 (39)	$\mu_a=1.409$ (2) $\mu_c=0.391$ (64)	81ALO2		2.5, 536a
258b	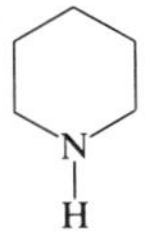 equatorial axial	$v=0$ $v=0$	0.820 (21) 1.189 (16)	$\mu_a=0.178$ (7) $\mu_c=0.800$ (20) $\mu_a=1.069$ (15) $\mu_c=0.521$ (7)	81PAR		2.5, 537
259	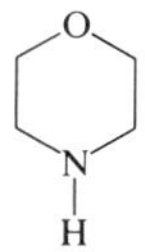	$v=0$	1.55 (2)	$\mu_a=1.53$ (1) $\mu_c=0.27$ (1)	72SLO	erratum	2.5, 539
260	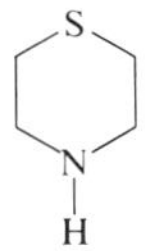	$v=0$	1.830 (16)	$\mu_a=1.766$ (6) $\mu_c=0.48$ (4)	80FAN		2.5, 540
261	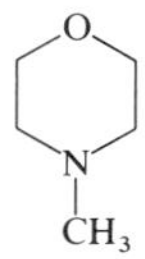	$v=0$	1.21 (2)	$\mu_a=1.21$ (2) $\mu_c=0.10$ (3)	74DAS		2.5, 541
262	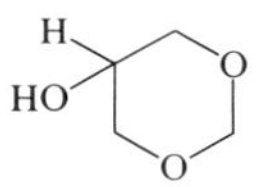	$v=0$	2.64 (3)	$\mu_a=1.61$ (1) $\mu_b\simeq 0$ $\mu_c=2.09$ (2)	80ALO1		2.5, 543
263	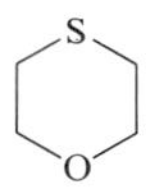	$v=0$	0.295 (3)	$\mu_a=0.290$ (3) $\mu_c=0.0537$ (4)	75KI2		2.5, 544

Nr.	Molecule	v	μ D	Components of μ D	Ref.	Varia, remarks Fig. in Chap. 3, subvol. b	See Tab., Nr.
264	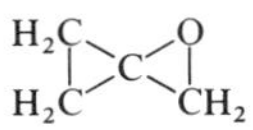	$v=0$	1.915 (9)	$\mu_a=0.840$ (12) $\mu_c=1.721$ (4)	75SLA	Fig. 149	2.5, 547
265	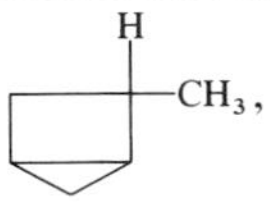, endo , exo	$v=0$ $v=0$	0.295 (10) 0.176 (1)	$\mu_a=0.050$ (11) *) $\mu_b=0.121$ (24) $\mu_c=0.264$ (1) $\mu_a=0.064$ (1) *) $\mu_b=0.038$ (3) $\mu_c=0.160$ (1)	75HA2	*) uncertainties are 3 σ Fig. 151	2.5, 549
266		$v=0$	2.139 (27)	$\mu_a=1.800$ (12) $\mu_c=1.155$ (24)	76MJO	Fig. 152	2.5, 550
267	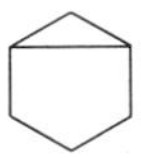	$v=0$	0.192 (3)	$\mu_a=0.093$ (1) $\mu_c=0.168$ (2)	74COO2	Fig. 153	2.5, 551
268	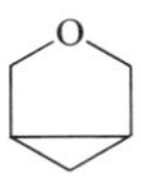	$v=0$	1.71 (3)	$\mu_a=1.36$ (2) $\mu_c=1.03$ (2)	74MAO		2.5, 552
269	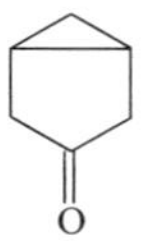	$v=0$	3.17 (7)	$\mu_a=2.93$ (7) $\mu_c=1.22$ (11)	78BEV	Fig. 154	2.5, 553
270	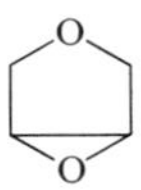	$v=0$	2.485 (40)	$\mu_a=0.276$ (10) $\mu_c=2.47$ (4)	73CRE	Fig. 156	2.5, 559
271	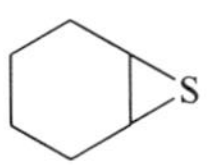	$v=0$	2.25 (36)	$\mu_a=1.92$ (40) $\mu_c=1.13$ (2)	73KEW	Fig. 157	2.5, 560
272	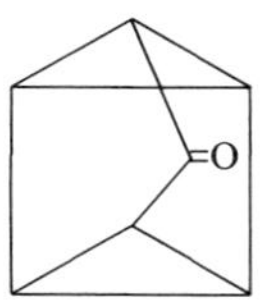	$v=0$	3.65 (2)	$\mu_a=3.65$ (2)	77COF2	Fig. 158	2.5, 561

Nr.	Molecule	v	μ D	Components of μ D	Ref.	Varia, remarks Fig. in Chap. 3, subvol. b	See Tab., Nr.
273	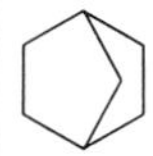	$v=0$	0.091 (8)	$\mu_c=0.091$ (8)	80CHP	Fig. 159	2.5, 562
273a	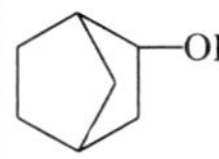 OH, exo	$v=0$	1.36 (9)	$\mu_a=0.53$ (9) $\mu_b=1.22$ (6) $\mu_c=0.294$ (4)	81TYB		2.5, 562a
274	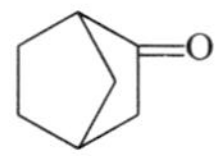	$v=0$	3.23 (2)	$\mu_a=3.14$ (2) $\mu_b=0.62$ (4) $\mu_c=0.42$ (4)	74COF		2.5, 563
275	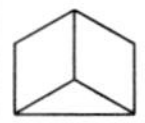	$v=0$	0.222 (10)	$\mu_b=0.090$ (10) $\mu_c=0.203$ (10)	75SUE	Fig. 160	2.5, 568
276	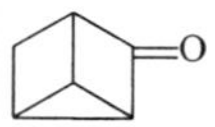	$v=0$	3.67 (2)	$\mu_a=3.61$ (1) $\mu_b=0.61$ (5) $\mu_c=0.28$ (17)	75COF		2.5, 569
277		$v=0$	1.621	$\mu_c=1.621$ (10)	75CR1	Fig. 161	2.5, 570
278	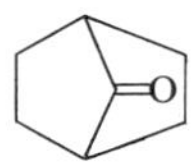	$v=0$	2.99 (3)	$\mu_a=2.99$ (3)	75DUR5	Fig. 162	2.5, 571
279		$v=0$	3.48 (5)	$\mu_a=2.76$ (1) $\mu_c=2.1$ (1)	76SUE	Fig. 163	2.5, 572
280	$CF_3COOH \cdot CH_3COOH$	$v=0$		$\mu_a=2.99$ (50)	75BEL		2.5, 573
281	$(CH_3)_3CCOOH \cdot CF_3COOH$	$v=0$		$\mu_a=2.28$ (50)	75BEL		2.5, 576
282	CH_2=CHCOOH · CF_3COOH	$v=0$		$\mu_a=1.30$ (50)	75BEL		2.5, 577
283	$(CH_3)_2C$=CHCOOH · CF_3COOH	$v=0$		$\mu_a=0.81$ (50)	75BEL		2.5, 578
284	H_2C/H_2C >CH—COOH · CF_3COOH	$v=0$		$\mu_a=2.53$ (50)	75BEL		2.5, 579

Nr.	Molecule	v	μ D	Components of μ D	Ref.	Varia, remarks Fig. in Chap. 3, subvol. b	See Tab., Nr.
284a	$CH_3CONH_2 \cdot CF_3COOH$	$v=0$		$\mu_a=3.09\,(50)$	75BEL		2.5, 582
285	$CH_3CH_2CONH_2 \cdot CF_3COOH$	$v=0$		$\mu_a=2.73\,(50)$	75BEL		2.5, 583
286	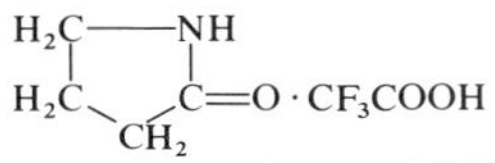	$v=0$		$\mu_a=1.75\,(50)$	75BEL		2.5, 584
287	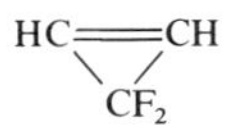	$v=0$	2.98 (2)	$\mu_a=2.98\,(2)$	76RAM		2.5, 586
288	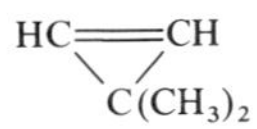	$v=0$	0.287 (3)	$\mu_a=0.287\,(3)$	78WIB		2.5, 587
289	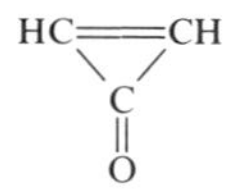	$v=0$	4.39 (6)	$\mu_a=4.39\,(6)$	73BEN1		2.5, 588
290	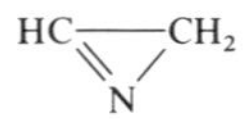	$v=0$	2.14 (5)	$\mu_a=0.54\,(2)$ $\mu_b=2.07\,(4)$	77FOR1		2.5, 589
291	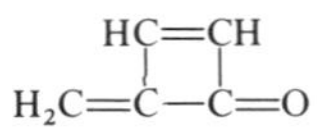	$v=0$	3.39 (5)	$\mu_a=2.04\,(2)$ $\mu_b=2.70\,(3)$	74CZ1	Fig. 164	2.5, 591
291a	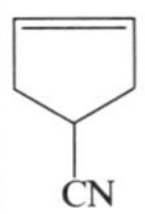 , equatorial axial	$v=0$ $v=0$	4.01 (5) 4.29 (22)	$\mu_a=3.85\,(2)$ $\mu_c=1.10\,(4)$ $\mu_a=3.48\,(3)$ $\mu_c=2.51\,(22)$	81CHO	Uncertainties are 2σ	2.5, 593a
292	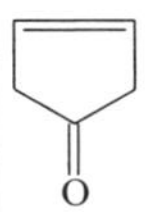	$v=0$	2.79 (3)	$\mu_a=2.79\,(3)$	73BEV1		2.5, 594
293	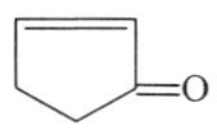	$v=0$		$\mu_a=3.64\,(1)$	79CHA		2.5, 595
294	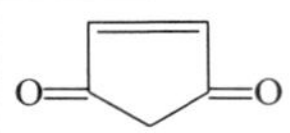	$v=0$	1.680 (9)	$\mu_b=1.680\,(9)$	78MJO		2.5, 596

Nr.	Molecule	v	μ D	Components of μ D	Ref.	Varia, remarks Fig. in Chap. 3, subvol. b	See Tab., Nr.
295	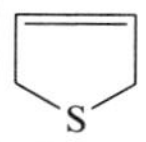	$v=0$	1.75 (1)	$\mu_a=1.75\,(1)$ $\mu_c=0.00\,(0)$	81DUR2		2.5, 597
295a	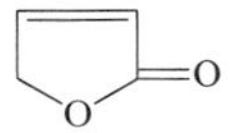	$v=0$	4.905 (11)	$\mu_a=4.726\,(10)$ $\mu_b=1.718\,(4)$ $\mu_c=0$	81ALO3		2.5, 597a
296	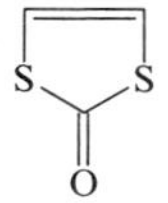	$v=0$	~0.3		80KUM		2.5, 600
297	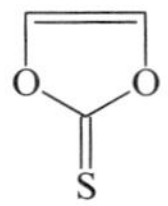	$v=0$	4.81 (6)	$\mu_a=4.81\,(6)$	80IQB		2.5, 601
298	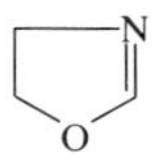	$v=0$	1.77 (1)	$\mu_a=1.14\,(1)$ $\mu_b=1.35\,(1)$ $\mu_c=0.00\,(0)$	74DUR12		2.5, 602
299		$v=0$	1.11 (2)	$\mu_a=1.10\,(2)$ $\mu_c=0.16\,(5)$	80DUR3		2.5, 604
300	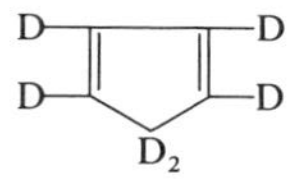	$v=0$	0.420 (3)	$\mu_a=0.420\,(3)$	76DAM		2.5, 605
301		$v=0$	4.25 (5)	$\mu_a=4.25\,(4)$ $\mu_b<0.3$	78FOR		2.5, 606
302	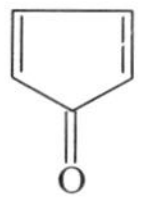	$v=0$	3.132 (7)	$\mu_a=3.132\,(7)$	73BAR		2.5, 607
303	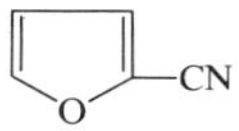	$v=0$		$\mu_a=4.51\,(2)$ $\mu_b=0.901\,(5)$	77WIE1, 76KOJ		2.5, 613

Nr.	Molecule	v	μ D	Components of μ D	Ref.	Varia, remarks Fig. in Chap. 3, subvol. b	See Tab., Nr.
304	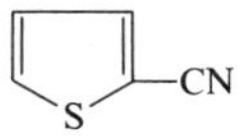	$v=0$	4.59 (2)	$\mu_a=4.51$ (2) $\mu_b=0.86$ (4)	75AVI4		2.5, 617
305	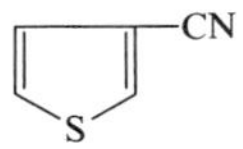	$v=0$	4.13 (2)	$\mu_a=4.09$ (2) $\mu_b=0.56$ (1)	77WIE2		2.5, 620
306	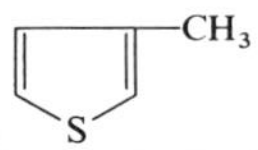	$v=0$	0.914 (15)	$\mu_a=0.888$ (4) $\mu_b=0.218$ (40)	73GUN		2.5, 619
307	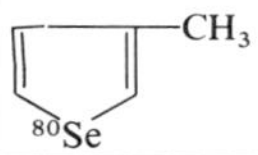	$v=0$	0.778 (5)	$\mu_a=0.760$ (3) $\mu_b=0.165$ (10)	73SHA	Fig. 166	2.5, 624
308	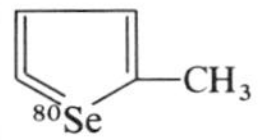	$v=0$	0.665 (4)	$\mu_a=0.665$ (2) $\mu_b=0.116$ (5)	78POZ2		2.5, 625
309	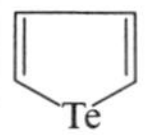	$v=0$	0.186	$\mu_a=0.186$	73BRO2		2.5, 626
310	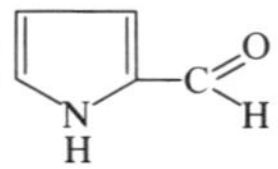	$v=0$	2.48 (2)	$\mu_a=2.47$ (2) $\mu_b=0.16$ (6)	74MAR3	Fig. 167	2.5, 628
311	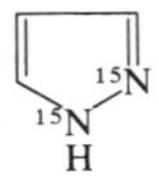	$v=0$	2.20 (1)	$\mu_a=1.320$ (15) $\mu_b=1.754$ (5)	74NAS	Fig. 169	2.5, 630
312	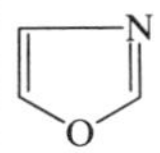	$v=0$	1.503 (28)	$\mu_a=1.365$ (20) $\mu_b=0.630$ (20)	78KUM2	for orientation of μ, see Fig. 171	2.5, 633
	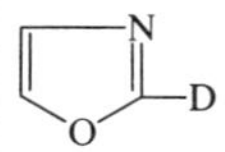	$v=0$	1.501 (64)	$\mu_a=0.755$ (20) $\mu_b=1.297$ (40)			
	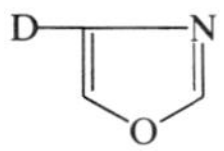	$v=0$	1.510 (53)	$\mu_a=0.532$ (50) $\mu_b=1.414$ (20)			
	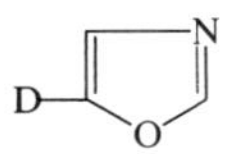	$v=0$	1.485 (63)	$\mu_a=1.459$ (20) $\mu_b=0.274$ (60)			

Nr.	Molecule	v	μ D	Components of μ D	Ref.	Varia, remarks Fig. in Chap. 3, subvol. b	See Tab., Nr.
313	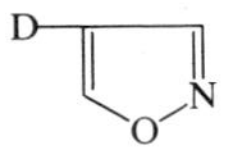	$v=0$		$\mu_a=2.93$ (4) $\mu_b=0.34$ (4)	78LOW	for orientation of μ, see Fig. 172	2.5, 634
314	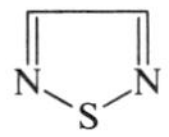	$v=0$	1.579 (7)	$\mu_a=1.579$ (7)	78STI2		2.5, 638
315	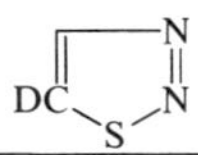	$v=0$	3.59	$\mu_a=1.04$ (10) $\mu_b=3.44$ (10)	76ST1	Fig. 177	2.5, 640
316	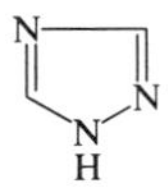	$v=0$	2.7 (1)	$\mu_a=0.81$ (9) $\mu_b=2.58$ (6)	75BOL	Fig. 179	2.5, 643
317	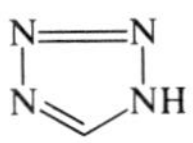	$v=0$	2.19 (5)	$\mu_a=1.91$ (5) $\mu_b=1.07$ (5)	74KRU	normal tetrazole and N—deuteriotetrazole are 2H—tetrazole	2.5, 644
	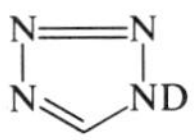	$v=0$	2.14 (5)	$\mu_a=1.82$ (5) $\mu_b=1.13$ (5)			
	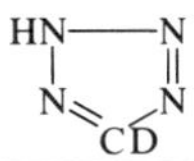	$v=0$	5.30 (5)	$\mu_a=3.30$ (5) $\mu_b=4.15$ (5)		C—deuteriotetrazole is 1H—tetrazole.	
318		$v=0$	1.283 (5)	$\mu_a=0.892$ (6) $\mu_b=0.507$ (10) $\mu_c=0.771$ (2)	74WE1		2.5, 645
319	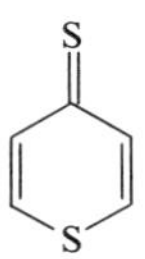	$v=0$	3.9	$\mu_a=3.9$ (2)	76COR		2.5, 648
320	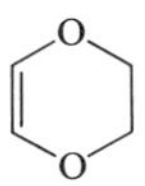	$v=0$	0.939 (8)	$\mu_b=0.939$ (8) $\mu_c=0.00$ (1)	74WE2		2.5, 649
321	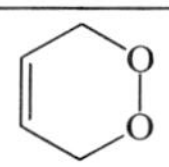	$v=0$	2.329 (1)	$\mu_b=2.329$ (1)	78KON		2.5, 650

Nr.	Molecule	v	μ D	Components of μ D	Ref.	Varia, remarks Fig. in Chap. 3, subvol. b	See Tab., Nr.
322	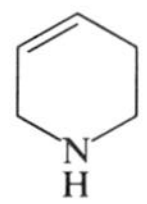 axial NH	$v=0$	1.007 (3)	$\mu_a=0.757$ (4)	76CH		2.5, 651
				$\mu_b=0.530$ (7)			
				$\mu_c=0.401$ (1)			
	equatorial NH	$v=0$	0.990 (3)	$\mu_a=0.293$ (2)			
				$\mu_b=0.428$ (3)			
				$\mu_c=0.843$ (3)			
323	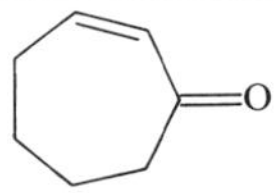	$v=0$	> 3.732	$\mu_a=3.626$ (23)	79KIT	*) no transition found with μ_b dependence	2.5, 656
				$\mu_b=$ *)			
				$\mu_c=0.882$ (40)			
324	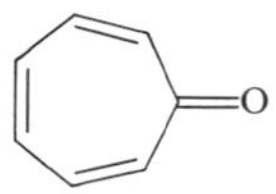	$v=0$	4.1 (3)	$\mu_a=4.1$ (3)	74CR3		2.5, 657
325		$v=0$	0.740	$\mu_b=0.734$ (3)	79AVI1		2.5, 658
				$\mu_c=0.0975$ (5)			
		$v_\alpha=1$		$\mu_b=0.732$ (23)			
				$\mu_c=0.083$ (2)			
		$v_\beta=1$		$\mu_b=0.675$ (18)			
				$\mu_c=0.076$ (1)			
326	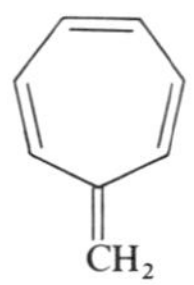	$v=0$	0.477 (5)	$\mu_a=0.477$ (5)	76BAU1		2.5, 659
327	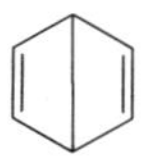	$v=0$	0.0437	$\mu_c=0.0437$	74GRI		2.5, 660
328		$v=0$		$\mu_a=0.327$ (5)	77CHI	Fig. 182	2.5, 663
				$\mu_b=0$			
				$\mu_c=0.00$ (1)			
330	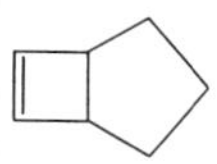	$v=0$	0.208 (1)	$\mu_a=0.201$ (1)	79AVI2		2.5, 666
				$\mu_c=0.052$ (1)			

Nr.	Molecule	v	μ D	Components of μ D	Ref.	Varia, remarks Fig. in Chap. 3, subvol. b	See Tab., Nr.
331	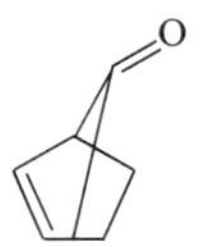	$v=0$	2.91 (4)	$\mu_a=2.88$ (3) $\mu_b=0.39$ (3) $\mu_c=0$	75DUR5		2.5, 667
332		$v=0$	0.299 (8)	$\mu_a=0.299$ (8)	76WAN		2.5, 668
333	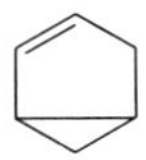	$v=0$	0.292 (12) *)	$\mu_a=0.166$ (9) *) $\mu_b=0.209$ (15) $\mu_c=0.119$ (1)	78MAT1	*) uncertainties are 3 σ.	2.5, 669
334	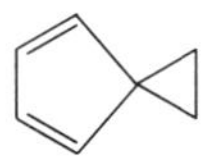	$v=0$	0.947 (1)	$\mu_a=0.947$ (1)	80STA		2.5, 670
335	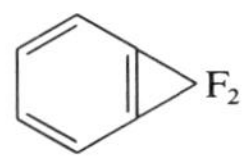	$v=0$ $v_x=1$	3.572 ± 0.02 3.54 (3)	$\mu_a=3.572 \pm 0.02$	75POZ		2.5, 671
336	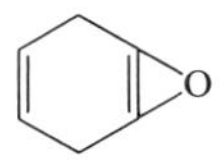	$v=0$	1.932 (5)	$\mu_a=0.775$ (2) $\mu_c=1.770$ (5)	78CHA		2.5, 672
337	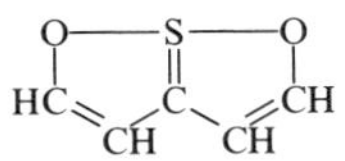	$v=0$	3.07 (4)	$\mu_b=3.07$ (4)	77PED		2.5, 673
338	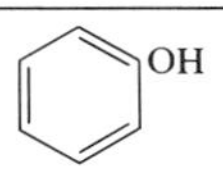	$v=0$	1.224	$\mu_a=-0.133$ (3) $\mu_b=$ 1.217 (8)	79LAR	orientation of μ determined	2.5, 677
	3-D	$v=0$		$\mu_a=0.1800$ (2) $\mu_b=1.220$ (7)			
	5-D	$v=0$		$\mu_a=0.0940$ (2) $\mu_b=1.210$ (6)			
	7-D	$v=0$		$\mu_a=0.1000$ (9) $\mu_b=1.202$ (5)			
	2-D	$v=0$		$\mu_a=0.105$ (2) $\mu_b=1.233$ (4)			
	6-D	$v=0$		$\mu_a=0.152$ (2) $\mu_b=1.204$ (4)			

Nr.	Molecule	v	μ D	Components of μ D	Ref.	Varia, remarks Fig. in Chap. 3, subvol. b	See Tab., Nr.
339	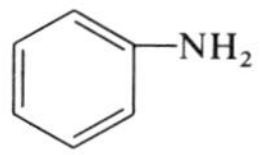	$v=0$		$\mu_a^+=1.129(1)$ *) $\mu_a^-=1.018(2)$	74LIS	*) for the 0^+ and 0^- states, respectively.	2.5, 679
	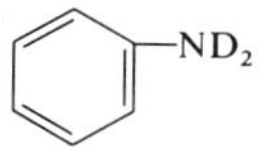	$v=0$	1.13 (1)	$\mu_a^2=1.27(2)$	73HAT		
340	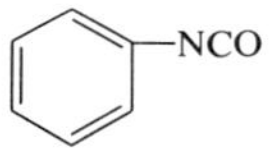	$v=0$	2.50 (3)	$\mu_a=2.50(2)$ $\mu_b\leqq 0.2$	77BOU4	Fig. 184	2.5, 681
	2-D trans	$v=0$	2.42 (9)	$\mu_a=2.41(9)$ $\mu_b=0.23(1)$	79MAS		
341	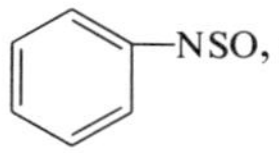 , cis	$v=0$	2.30 (6)	$\mu_a=2.20(6)$ $\mu_b=0.664(5)$	77CAM1		2.5, 683
342	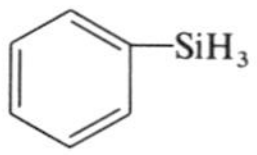	$v=0$	0.845 (12)	$\mu_a=0.845(12)$	75CA1		2.5, 689
343	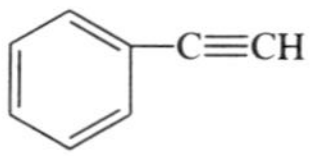	$v=0$	0.656 (5) *)	$\mu_a=0.656(5)$ *)	75COX	uncertainties are 2 σ	2.5, 690
	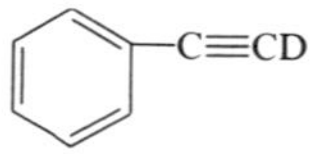	$v=0$	$\mu(H)-\mu(D)=0.010(3)$ *)				
344	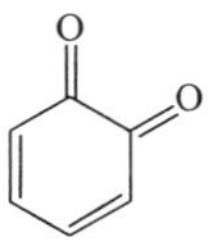	$v=0$	4.23 (2)	$\mu_a=4.23(2)$	75BL1		2.5, 701
345	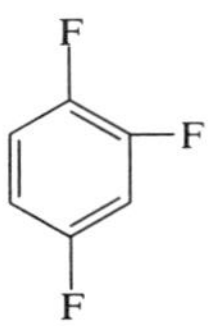	$v=0$	1.39 (3)	$\mu_a=0.95(2)$ $\mu_b=1.02(2)$	78DOR	Fig. 190	2.5, 724

Nr.	Molecule	v	μ D	Components of μ D	Ref.	Varia, remarks Fig. in Chap. 3, subvol. b	See Tab., Nr.
346	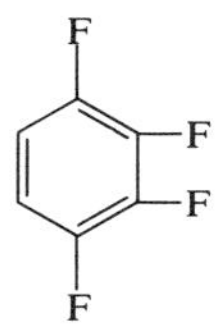	$v=0$	2.42 (5)	$\mu_b=2.42\,(5)$	74HAR3	Fig. 191	2.5, 725
347	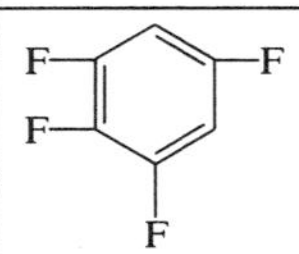	$v=0$	1.46 (6)	$\mu_a=1.46\,(6)$	77SHA		2.5, 726
348	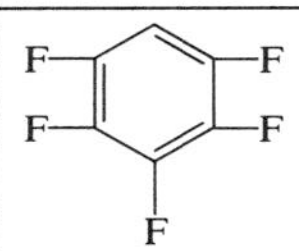	$v=0$	1.44 (5) *)	$\mu_b=1.44\,(5)$ *)	74DO2	*) uncertainties are 2 σ	2.5, 728
349		$v=0$	2.215 (10)	$\mu_a=2.215\,(10)$	74SOR, 76HAM1		2.5, 730
350	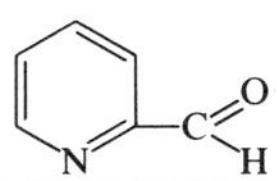	$v=0$	3.56 (7)	$\mu_a=3.48\,(5)$ $\mu_b=0.76\,(8)$	75KAW	Fig. 192	2.5, 736
351	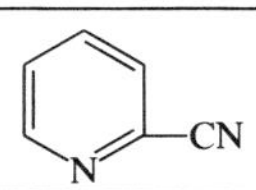	$v=0$	5.78 (11)	$\mu_a=5.47\,(10)$ $\mu_b=1.87\,(3)$	75FOR		2.5, 735
352	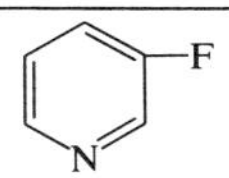	$v=0$	2.09 (26)	$\mu_a=0.39\,(10)$ $\mu_b=2.05\,(24)$	76SHA	Fig. 193	2.5, 737
353	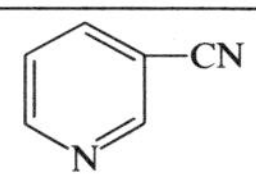	$v=0$	3.66 (11)	$\mu_a=3.13\,(8)$ $\mu_b=1.90\,(7)$	75FOR		2.5, 739
354	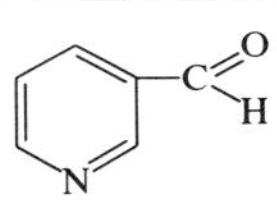	$v=0$		$\mu_a=1.35$ $\mu_b=0.5$	77HAE		2.5, 741

Nr.	Molecule	v	μ D	Components of μ D	Ref.	Varia, remarks Fig. in Chap. 3, subvol. b	See Tab., Nr.
355	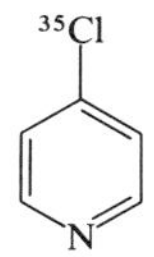	$v=0$	0.756 (5)	$\mu_a=0.756$ (5)	76CA3		2.5, 742
356	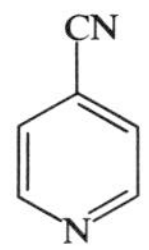	$v=0$	1.96 (3)	$\mu_a=1.96$ (3)	75FOR		2.5, 744
357	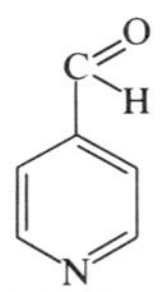	$v=0$	1.66	$\mu_a=0.37$ $\mu_b=1.62$	80GEO4	Fig. 194	2.5, 746
358	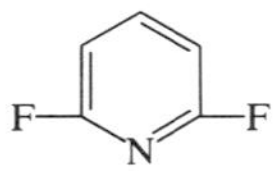	$v=0$	3.82 (6)	$\mu_b=3.82$ (6)	76ST2		2.5, 748
359	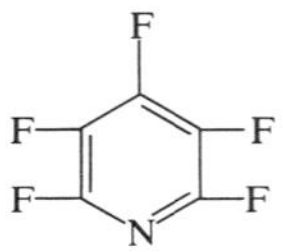	$v=0$	0.98 (8)	$\mu_b=0.98$ (8)	74DO1		2.5, 749
360	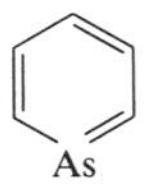	$v=0$	1.10 (4)	$\mu_a=1.10$ (4)	75LAT		2.5, 750
361	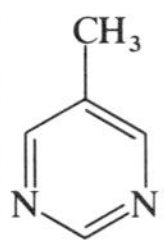	$v=0$	2.881 (6)	$\mu_a=2.881$ (6)	75CA2		2.5, 752
362	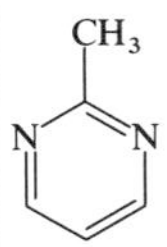	$v=0$	1.676 (10)	$\mu_a=1.676$ (10)	76CA4		2.5, 753

Nr.	Molecule	v	μ D	Components of μ D	Ref.	Varia, remarks Fig. in Chap. 3, subvol. b	See Tab., Nr.
363	Ar · CO_2	$v=0$	0.06793 (20)	$\mu_a=0.06793\ (20)$	79ST1		2.5, 755
364	Ar · OCS	$v=0$		$\mu_a=0.2146\ (10)$ $\mu_b=0.669\ (2)$	76HAR		2.5, 756
365	Ar · N_2O	$v=0$		$\mu_a=0.043$ $\mu_b=0.149$	79DIX		2.5, 757
366	Ar · H_2S Ar · D_2S	$v=0$ $v=0$		$\mu_a=0.186$ $(\mu_b^2+\mu_c^2)^{1/2}=0.8$ $\mu_a=0.17$	79VIS		2.5, 758
367	Ar · SO_2	$v=0$		$\mu_a=0.2674\ (1)$ $\mu_c=1.465\ (1)$	80DEL3		2.5, 759
368	Ar · O_3			$\mu_a=0.1010\ (1)$ $\mu_b\leqq 0.05$ $\mu_c=0.4643\ (6)$	80DEL1		2.5, 760
369	Ar · HC≡CH Ar · DC≡CD	$v=0$ $v=0$	0.02697 (3) 0.02697 (3)		80DEL2		2.5, 762
369 a	N_2O · HF N_2O · DF	$v=0$ $v=0$		$\mu_a=2.069\ (4)$ $\mu_b=0.69\ (1)$ $\mu_a=2.125\ (6)$ $\mu_b=0.63\ (2)$	81JOY		2.5, 763 a
369 b	 . HF			$\mu_a=3.85\ (2)$	81GEO		2.5, 764 a

2.7 Quadrupole coupling constants

2.7.1 Diatomic molecules

2.7.1.1 Preliminary remarks

1. Introduction

The electric hyperfine structure of rotational levels is observed if at least one of the nuclei in the molecule has a spin quantum number $I \geqq 1$. The energy splittings due to the nuclear quadrupole interaction are described by the molecular parameter eqQ, where e is the elementary charge, q the electric field gradient along the internuclear axis at the nucleus, and Q the nuclear quadrupole moment. If there are two nuclei with spin $I \geqq 1$ in the molecule the analysis of the hyperfine structure will yield two different coupling parameters $(eqQ)_i$, where i represents the nucleus under consideration.

More about the definition of the coupling parameters and the appropriate quantum numbers of the angular momenta can be found in reference [1]. Because the electric field gradient q is a function of the internuclear distance r, the observed coupling constant eqQ is expected to be dependent on the vibrational quantum number v and the rotational quantum number J. This is usually expressed by a power expansion:

$$eq_vQ = eq_eQ + eq_IQ(v+\tfrac{1}{2}) + eq_{II}Q(v+\tfrac{1}{2})^2 + \cdots + eq_JQJ(J+1) + \cdots . \quad (1)$$

The following table contains the information on the molecules in $^1\Sigma$-states and non $^1\Sigma$-states.

2. Explanation of the tables

Column	Parameters	Definitions
1	Nr. *Nr.*	serial number (as in Table 2.2) the following number represents the serial number of the molecule in 2.7.1 of Vol. II/6, Landolt-Börnstein, New Series
2	v J	vibrational quantum number of the state for which the value of eqQ was measured rotational quantum number of the state for which the value of eqQ was measured
3		quadrupole nucleus which produces hfs
4	eqQ $eq^{(a)}Q$ $eq^{(b)}Q$	nuclear quadrupole coupling constant for $^2\Pi$-states (compare part 4.1.5)
5	Ref.	references for columns 3 and 4; see 2.2.3, p. 34
6	Meth. MW MB EPR DR	method of measurement for the value in the preceding columns *) microwave spectroscopy molecular beam resonance experiment-electric or magnetic resonance electron paramagnetic resonance double resonance experiments (MODR, RF/DR)
7	eq_vQ eq_eQ eq_IQ $eq_{II}Q$ eq_JQ Q^m/Q^n	Varia, remarks compare eq. (1) (eq_eQ, eq_IQ, $eq_{II}Q$, eq_JQ) ratio of nuclear quadrupole moments of the nuclei m and n from eqQ-ratios
8		references to Tables 2.2.2, 2.6.1, 2.9.1, and in subvolume b, 4.1.

*) Results from other spectral regions are also included in the tables. The methods are defined by the abbreviations as described in 1.5.

3. Arrangement of substances

The molecules of this table are arranged according to increasing nuclear charge number. For each type of nucleus the ordering of the molecules in table 2.2 is retained. When the molecule consists of two different quadrupole nuclei, it is entered under both nuclei; however, the data is given only once.

4. References

[1] B. Schenk, E. Tiemann, J. Hoeft: Z. Naturforsch. **25a** (1970) 1827. A.R. Edmonds, Angular momentum in quantum mechanics Princeton Univ. Press, 1957.

References for 2.7.1.2 Data: in 2.2.3.

2.7.1.2 Data

Nr.	Molecule	J, v	Nuclear quadrupole coupling constants Nucleus	eqQ [MHz]	Ref.	Meth.	Varia, remarks	See Tab.
11 *6-14*	$^{16}O^{2}D$	$v=0$	^{2}D	$eq^{(a)}Q = 0.285\,(5)$	75MEE	MB	from fit using third order perturbation theory $X^{2}\Pi$	2.6.1
				$eq^{(b)}Q = -0.124\,(12)$	75MEE	MB		4.1.2
				$eq^{(a)}Q = 0.2850\,(11)$	78BEA	MB	from fit using direct matrix diagonalisation $X^{2}\Pi$	4.1.5
				$eq^{(b)}Q = -0.1189\,(42)$	78BEA	MB		
	$^{16}O^{2}D$	$v=0$	^{2}D	0.22 (3)	76GER	LC	$A^{2}\Sigma^{+}$	
		$v=1$	^{2}D	0.05 (4)	76GER	LC	$A^{2}\Sigma^{+}$	
12	$^{32}S^{2}D$	$v=0$	^{2}D	$eq^{(a)}Q = 0.149\,(4)$	75MEE	MB	$X^{2}\Pi_{3/2}$	2.6.1
				$eq^{(b)}Q = -0.050\,(24)$	75MEE	MB	$X^{2}\Pi_{3/2}$	4.1.5
17 *6-2a*	DBr		^{2}D	see ^{79}Br				
19 *6-5*	$^{6}Li^{19}F$	$J=1, 2, v=0$	^{6}Li	0.0085 (8)	76HEB	MB	same data as given in 2.7.1 of Vol. II/6 Landolt-Börnstein, New Series, from [H14]	2.2.2
		$J=1, 2, v=1$	^{6}Li	0.0086 (12)				2.6.1
		$J=1, 2, v=2$	^{6}Li	0.0071 (20)				2.9.1
	$^{7}Li^{19}F$	$J=1, 2, v=0$	^{7}Li	0.4156 (4)	76HEB	MB		
		$J=1, 2, v=1$	^{7}Li	0.4061 (6)	76HEB	MB		
		$J=1, 2, v=2$	^{7}Li	0.3965 (8)	76HEB	MB		
22 *6-8*	$^{7}Li^{127}I$	$J=1, v=0$	^{7}Li	0.171000 (30)	76JAC	MB	$\|ehH\|$ (^{127}I) < 0.075 kHz electric hexadecapole coupling for iodine	2.9.1
			^{127}I	−198.478627 (10)	76JAC	MB		
		$J=2, v=0$	^{7}Li	0.17108 (18)	76JAC	MB		
			^{127}I	−198.503955 (50)	76JAC	MB		
46 *6-10*	$^{12}C^{14}N$	$v=0$	^{14}N	−1.287 (41)	77DIX	MW	$X^{2}\Sigma^{+}$	2.6.1
		$v=1$	^{14}N	−1.27 (21)	77DIX	MW	$X^{2}\Sigma^{+}$	4.1.2
								4.1.5

Nr.	Molecule	J, v	Nuclear quadrupole coupling constants Nucleus	eqQ [MHz]	Ref.	Meth.	Varia, remarks	See Tab.
76 *6-11*	$^{14}N_2$	$v=0$	^{14}N	-2.518 (40)	73DeS	MB	$eq_eQ=-2.571$ (40) MHz; $eq_1Q=+0.0847$ (30) MHz; $A^3\Sigma_u^+$ (for $v=0$ to $v=12$)	4.1.3
		$v=1$	^{14}N	-2.446 (40)	73DeS	MB		
		$v=2$	^{14}N	-2.358 (40)	73DeS	MB		
		$v=3$	^{14}N	-2.290 (40)	73DeS	MB		
		$v=4$	^{14}N	-2.190 (40)	73DeS	MB		
		$v=5$	^{14}N	-2.107 (40)	73DeS	MB		
		$v=6$	^{14}N	-2.023 (40)	73DeS	MB		
		$v=7$	^{14}N	-1.945 (40)	73DeS	MB		
		$v=8$	^{14}N	-1.864 (40)	73DeS	MB		
		$v=9$	^{14}N	-1.771 (40)	73DeS	MB		
		$v=10$	^{14}N	-1.672 (40)	73DeS	MB		
		$v=11$	^{14}N	-1.610 (68)	73DeS	MB		
		$v=12$	^{14}N	-1.545 (68)	73DeS	MB		
80 *6-12*	$^{14}N^{16}O$	$v=0$	^{14}N	$eq^{(a)}Q=-1.857$ (1)	76MEE	MB	$X^2\Pi$	2.6.1
				$eq^{(b)}Q=$ 23.160 (24)	76MEE	MB		2.9.1
								4.1.2
								4.1.5
	$^{14}N^{16}O$	$v=0$	^{14}N	-2.88 (17)	74BER	RF/DR	$A^2\Sigma^+$	
82	$^{14}N^{19}F$	$v=0$	^{14}N	4.10 (23)	73CUR	EPR	excited state a $^1\Delta$	2.6.1 4.1.7
4 *6-17*	$^{23}Na_2$	$v=0$	^{23}Na	-0.4637 (9)	75ROS	MB		2.9.1
47 *6-22*	$^{27}Al^{19}F$	$v=0$	^{27}Al	-37.487 (43)	74HON3	MW		2.9.1
48	$^{27}Al^{35}Cl$	$v=0$	^{27}Al	-29.8 (10)	73HOE	MW		
		$v=0$	^{35}Cl	-8.6 (10)	73HOE	MW		
49	$^{27}Al^{79}Br$	$v=0$	^{79}Br	78.78 (20)	73HOE	MW		
		$v=0$	^{27}Al	-27.90 (50)	73HOE	MW		
88 *6-26*	$^{33}S^{16}O$	$v=0$	^{33}S	-15.96 (65)	82TIE	MW	reevaluation of data from [67AMA]	4.1.3 4.1.7

Nr.	Molecule	J, v	Nuclear quadrupole coupling constants Nucleus	eqQ [MHz]	Ref.	Meth.	Varia, remarks	See Tab.
16	$^{1}H^{35}Cl$	$J=1, v=0$	^{35}Cl	−67.61881 (15)	73DEL	MB		2.2.2
6–27								2.6.1
								2.9.1
27	$^{133}Cs^{35}Cl$	$v=0$	^{133}Cs	1.03090 (20)	77CED	MB	$eq_e Q(^{35}Cl)=1.83005$ (8) MHz,	2.2.2
6–57			^{35}Cl	1.76517 (6)	77CED	MB	$eq_I Q(^{35}Cl)=-130.18$ (9) kHz	2.9.1
		$v=1$	^{133}Cs	1.02288 (20)	77CED	MB	$eq_{II} Q(^{35}Cl)=0.87$ (2) kHz	
			^{35}Cl	1.63675 (6)	77CED	MB	$eq_e Q(^{133}Cs)=1.03497$ (13) MHz,	
		$v=2$	^{133}Cs	1.01521 (20)	77CED	MB	$eq_I Q(^{133}Cs)=-8.23$ (15) kHz	
			^{35}Cl	1.51002 (6)	77CED	MB	$eq_{II} Q(^{133}Cs)=0.12$ (4) kHz	
		$v=3$	^{133}Cs	1.00768 (20)	77CED	MB		
			^{35}Cl	1.38508 (6)	77CED	MB		
	$^{133}Cs^{37}Cl$	$v=0$	^{133}Cs	1.03115 (20)	77CED	MB		
			^{37}Cl	1.39230 (6)	77CED	MB		
31	$^{63}Cu^{35}Cl$	$v=0$	^{63}Cu	16.12 (20)	77TIE2	MW		2.2.2
		$v=0$	^{35}Cl	−32.28 (15)	77TIE2	MW		2.9.1
48	AlCl		^{35}Cl	see ^{27}Al				
84	$^{209}Bi^{35}Cl$	$v=0$	^{209}Bi	−1027 (12)	76KUI	MW	XO^+	4.1.3
			^{35}Cl	−34.5 (28)	76KUI	MW	XO^+	
	$^{209}Bi^{37}Cl$	$v=0$	^{209}Bi	−1025 (28)	76KUI	MW	XO^+	
			^{37}Cl	−27.1 (38)	76KUI	MW	XO^+	
92	$^{35}Cl^{16}O$	$v=0$	^{35}Cl	$eq^{(a)}Q=-87.95$ (19)	78KAK	MW	$X^2\Pi$	4.1.5
6–30				$eq^{(b)}Q=126.4$ (65)	78KAK	MW		
		$v=1$	^{35}Cl	$eq^{(a)}Q=-88.48$ (46)	78KAK	MW		
	$^{37}Cl^{16}O$	$v=0$	^{37}Cl	$eq^{(a)}Q=-69.50$ (34)	78KAK	MW	$X^2\Pi$	
				$eq^{(b)}Q^{der}=99.6$	78KAK	MW		
99	$^{79}Br^{35}Cl$	$v=0$	^{79}Br	875.32 (10)	76TIE5	MW		2.2.2
			^{35}Cl	−102.45 (15)	76TIE5	MW		2.6.1
		$v=1$	^{79}Br	875.77 (15)	76TIE5	MW		2.9.1
			^{35}Cl	−102.62 (20)	76TIE5	MW		
	$^{81}Br^{35}Cl$	$v=1$	^{81}Br	731.63 (10)	76TIE5	MW		
Cont.			^{35}Cl	−102.61 (15)	76TIE5	MW		

Nr.	Molecule	J, v	Nuclear quadrupole coupling constants Nucleus	eqQ [MHz]	Ref.	Meth.	Varia, remarks	See Tab.
99	$^{79}Br^{37}Cl$	$v=0$	^{79}Br	875.33 (15)	76TIE5	MW		
Cont.			^{37}Cl	−80.75 (15)	76TIE5	MW		
	$^{81}Br^{37}Cl$	$v=0$	^{81}Br	731.21 (15)	76TIE5	MW		
			^{37}Cl	−80.70 (15)	76TIE5	MW		
100	$^{127}I^{35}Cl$	$v=1$	^{127}I	−701 (1)	80KNÖ	MODR	$eq_e Q(^{127}I) = -597.6\,(35)$ MHz } $B^3\Pi_0^+$	2.2.2
6–54 a		$v=2$	^{127}I	−769.0 (4)	79KNÖ/80KNÖ	MODR	$eq_1 Q(^{127}I) = -68.5\,(14)$ MHz } $B^3\Pi_0^+$	2.6.1
			^{35}Cl	−27 (6)	79KNÖ/80KNÖ	MODR		4.1.6
	$^{127}I^{37}Cl$	$v=2$	^{127}I	−764.8 (4)	79KNÖ/80KNÖ	MODR		
23	$^{39}K^{127}I$	$v=0$	^{39}K	−4.12 (10)	73TIE1	MW	$eq_v Q(^{127}I) = [-85.32 - 2.93\,(v+\frac{1}{2}) \pm 0.12]$ MHz	2.6.1
			^{127}I	−86.79 (10)	73TIE1	MW		
		$v=1$	^{39}K	−4.06 (10)	73TIE1	MW		
			^{127}I	−89.80 (10)	73TIE1	MW		
		$v=2$	^{39}K	−3.98 (20)	73TIE1	MW		
			^{127}I	−92.62 (15)	73TIE1	MW		
		$v=3$	^{39}K	−4.13 (30)	73TIE1	MW		
			^{127}I	−95.59 (15)	73TIE1	MW		
30	$^{63}Cu^{19}F$	$v=0$	^{63}Cu	21.953 (66)	74HON1	MW		2.9.1
6–36								
31	CuCl		^{63}Cu	see ^{35}Cl				
32	$^{63}Cu^{79}Br$	$v=0$	^{79}Br	261.17 (20)	79HOE	MW		2.2.2
			^{63}Cu	12.76 (20)	79HOE	MW		
	$^{63}Cu^{81}Br$	$v=0$	^{81}Br	218.38 (20)	79HOE	MW		
			^{63}Cu	12.72 (20)	79HOE	MW		
33	$^{63}Cu^{127}I$	$v=0$	^{63}Cu	8.14 (20)	77NAI	MW		2.2.2
		$v=0$	^{127}I	−938.07 (20)	77NAI	MW		
50	$^{69}Ga^{19}F$	$v=0$	^{69}Ga	−106.485 (20)	74HON1	MW		2.2.2
6–37								2.9.1
51	$^{69}Ga^{79}Br$	$v=0$	^{79}Br	106.24 (20)	78PFA	MW	$eq_e Q(^{79}Br) = 105.78\,(34)$ MHz,	2.2.2
			^{69}Ga	−86.53 (23)	78PFA	MW	$eq_1 Q(^{79}Br) = 0.92\,(38)$ MHz	
		$v=1$	^{79}Br	107.16 (32)	78PFA	MW	$eq_e Q(^{69}Ga) = -86.68\,(30)$ MHz,	
Cont.			^{69}Ga	−86.07 (26)	78PFA	MW	$eq_1 Q(^{69}Ga) = 0.36\,(24)$ MHz	

Nr.	Molecule	J, v	Nuclear quadrupole coupling constants Nucleus	eqQ [MHz]	Ref.	Meth.	Varia, remarks	See Tab.
51	$^{69}Ga^{81}Br$	$v=0$	^{81}Br	88.85 (22)	78PFA	MW	$eq_eQ(^{81}Br)=88.55$ (35) MHz,	
Cont.			^{69}Ga	−86.46 (21)	78PFA	MW	$eq_1Q(^{81}Br)=0.60$ (32) MHz	
		$v=1$	^{81}Br	89.45 (23)	78PFA	MW		
			^{69}Ga	−86.21 (27)	78PFA	MW		
	$^{71}Ga^{79}Br$	$v=0$	^{79}Br	106.36 (23)	78PFA	MW		
			^{71}Ga	−54.55 (27)	78PFA	MW		
	$^{71}Ga^{81}Br$	$v=0$	^{81}Br	88.62 (22)	78PFA	MW		
			^{71}Ga	−54.61 (29)	78PFA	MW		
52	$^{69}Ga^{127}I$	$v=0$	^{69}Ga	−81.07 (8)	81NAI1	MW	$eq_eQ(^{69}Ga)=-81.29$ (25) MHz,	2.2.2
			^{127}I	−370.63 (9)	81NAI1	MW	$eq_1Q(^{69}Ga)=$ 0.43 (30) MHz	2.9.1
		$v=1$	^{69}Ga	−80.78 (12)	81NAI1	MW	$eq_eQ(^{127}I)=-369.35$ (10) MHz,	
			^{127}I	−373.26 (15)	81NAI1	MW	$eq_1Q(^{127}I)=-2.54$ (16) MHz	
		$v=2$	^{69}Ga	−80.05 (15)	81NAI1	MW		
			^{127}I	−375.91 (16)	81NAI1	MW		
		$v=3$	^{69}Ga	−79.77 (18)	81NAI1	MW		
			^{127}I	−378.26 (20)	81NAI1	MW		
	$^{71}Ga^{127}I$	$v=0$	^{71}Ga	−50.92 (12)	81NAI1	MW		
			^{127}I	−370.50 (15)	81NAI1	MW		
		$v=1$	^{71}Ga	−50.98 (30)	81NAI1	MW		
			^{127}I	−373.56 (40)	81NAI1	MW		
17	$^{1}H^{79}Br$	$J=1, v=0$	^{79}Br	532.30590 (24)	73DAB	MB		2.2.2
6-41	$^{1}H^{81}Br$	$J=1, v=0$	^{81}Br	444.68066 (15)	73DAB	MB		2.6.1
	$^{2}D^{79}Br$	$v=0$	^{79}Br	530.6315 (21)	74VAN	MW		2.9.1
			^{2}D	0.1469 (19)	74VAN	MW		
	$^{2}D^{81}Br$	$v=0$	^{81}Br	443.2799 (21)	74VAN	MW		
			^{2}D	0.1461 (19)	74VAN	MW		
24	$^{85}Rb^{79}Br$	$v=0$	^{85}Rb	−47.11 (17)	77TIE1	MW	$eq_eQ(^{85}Rb)=-47.20$ (25) MHz,	2.2.2
			^{79}Br	3.50 (29)	77TIE1	MW	$eq_1Q(^{85}Rb)=$ 0.28 (16) MHz	2.6.1
		$v=1$	^{85}Rb	−46.68 (21)	77TIE1	MW	$eq_eQ(^{79}Br)=3.08$ (39) MHz,	
			^{79}Br	4.10 (27)	77TIE1	MW	$eq_1Q(^{79}Br)=0.75$ (20) MHz	
		$v=2$	^{85}Rb	−46.56 (26)	77TIE1	MW		
			^{79}Br	5.00 (27)	77TIE1	MW		
	$^{85}Rb^{81}Br$	$v=0$	^{85}Rb	−47.00 (17)	77TIE1	MW		
Cont			^{81}Br	2.86 (27)	77TIE1	MW		

Nr.	Molecule	J, v	Nuclear quadrupole coupling constants Nucleus	eqQ [MHz]	Ref.	Meth.	Varia, remarks	See Tab.
24 Cont.	$^{87}Rb^{79}Br$	$v=0$	^{87}Rb	−22.67 (14)	77TIE1	MW		
			^{79}Br	3.62 (60)	77TIE1	MW		
	$^{87}Rb^{81}Br$	$v=0$	^{87}Rb	−22.78 (16)	77TIE1	MW		
			^{81}Br	3.41 (57)	77TIE1	MW		
32	CuBr		^{79}Br	see ^{63}Cu				
			^{81}Br	see ^{63}Cu				
49	AlBr		^{79}Br	see ^{27}Al				
51	GaBr		^{79}Br	see ^{69}Ga				
			^{81}Br	see ^{69}Ga				
53	$^{115}In^{79}Br$	$v=0$	^{115}In	−633.50 (26)	76TIE2	MW	$eq_vQ(^{115}In)=[-634.62+2.52\,(v+\frac{1}{2})\pm0.45]$ MHz	2.2.2
			^{79}Br	110.63 (27)	76TIE2	MW		
		$v=1$	^{115}In	−631.12 (44)	76TIE2	MW	$eq_vQ(^{79}Br)=[110.38+0.50\,(v+\frac{1}{2})\pm0.50]$ MHz	
			^{79}Br	111.13 (49)	76TIE2	MW		
	$^{115}In^{81}Br$	$v=0$	^{115}In	−633.20 (26)	76TIE2	MW	$eq_vQ(^{81}Br)=[91.88+0.45(v+\frac{1}{2})\pm0.50]$ MHz	
			^{81}Br	92.33 (26)	76TIE2	MW		
		$v=1$	^{115}In	−630.55 (45)	76TIE2	MW		
			^{81}Br	92.78 (53)	76TIE2	MW		
93 6–44	$^{79}Br^{16}O$	$v=0$	^{79}Br	$eq^{(a)}Q=$ 650.045 (32)	81COH	MW	$X^2\Pi_{3/2}$	4.1.5
			^{79}Br	$eq^{(b)}Q=-888$ (9)	81COH	MW	$X^2\Pi_{3/2}$	
		$v=1$	^{79}Br	$eq^{(a)}Q=$ 654.02 (24)	81COH	MW	$X^2\Pi_{3/2}$	
	$^{81}Br^{16}O$	$v=0$	^{81}Br	$eq^{(a)}Q=$ 543.080 (40)	81COH	MW	$X^2\Pi_{3/2}$	
			^{81}Br	$eq^{(b)}Q=-757$ (11)	81COH	MW	$X^2\Pi_{3/2}$	
		$v=1$	^{81}Br	$eq^{(a)}Q=$ 546.30 (26)	81COH	MW	$X^2\Pi_{3/2}$	
97	$^{79}Br^{19}F$	$v=0$	^{79}Br	1086.80 (30)	79NAI2	MW		2.2.2
		$v=1$	^{79}Br	1085.66 (60)	79NAI2	MW		2.6.1
	$^{81}Br^{19}F$	$v=0$	^{81}Br	908.09 (20)	79NAI2	MW		2.9.1
		$v=1$	^{81}Br	907.41 (60)	79NAI2	MW		
99	BrCl		^{79}Br	see ^{35}Cl				
			^{81}Br	see ^{35}Cl				

Nr.	Molecule	J, v	Nuclear quadrupole coupling constants Nucleus	eqQ [MHz]	Ref.	Meth.	Varia, remarks	See Tab.
101	$^{79}Br^{81}Br$	$v=1$	^{79}Br	810.0 (5)	81BET	LIF	ground state $X^1\Sigma_g^+$	2.9.1
		$v=2$	^{79}Br	810.0 (11)	81BET	LIF		4.1.6
		$v=13$	^{79}Br	179.1 (16)	81BET	LIF	$B^3\Pi_{0u}^+$ $\Delta eqQ = eq_{13}Q(B) - eq_1Q(X) = -630.9$ (11) MHz	
		$v=14$	^{79}Br	178.7 (16)	81BET	LIF	$\Delta eqQ = eq_{14}Q(B) - eq_1Q(X) = -631.3$ (11) MHz	
		$v=16$	^{79}Br	178.5 (12)	81BET	LIF	$\Delta eqQ = eq_{16}Q(B) - eq_2Q(X) = -631.5$ (7) MHz	
			^{79}Br		74ENG	SLS	$\Delta eqQ = eq_{17}Q(B) - eq_7Q(X) = -636.3$ (50) MHz	
			^{81}Br		74ENG	SLS	$\Delta eqQ = eq_{12}Q(B) - eq_4Q(X) = -531.3$ (51) MHz	
102	$^{127}I^{79}Br$	$v=0$	^{127}I	−2753.50 (10)	75TIE2	MW	$eq_eQ(^{127}I) = -2753.77$ (10) MHz	2.2.2
		$v=0$	^{79}Br	696.85 (15)	75TIE2	MW	$eq_IQ(^{127}I) = 0.57$ (30) MHz	2.6.1
		$v=1$	^{127}I	−2752.93 (25)	75TIE2	MW	$eq_eQ(^{79}Br) = 696.43$ (15) MHz	
		$v=1$	^{79}Br	697.66 (20)	75TIE2	MW	$eq_IQ(^{79}Br) = 0.82$ (30) MHz	
		$v=2$	^{127}I	−2752.36 (30)	75TIE2	MW	$eq_eQ(^{81}Br) = 581.71$ (15) MHz	
		$v=2$	^{79}Br	698.48 (25)	75TIE2	MW	$eq_IQ(^{81}Br) = 0.68$ (30) MHz	
	$^{127}I^{81}Br$	$v=0$	^{127}I	−2753.48 (10)	75TIE2	MW		
		$v=0$	^{81}Br	582.05 (20)	75TIE2	MW		
		$v=1$	^{127}I	−2752.90 (25)	75TIE2	MW		
		$v=1$	^{81}Br	582.73 (25)	75TIE2	MW		
		$v=2$	^{127}I	−2752.33 (40)	75TIE2	MW		
		$v=2$	^{81}Br	583.41 (35)	75TIE2	MW		
24	RbBr		^{85}Rb	see ^{79}Br				
			^{87}Rb	see ^{79}Br				
25	$^{85}Rb^{127}I$	$v=0$	^{85}Rb	−40.40 (30)	76TIE1	MW		2.6.1
		$v=0$	^{127}I	−59.89 (30)	76TIE1	MW		
53	InBr		^{115}In	see ^{79}Br				
18 *6–51*	$^1H^{127}I$	$v=1$	^{127}I	−1850 (13)	80STR	IR	$eq_IQ = -22$ (13) MHz	2.2.2
22 *6–8*	LiI		^{127}I	see ^7Li				
23	KI		^{127}I	see ^{39}K				
25	RbI		^{127}I	see ^{85}Rb				

Nr.	Molecule	J, v	Nuclear quadrupole coupling constants Nucleus	eqQ [MHz]	Ref.	Meth.	Varia, remarks	See Tab.
33	CuI		^{127}I	see ^{63}Cu				
52	GaI		^{127}I	see ^{69}Ga				
86	$^{209}Bi^{127}I$	$v=0$	^{209}Bi	−909.5 (20)	81TIS	MW	} XO^+	4.1.3
			^{127}I	−995.0 (20)	81TIS	MW		
94 *6-54*	$^{127}I^{16}O$	$v=0$	^{127}I	$eq^{(a)}Q=-1891.1$ (38)	73SAI	MW	$X^2\Pi_{3/2}$	4.1.5
100	ICl		^{127}I	see ^{35}Cl				
102	IBr		^{127}I	see ^{79}Br				
103 *6-54c*	I_2	$J=13, v=0$	^{127}I	−2452.5837 (16)	80YOK	MB	$X^1\Sigma_g^+$ ground state	2.6.1
	I_2	$J=12, v=43$	^{127}I	−558.665 (11)	80YOK	MB	$B^3\Pi_{0u}^+$ excited state	2.9.1
					80HAC	[1])		4.1.6
	I_2	$J=98, v=1$	^{127}I	−2374.0 (84) } [2])	79FOT	SLS	$X^1\Sigma_g^+$ ground state } $eqQ(B)-eqQ(X)=1880.198$ (90) MHz	
	I_2	$J=99, v=58$	^{127}I	−493.8 (84) }	79FOT	SLS	$B^3\Pi_{0u}^+$ excited state }	
	I_2	$J=127, v=6$	^{127}I	−2041 (139) } [2])	76LAN	SLS	$X^1\Sigma_g^+$ ground state } $eqQ(B)-eqQ(X)=1944.01$ (19) MHz	
		$J=128, v=11$	^{127}I	−98 (139) }	76LAN	SLS	$B^3\Pi_{0u}^+$ excited state }	
	I_2	$v=12$	^{127}I	−510 (5)	79CHU	LIF	$B^3\Pi_{0u}^+$ excited state [1]) heterodyne method [2]) strong correlation between both parameters	
27	CsCl		^{133}Cs	see ^{35}Cl				
44	$^{137}Ba^{35}Cl$	$v=0$	^{137}Ba	−39 (20)	81SCH2	MW	$X^2\Sigma^+$	4.1.2
83	$^{209}Bi^{19}F$	$v=0$	^{209}Bi	−1148.20 (12)	77KUI	MW	} XO^+ $eq_eQ(^{209}Bi)=-1150.28$ (12) MHz	4.1.3
		$v=1$	^{209}Bi	−1144.02 (17)	77KUI	MW	$eq_1Q(^{209}Bi)=4.20$ (12) MHz	
		$v=2$	^{209}Bi	−1139.68 (32)	77KUI	MW		
84	BiCl		^{209}Bi	see ^{35}Cl				
86	BiI		^{209}Bi	see ^{127}I				

2.7.2 Linear and symmetric top molecules

2.7.2.1 Preliminary remarks

1. Introduction

The electrostatic corrections to the point charge approximation due to the finite extent of the nucleus may alter the pure rotational energy of a molecule. Assuming that the electrons do not penetrate the extended nucleus–a good approximation in most cases–one can make a simple polynomial expansion for the interaction energy [53RAM]. The first relevant and also dominating term of this (multipole-) expansion which is centered on the nucleus in question, is the quadrupole coupling term which describes the interaction between the nuclear quadrupole tensor **Q** and the electric field gradient tensor **U** (see below) evaluated at the nucleus. The resulting splitting of the rotational lines is deemed quadrupole hyperfine structure (hfs). The corresponding energy operator $\mathscr{H}_Q$ can be conveniently written in terms of normalized second-order spherical harmonics Y_M^2 which favours the derivation of matrix elements by irreducible-tensor techniques [57EDM, 64TIN, 66ROS]:

$$\begin{aligned}\mathscr{H}_Q &= \tfrac{1}{8}\sum_{\mathrm{k},M}(-1)^M\left[\left(\frac{16\pi}{5}\right)^{\frac{1}{2}}\sum_{\mathrm{p}} e_{\mathrm{p}} r_{\mathrm{p}}^2\, Y_M^2(\omega_{\mathrm{p}})\right]_{\mathrm{k}}\left[\left(\frac{16\pi}{5}\right)^{\frac{1}{2}}\sum_{\mathrm{i}} e_{\mathrm{i}} r_{\mathrm{i}}^{-3}\, Y_{-M}^2(\omega_{\mathrm{i}})\right]_{\mathrm{k}}\\ &= \tfrac{1}{8}\sum_{\mathrm{k},M}(-1)^M Q_M^{\mathrm{k}}\, U_{-M}^{\mathrm{k}}\end{aligned} \tag{1}$$

where

k = the summation index for nuclei possessing a quadrupole moment
M = 0, ±1, ±2
e_{p} = the charge of the p-th proton in the nucleus in question
e_{i} = the charge of the i-th particle outside the nucleus
$r_{\mathrm{p}}, \omega_{\mathrm{p}}$; $r_{\mathrm{i}}, \omega_{\mathrm{i}}$ = the spherical coordinates of the p-th proton inside and the i-th particle outside measured from the center of the nucleus in question.

The formulation of energy and matrix element expressions depends on the coupling scheme which is adopted for nuclear spins I^{k} and rotational angular momentum J. The resulting expressions can always be written in terms of quadrupole coupling parameters $eQq_{JJ'}^{\mathrm{k}}$ which are defined as

$$eQq_{JJ'}^{\mathrm{k}} \equiv (eQ^{\mathrm{k}})q_{JJ'}^{\mathrm{k}} = (\langle I^{\mathrm{k}} M_{I^{\mathrm{k}}} = I^{\mathrm{k}} | Q_0^{\mathrm{k}} | I^{\mathrm{k}} M_{I^{\mathrm{k}}} = I^{\mathrm{k}}\rangle)\langle J\tau M_J = J | U_0^{\mathrm{k}} | J'\tau' M_{J'} = J\rangle \tag{2}$$

where

I^{k} = the nuclear spin quantum number of the k-th nucleus
J = the rotational angular momentum quantum number
$M_{I^{\mathrm{k}}}, M_J$ = the orientation quantum numbers
τ = the asymmetric rotor quantum number; can also be expressed as $K_{-1}K_{+1}$
eQ^{k} = the nuclear quadrupole moment; non-diagonal expressions in I^{k} do not appear because of the relatively large nuclear-energy separations.

Letting V be the potential at the nucleus in question due to all other charges, matrix elements taken of the electric field gradient tensor **U** are given by matrix elements of the second derivative of V directed along the space-fixed Z axis (= axis of quantization):

$$q_{JJ'}^{\mathrm{k}} = \langle J\tau M_J = J | U_0^{\mathrm{k}} | J'\tau' M_{J'} = J\rangle = \langle J\tau M_J = J | \left(\frac{\partial^2 V}{\partial Z^2}\right)_{r=0}^{\mathrm{k}} | J'\tau' M_{J'} = J\rangle. \tag{3}$$

The rotational dependence of $q_{JJ'}^{\mathrm{k}}$ can be calculated by transformation to the principal inertial axes system ($g = a, b, c$) yielding

$$eQq_{JJ'}^{\mathrm{k}} = \sum_{gg'}\langle\varphi_{Zg}\varphi_{Zg'}\rangle_{JJ'}\chi_{gg'}^{\mathrm{k}} \tag{4}$$

where

$\chi_{gg'}^{\mathrm{k}} = eQ^{\mathrm{k}}\left(\frac{\partial^2 V}{\partial g\,\partial g'}\right)_{r=0}^{\mathrm{k}}$ = the (g, g')-component of the quadrupole coupling tensor in the molecule-fixed axis system
φ_{Zg} = the direction cosine between space-fixed Z axis and g-th molecule-fixed axis.

The quadrupole-coupling tensor is symmetric, $\chi_{gg'}^{\mathrm{k}} = \chi_{g'g}^{\mathrm{k}}$, and its diagonal elements obey Laplace's equation,

$$\sum_g \chi_{gg}^{\mathrm{k}} = 0. \tag{5}$$

The first-order quadrupole coupling energy for *one* quadrupolar nucleus in the molecule can be written [55TOW, 64THA]

$$W_Q = \langle IJF | H_Q | IJF\rangle = \frac{1}{2}\,\frac{eQq_J}{I(2I-1)\,J(2J-1)}\left[\frac{3}{4}\,C(C+1) - I(I+1)\cdot J(J+1)\right] \tag{6a}$$

with

$$eQq_J \equiv eQq_{JJ} \equiv eQq(J\tau) = \frac{2}{(J+1)(2J+3)} \sum_g \langle P_g^2 \rangle \chi_{gg} \tag{6b}$$

where

F = the total-angular-momentum quantum number

C = $F(F+1) - I(I+1) - J(J+1)$

$\langle P_g^2 \rangle$ = the expectation value of the square of the g-th component of the angular momentum in the principal inertial axes system taken over in the unperturbed rotational state J_τ; P_g is in units of $\hbar = h/2\pi$, $g = a, b, c$.

The quadrupole coupling constants as defined in eqs. (2) and (4) have the dimension of energy. In this table, they are given in frequency units omitting the factor of Planck constant h.

Special remarks

For linear molecules and symmetric top molecules with a nuclear quadrupole located on the symmetry axis, the principal axes of the field gradient tensor and the inertia tensor coincide. Since two components of the field gradient tensor are equal because of symmetry, only one independent component of the quadrupole coupling tensor remains to be determined in an analysis of the quadrupole hfs:

$$eQq \equiv \chi_{aa} = -2\chi_{bb} = -2\chi_{cc} \tag{7}$$

(as shown for prolate symmetric top; for oblate symmetric top a and c should be interchanged).

It is sometimes necessary to take into account the effects of centrifugal distortion [72HOU]. The resulting formula for the quadrupole energy of a symmetric top molecule is

$$\begin{aligned} W_Q = \Bigg[& \left(\frac{3K^2}{J(J+1)} - D\right)\left(eqQ_0 + \chi_J J(J+1) + \chi_K K^2\right) \\ & + \chi_d \frac{K^2(4K^2-1)}{J(J+1)} \Bigg] \mathrm{Y}(I, J, F) + E_Q^{(2)} B^{-1} eqQ_0^2 + E_Q^{(3)} B^{-2} eqQ_0^3 \end{aligned} \tag{8}$$

where

$\mathrm{Y}(I, J, F)$ is the Casimir function,

$E_Q^{(2)}$ and $E_Q^{(3)}$ are the second-order and third-order perturbation contributions, respectively, B being the rotational constant,

$$D = \begin{cases} 1 & \text{for a non degenerate excited state} \\ 1 \pm 2\delta_{Kl,1} Kl\eta & \text{for a degenerate excited state} \end{cases} \tag{9}$$

η is a coefficient which describes the effective asymmetry produced by the degenerate vibration on the gradient of the molecular electric field around the molecular axis and has to be taken into account only for $K = l = \pm 1$. In the literature coefficients are also to be found which are two or four times larger. It is to be noted that only the relative signs of q_t (the l-type doubling constant) and η can be determined from microwave spectroscopy [76HER]. For a linear molecule in a first excited degenerate state $K = l$ in eq. (8).

2. Explanations of the columns in the tabulation

Column	Parameters		Definitions
1	Nr.		serial number of molecule in this table
2	v	v_j	vibrational quantum number of the j-th normal vibration In the vibrational ground state $v_1 = v_2 = \cdots = v_j = 0$, in short $v = 0$ vibrational state: $v_1 = \cdots = v_{j-1} = 0;\quad v_j = v;\quad v_{j+1} = \cdots = 0;$ in short $v_j = v$
		l	vibrational-angular-momentum quantum number in excited degenerate vibrational states

Column	Parameters		Definitions
3			quadrupolar nucleus which produces hfs
4	eQq [MHz]		nuclear quadrupole coupling constant, compare eq. (7)
		χ_{gg}, or $eQq(J_\tau)$ or $eQq(J_{K_{-1},K_{+1}})$	components of the nuclear quadrupole coupling tensor and diagonal quadrupole coupling parameters in the rotational state J_τ (or $J_{K_{-1},K_{+1}}$); see eqs. (4), (6); these notations apply here to asymmetries originating from isotopic substitution
		$(eqQ)_0$	nuclear quadrupole coupling constant independent of rotation, compare eq. (8)
		χ_J, χ_K, χ_d	centrifugal distortion terms of the quadrupole coupling, compare eq. (8)
		η	asymmetry parameter of the nuclear quadrupole coupling tensor, compare eq. (9)
5	Ref.		references for the previous column
6	Meth.		method of measurement
		BAS	beam absorption spectroscopy
		BMS	beam maser spectroscopy
		RFMWDR	radiofrequency-microwave double resonance
		IRMWDR	infrared-microwave double resonance
		IRRFDR	infrared-radiofrequency double resonance
		MBER	molecular beam electric resonance
		MW	microwave spectroscopy
		MWFT	microwave Fourier transform spectroscopy
		La	Laser
7	Varia, remarks		remarks concerning the previous columns.
8			references to the serial numbers in Tables 2.3, 2.4; figures in Chapter 3, subvolume b

3. Arrangement of substances

The molecules in this table are ordered according to the increasing nuclear charge number of their quadrupolar nuclei. When the molecule possesses two different quadrupolar nuclei, it is entered under both nuclei; however, the data is given only once. For each type of nucleus the ordering of the molecules in the tables 2.3 and 2.4 is retained.

2.7.2.2 Data

Nr.	Molecule	v	Nucleus	eqQ MHz	Ref.	Meth.	Varia, remarks	See Tab., Nr.
1	$DC^{15}N$	$v=0$	D	−0.207 (4)	80CAZ	LD *)	*) Lamb Dip Microwave Spectroscopy	2.3, 4
2	DCP	$v=0$	D	0.233 (40)	73HAR	MW		2.3, 10
	DCNO		D	see Nr. 12				
	Ne · DCl		D	see Nr. 26				
3	Ar · DF	$v=0$	D	0.1965 (8)	81DIX1	MBER		2.3, 39
	Ar · DCl		D	see Nr. 27				
	Ar · DBr		D	see Nr. 47				
	Kr · DF		D	see Nr. 55				
	Kr · DCl		D	see Nr. 29				
3a	CO · DF $C^{18}O$ · DF	$v=0$ $v=0$	D D	0.2786 (35) 0.2785 (35)	81LEG1	MWFT		2.3, 35
3b	CO_2 · DF	$v=0$	D	0.2639 (5)	81BAI	MBER		2.3, 37a
	DC≡CCN		D	see Nr. 14				
	ND_3		D	see Nr. 15				
	CD_3CN		D	see Nr. 16				
4	CF_3D	$v=0$	D	$eqQ\,(J=1, K=0)$ $= -0.03416\,(40)$	74REJ	MBER		2.4, 18
5	$H^{11}BS$ $D^{11}BS$ $D^{10}BS$	$v=0$ $v=0$ $v=0$	^{11}B ^{11}B ^{10}B	−3.71 (3) −3.74 (3) −7.91 (3)	74PEA	MW		2.3, 12
6	$CH_3{}^{11}BS$ $CH_3{}^{10}BS$	$v=0$ $v=0$	^{11}B ^{10}B	−3.714 (20) −7.743 (50)	80KIR1	MW		2.4, 40
7	$^{11}BH_3CO$ $^{10}BH_3CO$	$v=0$ $v=0$	^{11}B ^{10}B	1.60 (10) 3.45 (10)	77VEN	MW		2.4, 51

Nr.	Molecule	v	Nucleus	eqQ MHz	Ref.	Meth.	Varia, remarks	See Tab., Nr.
8	$^{11}BF_3CO$	$v=0$	^{11}B	2.75 (5)	78JAN	MBER		2.4, 52
	$^{10}BF_3CO$	$v=0$	^{10}B	5.74 (10)				
9	$Ar^{11}BF_3$	$v=0$	^{11}B	2.70 (5)	78JAN	MBER		2.4, 102
	$Ar^{10}BF_3$	$v=0$	^{10}B	5.64 (10)				
10	$^{35}ClCN$	$v=0$	^{14}N	−3.62277 (90)	80REI	MBER		2.3, 6
			^{35}Cl	−83.27519 (40)				
		$v_2=1^1$	^{14}N	$\chi_{zz}=-3.70219$ (40)				
				$\chi_{xx}-\chi_{yy}=0.25600$ (41)				
			^{35}Cl	$\chi_{zz}=-82.81554$ (55)				
				$\chi_{xx}-\chi_{yy}=-0.92784$ (49)				
	$^{37}ClCN$	$v=0$	^{14}N	−3.61 (8)	79COG	MW		
			^{37}Cl	−65.61 (6)				
11	HNC	$v=0$	^{14}N	−0.4 *)	77SNY	MW	*) [76BLA2] gives $eqQ=1.9$ (3) MHz	2.3, 9
12	HCNO	$v=0$	^{14}N	0.245 (5)	78HUT1	MW		2.3, 17
	DCNO	$v=0$	^{14}N	0.245 *)			*) fixed value	
			D	0.190 (25)				
13	$N_\alpha N_\beta O$	$v=0$	$^{14}N_\alpha$	−0.77376 (27)	78REI	MBER		2.3, 18
		$v_2=1^1$		$\chi_{zz}=-0.88653$ (12)				
				$\chi_{xx}-\chi_{yy}=0.31878$ (14)				
		$v=0$	$^{14}N_\beta$	−0.26758 (38)				
		$v_2=1^1$		$\chi_{zz}=-0.25814$ (18)				
				$\chi_{xx}-\chi_{yy}=-0.06084$ (18)				
14	HC≡CCN	$v=0$	^{14}N	−4.3188 (12)	78LAF	MW		2.3, 27
	$HC≡C^{13}CN$	$v=0$	^{14}N	−4.3001 (67)				
	$HC≡^{13}CCN$	$v=0$	^{14}N	−4.298 (23)				
	$H^{13}C≡CCN$	$v=0$	^{14}N	−4.273 (19)				
	$DC≡CC^{15}N$	$v=0$	D	0.2296 (48)	81FLI	MWFT		
14a	$HC^{14}N\cdots HC^{14}N$	$v=0$	^{14}N	$\chi_1=-4.0973$ (200)	81BUX1	MWFT	subscripts 1 and 2 indicate bonding and non-bonding nitrogen sites, respectively	2.3, 34
				$\chi_2=-4.4400$ (190)				
	$HC^{14}N\cdots HC^{15}N$	$v=0$	^{14}N	$\chi_1=-4.1059$ (10)				
	$HC^{15}N\cdots HC^{14}N$	$v=0$	^{14}N	$\chi_2=-4.4339$ (6)				

Nr.	Molecule	v	Nucleus	eqQ MHz	Ref.	Meth.	Varia, remarks	See Tab., Nr.
14b	NCCN···HF	$v=0$	^{14}N	$\chi_1=-4.56\,(5)$	81LEG2	MWFT		2.3, 34a
				$\chi_2=-4.28\,(5)$				
	^{15}NCCN···HF	$v=0$		$\chi_2=-4.21\,(1)$				
	NCC^{15}N···HF	$v=0$		$\chi_1=-4.65\,(1)$				
14c	CH_3CN···HF	$v=0$	^{14}N	$-3.727\,(3)$ *)	81SOP1	MWFT	*) from $J=0\rightarrow1$	2.4, 106
15	NH_3	$v=0$	^{14}N	$(eqQ)_0=-4.08965$ *)	72HOU	BMS	*) $eqQ=(eqQ)_0+1.048\cdot10^{-3}\,[J(J+1)-K^2]-0.471\cdot10^{-3}\,K^2$	2.4, 1
				$\chi_d=-1.070\cdot10^{-3}$				
				$(eqQ)_{J=5,\,K=3}=-0.39390\,(8)$	74RU1			
	ND_2H	$v=0$	^{14}N	$(eqQ)_{1_{11}}=-0.3594\,(5)$	76GAR			
				$(eqQ)_{1_{01}}=-0.4088\,(5)$				
			D	$(eqQ)_{1_{11}}=-0.0035\,(20)$				
				$(eqQ)_{1_{01}}=+0.0326\,(10)$				
	ND_3	$v=0$	^{14}N	$-4.080\,(3)$	79CED			
			D	$\chi_{xx}=0.207$				
				$\chi_{yy}=-0.111$				
				$\chi_{zz}=-0.096$				
	NT_3	$v=0$	^{14}N	$-4.170\,(49)$	74HEL1	MW		
16	CH_3CN	$v=0$	^{14}N	$-4.22534\,(73)$	77BOU1	BAS		2.4, 16
	CD_3CN	$v=0$	^{14}N	$-4.2293\,(15)$	73KUK3	BMS	*) 167.5 (40) kHz along the C–D bond direction	
			D	$0.0559\,(15)$ *)				
17	$^{28}SiH_3CN$	$v=0$	^{14}N	$-4.77\,(30)$	75CAE	MW		2.4, 43
		$v_7=1$		$-5.13\,(30)$				
		$v_8=1$		$-4.91\,(40)$				
18	SiH_3NCO	$v=0$	^{14}N	1.85 (5)	76DUC	MW		2.4, 44
19	SiH_3NCS	$v=0$	^{14}N	0.75	77DO2	MW		2.4, 45
20	^{17}OCS	$v=0$	^{17}O		80BES	MWFT		2.3, 2
	OC^{33}S	$v=0$	^{33}S	$-29.1184\,(12)$ *)	74REI	MBER	*) uncertainties are 2 σ	
		$v_2=1$		$eqQ_1=-28.6825\,(18)$ $^{+}$)			$^{+}$) $E_Q=Y(I,J,F)\left\{eqQ_1\left[\frac{3l^2}{J(J+1)}-1\right]\pm\frac{1}{2}\delta_{\lvert l\rvert 1}\,eqQ_2\right\}$	
				$eqQ_2=-1.1808\,(18)$ $^{+}$)				

Nr.	Molecule	v	Nucleus	eqQ MHz	Ref.	Meth.	Varia, remarks	See Tab., Nr.
21	$^{33}SPF_3$	$v=0$	^{33}S	−28.75 (40)	76SM2	MW		2.4, 32
	ClCN		Cl	see Nr. 10				
22	$^{35}Cl^{11}BS$	$v=0$	^{35}Cl	−42.54 (1)	80KIR2	MW		2.3, 13
		$v_2=1$		$\alpha_2=-0.428$ (34)				
		$v_3=1$		$\alpha_3=-0.62$ (23)				
	$^{35}Cl^{10}BS$	$v=0$	^{35}Cl	−42.56 (2)				
		$v_2=1$		$\alpha_2=-1.23$ (23)				
		$v_3=1$		$\alpha_3=-1.6$ (8)				
	$^{37}Cl^{11}BS$	$v=0$	^{37}Cl	−33.53 (1)				
		$v_2=1$		$\alpha_2=-1.46$ (23)				
		$v_3=1$		$\alpha_3=-1.6$ (4)				
	$^{37}Cl^{10}BS$	$v=0$	^{37}Cl	−33.58 (3)				
		$v_2=1$		$\alpha_2=-0.58$ (26)				
		$v_3=1$		$\alpha_3=-2.2$ (6)				
23	$HC{\equiv}C^{35}Cl$	$v=0$	^{35}Cl	−79.67 (8)	78JON	MW		2.3, 25
		$v_3=1$		−79.5 (2)				
		$v_4=1^1$		−79.15 (8)				
		$v_5=1^1$		−79.15 (8)				
		$v_5=2^2$		−78.9 (4)				
		$v_5=2^0$		−78.0 (4)				
	$DC{\equiv}C^{35}Cl$	$v=0$	^{35}Cl	−79.66				
		$v_3=1$		−79.9 (4)				
		$v_4=1^1$		−79.1 (4)				
		$v_5=1^1$		−79.15 (8)				
		$v_5=2^2$		−76.3 (4)				
		$v_5=2^0$		−78.3 (4)				
	$HC{\equiv}C^{37}Cl$	$v=0$	^{37}Cl	−62.75 (6)				
		$v_3=1$		−62.0 (3)				
		$v_4=1^1$		−62.5 (5)				
		$v_5=1^1$		−62.50 (6)				
		$v_5=2^2$		−62.1 (3)				
Cont.		$v_5=2^0$		−61.5 (3)				

Nr.	Molecule	v	Nucleus	eqQ MHz	Ref.	Meth.	Varia, remarks	See Tab., Nr.
23	$DC{\equiv}C^{37}Cl$	$v=0$	^{37}Cl	−63.12	78JON	MW		
Cont.		$v_3=1$		−62.7 (3)				
		$v_4=1^1$		−62.2 (5)				
		$v_5=1^1$		−62.25 (6)				
		$v_5=2^2$		−60.4 (3)				
		$v_5=2^0$		−61.7 (3)				
23a	Cl—C≡C—F	$v=0$	^{35}Cl	83.0 (1)	81GUA1	MW		2.3, 27a
			^{37}Cl	65.6 (1)				
24	$^{35}ClC{\equiv}CCN$	$v=0$	^{35}Cl	−75 (4)	74BJO	MW		2.3, 28
	$^{37}ClC{\equiv}CCN$	$v=0$	^{37}Cl	−62 (3)				
25	$OC\cdots H^{35}Cl$	$v=0$	^{35}Cl	−52.086 (9)	81SOP	MWFT		2.3, 36
	$OC\cdots D^{35}Cl$	$v=0$	^{35}Cl	−55.254 (7)				
	$O^{13}C\cdots H^{35}Cl$	$v=0$	^{35}Cl	−52.219 (29)				
	$OC\cdots H^{37}Cl$	$v=0$	^{37}Cl	−41.070 (38)				
26	$Ne\cdot D^{35}Cl$	$v=0$	^{35}Cl	−13.28692 (94)	80BAR2	MBER	$\chi_J=-4.93$ (22) kHz	2.3, 38
			D	0.0366 (13)				
	$Ne\cdot D^{37}Cl$	$v=0$	^{37}Cl	−10.522 (2)			$\chi_J=-3.8$ kHz fixed	
			D	0.0376 (12)				
27	$Ar\cdot H^{35}Cl$	$v=0$	^{35}Cl	−23.0297 (5)	81HUT	MBER		2.3, 40
				$\chi_J=-1.60(5)\cdot10^{-3}$				
	$Ar\cdot D^{35}Cl$	$v=0$	^{35}Cl	−36.1907 (7)				
				$\chi_J=-1.28(8)\cdot10^{-3}$				
			D	0.1000 (10)				
	$Ar\cdot D^{37}Cl$	$v=0$	^{37}Cl	−28.587	76NOV2			
28	$Ar\cdot{}^{35}ClF$	$v=0$	^{35}Cl	−140.869 (15)	74HAR1	MBER		2.3, 42
	$Ar\cdot{}^{37}ClF$	$v=0$	^{37}Cl	−111.053 (15)				
29	$^{82}Kr\cdot H^{35}Cl$	$v=0$	^{35}Cl	−29.268 (9)	80BAL	MWFT		2.3, 44
	$^{84}Kr\cdot H^{35}Cl$	$v=0$	^{35}Cl	−29.23757 (61)	80BAR1	MBER	$\chi_J=-0.730$ (70) kHz	
	$^{86}Kr\cdot H^{35}Cl$	$v=0$	^{35}Cl	−29.2451 (15)			$\chi_J=-0.46$ (13) kHz	
	$^{82}Kr\cdot D^{35}Cl$	$v=0$	^{35}Cl	−40.842 (14)	80BAL	MWFT		
	$^{84}Kr\cdot D^{35}Cl$	$v=0$	^{35}Cl	−40.80445 (97)	80BAR1	MBER	$\chi_J=-0.72$ (12) kHz	
Cont.			D	0.11317 (62)				

Nr.	Molecule	v	Nucleus	eqQ MHz	Ref.	Meth.	Varia, remarks	See Tab., Nr.
29	$^{86}Kr \cdot D^{35}Cl$	$v=0$	^{35}Cl	−40.849 (14)	80BAL	MWFT		
Cont.	$^{82}Kr \cdot H^{37}Cl$	$v=0$	^{37}Cl	−23.235 (33)				
	$^{84}Kr \cdot H^{37}Cl$	$v=0$	^{37}Cl	−23.0892 (25)	80BAR1	MBER	$\chi_J = -0.51$ (29) kHz	
	$^{86}Kr \cdot H^{37}Cl$	$v=0$	^{37}Cl	−23.139 (8)	80BAL	MWFT		
	$^{82}Kr \cdot D^{37}Cl$	$v=0$	^{37}Cl	−32.237 (122)				
	$^{84}Kr \cdot D^{37}Cl$	$v=0$	^{37}Cl	−32.211 (9)				
	$^{86}Kr \cdot D^{37}Cl$	$v=0$	^{37}Cl	−32.232 (119)				
	$^{83}Kr \cdot H^{35}Cl$	$v=0$	^{35}Cl	−29.238 (45)	81CAM5	MWFT		
			^{83}Kr	5.200 (100)				
	$^{83}Kr \cdot D^{35}Cl$	$v=0$	D	0.1135 *)			*) fixed value	
			^{35}Cl	−40.824 *)				
			^{83}Kr	7.192 (100)				
30	$^{82}Kr \cdot {}^{35}ClF$	$v=0$	^{35}Cl	−141.521 (20)	75NOV	MBER		2.3, 46
	$^{84}Kr \cdot {}^{35}ClF$	$v=0$	^{35}Cl	−141.519 (20)				
	$^{86}Kr \cdot {}^{35}ClF$	$v=0$	^{35}Cl	−141.517 (20)				
	$^{84}Kr \cdot {}^{37}ClF$	$v=0$	^{37}Cl	−111.544 (20)				
	$^{83}Kr \cdot {}^{35}ClF$	$v=0$	^{35}Cl	−141.519 *)	81BUX2	MWFT	*) fixed value	
			^{83}Kr	13.90 (25)				
31	$^{129}Xe \cdot H^{35}Cl$	$v=0$	^{35}Cl	−34.761 (42)	80KE2	MWFT		2.3, 47
	$^{129}Xe \cdot D^{35}Cl$	$v=0$	^{35}Cl	−44.816 (38)				
	$^{129}Xe \cdot H^{37}Cl$	$v=0$	^{37}Cl	−27.359 (94)				
	$^{131}Xe \cdot H^{35}Cl$	$v=0$	^{35}Cl	−34.76 *)	81CAM5		*) fixed value	
			^{131}Xe	−4.641 (50)				
	$^{131}Xe \cdot D^{35}Cl$	$v=0$	D	0.1245 *)				
			^{35}Cl	−44.780 (200)				
			^{131}Xe	−5.89 (20)				
	$^{132}Xe \cdot H^{35}Cl$	$v=0$	^{35}Cl	−34.670 (20)	80KE2			
	$^{132}Xe \cdot D^{35}Cl$	$v=0$	D	0.1245 *)	81CAM5			
			^{35}Cl	−44.800 (40)				
32	$N^{35}Cl_3$	$v=0$	^{35}Cl	−108 (3) *)	74CAZ	MW	*) for the N—Cl bond axis	2.4, 2
33	$CH_3{}^{35}Cl$	$v=0$	^{35}Cl	−74.7433 (14)	78EL1	MBER		2.4, 13
		$v_3=1$		−74.809 (45)	81DEM2	MW		
		$v_6=1$		−74.866 (99)	81DUB			
Cont.	$^{13}CH_3{}^{35}Cl$	$v=0$	^{35}Cl	−74.74 (10)	81DEM2			

Nr.	Molecule	v	Nucleus	eqQ MHz	Ref.	Meth.	Varia, remarks	See Tab., Nr.
33 Cont.	$CD_3{}^{35}Cl$	$v_3=1$	^{35}Cl	−74.9 (3)	76IM			
		$v_6=1$		−74.4 (4)				
	$CH_3{}^{37}Cl$	$v=0$	^{37}Cl	−58.9116 (34)	81DEM2			
		$v_3=1$		−59.058 (45)				
		$v_6=1$		−59.093 (49)	81DUB			
	$^{13}CH_3{}^{37}Cl$	$v=0$	^{37}Cl	−58.96 (12)	81DEM2			
	$CD_3{}^{37}Cl$	$v_3=1$	^{37}Cl	−58.7 (2)	76IM			
		$v_6=1$		−58.5 (5)				
34	$HSi^{35}Cl_3$	$v=0$	^{35}Cl	−37.0	77TA2	MW		2.4, 27
35	$^{35}ClO_3F$	$v=0$	^{35}Cl	−19.50 (15)	80DEL4	MBER		2.4, 35
36	$(CH_3)_3C^{35}Cl$	$v=0$	^{35}Cl	−67.56 (50)	74BRA	MW		2.4, 55
	$(CD_3)_2CH_3C^{35}Cl$	$v=0$	^{35}Cl	$\chi_{aa}=-65.815$ (755)				
				$\chi_{bb}=$ 34.10 (1602)				
				$\chi_{cc}=$ 31.708 (1914)				
37	$(CH_3)_3{}^{28}Si^{35}Cl$	$v=0$	^{35}Cl	−34.701	73ZEI	MW		2.4, 57
38	$SF_5{}^{35}Cl$	$v=0$	^{35}Cl	−85.35 (20)	79BEL	MW		2.4, 84
	$SF_5{}^{37}Cl$	$v=0$	^{37}Cl	−67.66 (100)				
	$^{34}SF_5{}^{35}Cl$	$v=0$	^{35}Cl	−91 (6)				
39	$^{35}ClF_5$	$v=0$	^{35}Cl	43.93 (8)	76BOD	MW		2.4, 88
	$^{37}ClF_5$	$v=0$	^{37}Cl	34.63 (6)				
40	$C_7H_9{}^{35}Cl$	$v=0$	^{35}Cl	−69.2 (2)	73LAU	MW		2.4, 95 3, 2
41	KOH	$v_2=2^2$	^{39}K	−8.3 (10)	77KUI	MW		2.3, 20
		$v_2=4^4$		−8.0 (10)				
		$v_2=6^4$		−8.0 (10)				
42	AsH_3	$v=0$		$eqQ_0=-162.651$ (28)	81SCA	IRRFDR		2.4, 6
				$\chi_J=0.0057$ (40)				
				$\chi_K=0.070$ (19)				
				$\chi_d=0.023$ (10)				
		$v_2=1$		−162.940 (40)				
		$v_4=1$		−160.87 (38)				

Nr.	Molecule	v	Nucleus	eqQ MHz	Ref.	Meth.	Varia, remarks	See Tab., Nr.
43	AsF_3	$v=0$		−236.128 (4)	78SMI	MW		2.4, 7
		$v_1=1$		−234.324 (80)				
		$v_2=1$		−235.140 (6)				
		$v_2=2$		−233.100 (54)				
		$v_3=1$		−234.79 (17)				
		$v_4=1$		−235.90 (15)				
44	$HC{\equiv}C^{79}Br$	$v=0$	^{79}Br	648.099 (21)	77POR	MW		2.3, 26
		$v_3=1$		646	77JON			
		$v_4=1^1$		644.9				
		$v_5=1^1$		644.90				
		$v_5=2^2$		644.9				
		$v_5=2^0$		646				
	$DC{\equiv}C^{79}Br$	$v=0$	^{79}Br	647.96				
		$v_5=1^1$		644.88				
	$HC{\equiv}C^{81}Br$	$v=0$	^{81}Br	541.470 (20)	77POR			
		$v_3=1$		540	77JON			
		$v_4=1^1$		538.5				
		$v_5=1^1$		538.54				
		$v_5=2^2$		540				
		$v_5=2^0$		540				
	$DC{\equiv}C^{81}Br$	$v=0$	^{81}Br	541.08				
		$v_5=1^1$		538.80				
45	$^{79}BrC{\equiv}CCN$	$v=0$	^{79}Br	660 (6)	74BJO	MW		2.3, 29
	$^{81}BrC{\equiv}CCN$	$v=0$	^{81}Br	550 (5)				
46	$OC\cdots H^{79}Br$	$v=0$	^{79}Br	409.7145 (42)	80KE3	MWFT		2.3, 37
	$OC\cdots D^{79}Br$	$v=0$	^{79}Br	437.12 (35)				
	$O^{13}C\cdots H^{79}Br$	$v=0$	^{79}Br	410.453 (89)				
	$^{18}OC\cdots H^{79}Br$	$v=0$	^{79}Br	410.44 (57)				
	$OC\cdots H^{81}Br$	$v=0$	^{81}Br	342.2854 (56)				
47	$Ar\cdot H^{79}Br$	$v=0$	^{79}Br	173.2257 (32)	80JAC,	MBER	$\chi_J=18.16\,(65)$ kHz	2.3, 41
	$Ar\cdot D^{79}Br$	$v=0$	^{79}Br	276.2235 (53)	80KE1		$\chi_J=15.43\,(91)$	
			D	0.07518 (87)				
Cont.	$Ar\cdot H^{81}Br$	$v=0$	^{81}Br	144.8081 (28)			$\chi_J=15.07\,(57)$	

Nr.	Molecule	v	Nucleus	eqQ MHz	Ref.	Meth.	Varia, remarks	See Tab., Nr.
47 Cont.	$Ar \cdot D^{81}Br$	$v=0$	^{81}Br	230.8766 (53)	80JAC,		$\chi_J = 12.94\,(92)$	
			D	0.07510 (98)	80KE1			
48	$^{82}Kr \cdot H^{79}Br$	$v=0$	^{79}Br	228.18 (9)	80KE1	MWFT		2.3, 45
	$^{84}Kr \cdot H^{79}Br$	$v=0$	^{79}Br	228.26 (14)				
	$^{86}Kr \cdot H^{79}Br$	$v=0$	^{79}Br	228.41 (36)				
	$^{84}Kr \cdot D^{79}Br$	$v=0$	^{79}Br	318.18 (14)				
	$^{82}Kr \cdot H^{81}Br$	$v=0$	^{81}Br	191.26 (30)				
	$^{84}Kr \cdot H^{81}Br$	$v=0$	^{81}Br	190.78 (14)				
	$^{86}Kr \cdot H^{81}Br$	$v=0$	^{81}Br	190.78 (14)				
	$^{84}Kr \cdot D^{81}Br$	$v=0$	^{81}Br	265.85 (19)				
49	$CH_3{}^{79}Br$	$v=0$	^{79}Br	$(eqQ)_0 = 577.143\,(8)$	80AR1	IRRFDR		2.4, 14
				$\chi_J = 0.66\,(5) \cdot 10^{-3}$				
				$\chi_K = 13.1\,(26) \cdot 10^{-3}$				
				$\chi_d = -8.5\,(20) \cdot 10^{-3}$				
		$v_3=1$		577.50 (10)				
		$v_3=2$		577.4 (1)	77MOS1	MW		
		$v_6=1$		$(eqQ)_0 = 578.847\,(48)$	81DUB			
				$eqQ\eta = 0.558\,(114)$ *)			*) the absolute sign has been determined from IRMWDR [76HER]	
				$\chi_J = 0.527\,(71) \cdot 10^{-3}$				
				$\chi_K = 12.5\,(29) \cdot 10^{-3}$				
		$v_6=2$		580.160 (95)	80AR1	IRRFDR		
		$v_3=1,\ v_6=1$		578.84 (12)				
	$^{13}CH_3{}^{79}Br$	$v=0$	^{79}Br	566.35 (553)	78MOS1	MW		
		$v_3=1$		577.44 (50)	80AR1	IRRFDR		
		$v_6=1$		579.44 (37)				
		$v_6=2$		580.8 (11)				
		$v_3=1,\ v_6=1$		580.6 (10)				
	$CD_3{}^{79}Br$	$v=0$	^{79}Br	575.66 (15)	79WIL2	MW		
		$v_3=1$		575.9 (10)	77MOS1			
		$v_3=2$		572.5 (20)				
	$^{13}CD_3{}^{79}Br$	$v=0$	^{79}Br	573.9 (20)	78MOS1			
	$CH_3{}^{81}Br$	$v=0$	^{81}Br	$(eqQ)_0 = 482.130\,(18)$	80AR1	IRRFDR		
				$\chi_J = 0.13\,(7) \cdot 10^{-3}$				
				$\chi_K = 15.4\,(17) \cdot 10^{-3}$				
Cont.				$\chi_d = -8.0\,(16) \cdot 10^{-3}$				

Nr.	Molecule	v	Nucleus	eqQ MHz	Ref.	Meth.	Varia, remarks	See Tab., Nr.
49	$CH_3{}^{81}Br$	$v_3=1$		482.43 (12)			*) the absolute sign has been	
Cont.		$v_3=2$		483.6 (10)	77MOS1	MW	determined from IRMWDR [76HER]	
		$v_6=1$		$(eqQ)_0=483.519$ (47)	81DUB			
				$eqQ\eta=0.391$ (42) *)				
				$\chi_J=$ 0.52 (18) $\cdot 10^{-3}$				
				$\chi_K=16.4$ (22) $\cdot 10^{-3}$				
		$v_6=2$		484.946 (56)	80AR1	IRRFDR		
		$v_3=1, v_6=1$		483.427 (60)				
		$v_1=1$		$\Delta(eqQ)=0.547$ (1) $^{+}$)	76HAL	La	$^{+}$) $\Delta(eqQ)=eqQ(v_1=1)-eqQ(v=0)$	
	$^{13}CH_3{}^{81}Br$	$v=0$	^{81}Br	482.23 (18)	81AR1	IRRFDR		
		$v_3=1$		481.58 (50)				
		$v_6=1$		483.93 (23)				
		$v_3=1, v_6=1$		484.25 (75)				
	$CD_3{}^{81}Br$	$v=0$	^{81}Br	480.89 (15)	79WIL2	MW		
		$v_3=1$		482.6 (10)	77MOS1			
		$v_3=2$		479.5				
	$^{13}CD_3{}^{81}Br$	$v=0$	^{81}Br	480.2 (20)	78MOS1			
50	$CF_3{}^{79}Br$	$v=0$	^{79}Br	618.2 (3) *)	80COX	RFMWDR	*) uncertainties are 2 σ	2.4, 20
	$CF_3{}^{81}Br$	$v=0$	^{81}Br	516.6 (3) *)				
51	$SiH_3{}^{79}Br$	$v=0$	^{79}Br	334.981 (8)	79DO1	MW		2.4, 24
	$SiD_3{}^{79}Br$	$v=0$	^{79}Br	334.158 (13)				
	$SiH_3{}^{81}Br$	$v=0$	^{81}Br	279.825 (5)	77DO3			
51 a	$^{74}GeHD_2{}^{79}Br$	$v=0$	^{79}Br	$\chi_{aa}=$ 383.0 (25)	81DUR3	MW		2.4, 45c
				$\chi_{bb}=-190.4$ (47)				
				$\chi_{cc}=-192.6$ (47)				
	$^{74}GeHD_2{}^{81}Br$	$v=0$	^{81}Br	$\chi_{aa}=$ 317.5 (27)				
				$\chi_{bb}=-159.5$ (50)				
				$\chi_{cc}=-158.0$ (50)				
	$^{74}GeH_2D^{79}Br$	$v=0$	^{79}Br	$\chi_{aa}=$ 381.1 (12)				
				$\chi_{bb}=-189.8$ (22)				
				$\chi_{cc}=-191.3$ (22)				
	$^{74}GeH_2D^{81}Br$	$v=0$	^{81}Br	$\chi_{aa}=$ 317.1 (23)				
				$\chi_{bb}=-158.9$ (43)				
Cont.				$\chi_{cc}=-158.2$ (43)				

Nr.	Molecule	v	Nucleus	eqQ MHz	Ref.	Meth.	Varia, remarks	See Tab., Nr.
51a	$^{74}GeD_3{}^{79}Br$	$v=0$	^{79}Br	383.0	81DUR3			
Cont.	$^{74}GeD_3{}^{81}Br$	$v=0$	^{81}Br	317.5				
52	$(CH_3)_3C^{79}Br$	$v=0$	^{79}Br	513.7 (8)	79LE2	MW		2.4, 56
	$(CH_3)_3{}^{13}C^{79}Br$	$v=0$	^{79}Br	512 (1)				
	$(CH_3)_3C^{81}Br$	$v=0$	^{81}Br	432 (2)				
	$(CH_3)_3{}^{13}C^{81}Br$	$v=0$	^{81}Br	429 (2)				
53	$SF_5{}^{81}Br$	$v=0$	^{81}Br	585 (8)	81VER	MW		2.4, 85
54	$^{79}BrF_5$	$v=0$	^{79}Br	−280.5 (4)	76BRI	MW		2.4, 89
		$v_5=1$		−277.7 (5)				
		$v_9=1$		−279.2				
	$^{81}BrF_5$	$v=0$	^{81}Br	−234.2 (4)				
		$v_5=1$		−233.5 (5)				
		$v_9=1$		−233.5				
55	$^{83}KrHF$	$v=0$	^{83}Kr	10.23 (8)	81BUX	MWFT		2.3, 43
	$^{83}KrDF$	$v=0$	^{83}Kr	13.83 (13)				
			D	0.213 *)			*) fixed value	
	$^{84}KrDF$	$v=0$	D	0.215 (8)				
	$^{86}KrDF$	$v=0$	D	0.206 (15)				
	Kr · HCl		^{83}Kr	see Nr. 29				
	Kr · ClF		^{83}Kr	see Nr. 30				
56	ICN	$v=0$	^{127}I	−2420.409 (75)	77BUR	MW		2.3, 8
57	IC≡CCN	$v=0$	^{127}I	−2310 (10)	74BJO	MW		2.3, 30
58	CH_3I	$v=0$	^{127}I	$(eqQ)_0 = -1934.136\,(5)$	78BOU	BAS	*) uncertainties are 3 σ	2.4, 15
				$\chi_J = -1.409\,(44) \cdot 10^{-3}$				
				$\chi_K = -38.1\,(10) \cdot 10^{-3}$				
				$\chi_d = 26.21\,(62) \cdot 10^{-3}$				
		$v_1=1$		−1937.06 (9) *)	76HAL	La		
		$v_2=1$		−1936.37 (184) *)	78ARI	IRRFDR		
		$v_3=1$		−1934.52 (34) *)				
		$v_3=2$		−1934.80 (61) *)				
Cont.		$v_3=3$		−1934.48 (39) *)				

Nr.	Molecule	v	Nucleus	eqQ MHz	Ref.	Meth.	Varia, remarks	See Tab., Nr.
58 Cont.	CH_3I	$v_6=1$		$(eqQ)_0=-1940.413$ (72)	81DUB	MW	*) uncertainties are 3 σ	
				$eqQ\eta=2.324$ (56) *)			the absolute sign has been	
				$\chi_J=-1.218$ (138)			determined from IRMWDR [80AR2]	
				$\chi_K=-47.9$ (24)				
				$\chi_d=$ 24.77 (85)				
		$v_6=2$		−1946.59 (114) *)	78ARI	IRRFDR		
		$v_2=1$, $v_6=1$		−1944.64 (432) *)				
		$v_3=1$, $v_6=1$		−1940.57 (87) *)				
		$v_3=2$, $v_6=1$		−1940.32 (144) *)				
	$^{13}CH_3I$	$v=0$	^{127}I	−1934.12 (80) *)				
		$v_6=1$		−1941.06 (69) *)				
	CD_3I	$v=0$	^{127}I	−1928.83 (26)	73KUC	MW		
		$v_2=1$		−1930.30 (26)				
		$v_3=1$		−1930.30 (26)				
		$v_3=2$		−1930.85 (46)				
		$v_5=1$		$(eqQ)_0=-1927.93$ (70)				
				$eqQ\eta=0.90$ (90)				
		$v_6=1$		$(eqQ)_0=-1932.80$ (60)				
				$eqQ\eta=2.33$ (40)				
		$v_3=1$, $v_6=1$		$(eqQ)_0=-1933.61$ (50)				
				$eqQ\eta=3.09$ (40)				
	$^{13}CD_3I$	$v=0$	^{127}I	−1929.08 (14)	75MAL			
	CH_2DI	$v=0$	^{127}I	$\chi_{aa}=-1928.746$ (81)			$\chi_{zz}=-1932.438$ (130)	
				$\chi_{bb}=$ 962.527 (90)			$\theta_{za}=2.05$ (7)°	
				$\chi_{cc}=$ 966.219 (90)				
	$^{13}CH_2DI$	$v=0$	^{127}I	$\chi_{aa}=-1929.290$ (73)			$\chi_{zz}=-1932.492$ (150)	
				$\chi_{bb}=$ 963.044 (80)			$\theta_{za}=1.90$ (7)°	
				$\chi_{cc}=$ 966.246 (80)				
59	CF_3I	$v=0$	^{127}I	−2144.7 (6) *)	80COX	RFMWDR	*) uncertainty is 2 σ	2.4, 21
60	IF_5	$v=0$	^{127}I	1069.07 (40)	81BAL1	MW		2.4, 90
		$v_5=1$		1066.3 (15)			$v_5(b_1)\sim200$ cm^{-1}	
		$v_9=1$		1067.6 (20)			$v_9(e)\sim189$ cm^{-1}	
61	$^{131}XeHF$	$v=0$	^{131}Xe	−8.589 (39)	81BAI1	MBER		2.3, 46a
	$^{131}XeDF$	$v=0$	^{131}Xe	−10.567 (27)				
	$^{131}XeHCl$		^{131}Xe	see Nr. 31				

2.7.3 Asymmetric top molecules

2.7.3.1 Preliminary remarks

1. Introduction

See 2.7.2.1, Preliminary remarks, page 654.

Special remarks:

For *asymmetric top molecules*, the principal axis systems of the inertia tensor and the field gradient tensor do not coincide in general. In the case of a completely non-symmetric position of the quadrupolar nucleus in the molecule, none of the components $\chi_{gg'}$ of the field gradient tensor equals zero. If a nucleus lies on a plane which contains the principal inertial axes g and g', and which is a symmetry plane of the molecule, then the off-diagonal elements $\chi_{gg''}$ and $\chi_{g'g''}$ vanish.

An analysis of the quadrupole hfs yields in first-order approximation the diagonal components $eQ\frac{\partial^2 V}{\partial g^2} \equiv \chi_{gg}$, $g = a, b, c$ (see eq. (6)). On account of the Laplace equation, only two of these three quantities are independent. The nondiagonal elements $\chi_{gg'}$ can be determined by analysis of the quadrupole hfs of an isotopically substituted molecule where the principal inertial axes are rotated or by analysis of quadrupole hfs in higher order (see eq. (3)).

In many cases the coupling tensor cannot be determined completely. In order to discuss bond properties, one frequently assumes that the bond direction coincides with one of the principal axes of the quadrupole coupling tensor (bond axis system), yielding information if the position of the principal inertial axes is known.

2. Explanations of the columns in the tabulation

Column	Parameters		Definitions
1	Nr.		serial number of molecule in this table
2	v	v_j	vibrational quantum number of the j-th normal vibration: $$j = 1, \ldots, 3N-6$$ where N is the number of atoms. In the vibrational ground state $$v_1 = v_2 = \cdots v_j \cdots = v_{3N-6} = 0,$$ in short $$v = 0$$ vibrational state: $$v_1 = \cdots v_{j-1} = 0; \quad v_j = v; \quad v_{j+1} = v_{3N-6} = 0,$$ in short $$v_j = v$$
3			quadrupolar nucleus which produces hfs
4	$\chi_{gg'}$ [MHz]		components of the nuclear quadrupole coupling tensor in the principal inertial axes system $$\chi_{gg'} = \chi_{g'g} = eQ\frac{\partial^2 V}{\partial g\, \partial g'}; \quad g, g' = a, b, c$$ with $$\sum_g \chi_{gg} = 0$$
		$eQq(J_\tau)$ or $eQq(J_{K_{-1},K_{+1}})$	diagonal quadrupole coupling parameter in the rotational state J_τ (or $J_{K_{-1},K_{+1}}$), see eqs. (2), (3), (6)
Cont.		eQq	nuclear quadrupole coupling constant for treatment of the asymmetric top molecule as the limiting case of a linear or symmetric top molecule

Column	Parameters	Definitions
4 Cont.	$\chi_{gg'}$ [MHz] η	asymmetry parameter of the nuclear quadrupole coupling tensor $$\eta=\frac{\frac{\partial^2 V}{\partial g^2}-\frac{\partial^2 V}{\partial g'^2}}{\frac{\partial^2 V}{\partial g''^2}}$$ where (g, g', g'') can be either defined axis system and g'' is the axis next to a symmetry axis
5	$\chi_{\alpha\beta}$, χ_{xy} [MHz]	components of the nuclear quadrupole coupling tensor relative to a coordinate system which is specified in column 6. Greek letters (α, β, γ) refer to the system of principal axes of the field gradient tensor, latin letters (x, y, z) to the system of bond axes or similarly defined coordinate systems
6	θ	angle between axes of the individual coordinate systems used description of the reference system for the components of the quadrupole coupling tensor in column 5
7	Ref.	references for the previous columns
8	Varia, remarks ν_j a, b	remarks concerning the previous columns: type or frequency of the j-th normal vibration symmetry species for classification of the vibration: a nondegenerate, symmetric; b nondegenerate, antisymmetric references to figures in Chapter 3, subvolume b
9		references to tables 2.5

2.7.3.2 Data

Nr.	Molecule	v	Nucleus	Nuclear quadrupole coupling tensor components: in system of principal axes of inertia χ_{gg}, [MHz]	in system of principal coupling tensor or bond axes $\chi_{\alpha\beta}$, χ_{xy} [MHz]	position of the axes θ	Ref.	Varia, remarks Fig. in Chap. 3, subvol. b	See Tab., Nr.
1	DNCO	$v=0$	D	See Nr. 12					
2	$(DF)_2$	$v=0$	D	$\chi_{aa}=0.110\,(8)$ *)			72DYK	*) MBER	2.5, 38
	HFDF	$v=0$	D	$\chi_{aa}=0.270\,(30)$ *)					
3	HDO	$v=0$	D	$\chi_{aa}=\ 0.27645\,(88)$ *) $\chi_{bb}=-0.11097\,(146)$ $\chi_{cc}=-0.16577\,(110)$	$\chi_{zz}=\ 0.3132$ $\chi_{yy}=-0.1477$ $\chi_{xx}=-0.1658$		82FRY	*) BMS z‖OD	2.5, 53
	D_2O	$v=0$	D	$eqQ(1_{10})$ $=0.03481\,(32)$ *) $eqQ(1_{01})$ $=-0.02960\,(55)$ *)	$\chi_{\alpha\alpha}=\ 0.3075\,(6)$ $\chi_{\beta\beta}=-0.1326\,(6)$ $\chi_{\gamma\gamma}=-0.1749\,(6)$	$\theta=1°16'\pm3''$	77GAR		
	$H_2{}^{17}O$	$v=0$	^{17}O	$\chi_{aa}=-8.899$ $\chi_{bb}=-1.269$ $\chi_{cc}=\ 10.169$			75LUC2		
4	CH_3OD	$v=0$			$\chi_{zz}=0.303\,(12)$ *)		73CAS	*) along the O—D bond direction, BMS	2.5, 65
5	CD_2F_2	$v=0$	D		$\chi_{zz}=\ 0.186\,(10)$ *) $\eta\ =-0.15\,(5)$		74NEL	*) along the C—D bond direction, BMS	2.5, 93
6	D_2CO	$v=0$	D	$\chi(1_{10})=\ 0.01694\,(3)$ *) $\chi(1_{11})=-0.01945\,(3)$	$V_{xx}=\ 8.04\,(6)$ **) $V_{yy}=-3.86\,(6)$ $V_{zz}=-4.180\,(5)$	$\theta_{ax}=58°31'\pm38'$	73TUC	*) BMS **) electric field gradients (in units of 10^{14} statvolt · cm^{-2}	2.5, 108
	HDCO	$v=0$	D	$\chi_{aa}=\ 0.00815\,(45)$ *) $\chi_{bb}=-0.07660\,(45)$			71TUC		
	$H_2C^{17}O$	$v=0$	^{17}O	$\chi_{aa}=-1.900\,(118)$ **) $\eta\ =-12.00\,(75)$			80COR1	**) uncertainties are 3 σ	
	$H_2{}^{13}C^{17}O$	$v=0$	^{17}O	$\chi_{aa}=-1.906\,(129)$ **) $\eta\ =-11.98\,(81)$					

Nr.	Molecule	v	Nucleus	Nuclear quadrupole coupling tensor components in system of principal axes of inertia χ_{gg}, [MHz]	in system of principal coupling tensor or bond axes $\chi_{\alpha\beta}$, χ_{xy} [MHz]	position of the axes θ	Ref.	Varia, remarks Fig. in Chap. 3, subvol. b	See Tab., Nr.
7	HCOOD trans DCOOH trans	$v=0$ $v=0$	D D	$\chi_{aa}=-0.1193$ (20) *) $\chi_{bb}=$ 0.2675 (30) $\chi_{cc}=-0.1482$ (20)	$\chi_{xx}=-0.148$ **) $\chi_{yy}=-0.124$ $\chi_{zz}=$ 0.272 (3) $\chi_{zz}=$ 0.166 (2)		74RU2	*) BMS **) the z axis is along the bond to D and the x axis is perpendicular to the plane of the molecule	2.5, 111
8	$D_2C{=}C{=}O$	$v=0$	D	$\chi_{aa}=-0.019793$ (132) *) $\chi_{bb}=$ 0.132370 (128)			77FAB	*) MBER	2.5, 299
9	Ar · DC≡CD	$v=0$	D	$\chi_{aa}+\chi_{bb}=$ 0.1020(27)*) $\chi_{aa}+\chi_{cc}=-0.1740$(30)	$\chi_{zz}=0.2044$ (10)		80DEL2	*) MBER	2.5, 764
9a	 DF	$v=0$	D	$\chi_{aa}=0.287$ (11)			81BUX3		2.5, 763b
	HFDCl		D	See Nr. 90					
9b	$^{11}BH_2OH$ $^{11}BH_2{}^{18}OH$ $^{11}BD_2OD$ $^{10}BH_2OH$	$v=0$ $v=0$ $v=0$ $v=0$	^{11}B ^{11}B ^{11}B ^{10}B	$\chi_{aa}=-1.1$ (7) $\chi_{bb}=-2.8$ (4) $\chi_{cc}=$ 3.9 (4) $\chi_{aa}=-0.9$ (8) $\chi_{bb}=-2.8$ (5) $\chi_{cc}=$ 3.7 (5) $\chi_{aa}=-1.4$ (9) $\chi_{bb}=-2.1$ (5) $\chi_{cc}=$ 3.5 (5) $\chi_{aa}=-2.2$ (8) $\chi_{bb}=-5.5$ (5) $\chi_{cc}=$ 7.9 (5)			81KAW		2.5, 47a
10	$CH_3{}^{11}BF_2$ $CH_3{}^{10}BF_2$	$v=0$ $v=0$	^{11}B ^{10}B	$\chi_{aa}=-2.713$ (21) $\chi_{bb}=-0.695$ (27) $\chi_{cc}=$ 3.408 (27) $\chi_{aa}=-5.654$ $\chi_{bb}=-1.448$ $\chi_{cc}=$ 7.102			79LAG		2.5, 412

Nr.	Molecule	v	Nucleus	Nuclear quadrupole coupling tensor components: in system of principal axes of inertia χ_{gg}, [MHz]	in system of principal coupling tensor or bond axes $\chi_{\alpha\beta}$, χ_{xy} [MHz]	position of the axes θ	Ref.	Varia, remarks Fig. in Chap. 3, subvol. b	See Tab., Nr.
11	HN=SOF$_2$	$v=0$	^{14}N	$\chi_{aa}=$ 0.46 (2)	$\chi_{\alpha\alpha}=-1.38$ (1)	$\theta_{\alpha a}=37.5$ (11)	77CAX1,	Fig. 9	2.5, 4
				$\chi_{bb}=$ 1.75 (2)	$\chi_{\beta\beta}=$ 3.59 32)		78CAX2		
				$\chi_{cc}=-2.210$ (5)	$\chi_{\gamma\gamma}=-2.21$ (1)				
	DN=SOF$_2$	$v=0$	^{14}N	$\chi_{aa}=$ 2.44 (25)			78CAX2		
				$\chi_{bb}=-0.67$ (85)					
				$\chi_{cc}=-1.77$ (67)					
12	HNCO	$v=0$	^{14}N	$\chi_{aa}=$ 2.0527 (10) *)			71KUK,	*) BMS data	2.5, 5
				$\chi_{bb}=-0.473$ (7)			75HO1		
				$\chi_{cc}=-1.583$ (7)					
		$v_4=1$	^{14}N	$\chi_{aa}=2.099$ (21)			77YAM4		
				$\chi_{bb}-\chi_{cc}=0.784$ (13)					
		$v_5=1$	^{14}N	$\chi_{aa}=1.991$ (24)					
				$\chi_{bb}-\chi_{cc}=0.728$ (11)					
		$v_6=1$	^{14}N	$\chi_{aa}=1.982$ (19)					
				$\chi_{bb}-\chi_{cc}=0.445$ (16)					
	DNCO	$v=0$	D	$\chi_{aa}=$ 0.0536 (2) *)	$\chi_{zz}=0.345$ (2)		71KUK		
			^{14}N	$\chi_{aa}=$ 2.1230 (10) *)			71KUK,		
				$\chi_{bb}=-0.540$ (12)			75HO1		
				$\chi_{cc}=-1.570$ (12)					
	HN^{13}CO	$v=0$	^{14}N	$\chi_{aa}=$ 2.067 (20)			75HO1		
				$\chi_{bb}=-0.482$ (11)					
				$\chi_{cc}=-1.585$ (11)					
	HNC^{18}O	$v=0$	^{14}N	$\chi_{aa}=$ 2.060 (13)					
				$\chi_{bb}=-0.472$ (8)					
				$\chi_{cc}=-1.588$ (8)					
13	HNCS	$v=0$	^{14}N	$\chi_{aa}=1.114$ (26)			79YA	Fig. 11	2.5, 6
				$\chi_{bb}-\chi_{cc}=0.055$ (115)					
		$v_4=1$	^{14}N	$\chi_{aa}=1.206$ (28)			78YAM		
		$v_5=1$	^{14}N	$\chi_{aa}=1.080$ (22)					
		$v_6=1$	^{14}N	$\chi_{aa}=1.016$ (18)					
	HN^{13}CS	$v=0$	^{14}N	$\chi_{aa}=1.114$ (17)			80YA3		
	HNC^{34}S	$v=0$	^{14}N	$\chi_{aa}=1.121$ (22)					
	DNCS	$v=0$	^{14}N	$\chi_{aa}=1.189$ (19)			78SZA		
				$\chi_{bb}-\chi_{cc}=0.014$ (140)					

Nr.	Molecule	v	Nucleus	Nuclear quadrupole coupling tensor components: in system of principal axes of inertia χ_{gg}, [MHz]	in system of principal coupling tensor or bond axes $\chi_{\alpha\beta}$, χ_{xy} [MHz]	position of the axes θ	Ref.	Varia, remarks Fig. in Chap. 3, subvol. b	See Tab., Nr.
14	HNCSe	$v=0$	^{14}N	$\chi_{aa}=1.17$			79LAD		2.5, 7
15	HNO	$v=0$	^{14}N	$\chi_{aa}=$ 0.36 (56) $\chi_{bb}=-5.46$ (30) $\chi_{cc}=$ 5.10 (26)			73SAI	Fig. 13	2.5, 12
16	$NOC^{15}N$ $^{15}NOCN$	$v=0$ $v=0$	^{14}N ^{14}N	$\chi_{aa}=$ 1.52 $\chi_{bb}=-5.68$ $\chi_{aa}=-3.92$ $\chi_{bb}=$ 1.08			78DIC	Fig. 14	2.5, 16
17	$N_\delta CN_\alpha N_\beta N_\gamma$	$v=0$	$^{14}N_\alpha$ $^{14}N_\beta$ $^{14}N_\gamma$ $^{14}N_\delta$	$\chi_{aa}=$ 4.82 (2) $\chi_{bb}=-0.70$ (8) $\chi_{cc}=-4.12$ (10) $\chi_{aa}=-0.85$ (7) $\chi_{bb}=$ 0.70 (8) $\chi_{cc}=$ 0.15 (15) $\chi_{aa}=-0.75$ (7) $\chi_{bb}=$ 1.55 (5) $\chi_{cc}=-0.80$ (12) $\chi_{aa}=-2.27$ (4) $\chi_{bb}=$ 1.15 (7) $\chi_{cc}=$ 1.12 (11)			73BLA	Fig. 16	2.5, 22
18	HNSO	$v=0$	^{14}N		$\chi_{\alpha\alpha}=-3.21$ (50) $\chi_{\beta\beta}=$ 1.64 (75) $\chi_{\gamma\gamma}=$ 1.57 (14)	$\theta_{\alpha a}=36.4$ (40)°	78CAX2		2.5, 23
19	N_2H_4 N_2D_4	$v=0$ $v=0$	^{14}N ^{14}N	$\chi_{aa}=$ 4.09 $\chi_{aa}=$ 4.234 (36) $\chi_{bb}=-1.978$ (51) $\chi_{cc}=-2.255$ (51)			77HAR2	Fig. 17	2.5, 26
	ClNCO		^{14}N	See Nr. 81					
	CH_3NHCl		^{14}N	See Nr. 94					

Nr.	Molecule	v	Nucleus	Nuclear quadrupole coupling tensor components in system of principal axes of inertia χ_{gg}, [MHz]	in system of principal coupling tensor or bond axes $\chi_{\alpha\beta}$, χ_{xy} [MHz]	position of the axes θ	Ref.	Varia, remarks Fig. in Chap. 3, subvol. b	See Tab., Nr.
20	CH_3NHOH	$v=0$	^{14}N	$\chi_{aa}=$ 4.35 (1) $\chi_{bb}=$ 0.67 (1) $\chi_{cc}=-5.02$ (2) $\chi_{ab}=$ 0.2 (1) $\chi_{ac}=-0.4$ (4) $\chi_{bc}=-3.03$ (20)	$\chi_{\alpha\alpha}=$ 4.41 (30) $\chi_{\beta\beta}=$ 1.93 (45) $\chi_{\gamma\gamma}=-6.34$ (30)		79SUN		2.5, 71
	CH_3NDOH	$v=0$	^{14}N	$\chi_{aa}=$ 4.34 (1) $\chi_{bb}=$ 1.26 (1) $\chi_{cc}=-5.60$ (2)					
	CH_3NHOD	$v=0$	^{14}N	$\chi_{aa}=$ 4.35 (2) $\chi_{bb}=$ 0.75 (1) $\chi_{cc}=-5.10$ (3)					
	CH_3NDOD	$v=0$	^{14}N	$\chi_{aa}=$ 4.32 (1) $\chi_{bb}=$ 1.30 (1) $\chi_{cc}=-5.62$ (2) $\chi_{ab}=$ 0.2 (1) $\chi_{ac}=-0.6$ (4) $\chi_{bc}=-2.29$ (15)					
21	CH_3ONO, cis	$v=0$	^{14}N	$\chi_{aa}=$ 1.443 (81) $\chi_{bb}=-4.874$ (76) $\chi_{cc}=$ 3.431 (80) $\chi_{ab}=$ 2.8 (6)	$\chi_{\alpha\alpha}=-5.98$ (50) $\chi_{\beta\beta}=$ 2.55 (50) $\chi_{\gamma\gamma}=$ 3.43 (8)	$\theta_{a\beta}=20.9$ (30)°	80GHO 79TUR1	Fig. 29	2.5, 82
	cis CD_3ONO	$v=0$	^{14}N	$\chi_{aa}=$ 0.97 (20) $\chi_{bb}=-4.47$ (3) $\chi_{cc}=$ 3.50 (3)	$\chi_{\alpha\alpha}=-5.9$ (5) $\chi_{\beta\beta}=$ 2.4 (5) $\chi_{\gamma\gamma}=$ 3.48 (3)	$\theta_{\alpha b}=20.7$ (30)°			
22	CH_3ONH_2	$v=0$	^{14}N	$\chi_{aa}=$ 3.63 (3) $\chi_{bb}=-3.69$ (7) $\chi_{cc}=$ 0.06 (7)			74FON		2.5, 84
	CH_3ONHD	$v=0$	^{14}N	$\chi_{aa}=$ 3.73 (3) $\chi_{bb}=-3.74$ (8) $\chi_{cc}=$ 0.01 (8)					
	CH_3OND_2	$v=0$	^{14}N	$\chi_{aa}=$ 3.68 (5) $\chi_{bb}=-3.71$ (14) $\chi_{cc}=$ 0.03 (14)					

Nr.	Molecule	v	Nucleus	Nuclear quadrupole coupling tensor components in system of principal axes of inertia χ_{gg}, [MHz]	in system of principal coupling tensor or bond axes $\chi_{\alpha\beta}$, χ_{xy} [MHz]	position of the axes θ	Ref.	Varia, remarks Fig. in Chap. 3, subvol. b	See Tab., Nr.
23	CH_2OHCN	$v=0$	^{14}N	$\chi_{aa}=-3.63$			73CAZ3	Fig. 32	2.5, 100
				$\chi_{bb}=1.64$					
		$v_\tau=1$	^{14}N	$\chi_{aa}=-3.35$				OH torsion	
				$\chi_{bb}=1.39$					
	CH_2ODCN	$v=0$	^{14}N	$\chi_{aa}=-3.64$					
				$\chi_{bb}=1.56$					
		$v_\tau=1$	^{14}N	$\chi_{aa}=-3.41$					
24	$N_\alpha H_2CH_2CN_\beta$	$v=0$	$^{14}N_\alpha$	$\chi_{aa}=-2.77(4)$			77BRO		2.5, 101
				$\chi_{bb}=1.20(9)$					
			$^{14}N_\beta$	$\chi_{aa}=-3.48(3)$					
				$\chi_{bb}=1.50(6)$					
25	CH_2NH	$v=0$	^{14}N	$\chi_{aa}=-0.899(28)$			77PEA		2.5, 103
				$\chi_{bb}=-2.653(19)$					
				$\chi_{cc}=3.553(21)$					
	CH_2ND	$v=0$	^{14}N	$\chi_{aa}=-1.18(20)$					
				$\chi_{bb}=-2.76(9)$					
				$\chi_{cc}=3.95(13)$					
26	$HN(CHO)_2$	$v=0$	^{14}N	$\chi_{aa}=1.88(11)$			73STE1		2.5, 122
				$\chi_{bb}=1.53(7)$					
				$\chi_{cc}=-3.41(7)$					
	$DN(CHO)_2$	$v=0$	^{14}N	$\chi_{aa}=1.65(28)$					
				$\chi_{bb}=1.69(18)$					
				$\chi_{cc}=-3.34(18)$					
27	$OC(NH_2)_2$	$v=0$	^{14}N	$\chi_{aa}=2.16(2)$			75BRW1		2.5, 129
				$\chi_{bb}=1.88(2)$					
				$\chi_{cc}=-4.04(4)$					
28	$HC(=S)NH_2$	$v=0$	^{14}N	$\chi_{aa}=2.9(12)$			74SUG		2.5, 149
				$\chi_{cc}-\chi_{bb}=-5.39(15)$					

Nr.	Molecule	v	Nucleus	Nuclear quadrupole coupling tensor components in system of principal axes of inertia χ_{gg}, [MHz]	in system of principal coupling tensor or bond axes $\chi_{\alpha\beta}$, χ_{xy} [MHz]	position of the axes θ	Ref.	Varia, remarks Fig. in Chap. 3, subvol. b	See Tab., Nr.
29	CH_3CD_2CN	$v=0$	^{14}N	$\chi_{aa}=-3.449$ (16) $\chi_{bb}=1.399$ (15) $\chi_{cc}=2.050$ (17) $\chi_{ab}=-2.01$ (18)	$\chi_{\alpha\alpha}=-4.173\pm0.16$ $\chi_{\beta\beta}=2.119\pm0.16$ $\chi_{\gamma\gamma}=2.055\pm0.018$ $\eta=0.02$ (4)	$\theta_{\alpha a}=19.8$ (17)°	74MAD, 74HEI		2.5, 155
	CD_3CH_2CN	$v=0$	^{14}N	$\chi_{aa}=-3.209$ (17) $\chi_{bb}=1.150$ (16) $\chi_{cc}=2.059$ (18) $\chi_{ab}=-2.27$ (12)			74HEI		
	$^{13}CH_3CH_2CN$	$v=0$	^{14}N	$\chi_{aa}=-3.290$ (11) $\chi_{bb}=1.242$ (10) $\chi_{cc}=2.048$ (12)					
	$CH_3{}^{13}CH_2CN$	$v=0$	^{14}N	$\chi_{aa}=-3.377$ (21) $\chi_{bb}=1.334$ (17) $\chi_{cc}=2.043$ (19)					
	$CH_3CH_2{}^{13}CN$	$v=0$	^{14}N	$\chi_{aa}=-3.349$ (11) $\chi_{bb}=1.297$ (10) $\chi_{cc}=2.052$ (12)					
	CH_2DCH_2CN symmetric	$v=0$	^{14}N	$\chi_{aa}=-3.326$ (15) $\chi_{bb}=1.319$ (14) $\chi_{cc}=2.007$ (16)			74MAD		
	CH_2DCH_2CN asymmetric	$v=0$	^{14}N	$\chi_{aa}=-3.290$ (19) $\chi_{bb}=1.180$ (18) $\chi_{cc}=2.110$ (22)					
	CD_3CD_2CN	$v=0$	^{14}N	$\chi_{aa}=-3.213$ $\chi_{bb}=1.168$ $\chi_{cc}=2.045$			74LI		
30	$CH_3CH_2NHCH_3$*)	$v=0$	^{14}N	$\chi_{aa}=2.82$ (9) $\chi_{bb}=0.88$ (13) $\chi_{cc}=-3.70$ (9)			73PEN1	*) trans form	2.5, 165
	$CH_3CH_2NDCH_3$*)	$v=0$	^{14}N	$\chi_{aa}=2.82$ (3) $\chi_{bb}=1.33$ (4) $\chi_{cc}=-4.15$ (2)					

Nr.	Molecule	v	Nucleus	Nuclear quadrupole coupling tensor components: in system of principal axes of inertia χ_{gg}, [MHz]	in system of principal coupling tensor or bond axes $\chi_{\alpha\beta}$, χ_{xy} [MHz]	position of the axes θ	Ref.	Varia, remarks Fig. in Chap. 3, subvol. b	See Tab., Nr.
31	$CH_3CH{=}NH$ cis	$v=0$	^{14}N	$\chi_{aa} = 0.70\,(45)$ *) $\chi_{bb} = -3.90\,(24)$ *) $\chi_{cc} = 3.20\,(37)$			80LOV	*) uncertainties are 3 σ	2.5, 171
	$CH_3CH{=}NH$ trans	$v=0$	^{14}N	$\chi_{aa} = -3.51\,(20)$ *) $\chi_{bb} = 1.01\,(61)$ $\chi_{cc} = 2.50\,(51)$					
32	$CH_2NH_2CH_2F$ conformer I *)	$v=0$	^{14}N	$\chi_{aa} = 2.02\,(13)$ $\chi_{bb} = 2.49\,(10)$			80MAR1	*) see Fig. 52	2.5, 177
	$CH_2NH_2CH_2F$ conformer II *)	$v=0$	^{14}N	$\chi_{aa} = -2.78\,(14)$ $\chi_{bb} = 2.34\,(8)$					
33	$CH_3OCH_2CH_2NH_2$	$v=0$	^{14}N	$\chi_{aa} = -3.33\,(3)$ $\chi_{bb} = 2.50\,(2)$ $\chi_{cc} = 0.83\,(5)$			80CAM1	Fig. 55	2.5, 182
34	$CH_3C(=O)CN$	$v=0$	^{14}N	$\chi_{aa} = -4.34\,(7)$ $\chi_{bb} = 2.03\,(31)$ $\chi_{cc} = 2.31\,(31)$	$\chi_{xx} = 2.06$ $\chi_{yy} = 2.31$ $\chi_{zz} = -4.37$		76SCA2		2.5, 186
		$v_\tau=1$	^{14}N	$\chi_{aa} = -4.72\,(7)$ $\chi_{bb} = 1.14\,(15)$ $\chi_{cc} = 3.58\,(15)$					
		$v_\delta=1$	^{14}N	$\chi_{aa} = -4.15\,(9)$ $\chi_{bb} = 2.01\,(24)$ $\chi_{cc} = 2.14\,(24)$					
	$CH_2DC(=O)CN$ symmetric	$v=0$	^{14}N	$\chi_{aa} = -4.72\,(7)$ $\chi_{bb} = 1.93\,(29)$ $\chi_{cc} = 2.79\,(29)$			77PAN1		
	$CH_2DC(=O)CN$ asymmetric	$v=0$	^{14}N	$\chi_{aa} = -3.88\,(9)$ $\chi_{bb} = 2.38\,(31)$ $\chi_{cc} = 1.50\,(31)$					

Nr.	Molecule	v	Nucleus	Nuclear quadrupole coupling tensor components: in system of principal axes of inertia χ_{gg}, [MHz]	in system of principal coupling tensor or bond axes $\chi_{\alpha\beta}$, χ_{xy} [MHz]	position of the axes θ	Ref.	Varia, remarks Fig. in Chap. 3, subvol. b	See Tab., Nr.
35	$CH_2FC(=O)NH_2$	$v=0$	^{14}N	$\chi_{aa}=1.6(2)$ $\chi_{bb}=1.6(2)$ $\chi_{cc}=-3.2(3)$			74MAR2		2.5, 205
36	$(CH_3)_2CHNH_2$ $(CH_3)_2CHND_2$	$v=0$ $v=0$	^{14}N ^{14}N	$\chi_{aa}=2.160(30)$ *) $\chi_{bb}=2.160(30)$ $\chi_{cc}=-4.377(70)$ $\chi_{aa}=2.205(30)$ *) $\chi_{bb}=2.205(30)$ $\chi_{cc}=-4.523(25)$			77MEH	*) symmetry plane assumed (trans form)	2.5, 239
37	$CH_2=CHNH_2$	$v=0$ $v_i=1$	^{14}N ^{14}N	$\chi_{aa}=2.20(22)$ $\chi_{bb}=1.95(18)$ $\chi_{cc}=-4.13(20)$ $\chi_{aa}=1.64(35)$ $\chi_{bb}=2.17(31)$ $\chi_{cc}=-3.81(27)$			75LOV		2.5, 287
38	$CH_2=CHNCO$ trans $CH_2=CHNCO$ cis	$v=0$ $v=0$	^{14}N ^{14}N	$\chi_{aa}=2.52(2)$ $\chi_{bb}=-1.56(1)$ $\chi_{cc}=-0.96(2)$ $\chi_{aa}=2.76(2)$ $\chi_{bb}=-1.88(1)$ $\chi_{cc}=-0.88(1)$	$\chi_{\alpha\alpha}=2.83$ $\chi_{\beta\beta}=-1.87$ $\chi_{\gamma\gamma}=-0.96$ $\chi_{\alpha\alpha}=2.80$ $\chi_{\beta\beta}=-1.91$ $\chi_{\gamma\gamma}=-0.88$	$\theta_{a\alpha}=15.0°$ $\theta_{a\alpha}=-5.0°$	77BOU3		2.5, 289
39	$CH_2=CHCH_2NH_2$ N cis lone pair gauche	$v=0$	^{14}N	$\chi_{aa}=1.05(6)$ $\chi_{bb}-\chi_{cc}=1.81(16)$			78BOT	Fig. 92	2.5, 311
40	$CH_2=CHCH=NH$ trans	$v=0$	^{14}N	$\chi_{aa}=-4.0(4)$			78PEN1	Fig. 94	2.5, 322
41	$CH_2=C=CHCN$	$v=0$	^{14}N	$\chi_{aa}=-2.71(5)$ $\chi_{bb}=0.63(15)$ $\chi_{cc}=2.08(15)$			73BOU	Fig. 99	2.5, 347

Nr.	Molecule	v	Nucleus	Nuclear quadrupole coupling tensor components: in system of principal axes of inertia χ_{gg}, [MHz]	in system of principal coupling tensor or bond axes $\chi_{\alpha\beta}$, χ_{xy} [MHz]	position of the axes θ	Ref.	Varia, remarks Fig. in Chap. 3, subvol. b	See Tab., Nr.
42	$HC{\equiv}C{-}C(=O)NH_2$	$v=0$	^{14}N	$\chi_{aa}=1.85\,(14)$ $\chi_{bb}-\chi_{cc}=5.79\,(7)$			78LIT	Fig. 103	2.5, 368
		$v_{in}=1$	^{14}N	$\chi_{aa}=1.58\,(34)$ $\chi_{bb}-\chi_{cc}=5.84\,(11)$				$\nu_{in}=273\,(75)\ cm^{-1}$, C≡C—C in-plane bending.	
		$v_{out}=1$	^{14}N	$\chi_{aa}=2.52\,(75)$ $\chi_{bb}-\chi_{cc}=5.63\,(20)$				$\nu_{out}=275\ cm^{-1}$	
		$v_{inv}=1$	^{14}N	$\chi_{aa}=1.91\,(47)$ $\chi_{bb}-\chi_{cc}=5.46\,(13)$				$\nu_{inv}=333\,(75)\ cm^{-1}$, NH_2 out-of-plane wagging.	
	$HC{\equiv}C{-}C(=O)ND_2$	$v=0$	^{14}N	$\chi_{aa}=16.11\,(4.86)$ $\chi_{bb}-\chi_{cc}=4.92\,(22)$					
	$DC{\equiv}C{-}C(=O)ND_2$	$v=0$	^{14}N	$\chi_{aa}=1.66\,(2.01)$ $\chi_{bb}-\chi_{cc}=5.81\,(37)$					
43	$(CH_3)_2{}^{14}N^{15}NO$	$v=0$	^{14}N	$\chi_{aa}=1.92\,(10)$ $\chi_{bb}=2.09\,(10)$ $\chi_{cc}=-4.11\,(10)$			75GUA	Fig. 129	2.5, 440
	$(CH_3)_2{}^{15}N^{14}NO$	$v=0$	^{14}N	$\chi_{aa}=1.83\,(10)$ $\chi_{bb}=-5.55\,(10)$ $\chi_{cc}=3.72\,(10)$					
44	$B_2H_5NH_2$	$v=0$	^{14}N	$\chi_{aa}=-2.71\,(3)$ $\chi_{bb}=0.84\,(5)$ $\chi_{cc}=1.87\,(8)$			74LAU	Fig. 115	2.5, 415
45	$H_2C{-}CHCN$ (ring closed by CH_2)	$v=0$	^{14}N	$\chi_{aa}=-3.453\,(11)$ $\chi_{bb}=1.759\,(16)$			80BRO1		2.5, 458
46	$H_2C{-}CHNO$ (ring closed by CH_2)	$v=0$	^{14}N	$\chi_{aa}=0.80\,(23)$ $\chi_{bb}=5.25\,(13)$ $\chi_{cc}=-6.05\,(13)$			78COR		2.5, 463
47	$H_2C{-}CHCH_3$, cis (ring closed by NH)	$v=0$	^{14}N	$\chi_{aa}=-1.16$ $\chi_{bb}=1.80$ $\chi_{cc}=-0.64$			73LI3, 73SCH, 75LI1		2.5, 491

Nr.	Molecule	v	Nucleus	Nuclear quadrupole coupling tensor components in system of principal axes of inertia χ_{gg}, [MHz]	in system of principal coupling tensor or bond axes $\chi_{\alpha\beta}$, χ_{xy} [MHz]	position of the axes θ	Ref.	Varia, remarks Fig. in Chap. 3, subvol. b	See Tab., Nr.
48	H_2C—$CHCN_\beta$, cis (ring with $N_\alpha H$)	$v=0$	$^{14}N_\alpha$ $^{14}N_\beta$	$\chi_{aa} = -1.249\,(14)$ $\chi_{bb} = 1.407\,(12)$ $\chi_{aa} = -3.547\,(6)$ $\chi_{bb} = 1.865\,(8)$			80BRO2	Fig. 142	2.5, 492
49	H_2C—$CHCN$ H_2C—CH_2	$v=0$	^{14}N	$\chi_{aa} = -3.663\,(5)$ $\chi_{bb} = 2.086\,(6)$ $\chi_{cc} = 1.577\,(6)$			73FON		2.5, 494
50	H_2C—$CHCN$ H_2C=C—CH_2	$v=0$	^{14}N	$\chi_{aa} = -3.49\,(15)$			74DUR9		2.5, 498
51	H_2C—NCH_3 H_2C—NCH_3	$v=0$	^{14}N	$\chi_{aa} = 2.939\,(15)$ $\chi_{bb} = 1.400\,(12)$ $\chi_{cc} = -4.339\,(10)$			80GEB		2.5, 509
51 a	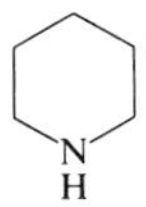, equatorial axial	$v=0$ $v=0$	^{14}N ^{14}N	$\chi_{aa} = 2.82\,(398)$ $\chi_{bb} = 2.01\,(195)$ $\chi_{cc} = -4.83\,(4)$ $\chi_{aa} = -3.80\,(10)$ $\chi_{bb} = 2.97\,(10)$ $\chi_{cc} = 0.83\,(26)$			81PAR		2.5, 537
52	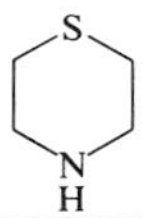	$v=0$	^{14}N	$\chi_{aa} = 1.74\,(9)$ $\chi_{bb} = 2.89\,(26)$ $\chi_{cc} = -4.63\,(35)$			80FAN		2.5, 540
53	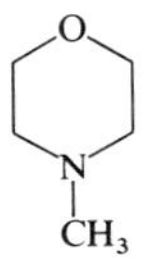	$v=0$	^{14}N	$\chi_{aa} = 2.4\,(1)$ $\chi_{bb} = 2.6\,(3)$ $\chi_{cc} = -5.0\,(2)$			74DAS		2.5, 541

Nr.	Molecule	v	Nucleus	Nuclear quadrupole coupling tensor components: in system of principal axes of inertia χ_{gg}, [MHz]	in system of principal coupling tensor or bond axes $\chi_{\alpha\beta}$, χ_{xy} [MHz]	position of the axes θ	Ref.	Varia, remarks Fig. in Chap. 3, subvol. b	See Tab., Nr.
54	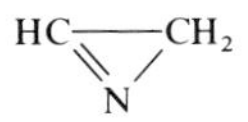	$v=0$	^{14}N	$\chi_{aa}=1.04$ $\chi_{bb}=-3.43$ $\chi_{cc}=2.39$			77FOR1		2.5, 589
55	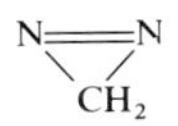	$v=0$	^{14}N	$\chi_{aa}=0.196\,(150)$ $\chi_{bb}=-3.010\,(150)$ $\chi_{cc}=2.814\,(150)$			72POC		2.5, 590
56	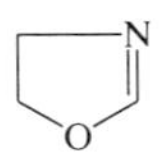	$v=0$	^{14}N	$\chi_{aa}=-3.54$ $\chi_{bb}=2.03$ $\chi_{cc}=1.51$			74DUR12		2.5, 602
		$v_R=1$		$\chi_{aa}=-3.52$ $\chi_{bb}=2.07$ $\chi_{cc}=1.45$				v_R, ring-puckering vibration	
		$v_R=2$		$\chi_{aa}=-3.52$ $\chi_{bb}=2.02$ $\chi_{cc}=1.50$					
57		$v=0$	^{14}N	$\chi_{aa}=-3.36\,(9)$ $\chi_{bb}=-0.26\,(18)$			80HAL		2.5, 603
58	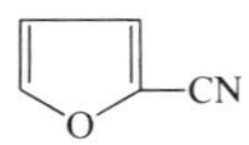	$v=0$	^{14}N	$\chi_{bb}+\chi_{cc}=4.26\,(5)$ $\chi_{bb}-\chi_{cc}=0.96\,(7)$			76EN		2.5, 613
59	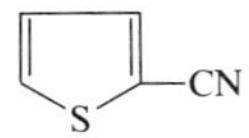	$v=0$	^{14}N	$\chi_{bb}+\chi_{cc}=4.04\,(16)$ $\chi_{bb}-\chi_{cc}\sim 0.96$			77WIE1		2.5, 617
60	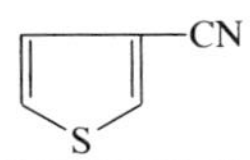	$v=0$	^{14}N	$\chi_{bb}+\chi_{cc}=4.12\,(6)$			77WIE2		2.5, 618
61	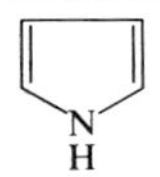	$v=0$	^{14}N	$\chi_{aa}=1.4057\,(11)$ $\chi_{bb}=1.2945\,(10)$			73GAI, 74BOL1		2.5, 627

Nr.	Molecule	v	Nucleus	Nuclear quadrupole coupling tensor components: in system of principal axes of inertia χ_{gg}, [MHz]	in system of principal coupling tensor or bond axes $\chi_{\alpha\beta}$, χ_{xy} [MHz]	position of the axes θ	Ref.	Varia, remarks Fig. in Chap. 3, subvol. b	See Tab., Nr.
62	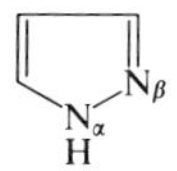	$v=0$	$^{14}N_\alpha$	$\chi_{aa}=$ 1.49 (6) $\chi_{bb}=$ 1.57 (10) $\chi_{cc}=-3.06$ (10) $\chi_{ab}=$ 0.562 (100)	$\chi_{\alpha\alpha}=$ 0.97 $\chi_{\beta\beta}=$ 2.09 $\chi_{\gamma\gamma}=-3.06$		73NAS		2.5, 630
			$^{14}N_\beta$	$\chi_{aa}=-3.89$ (4) $\chi_{bb}=$ 3.18 (6) $\chi_{cc}=$ 0.71 (6) $\chi_{ab}=$ 2.11 (6)	$\chi_{\alpha\alpha}=-4.47$ $\chi_{\beta\beta}=$ 3.76 $\chi_{\gamma\gamma}=$ 0.71				
	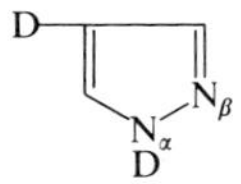	$v=0$	$^{14}N_\alpha$	$\chi_{aa}=$ 0.97 $\chi_{bb}=$ 2.09 $\chi_{cc}=-3.06$					
			$^{14}N_\beta$	$\chi_{aa}=-2.64$ $\chi_{bb}=$ 1.93 $\chi_{cc}=$ 0.71					
	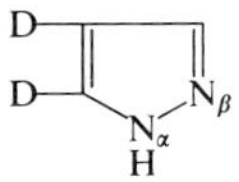	$v=0$	$^{14}N_\alpha$	$\chi_{aa}=$ 1.52 $\chi_{bb}=$ 1.54 $\chi_{cc}=-3.06$					
			$^{14}N_\beta$	$\chi_{aa}=-3.78$ $\chi_{bb}=$ 3.07 $\chi_{cc}=$ 0.71					
	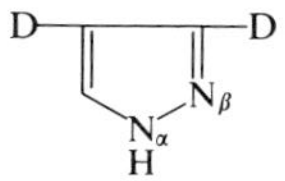	$v=0$	$^{14}N_\alpha$	$\chi_{aa}=$ 0.98 $\chi_{bb}=$ 2.08 $\chi_{cc}=-3.06$					
			$^{14}N_\beta$	$\chi_{aa}=-1.30$ $\chi_{bb}=$ 0.59 $\chi_{cc}=$ 0.71					
	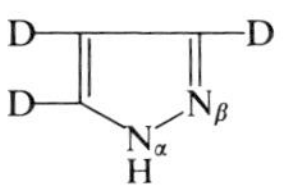	$v=0$	$^{14}N_\alpha$	$\chi_{aa}=$ 2.07 $\chi_{bb}=$ 0.99 $\chi_{cc}=-3.06$					
			$^{14}N_\beta$	$\chi_{aa}=$ 1.95 $\chi_{bb}=-2.66$ $\chi_{cc}=$ 0.71					
Cont.									

Nr.	Molecule	v	Nucleus	Nuclear quadrupole coupling tensor components in system of principal axes of inertia χ_{gg}, [MHz]	in system of principal coupling tensor or bond axes $\chi_{\alpha\beta}$, χ_{xy} [MHz]	position of the axes θ	Ref.	Varia, remarks Fig. in Chap. 3, subvol. b	See Tab., Nr.
62 Cont.	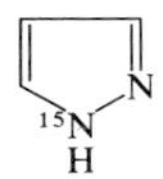	$v=0$	^{14}N	$\chi_{aa}=-3.66$			73NAS		
				$\chi_{bb}=2.95$					
				$\chi_{cc}=0.71$					
	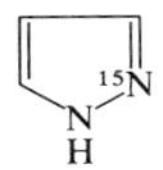	$v=0$	^{14}N	$\chi_{aa}=1.53$					
				$\chi_{bb}=1.53$					
				$\chi_{cc}=-3.06$					
	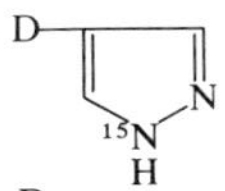	$v=0$	^{14}N	$\chi_{aa}=-3.71$					
				$\chi_{bb}=3.00$					
				$\chi_{cc}=0.71$					
	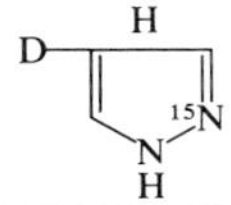	$v=0$	^{14}N	$\chi_{aa}=1.11$					
				$\chi_{bb}=1.95$					
				$\chi_{cc}=-3.06$					
63	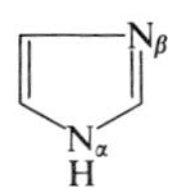	$v=0$	$^{14}N_\alpha$	$\chi_{aa}=1.254(29)$	$\chi_{\alpha\alpha}=1.043$		76BLA1	Fig. 170: $N_\alpha=N(1)$, $N_\beta=N(2)$	2.5, 632
				$\chi_{bb}=1.283(37)$	$\chi_{\beta\beta}=1.494$				
				$\chi_{cc}=-2.537$	$\chi_{\gamma\gamma}=-2.537$				
			$^{14}N_\beta$	$\chi_{aa}=-3.952(24)$	$\chi_{\alpha\alpha}=1.774$				
				$\chi_{bb}=1.694(36)$	$\chi_{\beta\beta}=2.258$				
				$\chi_{cc}=2.258$	$\chi_{\gamma\gamma}=-4.032$				
	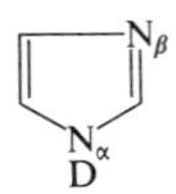	$v=0$	$^{14}N_\alpha$	$\chi_{aa}=1.407(39)$					
				$\chi_{bb}=1.146(42)$					
				$\chi_{cc}=-2.553$					
			$^{14}N_\beta$	$\chi_{aa}=-2.861(31)$					
				$\chi_{bb}=0.676(41)$					
				$\chi_{cc}=2.285$					
64	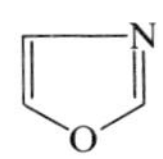	$v=0$	^{14}N	$\chi_{aa}=-3.92(2)$	$\chi_{\alpha\alpha}=-4.04(2)$	$\theta_{\alpha a}=8.3°$	78KUM2	Fig. 171	2.5, 633
				$\chi_{bb}=1.54(2)$	$\chi_{\beta\beta}=1.66(2)$				
				$\chi_{cc}=2.38$	$\chi_{\gamma\gamma}=2.38$				
	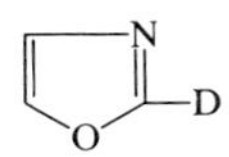	$v=0$	^{14}N	$\chi_{aa}=1.22(2)$					
				$\chi_{bb}=-3.60(2)$					
Cont.				$\chi_{cc}=2.38$					

Nr.	Molecule	v	Nucleus	Nuclear quadrupole coupling tensor components in system of principal axes of inertia χ_{gg}, [MHz]	in system of principal coupling tensor or bond axes $\chi_{\alpha\beta}$, χ_{xy} [MHz]	position of the axes θ	Ref.	Varia, remarks Fig. in Chap. 3, subvol. b	See Tab., Nr.
64 Cont.	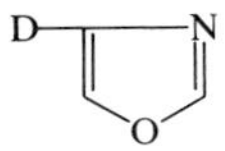	$v=0$	^{14}N	$\chi_{aa} = -0.37\ (5)$ $\chi_{bb} = -2.01\ (5)$ $\chi_{cc} = 2.38$			78KUM2		
	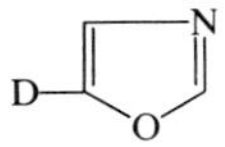	$v=0$	^{14}N	$\chi_{aa} = -2.98\ (5)$ $\chi_{bb} = 0.60\ (5)$ $\chi_{cc} = 2.38$					
65	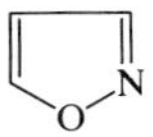	$v=0$	^{14}N	$\chi_{aa} = 0.526$ $\chi_{bb} = -0.609$ $\chi_{cc} = 0.083$	$\chi_{\alpha\alpha} = -5.42\ (7)$ $\chi_{\beta\beta} = 5.34\ (7)$ $\chi_{\gamma\gamma} = 0.08\ (5)$		75ST5, 78LOW	Fig. 172	2.5, 634
	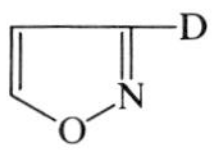	$v=0$	^{14}N	$\chi_{aa} \simeq -4.5$ $\chi_{bb} \simeq 5.0$			75ST3		
	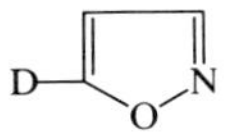	$v=0$	^{14}N	$\chi_{aa} \simeq 2.4$ $\chi_{bb} \simeq -2.4$					
	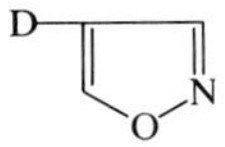	$v=0$	^{14}N	$\chi_{aa} = -5.33\ (4)$ $\chi_{bb} = 5.31\ (7)$ $\chi_{cc} = 0.02\ (11)$			78LOW		
66	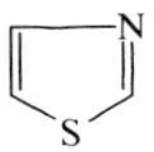	$v=0$	^{14}N	$\chi_{aa} = -2.708\ (22)$ $\chi_{bb} = 0.113\ (34)$ $\chi_{cc} = 2.595\ (34)$			80WI2	Fig. 174	2.5, 636
67	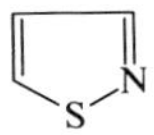	$v=0$	^{14}N	$\chi_{aa} = 1.037\ (23)$ $\chi_{bb} = -2.404\ (27)$ $\chi_{cc} = 1.367\ (27)$			80WI2	Fig. 175	2.5, 637
68	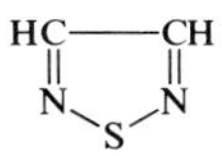	$v=0$	^{14}N	$\chi_{aa} = 0.90\ (1)$ *) $\chi_{bb} = -3.13\ (6)$ $\chi_{cc} = 2.22\ (6)$			78STI2	*) uncertainties are 3 σ	2.5, 638
69 Cont.	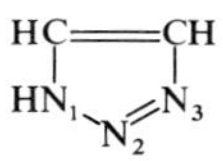	$v=0$	$^{14}N_1$	$\chi_{aa} = 0.64\ (28)$ $\chi_{bb} = 3.22\ (44)$ $\chi_{cc} = -3.86$ $\chi_{ab} = -1.04$	$\chi_{\alpha\alpha} = 0.27$ $\chi_{\beta\beta} = 3.58$ $\chi_{\gamma\gamma} = -3.86$		77BLA1	Fig. 178	2.5, 642

Nr.	Molecule	v	Nucleus	Nuclear quadrupole coupling tensor components: in system of principal axes of inertia χ_{gg}, [MHz]	in system of principal coupling tensor or bond axes $\chi_{\alpha\beta}$, χ_{xy} [MHz]	position of the axes θ	Ref.	Varia, remarks Fig. in Chap. 3, subvol. b	See Tab., Nr.
69 Cont.	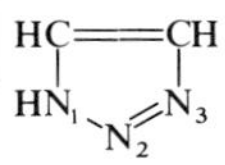		$^{14}N_2$	$\chi_{aa}=3.62(14)$ $\chi_{bb}=-4.15(17)$ $\chi_{cc}=0.53$ $\chi_{ab}=1.17$	$\chi_{\alpha\alpha}=0.53$ $\chi_{\beta\beta}=3.79$ $\chi_{\gamma\gamma}=-4.33$		77BLA1		
			$^{14}N_3$	$\chi_{aa}=-4.77(14)$ $\chi_{bb}=2.57(14)$ $\chi_{cc}=2.19$ $\chi_{ab}=0.90$	$\chi_{\alpha\alpha}=2.19$ $\chi_{\beta\beta}=2.68$ $\chi_{\gamma\gamma}=-4.87$				
	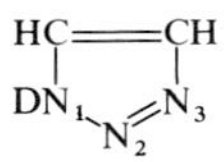	$v=0$	$^{14}N_1$	$\chi_{aa}=0.55(16)$ $\chi_{bb}=3.32(88)$ $\chi_{cc}=-3.86$ $\chi_{ab}=0.91$					
			$^{14}N_2$	$\chi_{aa}=2.04(12)$ $\chi_{bb}=-2.58(43)$ $\chi_{cc}=0.53$ $\chi_{ab}=-3.33$					
			$^{14}N_3$	$\chi_{aa}=-1.37(13)$ $\chi_{bb}=-0.82(41)$ $\chi_{cc}=2.19$ $\chi_{ab}=3.76$					
70	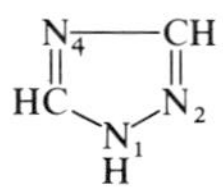	$v=0$	$^{14}N_1$	$\chi_{aa}=0.82(4)$ $\chi_{bb}=2.13(4)$ $\chi_{cc}=-2.95$ $\chi_{ab}=0.49$	$\chi_{\alpha\alpha}=0.66$ $\chi_{\beta\beta}=2.29$ $\chi_{\gamma\gamma}=-2.95$	$\theta_{a\alpha}=-18.4°$	75BL2	Fig. 179	2.5, 643
			$^{14}N_2$	$\chi_{aa}=-4.39(4)$ $\chi_{bb}=3.79(4)$ $\chi_{cc}=0.60$ $\chi_{ab}=-0.42$	$\chi_{\alpha\alpha}=0.60$ $\chi_{bb}=3.81$ $\chi_{\gamma\gamma}=-4.41$	$\theta_{a\gamma}=2.9°$			
Cont.			$^{14}N_4$	$\chi_{aa}=-3.55(4)$ $\chi_{bb}=1.35(4)$ $\chi_{cc}=2.20$ $\chi_{ab}=1.56$	$\chi_{\alpha\alpha}=1.80$ $\chi_{\beta\beta}=2.21$ $\chi_{\gamma\gamma}=-4.01$	$\theta_{a\gamma}=-16.2°$			

Nr.	Molecule	v	Nucleus	Nuclear quadrupole coupling tensor components: in system of principal axes of inertia χ_{gg}, [MHz]	in system of principal coupling tensor or bond axes $\chi_{\alpha\beta}$, χ_{xy} [MHz]	position of the axes θ	Ref.	Varia, remarks Fig. in Chap. 3, subvol. b	See Tab., Nr.
70 Cont.	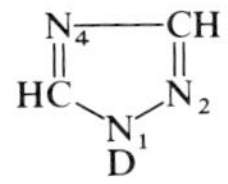	$v=0$	$^{14}N_1$ $^{14}N_2$ $^{14}N_4$	$\chi_{aa}=0.80(2)$ $\chi_{bb}=2.15(6)$ $\chi_{cc}=-2.95$ $\chi_{ab}=-0.46$ $\chi_{aa}=-1.23(4)$ $\chi_{bb}=0.63(7)$ $\chi_{cc}=0.60$ $\chi_{ab}=-4.00$ $\chi_{aa}=-3.38(7)$ $\chi_{bb}=1.18(7)$ $\chi_{cc}=2.20$ $\chi_{ab}=-1.81$			75BL2		
71	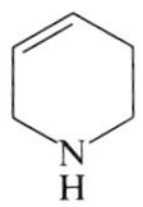 axial NH equatorial NH, equatorial ND,	$v=0$ $v=0$ $v=0$	^{14}N	$\chi_{cc}=1.32(2)$ $\chi_{bb}-\chi_{aa}=0.72(2)$ $\chi_{cc}=-4.48(4)$ $\chi_{bb}-\chi_{aa}=0.09(9)$ $\chi_{cc}=-4.51(6)$			76CH		2.5, 651
71a	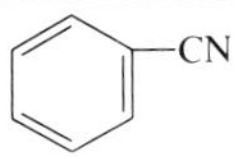	$v=0$	^{14}N	$\chi_{aa}=-4.187(160)$ $\chi_{bb}=2.301(80)$ $\chi_{cc}=1.886(80)$			81FLI1		2.5, 675a
72	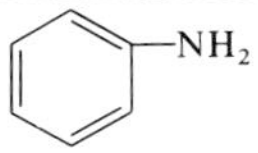	$v=0$	^{14}N	$\chi_{aa}=2.34(6)$ $\chi_{bb}=1.86(6)$ $\chi_{cc}=-4.20(6)$			73HAT		2.5, 679
72a	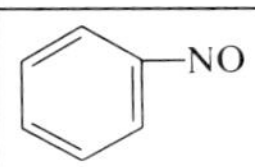	$v=0$	^{14}N	$\chi_{aa}=-0.037(60)$ $\chi_{bb}=-4.968(60)$ $\chi_{cc}=5.005(60)$			81FLI2		2.5, 680a

Nr.	Molecule	v	Nucleus	Nuclear quadrupole coupling tensor components in system of principal axes of inertia χ_{gg}, [MHz]	in system of principal coupling tensor or bond axes $\chi_{\alpha\beta}$, χ_{xy} [MHz]	position of the axes θ	Ref.	Varia, remarks Fig. in Chap. 3, subvol. b	See Tab., Nr.
73	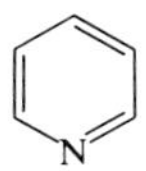	$v=0$	^{14}N	$\chi_{aa}=-4.87(4)$ **) $\chi_{bb}=1.40(6)$ $\chi_{cc}=3.47(3)$			74SOR	**) uncertainties are 3 σ	2.5, 730
	2-^{13}C	$v=0$		$\chi_{aa}=-4.50(8)$ **) $\chi_{bb}=1.04(8)$ $\chi_{cc}=3.47(3)$					
	3-^{13}C	$v=0$		$\chi_{aa}=-4.31(6)$ **) $\chi_{bb}=0.84(7)$ $\chi_{cc}=3.47(3)$					
	4-D	$v=0$		$\chi_{aa}=-4.894(20)$ $\chi_{bb}=1.400(40)$ $\chi_{cc}=3.50(6)$			76HAM1		
74	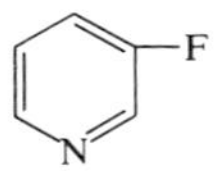	$v=0$	^{14}N	$\chi_{aa}=0.06(3)$ $\chi_{bb}=-3.52(8)$ $\chi_{cc}=3.46(11)$			76SHA	Fig. 193	2.5, 737
75	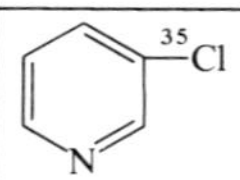	$v=0$	^{14}N	$\chi_{aa}=-0.08$ $\chi_{bb}=-3.41(10)$ $\chi_{cc}=3.49(10)$			75BRW2		2.5, 738
			^{35}Cl	$\chi_{aa}=-72.15$ $\chi_{bb}=38.49(10)$ $\chi_{cc}=33.66(10)$					
	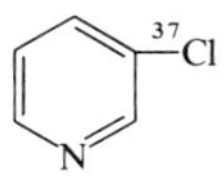	$v=0$	^{14}N	$\chi_{aa}=-0.08$ $\chi_{bb}=-3.41(10)$ $\chi_{cc}=3.49(10)$					
			^{37}Cl	$\chi_{aa}=-56.30$ $\chi_{bb}=30.01(20)$ $\chi_{cc}=26.29(20)$					
76	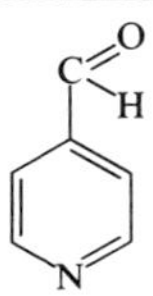	$v=0$	^{14}N	$\chi_{aa}=-4.67(9)$ $\chi_{bb}=1.19(26)$ $\chi_{cc}=3.48(26)$			80GEO4	Fig. 194	2.5, 746

Nr.	Molecule	v	Nucleus	Nuclear quadrupole coupling tensor components: in system of principal axes of inertia χ_{gg}, [MHz]	in system of principal coupling tensor or bond axes $\chi_{\alpha\beta}$, χ_{xy} [MHz]	position of the axes θ	Ref.	Varia, remarks Fig. in Chap. 3, subvol. b	See Tab., Nr.
77	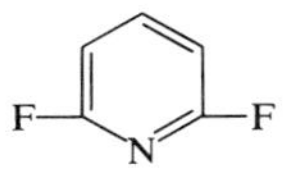	$v=0$	^{14}N	$\chi_{aa}=\ 1.82\,(5)$ $\chi_{bb}=-4.16\,(2)$ $\chi_{cc}=\ 2.34\,(5)$			76ST2, 76DOR		2.5, 748
78	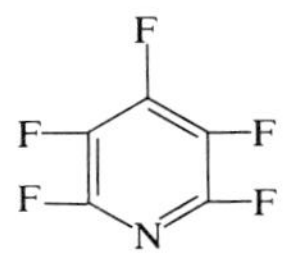	$v=0$	^{14}N	$\chi_{aa}=\ 1.94\,(25)$ $\chi_{bb}=-4.08\,(6)$ $\chi_{cc}=\ 2.14\,(25)$			74DO1		2.5, 749
	ArClCN		^{14}N	see Nr. 144a					
	H_2O		^{17}O	see Nr. 3					
	H_2CO		^{17}O	see Nr. 6					
79	HOCH=CHC(=O)H mono ^{17}O	$v=0$	^{17}O	$\chi_{bb}=-9.6$			78BAU		2.5, 256
80	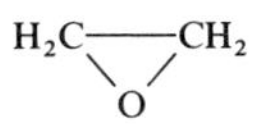	$v=0$	^{17}O	$\chi_{xx}=\ 12.6\,(1)$ *) $\chi_{yy}=-5.2\,(1)$ $\chi_{zz}=-7.4\,(1)$			74CR2	*) z is the C_2 axis, x the out of plane axis	2.5, 478
81	$^{35}Cl^{15}NCO$	$v=0$	^{35}Cl	$\chi_{aa}=-71.25\,(8)$ $\chi_{bb}-\chi_{cc}=-42.67\,(4)$			75HO2		2.5, 8
	$^{37}Cl^{15}NCO$	$v=0$	^{37}Cl	$\chi_{aa}=-56.87\,(12)$ $\chi_{bb}-\chi_{cc}=-33.09\,(6)$					
	$^{35}ClN^{13}CO$	$v=0$	^{35}Cl	$\chi_{aa}=-70.57\,(13)$ $\chi_{bb}-\chi_{cc}=-42.38\,(5)$					
			^{14}N	$\chi_{aa}=4.13\,(7)$ $\chi_{bb}-\chi_{cc}=2.04\,(3)$					
	$^{35}ClNC^{18}O$	$v=0$	^{35}Cl	$\chi_{aa}=-69.82\,(9)$ $\chi_{bb}-\chi_{cc}=-43.81\,(5)$					
			^{14}N	$\chi_{aa}=4.02\,(5)$ $\chi_{bb}-\chi_{cc}=2.00\,(3)$					

Nr.	Molecule	v	Nucleus	Nuclear quadrupole coupling tensor components: in system of principal axes of inertia χ_{gg}, [MHz]	in system of principal coupling tensor or bond axes $\chi_{\alpha\beta}$, χ_{xy} [MHz]	position of the axes θ	Ref.	Varia, remarks Fig. in Chap. 3, subvol. b	See Tab., Nr.
82	$^{35}ClSCN$	$v=0$	^{35}Cl	$\chi_{aa} = -36\,(4)$ $\chi_{bb}-\chi_{cc} = -65\,(1)$			80RIC		2.5, 11
83	$^{15}NO^{35}Cl$	$v=0$	^{35}Cl	$\chi_{aa} = -49.3\,(7)$ $\chi_{bb} = 29.4\,(6)$ $\chi_{cc} = 19.9\,(6)$	$\chi_{zz} = -57.8\,(16)$ $\chi_{xx} = 37.9\,(14)$ $\chi_{yy} = 19.9\,(30)$		79EN3		2.5, 14
	$^{15}NO^{37}Cl$	$v=0$	^{37}Cl	$\chi_{aa} = -38.6\,(6)$ $\chi_{bb} = 23.4\,(8)$ $\chi_{cc} = 15.3\,(14)$	$\chi_{zz} = -45.3\,(14)$ $\chi_{xx} = 30.1\,(18)$ $\chi_{yy} = 15.2\,(32)$				
84	$^{15}NO_2{}^{35}Cl$	$v=0$	^{35}Cl	$\chi_{aa} = -94.8\,(5)$ $\chi_{bb} = 52.5\,(15)$ $\chi_{cc} = 42.3\,(20)$			79EN1		2.5, 17
	$^{15}NO_2{}^{37}Cl$	$v=0$	^{37}Cl	$\chi_{aa} = -74.7\,(8)$ $\chi_{bb} = 41.4\,(20)$ $\chi_{cc} = 33.3\,(30)$					
85	$^{35}ClONO$ cis	$v=0$	^{35}Cl	$\chi_{aa} = -39.5\,(10)$ $\chi_{bb} = -18.7\,(8)$ $\chi_{cc} = 58.2\,(13)$			79KAW1		2.5, 18
	$^{37}ClONO$ cis	$v=0$	^{37}Cl	$\chi_{aa} = -31.1\,(21)$ $\chi_{bb} = -14.7\,(17)$ $\chi_{cc} = 45.8\,(25)$					
86	$^{35}ClNO_3$	$v=0$	^{35}Cl	$\chi_{aa} = -83.9\,(3)$ $\chi_{bb} = 18.8\,(10)$ $\chi_{cc} = 65.1\,(11)$			77SUE		2.5, 24
		$v=1$		$\chi_{aa} = -82.3\,(5)$ $\chi_{bb} = 21.1\,(10)$ $\chi_{cc} = 61.2\,(11)$					
	$^{37}ClNO_3$	$v=0$	^{37}Cl	$\chi_{aa} = -65.4\,(4)$ $\chi_{bb} = 14.7\,(10)$ $\chi_{cc} = 50.7\,(10)$					
		$v=1$		$\chi_{aa} = -67.2\,(6)$ $\chi_{bb} = 11.6\,(32)$ $\chi_{cc} = 55.6\,(35)$					

Nr.	Molecule	v	Nucleus	Nuclear quadrupole coupling tensor components in system of principal axes of inertia χ_{gg}, [MHz]	in system of principal coupling tensor or bond axes $\chi_{\alpha\beta}$, χ_{xy} [MHz]	position of the axes θ	Ref.	Varia, remarks Fig. in Chap. 3, subvol. b	See Tab., Nr.
87	$SO^{35}Cl_2$	$v=0$	^{35}Cl	$\chi_{aa}=-25.02\,(4)$ $\chi_{bb}=-0.25\,(4)$ $\chi_{cc}=25.27\,(8)$	$\chi_{zz}=-96.75$ $\chi_{xx}=65.62$ $\chi_{yy}=31.13$		81SUZ1, 73WEN		2.5, 31
88	$S^{35}Cl_2$	$v=0$	^{35}Cl	$\chi_{aa}=-38.98\,(10)$ $\chi_{bb}=-8.87\,(10)$ $\chi_{cc}=47.95\,(10)$ $\lvert\chi_{ab}\rvert=64\,(10)$	$\chi_{\alpha\alpha}=47.85\,(10)$ *) $\chi_{\beta\beta}=42.09\,(10)$ $\chi_{\gamma\gamma}=-89.94\,(10)$	$\theta_{b\gamma}=51.6°$	76MUR	*) the bond axis is assumed to be a principal axis of the field gradient	2.5, 37
	$S^{35}Cl^{37}Cl$	$v=0$	^{35}Cl ^{37}Cl	$\chi_{aa}=-37.85$ $\eta=1.529$ $\chi_{aa}=-31.61$ $\eta=1.386$			77DAV		
89	$HF^{35}ClF$ $HF^{37}ClF$ $DF^{35}ClF$	$v=0$ $v=0$ $v=0$	^{35}Cl ^{37}Cl ^{35}Cl	$\chi_{aa}=-144.746\,(25)$ *) $\chi_{aa}=-114.067\,(25)$ *) $\chi_{aa}=-144.548\,(25)$ *)			76NOV1	Fig. 19 *) MBER	2.5, 39
90	$HFH^{35}Cl$ $HFH^{37}Cl$ $HFD^{35}Cl$	$v=0$ $v=0$ $v=0$	^{35}Cl ^{37}Cl ^{35}Cl D	$\chi_{aa}=-52.291\,(25)$ *) $\chi_{aa}=-41.225\,(25)$ *) $\chi_{aa}=-55.600\,(10)$ *) $\chi_{aa}=0.145\,(5)$ *)			77JAN	Fig. 20 *) MBER	2.5, 40
91	$F^{35}ClO_2$ $F^{37}ClO_2$ $F^{35}Cl^{16}O^{18}O$	$v=0$ $v=0$ $v=0$	^{35}Cl ^{37}Cl ^{35}Cl	$\chi_{cc}=51.82\,(16)$ $\chi_{bb}-\chi_{aa}=17.67\,(21)$ $\chi_{cc}=40.63\,(16)$ $\chi_{bb}-\chi_{aa}=14.12\,(23)$ $\chi_{cc}=52.56\,(36)$ $\chi_{bb}-\chi_{aa}=15.96\,(45)$			74PAR		2.5, 42
92	$^{11}B^{35}ClF_2$ $^{11}B^{37}ClF_2$	$v=0$ $v=0$	^{35}Cl ^{37}Cl	$\chi_{aa}=-42.8\,(10)$ $\chi_{bb}=30.2\,(15)$ $\chi_{cc}=12.6\,(15)$ $\chi_{aa}=-33.8\,(10)$ $\chi_{bb}=24.6\,(15)$ $\chi_{cc}=9.2\,(15)$			77KRO		2.5, 45

Nr.	Molecule	v	Nucleus	Nuclear quadrupole coupling tensor components: in system of principal axes of inertia χ_{gg}, [MHz]	in system of principal coupling tensor or bond axes $\chi_{\alpha\beta}$, χ_{xy} [MHz]	position of the axes θ	Ref.	Varia, remarks Fig. in Chap. 3, subvol. b	See Tab., Nr.
93	S_2Cl_2	$v=0$	^{35}Cl	$\chi_{aa}= -7.35(10)$ $\chi_{bb}= -16.05(5)$			79MAR		2.5, 59
94	$CH_3NH^{35}Cl$	$v=0$	^{35}Cl	$\chi_{aa}= -67.02(6)$ $\chi_{bb}= 18.90(7)$	$\chi_{\alpha\alpha}=100(8)$ $\chi_{\beta\beta}= 54.5(10)$	$\theta_{\alpha a}=29(1)°$ $\theta_{\alpha b}=118°50' \pm 1°20'$	73MIR, 74CA2	∢(NCl, a) $=26°08' \pm 9'$ ∢(NCl, b)$=116°2' \pm 8'$	2.5, 69
			^{14}N	$\chi_{aa}= 4.63(9)$ $\chi_{bb}= -0.20(5)$					
	$CH_3NH^{35}Cl$	$v_\tau=1$	^{35}Cl	$\chi_{aa}= -67.8(11)$ $\chi_{bb}= 18.0(12)$					
			^{14}N	$\chi_{aa}= 5.1(15)$ $\chi_{bb}= 0.9(15)$					
	$CH_3ND^{35}Cl$	$v=0$	^{35}Cl	$\chi_{aa}= -67.34(12)$ $\chi_{bb}= 19.26(8)$					
			^{14}N	$\chi_{aa}= 4.70(10)$ $\chi_{bb}= -0.25(10)$					
	$CH_3NH^{37}Cl$	$v=0$	^{37}Cl	$\chi_{aa}= -53.3(4)$ $\chi_{bb}= 17.7(4)$					
			^{14}N	$\chi_{aa}= 5.1(8)$					
	$CH_3ND^{37}Cl$	$v=0$	^{37}Cl	$\chi_{aa}= -53.7(2)$ $\chi_{bb}= 16.0(2)$					
			^{14}N	$\chi_{aa}= 4.8(2)$ $\chi_{bb}= -0.16(14)$					
	$CH_2D(1)NH^{35}Cl$	$v=0$	^{35}Cl	$\chi_{aa}= -64.91(16)$ $\chi_{bb}= 14.78(15)$			74CA2	see Fig. 25 for the numbering of the 3 different positions of D	
			^{14}N	$\chi_{aa}= 4.53(19)$ $\chi_{bb}= 0.25(11)$					
	$CH_2D(2)NH^{35}Cl$	$v=0$	^{35}Cl	$\chi_{aa}= -64.78(11)$ $\chi_{bb}= 15.78(12)$					
			^{14}N	$\chi_{aa}= 4.42(18)$ $\chi_{bb}= -1.01(12)$					
	$CH_2D(3)NH^{35}Cl$	$v=0$	^{35}Cl	$\chi_{aa}= -68.08(16)$ $\chi_{bb}= 18.52(16)$					
			^{14}N	$\chi_{aa}= 4.64(20)$ $\chi_{bb}= -0.08(12)$					

Nr.	Molecule	v	Nucleus	Nuclear quadrupole coupling tensor components in system of principal axes of inertia χ_{gg}, [MHz]	in system of principal coupling tensor or bond axes $\chi_{\alpha\beta}$, χ_{xy} [MHz]	position of the axes θ	Ref.	Varia, remarks Fig. in Chap. 3, subvol. b	See Tab., Nr.
95	$CH_3O^{35}Cl$	$v=0$	^{35}Cl	$\chi_{aa}=-84.456\ (26)$ $\chi_{bb}=25.361\ (22)$ $\chi_{cc}=59.095$ $\chi_{ab}=68.2\ (13)$	$\chi_{\alpha\alpha}=-117.1\ (20)$ $\chi_{\beta\beta}=58.0\ (20)$ $\chi_{\gamma\gamma}=59.1$ $\eta=0.009\ (17)$	$\theta_{\alpha a}=25.6\ (3)°$	76SUZ	Fig. 27	2.5, 75
	$CD_3O^{35}Cl$	$v=0$	^{35}Cl	$\chi_{aa}=-80.045\ (11)$ $\chi_{bb}=21.072\ (11)$ $\chi_{cc}=58.973$ $\chi_{ab}=70\ (3)$					
96	$CH_3S^{35}Cl$	$v=0$	^{35}Cl	$\chi_{aa}=-58.31\ (5)$ $\chi_{bb}=14.80\ (5)$ $\chi_{cc}=43.51\ (5)$ $\chi_{ab}=43.94\pm1.8$	$\chi_{\alpha\alpha}=-78.91\pm1.4$ $\chi_{\beta\beta}=35.40\pm0.6$ $\chi_{\gamma\gamma}=43.51\pm0.05$	$\theta_{\alpha a}=25.12\ (60)°$	73GUA		2.5, 76
	$CH_3{}^{34}S^{35}Cl$	$v=0$	^{35}Cl	$\chi_{aa}=-59.09\ (8)$ $\chi_{bb}=15.63\ (8)$ $\chi_{cc}=43.46\ (8)$					
	$CD_3S^{35}Cl$	$v=0$	^{35}Cl	$\chi_{aa}=-52.79\ (5)$ $\chi_{bb}=9.27\ (5)$ $\chi_{cc}=43.52\ (5)$ $\chi_{ab}=48.00\pm2.0$	$\chi_{\alpha\alpha}=-78.92\pm1.4$ $\chi_{\beta\beta}=35.40\pm0.6$ $\chi_{\gamma\gamma}=43.52\pm0.05$	$\theta_{\alpha a}=28.56\ (60)°$			
	$CH_3S^{37}Cl$	$v=0$	^{37}Cl	$\chi_{aa}=-46.29\ (5)$ $\chi_{bb}=12.15\ (5)$ $\chi_{cc}=34.14\ (5)$ $\chi_{ab}=33.59\pm1.4$	$\chi_{\alpha\alpha}=-61.59\pm1.1$ $\chi_{\beta\beta}=27.45\pm0.5$ $\chi_{\gamma\gamma}=34.14\pm0.05$	$\theta_{\alpha a}=24.49\ (60)°$			
	$CD_3S^{37}Cl$	$v=0$	^{37}Cl	$\chi_{aa}=-42.12\ (5)$ $\chi_{bb}=7.83\ (5)$ $\chi_{cc}=34.29\ (5)$ $\chi_{ab}=36.69\pm1.5$	$\chi_{\alpha\alpha}=-61.53\pm1.1$ $\chi_{\beta\beta}=27.24\pm0.5$ $\chi_{\gamma\gamma}=34.28\pm0.05$	$\theta_{\alpha a}=27.88\ (60)°$			
97	$CF_3OO^{35}Cl$	$v=0$	^{35}Cl	$\chi_{aa}=47.6\ (13)$ $\chi_{bb}=-3.3\ (35)$ $\chi_{cc}=-44.3\ (35)$			80DUR2		2.5, 79

Nr.	Molecule	v	Nucleus	Nuclear quadrupole coupling tensor components: in system of principal axes of inertia $\chi_{gg'}$ [MHz]	in system of principal coupling tensor or bond axes $\chi_{\alpha\beta}$, χ_{xy} [MHz]	position of the axes θ	Ref.	Varia, remarks Fig. in Chap. 3, subvol. b	See Tab., Nr.
98	$CH_3OCH_2{}^{35}Cl$	$v=0$	^{35}Cl	$\chi_{aa} = -27.4\ (2)$ $\chi_{bb} = -3.1\ (1)$ $\chi_{cc} = 30.5\ (2)$	$\chi_{zz} = -71.3\ (4)$ $\chi_{xx} = 35.6$ $\chi_{yy} = 35.6$	$\theta_{za} = 39°54'$ $\theta_{zb} = 127°35'$ $\theta_{zc} = 78°32'$	74IKE, 74HAY4	Fig. 30	2.5, 86
	$CH_3OCH_2{}^{37}Cl$	$v=0$	^{37}Cl	$\chi_{aa} = -21.7\ (3)$ $\chi_{bb} = -2.2\ (2)$ $\chi_{cc} = 23.9\ (3)$	$\chi_{zz} = -57.0\ (4)$ $\chi_{xx} = 28.5$ $\chi_{yy} = 28.5$	$\theta_{za} = 39°34'$ $\theta_{zb} = 127°16'$ $\theta_{zc} = 78°37'$			
	$CD_3OCH_2{}^{35}Cl$	$v=0$	^{35}Cl	$\chi_{aa} = -23.0\ (9)$ *) $\eta = 1.61\ (9)$			74HAY4	*) uncertainties are 2.5 σ	
	$CD_3OCH_2{}^{37}Cl$	$v=0$	^{37}Cl	$\chi_{aa} = -19.2\ (20)$ *) $\eta = 1.56\ (19)$					
	$CH_3OCD_2{}^{35}Cl$	$v=0$	^{35}Cl	$\chi_{aa} = -27.7\ (10)$ *) $\eta = 1.31\ (7)$					
	$CH_3OCD_2{}^{37}Cl$	$v=0$	^{37}Cl	$\chi_{aa} = -22.2\ (6)$ *) $\eta = 1.26\ (5)$					
	$CH_3O^{13}CH_2{}^{35}Cl$	$v=0$	^{35}Cl	$\chi_{aa} = -26.5\ (40)$ *) $\eta\chi_{aa} = -33.8\ (29)$			79HAY2		
	$CH_3OCHD^{35}Cl$-1	$v=0$	^{35}Cl	$\chi_{aa} = -26.4\ (47)$ *) $\eta\chi_{aa} = -35.8\ (29)$					
	$CH_3OCHD^{35}Cl$-2	$v=0$	^{35}Cl	$\chi_{aa} = -26.5\ (47)$ *) $\eta\chi_{aa} = -33.3\ (30)$					
	$^{13}CH_3OCH_2{}^{35}Cl$	$v=0$	^{35}Cl	$\chi_{aa} = -26.3\ (49)$ *) $\eta\chi_{aa} = -34.2\ (31)$					
	$CH_2DOCH_2{}^{35}Cl$-3	$v=0$	^{35}Cl	$\chi_{aa} = -24.1\ (54)$ *) $\eta\chi_{aa} = -35.3\ (30)$					
	$CH_2DOCH_2{}^{35}Cl$-4	$v=0$	^{35}Cl	$\chi_{aa} = -24.8\ (56)$ *) $\eta\chi_{aa} = -34.5\ (37)$					
	$CH_2DOCH_2{}^{35}Cl$-5	$v=0$	^{35}Cl	$\chi_{aa} = -25.9\ (49)$ *) $\eta\chi_{aa} = -32.8\ (29)$				*) uncertainties are 2.5 σ	
	$CD_3OCD_2{}^{35}Cl$	$v=0$	^{35}Cl	$\chi_{aa} = -22.7\ (43)$ *) $\eta\chi_{aa} = -39.3\ (34)$					
	$CH_3O^{13}CH_2{}^{37}Cl$	$v=0$	^{37}Cl	$\chi_{aa} = -23.1\ (76)$ *) $\eta\chi_{aa} = -26.9\ (46)$					
	$CH_3OCHD^{37}Cl$-1	$v=0$	^{37}Cl	$\chi_{aa} = -20.1\ (64)$ *) $\eta\chi_{aa} = -26.4\ (43)$					
Cont.									

Nr.	Molecule	v	Nucleus	Nuclear quadrupole coupling tensor components: in system of principal axes of inertia χ_{gg}, [MHz]	in system of principal coupling tensor or bond axes $\chi_{\alpha\beta}$, χ_{xy} [MHz]	position of the axes θ	Ref.	Varia, remarks Fig. in Chap. 3, subvol. b	See Tab., Nr.
98 Cont.	$CH_3OCHD^{37}Cl$-2	$v=0$	^{37}Cl	$\chi_{aa} = -19.8\ (64)$ *) $\eta\chi_{aa} = -24.0\ (45)$			79HAY2		
	$^{13}CH_3OCH_2{}^{37}Cl$	$v=0$	^{37}Cl	$\chi_{aa} = -20.5\ (80)$ *) $\eta\chi_{aa} = -25.8\ (49)$					
	$CH_2DOCH_2{}^{37}Cl$-3	$v=0$	^{37}Cl	$\chi_{aa} = -17.7\ (122)$ *) $\eta\chi_{aa} = -26.4\ (82)$					
	$CH_2DOCH_2{}^{37}Cl$-4	$v=0$	^{37}Cl	$\chi_{aa} = -20.1\ (83)$ *) $\eta\chi_{aa} = -26.7\ (67)$					
	$CH_2DOCH_2{}^{37}Cl$-5	$v=0$	^{37}Cl	$\chi_{aa} = -20.6\ (90)$ *) $\eta\chi_{aa} = -24.6\ (56)$					
	$CD_3OCD_2{}^{37}Cl$	$v=0$	^{37}Cl	$\chi_{aa} = -16.6\ (45)$ *) $\eta\chi_{aa} = -30.0\ (33)$					
	$CH_3{}^{18}OCH_2{}^{35}Cl$	$v=0$	^{35}Cl	$\chi_{aa} = -27.7\ (35)$ *) $\eta\chi_{aa} = -31.5\ (21)$			80HAY3		
99	$CH_3SO_2{}^{35}Cl$	$v=0$	^{35}Cl	$\chi_{aa} = -65.7\ (4)$ $\chi_{bb} = 29.9\ (3)$ $\chi_{cc} = 35.8\ (3)$			75VAN		2.5, 91
	$CH_3SO_2{}^{37}Cl$	$v=0$	^{37}Cl	$\chi_{aa} = -51.6\ (5)$ $\chi_{bb} = 23.8\ (4)$ $\chi_{cc} = 27.8\ (4)$					
100	$CH_2F^{35}Cl$	$v=0$	^{35}Cl	$\chi_{aa} = -50.86\ (21)$ $\chi_{bb} = 12.91\ (42)$ $\chi_{cc} = 37.95\ (21)$			75NAN		2.5, 97
	$CH_2F^{37}Cl$	$v=0$	^{37}Cl	$\chi_{aa} = -38.59\ (36)$ $\chi_{bb} = 9.18\ (72)$ $\chi_{cc} = 29.41\ (36)$					
101	$CF_2{}^{35}Cl_2$	$v=0$	^{35}Cl	$\chi_{aa} = -43.42\ (16)$ $\chi_{bb} = 4.70\ (20)$ $\chi_{cc} = 38.71\ (22)$ $\chi_{ab} = -56.1\ (40)$	$\chi_{\alpha\alpha} = -80.37 \pm 3.0$ $\chi_{\beta\beta} = 41.65 \pm 2.5$ $\chi_{\gamma\gamma} = 38.71 \pm 0.22$		77SU		2.5, 107

Nr.	Molecule	v	Nucleus	Nuclear quadrupole coupling tensor components in system of principal axes of inertia χ_{gg}, [MHz]	in system of principal coupling tensor or bond axes $\chi_{\alpha\beta}$, χ_{xy} [MHz]	position of the axes θ	Ref.	Varia, remarks Fig. in Chap. 3, subvol. b	See Tab., Nr.
102	$HCO^{35}Cl$	$v=0$	^{35}Cl	$\chi_{aa}=-50.9_0$ $\chi_{bb}=30.0_7$ $\chi_{cc}=20.8_3$	$\chi_{zz}=-59.8$ $\eta=-0.303$		76TAK	Estimated errors: ± 0.2 MHz for χ_{aa}, χ_{bb}, χ_{cc} ± 0.7 MHz for χ_{zz} ± 0.002 for η	2.5, 109
	$HCO^{37}Cl$	$v=0$	^{37}Cl	$\chi_{aa}=-40.2_6$ $\chi_{bb}=23.9_4$ $\chi_{cc}=16.3_2$	$\chi_{zz}=-47.1$ $\eta=-0.307$				
	$DCO^{35}Cl$	$v=0$	^{35}Cl	$\chi_{aa}=-52.0_5$ $\chi_{bb}=31.2_6$ $\chi_{cc}=20.7_9$	$\chi_{zz}=-60.2$ $\eta=-0.309$				
	$DCO^{37}Cl$	$v=0$	^{37}Cl	$\chi_{aa}=-40.8_3$ $\chi_{bb}=24.5_3$ $\chi_{cc}=16.3_0$	$\chi_{zz}=-46.6$ $\eta=-0.301$				
103	$OC^{35}Cl_2$	$v=0$	^{35}Cl	$\chi_{aa}=-37.641(25)$ *) $\chi_{bb}=10.337(29)$ $\chi_{cc}=27.304(16)$ $\chi_{ab}=51.0(16)$	$\chi_{\alpha\alpha}=-70.1(14)$ *) $\chi_{\beta\beta}=42.7(14)$ $\chi_{\gamma\gamma}=27.304(16)$ $\chi_{zz}=-69.9(15)$ *) $\chi_{yy}=42.6(15)$ $\chi_{xx}=27.304(16)$		79FER	*) uncertainties are 2 σ α axis nearly coincident with the C—Cl bond, γ axis perpendicular to the plane of the molecule	2.5, 126
	$OC^{35}Cl^{37}Cl$	$v=0$	^{35}Cl	$\chi_{aa}=-36.009(36)$ *) $\chi_{bb}=8.705(39)$ $\chi_{cc}=27.304(16)$ $\chi_{ab}=-51.8(16)$					
		$v=0$	^{37}Cl	$\chi_{aa}=-30.899(67)$ *) $\chi_{bb}=9.379(69)$ $\chi_{cc}=21.520(13)$ $\chi_{ab}=39.6(12)$					
	$OC^{37}Cl_2$	$v=0$	^{37}Cl	$\chi_{aa}=-29.719(34)$ $\chi_{bb}=8.190(22)$ $\chi_{cc}=21.529(19)$			78CAR		
104	$OCF^{35}Cl$	$v=0$	^{35}Cl	$\chi_{aa}=-73.04(2)$ $\chi_{bb}=44.71(2)$ $\chi_{cc}=28.33(3)$	$\chi_{zz}=-74.78$ $\chi_{xx}=46.46$ $\chi_{yy}=28.32$		76SCA1		2.5, 128

Nr.	Molecule	v	Nucleus	Nuclear quadrupole coupling tensor components: in system of principal axes of inertia χ_{gg}, [MHz]	in system of principal coupling tensor or bond axes $\chi_{\alpha\beta}$, χ_{xy} [MHz]	position of the axes θ	Ref.	Varia, remarks Fig. in Chap. 3, subvol. b	See Tab., Nr.
105	$^{35}ClC(=O)OCH_3$	$v=0$	^{35}Cl	$\chi_{aa} = -59.8\,(5)$ $\chi_{bb} = 31.2\,(10)$ $\chi_{cc} = 28.6\,(10)$	$\chi_{xx} = -73.7$ $\chi_{yy} = 45.1$ $\chi_{zz} = 28.6$	$\theta_{xa} = 20°$	77DUR2	x axis along C—Cl bond	2.5, 131
	$^{37}ClC(=O)OCH_3$	$v=0$	^{37}Cl	$\chi_{aa} = -46.4\,(5)$ $\chi_{bb} = 25.4\,(10)$ $\chi_{cc} = 21.0\,(10)$					
	$^{35}ClC(=O)OCD_3$	$v=0$	^{35}Cl	$\chi_{aa} = -56.7\,(5)$ $\chi_{bb} = 29.8\,(10)$ $\chi_{cc} = 26.9\,(10)$					
	$^{35}ClC(=O)O^{13}CH_3$	$v=0$	^{35}Cl	$\chi_{aa} = -57.2\,(5)$ $\chi_{bb} = 30.9\,(10)$ $\chi_{cc} = 26.3\,(10)$					
	$^{35}ClC(=O)OCH_2D$ out of plane	$v=0$	^{35}Cl	$\chi_{aa} = -58.6\,(5)$ $\chi_{bb} = 29.7\,(10)$ $\chi_{cc} = 28.9\,(10)$					
	$^{35}ClC(=O)OCH_2D$ in plane	$v=0$	^{35}Cl	$\chi_{aa} = -56.2\,(5)$ $\chi_{bb} = 29.6\,(10)$ $\chi_{cc} = 26.6\,(10)$					
106	$^{35}ClC(=O)SCH_3$	$v=0$	^{35}Cl	$\chi_{aa} = -49.2\,(4)$ $\chi_{bb} = 22.4\,(8)$			80CAM3		2.5, 132
	$^{37}ClC(=O)SCH_3$	$v=0$	^{37}Cl	$\chi_{aa} = -39.0\,(6)$ $\chi_{bb} = 18.3\,(11)$					
107	$SC^{35}Cl_2$	$v=0$	^{35}Cl	$\chi_{aa} = 8.89\,(3)$ $\chi_{bb} = -36.58\,(3)$ $\chi_{cc} = 27.68\,(5)$ $\chi_{ab} = \pm 55.36\,(7)$	$\chi_{\gamma\gamma} = -73.70\,(7)$ $\chi_{\alpha\alpha} = 27.69\,(10)$ $\chi_{\beta\beta} = 46.01\,(7)$	$\theta_{\gamma a} = 56.16\,(2)°$	75CAR	α perpendicular to the plane of the molecule	2.5, 147
	$SC^{35}Cl^{37}Cl$	$v=0$	^{35}Cl	$\chi_{aa} = 45.46\,(4)$ $\chi_{bb} = -73.06\,(4)$ $\chi_{cc} = 27.60\,(4)$ $\chi_{ab} = -8.38\,(9)$					
Cont.									

Nr.	Molecule	v	Nucleus	Nuclear quadrupole coupling tensor components: in system of principal axes of inertia χ_{gg}, [MHz]	in system of principal coupling tensor or bond axes $\chi_{\alpha\beta}$, χ_{xy} [MHz]	position of the axes θ	Ref.	Varia, remarks Fig. in Chap. 3, subvol. b	See Tab., Nr.
107 Cont.	$SC^{35}Cl^{37}Cl$	$v=0$	^{37}Cl	$\chi_{aa}=-48.84$ (4) $\chi_{bb}=$ 26.89 (4) $\chi_{cc}=$ 21.95 (4) $\chi_{ab}=-28.12$ (7)			75CAR		
108	$SCF^{35}Cl$	$v=0$	^{35}Cl	$\chi_{aa}=-51.298$ (42) $\chi_{bb}=$ 23.587 (58) $\chi_{cc}=$ 27.711 (52)	$\chi_{xx}=-73.320$ (64) $\chi_{yy}=$ 45.609 (81) $\chi_{zz}=$ 27.711 (52)	$\theta_{xa}=25.487°$	75KOH		2.5, 148
	$SCF^{37}Cl$	$v=0$	^{37}Cl	$\chi_{aa}=-41.566$ (52) $\chi_{bb}=$ 19.544 (68) $\chi_{cc}=$ 22.021 (71)	$\chi_{xx}=-57.751$ (76) $\chi_{yy}=$ 35.729 (94) $\chi_{zz}=$ 22.021 (71)	$\theta_{xa}=24.488°$			
	$^{34}SCF^{35}Cl$	$v=0$	^{35}Cl	$\chi_{aa}=-49.893$ (68) $\chi_{bb}=$ 22.087 (131) $\chi_{cc}=$ 27.806 (120)	$\chi_{xx}=-73.546$ (109) $\chi_{yy}=$ 45.740 (179) $\chi_{zz}=$ 27.806 (120)	$\theta_{xa}=26.443°$			
109	$CH_3CH_2{}^{35}Cl$	$v=0$	^{35}Cl	$\chi_{aa}=-49.16$ (6) $\chi_{bb}=$ 13.46 (9) $\chi_{cc}=$ 35.70 (7) $\chi_{ab}=-42.10$ (15)	$\chi_{\alpha\alpha}=-70.28\pm0.1$ $\chi_{\beta\beta}=$ 34.63 $\pm$ 0.1 $\chi_{\gamma\gamma}=$ 35.65 $\pm$ 0.03 η = 0.015 $\pm$ 0.0017	$\theta=0.7°$ *) $\phi=25.93$ **)	78BRA	*) angle between the field gradient axis and the C—Cl bond **) angle between the a axis and the C—Cl bond	2.5, 152
	$CD_3CD_2{}^{35}Cl$	$v=0$	^{35}Cl	$\chi_{aa}=-47.78$ (3) $\chi_{bb}=$ 12.16 (8) $\chi_{cc}=$ 35.62 (6)		$\phi=26.85°$ **)			
	$CD_2HCD_2{}^{35}Cl$, symmetric	$v=0$	^{35}Cl	$\chi_{aa}=-47.38$ (5) $\chi_{bb}=$ 11.72 (8) $\chi_{cc}=$ 35.66 (8)		$\phi=27.21°$ **)			
	$CH_2DCH_2{}^{35}Cl$, symmetric	$v=0$	^{35}Cl	$\chi_{aa}=-49.59$ (3) $\chi_{bb}=$ 13.93 (5) $\chi_{cc}=$ 35.66 (3)		$\phi=25.65°$ **)			
	$CD_3CH_2{}^{35}Cl$	$v=0$	^{35}Cl	$\chi_{aa}=-46.43$ (3) $\chi_{bb}=$ 10.78 (7) $\chi_{cc}=$ 35.65 (7)		$\phi=27.77°$ **)			
Cont.	$CH_3CD_2{}^{35}Cl$	$v=0$	^{35}Cl	$\chi_{aa}=-51.07$ (3) $\chi_{bb}=$ 15.45 (7) $\chi_{cc}=$ 35.62 (7)		$\phi=24.66°$ **)			

Nr.	Molecule	v	Nucleus	Nuclear quadrupole coupling tensor components: in system of principal axes of inertia χ_{gg}, [MHz]	in system of principal coupling tensor or bond axes $\chi_{\alpha\beta}$, χ_{xy} [MHz]	position of the axes θ	Ref.	Varia, remarks Fig. in Chap. 3, subvol. b	See Tab., Nr.
109 Cont.	$CD_2HCD_2{}^{35}Cl$, asymmetric	$v=0$	^{35}Cl	$\chi_{aa}=-49.55(4)$ $\chi_{bb}=13.92(5)$ $\chi_{cc}=35.63(5)$		$\phi=25.74°$ **)	78BRA	**) angle between the *a* axis and the C—Cl bond	
	$CH_2DCH_2{}^{35}Cl$, asymmetric	$v=0$	^{35}Cl	$\chi_{aa}=-47.42(2)$ $\chi_{bb}=11.82(3)$ $\chi_{cc}=35.60(3)$		$\phi=27.14°$ **)			
	$CH_3CHD^{35}Cl$	$v=0$	^{35}Cl	$\chi_{aa}=-50.33(4)$ $\chi_{bb}=14.86(7)$ $\chi_{cc}=35.47(5)$		$\phi=25.14°$ **)			
110	$CH_3CH_2O^{35}Cl$	$v=0$	^{35}Cl	$\chi_{aa}=-93.96(62)$ $\chi_{bb}=35.76(39)$ $\chi_{cc}=58.20(28)$	$\chi_{zz}=-114.2(20)$ $\chi_{xx}=56.0(23)$ $\chi_{yy}=58.20(28)$	$\theta_{za}=20.1(10)°$	78SU2		2.5, 166
		$v_\tau=1$	^{35}Cl	$\chi_{aa}=-91.2(60)$ $\chi_{bb}=35.1(37)$ $\chi_{cc}=56.1(24)$				torsional motion around the C—O bond	
	$CH_3CH_2O^{37}Cl$	$v=0$	^{37}Cl	$\chi_{aa}=-67.4(25)$ $\chi_{bb}=24.8(13)$ $\chi_{cc}=42.6(12)$					
111	$CH_3CH_2OCH_2{}^{35}Cl$	$v=0$	^{35}Cl	$\chi_{aa}=-7.9(60)$ *) $\chi_{bb}-\chi_{cc}=-54.3(48)$			79HAY1	Fig. 49 *) uncertainties are 2.5 σ	2.5, 168
	$CH_3CH_2O^{13}CH_2{}^{35}Cl$	$v=0$	^{35}Cl	$\chi_{aa}=-5.8(95)$ *) $\chi_{bb}-\chi_{cc}=-54.4(50)$					
	$CH_3CH_2OCHD^{35}Cl$-1	$v=0$	^{35}Cl	$\chi_{aa}=-6.6(61)$ *) $\chi_{bb}-\chi_{cc}=-56.4(42)$					
	$CH_3CH_2OCHD^{35}Cl$-2	$v=0$	^{35}Cl	$\chi_{aa}=-7.9(77)$ *) $\chi_{bb}-\chi_{cc}=-54.2(51)$					
	$CH_3{}^{13}CH_2OCH_2{}^{35}Cl$	$v=0$	^{35}Cl	$\chi_{aa}=-6.4(65)$ *) $\chi_{bb}-\chi_{cc}=-53.8(48)$					
	$CH_3CHDOCH_2{}^{35}Cl$-3	$v=0$	^{35}Cl	$\chi_{aa}=-7.0(65)$ *) $\chi_{bb}-\chi_{cc}=-53.4(45)$					
Cont.	$CH_3CHDOCH_2{}^{35}Cl$-4	$v=0$	^{35}Cl	$\chi_{aa}=-7.2(77)$ *) $\chi_{bb}-\chi_{cc}=-54.5(51)$					

Nr.	Molecule	v	Nucleus	Nuclear quadrupole coupling tensor components: in system of principal axes of inertia χ_{gg}, [MHz]	in system of principal coupling tensor or bond axes $\chi_{\alpha\beta}$, χ_{xy} [MHz]	position of the axes θ	Ref.	Varia, remarks Fig. in Chap. 3, subvol. b	See Tab., Nr.
111 Cont.	$^{13}CH_3CH_2OCH_2{}^{35}Cl$	$v=0$	^{35}Cl	$\chi_{aa}=-5.8\,(62)$ *) $\chi_{bb}-\chi_{cc}=-54.7\,(40)$			79HAY1	*) uncertainties are 2.5 σ	
	$CH_2DCH_2OCH_2{}^{35}Cl$-5	$v=0$	^{35}Cl	$\chi_{aa}=-8.7\,(72)$ *) $\chi_{bb}-\chi_{cc}=-53.0\,(48)$					
	$CH_2DCH_2OCH_2{}^{35}Cl$-6	$v=0$	^{35}Cl	$\chi_{aa}=-8.4\,(91)$ *) $\chi_{bb}-\chi_{cc}=-54.6\,(51)$					
	$CH_2DCH_2OCH_2{}^{35}Cl$-7	$v=0$	^{35}Cl	$\chi_{aa}=-6.0\,(87)$ *) $\chi_{bb}-\chi_{cc}=-54.5\,(61)$					
	$CH_3CH_2OCD_2{}^{35}Cl$	$v=0$	^{35}Cl	$\chi_{aa}=-8.2\,(82)$ *) $\chi_{bb}-\chi_{cc}=-57.0\,(55)$					
	$CH_3CD_2OCH_2{}^{35}Cl$	$v=0$	^{35}Cl	$\chi_{aa}=-7.9\,(54)$ *) $\chi_{bb}-\chi_{cc}=-54.2\,(33)$					
	$CH_3CH_2{}^{18}OCH_2{}^{35}Cl$	$v=0$	^{35}Cl	$\chi_{aa}=-5.7\,(77)$ *) $\chi_{bb}-\chi_{cc}=-50.7\,(54)$			80HAY3		
112	$CH_3CHF^{35}Cl$	$v=0$	^{35}Cl	$\chi_{aa}=-62.26$ $\chi_{bb}=34.18$ $\chi_{cc}=28.08$	$\chi_{zz}=-68.21\pm1.0$ *) $\chi_{yy}=38.49\pm1.0$ $\chi_{xx}=29.72\pm1.0$		74THO	*) z axis along C—Cl bond, x axis in the ClCF plane	2.5, 170
113	$CH_2{}^{35}ClC(=O)OH$	$v=0$	^{35}Cl	$\chi_{aa}=-47.5\,(14)$ $\chi_{bb}=10.4\,(7)$ $\chi_{cc}=37.1\,(12)$	$\chi_{zz}=-73.3\,(23)$ *) $\eta=0.012\,(13)$		74VAN2	*) z axis along C—Cl bond	2.5, 200
114	$CH_2FC(=O){}^{35}Cl$	$v=0$	^{35}Cl	$\chi_{aa}=-47.7$ $\chi_{bb}=23.7$ $\chi_{cc}=24.1$	$\chi_{xx}=-61.7$ $\chi_{yy}=37.7$ $\chi_{zz}=24.1$	$\theta=22°2'$ *)	74SZA	*) angle between a and x axes	2.5, 202
		$v_\tau=1$	^{35}Cl	$\chi_{aa}=-46.3$ $\chi_{bb}=22.8$ $\chi_{cc}=23.5$					
	$CH_2FC(=O){}^{37}Cl$	$v=0$	^{37}Cl	$\chi_{aa}=-38.0$ $\chi_{bb}=18.9$ $\chi_{cc}=19.1$	$\chi_{xx}=-48.7$ $\chi_{yy}=29.6$ $\chi_{zz}=19.1$	$\theta=21°39'$ *)			

Nr.	Molecule	v	Nucleus	Nuclear quadrupole coupling tensor components in system of principal axes of inertia χ_{gg}, [MHz]	in system of principal coupling tensor or bond axes $\chi_{\alpha\beta}$, χ_{xy} [MHz]	position of the axes θ	Ref.	Varia, remarks Fig. in Chap. 3, subvol. b	See Tab., Nr.
115	$CH_2{}^{35}ClC(=O)H$ cis *)	$v=0$	^{35}Cl	$\chi_{aa}=-18.03$ $\chi_{bb}=-18.20$ $\chi_{cc}=36.24$			76FOR	*) Cl and O cis	2.5, 203
	$CH_2{}^{35}ClC(=O)H$ trans	$v=0$	^{35}Cl	$\chi_{aa}=-50.09$ $\chi_{bb}=12.56$ $\chi_{cc}=37.53$					
		$v_\tau=1$	^{35}Cl	$\chi_{aa}=-48.03$ $\chi_{bb}=11.61$ $\chi_{cc}=36.42$					
		$v_\tau=2$	^{35}Cl	$\chi_{aa}=-47.88$ $\chi_{bb}=11.33$ $\chi_{cc}=36.55$					
		$v_\tau=3$	^{35}Cl	$\chi_{aa}=-46.39$ $\chi_{bb}=11.26$ $\chi_{cc}=35.13$					
116	$CH_2{}^{35}Cl_\alpha C(=O){}^{35}Cl_\beta$	$v=0$	$^{35}Cl_\alpha$	$\chi_{aa}=-64.1\ (13)$ $\chi_{bb}=25.3\ (7)$ $\chi_{cc}=38.8\ (6)$	$\chi_{zz}=-77.9\ (17)$ *) $\chi_{xx}=39.1\ (10)$ $\eta=-0.004\ (6)$	$\theta_{az}=20°4'$	76VAN	*) z along C—Cl bond	2.5, 204
			$^{35}Cl_\beta$	$\chi_{aa}=-46.5\ (14)$ $\chi_{bb}=23.5\ (8)$ $\chi_{cc}=23.0\ (7)$	$\chi_{zz}=-62.4\ (19)$ *) $\chi_{xx}=39.4\ (13)$ $\eta=-0.263\ (10)$	$\theta_{az}=23°16'$			
117	$CH_3CH_2C(=O){}^{35}Cl$ cis	$v=0$	^{35}Cl	$\chi_{aa}=-48.8\ (1)$ $\chi_{bb}=27.3\ (6)$ $\chi_{cc}=21.5\ (6)$	$\chi_{xx}=-60.1$ *) $\chi_{yy}=38.6$ $\chi_{zz}=21.5$	$\theta_{xa}=19.8°$	76KAR1, 79MAT	*) x along C—Cl bond $z \perp$ to plane of framework	2.5, 259
	$CH_3CH_2C(=O){}^{37}Cl$ cis	$v=0$	^{37}Cl	$\chi_{aa}=-38.3\ (2)$ $\chi_{bb}=23.1\ (8)$ $\chi_{cc}=15.2\ (8)$	$\chi_{xx}=-47.1$ *) $\chi_{yy}=31.9$ $\chi_{zz}=15.2$	$\theta_{xa}=19.5°$			

Nr.	Molecule	v	Nucleus	Nuclear quadrupole coupling tensor components in system of principal axes of inertia $\chi_{gg'}$ [MHz]	in system of principal coupling tensor or bond axes $\chi_{\alpha\beta}, \chi_{xy}$ [MHz]	position of the axes θ	Ref.	Varia, remarks Fig. in Chap. 3, subvol. b	See Tab., Nr.
118	$CH_2{=}CH^{35}Cl$	$v=0$	^{35}Cl ^{37}Cl	$\chi_{aa} = -57.13\,(11)$ *) $\eta = 0.1034\,(9)$ $\chi_{aa} = -45.15\,(11)$ *) $\eta = 0.1039\,(14)$	$\chi_{xx} = 38.37$ **) $\chi_{yy} = 31.52$ $\chi_{zz} = -69.89$		71GER	*) confidence limits of 95%. **) $z \parallel$C—Cl bond, $y \perp$molecular plane	2.5, 282
119	$CH_2{=}C^{35}ClCN$	$v=0$	^{35}Cl	$\chi_{aa} = -31.16\,(33)$ $\chi_{bb} - \chi_{cc} = -38.26\,(17)$	$\chi_{xx} = 44.9\,(8)$ *) $\chi_{yy} = 34.7\,(2)$ $\chi_{zz} = -79.6\,(10)$		75AVI3	*) z along C—Cl bond	2.5, 297
119a	$(CH_3)_2C{=}CHCl$	$v=0$	^{35}Cl ^{37}Cl	$\chi_{aa} = -49.2$ $\chi_{bb} = 19.5$ $\chi_{cc} = 29.7$ $\chi_{aa} = -38.8$ $\chi_{bb} = 15.4$ $\chi_{cc} = 23.4$			81GUI		2.5, 324a
120	$CH_2{=}CH{-}CH{=}CH^{35}Cl$ cis $CH_2{=}CH{-}CH{=}CH^{35}Cl$ trans	$v=0$ $v=0$	^{35}Cl ^{35}Cl	$\chi_{aa} = -2.62\,(28)$ $\chi_{bb} = -27.85\,(14)$ $\chi_{aa} = -51.603$ $\chi_{bb} = 21.103$	$\chi_{xx} = 37.6\,(8)$ *) $\chi_{yy} = 30.5\,(8)$ $\chi_{xx} = 37.6\,(8)$ *) $\chi_{yy} = 30.5\,(8)$		80KAR 80CED2	$z \parallel$C—Cl bond $y \perp$molecular plane Fig. 96	2.5, 341
121	$CH_2{=}CH{-}C^{35}Cl{=}CH_2$	$v=0$	^{35}Cl	$\chi_{aa} = 5.2\,(2)$ $\chi_{bb} = -37.5\,(2)$	$\chi_{zz} = -67.6\,(6)$ *) $\chi_{yy} = 32.3\,(8)$		79KAR, 81CED	*) $z \parallel$C—Cl bond. $y \perp$molecular plane	2.5, 344
122	$CH_2{=}C{=}CHCH_2{}^{35}Cl$ skew $CH_2{=}C{=}CHCH_2{}^{37}Cl$ skew	$v=0$ $v=0$	^{35}Cl ^{37}Cl	$\chi_{aa} = -30$ $\chi_{aa} = -24$			77KAR	Fig. 100	2.5, 348
123	$CH_2{=}C{=}C{=}CH^{35}Cl$	$v=0$	^{35}Cl	$\chi_{aa} = -25.8\,(5)$			74KAR2		2.5, 354

Nr.	Molecule	v	Nucleus	Nuclear quadrupole coupling tensor components: in system of principal axes of inertia χ_{gg}, [MHz]	in system of principal coupling tensor or bond axes $\chi_{\alpha\beta}$, χ_{xy} [MHz]	position of the axes θ	Ref.	Varia, remarks Fig. in Chap. 3, subvol. b	See Tab., Nr.
124	$HC{\equiv}C{-}C({=}O)^{35}Cl$	$v=0$	^{35}Cl	$\chi_{aa} = -20.6\,(3)$ $\chi_{bb} = -3.2\,(3)$ $\chi_{cc} = 23.8\,(3)$			74DAV3		2.5, 366
	$HC{\equiv}C{-}C({=}O)^{37}Cl$	$v=0$	^{37}Cl	$\chi_{aa} = -16.7\,(3)$ $\chi_{bb} = -1.7\,(3)$ $\chi_{cc} = 18.4\,(3)$					
125	$^{35}Cl_{\alpha}C{\equiv}CCH_2{}^{35}Cl_{\beta}$	$v=0$	$^{35}Cl_{\alpha}$	$\chi_{aa} = -73.5\,(6)$ $\chi_{bb} = 31.5\,(3)$ $\chi_{cc} = 42.0\,(6)$	$\chi_{zz} = -79.3\,(2)$		79GUN		2.5, 371
			$^{35}Cl_{\beta}$	$\chi_{aa} = -7.5\,(3)$ $\chi_{bb} = -32.4\,(2)$ $\chi_{cc} = 39.9\,(3)$	$\chi_{zz} = -74.8\,(2.7)$				
		$v=1$	$^{35}Cl_{\alpha}$	$\chi_{aa} = -67.8\,(6)$ $\chi_{bb} = 28.4\,(3)$ $\chi_{cc} = 39.4\,(3)$					
			$^{35}Cl_{\beta}$	$\chi_{aa} = -5.6\,(2)$ $\chi_{bb} = -33.3\,(1)$ $\chi_{cc} = 38.9\,(2)$					
	$^{37}ClC{\equiv}CCH_2{}^{35}Cl$	$v=0$	^{37}Cl	$\chi_{aa} = -53.1\,(3)$ $\chi_{bb} = 22.5\,(2)$ $\chi_{cc} = 30.6\,(3)$					
			^{35}Cl	$\chi_{aa} = -5.4\,(3)$ $\chi_{bb} = -33.5\,(1)$ $\chi_{cc} = 38.9\,(2)$					
	$^{35}ClC{\equiv}CCH_2{}^{37}Cl$	$v=0$	^{35}Cl	$\chi_{aa} = -72.8\,(7)$ $\chi_{bb} = 31.0\,(4)$ $\chi_{cc} = 41.8\,(7)$					
			^{37}Cl	$\chi_{aa} = -5.2\,(3)$ $\chi_{bb} = -25.7\,(1)$ $\chi_{cc} = 30.9\,(3)$					

Nr.	Molecule	v	Nucleus	Nuclear quadrupole coupling tensor components in system of principal axes of inertia χ_{gg}, [MHz]	in system of principal coupling tensor or bond axes $\chi_{\alpha\beta}$, χ_{xy} [MHz]	position of the axes θ	Ref.	Varia, remarks Fig. in Chap. 3, subvol. b	See Tab., Nr.
126	$HC{\equiv}C{-}C^{35}Cl{=}CH_2$	$v=0$	^{35}Cl	$\chi_{aa}=-27.1\,(1)$ $\chi_{bb}=-6.4\,(1)$			74KAR1	Fig. 106	2.5, 374
	$HC{\equiv}C{-}C^{37}Cl{=}CH_2$	$v=0$	^{37}Cl	$\chi_{aa}=-21.8\,(3)$ $\chi_{bb}=-4.0\,(3)$					
127	$HC{\equiv}C{-}CH{=}CH^{35}Cl$ trans	$v=0$	^{35}Cl	$\chi_{aa}=-63.6\,(4)$			73KAR2	Fig. 107	2.5, 375
	$HC{\equiv}C{-}CH{=}CH^{37}Cl$ trans	$v=0$	^{37}Cl	$\chi_{aa}=-51.0\,(4)$					
	$HC{\equiv}C{-}CH{=}CH^{35}Cl$ cis	$v=0$	^{35}Cl	$\chi_{aa}=16.3\,(4)$ $\chi_{bb}=-46.8\,(2)$	$\chi_{xx}=39.3\,(7)$ *) $\chi_{yy}=30.5\,(7)$		75KA2	*) $z \parallel$ C—Cl bond, $y \perp$ molecular plane	
	$HC{\equiv}C{-}CH{=}CH^{37}Cl$ cis	$v=0$	^{37}Cl	$\chi_{aa}=11.8\,(8)$ $\chi_{bb}=-36.2\,(4)$					
128	$^{35}ClC{\equiv}C{-}CH{=}CH_2$	$v=0$	^{35}Cl	$\chi_{aa}=-76.9\,(4)$			75KA1	Fig. 108	2.5, 376
	$^{37}ClC{\equiv}C{-}CH{=}CH_2$	$v=0$	^{37}Cl	$\chi_{aa}=-60.9\,(4)$					
129	$^{28}SiH_2{}^{35}Cl_2$	$v=0$	^{35}Cl	$\chi_{aa}=-21.13\,(35)$ $\chi_{bb}-\chi_{cc}=-20.88\,(30)$	$\chi_{zz}=-42.0\,(7)$ *) $\chi_{xx}=21.0\,(6)$ $\chi_{yy}=21.00\,(23)$		76DAV2	*) z parallel to Si—Cl bond x in molecular plane perpendicular to Si—Cl bond	2.5, 379
130	$CH_3SiH_2{}^{35}Cl$	$v=0$	^{35}Cl	$\chi_{aa}=-27.76\,(8)$ $\chi_{bb}=8.89\,(15)$ $\chi_{cc}=18.87\,(12)$			75ZEI		2.5, 383
	$CH_3SiD_2{}^{35}Cl$	$v=0$	^{35}Cl	$\chi_{aa}=-28.57\,(4)$ $\chi_{bb}=9.51\,(8)$ $\chi_{cc}=19.06\,(7)$					
	$CD_3SiD_2{}^{35}Cl$	$v=0$	^{35}Cl	$\chi_{aa}=-26.01\,(7)$ $\chi_{bb}=7.12\,(10)$ $\chi_{cc}=18.89\,(9)$					
	$CH_3SiH_2{}^{37}Cl$	$v=0$	^{37}Cl	$\chi_{aa}=-22.09\,(3)$ $\chi_{bb}=7.21\,(6)$ $\chi_{cc}=14.88\,(7)$					
Cont.									

Nr.	Molecule	v	Nucleus	Nuclear quadrupole coupling tensor components: in system of principal axes of inertia χ_{gg}, [MHz]	in system of principal coupling tensor or bond axes $\chi_{\alpha\beta}$, χ_{xy} [MHz]	position of the axes θ	Ref.	Varia, remarks Fig. in Chap. 3, subvol. b	See Tab., Nr.
130 Cont.	$CH_3SiD_2{}^{37}Cl$	$v=0$	^{37}Cl	$\chi_{aa}= -22.61$ (6) $\chi_{bb}=$ 7.62 (9) $\chi_{cc}=$ 14.99 (7)			75ZEI		
	$CD_3SiD_2{}^{37}Cl$	$v=0$	^{37}Cl	$\chi_{aa}= -20.90$ (5) $\chi_{bb}=$ 6.11 (11) $\chi_{cc}=$ 14.79 (11)					
131	$CH_3{}^{28}SiH^{35}Cl_2$	$v=0$	^{35}Cl	$\chi_{aa}= -19.1$ (3) $\chi_{bb}=$ 5.4 (13) $\chi_{cc}=$ 13.7 (14)	$\chi_{zz}= -41.2$ (20) *) η = −0.16 (3)		77EN	z along Si—Cl bond.	2.5, 384
132	$(CH_3)_2Si^{35}Cl_2$	$v=0$	^{35}Cl	$\chi_{aa}= -19.6$ (3) $\chi_{bb}= -3.7$ (1.4) $\chi_{cc}=$ 23.3 (1.4)	$\chi_{zz}= -38.0$ (1.6) η = 0.22 (8)		80NA3		2.5, 393
133	$CH_3CH_2SiH_2{}^{35}Cl$ trans	$v=0$	^{35}Cl	$\chi_{bb}+\chi_{cc}=$ 26.94 (39) $\chi_{bb}-\chi_{cc}= -11.2$ (34)	$\chi_{zz}= -35.54$ (79) $\chi_{xx}=$ 13.3 (31) $\chi_{yy}=$ 22.2 (24)		74TYP	Fig. 110	2.5, 396
	$CH_3CH_2SiH_2{}^{37}Cl$ trans	$v=0$	^{37}Cl	$\chi_{bb}+\chi_{cc}=$17.35 (84) $\chi_{bb}-\chi_{cc}=$ 5.8 (108)					
	$CH_3CH_2SiD_2{}^{35}Cl$ trans	$v=0$	^{35}Cl	$\chi_{bb}+\chi_{cc}=$ 25.9 (12) $\chi_{bb}=\chi_{cc}= -24.4$ (91)					
	$CH_3CD_2SiH_2{}^{35}Cl$ trans	$v=0$	^{35}Cl	$\chi_{bb}+\chi_{cc}=$ 25.87 (72) $\chi_{bb}-\chi_{cc}= -28.4$ (91)					
	$CD_3CH_2SiH_2{}^{35}Cl$ trans	$v=0$	^{35}Cl	$\chi_{bb}+\chi_{cc}=$ 24.98 (88) $\chi_{bb}-\chi_{cc}= -19.4$ (216)					
	$CH_3CH_2SiH_2{}^{35}Cl$ gauche	$v=0$	^{35}Cl	$\chi_{bb}+\chi_{cc}=$ 8.989 (82) $\chi_{bb}-\chi_{cc}= -27.289$ (71)	$\chi_{zz}= -36.3$ (27) $\chi_{xx}=$ 16.7 (35) $\chi_{yy}=$ 19.61 (79)				
	$CH_3CH_2SiH_2{}^{37}Cl$ gauche	$v=0$	^{37}Cl	$\chi_{bb}+\chi_{cc}=$ 7.08 (60) $\chi_{bb}-\chi_{cc}= -20.17$ (39)					
	$CH_3CH_2SiD_2{}^{35}Cl$ gauche	$v=0$	^{35}Cl	$\chi_{bb}+\chi_{cc}=$ 8.14 (40) $\chi_{bb}-\chi_{cc}= -27.61$ (23)					
	$CH_3CD_2SiH_2{}^{35}Cl$ gauche	$v=0$	^{35}Cl	$\chi_{bb}+\chi_{cc}=$ 9.85 (32) $\chi_{bb}-\chi_{cc}= -25.10$ (23)					
	$CD_3CH_2SiH_2{}^{35}Cl$ gauche	$v=0$	^{35}Cl	$\chi_{bb}+\chi_{cc}=$ 5.24 (19) $\chi_{bb}-\chi_{cc}= -29.56$ (18)					

Nr.	Molecule	v	Nucleus	Nuclear quadrupole coupling tensor components: in system of principal axes of inertia χ_{gg}, [MHz]	in system of principal coupling tensor or bond axes $\chi_{\alpha\beta}$, χ_{xy} [MHz]	position of the axes θ	Ref.	Varia, remarks Fig. in Chap. 3, subvol. b	See Tab., Nr.
134	$^{74}GcH_3CH_2{}^{35}Cl$	$v=0$	^{35}Cl	$\chi_{aa}=-29.7$ (4) *) $\eta\chi_{aa}=-45.0$ (4)	$\chi_{zz}=-74.5$ $\eta\chi_{zz}=-0.2$		76NAK1	*) uncertainties are 2.5 σ	2.5, 407
		$v_\tau=1$	^{35}Cl	$\chi_{aa}=-29.7$ (15) *) $\eta\chi_{aa}=-44.3$ (8)					
	$^{72}GeH_3CH_2{}^{35}Cl$	$v=0$	^{35}Cl	$\chi_{aa}=-29.2$ (5) $\eta\chi_{aa}=-45.1$ (4)	$\chi_{zz}=-72.0$ $\eta\chi_{zz}=-2.2$				
		$v_\tau=1$	^{35}Cl	$\chi_{aa}=-29.7$ (20) $\eta\chi_{aa}=-44.2$ (10)					
	$^{70}GeH_3CH_2{}^{35}Cl$	$v=0$	^{35}Cl	$\chi_{aa}=-29.2$ (5) $\eta\chi_{aa}=-45.3$ (5)	$\chi_{zz}=-71.1$ $\eta\chi_{zz}=-3.4$				
	$^{74}GeH_3CH_2{}^{37}Cl$	$v=0$	^{37}Cl	$\chi_{aa}=-23.5$ (6) $\eta\chi_{aa}=-35.0$ (4)	$\chi_{zz}=-57.7$ $\eta\chi_{zz}=-0.9$				
	$^{74}GeD_3CH_2{}^{35}Cl$	$v=0$	^{35}Cl	$\chi_{aa}=-29.7$ (7) $\eta\chi_{aa}=-44.9$ (4)	$\chi_{zz}=-74.4$ $\eta\chi_{zz}=-0.2$				
135	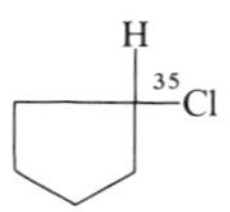	$v=0$	^{35}Cl	$\chi_{aa}=-23.70$ (10) $\chi_{bb}=32.33$ (36) $\chi_{cc}=-8.63$ (37)	$\chi_{zz}=-64.67$ (72)	$\theta_{az}=40.5$ (1.6)°	78LOY	Fig. 144	2.5, 511
136	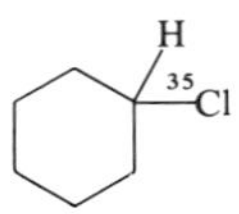	$v=0$	^{35}Cl	$\chi_{aa}=-61.4$ (2) $\chi_{bb}=30.5$ (7) $\chi_{cc}=30.9$ (9)			73DAM	equatorial	2.5, 524
137	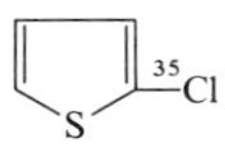	$v=0$	^{35}Cl	$\chi_{aa}=-74.77$ (5) $\chi_{bb}=37.51$ (17) $\chi_{cc}=37.25$ (18)			73MJO		2.5, 615
	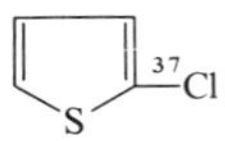	$v=0$	^{37}Cl	$\chi_{aa}=-58.98$ (9) $\chi_{bb}=29.55$ (26) $\chi_{cc}=29.43$ (28)					

Nr.	Molecule	v	Nucleus	Nuclear quadrupole coupling tensor components: in system of principal axes of inertia χ_{gg}, [MHz]	in system of principal coupling tensor or bond axes $\chi_{\alpha\beta}$, χ_{xy} [MHz]	position of the axes θ	Ref.	Varia, remarks Fig. in Chap. 3, subvol. b	See Tab., Nr.
138	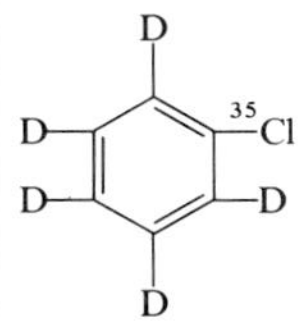	$v=0$	^{35}Cl	$\chi_{aa}=-71.10$ $\eta=0.0$			77CHA		2.5, 675
	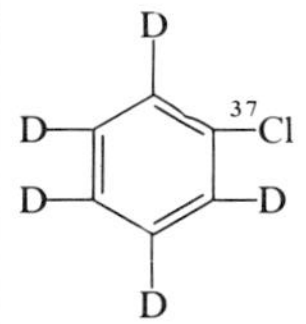	$v=0$	^{37}Cl	$\chi_{aa}=-56.10$ $\eta=0.0$					
139	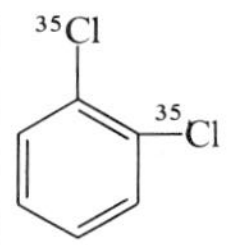	$v=0$	^{35}Cl	$\chi_{aa}=-37.9\ (13)$ *) $\chi_{bb}=7.0\ (7)$ $\chi_{cc}=30.9\ (20)$			76OND1	*) uncertainties are 2.5 σ	2.5, 698
140	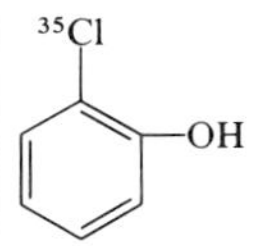	$v=0$	^{35}Cl	$\chi_{aa}=-67.6\ (3)$ $\chi_{bb}=33.3\ (1)$ $\chi_{cc}=34.3\ (4)$	$\chi_{zz}=-68.7\ (3)$ $\chi_{yy}=34.4\ (1)$ $\chi_{xx}=34.3\ (4)$	$\theta_{za}=5.87°$	78OND1	z=C—Cl bond axis $x\perp$molecular plane	2.5, 699
	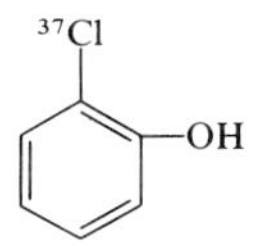	$v=0$	^{37}Cl	$\chi_{aa}=-55.9\ (20)$ $\chi_{bb}=26.5\ (3)$ $\chi_{cc}=29.4\ (23)$	$\chi_{zz}=-56.5\ (45)$ $\chi_{yy}=27.1\ (11)$ $\chi_{xx}=29.4\ (23)$	$\theta_{za}=4.97°$			
	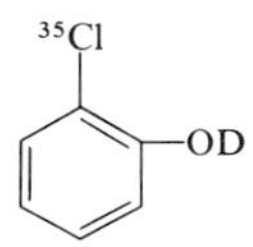	$v=0$	^{35}Cl	$\chi_{aa}=-67.1\ (20)$ $\chi_{bb}=33.1\ (2)$ $\chi_{cc}=34.0\ (22)$	$\chi_{zz}=-68.6\ (20)$ $\chi_{yy}=34.6\ (2)$ $\chi_{xx}=34.0\ (22)$	$\theta_{za}=6.84°$	78OND2		

Nr.	Molecule	v	Nu-cleus	Nuclear quadrupole coupling tensor components: in system of principal axes of inertia χ_{gg}, [MHz]	in system of principal coupling tensor or bond axes $\chi_{\alpha\beta}$, χ_{xy} [MHz]	position of the axes θ	Ref.	Varia, remarks Fig. in Chap. 3, subvol. b	See Tab., Nr.
141	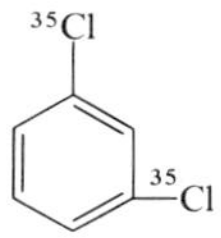	$v=0$	^{35}Cl	$\chi_{aa}=-43.3$ (8) $\chi_{bb}=$ 10.5 (3) $\chi_{cc}=$ 32.8 (11)	$\chi_{zz}=-70.2$ (11) $\chi_{yy}=$ 37.4 (1) $\chi_{xx}=$ 32.8 (12)	$\theta_{za}=30°$	77OND	z=C—Cl bond axis $x\perp$molecular plane	2.5, 707
142	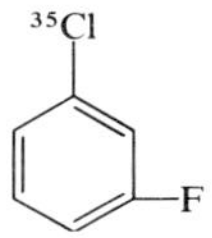	$v=0$	^{35}Cl	$\chi_{aa}=-64.8$ (6) $\chi_{bb}=$ 23.2 (3) $\chi_{cc}=$ 41.6 (3)	$\chi_{\alpha\alpha}=$ 34.4 (10) $\chi_{\beta\beta}=$ 41.6 (3) $\chi_{\gamma\gamma}=-76.0$ (7)		73KIR5		2.5, 708
143	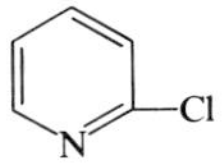	$v=0$ $v=0$	^{35}Cl ^{37}Cl	$\chi_{aa}=-71.9$ (15) $\chi_{aa}=-54.9$ (10)			74WAL		2.5, 733
	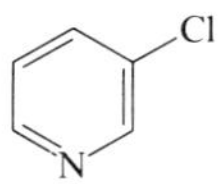		^{35}Cl	see Nr. 75					
144	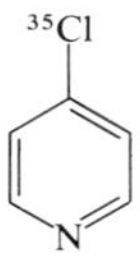	$v=0$	^{35}Cl	$\chi_{aa}=-72.0$ (2) *) $\chi_{bb}=$ 40.2 (12) η = 0.117 (35)			76CA3	*) N quadrupole constants of pyridine assumed [Nr. 73]	2.5, 742
144a	$Ar^{35}ClCN$	$v=0$	^{35}Cl ^{14}N	$\chi_{aa}=$ 37.9468 (23) $\chi_{bb}=-79.5239$ (20) $\chi_{cc}=$ 41.5771 (43) $\chi_{ab}=$ 21.9 *) $\chi_{aa}=$ 1.6403 (22) $\chi_{bb}=-3.4571$ (20) $\chi_{cc}=$ 1.8168 (42) $\chi_{ab}=$ 0.95 *)			81KEE1	*) assumed value	2.5, 756a

Nr.	Molecule	v	Nu-cleus	Nuclear quadrupole coupling tensor components: in system of principal axes of inertia χ_{gg}, [MHz]	in system of principal coupling tensor or bond axes $\chi_{\alpha\beta}$, χ_{xy} [MHz]	position of the axes θ	Ref.	Varia, remarks Fig. in Chap. 3, subvol. b	See Tab., Nr.
144b	HC≡CH · HCl		^{35}Cl	$\chi_{aa} = -54.342\,(4)$ $\chi_{bb} = 26.862\,(5)$ $\chi_{cc} = 27.480\,(9)$			81LEG		2.5, 762a
			^{37}Cl	$\chi_{aa} = -42.832\,(3)$ $\chi_{bb} = 21.173\,(4)$ $\chi_{cc} = 21.659\,(7)$					
	HC≡CH · DCl		^{35}Cl	$\chi_{aa} = -56.587\,(9)$ $\chi_{bb} = 27.91\,(1)$ $\chi_{cc} = 28.68\,(2)$					
145	$H^{35}Cl \cdot C_3H_6$	$v=0$	^{35}Cl	$\chi_{aa} = -54.41\,(2)$ $\chi_{bb} = 27.26\,(7)$ $\chi_{cc} = 27.15\,(7)$			80LG3	Fig. 195	2.5, 764
145a	$CH_2{=}CH_2 \cdot HCl$		^{35}Cl	$\chi_{aa} = -54.076\,(4)$ $\chi_{bb} = 27.091\,(6)$ $\chi_{cc} = 26.985\,(10)$			81ALD		2.5, 766
			^{37}Cl	$\chi_{aa} = -42.633\,(7)$ $\chi_{bb} = 21.399\,(13)$ $\chi_{cc} = 21.235\,(19)$					
	$CH_2{=}CH_2 \cdot DCl$		^{35}Cl	$\chi_{aa} = -56.331\,(14)$ $\chi_{bb} = 28.218\,(17)$ $\chi_{cc} = 28.113\,(31)$					
146		$v=0$	^{75}As	$\chi_{aa} = -186.4\,(1)$ $\chi_{bb} = 43.5\,(2)$ $\chi_{cc} = 142.9\,(2)$			75LAT		2.5, 750
146a	$GeH_3CH{=}CH_2$	$v=0$	^{73}Ge	$\chi_{aa} < 2.8$			74DUR7	Fig. 113	2.5, 409
147	$SO_2{}^{79}BrF$	$v=0$	^{79}Br	$\chi_{aa} = 626.6$ $\eta = -0.011$			73RAL1		2.5, 33
	$SO_2{}^{81}BrF$	$v=0$	^{81}Br	$\chi_{aa} = 524.1$ $\eta = -0.010$					

Nr.	Molecule	v	Nucleus	Nuclear quadrupole coupling tensor components: in system of principal axes of inertia χ_{gg}, [MHz]	in system of principal coupling tensor or bond axes $\chi_{\alpha\beta}$, χ_{xy} [MHz]	position of the axes θ	Ref.	Varia, remarks Fig. in Chap. 3, subvol. b	See Tab., Nr.
148	$CH_3OCH_2{}^{79}Br$	$v=0$	^{79}Br	$\chi_{aa}=284.6\,(76)$ *) η = 0.6701 (296) $\chi_{ab}=320.4\,(168)$ $\chi_{ac}=109.3\,(232)$ $\chi_{bc}=70.6\,(128)$			74HAY4	*) uncertainties are 2.5 σ	2.5, 87
	$CH_3OCH_2{}^{81}Br$	$v=0$	^{81}Br	$\chi_{aa}=237.1\,(61)$ η = 0.6665 (287) $\chi_{ab}=270.1\,(188)$ $\chi_{ac}=91.4\,(57)$ $\chi_{bc}=63.9\,(103)$					
	$CD_3OCH_2{}^{79}Br$	$v=0$	^{79}Br	$\chi_{aa}=258.0\,(110)$ η = 0.8258 (500) $\chi_{ab}=350.6\,(177)$					
	$CD_3OCH_2{}^{81}Br$	$v=0$	^{81}Br	$\chi_{aa}=218.0\,(93)$ η = 0.8103 (498) $\chi_{ab}=299.3\,(108)$					
	$CH_3OCD_2{}^{79}Br$	$v=0$	^{79}Br	$\chi_{aa}=294.2\,(100)$ η = 0.6657 (340) $\chi_{ab}=321.2\,(178)$					
	$CH_3OCD_2{}^{81}Br$	$v=0$	^{81}Br	$\chi_{aa}=245.0\,(80)$ η = 0.6688 (333) $\chi_{ab}=272.8\,(179)$					
149	$CH_2{}^{79}BrCN$	$v=0$	^{79}Br	$\chi_{aa}=342.4$ $\chi_{bb}=-25.6$ $\chi_{cc}=-316.8$ $\chi_{ab}=-438.12$	$\chi_{\alpha\alpha}=633.6$	$\theta_{\alpha a}=33.62°$ $\theta_{\alpha b}=56.38°$	76GUM		2.5, 98
	$CH_2{}^{81}BrCN$	$v=0$	^{81}Br	$\chi_{aa}=285.2$ $\chi_{bb}=-21.9$ $\chi_{cc}=-263.2$ $\chi_{ab}=-363.7$	$\chi_{\alpha\alpha}=526.4$	$\theta_{\alpha a}=33.55°$ $\theta_{\alpha b}=56.43°$			

Nr.	Molecule	v	Nucleus	Nuclear quadrupole coupling tensor components in system of principal axes of inertia χ_{gg}, [MHz]	in system of principal coupling tensor or bond axes $\chi_{\alpha\beta}$, χ_{xy} [MHz]	position of the axes θ	Ref.	Varia, remarks Fig. in Chap. 3, subvol. b	See Tab., Nr.
150	$OC^{79}Br_2$	$v=0$	^{79}Br	$\chi_{aa}=$ 316.9 (21)	$\chi_{zz}=$ 626.5		77CAR		2.5, 127
				$\chi_{bb}=$ −83.5 (17)	$\chi_{xx}=$ −229.4				
				$\chi_{cc}=$ −233.4 (11)	$\chi_{yy}=$ −397.1				
				$\chi_{ab}=$ 490 (10)					
	$OC^{79}Br^{81}Br$	$v=0$	^{79}Br	$\chi_{aa}=$ 308.4 (19)					
				$\chi_{bb}=$ −78.7 (15)					
				$\chi_{cc}=$ −229.7 (10)					
		$v=0$	^{81}Br	$\chi_{aa}=$ 266.7 (19)					
				$\chi_{bb}=$ −71.0 (15)					
				$\chi_{cc}=$ −195.7 (10)					
	$OC^{81}Br_2$	$v=0$	^{81}Br	$\chi_{aa}=$ 248.6 (18)					
				$\chi_{bb}=$ −62.3 (15)					
				$\chi_{cc}=$ −186.3 (10)					
151	$CH_2{}^{79}BrC(=O)OH$	$v=0$	^{79}Br	$\chi_{aa}=$ 421 (4)	$\chi_{\alpha\alpha}=$ 604 (3)	$\theta_{\alpha a}=26.7\ (2)°$	78VAN1		2.5, 201
				$\chi_{bb}=$ −121 (2)	$\chi_{\beta\beta}=$ −304 (2)				
				$\chi_{cc}=$ −300 (2)	$\chi_{\gamma\gamma}=$ −300 (2)				
				$\chi_{ab}=$ 365 (2)					
		$v_\tau=1$	^{79}Br	$\chi_{aa}=$ 419 (7)					
				$\chi_{bb}=$ −125 (4)					
				$\chi_{cc}=$ −294 (4)					
				$\chi_{ab}=$ 363 (3)					
	$CH_2{}^{79}BrC(=O)OD$	$v=0$	^{79}Br	$\chi_{aa}=$ 422 (5)					
				$\chi_{bb}=$ −122 (3)					
				$\chi_{cc}=$ −301 (2)					
				$\chi_{ab}=$ 366 (4)					
	$CH_2{}^{81}BrC(=O)OH$	$v=0$	^{81}Br	$\chi_{aa}=$ 354 (4)					
				$\chi_{bb}=$ −102 (2)					
				$\chi_{cc}=$ −251 (2)					
				$\chi_{ab}=$ 306 (4)					

Nr.	Molecule	v	Nucleus	Nuclear quadrupole coupling tensor components: in system of principal axes of inertia χ_{gg}, [MHz]	in system of principal coupling tensor or bond axes $\chi_{\alpha\beta}$, χ_{xy} [MHz]	position of the axes θ	Ref.	Varia, remarks Fig. in Chap. 3, subvol. b	See Tab., Nr.
151 a	$CH_3CH_2CH_2Br$ gauche	$v=0$	^{79}Br	$\chi_{aa}=$ 256.9 (11)	$\chi_{\alpha\alpha}=$ 526.4	$\theta_{\alpha a}=35.34°$	81NII		2.5, 221 a
				$\chi_{bb}=$ −9.5 (5)	$\chi_{\beta\beta}=-247.4$				
				$\chi_{cc}=-247.4$ (13)	$\chi_{\gamma\gamma}=-279.0$				
				$\chi_{ab}=$ 380.0 (43)					
		$v_\tau=1$		$\chi_{aa}=$ 250.1 (14)	$\chi_{\alpha\alpha}=$ 536.5			$v_\tau=127\ cm^{-1}$, central C—C torsion	
				$\chi_{bb}=$ −6.9 (7)	$\chi_{\beta\beta}=-243.2$				
				$\chi_{cc}=-243.2$ (16)	$\chi_{\gamma\gamma}=-293.2$				
				$\chi_{ab}=$ 394.5 (36)					
		$v_\tau=2$		$\chi_{aa}=$ 247.1 (27)	$\chi_{\alpha\alpha}=$ 533.3				
				$\chi_{bb}=$ −4.9 (31)	$\chi_{\beta\beta}=-242.2$				
				$\chi_{cc}=-242.2$ (41)	$\chi_{\gamma\gamma}=-291.1$				
				$\chi_{ab}=$ 392.4 (64)					
		$v=0$	^{81}Br	$\chi_{aa}=$ 214.1 (11)	$\chi_{\alpha\alpha}=$ 434.1	$\theta_{\alpha a}=35.20°$			
				$\chi_{bb}=$ −8.1 (5)	$\chi_{\beta\beta}=-206.0$				
				$\chi_{cc}=-206.0$ (12)	$\chi_{\gamma\gamma}=-228.1$				
				$\chi_{ab}=$ 311.9 (69)					
		$v_\tau=1$		$\chi_{aa}=$ 209.9 (12)	$\chi_{\alpha\alpha}=$ 456.0				
				$\chi_{bb}=$ −6.8 (6)	$\chi_{\beta\beta}=-203.1$				
				$\chi_{cc}=-203.1$ (13)	$\chi_{\gamma\gamma}=-252.9$				
				$\chi_{ab}=$ 337.5 (32)					
		$v_\tau=2$		$\chi_{aa}=$ 213.5 (34)	$\chi_{\alpha\alpha}=$ 452.9				
				$\chi_{bb}=$ −9.4 (37)	$\chi_{\beta\beta}=-204.1$				
				$\chi_{cc}=-204.1$ (50)	$\chi_{\gamma\gamma}=-248.8$				
				$\chi_{ab}=$ 332.7 (34)					
152	CH_2=$CH^{79}Br$	$v_{10}=1$	^{79}Br	$\chi_{aa}=$ 472.13	$\chi_{xx}=$ 558.91	$\theta=18.45°$	72MAR		2.5, 283
				$\chi_{bb}=-220.72$	$\chi_{yy}=-307.50$				
				$\chi_{cc}=-251.41$	$\chi_{zz}=-251.41$				
	CH_2=$CH^{81}Br$	$v_{10}=1$	^{81}Br	$\chi_{aa}=$ 393.95	$\chi_{xx}=$ 474.95	$\theta=19.34°$			
				$\chi_{bb}=-182.32$	$\chi_{yy}=-263.22$				
				$\chi_{cc}=-211.62$	$\chi_{zz}=-211.62$				
	CD_2=$CD^{79}Br$	$v=0$	^{79}Br	$\chi_{aa}=$ 470.60	$\chi_{\alpha\alpha}=$ 533.94	$\theta=16.22°$	73MAR		
				$\chi_{bb}=-216.02$	$\chi_{\beta\beta}=-279.42$				
Cont.				$\chi_{cc}=-254.58$	$\chi_{\gamma\gamma}=-254.63$				

Nr.	Molecule	v	Nucleus	Nuclear quadrupole coupling tensor components in system of principal axes of inertia χ_{gg}, [MHz]	in system of principal coupling tensor or bond axes $\chi_{\alpha\beta}$, χ_{xy} [MHz]	position of the axes θ	Ref.	Varia, remarks Fig. in Chap. 3, subvol. b	See Tab., Nr.
152 Cont.	$CD_2{=}CD^{81}Br$	$v=0$	^{81}Br	$\chi_{aa}=$ 395.07 $\chi_{bb}=-194.47$ $\chi_{cc}=-165.60$	$\chi_{\alpha\alpha}=$ 454.21 $\chi_{\beta\beta}=-241.74$ $\chi_{\gamma\gamma}=-212.47$	$\theta=16.95°$			
153	$CH_2{=}CHCH_2{}^{79}Br$	$v=0$	^{79}Br	$\chi_{aa}=$ 348.2 (46) $\chi_{bb}=-71.9$ (27) $\chi_{cc}=-276.3$ (53) $\chi_{ab}=$ 358.7 (54)	$\chi_{\alpha\alpha}=$ 553.8 *)		76NI	*) α axis along C—Br bond Fig. 91	2.5, 309
		$v_\tau=1$	^{79}Br	$\chi_{aa}=$ 345.8 (19) $\chi_{bb}=-71.5$ (10) $\chi_{cc}=-274.3$ (21) $\chi_{ab}=$ 369.0 (17)	$\chi_{\beta\beta}=-286.8$ $\chi_{\gamma\gamma}=-274.3$ $\chi_{\alpha\alpha}=$ 561.1		79SAS	v_τ, C—C torsion	
		$v_\tau=2$	^{79}Br	$\chi_{aa}=$ 344.4 (40) $\chi_{bb}=-71.5$ (22) $\chi_{cc}=-272.9$ (46) $\chi_{ab}=$ 371.3 (27)	$\chi_{\beta\beta}=-289.1$ $\chi_{\gamma\gamma}=-272.9$ $\chi_{\alpha\alpha}=$ 562.0				
	$CH_2{=}CHCH_2{}^{81}Br$	$v=0$	^{81}Br	$\chi_{aa}=$ 283.2 (49) $\chi_{bb}=-55.6$ (29) $\chi_{cc}=-227.6$ (57) $\chi_{ab}=$ 296.0 (68)	$\chi_{\alpha\alpha}=$ 448.6		76NI		
		$v_\tau=1$	^{81}Br	$\chi_{aa}=$ 284.0 (19) $\chi_{bb}=-57.8$ (11) $\chi_{cc}=-226.2$ (22) $\chi_{ab}=$ 303.2 (22)	$\chi_{\beta\beta}=-234.9$ $\chi_{\gamma\gamma}=-226.2$ $\chi_{\alpha\alpha}=$ 461.1		79SAS		
		$v_\tau=2$	^{81}Br	$\chi_{aa}=$ 282.1 (24) $\chi_{bb}=-57.5$ (13) $\chi_{cc}=-224.6$ (27) $\chi_{ab}=$ 298.6 (20)	$\chi_{\beta\beta}=-231.2$ $\chi_{\gamma\gamma}=-224.6$ $\chi_{\alpha\alpha}=$ 455.8				
154	$CH_2{=}C{=}CH^{79}Br$	$v=0$	^{79}Br	$\chi_{aa}=377$ (2) $\chi_{bb}-\chi_{cc}=183$ (5)			76DEM2		2.5, 346
	$CH_2{=}C{=}CH^{81}Br$	$v=0$	^{81}Br	$\chi_{aa}=316$ (3) $\chi_{bb}-\chi_{cc}=154$ (6)					

Nr.	Molecule	v	Nucleus	Nuclear quadrupole coupling tensor components: in system of principal axes of inertia χ_{gg}, [MHz]	in system of principal coupling tensor or bond axes $\chi_{\alpha\beta}$, χ_{xy} [MHz]	position of the axes θ	Ref.	Varia, remarks Fig. in Chap. 3, subvol. b	See Tab., Nr.
155	$SiH_3CH_2{}^{79}Br$	$v=0$	^{79}Br	$\chi_{aa}=339.6(106)$ $\chi_{bb}-\chi_{cc}=244.8(24)$ $\chi_{ab}=381.4(106)$	$\chi_{\alpha\alpha}=573.7(11)$ $\chi_{\beta\beta}=-281.6(183)$ $\chi_{\gamma\gamma}=-292.2(65)$	$\theta_{\alpha a}=31°33'(40')$	74HAY2		2.5, 380
	$SiH_3CH_2{}^{81}Br$	$v=0$	^{81}Br	$\chi_{aa}=283.2(80)$ $\chi_{bb}-\chi_{cc}=204.6(17)$ $\chi_{ab}=319.1(101)$	$\chi_{\alpha\alpha}=479.5(28)$ $\chi_{\beta\beta}=-235.6(156)$ $\chi_{\gamma\gamma}=-243.9(48)$	$\theta_{\alpha a}=31°36'(41')$			
	$SiD_3CH_2{}^{79}Br$	$v=0$	^{79}Br	$\chi_{aa}=332.0(176)$ $\chi_{bb}-\chi_{cc}=254.1(31)$ $\chi_{ab}=399.1(111)$	$\chi_{\alpha\alpha}=586.6(39)$ $\chi_{\beta\beta}=-293.5(318)$ $\chi_{\gamma\gamma}=-293.1(103)$	$\theta_{\alpha a}=32°32'(1°5')$			
	$SiD_3CH_2{}^{81}Br$	$v=0$	^{81}Br	$\chi_{aa}=275.5(133)$ $\chi_{bb}-\chi_{cc}=211.9(22)$ $\chi_{ab}=332.6(137)$	$\chi_{\alpha\alpha}=488.3(19)$ $\chi_{\beta\beta}=-244.6(230)$ $\chi_{\gamma\gamma}=-243.7(78)$	$\theta_{\alpha a}=32°36'(56')$			
155a	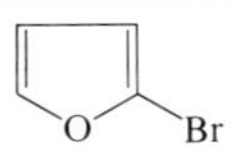	$v=0$	^{79}Br	$\chi_{aa}=592.8(3)$ $\chi_{bb}-\chi_{cc}=-11.1(2)$ $\chi_{ab}=-60.6(18)$			81STO		2.5, 612a
		$v=0$	^{81}Br	$\chi_{aa}=495.5(3)$ $\chi_{bb}-\chi_{cc}=-9.3(2)$ $\chi_{ab}=-44.2(25)$					
155b	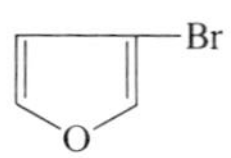	$v=0$	^{79}Br	$\chi_{aa}=589.5(4)$ $\chi_{bb}-\chi_{cc}=-7.6(3)$ $\chi_{ab}=19.9(65)$			81STO		2.5, 612b
		$v=0$	^{81}Br	$\chi_{aa}=492.3(4)$ $\chi_{bb}-\chi_{cc}=-6.8(3)$ $\chi_{ab}=16.7(20)$					
156	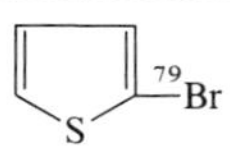	$v=0$	^{79}Br	$\chi_{aa}=592.7(15)$ $\chi_{bb}=-295.3(6)$ $\chi_{cc}=-297.4(16)$ $\chi_{ab}=80(9)$	$\chi_{\beta\beta}=-302.4(27)$ $\chi_{\gamma\gamma}=-297.4(16)$ $\chi_{\alpha\alpha}=599.8(22)$	$\theta_{\alpha a}=5.11(56)$	75MJ1		2.5, 616
	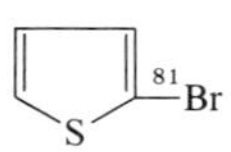	$v=0$	^{81}Br	$\chi_{aa}=493.7(15)$ $\chi_{bb}=-245.6(7)$ $\chi_{cc}=-248.1(16)$ $\chi_{ab}=64(8)$	$\chi_{\beta\beta}=-251.1(26)$ $\chi_{\gamma\gamma}=-248.1(16)$ $\chi_{\alpha\alpha}=499.2(20)$	$\theta_{\alpha a}=4.91(60)$			

Nr.	Molecule	v	Nucleus	Nuclear quadrupole coupling tensor components in system of principal axes of inertia $\chi_{gg'}$ [MHz]	in system of principal coupling tensor or bond axes $\chi_{\alpha\beta}$, χ_{xy} [MHz]	position of the axes θ	Ref.	Varia, remarks Fig. in Chap. 3, subvol. b	See Tab., Nr.
156a	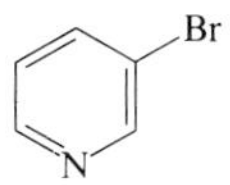	$v=0$	^{79}Br ^{81}Br	$\chi_{aa}=561.3(29)$ $\chi_{bb}=-293.6(15)$ $\chi_{cc}=-267.7(29)$ $\chi_{aa}=475.8(32)$ $\chi_{bb}=-248.6(16)$ $\chi_{cc}=-227.2(32)$			81DOR		2.5, 738a
157	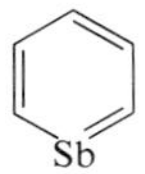	$v=0$	^{121}Sb ^{123}Sb	$\chi_{aa}=456.4(41)$ $\chi_{bb}=-137.8(31)$ $\chi_{cc}=-318.6(47)$ $\chi_{aa}=583.0(53)$ $\chi_{bb}=-175.2(39)$ $\chi_{cc}=-407.1(60)$			78FO1		2.5, 751
158	CH_2ICN	$v=0$	^{127}I	$\chi_{aa}=-1276$ $\chi_{bb}=233$ $\chi_{ab}=1370$	$\chi_{\alpha\alpha}=-2086$	$\theta_{\alpha a}=30.6°$	75AUL		2.5, 99
159	CH_3CH_2I	$v=0$	^{127}I	$\chi_{aa}=-1478.111(68)$ $\chi_{bb}=564.464(76)$ $\chi_{cc}=913.648(73)$ $\chi_{ab}=896.38(47)$	$\chi_{\alpha\alpha}=-1815.693(210)$ $\chi_{\beta\beta}=902.046(138)$ $\chi_{\gamma\gamma}=913.648(73)$ $\eta=0.00639$	$\theta_{\alpha a}=20.637°$	80BOU2		2.5, 154
160	$(CH_3)_2CHI$	$v=0$	^{127}I	$\chi_{aa}=-1626.6088\pm2.1$ *) $\chi_{bb}=883.6775\pm1.9$ $\chi_{cc}=742.9313\pm2.5$ $\chi_{ac}=535.6870\pm1.3$	$\chi_{\alpha\alpha}=-1742.1$ $\chi_{\beta\beta}=883.7$ $\chi_{\gamma\gamma}=858.4$ $\eta=-0.0145$		78GRO	*) uncertainties are 2 σ	2.5, 237
161	$CH_3CI{=}CH_2$	$v=0$	^{127}I	$\chi_{aa}=-1820.7832(331)$ $\chi_{bb}=957.0180(406)$ $\chi_{cc}=863.7652(402)$ $\chi_{ab}=147.48(102)$	$\chi_{\alpha\alpha}=-1828.59(15)$ *) $\chi_{\beta\beta}=964.83(16)$ $\chi_{\gamma\gamma}=863.77(4)$	$\theta_{\alpha a}=3.03(2)°$	79GRO	*) α axis approximately along C—I bond, β axis in the molecular symmetry plane. v_{21}, torsion	2.5, 306
		$v_{21}=1$	^{127}I	$\chi_{aa}=-1821.724(151)$ $\chi_{bb}=959.049(202)$ $\chi_{cc}=862.675(123)$ $\chi_{ab}=137.81(582)$	$\chi_{\alpha\alpha}=-1828.54(72)$ *) $\chi_{\beta\beta}=965.86(77)$ $\chi_{\gamma\gamma}=862.68(12)$	$\theta_{\alpha a}=2.83(12)°$			
Cont.									

Nr.	Molecule	v	Nucleus	Nuclear quadrupole coupling tensor components: in system of principal axes of inertia χ_{gg}, [MHz]	in system of principal coupling tensor or bond axes $\chi_{\alpha\beta}$, χ_{xy} [MHz]	position of the axes θ	Ref.	Varia, remarks, Fig. in Chap. 3, subvol. b	See Tab., Nr.
161 Cont.		$v_{14}=1$	^{127}I	$\chi_{aa}=-1821.367\ (101)$ $\chi_{bb}=955.457\ (118)$ $\chi_{cc}=865.911\ (100)$ $\chi_{ab}=145.23\ (434)$	$\chi_{\alpha\alpha}=-1828.94\ (56)$ *) $\chi_{\beta\beta}=963.03\ (58)$ $\chi_{\gamma\gamma}=865.91\ (10)$	$\theta_{\alpha a}=2.99\ (9)°$	79GRO	v_{14}, in plane vibration. *) α axis approximately along C—I bond, β axis in the molecular symmetry plane.	
162	$CH_2{=}CHCH_2I$ skew	$v=0$	^{127}I	$\chi_{aa}=-1337\ (12)$ $\chi_{bb}=387\ (6)$ $\chi_{cc}=950\ (9)$ $\chi_{ab}=1160\ (5)$	$\chi_{\alpha\alpha}=-1920$		80OHK		2.5, 310
		$v_{\tau}=1$	^{127}I	$\chi_{aa}=-1333\ (8)$ $\chi_{bb}=386\ (4)$ $\chi_{cc}=947\ (6)$ $\chi_{ab}=1086\ (2)$	$\chi_{\alpha\alpha}=-1859$ $\chi_{\beta\beta}=912$ $\chi_{\gamma\gamma}=947$	$\theta_{\alpha a}=25.8°$	79TAK2	v_{τ}, CH_2I torsion	
		$v_{\tau}=2$	^{127}I	$\chi_{aa}=-1319\ (10)$ $\chi_{bb}=383\ (5)$ $\chi_{cc}=936\ (8)$ $\chi_{ab}=1073\ (3)$	$\chi_{\alpha\alpha}=-1839$ $\chi_{\beta\beta}=903$ $\chi_{\gamma\gamma}=936$	$\theta_{\alpha a}=25.8°$			
163	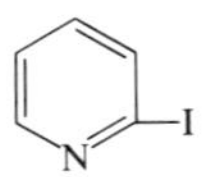	$v=0$	^{127}I	$\chi_{aa}=-1881\ (2)$ $\chi_{bb}=1017\ (2)$			74CA1		2.5, 734
164	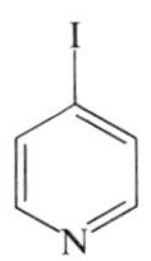	$v=0$	^{127}I	$\chi_{aa}=-1938\ (2)$ $\chi_{bb}=1008\ (2)$			74CA1		2.5, 743

2.8 Hindered rotation

2.8.1 Preliminary remarks

1. Introduction

The rotational spectrum of a molecule with internal rotation (torsion) is modified due to torsion-rotation interaction. The interpretation of this modification allows the determination of the internal rotation potential barrier [59LIN, 68DRE, 70GOR]. The molecule is generally taken to be rigid except for internal rotation. However, special methods have been developed to include the interaction with molecular vibrations.

For higher potential barriers, a torsion fine structure appears in the rotational spectrum. The PAM (principal axis method) or IAM (internal axis method) is used in the analysis for potential barriers greater than approximately 0.5 kcal/mole. This limit depends on the molecular geometry and mass distribution. IAM(W) is a method introduced by Woods [66WOO, 67WOO] and related to IAM. For lower potentials, line bands appear instead of a clear multiplet fine structure. In this case, the rotational spectrum modified by torsion is analysed by proceeding from the limiting case of free internal rotation using the "low barrier" method.

If no torsion fine structure appears in the rotational spectrum, but it is possible to identify rotational lines in higher torsional states, then the magnitude of the potential barrier can be determined by comparing intensities of rotational lines which have been assigned to different torsional states: "Int." (= intensity comparison) method.

2. Explanations of the columns in the tabulation

Column	Parameters		Definitions
1	Nr.		serial number of the molecule in this table
2	v	v_τ	quantum number of the torsional vibration. Molecular torsional state data corresponding to the evaluation method: For the methods PAM and IAM quantum number of the torsional state to which the analysed rotational lines correspond. For molecules with only *one* torsional vibration (one-top molecules), 0 or 1 means $v_\tau=0$ or $v_\tau=1$; for molecules with *two* torsional vibrations (two-top molecules) 0 0 means $v_\tau v_{\tau'}=00$. For two-top molecules with two equivalent internal rotors, a double designation is used, e.g. the singly excited state which is (nearly) doubly degenerate is designated by (01, 10) meaning that $v_\tau v_{\tau'}=01$ is degenerate with $v_\tau v_{\tau'}=10$. The degeneracy is usually removed by top-top coupling, yielding two distinct singly excited states, $(01,10)_1$ and $(01,10)_2$. For the "Int." method data of torsional states to which the rotational lines used for intensity comparison correspond, e.g. $(v_\tau=0):(v_\tau=1):(v_\tau=2)$, i.e., 0, 1, 2
3	N		N-fold axis of the hindering potential
4	V		hindering potential
		V_N [cal/mol] [cm^{-1}]	Fourier expansion coefficient of the hindering potential $V(\alpha)$. (In most cases simply called the hindering potential when only the coefficient with the smallest N is given) [59LIN, 68DRE, 70GOR]
		$V(\alpha)$	hindering potential (torsional potential) as a function of the torsional angle α [59LIN, 68DRE] $$V(\alpha)=\sum_N \frac{V_N}{2}(1-\cos N\alpha)$$
Cont.		s	reduced hindering potential $$s=\frac{4V_N}{N^2 F}$$

Column	Parameters		Definitions
4 Cont.	V	Δ_0 [MHz]	Wood' IAM Parameter $\Delta_0 = -\frac{27}{8} F w_1^{(0)}$ where the Fourier coefficient $w_1^{(0)}$ is a tabulated function of the barrier height [61HAY]
5			structural parameters
		F [GHz] [cm^{-1}]	inverse reduced moment of inertia for one-top molecules $F = \frac{h}{8\pi^2} \frac{1}{r I_\alpha}$
		I_α [amu Å^2] [g cm^2]	moment of inertia of the internal top (most often methyl group) about its symmetry axis
		$\sphericalangle(i, g) = \theta_g$	angle between the symmetry axis of the internal top and the principal axis $g (= a, b, c)$
		$\lambda_a, \lambda_b, \lambda_c$	direction cosine between symmetry axis of the internal top and the "ordered" principal inertia axes a, b, c ($I_a \leqq I_b \leqq I_c$)
		$\lambda_x, \lambda_y, \lambda_z$	direction cosine between symmetry axis of the internal top and the principal axes x, y, z
		r	reduction factor $r = 1 - \frac{\lambda_x^2 I_\alpha}{I_x} - \frac{\lambda_y^2 I_\alpha}{I_y} - \frac{\lambda_z^2 I_\alpha}{I_z}$
		I_x, I_y, I_z [amu Å^2]	principal moments of inertia
		I_a, I_b, I_c	principal moments of inertia in order $I_a \leqq I_b \leqq I_c$
		α, β, γ [amu Å^2]	$\alpha = \lambda_a I_\alpha / I_a$; $\beta = \lambda_b I_\alpha / I_b$; $\gamma = \lambda_c I_\alpha / I_c$
		$\varrho_a, \varrho_b, \varrho_c$	$\varrho_a = \alpha$; $\varrho_b = \beta$; $\varrho_c = \gamma$
		$\varrho, \beta_w, \gamma_w$	$\varrho = \sqrt{\alpha^2 + \beta^2 + \gamma^2}$; $\beta_w = \arccos(\alpha/\varrho)$; $\gamma_w = \arccos(\beta/\sqrt{\beta^2 + \gamma^2})$ parameters occurring in IAM (W)
6	Meth.		method of measurement for the values in the previous columns PAM: Principal Axis Method IAM: Internal Axis Method IAM (W): Internal Axis Method, Wood's version Int.: Intensity RTV: Rotation-Torsion-Vibration [72DRE, 76BAU]
7	Ref.		references for the previous columns
8	Varia, remarks		remarks concerning the previous columns references to figures in Chapter 3, subvolume b
9			reference to Tables 2.4 and 2.5

3. Conversion factors

1 cal/mol = 4.1868 J/mol ≙ 0.349989 cm^{-1} ≙ $1.04924 \cdot 10^{10}$ Hz

See also Introduction, Chapter 1.

4. Arrangement of the substances

a) C—C bonds	Nr. 1 ··· 59
b) C—Si, C—Ge bonds	Nr. 60 ··· 72
c) C—O bonds	Nr. 73 ··· 92
d) C—S, C—Se bonds	Nr. 93 ··· 107
e) C—N, C—P bonds	Nr. 108 ··· 126
f) other bonds	Nr. 127 ··· 140.

2.8.2 Data

Nr.	Molecule	v	N	Potential	Structural parameters	Meth.	Ref.	Varia, remarks Fig. in 3	See Tab., Nr.
a) C—C bonds									
1	CH_3CHD_2	0	3	$V_3=2878.3\ (2.8)$	$F=257.70\ (35)$	IAM	79HIR1	uncertainties: 2.5 σ	2.4, 72
		1	3	$V_3=2865.9\ (18)$	$F=257.70$ *)	IAM	81HIR1	*) ground state value used	
	CH_3CH_2D	0	3	$V_3=2888.38\ (36)$	$F=282.795\ (49)$				
	CD_3CH_2D	0	3	$V_3=2862\ (11)$	$F=203.49\ (65)$				
	CH_3CH_3	0	3	$V_3=2882\ (10)$ ⁺)				⁺) by extrapolation	
				$V_6=20\ (10)$					
2	CH_3CF_3	0	3	$V_3=3160\ (110)$	$I_\alpha=3.17\ (11)$	MBER	78MEE		2.4, 73
3	CH_3CCl_3		3	$V_3=5100\ (300)$		Int.	73DUR2		2.4, 74
4	CH_3CH_2F	$v_\tau=1$	3	$s=79.7\ (8)$	$F=192.34$	PAM	80SH2		2.5, 151
					$\alpha=0.1846$				
					$\beta=0.0336$				
		$v_\tau=2$		$s=80.1\ (1)$	$F=192.47$				
					$\alpha=0.1852$				
					$\beta=0.0334$				
		$v_\delta=1$		$s=63.4\ (1.5)$	$F=191.87$			v_δ skeletal bending state	
					$\alpha=0.1835$				
					$\beta=0.0348$				
	CD_3CH_2F	$v_\tau=2$		$s=183.6\ (9)$	$F=110.60$				
					$\alpha=0.2995$				
					$\beta=0.0571$				
		$v_\delta=1$		$s=64.4\ (10)$	$F=112.17$				
					$\alpha=0.3084$				
					$\beta=0.0543$				
5	CH_3CH_2I	$v_\tau=1$	3	$V_3=3623\ (150)$	$I_\alpha=3.18$	PAM	80BOU3		2.5, 154
				$s=96.46\ (12)$	$\measuredangle(i,a)=48.5°$				
6	CH_3CH_2CN	0	3	$V_3=3007\ (90)$	$I_\alpha=3.167$	PAM	80BOU4	Fig. 42	2.5, 155
					$\Theta=48.65°$				
		$v_\tau=1$		$V_3=3226\ (10)$	$I_\alpha=3.159$	RTV	76HEI		
				$V_6=-172.0\ (0.5)$	$\lambda_x=0.678$				
Cont.				$V'_{3c}=-1421\ (134)$ *)	$\lambda_v=-0.7351$			*) in cal/molrad	

Nr.	Molecule	v	N	Potential	Structural parameters	Meth.	Ref.	Varia, remarks Fig. in 3	See Tab., Nr.
6 Cont.	CD_3CD_2CN	$v_\tau=3$		$V_3 =$ 3002 (150)	$I'_\alpha =$ 6.389 $\alpha =$ 0.1528 $\beta = -0.0368$	PAM	74LI		
7	CH_3CH_2OH gauche	0	3	$V_3=1331\ cm^{-1}$	$F=193.2$ $\varrho_z=0.1754$ $\lambda_z=0.8310$	PAM	80KA2	methyl internal rotation analysis	2.5, 156
			1	$V_1=$ 57.0 cm^{-1} $V_2=$ 0.8 cm^{-1} $V_3=$ 395.0 cm^{-1}				hydroxyl internal rotation analysis	
	CH_3CH_2OH trans	0	3	$V_3=1158\ cm^{-1}$					
8	CH_3CH_2SH trans	0	3	$V_3=1156\ (30)\ cm^{-1}$	$F=177.4$ $\lambda_a=0.7215$	PAM	75SC	Fig. 43	2.5, 157
		1	3	$V_3=3260\ (30)$	$I_\alpha=3.1773$ $\lambda_a=0.7328$ $\lambda_b=0.6805$		76NAK2		
9	CH_3CH_2SeH	0	3	$V_3=3590\ (10)$ $V_2=-96.4\ (1)\ cm^{-}1$ $V_3=$ 432 (4) cm^{-1} $V_6=-20\ (2)\ cm^{-1}$	$F=5.80\ cm^{-1}$	PAM	77DUR3	methyl torsion selenol torsion	2.5, 158
	CH_3CH_2SeD			$V_2=-91.7\ (4)\ cm^{-1}$ $V_3=$ 408.6 (2) cm^{-1} $V_6=$ $-7.8\ (2)\ cm^{-1}$		PAM		selenol torsion	
10	$CH_3CH_2OCH_3$ trans	0	3	$V_3=2554\ (20)$ $s=61.54\ (0.54)$	$F=193.452$ $\lambda_a=0.9344$ $\lambda_b=0.3562$ $\lambda_c=0$	PAM	75HAY	one-top approximation uncertainties are 2σ OCH_3 group Fig. 44	2.5, 159
	$^{13}CH_3CH_2OCH_3$	0	3	$V_3=2554\ (30)$ $s=61.54\ (0.81)$	$F=193.445$ $\lambda_a=0.9354$ $\lambda_b=0.3537$ $\lambda_c=0$			OCH_3 group	
	$CH_3{}^{13}CH_2OCH_3$	0	3	$V_3=2545\ (35)$ $s=61.59\ (0.95)$	$F=192.625$ $\lambda_a=0.9335$ $\lambda_b=0.3586$ $\lambda_c=0$			OCH_3 group	

Nr.	Molecule	v	N	Potential	Structural parameters	Meth.	Ref.	Varia, remarks Fig. in 3	See Tab., Nr.
10 Cont.	$CH_3CH_2O^{13}CH_3$	0	3	$V_3=2549$ (25) $s=61.42$ (0.68)	$F=193.450$ $\lambda_a=0.9349$ $\lambda_b=0.3549$ $\lambda_c=0$		75HAY	OCH_3 group	
	$CD_3CH_2OCH_3$	0	3	$V_3=2554$ (90) $s=61.84$ (2.43)	$F=192.530$ $\lambda_a=0.9387$ $\lambda_b=0.3446$ $\lambda_c=0$			OCH_3 group	
		1	3	$V_3=2702$ (7) $s=65.44$ (0.19)	$F=192.457$ $\lambda_a=0.9387$ $\lambda_b=0.3446$ $\lambda_c=0$			OCH_3 group	
	$CH_3CHDOCH_3$	0	3	$V_3=2525$ (35) $s=62.52$ (0.95)	$F=188.251$ $\lambda_a=0.9328$ $\lambda_b=0.3584$ $\lambda_c=0.0378$			OCH_3 group	
	$CH_3CD_2OCH_3$	0	3	$V_3=2494$ (40) $s=63.05$ (108)	$F=184.396$ $\lambda_a=0.9313$ $\lambda_b=0.3643$ $\lambda_c=0$			OCH_3 group	
	s—$CH_2DCH_2OCH_3$	0	3	$V_3=2574$ (40) $s=61.70$ (1.08)	$F=194.477$ $\lambda_a=0.9293$ $\lambda_b=0.3693$ $\lambda_c=0$			OCH_3 group	
	a—$CH_2DCH_2OCH_3$	0	3	$V_3=2564$ (35) $s=62.03$ (0.95)	$F=192.691$ $\lambda_a=0.9393$ $\lambda_b=0.3404$ $\lambda_c=0.0417$			OCH_3 group	
	$CH_3CH_2OCD_3$	0	3	$V_3=3294$ (100) $s=79.74$ (2.70)	$F=192.486$ $\lambda_a=-0.8962$ $\lambda_b=-0.4436$ $\lambda_c=0$			CH_3C group, one-top approximation	
		1	3	$V_3=3300$ (25) $s=79.95$ (0.68)	$F=192.395$ $\lambda_a=-0.8962$ $\lambda_b=-0.4436$ $\lambda_c=0$			CH_3C group, one-top approximation	

Nr.	Molecule	v	N	Potential	Structural parameters	Meth.	Ref.	Varia, remarks Fig. in 3	See Tab., Nr.
11	$CH_3CH_2OCH_2F$	0	3	$V_3=3150\,(50)$	$F=174.12$ $\alpha=0.08655$ $\beta=0.00227$ $\gamma=0.00534$	PAM	80HAY1	Fig. 48	2.5, 167
12	CH_3CHF_2	$v_\tau=1$ $v_\tau=2$	3	$V_3=3199\,(8)$ $s=88.44$ $V_3=3193\,(10)$ $s=88.34$	$F=168.64$ $\lambda_a=0.9602$ $F=168.50$ $\lambda_a=0.9455$	PAM	74AGE		2.5, 169
13	$CH_3CHF^{35}Cl$	0, 1	3	$V_3=4300\,(300)$		Int.	74THO		2.5, 170
14	$CH_3CH{=}NH$ cis $CH_3CH{=}NH$ trans	0	3	$V_3=557.8\,(19)\ cm^{-1}$ $s=33.82$ $V_3=515.1\,(63)\ cm^{-1}$ $s=31.49$	$\theta_{i,a}=25.69\,(40)°$ $I_\alpha=3.225\,(9)$ $\theta_{i,a}=24.79\,(66)°$ $I_\alpha=3.174\,(60)$	IAM(W)	80LOV	uncertainties are 3σ see [76BAU]	2.5, 171
15	$CH_3CH{=}NCH_3$ trans $CH_3CD{=}NCH_3$ trans $CH_3CH{=}NCD_3$ trans $CD_3CD{=}NCH_3$ trans	0 0 0 0	3	$V_{30}=537.80\,(16)\ cm^{-1}$ $V_{03}=693.45\,(16)\ cm^{-1}$ $s_1=32.992$ $s_2=42.588$ $V_{30}=540.39\,(16)\ cm^{-1}$ $V_{03}=701.7\,(32)\ cm^{-1}$ $s_1=35.365$ $s_2=45.581$ $V_{30}=539.58\,(29)\ cm^{-1}$ $V_{03}=688.3\,(77)\ cm^{-1}$ $s_1=33.160$ $s_2=66.159$ $V_{30}=542.0\,(20)\ cm^{-1}$ $V_{03}=707.47\,(23)\ cm^{-1}$ $s_1=56.681$ $s_2=46.127$	$I_{\alpha_1}=3.22558\,(36)$ $I_{\alpha_2}=3.230\,(10)$	IAM(W) *)	79BOS	*) rigorous analysis non-rigid model $V_{30}=C{-}CH_3$ top $V_{03}=N{-}CH_3$ top	2.5, 172
16	$CH_3CH{=}SO$	0	3	$V_3=400$			78PEN3		2.5, 173
17	$CH_3C(=O)H$	0	3	$\Delta_0=-2074.889\,(1409)$ MHz $V_3=399.9\ cm^{-1}$ $s=23.445$	$F=227.262$ $\lambda_a=0.90943$ $\lambda_b=0.41585$	IAM(W)	76BAU	Fig. 57	2.5, 185

Nr.	Molecule	v	N	Potential	Structural parameters	Meth.	Ref.	Varia, remarks Fig. in 3	See Tab., Nr.
18	$CH_3C(=O)CN$	0	3	$V_3 = 1198\,(1)$ $V_6 = -13\,(1)$		RTV	78PAN2	interaction parameter with vibration also determined	2.5, 186
	$CH_3C(=O)C^{15}N$			$V_3 = 1200\,(1)$ $V_6 = -6\,(7)$					
19	$CH_3C(=O)NCO$	0	3	$V_3 = 1018\,(15)$ $s = 28.11$	$F = 168.828$ $\lambda_a = 0.8205$ $\lambda_b = 0.5717$	PAM	80LAD		2.5, 187
19a	$CH_3C(=O)OH$	0	3	$V_3 = 168.238\,(17)\ cm^{-1}$	$I_\alpha = 3.2116\,(4)$ $\theta = 4.402\,(19)$	IAM	82DEM		2.5, 188
	$CD_3C(=O)OH$	0		$V_3 = 163.64\,(9)\ cm^{-1}$	$I_\alpha = 6.3906\,(40)$ $\theta = 3.26\,(10)$	IAM	81VAN1		
	$CH_3C(=O)OD$	0		$V_3 = 165.82\,(8)\ cm^{-1}$	$I_\alpha = 3.2216\,(14)$ $\theta = 12.39\,(6)$				
	$CH_3C(=O)^{18}OH$	0		$V_3 = 167.73\,(9)\ cm^{-1}$	$I_\alpha = 3.2197\,(14)$ $\theta = 15.51\,(4)$				
	$CH_3C(={}^{18}O)OH$	0		$V_3 = 168.19\,(3)\ cm^{-1}$	$I_\alpha = 3.2074\,(5)$ $\theta = (-)\,6.03\,(4)$				
	$^{13}CH_3C(=O)OH$	0		$V_3 = 168.03\,(10)\ cm^{-1}$	$I_\alpha = 3.2154\,(14)$ $\theta = 4.22\,(11)$				
	$^{13}CH_3C(=O)OD$	0		$V_3 = 166.12\,(8)$	$I_\alpha = 3.2168\,(13)$ $\theta = 10.62\,(5)$				
19b	$CH_3C(=O)OCH_3$	0	3	$V_3 = 99.559\,(83)\ cm^{-1}$ $V_3 = 425\,(31)\ cm^{-1}$	$I_\alpha = 3.2085\,(26)$ $\theta = 27.982\,(23)$ $I_\alpha = 3.16\,(24)$ $\theta = 25.3\,(30)$	low barrier *)	80SHE	CH_3C group *) see original paper OCH_3 group	2.5, 188a
20 Cont.	$CH_3C(=O)OOH$	0	3	$V_3 = 76.70\,(45)\ cm^{-1}$ $s = 6.10$	$\lambda_a = 0.9096\,(50)$ $\lambda_b = 0.4313\,(20)$ $I_\alpha = 3.217\,(3)$	*)	76CUG	Fig. 59 *) semirigid model, see original paper	2.5, 189

Nr.	Molecule	v	N	Potential	Structural parameters	Meth.	Ref.	Varia, remarks Fig. in 3	See Tab., Nr.
20 Cont.	$CH_3C(=O)OOD$	0	3	$V_3 = 76.70$ (ass.) cm^{-1} $s = 6.12$	$\lambda_a = 0.9057$ (140) $\lambda_b = 0.4179$ (56) $I_\alpha = 3.223$ (23)		76CUG		
21	$CH_3C(=O)SH$	0	3	$V_3 = 222.6$ (14) $s = 6.2980$ (12)	$I_\alpha = 3.198$ (20) $\theta_{i,a} = 57.080$ (30)	PAM	77NAI	Fig. 60	2.5, 190
	$CH_3C(=O)SD$			$V_3 = 212.9$ (14) $s = 5.9742$ (16)	$I_\alpha = 3.171$ (20) $\theta_{i,a} = 55.097$ (24)				
22	$CH_3C(=O)C(=O)OH$	0	3	$V_3 = 339.4$ (5) cm^{-1}		IAM(W)	77DYL	Fig. 71	2.5, 214
23	$CH_3C(=O)C(=O)H$ trans	0	3	$V_3 = 269.1$ (3) cm^{-1} $s = 22.05$ (3)		IAM(W)	78DYL	Fig. 72	2.5, 215
24	$CH_3C(=NOD)H$ syn. anti	0	3	$V_3 = 1948$ (100) $V_3 = 373$ (30)	$\lambda_z = 0.9473$ $F = 182.457$	PAM	76OH 77OH1	Fig. 73	2.5, 216
25	$CH_3C(=S)H$	0	3	$V_3 = 1572$ (30) $s = 34.94$	$F = 209.705$ $\lambda_a = 0.8535$	PAM	76KRO		2.5, 217
		1		$V_3 = 1646$ $s = 36.63$					
	$CH_3C(=S)D$	0		$V_3 = 1600$ $s = 37.74$	$F = 197.000$				
		1		$V_3 = 1654$ $s = 39.02$					
	$CD_3C(=S)H$	0		$V_3 = 1501$ $s = 35.88$	$F = 125.201$				
		1		$V_3 = 1555$ $s = 57.89$					
	$CH_3C(={}^{34}S)H$	0		$V_3 = 1580$ $s = 35.18$	$F = 209.339$				
	${}^{13}CH_3C(=S)D$	0		$V_3 = 1580$ $s = 35.23$	$F = 209.071$				
	$CH_3{}^{13}C(=S)H$	0		$V_3 = 1580$ $s = 35.32$	$F = 208.529$				

Nr.	Molecule	v	N	Potential	Structural parameters	Meth.	Ref.	Varia, remarks Fig. in 3	See Tab., Nr.
26	$CH_3C(={}^{80}Se)H$	0	3	$V_3=1602\,(30)$ $s=37.01$	$F=201.687$ $\lambda_a=0.823$	PAM	78HUT		2.5, 218
		1		$V_3=1648\,(30)$ $s=38.63$					
	$CH_3C(={}^{78}Se)H$	0		$V_3=1604\,(30)$ $s=37.06$	$F=201.716$				
		1		$V_3=1648\,(30)$ $s=38.64$					
	$CH_3C(={}^{76}Se)H$	0		$V_3=1602\,(30)$ $s=37.00$	$F=201.746$				
27	$CH_3CH_2CH_2OH$ trans	0	3	$V_3=2730\,(60)$ $s=69.6\,(15)$	$I_\alpha=3.193$ $\sphericalangle(i,a)=29\,(1)$	PAM	81DRE1		2.5, 225
		1		$V_3=3084\,(50)$ $s=77.8\,(10)$	$I_\alpha=3.19312$ $\alpha^2=0.022717$		72AB		
	$CH_3CH_2CH_2OH$ gauche	1		$V_3=2876\,(50)$ $s=78.9\,(10)$	$I_\alpha=3.13024$ $\alpha^2=0.002515$				
		$v_t=1$		$V_3=2620\,(50)$ $s=71.7$	$I_\alpha=3.13$ $\lambda_a=-0.5739$ $\lambda_b=-0.7642$ $\lambda_c=0.2907$		81AB	v_t, methyl group twisting	
28	$CH_3CH_2CH_2OCH_3$ trans-trans	1	3	$V_3=3300\,(60)$ $s=86.5$	$I_\alpha=3.1202$ $\lambda_a=0.8407$ $\lambda_b=0.5415$	PAM	80KAT	barrier to internal rotation of the CH_3C group Fig. 74	2.5, 226
29	$CH_3CH(OH)CH_2NH_2$	1	3	$V_3=3173\,(100)$ $s=89.37$	$I_\alpha=3.20$ $\lambda_a=0.868$ $\lambda_b=0.450$	PAM	76MAR1	Fig. 77	2.5, 234
30	$CH_3CH(OH)CH_2F$	1	3	$V_3=2796\,(50)$ $s=78.78$	$I_\alpha=3.20$ $\lambda_a=0.8544$ $\lambda_b=0.4641$ $\lambda_c=0.2336$	PAM	77MAR	Fig. 78	2.5, 235

Nr.	Molecule	v	N	Potential	Structural parameters	Meth.	Ref.	Varia, remarks Fig. in 3	See Tab., Nr.
31	CH_3CHFCH_3	1	3	$V_3=3560\,(300)$		PAM	73GRI		2.5, 236
32	$(CH_3)_2CHCN$		3	$V_3=3300$	$F=5.45\ cm^{-1}$	Int.	74DUR8		2.5, 238
32a	$CH_3C(=O)CH_2OH$	0	3	$V_3=193\,(10)$	$I_\alpha=\ 3.20$ *) $\theta_a=25.8\,(4)$	low-barrier	80KAT1	*) assumed value	2.5, 242a
33	$(CH_3)_2CS$		3	$V_3=1300\,(50)$		PAM	74KRO		2.5, 253
34	$CHOC(CH_3)CHOH$						81SAN	coupling of methyl torsion with tunneling of the hydrogen-bond hydrogen analysed	2.5, 257
34a	$CH_3CH_2C(=O)F$ cis	0	3	$s=65.2\,(3)$ $V_3=2350\,(11)$	$I_\alpha=3.18\,(3)$ $\theta_a=32.7\,(20)$	PAM	81SCA1		2.5, 258a
35	$CH_3CH_2C(=O){}^{35}Cl$ cis	1, 0	3	$V_3=2500\,(50)$ $s=68.6\,(7)$	$F=169.69\,(155)$ $I_\alpha=3.136\,(20$ *)) $\lambda_a=0.9421\,(20)$ $\lambda_b=0.3355\,(20)$	PAM	79MAT	*) assumed uncertainties	2.5, 259
36	$CH_3CH_2C(=O)OH$ cis	1	3	$V_3^{(a)}=2339$ $s=64.2$	$F=169.486$ $I_\alpha=3.136$ $\lambda_a=0.8377$ $\lambda_b=0.5462$	PAM	75ST1	[a] $V_3=2340\,(30)$ cal/mol for CH_3CH_2COOH and CH_3CH_2COOD Fig. 82	2.5, 260
	$CH_3CH_2C(=O)OD$ cis			$V_3^{(a)}=2342$ $s=64.3$	$F=169.794$ $I_\alpha=3.136$ $\lambda_a=0.8377$ $\lambda_b=0.5462$				
	$CH_3CD_2C(=O)OH$ cis			$V_3^{(b)}=2370$ $s=65.3$	$F=169.178$ $I_\alpha=3.136$ $\lambda_a=0.8508$ $\lambda_b=0.5255$			[b] $V_3=2370\,(30)$ cal/mol for CH_3CD_2COOH and CH_3CD_2COOD	
Cont.									

Nr.	Molecule	v	N	Potential	Structural parameters	Meth.	Ref.	Varia, remarks Fig. in 3	See Tab., Nr.
36 Cont.	$CH_3CD_2C(=O)OD$			$V_3^{(b)}=2369$ $s=65.3$	$F=169.132$ $I_\alpha=3.136$ $\lambda_a=0.8508$ $\lambda_b=0.5255$		75ST1		
37	$CH_2{=}CHOCH_2CH_3$	0, 2, 3	3	$V_3=3070\,(25)$		PAM	79OW	Fig. 88	2.5, 292
38	$(CH_3)_2CHC{\equiv}CH$		3	$V_3=3650\,(500)$		Int.	73MOC2		2.5, 355
39	$CH_3CH_2SiH_2F$ gauche	$v_\tau=1$	3	$V_3=2890\,(60)$		PAM	77HAY1	uncertainties: 2.5 σ	
40	$CH_3CH_2GeH_3$	0, 1, 2	3	$V_{03}=2850\,(30)$ $V_{30}=1410\,(30)$		*)	77DUR4	V_{03} for methyl top V_{30} for germyl top *) see [77DUR4]	2.5, 408
41	$(CH_3)_2C{=}CH_2$	0	3	$\Delta_0=-9.659\,(58)$ MHz $V_3=2170\,(9)$	$I_\alpha=3.18\,(1)$ $\varrho=0.05620\,(20)$ $\beta=0.5172\,(12)$ $\theta=58.21\,(6)$	IAM	75DEM	uncertainties: 2 σ	2.5, 298
		1		$V_3=2194\,(10)$ $V'_{12}=-210\,(30)$ GHz		PAM			
42	$CH_3CH{=}C{=}S$		3	$V_3=435\ cm^{-1}$			80BAK3		2.5, 302
43	$CH_3CH{=}CHNO_2$ trans		3	$V_3>2250$			74FOR1		2.5, 305
44	$CH_3CI{=}CH_2$	1	3	$\Delta_0^{(1)}=117.79\,(17)$ MHz $V_3=2590.2\,(5)$ $s=73.91$	$\varrho=0.031306$ $I_\tau=3.1767$ $\beta=0.4209\,(28)$ $F=163.192$	IAM(W)	79GRO		2.5, 306
45	$CF_3CH{=}CH_2$	0, 1	3	$V_3=1640\,(800)\ cm^{-1}$		Int.	74SAI	Fig. 90	2.5, 307
46	$CF_3CF{=}CF_2$	0, 1, 2, 3, 4	3	$V_3=350\ cm^{-1}$		Int.	73JAC		2.5, 308
47	$CH_2{=}C(CH_3)C(=O)F$ s. trans	0	3	$V_3=596.3\ cm^{-1}$ $s=49.42$	$I_\alpha=3.23$ $\lambda_a=0.690\,(8)$ $F=160.96$	PAM	76LAT		2.5, 321

Nr.	Molecule	v	N	Potential	Structural parameters	Meth.	Ref.	Varia, remarks Fig. in 3	See Tab., Nr.
48	$CD_3CH{=}CHCD_3$ cis $CD_3CD{=}CDCD_3$ cis	0 0	3	$V_3=777$ (30) $s=39.31$ (8) $V_3=811$ (30) $s=40.34$ (10)	$I_\alpha=5.98$ (20) $\theta_{i,a}=54.27$ (1) $I_\alpha=5.83$ (20) $\theta_{i,a}=54.08$ (1)	PAM	79GUY		2.5, 324
48a	$(CH_3)_2C{=}CH^{35}Cl$ $(CH_3)_2C{=}CH^{37}Cl$	0 0	3	$s=22.5$ $V_3=283$ (14) cm^{-1} $s=23.2$ $V_3=292$ (15) cm^{-1}	$F=167.8$ $\alpha=-0.00126$ $\beta=\ \ 0.01364$ $F=167.7$ $\alpha=-0.00124$ $\beta=\ \ 0.01329$	PAM	81GUI		2.5, 324a
49	$CH_3CH{=}CHCH_2CH_3$ cis	0	3	$V_3=277$ (1) cm^{-1} $s=22.649$ (16)	$I_\alpha=3.189$ (10) $F=163.2$ (5) $\varrho_a=0.03349$ (12) $\varrho_b=0.01430$ (4) $\varrho_c=0.00125$ (8)	PAM	81VAN	V_3: barrier to internal rotation about C_1-C_2 axis	2.5, 326
49a	$CH_3CH_2SiH_2F$ gauche	1	3	$V_3=2775$ (25)	$I_\alpha=3.2599$ $\lambda_a=0.5594$ $\lambda_b=0.7113$ $\lambda_c=0.4257$	PAM	81HAY1		2.5, 395
	$CH_3CH_2SiH_3$ see Nr. 60								
50	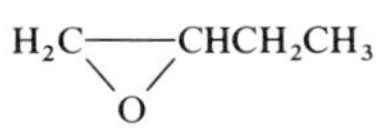 $v_\tau=1$	0	3	$s=84.0$ (1.0) $V_3=3020$ (40) $s=88.15$ (8) $V_3=3166$ (3)	$I_\alpha=3.20$ *) $\lambda_a=0.760$ $\lambda_b=0.636$ $\lambda_c=0.133$ $I_\alpha=3.20$ *) $\lambda_a=0.760$ $\lambda_b=0.636$ $\lambda_c=0.133$	IAM(W)	78LJ	*) assumed	2.5, 487
50a	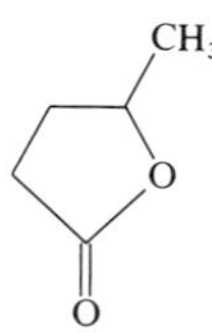	0	3	$V_3=904$ (19)	$I_\alpha=3.25$ $F=159.67$		81KEW		2.5, 520

Nr.	Molecule	v	N	Potential	Structural parameters	Meth.	Ref.	Varia, remarks Fig. in 3	See Tab., Nr.
51	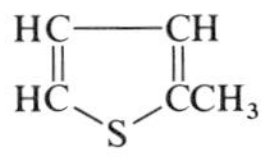	$v_\tau=1$	3	$V_3=208.9\ cm^{-1}$ $V_6=\ 1.5\ cm^{-1}$	$F=174700$ $\alpha=0.0312$	PAM	76POZ1		2.5, 619
52	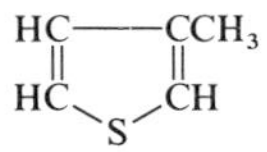	$v_\tau=0,1$	3	$V_3=266\ cm^{-1}$ $V_6=-7.4\ cm^{-1}$	$I_\alpha=3.15$ $\lambda_a=0.989$	PAM	73GUN		2.5, 620
53	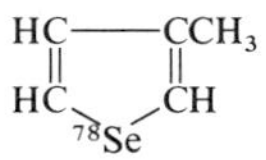 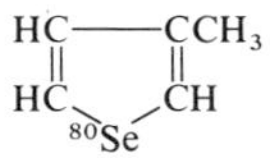	$v_\tau=0,1$ $v_\tau=0,1$	3	$V_3=\ 333\ cm^{-1}$ $V_6=-11.6\ cm^{-1}$ $V_3=\ 333\ cm^{-1}$ $V_6=-11.6\ cm^{-1}$	$I_\alpha=3.07$ $\lambda_a=0.974$ $\lambda_b=0.226$ $I_\alpha=3.07$ $\lambda_a=0.974$ $\lambda_b=0.226$	PAM	73SHA		2.5, 624
54	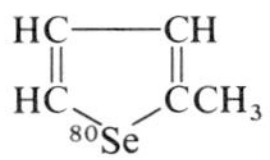	$v_\tau=0,1$	3	$V_3=\ 287\ cm^{-1}$ $V_6=-12.3\ cm^{-1}$	$I_\alpha=3.15$ *) $\lambda_a=0.865\ (5)$ $\lambda_b=0.5018\ (29)$	PAM	78POZ2	*) assumed	2.5, 625
55	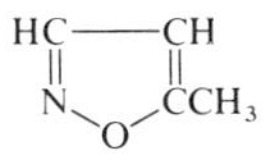	0	3	$s=21.638\ (23)$ $V_3=777.2\ (55)$	$I_\alpha=3.204\ (16)$ $\lambda_a=0.990914$ $\lambda_b=0.134499$	PAM	75MJ2, 77SRI		2.5, 635
56	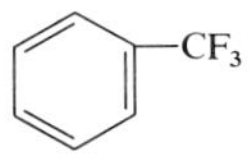		6	$V_6=10.2\ (3)$	$A(CF_3)=5700$ MHz *)	low barrier	76OGA	*) assumed	2.5, 685
57	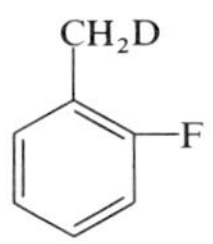 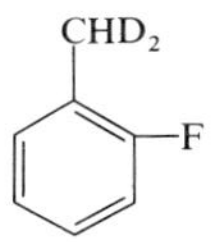	0 0	3	$V_3=567\ (48)$ $V_3=711\ (40)$		*)	75SCH	*) see original paper	2.5, 694

Nr.	Molecule	v	N	Potential	Structural parameters	Meth.	Ref.	Varia, remarks Fig. in 3	See Tab., Nr.
58	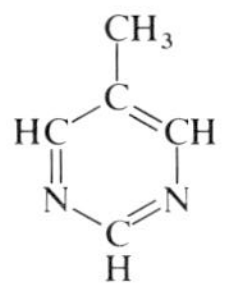		6	$V_6 = 11.73\,(1)$	$I_\alpha = 3.20$ *)	low barrier	75CA2	*) assumed	2.5, 752
59	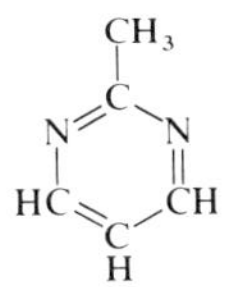		6	$V_6 = 4.454\,(4)$	$F = 164087$ *)	low barrier	76CA4	*) assumed	2.5, 753
b) C—Si and C—Ge bonds									
60	$CH_3CH_2SiH_3$	0, 1	3	$V_{30} = 2624.8\,(104)$ $V_{03} = 1979.1\,(68)$ $V'_{33} = -185.0\,(579)$	$F_1 = 6.04213\ cm^{-1}$ $F_2 = 3.85851\ cm^{-1}$ $F' = -0.70242\ cm^{-1}$		77DUR1	V_{30}: CH_3 group V_{03}: SiH_3 group	2.5, 381
61	CH_3SiHCl_2	0	3	$V_3 = 1690\,(50)$ $s = 49.15$	$I_\alpha = 3.2$ $\lambda_b = 0.902585$ $\lambda_c = 0.430511$ $F = 160.26$	PAM	77EN	V_3, s: average value	2.5, 384
62	$(CH_3)_2SiH_2$	1	3	$V_3 = 1646\,(3)$ $V'_{12} = -38\,(2)$ $s = 43.83\,(8)$	$F = 175.12$ $F' = 11.18$ $I_\alpha = 3.16$ $\theta = 55.22\,(4)$	PAM	73TRI		2.5, 391
63	$(SiH_3)_2CH_2$	0	3	$V_3 = 1210\,(40)$ $s = 53.11$	$I_\alpha = 5.7903$ $\lambda_a = 0.8311$ $\lambda_b = -0.5562$ $F = 106.21$ $F' = -17.07$	PAM	78SHI	barrier to internal rotation of the silyl group	2.5, 392
	$^{29}SiH_3CH_2SiH_3$			$V_3 = 1210\,(40)$ $s = 53.16$	$I_\alpha = 5.7903$ $\lambda_a = 0.8300$ $\lambda_b = -0.5578$ $F = 106.10$ $F' = -16.98$			F': coefficient of the kinetic coupling term	
Cont.									

Nr.	Molecule	v	N	Potential	Structural parameters	Meth.	Ref.	Varia, remarks Fig. in 3	See Tab., Nr.
63 Cont.	$(SiH_3)CHD$			$V_3=1200\,(40)$ $s=53.88$	$I_\alpha=\ 5.7903$ $\lambda_a=\ 0.8311$ $\lambda_b=-0.5539$ $\lambda_c=\ 0.0505$ $F=\ 103.82$ $F'=-14.69$		78SHI		
	$SiH_3CH_2SiH_2D$—s			$V_3=1160\,(40)$ $s=52.32$	$I_\alpha=\ 5.7903$ $\lambda_a=\ 0.8379$ $\lambda_b=-0.5459$ $F=103.36$			for these species, the one-top approximation was applied	
	$SiH_3CH_2SiH_2D$—a			$V_3=1160\,(20)$ $s=53.13$	$I_\alpha=\ 5.7903$ $\lambda_a=\ 0.8245$ $\lambda_b=-0.5641$ $\lambda_c=-0.0451$ $F=101.77$				
64	$CH_2{=}CHSiH_2CH_3$ cis	0	3	$s=52.0\,(11)$ $V_3=1866\,(39)$	$I_\alpha=3.1206$ $\lambda_a=0.4425$ $\lambda_b=0.8968$ $\lambda_c=0$	PAM	81IMA		2.5, 398
	$CH_2{=}CHSiD_2CH_3$ cis	0		$s=52.2\,(13)$ $V_3=1868\,(47)$	$I_\alpha=3.1206$ $\lambda_a=0.4059$ $\lambda_b=0.9139$ $\lambda_c=0$				
	$CH_2{=}CHSiHDCH_3$ cis	0		$s=52.1\,(28)$ $V_3=1867\,(99)$	$I_\alpha=3.1206$ $\lambda_a=0.4240$ $\lambda_b=0.9052$ $\lambda_c=0.0285$				
	$CH_2{=}CHSiH_2{}^{13}CH_3$ cis	0		$s=52.1\,(22)$ $V_3=1868\,(78)$	$I_\alpha=3.1206$ $\lambda_a=0.4606$ $\lambda_b=0.8876$ $\lambda_c=0$				
	$CH_2{=}CHSiH_2CH_3$ skew	0		$s=45.6\,(6)$ $V_3=1698\,(22)$	$I_\alpha=3.0732$ $\lambda_a=0.7311$ $\lambda_b=0.6755$ $\lambda_c=0.0959$				
Cont.									

Nr.	Molecule	v	N	Potential	Structural parameters	Meth.	Ref.	Varia, remarks Fig. in 3	See Tab., Nr.
64 Cont.	$CH_2{=}CHSiD_2CH_3$ skew	0		$s=45.8\,(17)$ $V_3=1692\,(61)$	$I_\alpha=3.0732$ $\lambda_a=0.7213$ $\lambda_b=0.6861$ $\lambda_c=0.0953$			for atom numbering see Fig. 110a: e.g. $4=D$ in 4	
	$CH_2{=}CHSiHDCH_3$-4 skew	0		$s=45.7\,(14)$ $V_3=1694\,(52)$	$I_\alpha=3.0732$ $\lambda_a=0.7233$ $\lambda_b=0.6901$ $\lambda_c=0.0212$				
	$CH_2{=}CHSiHDCH_3$-5 skew	0		$s=45.6\,(14)$ $V_3=1693\,(53)$	$I_\alpha=3.0732$ $\lambda_a=0.7289$ $\lambda_b=0.6622$ $\lambda_c=0.1738$				
	$CH_2{=}CHSiH_2{}^{13}CH_3$ skew	0		$s=45.2\,(8)$ $V_3=1682\,(30)$	$I_\alpha=3.0732$ $\lambda_a=0.7380$ $\lambda_b=0.6688$ $\lambda_c=0.0895$				
65	$CH_2{=}CHCH_2SiH_3$	1	3	$V_3=2130\,(70)$	$I_\alpha=5.868$ $\lambda_a=0.8236$ $\lambda_b=0.5526$ $\lambda_c=0.1281$	PAM	77HAY2		2.5, 399
66	$CF_3{}^{72}GeH_3$	0, 1, 2, 3	3	$V_3=1280\,(150)$ $s=67\,(10)$	$I_\alpha=6.292$ $F=85.500$	PAM	74KR2		2.4, 77
	$CF_3{}^{74}GeH_3$			$V_3=1280\,(150)$ $s=70\,(10)$	$I_\alpha=6.292$ $F=85.500$				
67	CH_3GeCl_3		3	$V_3=1450$	$F=5.457\ cm^{-1}$	Int.	75DUR2		2.4, 76
68	$^{74}GeH_3CH_2F$	0	3	$V_3=1390\,(40)$ $s=57.60$	$I_\alpha=6.292$ $\lambda_a=0.92859$ $\lambda_b=0.37110$ $F=112.658$	PAM	74KR2	Fig. 112	2.5, 406
Cont.	$^{72}GeH_3CH_2F$			$V_3=1390\,(40)$ $s=57.60$	$I_\alpha=6.292$ $\lambda_a=0.92842$ $\lambda_b=0.37154$ $F=112.658$				

Nr.	Molecule	v	N	Potential	Structural parameters	Meth.	Ref.	Varia, remarks Fig. in 3	See Tab., Nr.
68 Cont.	$^{70}GeH_3CH_2F$			$V_3=1390\,(40)$ $s=57.60$	$I_\alpha=6.292$ $\lambda_a=0.92823$ $\lambda_b=0.37201$ $F=112.657$		74KR2		
69	$^{74}GeH_3CH_2{}^{35}Cl$	$v_\tau=1$	3	$V_3=1740\,(30)$ $s=81.255$	$I_\alpha=6.3133$ $\lambda_a=0.86162$ $\lambda_b=0.50759$ $F=99.819$	PAM	76NAK1		2.5, 407
	$^{72}GeH_3CH_2{}^{35}Cl$			$V_3=1741\,(30)$ $s=81.301$	$I_\alpha=6.3133$ $\lambda_a=0.86123$ $\lambda_b=0.50821$ $F=99.820$				
	$CH_3CH_2GeH_3$ see Nr. 40								
70	$GeH_3CH{=}CH_2$	0, 1	3	$V_3=1238\,(57)$	$F=4.320\ cm^{-1}$ $\alpha=0.3829$		74DUR7	Fig. 113	2.5, 409
71	$CH_3{}^{74}GeH_2F$	0	3	$V_3=939.8$ $s=25.80$	$I_\alpha=3.197$ $\lambda_a=0.804888$ $F=169.804$	IAM (W)	76ROB1	Fig. 114	2.5, 410
	$CH_3{}^{72}GeH_2F$			$V_3=942.4$ $s=25.86$	$\lambda_a=0.804989$ $F=169.864$				
	$CH_3{}^{70}GeH_2F$			$V_3=941.6$ $s=25.83$	$\lambda_a=0.805098$ $F=169.927$				
	$^{13}CH_3{}^{74}GeH_2F$			$V_3=941.0$ $s=25.83$	$\lambda_a=0.821638$ $F=169.808$				
	$^{13}CH_3{}^{72}GeH_2F$			$V_3=941.2$ *) $s=25.83$	$\lambda_a=0.821525$ $F=169.868$			*) assumed	
	$^{13}CH_3{}^{70}GeH_2F$			$V_3=941.2$ *) $s=25.82$	$\lambda_a=0.821401$ $F=169.932$				
	$CD_3{}^{74}GeH_2F$			$V_3=924.8$ $s=47.52$	$I_\alpha=6.388$ $\lambda_a=0.857871$ $F=90.174$				
	$CD_3{}^{72}GeH_2F$			$V_3=919.3$ $s=47.21$	$\lambda_a=0.857342$ $F=90.773$				
	$CD_3{}^{70}GeH_2F$			$V_3=920.0$ $s=47.21$	$\lambda_a=0.856795$ $F=90.836$				

Nr.	Molecule	v	N	Potential	Structural parameters	Meth.	Ref.	Varia, remarks Fig. in 3	See Tab., Nr.
72	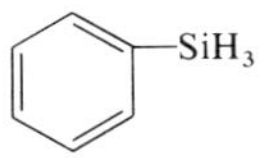		6	$V_6=17.78$ (2)	$F=88756$ MHz *)	low barrier	75CA1	*) assumed	2.5, 689
	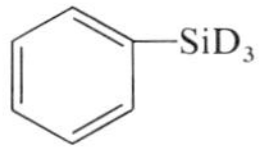			$V_6=15.93$ (3)	$F=48377$ MHz *)		76CA2		
c) C—O bonds									
73	CH_3OH	0	3	$V_3=373.1$ (3.0) cm^{-1} $V_6=-0.80$ (1.0) cm^{-1}	$I_{a_1}=0.7536$ (110) *) $I_\alpha=3.2063$ (130)	IAM $^+$)	73LEE	*) moment of inertia of the framework: $I_a=I_{a_1}+I_\alpha$ $^+$) see [68LEE]	2.5, 65
	$^{13}CH_3OH$	0		$V_3=374.99$ (8) cm^{-1}	$I_{a_1}=0.75077$ (13) *) $I_\alpha=3.21164$ (21)	IAM $^+$)	76GER		
		$v_x=1$		$V_3=392.35$ (30) cm^{-1}	$I_{a_1}=0.7542$ (5) $I_\alpha=3.2119$ (5)	IAM $^+$)	81HEN, 81AMA	v_x=C—O stretch	
	$CH_3{}^{18}OH$	0		$V_3=374.91$ (18) cm^{-1}	$I_{a_1}=0.75929$ (27) *) $I_\alpha=3.21151$ (32)	IAM $^+$)	76GER		
	CH_3OD	0		$V_3=365.88$ (9) cm^{-1}	$I_{a_1}=1.41237$ (26) *) $I_\alpha=3.21461$ (46)	IAM $^+$)	80KAU		
	CH_2DOH	0		$V_1=12$ (1) cm^{-1} $V_3=374.8$ cm^{-1}		FFAM **)	80QUA	**) framework fixed-axis method, see original work	
74	CF_3OF		3	$V_3=3900$		Int.	74BUC		2.5, 77
75	CF_3OOF		3	$V_3=2800$ (400)		Int.	80DES		2.5,78
76	CH_3ONO, cis	0	3	$V_3=2099$ (6)	$\lambda_a=0.589$ *) $I_\alpha=3.18$ $^+$)	IAM (W)	80GHO	*) from structure $^+$) assumed Fig. 29	2.5, 82
		$v_\tau=1$		$V_3=2061$ (10)	$\lambda_a=0.586$ $I_\alpha=2.90$				
		$v_\tau=2$		$V_3=2084$ (5)	$\lambda_a=0.605$ $I_\alpha=3.27$				
		$v_\tau=1, v_{14}=1$		$V_3=1896$ (14)	$\lambda_a=0.582$ $I_\alpha=3.206$			v_{14}=N—O torsion	
	$CH_3O^{15}NO$, cis	0		$V_3=2090$	$\lambda_a=0.5892$ *) $I_\alpha=3.18$ $^+$)	IAM (W)	79TUR1		
	CH_3ONO, trans	0		$V_3=29$	$\theta_{i,a}=28.50$ *) $I_\alpha=3.18$ $^+$)	low barrier			
Cont.									

Nr.	Molecule	v	N	Potential	Structural parameters	Meth.	Ref.	Varia, remarks Fig. in 3	See Tab., Nr.
76 Cont.	$CH_3O^{15}NO$, trans	0		$V_3 =$ 29	$\theta_{i,a} = 28.66$ *) $I_\alpha = 3.18$ $^+$)		79TUR1		
	$CH_3ON^{18}O$, trans	0		$V_3 =$ 27	$\theta_{i,a} = 28.34$ *) $I_\alpha = 3.18$ $^+$)				
	CD_3ONO, trans	0		$V_3 =$ 46	$\theta_{i,a} = 27.73$ *) $I_\alpha = 6.36$ $^+$)				
77	CH_3ONH_2	0	3	$V_3 = 880\ cm^{-1}$	$\lambda_a = 0.8241$ $\lambda_b = 0.5664$ $I_\alpha = 3.195$	PAM	74FON		2.5, 84
	CH_3ONHD	0		$V_3 = 873\ cm^{-1}$					
	CH_3OND_2	0		$V_3 = 867\ cm^{-1}$					
78	CH_3OCH_2F	1	3	$s = 39.64\ (105)$ $V_3 = 1538\ (40)$	$I_\alpha = 3.250$ $\lambda_a = 0.63011$ $\lambda_b = 0.73175$ $\lambda_c = 0.25983$	PAM	81NA1		2.5, 85
79	CH_3OCH_2Cl	0	3	$V_3 = 1819\ (95)$ $V_s = -38\ (250)$	$I_\alpha = 3.312$ $\lambda_a = 0.433700$ $\lambda_b = 0.835684$ $\lambda_c = 0.336952$	PAM *)	79HAY2	*) coupling effect with the skeletal torsion v_x (CH_2Cl) taken into account: V_s = interaction cofficient, see [74IKE] Fig. 30	2.5, 86
		$v_x = 1$		$V_3 = 1816$					
		$v_\tau = 1$		$V_3 = 1820$					
80	CH_3OCH_2CN	0	3	$V_3 = 2280\ (60)$	$I_\alpha = 3.207$ $\lambda_a = 0.2856$ $\lambda_b = 0.8592$ $\lambda_c = 0.4245$	PAM	76KAR2		2.5, 88
		$v_x = 1$		$V_3 = 1920\ (60)$				$v_x = CH_2$—CN torsion	
		$v_\tau = 1$		$V_3 = 2470\ (60)$					
81	HC(=O)OCH_3	0	3	$\Delta = -421.29\ (198)$ MHz $V_3 = 406.1\ cm^{-1}$	$\varrho = 0.083259\ (222)$ $\beta = 0.430600\ (180)$	IAM(W)	79BAU1		2.5, 113

Nr.	Molecule	v	N	Potential	Structural parameters	Meth.	Ref.	Varia, remarks Fig. in 3	See Tab., Nr.
82	$^{35}ClC(=O)OCH_3$, cis	0	3	$\Delta = -294.45$ MHz $V_3 = 1246\,(30)$	$\varrho = 0.05930$ $\beta = 0.06632$	IAM(W)	73LIS1, 77DUR2	CH_3 cis to C=O group	2.5, 131
	$^{37}ClC(=O)OCH_3$, cis	0		$\Delta = -294.43$ MHz $V_3 = 1253\,(30)$	$\varrho = 0.05921$ $\beta = 0.06161$				
83	$N_3C(=O)OCH_3$	$v_x = 1$	3	$V_3 = 370\,(10)\ cm^{-1}$	$F = 5.68\ cm^{-1}$ $F\varrho_a^2 = 408$ MHz $F\varrho_b^2 = 6.97$ MHz	PAM	80KA1	v_x = vibrational mode	2.5, 119
		$v_\tau = 1$		$V_3 = 400\,(10)\ cm^{-1}$					
	$CH_3OCH_2CH_3$ see Nr. 10								
84	$CH_3OCH_2CH_2OH$	$v_\tau = 1$	3	$s = 84.8\,(1.0)$ $V_3 = 3100\,(100)$	$\varrho = 0.07836$ $\beta = 0.0702$ $\gamma = 1.507$	IAM(W)	80CAM1		2.5, 181
85	$CH_3OCH_2CH_2NH_2$	$v_\tau = 1$	3	$s = 86.5\,(5)$ $V_3 = 3150\,(50)$	$\varrho = 0.07644$ $\beta = 0.0174$ $\gamma = 1.512$	IAM(W)	80CAM1		2.5, 182
86	$HC\equiv CCH_2OCH_3$	0	3	$V_3 = 2314\,(120)$ $V_s = -284\,(320)$	$I_\alpha = 3.3872$ $\lambda_a = -0.3290$ $\lambda_b = 0.8637$ $\lambda_c = -0.3818$	PAM *)	80HAY4	*) coupling effect of the skeletal torsion taken into account [74IKE]	2.5, 359
		$v_x = 1$ $v_\tau = 1$		$V_3 = 2386\,(50)$ $V_3 = 2398\,(40)$					
87	CH_3OSiH_3	0	3	$V_{31} = 557$ $V_{32} = 1110$		PAM	74LEC	1 = methyl-group 2 = silyl-group	2.5, 385
	CD_3OSiH_3	0		$V_{31} = 510$ $V_{32} = 1108$					
	CH_3OSiD_3	0		$V_{31} = 550$ $V_{32} = 1045$					
	SiD_3OCD_3	0		$V_{31} = 586$ $V_{32} = 1030$					

Nr.	Molecule	v	N	Potential	Structural parameters	Meth.	Ref.	Varia, remarks Fig. in 3	See Tab., Nr.
88	$SiH_3CH_2OCH_3$	0	3	$s=64.41$ $V_3=2520\,(45)$	$I_\alpha=3.2064$ $\lambda_a=0.9603$ $\lambda_b=0.2788$ $\lambda_c=0$	PAM	81SHI	CH_3 group	2.5, 390
	s-$SiH_2DCH_2OCH_3$	0	3	$s=65.12$ $V_3=2528\,(120)$	$I_\alpha=3.2064$ $\lambda_a=0.9562$ $\lambda_b=0.2927$ $\lambda_c=0$			CH_3 group	
	a-$SiH_2DCH_2OCH_3$	0	3	$s=65.43$ $V_3=2525\,(25)$	$I_\alpha=3.2064$ $\lambda_a=0.9629$ $\lambda_b=0.2625$ $\lambda_c=0$			CH_3 group	
	$SiH_3CD_2OCH_3$	0	3	$s=66.31$ $V_3=2523\,(40)$	$I_\alpha=3.2064$ $\lambda_a=0.9612$ $\lambda_b=0.2759$ $\lambda_c=0$			CH_3 group	
	$SiH_3CH_2OCD_3$	1	3	$s=104.40$ $V_3=2319\,(50)$	$I_\alpha=6.4079$ $\lambda_a=0.9610$ $\lambda_b=0.2765$ $\lambda_c=0$			CD_3 group	
	$SiH_3CH_2OCD_3$	1	3	$s=91.67$ $V_3=2128\,(10)$	$I_\alpha=5.8738$ $\lambda_a=-0.9309$ $\lambda_b=-0.3654$ $\lambda_c=\ 0$			SiH_3 group	
89	$(CH_3)_2O$	0	3	$s=60.32\,(16)$ $V_3=2630\,(7)$	$\theta_{b,i}=57.7\,(8)$ $I_\alpha=3.2084$ $F'=-39.7$	PAM	78LUT	uncertainties are 2 σ	2.5, 434
		$v_\tau=1_n$		$s=59.96\,(2)$ $V_3=2663\,(2)$ $V'_{12}=13\,(2)$	$\theta_{b,i}=60.6\,(2)$ $I_\alpha=3.2084$ $F'=-44.44$				
		$v_\tau=2_n$		$s=59.76\,(90)$ $V_3=2618\,(4)$ $V'_{12}=18\,(8)$	$\theta_{b,i}=58.4$ $I_\alpha=3.2084$ $F'=-40.91$				
Cont.									

Nr.	Molecule	v	N	Potential	Structural parameters	Meth.	Ref.	Varia, remarks Fig. in 3	See Tab., Nr.
89 Cont.	$(CD_3)_2O$	$v_\tau=1_n$ $v_\tau=2_n$		$s=103.39\,(80)$ $V_3=2584\,(20)$ $V'_{12}=21\,(50)$ $s=103.14\,(19)$ $V_3=2572\,(6)$ $V'_{12}=29\,(12)$	$\theta_{b,i}=58.6\,(20)$ $I_\alpha=6.4294$ $F'=-33.60$ $\theta_{b,i}=58.4$ $I_\alpha=6.4294$ $F'=-33.28$		78LUT		
90	CH_3OPF_2 CD_3OPF_2	0 0	3	$V_3=422\,(5)$ $V_3=404\,(5)$	$I_\alpha=\ 3.179$ $\theta\ =51.0$ $I_\alpha=\ 6.425$ $\theta\ =47.8$	PAM	74CO1		2.5, 426
91	H_2C—$CHOCH_3$ (ring with CH_2)	0 $v_\tau=1$		$s=63.15$ $V_3=2276\,(80)$ $s=63.55\,(80)$ $V_3=2290$	$I_\alpha=3.2$ *) $\lambda_a=0.800$ $\lambda_b=0.589$ $\lambda_c=0.113$ $I_\alpha=3.2$ *) $\lambda_a=0.814$ $\lambda_b=0.570$ $\lambda_c=0.113$	PAM	73PEN2	*) assumed Fig. 132	2.5, 470
92	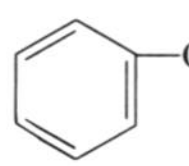—OD	$v_\tau=0, 1$	2	$V_2=1353.6\,(1000)\ cm^{-1}$ $V_4=-20.3\,(200)\ cm^{-1}$		*)	73LAR	see original paper	2.5, 677
d) C—S and C—Se bonds									
93	CH_3SH	0	3	$V_3=444.763\,(87)\ cm^{-1}$ $V_6=-2.070\,(66)\ cm^{-1}$	$I_{a_1}=1.71439\,(47)$ *) $I_\alpha\ =3.20797\,(43)$	IAM $^+$)	80LEE	*) moment of inertia of the framework: $I_a=I_{a_1}+I_\alpha$ $^+$) see [68LEE]	2.5, 66
94	$CH_3{}^{80}SeH$ $CH_3{}^{80}SeD$ $CD_3{}^{80}SeH$ $CD_3{}^{80}SeD$	$v_\tau=0, 1, 2$	3	$s=11.4\,(4)$ $V_3=957\,(50)$ $s=16.8\,(4)$ $V_3=1022\,(50)$ $s=14.4\,(4)$ $V_3=1012\,(50)$ $s=22.9\,(4)$ $V_3=1008\,(50)$	$I_\alpha=3.218\,(20)$ $\lambda_b=0.001\,(5)$ $I_\alpha=3.218\,(20)$ $\lambda_b=0.0065\,(50)$ $I_\alpha=6.396\,(20)$ $\lambda_b=0.0001\,(50)$ $I_\alpha=6.396\,(20)$ $\lambda_b=0.016\,(5)$	PAM	73THO		2.5, 67

Nr.	Molecule	v	N	Potential	Structural parameters	Meth.	Ref.	Varia, remarks Fig. in 3	See Tab., Nr.
95	CH_3SCN		3	$V_3=1618$		RTV	74AND		2.5, 80
96	$CH_3{}^{80}SeCN$	0	3	$s=35.83$ $V_3=1243.7\ (10)$	$I_\alpha=3.212$ $\theta_{z,i}=69°$	PAM	78LAN		2.5, 81
		$v_{10}=1$		$s=33.93$ $V_3=1178.5\ (42)$				$\nu_{10}=168\ cm^{-1}$, C—Se—C bend	
		$v_{10}=2$		$s=31.17$ $V_3=1082.1\ (44)$					
	$CH_3{}^{82}SeCN$	0		$s=36.1\ (10)$ $V_3=1234$	$\lambda_a=0.316087$ *)		78SAK	*) $I_\alpha(CH_3)=3.267$ amu Å^2 $I_\alpha(CD_3)=6.529$ amu Å^2	
	$CH_3{}^{78}SeCN$	0		$s=36.3\ (6)$ $V_3=1238$	$\lambda_a=0.322314$ *)				
	$CH_3{}^{77}SeCN$	0		$s=36.4\ (5)$ $V_3=1242$	$\lambda_a=0.323935$ *)				
	$CH_3{}^{76}SeCN$	0		$s=36.3\ (4)$ $V_3=1239$	$\lambda_a=0.325592$ *)				
	$CD_3{}^{80}SeCN$	0		$s=70.1\ (6)$ $V_3=1229$	$\lambda_a=0.376662$ *)				
	$CD_3{}^{78}SeCN$	0		$s=70.0\ (14)$ $V_3=1227$	$\lambda_a=0.380020$ *)				
97	HC(=O)SCH_3	0		$V_3>10.5\ kJ\cdot mol^{-1}$			76JON1		2.5, 114
97a	FC(=O)SCH_3	0	3	$s=26.10\ (5)$ $V_3=917$	$I_\alpha=3.198$ $\theta=44.1\ (4)$	PAM	81CAM2		2.5, 131 a
		$v_\tau=1$		$s=25.7\ (2)$ $V_3=903$	$I_\alpha=3.198$ $\theta=45.2\ (7)$				
		$v_{SK}=1$		$s=41.08\ (9)$ $V_3=1443$	$I_\alpha=3.198$ $\theta=48.4\ (7)$			Skeletal torsion	
98	^{35}ClC(=O)SCH_3	0	3	$s=25.41\ (3)$ $V_3=893.0\ (20)$	$I_\alpha=3.2$ $\theta_{a,i}=32.4°$	PAM	80CAM3		2.5, 132
	^{37}ClC(=O)SCH_3	0		$s=25.34\ (8)$ $V_3=890.0\ (20)$	$I_\alpha=3.2$ $\theta_{a,i}=32.0\ (12)°$				

Nr.	Molecule	v	N	Potential	Structural parameters	Meth.	Ref.	Varia, remarks Fig. in 3	See Tab., Nr.
98a	NCC(=O)SCH_3	0	3	$s=20.04(3)$ $V_3=705(20)$	$I_\alpha=3.198$ $\theta=29.5(4)$	PAM	81CAM3		2.5, 132a
99	CH_3SOH	0	3	$V_3=2400(100)$			78PEN4	OH internal rotation: $V_{cis}=5.4$ kcal · mol^{-1}; $V_{trans}=4.1$ kcal · mol^{-1}	
100	$CH_3CH_2SCH_3$, trans	0	3	$s=53.91(150)$ $V_3=1980(55)$	$I_\alpha=3.191$ $\lambda_a=0.8236$ $\lambda_b=0.5671$	PAM	81HAY	SCH_3 group	2.5, 160
101	CH_3SSiH_3	0	3	$s=30.92$ $V_3=1100(7)$	$F=165.80$ $\lambda_a=0.5875$ $\lambda_b=-0.8092$	PAM	80SH1	CH_3 group	2.5, 389
	CH_3SSiD_3	0	3	$s=30.94$ $V_3=1093(8)$	$F=164.67$ $\lambda_a=0.5794$ $\lambda_b=-0.8150$				
102	$(CH_3)_2S$	0	3	$\Delta_0=13.7791(14)$ MHz $V_3=2099.2(32)$	$\varrho=0.094427(114)$ $\beta=0.32804(60)$ $I_\alpha=3.2421(43)$ $\theta_{b,i}=51.504(54)°$	IAM(W)	80DEM	uncertainties are 2σ	2.5, 435
		$v_\tau=1_n$		$s=57.302(50)$ $V_3=2136(2)$ $V'_{12}=34(4)$	$I_\alpha=3.204$ $\theta_{b,i}=52.11°$ $F'=-9.964$	PAM	81DEM1		
	$(CD_3)_2S$	$v_\tau=1_n, 2_n$		$s=106.195(22)$ $V_3=2097.3(4)$ $V'_{12}=32.2(5)$	$I_\alpha=6.403$ $\theta_{b,i}=52.183$ $F'=-8.4978$				
103	$(CH_3)_2{}^{76}Se$	0	3	$s=42.06(3)$ $V_3=1498.0(13)$	$I_\alpha=3.244$ $\theta_{i,a}=39.87(11)$	IAM(W)	80DRE	*) $I_\alpha=3.244$ amu Å^2 assumed	2.5, 436
	$(CH_3)_2{}^{77}Se$	0		$s=42.09(1)$ $V_3=1498.8(8)$	$\theta_{i,a}=39.90(7)$ *)				
	$(CH_3)_2{}^{78}Se$	0		$s=42.09(2)$ $V_3=1498.4(9)$	$\theta_{i,a}=39.90(7)$ *)				
	$(CH_3)_2{}^{80}Se$	0		$s=42.10(2)$ $V_3=1498.4(10)$	$\theta_{i,a}=39.91(9)$ *)				
Cont.	$(CH_3)_2{}^{82}Se$	0		$s=42.12(2)$ $V_3=1498.8(9)$	$\theta_{i,a}=39.80(8)$ *)				

Nr.	Molecule	v	N	Potential	Structural parameters	Meth.	Ref.	Varia, remarks Fig. in 3	See Tab., Nr.
103 Cont.	$(CD_3)_2{}^{80}Se$	$v_\tau=1_n$		$s=79.2$ (5) $V_3=1493$ (9) $V'_{12}=28.4$ (3)	$I_\alpha=6.35$ $\theta=50°1'\pm7'$ $F'=-3.605$	PAM	76PAN2		
104	CH_3SSCD_3		3	$s=42.6$ $V_3=1469$		RTV *)	74KUH	*) coupling of CH_3 torsion and S—S torsion analyzed	2.5, 444
105	$(CH_3)_3CSH$	0	3	$\Delta_0=1414.2607\pm1$ MHz $V_3=1742.0$	$\varrho=0.984819$ (50) $\beta=0.03209$ (10) $I_\alpha=111.21076$ $\lambda_a=0.998853$ $\lambda_b=0.0478756$	IAM *)	74VAL	internal rotation about C—S bond. *) suited for the intermediate barrier case, see [74VAL]	2.5, 446
106	$(SiH_3)_2S$	0	3	$s=27.339$ (7) $V_3=522.6$ (4)	$I_\alpha=6.219$ (6) $\theta=39.176$ (28)	PAM	78DOS2		2.5, 388
107	H_2C—$CHSCH_3$ (ring via CH_2)	0	3	$s=54.7$ (7) $V_3=1913$ (40)	$I_\alpha=3.2$ *) $\lambda_a=0.6031$ $\lambda_b=0.7939$ $\lambda_c=0.0773$	PAM	75PEN1	*) assumed Fig. 133	2.5, 471
e)	C—N and C—P bonds								
108	$(CH_3)_3N$		3	$V_3=4.4\ \mathrm{kcal\cdot mol^{-1}}$		Int.	73RIN		2.4, 65
109	$CH_3NH^{35}Cl$	$v_\tau=1$	3	$V_3=3781$ (13)	$\varrho=0.17321$ $\beta=0.14508$ $\gamma=0.05513$	IAM	74CA2	Fig. 25	2.5, 69
	$CH_3ND^{35}Cl$	$v_\tau=1$		$V_3=3782$ (13)	$\varrho=0.15032$ $\beta=0.16742$ $\gamma=0.12289$				
110	CH_3NHCN		3	$V_3=260\ \mathrm{cm^{-1}}$			80BAK2		2.5, 70
111	CH_3NHOH, trans	0	3	$V_3=3553$ (36)	$I_\alpha=3.195$ $\lambda_a=0.8042$ $\lambda_b=0.5920$ $\lambda_c=0.0531$	PAM	79SUN		2.5, 71

Nr.	Molecule	v	N	Potential	Structural parameters	Meth.	Ref.	Varia, remarks Fig. in 3	See Tab., Nr.
112	CH_3NO CD_3NO	0 0	3	$s=23.32$ (2) $V_3=1143$ (4) $s=34.19$ (8) $V_3=1095$ (9)	$I_\alpha=3.28$ (1) $\theta_a=27.79$ (5)° $I_\alpha=6.47$ (4) $\theta_a=26.24$ (6)°	PAM	78TUR	Fig. 26	2.5, 72
113	CF_3NO	0	3	$V_3=770$ (50)		IAM(W)	76TUR2		2.5, 73
114	CH_3NO_2	0	6	$V_6=62500$ (60) MHz		low barrier *)	75ROH	*) see original paper	2.5, 74
115	$CH_3N{=}NCH_3$ cis	0	3	$s=36.1$ (2) $V_3=1320$	$I_\alpha=3.16$ $\lambda_a=0.59$ (4)		79STE	top-top coupling neglected	2.5, 106
116	$HC(=O)NHCH_3$		3	$V_3=205$	$I_\alpha=3.04$ $\theta_a=50.13°$		73ELZ2		2.5, 115
117	$HC(=O)N(CH_3)_2$		3	$V_3=1079$			73ELZ1	cis CH_3 group	2.5, 116
	$CH_3CH{=}NCH_3$ see Nr. 15								
118	$(CH_3)_3CNO$	0 $v_\tau=1$	3	$V_3=920$ (25) $V_3=1075$ (25)			78COX	internal rotation around the C—N bond	2.5, 454
119	$(CH_3)_3CNO_2$ $(CH_3)_3C^{15}NO_2$	0 0	6	$V_6=202.467$ (20) $V_6=202.473$ (10)	$F=17.87495$ $F=17.87715$	low barrier	80LAN		2.5, 455
120	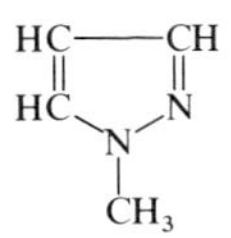	0	3	$V_3=530.1$ (7)	$I_\alpha=3.155$ (4) $\lambda_z=0.99962$	IAM	77KRI		2.5, 631
121	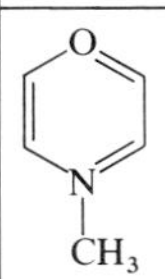	$v_\tau=1$	3	$s=130$ *) $V_3=4560$ (250)	$I_\alpha=3.2$	Int.	74DAS	*) other possibility: $s=98$ $V_3=3440$ (150)	2.5, 541
122	$CH_3PH_2BH_3$ *)	$v_\tau=1$ $v_{\tau'}=1$	3	$V_3(CH_3)=2490$ (60) $V_3(BH_3)=1570$ (60)			75DUR4	*) $CH_3PD_2BH_3$ also measured	2.5, 432

Nr.	Molecule	v	N	Potential	Structural parameters	Meth.	Ref.	Varia, remarks Fig. in 3	See Tab., Nr.
123	$CH_3PF_2BH_3$	0	3	$V_3(CH_3)>2000$ $V_3(BH_3)>2000$			75CR2		2.5, 433
124	CH_3PF_2	$v_\tau=0,1$	3	$V_3=2300\,(6)$	$I_\alpha=3.16$ $\theta=25.0°$	PAM	74CO2	Fig. 124	2.5, 424
125	CH_3POF_2	$v_\tau=1,2$	3	$V_3=3580\,(40)$	$F=5.65\ cm^{-1}$		76DUR1		2.5, 427
126	$(CH_3)_2PH$	$v_\tau=0,1_n,2_n$	3	$V_3=2319\,(9)$ $V'_{33}=83.2\,(80)$	$I_\alpha=3.18$ $\lambda_a=0.764$ $\lambda_b=0.644$ $F'=-0.255\ cm^{-1}$	*)	77DUR5	*) "infinite matrix" diagonalization	2.5, 437
	CH_3PHCD_3	1		$V_3=2340$ $s=63.46$	$I_\alpha=3.15$ $\lambda_a=0.718$ $\lambda_b=0.696$	**)	81DUR1	**) semirigid model, see original paper	
f)	Other bonds								
127	SiF_3BF_2		6	$V_6=1.9\,(8)$		IAM	74OGA		2.5, 400
128	$CH_3CH_2BF_2$		2	$V_2=1170$	$F=1\ 25\ cm^{-1}$	Int.	78OD1	BF_2 torsion	2.5, 413
	$CH_3PH_2BH_3$ see Nr. 122								
	$CH_3PF_2BH_3$ see Nr. 123								
129	$CH_3B_2H_5$	0	3	$s=105.1$ $V_3=4280\,(90)$	$I_\alpha=3.16$ $\lambda_a=0.8632$ $\lambda_b=0.5050$	PAM	77PEN1	Fig. 117	2.5, 417
130	$CH_2{=}CHBF_2$		2	$V_2=4170$		Int.	74DUR3		2.5, 414
131	SiH_3PH_2	0	3	$s=33.29$ $V_3=1535\,(40)$	$I_\alpha=5.852$ $\lambda_a=1.0$ $\lambda_c=0.0$	PAM and IAM(W)	75VAR	SiH_3 group	2.5, 401
	SiH_3PD_2	0	3	$V_3=1512\,(26)$	$I_\alpha=5.928$ $\alpha=0.4408$ $\beta=0$ $\gamma=0.0011$	PAM	76DUR2		

Nr.	Molecule	v	N	Potential	Structural parameters	Meth.	Ref.	Varia, remarks Fig. in 3	See Tab., Nr.
132	$^{11}BD_2HPH_3$	$v_\tau=1$	3	$s=63.4$ $V_3=2467\ (50)$	$I_\alpha=4.71$ $\lambda_a=0.9988$ $\lambda_b=0$ $\lambda_c=0.0483$	PAM	73DUR5		2.4, 81
133	BF_3PH_3 BF_3PD_3		3	$V_3=3340\ (350)$ $V_3=3440\ (300)$	$F=3.756\ cm^{-1}$ $F=1.965\ cm^{-1}$	Int. Int.	75ODO		2.4, 80
134	$CH_3NC \cdot BH_3$		3	$V_3<4.2$		Int.	77STE		2.4, 107
135	$(CH_3)_3N \cdot BH_3$		3	$V_3=3400$		Int.	73DUR6	BH_3 group	2.4, 68
136	$(CH_3)_3NGaD_3$		3	$V_3=1200$		Int.	80DUR1	GaD_3 group	2.4, 70
137	N_2H_4	0	2	$V_2=1240\ (50)\ cm^{-1}$			80TSU	Fig. 17	2.5, 25
138	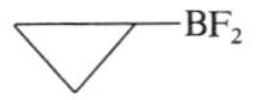		2	$V_2=4230$		Int.	80ODO		2.5, 477
139	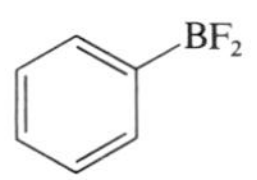	$v_\tau=0,1$	2	$V_2=1114\ (200)\ cm^{-1}$		Int.	74CHR		2.5, 687
140	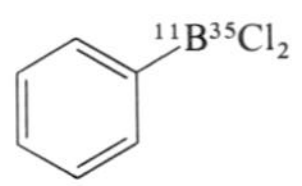 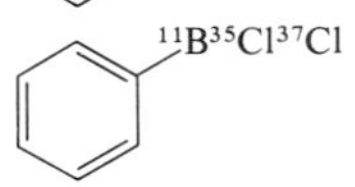	$v_\tau=0,1$ $v_\tau=0,1$	2	$V_2=1151\ (70)\ cm^{-1}$ $V_2=1161\ (100)\ cm^{-1}$		*)	78CAM	*) from variation of inertial defect with v_τ	2.5, 688

2.9 Magnetic constants

2.9.1 Diatomic molecules

2.9.1.1 Preliminary remarks

1. Introduction

This table contains the molecular parameters which describe the interaction of the molecule with any internal or external magnetic field. These parameters are: the spin-rotation constants C_i for the nucleus (i), the tensor (C_3) and scalar (C_4) part of the nuclear spin-spin interaction, the scalar (σ_{iS}) and tensor (σ_{iT}) part of the magnetic shielding of the nuclear magnetic moment of nucleus (i) for the nuclear Zeeman effect, the rotational g_J-factor and the scalar (ξ_s) and tensor (ξ_T) part of the magnetic susceptibility describing the induced magnetic moment.

These molecular parameters are defined in 2.9.1 of Vol. II/6, Landolt-Börnstein, New Series, using a phenomenological Hamiltonian to describe the spectra. Some remarks about derived constants like the molecular quadrupole moment Θ are also to be found there.

2. Explanations of the columns in the tabulation

Column	Parameters	Definitions
1	Nr. *Nr.*	serial number of the molecule of Table 2.2, the following number represents the serial number of the molecule in 2.9.1 of Vol. II/6, of Landolt-Börnstein, New Series
2	v J un.	quantum number of vibrational state and quantum number of rotational state for the constants in columns 3···6 unassigned to a particular state v, J; values from molecular beam magnetic resonance experiments
3	c_1 [kHz]	spin-rotation constant of atom *A* in the molecule *AB*
4	c_2 [kHz]	spin-rotation constant of atom *B* in the molecule *AB*
5	c_3 [kHz]	tensor part of the nuclear spin-spin interaction constant
6	c_4 [kHz]	scalar part of the nuclear spin-spin interaction constant
3···6	σ_{iS}, σ_{iT}	scalar and tensor part of the magnetic shielding of the nuclear magnetic moment of the nucleus i
7	Ref.	references for columns 3···6 (compiled in 2.2.3)
8	g_J	rotational g-factor
9	$\xi_T \left[\frac{Hz}{kG^2}\right]$	tensor part of the magnetic susceptibility constant
10	Ref.	references for columns 8 and 9 (compiled in 2.2.3)
11	Meth. MW MB NMR	method of measurement for the values in the corresponding line of the table on the left side of this column: microwave spectroscopy molecular beam resonance experiment-electric or magnetic resonance nuclear magnetic resonance
12	(Θ/e) [Å^2] (Θ^{el}/e) [Å^2] (Θ^{n}/e) [Å^2]	molecular quadrupole moment with reference to the center of mass of the molecule in units of the fundamental charge of the electron electronic part of the molecular quadrupole moment... nuclear part of the molecular quadrupole moment...
13	Ref.	references for column 12 (compiled in 2.2.3)
14		references to Tables 2.2.2, 2.6.1, 2.7.1; and in subvolume b, 4.1

Some results from other spectral regions are included. The experimental methods are indicated by the abbreviations as explained in 1.5, p. 3.

List of abbreviations for indexes:

der	derived	= derived from the measured values by simple interrelations
dia	diamagnetic	
el	electronic	
ev	evaluated	= evaluated with the use of experimental values from other spectroscopic regions
hf	high-frequency	
n	nuclear	
para	paramagnetic	
S	scalar (part)	
T	tensor (part)	

3. Conversion factors

In some cases conversion factors had to be used to give the tabulated parameters in the quoted units. These factors have been calculated with the fundamental constants given in 1. But the accuracy of the measured magnetic constants is not sufficient to differentiate between earlier used fundamental constants.

4. Arrangement of substances

Compare 2.2.1.

5. References: in 2.2.3, p. 34.

2.9.1.2 Data

Nr.	Molecule v, J	c_1 kHz	c_2 kHz	c_3 kHz	c_4 kHz	Ref.	g_J	ξ_T $\frac{Hz}{kG^2}$	Ref.	Meth.	(Θ/e) Å^2	Ref.	See Tab., Nr.
1 *6-1*	$^1H_2$ (ortho) G(3d)$^1\Sigma_g^+$ $J=1, v=1$	1.00 (17) MHz				74MEL							2.2.2 4.1.3 4.1.6
2 *6-2*	$^7Li^1H$ $J=1$ $v=0$	10.054 (7)	−9.345 (27)	11.346 (7)	0.135 (10)	75FRE	−0.65842 (17)	213 (15)	75FRE	MB			2.6.1
	$J=1$ $v=1$	9.88 (4)	−8.87 (20)	11.03 (8)	0.17 (4)	75FRE							
	$J=2$ $v=0$	10.034 (3)	−9.31 (2)	11.329 (12)	0.160 (5)	75FRE							
		$\sigma_{1T}=-7.3$ (20) ppm, $\sigma_{2T}=-1.7$ (17) ppm				75FRE							
	$^7Li^2D$ $J=1$ $v=0$	5.682 (5)	−0.819 (6)	1.7430 (70)	0.005 (10)	75FRE	−0.27674 (11)	218 (24)	75FRE	MB			
		$\sigma_{1T}=-6.3$ (50) ppm, $\sigma_{2T}=-13$ (21) ppm				75FRE							
4 *6-4*	$^{23}Na_2$ $v=0$	0.17 (3)				75ROS				MB			2.7.1
15 *6-8*	$^1H^{19}F$ $J=1$ $v=0$						0.74104 (15)	−44 (2)	73DEL	MB	0.4913 (62)	73DEL	2.2.2
		$\sigma_{1T}=-8$ (3) ppm, $\sigma_{2T}=-36$ (3) ppm				73DEL							
16 *6-9*	$^1H^{35}Cl$ $J=1$ $v=0$	−41.70 (8)	53.841 (18)	5.59 (9)	0.02 (6)	73DEL	0.45935 (9)	+16 (6)	73DEL	MB	0.779 (25)	73DEL	2.2.2 2.6.1 2.7.1
		$\sigma_{1T}=-7.0$ (17) ppm; $\sigma_{2T}=-100$ (8) ppm				73DEL							
17 *6-10*	$^1H^{79}Br$ $J=1$ $v=0$	−41.97 (26)	290.796 (28)	10.296 (70)	−0.15 (16)	73DAB				MB			2.2.2 2.6.1 2.7.1
	$^1H^{81}Br$ $J=1$ $v=0$	−42.14 (12)	313.370 (21)	11.138 (51)	−0.063 (72)	73DAB	0.37122 (8)	+46.3 (23)	73DAB	MB	0.862 (27)	73DAB	
Cont.		$\sigma_{1T}=-8$ (4) ppm, $\sigma_{2T}=-244.0$ (47) ppm				73DAB							

Nr.	Molecule v, J	c_1 kHz	c_2 kHz	c_3 kHz	c_4 kHz	Ref.	g_J	ξ_T $\frac{Hz}{kG^2}$	Ref.	Meth.	(Θ/e) $Å^2$	Ref.	See Tab., Nr.
17 6–10 Cont.	$^2D^{79}Br$ $v=0$	−3.25 (57)	145.82 (24)	1.59 (24)		74VAN				MW			
	$^2D^{81}Br$ $v=0$	−3.55 (57)	157.26 (24)	1.74 (24)		74VAN				MW			
19 *6–12*	$^6Li^{19}F$					76HEB[1])				MB			2.2.2 2.6.1 2.7.1
	$^7Li^{19}F$					76HEB[1])				MB			
20 *6–13*	$^7Li^{35}Cl$ $J=1$ $v=0$						0.100419 (31)	127 (13)	74FRE	MB	−2.72 (14)	74FRE	2.2.2
		$\sigma_{1T}=-12\,(5)$ ppm; $\sigma_{2T}=-61\,(9)$ ppm				74FRE				MB			
	$^7Li^{35}Cl$ $J=1$ $v=1$						0.100642 (80)	103 (14)	74FRE	MB	−2.19 (21)	74FRE	
		$\sigma_{1T}=-14\,(17)$ ppm; $\sigma_{2T}=-100\,(28)$ ppm				74FRE				MB			
	$^7Li^{37}Cl$ $J=1$ $v=0$						0.101336 (40)	157 (15)	74FRE	MB	−2.86 (17)	74FRE	
		$\sigma_{1T}=-25\,(10)$ ppm; $\sigma_{2T}=-84\,(33)$ ppm				74FRE				MB			
21 *6–14*	$^6Li^{79}Br$ $v=0$ $J=1$						0.13184 (24)	176 (76)	73CEC	MB	$(\Theta^{el}/e)_{der}$ = −9.8 (10)	73CEC	
		$\sigma_{1T}=-16\,(6)$ ppm; $\sigma_{2T}=-173\,(87)$ ppm				73CEC				MB			
	$^6Li^{81}Br$ $v=0$ $J=1$						0.13172 (14)	156 (65)	73CEC	MB	$(\Theta^{el}/e)_{der}$ = −9.9 (10)	73CEC	
		$\sigma_{1T}=-21\,(5)$ ppm; $\sigma_{2T}=-140\,(90)$ ppm				73CEC				MB			
	$^7Li^{79}Br$ $v=0$ $J=1$						0.112056 (64)	118 (25)	73CEC	MB	$(\Theta^{el}/e)_{der}$ = −10.0 (10)	73CEC	
Cont.		$\sigma_{1T}=-16.0\,(14)$ ppm; $\sigma_{2T}=-107\,(27)$ ppm				73CEC				MB			

[1]) Same data as in [H14] of Vol. II/6 Landolt-Börnstein, New Series.

Nr.	Molecule v, J	c_1 kHz	c_2 kHz	c_3 kHz	c_4 kHz	Ref.	g_J	ξ_T $\frac{Hz}{kG^2}$	Ref.	Meth.	(Θ/e) Å^2	Ref.	See Tab., Nr.
21 6–14 Cont.	$^7Li^{81}Br$ $v=0$ $J=1$						0.112217 (64)	128 (25)	73CEC	MB	$(\Theta^{el}/e)_{der}$ $=-10.0$ (10)	73CEC	
		$\sigma_{1T}=-11.3$ (17) ppm; $\sigma_{2T}=-114$ (29) ppm				73CEC				MB			
22 *6–15*	$^7Li^{127}I$ $J=1$ $v=0$	0.754 (3)	6.8039 (8)	0.627 (2)	0.062 (1)	76JAC [1]				MB			2.7.1
27 *6–26*	$^{133}Cs^{35}Cl$ $v=0$ $J=2$	0.6790 (15)	0.5872 (37)	0.0280 (15)	0.0605 (16)	77CED [2]				MB			2.2.2 2.9.1
	$v=1$ $J=2$	0.6735 (15)	0.5736 (37)	0.0273 (15)	0.0621 (16)	77CED				MB			
	$v=2$ $J=2$	0.6690 (15)	0.5703 (37)	0.0286 (15)	0.0604 (16)	77CED				MB			
	$v=3$ $J=2$	0.6621 (15)	0.5684 (37)	0.0286 (15)	0.0597 (16)	77CED				MB			
	$^{133}Cs^{37}Cl$ $v=0$ $J=2$	0.6505 (15)	0.4630 (37)	0.0318 (15)	0.0554 (16)	77CED				MB			
30 *6–27*	$^{63}Cu^{19}F$ $v=0$	36.1 (74)				74HON1	−0.0628 (2)	540 (60)	74HON1	MW	−1.26 (17)	74HON1	2.7.1
		$\sigma_{1T}=-1760$ (830) ppm				74HON1				MW			
31	$^{63}Cu^{35}Cl$ $v=0$	10 (7)				77TIE2				MW			2.2.2 2.7.1
47 *6–29*	$^{27}Al^{19}F$ $v=0$	8.2 (13)				74HON3	−0.08051 (8)	434 (37)	74HON3	MW	−1.23 (10)	74HON3	2.7.1
		$\sigma_{1T}=-224$ (71) ppm				74HON3				MW			

[1]) 76JAC: $|e\omega\Omega|<0.002$ kHz; magnetic octupole coupling for iodine nucleus.
$|ehH|<0.075$ kHz; electric hexadecapole coupling for iodine nucleus.

[2]) 77CED: $c_1=[0.681\,(3)-0.004\,(3)\,(v+\frac{1}{2})+0.0004\,(8)\,(v+\frac{1}{2})^2]$ kHz.
$c_2=[0.595\,(4)-0.018\,(5)\,(v+\frac{1}{2})+0.003\,(1)\,(v+\frac{1}{2})^2]$ kHz

Nr.	Molecule v, J	c_1 kHz	c_2 kHz	c_3 kHz	c_4 kHz	Ref.	g_J	ξ_T $\frac{\text{Hz}}{\text{kG}^2}$	Ref.	Meth.	(Θ/e) Å^2	Ref.	See Tab., Nr.
50 *6-30*	$^{69}Ga^{19}F$ $v=0$	15.3 (18)				74HON1	−0.06012 (12)	521 (17)	74HON1	MW	−1.34 (5)	74HON1	2.2.2 2.7.1
		$\sigma_{1T}=-386$ (165) ppm				74HON1				MW			
52	$^{69}Ga^{127}I$ $v=0$	2.9 (28)	6.8 (28)			81NAI1							2.2.2 2.7.1
59	$^{12}C^{16}O$ $J=1$ $v=0$						−0.26895 (5)	690.2 (10)	77MEE	MB	−0.4039 (83)	77MEE	2.2.2 2.6.1 4.1.6
	$^{12}C^{18}O$ $J=1$ $v=0$						−0.25625 (5)	688.5 (17)	77MEE	MB			
	$^{13}C^{16}O$ $J=1$ $v=0$	32.70 (12)				77MEE				MB			
60 *6-37*	$^{12}C^{32}S$ $v=0$						−0.2702 (4)	2022 (100)	73McG1	MW	+0.17 (29)	73McG1	2.2.2
61	$^{12}C^{80}Se$ $v=0$						−0.2431 (16)	2322 (117)	73McG1	MW	−0.54 (33)	73McG1	2.2.2 2.6.1
62	$^{28}Si^{16}O$ $v=0$						−0.1546 (3)	930 (80)	74HON4	MW	−0.95 (23)	74HON4	2.2.2
	$^{28}Si^{16}O$ $v=0$ $J=1$						−0.15359 (12)	1177 (33)	77HON 74DAV [1]	MB	−0.23 (10)	74DAV	
	$v=1$ $J=1$						−0.15450 (11)	1220 (27)	74DAV	MB			
	$v=2$ $J=1$						−0.15539 (12)	1267 (33)	74DAV	MB			

[1]) Compare GeO about differences between [74HON4] and [74DAV].

Nr.	Molecule v, J	c_1 kHz	c_2 kHz	c_3 kHz	c_4 kHz	Ref.	g_J	ξ_T $\frac{Hz}{kG^2}$	Ref.	Meth.	(Θ/e) $Å^2$	Ref.	See Tab., Nr.
63 *6–37a*	$^{28}Si^{32}S$ $v=0$						−0.091131 (108)	1710 (11)	73HON2 [1]	MW	$\Theta/e = -0.214\,(20) + 0.012\,(11) \cdot (v+\frac{1}{2})$	73HON2	2.2.2
	$v=1$						−0.091451 (132)	1763 (15)	73HON2 [1]	MW			
	$v=2$						−0.091632 (117)	1764 (13)	73HON2 [1]	MW			
	$v=3$						−0.092176 (220)	1795 (28)	73HON2 [1]	MW			
									77HON	MW			
64 *6–38*	$^{74}Ge^{16}O$ $v=0$						−0.141369 (151)	1254 (24)	73HON2 [2]	MW	$\Theta/e = -1.248\,(41) - 0.015\,(10) \cdot (v+\frac{1}{2})$	73HON2	2.2.2 2.6.1
	$v=1$						−0.142034 (203)	1289 (24)	73HON2	MW			
	$v=2$						−0.142686 (292)	1300 (28)	73HON2	MW			
									77HON	MW			
	$^{70}Ge^{16}O$ $J=1$ $v=0$						−0.14179 (10)	1393 (33)	74DAV [3]	MB			
Cont.	$J=1$ $v=1$						−0.14254 (12)	1420 (40)	74DAV	MB			

[1]) [73HON2]: $g_J(v) = -0.090974\,(65) - 0.000296\,(24)\,(v+\frac{1}{2})$.
$\xi_T(v) = [1702.3\,(69) + 27.3\,(39)\,(v+\frac{1}{2})]\,\frac{Hz}{kG^2}$.

[2]) [73HON2]: $g_J(v) = -0.141040\,(112) - 0.000660\,(6)\,(v+\frac{1}{2})$.
$\xi_T(v) = [1245.5\,(145) + 23.9\,(35)\,(v+\frac{1}{2})]\,\frac{Hz}{kG^2}$.

[3]) [74DAV] includes a remark about the descrepancy between the data of [73HON2] and [74DAV].

Nr.	Molecule v, J	c_1 kHz	c_2 kHz	c_3 kHz	c_4 kHz	Ref.	g_J	ξ_T $\frac{Hz}{kG^2}$	Ref.	Meth.	(Θ/e) $Å^2$	Ref.	See Tab., Nr.
64 6-38 Cont.	$^{72}Ge^{16}O$ $J=1$ $v=0$						−0.14134 (11)	1390 (37)	74DAV	MB			
	$J=1$ $v=1$						−0.14208 (10)	1417 (37)	74DAV	MB			
	$^{74}Ge^{16}O$ $J=1$ $v=0$						−0.14089 (10)	1393 (30)	74DAV	MB	−0.94 (13)	74DAV	
	$J=1$ $v=1$						−0.14164 (12)	1420 (37)	74DAV	MB			
65 *6-39*	$^{72}Ge^{32}S$ $v=0$						−0.06828 (11)		77HON	MW			2.2.2
66	$^{74}Ge^{80}Se$ $v=0$						−0.03765 (6)		77HON	MW			2.2.2
67	$^{74}Ge^{130}Te$ $v=0$						−0.03193 (10)		77HON	MW			
68	$^{120}Sn^{16}O$ $v=0$						−0.14631 (38)	1844 (18)	73HON1	MW	−1.70 (5)	73HON1	
									77HON	MW			
69	$^{120}Sn^{32}S$ $v=0$						−0.06120 (7)		77HON	MW			2.2.2
70	$^{120}Sn^{80}Se$ $v=0$						−0.03007 (10)		77HON	MW			2.2.2
71	$^{120}Sn^{130}Te$ $v=0$						−0.02212 (13)		77HON	MW			2.2.2
72	$^{208}Pb^{16}O$ $v=0$						−0.16233 (39)	2598 (18)	73HON1	MW	−1.54 (6)	73HON1	
									77HON	MW			
73	$^{208}Pb^{32}S$ $v=0$						−0.06440 (4)		77HON	MW			2.2.2

Nr.	Molecule v, J	c_1 kHz	c_2 kHz	c_3 kHz	c_4 kHz	Ref.	g_J	ξ_T $\frac{Hz}{kG^2}$	Ref.	Meth.	(Θ/e) Å^2	Ref.	See Tab., Nr.
74	$^{208}Pb^{80}Se$ $v=0$						−0.02764 (7)		77HON	MW			2.2.2
75	$^{208}Pb^{130}Te$ $v=0$						−0.01800 (18)	4 (1) $\cdot 10^{3\mathrm{der}}$	74HON4 77HON	MW MW			2.2.2
80	$^{14}N^{16}O$ $v=0$ $X^2\Pi$							27 (24)	80MEE	MB	Θ^{el}/e $= -7.50$ (7)	80MEE	2.6.1 2.7.1 4.1.2 4.1.5
97 *6–43*	$^{79}Br^{19}F$ $v=0$ $^{81}Br^{19}F$ $v=0$	75 (50) 83 (50)				79NAI2 79NAI2				MW MW			2.2.2 2.6.1 2.7.1
98 *6–43a*	$^{127}I^{19}F$ $v=0$ $v=1$	91 (10) 89 (10)				73TIE2				MW			2.2.2 2.6.1
99	$^{79}Br^{35}Cl$ $v=0$ $v=1$ $^{81}Br^{35}Cl$ $v=1$ $^{79}Br^{37}Cl$ $v=0$ $^{81}Br^{37}Cl$ $v=0$	 23.4 (40) 16.9 (50) 18.9 (50) 15.2 (70) 18.9 (50)				 76TIE5 76TIE5 76TIE5 76TIE5 76TIE5				 MW MW MW MW MW			2.2.2 2.6.1 2.7.1
101	$^{79}Br_2$ $v=17$ $B^3\Pi^+_{0u}$ $^{81}Br_2$ $v=12$ $B^3\Pi^+_{0u}$	 87 (10) 66 (44)	 $c_1=c_2$ $c_1=c_2$			 74ENG 74ENG				 SLS SLS			2.7.1 4.1.6

Nr.	Molecule v, J	c_1 kHz	c_2 kHz	c_3 kHz	c_4 kHz	Ref.	g_J	ξ_T $\frac{Hz}{kG^2}$	Ref.	Meth.	(Θ/e) Å²	Ref.	See Tab., Nr.
103 *6–45*	$^{127}I_2$ $J=13$ $v=0$ $X^1\Sigma_g^+$	3.162 (8)	$c_1=c_2$	1.58 (5)	3.66 (3)	80YOK				MB			2.6.1 2.7.1 4.1.6
	$^{127}I_2$ $J=12$ $v=43$ $B^3\Pi_{0u}^+$	190.08 (17)	$c_1=c_2$	−100.3 (7)	−0.2 (4)	80YOK 80HAC							
	$^{127}I_2$ [1]	$c_1'-c_1''$ =829.20 (4)		$c_3'-c_3''$ =−500.4 (35)	$c_4'-c_4''$ =356.8 (11)	79FOT	−0.4 (1)		75BRO	LC SLS			
	$^{127}I_2$ [2]	$c_1'-c_1''$ =28.81 (15)		$c_3'-c_3''$ =−24.2 (28)	$c_4'-c_4''$ =−11.7 (16)	76LAN				SLS			
	$^{127}I_2$ $v=17$						−0.064 (22)		75BRO	LC			
	$v=18$ $B^3\Pi_{0u}^+$						−0.059 (17)		75BRO	LC			
	$v=21$						−0.10 (6)		75BRO	LC			
	$v=32$						−0.21 (3)		75BRO	LC			
	$v=40$						−0.32 (4)		75BRO	LC			
	$v=62$						−1.82 (6)		76GUE2	[3]			
	$v=64$						−2.7 (3)		76GUE2	[3]			
	$v=67$						−3.6 (6)		76GUE2	[3]			
	$v=70$						−5.45 (15)		76GUE2	[3]			

[1]) Difference between the molecular parameters of the two states: $' \triangleq B^3\Pi_{0u}^+$, $v'=58$, $J'=99$; $'' \triangleq X^1\Sigma_g^+$, $v''=1$, $J''=98$.

[2]) Difference between the molecular parameters of the two states: $' \triangleq B^3\Pi_{0u}^+$, $v'=11$, $J'=128$; $'' \triangleq X^1\Sigma_g^+$, $v''=6$, $J''=127$.

[3]) Zeeman-resonances in a modulated laser light beam.

2.9.2 Linear and symmetric top molecules

Rotational g-values, magnetic susceptibilities and anisotropies, paramagnetic and diamagnetic contributions, molecular quadrupole moments, electronic charge distributions, spin-rotation and spin-spin coupling parameters, nuclear g-values from the rotational Zeeman effect and nuclear magnetic shielding parameters from the rotational Zeeman effect as well as indirectly via the paramagnetic contribution to the spin-rotation tensor components

2.9.2.1 Preliminary remarks

1. Introduction

The spectroscopic determination of the spin-rotation-, spin-spin-, and nuclear-shielding parameters of diamagnetic molecules compiled here together with the energetically more effective rotational Zeeman parameters requires a resolving power of the order of 1 kHz.

Several high-resolution techniques

molecular beam radiofrequency spectroscopy (MBRF)
molecular beam microwave and mm-wave spectroscopy (MBMW)
molecular beam electric and magnetic resonance (MBE, MBM), beam maser spectroscopy (BMS)
saturation or Lamb-dip spectroscopy
radiofrequency infrared double resonance (RFIRDR), and molecular beam Fourier transform microwave spectroscopy

have been developed to meet this demand, and are now widely used in order to measure and analyse rotational hyperfine structures (hfs) of polyatomic molecules. Rotational molecular g-factors and susceptibility anisotropies, on the other hand, can frequently be determined accurately enough with the aid of the conventional rotational-Zeeman microwave spectroscopy which is more easy to handle. Nevertheless, the number of polyatomic molecules studied in external magnetic fields by beam techniques as compared to the lists in the foregoing table in Volume II/6 is increasing.

Basic changes in theory, definition and designation of internal and external magnetic coupling parameters have not occurred in the past seven years. Therefore the overall-disposal and arrangement of the first table of magnetic constants of polyatomic molecules in Vol. II/6 is retained.

The entries in the table below are drawn completely from the excellent documentation material available in the

Sektion für Strukturdokumentation der Universität Ulm.

The author is grateful for the help of B. Starck and her coworkers, especially R. Mutter and Ch. Spreter.

2. Basic formulae and definition of parameters

The theoretical formulae describing the hyperfine and external-field splittings of polyatomic molecules was already compiled for Vol. II/6, 2.9.2. For the convenience of the user it is given here again with some minor changes and additions.

The hfs plus external-field Hamiltonian may be written in tensor notation as [64T1, 70V1, 67H1] [1])

$$\mathcal{H}' = \tfrac{1}{2}\sum_{k}(\boldsymbol{I}^k \cdot \mathsf{M}^k \cdot \boldsymbol{J} + (\boldsymbol{I}^k \cdot \mathsf{M}^k \cdot \boldsymbol{J})^\dagger) \quad (1a)$$

$$+ \sum_{k \neq l} \boldsymbol{I}^k \cdot \mathsf{D}^{kl} \cdot \boldsymbol{I}^l \quad (1b)$$

$$- \tfrac{1}{2}\mu_n(\boldsymbol{H} \cdot \mathsf{g} \cdot \boldsymbol{J} + (\boldsymbol{H} \cdot \mathsf{g} \cdot \boldsymbol{J})^\dagger) \quad (1c)$$

$$- \tfrac{1}{2}\boldsymbol{H} \cdot \boldsymbol{\xi} \cdot \boldsymbol{H} \quad (1d)$$

$$- \sum_{k} \mu_n\, g_I^k\, \boldsymbol{I}^k \cdot (\boldsymbol{1} - \boldsymbol{\sigma}^k) \cdot \boldsymbol{H} \quad (1e)$$

[1]) Concerning completeness compare footnote on page 383 of Vol. II/6. In polyatomic molecules, ΔB corrections and pseudo-quadrupolic coupling are normally not considered. However, see for example [53B1].

where

k, l = the summation indices for nuclei with non-zero spin

μ_n = the nuclear magneton

g_I^k = the magnetogyric ratio (nuclear g-factor) of the k-th nucleus

$\boldsymbol{I}^k$ = the nuclear spin of the k-th nucleus, with quantum number I^k

$\boldsymbol{J}$ = the rotational angular momentum with quantum number J

$\boldsymbol{H}$ = the magnetic field strength

M^k = the spin-rotation coupling tensor of the k-th nucleus

D^{kl} = the spin-spin coupling tensor of nuclei k and l

g = the molecular magnetogyric-ratio tensor (rotationally induced molecular g-values)

ξ = the magnetic-susceptibility tensor

$\boldsymbol{\sigma}^k$ = the nuclear-shielding tensor of the k-th nucleus

† = conjugate transpose

It is appropriate to refer these quantities to the molecular-principal-axis system (a, b, c). In this way, the tensor components $M^k_{gg'}$, $D^k_{gg'}$, $g_{gg'}$, $\xi_{gg'}$, and $\sigma^k_{gg'}$ become uniquely defined molecular properties ($g, g' = a, b, c$). However, only the diagonal tensor components are useful in analysing hfs spectra as non-diagonal terms, $g \neq g'$, do not enter matrix elements diagonal in J, and higher-than-first-order rotational perturbations of $\mathscr{H}'$ are barely observable. Thus it is sufficient to consider an effective Hamiltonian diagonal in J. This Hamiltonian is customarily written in terms of rotationally dependent interaction strengths $C^k(J_{K_{-1},K_{+1}})$, $D^{ki}(J_{K_{-1},K_{+1}})$, $g(J_{K_{-1},K_{+1}})$, $\xi(J_{K_{-1},K_{+1}})$, and $\sigma^k(J_{K_{-1},K_{+1}})$, in the following way:

$$\mathscr{H} = \sum_k C^k(J_{K_{-1},K_{+1}})\, \boldsymbol{I}^k \cdot \boldsymbol{J} \tag{2a}$$

$$+ \sum_{k \neq 1} \sum_{l} D^{kl}(J_{K_{-1},K_{+1}}) \frac{1}{J(2J-1)} \{\tfrac{3}{2}[(\boldsymbol{I}^l \cdot \boldsymbol{J})(\boldsymbol{I}^k \cdot \boldsymbol{J}) + (\boldsymbol{I}^k \cdot \boldsymbol{J})(\boldsymbol{I}^l \cdot \boldsymbol{J})] - (\boldsymbol{I}^k \cdot \boldsymbol{I}^l)\, J(J+1)\} \tag{2b}$$

$$- \mu_n g(J_{K_{-1},K_{+1}})\, \boldsymbol{H} \cdot \boldsymbol{J} \tag{2c}$$

$$- \tfrac{1}{2} \xi \boldsymbol{H}^2 \tag{2d}$$

$$- \xi(J_{K_{-1},K_{+1}}) \frac{1}{(2J+3)(2J-1)} [3(\boldsymbol{H} \cdot \boldsymbol{J})^2 - \boldsymbol{H}^2 J(J+1)] \tag{2e}$$

$$+ \sum_k \mu_n g_I^k \sigma^k(J_{K_{-1},K_{+1}}) \frac{1}{(2J+3)(2J-1)} \{3[(\boldsymbol{I}^k \cdot \boldsymbol{J})(\boldsymbol{H} \cdot \boldsymbol{J}) + (\boldsymbol{H} \cdot \boldsymbol{J})(\boldsymbol{I}^k \cdot \boldsymbol{J})] - (\boldsymbol{I}^k \cdot \boldsymbol{H}) J(J+1)\} \tag{2f *)}$$

$$- \mu_n g_I^k (1 - \sigma^k)\, \boldsymbol{I}^k \cdot \boldsymbol{H} \tag{2g}$$

with

$$C^k(J_{K_{-1},K_{+1}}) = \sum_g M^k_{gg} \frac{1}{J(J+1)} \langle J_g^2 \rangle, \tag{3a}$$

$$D^{kl}(J_{K_{-1},K_{+1}}) = 2 \sum_g D^{kl}_{gg} \frac{1}{(J+1)(2J+3)} \langle J_g^2 \rangle, \tag{3b}$$

$$g(J_{K_{-1},K_{+1}}) = \sum_g g_{gg} \frac{1}{J(J+1)} \langle J_g^2 \rangle, \tag{3c}$$

$$\xi(J_{K_{-1},K_{+1}}) = \sum_g (\xi_{gg} - \xi) \frac{1}{J(J+1)} \langle J_g^2 \rangle, \tag{3d}$$

$$\sigma^k(J_{K_{-1},K_{+1}}) = \sum_g (\sigma^k_{gg} - \sigma^k) \frac{1}{J(J+1)} \langle J_g^2 \rangle, \tag{3e}$$

$$\xi = \tfrac{1}{3} \sum_g \xi_{gg}, \tag{3f}$$

$$\sigma^k = \tfrac{1}{3} \sum_g \sigma^k_{gg}. \tag{3g}$$

$\langle J_g^2 \rangle$ = the expectation value of J_g^2 over the state $J_{K_{-1},K_{+1}}$. For nomenclature regarding rotational-energy states, see Table 2.5.

*) The sign of this expression was erroneously given as negative in Vol. II/6, 2.9.2.1.

A derivation of the well-known expressions (2a)···(2g) using irreducible-tensor techniques is given in [77H1].

The scalar spin-spin coupling term is omitted since the electron-coupled spin-spin interaction seems to be beyond the resolution achievable with techniques reported here. ξ as defined in (3f) is the bulk magnetic susceptibility, and σ^k as defined in (3g) is the average magnetic shielding at the k-th nucleus. ξ cannot be obtained spectroscopically since the term (2d) cancels out in subtracting energy levels.

Matrix element expressions depend on the coupling scheme which is adopted for nuclear spins $\boldsymbol{I}^k$ and rotational angular momentum $\boldsymbol{J}$, and are developed conveniently in terms of vector-coupling coefficients [64T1, 67H1].

The theoretical expressions for the molecular tensor quantities introduced above are as follows (principal-axis-of-inertia system throughout):

$$M^{k}_{gg'} = M^{nk}_{gg'} + M^{elk}_{gg'} = -\frac{2|e|\mu_n g_I^k G_g}{\hbar c} \sum_{l \neq k} Z_l \langle r_{lk}^{-3}(r_{kl}^2 \delta_{gg'} - (r_{kl})_g (r_{kl})_{g'})\rangle$$
$$+\frac{2|e|\mu_n g_I^k G_g}{\hbar c\, m_{el}} \sum_{n>0} (W_n - W_0)^{-1}(\langle 0|\sum_i (L_{ik})_g r_{ik}^{-3}|n\rangle \langle n|(L_k)_{g'}|0\rangle$$
$$+\langle 0|(L_k)_{g'}|n\rangle \langle n|\sum_i (L_{ik})_g r_{ik}^{-3}|0\rangle) \quad [64F1], \tag{4a}$$

$$D^{kl}_{gg'} = \mu_n^2 g_I^k g_I^l r_{kl}^{-5}(r_{kl}^2 \delta_{gg'} - 3(r_{lk})_g (r_{lk})_{g'}) \quad [64T1], \tag{4b}$$

$$g_{gg'} = g^{n}_{gg'} + g^{el}_{gg'} = \frac{4\pi m_p G_g}{\hbar} \sum_k Z_k (r_k^2 \delta_{gg'} - (r_k)_g (r_k)_{g'})$$
$$-\frac{4\pi m_p G_g}{\hbar m_{el}} \sum_{n>0} (W_n - W_0)^{-1}(\langle 0|L_g|n\rangle \langle n|L_{g'}|0\rangle + \langle 0|L_{g'}|n\rangle \langle n|L_g|0\rangle) \quad [52E1], \tag{4c}$$

$$\xi_{gg'} = \xi^{dia}_{gg'} + \xi^{para}_{gg'} = -\frac{e^2 N}{4 m_{el} c^2} \langle 0|\sum_i (r_i^2 \delta_{gg'} - (r_i)_g (r_i)_{g'})|0\rangle$$
$$+\frac{e^2 N}{4 m_{el}^2 c^2} \sum_{n>0} (W_n - W_0)^{-1}(\langle 0|L_g|n\rangle \langle n|L_{g'}|0\rangle + \langle 0|L_{g'}|n\rangle \langle n|L_g|0\rangle) \quad [32V1], \tag{4d}$$

$$\sigma^k_{gg'} = \sigma^{dia\,k}_{gg'} + \sigma^{para\,k}_{gg'} = \frac{e^2}{2 m_{el} c^2} \langle 0|\sum_i (r_{ik}^2 \delta_{gg'} - (r_{ki})_g (r_{ki})_{g'})\, r_{ik}^{-3}|0\rangle$$
$$-\frac{e^2}{2 m_{el}^2 c^2} \sum_{n>0} (W_n - W_0)^{-1}(\langle 0|\sum_i (L_{ik})_g r_{ik}^{-3}|n\rangle \langle n|(L_k)_{g'}|0\rangle$$
$$+\langle 0|(L_k)_{g'}|n\rangle \langle n|\sum_i L_{ik})_g r_{ik}^{-3}|0\rangle) \quad [50R1] \tag{4e}$$

where

m_p	= the proton mass
$\hbar$	= Planck's constant divided by 2π
c	= the speed of light
e	= the electron charge
m_{el}	= the electron mass
N	= Avogadro's number
G_g	= the rotational constant around axis g
Z_k	= the atomic number of the k-th nucleus
$\delta_{gg'}$	= the Kronecker symbol
$r_{lk} = [\sum_g (r_{lk})_g^2]^{\frac{1}{2}}$	= the distance between nucleus l and nucleus k
$(r_{lk})_g = (r_k)_g - (r_l)_g$	= the projection of r_{lk} on axis g, $(r_k)_g$ being the g-th component of the position vector of the k-th nucleus measured from the center of mass
$r_{ik}, (r_{ik})_g$	= the corresponding quantities involving electron i and nucleus k
$W_n - W_0$	= the energy difference between n-th excited and electronic ground state
$L_g = \sum_i (L_i)_g$	= the g-th component of total electronic orbital angular momentum referred to the center of mass
$(L_k)_g = \sum_i (L_{ik})_g$	= the g-th component of total electronic orbital angular momentum referred to nucleus k
$(L_{ik})_g$	= the g-th component of the orbital angular momentum of the i-th electron referred to nucleus k
$\langle 0\| \|n\rangle$	= the matrix element connecting the electronic ground state and the n-th excited state.

Higher-order contributions to (4a)–(4e) have been omitted because the accuracy of the data presented is barely sufficient to resolve smaller terms. For instance, the nuclear diamagnetic contribution is not included in eq. (4d) though it can easily be taken into account from structural information if necessary.

In the case of weakly coupling nuclei in an asymmetric-top molecule interacting with a reasonably high magnetic field of strength H, the hfs and external magnetic field energies described by (2a)···(2g) read as follows:

$$E(J_{K-1,K+1})=\sum_k C^K(J_{K-1,K+1})M_{I^k}M_J \quad (5a)$$

$$+\sum_{k\neq 1}\sum_{1} D^{kl}(J_{K-1,K+1})\frac{1}{J(2J-1)}M_{I^l}M_{I^k}[3M_J^2-J(J+1)] \quad (5b)$$

$$-\mu_n g(J_{K-1,K+1})M_J H \quad (5c)$$

$$-\tfrac{1}{2}\xi H^2 \quad (5d)$$

$$-\xi(J_{K-1,K+1})\frac{H^2}{(2J+3)(2J-1)}[3M_J^2-J(J+1)] \quad (5e)$$

$$+\sum_k \mu_n g_I^k \sigma^k(J_{K-1,K+1})\frac{2}{(2J+3)(2J-1)}M_{I^k}H[3M_J^2-J(J+1)] \quad (5f)$$

$$-\sum_k \mu_n g_I^k(1-\sigma^k)M_{I^k}H \quad (5g)$$

In the case of a prolate symmetric-top molecule with one interacting nucleus on the symmetry axis, the expressions (5a)···(5g) are simplified to

$$E(J_K)=\frac{1}{J(J+1)}\{M_{\parallel}K^2+M_{\perp}(J(J+1)-K^2)\}M_I M_J \quad (6a)$$

$$-\frac{\mu_n}{J(J+1)}\{g_{\parallel}K^2+g_{\perp}(J(J+1)-K^2)\}M_J H \quad (6c)$$

$$-\tfrac{1}{2}\xi H^2 \quad (6d)$$

$$-\frac{1}{3}\frac{1}{J(J+1)}(\xi_{\parallel}-\xi_{\perp})[3K^2-J(J+1)]\frac{H^2}{(2J+3)(2J-1)}[3M_J^2-J(J+1)] \quad (6e)$$

$$+\mu_n g_I\frac{1}{3}\frac{1}{J(J+1)}(\sigma_{\parallel}-\sigma_{\perp})[3K^2-J(J+1)]\frac{2M_I H}{(2J+3)(2J-1)}[3M_J^2-J(J+1)] \quad (6f)$$

$$-\mu_n g_I(1-\sigma)M_I H \quad (6g)$$

In the zero-field case the coupled basis is appropriate, and eqs. (6) collapse to

$$E(J_K)=\frac{1}{J(J+1)}\{M_{\parallel}K^2+M_{\perp}(J(J+1)-K^2)\}\tfrac{1}{2}(F(F+1)-I(I+1)-J(J+1)) \quad (7).$$

For a linear molecule $K=0$ in eqs. (6a)···(7). The newly used symbols mean

M_J = magnetic quantum number to J_z, z = magnetic-field direction
M_{I^k} = magnetic quantum number to I_z^k
H = magnetic field strength
K = symmetric-rotor quantum number to J_a, see Table 2.4
$\parallel, \perp$ = indices designating directions parallel and perpendicular to the figure axis
F = angular-momentum quantum number to $\boldsymbol{J}+\boldsymbol{I}$.

The quantities (4a)···(4e) above are averages over the vibrational state in question. Vibrational ground-state data are normally handled in the rigid-rotor approximation. The error introduced thereby is considered to be less than 1% in the measured quantities. Thus, in computing $M_{gg}^{\text{n}k}$ from structural information and observing that $M_{gg}^{\text{el}k}$ and $\sigma_{gg}^{\text{para}k}$ in eqs. (4a) and (4e) are the same but for a constant one obtains the diamagnetic and paramagnetic shielding, $\sigma_{gg}^{\text{dia}k}$ and $\sigma_{gg}^{\text{para}k}$, from the measurement of M_{gg}^k and σ_{gg}^k. Moreover, $\sigma_{gg}^{\text{dia}k}$ can be determined by ab-initio Hartree-Fock calculations, and even an accurate and convenient semi-empirical method [72G2] is available. Nuclear-shielding tensors have been determined this way from zero-field spectra and are included in the tables below.

The diamagnetic and paramagnetic susceptibilities, ξ_{gg}^{dia} and ξ_{gg}^{para}, can be obtained from measurements of g_{gg} and ξ_{gg}. However, the bulk susceptibility is needed in order to determine ξ_{gg}, $g=a, b, c$ from the spectroscopically measured anisotropies (eqs. (2e), (3d)). Lacking gas-phase data, bulk susceptibilities measured in the liquid state are frequently used in order to obtain the desired information. The error introduced is normally not known.

The molecular electric quadrupole moments can be extracted from g-values and susceptibility anisotropies via [68H1]

$$\begin{aligned}\Theta_{aa}&=\tfrac{1}{2}|e|\sum_{k} Z_k(3(r_k)_a^2-r_k^2)-\tfrac{1}{2}|e|\,\langle 0|\sum_{i}(3(r_i)_a^2-r_i^2)\,|0\rangle\\&=-\frac{\hbar}{8\pi}\frac{|e|}{m_p}\left(\frac{2g_{aa}}{G_a}-\frac{g_{bb}}{G_b}-\frac{g_{cc}}{G_c}\right)-\frac{4m_{el}c^2}{|e|N}(2\xi_{aa}-\xi_{bb}-\xi_{cc}).\end{aligned} \tag{8}$$

The expectation values

$$\langle 0|\sum_{i}((r_i)_g^2-(r_i)_{g'}^2)\,|0\rangle=\langle g^2\rangle-\langle g'^2\rangle,\qquad g, g'=a, b, c \tag{9}$$

follow directly from diamagnetic-susceptibility anisotropies, and the individual-coordinate values

$$\langle 0|\sum_{i}(r_i)_g^2\,|0\rangle=\langle g^2\rangle,\qquad g=a, b, c \tag{9a}$$

can be extracted if bulk susceptibility information is available.

Further magnetic interactions are those of a vibrationally induced moment either internally with a nucleus $I_k>\frac{1}{2}$ or externally with a magnetic field. They can be observed in degenerate vibrational states v of linear and symmetric-top molecules. The effective Hamiltonian describing spectra appropriately in *linear* molecules is based upon a slightly-bent-rigid-rotor model. The magnetic (x, y)-anisotropy in the two directions perpendicular to the figure axis seems negligible [78R1, 78H4], and therefore eqs. (6a)···(6g) can be applied by setting $M_\parallel^v=M_{zz}^v$, $M_\perp^v=\frac{1}{2}(M_{xx}^v+M_{yy}^v)$, $g_\parallel^v=g_{zz}^v$, $g_\perp^v=\frac{1}{2}(g_{xx}^v+g_{yy}^v)$, $(\xi_\parallel-\xi_\perp)^v=\xi_{zz}^v-\frac{1}{2}(\xi_{xx}^v+\xi_{yy}^v)$, and $(\sigma_\parallel-\sigma_\perp)^v=\sigma_{zz}^v-\frac{1}{2}(\sigma_{xx}^v+\sigma_{yy}^v)$. In the ground vibrational state $M_\parallel^0=g_\parallel^0=0$ while all the other quantities do not change significantly. In *symmetric*-top molecules, the vector addition of the vibrational, and the skeleton-rotational induced magnetic moments along the figure axis must be considered. There are four possible relative orientations of the total and vibrationally induced angular momenta, J_z and p_α, which obey the relations

$$J_z|J M_J K\rangle=K|J M_J K\rangle,\quad K=\pm J,\ \pm(J-1),\ldots,0$$

as well as

$$p_\alpha|v\,l\rangle=\zeta\,l|v\,l\rangle,\quad l=\pm v,\ \pm(v-2),\ldots,\pm 1\ \text{or}\ 0,\quad l\leqq J,$$

and which belong to a set of four almost degenerate rovibrational energy levels [59N1]. $|J M_J K\rangle$ and $|v\,l\rangle$ are the symmetric-rotor and twodimensional-isotropic-oscillator functions, and ζ is the Coriolis coupling constant. It has been shown experimentally [77N1] that the vibrational-rotational Zeeman effect is properly described by the Hamiltonian

$$\mathscr{H}^v=-\mu_n H\{g_{xx}^v\phi_{xZ}J_x+g_{yy}^v\phi_{yZ}J_y+\phi_{zZ}[g_{zz}^v J_z+(g_\alpha^v-g_{zz}^v)p_\alpha]\}-\tfrac{1}{2}H^2\sum_{g}\xi_{gg}^v(\phi_{gZ}^2-\tfrac{1}{3}), \tag{10}$$

and the corresponding first order energy expression,

$$\begin{aligned}E_\pm^v(J M_J\sigma|K|\,|l|)&=\langle v J M_J\sigma|K|\,|l|\pm|\mathscr{H}^v|v J M_J\sigma|K|\,|l|\pm\rangle\\&=-\mu_n H\frac{M_J}{J(J+1)}[g_\perp^v(J(J+1)-K^2)+K^2 g_{zz}^v+\sigma|K|\,|l|\,\zeta(g_\alpha^v-g_{zz}^v)]\\&\quad-H^2\frac{3M_J^2-J(J+1)}{(2J-1)(2J+3)J(J+1)}[K^2-\tfrac{1}{3}J(J+1)]\,(\xi_\parallel-\xi_\perp)^v\end{aligned} \tag{11}$$

where

ϕ_{gZ} = directional cosine operator connecting external-field axis Z and molecular axis g

p_α = vibrational angular momentum operator

J_z = operator of total angular momentum along the figure axis

ζ = Coriolis coupling constant

σ = ± 1, quantum number to distinguish between states $Kl>0$ and $Kl<0$, and

$$|v J M_J\sigma|K|\,|l|\pm\rangle=\frac{1}{\sqrt{2}}\{|J M_J|K|\rangle\,|v|l|\rangle\pm|J M_J-|K|\rangle\,|v-|l|\rangle\},\qquad \sigma=+1,$$

or

$$=\frac{1}{\sqrt{2}}\{|J M_J - |K|\rangle\,|v|l|\rangle \pm |J M_J |K|\rangle\,|v - |l|\rangle\}, \qquad \sigma = -1,$$

properly adapted zero-order basis functions

g^v_α = vibrationally induced g-value

g^v_{zz} = pure rotational g-value which would be measured if no vibrational angular momentum were present

(x, y)-anisotropies have again been neglected.

With $\sigma = +1$, $K = l$, $\zeta = 1$, $g^v_{zz} = 0$, $g^v_\alpha = g^v_\parallel$, eq. (11) applies to the linear molecule. The spin-vibration coupling in symmetric-top molecules can be treated analogously replacing ϕ_{gZ} by I^k_g, and g^v_{gg}, g^v_α by M^v_{gg}, M^v_α.

3. Explanation of the columns in the tabulations 2.9.2.2 and 2.9.3.2

Column	Parameters		Definitions
1	Nr.		serial number of the molecule in this table
2			chemical formula of the considered molecule; information about its conformation and isotopic composition
3	g		molecular g-value
		g_{aa}, g_{bb}, g_{cc}	molecular g-values induced by rotation around the principal axis $g = a, b$, or c. The corresponding magnetic moment is $\mu_g = \mu_n g_{gg} J_g$
		$g_\parallel$, $g_\perp$	the same quantity in linear and symmetric-top molecules referred to directions along and perpendicular to the molecular axis
		$g(J_{K_{-1}, K_{+1}})$	effective g-value of the rotational state $J_{K_{-1}, K_{+1}}$; see eq. (3c)
			the sign of the g-values is always given when strong arguments exclude the other possibility. Cases where the sign is determined experimentally are mentioned in column 10
4	$\Delta\xi$ [10^{-6} erg/G^2 mol]		magnetic-susceptibility anisotropy (per mole)
		$\Delta\xi = \xi_\parallel - \xi_\perp$	magnetic-susceptibility anisotropy (per mole) in symmetric-top and linear molecules
		$\Delta\xi_1 = 2\xi_{aa} - \xi_{bb} - \xi_{cc}$ $\Delta\xi_2 = 2\xi_{bb} - \xi_{cc} - \xi_{aa}$	independent anisotropies of the magnetic susceptibility (per mole). $3\Delta\xi_1 = \xi_{aa} - \xi$, $3\Delta\xi_2 = \xi_{bb} - \xi$, see eqs. (3d), (3f)
5	ξ^{para} [10^{-6} erg/G^2 mol]		paramagnetic susceptibility (per mole)
		$\xi^{para}_\perp$	paramagnetic susceptibility (per mole), perpendicular to the molecular axis in linear and symmetric-top molecules
		$\xi^{para}_\parallel$	paramagnetic susceptibility (per mole), parallel to the molecular axis in symmetric-top molecules
		ξ^{para}_{aa}, ξ^{para}_{bb}, ξ^{para}_{cc}	paramagnetic susceptibilities (per mole), see eq. (4d)
6	Θ[10^{-26} esu cm^2]		molecular electric quadrupole moment
		Θ_{aa}, Θ_{bb}, Θ_{cc}	molecular electric quadrupole moments, see eq. (8)
		$\Theta_\parallel$, $\Theta_\perp$	molecular electric quadrupole moments in symmetric-top and linear molecules, see eq. (8)

Column	Parameters		Definitions
7	$\langle a^2\rangle, \langle b^2\rangle, \langle c^2\rangle$, [$10^{-16}$ cm^2]	or linear combinations thereof	expectation values of the sums of the squares of the electronic center-of-mass coordinates, see eqs. (9), (9a)
	$\langle \parallel^2\rangle, \langle \perp^2\rangle$ [10^{-16} cm^2]	or the difference of both	the same quantities in linear and symmetric-top molecules referred to directions along and perpendicular to the molecular axis, see eqs. (9), (9a)
8	Meth.		experimental method yielding the values listed to the left or/and to the right of this column. Explanations of abbreviations see 1.5, p. 3
9	Ref.		references for the values in columns 3···8 and for column 10
10			Varia, remarks: (See footnotes)
		$g^n_{aa}, g^n_{bb}, g^n_{cc}$	nuclear contributions to the molecular g-values
		$M^k_{aa}, M^k_{bb}, M^k_{cc}$ [1]) [2])	spin-rotation coupling tensor elements, see eqs. (1a), (4a)
		$M^k_{\parallel}, M^k_{\perp}$ [1]) [2])	spin-rotation coupling tensor elements of linear and symmetric-top molecules, nucleus k on the molecular axis, see eqs. (1a), (4a)
		$C^k(J_{K_{-1},K_{+1}})$ [1]) [2])	spin-rotation interaction strength of the rotational state $J_{K_{-1},K_{+1}}$, see eqs. (2a), (3a)
		g^k_I [2]) (or simply g_k)	nuclear g-value; the corresponding nuclear magnetic moment is $\mu^k_I = \mu_n g^k_I I^k$
		$D^{kl}_{aa}, D^{kl}_{bb}, D^{kl}_{cc}$ [3])	spin-spin interaction tensor elements involving nuclei k and l, see eqs. (1b), (4b)
		$D^{kl}(J_{K_{-1},K_{+1}})$ [3])	spin-spin interaction strength of the rotational state $J_{K_{-1},K_{+1}}$, see eq. (3b)
		$\mu_n^2 g^k_I g^l_I / r^3_{kl}, r^{-3}_{kl}$, or similar [3])	spin-spin interaction parameters related to D^{kl}_{gg} or $D^{kl}(J_{K_{-1},K_{+1}})$ above, see eqs. (3b), (4b)
		$\xi_{aa}, \xi_{bb}, \xi_{cc}$ $\xi^{dia}_{aa}, \xi^{dia}_{bb}, \xi^{dia}_{cc}$	individual tensor components of the total and diamagnetic susceptibilities, see eq. (4d)
		$\xi_{\parallel}, \xi_{\perp}$ [4]) $\xi^{dia}_{\parallel}, \xi^{dia}_{\perp}$	the same quantities in linear and symmetric-top molecules referred to directions along and perpendicular to the molecular axis
		ξ	bulk susceptibility used to obtain $\xi_{aa}, \xi_{bb}, \xi_{cc}(\xi_{\parallel}, \xi_{\perp})$ $\xi^{dia}_{aa}, \xi^{dia}_{bb}, \xi^{dia}_{cc}$ $(\xi^{dia}_{\parallel}, \xi^{dia}_{\perp})$, and $\langle a^2\rangle, \langle b^2\rangle, \langle c^2\rangle$, $(\langle\parallel^2\rangle, \langle\perp^2\rangle)$ from the corresponding anisotropies, see text above
		σ^k [2])	nuclear-magnetic shielding (ppm), see eqs. (2g), (3g)
		$\sigma^k_{aa}, \sigma^k_{bb}, \sigma^k_{cc}$ [2])	nuclear-magnetic shielding tensor components (ppm) see eqs. (1e), (4e)
		$\sigma^k_{\parallel}, \sigma^k_{\perp}$ [1])	nuclear-magnetic shielding tensor components (ppm) of linear and symmetric-top molecules, nucleus k on the molecular axis
11			references to Tables 2.3, 2.4 and 2.5

[1]) Different sign conventions are used in the literature for spin-rotation coupling constants. The signs have been converted to comply with the definition given in eqs. (1a), (2a), and (4a) if necessary. Thus, a positive value of M^k_{gg} indicates that the direction of the magnetic field at the site of the k-th nucleus as induced by molecular rotation around the g-th axis is determined by the electronic charge distribution (assuming a positive nuclear magnetic moment).

[2]) The suffix k is sometimes left omitted if only one nucleus with $I>0$ is present in the molecule.

[3]) Spin-spin interaction parameters are given only if they are used in the spectral fit. In most cases authors obtain them from the known structure.

A suffix v is used (M^v_{gg}, g^v_{gg}, ξ^v_{gg} etc.) for parameters which have been determined in an excited vibrational state. The type of vibration is specified.

Uncertainties given behind each entry in parenthesis are in units of the last digit. They mean in most cases one, two or sometimes three standard deviations. The original literature should be consulted if more detailed information is desired. Molecular constants without any error estimates have not been recorded.

Individual second-moment electronic charge distributions $\langle a^2\rangle$, $\langle b^2\rangle$, $\langle c^2\rangle$ (or $\langle\|^2\rangle$, $\langle\perp^2\rangle$) are listed in column 7 if bulk susceptibility information was available to the authors and they could be obtained from the anisotropies, $\langle a^2\rangle-\langle b^2\rangle$, $\langle b^2\rangle-\langle c^2\rangle$ (or $\langle\|^2\rangle-\langle\perp^2\rangle$). Observe, however, that the latter are more accurate as they follow from spectroscopical information alone.

4. Units and conversion factors

In agreement with common practice in the literature, cgs units are used in the tabulations below.

Conversion of the entries to SI units is as follows.

Susceptibilities: $\xi(\mathrm{SI})/\mathrm{m}^3 = 4\pi \cdot 10^{-6}\, \xi(\mathrm{cgs})/(\mathrm{erg/G^2\ mol})$

quadrupole moment: $\Theta(\mathrm{SI})/(\mathrm{C\ m^2}) = 3.3357 \cdot 10^{-14}\, \Theta(\mathrm{cgs})/(\mathrm{esu\ cm^2})$

second-order electronic moment: $\langle a^2\rangle\,(\mathrm{SI})/\mathrm{m}^2 = 10^{-4}\, \langle a^2\rangle\,(\mathrm{cgs})/\mathrm{cm}^2$

5. Arrangement of substances

a) Linear molecules	1…14
b) Symmetric top molecules:	
Pyramidal molecules, XY_3	15…18
Five-atom molecules, XY_4 and XY_3Z, X=C, Si, Ge	19…30
Other five atom molecules, XYZ_3	31
CH_3 and SiH_3 derivatives of linear molecules	32…34
others	35…37

2.9.2.2 Data

Nr.	Molecule	g	$\Delta\xi=\xi_\parallel-\xi_\perp$	$\xi_\parallel^{para}, \xi_\perp^{para}$	$\Theta_\parallel, \Theta_\perp$	$\langle\parallel^2\rangle, \langle\perp^2\rangle$	Meth.	Ref.	Varia, remarks	See Tab., Nr.
			10^{-6} erg/Gauss^2 mol		10^{-26} esu cm^2	10^{-16} cm^2				
a) Linear molecules										
1	HBS DBS $D^{10}BS$	−0.0414 (2) −0.0356 (2)	−7.2 (5) −9.8 (21)	110.2 (5)	2.7 (6)	$\langle\parallel^2\rangle-\langle\perp^2\rangle=14.6$	MW	74P1	[1] [2] [3]	2.3, 12
2	OCS	see Vol. II/6, 2.9.2, Nr. 2 $g_\perp^v=-0.02930\,(4)$ *) $g_\parallel^v-g_\perp^v=0.0905\,(5)$ *) $g_{xx}^v-g_{yy}^v=0.00023\,(2)$ *)	$\xi_\parallel^v-\xi_\perp^v=-9.52\,(40)$ *) $\xi_{xx}^v-\xi_{yy}^v=-0.12\,(12)$ *)				MBE MW MBE	74D2 73T1 72R1	[1] [2]	
	$OC^{33}S$						MBE	74R2	[3]	
3	HCN	−0.0980 (20)					MW BMS	73G1 74G1, 69L1		2.3, 9
	$D^{13}CN$ $HC^{15}N$	−0.0917 (7)	−7.6 (8)		3.5 (9)		MW	73G1, 73H1	[1] [2] [3]	
4	$FC^{15}N$	−0.0504 (8)	−7.2 (8)		−3.7 (10)		MW	73R1		

to Nr.:

1 [1]) $M_\perp^B=-8.1\,(30)$ kHz.
[2]) $M_\perp^B=-6.4(30)$ kHz; the polarity from the isotopic effect in the g-values is (+HBS−).
[3]) $M_\perp^{^{10}B}=-2.6(10)$ kHz.

2 [1]) $g_\perp$ has been measured in the rotational states $J=1$ through $J=5$, the values being identical to within $\pm 0.01\%$ (no centrifugal distortion effects) as well as in agreement with a previous result for $J=1$ [70 L1, Vol. II/6, Tab. 2.9.2, Nr. 2].
[2]) Cavity work at 62 kG, results in agreement with MBE [70L1, 74 D2].

to Nr.:

2 [3]) $M_\perp^{^{33}S}=0.87(5)$ kHz; $(M_\perp^v)^{^{33}S}=1.16(8)$ kHz*); $(M_\parallel^v-M_\perp^v)^{^{33}S}=11.5(5)$ kHz*); $(M_{xx}^v-M_{yy}^v)^{^{33}S}=-0.12(11)$ kHz*).
*) Measured in the $v=(01^10)$ bending mode by observation of direct l-type transitions
$g_\perp^v=\frac{1}{2}(g_{xx}^v+g_{yy}^v)$; $g_\parallel^v=g_{zz}^v$; $\xi_\perp^v=\frac{1}{2}(\xi_{xx}^v+\xi_{yy}^v)$; $\xi_\parallel^v=\xi_{zz}^v$; $M_\perp^v=\frac{1}{2}(M_{xx}^v+M_{yy}^v)$; $M_\parallel^v=M_{zz}^v$, (z=molecular axis).

3 [1]) $M_\perp^H=-3.7(3)$ kHz; $M_\perp^N=10.4(3)$ kHz.
[2]) $M_\perp^D=-0.6(3)$ kHz; $M_\perp^{^{13}C}=15.0(10)$ kHz.
[3]) The polarity from the isotopic effect in the g-values is (+HCN−).

Nr.	Molecule	g	$\Delta\xi=\xi_{\parallel}-\xi_{\perp}$	$\xi_{\parallel}^{para}, \xi_{\perp}^{para}$	$\Theta_{\parallel}, \Theta_{\perp}$	$\langle\parallel^2\rangle, \langle\perp^2\rangle$	Meth.	Ref.	Varia, remarks	See Tab., Nr.
			10^{-6} erg/Gauss² mol		10^{-26} esu cm²	10^{-16} cm²				
5	ICN						MW	77B3	1)	2.3, 8
6	HCP DCP	−0.0430 (10) −0.0353 (10)	−8.4 (9)		4.4 (12)		MW	73H1		2.3, 10
7	N_2O ($N^{end}N^{cent}O$)	−0.07887 (8) $g_{\perp}^{v}+g_{\parallel}^{v}$ = −0.031 (8) *) $g_{\parallel}^{v}\approx g_{\perp}^{v}+g_{\parallel}^{v}-g_{\perp}$ = 0.048 (8) *)	−10.43 (6)		−3.296 (15)		BMS MBE	75C1 78R1	1) 2)	2.3, 18
	$^{15}N^{15}NO$	see Vol. II/6, 2.9.2, Nr. 7					3)	73B4		
8	HCNO	−0.0356 (4)	−9.2 (2)	110.8 (3)	3.0 (3)	$\langle\parallel^2\rangle-\langle\perp^2\rangle$ = 23.9 (1)	MW	78H3	1)	2.3, 17
	DCNO	−0.0328 (5)	−9.2 (2)							
9	HC≡CH						1) 2)	73H1 73H2		2.3, 23

to Nr.:

5 1) $M_{\perp}^{I}=-11.2(12)$ kHz;
$\sigma_{\perp}^{I}=2.6(6)\cdot 10^{-3}$ using semiempirical methods [74 F1, 72 G2].

7 1) $M_{\perp}^{end}=2.35(20)$ kHz; $M_{\perp}^{cent}=2.90(26)$ kHz.
2) $(\sigma_{\parallel}-\sigma_{\perp})^{end}=322(400)\cdot 10^{-6}$; $(\sigma_{\parallel}-\sigma_{\perp})^{cent}=458(300)\cdot 10^{-6}$
$M_{\perp}^{end}=1.83(7)$ kHz; $M_{\perp}^{cent}=3.06(12)$ kHz
$(\sigma_{\perp}^{para})^{end}=-561(36)\cdot 10^{-6}$; $(\sigma_{\perp}^{para})^{cent}=-707(47)\cdot 10^{-6}$
Using the accurate shielding anisotropies of $^{15}N^{15}NO$ below [73B4]:
$(\sigma_{\parallel}^{dia}-\sigma_{\perp}^{dia})^{end}=-194(18)\cdot 10^{-6}$; $(\sigma_{\parallel}^{dia}-\sigma_{\perp}^{dia})^{cent}=-199(48)\cdot 10^{-6}$
$(M_{\perp}^{v})^{end}=1.904(15)$ kHz *); $(M_{\parallel}^{v}-M_{\perp}^{v})^{end}=-1.78(7)$ kHz *)
$(M_{\perp}^{v})^{cent}=2.60(2)$ kHz *); $(M_{\parallel}^{v}-M_{\perp}^{v})^{end}=1.2(1)$ kHz *).
*) Measured in the $v=(01^10)$ bending mode, compare OCS data of this table.
3) Nuclear Magnetic Resonance in liquid-crystal solution
$(\sigma_{\parallel}-\sigma_{\perp})^{end}=367(12)\cdot 10^{-6}$; $(\sigma_{\parallel}-\sigma_{\perp})^{cent}=508(10)\cdot 10^{-6}$.

to Nr.:

8 1) The sign of $g_{\perp}$ is from the isotopic effect assuming the polarity (+HCNO−).

9 1) Semiempirical arguments show conclusively that the sign of $g_{\perp}$ must be positive, in contradiction to a previous MBM measurement [compare Vol. II, Tab. 2.9.2, Nr. 8].
2) The measurement of the Senftleben-Beenacker effect also reveals that $g_{\perp}$ is positive (see 1) above.)

Nr.	Molecule	g	$\Delta\xi = \xi_{\parallel} - \xi_{\perp}$	$\xi_{\parallel}^{para}, \xi_{\perp}^{para}$	$\Theta_{\parallel}, \Theta_{\perp}$	$\langle \parallel^2 \rangle, \langle \perp^2 \rangle$	Meth.	Ref.	Varia, remarks	See Tab., Nr.
			10^{-6} erg/Gauss² mol		10^{-26} esu cm²	10^{-16} cm²				
10	HC≡CF	−0.00767 (5) $g_{\perp}^{v} = -0.0087$ (6) ²) $g_{\parallel}^{v} = 0.4307$ (14) ²) $g_{\perp}^{v} = -0.0081$ (9) ³) $g_{\parallel}^{v} = 0.0472$ (20) ³)	−5.2 (1) $\xi_{\parallel}^{v} - \xi_{\perp}^{v}$ = −5.4 (9) ²) $\xi_{\parallel}^{v} - \xi_{\perp}^{v}$ = −5.0 (8) ³)				MW	78H4	¹)	2.3, 24
	DC≡CF	−0.00709 (8) $g_{\perp}^{v} = -0.0078$ (6) ²) $g_{\parallel}^{v} = 0.2193$ (13) ²) $g_{\perp}^{v} = -0.0073$ (6) ³) $g_{\parallel}^{v} = 0.0462$ (19) ³) $g_{\parallel}^{v} - g_{\perp}^{v} = 0.050$ (3) ⁴)	−5.2 (3) $\xi_{\parallel}^{v} - \xi_{\perp}^{v}$ = −5.2 (8) ²) $\xi_{\parallel}^{v} - \xi_{\perp}^{v}$ = −5.3 (8) ³)							
11	HC≡C⁷⁹Br HC≡C⁸¹Br	−0.00395 (32) −0.00388 (14)	−9.5 (9)		8.5 (11)		MW	73H1	¹)	2.3, 26
12	ArH⁷⁹Br						MW⁹)	80K1	¹)	2.3, 41
							MBE	80J1	²)	
	ArH⁸¹Br						MW⁹)	80K1	³)	
							MBE	80J1	⁴)	
	ArD⁷⁹Br						MW⁹)	80K1	⁵)	
							MBE	80J1	⁶)	
	ArD⁸¹Br						MW⁹)	80K1	⁷)	
							MBE	80J1	⁸)	

to Nr.:

10 ¹) Compare Vol. II.6, Tab. 2.9.2, Nr. 9. The sign of the vibrational-ground-state g-values is from the isotopic effect assuming the polarity (+HC≡CF−).
²) Measured in the $v=(0001^10)$ bending mode.
³) Measured in the $v=(00001^1)$ bending mode.
⁴) Measured in the $v=(00002^2_{1,2})$ bending states.

11 ¹) The sign of the g-values is determined relative to g_I^{Br}.

12 ¹) $M_{\perp}^{^{79}Br} = 1.09(13)$ *).
²) $M_{\perp}^{^{79}Br} = 0.63$ *).
³) $M_{\perp}^{^{81}Br} = 0.76(12)$ *).
⁴) $M_{\perp}^{^{81}Br} = 0.66$ *).
⁵) $M_{\perp}^{^{79}Br} = 0.94(20)$ *).
⁶) $M_{\perp}^{^{79}Br} = 0.88$ *); $D_{\perp}^{D^{79}Br} = 0.82(12)$ *).
⁷) $M_{\perp}^{^{81}Br} = 1.02(16)$ *).
⁸) $M_{\perp}^{^{81}Br} = 0.94$ *); $D_{\perp}^{D^{81}Br} = 0.91(13)$ *).
*) All values in kHz. Spin-spin interactions are ignored in [80K1]. In [80J1], they have been partly determined from the spectral fit while the Br spin-rotation parameters were indirectly concluded and held fixed.
⁹) Fabry-Perot Fourier-transform technique.

Nr.	Molecule	g	$\Delta\xi=\xi_{\parallel}-\xi_{\perp}$ 10⁻⁶ erg/Gauss² mol	$\xi_{\parallel}^{para}, \xi_{\perp}^{para}$	$\Theta_{\parallel}, \Theta_{\perp}$ 10⁻²⁶ esu cm²	$\langle\parallel^2\rangle, \langle\perp^2\rangle$ 10⁻¹⁶ cm²	Meth.	Ref.	Varia, remarks	See Tab., Nr.
13	$^{84}KrH^{35}Cl$ $^{84}KrH^{37}Cl$ $^{86}KrH^{35}Cl$						MBE	80B1	[1] [2] [3]	2.3, 44
14	$OCH^{79}Br$ $OCH^{81}Br$ $OCD^{79}Br$ $O^{13}CH^{79}Br$						MW [5]	80K2	[1] [2] [3] [4]	2.3, 37
b) symmetric top molecules										
15	$^{14}NH_3$	see Vol. II/6, 2.9.2, Nr. 10	1.3 (3) [1]		$\Theta_{\parallel}=-3.3\,(4)$ [1]	see Vol. II/6, 2.9.2, Nr. 10	BMS	71K1		2.4, 1
		$g^+(3_2)-g^-(3_2)=1.38\,(8)\cdot10^{-4}$ [2]					BMS	73K1		
							BMS Theory Theory Theory	74R3 75K1 76G1 75K1	[3] [4] [5] [6]	
Cont.	$^{15}NH_3$	see Vol. II/6, 2.9.2, Nr. 10								

to Nr.:

13 [1]) $M_{\perp}^{35Cl}=0.132(24)$ kHz; $D_{\perp}^{35ClH}=2.43(29)$ kHz.
[2]) $M_{\perp}^{37Cl}=0.124(59)$ kHz; $D_{\perp}^{37ClH}=1.95(11)$ kHz.
[3]) $M_{\perp}^{35Cl}=0.13$ kHz *); $D_{\perp}^{35ClH}=2.40(11)$ kHz.
*) Held fixed in the spectral fit.

14 [1]) $M_{\perp}^{79Br}=1.19(15)$ kHz.
[2]) $M_{\perp}^{81Br}=1.39(16)$ kHz.
[3]) $M_{\perp}^{79Br}=2.60(157)$ kHz.
[4]) $M_{\perp}^{79Br}=1.42(51)$ kHz.
[5]) Fabry-Perot Fourier transform technique.

15 [1]) Erratum, replace entries [70K1] in Vol. II/6, 2.9.2, Nr. 10.
[2]) Difference of g-values in the upper (+) and lower (−) inversion states $J=3$, $K=2$.
[3]) $C^N(5_3)=-6.746(3)$ kHz; $C^H(5_3)=18.164(8)$ kHz.
[4]) The following spin-rotation tensor elements are obtained by least-squares fitting the above [74R3], and earlier [65K1, 67K1, 68K1, 70K2]

to Nr.:

15 BMS data:
$M_{\parallel}^{N}=6.695(5)$ kHz; $M_{\perp}^{N}=6.764(5)$ kHz; $M_{\parallel}^{H}=-19.05(2)$ kHz;
$\frac{1}{2}(M_{xx}^{H}+M_{yy}^{H})=-17.73(2)$ kHz; $M_{xx}^{H}-M_{yy}^{H}=23.98(3)$ kHz;
z along symmetry axis, x along projection of NH bond on hydrogen plane. With the aid of Hartree-Fock diamagnetic-shielding calculations [72L1] follow: $\sigma^{H}=31.2(1)\cdot10^{-6}$; $\sigma^{14N}=264.5(6)\cdot10^{-6}$.
[5]) Λ_{gg}^{k} as defined by $M_{gg}^{k}=g_I^{k}G_g\Lambda_{gg}^{k}$ is independent of isotopic substitution, within the rigid-rotor model. Fitting to $^{14}NH_3$, $^{15}NH_3$, and $^{14}ND_3$ hydrogen spin-rotation data yields (definition of axes see above):
$\Lambda_{xx}^{H}=-19.84(35)\cdot10^{-10}$; $\Lambda_{yy}^{H}=-193.93(35)\cdot10^{-10}$;
$\Lambda_{zz}^{H}=-181.36(29)\cdot10^{-10}$.
[6]) The following spin-rotation tensor elements are obtained by least-squares fitting earlier [65K1, 67K1, 68K1, 70K2] BMS data:
$M_{\parallel}^{N}=-9.37(2)$ kHz; $M_{\perp}^{N}=-9.60(2)$ kHz;
$M_{\parallel}^{H}=-18.97(3)$ kHz; $\frac{1}{2}(M_{xx}^{H}+M_{yy}^{H})=-17.80(3)$ kHz;
for definition of axes see $^{14}NH_3$ above.

Nr.	Molecule	g	$\Delta\xi=\xi_{\parallel}-\xi_{\perp}$ (10^{-6} erg/Gauss^2 mol)	$\xi_{\parallel}^{para}$, $\xi_{\perp}^{para}$ (10^{-6} erg/Gauss^2 mol)	$\Theta_{\parallel}$, $\Theta_{\perp}$ (10^{-26} esu cm^2)	$\langle\parallel^2\rangle$, $\langle\perp^2\rangle$ (10^{-16} cm^2)	Meth.	Ref.	Varia, remarks	See Tab., Nr.
15 Cont.	ND_2H ND_3						BMS BMS BMS MBE	76G1 76G1 77G1 79C1	[7] [8] [9] [10]	
16	PH_3						MBE	71D1	[1]	2.4, 3
17	PF_3	$g_{\parallel}=-0.0470(40)$ [1] (compare Vol. II/6, 2.9.2, Nr. 13) $g_{\perp}^{v}=-0.0659(53)$ [2] $g_{zz}^{v}=-0.0439(31)$ [2] $g_{\alpha}^{v}=-0.0735(99)$ [2]	see Vol. II/6, 2.9.2, Nr. 13 $\xi_{\parallel}^{v}-\xi_{\perp}^{v}$ $=-1.03(47)$ [2]				MW	77N1		2.4, 4
18	AsH_3						MW [2]	78H1	[1]	2.4, 6
19	CH_4	$g=0.31338(40)$ (compare Vol. II/6, 2.9.2, Nr. 17)		$\xi^{para}=9.295(12)$			MBM [2]	76O1	[1]	2.4, 11
	$^{13}CH_4$						MBM	72O1	[3]	

to Nr.:

15 7) $C^N(1_{11})=4.5(5)$ kHz; $C^N(1_{01})=3.4(5)$ kHz; $C^H(1_{11})=-16.9(5)$ kHz; $C^H(1_{01})=-12.5(20)$ kHz; $C^D(1_{11})=-1.5$ kHz, and $C^D(1_{01})=-1.9$ kHz calculated from hydrogen spin-rotation results.

8) $C^D(6_6)=-1.437(1)$ kHz; $D^{DD}(6_6)=0.72(8)$ kHz; $D^{ND}(6_6)=0.79(15)$ kHz; obtained by analyzing the data in [62Z1].

9) $M_{\perp}^{N}=2.0(15)$ kHz; $\frac{1}{2}(M_{xx}^{D}+M_{yy}^{D})=-2.2(8)$ kHz.

10) $\frac{1}{2}(M_{xx}^{D}-M_{yy}^{D})=1.16(12)$ kHz; definition of axes see $^{14}NH_3$ above.

16 1) $M_{\perp}^{P}=114.90(13)$ kHz; $M_{\parallel}^{P}=116.38(32)$ kHz; $\frac{1}{2}(M_{bb}^{H}+M_{cc}^{H})=-8.01(8)$ kHz; $M_{\parallel}=-7.69(19)$ kHz

17 1) Reanalysis of a previous investigation [69S2].

2) $v=1$, measured in the ν_4 mode (degenerate PF_3 bending vibration, 344 cm^{-1}); for the designations, see Preliminary remarks, eq. (10), p. 759 and following.

to Nr.:

18 1) $M_{\perp}^{As}=111(16)$ kHz; $M_{\parallel}^{As}=113(17)$ kHz

2) Allowed higher-order mm-wave transitions; $M_{\parallel}^{As}$ disagrees with an earlier result, see Vol. II/6, 2.9.2., Nr. 14.

19 1) The positive sign of g was experimentally confirmed by the rotating-field method [56R1]. $\xi=-17.4(8)\cdot10^{-6}$ erg/G^2 mol [60B1] yields $\xi^{dia}=-26.7(8)\cdot10^{-6}$ erg/G^2 mol.

2) Rotational-moment spectrum; $\frac{1}{3}(M_z^H+2M_x^H)=-10.5(5)$ kHz agrees in sign and magnitude with an earlier nuclear-moment spectrum (Vol. II/6, 2.9.2, Nr. 17, z along CH-bond), $\sigma^{para}=-355(4)\cdot10^{-6}$.

3) $M^{^{13}C}=-15.94(35)$ kHz.

Nr.	Molecule	g	$\Delta\xi = \xi_\parallel - \xi_\perp$	$\xi_\parallel^{para}, \xi_\perp^{para}$	$\Theta_\parallel, \Theta_\perp$	$\langle\parallel^2\rangle, \langle\perp^2\rangle$	Meth.	Ref.	Varia, remarks	See Tab., Nr.
			10^{-6} erg/Gauss^2 mol		10^{-26} esu cm^2	10^{-16} cm^2				
20	CH_3F	$g_\parallel = 0.245\,(2)$ [1]) (compare Vol. II/6, 2.9.2, Nr. 18)	see Vol. II/6, 2.9.2, Nr. 18				MW	75E1		2.4, 12
21	CH_3Cl	$g_\perp = -0.016672\,(91)$ (compare Vol. II/6, 2.9.2, Nr. 19)	$-8.371\,(17)$				BMS	78E2	[1])	2.4, 13
							BMS, MBMW	78E2, 73K3, 77D3	[2])	
							MBMW	77D3	[3])	
22	$CH_3{}^{79}Br$	see Vol. II/6, 2.9.2, Nr. 20					RF [2])	80A1	[1])	2.4, 14
							MBMW	77D1, 76D1	[3])	
	$CH_3{}^{81}Br$	$g_\perp = -0.0057\,(3)$	$-8.8\,(5)$				MW	79D2	[4])	
							RF [6])	80A1	[5])	
							MBMW	77D1, 76D1	[7])	

to Nr.:

20 [1]) Reanalysis of previous data [56C1, 74N2], uncertainty estimated.

21 [1]) $(1-\sigma)\,g_I(^{35}Cl) = 0.547409(55)$; $(\sigma_\parallel - \sigma_\perp)^{35Cl} = 1441(130)\cdot 10^{-6}$
$\sigma_\parallel^{35Cl} = 1796(117)\cdot 10^{-6}$; $\sigma_\perp^{35Cl} = 356(90)\cdot 10^{-6}$
$(\sigma_\parallel - \sigma_\perp)^H < 2.2\cdot 10^{-6}$; $\frac{1}{2}(M_{aa}^H + M_{bb}^H) = -0.19\,(16)$ kHz.

[2]) $M_\perp^{35Cl} = 1.83(20)$ kHz.

[3]) $M_\parallel^{35Cl} = 7.0(16)$ kHz.

22 [1]) $M_\perp^{79Br} = 12.49(11)$ kHz; $M_\parallel^{79Br} = 13.6(13)$ kHz.

[2]) Radiofrequency Spectroscopy inside a Laser cavity.

to Nr.: 0

22 [3]) $M_\perp^{79Br} = 12.63(10)$ kHz; $M_\parallel^{79Br} = 18.8(17)$ kHz
$(\sigma_\perp^{para})^{79Br} = -931(23)\cdot 10^{-6}$; $(\sigma_\perp^{dia})^{79Br} = 3177\cdot 10^{-6}$
$(\sigma_\parallel^{para})^{79Br} = -83(15)\cdot 10^{-6}$; $(\sigma_\parallel^{dia})^{79Br} = 3130\cdot 10^{-6}$
$(\sigma^{para})^{79Br} = -648(21)\cdot 10^{-6}$; $(\sigma^{dia})^{79Br} = 3161\cdot 10^{-6}$
$\sigma_\perp^{79Br} = 2246(100)\cdot 10^{-6}$; $\sigma_\parallel^{79Br} = 3047(100)\cdot 10^{-6}$; $\sigma^{79Br} = 2513(100)\cdot 10^{-6}$;
the diamagnetic parts ar obtained by the atom-dipole method [72G2] and are believed accurate better 3%.

[4]) $(1-\sigma)\,g_I^{81Br} = 1.5093(2)$; $\sigma^{81Br} = 2467(572)\cdot 10^{-6}$;
$(\sigma_\parallel - \sigma_\perp)^{81Br} = 1110(956)\cdot 10^{-6}$.

[5]) $M_\perp^{81Br} = 13.46(17)$ kHz; $M_\parallel^{81Br} = 19.9(13)$ kHz.

[6]) Radiofrequency Spectroscopy inside a Laser cavity; MBMW frequencies [77D1] were included in the fit.

[7]) $M_\perp^{81Br} = 13.78(19)$ kHz; $M_\parallel^{81Br} = 25.4(20)$ kHz.

Nr.	Molecule	g	$\Delta\xi=\xi_\parallel-\xi_\perp$ 10^{-6} erg/Gauss^2 mol	$\xi_\parallel^{para}, \xi_\perp^{para}$	$\Theta_\parallel, \Theta_\perp$ 10^{-26} esu cm^2	$\langle\parallel^2\rangle, \langle\perp^2\rangle$ 10^{-16} cm^2	Meth.	Ref.	Varia, remarks	See Tab., Nr.
23	CH_3I	see Vol. II/6, 2.9.2, Nr. 21					RF ²)	78A1	¹)	2.4, 15
							MBMW ⁴)	78B1, 76B1	³)	
24	CHF_3	$g_\perp=-0.036266\,(93)$ $g_\parallel=-0.033330\,(93)$	1.727 (35)		$\Theta_\parallel=3.867\,(24)$		BMS	78E2	¹)	2.4, 18
								78E2, 74R1	²)	
	CDF_3							72K1 74R1	³) ⁴)	
25	CF_4	$g=-0.03124\,(5)$		$\xi^{para}=188\,(3)$			MBM	76O1	¹)	2.4, 19
26	SiH_4	$g=-0.27036\,(50)$		$\xi^{para}=31.54\,(6)$			MBM	76O1	¹)	2.4, 22
27	$SiH_3{}^{81}Br$	$g_\perp=-0.02185\,(13)$ $g_\parallel=-0.3185\,(5)$	−2.7 (3)	$\xi_\perp^{para}=284.7\,(5)$ $\xi_\parallel^{para}=33.93\,(5)$	$\Theta_\parallel=-0.1\,(4)$		MW	77D2	¹)	2.4, 24
	$SiD_3{}^{81}Br$	$g_\perp=-0.02020\,(20)$ $g_\parallel=-0.1615\,(15)$	−2.5 (9)						²)	

to Nr.:

23 ¹) $M_\perp^{127I}=17.6(10)$ kHz; $M_\parallel^{127I}=17.0(46)$ kHz; obtained in a simultaneous fit with RFMWDR [60O1] data
$M_\perp^{127I}=17.3(21)$ kHz *); $M_\parallel^{127I}=17.6(10)$ kHz *)
*) In the $v_6=1$ vibrational state.

²) Radiofrequency Spectroscopy inside a Laser cavity, errors are three times the standard deviations.

³) $M_\perp^{127I}=17.405(58)$ kHz; $M_\parallel^{127I}=19.41(54)$ kHz; obtained in a simultaneous fit with the RFIRDR data above [78A1], errors are standard deviations. $(\sigma_\perp^{para})^{127I}=-1950(13)\cdot10^{-6}$; $(\sigma_\parallel^{para})^{127I}=-105(6)\cdot10^{-6}$.

⁴) Including several Fabry-Perot Lamb dip measurements.

24 ¹) $(2\sigma_{cc}-\sigma_{aa}-\sigma_{bb})^F=-156.0(78)\cdot10^{-6}$; $(\sigma_\parallel-\sigma_\perp)^H=66(46)\cdot10^{-6}$.
²) $\frac{1}{2}(M_{aa}^F+M_{bb}^F)=10.37(13)$ kHz.
³) $M_\perp^H=-3.0(6)$ kHz.
⁴) $\frac{1}{2}(M_{aa}^F+M_{bb}^F)=9.67(30)$ kHz; $M_\perp^D=-0.16(40)$ kHz.

to Nr.:

25 ¹) $\frac{1}{3}(M_z^F+2M_x^F)=6.78(6)$ kHz, the positive sign is from theoretical arguments involving nuclear shielding [70T1]; the negative sign of the g-value is determined experimentally relative to the average spin-rotation parameter.

26 ¹) The negative sign of g was determined experimentally by the rotating-field method [56R1].
$\xi=-20.4(8)\cdot10^{-6}$ erg/G^2 mol yields $\xi^{dia}=-51.9(8)\cdot10^{-6}$ erg/G^2 mol.
The sign of $\frac{1}{3}(M_z^H+2M_x^H)=-3.6(6)$ kHz, z along SiH bond, is determined relative to g above.

27 ¹) The absolute sign of the g-values is measured relative to g_I^{81Br}.
$(1-\sigma)\,g_I^{81Br}=1.5077(3)$; $\sigma^{81Br}=3569(616)\cdot10^{-6}$; $(\sigma_\parallel-\sigma_\perp)^{81Br}<|200|\cdot10^{-6}$
$M_\perp^{81Br}=2.32(40)$ kHz; $M_\parallel^{81Br}=34.2(11)$ kHz
$(\sigma_\perp^{para})^{81Br}=-430(55)\cdot10^{-6}$; $(\sigma_\parallel^{para})^{81Br}=-277(8)\cdot10^{-6}$.
²) The polarity as obtained from the isotopic effect in the g-values is $(+H_3SiBr-)$.

Nr.	Molecule	g	$\Delta\xi=\xi_{\parallel}-\xi_{\perp}$ (10^{-6} erg/Gauss^2 mol)	$\xi_{\parallel}^{para}, \xi_{\perp}^{para}$ (10^{-6} erg/Gauss^2 mol)	$\Theta_{\parallel}, \Theta_{\perp}$ (10^{-26} esu cm^2)	$\langle\parallel^2\rangle, \langle\perp^2\rangle$ (10^{-16} cm^2)	Meth.	Ref.	Varia, remarks	See Tab., Nr.
28	SiF_4	$g=-0.03193\,(4)$		$\xi^{para}=258\,(7)$			MBM	76O1	1)	2.4, 25
29	GeH_4	$g=-0.10815\,(14)$		$\xi^{para}=29.16\,(4)$			MBM	76O1	1)	2.4, 28
30	GeF_4	$g=-0.02658\,(5)$		$\xi^{para}=300\,(11)$			MBM	76O1	1)	2.4, 30
31	POF_3	$g_{\perp}=-0.04460\,(4)$ $g_{\parallel}=-0.03370\,(30)$ the signs of the g-values were obtained in an avoided-crossing experiment	$-1.7\,(3)$		$\Theta_{\parallel}=-4.59\,(33)$		MBE MBM	79M1, 78O1 70F1	1) 2)	2.4, 31
32	CH_3CN CD_3CN	see Vol. II/6, 2.9.2, Nr. 30					MBMW BMS	77B1 73K2	1) 2) 3)	2.4, 16

to Nr.:

28 1) The sign of $\frac{1}{3}(M_z^F+2M_x^F)=2.45(8)$ kHz, z along SiH bond, is from theoretical arguments involving nuclear shielding [70T1]; the sign of g is determined experimentally relative to the former.

29 1) The negative sign of g was determined experimentally by the rotating-field method [56R1].
$\xi=-27.7(8)\cdot10^{-6}$ erg/G^2 mol yields $\xi^{dia}=58.9(8)\cdot10^{-6}$ erg/G^2 mol.
The sign of $\frac{1}{3}(M_z^H+2M_x^H)=-4.0(3)$ kHz, with z along GeH bond, is determined relative to g above.

30 1) The sign of $\frac{1}{3}(M_z^F+2M_x^F)=1.85(5)$ kHz, with z along GeF bond, is from theoretical arguments involving nuclear shielding [70T1]; the sign of g is determined experimentally relative to it.

to Nr.:

31 1) $M_{\perp}^P=7.7(8)$ kHz *); $M_{\parallel}^P=21(6)$ kHz **)
$\frac{1}{2}(M_{aa}^F+M_{bb}^F)=4.6(4)$ kHz *); $M_{\parallel}^F=4.0(25)$ kHz **)
$\mu_n^2(g_I^F)^2/r_{FF}^3=8.2(18)$ kHz; $\mu_n^2 g_I^F g_I^P(1-\frac{3}{2}\sin^2(\sphericalangle OPF))/r_{PF}^3=-2.3(9)$ kHz.
*) Measured with relative signs; absolute signs from indirect reasoning.
**) Obtained via the average spin-rotation parameters below [78O1].
2) $|\frac{1}{3}(2M_{\perp}^P+M_{\parallel}^P)|=12(2)$ kHz; $|\frac{1}{3}(M_{aa}^F+M_{bb}^F+M_{\parallel}^F)|=4.4(8)$ kHz.

32 1) $M_{\perp}^N=1.85(19)$ kHz; $M_{\parallel}^N=0.7$ kHz (held fixed)
$\sigma_{\parallel}^N=341(10)\cdot10^{-6}$; $\sigma_{\perp}^N=-163(10)\cdot10^{-6}$.
2) $M_{\perp}^H=-0.7(5)$ kHz.
3) $M_{\perp}^N=1.2(5)$ kHz.

Nr.	Molecule	g	$\Delta\xi = \xi_\parallel - \xi_\perp$ (10⁻⁶ erg/Gauss² mol)	$\xi_\parallel^{para}, \xi_\perp^{para}$ (10⁻⁶ erg/Gauss² mol)	$\Theta_\parallel, \Theta_\perp$ (10⁻²⁶ esu cm²)	$\langle \parallel^2 \rangle, \langle \perp^2 \rangle$ (10⁻¹⁶ cm²)	Meth.	Ref.	Varia, remarks	See Tab., Nr.
33	$CH_3C{\equiv}C{-}H$	$g_\parallel = 0.2907(30)$ [1]) (compare Vol. II/6, 2.9.2, Nr. 33) $g_\perp^v = 0.00314(29)$ [2]) $g_{zz}^v = 0.2914(79)$ [2]) $g_\alpha^v = 0.0628(18)$ [2]) $g_\alpha^v = 0.433(30)$ [3])	see Vol. II/6, 2.9.2, Nr. 33 $\xi_\parallel^v - \xi_\perp^v = -7.46(38)$ [2])				MW	77N1		2.4, 47
34	$^{28}SiH_3{}^{14}NC^{32}S$	$g_\perp = -0.01521(10)$ $g_\parallel = -0.315(5)$	$-28.0(30)$	$\xi_\perp^{para} = 796.1(8)$ $\xi_\parallel^{para} = 33.61(16)$	$\Theta_\parallel = 16.6(37)$		MW	79D1		2.4, 45
	$^{28}SiH_3{}^{14}NC^{34}S$	$g_\perp = -0.01453(15)$							[1])	
	$^{28}SiD_3{}^{14}NC^{32}S$	$g_\perp = -0.01445(15)$								
	$^{30}SiD_3{}^{14}NC^{32}S$	$g_\perp = -0.01419(25)$								
35	$(CH_3)_3CH$	$g_\perp = 0.0131(8)$ $g_\parallel = 0.0264(15)$	$2.72(29)$	$\xi_\perp^{para} = 38.7(4)$ $\xi_\parallel^{para} = 61.2(8)$	$\Theta_\parallel = -13.2(9)$	$\langle \perp^2 \rangle = 62.2(6)$ $\langle \parallel^2 \rangle = 39.8(12)$	MW	82G1	[1])	2.4, 54
36	SF_6	$g = -0.03732(5)$		$\xi^{para} = 403(6)$			MBM	77O1	[1])	2.4, 83
37	$(CH_2O)_3$	$g_\perp = -0.0336(4)$ $g_\parallel = g_\perp \pm 0.0030$	$16.6(30)$		$\Theta_\parallel = -6.5(28)$		MW	80H2		2.4, 9?

to Nr.:

33 [1]) Reanalysis of previous investigations [56C1, 69S2].

[2]) $v=1$, measured in the ν_{10} mode ($C-C{\equiv}C$ bending vibration, 327 cm⁻¹); $g_\perp^v = \frac{1}{2}(g_{xx}^v + g_{yy}^v)$; g_α^v = vibrationally induced g value; $\xi_\parallel^v - \xi_\perp^v = \frac{1}{2}(2\,\xi_{zz}^v - \xi_{xx}^v - \xi_{vy}^v)$; see Preliminary remarks, e.q. (10), p. 759 and following.

[3]) $v=1$, measured in the ν_9 mode ($C{\equiv}C-H$ bending vibration, 642 cm⁻¹); all other parameters held fixed on the ν_{10} values above in the data fit.

34 [1]) The polarity as obtained from the isotopic effect in the g-values is $(+H_3SiNCS-)$.

to Nr.:

35 [1]) The gaseous-state bulk value $\xi = -50.5(8) \cdot 10^{-6}$ erg/G² mol [60B1] yields (same units) $\xi_\parallel = -48.7(6)$, $\xi_\perp = -51.4(9)$, $\xi_\parallel^{dia} = -109.9(10)$, $\xi_\perp^{dia} = -90.1(10)$.

36 [1]) $\frac{1}{3}(M_z^F + 2M_x^F) = 5.27(40)$ kHz, z along S−F bond, the sign is determined from theoretical arguments [70T1] involving nuclear shielding; the sign of g is in turn determined relative to the spin-rotation parameter.

2.9.3 Asymmetric top molecules

Rotational *g*-values, magnetic susceptibilities and anisotropies, paramagnetic and diamagnetic contributions, molecular quadrupole moments, electronic charge distributions, spin-rotation and spin-spin coupling parameters, nuclear *g*-values from the rotational Zeeman effect and nuclear magnetic shielding parameters from the rotational Zeeman effect as well as indirectly via the paramagnetic contribution to the spin-rotation tensor components

2.9.3.1 Preliminary remarks

For 1···3, see 2.9.2.1, p. 755.

4. Arrangement of substances

Inorganic molecules	1··· 8
Methane, ethane derivatives	9···22
Aliphatic molecules with double carbon bonds	23···31
Saturated cyclic compounds	32···34
Unsaturated cyclic and aromatic compounds	35···54

2.9.3.2 Data

Nr.	Molecule	g_{aa}, g_{bb}, g_{cc}	$\Delta\xi_1, \Delta\xi_2$ 10^{-6} erg/G^2 mol	$\xi_{aa}^{para}, \xi_{bb}^{para}, \xi_{cc}^{para}$ 10^{-6} erg/G^2 mol	$\Theta_{aa}, \Theta_{bb}, \Theta_{cc}$ 10^{-26} esu cm^2	$\langle a^2\rangle, \langle b^2\rangle, \langle c^2\rangle$ 10^{-16} cm^2	Meth.	Ref.	Varia, remarks	See Tab., Nr.
1	NOF	$g_{aa}=-0.9542\,(14)$ $g_{bb}=-0.1100\,(13)$ $g_{cc}=-0.0640\,(18)$	$\Delta\xi_1=11.8\,(9)$ $\Delta\xi_2=3.2\,(9)$	$\xi_{aa}^{para}=32.6\,(3)$ $\xi_{bb}^{para}=107.8\,(17)$ $\xi_{cc}^{para}=112.2\,(18)$	$\Theta_{aa}=-1.1\,(9)$ $\Theta_{bb}=1.2\,(10)$ $\Theta_{cc}=-0.1\,(16)$	$\langle a^2\rangle=24.19\,(30)$ $\langle b^2\rangle=5.79\,(28)$ $\langle c^2\rangle=3.32\,(33)$	MW	80R1	[1])	2.5, 13
2	SO_2	$g_{aa}=-0.6043\,(3)$ $g_{bb}=-0.11634\,(12)$ $g_{cc}=-0.08865\,(10)$	$\Delta\xi_1=5.58\,(12)$ $\Delta\xi_2=3.60\,(9)$		$\Theta_{aa}=-4.91\,(10)$ $\Theta_{bb}=1.02\,(3)$ $\Theta_{cc}=-3.86\,(6)$		BMS, MBE MBE	76E1, 80M1 80M1	[1]) [2])	2.5, 28
3	O_3	$g_{aa}=-2.9877\,(9)$ $g_{bb}=-0.2295\,(3)$ $g_{cc}=-0.0760\,(3)$	$\Delta\xi_1=71.55\,(33)$ $\Delta\xi_2=1{,}23\,(24)$	$\xi_{aa}^{para}=69.80\,(2)$ $\xi_{bb}^{para}=117.07\,(3)$ $\xi_{cc}^{para}=104.30\,(5)$	$\Theta_{aa}=-1.37\,(18)$ $\Theta_{bb}=-0.80\,(13)$ $\Theta_{cc}=2.17\,(22)$	$\langle a^2\rangle-\langle b^2\rangle = 16.67\,(3)$ $\langle b^2\rangle-\langle c^2\rangle = 2.80\,(3)$ $\langle c^2\rangle-\langle a^2\rangle = -19.47\,(3)$	MBE	77M2, 77M1	[1])	2.5, 52
4	HOF	$g_{aa}=0.642\,(1)$ $g_{bb}=-0.119\,(1)$ $g_{cc}=-0.061\,(1)$	$\Delta\xi_1=-19.6\,(6)$ $\Delta\xi_2=12.8\,(12)$	$\xi_{aa}^{para}=1.1\,(1)$ $\xi_{bb}^{para}=50.3\,(1)$ $\xi_{cc}^{para}=49.5\,(1)$	$\Theta_{aa}=0.2\,(4)$ $\Theta_{bb}=1.9\,(8)$ $\Theta_{cc}=-2.1\,(11)$	$\langle a^2\rangle=11.6\,(4)$ [1]) $\langle b^2\rangle=2.6\,(4)$ [1]) $\langle c^2\rangle=2.3\,(3)$ [1])	MW	73R2		2.5, 41

to Nr.:

1 [1]) Signs of g-values determined to be negative relative to the known nitrogen nuclear magnetic moment. An estimated bulk value $\xi=-10(1)\cdot10^{-6}$ erg/G^2 mol was used to compute $\langle a^2\rangle$, $\langle b^2\rangle$, $\langle c^2\rangle$, and in 10^{-6} erg/G^2 mol:
$\xi_{aa}^{dia}=-38.7\,(10)$; $\xi_{bb}^{dia}=-116.7\,(20)$; $\xi_{cc}^{dia}=-127.2\,(21)$.

2 [1]) Essential agreement with the less accurate MW results, Vol. II/6, 2.9.3, Nr. 1.
[2]) The absolute sign of the g-values was confirmed using circularly polarized radiation.

to Nr.:

3 [1]) Earlier measurements of $\Delta\xi_1$ and $\Delta\xi_2$ (see Vol. II/6, 2.9.3, Nr. 5) seem to be in error.

4 [1]) Computed relative to the semi-empirically [72G1] obtained $\langle c^2\rangle$-value.

Nr.	Molecule	g_{aa}, g_{bb}, g_{cc}	$\Delta\xi_1, \Delta\xi_2$ 10^{-6} erg/G² mol	$\xi_{aa}^{para}, \xi_{bb}^{para}, \xi_{cc}^{para}$	$\Theta_{aa}, \Theta_{bb}, \Theta_{cc}$ 10^{-26} esu cm²	$\langle a^2\rangle, \langle b^2\rangle, \langle c^2\rangle$ 10^{-16} cm²	Meth.	Ref.	Varia, remarks	See Tab., Nr.
5	HOCl	see Vol. II/6; 2.5, 27 $g_{aa}=$ 0.6390 (55) $g_{bb}=-0.0752$ (9) $g_{cc}=-0.0616$ (10)	$\Delta\xi_1=-24.8$ (3) $\Delta\xi_2=$ 15.8 (7)	$\xi_{aa}^{para}=$ 1.2 (2) $\xi_{bb}^{para}=84.4$ (3) $\xi_{cc}^{para}=86.2$ (4)	$\Theta_{aa}=$ 0.50 (35) $\Theta_{bb}=$ 0.74 (67) $\Theta_{cc}=-1.24$ (102)	$\langle a^2\rangle=21.3$ (5) $\langle b^2\rangle=$ 4.8 (5) $\langle c^2\rangle=$ 3.9 (3)	MW	75S1	[1]	
6	HDO	see Vol. II/6, 2.9.3, Nr. 6					BMS [2]	77K1	[1]	2.5, 53
	$H_2{}^{17}O$						MW	75L1	[3]	
7	H_2O_2	$g_{aa}=-0.0532$ (5) $g_{bb}=-0.0828$ (8) $g_{cc}=$ 0.647 (6)	$\Delta\xi_1=$ 8.40 (36) $\Delta\xi_2=11.52$ (66)		$\Theta_{aa}=-1.24$ (30) $\Theta_{bb}=$ 1.55 (35) $\Theta_{cc}=-0.31$ (46)		BMS	78E1	[1]	2.5, 62
8	H–NCO	see Vol. II/6; 2.5, 3					BMS	71K3	[1]	
	D–NCO								[2]	
9	CH_3OH						BMS	73H3	[1]	2.5, 65

to Nr.:

5 [1]) $(1-\sigma)\,g_I^{Cl}=0.5490(14)$; the sign of the g-values is measured relative to the known sign of g_I^{Cl} [65S1].
$\langle a^2\rangle, \langle b^2\rangle$ are computed relative to $\langle c^2\rangle$ which is obtained with a semi-empirical additivity rule [72G1], and in turn the following quantities in 10^{-6} erg/G² mol:
$\xi=-27.2(37)$; $\xi_{aa}=-35.5(34)$; $\xi_{bb}=-21.9(35)$; $\xi_{cc}=-24.2(43)$
$\xi_{aa}^{dia}=-36.7(32)$; $\xi_{bb}^{dia}=-106.3(32)$; $\xi_{cc}^{dia}=-110.4(39)$.

6 [1]) $C^H(2_{20})=-43.1(2)$ kHz; $C^H(2_{21})=-43.5(2)$ kHz
$C^D(2_{20})=-2.31(9)$ kHz; $C^D(2_{21})=-1.96(9)$ kHz.
[2]) Two-cavity double-resonance technique with magnetic dipole transitions; agreement with differently obtained results, see Vol. II/6, 2.9.3, Nr. 6.
[3]) $C^{^{17}O}(3_{13})=-12(3)$ kHz; $C^{^{17}O}(2_{20})=-17(3)$ kHz.

to Nr.:

7 [1]) The tensors are defined in a coordinate system where internal and overall rotation are decoupled [65H2]. The g-values are obtained assuming 1% deviation in different torsional states; the singly measured coupling strengths not given here are more accurate.

8 [1]) $C^N(1_{01})=-0.50(20)$ kHz; $C^H(1_{01})$ negligible.
[2]) $C^N(1_{01})=-0.55(20)$ kHz; $C^D(1_{01})=-0.75(10)$ kHz.

9 [1]) Suitably defined spin-rotation and spin-spin coupling strengths within a rotor-internal rotor model [73H4] have been measured.

Nr.	Molecule	g_{aa}, g_{bb}, g_{cc}	$\Delta\xi_1, \Delta\xi_2$ 10^{-6} erg/G² mol	$\xi_{aa}^{para}, \xi_{bb}^{para}, \xi_{cc}^{para}$	$\Theta_{aa}, \Theta_{bb}, \Theta_{cc}$ 10^{-26} esu cm²	$\langle a^2\rangle, \langle b^2\rangle, \langle c^2\rangle$ 10^{-16} cm²	Meth.	Ref.	Varia, remarks	See Tab., Nr.
10	CH_3NO_2	$g_{aa} = -0.1108\,(10)$ $g_{bb} = -0.1206\,(3)$ $g_{cc} = -0.0420\,(3)$ $g_\alpha = 0.347\,(2)$	$\Delta\xi_1 = -6.0\,(7)$ $\Delta\xi_2 = 23.8\,(9)$				MW	73E1	¹)	2.5, 74
11	$H_3C-O-Cl$	$g_{aa} = -0.0593\,(8)$ $g_{bb} = -0.0249\,(9)$ $g_{cc} = -0.0379\,(6)$	$\Delta\xi_1 = -24.7\,(11)$ $\Delta\xi_2 = 1.8\,(24)$		$\Theta_{aa} = 4.6\,(10)$ $\Theta_{bb} = -0.9\,(19)$ $\Theta_{cc} = -3.5\,(29)$	$\langle a^2\rangle - \langle b^2\rangle = 37.7(4)$ $\langle b^2\rangle - \langle c^2\rangle = 3.7\,(4)$ $\langle c^2\rangle - \langle a^2\rangle = -41.4\,(8)$	MW	76S3	¹)	2.5, 75
	$D_3C-O-Cl$	$g_{aa} = -0.0478\,(10)$ $g_{bb} = -0.0215\,(10)$ $g_{cc} = -0.0330\,(9)$	$\Delta\xi_1 = -23.2\,(11)$ $\Delta\xi_2 = 0.5\,(20)$		$\Theta_{aa} = 4.0\,(10)$ $\Theta_{bb} = -0.8\,(17)$ $\Theta_{cc} = -3.2\,(27)$	$\langle a^2\rangle - \langle b^2\rangle = 37.6\,(3)$ $\langle b^2\rangle - \langle c^2\rangle = 3.7\,(4)$ $\langle c^2\rangle - \langle a^2\rangle = -41.3\,(7)$	MW	76S3	²)	
12	CH_2F_2	see Vol. II/6, 2.9.3, Nr. 10					BMS	73K4, 74N1	¹)	2.5, 93
13	CH_3CH_2Cl	$g_{aa} = 0.0372\,(25)$ $g_{bb} = -0.0042\,(5)$ $g_{cc} = -0.0113\,(7)$	$\Delta\xi_1 = -9.6\,(22)$ $\Delta\xi_2 = -1.7\,(15)$		$\Theta_{aa} = -1.1\,(12)$ $\Theta_{bb} = -0.4\,(10)$ $\Theta_{cc} = 1.5\,(16)$		MW	80H1		2.5, 152

to Nr.:

10 ¹) g_α is the g-value of the internal methyl rotor, the other g-values hold for the rigid molecule. The underlying effective Zeeman Hamiltonian was given in [69H1].
Using the liquid-state bulk value $\xi = -21.1(21)\cdot 10^{-6}$ erg/G² mol [76W1] yields (same units) $\xi_{aa} = -23.1(25)$; $\xi_{bb} = -13.2(25)$; $\xi_{cc} = -27.0(25)$.

11 ¹) $(1-\sigma)\,g_I^{35Cl} = 0.5478(5)$; $(2\sigma_{aa} - \sigma_{bb} - \sigma_{cc})^{35Cl} = 5.4(51)$; $(2\sigma_{bb} - \sigma_{cc} - \sigma_{aa})^{35Cl} = -4.5(49)$.
The signs of the molecular g-values were determined relative to g_I^{35Cl}.

11 ²) $(1-\sigma)\,g_I^{35Cl} = 0.5480(9)$; $(2\sigma_{aa} - \sigma_{bb} - \sigma_{cc})^{35Cl} = 8.2(64)$; $(2\sigma_{bb} - \sigma_{cc} - \sigma_{aa})^{35Cl} = -8.3(91)$.
The signs of the molecular g-values were determined relative to g_I^{35Cl}.

12 ¹) $M_{aa}^F = 20.4(10)$ kHz; $M_{bb}^F = 4.9(5)$ kHz; $M_{cc}^F = 13.1(5)$ kHz
$M_{aa}^H = -6.5(30)$ kHz; $M_{bb}^H = -1.0(14)$ kHz; $M_{cc}^H = -0.3(13)$ kHz
Fluorine nuclear shielding (ppm):
$\sigma_{aa} = 398(22)$; $\sigma_{aa}^{para} = -104(5)$; $\sigma_{aa}^{dia} = 502(27)$
$\sigma_{bb} = 388(22)$; $\sigma_{bb}^{para} = -186(20)$; $\sigma_{bb}^{dia} = 574(42)$
$\sigma_{cc} = 255(22)$; $\sigma_{cc}^{para} = -378(20)$; $\sigma_{cc}^{dia} = 603(42)$
$\sigma = 337(5)$; $\sigma^{para} = -223(15)$; $\sigma^{dia} = 560(37)$

Nr.	Molecule	g_{aa}, g_{bb}, g_{cc}	$\Delta\xi_1, \Delta\xi_2$	$\xi^{para}_{aa}, \xi^{para}_{bb}, \xi^{para}_{cc}$	$\Theta_{aa}, \Theta_{bb}, \Theta_{cc}$	$\langle a^2\rangle, \langle b^2\rangle, \langle c^2\rangle$	Meth.	Ref.	Varia, remarks	See Tab., Nr.
			10^{-6} erg/G^2 mol		10^{-26} esu cm^2	10^{-16} cm^2				
14	$HC(=O)H$	see Vol. II/6, 2.9.3, Nr. 11					MBE, BMS, MBRF	77F1, 75K1, 72T1, 74C1, 73C1	1)	2.5, 108
							MBE	77F1	2)	
							BMS	75K1	3)	
							MBRF	77C1	4)	
	$HC(=O)D$	see Vol. II/6, 2.9.3, Nr. 11					BMS	71T1	5)	
	$DC(=O)D$	see Vol. II/6, 2.9.3, Nr. 11					BMS, MBRF	73T3, 74C1	6)	
	$H^{13}C(=O)H$						BMS	71T1	7)	
							BMS, MBRF	72T1, 74C1	8)	
Cont.							MBE	77F1	9)	

to Nr.:

14 1) $M^H_{aa}+4M^H_{bb}+M^H_{cc}=0.375(7)$ kHz from the 2_{11} level; combining this with previous BMS results [71K2, 71T1] on $1_{10}\rightarrow 1_{11}$ yields: $M^H_{aa}=-3.71(16)$ kHz; $M^H_{bb}=2.09(4)$ kHz; $M^H_{cc}=-2.41(22)$ kHz.

2) $r_{HH}=1.894(4)$ Å from spin-spin interaction agrees with structure from rotational constants.

3) $(\sigma^{para}_{aa})^H=-76.7(2)\cdot 10^{-6}$; $(\sigma^{para}_{bb})^H=-54.0(14)\cdot 10^{-6}$; $(\sigma^{para}_{cc})^H=-148.8(16)\cdot 10^{-6}$.
Hartree-Fock diamagnetic-shielding calculations [69N1] yield (uncertainties estimated):
$\sigma^H_{aa}=\ 15.9(30)\cdot 10^{-6}$; $\sigma^H_{bb}=\ 40.7(30)\cdot 10^{-6}$;
$\sigma^H_{cc}=-1.7(30)\cdot 10^{-6}$; $\sigma^H=\ 18.3(30)\cdot 10^{-6}$.

4) $v_6=1$ vibrational state:
$M^H_{aa}=-5.90(5)$ kHz; $M^H_{bb}+M^H_{cc}=0.95(5)$ kHz;
$\mu_n^2(g^H_I)^2\langle r^{-3}_{HH}\rangle=16.5$ kHz.

to Nr.:

14 5) $C^D(1_{10})=0.11(4)$ kHz; $C^D(1_{11})=-0.12(4)$ kHz;
$C^H(1_{10})=-1.43(8)$ kHz; $C^H(1_{11})=-3.70(12)$ kHz;
$\mu_n^2\, g^H_I\, g^D_I\langle r^{-3}_{HD}\rangle=2.75(10)$ kHz.

6) $C^D(1_{10})=-0.02(2)$ kHz; $C^D(1_{11})=-0.33(2)$ kHz

7) $C^{^{13}C}(1_{10})=73.87(14)$ kHz; $C^{^{13}C}(1_{11})=67.79(14)$ kHz;
$C^H(1_{10})=-0.77(6)$ kHz, $C^H(1_{11})=-3.03(6)$ kHz
$\mu_n^2(g^H_I)^2\langle r^{-3}_{HH}\rangle=17.52(17)$ kHz; $\mu_n^2(g^H_I)(g^C_I)\langle r^{-3}_{HC}\rangle=21.65(18)$ kHz

8) $C^{^{13}C}(2_{11})=36.01(15)$ kHz; $C^{^{13}C}(2_{12})=29.71(15)$ kHz;
$C^H(2_{11})=0.30(9)$ kHz; $C^H(2_{12})=-1.94(9)$ kHz
$M^H_{aa}=-3.52(16)$ kHz; $M^H_{bb}=\ 1.97(11)$ kHz; $M^H_{cc}=-2.53(11)$ kHz
$M^{^{13}C}_{aa}=127.86(31)$ kHz; $M^{^{13}C}_{bb}=19.99(23)$ kHz; $M^{^{13}C}_{cc}=7.61\ (23)$ kHz

9) $|M^{^{13}C}_{bb}+M^{^{13}C}_{cc}|=27.490(68)$ kHz.

Nr.	Molecule	g_{aa}, g_{bb}, g_{cc}	$\Delta\xi_1, \Delta\xi_2$ 10^{-6} erg/G² mol	$\xi_{aa}^{para}, \xi_{bb}^{para}, \xi_{cc}^{para}$	$\Theta_{aa}, \Theta_{bb}, \Theta_{cc}$ 10^{-26} esu cm²	$\langle a^2\rangle, \langle b^2\rangle, \langle c^2\rangle$ 10^{-16} cm²	Meth.	Ref.	Varia, remarks	See Tab., Nr.
14 Cont.	$HC(=^{18}O)H$						BMS	71T1 72T1	[10] [11]	
	$H^{13}C(=^{17}O)H$						MW	80C1	[12]	
15	$H_{(2)}C(=O)OH_{(1)}$	see Vol. II/6, 2.9.3, Nr. 14					MBE	76C2	[1]	2.5, 111
							MBRF	75A1	[2]	
16	$HC(=O)OCH{=}CH_2$	$g_{aa}=-0.1759\,(10)$ $g_{bb}=-0.0342\,(6)$ $g_{cc}=-0.0059\,(6)$	$\Delta\xi_1=9.5\,(13)$ $\Delta\xi_2=20.8\,(11)$	$\xi_{aa}^{para}=85.6\,(2)$ $\xi_{bb}^{para}=397.8\,(8)$ $\xi_{cc}^{para}=446.6\,(9)$	$\Theta_{aa}=-0.1\,(14)$ $\Theta_{bb}=1.2\,(14)$ $\Theta_{cc}=-1.1\,(22)$	$\langle a^2\rangle=94.2\,(8)$ $\langle b^2\rangle=21.6\,(8)$ $\langle c^2\rangle=6.0\,(8)$	MW	78V1	[1]	2.5, 117
17	$HC(=O)Cl$	$g_{aa}=-0.5111\,(48)$ $g_{bb}=-0.0762\,(51)$ $g_{cc}=-0.0288\,(36)$	$\Delta\xi_1=4.1\,(13)$ $\Delta\xi_2=4.1$ [2]	$\xi_{aa}^{para}=32.8\,(2)$ $\xi_{bb}^{para}=199.8\,(22)$ $\xi_{cc}^{para}=202.9\,(18)$	$\Theta_{aa}=-7.6\,(27)$ $\Theta_{bb}=13.6\,(57)$ *) $\Theta_{cc}=-6.0\,(63)$ *)	$\langle a^2\rangle=47.1\,(20)$ *) $\langle b^2\rangle=7.6\,(20)$ *) $\langle c^2\rangle=6.1\,(20)$ *)	MW	78D1	[1]	2.5, 109
18	$F{-}C(=O){-}F$	see Vol. II/6, 2.9.3, Nr. 16					BMS	73W1	[1]	2.5, 125

to Nr.:

14 [10]) $C^H(1_{10})=-0.98(10)$ kHz; $C^H(1_{11})=-2.98(10)$ kHz; $\mu_n^2(g_I^H)^2\langle r_{HH}^{-3}\rangle=17.90(25)$ kHz.

[11]) $C^H(2_{11})=0.03(15)$ kHz; $C^H(2_{12})=-2.23(15)$ kHz $M_{aa}^H=-3.30(27)$ kHz; $M_{bb}^H=1.47(20)$ kHz; $M_{cc}^H=-2.79(20)$ kHz.

[12]) $M_{aa}^{^{17}O}=-358(24)$ kHz; $M_{bb}^{^{17}O}=-28(4)$ kHz.

15 [1]) $M_{aa}^{H(1)}=-7.50(20)$ kHz; $M_{bb}^{H(1)}+M_{cc}^{H(1)}=0.3(4)$ kHz; $M_{bb}^{H(1)}=-7.2(40)$ kHz $M_{aa}^{H(2)}=-6.55(20)$ kHz; $M_{bb}^{H(2)}+M_{cc}^{H(2)}=-0.4(4)$ kHz; $M_{bb}^{H(2)}=8.2(40)$ kHz.

[2]) Experimental verification of the negative sign of the g-values, compare Vol. II/6, 2.9.3, Nr. 14.

to Nr.:

16 [1]) The liquid-state bulk value $\xi=-34.7$ [77B2] was used to compute $\langle a^2\rangle$, $\langle b^2\rangle$, $\langle c^2\rangle$, and also the following quantities in 10^{-6} erg/G² mol: $\xi_{aa}^{dia}=-117.1(17)$; $\xi_{ba}^{dia}=-425.6(22)$; $\xi_{cc}^{dia}=-491.4(27)$ $\xi_{aa}=-31.5(14)$; $\xi_{bb}=-27.8(14)$; $\xi_{cc}=-44.8(18)$

17 [1]) Signs of g-values determined to be negative relative to the known chlorine nuclear g-value. An estimated [65H1] bulk value $\xi=-27.3(30)$ was used to compute $\langle a^2\rangle$, $\langle b^2\rangle$, $\langle c^2\rangle$ *), and in 10^{-6} erg/G² mol: $\xi_{aa}^{dia}=-58.6(37)$ *); $\xi_{bb}^{dia}=-225.6(69)$ *); $\xi_{cc}^{dia}=-232.8(69)$ *);

*) Computed assuming an error of $\pm 5\cdot 10^{-6}$ erg/G² mol for $\Delta\xi_2$.

[2]) Since data were insensitive against $\Delta\xi_2$ the fit was constrained to the condition $\Delta\xi_1=\Delta\xi_2$.

18 [1]) $M_{aa}^F=19.77(21)$ kHz; $M_{bb}^F=13.46(14)$ kHz; $M_{cc}^F=7.80(26)$ kHz.

Nr.	Molecule	g_{aa}, g_{bb}, g_{cc}	$\Delta\xi_1, \Delta\xi_2$ 10^{-6} erg/G^2 mol	$\xi_{aa}^{para}, \xi_{bb}^{para}, \xi_{cc}^{para}$	$\Theta_{aa}, \Theta_{bb}, \Theta_{cc}$ 10^{-26} esu cm^2	$\langle a^2\rangle, \langle b^2\rangle, \langle c^2\rangle$ 10^{-16} cm^2	Meth.	Ref.	Varia, remarks	See Tab., Nr.
19	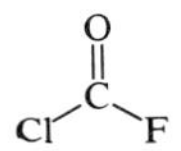	$g_{aa}=-0.056\,(11)$ $g_{bb}=-0.021\,(12)$ $g_{cc}=-0.044\,(10)$	$\Delta\xi_1=-1.6\,(13)$ $\Delta\xi_2=3.0\,(6)$	$\xi_{aa}^{para}=99\,(2)$ $\xi_{bb}^{para}=205\,(4)$ $\xi_{cc}^{para}=311\,(5)$	$\Theta_{aa}=-7\,(3)$ $\Theta_{bb}=-12\,(7)$ $\Theta_{cc}=19\,(7)$	$\langle a^2\rangle=55.9\,(23)$ $\langle b^2\rangle=31.1\,(17)$ $\langle c^2\rangle=5.8\,(6)$ [2]	MW	81S1, 76S1	[1]	2.5, 128
20	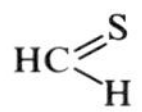	see Vol. II/6, 2.9.3, Nr. 16a					MBE	77F1	[1]	2.5, 145
21	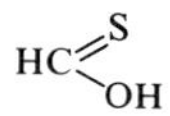 cis	$g_{aa}=-0.4017\,(12)$ $g_{bb}=-0.0711\,(4)$ $g_{cc}=-0.0318\,(4)$	$\Delta\xi_1=-7.3\,(12)$ $\Delta\xi_2=16.5\,(8)$	$\xi_{aa}^{para}=38.3\,(7)$ $\xi_{bb}^{para}=205.9\,(7)$ $\xi_{cc}^{para}=218.1\,(7)$	$\Theta_{aa}=-1.3\,(8)$ $\Theta_{bb}=4.1\,(8)$ $\Theta_{cc}=-2.8\,(12)$	$\langle a^2\rangle=48.0\,(23)$ $\langle b^2\rangle=10.4\,(23)$ $\langle c^2\rangle=5.5\,(23)$	MW	78P1	[1]	
	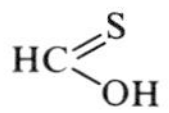 trans	$g_{aa}=-0.4091\,(9)$ $g_{bb}=-0.0703\,(3)$ $g_{cc}=-0.0345\,(3)$	$\Delta\xi_1=-0.7\,(8)$ $\Delta\xi_2=11.4\,(5)$	$\xi_{aa}^{para}=40.3\,(5)$ $\xi_{bb}^{para}=201.9\,(5)$ $\xi_{cc}^{para}=216.9\,(5)$	$\Theta_{aa}=-5.0\,(6)$ $\Theta_{bb}=-5.8\,(5)$ $\Theta_{cc}=-0.8\,(9)$	$\langle a^2\rangle=47.7\,(21)$ $\langle b^2\rangle=10.5\,(21)$ $\langle c^2\rangle=5.3\,(21)$	MW	78P1	[2]	
22	$CH_3CH_2CH_3$	$g_{aa}=0.0605\,(4)$ $g_{bb}=0.0061\,(4)$ $g_{cc}=-0.0001\,(2)$	$\Delta\xi_1=-6.75\,(78)$ $\Delta\xi_2=1.74\,(84)$		$\Theta_{aa}=-0.32\,(44)$ $\Theta_{bb}=-0.25\,(50)$ $\Theta_{cc}=0.56\,(36)$		MW	80H1		2.5, 219
23	CH_2CF_2	see Vol. II/6, 2.9.3, Nr. 22					BMS	76C1	[1]	2.5, 295

to Nr.:

19 [1]) $(1-\sigma)\,g_I^{35Cl}=0.5461(10)$; the signs of the molecular g-values were determined relative to g_I^{35Cl}. $\xi_{aa}^{dia}=-157(8)$; $\xi_{bb}^{dia}=-262(10)$; $\xi_{cc}^{dia}=-369(21)$.

[2]) Estimated with an additivity rule [72G1].

20 [1]) $M_{aa}^{H}=0.197(93)$ kHz; $M_{bb}^{H}=1.682(63)$ kHz; $M_{cc}^{H}=-1.236(112)$ kHz $r_{HH}=1.860(2)$ Å from spin-spin interaction agrees with structure from rotational constants.

21 [1]) An estimated [76S2] bulk value $\xi=-26.5(50)$ was used to compute $\langle a^2\rangle, \langle b^2\rangle, \langle c^2\rangle$, and also the following quantities in 10^{-6} erg/G^2 mol: $\xi_{aa}^{dia}=-42.5(3)$; $\xi_{bb}^{dia}=-67.2(60)$; $\xi_{cc}^{dia}=-226.9(60)$ $\xi_{aa}=-28.9(53)$; $\xi_{bb}=-21.0(52)$; $\xi_{cc}=-29.5(54)$

to Nr.:

21 [2]) An estimated [76S2] bulk value $\xi=-26.5(50)$ was used to compute $\langle a^2\rangle, \langle b^2\rangle, \langle c^2\rangle$, and also the following quantities in 10^{-6} erg/G^2 mol: $\xi_{aa}^{dia}=-67.0(57)$; $\xi_{bb}^{dia}=-224.6(57)$; $\xi_{cc}^{dia}=-247.0(60)$ $\xi_{aa}=-26.7(53)$; $\xi_{bb}=-22.7(52)$; $\xi_{cc}=-30.0(54)$

23 [1]) $M_{aa}^{F}=13.4(36)$ kHz; $M_{bb}^{F}=10.9(7)$ kHz; $M_{cc}^{F}=6.1(7)$ kHz $(\sigma_{aa}^{para})^F=-323(57)\cdot 10^{-6}$; $(\sigma_{bb}^{para})^F=-241(11)\cdot 10^{-6}$; $(\sigma_{cc}^{para})^F=-367(22)\cdot 10^{-6}$; $(\sigma^{para})^F=-310(21)\cdot 10^{-6}$.

Nr.	Molecule	g_{aa}, g_{bb}, g_{cc}	$\Delta\xi_1, \Delta\xi_2$ 10^{-6} erg/G^2 mol	$\xi_{aa}^{para}, \xi_{bb}^{para}, \xi_{cc}^{para}$	$\Theta_{aa}, \Theta_{bb}, \Theta_{cc}$ 10^{-26} esu cm^2	$\langle a^2\rangle, \langle b^2\rangle, \langle c^2\rangle$ 10^{-16} cm^2	Meth.	Ref.	Varia, remarks	See Tab., Nr.
24	CH_3—O—CH=CH_2	$g_{aa} = -0.0868$ (5) $g_{bb} = -0.0224$ (2) $g_{cc} = 0.0000$ (5)	$\Delta\xi_1 = 8.0$ (6) $\Delta\xi_2 = 3.4$ (5)	$\xi_{aa}^{para} = 88.4$ (2) $\xi_{bb}^{para} = 215.6$ (3) $\xi_{cc}^{para} = 276.9$ (4)	$\Theta_{aa} = 2.8$ (5) $\Theta_{bb} = 0.8$ (5) $\Theta_{cc} = -3.6$ (9)	$\langle a^2\rangle - \langle b^2\rangle = 30.3$ (2) $\langle b^2\rangle - \langle c^2\rangle = 14.7$ (2) $\langle c^2\rangle - \langle a^2\rangle = -45.0$ (2)	MW	76D2		2.5, 291
25	CH_2=C=O	$g_{aa} = -0.4322$ (4) $g_{bb} = -0.0362$ (7) $g_{cc} = -0.0252$ (5)	$\Delta\xi_1 = -5.53$ (40) $\Delta\xi_2 = 0.03$ (30)				MW MBE	74S1 77F1	[1]) [2])	2.5, 299
	CD_2=C=O	$g_{aa} = -0.2195$ (5) $g_{bb} = -0.0329$ (5) $g_{cc} = -0.0230$ (5)	$\Delta\xi_1 = -5.15$ (50) $\Delta\xi_2 = 0.13$ (40)				MW MBE	74S1 77F1	 [3])	
26	CH_2=CH—C(=O)—H s-trans	$g_{aa} = -0.5512$ (19) $g_{bb} = -0.0567$ (10) $g_{cc} = -0.0080$ (10)	$\Delta\xi_1 = 24.1$ (9) $\Delta\xi_2 = 17.1$ (15)	$\xi_{aa}^{para} = 60.9$ (3) $\xi_{bb}^{para} = 278.4$ (17) $\xi_{cc}^{para} = 292.6$ (18)	$\Theta_{aa} = -2.5$ (11) $\Theta_{bb} = 3.3$ (17) $\Theta_{cc} = -0.8$ (22)	$\langle a^2\rangle - \langle b^2\rangle = 51.8$ (7) $\langle b^2\rangle - \langle c^2\rangle = 7.9$ (6) $\langle c^2\rangle - \langle a^2\rangle = -59.7$ (7)	MW	73B3		2.5, 318
27	CH_3—CH=CH—CH_3 cis	$g_{aa} = -0.0485$ (18) $g_{bb} = -0.0148$ (16) $g_{cc} = 0.0044$ (16)	$\Delta\xi_1 = 2.6$ (14) $\Delta\xi_2 = 8.0$ (5)		$\Theta_{aa} = 3.6$ (18) $\Theta_{bb} = 0.1$ (11) $\Theta_{cc} = -3.7$ (14)		MW	80H1		2.5, 324

to Nr.:

25 [1]) Polarity from the isotopic effect in the g-values (see CD_2CO below) is
$\overset{+}{H_2}$C=C=O^{-} .
The sign of the g-values was obtained via INDO calculations of $\langle a^2\rangle$, $\langle b^2\rangle$, $\langle c^2\rangle$.

to Nr.:

25 [2]) $M_{aa}^{H} = -18.296(76)$ kHz; $M_{bb}^{H} = -0.087(49)$ kHz; $M_{cc}^{H} = -1.156(51)$ kHz.
$r_{HH} = 1.890(2)$ Å from spin-spin interaction agrees with structure from rotational constants.
[3]) $M_{aa}^{D} = -1.386(55)$ kHz; $M_{bb}^{D} = -0.032(22)$ kHz; $M_{cc}^{D} = -0.176(37)$ kHz.

Nr.	Molecule	g_{aa}, g_{bb}, g_{cc}	$\Delta\xi_1, \Delta\xi_2$	$\xi_{aa}^{para}, \xi_{bb}^{para}, \xi_{cc}^{para}$	$\Theta_{aa}, \Theta_{bb}, \Theta_{cc}$	$\langle a^2\rangle, \langle b^2\rangle, \langle c^2\rangle$	Meth.	Ref.	Varia, remarks	See Tab., Nr.
			10^{-6} erg/G^2 mol		10^{-26} esu cm^2	10^{-16} cm^2				
28	$CH_2{=}CH{-}CH_2{-}CH_3$ cis	$g_{aa} = -0.0454(4)$ $g_{bb} = -0.0203(2)$ $g_{cc} = 0.0080(2)$	$\Delta\xi_1 = 7.80(30)$ $\Delta\xi_2 = 10.26(30)$		$\Theta_{aa} = 0.59(18)$ $\Theta_{bb} = 1.63(20)$ $\Theta_{cc} = -2.22(18)$		MW	80H1		2.5, 323
	$CH_2{=}CH{-}CH_2{-}CH_3$ skew	$g_{aa} = 0.0340(6)$ $g_{bb} = 0.0064(6)$ $g_{cc} = 0.0096(6)$	$\Delta\xi_1 = -2.28(12)$ $\Delta\xi_2 = -0.78(30)$		$\Theta_{aa} = 2.36(26)$ $\Theta_{bb} = 1.40(42)$ $\Theta_{cc} = -3.76(42)$					
29	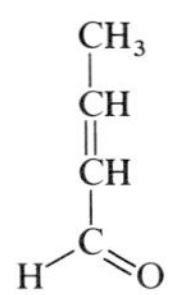	$g_{aa} = -0.3199(14)$ $g_{bb} = -0.0327(7)$ $g_{cc} = -0.0050(7)$	$\Delta\xi_1 = 17.9(13)$ $\Delta\xi_2 = 16.5(13)$	$\xi_{aa}^{para} = 73.8(2)$ $\xi_{bb}^{para} = 583.8(15)$ $\xi_{cc}^{para} = 596.5(16)$	$\Theta_{aa} = -7.5(16)$ $\Theta_{bb} = 12.1(20)$ $\Theta_{cc} = -4.6(27)$	$\langle a^2\rangle = 136.4(10)$ $\langle b^2\rangle = 15.9(10)$ $\langle c^2\rangle = 8.9(10)$	MW	78H2	[1]	2.5, 142
30	$HC{\equiv}CCH_2CH_2CH_3$ trans	$g_{aa} = 0.0450(6)$ $g_{bb} = -0.0004(4)$ $g_{cc} = -0.0022(4)$	$\Delta\xi_1 = -15.27(60)$ $\Delta\xi_2 = 6.75(78)$		$\Theta_{aa} = 2.53(46)$ $\Theta_{bb} = -2.32(66)$ $\Theta_{cc} = -0.21(58)$		MW	80H1		2.5, 155
	$HC{\equiv}CCH_2CH_2CH_3$ gauche	$g_{aa} = 0.0088(4)$ $g_{bb} = 0.0036(10)$ $g_{cc} = -0.0015(4)$	$\Delta\xi_1 = -\ 4.35(18)$ $\Delta\xi_2 = -\ 1.98(30)$		$\Theta_{aa} = 1.00(44)$ $\Theta_{bb} = -1.23(80)$ $\Theta_{cc} = 0.23(54)$					
31	$(CH_3)_2C{=}CH_2$	$g_{aa} = -0.0288(2)$ $g_{bb} = -0.0232(4)$ $g_{cc} = 0.0045(2)$	$\Delta\xi_1 = 7.29(18)$ $\Delta\xi_2 = 5.82(6)$		$\Theta_{aa} = 1.31(14)$ $\Theta_{bb} = 0.75(14)$ $\Theta_{cc} = -2.07(14)$		MW	80H1		2.5, 326

to Nr.:

29 [1]) An estimated [77B2] bulk value $\xi = -37.1$ was used to compute $\langle a^2\rangle$, $\langle b^2\rangle$, $\langle c^2\rangle$, and also the following quantities in 10^{-6} erg/G^2 mol:
$\xi_{aa}^{dia} = -104.9(16)$; $\xi_{bb}^{dia} = -615.4(29)$; $\xi_{cc}^{dia} = -645.1(35)$
$\xi_{aa} = -31.1(14)$; $\xi_{bb} = -31.6(14)$; $\xi_{cc} = -48.6(19)$

Nr.	Molecule	g_{aa}, g_{bb}, g_{cc}	$\Delta\xi_1, \Delta\xi_2$ (10^{-6} erg/G^2 mol)	$\xi_{aa}^{para}, \xi_{bb}^{para}, \xi_{cc}^{para}$ (10^{-6} erg/G^2 mol)	$\Theta_{aa}, \Theta_{bb}, \Theta_{cc}$ (10^{-26} esu cm^2)	$\langle a^2\rangle, \langle b^2\rangle, \langle c^2\rangle$ (10^{-16} cm^2)	Meth.	Ref.	Varia, remarks	See Tab., Nr.
32	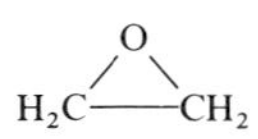	$g_{aa} = -0.09692\,(4)$ $g_{bb} = 0.01848\,(5)$ $g_{cc} = 0.03361\,(7)$	$\Delta\xi_1 = 18.46\,(7)$ $\Delta\xi_2 = -0.05\,(10)$	$\xi_{aa}^{para} = 60.8\,(3)$ $\xi_{bb}^{para} = 67.3\,(5)$ $\xi_{cc}^{para} = 87.9\,(5)$	$\Theta_{aa} = 2.6\,(1)$ $\Theta_{bb} = -3.7\,(1)$ $\Theta_{cc} = 1.1\,(1)$	$\langle a^2\rangle = 16.2\,(5)$ $\langle b^2\rangle = 13.2\,(6)$ $\langle c^2\rangle = 6.9\,(5)$	MW	76H1	[1])	2.5, 478
	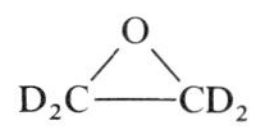	$g_{aa} = -0.07794\,(9)$ $g_{bb} = 0.01436\,(17)$ $g_{cc} = 0.02683\,(11)$	$\Delta\xi_1 = 18.45\,(16)$ $\Delta\xi_1 = 0.24\,(22)$	$\xi_{aa}^{para} = 60.6\,(3)$ $\xi_{bb}^{para} = 67.1\,(5)$ $\xi_{cc}^{para} = 87.8\,(5)$	$\Theta_{aa} = 2.7\,(1)$ $\Theta_{bb} = -4.2\,(2)$ $\Theta_{cc} = 1.5\,(3)$	$\langle a^2\rangle = 16.2\,(6)$ $\langle b^2\rangle = 13.2\,(6)$ $\langle c^2\rangle = 6.9\,(6)$			[2])	
33	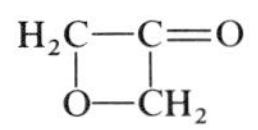	$g_{aa} = -0.1059\,(8)$ $g_{bb} = -0.0581\,(4)$ $g_{cc} = -0.0437\,(4)$	$\Delta\xi_1 = 9.6\,(5)$ $\Delta\xi_2 = -7.8\,(6)$	$\xi_{aa}^{para} = 137.7\,(4)$ $\xi_{bb}^{para} = 260.7\,(32)$ $\xi_{cc}^{para} = 353.1\,(32)$	$\Theta_{aa} = -12.8\,(8)$ $\Theta_{bb} = 7.9\,(8)$ $\Theta_{cc} = 4.9\,(8)$	$\langle a^2\rangle = 60.4\,(14)$ $\langle b^2\rangle = 30.1\,(14)$ $\langle c^2\rangle = 8.8\,(14)$	MW	73N1	[1])	2.5, 503
34	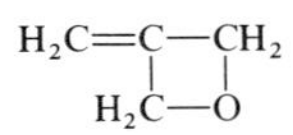	$g_{aa} = -0.0510\,(18)$ $g_{bb} = -0.0435\,(10)$ $g_{cc} = -0.0313\,(10)$	$\Delta\xi_1 = -10.9\,(7)$ $\Delta\xi_2 = 2.3\,(9)$	$\xi_{aa}^{para} = 132.7\,(21)$ $\xi_{bb}^{para} = 284.0\,(31)$ $\xi_{cc}^{para} = 378.4\,(32)$	$\Theta_{aa} = -5.4\,(10)$ $\Theta_{bb} = 5.1\,(12)$ $\Theta_{cc} = 0.2\,(15)$	$\langle a^2\rangle - \langle b^2\rangle = 34.6\,(13)$ $\langle b^2\rangle - \langle c^2\rangle = 21.8\,(8)$ $\langle c^2\rangle - \langle a^2\rangle = -56.4\,(10)$	MW	73N1		2.5, 505
35	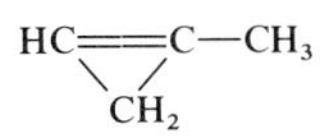	$g_{aa} = -0.0813\,(7)$ $g_{bb} = -0.0261\,(4)$ $g_{cc} = 0.0166\,(3)$	$\Delta\xi_1 = 13.9\,(3)$ $\Delta\xi_2 = 16.4\,(6)$	$\xi_{aa}^{para} = 83.5\,(26)$ $\xi_{bb}^{para} = 223.9\,(27)$ $\xi_{cc}^{para} = 255.5\,(27)$	$\Theta_{aa} = 0.6\,(4)$ $\Theta_{bb} = -0.3\,(6)$ $\Theta_{cc} = -0.3\,(8)$	$\langle a^2\rangle - \langle b^2\rangle = 32.9\,(11)$ $\langle b^2\rangle - \langle c^2\rangle = 11.1\,(8)$ $\langle c^2\rangle - \langle a^2\rangle = -44.0\,(7)$	MW	73B1		

to Nr.:

32 [1]) The liquid-state bulk value $\xi = -30.7(10)$ [76W1] was used to compute $\langle a^2\rangle, \langle b^2\rangle, \langle c^2\rangle$, and in 10^{-6} erg/G^2 mol:
$\xi_{aa}^{dia} = -85.3(1)$; $\xi_{bb}^{dia} = -98.0(2)$; $\xi_{cc}^{dia} = -124.7(2)$.
See Vol. II.6, Tab. 2.9.3, Nr. 33 for earlier results on C_2H_4O.

[2]) The same ξ as in the main isotopic species above was used to compute $\langle a^2\rangle, \langle b^2\rangle, \langle c^2\rangle$, and in 10^{-6} erg/G^2 mol:
$\xi_{aa}^{dia} = -84.1(1)$; $\xi_{bb}^{dia} = -97.7(2)$; $\xi_{cc}^{dia} = -124.7(2)$.
The polarity as determined from the isotopic effect in the g-values is
$^{+}{}^{CH_2}_{CH_2}\!\rhd O^{-}$

to Nr.:

33 [1]) The liquid-phase bulk value $\xi = -30.6(15) \cdot 10^{-6}$ erg/G^2 mol was measured and used to calculate $\langle a^2\rangle, \langle b^2\rangle, \langle c^2\rangle$, and in 10^{-6} erg/G^2 mol:
$\xi_{aa}^{dia} = -165.1(20)$; $\xi_{bb}^{dia} = -294.0(48)$; $\xi_{cc}^{dia} = -384.3(5)$.

Nr.	Molecule	g_{aa}, g_{bb}, g_{cc}	$\Delta\xi_1, \Delta\xi_2$ 10^{-6} erg/G^2 mol	$\xi_{aa}^{para}, \xi_{bb}^{para}, \xi_{cc}^{para}$	$\Theta_{aa}, \Theta_{bb}, \Theta_{cc}$ 10^{-26} esu cm^2	$\langle a^2\rangle, \langle b^2\rangle, \langle c^2\rangle$ 10^{-16} cm^2	Meth.	Ref.	Varia, remarks	See Tab., Nr.
36	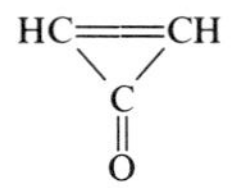	$g_{aa} = -0.2900\,(13)$ $g_{bb} = -0.0963\,(4)$ $g_{cc} = -0.0121\,(4)$	$\Delta\xi_1 = 13.6\,(11)$ $\Delta\xi_2 = 22.0\,(8)$	$\xi_{aa}^{para} = 61.2\,(3)$ $\xi_{bb}^{para} = 173.9\,(6)$ $\xi_{cc}^{para} = 193.8\,(6)$	$\Theta_{aa} = -3.0\,(9)$ $\Theta_{bb} = 4.0\,(7)$ $\Theta_{cc} = -1.0\,(13)$	$\langle a^2\rangle - \langle b^2\rangle = 25.9\,(4)$ $\langle b^2\rangle - \langle c^2\rangle = 9.2\,(3)$ $\langle c^2\rangle - \langle a^2\rangle = -35.1\,(4)$	MW	73B2		2.5, 588
37	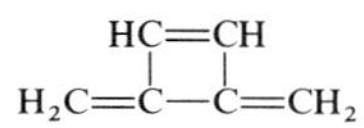	$g_{aa} = -0.0703\,(7)$ $g_{bb} = -0.0532\,(7)$ $g_{cc} = 0.0023\,(7)$	$\Delta\xi_1 = 22.1\,(7)$ $\Delta\xi_2 = 21.2\,(6)$	$\xi_{aa}^{para} = 271.3\,(28)$ $\xi_{bb}^{para} = 329.2\,(37)$ $\xi_{cc}^{para} = 545.0\,(40)$	$\Theta_{aa} = 4.0\,(12)$ $\Theta_{bb} = 4.0\,(12)$ $\Theta_{cc} = -8.0\,(18)$	$\langle a^2\rangle - \langle b^2\rangle = 43.7\,(10)$ $\langle b^2\rangle - \langle c^2\rangle = 55.9\,(10)$ $\langle c^2\rangle - \langle a^2\rangle = -69.6\,(10)$	MW	73B3, 72B1		2.5, 592
38	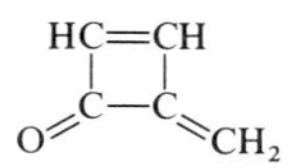	$g_{aa} = -0.0925\,(3)$ $g_{bb} = -0.0729\,(3)$ $g_{cc} = -0.0086\,(3)$	$\Delta\xi_1 = 23.0\,(5)$ $\Delta\xi_2 = 25.5\,(5)$	$\xi_{aa}^{para} = 247.7\,(22)$ $\xi_{bb}^{para} = 316.3\,(27)$ $\xi_{cc}^{para} = 501.4\,(28)$	$\Theta_{aa} = 1.0\,(6)$ $\Theta_{bb} = 2.8\,(6)$ $\Theta_{cc} = -3.8\,(10)$	$\langle a^2\rangle - \langle b^2\rangle = 16.0\,(12)$ $\langle b^2\rangle - \langle c^2\rangle = 49.4\,(7)$ $\langle c^2\rangle - \langle a^2\rangle = -65.4\,(8)$	MW	74C2	[1])	2.5, 591
39	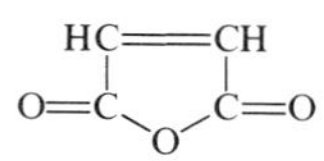	$g_{aa} = -0.1131\,(10)$ $g_{bb} = -0.0499\,(14)$ $g_{cc} = -0.0150\,(12)$	$\Delta\xi_1 = 30.3\,(20)$ $\Delta\xi_2 = 22.9\,(15)$	$\xi_{aa}^{para} = 213.9\,(20)$ $\xi_{bb}^{para} = 486.1\,(63)$ $\xi_{cc}^{para} = 639.4\,(65)$	$\Theta_{aa} = -11.6\,(30)$ $\Theta_{bb} = 5.9\,(32)$ $\Theta_{cc} = 5.7\,(44)$	$\langle a^2\rangle = 114.0\,(24)$ $\langle b^2\rangle = 49.3\,(24)$ $\langle c^2\rangle = 7.2\,(24)$	MW	73B3	[1])	2.5, 598
40	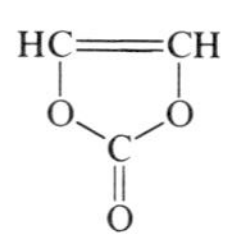	$g_{aa} = -0.0856\,(16)$ $g_{bb} = -0.0502\,(9)$ $g_{cc} = -0.0112\,(10)$	$\Delta\xi_1 = 7.2\,(12)$ $\Delta\xi_2 = 21.7\,(14)$	$\xi_{aa}^{para} = 143.0\,(6)$ $\xi_{bb}^{para} = 301.8\,(9)$ $\xi_{cc}^{para} = 408.0\,(12)$	$\Theta_{aa} = -1.1\,(18)$ $\Theta_{bb} = 0.9\,(19)$ $\Theta_{cc} = 0.2\,(28)$	$\langle a^2\rangle - \langle b^2\rangle = 36.3\,(6)$ $\langle b^2\rangle - \langle c^2\rangle = 29.0\,(7)$ $\langle c^2\rangle - \langle a^2\rangle = -65.3\,(7)$	MW	73B3		2.5, 599

to Nr.:

38 [1]) A tentative structure [74C3] was used to compute the values in columns 4 and 6.

to Nr.:

39 [1]) The liquid-state bulk value $\xi = -35.8(5) \cdot 10^{-6}$ erg/G^2 mol [14G1] is used to compute $\langle a^2\rangle$, $\langle b^2\rangle$, $\langle c^2\rangle$, and also the following quantities in 10^{-6} erg/G^2 mol: $\xi_{aa}^{dia} = -239.6(32)$; $\xi_{bb}^{dia} = -514.3(73)$; $\xi_{cc}^{dia} = -692.9(82)$ $\xi_{aa} = -25.7(12)$; $\xi_{bb} = -28.2(10)$; $\xi_{cc} = -53.5(17)$.

Nr.	Molecule	g_{aa}, g_{bb}, g_{cc}	$\Delta\xi_1, \Delta\xi_2$ 10^{-6} erg/G^2 mol	$\xi_{aa}^{para}, \xi_{bb}^{para}, \xi_{cc}^{para}$	$\Theta_{aa}, \Theta_{bb}, \Theta_{cc}$ 10^{-26} esu cm^2	$\langle a^2\rangle, \langle b^2\rangle, \langle c^2\rangle$ 10^{-16} cm^2	Meth.	Ref.	Varia, remarks	See Tab., Nr.
41	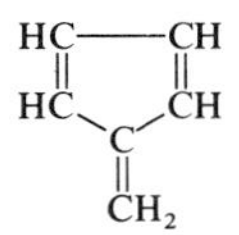	$g_{aa}=-0.1059\,(14)$ $g_{bb}=-0.0482\,(7)$ $g_{cc}=\ 0.0219\,(7)$	$\Delta\xi_1=35.9\,(7)$ $\Delta\xi_2=38.1\,(11)$	$\xi_{aa}^{para}=190.5\,(8)$ $\xi_{bb}^{para}=361.9\,(6)$ $\xi_{cc}^{para}=479.9\,(7)$	$\Theta_{aa}=\ 5.8\,(14)$ $\Theta_{bb}=\ 3.6\,(16)$ $\Theta_{cc}=-9.4\,(21)$	$\langle a^2\rangle=84.6\,(4)$ $\langle b^2\rangle=44.4\,(8)$ $\langle c^2\rangle=\ 7.8\,(8)$	MW	73B3, 72B1	[1])	2.5, 611
42	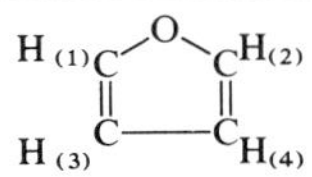	see Vol. II/6, 2.9.3, Nr. 43					BMS	73T2	[1])	2.5, 610
43	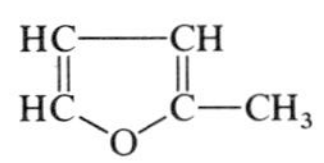	$g_{aa}=-0.0704\,(5)$ [2]) $g_{bb}=-0.0335\,(4)$ [2]) $g_{cc}=\ 0.0253\,(5)$ [2])	$\Delta\xi_1=23.6\,(9)$ [2]) $\Delta\xi_2=44.2\,(7)$ [2])	$\xi_{aa}^{para}=164.3\,(22)$ $\xi_{bb}^{para}=387.9\,(49)$ $\xi_{cc}^{para}=480.5\,(51)$	$\Theta_{aa}=\ 6.4\,(10)$ $\Theta_{bb}=\ 0.0\,(10)$ $\Theta_{cc}=-6.4\,(10)$	$\langle a^2\rangle=90.8\,(34)$ $\langle b^2\rangle=39.7\,(34)$ $\langle c^2\rangle=\ 9.1\,(34)$	MW	73C2	[1])	2.5, 611
44	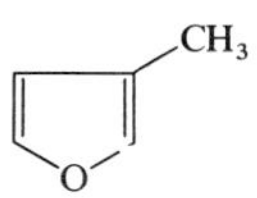	$g_{aa}=-0.0716\,(6)$ $g_{bb}=-0.0370\,(3)$ $g_{cc}=\ 0.0222\,(3)$	$\Delta\xi_1=25.7\,(6)$ $\Delta\xi_2=43.1\,(5)$	$\xi_{aa}^{para}=171.6\,(15)$ $\xi_{bb}^{para}=400.3\,(37)$ $\xi_{cc}^{para}=498.8\,(37)$	$\Theta_{aa}=\ 2.4\,(8)$ $\Theta_{bb}=\ 3.1\,(7)$ $\Theta_{cc}=-5.5\,(11)$	$\langle a^2\rangle=94.0\,(30)$ $\langle b^2\rangle=41.5\,(30)$ $\langle c^2\rangle=\ 9.5\,(32)$	MW	76C3	[1])	2.5, 612
45	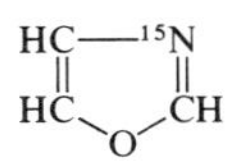	$g_{aa}=-0.0963\,(24)$ $g_{bb}=-0.1214\,(21)$ $g_{cc}=\ 0.0394\,(16)$	$\Delta\xi_1=39.2\,(30)$ $\Delta\xi_2=35.2\,(34)$		$\Theta_{aa}=-4.9\,(29)$ $\Theta_{bb}=\ 9.5\,(31)$ $\Theta_{cc}=-4.6\,(49)$		MW	74D1		2.5, 633

to Nr.:

41 [1]) The liquid-state bulk value $\xi=-42.9(10)\cdot 10^{-6}$ erg/G^2 mol [67L1] was used to compute $\langle a^2\rangle$, $\langle b^2\rangle$, $\langle c^2\rangle$, and also the following quantities in 10^{-6} erg/G^2 mol:
$\xi_{aa}^{dia}=-221.5(21)$; $\xi_{bb}^{dia}=-392.1(19)$; $\xi_{cc}^{dia}=-547.5(23)$
$\xi_{aa}=-30.9(12)$; $\xi_{bb}=-30.2(14)$; $\xi_{cc}=-67.6(16)$.

42 [1]) $M_{aa}^{H(1)}=-537(3)$ Hz; $M_{bb}^{H(1)}=611(5)$ Hz; $M_{cc}^{H(1)}=-548(13)$ Hz
$M_{aa}^{H(3)}=253(4)$ Hz; $M_{bb}^{H(3)}=-189(5)$ Hz; $M_{cc}^{H(3)}=-512(17)$ Hz.
From spin-spin interactions [10^{-10} m]: $r_{12}=4.116(8)$, $r_{34}=2.762(2)$, a_{13} (=projection of r_{13} on the symmetry axis)=2.663(3).

to Nr.:

43 [1]) The other choice of the g-value signs yields $\langle c^2\rangle<0$; the liquid-state bulk value $\xi=-50.9(16)$ [H. Frieske, private communication to the authors] was used to compute $\langle a^2\rangle$, $\langle b^2\rangle$, $\langle c^2\rangle$, and also the following quantities (in 10^{-6} erg/G^2 mol; the uncertainty of ξ was enhanced to ± 5.0 to account for a possible difference between gas and liquid):
$\xi_{aa}^{dia}=-207.3(76)$; $\xi_{bb}^{dia}=-424.0(102)$; $\xi_{cc}^{dia}=-554.1(106)$
$\xi_{aa}=-43.0(53)$; $\xi_{bb}=-36.2(52)$; $\xi_{cc}=-73.5(55)$

[2]) From torsional A species only; E species values agree within two standard deviations.

44 [1]) The liquid-state bulk value $\xi=-53.2(50)$ was used to compute $\langle a^2\rangle$, $\langle b^2\rangle$, $\langle c^2\rangle$, and also the following quantities in 10^{-6} erg/G^2 mol:
$\xi_{aa}^{dia}=-216.3(67)$; $\xi_{bb}^{dia}=-439.1(88)$; $\xi_{cc}^{dia}=-574.9(91)$.

Nr.	Molecule	g_{aa}, g_{bb}, g_{cc}	$\Delta\xi_1, \Delta\xi_2$ (10^{-6} erg/G^2 mol)	$\xi_{aa}^{para}, \xi_{bb}^{para}, \xi_{cc}^{para}$ (10^{-6} erg/G^2 mol)	$\Theta_{aa}, \Theta_{bb}, \Theta_{cc}$ (10^{-26} esu cm^2)	$\langle a^2\rangle, \langle b^2\rangle, \langle c^2\rangle$ (10^{-16} cm^2)	Meth.	Ref.	Varia, remarks	See Tab., Nr.
46	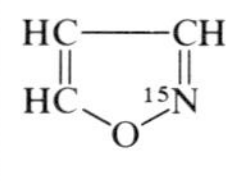	$g_{aa}=-0.1044\,(6)$ $g_{bb}=-0.1220\,(9)$ $g_{cc}=\ \ 0.0406\,(7)$	$\Delta\xi_1=35.7\,(13)$ $\Delta\xi_2=45.6\,(14)$	$\xi_{aa}^{para}=148.3\,(2)$ $\xi_{bb}^{para}=156.0\,(3)$ $\xi_{cc}^{para}=235.7\,(1)$	$\Theta_{aa}=\ \ 0.1\,(11)$ $\Theta_{bb}=\ \ 3.0\,(13)$ $\Theta_{cc}=-3.1\,(16)$	$\langle a^2\rangle=36.0\,(4)$ $\langle b^2\rangle=34.9\,(5)$ $\langle c^2\rangle=\ \ 6.2\,(4)$	MW	74D1	[1])	2.5, 634
47	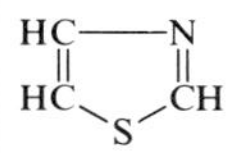	$g_{aa}=-0.1057\,(3)$ $g_{bb}=-0.0847\,(3)$ $g_{cc}=\ \ 0.0404\,(3)$	$\Delta\xi_1=51.3\,(5)$ $\Delta\xi_2=50.7\,(5)$		$\Theta_{aa}=-3.2\,(5)$ $\Theta_{bb}=\ \ 8.0\,(6)$ $\Theta_{cc}=-4.8\,(9)$	$\langle a^2\rangle=55.9\,(10)$ $\langle b^2\rangle=41.5\,(10)$ $\langle c^2\rangle=\ \ 7.9\,(10)$	MW	80W1	[1])	2.5, 636
48	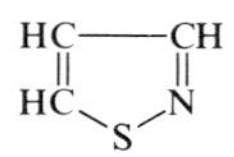	$g_{aa}=-0.1287\,(3)$ $g_{bb}=-0.0785\,(3)$ $g_{cc}=\ \ 0.0433\,(3)$	$\Delta\xi_1=55.6\,(6)$ $\Delta\xi_2=51.0\,(6)$		$\Theta_{aa}=\ \ 5.1\,(6)$ $\Theta_{bb}=\ \ 0.0\,(6)$ $\Theta_{cc}=-5.1\,(10)$	$\langle a^2\rangle=54.4\,(22)$ $\langle b^2\rangle=42.0\,(22)$ $\langle c^2\rangle=\ \ 8.0\,(22)$	MW	80W1	[1])	2.5, 637
49	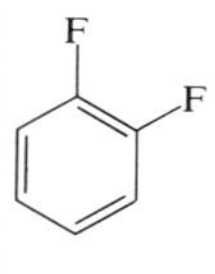	see Vol. II/6, 2.5, 284 $g_{aa}=-0.0412\,(12)$ $g_{bb}=-0.0371\,(8)$ $g_{cc}=\ \ 0.0163\,(7)$	$\Delta\xi_1=50.8\,(15)$ $\Delta\xi_2=57.9\,(10)$	$\xi_{aa}^{para}=380.0\,(29)$ $\xi_{bb}^{para}=542.6\,(29)$ $\xi_{cc}^{para}=834.1\,(31)$	$\Theta_{aa}=-3.6\,(26)$ $\Theta_{bb}=\ \ 7.0\,(24)$ $\Theta_{cc}=-3.4\,(33)$	$\langle a^2\rangle=129.0\,(46)$ $\langle b^2\rangle=\ \ 91.2\,(46)$ $\langle c^2\rangle=\ \ \ 9.5\,(46)$	MW	74S2	[1])	
50	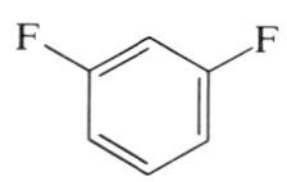	$g_{aa}=-0.0486\,(2)$ $g_{bb}=-0.0315\,(3)$ $g_{cc}=\ \ 0.0116\,(3)$	$\Delta\xi_1=46.6\,(4)$ $\Delta\xi_2=55.7\,(4)$	$\xi_{aa}^{para}=330\,(9)$ $\xi_{bb}^{para}=646\,(9)$ $\xi_{cc}^{para}=890\,(9)$	$\Theta_{aa}=-5.0\,(9)$ $\Theta_{bb}=\ \ 7.6\,(10)$ $\Theta_{cc}=-2.6\,(13)$	$\langle a^2\rangle=153.2\,(49)$ $\langle b^2\rangle=\ \ 79.4\,(49)$ $\langle c^2\rangle=\ \ \ 9.5\,(49)$	MW	76C3	[1])	2.5, 702

to Nr.:

46 [1]) The liquid-state bulk value of $\xi=-38.0(5)$ [74G2] is used to compute $\langle a^2\rangle$, $\langle b^2\rangle$, $\langle c^2\rangle$, and also the following quantities in 10^{-6} erg/G^2 mol:
$\xi_{aa}=-26.1(9)$; $\xi_{bb}=-22.8(10)$; $\xi_{cc}=-65.1(14)$
$\xi_{aa}^{dia}=-174.4(11)$; $\xi_{bb}^{dia}=-178.8(12)$; $\xi_{cc}^{dia}=-300.8(15)$.

47 [1]) The liquid-state bulk value $\xi=-50.6(20)$ [76W1] was used to compute $\langle a^2\rangle$, $\langle b^2\rangle$, $\langle c^2\rangle$, and in 10^{-6} erg/G^2 mol:
$\xi_{aa}=-33.5(22)$; $\xi_{bb}=-33.7(22)$; $\xi_{cc}=-84.6(23)$.

48 [1]) The liquid-state bulk value of thiazole, $\xi=-50.6(30)$ [76W1], as well as an estimated structure based on the rotational constants were used to compute $\langle a^2\rangle$, $\langle b^2\rangle$, $\langle c^2\rangle$, and in 10^{-6} erg/G^2 mol:
$\xi_{aa}=-32.1(32)$; $\xi_{bb}=-33.6(32)$; $\xi_{cc}=-86.1(34)$.

to Nr.:

49 [1]) A partly assumed structure was used to calculate the paramagnetic susceptibilities; in addition, the liquid-state bulk value $\xi=-64.5(4)$ served to compute $\langle a^2\rangle$, $\langle b^2\rangle$, $\langle c^2\rangle$, and also the following quantities in 10^{-6} erg/G^2 mol:
$\xi_{aa}=-47.6(65)$; $\xi_{bb}=-45.2(63)$; $\xi_{cc}=-100.7(68)$
$\xi_{aa}^{dia}=-427.4(115)$; $\xi_{bb}^{dia}=-587.7(113)$; $\xi_{cc}^{dia}=-932.1(122)$.
Assuming g_{aa}, g_{bb} positive and g_{cc} negative yields
$\langle c^2\rangle=-11.4(46)\cdot 10^{-16}$ cm^2.

50 [1]) The liquid-state bulk value $\xi=-62.9(20)$ was used to compute $\langle a^2\rangle$, $\langle b^2\rangle$, $\langle c^2\rangle$, and in 10^{-6} erg/G^2 mol:
$\xi_{aa}^{dia}=-378(11)$; $\xi_{bb}^{dia}=-691(11)$; $\xi_{cc}^{dia}=-987(11)$.

Nr.	Molecule	g_{aa}, g_{bb}, g_{cc}	$\Delta\xi_1, \Delta\xi_2$	$\xi^{para}_{aa}, \xi^{para}_{bb}, \xi^{para}_{cc}$	$\Theta_{aa}, \Theta_{bb}, \Theta_{cc}$	$\langle a^2\rangle, \langle b^2\rangle, \langle c^2\rangle$	Meth.	Ref.	Varia, remarks	See Tab., Nr.
			10^{-6} erg/G^2 mol		10^{-26} esu cm^2	10^{-16} cm^2				
51	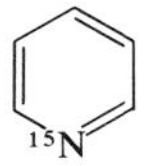	$g_{aa} = -0.08086$ (19) $g_{bb} = -0.09974$ (16) $g_{cc} = 0.04101$ (17)	$\Delta\xi_1 = 54.1$ (4) $\Delta\xi_2 = 62.0$ (4)	$\xi^{para}_{aa} = 242.7$ (2) $\xi^{para}_{bb} = 247.9$ (2) $\xi^{para}_{cc} = 394.0$ (3)	$\Theta_{aa} = -2.6$ (4) $\Theta_{bb} = 8.0$ (4) $\Theta_{cc} = -5.4$ (6)	$\langle a^2\rangle = 57.2$ (3) $\langle b^2\rangle = 56.6$ (3) $\langle c^2\rangle = 8.0$ (3)	MW	76H1	[1])	2.5, 730
	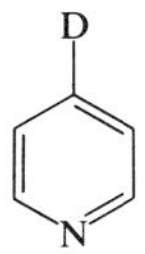	$g_{aa} = -0.07935$ (7) $g_{bb} = -0.09438$ (6) $g_{cc} = 0.04063$ (6) see Vol. II/6, 2.9.3, Nr. 49 for corresponding data of the main isotopic species	$\Delta\xi_1 = 53.3$ (10) $\Delta\xi_2 = 60.1$ (16)	$\xi^{para}_{aa} = 242.1$ (4) $\xi^{para}_{bb} = 247.3$ (4) $\xi^{para}_{cc} = 394.0$ (5)	$\Theta_{aa} = -2.3$ (1) $\Theta_{bb} = 9.3$ (2) $\Theta_{cc} = -6.9$ (2)	$\langle a^2\rangle = 57.0$ (7) $\langle b^2\rangle = 56.3$ (7) $\langle c^2\rangle = 8.1$ (7)			[2])	
52	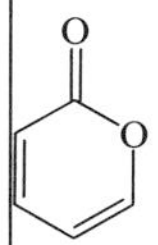	$g_{aa} = -0.0753$ (16) $g_{bb} = -0.0488$ (11) $g_{cc} = -0.0059$ (9)	$\Delta\xi_1 = 22.1$ (12) $\Delta\xi_2 = 27.4$ (14)	$\xi^{para}_{aa} = 247.8$ (63) $\xi^{para}_{bb} = 437.1$ (86) $\xi^{para}_{cc} = 627.2$ (87)	$\Theta_{aa} = -4.7$ (27) $\Theta_{bb} = 5.6$ (26) $\Theta_{cc} = -0.9$ (34)	$\langle a^2\rangle - \langle b^2\rangle = 44.2$ (37) $\langle b^2\rangle - \langle c^2\rangle = 50.8$ (21) $\langle c^2\rangle - \langle a^2\rangle = -95.0$ (25)	MW	73N2		
53	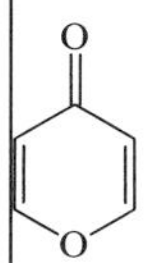	$g_{aa} = -0.0771$ (21) $g_{bb} = -0.0542$ (10) $g_{cc} = -0.0093$ (9)	$\Delta\xi_1 = 23.9$ (14) $\Delta\xi_2 = 21.9$ (20)	$\xi^{para}_{aa} = 247.1$ (33) $\xi^{para}_{bb} = 448.1$ (43) $\xi^{para}_{cc} = 635.4$ (46)	$\Theta_{aa} = -11.9$ (27) $\Theta_{bb} = 13.8$ (30) $\Theta_{cc} = -1.9$ (40)	$\langle a^2\rangle - \langle b^2\rangle = 47.5$ (21) $\langle b^2\rangle - \langle c^2\rangle = 49.4$ (14) $\langle c^2\rangle - \langle a^2\rangle = -97.0$ (16)	MW	73N2		2.5, 646
54	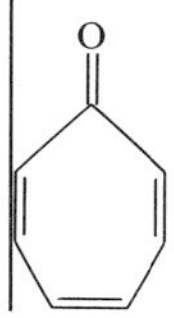		$\Delta\xi_1 = 32.1$ (50) $\Delta\xi_2 = 39.8$ (70)				MW	73N2		2.5, 657

to Nr.:

51 [1]) The liquid-state bulk value $\xi = -49.2(9)$ [76W1] was used to compute $\langle a^2\rangle$, $\langle b^2\rangle$, $\langle c^2\rangle$, and in 10^{-6} erg/G^2 mol: $\xi^{dia}_{aa} = -273.8(7)$; $\xi^{dia}_{bb} = -276.4(7)$; $\xi^{dia}_{cc} = -482.8(9)$.

to Nr.:

51 [2]) The same ξ as in the ^{15}N isotopic species above was used to compute $\langle a^2\rangle$, $\langle b^2\rangle$, $\langle c^2\rangle$, and in 10^{-6} erg/G^2 mol: $\xi^{dia}_{aa} = -273.6(12)$; $\xi^{dia}_{bb} = -276.4(13)$; $\xi^{dia}_{cc} = -481.0(18)$.

2.9.4 References for 2.9.2 and 2.9.3

14G1 Gray, F.W., Birse, W.M.: J. Chem. Soc. **105** (1914) 2707.
32V1 Van Vleck, J.H.: "The Theory of Electric and Magnetic Susceptibilities", Oxford, England: Oxford University Press, **1932**.
50R1 Ramsey, N.F.: Phys. Rev. **78** (1950) 699.
52E1 Eshbach, J.R., Strandberg, M.W.P.: Phys. Rev. **85** (1952) 24.
53B1 Burke, B.F., Strandberg, M.W.P.: Phys. Rev. **90** (1953) 303.
56C1 Cox, J.T., Gordy, W.: Phys. Rev. **101** (1956) 1298.
56R1 Ramsey, N.F.: Molecular Beams, Oxford, **1956**.
59N1 Nielsen, H.H.: In Encyclopedia of Physics, Vol. XXXVII/1, (S. Flügge, ed.) Springer-Verlag Berlin, Göttingen, Heidelberg, 1959.
60B1 Barter, C., Meisenheimer, R.G., Stevenson, D.P.: J. Phys. Chem. **64** (1960) 1312.
60O1 Osipov, B.D.: Opt. Spektrosk. **8** (1960) 581 (Opt. Spectrosc. **8** (1960) 309, Engl. Transl.).
62Z1 Zuev, V.S.: Opt. Spektrosk. **12** (1962) 641 (Opt. Spectrosc. **12** (1962) 358, Engl. Transl.).
64F1 Flygare, W.H.: J. Chem. Phys. **41** (1964) 793.
64T1 Thaddeus, P., Krisher, L.C., Loubser, J.H.N.: J. Chem. Phys. **40** (1964) 257.
65H1 Hameka, H.F.: Advanced Quantum Chemistry, Addison-Wesley, Reading Ma., **1965**.
65H2 Hunt, R.H., Leacock, R.A., Peters, C.W., Hecht, K.T.: J. Chem. Phys. **42** (1965) 1931.
65K1 Kukolich, S.G.: Phys. Rev. **138** (1965) A 1322.
65S1 Saito, Y.: Can. J. Chem. **43** (1965) 2530.
67H1 Hüttner, W., Flygare, W.H.: J. Chem. Phys. **47** (1967) 4137.
67K1 Kukolich, S.G.: Phys. Rev. **156** (1967) 83.
67L1 Landolt-Börnstein, Critical Tables (K.H. Hellwege, ed.) Vol. II/10, p. 85, Springer-Verlag, Berlin, **1967**.
68H1 Hüttner, W., Lo, M.-K., Flygare, W.H.: J. Chem. Phys. **48** (1968) 1206.
68K1 Kukolich, S.G.: Phys. Rev. **172** (1968) 59.
69H1 Hüttner, W., Flygare, W.H.: Trans. Faraday Soc. **65** (1969) 1953.
69L1 De Lucia, F.C., Gordy, W.: Phys. Rev. **187** (1969) 58.
69N1 Neumann, D.B., Moskowitz, J.W.: J. Chem. Phys. **50** (1969) 2216.
69S1 Shoemaker, R.L., Flygare, W.H.: J. Am. Chem. Soc. **91** (1969) 5417.
69S2 Stone, R.G., Pochan, J.M., Flygare, W.H.: Inorg. Chem. **8** (1969) 2647.
70F1 Follett, T.L.: PhD. Thesis, Harvard University, Cambridge, MA., **1970**.
70K1 Kukolich, S.G.: Chem. Phys. Lett. **5** (1970) 401, Erratum: Chem. Phys. Lett. **12** (1971) 216.
70K2 Kukolich, S.G., Wofsy, S.C.: J. Chem. Phys. **52** (1970) 5477.
70L1 De Leeuw, F.H., Dymanus, A.: Chem. Phys. Lett. **7** (1970) 288.
70T1 Tigelaar, H.L., Flygare, W.H.: Chem. Phys. Lett. **7** (1970) 254.
70V1 Verhoeven, J., Dymanus, A.: J. Chem. Phys. **52** (1970) 3222.
71D1 Davies, P.B., Neumann, R.M., Wofsy, S.C., Klemperer, W.: J. Chem. Phys. **55** (1971) 3564.
71K1 Kukolich, S.G.: Chem. Phys. Lett. **12** (1971) 216.
71K2 Kukolich, S.G., Ruben, D.J.: J. Mol. Spectry. **38** (1971) 130.
71K3 Kukolich, S.G., Nelson, A.C., Yamanaski, B.S.: J. Am. Chem. Soc. **93** (1971) 6769.
71T1 Tucker, K.D., Tomasevich, G.R., Thaddeus, P.: Astrophys. J. **169** (1971) 429.
72B1 Benson, R.C.: Ph. D. Thesis, University of Illinois, Urbana, Il., **1972**.
72G1 Gierke, T.D., Tigelaar, H.L., Flygare, W.H.: J. Am. Chem. Soc. **94** (1972) 330.
72G2 Gierke, T.D., Flygare, W.H.: J. Am. Chem. Soc. **94** (1972) 7277.
72K1 Kukolich, S.G., Ruben, D.J.: J. Mol. Spectry. **44** (1972) 607.
72L1 Laws, E.A., Stevens, R.M., Libscomb, W.N.: J. Chem. Phys. **56** (1972) 2029.
72O1 Ozier, I., Vitkevich, J.A., Ramsey, N.F.: Symp. Mol. Struct. and Spectr. Ohio, **1972**, Z2.
72R1 Reinartz, J.M.L.J., Meerts, W.L., Dymanus, A.: Chem. Phys. Lett. **16** (1972) 576.
72T1 Tucker, K.D., Tomasevich, G.R., Thaddeus, P.: Astrophys. J. **174** (1972) 463.
73B1 Benson, R.C., Flygare, W.H.: J. Chem. Phys. **58** (1973) 2651.
73B2 Benson, R.C., Flygare, W.H., Oda, M., Breslow, R.: J. Am. Chem. Soc. **95** (1973) 2772.
73B3 Benson, R.C., Flygare, W.H.: J. Chem. Phys. **58** (1973) 2366.
73B4 Bhattacharyya, P.K., Dailey, B.P.: J. Chem. Phys. **59** (1973) 5820.
73C1 Chardon, J.C., Guichon, D.: J. Phys. (Paris) **34** (1973) 791.
73C2 Czieslik, W., Andresen, U., Dreizler, H.: Z. Naturforsch. **28a** (1973) 1906.

73E1 Engelbrecht, L., Sutter, D., Dreizler, H.: Z. Naturforsch. **28a** (1973) 709.
73G1 Gustafson, S., Gordy, W.: J. Chem. Phys. **58** (1973) 5181.
73H1 Hartford, S.L, Allen, Wm.C., Norris, C.L., Pearson, E.F., Flygare, W.H.: Chem. Phys. Lett. **18** (1973) 153.
73H2 Heemskerk, J.P.J., Boogaard, A., Knaap, H.F.P.: Chem. Phys. Lett. **18** (1973) 77.
73H3 Heuvel, J.E.M., Dymanus, A.: J. Mol. Spectrosc. **45** (1973) 282.
73H4 Heuvel, J.E.M., Dymanus, A.: J. Mol. Spectrosc. **47** (1973) 363.
73K1 Kukolich, S.G., Casleton, K.H.: Chem. Phys. Lett. **18** (1973) 408.
73K2 Kukolich, S.G., Ruben, D.J., Wang, J.H.S., Williams, J.R.: J. Chem. Phys. **58** (1973) 3155.
73K3 Kukolich, S.G., Nelson, A.C.: J. Am. Chem. Soc. **95** (1973) 680.
73K4 Kukolich, S.G., Wang, J.H.S., Ruben, D.J.: J. Chem. Phys. **58** (1973) 5474.
73N1 Norris, C.L., Tigelaar, H.L., Flygare, W.H.: Chem. Phys. **1** (1973) 1.
73N2 Norris, C.L., Benson, R.C., Beak, P., Flygare, W.H.: J. Am. Chem. Soc. **95** (1973) 2766.
73R1 Rock, S.L., McGurk, J.C., Flygare, W.H.: Chem. Phys. Lett. **19** (1973) 153.
73R2 Rock, S.L., Pearson, E.F., Appleman, E.H., Norris, C.L., Flygare, W.H.: J. Chem. Phys. **59** (1973) 3940.
73T1 Taft, H., Bhattacharyya, P., Smith, N., Dailey, B.P.: Chem. Phys. Lett. **22** (1973) 113.
73T2 Tomasevich, G.R., Tucker, K.D., Thaddeus, P.: J. Chem. Phys. **59** (1973) 131.
73T3 Tucker, K.D., Tomasevich, G.R.: J. Mol. Spectrosc. **48** (1973) 475.
73W1 Wang, J.H.S., Kukolich, S.G.: J. Am. Chem. Soc. **95** (1973) 4138.
74C1 Chardon, J.C., Genty, C., Guichon, D., Sungur, N., Theobald, J.G.: Rev. Phys. Appl. **9** (1974) 961.
74C2 Czieslik, W., Sutter, D.H.: Z. Naturforsch. **29a** (1974) 1820.
74C3 Czieslik, W., Charpentier, L., Sutter, D.H.: Z. Naturforsch. **29a** (1974) 1498.
74D1 Davidson, J.R., Burnham, A.K., Siegel, B., Beak, P., Flygare, W.H.: J.A.C.S. **96** (1974) 7394.
74D2 Davis, R.E., Muenter, J.S.: Chem. Phys. Lett. **24** (1974) 343.
74F1 Flygare, W.H.: Chem. Rev. **74** (1974) 653.
74G1 Garvey, R.M., de Lucia, F.C.: J. Mol. Spectrosc. **50** (1974) 38.
74G2 Gierke, T.D.: Ph. D. Thesis, **1974**, University of Illinois.
74N1 Nelson, A.C., Kukolich, S.G., Ruben, D.J.: J. Mol. Spectrosc. **51** (1974) 107.
74N2 Norris, C.L., Pearson, E.F., Flygare, W.H.: J. Chem. Phys. **60** (1974) 1758.
74P1 Pearson, E.F., Norris, C.L., Flygare, W.H.: J. Chem. Phys. **60** (1974) 1761.
74R1 Reijnders, J.M.H., Ellenbroek, A.W., Dymanus, A.: Chem. Phys. Lett. **26** (1974) 470.
74R2 Reinartz, J.M.L.J., Dymanus, A.: Chem. Phys. Lett. **24** (1974) 346.
74R3 Ruben, D.J., Kukolich, S.G.: J. Chem. Phys. **61** (1974) 3780.
74S1 Sastry, K.V.L.N., Guarnieri, A.: Z. Naturforsch. **29a** (1974) 1495.
74S2 Sutter, D.H.: Z. Naturforsch. **29a** (1974) 786.
75A1 Annabi, M., Gillet, D., Touré, C.: C. R. Acad. Sci. **281C** (1975) 663.
75C1 Casleton, K.H., Kukolich, S.G.: J. Chem. Phys. **62** (1975) 2696.
75E1 Engelbrecht, L., Sutter, D.H.: Z. Naturforsch. **30a** (1975) 1265.
75K1 Kukolich, S.G.: J. Am. Chem. Soc. **97** (1975) 5704.
75L1 de Lucia, F.C., Helminger, P.: J. Mol. Spectrosc. **56** (1975) 138.
75S1 Suzuki, M., Guarnieri, A.: Z. Naturforsch. **30a** (1975) 497.
76B1 Burie, J., Boucher, D., Demaison, J., Dubrulle, A.: Mol. Phys. **32** (1976) 289.
76C1 Casleton, K.H., Gierke, T.D., Wang, J.H.-S., Kukolich, S.G.: J. Chem. Phys. **64** (1976) 471.
76C2 Chardon, J.-C., Genty, C., Guichon, D., Theobald, J.-G.: J. Chem. Phys. **64** (1976) 1437.
76C3 Czieslik, W., Wiese, J., Sutter, D.H.: Z. Naturforsch. **31a** (1976) 1210.
76D1 Demaison, J., Boucher, D., Burie, J., Dubrulle, A.: J. Mol. Spectrosc. **61** (1976) 141.
76D2 Duda, L.E., Flygare, W.H.: J. Chem. Phys. **65** (1976) 2841.
76E1 Ellenbroek, A.W., Dymanus, A.: Chem. Phys. Lett. **42** (1976) 303.
76G1 Garvey, R.M., de Lucia, F.C., Cederberg, J.W.: Mol. Phys. **31** (1976) 265.
76H1 Hamer, E., Sutter, D.H.: Z. Naturforsch. **31a** (1976) 265.
76O1 Ozier, I., Lee, S.S., Ramsey, N.F.: J. Chem. Phys. **65** (1976) 3985.
76S1 Scappini, F., Guarnieri, A.: Z. Naturforsch. **31a** (1976) 369, Erratum: Z. Naturforsch. **36a** (1981) 1393.
76S2 Sutter, D.H., Flygare, W.H.: Top. Current Chem. **63** (1976) 89.
76S3 Suzuki, M., Guarnieri, A.: Z. Naturforsch. **31a** (1976) 1242.

76W1	Weast, R.C. (ed.): Handbook of Chemistry and Physics, CRC Press, Cleveland, Ohio, 56-th edition.
77B1	Boucher, D., Burie, J., Demaison, J., Dubrulle, A., Legrand, J., Segard, B.: J. Mol. Spectrosc. **64** (1977) 290.
77B2	Burnham, A.K., Jaekeum Lee, Schmalz, T.G., Beak, P., Flygare, W.H.: J. Am. Chem. Soc. **99** (1977) 1836.
77B3	Burie, J., Demaison, J., Dubrulle, A., Boucher, D.: J. Mol. Spectrosc. **68** (1977) 495.
77C1	Chardon, J.C., Guichon, D.: J. Phys. (Paris) **38** (1977) 113.
77D1	Demaison, J., Dubrulle, A., Boucher, D., Burie, J.: J. Chem. Phys. **67** (1977) 254.
77D2	Dössel, K.-F., Sutter, D.H.: Z. Naturforsch. **32a** (1977) 1444.
77D3	Dubrulle, A., Boucher, D., Burie, J., Demaison, J.: Chem. Phys. Lett. **45** (1977) 559.
77F1	Fabricant, B., Krieger, D., Muenter, J.S.: J. Chem. Phys. **67** (1977) 1576.
77G1	Garvey, R.M., de Lucia, F.C.: Can. J. Phys. **55** (1977) 1115.
77H1	Hüttner, W.: Chem. Phys. **23** (1977) 71.
77K1	Kukolich, S.G.: J. Chem. Phys. **66** (1977) 4345.
77M1	Mack, K.M., Muenter, J.S.: J. Chem. Phys. **66** (1977) 5278.
77M2	Meerts, W.L., Stolte, S., Dynamus, A.: Chem. Phys. **19** (1977) 467.
77N1	Nowicki, P., Hüttner, W.: to be published; Nowicki, P., Dissertation, Ulm, **1977**.
77O1	Ozier, I., Yi, P.N., Ramsey, N.F.: J. Chem. Phys. **66** (1977) 143.
78A1	Arimondo, E., Glorieux, P., Oka, T.: Phys. Rev. **A 17** (1978) 1375.
78B1	Boucher, D., Burie, J., Dangoisse, D., Demaison, J., Dubrulle, A.: Chem. Phys. **29** (1978) 323.
78D1	Dössel, K.-F., Wiese, J., Sutter, D.H.: Z. Naturforsch. **33a** (1978) 21.
78E1	Ellenbroek, A.W., Dymanus, A.: Chem. Phys. **33** (1978) 107.
78E2	Ellenbroek, A.W., Dymanus, A.: Chem. Phys. **35** (1978) 227, Erratum: Chem. Phys. **55** (1981) 428.
78H1	Helms, D.A., Gordy, W.: J. Mol. Spectrosc. **69** (1978) 473.
78H2	Hoke, W.E., Voss, H.L., Campbell, E.J., Flygare, W.H.: Chem. Phys. Lett. **58** (1978) 441.
78H3	Hüttner, W., Bodenseh, H.K., Nowicki, P., Morgenstern, K.: J. Mol. Spectrosc. **71** (1978) 246.
78H4	Hüttner, W., Bodenseh, H.K., Nowicki, P.: Mol Phys. **35** (1978) 729.
78O1	Ozier, I., Meerts, W.L.: Phys. Rev. Lett. **40** (1978) 226.
78P1	Pandey, G.K., Sutter, D.H.: Z. Naturforsch. **33a** (1978) 29.
78R1	Reinartz, J.M.L.J., Meerts, W.L., Dymanus, A.: Chem. Phys. **31** (1978) 19.
78V1	Voss, H.L., Krajnovich, D., Hoke, W.E., Flygare, W.H.: J. Chem. Phys. **68** (1978) 1439.
79C1	Cederberg, J.: J. Mol. Spectrosc. **77** (1979) 102.
79D1	Dössel, K.-F., Sutter, D.H.: Z. Naturforsch. **34a** (1979) 482.
79D2	Dössel, K.-F., Sutter, D.H.: Z. Naturforsch. **34a** (1979) 469.
79M1	Meerts, W.L., Ozier, I., Dymanus, A.: Can. J. Phys. **57** (1979) 1163.
80A1	Arimondo, E., Baker, J.G., Glorieux, P., Oka, T.: J. Mol. Spectrosc. **82** (1980) 54.
80B1	Barton, A.E., Henderson, T.J., Langridge-Smith, P.R.R., Howard, B.J.: Chem. Phys. **45** (1980) 429.
80C1	Cornet, R., Landsberg, B.M., Winnewisser, G.: J. Mol. Spectrosc. **82** (1980) 253.
80H1	Häußler, H., Hüttner, W.: unpublished.
80J1	Jackson, K.C., Langridge-Smith, P.R.R., Howard, B.J.: Mol. Phys. **39** (1980) 817.
80K1	Keenan, M.R., Campbell, E.J., Balle, T.J., Buxton, L.W., Minton, T.K., Soper, P.D., Flygare, W.H.: J. Chem. Phys. **72** (1980) 3070.
80K2	Keenan, M.R., Minton, T.K., Legon, A.C., Balle, T.J., Flygare, W.H.: Proc. Natl. Acad. Sci. USA **77** (1980) 5583.
80M1	Miller, D.J., De Leon, R., Muenter, J.S.: J. Mol. Spectrosc. **83** (1980) 283.
80R1	Rohwer, F., Guarnieri, A.: Z. Naturforsch. **35a** (1980) 336.
80W1	Wiese, J., Sutter, D.H.: Z. Naturforsch. **35a** (1980) 712.
81S1	Scappini, F., Guarnieri, A.: Z. Naturforsch. **36a** (1981) 1393.
82G1	Geschka, H., Pferrer, S., Häußler, H., Hüttner, W.: Ber. Bunsengesellschaft Phys. Chem. **86** (1982) 790.